Arithmetic Operations:

$$ab + ac = a(b + c)$$

$$\frac{a}{b} + \frac{c}{d} = \frac{ad + bc}{bd}$$

$$\frac{a + b}{c} = \frac{a}{c} + \frac{b}{c}$$

$$\frac{\left(\dfrac{a}{b}\right)}{\left(\dfrac{c}{d}\right)} = \frac{ad}{bc}$$

$$a\left(\frac{b}{c}\right) = \frac{ab}{c}$$

$$\frac{a - b}{c - d} = \frac{b - a}{d - c}$$

$$\frac{ab + ac}{a} = b + c,\ a \neq 0$$

$$\frac{\left(\dfrac{a}{b}\right)}{c} = \frac{a}{bc}$$

$$\frac{a}{\left(\dfrac{b}{c}\right)} = \frac{ac}{b}$$

Exponents and Radicals:

$$a^0 = 1,\ a \neq 0$$

$$\frac{a^x}{a^y} = a^{x-y}$$

$$\left(\frac{a}{b}\right)^x = \frac{a^x}{b^x}$$

$$\sqrt[n]{a^m} = a^{m/n} = \left(\sqrt[n]{a}\right)^m$$

$$a^{-x} = \frac{1}{a^x}$$

$$(a^x)^y = a^{xy}$$

$$\sqrt{a} = a^{1/2}$$

$$\sqrt[n]{ab} = \sqrt[n]{a}\,\sqrt[n]{b}$$

$$a^x a^y = a^{x+y}$$

$$(ab)^x = a^x b^x$$

$$\sqrt[n]{a} = a^{1/n}$$

$$\sqrt[n]{\left(\frac{a}{b}\right)} = \frac{\sqrt[n]{a}}{\sqrt[n]{b}}$$

Algebraic Errors to Avoid:

$$\frac{a}{x + b} \neq \frac{a}{x} + \frac{a}{b}$$

(To see this error, let $a = b = x = 1$.)

$$\sqrt{x^2 + a^2} \neq x + a$$

(To see this error, let $x = 3$ and $a = 4$.)

$$a - b(x - 1) \neq a - bx - b$$

[Remember to distribute negative signs. The equation should be $a - b(x - 1) = a - bx + b$.]

$$\frac{\left(\dfrac{x}{a}\right)}{b} \neq \frac{bx}{a}$$

[To divide fractions, invert and multiply. The equation should be

$$\frac{\left(\dfrac{x}{a}\right)}{b} = \frac{\left(\dfrac{x}{a}\right)}{\left(\dfrac{b}{1}\right)} = \left(\frac{x}{a}\right)\left(\frac{1}{b}\right) = \frac{x}{ab}.]$$

$$\sqrt{-x^2 + a^2} \neq -\sqrt{x^2 - a^2}$$

(The negative sign cannot be factored out of the square root.)

$$\frac{a + bx}{a} \neq 1 + bx$$

(This is one of many examples of incorrect cancellation. The equation should be $\dfrac{a + bx}{a} = \dfrac{a}{a} + \dfrac{bx}{a} = 1 + \dfrac{bx}{a}$.)

$$\frac{1}{x^{1/2} - x^{1/3}} \neq x^{-1/2} - x^{-1/3}$$

(This error is a more complex version of the first error.)

$$(x^2)^3 \neq x^5$$

[This equation should be $(x^2)^3 = x^2 x^2 x^2 = x^6$.]

Conversion Table:

1 centimeter ≈ 0.394 inch	1 joule ≈ 0.738 foot-pound	1 mile ≈ 1.609 kilometers
1 meter ≈ 39.370 inches	1 gram ≈ 0.035 ounce	1 gallon ≈ 3.785 liters
≈ 3.281 feet	1 kilogram ≈ 2.205 pounds	1 pound ≈ 4.448 newtons
1 kilometer ≈ 0.621 mile	1 inch ≈ 2.540 centimeters	1 foot-lb ≈ 1.356 joules
1 liter ≈ 0.264 gallon	1 foot ≈ 30.480 centimeters	1 ounce ≈ 28.350 grams
1 newton ≈ 0.225 pound	≈ 0.305 meter	1 pound ≈ 0.454 kilogram

Trigonometry

Sixth Edition

Instructor's Annotated Edition

Trigonometry

Sixth Edition

▶ **Ron Larson**
▶ **Robert P. Hostetler**

The Pennsylvania State University
The Behrend College

▶ **With the assistance of David C. Falvo**

The Pennsylvania State University
The Behrend College

Houghton Mifflin Company   **Boston   New York**

Publisher : Jack Shira
Managing Editor: Cathy Cantin
Development Manager: Maureen Ross
Development Editor: Laura Wheel
Assistant Editor: Jennifer King
Assistant Editor: James Cohen
Supervising Editor: Karen Carter
Senior Project Editor: Patty Bergin
Production Technology Supervisor: Gary Crespo
Senior Marketing Manager: Danielle Potvin
Marketing Associate: Nicole Mollica
Senior Manufacturing Coordinator: Jane Spelman
Composition and Art: Meridian Creative Group
Cover Design Manager: Diana Coe

Cover Image: © Ralph Mercer

Printed in the U.S.A.

Library of Congress Catalog Card Number: 2002109502

ISBN: 0-618-31796-1

123456789–DOW–07 06 05 04 03

Contents

Contents ▶

CONTENTS

A Word from the Authors

Welcome to *Trigonometry*, Sixth Edition. In this revision we continue to focus on promoting student success, while providing an accessible text that offers flexible teaching and learning options.

In keeping with our philosophy that students learn best when they know what they are expected to learn, we have retained the thematic study thread from the Fifth Edition. We first introduce this study thread in the Chapter Opener. Each chapter begins with a study guide that contains a comprehensive overview of the chapter concepts (*What you should learn*), a list of *Important Vocabulary* integral to learning the chapter concepts, a list of additional chapter-specific *Study Tools*, and additional text-specific resources. The study guide allows students to get organized and prepare for the chapter. Then, each section opens with a a set of learning objectives outlining the concepts and skills students are expected to learn (*What you should learn*), followed by an interesting real-life application used to illustrate why it is important to learn the concepts in that section (*Why you should learn it*). *Study Tips* at point-of-use provide support as students read through the section. And finally, to provide study support and a comprehensive review of the chapter, each chapter concludes with a chapter summary (*What did you learn?*), which reinforces the section objectives, and chapter *Review Exercises*, which are correlated to the chapter summary.

In addition to providing in-text study support, we have taken care to write a text for the student. We paid careful attention to the presentation, using precise mathematical language and clear writing, to create an effective learning tool. We are committed to providing a text that makes the mathematics within it accessible to all students. In the Sixth Edition, we have revised and improved upon many text features designed for this purpose. The *Technology, Exploration* features have been expanded. *Chapter Tests*, which gave students an opportunity for self-assessment, are included in every chapter. We have retained the *Synthesis* exercises, which check students' conceptual understanding, and the *Review* exercises, which reinforce skills learned in previous sections within each section exercise set. Also, students have access to several media resources that offer additional text-specific resources to enhance the learning process.

From the time we first began writing in the early 1970s, we have always viewed part of our authoring role as that of providing instructors with flexible teaching programs. The optional features within the text allow instructors with different pedagogical approaches to design their course to meet both their instructional needs and the needs of their students. Instructors who stress applications and problem solving, or exploration and technology, and more traditional methods will be able to use this text successfully. We hope you enjoy the Sixth Edition.

Ron Larson

Ron Larson

Robert P. Hostetler

Robert P. Hostetler

Acknowledgments

We would like to thank the many people who helped us at various stages of this project. Their encouragement, criticisms, and suggestions have been invaluable to us.

Sixth Edition Reviewers

Ahmad Abusaid, Southern Polytechnic University; Catherine Banks, Texas Woman's College; Jared Burch, College of the Sequoias; Dr. Michelle R. DeDeo, University of North Florida; Gangadhar R. Hiremath, Miles College; Dr. Kevin W. Hopkins, Southwest Baptist University; Charles W. Johnson, South Georgia College; Gary S. Kersting, North Central Michigan College; Namyong Lee, Minnesota State University; Mary Leeseberg, Manatee Community College; Tristan Londré, Blue River Community College; Bruce N. Lundberg, University of Southern Colorado; Rudy Maglio, Oakton Community College; Steve O'Donnell, Rogue Community College; Armando I. Perez, Laredo Community College; Rita Randolfi, Brevard Community College; David Ray, The University of Tennessee at Martin; Miguel San Miguel Gonzalez, Texas A&M International University; Scott Satake, North Idaho College; Jed Soifer, Atlantic Cape Community College; Dr. Roy N. Tucker, Palo Alto College and The University of Texas at San Antonio; Karen Villarreal, Xavier University of Louisiana; Carol Walker, Hinds Community College; J. Lewis Walston, Methodist College; Jun Wang, Alabama State University; Ibrahim Wazir, American International School; Robert Wylie, Carl Albert State College

We would like to extend a special thanks to all of the instructors who took time to participate in our phone interviews.

We would like to thank the staff of Larson Texts, Inc. and the staff of Meridian Creative Group, who assisted in proofreading the manuscript, preparing and proofreading the art package, and typesetting the supplements.

We are grateful to our wives, Deanna Gilbert Larson and Eloise Hostetler, for their love, patience, and support. Also, a special thanks goes to R. Scott O'Neil.

If you have suggestions for improving this text, please feel free to write to us. Over the years we have received many useful comments from both instructors and students, and we value these comments very much.

Ron Larson
Robert P. Hostetler

How can this book help you

Support for Student Success

- Larson provides clear, easy-to-read examples that include all the steps needed to understand a new concept.
- Numerous examples are provided throughout the book that correspond to the exercise sets, giving students support with the key concepts in their homework assignments.
- Additional resources are also available, such as SMARTHINKING's live, one-on-one online tutoring service. This enables students to receive tutorial help from the comfort and privacy of their own home.
- Key course material is also presented on a DVD by a qualified instructor, making it easy to review content or material missed due to an absence.

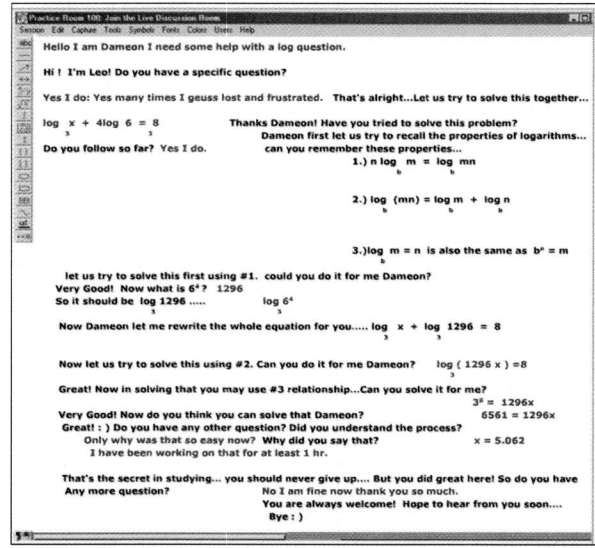

Options for Students and Instructors

- Concepts are presented through examples, applications, technology, or explorations to adapt the course to the curriculum needs or student learning styles.
- A variety of exercises that increase in difficulty allows professors the flexibility to assign homework to students with various learning styles. Exercise options include skills, technology, critical thinking, writing, applications, modeling data, true/false, proofs, and theoretical questions.
- The P.S. Problem Solving section at the end of every chapter offers more challenging exercises for advanced students.
- This text provides a solid mathematical foundation by foreshadowing concepts that will be used in future courses. Topics that will be especially helpful to students in Calculus are labeled with an "Algebra of Calculus" $\int$ icon.

Exploration

Graph each of the functions with a graphing utility. Determine whether the function is *even*, *odd*, or *neither*.

Technology

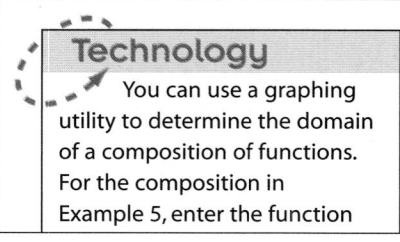

You can use a graphing utility to determine the domain of a composition of functions. For the composition in Example 5, enter the function

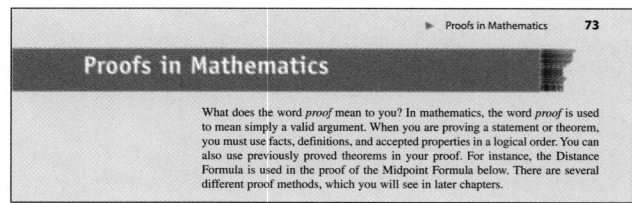

Proofs in Mathematics

What does the word *proof* mean to you? In mathematics, the word *proof* is used to mean simply a valid argument. When you are proving a statement or theorem, you must use facts, definitions, and accepted properties in a logical order. You can also use previously proved theorems in your proof. For instance, the Distance Formula is used in the proof of the Midpoint Formula below. There are several different proof methods, which you will see in later chapters.

succeed in your math course?

For more information, see pages xii–xvi.

Applications That Motivate Students

- Applications in the exposition, examples, and exercises use real life data for students to see the relevance of what they are learning.
- Interesting topics are included throughout the book to help students see the practical, as well as theoretical, side of mathematics.
- Sourced data sets are included throughout the text, allowing students the opportunity to generate mathematical models that represent real data.

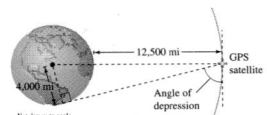

Readable and Understandable Text for Students

- Examples, explanations, and proofs begin and end on the same page to allow students to see concepts as a whole, without page-turning distractions. This unique design is one more example of the carefully developed texts created by the Larson Team.
- Examples include detailed solutions that show all steps to make it easy for students to understand the material being presented.
- Many examples include numerical, algebraic, and/or graphical presentations to provide students an opportunity to see the solution represented in a way that is most clear to them.

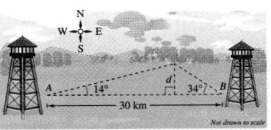

FEATURES

Student Success Tools

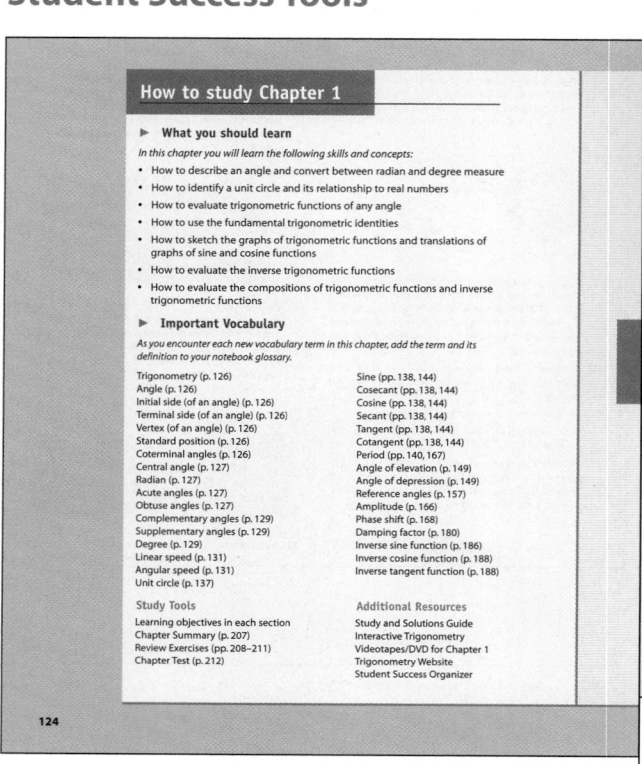

How to study Chapter 1

▶ **What you should learn**

In this chapter you will learn the following skills and concepts:

- How to describe an angle and convert between radian and degree measure
- How to identify a unit circle and its relationship to real numbers
- How to evaluate trigonometric functions of any angle
- How to use the fundamental trigonometric identities
- How to sketch the graphs of trigonometric functions and translations of graphs of sine and cosine functions
- How to evaluate the inverse trigonometric functions
- How to evaluate the compositions of trigonometric functions and inverse trigonometric functions

▶ **Important Vocabulary**

As you encounter each new vocabulary term in this chapter, add the term and its definition to your notebook glossary.

Trigonometry (p. 126)
Angle (p. 126)
Initial side (of an angle) (p. 126)
Terminal side (of an angle) (p. 126)
Vertex (of an angle) (p. 126)
Standard position (p. 126)
Coterminal angles (p. 126)
Central angle (p. 127)
Radian (p. 127)
Acute angles (p. 127)
Obtuse angles (p. 127)
Complementary angles (p. 129)
Supplementary angles (p. 129)
Degree (p. 129)
Linear speed (p. 131)
Angular speed (p. 131)
Unit circle (p. 137)

Sine (pp. 138, 144)
Cosecant (pp. 138, 144)
Cosine (pp. 138, 144)
Secant (pp. 138, 144)
Tangent (pp. 138, 144)
Cotangent (pp. 138, 144)
Period (pp. 140, 167)
Angle of elevation (p. 149)
Angle of depression (p. 149)
Reference angles (p. 157)
Amplitude (p. 166)
Phase shift (p. 168)
Damping factor (p. 180)
Inverse sine function (p. 186)
Inverse cosine function (p. 188)
Inverse tangent function (p. 188)

Study Tools

Learning objectives in each section
Chapter Summary (p. 207)
Review Exercises (pp. 208–211)
Chapter Test (p. 212)

Additional Resources

Study and Solutions Guide
Interactive Trigonometry
Videotapes/DVD for Chapter 1
Trigonometry Website
Student Success Organizer

124

Trigonometry

1.1 **Radian and Degree Measure**
1.2 **Trigonometric Functions: The Unit Circle**
1.3 **Right Triangle Trigonometry**
1.4 **Trigonometric Functions of Any Angle**
1.5 **Graphs of Sine and Cosine Functions**
1.6 **Graphs of Other Trigonometric Functions**
1.7 **Inverse Trigonometric Functions**
1.8 **Applications and Models**

"How to Study This Chapter"

The chapter-opening study guide includes: *What you should learn*, an objective-based overview of the main concepts of the chapter, *Important Vocabulary*, key mathematical terms integral to learning the concepts outlined in *What you should learn*, a list of *Study Tools*, additional study resources within the text chapter, and *Additional Resources*, text-specific supplemental resources available for each chapter.

Section Openers include: "What you should learn"

A list of section objectives outlining the main concepts to help students focus while reading through the section.

"Why you should learn it"

A real-life application or a reference to other branches of mathematics illustrates the relevance of the section's content. The real-life application is showcased in *Model It* found in the section exercise set.

---

144   Chapter 1   ▶   Trigonometry

1.3   Right Triangle Trigonometry

▶ **What you should learn**

- How to evaluate trigonometric functions of acute angles
- How to use the fundamental trigonometric identities
- How to use a calculator to evaluate trigonometric functions
- How to use trigonometric functions to model and solve real-life problems

▶ **Why you should learn it**

Trigonometric functions are often used to analyze real-life situations. For instance, in Exercise 63 on page 153, you are asked to use trigonometric functions to find the height of a helium-filled balloon.

The Six Trigonometric Functions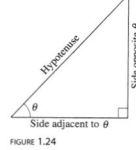

Our second look at the trigonometric functions is from a *right triangle* perspective. Consider a right triangle, with one acute angle labeled θ, as shown in Figure 1.24. Relative to the angle θ, the three sides of the triangle are the **hypotenuse,** the **opposite side** (the side opposite the angle θ), and the **adjacent side** (the side adjacent to the angle θ).

FIGURE 1.24

Using the lengths of these three sides, you can form six ratios that define the six trigonometric functions of the acute angle θ.

sine   cosecant   cosine   secant   tangent   cotangent

In the following definition, it is important to see that $0° < \theta < 90°$ and that for such angles the value of each trigonometric function is *positive*.

Right Triangle Definitions of Trigonometric Functions

Let θ be an *acute* angle of a right triangle. The six trigonometric functions of the angle θ are defined as follows. (Note that the functions in the second row are the *reciprocals* of the corresponding functions in the first row.)

$$\sin \theta = \frac{\text{opp}}{\text{hyp}} \qquad \cos \theta = \frac{\text{adj}}{\text{hyp}} \qquad \tan \theta = \frac{\text{opp}}{\text{adj}}$$

$$\csc \theta = \frac{\text{hyp}}{\text{opp}} \qquad \sec \theta = \frac{\text{hyp}}{\text{adj}} \qquad \cot \theta = \frac{\text{adj}}{\text{opp}}$$

The abbreviations opp, adj, and hyp represent the lengths of the three sides of a right triangle.

opp = the length of the side *opposite* θ

adj = the length of the side *adjacent to* θ

hyp = the length of the *hypotenuse*

(Reproduction of page 153)

Section 1.3 ▶ Right Triangle Trigonometry   **153**

▶ **Model It**

63. Height A 20-meter line is used to tether a helium-filled balloon. Because of a breeze, the line makes an angle of approximately 85° with the ground.

(a) Draw a right triangle that gives a visual representation of the problem. Show the known quantities of the triangle and use a variable to indicate the height of the balloon.

(b) Use a trigonometric function to write an equation involving the unknown quantity.

(c) What is the height of the balloon?

(d) The breeze becomes stronger and the angle the balloon makes with the ground decreases. How does this affect the triangle you drew in part (a)?

(e) Complete the table, which shows the height (in meters) of the balloon for decreasing angle measures θ.

Angle, θ	80°	70°	60°	50°
Height				

Angle, θ	40°	30°	20°	10°
Height				

(f) As the angle the balloon makes with the ground approaches 0°, how does this affect the height of the balloon? Draw a right triangle to explain your reasoning.

64. Angle of Elevation A ramp 20 feet in length rises to a loading platform that is 3⅓ feet off the ground.

(a) Draw a right triangle that gives a visual representation of the problem. Show the known quantities of the triangle and use a variable to indicate the angle of elevation of the ramp.

(b) Use a trigonometric function to write an equation involving the unknown quantity.

(c) What is the angle of elevation of the ramp?

65. Width of a River A biologist wants to know the width *w* of a river in order to set instruments for studying the pollutants in the water. From point *A*, the biologist walks downstream 100 feet and sights to point *C* (see figure). From this sighting, it is determined that θ = 54°. How wide is the river?

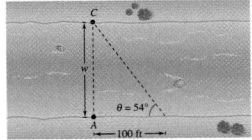

66. Height of a Mountain In traveling across flat land, you notice a mountain directly in front of you. Its angle of elevation (to the peak) is 3.5°. After you drive 13 miles closer to the mountain, the angle of elevation is 9°. Approximate the height of the mountain.

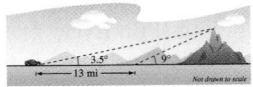

67. Machine Shop Calculations A steel plate has the form of one-fourth of a circle with a radius of 60 centimeters. Two two-centimeter holes are to be drilled in the plate positioned as shown in the figure. Find the coordinates of the center of each hole.

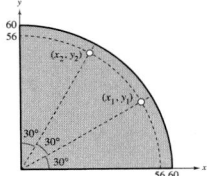

(Reproduction of page 207)

▶ Chapter Summary   **207**

Chapter Summary

▶ **What did you learn?**

	Review Exercises
Section 1.1	
☐ How to describe angles	1–4
☐ How to use radian and degree measure	5–20
☐ How to use angles to model and solve real-life problems	21, 22
Section 1.2	
☐ How to identify a unit circle and its relationship to real numbers	23–26
☐ How to evaluate trigonometric functions using the unit circle	27–30
☐ How to use the domain and period to evaluate sine and cosine functions	31–34
☐ How to use a calculator to evaluate trigonometric functions	35–38
Section 1.3	
☐ How to evaluate trigonometric functions of acute angles	39–42
☐ How to use the fundamental trigonometric identities	43–46
☐ How to use a calculator to evaluate trigonometric functions	47–52
☐ How to use trigonometric functions to model and solve real-life problems	53, 54
Section 1.4	
☐ How to evaluate trigonometric functions of any angle	55–68
☐ How to use reference angles to evaluate trigonometric functions	69–76
☐ How to evaluate trigonometric functions of real numbers	77–82
Section 1.5	
☐ How to use amplitude and period to sketch the graphs of sine and cosine functions	83–86
☐ How to sketch translations of graphs of sine and cosine functions	87–90
☐ How to use sine and cosine functions to model real-life data	91, 92
Section 1.6	
☐ How to sketch the graphs of tangent and cotangent functions	93–96
☐ How to sketch the graphs of secant and cosecant functions	97–100
☐ How to sketch the graphs of damped trigonometric functions	101, 102
Section 1.7	
☐ How to evaluate the inverse sine function	103–108
☐ How to evaluate the other inverse trigonometric functions	109–120
☐ How to evaluate the compositions of trigonometric functions	121–128
Section 1.8	
☐ How to solve real-life problems involving right triangles	129, 130
☐ How to solve real-life problems involving directional bearings	131
☐ How to solve real-life problems involving harmonic motion	132

(Reproduction of page 208)

208   Chapter 1 ▶ Trigonometry

Review Exercises

1.1 In Exercises 1–4, estimate the angle to the nearest one-half radian.

1. 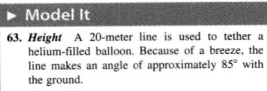   2.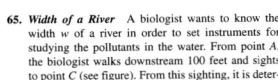

3.   4.

In Exercises 5–12, sketch the angle in standard position. List one positive and one negative coterminal angle.

5. $\frac{11\pi}{4}$ 6. $\frac{2\pi}{9}$

7. $-\frac{4\pi}{3}$ 8. $-\frac{23\pi}{3}$

9. 70°   10. 280°

11. −110°   12. −405°

In Exercises 13–16, convert the measure from radians to degrees. Round your answer to two decimal places.

13. $\frac{5\pi}{7}$ 14. $-\frac{11\pi}{6}$

15. −3.5   16. 5.7

In Exercises 17–20, convert the measure from degrees to radians. Round your answer to four decimal places.

17. 480°   18. −127.5°

19. −33° 45′   20. 196° 77′

21. Phonograph Compact discs have all but replaced phonograph records. Phonograph records are vinyl discs that rotate on a turntable. A typical record album is 12 inches in diameter and plays at 33⅓ revolutions per minute.

(a) What is the angular speed of a record album?

(b) What is the linear speed of the outer edge of a record album?

22. Bicycle At what speed is a bicyclist traveling when his 27-inch-diameter tires are rotating at an angular speed of 5π radians per second?

1.2 In Exercises 23–26, find the point (x, y) on the unit circle that corresponds to the real number *t*.

23. $t = \frac{2\pi}{3}$ 24. $t = \frac{3\pi}{4}$

25. $t = \frac{5\pi}{6}$ 26. $t = -\frac{4\pi}{3}$

In Exercises 27–30, evaluate (if possible) the six trigonometric functions of the real number.

27. $t = \frac{7\pi}{6}$ 28. $t = \frac{\pi}{4}$

29. $t = -\frac{2\pi}{3}$ 30. $t = 2\pi$

In Exercises 31–34, evaluate the trigonometric function using its period as an aid.

31. $\sin \frac{11\pi}{4}$ 32. $\cos 4\pi$

33. $\sin\left(-\frac{17\pi}{6}\right)$ 34. $\cos\left(-\frac{13\pi}{6}\right)$

In Exercises 35–38, use a calculator to evaluate the trigonometric function. Round your answer to two decimal places.

35. tan 33   36. csc 10.5

37. $\sec \frac{12\pi}{5}$ 38. $\sin\left(-\frac{\pi}{9}\right)$

1.3 In Exercises 39–42, find the exact values of the six trigonometric functions of the angle θ shown in the figure.

39. 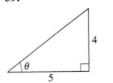   40.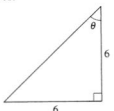

NEW! Model It

Often involving real-life data, these multi-part applications, referenced in *Why you should learn it*, offer students the opportunity to generate and analyze mathematical models.

"What did you learn?" Chapter Summary

The chapter summary provides a concise, section-by-section review of the section objectives. These objectives are correlated to the chapter Review Exercises allowing students to identify sections and concepts needing further review and study.

Review Exercises

Following the chapter summary, the Review Exercises provide additional practice and review of chapter concepts. The Review Exercises are organized by section and keyed directly to the section objectives listed in the chapter summary.

Additional **Student Success Tools** include point-of-use *Study Tips* and *Chapter* and *Cumulative Tests*.

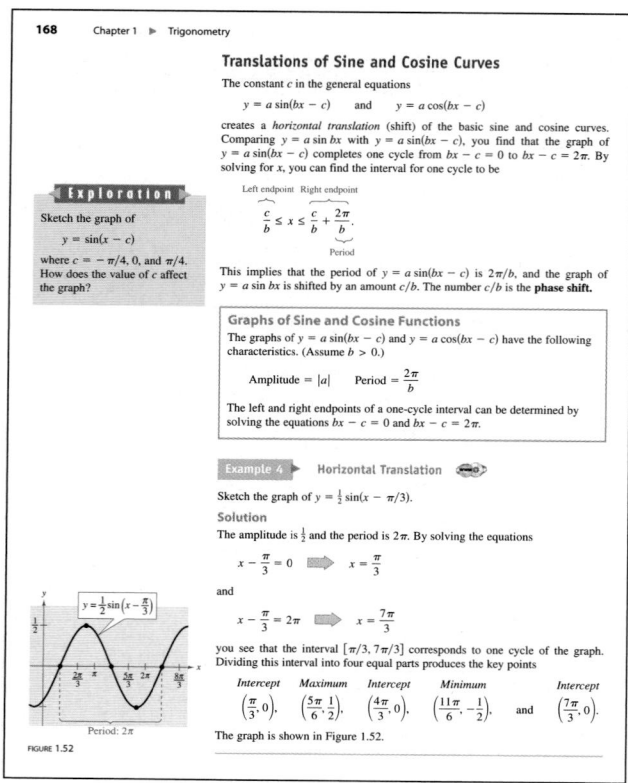

Exploration

Before introducing selected topics, *Explorations* engage students in active discovery of mathematical concepts and relationships, often through the power of technology, while strengthening their critical thinking skills and developing an intuitive understanding of theoretical concepts.

Examples

Each example was carefully chosen to illustrate a particular mathematical concept or problem solving skill. Every example contains step-by-step solutions, most with side-by-side explanations that lead students through the solution process.

Technology

Point-of-use instructions for graphing utilities appear in the margin. Emphasis is placed on using technology as a tool for visualizing mathematical concepts, for verifying solutions, and for facilitating mathematical computation. The use of technology is optional and this feature and related exercises, identified by the icon , can be omitted without loss of continuity in coverage of topics.

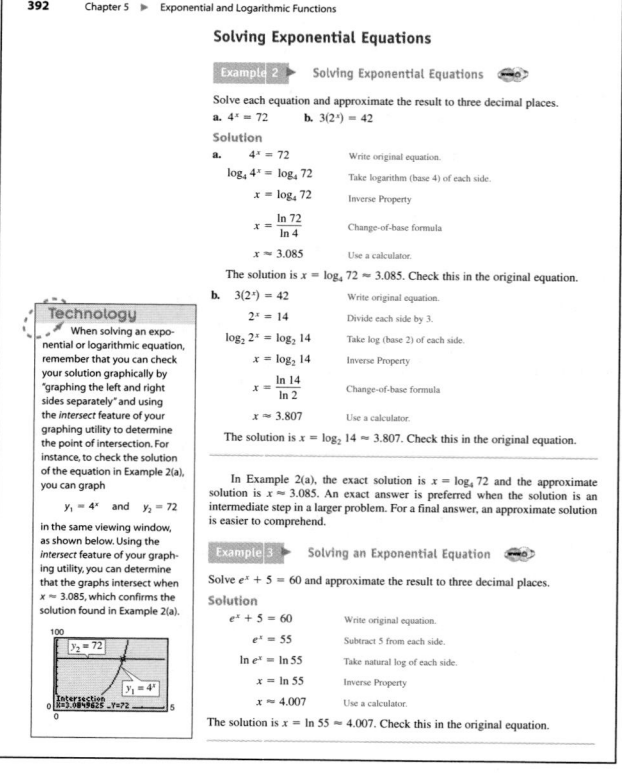

P.S. Problem Solving

1. The restaurant at the top of the Space Needle in Seattle, Washington is circular and has a radius of 47.25 feet. The dining part of the restaurant revolves, making about one complete revolution every 48 minutes. A dinner party was seated at the edge of the revolving restaurant at 6:45 P.M. and was finished at 8:57 P.M.
   (a) Find the angle through which the dinner party rotated.
   (b) Find the distance the party traveled during dinner.

2. A bicycle's gear ratio is the number of times the freewheel turns for every one turn of the chainwheel (see figure). The table shows the numbers of teeth in the freewheel and chainwheel for the first five gears of an 18-speed touring bicycle. The chainwheel completes one rotation for each gear. Find the angle through which the freewheel turns for each gear. Give your answers in both degrees and radians.

Gear number	Number of teeth in freewheel	Number of teeth in chainwheel
1	32	24
2	26	24
3	22	24
4	32	40
5	19	24

Freewheel

Chainwheel

3. A surveyor in a helicopter is trying to determine the width of an island, as shown in the figure.
   (a) What is the shortest distance d the helicopter would have to travel to land on the island?
   (b) What is the horizontal distance x that the helicopter would have to travel before it would be directly over the nearer end of the island?
   (c) Find the width w of the island. Explain how you obtained your answer.

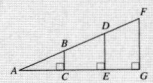

FIGURE FOR 3

Not drawn to scale

4. Use the figure below.

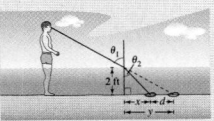

(a) Explain why $\triangle ABC$, $\triangle ADE$, and $\triangle AFG$ are similar triangles.
(b) What does similarity imply about the ratios
$$\frac{BC}{AB}, \frac{DE}{AD}, \text{ and } \frac{FG}{AF}?$$
(c) Does the value of $\sin A$ depend on which triangle from part (a) is used to calculate it? Would the value of $\sin A$ change if it were found using a different right triangle that was similar to the three given triangles?
(d) Do your conclusions from part (c) apply to the other five trigonometric functions? Explain.

5. Use a graphing utility to graph h, and use the graph to decide whether h is even, odd, or neither.
   (a) $h(x) = \cos^2 x$ (b) $h(x) = \sin^2 x$

6. If f is an even function and g is an odd function, use the results of Exercise 5 to make a conjecture about h where
   (a) $h(x) = [f(x)]^2$ (b) $h(x) = [g(x)]^2$.

7. The model for the height h of a Ferris wheel car is
$$h = 50 + 50 \sin 8\pi t$$
where t is the time in minutes. (The Ferris wheel has a radius of 50 feet.) This model yields a height of 50 feet when $t = 0$. Alter the model so that the height of the car is 1 foot when $t = 0$.

8. A popular theory that attempts to explain the ups and downs of everyday life states that each of us has three cycles, called biorhythms, which begin at birth. These three cycles can be modeled by sine waves.

Physical (23 days): $P = \sin \frac{2\pi t}{23}$, $t \geq 0$

Emotional (28 days): $E = \sin \frac{2\pi t}{28}$, $t \geq 0$

Intellectual (33 days): $I = \sin \frac{2\pi t}{33}$, $t \geq 0$

where t is the number of days since birth. Consider a person who was born on July 20, 1984.
(a) Use a graphing utility to graph the three models in the same viewing window for $7300 \leq t \leq 7380$.
(b) Describe the person's biorhythms during the month of September 2004.
(c) Calculate the person's three energy levels on September 22, 2004.

9. (a) Use a graphing utility to graph the functions
$$f(x) = 2 \cos 2x + 3 \sin 3x$$
and
$$g(x) = 2 \cos 2x + 3 \sin 4x.$$
(b) Use the graphs from part (a) to find the period of each function.
(c) If α and β are positive integers, is the function
$$h(x) = A \cos \alpha x + B \sin \beta x$$
periodic? Explain your reasoning.

10. Two trigonometric functions f and g have periods of 2, and their graphs intersect at $x = 5.35$.
(a) Give one smaller and one larger positive value of x at which the functions have the same value.
(b) Determine one negative value of x at which the graphs intersect.
(c) Is it true that $f(13.35) = g(-4.65)$? Explain your reasoning.

11. The function f is periodic, with period c. So, $f(t + c) = f(t)$. Are the following equal? Explain.
(a) $f(t - 2c) = f(t)$ (b) $f(t + \frac{1}{2}c) = f(\frac{1}{2}t)$
(c) $f(\frac{1}{2}(t + c)) = f(\frac{1}{2}t)$

12. If you stand in shallow water and look at an object below the surface of the water, the object will look farther away from you than it really is. This is because when light rays pass between air and water, the water refracts, or bends, the light rays. The index of refraction for water is 1.333. This is the ratio of the sine of θ_1 and the sine of θ_2 (see figure).
(a) You are standing in water that is 2 feet deep and are looking at a rock at angle $\theta_1 = 60°$ (measured from a line perpendicular to the surface of the water). Find θ_2.
(b) Find the distances x and y.
(c) Find the distance d between where the rock is and where it appears to be.
(d) What happens to d as you move closer to the rock? Explain your reasoning.

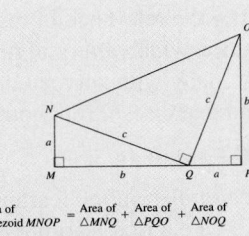

13. In calculus it can be shown that the arctangent function can be approximated by the polynomial
$$\arctan x \approx x - \frac{x^3}{3} + \frac{x^5}{5} - \frac{x^7}{7}$$
where x is in radians.
(a) Use a graphing utility to graph the arctangent function and its polynomial approximation in the same viewing window. How do the graphs compare?
(b) Study the pattern in the polynomial approximation of the arctangent function and guess the next term. Then repeat part (a). How does the accuracy of the approximation change when additional terms are added?

NEW! P.S. Problem Solving

Each chapter concludes with a collection of thought-provoking and challenging exercises that further explore and expand upon the chapter concepts. These exercises have unusual characteristics that set them apart from traditional text exercises.

NEW! Proofs in Mathematics

At the end of every chapter, Proofs in Mathematics emphasizes the importance of proofs in mathematics. Proofs of important mathematical properties and theorems are presented as well as discussions of various proof techniques.

▶ Proofs in Mathematics   213

Proofs in Mathematics

The Pythagorean Theorem

The Pythagorean Theorem is one of the most famous theorems in mathematics. More than 100 different proofs now exist. James A. Garfield, the twentieth president of the United States, developed a proof of the Pythagorean Theorem in 1876. His proof, shown below, involved the fact that a trapezoid can be formed from two congruent right triangles and an isosceles right triangle.

The Pythagorean Theorem

In a right triangle, the sum of the squares of the lengths of the legs is equal to the square of the length of the hypotenuse, where a and b are the legs and c is the hypotenuse.

$$a^2 + b^2 = c^2$$

Proof

$$\text{Area of trapezoid } MNOP = \text{Area of } \triangle MNQ + \text{Area of } \triangle PQO + \text{Area of } \triangle NOQ$$

$$\frac{1}{2}(a + b)(a + b) = \frac{1}{2}ab + \frac{1}{2}ab + \frac{1}{2}c^2$$

$$\frac{1}{2}(a + b)(a + b) = ab + \frac{1}{2}c^2$$

$$(a + b)(a + b) = 2ab + c^2$$

$$a^2 + 2ab + b^2 = 2ab + c^2$$

$$a^2 + b^2 = c^2$$

1.3 Exercises

In Exercises 1–4, find the exact values of the six trigonometric functions of the angle θ shown in the figure. (Use the Pythagorean Theorem to find the third side of the triangle.)

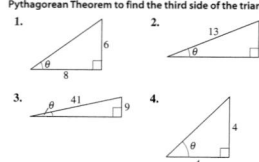

In Exercises 5–8, find the exact values of the six trigonometric functions of the angle θ for each of the two triangles. Explain why the function values are the same.

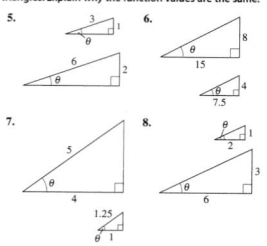

In Exercises 9–16, sketch a right triangle corresponding to the trigonometric function of the acute angle θ. Use the Pythagorean Theorem to determine the third side and then find the other five trigonometric functions of θ.

9. $\sin \theta = \frac{3}{4}$ 10. $\cos \theta = \frac{5}{7}$
11. $\sec \theta = 2$ 12. $\cot \theta = 5$
13. $\tan \theta = 3$ 14. $\sec \theta = 6$
15. $\cot \theta = \frac{3}{2}$ 16. $\csc \theta = \frac{17}{4}$

In Exercises 17–22, use the given function value(s), and trigonometric identities (including the cofunction identities), to find the indicated trigonometric functions.

17. $\sin 60° = \frac{\sqrt{3}}{2},\quad \cos 60° = \frac{1}{2}$
   (a) tan 60°   (b) sin 30°
   (c) cos 30°   (d) cot 60°

18. $\sin 30° = \frac{1}{2},\quad \tan 30° = \frac{\sqrt{3}}{3}$
   (a) csc 30°   (b) cot 60°
   (c) cos 30°   (d) cot 30°

19. $\csc \theta = \frac{\sqrt{13}}{2},\quad \sec \theta = \frac{\sqrt{13}}{3}$
   (a) sin θ   (b) cos θ
   (c) tan θ   (d) sec(90° − θ)

20. $\sec \theta = 5,\quad \tan \theta = 2\sqrt{6}$
   (a) cos θ   (b) cot θ
   (c) cot(90° − θ)   (d) sin θ

21. $\cos \alpha = \frac{1}{3}$
   (a) sec α   (b) sin α
   (c) cot α   (d) sin(90° − α)

22. $\tan \beta = 5$
   (a) cot β   (b) cos β
   (c) tan(90° − β)   (d) csc β

In Exercises 23–26, evaluate the trigonometric function by memory or by constructing an appropriate triangle for the given special angle.

23. (a) cos 60°   (b) csc 30°   (c) tan 60°
24. (a) cot 45°   (b) cos 45°   (c) csc 45°
25. (a) sin 45°   (b) cos 30°   (c) tan 30°
26. (a) sin 60°   (b) tan 45°   (c) sec 30°

In Exercises 27–36, use a calculator to evaluate each function. Round your answers to four decimal places. (Be sure the calculator is in the correct angle mode.)

27. (a) sin 10°   (b) cos 80°
28. (a) tan 23.5°   (b) cot 66.5°
29. (a) sin 16.35°   (b) csc 16.35°

---

75. *Fuel Consumption* The daily consumption C (in gallons) of diesel fuel on a farm is modeled by

$$C = 30.3 + 21.6 \sin\left(\frac{2\pi t}{365} + 10.9\right)$$

where t is the time in days, with t = 1 corresponding to January 1.
   (a) What is the period of the model? Is it what you expected? Explain.
   (b) What is the average daily fuel consumption? Which term of the model did you use? Explain.
   (c) Use a graphing utility to graph the model. Use the graph to approximate the time of the year when consumption exceeds 40 gallons per day.

Synthesis

True or False? In Exercises 76–78, determine whether the statement is true or false. Justify your answer.

76. The graph of the function $f(x) = \sin(x + 2\pi)$ translates the graph of $f(x) = \sin x$ exactly one period to the right so that the two graphs look identical.

77. The function $y = \frac{1}{2}\cos 2x$ has an amplitude that is twice that of the function $y = \cos x$.

78. The graph of $y = -\cos x$ is a reflection of the graph of $y = \sin(x + \pi/2)$ in the x-axis.

Conjecture In Exercises 79 and 80, graph f and g on the same set of coordinate axes. Include two full periods. Make a conjecture about the functions.

79. $f(x) = \sin x,\quad g(x) = \cos\left(x - \frac{\pi}{2}\right)$

80. $f(x) = \sin x,\quad g(x) = -\cos\left(x + \frac{\pi}{2}\right)$

81. *Writing* Use a graphing utility to graph the function $y = a \sin x$ for $a = \frac{1}{2}$, $a = 2$, and $a = -3$. Write a paragraph describing the changes in the graph corresponding to the specified changes in a.

82. *Writing* Use a graphing utility to graph the function $y = d + \sin x$ for $d = 2$, $d = 3.5$, and $d = -2$. Write a paragraph describing the changes in the graph corresponding to the specified changes in d.

83. *Writing* Use a graphing utility to graph the function $y = \sin bx$ for $b = \frac{1}{2}$, $b = \frac{3}{2}$, and $b = 4$. Write a paragraph describing the changes in the graph corresponding to the specified changes in b.

84. *Writing* Use a graphing utility to graph the function $y = \sin(x - c)$ for $c = 1$, $c = 3$, and $c = -2$. Write a paragraph describing the changes in the graph corresponding to the specified changes in c.

85. *Exploration* Using calculus, it can be shown that the sine and cosine functions can be approximated by the polynomials

$$\sin x \approx x - \frac{x^3}{3!} + \frac{x^5}{5!} \quad\text{and}\quad \cos x \approx 1 - \frac{x^2}{2!} + \frac{x^4}{4!}$$

where x is in radians.
   (a) Use a graphing utility to graph the sine function and its polynomial approximation in the same viewing window. How do the graphs compare?
   (b) Use a graphing utility to graph the cosine function and its polynomial approximation in the same viewing window. How do the graphs compare?
   (c) Study the patterns in the polynomial approximations of the sine and cosine functions and guess the next term in each. Then repeat parts (a) and (b). How did the accuracy of the approximations change when additional terms were added?

86. *Exploration* Use the polynomial approximations for the sine and cosine functions from Exercise 85 to approximate the following functional values. Compare the results with those given by a calculator. Is the error in the approximation the same in each case? Explain.
   (a) $\sin \frac{1}{2}$ (b) sin 1   (c) $\sin \frac{\pi}{6}$
   (d) cos(−0.5)   (e) cos 1   (f) $\cos \frac{\pi}{4}$

Review

In Exercises 87–90, perform the operations and simplify.

87. $\frac{4}{x} + \frac{4}{1-x}$ 88. $\frac{2}{x+5} - \frac{2}{x-5}$

89. $\frac{3}{x-1} - \frac{2}{x(x-1)}$ 90. $\frac{x}{x-5} + \frac{1}{2}$

In Exercises 91–94, find the domain of the function.

91. $f(x) = \frac{2}{11-x}$ 92. $f(x) = \frac{\sqrt{x-3}}{x-8}$

93. $f(x) = \sqrt{81 - x^2}$

94. $f(x) = \sqrt[3]{4 - x^2}$

Exercises

A hallmark feature of the text, the exercise sets contain a variety of computational, conceptual, and applied problems. Each section exercise set contains *Synthesis* exercises, which promote further exploration of mathematical concepts, critical thinking skills, and writing about mathematics and *Review* exercises, which provide continuous review of previously learned skills and concepts.

Applications

Demonstrating the relevance of mathematics to the real world, a wide variety of practical, real-life applications, many with sourced data, are found in examples and exercises throughout the text.

Additional Features

Additional carefully crafted learning tools designed to create a rich learning environment for all students can be found throughout the text. These learning tools include Historical Notes, Writing About Mathematics, Algebra of Calculus, and an extensive art program.

---

24. *Angle of Depression* A Global Positioning System satellite orbits 12,500 miles above Earth's surface. Find the angle of depression from the satellite to the horizon. Assume the radius of Earth is 4000 miles.

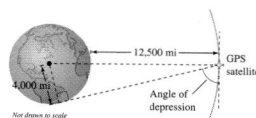

Not drawn to scale

25. *Angle of Depression* A cellular telephone tower that is 150 feet tall is placed on top of a mountain that is 1200 feet above sea level. What is the angle of depression from the top of the tower to a cell phone user who is 5 horizontal miles away and 400 feet above sea level?

26. *Airplane Ascent* During takeoff, an airplane's angle of climb is 18° and its speed is 275 feet per second.
   (a) Find the plane's altitude after 1 minute.
   (b) How long will it take the plane to climb to an altitude of 10,000 feet?

27. *Mountain Descent* A sign on a roadway at the top of a mountain indicates that for the next 4 miles the grade is 10.5° (see figure). Find the change in elevation for a car descending the mountain.

Not drawn to scale

28. *Mountain Descent* A roadway sign at the top of a mountain indicates that for the next 4 miles the grade is 12%. Find the angle of the grade and the change in elevation for a car descending the mountain.

29. *Navigation* An airplane flying at 600 miles per hour has a bearing of 52°. After flying for 1.5 hours, how far north and how far east will the plane have traveled from its point of departure?

30. *Surveying* A surveyor wishes to find the distance across a swamp (see figure). The bearing from A to B is N 32° W. The surveyor walks 50 meters from A, and at the point C the bearing to B is N 68° W. Find (a) the bearing from A to C and (b) the distance from A to B.

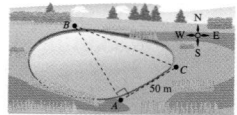

31. *Location of a Fire* Two fire towers are 30 kilometers apart, where tower A is due west of tower B. A fire is spotted from the towers, and the bearings from A and B are E 14° N and W 34° N, respectively (see figure). Find the distance d of the fire from the line segment AB.

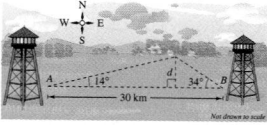

Not drawn to scale

32. *Navigation* A ship is 45 miles east and 30 miles south of port. The captain wants to sail directly to port. What bearing should be taken?

33. *Distance* An observer in a lighthouse 350 feet above sea level observes two ships directly offshore. The angles of depression to the ships are 4° and 6.5° (see figure). How far apart are the ships?

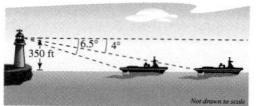

Not drawn to scale

Program Components

Trigonometry, Student Edition
Trigonometry, Instructor's Annotated Edition

Interactive Trigonometry 3.0 CD-ROM (can be used alone or with the printed textbook)
Internet Trigonometry 3.0 (can be used alone or with the printed textbook)

Additional Resources

Student Resources

Student Success Organizer

Study and Solutions Guide
   by Dianna L. Zook, (Indiana University/Purdue
   University–Fort Wayne)

Instructor Resources

Instructor Success Organizer

Complete Solutions Guide
   by Dianna L. Zook, (Indiana University/Purdue
   University–Fort Wayne)

Instructor's Annotated Edition

Test Item File

Student Technology Resources

Instructional Videotapes for Graphing Calculators
   by Dana Mosley

Learning Tools Student CD-ROM

Smarthinking^{TM}*.com* live online tutoring

Instructional DVDs by Dana Mosley

Instructional Videotapes for Graphing Calculators
   by Dana Mosley

Interactive Trigonometry 3.0 CD-ROM

Internet Trigonometry 3.0

HM eduSpace website

BlackBoard Course Cartridge

WebCT e-pack

Instructor Technology Resources

HMClassPrep^{TM} Instructor's CD-ROM

HM Testing 6.03

PowerPoint Presentations

Instructional Videotapes by Dana Mosley
   (ideal for libraries and resource centers)

Interactive Trigonometry 3.0 CD-ROM

Internet Trigonometry 3.0

HM eduSpace website

BlackBoard Course Cartridge

WebCT e-pack

Textbook website (math.college.hmco.com)

Textbook website (math.college.hmco.com)

*For more information on these and other resources available,
visit our website at* **math.college.hmco.com.**

Trigonometry, Sixth Edition
Learning Tools Student CD-ROM

The Learning Tools Student CD-ROM that accompanies the text provides students with an unprecedented quantity of support materials and resources that help bring mathematics to life with motion and sound. These electronic learning tools are separated into three components described below. The CD-ROM also provides access to MathGraphs, ACE Practice Tests, and SMARTHINKING, the online tutoring center.

Study the Lesson

The Glossary of Terms provides a comprehensive list of important mathematical terms for each chapter with a short definition of each term.

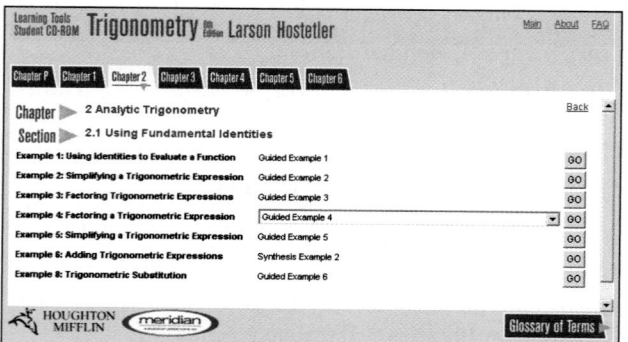

Review and Practice

- **Guided Examples** provide a full range of support by walking students step-by-step through problems that relate to a specific concept in the text.
- **Synthesis Examples** require the use of more than one concept from a section and encourage students to work through a solution of a problem one step at a time.

Visualize and Extend the Concepts

- **Animations** use motion and sound to explain concepts and can be played, paused, stopped, and replayed as many times as the student desires.
- **Simulations** encourage students to explore mathematical concepts experimentally.
- **Editable Graph Explorations** engage students in active discovery of mathematical concepts and relationships through the use of technology.

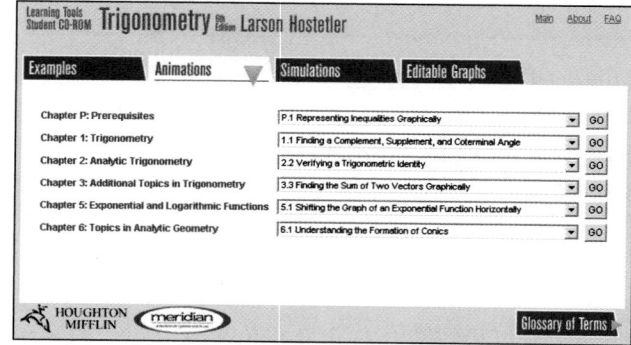

Trigonometry, Sixth Edition
Learning Tools Student CD-ROM

Selected examples and concepts throughout the text are identified by the Learning Tools Student CD-ROM icon ⬤www⬤. The chart on this and the following pages indicates the feature(s) of the CD—Guided Example, Synthesis Example, Animation, Simulation, and Editable Graph Exploration—that corresponds to the example or concept.

Chapter	Section	Example/Concept	Guided Example	Synthesis Example	Animation	Simulation	Editable Graph
P	1	Real Numbers	✓				
P	1	Ordering Real Numbers	✓				
P	1	Example 1	✓		✓		
P	1	Example 2	✓				
P	1	Absolute Value and Distance		✓			
P	1	Examples 4, 5	✓	✓			
P	1	Algebraic Expressions	✓		✓		
P	1	Basic Rules of Algebra	✓				
P	2	Solutions of Equations	✓	✓			
P	2	Examples 1, 4, 5, 10	✓	✓			
P	2	Quadratic Equations	✓				
P	2	Examples 6, 8, 9, 11, 12	✓				
P	2	Example 7			✓		
P	2	Example 10	✓	✓			
P	3	Examples 1, 3, 5, 8, 11	✓				
P	3	Example 2			✓		
P	3	Example 4			✓		
P	3	Example 6	✓		✓		
P	3	Example 7	✓				✓
P	3	Symmetry			✓		
P	3	Example 9		✓	✓		
P	3	Example 10		✓			✓
P	4	Using Slope	✓			✓	
P	4	Example 1		✓			
P	4	Examples 4, 5	✓	✓			
P	4	Examples 6, 7	✓				
P	4	Parallel and Perpendicular Lines	✓				
P	5	Examples 2, 6, 7, 9	✓				
P	5	Examples 3, 5	✓	✓			
P	6	Examples 1, 2	✓				
P	6	Examples 4, 6	✓	✓			
P	6	Even and Odd Functions				✓	
P	7	Example 2	✓				

Trigonometry, Sixth Edition
Learning Tools Student CD-ROM

Chapter	Section	Example/Concept	Guided Example	Synthesis Example	Animation	Simulation	Editable Graph
P	7	Example 3		✓			
P	8	Shifting Graphs	✓		✓		
P	8	Example 1	✓				
P	8	Reflecting Graphs			✓		
P	8	Examples 2, 4		✓			
P	8	Nonrigid Transformations			✓		
P	9	Combinations of Functions	✓		✓		
P	9	Examples 2, 4, 5	✓				
P	9	Composition of Functions			✓		
P	9	Example 6		✓			
P	10	Examples 1, 2, 7	✓				
P	10	Graph of an Inverse Function			✓		
P	10	Examples 5, 6	✓	✓			
P	10	Finding Inverse Functions	✓				
1	1	Examples 1, 3, 4, 7	✓				
1	1	Example 2	✓		✓		
1	1	Degree Measure	✓	✓			
1	1	Example 5	✓			✓	
1	2	The Unit Circle		✓	✓		
1	2	The Trigonometric Functions			✓		
1	2	Examples 1–4	✓				
1	2	Domain and Period of Sine and Cosine	✓				
1	3	The Six Trigonometric Functions			✓		
1	3	Examples 1, 4, 7	✓	✓			
1	3	Example 3	✓		✓		
1	3	Example 6	✓				
1	4	Examples 1, 2, 5	✓				
1	4	Reference Angles				✓	
1	4	Examples 4, 7	✓	✓			
1	5	Amplitude and Period	✓				
1	5	Example 2		✓			✓
1	5	Example 4					✓
1	5	Example 5	✓		✓		
1	5	Example 6	✓				✓
1	5	Mathematical Modeling	✓				
1	6	Examples 1, 3–6	✓				
1	6	Example 2			✓		

Trigonometry, Sixth Edition
Learning Tools Student CD-ROM

Chapter	Section	Example/Concept	Guided Example	Synthesis Example	Animation	Simulation	Editable Graph
1	6	Graphs of Reciprocal Functions					✓
1	7	Example 2			✓		
1	7	Other Inverse Trigonometric Functions		✓	✓		
1	7	Examples 3, 5, 6	✓				
1	7	Examples 4, 7	✓	✓			
1	8	Examples 1, 3, 5, 6	✓				
1	8	Example 4	✓	✓			
2	1	Examples 1–3, 5, 8	✓				
2	1	Example 4	✓	✓			
2	1	Example 6		✓			
2	2	Introduction			✓		
2	2	Examples 1, 3, 5	✓				
2	3	Examples 3, 4, 6, 8	✓				
2	3	Equations of Quadratic Type	✓				
2	3	Example 5		✓			
2	3	Functions Involving Multiple Angles		✓			
2	4	Sum and Difference Formulas		✓			
2	4	Examples 1, 3, 4, 8	✓				
2	4	Example 2		✓			
2	5	Multiple-Angle Formulas			✓		
2	5	Examples 1, 9	✓				
2	5	Example 3		✓			
2	5	Power-Reducing Formulas			✓		
2	5	Example 5	✓	✓			
2	5	Half-Angle Formulas			✓		
2	5	Example 6	✓	✓			
2	5	Product-To-Sum Formulas	✓				
3	1	Examples 1, 3, 5–7	✓				
3	1	The Ambiguous Case (SSA)				✓	
3	2	Example 1	✓	✓			
3	2	Examples 2, 4, 5	✓				
3	2	Example 3		✓			
3	3	Examples 2, 9	✓				
3	3	Vector Operations			✓		
3	3	Example 3	✓	✓	✓		
3	3	Unit Vectors	✓				

Trigonometry, Sixth Edition
Learning Tools Student CD-ROM

Chapter	Section	Example/Concept	Guided Example	Synthesis Example	Animation	Simulation	Editable Graph
3	3	Examples 6, 7		✓			
3	3	Direction Angles		✓			
3	4	Examples 1, 2, 6, 8	✓				
3	4	Example 4	✓	✓			
3	4	Finding Vector Components				✓	
4	1	The Imaginary Unit *i*	✓				
4	1	Examples 1, 3, 4, 6	✓				
4	1	Examples 2, 5	✓	✓			
4	2	Examples 2, 3, 6	✓				
4	2	Example 5	✓	✓			
4	3	Examples 1, 2, 4	✓				
4	3	Example 5		✓			
4	3	Example 6	✓	✓			
4	4	Example 1	✓				
4	4	Example 3	✓	✓			
5	1	Example 2		✓			✓
5	1	Example 3	✓				✓
5	1	Example 4	✓	✓	✓		
5	1	Examples 5, 7, 8	✓				
5	1	Applications				✓	
5	2	Example 1	✓	✓			
5	2	Examples 2, 7, 8, 10	✓				
5	2	Example 4			✓		
5	2	Example 6	✓	✓	✓		
5	3	Examples 1, 2, 6	✓				
5	3	Example 4	✓	✓			
5	3	Example 5		✓			
5	4	Examples 1, 2, 4–8, 10	✓				
5	4	Example 3	✓	✓			
5	5	Example 1	✓	✓			
5	5	Examples 2, 3, 5, 6	✓				
6	1	Inclination of a Line				✓	
6	1	Example 1	✓	✓			
6	1	The Angle Between Two Lines				✓	
6	1	Examples 2–4	✓				
6	2	Conics			✓		
6	2	Examples 1, 2, 4	✓				
6	2	Example 3	✓	✓			

Trigonometry, Sixth Edition
Learning Tools Student CD-ROM

Chapter	Section	Example/Concept	Guided Example	Synthesis Example	Animation	Simulation	Editable Graph
6	3	Introduction			✓		
6	3	Example 1	✓	✓			
6	3	Example 2		✓			
6	3	Application		✓			
6	3	Eccentricity	✓				
6	4	Example 1	✓	✓			
6	4	Examples 2, 4, 6	✓				
6	5	Examples 1, 3, 4	✓				
6	6	Plane Curves			✓		
6	6	Example 1			✓		
6	6	Example 2	✓				
6	6	Example 3	✓	✓			
6	6	Finding Parametric Equations for a Graph		✓			
6	6	Example 4		✓			
6	6	Example 5	✓		✓		
6	7	Introduction			✓		
6	7	Examples 3–5	✓				
6	7	Equation Conversion	✓				
6	8	Introduction		✓			
6	8	Symmetry	✓				
6	8	Example 2		✓			
6	8	Examples 3, 5, 6	✓				
6	8	Example 4			✓		
6	8	Special Polar Graphs		✓			
6	9	Polar Equations of Conics				✓	
6	9	Examples 1, 2, 4	✓				
6	9	Example 3	✓	✓			

To accommodate a wide variety of teaching and learning styles, *Trigonometry* is also available as *Interactive Trigonometry* 3.0 on an interactive CD-ROM and *Internet Trigonometry* 3.0. Students using the interactive CD-ROM or those with internet access will benefit from a wide range of compelling, interactive pedagogy, plus solutions to all odd exercises in the text. For instructors who conduct part of their course online, the internet version is an ideal solution, offering the additional advantages of online interaction with instructors and course management tools.

Hands-on Interaction

- The graphing calculator emulator provides students with an onscreen graphing utility that can be used for computation and exploration.
- The animations, simulations, and editable graph explorations make mathmatical concepts come alive.
- Guided and synthesis examples are designed to have students work through a solution one step at a time.
- Section quizzes require students to enter free-response answers, click and drag answers into place, or click on correct answers.

Classroom and Syllabus Management Systems

All of the content of the Sixth Edition—a wealth of applications, exercises, worked-out examples, and detailed explanations—is included in *Interactive Trigonometry* 3.0 on CD-ROM and *Internet Trigonometry* 3.0. Instructors have the flexibility of customizing content and interactive features for students as desired. Instructors may simply add dates to a default syllabus or may modify the order of topics. Either way, a customized syllabus is easy to distribute electronically and update instantly. This tool is particularly useful for managing distance learning courses.

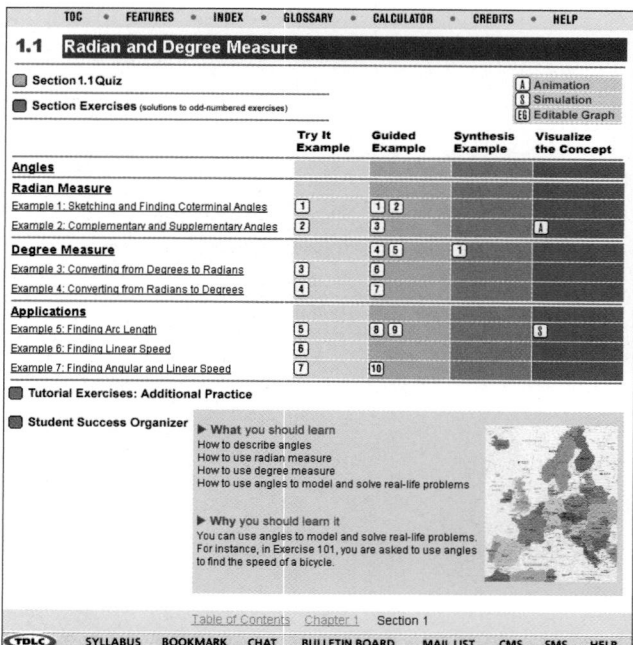

Features

Interactive Trigonometry **3.0 CD-ROM and**
Internet Trigonometry **3.0**

Exercises with full worked-out solutions to all of the odd exercises in the text provide immediate feedback for students.

Try Its allow students to try problems similar to the examples and to check their work using the worked-out solutions provided.

Guided Examples provide a full range of support by walking students step-by-step through problems that relate to a specific concept in the text.

Synthesis Examples require the use of more than one concept from a section and encourage students to work through a solution of a problem one step at a time.

Animations, which use motion and sound to explain concepts, can be played, paused, stopped, and replayed as many times as the student desires.

Simulations are interactive activities that encourage exploration and hands-on use of mathematical concepts.

Editable Graphs encourage students to explore concepts by graphing "editable" graphs. Students can also change the viewing window and use *zoom* and *trace* features.

MathGraphs are enlarged, printable versions of graphs from exercises in the book in which students are asked to draw on the graphs.

Tutorial Exercises are *additional* exercises that furnish students with much needed guided practice and refer back to a **Guided Example** for help if necessary.

Graphing Calculator Emulator is a powerful tool built into the program for convenient computation and exploration, and is also useful for working exercises that require the use of a graphing calculator.

Section Quizzes with responses and **Chapter Tests** with answers help students assess their mastery of the material.

Chapter Pre-Tests and **Post-Tests** offer added practice and assessment opportunities.

Glossary of Terms provides a comprehensive list of important mathematical terms, which students can quickly and easily access at any time.

Index and **Features Index** facilitate cross-referencing by providing complete searchable text-specific content.

Syllabus Builder enables instructors to save administrative time and to convey important information online.

How to study Chapter P

▶ What you should learn

In this chapter you will learn the following skills and concepts:

- How to represent, classify, and order real numbers
- How to evaluate algebraic expressions and solve linear and nonlinear equations
- How to plot points in the coordinate plane, use the Distance and Midpoint Formulas, and sketch graphs of equations
- How to find and use the slopes of lines to write and graph linear equations in two variables
- How to evaluate functions and find their domains
- How to analyze graphs of functions, identify rigid and nonrigid transformations of functions, and find inverse functions
- How to find arithmetic combinations and compositions of functions

▶ Important Vocabulary

As you encounter each new vocabulary term in this chapter, add the term and its definition to your notebook glossary.

Real numbers (p. 2)
Inequality (p. 3)
Absolute value (p. 5)
Algebraic expressions (p. 6)
Equation (p. 12)
Extraneous solution (p. 14)
Quadratic equation (p. 15)
Rectangular coordinate system (p. 25)
Ordered pair (p. 25)
Graph of an equation (p. 29)
Intercepts (p. 30)
Symmetry (p. 31)
Slope (pp. 38, 41)
Slope-intercept form (p. 38)
Point-slope form (p. 43)
Parallel (p. 45)

Perpendicular (p. 45)
Function (p. 53)
Domain (p. 53)
Range (p. 53)
Independent variable (p. 55)
Dependent variable (p. 55)
Implied domain (p. 57)
Vertical Line Test (p. 68)
Zeros of a function (p. 69)
Relative minimum (p. 71)
Relative maximum (p. 71)
Even function (p. 72)
Odd function (p. 72)
Inverse function (p. 103)
Horizontal Line Test (p. 106)
One-to-one function (p. 106)

Study Tools

Learning objectives in each section
Chapter Summary (pp. 113–114)
Review Exercises (pp. 115–119)
Chapter Test (p. 120)

Additional Resources

Study and Solutions Guide
Interactive Trigonometry
Videotapes/DVD for Chapter P
Trigonometry Website
Student Success Organizer

Andreas Stirnberg/Getty Images

P

Prerequisites

Review of Real Numbers and Their Properties

▶ **What you should learn**
- How to represent and classify real numbers
- How to order real numbers and use inequalities
- How to find the absolute values of real numbers and find the distance between two real numbers
- How to evaluate algebraic expressions
- How to use the basic rules and properties of algebra

▶ **Why you should learn it**

Real numbers are used to represent many real-life quantities. For example, in Exercise 69 on page 10, you will use real numbers to represent the federal deficit.

Real Numbers

Real numbers are used in everyday life to describe quantities such as age, miles per gallon, container size, and population. Real numbers are represented by symbols such as

$$-5, 9, 0, \frac{4}{3}, 0.666\ldots, 28.21, \sqrt{2}, \pi, \text{ and } \sqrt[3]{-32}.$$

Here are some important subsets of the real numbers.

$$\{1, 2, 3, 4, \ldots\} \qquad \text{Set of natural numbers}$$

$$\{0, 1, 2, 3, 4, \ldots\} \qquad \text{Set of whole numbers}$$

$$\{\ldots, -3, -2, -1, 0, 1, 2, 3, \ldots\} \qquad \text{Set of integers}$$

A real number is **rational** if it can be written as the ratio p/q of two integers, where $q \neq 0$. For instance, the numbers

$$\frac{1}{3} = 0.3333\ldots = 0.\overline{3}, \frac{1}{8} = 0.125, \text{ and } \frac{125}{111} = 1.126126\ldots = 1.\overline{126}$$

are rational. The decimal representation of a rational number either repeats $\left(\text{as in } \frac{173}{55} = 3.1\overline{45}\right)$ or terminates $\left(\text{as in } \frac{1}{2} = 0.5\right)$. A real number that cannot be written as the ratio of two integers is called **irrational.** Irrational numbers have infinite nonrepeating decimal representations. For instance, the numbers

$$\sqrt{2} \approx 1.4142136 \quad \text{and} \quad \pi \approx 3.1415927$$

are irrational. (The symbol $\approx$ means "is approximately equal to.")

Real numbers are represented graphically by a **real number line.** The point 0 on the real number line is the **origin.** Numbers to the right of 0 are positive, and numbers to the left of 0 are negative, as shown in Figure P.1. The term **nonnegative** describes a number that is either positive or zero.

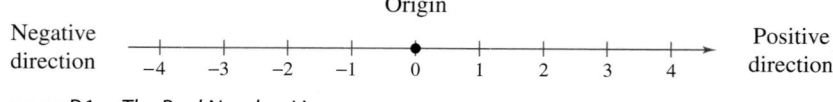

FIGURE P.1 *The Real Number Line*

As illustrated in Figure P.2, there is a *one-to-one correspondence* between real numbers and points on the real number line.

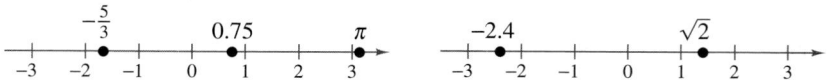

Every real number corresponds to exactly one point on the real number line.

Every point on the real number line corresponds to exactly one real number.

FIGURE P.2 *One-to-One Correspondence*

The icon identifies examples and concepts related to features of the Learning Tools CD-ROM and the *Interactive* and *Internet* versions of this text. For more details see the chart on pages *xix-xxiii.*

Ordering Real Numbers

One important property of real numbers is that they are *ordered*.

Definition of Order on the Real Number Line

If a and b are real numbers, a is less than b if $b - a$ is positive. The **order** of a and b is denoted by the **inequality**

$$a < b.$$

This relationship can also be described by saying that b is *greater than a* and writing $b > a$. The inequality $a \leq b$ means that a is *less than or equal to b*, and the inequality $b \geq a$ means that b is *greater than or equal to a*. The symbols $<$, $>$, $\leq$, and $\geq$ are *inequality symbols*.

Geometrically, this definition implies that $a < b$ if and only if a lies to the *left* of b on the real number line, as shown in Figure P.3.

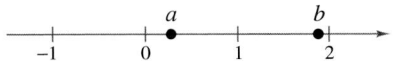

FIGURE P.3 $a < b$ *if and only if a lies to the left of b.*

Example 1 ▶ **Interpreting Inequalities**

Describe the subset of real numbers represented by each inequality.

a. $x \leq 2$ **b.** $-2 \leq x < 3$

Solution

a. The inequality $x \leq 2$ denotes all real numbers less than or equal to 2, as shown in Figure P.4.

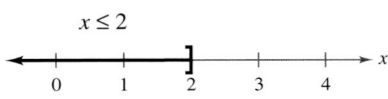

FIGURE P.4

b. The inequality $-2 \leq x < 3$ means that $x \geq -2$ and $x < 3$. This "double inequality" denotes all real numbers between -2 and 3, including -2 but not including 3, as shown in Figure P.5.

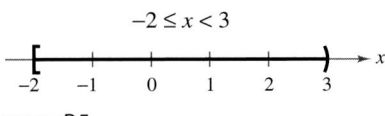

FIGURE P.5

Inequalities can be used to describe subsets of real numbers called **intervals**. In the bounded intervals below, the real numbers a and b are the **endpoints** of each interval.

The *Interactive* CD-ROM and *Internet* versions of this text offer a Try It for each example in the text.

Bounded Intervals on the Real Number Line

Notation	Interval Type	Inequality	Graph
$[a, b]$	Closed	$a \leq x \leq b$	
(a, b)	Open	$a < x < b$	
$[a, b)$		$a \leq x < b$	
$(a, b]$		$a < x \leq b$	

The symbols ∞, **positive infinity,** and −∞, **negative infinity,** do not represent real numbers. They are simply convenient symbols used to describe the unboundedness of an interval such as $(1, \infty)$ or $(-\infty, 3]$.

Unbounded Intervals on the Real Number Line

Notation	Interval Type	Inequality	Graph
$[a, \infty)$		$x \geq a$	
(a, ∞)	Open	$x > a$	
$(-\infty, b]$		$x \leq b$	
$(-\infty, b)$	Open	$x < b$	
$(-\infty, \infty)$	Entire real line	$-\infty \leq x \leq \infty$	

Example 2 ▶   Using Inequalities to Represent Intervals

Use inequality notation to describe each of the following.

a. c is at most 2.

b. m is at least -3.

c. All x in the interval $(-3, 5]$

Solution

a. The statement "c is at most 2" can be represented by $c \leq 2$.

b. The statement "m is at least -3" can be represented by $m \geq -3$.

c. "All x in the interval $(-3, 5]$" can be represented by $-3 < x \leq 5$.

Example 3 ▶   Interpreting Intervals

Give a verbal description of each interval.

a. $(-1, 0)$ **b.** $[2, \infty)$ **c.** $(-\infty, 0)$

Solution

a. This interval consists of all real numbers that are greater than -1 and less than 0.

b. This interval consists of all real numbers that are greater than or equal to 2.

c. This interval consists of all real numbers that are less than zero (the negative real numbers).

The **Law of Trichotomy** states that for any two real numbers a and b, *precisely* one of three relationships is possible:

$$a = b, \quad a < b, \quad \text{or} \quad a > b. \qquad \text{Law of Trichotomy}$$

Additional Examples

Use inequality notation to describe each of the following.

a. All x in the interval $(-2, 2]$

b. w is at least 1 and at most 5.

c. All q in the interval $(-12, 0)$

Solution

a. $-2 < x \leq 2$

b. $1 \leq w \leq 5$

c. $-12 < q < 0$

◀ E x p l o r a t i o n ▶

Absolute value expressions can be evaluated on a graphing utility. When an expression such as $|3 - 8|$ is evaluated, parentheses should surround the expression, as in abs$(3 - 8)$. Evaluate each expression below. What can you conclude?

a. $|6|$ **b.** $|-1|$

c. $|5 - 2|$ **d.** $|2 - 5|$

Absolute Value and Distance

The **absolute value** of a real number is its *magnitude*, or the distance between the origin and the point representing the real number on the real number line.

Definition of Absolute Value

If a is a real number, then the absolute value of a is

$$|a| = \begin{cases} a, & \text{if } a \geq 0 \\ -a, & \text{if } a < 0. \end{cases}$$

Notice in this definition that the absolute value of a real number is never negative. For instance, if $a = -5$, then $|-5| = -(-5) = 5$. The absolute value of a real number is either positive or zero. Moreover, 0 is the only real number whose absolute value is 0. So, $|0| = 0$.

Example 4 ▶   Evaluating the Absolute Value of a Number

Evaluate $\dfrac{|x|}{x}$ for (a) $x > 0$ and (b) $x < 0$.

Solution

a. If $x > 0$, then $|x| = x$ and $\dfrac{|x|}{x} = \dfrac{x}{x} = 1$.

b. If $x < 0$, then $|x| = -x$ and $\dfrac{|x|}{x} = \dfrac{-x}{x} = -1$.

Properties of Absolute Values

1. $|a| \geq 0$ **2.** $|-a| = |a|$

3. $|ab| = |a||b|$ **4.** $\left|\dfrac{a}{b}\right| = \dfrac{|a|}{|b|}, \qquad b \neq 0$

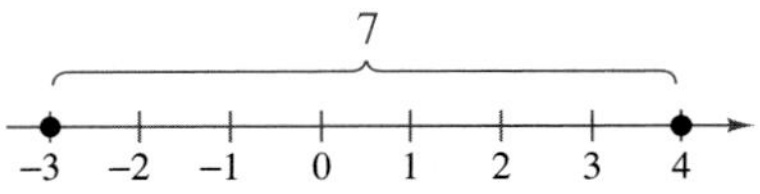

Absolute value can be used to define the distance between two points on the real number line. For instance, the distance between -3 and 4 is

$$|-3 - 4| = |-7|$$
$$= 7$$

as shown in Figure P.6.

$$\overbrace{}^{7}$$

-3 -2 -1 0 1 2 3 4

FIGURE P.6   *The distance between -3 and 4 is 7.*

Distance Between Two Points on the Real Line

Let a and b be real numbers. The **distance between a and b** is

$$d(a, b) = |b - a| = |a - b|.$$

Algebraic Expressions

One characteristic of algebra is the use of letters to represent numbers. The letters are **variables,** and combinations of letters and numbers are **algebraic expressions.** Here are a few examples of algebraic expressions.

$$5x, \qquad 2x - 3, \qquad \frac{4}{x^2 + 2}, \qquad 7x + y$$

Definition of an Algebraic Expression

An **algebraic expression** is a collection of letters **(variables)** and real numbers **(constants)** combined using the operations of addition, subtraction, multiplication, division, and exponentiation.

The **terms** of an algebraic expression are those parts that are separated by *addition.* For example,

$$x^2 - 5x + 8 = x^2 + (-5x) + 8$$

has three terms: x^2 and $-5x$ are the **variable terms** and 8 is the **constant term.** The numerical factor of a variable term is the **coefficient** of the variable term. For instance, the coefficient of $-5x$ is -5, and the coefficient of x^2 is 1.

To **evaluate** an algebraic expression, substitute numerical values for each of the variables in the expression. Here are two examples.

Expression	Value of Variable	Substitute	Value of Expression
$-3x + 5$	$x = 3$	$-3(3) + 5$	$-9 + 5 = -4$
$3x^2 + 2x - 1$	$x = -1$	$3(-1)^2 + 2(-1) - 1$	$3 - 2 - 1 = 0$

When an algebraic expression is evaluated, the **Substitution Principle** is used. It states that "If $a = b$, then a can be replaced by b in any expression involving a." In the first evaluation shown above, for instance, 3 is *substituted* for x in the expression $-3x + 5$.

Basic Rules of Algebra

There are four arithmetic operations with real numbers: *addition, multiplication, subtraction,* and *division,* denoted by the symbols $+$, $\times$ or $\cdot$, $-$, and $\div$. Of these, addition and multiplication are the two primary operations. Subtraction and division are the inverse operations of addition and multiplication, respectively.

Subtraction: Add the opposite.   *Division:* Multiply by the reciprocal.

$$a - b = a + (-b) \qquad \text{If } b \neq 0, \text{ then } a/b = a\left(\frac{1}{b}\right) = \frac{a}{b}.$$

In these definitions, $-b$ is the **additive inverse** (or opposite) of b, and $1/b$ is the **multiplicative inverse** (or reciprocal) of b. In the fractional form a/b, a is the **numerator** of the fraction and b is the **denominator.**

Because the properties of real numbers on page 7 are true for variables and algebraic expressions as well as for real numbers, they are often called the **Basic Rules of Algebra.**

Basic Rules of Algebra

Let a, b, and c be real numbers, variables, or algebraic expressions.

Property		*Example*
Commutative Property of Addition:	$a + b = b + a$	$4x + x^2 = x^2 + 4x$
Commutative Property of Multiplication:	$ab = ba$	$(4 - x)x^2 = x^2(4 - x)$
Associative Property of Addition:	$(a + b) + c = a + (b + c)$	$(x + 5) + x^2 = x + (5 + x^2)$
Associative Property of Multiplication:	$(ab)c = a(bc)$	$(2x \cdot 3y)(8) = (2x)(3y \cdot 8)$
Distributive Properties:	$a(b + c) = ab + ac$	$3x(5 + 2x) = 3x \cdot 5 + 3x \cdot 2x$
	$(a + b)c = ac + bc$	$(y + 8)y = y \cdot y + 8 \cdot y$
Additive Identity Property:	$a + 0 = a$	$5y^2 + 0 = 5y^2$
Multiplicative Identity Property:	$a \cdot 1 = a$	$(4x^2)(1) = 4x^2$
Additive Inverse Property:	$a + (-a) = 0$	$5x^3 + (-5x^3) = 0$
Multiplicative Inverse Property:	$a \cdot \dfrac{1}{a} = 1, \quad a \neq 0$	$(x^2 + 4)\left(\dfrac{1}{x^2 + 4}\right) = 1$

Because subtraction is defined as "adding the opposite," the Distributive Properties are also true for subtraction. So, the first Distributive Property can be applied to an expression of the form $a(b - c)$ as follows.

$$a(b - c) = ab - ac$$

STUDY TIP

Be sure you see the difference between the *opposite of a number* and a *negative number*. If a is already negative, then its opposite, $-a$, is positive. For instance, if $a = -5$, then

$$-a = -(-5) = 5.$$

Properties of Negation

Let a and b be real numbers, variables, or algebraic expressions.

Property	*Example*
1. $(-1)a = -a$	$(-1)7 = -7$
2. $-(-a) = a$	$-(-6) = 6$
3. $(-a)b = -(ab) = a(-b)$	$(-5)3 = -(5 \cdot 3) = 5(-3)$
4. $(-a)(-b) = ab$	$(-2)(-x) = 2x$
5. $-(a + b) = (-a) + (-b)$	$-(x + 8) = (-x) + (-8)$
	$= -x - 8$

Properties of Equality

Let a, b, and c be real numbers, variables, or algebraic expressions.

1. If $a = b$, then $a + c = b + c$.	Add c to each side.
2. If $a = b$, then $ac = bc$.	Multiply each side by c.
3. If $a + c = b + c$, then $a = b$.	Subtract c from each side.
4. If $ac = bc$ and $c \neq 0$, then $a = b$.	Divide each side by c.

Activities

1. Evaluate: $4 - |-3|$.

   Answer: 1

2. Find the distance between -41 and 16.

   Answer: 57

3. Use inequality notation to describe the set of nonnegative numbers.

   Answer: $x \geq 0$

4. Use interval notation to describe the inequality $-6 < x \leq 13$.

   Answer: $(-6, 13]$

Point out to students that to add or subtract fractions with unlike denominators, they can use Property 5 of fractions, as in Example 5(c), or they can rewrite the fractions with like denominators using the least common denominator (LCD) of the fractions.

Properties of Zero

Let a and b be real numbers, variables, or algebraic expressions.

1. $a + 0 = a$ and $a - 0 = a$ 2. $a \cdot 0 = 0$

3. $\dfrac{0}{a} = 0, \qquad a \neq 0$ 4. $\dfrac{a}{0}$ is undefined.

5. **Zero-Factor Property:** If $ab = 0$, then $a = 0$ or $b = 0$.

Properties and Operations of Fractions

Let a, b, c, and d be real numbers, variables, or algebraic expressions such that $b \neq 0$ and $d \neq 0$.

1. **Equivalent Fractions:** $\dfrac{a}{b} = \dfrac{c}{d}$ if and only if $ad = bc$.

2. **Rules of Signs:** $-\dfrac{a}{b} = \dfrac{-a}{b} = \dfrac{a}{-b}$ and $\dfrac{-a}{-b} = \dfrac{a}{b}$

3. **Generate Equivalent Fractions:** $\dfrac{a}{b} = \dfrac{ac}{bc}, \qquad c \neq 0$

4. **Add or Subtract with Like Denominators:** $\dfrac{a}{b} \pm \dfrac{c}{b} = \dfrac{a \pm c}{b}$

5. **Add or Subtract with Unlike Denominators:** $\dfrac{a}{b} \pm \dfrac{c}{d} = \dfrac{ad \pm bc}{bd}$

6. **Multiply Fractions:** $\dfrac{a}{b} \cdot \dfrac{c}{d} = \dfrac{ac}{bd}$

7. **Divide Fractions:** $\dfrac{a}{b} \div \dfrac{c}{d} = \dfrac{a}{b} \cdot \dfrac{d}{c} = \dfrac{ad}{bc}, \qquad c \neq 0$

Example 5 ▶ **Properties and Operations of Fractions**

a. Equivalent fractions: $\dfrac{x}{5} = \dfrac{3 \cdot x}{3 \cdot 5} = \dfrac{3x}{15}$ **b.** Divide fractions: $\dfrac{7}{x} \div \dfrac{3}{2} = \dfrac{7}{x} \cdot \dfrac{2}{3} = \dfrac{14}{3x}$

c. Add fractions with unlike denominators: $\dfrac{x}{3} + \dfrac{2x}{5} = \dfrac{5 \cdot x + 3 \cdot 2x}{3 \cdot 5} = \dfrac{11x}{15}$

If a, b, and c are integers such that $ab = c$, then a and b are **factors** or **divisors** of c. A **prime number** is an integer that has exactly two positive factors:— itself and 1—such as 2, 3, 5, 7, and 11. The numbers 4, 6, 8, 9, and 10 are **composite** because they can be written as the product of two or more prime numbers. The number 1 is neither prime nor composite. The **Fundamental Theorem of Arithmetic** states that every positive integer greater than 1 can be written as the product of prime numbers in precisely one way (disregarding order). For instance, the *prime factorization* of 24 is $24 = 2 \cdot 2 \cdot 2 \cdot 3$.

P.1 Exercises

The *Interactive* CD-ROM and *Internet* versions of this text contain step-by-step solutions to all odd-numbered exercises. They also provide Tutorial Exercises for additional help.

In Exercises 1–6, determine which numbers are (a) natural numbers, (b) integers, (c) rational numbers, and (d) irrational numbers.

1. $-9, -\frac{7}{2}, 5, \frac{2}{3}, \sqrt{2}, 0, 1, -4, 2, -11$

2. $\sqrt{5}, -7, -\frac{7}{3}, 0, 3.12, \frac{5}{4}, -3, 12, 5$

3. $2.01, 0.666\ldots, -13, 0.010110111\ldots, 1, -6$

4. $2.3030030003\ldots, 0.7575, -4.63, \sqrt{10}, -75, 4$

5. $-\pi, -\frac{1}{3}, \frac{6}{3}, \frac{1}{2}\sqrt{2}, -7.5, -1, 8, -22$

6. $25, -17, -\frac{12}{5}, \sqrt{9}, 3.12, \frac{1}{2}\pi, 7, -11.1, 13$

In Exercises 7–10, use a calculator to find the decimal form of the rational number. If it is a nonterminating decimal, write the repeating pattern.

7. $\frac{5}{8}$ **8.** $\frac{1}{3}$

9. $\frac{41}{333}$ **10.** $\frac{6}{11}$

In Exercises 11 and 12, approximate the numbers and place the correct symbol (< or >) between them.

11.

12.

In Exercises 13–18, plot the two real numbers on the real number line. Then place the appropriate inequality symbol (< or >) between them.

13. $-4, -8$ **14.** $-3.5, 1$

15. $\frac{3}{2}, 7$ **16.** $1, \frac{16}{3}$

17. $\frac{5}{6}, \frac{2}{3}$ **18.** $-\frac{8}{7}, -\frac{3}{7}$

In Exercises 19–28, verbally describe the subset of real numbers represented by the inequality. Then sketch the subset on the real number line. State whether the interval is bounded or unbounded.

19. $x \le 5$ **20.** $x \ge -2$

21. $x < 0$ **22.** $x > 3$

23. $x \ge 4$ **24.** $x < 2$

25. $-2 < x < 2$ **26.** $0 \le x \le 5$

27. $-1 \le x < 0$ **28.** $0 < x \le 6$

In Exercises 29–36, use inequality notation to describe the set.

29. All x in the interval $(-2, 4]$

30. All y in the interval $[-6, 0)$

31. y is nonnegative.

32. y is no more than 25.

33. t is at least 10 and at most 22.

34. k is less than 5 but no less than -3.

35. The dog's weight W is more than 65 pounds.

36. The annual rate of inflation r is expected to be at least 2.5% but no more than 5%.

In Exercises 37–40, give a verbal description of the interval.

37. $[0, 8)$ **38.** $[-5, 7]$

39. $(-6, \infty)$ **40.** $(-\infty, 4]$

In Exercises 41–50, evaluate the expression.

41. $|-10|$ **42.** $|0|$

43. $|3 - 8|$ **44.** $|4 - 1|$

45. $|-1| - |-2|$ **46.** $-3 - |-3|$

47. $\dfrac{-5}{|-5|}$ **48.** $-3|-3|$

49. $\dfrac{|x + 2|}{x + 2}, \quad x < -2$ **50.** $\dfrac{|x - 1|}{x - 1}, \quad x > 1$

In Exercises 51–56, place the correct symbol (<, >, or =) between the pair of real numbers.

51. $|-3| \quad -|-3|$ **52.** $|-4| \quad |4|$

53. $-5 \quad -|5|$ **54.** $-|-6| \quad |-6|$

55. $-|-2| \quad -|2|$ **56.** $-(-2) \quad -2$

In Exercises 57–64, find the distance between a and b.

57. $a = -1$ $b = 3$

58. $a = -4$ $b = -\frac{3}{2}$

59. $a = 126, b = 75$ **60.** $a = -126, b = -75$

61. $a = -\frac{5}{2}, b = 0$ **62.** $a = \frac{1}{4}, b = \frac{11}{4}$

63. $a = \frac{16}{5}, b = \frac{112}{75}$ **64.** $a = 9.34, b = -5.65$

Budget Variance In Exercises 65–68, the accounting department of a sports drink bottling company is checking to see whether the actual expenses of a department differ from the budgeted expenses by more than $500 or by more than 5%. Fill in the missing parts of the table, and determine whether each actual expense passes the "budget variance test."

		Budgeted Expense, b	Actual Expense, a	$\|a - b\|$	$0.05b$
65.	Wages	$112,700	$113,356		
66.	Utilities	$9,400	$9,772		
67.	Taxes	$37,640	$37,335		
68.	Insurance	$2,575	$2,613		

▶ Model It

69. *Federal Deficit* The bar graph shows the federal government receipts (in billions of dollars) for selected years from 1960 through 2000. (Source: U.S. Office of Management and Budget)

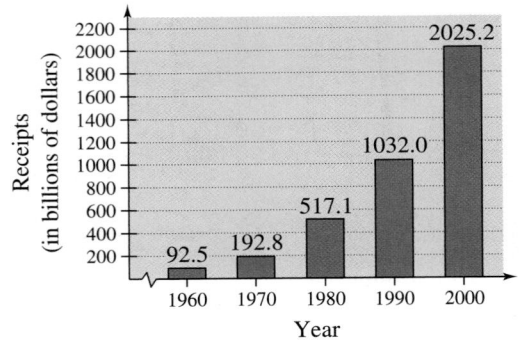

(a) Complete the table. (*Hint:* Find |Receipts – Expenditures|.)

Year	Expenditures (in billions)	Surplus or deficit (in billions)
1960	$92.2	
1970	$195.6	
1980	$590.9	
1990	$1253.2	
2000	$1788.8	

(b) Use the table in part (a) to construct a bar graph showing the magnitude of the surplus or deficit for each year.

70. *Veterans* The table shows the number of surviving spouses of deceased veterans of United States wars (as of May 2001). Construct a circle graph showing the percent of surviving spouses for each war as a fraction of the total number of surviving spouses of deceased war veterans. (Source: Department of Veteran Affairs)

War	Number of surviving spouses
Civil War	1
Indian Wars	0
Spanish-American War	386
Mexican Border War	181
World War I	25,573
World War II	272,793
Korean War	63,579
Vietnam War	114,514
Gulf War	6,261

In Exercises 71–78, use absolute value notation to describe the situation.

71. While traveling on the Pennsylvania Turnpike, you pass milepost 57 near Pittsburgh, then milepost 236 near Gettysburg. How far do you travel during that time period?

72. While traveling on the Pennsylvania Turnpike, you pass milepost 326 near Valley Forge, then milepost 351 near Philadelphia. How far do you travel during that time period?

73. The temperature in Bismarck, North Dakota, was 60° at noon, then 23° at midnight. What was the change in temperature over the 12-hour period?

74. The temperature in Chicago, Illinois was 48° last night at midnight, then 82° at noon today. What was the change in temperature over the 12-hour period?

75. The distance between x and 5 is no more than 3.

76. The distance between x and -10 is at least 6.

77. y is at least six units from 0.

78. y is at most two units from a.

In Exercises 79–84, identify the terms. Then identify the coefficients of the variable terms of the expression.

79. $7x + 4$

80. $6x^3 - 5x$

81. $\sqrt{3}x^2 - 8x - 11$

82. $3\sqrt{3}x^2 + 1$

83. $4x^3 + \dfrac{x}{2} - 5$

84. $3x^4 - \dfrac{x^2}{4}$

In Exercises 85–90, evaluate the expression for each value of x. (If not possible, state the reason.)

	Expression		Values	
85.	$4x - 6$	(a) $x = -1$	(b) $x = 0$
86.	$9 - 7x$	(a) $x = -3$	(b) $x = 3$
87.	$x^2 - 3x + 4$	(a) $x = -2$	(b) $x = 2$
88.	$-x^2 + 5x - 4$	(a) $x = -1$	(b) $x = 1$
89.	$\dfrac{x+1}{x-1}$	(a) $x = 1$	(b) $x = -1$
90.	$\dfrac{x}{x+2}$	(a) $x = 2$	(b) $x = -2$

In Exercises 91–100, identify the rule(s) of algebra illustrated by the statement.

91. $x + 9 = 9 + x$

92. $2\left(\frac{1}{2}\right) = 1$

93. $\dfrac{1}{h+6}(h + 6) = 1, \quad h \neq -6$

94. $(x + 3) - (x + 3) = 0$

95. $2(x + 3) = 2x + 6$

96. $(z - 2) + 0 = z - 2$

97. $1 \cdot (1 + x) = 1 + x$

98. $x + (y + 10) = (x + y) + 10$

99. $x(3y) = (x \cdot 3)y = (3x)y$

100. $\frac{1}{7}(7 \cdot 12) = \left(\frac{1}{7} \cdot 7\right)12 = 1 \cdot 12 = 12$

In Exercises 101–108, perform the operation(s). (Write fractional answers in simplest form.)

101. $\frac{3}{16} + \frac{5}{16}$

102. $\frac{6}{7} - \frac{4}{7}$

103. $\frac{5}{8} - \frac{5}{12} + \frac{1}{6}$

104. $\frac{10}{11} + \frac{6}{33} - \frac{13}{66}$

105. $12 \div \frac{1}{4}$

106. $-\left(6 \cdot \frac{4}{8}\right)$

107. $\dfrac{2x}{3} - \dfrac{x}{4}$

108. $\dfrac{5x}{6} \cdot \dfrac{2}{9}$

109. (a) Use a calculator to complete the table.

n	1	0.5	0.01	0.0001	0.000001
$5/n$					

 (b) Use the result from part (a) to make a conjecture about the value of $5/n$ as n approaches 0.

110. (a) Use a calculator to complete the table.

n	1	10	100	10,000	100,000
$5/n$					

 (b) Use the result from part (a) to make a conjecture about the value of $5/n$ as n increases without bound.

Synthesis

True or False? **In Exercises 111 and 112, determine whether the statement is true or false. Justify your answer.**

111. If $a < b$, then $\dfrac{1}{a} < \dfrac{1}{b}$, where $a \neq b \neq 0$.

112. Because $\dfrac{a+b}{c} = \dfrac{a}{c} + \dfrac{b}{c}$, then $\dfrac{c}{a+b} = \dfrac{c}{a} + \dfrac{c}{b}$.

113. ***Exploration*** Consider $|u + v|$ and $|u| + |v|$.

 (a) Are the values of the expressions always equal? If not, under what conditions are they unequal?

 (b) If the two expressions are not equal for certain values of u and v, is one of the expressions always greater than the other? Explain.

114. ***Think About It*** Is there a difference between saying that a real number is positive and saying that a real number is nonnegative? Explain.

115. ***Think About It*** Because every even number is divisible by 2, is it possible that there exist any even prime numbers? Explain.

116. ***Writing*** Describe the differences among the sets of natural numbers, integers, rational numbers, and irrational numbers.

In Exercises 117 and 118, use the real numbers A, B, and C shown on the number line. Determine the sign of each expression.

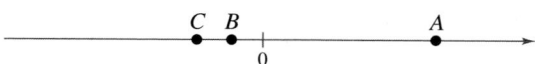

117. (a) $-A$

 (b) $B - A$

118. (a) $-C$

 (b) $A - C$

119. ***Writing*** You may hear it said that to take the absolute value of a real number you simply remove any negative sign and make the number positive. Can it ever be true that $|a| = -a$ for a real number a? Explain.

P.2 Solving Equations

▶ What you should learn

- How to identify different types of equations
- How to solve linear equations in one variable and equations that lead to linear equations
- How to solve quadratic equations by factoring, extracting square roots, completing the square, and using the Quadratic Formula
- How to solve polynomial equations of degree three or greater
- How to solve equations involving radicals
- How to solve equations involving absolute values

▶ Why you should learn it

Linear equations are used in many real-life applications. For example, in Exercise 185 on page 23, linear equations can be used to model the relationship between the length of a thigh bone and the height of a person, helping researchers learn about ancient cultures.

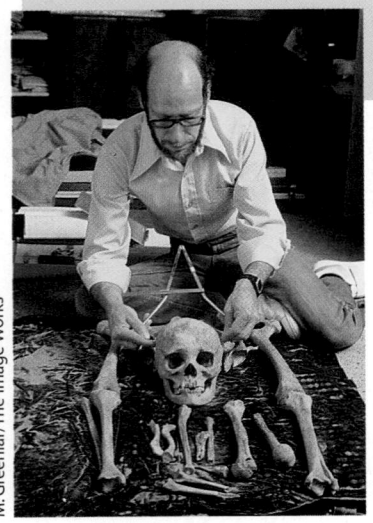

M. Greenlar/The Image Works

Equations and Solutions of Equations

An **equation** in x is a statement that two algebraic expressions are equal. For example,

$$3x - 5 = 7,\ x^2 - x - 6 = 0,\ \text{and } \sqrt{2x} = 4$$

are equations. To **solve** an equation in x means to find all values of x for which the equation is true. Such values are **solutions.** For instance, $x = 4$ is a solution of the equation

$$3x - 5 = 7$$

because $3(4) - 5 = 7$ is a true statement.

The solutions of an equation depend on the kinds of numbers being considered. For instance, in the set of rational numbers, $x^2 = 10$ has no solution because there is no rational number whose square is 10. However, in the set of real numbers, the equation has the two solutions $\sqrt{10}$ and $-\sqrt{10}$.

An equation that is true for *every* real number in the domain of the variable is called an **identity.** For example,

$$x^2 - 9 = (x + 3)(x - 3) \qquad \text{Identity}$$

is an identity because it is a true statement for any real value of x, and

$$\frac{x}{3x^2} = \frac{1}{3x} \qquad \text{Identity}$$

where $x \neq 0$, is an identity because it is true for any nonzero real value of x.

An equation that is true for just *some* (or even none) of the real numbers in the domain of the variable is called a **conditional equation.** For example, the equation

$$x^2 - 9 = 0 \qquad \text{Conditional equation}$$

is conditional because $x = 3$ and $x = -3$ are the only values in the domain that satisfy the equation. The equation $2x - 4 = 2x + 1$ is conditional because there are no real values of x for which the equation is true. Learning to solve conditional equations is the primary focus of this chapter.

Linear Equations in One Variable

> **Definition of Linear Equation**
>
> A **linear equation in one variable** x is an equation that can be written in the standard form
>
> $$ax + b = 0$$
>
> where a and b are real numbers with $a \neq 0$.

A linear equation has exactly one solution. To see this, consider the following steps. (Remember that $a \neq 0$.)

$$ax + b = 0 \qquad \text{Write original equation.}$$

$$ax = -b \qquad \text{Subtract } b \text{ from each side.}$$

$$x = -\frac{b}{a} \qquad \text{Divide each side by } a.$$

To solve a conditional equation in x, isolate x on one side of the equation by a sequence of **equivalent** (and usually simpler) **equations,** each having the same solution(s) as the original equation. The operations that yield equivalent equations come from the Substitution Principle and the Properties of Equality studied in Section P.1.

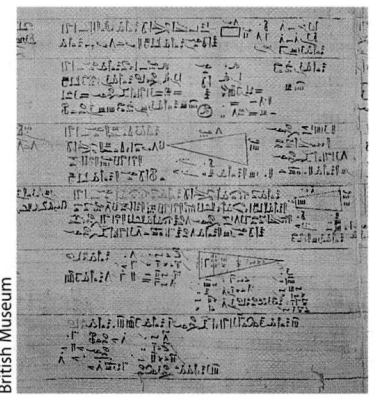

British Museum

Historical Note
This ancient Egyptian papyrus, discovered in 1858, contains one of the earliest examples of mathematical writing in existence. The papyrus itself dates back to around 1650 B.C., but it is actually a copy of writings from two centuries earlier. The algebraic equations on the papyrus were written in words. Diophantus, a Greek who lived around A.D. 250, is often called the Father of Algebra. He was the first to use abbreviated word forms in equations.

Generating Equivalent Equations

An equation can be transformed into an *equivalent equation* by one or more of the following steps.

	Given Equation	*Equivalent Equation*
1. Remove symbols of grouping, combine like terms, or simplify fractions on one or both sides of the equation.	$2x - x = 4$	$x = 4$
2. Add (or subtract) the same quantity to (from) *each* side of the equation.	$x + 1 = 6$	$x = 5$
3. Multiply (or divide) *each* side of the equation by the same *nonzero* quantity.	$2x = 6$	$x = 3$
4. Interchange the two sides of the equation.	$2 = x$	$x = 2$

Example 1 ► **Solving a Linear Equation**

a. $3x - 6 = 0$ Original equation

$\qquad 3x = 6$ Add 6 to each side.

$\qquad\quad x = 2$ Divide each side by 3.

b. $5x + 4 = 3x - 8$ Original equation

$\quad 2x + 4 = -8$ Subtract $3x$ from each side.

$\qquad 2x = -12$ Subtract 4 from each side.

$\qquad\quad x = -6$ Divide each side by 2.

To solve an equation involving fractional expressions, find the least common denominator (LCD) of all terms and multiply every term by the LCD.

Example 2 ▶ **An Equation Involving Fractional Expressions**

Solve $\dfrac{x}{3} + \dfrac{3x}{4} = 2$.

Solution

$$\frac{x}{3} + \frac{3x}{4} = 2 \qquad \text{Write original equation.}$$

$$(12)\frac{x}{3} + (12)\frac{3x}{4} = (12)2 \qquad \text{Multiply each term by the LCD of 12.}$$

$$4x + 9x = 24 \qquad \text{Divide out and multiply.}$$

$$13x = 24 \qquad \text{Combine like terms.}$$

$$x = \frac{24}{13} \qquad \text{Divide each side by 13.}$$

The solution is $x = \frac{24}{13}$. Check this in the original equation.

When multiplying or dividing an equation by a *variable* quantity, it is possible to introduce an extraneous solution. An **extraneous solution** is one that does not satisfy the original equation.

Example 3 ▶ **An Equation with an Extraneous Solution**

Solve $\dfrac{1}{x - 2} = \dfrac{3}{x + 2} - \dfrac{6x}{x^2 - 4}$.

Solution

The LCD is $x^2 - 4$, or $(x + 2)(x - 2)$. Multiply each term by this LCD.

$$\frac{1}{x - 2}(x + 2)(x - 2) = \frac{3}{x + 2}(x + 2)(x - 2) - \frac{6x}{x^2 - 4}(x + 2)(x - 2)$$

$$x + 2 = 3(x - 2) - 6x, \qquad x \neq \pm 2$$

$$x + 2 = 3x - 6 - 6x$$

$$x + 2 = -3x - 6$$

$$4x = -8$$

$$x = -2 \qquad \text{Extraneous solution}$$

In the original equation, $x = -2$ yields a denominator of zero. So, $x = -2$ is an extraneous solution, and the original equation has *no solution*.

STUDY TIP

An equation with a *single fraction* on each side can be cleared of denominators by **cross multiplying,** which is equivalent to multiplying by the LCD and then dividing out.

$$\frac{a}{b} = \frac{c}{d} \qquad \text{LCD is } bd.$$

$$\frac{a}{b} \cdot bd = \frac{c}{d} \cdot bd \qquad \text{Multiply by LCD.}$$

$$ad = cb \qquad \begin{array}{l}\text{Divide out}\\\text{common factors.}\end{array}$$

By comparing the last equation with the original equation, you can see that the left numerator was multiplied by the right denominator and the right numerator was multiplied by the left denominator. Try cross multiplying the following equation to clear the equation of denominators.

$$\frac{2}{x - 3} = \frac{3}{x + 1}$$

Quadratic Equations

A **quadratic equation** in x is an equation that can be written in the general form

$$ax^2 + bx + c = 0$$

where a, b, and c are real numbers with $a \neq 0$. A quadratic equation in x is also known as a **second-degree polynomial equation in x.**

You should be familiar with the following four methods for solving quadratic equations.

Additional Example: Completing the Square Within an Algebraic Expression

Rewrite the denominator as the sum or difference of two squares.

$$\frac{1}{x^2 - 2x - 3}$$

Solution

Complete the square of the denominator.

$$x^2 - 2x - 3 = x^2 - 2x + (-1)^2$$
$$- 3 - (-1)^2$$
$$= (x^2 - 2x + 1) - 4$$
$$= (x - 1)^2 - 2^2$$

The original expression can be written as

$$\frac{1}{(x - 1)^2 - 2^2}.$$

Solving a Quadratic Equation

Factoring: If $ab = 0$, then $a = 0$ or $b = 0$.

Example:
$$x^2 - x - 6 = 0$$
$$(x - 3)(x + 2) = 0$$
$$x - 3 = 0 \quad \Longrightarrow \quad x = 3$$
$$x + 2 = 0 \quad \Longrightarrow \quad x = -2$$

Square Root Principle: If $u^2 = c$, where $c > 0$, then $u = \pm\sqrt{c}$.

Example:
$$(x + 3)^2 = 16$$
$$x + 3 = \pm 4$$
$$x = -3 \pm 4$$
$$x = 1 \quad \text{or} \quad x = -7$$

Completing the Square: If $x^2 + bx = c$, then

$$x^2 + bx + \left(\frac{b}{2}\right)^2 = c + \left(\frac{b}{2}\right)^2$$
$$\left(x + \frac{b}{2}\right)^2 = c + \frac{b^2}{4}.$$

Example:
$$x^2 + 6x = 5$$
$$x^2 + 6x + 3^2 = 5 + 3^2$$
$$(x + 3)^2 = 14$$
$$x + 3 = \pm\sqrt{14}$$
$$x = -3 \pm \sqrt{14}$$

Quadratic Formula: If $ax^2 + bx + c = 0$, then $x = \dfrac{-b \pm \sqrt{b^2 - 4ac}}{2a}$.

Example:
$$2x^2 + 3x - 1 = 0$$
$$x = \frac{-3 \pm \sqrt{3^2 - 4(2)(-1)}}{2(2)}$$
$$= \frac{-3 \pm \sqrt{17}}{4}$$

Example 4 ▶ Solving a Quadratic Equation by Factoring

a.

$2x^2 + 9x + 7 = 3$	Original equation
$2x^2 + 9x + 4 = 0$	Write in general form.
$(2x + 1)(x + 4) = 0$	Factor.
$2x + 1 = 0$ ⟹ $x = -\dfrac{1}{2}$	Set 1st factor equal to 0.
$x + 4 = 0$ ⟹ $x = -4$	Set 2nd factor equal to 0.

The solutions are $x = -\frac{1}{2}$ and $x = -4$. Check these in the original equation.

A common error occurs when there are only two terms in an equation, as in Example 4(b).

Sample Error:

$6x^2 - 3x = 0$

$6x^2 = 3x$

$2x = 1$

$x = \dfrac{1}{2}$

The solution $x = 0$ is lost when $6x^2 = 3x$ is divided by $3x$.

b.

$6x^2 - 3x = 0$	Original equation
$3x(2x - 1) = 0$	Factor.
$3x = 0$ ⟹ $x = 0$	Set 1st factor equal to 0.
$2x - 1 = 0$ ⟹ $x = \dfrac{1}{2}$	Set 2nd factor equal to 0.

The solutions are $x = 0$ and $x = \frac{1}{2}$. Check these in the original equation.

Be sure you see that the Zero-Factor Property works *only* for equations written in general form (in which the right side of the equation is zero). So, all terms must be collected on one side *before* factoring. For instance, in the equation

$$(x - 5)(x + 2) = 8$$

it is *incorrect* to set each factor equal to 8. Try to solve this equation correctly.

Example 5 ▶ Extracting Square Roots

Solve each equation by extracting square roots.

a. $4x^2 = 12$ **b.** $(x - 3)^2 = 7$

Solution

a.

$4x^2 = 12$	Write original equation.
$x^2 = 3$	Divide each side by 4.
$x = \pm\sqrt{3}$	Extract square roots.

The solutions are $x = \sqrt{3}$ and $x = -\sqrt{3}$. Check these in the original equation.

b.

$(x - 3)^2 = 7$	Write original equation.
$x - 3 = \pm\sqrt{7}$	Extract square roots.
$x = 3 \pm \sqrt{7}$	Add 3 to each side.

The solutions are $x = 3 \pm \sqrt{7}$. Check these in the original equation.

Example 6 ▶ **The Quadratic Formula: Two Distinct Solutions**

Use the Quadratic Formula to solve

$$x^2 + 3x = 9.$$

Solution

$x^2 + 3x = 9$	Write original equation.
$x^2 + 3x - 9 = 0$	Write in general form.
$x = \dfrac{-b \pm \sqrt{b^2 - 4ac}}{2a}$	Quadratic Formula
$x = \dfrac{-3 \pm \sqrt{(3)^2 - 4(1)(-9)}}{2(1)}$	Substitute $a = 1$, $b = 3$, and $c = -9$.
$x = \dfrac{-3 \pm \sqrt{45}}{2}$	Simplify.
$x = \dfrac{-3 \pm 3\sqrt{5}}{2}$	Simplify.

The equation has two solutions:

$$x = \frac{-3 + 3\sqrt{5}}{2} \quad \text{and} \quad x = \frac{-3 - 3\sqrt{5}}{2}.$$

Check these in the original equation.

Example 7 ▶ **The Quadratic Formula: One Solution**

Use the Quadratic Formula to solve

$$8x^2 - 24x + 18 = 0.$$

Solution

$8x^2 - 24x + 18 = 0$	Write original equation.
$4x^2 - 12x + 9 = 0$	Divide out common factor of 2.
$x = \dfrac{-b \pm \sqrt{b^2 - 4ac}}{2a}$	Quadratic Formula
$x = \dfrac{-(-12) \pm \sqrt{(-12)^2 - 4(4)(9)}}{2(4)}$	Substitute.
$x = \dfrac{12 \pm \sqrt{0}}{8}$	Simplify.
$x = \dfrac{3}{2}$	Simplify.

This quadratic equation has only one solution: $x = \frac{3}{2}$. Check this in the original equation.

Activities

1. Solve $3x^2 = 5x$.

   Answer: $x = 0, \dfrac{5}{3}$

2. Solve $(x + 3)^2 = 12$.

   Answer: $x = -3 \pm 2\sqrt{3}$

3. Complete the square for the quadratic portion of $\dfrac{1}{\sqrt{x^2 - 6x + 13}}$.

   Answer: $\dfrac{1}{\sqrt{(x - 3)^2 + 4}}$

4. Use the Quadratic Formula to solve $3x^2 = 3x + 1$.

   Answer: $x = \dfrac{3 \pm \sqrt{21}}{6}$

5. If the perimeter of a right triangle is 42 units and the length of the hypotenuse is 20 units, find the lengths of the other two sides of the triangle.

   Answer: 19.888; 2.112

Polynomial Equations of Higher Degree

The methods used to solve quadratic equations can sometimes be extended to polynomials of higher degree

Example 8 ▶ **Solving a Polynomial Equation by Factoring**

Solve $3x^4 = 48x^2$.

Solution

First write the polynomial equation in general form with zero on one side, factor the other side, and then set each factor equal to zero and solve.

$3x^4 = 48x^2$	Write original equation.
$3x^4 - 48x^2 = 0$	Write in general form.
$3x^2(x^2 - 16) = 0$	Factor out common factor.
$3x^2(x + 4)(x - 4) = 0$	Write in factored form.
$3x^2 = 0 \implies x = 0$	Set 1st factor equal to 0.
$x + 4 = 0 \implies x = -4$	Set 2nd factor equal to 0.
$x - 4 = 0 \implies x = 4$	Set 3rd factor equal to 0.

You can check these solutions by substituting in the original equation, as follows.

Check

$3(0)^4 = 48(0)^2$	0 checks. ✓
$3(-4)^4 = 48(-4)^2$	-4 checks. ✓
$3(4)^4 = 48(4)^2$	4 checks. ✓

So, you can conclude that the solutions are $x = 0$, $x = -4$, and $x = 4$.

Example 9 ▶ **Solving a Polynomial Equation by Factoring**

Solve $x^3 - 3x^2 - 3x + 9 = 0$.

Solution

$x^3 - 3x^2 - 3x + 9 = 0$	Write original equation.
$x^2(x - 3) - 3(x - 3) = 0$	Factor by grouping.
$(x - 3)(x^2 - 3) = 0$	Distributive Property
$x - 3 = 0 \implies x = 3$	Set 1st factor equal to 0.
$x^2 - 3 = 0 \implies x = \pm\sqrt{3}$	Set 2nd factor equal to 0.

The solutions are $x = 3$, $x = \sqrt{3}$, and $x = -\sqrt{3}$. Check these in the original equation.

Equations Involving Radicals

The steps involved in solving the remaining equations in this section will often introduce *extraneous solutions*. Operations such as squaring each side of an equation, raising each side of an equation to a rational power, and multiplying each side of an equation by a variable quantity all can introduce extraneous solutions. So, when you use any of these operations, checking is crucial.

Example 10 ► Solving Equations Involving Radicals

a.

$\sqrt{2x + 7} - x = 2$	Original equation
$\sqrt{2x + 7} = x + 2$	Isolate radical.
$2x + 7 = x^2 + 4x + 4$	Square each side.
$0 = x^2 + 2x - 3$	Write in general form.
$0 = (x + 3)(x - 1)$	Factor.
$x + 3 = 0 \implies x = -3$	Set 1st factor equal to 0.
$x - 1 = 0 \implies x = 1$	Set 2nd factor equal to 0.

By checking these values, you can determine that the only solution is $x = 1$.

b.

$\sqrt{2x - 5} - \sqrt{x - 3} = 1$	Original equation
$\sqrt{2x - 5} = \sqrt{x - 3} + 1$	Isolate $\sqrt{2x - 5}$.
$2x - 5 = x - 3 + 2\sqrt{x - 3} + 1$	Square each side.
$2x - 5 = x - 2 + 2\sqrt{x - 3}$	Combine like terms.
$x - 3 = 2\sqrt{x - 3}$	Isolate $2\sqrt{x - 3}$.
$x^2 - 6x + 9 = 4(x - 3)$	Square each side.
$x^2 - 10x + 21 = 0$	Write in general form.
$(x - 3)(x - 7) = 0$	Factor.
$x - 3 = 0 \implies x = 3$	Set 1st factor equal to 0.
$x - 7 = 0 \implies x = 7$	Set 2nd factor equal to 0.

The solutions are $x = 3$ and $x = 7$. Check these in the original equation.

To show why the radical should be isolated, have students square each side of the equation before isolating the radical. Compare problem-solving strategies to help convince students of the need to isolate the radical.

STUDY TIP

The essential operations in Example 10 are isolating the square root and squaring each side. In Example 11, this is equivalent to isolating the factor with the rational exponent and raising each side to the *reciprocal power*.

Example 11 ► Solving an Equation Involving a Rational Exponent

$(x - 4)^{2/3} = 25$	Original equation
$x - 4 = 25^{3/2}$	Raise each side to the $\frac{3}{2}$ power.
$x - 4 = 125$	Simplify.
$x = 129$	Add 4 to each side.

The solution is $x = 129$. Check this in the original equation.

Equations Involving Absolute Values

To solve an equation involving an absolute value, remember that the expression inside the absolute value signs can be positive or negative. This results in *two* separate equations, each of which must be solved. For instance, the equation $|x - 2| = 3$ results in the two equations $x - 2 = 3$ and $-(x - 2) = 3$, which implies that the equation has two solutions: $x = 5$ and $x = -1$.

Example 12 ▶ Solving an Equation Involving Absolute Value

Solve $|x^2 - 3x| = -4x + 6$.

Solution

Because the variable expression inside the absolute value signs can be positive or negative, you must solve the following two equations.

First Equation

$x^2 - 3x = -4x + 6$	Use positive expression.
$x^2 + x - 6 = 0$	Write in general form.
$(x + 3)(x - 2) = 0$	Factor.
$x + 3 = 0 \implies x = -3$	Set 1st factor equal to 0.
$x - 2 = 0 \implies x = 2$	Set 2nd factor equal to 0.

Second Equation

$-(x^2 - 3x) = -4x + 6$	Use negative expression.
$x^2 - 7x + 6 = 0$	Write in general form.
$(x - 1)(x - 6) = 0$	Factor.
$x - 1 = 0 \implies x = 1$	Set 1st factor equal to 0.
$x - 6 = 0 \implies x = 6$	Set 2nd factor equal to 0.

Check

$\left|(-3)^2 - 3(-3)\right| \overset{?}{=} -4(-3) + 6$	Substitute -3 for x.
$18 = 18$	-3 checks. ✓
$\left|(2)^2 - 3(2)\right| \overset{?}{=} -4(2) + 6$	Substitute 2 for x.
$2 \neq -2$	2 does not check.
$\left|(1)^2 - 3(1)\right| \overset{?}{=} -4(1) + 6$	Substitute 1 for x.
$2 = 2$	1 checks. ✓
$\left|(6)^2 - 3(6)\right| \overset{?}{=} -4(6) + 6$	Substitute 6 for x.
$18 \neq -18$	6 does not check.

The solutions are $x = -3$ and $x = 1$.

P.2 Exercises

In Exercises 1–10, determine whether the equation is an identity or a conditional equation.

1. $2(x - 1) = 2x - 2$

2. $3(x + 2) = 5x + 4$

3. $-6(x - 3) + 5 = -2x + 10$

4. $3(x + 2) - 5 = 3x + 1$

5. $4(x + 1) - 2x = 2(x + 2)$

6. $-7(x - 3) + 4x = 3(7 - x)$

7. $x^2 - 8x + 5 = (x - 4)^2 - 11$

8. $x^2 + 2(3x - 2) = x^2 + 6x - 4$

9. $3 + \dfrac{1}{x + 1} = \dfrac{4x}{x + 1}$ **10.** $\dfrac{5}{x} + \dfrac{3}{x} = 24$

In Exercises 11–26, solve the equation and check your solution.

11. $x + 11 = 15$ **12.** $7 - x = 19$

13. $7 - 2x = 25$ **14.** $7x + 2 = 23$

15. $8x - 5 = 3x + 20$ **16.** $7x + 3 = 3x - 17$

17. $2(x + 5) - 7 = 3(x - 2)$

18. $3(x + 3) = 5(1 - x) - 1$

19. $x - 3(2x + 3) = 8 - 5x$

20. $9x - 10 = 5x + 2(2x - 5)$

21. $\dfrac{5x}{4} + \dfrac{1}{2} = x - \dfrac{1}{2}$ **22.** $\dfrac{x}{5} - \dfrac{x}{2} = 3 + \dfrac{3x}{10}$

23. $\frac{3}{2}(z + 5) - \frac{1}{4}(z + 24) = 0$

24. $\dfrac{3x}{2} + \dfrac{1}{4}(x - 2) = 10$

25. $0.25x + 0.75(10 - x) = 3$

26. $0.60x + 0.40(100 - x) = 50$

In Exercises 27–48, solve the equation and check your solution. (If not possible, explain why.)

27. $x + 8 = 2(x - 2) - x$

28. $8(x + 2) - 3(2x + 1) = 2(x + 5)$

29. $\dfrac{100 - 4x}{3} = \dfrac{5x + 6}{4} + 6$

30. $\dfrac{17 + y}{y} + \dfrac{32 + y}{y} = 100$

31. $\dfrac{5x - 4}{5x + 4} = \dfrac{2}{3}$ **32.** $\dfrac{10x + 3}{5x + 6} = \dfrac{1}{2}$

33. $10 - \dfrac{13}{x} = 4 + \dfrac{5}{x}$ **34.** $\dfrac{15}{x} - 4 = \dfrac{6}{x} + 3$

35. $3 = 2 + \dfrac{2}{z + 2}$ **36.** $\dfrac{1}{x} + \dfrac{2}{x - 5} = 0$

37. $\dfrac{x}{x + 4} + \dfrac{4}{x + 4} + 2 = 0$

38. $\dfrac{7}{2x + 1} - \dfrac{8x}{2x - 1} = -4$

39. $\dfrac{2}{(x - 4)(x - 2)} = \dfrac{1}{x - 4} + \dfrac{2}{x - 2}$

40. $\dfrac{4}{x - 1} + \dfrac{6}{3x + 1} = \dfrac{15}{3x + 1}$

41. $\dfrac{1}{x - 3} + \dfrac{1}{x + 3} = \dfrac{10}{x^2 - 9}$

42. $\dfrac{1}{x - 2} + \dfrac{3}{x + 3} = \dfrac{4}{x^2 + x - 6}$

43. $\dfrac{3}{x^2 - 3x} + \dfrac{4}{x} = \dfrac{1}{x - 3}$

44. $\dfrac{6}{x} - \dfrac{2}{x + 3} = \dfrac{3(x + 5)}{x^2 + 3x}$

45. $(x + 2)^2 + 5 = (x + 3)^2$

46. $(x + 1)^2 + 2(x - 2) = (x + 1)(x - 2)$

47. $(x + 2)^2 - x^2 = 4(x + 1)$

48. $(2x + 1)^2 = 4(x^2 + x + 1)$

In Exercises 49–54, write the quadratic equation in general form.

49. $2x^2 = 3 - 8x$ **50.** $x^2 = 16x$

51. $(x - 3)^2 = 3$ **52.** $13 - 3(x + 7)^2 = 0$

53. $\frac{1}{5}(3x^2 - 10) = 18x$ **54.** $x(x + 2) = 5x^2 + 1$

In Exercises 55–68, solve the quadratic equation by factoring.

55. $6x^2 + 3x = 0$ **56.** $9x^2 - 1 = 0$

57. $x^2 - 2x - 8 = 0$ **58.** $x^2 - 10x + 9 = 0$

59. $x^2 + 10x + 25 = 0$ **60.** $4x^2 + 12x + 9 = 0$

61. $3 + 5x - 2x^2 = 0$ **62.** $2x^2 = 19x + 33$

63. $x^2 + 4x = 12$ **64.** $-x^2 + 8x = 12$

65. $\frac{3}{4}x^2 + 8x + 20 = 0$ **66.** $\frac{1}{8}x^2 - x - 16 = 0$

67. $x^2 + 2ax + a^2 = 0$ **68.** $(x + a)^2 - b^2 = 0$

In Exercises 69–82, solve the equation by extracting square roots. List both the exact solution and the decimal solution rounded to two decimal places.

69. $x^2 = 49$

70. $x^2 = 169$

71. $x^2 = 11$

72. $x^2 = 32$

73. $3x^2 = 81$

74. $9x^2 = 36$

75. $(x - 12)^2 = 16$

76. $(x + 13)^2 = 25$

77. $(x + 2)^2 = 14$

78. $(x - 5)^2 = 30$

79. $(2x - 1)^2 = 18$

80. $(4x + 7)^2 = 44$

81. $(x - 7)^2 = (x + 3)^2$

82. $(x + 5)^2 = (x + 4)^2$

In Exercises 83–92, solve the quadratic equation by completing the square.

83. $x^2 - 2x = 0$

84. $x^2 + 4x = 0$

85. $x^2 + 4x - 32 = 0$

86. $x^2 - 2x - 3 = 0$

87. $x^2 + 6x + 2 = 0$

88. $x^2 + 8x + 14 = 0$

89. $9x^2 - 18x = -3$

90. $9x^2 - 12x = 14$

91. $8 + 4x - x^2 = 0$

92. $4x^2 - 4x - 99 = 0$

In Exercises 93–116, use the Quadratic Formula to solve the equation.

93. $2x^2 + x - 1 = 0$

94. $2x^2 - x - 1 = 0$

95. $16x^2 + 8x - 3 = 0$

96. $25x^2 - 20x + 3 = 0$

97. $2 + 2x - x^2 = 0$

98. $x^2 - 10x + 22 = 0$

99. $x^2 + 14x + 44 = 0$

100. $6x = 4 - x^2$

101. $x^2 + 8x - 4 = 0$

102. $4x^2 - 4x - 4 = 0$

103. $12x - 9x^2 = -3$

104. $16x^2 + 22 = 40x$

105. $9x^2 + 24x + 16 = 0$

106. $36x^2 + 24x - 7 = 0$

107. $4x^2 + 4x = 7$

108. $16x^2 - 40x + 5 = 0$

109. $28x - 49x^2 = 4$

110. $3x + x^2 - 1 = 0$

111. $8t = 5 + 2t^2$

112. $25h^2 + 80h + 61 = 0$

113. $(y - 5)^2 = 2y$

114. $(z + 6)^2 = -2z$

115. $\frac{1}{2}x^2 + \frac{3}{8}x = 2$

116. $\left(\frac{5}{7}x - 14\right)^2 = 8x$

In Exercises 117–124, use the Quadratic Formula to solve the equation. (Round your answers to three decimal places.)

117. $5.1x^2 - 1.7x - 3.2 = 0$

118. $2x^2 - 2.50x - 0.42 = 0$

119. $-0.067x^2 - 0.852x + 1.277 = 0$

120. $-0.005x^2 + 0.101x - 0.193 = 0$

121. $422x^2 - 506x - 347 = 0$

122. $1100x^2 + 326x - 715 = 0$

123. $12.67x^2 + 31.55x + 8.09 = 0$

124. $-3.22x^2 - 0.08x + 28.651 = 0$

In Exercises 125–134, solve the equation using any convenient method.

125. $x^2 - 2x - 1 = 0$

126. $11x^2 + 33x = 0$

127. $(x + 3)^2 = 81$

128. $x^2 - 14x + 49 = 0$

129. $x^2 - x - \frac{11}{4} = 0$

130. $x^2 + 3x - \frac{3}{4} = 0$

131. $(x + 1)^2 = x^2$

132. $a^2x^2 - b^2 = 0$

133. $3x + 4 = 2x^2 - 7$

134. $4x^2 + 2x + 4 = 2x + 8$

In Exercises 135–152, find all real number solutions of the equation. Check your solutions in the original equation.

135. $4x^4 - 18x^2 = 0$

136. $20x^3 - 125x = 0$

137. $x^4 - 81 = 0$

138. $x^6 - 64 = 0$

139. $x^3 + 216 = 0$

140. $27x^3 - 512 = 0$

141. $5x^3 + 30x^2 + 45x = 0$

142. $9x^4 - 24x^3 + 16x^2 = 0$

143. $x^3 - 3x^2 - x + 3 = 0$

144. $x^3 + 2x^2 + 3x + 6 = 0$

145. $x^4 - x^3 + x - 1 = 0$

146. $x^4 + 2x^3 - 8x - 16 = 0$

147. $x^4 - 4x^2 + 3 = 0$ **148.** $x^4 + 5x^2 - 36 = 0$

149. $4x^4 - 65x^2 + 16 = 0$

150. $36t^4 + 29t^2 - 7 = 0$

151. $x^6 + 7x^3 - 8 = 0$ **152.** $x^6 + 3x^3 + 2 = 0$

In Exercises 153–170, find all solutions of the equation. Check your solutions in the original equation.

153. $\sqrt{2x} - 10 = 0$ **154.** $4\sqrt{x} - 3 = 0$

155. $\sqrt{x - 10} - 4 = 0$ **156.** $\sqrt{5 - x} - 3 = 0$

157. $\sqrt[3]{2x + 5} + 3 = 0$ **158.** $\sqrt[3]{3x + 1} - 5 = 0$

159. $-\sqrt{26 - 11x} + 4 = x$

160. $x + \sqrt{31 - 9x} = 5$

161. $\sqrt{x + 1} = \sqrt{3x + 1}$ **162.** $\sqrt{x + 5} = \sqrt{x - 5}$

163. $(x - 5)^{3/2} = 8$ **164.** $(x + 3)^{3/2} = 8$

165. $(x + 3)^{2/3} = 8$ **166.** $(x + 2)^{2/3} = 9$

167. $(x^2 - 5)^{3/2} = 27$

168. $(x^2 - x - 22)^{3/2} = 27$

169. $3x(x - 1)^{1/2} + 2(x - 1)^{3/2} = 0$

170. $4x^2(x - 1)^{1/3} + 6x(x - 1)^{4/3} = 0$

In Exercises 171–184, find all solutions of the equation. Check your solutions in the original equation.

171. $x = \dfrac{3}{x} + \dfrac{1}{2}$

172. $\dfrac{4}{x} - \dfrac{5}{3} = \dfrac{x}{6}$

173. $\dfrac{1}{x} - \dfrac{1}{x+1} = 3$

174. $\dfrac{4}{x+1} - \dfrac{3}{x+2} = 1$

175. $\dfrac{20 - x}{x} = x$

176. $4x + 1 = \dfrac{3}{x}$

177. $\dfrac{x}{x^2 - 4} + \dfrac{1}{x+2} = 3$

178. $\dfrac{x+1}{3} - \dfrac{x+1}{x+2} = 0$

179. $|2x - 1| = 5$

180. $|3x + 2| = 7$

181. $|x| = x^2 + x - 3$

182. $|x^2 + 6x| = 3x + 18$

183. $|x + 1| = x^2 - 5$

184. $|x - 10| = x^2 - 10x$

▶ Model It

185. *Anthropology* The relationship between the length of an adult's femur (thigh bone) and the height of the adult can be approximated by the linear equations

$$y = 0.432x - 10.44 \qquad \text{Female}$$

$$y = 0.449x - 12.15 \qquad \text{Male}$$

where y is the length of the femur in inches and x is the height of the adult in inches (see figure).

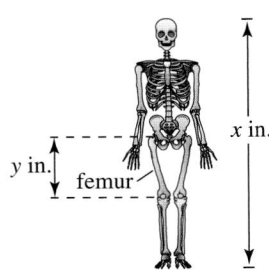

x in.

y in.
femur

(a) An anthropologist discovers a femur belonging to an adult human female. The bone is 16 inches long. Estimate the height of the female.

(b) From the foot bones of an adult human male, an anthropologist estimates that the person's height was 69 inches. A few feet away from the site where the foot bones were discovered, the anthropologist discovers a male adult femur that is 19 inches long. Is it likely that the foot bones and the femur came from the same person?

▶ Model It *(continued)*

(c) Complete the table to determine if there is a height of an adult for which an anthropologist would not be able to determine whether the femur belonged to a male or a female.

Height, x	Female femur length, y	Male femur length, y
60		
70		
80		
90		
100		
110		

(d) Solve part (c) algebraically by setting the two equations equal to each other and solving for x. Compare your solutions. Do you believe an anthropologist would ever have the problem of not being able to determine whether a femur belonged to a male or female? Why or why not?

---

186. *Operating Cost* A delivery company has a fleet of vans. The annual operating cost C per van is $C = 0.32m + 2500$, where m is the number of miles traveled by a van in a year. What number of miles will yield an annual operating cost of $10,000?

187. *Flood Control* A river has risen 8 feet above its flood stage. The water begins to recede at a rate of 3 inches per hour. Write a mathematical model that shows the number of feet above flood stage after t hours. If the water continually recedes at this rate, when will the river be 1 foot above its flood stage?

188. *Floor Space* The floor of a one-story building is 14 feet longer than it is wide. The building has 1632 square feet of floor space.

(a) Draw a diagram that gives a visual representation of the floor space. Represent the width as w and show the length in terms of w.

(b) Write a quadratic equation in terms of w.

(c) Find the length and width of the floor of the building.

189. *Geometry* The hypotenuse of an isosceles right triangle is 5 centimeters long. How long are its sides?

190. *Geometry* An equilateral triangle has a height of 10 inches. How long is one of its sides? (*Hint:* Use the height of the triangle to partition the triangle into two congruent right triangles.)

191. *Flying Speed* Two planes leave simultaneously from Chicago's O'Hare Airport, one flying due north and the other due east (see figure). The northbound plane is flying 50 miles per hour faster than the eastbound plane. After 3 hours, the planes are 2440 miles apart. Find the speed of each plane.

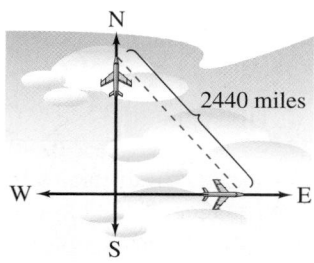

192. *Airline Passengers* An airline offers daily flights between Chicago and Denver. The total monthly cost C (in millions of dollars) of these flights is

$$C = \sqrt{0.2x + 1}$$

where x is the number of passengers (in thousands). The total cost of the flights for June is 2.5 million dollars. How many passengers flew in June?

193. *Demand* The demand equation for a video game is modeled by

$$p = 40 - \sqrt{0.01x + 1}$$

where x is the number of units demanded per day and p is the price per unit. Approximate the demand when the price is \$37.55.

194. *Demand* The demand equation for a coffee maker is modeled by

$$p = 40 - \sqrt{0.0001x + 1}$$

where x is the number of units demanded per day and p is the price per unit. Approximate the demand when the price is \$34.70.

Synthesis

True or False? In Exercises 195–197, determine whether the statement is true or false. Justify your answer.

195. The equation $x(3 - x) = 10$ is a linear equation.

196. The equation $x^2 + 9x - 5 = 4 - x^3$ has no real solution.

197. When solving an absolute value equation, you will always have to check more than one solution.

198. To solve the equation

$$2x^2 + 3x = 15x$$

a student divides each side by x and solves the equation $2x + 3 = 15$. The resulting solution ($x = 6$) satisfies the original equation. Is there an error? Explain.

199. *Think About It* What is meant by "equivalent equations"? Give an example of two equivalent equations.

200. *Writing* Describe the steps used to transform an equation into an equivalent equation.

201. Solve $3(x + 4)^2 + (x + 4) - 2 = 0$ in two ways.

(a) Let $u = x + 4$, and solve the resulting equation for u. Then solve the u-solution for x.

(b) Expand and collect like terms in the equation, and solve the resulting equation for x.

(c) Which method is easier? Explain.

202. Solve the equations, given that a and b are not zero.

(a) $ax^2 + bx = 0$ (b) $ax^2 - ax = 0$

In Exercises 203 and 204, consider an equation of the form $x + |x - a| = b$, where a and b are constants.

203. Find a and b when the solution to the equation is $x = 9$. (There are many correct answers.)

204. *Writing* Write a short paragraph listing the steps required to solve this equation involving absolute values.

In Exercises 205 and 206, consider an equation of the form $x + \sqrt{x - a} = b$, where a and b are constants.

205. Find a and b when the solution to the equation is $x = 20$. (There are many correct answers.)

206. *Writing* Write a short paragraph listing the steps required to solve this equation involving radicals.

The Cartesian Plane and Graphs of Equations

▶ **What you should learn**

- How to plot points in the Cartesian plane
- How to use the Distance Formula to find the distance between two points
- How to use the Midpoint Formula to find the midpoint of a line segment
- How to sketch graphs of equations
- How to find *x*- and *y*-intercepts of graphs of equations
- How to use symmetry to sketch graphs of equations
- How to find equations and sketch graphs of circles

▶ **Why you should learn it**

The graph of an equation can help you see relationships between real-life quantities. For example, in Exercise 93 on page 37, a graph can be used to estimate the life expectancies of children who are born in the years 2005 and 2010.

The Cartesian Plane

Just as you can represent real numbers by points on a real number line, you can represent ordered pairs of real numbers by points in a plane called the **rectangular coordinate system,** or the **Cartesian plane,** named after the French mathematician René Descartes (1596–1650).

The Cartesian plane is formed by using two real number lines intersecting at right angles, as shown in Figure P.7. The horizontal real number line is usually called the **x-axis,** and the vertical real number line is usually called the **y-axis.** The point of intersection of these two axes is the **origin,** and the two axes divide the plane into four parts called **quadrants.**

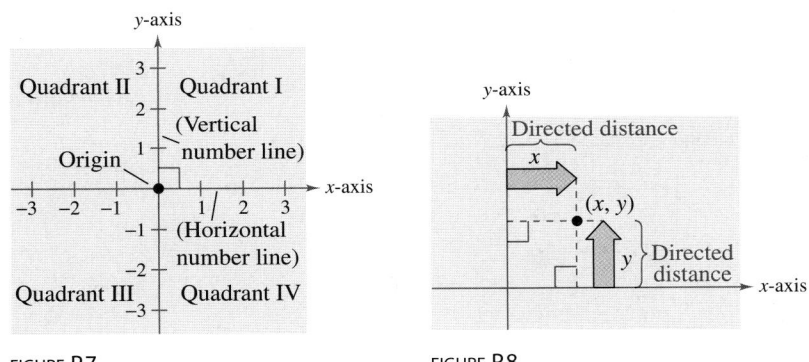

FIGURE P.7   FIGURE P.8

Each point in the plane corresponds to an **ordered pair** (x, y) of real numbers x and y, called **coordinates** of the point. The **x-coordinate** represents the directed distance from the y-axis to the point, and the **y-coordinate** represents the directed distance from the x-axis to the point, as shown in Figure P.8.

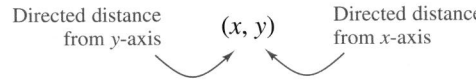

Directed distance   Directed distance
from y-axis (x, y) from x-axis

The notation (x, y) denotes both a point in the plane and an open interval on the real number line. The context will tell you which meaning is intended.

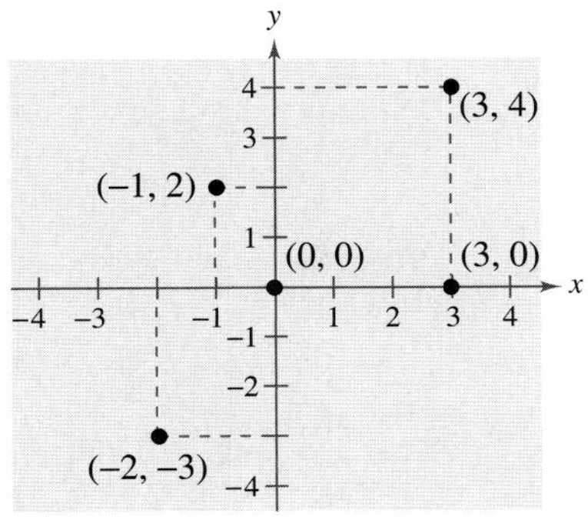

Example 1 ▶ **Plotting Points in the Cartesian Plane**   www ⊙

Plot the points $(-1, 2)$, $(3, 4)$, $(0, 0)$, $(3, 0)$, and $(-2, -3)$.

Solution

To plot the point $(-1, 2)$, imagine a vertical line through -1 on the x-axis and a horizontal line through 2 on the y-axis. The intersection of these two lines is the point $(-1, 2)$. The other four points can be plotted in a similar way, as shown in Figure P.9.

FIGURE P.9

The beauty of a rectangular coordinate system is that it allows you to *see* relationships between two variables. It would be difficult to overestimate the importance of Descartes's introduction of coordinates in the plane. Today, his ideas are in common use in virtually every scientific and business-related field.

| Example 2 ▶ | Sketching a Scatter Plot | |

From 1990 through 1999, the amount *A* (in millions of dollars) spent on skiing equipment in the United States is shown in the table, where *t* represents the year. Sketch a scatter plot of the data. (Source: National Sporting Goods Association)

Solution

To sketch a *scatter plot* of the data shown in the table, you simply represent each pair of values by an ordered pair (t, A) and plot the resulting points, as shown in Figure P.10. For instance, the first pair of values is represented by the ordered pair (1990, 475). Note that the break in the *t*-axis indicates that the numbers between 0 and 1990 have been omitted.

Year, *t*	Amount, *A*
1990	475
1991	577
1992	521
1993	569
1994	609
1995	562
1996	707
1997	723
1998	718
1999	739

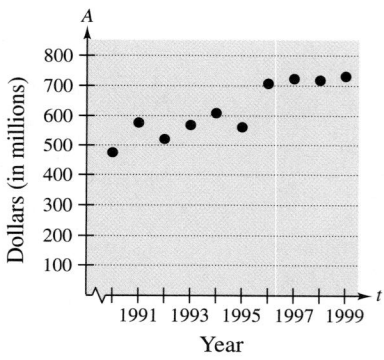

FIGURE P.10

STUDY TIP

In Example 2, you could have let $t = 1$ represent the year 1990. In that case, the horizontal axis would not have been broken, and the tick marks would have been labeled 1 through 10 (instead of 1990 through 1999).

Technology

The scatter plot in Example 2 is only one way to represent the data graphically. Two other techniques are shown at the right. The first is a bar graph and the second is a line graph. All three graphical representations were created with a computer. If you have access to a graphing utility, try using it to represent graphically the data given in Example 2.

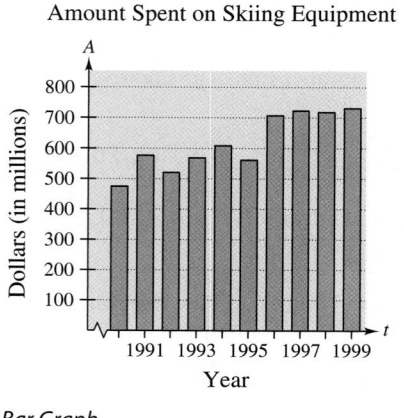

Bar Graph

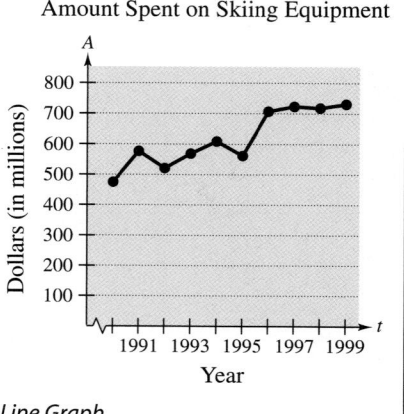

Line Graph

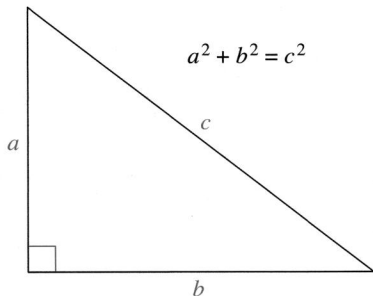

$$a^2 + b^2 = c^2$$

a

c

b

FIGURE **P.11**

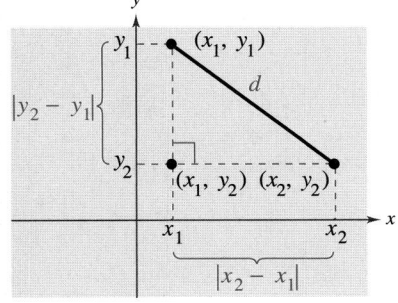

FIGURE **P.12**

The Distance Formula

Recall from the Pythagorean Theorem that, for a right triangle with hypotenuse of length c and sides of lengths a and b, you have

$$a^2 + b^2 = c^2 \qquad \text{Pythagorean Theorem}$$

as shown in Figure P.11. (The converse is also true. That is, if $a^2 + b^2 = c^2$, then the triangle is a right triangle.)

Suppose you want to determine the distance d between two points (x_1, y_1) and (x_2, y_2) in the plane. With these two points, a right triangle can be formed, as shown in Figure P.12. The length of the vertical side of the triangle is $|y_2 - y_1|$, and the length of the horizontal side is $|x_2 - x_1|$. By the Pythagorean Theorem, you can write

$$d^2 = |x_2 - x_1|^2 + |y_2 - y_1|^2$$
$$d = \sqrt{|x_2 - x_1|^2 + |y_2 - y_1|^2}$$
$$d = \sqrt{(x_2 - x_1)^2 + (y_2 - y_1)^2}.$$

This result is the **Distance Formula.**

The Distance Formula

The distance d between the points (x_1, y_1) and (x_2, y_2) in the plane is

$$d = \sqrt{(x_2 - x_1)^2 + (y_2 - y_1)^2}.$$

Example 3 ▶ **Finding a Distance**

Find the distance between the points $(-2, 1)$ and $(3, 4)$.

Solution

Let $(x_1, y_1) = (-2, 1)$ and $(x_2, y_2) = (3, 4)$. Then apply the Distance Formula.

$$d = \sqrt{(x_2 - x_1)^2 + (y_2 - y_1)^2} \qquad \text{Distance Formula}$$
$$= \sqrt{[3 - (-2)]^2 + (4 - 1)^2} \qquad \text{Substitute for } x_1, y_1, x_2, \text{ and } y_2.$$
$$= \sqrt{(5)^2 + (3)^2} \qquad \text{Simplify.}$$
$$= \sqrt{34} \qquad \text{Simplify.}$$
$$\approx 5.83 \qquad \text{Use a calculator.}$$

Note in Figure P.13 that a distance of 5.83 looks about right. You can use the Pythagorean Theorem to check that the distance is correct.

$$d^2 \stackrel{?}{=} 3^2 + 5^2 \qquad \text{Pythagorean Theorem}$$
$$\left(\sqrt{34}\right)^2 \stackrel{?}{=} 3^2 + 5^2 \qquad \text{Substitute for } d.$$
$$34 = 34 \qquad \text{Distance checks. } \checkmark$$

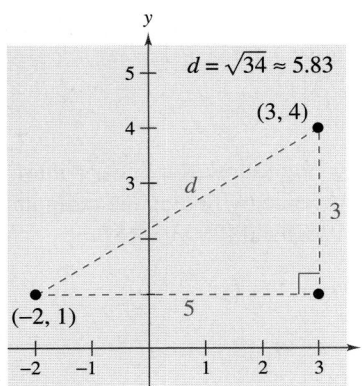

$d = \sqrt{34} \approx 5.83$

$(3, 4)$

d

3

$(-2, 1)$

5

FIGURE **P.13**

The Midpoint Formula

To find the **midpoint** of the line segment that joins two points in a coordinate plane, you can simply find the average values of the respective coordinates of the two endpoints using the **Midpoint Formula.**

The Midpoint Formula

The midpoint of the line segment joining the points (x_1, y_1) and (x_2, y_2) is given by the Midpoint Formula

$$\text{Midpoint} = \left(\frac{x_1 + x_2}{2}, \frac{y_1 + y_2}{2} \right).$$

For a proof of the Midpoint Formula, see Proofs in Mathematics on page 121.

Example 4 ▶ **Finding a Line Segment's Midpoint**

Find the midpoint of the line segment joining the points $(-5, -3)$ and $(9, 3)$, as shown in Figure P.14.

Solution

Let $(x_1, y_1) = (-5, -3)$ and $(x_2, y_2) = (9, 3)$.

$$\text{Midpoint} = \left(\frac{x_1 + x_2}{2}, \frac{y_1 + y_2}{2} \right) \qquad \text{Midpoint Formula}$$

$$= \left(\frac{-5 + 9}{2}, \frac{-3 + 3}{2} \right) \qquad \text{Substitute for } x_1, y_1, x_2, \text{ and } y_2.$$

$$= (2, 0) \qquad \text{Simplify.}$$

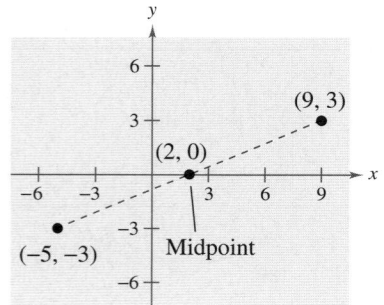

FIGURE P.14

Example 5 ▶ **Estimating Annual Revenue**

The United Parcel Service had annual revenues of $24.8 billion in 1998 and $29.8 billion in 2000. Without knowing any additional information, what would you estimate the 1999 revenue to have been? (Source: United Parcel Service of America Corp.)

Solution

One solution to the problem is to assume that revenue followed a linear pattern. With this assumption, you can estimate the 1999 revenue by finding the midpoint of the line segment connecting the points (1998, 24.8) and (2000, 29.8).

$$\text{Midpoint} = \left(\frac{1998 + 2000}{2}, \frac{24.8 + 29.8}{2} \right)$$

$$= (1999, 27.3)$$

So, you would estimate the 1999 revenue to have been about $27.3 billion, as shown in Figure P.15. (The actual 1999 revenue was $27.1 billion.)

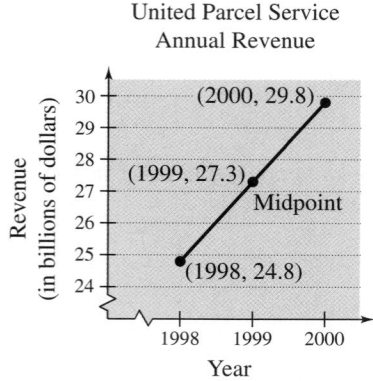

United Parcel Service
Annual Revenue

FIGURE P.15

The Graph of an Equation

Earlier in this section, you used a coordinate system to represent graphically the relationship between two quantities. There, the graphical picture consisted of a collection of points in a coordinate plane (see Example 2).

Frequently, a relationship between two quantities is expressed as an **equation in two variables.** For instance, $y = 7 - 3x$ is an equation in x and y. An ordered pair (a, b) is a **solution** or **solution point** of an equation in x and y if the equation is true when a is substituted for x and b is substituted for y. For instance, $(1, 4)$ is a solution of $y = 7 - 3x$ because $4 = 7 - 3(1)$ is a true statement.

In the remainder of this section, you will review some basic procedures for sketching the graph of an equation in two variables. The **graph of an equation** is the set of all points that are solutions of the equation.

Example 6 ▶ Sketching the Graph of an Equation

Sketch the graph of $y = x^2 - 2$.

Solution

Begin by constructing a table of values.

x	$y = x^2 - 2$	(x, y)
-2	2	$(-2, 2)$
-1	-1	$(-1, -1)$
0	-2	$(0, -2)$
1	-1	$(1, -1)$
2	2	$(2, 2)$
3	7	$(3, 7)$

Next, plot the points given in the table, as shown in Figure P.16. Finally, connect the points with a smooth curve, as shown in Figure P.17.

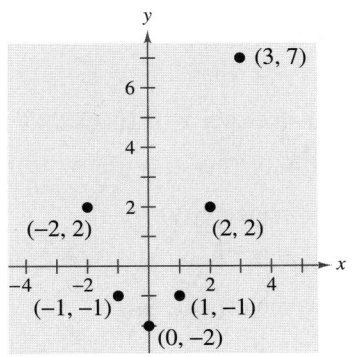

FIGURE P.16

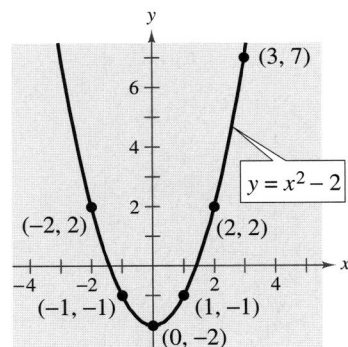

FIGURE P.17

Intercepts of a Graph

It is often easy to determine the solution points that have zero as either the *x*-coordinate or the *y*-coordinate. These points are called **intercepts** because they are the points at which the graph intersects the *x*- or *y*-axis. It is possible for a graph to have no intercepts, one intercept, or several intercepts, as shown in Figure P.18.

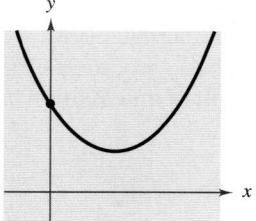

No x-intercept
One y-intercept
FIGURE P.18

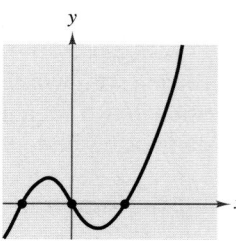

Three x-intercepts
One y-intercept

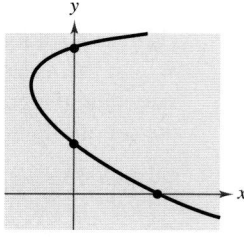

One x-intercept
Two y-intercepts

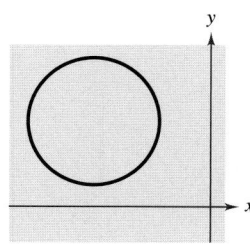

No intercepts

Note that an *x*-intercept is written as the ordered pair $(x, 0)$ and a *y*-intercept is written as the ordered pair $(0, y)$.

Finding Intercepts

1. To find *x*-intercepts, let *y* be zero and solve the equation for *x*.

2. To find *y*-intercepts, let *x* be zero and solve the equation for *y*.

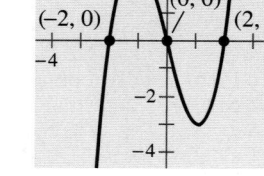

FIGURE P.19

Example 7 ▶ Finding *x*- and *y*-Intercepts

Find the *x*- and *y*-intercepts of the graph of each equation.

a. $y = x^3 - 4x$

b. $y^2 = x + 4$

Solution

a. Let $y = 0$. Then $0 = x^3 - 4x = x(x^2 - 4)$ has solutions $x = 0$ and $x = \pm 2$.

   x-intercepts: $(0, 0), (2, 0), (-2, 0)$

   Let $x = 0$. Then $y = (0)^3 - 4(0) = 0$.

   y-intercept: $(0, 0)$ (See Figure P.19.)

b. Let $y = 0$. Then $(0)^2 = x + 4$, and $-4 = x$.

   x-intercept: $(-4, 0)$

   Let $x = 0$. Then $y^2 = 0 + 4 = 4$ has solutions $y = \pm 2$.

   y-intercepts: $(0, 2), (0, -2)$ (See Figure P.20.)

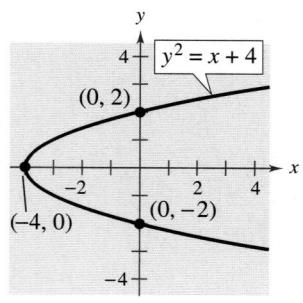

FIGURE P.20

Symmetry

Graphs of equations can have **symmetry** with respect to one of the coordinate axes or with respect to the origin. Symmetry with respect to the *x*-axis means that if the Cartesian plane were folded along the *x*-axis, the portion of the graph above the *x*-axis would coincide with the portion below the *x*-axis. Symmetry with respect to the *y*-axis or the origin can be described in a similar manner, as shown in Figure P.21.

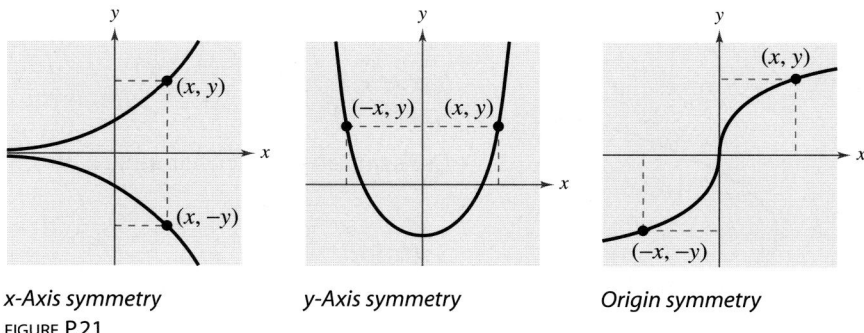

x-Axis symmetry   *y*-Axis symmetry   Origin symmetry
FIGURE P.21

Knowing the symmetry of a graph *before* attempting to sketch it is helpful, because then you need only half as many solution points to sketch the graph. There are three basic types of symmetry, described as follows.

Graphical Tests for Symmetry

1. A graph is **symmetric with respect to the *x*-axis** if, whenever (x, y) is on the graph, $(x, -y)$ is also on the graph.

2. A graph is **symmetric with respect to the *y*-axis** if, whenever (x, y) is on the graph, $(-x, y)$ is also on the graph.

3. A graph is **symmetric with respect to the origin** if, whenever (x, y) is on the graph, $(-x, -y)$ is also on the graph.

Example 8 ► **Testing for Symmetry**

The graph of

$$y = x^2 - 2$$

is symmetric with respect to the *y*-axis because the point $(-x, y)$ is also on the graph of $y = x^2 - 2$. (See Figure P.22.) The table below confirms that the graph is symmetric with respect to the *y*-axis.

x	-3	-2	-1	1	2	3
y	7	2	-1	-1	2	7
(x, y)	$(-3, 7)$	$(-2, 2)$	$(-1, -1)$	$(1, -1)$	$(2, 2)$	$(3, 7)$

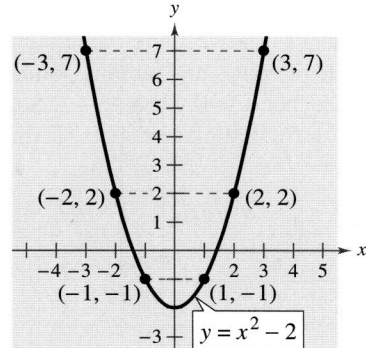

FIGURE P.22 *y-Axis symmetry*

Algebraic Tests for Symmetry

1. The graph of an equation is symmetric with respect to the x-axis if replacing y with $-y$ yields an equivalent equation.

2. The graph of an equation is symmetric with respect to the y-axis if replacing x with $-x$ yields an equivalent equation.

3. The graph of an equation is symmetric with respect to the origin if replacing x with $-x$ and y with $-y$ yields an equivalent equation.

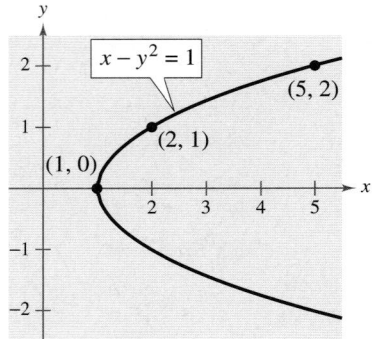

FIGURE P.23

Example 9 ▶ **Using Symmetry as a Sketching Aid**

Use symmetry to sketch the graph of

$$x - y^2 = 1.$$

Solution

Of the three tests for symmetry, the only one that is satisfied is the test for x-axis symmetry because $x - (-y)^2 = 1$ is equivalent to $x - y^2 = 1$. So, the graph is symmetric with respect to the x-axis. Using symmetry, you need only to find the solution points above the x-axis and then reflect them to obtain the graph, as shown in Figure P.23.

y	$x = y^2 + 1$	(x, y)
0	1	$(1, 0)$
1	2	$(2, 1)$
2	5	$(5, 2)$

STUDY TIP

Notice that when creating the table in Example 9, it is easier to choose y-values and then find the corresponding x-values of the ordered pairs.

Example 10 ▶ **Sketching the Graph of an Equation**

Sketch the graph of

$$y = |x - 1|.$$

Solution

This equation fails all three tests for symmetry and consequently its graph is not symmetric with respect to either axis or to the origin. The absolute value sign indicates that y is always nonnegative. Create a table of values and plot the points as shown in Figure P.24. From the table, you can see that $x = 0$ when $y = 1$. So, the y-intercept is $(0, 1)$. Similarly, $y = 0$ when $x = 1$. So, the x-intercept is $(1, 0)$.

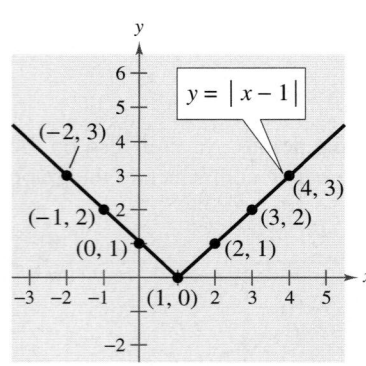

FIGURE P.24

x	-2	-1	0	1	2	3	4
$y = |x - 1|$	3	2	1	0	1	2	3
(x, y)	$(-2, 3)$	$(-1, 2)$	$(0, 1)$	$(1, 0)$	$(2, 1)$	$(3, 2)$	$(4, 3)$

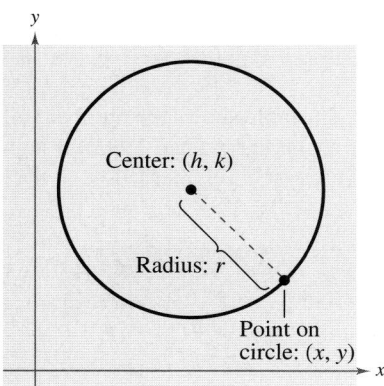

FIGURE P.25

Throughout this course, you will learn to recognize several types of graphs from their equations. For instance, you will learn to recognize that the graph of a second-degree equation of the form

$$y = ax^2 + bx + c$$

is a parabola (see Example 6). Another easily recognized graph is that of a **circle.**

Circles

Consider the circle shown in Figure P.25. A point (x, y) is on the circle if and only if its distance from the center (h, k) is r. By the Distance Formula,

$$\sqrt{(x - h)^2 + (y - k)^2} = r.$$

By squaring each side of this equation, you obtain the **standard form of the equation of a circle.**

Standard Form of the Equation of a Circle

The point (x, y) lies on the circle of radius r and center (h, k) if and only if

$$(x - h)^2 + (y - k)^2 = r^2.$$

From this result, you can see that the standard form of the equation of a circle *with its center at the origin*, $(h, k) = (0, 0)$, is simply

$$x^2 + y^2 = r^2. \qquad \text{Circle with center at origin}$$

Example 11 ▶ Finding the Equation of a Circle

The point $(3, 4)$ lies on a circle whose center is at $(-1, 2)$, as shown in Figure P.26. Write the standard form of the equation of this circle.

Solution

The radius of the circle is the distance between $(-1, 2)$ and $(3, 4)$.

$$r = \sqrt{(x - h)^2 + (y - k)^2} \qquad \text{Distance Formula}$$
$$r = \sqrt{[3 - (-1)]^2 + (4 - 2)^2} \qquad \text{Substitute for } x, y, h, \text{ and } k.$$
$$= \sqrt{4^2 + 2^2} \qquad \text{Simplify.}$$
$$= \sqrt{16 + 4} \qquad \text{Simplify.}$$
$$= \sqrt{20} \qquad \text{Radius}$$

Using $(h, k) = (-1, 2)$ and $r = \sqrt{20}$, the equation of the circle is

$$(x - h)^2 + (y - k)^2 = r^2 \qquad \text{Equation of circle}$$
$$[x - (-1)]^2 + (y - 2)^2 = \left(\sqrt{20}\right)^2 \qquad \text{Substitute for } h, k, \text{ and } r.$$
$$(x + 1)^2 + (y - 2)^2 = 20. \qquad \text{Standard form}$$

STUDY TIP

To find the correct h and k, in Example 11, it may be helpful to rewrite the quantities $(x + 1)^2$ and $(y - 2)^2$, using subtraction.

$$(x + 1)^2 = [x - (-1)]^2,$$
$$h = -1$$
$$(y - 2)^2 = [y - (2)]^2,$$
$$k = 2$$

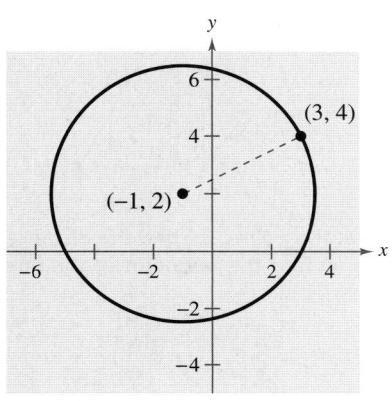

FIGURE P.26

P.3 Exercises

In Exercises 1 and 2, approximate the coordinates of the points.

1.   **2.**

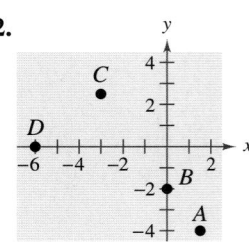

In Exercises 3 and 4, find the coordinates of the point.

3. The point is located three units to the left of the y-axis and four units above the x-axis.

4. The point is located eight units below the x-axis and four units to the right of the y-axis.

In Exercises 5–14, determine the quadrant(s) in which (x, y) is located so that the condition(s) is (are) satisfied.

5. $x > 0$ and $y < 0$ **6.** $x < 0$ and $y < 0$

7. $x = -4$ and $y > 0$ **8.** $x > 2$ and $y = 3$

9. $y < -5$ **10.** $x > 4$

11. $(x, -y)$ is in the second quadrant.

12. $(-x, y)$ is in the fourth quadrant.

13. $xy > 0$ **14.** $xy < 0$

In Exercises 15 and 16, sketch a scatter plot of the data given in the table.

15. *Meteorology*   The table shows the lowest temperature on record y (in degrees Fahrenheit) in Duluth, Minnesota, for each month x, where $x = 1$ represents January. (Source: NOAA)

x	y	x	y
1	-39	7	35
2	-33	8	32
3	-29	9	22
4	-5	10	8
5	17	11	-23
6	27	12	-34

16. *Number of Stores*   The table shows the number y of Wal-Mart stores for each year x from 1993 through 2000. (Source: Wal-Mart Stores, Inc.)

Year, x	Number of stores, y
1993	2440
1994	2759
1995	2943
1996	3054
1997	3406
1998	3599
1999	3985
2000	4190

In Exercises 17–20, (a) find the length of each side of the right triangle, and (b) show that these lengths satisfy the Pythagorean Theorem.

17.

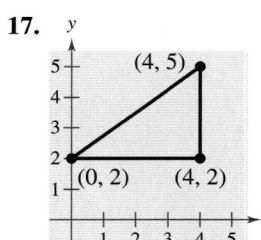

18.

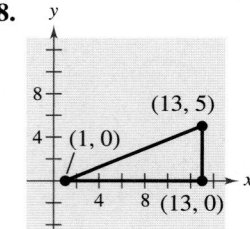

19.

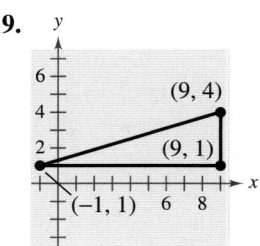

20.

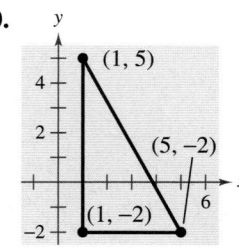

In Exercises 21–28, (a) plot the points, (b) find the distance between the points, and (c) find the midpoint of the line segment joining the points.

21. $(-4, 10), (4, -5)$ **22.** $(-7, -4), (2, 8)$

23. $(-1, 2), (5, 4)$ **24.** $(2, 10), (10, 2)$

25. $\left(\frac{1}{2}, 1\right), \left(-\frac{5}{2}, \frac{4}{3}\right)$ **26.** $\left(-\frac{1}{3}, -\frac{1}{3}\right), \left(-\frac{1}{6}, -\frac{1}{2}\right)$

27. $(6.2, 5.4), (-3.7, 1.8)$

28. $(-6.8, 1.3), (5.6, 4.9)$

29. *Music* The graph shows the numbers of recording artists who were elected to the Rock and Roll Hall of Fame from 1986 to 2001.

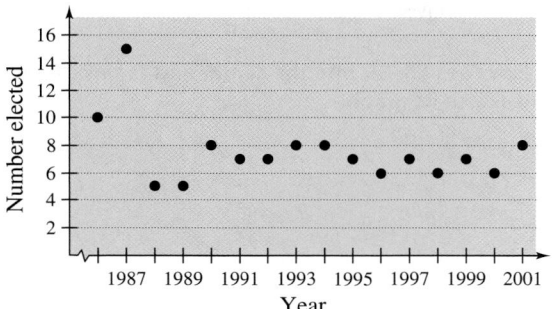

(a) Describe any trends in the data. From these trends, predict the number of artists elected in 2004.

(b) Why do you think the numbers elected in 1986 and 1987 were greater than in other years?

30. *Make a Conjecture* Plot the points $(2, 1)$, $(-3, 5)$, and $(7, -3)$ on a rectangular coordinate system. Then change the sign of the x-coordinate of each point and plot the three new points on the same rectangular coordinate system. Make a conjecture about the location of a point when each of the following occurs.

(a) The sign of the x-coordinate is changed.

(b) The sign of the y-coordinate is changed.

(c) The signs of both the x- and y-coordinates are changed.

31. *Sports* In a football game, a quarterback throws a pass from the 15-yard line, 10 yards from the sideline, as shown in the figure. The pass is caught on the 40-yard line, 45 yards from the same sideline. How long is the pass?

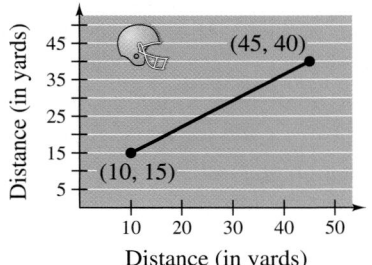

32. *Flying Distance* A jet plane flies from Naples, Italy, in a straight line to Rome, Italy, which is 120 kilometers west and 150 kilometers north of Naples. How far does the plane fly?

33. *Revenue* Polo Ralph Lauren Corp. had annual revenues of $1713.1 million in 1998 and $2225.8 million in 2000. Use the Midpoint Formula to estimate the revenue in 1999. (Source: Polo Ralph Lauren Corp.)

34. *Revenue* Zale Corp. had annual revenues of $1428.9 million in 1999 and $2068.2 million in 2001. Use the Midpoint Formula to estimate the sales in 2000. (Source: Zale Corp.)

In Exercises 35–38, determine whether each point lies on the graph of the equation.

Equation	*Points*	
35. $y = \sqrt{x + 4}$	(a) $(0, 2)$	(b) $(5, 3)$
36. $y = x^2 - 3x + 2$	(a) $(2, 0)$	(b) $(-2, 8)$
37. $y = 4 - |x - 2|$	(a) $(1, 5)$	(b) $(6, 0)$
38. $y = \frac{1}{3}x^3 - 2x^2$	(a) $\left(2, -\frac{16}{3}\right)$	(b) $(-3, 9)$

In Exercises 39 and 40, complete the table. Use the resulting solution points to sketch the graph of the equation.

39. $y = \frac{3}{4}x - 1$

x	-2	0	1	$\frac{4}{3}$	2
y					
(x, y)					

40. $y = x^2 - 3x$

x	-1	0	1	2	3
y					
(x, y)					

In Exercises 41–52, find the x- and y-intercepts of the graph of the equation.

41. $y = 16 - 4x^2$ **42.** $y = (x + 3)^2$

43. $y = 5x - 6$ **44.** $y = 8 - 3x$

45. $y = \sqrt{x + 4}$ **46.** $y = \sqrt{2x - 1}$

47. $y = |3x - 7|$ **48.** $y = -|x + 10|$

49. $y = 2x^3 - 4x^2$ **50.** $y = x^4 - 25$

51. $y^2 = 6 - x$ **52.** $y^2 = x + 1$

In Exercises 53–60, use the algebraic tests to check for symmetry with respect to both axes and the origin.

53. $x^2 - y = 0$

54. $x - y^2 = 0$

55. $y = x^3$

56. $y = x^4 - x^2 + 3$

57. $y = \dfrac{x}{x^2 + 1}$

58. $y = \dfrac{1}{x^2 + 1}$

59. $xy^2 + 10 = 0$

60. $xy = 4$

In Exercises 61–64, assume that the graph has the indicated type of symmetry. Sketch the complete graph of the equation. To print an enlarged copy of the graph, go to the website www.mathgraphs.com.

61.

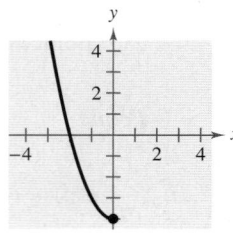

y-Axis symmetry

62.

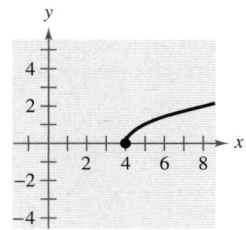

x-Axis symmetry

63.

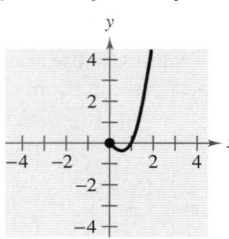

Origin symmetry

64.

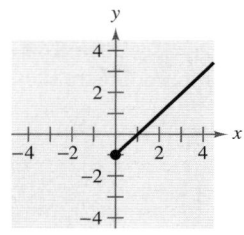

y-Axis symmetry

In Exercises 65–76, use symmetry to sketch the graph of the equation.

65. $y = -3x + 1$

66. $y = 2x - 3$

67. $y = x^2 - 2x$

68. $y = -x^2 - 2x$

69. $y = x^3 + 3$

70. $y = x^3 - 1$

71. $y = \sqrt{x - 3}$

72. $y = \sqrt{1 - x}$

73. $y = |x - 6|$

74. $y = 1 - |x|$

75. $x = y^2 - 1$

76. $x = y^2 - 5$

In Exercises 77–82, write the standard form of the equation of the specified circle.

77. Center: $(2, -1)$; Radius: 4

78. Center: $(-7, -4)$; Radius: 7

79. Center: $(-1, 2)$; Solution point: $(0, 0)$

80. Center: $(3, -2)$; Solution point: $(-1, 1)$

81. Endpoints of a diameter: $(0, 0)$, $(6, 8)$

82. Endpoints of a diameter: $(-4, -1)$, $(4, 1)$

In Exercises 83–88, find the center and radius of the circle, and sketch its graph.

83. $x^2 + y^2 = 25$

84. $x^2 + y^2 = 16$

85. $(x - 1)^2 + (y + 3)^2 = 9$

86. $x^2 + (y - 1)^2 = 1$

87. $\left(x - \tfrac{1}{2}\right)^2 + \left(y - \tfrac{1}{2}\right)^2 = \tfrac{9}{4}$

88. $(x - 2)^2 + (y + 1)^2 = 3$

89. *Depreciation*   A manufacturing plant purchases a new molding machine for $225,000. The depreciated value y after t years is

$$y = 225{,}000 - 20{,}000t, \qquad 0 \le t \le 8.$$

Sketch the graph of the equation.

90. *Consumerism*   You purchase a jet ski for $8100. The depreciated value y after t years is

$$y = 8100 - 929t, \qquad 0 \le t \le 6.$$

Sketch the graph of the equation.

91. *Geometry*   A rectangle of length x and width w has a perimeter of 12 meters.

(a) Draw a rectangle that gives a visual representation of the problem. Use the specified variables to label the sides of the rectangle.

(b) Show that the width of the rectangle is $w = 6 - x$ and its area is $A = x(6 - x)$.

(c) Use a graphing utility to graph the area equation.

(d) From the graph in part (c), estimate the dimensions of the rectangle that yield a maximum area.

92. *Geometry*   A rectangle of length x and width w has a perimeter of 22 yards.

(a) Draw a rectangle that gives a visual representation of the problem. Use the specified variables to label the sides of the rectangle.

(b) Show that the width of the rectangle is $w = 11 - x$ and its area is $A = x(11 - x)$.

(c) Use a graphing utility to graph the area equation. Be sure to adjust your window settings.

(d) From the graph in part (c), estimate the dimensions of the rectangle that yield a maximum area.

The symbol ⚙ indicates an exercise or parts of an exercise in which you are instructed to use a graphing utility.

▶ Model It

93. *Population Statistics* The table shows the life expectancy of a child (at birth) in the United States for selected years from 1920 to 2000. (Source: U.S. National Center for Health Statistics, U.S. Census Bureau)

Year, t	Life expectancy, y
1920	54.1
1930	59.7
1940	62.9
1950	68.2
1960	69.7
1970	70.8
1980	73.7
1990	75.4
2000	77.1

A model for the life expectancy during this period is

$$y = -0.0025t^2 + 0.572t + 44.31$$

where y represents the life expectancy and t is the time in years, with $t = 20$ corresponding to 1920.

(a) Sketch a scatter plot of the data.

(b) Graph the model for the data and compare the scatter plot and the graph.

(c) Use the graph of the model to estimate the life expectancy of a child for the years 2005 and 2010.

(d) Do you think this model can be used to predict the life expectancy of a child 50 years from now? Explain.

94. *Electronics* The resistance y (in ohms) of 1000 feet of solid copper wire at 77 degrees Fahrenheit can be approximated by the model

$$y = \frac{10{,}770}{x^2} - 0.37, \qquad 5 \le x \le 100$$

where x is the diameter of the wire in mils (0.001 in.). Use a graphing utility to graph the model and estimate the resistance when $x = 50$. (Source: American Wire Gage)

Synthesis

True or False? In Exercises 95–98, determine whether the statement is true or false. Justify your answer.

95. In order to divide a line segment into 16 equal parts, you would have to use the Midpoint Formula 16 times.

96. The points $(-8, 4)$, $(2, 11)$, and $(-5, 1)$ represent the vertices of an isosceles triangle.

97. In order to find the y-intercepts of the graph of an equation, let $y = 0$ and solve the equation for x.

98. The graph of a linear equation of the form $y = mx + b$ has one y-intercept.

99. *Think About It* What is the y-coordinate of any point on the x-axis? What is the x-coordinate of any point on the y-axis?

100. *Think About It* When plotting points on the rectangular coordinate system, is it true that the scales on the x- and y-axes must be the same? Explain.

101. *Proof* Prove that the diagonals of the parallelogram in the figure intersect at their midpoints.

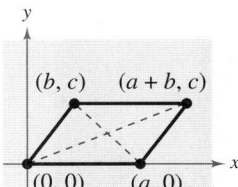

102. *Think About It* Suppose you correctly enter an expression for the variable y on a graphing utility. However, no graph appears on the display when you graph the equation. Give a possible explanation and the steps you could take to remedy the problem. Illustrate your explanation with an example.

103. *Think About It* Find a and b if the graph of $y = ax^2 + bx^3$ is symmetric with respect to (a) the y-axis and (b) the origin. (There are many correct answers

P.4 Linear Equations in Two Variables

▶ **What you should learn**

- How to use slope to graph linear equations in two variables
- How to find slopes of lines
- How to write linear equations in two variables
- How to use slope to identify parallel and perpendicular lines
- How to use linear equations in two variables to model and solve real-life problems

▶ **Why you should learn it**

Linear equations in two variables can be used to model and solve real-life problems. For instance, in Exercise 119 on page 51, a linear equation is used to model the average monthly cellular phone bills for subscribers in the United States.

Dick Luria/Getty Images

Using Slope

The simplest mathematical model for relating two variables is the **linear equation in two variables** $y = mx + b$. The equation is called *linear* because its graph is a line. (In mathematics, the term *line* means *straight line*.) By letting $x = 0$, you can see that the line crosses the y-axis at $y = b$, as shown in Figure P.27. In other words, the y-intercept is $(0, b)$. The steepness or slope of the line is m.

$$y = mx + b$$

Slope ⎯⎯ ⎿⎯ y-Intercept

The **slope** of a nonvertical line is the number of units the line rises (or falls) vertically for each unit of horizontal change from left to right, as shown in Figure P.27 and Figure P.28.

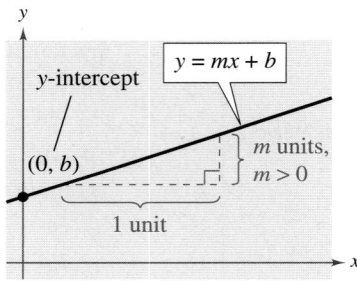

Positive slope, line rises.
FIGURE P.27

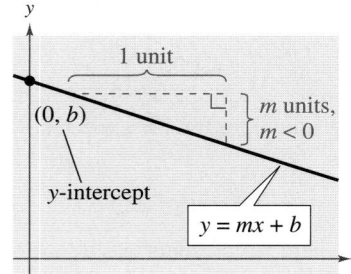

Negative slope, line falls.
FIGURE P.28

A linear equation that is written in the form $y = mx + b$ is said to be written in **slope-intercept form.**

The Slope-Intercept Form of the Equation of a Line

The graph of the equation

$$y = mx + b$$

is a line whose slope is m and whose y-intercept is $(0, b)$.

◀ **Exploration** ▶

Use a graphing utility to compare the slopes of the lines $y = mx$ where $m = 0.5, 1, 2,$ and 4. Which line rises most quickly? Now, let $m = -0.5, -1, -2,$ and -4. Which line falls most quickly? Use a square setting to obtain a true geometric perspective. What can you conclude about the slope and the "rate" at which the line rises or falls?

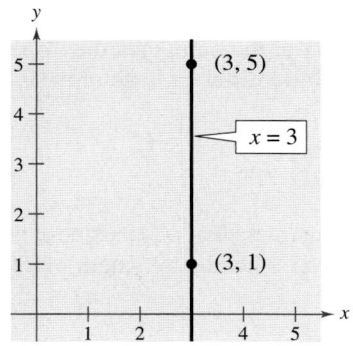

FIGURE P.29 *Slope is undefined.*

Once you have determined the slope and the y-intercept of a line, it is a relatively simple matter to sketch its graph. In the next example, note that none of the lines is vertical. A vertical line has an equation of the form

$$x = a. \qquad \text{Vertical line}$$

The equation of a vertical line cannot be written in the form $y = mx + b$ because the slope of a vertical line is undefined, as indicated in Figure P.29.

Example 1 ▶ **Graphing a Linear Equation**

Sketch the graph of each linear equation.

a. $y = 2x + 1$

b. $y = 2$

c. $x + y = 2$

Solution

a. Because $b = 1$, the y-intercept is $(0, 1)$. Moreover, because the slope is $m = 2$, the line *rises* two units for each unit the line moves to the right, as shown in Figure P.30.

b. By writing this equation in the form $y = (0)x + 2$, you can see that the y-intercept is $(0, 2)$ and the slope is zero. A zero slope implies that the line is horizontal—that is, it doesn't rise *or* fall, as shown in Figure P.31.

c. By writing this equation in slope-intercept form

$$x + y = 2 \qquad \text{Write original equation.}$$

$$y = -x + 2 \qquad \text{Subtract } x \text{ from each side.}$$

$$y = (-1)x + 2 \qquad \text{Write in slope-intercept form.}$$

you can see that the y-intercept is $(0, 2)$. Moreover, because the slope is $m = -1$, the line *falls* one unit for each unit the line moves to the right, as shown in Figure P.32.

Common Error

Many students confuse the line $x = a$ with the point $x = a$ on the real number line, or the line $y = b$ with the point $y = b$. Point out to students that they need to be aware of the context in which $x = a$ or $y = b$ is presented to know whether it refers to the line in the plane or the point on the number line.

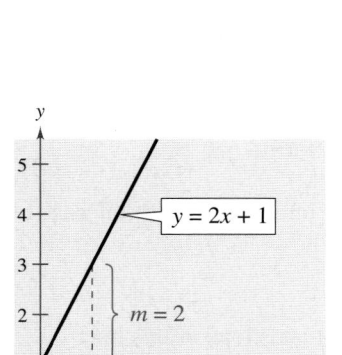

When m is positive, the line rises.

FIGURE P.30

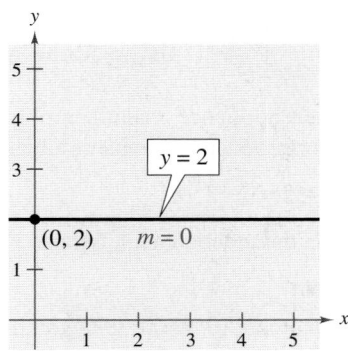

When m is 0, the line is horizontal.

FIGURE P.31

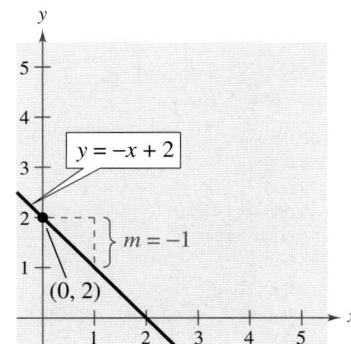

When m is negative, the line falls.

FIGURE P.32

In real-life problems, the slope of a line can be interpreted as either a *ratio* or a *rate*. If the *x*-axis and *y*-axis have the same unit of measure, then the slope has no units and is a **ratio.** If the *x*-axis and *y*-axis have different units of measure, then the slope is a **rate** or **rate of change.**

Example 2 ▶ Using Slope as a Ratio

The maximum recommended slope of a wheelchair ramp is $\frac{1}{12}$. A business is installing a wheelchair ramp that rises 22 inches over a horizontal length of 24 feet. Is the ramp steeper than recommended? (Source: Americans with Disabilities Act Handbook)

Solution

The horizontal length of the ramp is 24 feet or $12(24) = 288$ inches, as shown in Figure P.33. So, the slope of the ramp is

$$\text{Slope} = \frac{\text{vertical change}}{\text{horizontal change}}$$

$$= \frac{22 \text{ in.}}{288 \text{ in.}}$$

$$\approx 0.076.$$

Because $\frac{1}{12} \approx 0.083$, the slope of the ramp is not steeper than recommended.

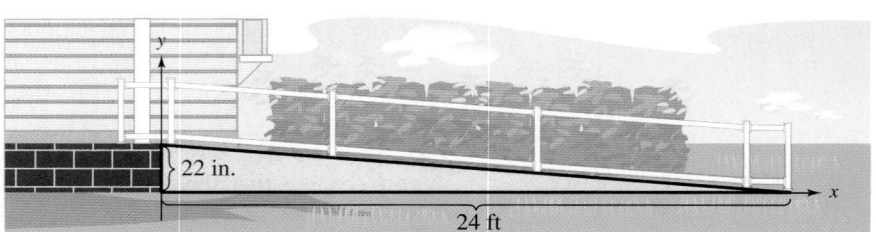

22 in.

24 ft

FIGURE P.33

Example 3 ▶ Using Slope as a Rate of Change

A kitchen appliance manufacturing company determines that the total cost in dollars of producing *x* units of a blender is

$$C = 25x + 3500. \qquad \text{Cost equation}$$

Describe the practical significance of the *y*-intercept and slope of this line.

Solution

The *y*-intercept $(0, 3500)$ tells you that the cost of producing zero units is $3500. This is the *fixed cost* of production—it includes costs that must be paid regardless of the number of units produced. The slope of $m = 25$ tells you that the cost of producing each unit is $25, as shown in Figure P.34. Economists call the cost per unit the *marginal cost*. If the production increases by one unit, then the "margin," or extra amount of cost, is $25. So, the cost increases at a rate of $25 per unit.

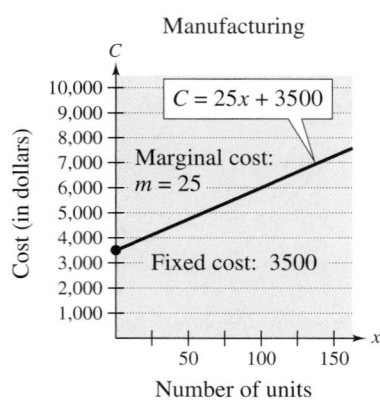

FIGURE P.34   *Production cost*

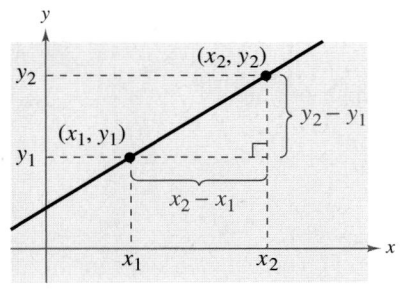

FIGURE P.35

Finding the Slope of a Line

Given an equation of a line, you can find its slope by writing the equation in slope-intercept form. If you are not given an equation, you can still find the slope of a line. For instance, suppose you want to find the slope of the line passing through the points (x_1, y_1) and (x_2, y_2), as shown in Figure P.35. As you move from left to right along this line, a change of $(y_2 - y_1)$ units in the vertical direction corresponds to a change of $(x_2 - x_1)$ units in the horizontal direction.

$$y_2 - y_1 = \text{the change in } y = \text{rise}$$

and

$$x_2 - x_1 = \text{the change in } x = \text{run}$$

The ratio of $(y_2 - y_1)$ to $(x_2 - x_1)$ represents the slope of the line that passes through the points (x_1, y_1) and (x_2, y_2).

$$\begin{aligned}
\text{Slope} &= \frac{\text{change in } y}{\text{change in } x} \\[1em]
&= \frac{\text{rise}}{\text{run}} \\[1em]
&= \frac{y_2 - y_1}{x_2 - x_1}
\end{aligned}$$

> ### The Slope of a Line Passing Through Two Points
>
> The **slope** m of the nonvertical line through (x_1, y_1) and (x_2, y_2) is
>
> $$m = \frac{y_2 - y_1}{x_2 - x_1}$$
>
> where $x_1 \neq x_2$.

When this formula is used for slope, the *order of subtraction* is important. Given two points on a line, you are free to label either one of them as (x_1, y_1) and the other as (x_2, y_2). However, once you have done this, you must form the numerator and denominator using the same order of subtraction.

$$m = \frac{y_2 - y_1}{x_2 - x_1} \qquad m = \frac{y_1 - y_2}{x_1 - x_2} \qquad m = \frac{y_2 - y_1}{x_1 - x_2}$$

Correct　　　　　　Correct　　　　　　Incorrect

For instance, the slope of the line passing through the points $(3, 4)$ and $(5, 7)$ can be calculated as

$$m = \frac{7 - 4}{5 - 3} = \frac{3}{2}$$

or, reversing the subtraction order in both the numerator and denominator, as

$$m = \frac{4 - 7}{3 - 5} = \frac{-3}{-2} = \frac{3}{2}.$$

Example 4 ▶ **Finding the Slope of a Line Through Two Points**

Find the slope of the line passing through each pair of points.

a. $(-2, 0)$ and $(3, 1)$ **b.** $(-1, 2)$ and $(2, 2)$

c. $(0, 4)$ and $(1, -1)$ **d.** $(3, 4)$ and $(3, 1)$

Solution

a. Letting $(x_1, y_1) = (-2, 0)$ and $(x_2, y_2) = (3, 1)$, you obtain a slope of

$$m = \frac{y_2 - y_1}{x_2 - x_1} = \frac{1 - 0}{3 - (-2)} = \frac{1}{5}.$$ See Figure P.36.

b. The slope of the line passing through $(-1, 2)$ and $(2, 2)$ is

$$m = \frac{2 - 2}{2 - (-1)} = \frac{0}{3} = 0.$$ See Figure P.37.

c. The slope of the line passing through $(0, 4)$ and $(1, -1)$ is

$$m = \frac{-1 - 4}{1 - 0} = \frac{-5}{1} = -5.$$ See Figure P.38.

d. The slope of the line passing through $(3, 4)$ and $(3, 1)$ is

$$m = \frac{1 - 4}{3 - 3} = \frac{-3}{0}.$$ See Figure P.39.

Because division by 0 is undefined, the slope is undefined and the line is vertical.

Common Error

A common error when finding the slope of a line is combining x- and y-coordinates in either the numerator or denominator, or both, as in

$$m = \frac{y_2 - x_1}{x_2 - y_1}.$$

STUDY TIP

In Figures P.36–P.39, note the relationships between slope and the orientation of the line.

a. Positive slope; line rises from left to right

b. Zero slope; line is horizontal

c. Negative slope; line falls from left to right

d. Undefined slope; line is vertical

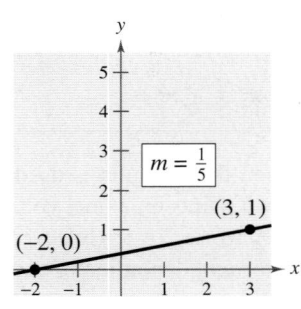

FIGURE **P.36**

FIGURE **P.37**

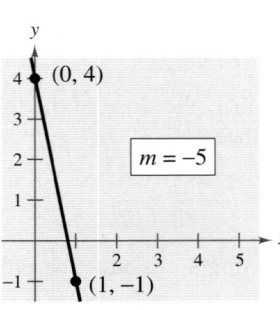

FIGURE **P.38**

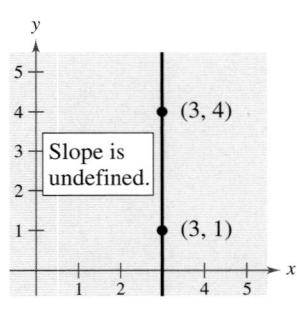

FIGURE **P.39**

Writing Linear Equations in Two Variables

If (x_1, y_1) is a point on a line of slope m and (x, y) is *any other* point on the line, then

$$\frac{y - y_1}{x - x_1} = m.$$

This equation, involving the variables x and y, can be rewritten in the form

$$y - y_1 = m(x - x_1)$$

which is the **point-slope form** of the equation of a line.

Point-Slope Form of the Equation of a Line

The equation of the line with slope m passing through the point (x_1, y_1) is

$$y - y_1 = m(x - x_1).$$

The point-slope form is most useful for *finding* the equation of a line. You should remember this form.

Example 5 ▶ Using the Point-Slope Form

Find the slope-intercept form of the equation of the line that has a slope of 3 and passes through the point $(1, -2)$.

Solution

Use the point-slope form with $m = 3$ and $(x_1, y_1) = (1, -2)$.

$$y - y_1 = m(x - x_1) \qquad \text{Point-slope form}$$
$$y - (-2) = 3(x - 1) \qquad \text{Substitute for } m, x_1, \text{ and } y_1.$$
$$y + 2 = 3x - 3 \qquad \text{Simplify.}$$
$$y = 3x - 5 \qquad \text{Write in slope-intercept form.}$$

The slope-intercept form of the equation of the line is $y = 3x - 5$. The graph of this line is shown in Figure P.40.

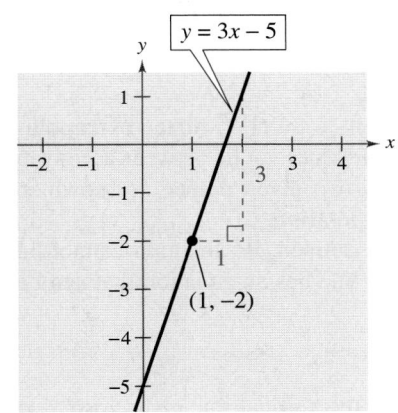

$y = 3x - 5$

$(1, -2)$

FIGURE P.40

The point-slope form can be used to find an equation of the line passing through two points (x_1, y_1) and (x_2, y_2). To do this, first find the slope of the line

$$m = \frac{y_2 - y_1}{x_2 - x_1}, \qquad x_1 \neq x_2$$

and then use the point-slope form to obtain the equation

$$y - y_1 = \frac{y_2 - y_1}{x_2 - x_1}(x - x_1). \qquad \text{Two-point form}$$

This is sometimes called the **two-point form** of the equation of a line.

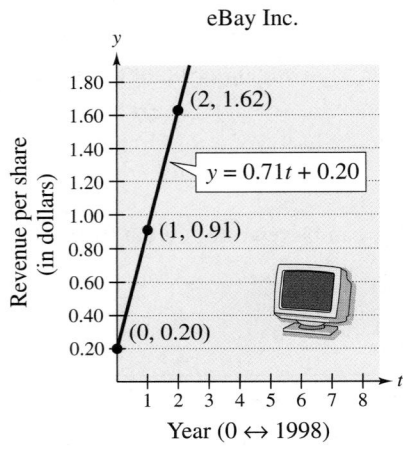

FIGURE P.41

Example 6 ▶   **Predicting Revenue per Share**

The revenue per share for eBay Inc. was \$0.20 in 1998 and \$0.91 in 1999. Using only this information, write a linear equation that gives the revenue per share in terms of the year. Then predict the revenue per share for 2000.   (Source: eBay Inc.)

Solution

Let $t = 0$ represent 1998. Then the two given values are represented by the data points $(0, 0.20)$ and $(1, 0.91)$. The slope of the line through these points is

$$m = \frac{0.91 - 0.20}{1 - 0}$$

$$= 0.71.$$

Using the point-slope form, you can find the equation that relates the revenue per share y and the year t to be

$$y - 0.20 = 0.71(t - 0) \qquad \text{Write in point-slope form.}$$

$$y = 0.71t + 0.20. \qquad \text{Write in slope-intercept form.}$$

According to this equation, the revenue per share in 2000 was \$1.62, as shown in Figure P.41. (In this case, the prediction is quite good—the actual revenue per share in 2000 was \$1.60.)

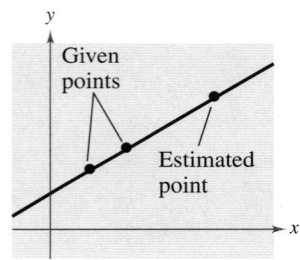

Linear extrapolation
FIGURE P.42

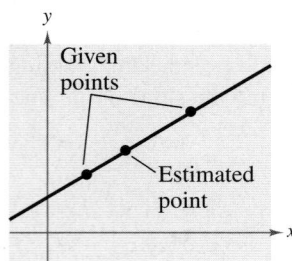

Linear interpolation
FIGURE P.43

The prediction method illustrated in Example 6 is called **linear extrapolation.** Note in Figure P.42 that an extrapolated point does not lie between the given points. When the estimated point lies between two given points, as shown in Figure P.43, the procedure is called **linear interpolation.**

Because the slope of a vertical line is not defined, its equation cannot be written in slope-intercept form. However, every line has an equation that can be written in the **general form**

$$Ax + By + C = 0 \qquad \text{General form}$$

where A and B are not both zero. For instance, the vertical line given by $x = a$ can be represented by the general form $x - a = 0$.

Equations of Lines

1. General form: $Ax + By + C = 0$

2. Vertical line: $x = a$

3. Horizontal line: $y = b$

4. Slope-intercept form: $y = mx + b$

5. Point-slope form: $y - y_1 = m(x - x_1)$

6. Two-point form: $y - y_1 = \dfrac{y_2 - y_1}{x_2 - x_1}(x - x_1)$

Exploration

Find d_1 and d_2 in terms of m_1 and m_2, respectively (see figure). Then use the Pythagorean Theorem to find a relationship between m_1 and m_2.

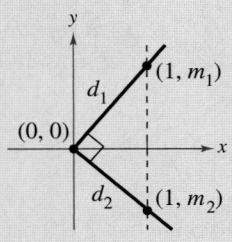

Parallel and Perpendicular Lines

Slope can be used to decide whether two nonvertical lines in a plane are parallel, perpendicular, or neither.

> **Parallel and Perpendicular Lines**
>
> 1. Two distinct nonvertical lines are **parallel** if and only if their slopes are equal. That is, $m_1 = m_2$.
>
> 2. Two nonvertical lines are **perpendicular** if and only if their slopes are negative reciprocals of each other. That is, $m_1 = -1/m_2$.

Example 7 ▶ Finding Parallel and Perpendicular Lines

Find the slope-intercept forms of the equations of the lines that pass through the point $(2, -1)$ and are (a) parallel to and (b) perpendicular to the line $2x - 3y = 5$.

Solution

By writing the equation of the given line in slope-intercept form

$$2x - 3y = 5 \qquad \text{Write original equation.}$$
$$-3y = -2x + 5 \qquad \text{Subtract } 2x \text{ from each side.}$$
$$y = \tfrac{2}{3}x - \tfrac{5}{3} \qquad \text{Write in slope-intercept form.}$$

you can see that it has a slope of $m = \tfrac{2}{3}$, as shown in Figure P.44.

a. Any line parallel to the given line must also have a slope of $\tfrac{2}{3}$. So, the line through $(2, -1)$ that is parallel to the given line has the following equation.

$$y - (-1) = \tfrac{2}{3}(x - 2) \qquad \text{Write in point-slope form.}$$
$$3(y + 1) = 2(x - 2) \qquad \text{Multiply each side by 3.}$$
$$3y + 3 = 2x - 4 \qquad \text{Distributive Property}$$
$$2x - 3y - 7 = 0 \qquad \text{Write in general form.}$$
$$y = \tfrac{2}{3}x - \tfrac{7}{3} \qquad \text{Write in slope-intercept form.}$$

b. Any line perpendicular to the given line must have a slope of $-1/(2/3)$ or $-3/2$. So, the line through $(2, -1)$ that is perpendicular to the given line has the following equation.

$$y - (-1) = -\tfrac{3}{2}(x - 2) \qquad \text{Write in point-slope form.}$$
$$2(y + 1) = -3(x - 2) \qquad \text{Multiply each side by 2.}$$
$$2y + 2 = -3x + 6 \qquad \text{Distributive Property}$$
$$3x + 2y - 4 = 0 \qquad \text{Write in general form.}$$
$$y = -\tfrac{3}{2}x + 2 \qquad \text{Write in slope-intercept form.}$$

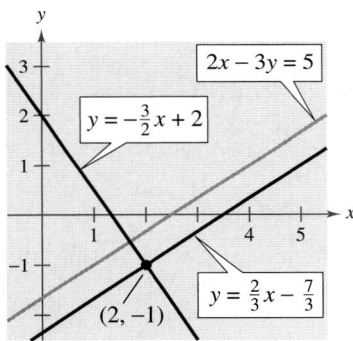

FIGURE P.44

Technology

On a graphing utility, lines will not appear to have the correct slope unless you use a viewing window that has a square setting. For instance, try graphing the lines in Example 7 using the standard setting $-10 \le x \le 10$ and $-10 \le y \le 10$. Then reset the viewing window with the square setting $-9 \le x \le 9$ and $-6 \le y \le 6$. On which setting do the lines $y = \tfrac{2}{3}x - \tfrac{5}{3}$ and $y = -\tfrac{3}{2}x + 2$ appear perpendicular?

Notice in Example 7 how the slope-intercept form is used to obtain information about the graph of a line, whereas the point-slope form is used to write the equation of a line.

Application

Most business expenses can be deducted in the same year they occur. One exception is the cost of property that has a useful life of more than 1 year. Such costs must be *depreciated* over the useful life of the property. If the *same amount* is depreciated each year, the procedure is called *linear* or *straight-line depreciation*. The *book value* is the difference between the original value and the total amount of depreciation accumulated to date.

Example 8 ▶ **Straight-Line Depreciation**

Your publishing company has purchased a $12,000 machine that has a useful life of 8 years. The salvage value at the end of 8 years is $2000. Write a linear equation that describes the book value of the machine each year.

Solution

Let V represent the value of the machine at the end of year t. You can represent the initial value of the machine by the data point $(0, 12{,}000)$ and the salvage value of the machine by the data point $(8, 2000)$. The slope of the line is

$$m = \frac{2000 - 12{,}000}{8 - 0} = -\$1250$$

which represents the annual depreciation in *dollars per year*. Using the point-slope form, you can write the equation of the line as follows.

$$V - 12{,}000 = -1250(t - 0) \qquad \text{Write in point-slope form.}$$

$$V = -1250t + 12{,}000 \qquad \text{Write in slope-intercept form.}$$

The table shows the book value at the end of each year, and the graph of the equation is shown in Figure P.45.

> **STUDY TIP**
>
> In many real-life applications, the two data points that determine the line are often given in a disguised form. Note how the data points are described in Example 8.

Useful Life of a Machine

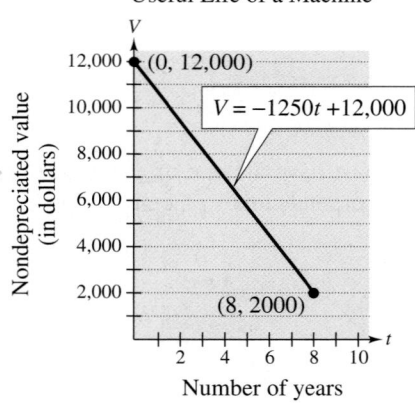

FIGURE P.45 *Straight-line depreciation*

Year, t	Value, V
0	12,000
1	10,750
2	9,500
3	8,250
4	7,000
5	5,750
6	4,500
7	3,250
8	2,000

P.4 Exercises

In Exercises 1 and 2, identify the line that has each slope.

1. (a) $m = \frac{2}{3}$

   (b) m is undefined.

   (c) $m = -2$

2. (a) $m = 0$

   (b) $m = -\frac{3}{4}$

   (c) $m = 1$

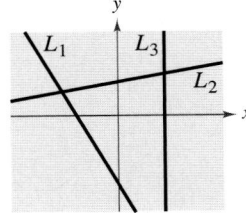   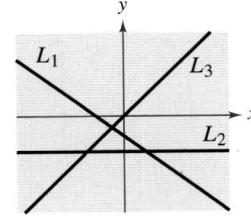

In Exercises 3 and 4, sketch the lines through the point with the indicated slopes on the same set of coordinate axes.

	Point	Slopes
3.	$(2, 3)$	(a) 0 (b) 1 (c) 2 (d) -3
4.	$(-4, 1)$	(a) 3 (b) -3 (c) $\frac{1}{2}$ (d) Undefined

In Exercises 5–8, estimate the slope of the line.

5.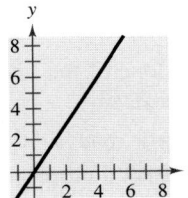

6.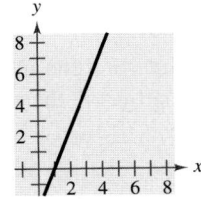

7.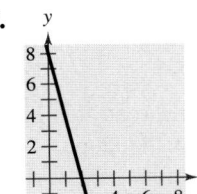

8.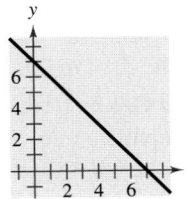

In Exercises 9–20, find the slope and y-intercept (if possible) of the equation of the line. Sketch the line.

9. $y = 5x + 3$

10. $y = x - 10$

11. $y = -\frac{1}{2}x + 4$

12. $y = -\frac{3}{2}x + 6$

13. $5x - 2 = 0$

14. $3y + 5 = 0$

15. $7x + 6y = 30$

16. $2x + 3y = 9$

17. $y - 3 = 0$

18. $y + 4 = 0$

19. $x + 5 = 0$

20. $x - 2 = 0$

In Exercises 21–28, plot the points and find the slope of the line passing through the pair of points.

21. $(-3, -2), (1, 6)$

22. $(2, 4), (4, -4)$

23. $(-6, -1), (-6, 4)$

24. $(0, -10), (-4, 0)$

25. $\left(\frac{11}{2}, -\frac{4}{3}\right), \left(-\frac{3}{2}, -\frac{1}{3}\right)$

26. $\left(\frac{7}{8}, \frac{3}{4}\right), \left(\frac{5}{4}, -\frac{1}{4}\right)$

27. $(4.8, 3.1), (-5.2, 1.6)$

28. $(-1.75, -8.3), (2.25, -2.6)$

In Exercises 29–38, use the point on the line and the slope of the line to find three additional points through which the line passes. (There are many correct answers.)

	Point	Slope
29.	$(2, 1)$	$m = 0$
30.	$(-4, 1)$	m is undefined.
31.	$(5, -6)$	$m = 1$
32.	$(10, -6)$	$m = -1$
33.	$(-8, 1)$	m is undefined.
34.	$(-3, -1)$	$m = 0$
35.	$(-5, 4)$	$m = 2$
36.	$(0, -9)$	$m = -2$
37.	$(7, -2)$	$m = \frac{1}{2}$
38.	$(-1, -6)$	$m = -\frac{1}{2}$

In Exercises 39–42, determine whether the lines L_1 and L_2 passing through the pairs of points are parallel, perpendicular, or neither.

39. L_1: $(0, -1), (5, 9)$

    L_2: $(0, 3), (4, 1)$

40. L_1: $(-2, -1), (1, 5)$

    L_2: $(1, 3), (5, -5)$

41. L_1: $(3, 6), (-6, 0)$

    L_2: $(0, -1), \left(5, \frac{7}{3}\right)$

42. L_1: $(4, 8), (-4, 2)$

    L_2: $(3, -5), \left(-1, \frac{1}{3}\right)$

43. **Sales** The following are the slopes of lines representing annual sales y in terms of time x in years. Use the slopes to interpret any change in annual sales for a one-year increase in time.

    (a) The line has a slope of $m = 135$.

    (b) The line has a slope of $m = 0$.

    (c) The line has a slope of $m = -40$.

44. *Revenue* The following are the slopes of lines representing daily revenues y in terms of time x in days. Use the slopes to interpret any change in daily revenues for a one-day increase in time.

(a) The line has a slope of $m = 400$.

(b) The line has a slope of $m = 100$.

(c) The line has a slope of $m = 0$.

45. *Earnings per Share* The graph shows the earnings per share of stock for Auto Zone, Inc. for the years 1991 through 2001. (Source: Auto Zone, Inc.)

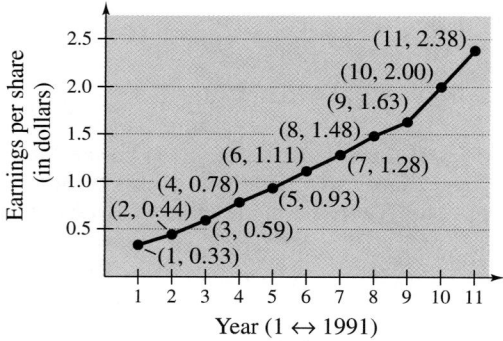

Year (1 ↔ 1991)

(a) Use the slopes to determine the years in which the earnings per share showed the greatest increase and the smallest increase.

(b) Find the slope of the line segment connecting the years 1991 and 2001.

(c) Interpret the meaning of the slope in part (b) in the context of the problem.

46. *Net Profit* The graph shows the net profit (in millions of dollars) for Outback Steakhouse for the years 1991 through 2001. (Source: Outback Steakhouse, Inc.)

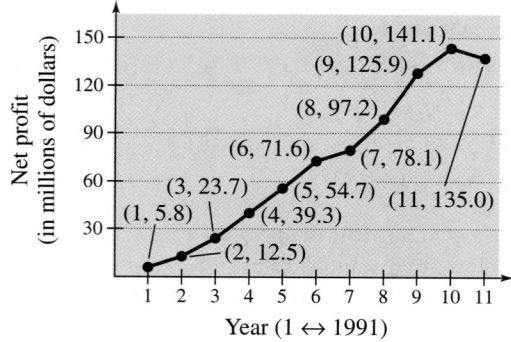

Year (1 ↔ 1991)

(a) Use the slopes to determine the years in which the net profit showed the greatest increase and the smallest increase.

(b) Find the slope of the line segment connecting the years 1991 and 2001.

(c) Interpret the meaning of the slope in part (b) in the context of the problem.

47. *Road Grade* From the top of a mountain road, a surveyor takes several horizontal measurements x and several vertical measurements y, as shown in the table (x and y are measured in feet).

x	y
300	-25
600	-50
900	-75
1200	-100
1500	-125
1800	-150
2100	-175

(a) Sketch a scatter plot of the data.

(b) Use a straightedge to sketch the best-fitting line through the points.

(c) Find an equation for the line you sketched in part (b).

(d) Interpret the meaning of the slope of the line in part (c) in the context of the problem.

(e) The surveyor needs to put up a road sign that indicates the steepness of the road. For instance, a surveyor would put up a sign that states "8% grade" on a road with a downhill grade that has a slope of $-\frac{8}{100}$. What should the sign state for the road in this problem?

48. *Road Grade* You are driving on a road that has a 6% uphill grade. This means that the slope of the road is $\frac{6}{100}$. Approximate the amount of vertical change in your position if you drive 200 feet.

Rate of Change In Exercises 49 and 50, you are given the dollar value of a product in 2003 and the rate at which the value of the product is expected to change during the next 5 years. Use this information to write a linear equation that gives the dollar value V of the product in terms of the year t. (Let $t = 3$ represent 2003.)

2003 Value	Rate
49. $2540	$125 increase per year
50. $156	$4.50 increase per year

Graphical Interpretation In Exercises 51–54, match the description of the situation with its graph. Also determine the slope of each graph and interpret the slope in the context of the situation. [The graphs are labeled (a), (b), (c), and (d).]

(a)
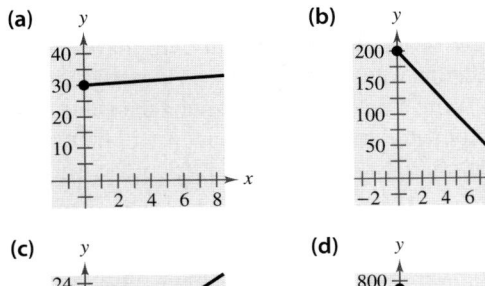

(b)

(c)
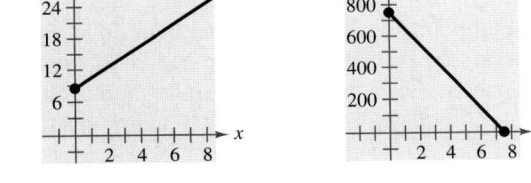

(d)

51. A person is paying $20 per week to a friend to repay a $200 loan.

52. An employee is paid $8.50 per hour plus $2 for each unit produced per hour.

53. A sales representative receives $30 per day for food plus $0.32 for each mile traveled.

54. A word processor that was purchased for $750 depreciates $100 per year.

In Exercises 55–66, find the slope-intercept form of the equation of the line that passes through the given point and has the indicated slope. Sketch the line.

Point	Slope
55. $(0, -2)$	$m = 3$
56. $(0, 10)$	$m = -1$
57. $(-3, 6)$	$m = -2$
58. $(0, 0)$	$m = 4$

59. $(4, 0)$	$m = -\frac{1}{3}$
60. $(-2, -5)$	$m = \frac{3}{4}$
61. $(6, -1)$	m is undefined.
62. $(-10, 4)$	m is undefined.
63. $\left(4, \frac{5}{2}\right)$	$m = 0$
64. $\left(-\frac{1}{2}, \frac{3}{2}\right)$	$m = 0$
65. $(-5.1, 1.8)$	$m = 5$
66. $(2.3, -8.5)$	$m = -\frac{5}{2}$

In Exercises 67–80, find the slope-intercept form of the equation of the line passing through the points. Sketch the line.

67. $(5, -1), (-5, 5)$ **68.** $(4, 3), (-4, -4)$

69. $(-8, 1), (-8, 7)$ **70.** $(-1, 4), (6, 4)$

71. $\left(2, \frac{1}{2}\right), \left(\frac{1}{2}, \frac{5}{4}\right)$ **72.** $\left(1, 1\right), \left(6, -\frac{2}{3}\right)$

73. $\left(-\frac{1}{10}, -\frac{3}{5}\right), \left(\frac{9}{10}, -\frac{9}{5}\right)$

74. $\left(\frac{3}{4}, \frac{3}{2}\right), \left(-\frac{4}{3}, \frac{7}{4}\right)$

75. $(1, 0.6), (-2, -0.6)$

76. $(-8, 0.6), (2, -2.4)$

77. $(2, -1), \left(\frac{1}{3}, -1\right)$

78. $\left(\frac{1}{5}, -2\right), (-6, -2)$

79. $\left(\frac{7}{3}, -8\right), \left(\frac{7}{3}, 1\right)$

80. $(1.5, -2), (1.5, 0.2)$

In Exercises 81–86, use the *intercept form* to find the equation of the line with the given intercepts. The intercept form of the equation of a line with intercepts $(a, 0)$ and $(0, b)$ is

$$\frac{x}{a} + \frac{y}{b} = 1, \quad a \neq 0, \quad b \neq 0.$$

81. x-intercept: $(2, 0)$ **82.** x-intercept: $(-3, 0)$
  y-intercept: $(0, 3)$ y-intercept: $(0, 4)$

83. x-intercept: $\left(-\frac{1}{6}, 0\right)$ **84.** x-intercept: $\left(\frac{2}{3}, 0\right)$
  y-intercept: $\left(0, -\frac{2}{3}\right)$ y-intercept: $(0, -2)$

85. Point on line: $(1, 2)$
  x-intercept: $(c, 0)$
  y-intercept: $(0, c), \quad c \neq 0$

86. Point on line: $(-3, 4)$
  x-intercept: $(d, 0)$
  y-intercept: $(0, d), \quad d \neq 0$

In Exercises 87–96, write the slope-intercept forms of the equations of the lines through the given point (a) parallel to the given line and (b) perpendicular to the given line.

Point	Line
87. $(2, 1)$	$4x - 2y = 3$
88. $(-3, 2)$	$x + y = 7$
89. $\left(-\frac{2}{3}, \frac{7}{8}\right)$	$3x + 4y = 7$
90. $\left(\frac{7}{8}, \frac{3}{4}\right)$	$5x + 3y = 0$
91. $(-1, 0)$	$y = -3$
92. $(4, -2)$	$y = 1$
93. $(2, 5)$	$x = 4$
94. $(-5, 1)$	$x = -2$
95. $(2.5, 6.8)$	$x - y = 4$
96. $(-3.9, -1.4)$	$6x + 2y = 9$

 Graphical Interpretation   In Exercises 97–100, identify any relationships that exist among the lines, and then use a graphing utility to graph the three equations in the same viewing window. Adjust the viewing window so that the slope appears visually correct—that is, so that parallel lines appear parallel and perpendicular lines appear to intersect at a right angle.

97. (a) $y = 2x$ (b) $y = -2x$ (c) $y = \frac{1}{2}x$

98. (a) $y = \frac{2}{3}x$ (b) $y = -\frac{3}{2}x$ (c) $y = \frac{2}{3}x + 2$

99. (a) $y = -\frac{1}{2}x$ (b) $y = -\frac{1}{2}x + 3$ (c) $y = 2x - 4$

100. (a) $y = x - 8$ (b) $y = x + 1$ (c) $y = -x + 3$

In Exercises 101–104, find a relationship between x and y such that (x, y) is equidistant from the two points.

101. $(4, -1), (-2, 3)$

102. $(6, 5), (1, -8)$

103. $\left(3, \frac{5}{2}\right), (-7, 1)$

104. $\left(-\frac{1}{2}, -4\right), \left(\frac{7}{2}, \frac{5}{4}\right)$

105. *Cash Flow per Share*   The cash flow per share for Timberland Co. was $0.18 in 1995 and $3.65 in 2000. Write a linear equation that gives the cash flow per share in terms of the year. Let $t = 0$ represent 1995. Then predict the cash flows for the years 2005 and 2010.   (Source: Timberland Co.)

106. *Number of Stores*   In 1996 there were 3927 J.C. Penney stores and in 2000 there were 3800 stores. Write a linear equation that gives the number of stores in terms of the year. Let $t = 0$ represent 1996. Then predict the numbers of stores for the years 2005 and 2010.   (Source: J.C. Penney Co.)

107. *Annual Salary*   A jeweler's salary was $28,500 in 1998 and $32,900 in 2000. The jeweler's salary follows a linear growth pattern. What will the jeweler's salary be in 2005?

108. *College Enrollment*   Ohio University had 27,913 students in 1999 and 28,197 students in 2001. The enrollment appears to follow a linear growth pattern. How many students will Ohio University have in 2005?   (Source: Ohio University)

109. *Depreciation*   A sub shop purchases a used pizza oven for $875. After 5 years, the oven will have to be replaced. Write a linear equation giving the value V of the equipment during the 5 years it will be in use.

110. *Depreciation*   A school district purchases a high-volume printer, copier, and scanner for $25,000. After 10 years, the equipment will have to be replaced. Its value at that time is expected to be $2000. Write a linear equation giving the value V of the equipment during the 10 years it will be in use.

111. *Sales*   A discount outlet is offering a 15% discount on all items. Write a linear equation giving the sale price S for an item with a list price L.

112. *Hourly Wage*   A microchip manufacturer pays its assembly line workers $11.50 per hour. In addition, workers receive a piecework rate of $0.75 per unit produced. Write a linear equation for the hourly wage W in terms of the number of units x produced per hour.

113. *Cost, Revenue, and Profit*   A roofing contractor purchases a shingle delivery truck with a shingle elevator for $36,500. The vehicle requires an average expenditure of $5.25 per hour for fuel and maintenance, and the operator is paid $11.50 per hour.

(a) Write a linear equation giving the total cost C of operating this equipment for t hours. (Include the purchase cost of the equipment.)

(b) Assuming that customers are charged $27 per hour of machine use, write an equation for the revenue R derived from t hours of use.

(c) Use the formula for profit $(P = R - C)$ to write an equation for the profit derived from t hours of use.

(d) Use the result of part (c) to find the break-even point—that is, the number of hours this equipment must be used to yield a profit of 0 dollars.

114. *Rental Demand* A real estate office handles an apartment complex with 50 units. When the rent per unit is $580 per month, all 50 units are occupied. However, when the rent is $625 per month, the average number of occupied units drops to 47. Assume that the relationship between the monthly rent p and the demand x is linear.

(a) Write the equation of the line giving the demand x in terms of the rent p.

(b) Use this equation to predict the number of units occupied when the rent is $655.

(c) Predict the number of units occupied when the rent is $595.

115. *Geometry* The length and width of a rectangular garden are 15 meters and 10 meters, respectively. A walkway of width x surrounds the garden.

(a) Draw a diagram that gives a visual representation of the problem.

(b) Write the equation for the perimeter y of the walkway in terms of x.

(c) Use a graphing utility to graph the equation for the perimeter.

(d) Determine the slope of the graph in part (c). For each additional one-meter increase in the width of the walkway, determine the increase in its perimeter.

116. *Monthly Salary* A pharmaceutical salesperson receives a monthly salary of $2500 plus a commission of 7% of sales. Write a linear equation for the salesperson's monthly wage W in terms of monthly sales S.

117. *Business Costs* A sales representative of a company using a personal car receives $120 per day for lodging and meals plus $0.35 per mile driven. Write a linear equation giving the daily cost C to the company in terms of x, the number of miles driven.

118. *Sports* The average annual salaries of major league baseball players (in thousands of dollars) from 1995 to 2002 are shown in the scatter plot. Find the equation of the line that you think best fits this data. (Let y represent the average salary and let t represent the year, with $t = 5$ corresponding to 1995.) (Source: The Associated Press)

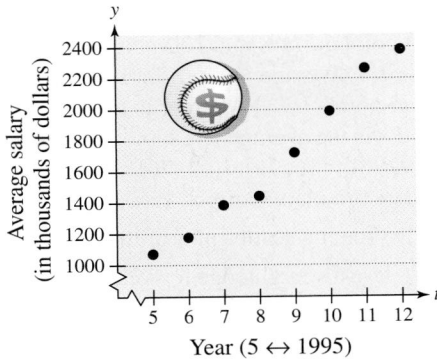

FIGURE FOR **118**

▶ **Model It**

119. *Data Analysis* The average monthly cellular phone bills y (in dollars) for subscribers in the United States from 1990 through 1999, where x is the year, are shown as data points (x, y). (Source: Cellular Telecommunications Industry Association)

(1990, 80.90)	(1995, 51.00)
(1991, 72.74)	(1996, 47.70)
(1992, 68.68)	(1997, 42.78)
(1993, 61.48)	(1998, 39.43)
(1994, 56.21)	(1999, 41.24)

(a) Sketch a scatter plot of the data. Let $x = 0$ correspond to 1990.

(b) Sketch the best-fitting line through the points.

(c) Find the equation of the line from part (b). Explain the procedure you used.

(d) Write a short paragraph explaining the meaning of the slope and y-intercept of the line in terms of the data.

(e) Compare the values obtained using your model with the actual values.

(f) Use your model to estimate the average monthly cellular phone bill in 2005.

120. *Data Analysis* An instructor gives regular 20-point quizzes and 100-point exams in an algebra course. Average scores for six students, given as data points (x, y) where x is the average quiz score and y is the average test score, are (18, 87), (10, 55), (19, 96), (16, 79), (13, 76), and (15, 82). [*Note:* There are many correct answers for parts (b)–(d).]

(a) Sketch a scatter plot of the data.

(b) Use a straightedge to sketch the best-fitting line through the points.

(c) Find an equation for the line sketched in part (b).

(d) Use the equation in part (c) to estimate the average test score for a person with an average quiz score of 17.

(e) The instructor adds 4 points to the average test score of everyone in the class. Describe the changes in the positions of the plotted points and the change in the equation of the line.

Synthesis

True or False? **In Exercises 121 and 122, determine whether the statement is true or false. Justify your answer.**

121. A line with a slope of $-\frac{5}{7}$ is steeper than a line with a slope of $-\frac{6}{7}$.

122. The line through $(-8, 2)$ and $(-1, 4)$ and the line through $(0, -4)$ and $(-7, 7)$ are parallel.

123. Explain how you could show that the points $A\,(2, 3)$, $B\,(2, 9)$, and $C\,(4, 3)$ are the vertices of a right triangle.

124. Explain why the slope of a vertical line is said to be undefined.

125. With the information given in the graphs, is it possible to determine the slope of each line? Is it possible that the lines could have the same slope? Explain.

(a)   (b)

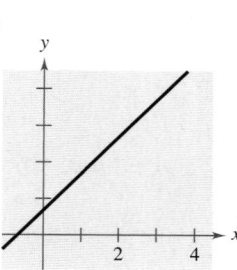

126. The slopes of two lines are -4 and $\frac{5}{2}$. Which is steeper? Explain.

127. The value V of a molding machine t years after it is purchased is

$$V = -4000t + 58{,}500, \quad 0 \le t \le 5.$$

Explain what the V-intercept and slope measure.

128. *Think About It* Is it possible for two lines with positive slopes to be perpendicular? Explain.

▶ What you should learn

- How to determine whether relations between two variables are functions
- How to use function notation and evaluate functions
- How to find the domains of functions
- How to use functions to model and solve real-life problems

▶ Why you should learn it

Functions can be used to model and solve real-life problems. For instance, in Exercise 99 on page 66, you will use a function to find the number of threatened and endangered fish in the world.

B&C Alexander

Introduction to Functions

Many everyday phenomena involve two quantities that are related to each other by some rule of correspondence. The mathematical term for such a rule of correspondence is a **relation.** In mathematics, relations are often represented by mathematical equations and formulas. For instance, the simple interest I earned on $1000 for 1 year is related to the annual interest rate r by the formula $I = 1000r$.

The formula $I = 1000r$ represents a special kind of relation that matches each item from one set with exactly one item from a different set. Such a relation is called a **function.**

Definition of a Function

A **function** f from a set A to a set B is a relation that assigns to each element x in the set A exactly one element y in the set B. The set A is the **domain** (or set of inputs) of the function f, and the set B contains the **range** (or set of outputs).

To help understand this definition, look at the function that relates the time of day to the temperature in Figure P.46.

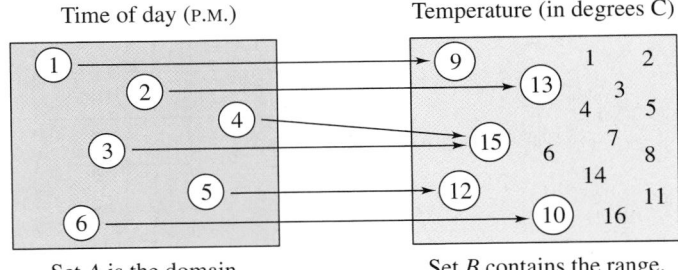

Set A is the domain.
Inputs: 1, 2, 3, 4, 5, 6

Set B contains the range.
Outputs: 9, 10, 12, 13, 15

FIGURE P.46

This function can be represented by the following ordered pairs, in which the first coordinate is the input and the second coordinate is the output.

$$\{(1, 9°), (2, 13°), (3, 15°), (4, 15°), (5, 12°), (6, 10°)\}$$

Characteristics of a Function from Set A to Set B

1. Each element in A must be matched with an element in B.

2. Some elements in B may not be matched with any element in A.

3. Two or more elements in A may be matched with the same element in B.

4. An element in A (the domain) cannot be matched with two different elements of B.

Functions are commonly represented in four ways.

Four Ways to Represent a Function

1. *Verbally* by a sentence that describes how the input variable is related to the output variable

2. *Numerically* by a table or a list of ordered pairs that matches input values with output values

3. *Graphically* by points on a graph in a coordinate plane in which the input values are represented by the horizontal axis and the output values are represented by the vertical axis

4. *Algebraically* by an equation in two variables

To determine whether or not a relation is a function, you must decide whether each input value is matched with exactly one output value. If any input value is matched with two or more output values, the relation is not a function.

Example 1 ▶ Testing for Functions

Determine whether the relation represents y as a function of x.

a. The input value x is the number of representatives from a state, and the output value y is the number of senators.

b.

Input x	Output y
2	11
2	10
3	8
4	5
5	1

c.

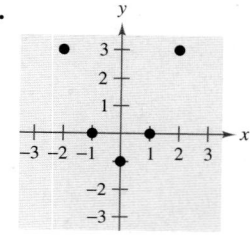

FIGURE P.47

Solution

a. This verbal description *does* describe y as a function of x. Regardless of the value of x, the value of y is always 2. Such functions are called *constant functions*.

b. This table *does not* describe y as a function of x. The input value 2 is matched with two different y-values.

c. The graph in Figure P.47 *does* describe y as a function of x. Each input value is matched with exactly one output value.

Representing functions by sets of ordered pairs is common in *discrete mathematics*. In algebra, however, it is more common to represent functions by equations or formulas involving two variables. For instance, the equation

$$y = x^2 \qquad \text{\small y is a function of x.}$$

represents the variable y as a function of the variable x. In this equation, x is the **independent variable** and y is the **dependent variable.** The domain of the function is the set of all values taken on by the independent variable x, and the range of the function is the set of all values taken on by the dependent variable y.

The Granger Collection

Historical Note

Leonhard Euler (1707–1783), a Swiss mathematician, is considered to have been the most prolific and productive mathematician in history. One of his greatest influences on mathematics was his use of symbols, or notation. The function notation $y = f(x)$ was introduced by Euler.

Example 2 ▶ Testing for Functions Represented Algebraically

Which of the equations represent(s) y as a function of x?

a. $x^2 + y = 1$ **b.** $-x + y^2 = 1$

Solution

To determine whether y is a function of x, try to solve for y in terms of x.

a. Solving for y yields

$$x^2 + y = 1 \qquad \text{Write original equation.}$$
$$y = 1 - x^2. \qquad \text{Solve for } y.$$

To each value of x there corresponds exactly one value of y. So, y is a function of x.

b. Solving for y yields

$$-x + y^2 = 1 \qquad \text{Write original equation.}$$
$$y^2 = 1 + x \qquad \text{Add } x \text{ to each side.}$$
$$y = \pm\sqrt{1 + x}. \qquad \text{Solve for } y.$$

The $\pm$ indicates that to a given value of x there correspond two values of y. So, y is not a function of x.

Function Notation

Understanding the concept of functions is essential. Be sure that students understand function notation. Frequently $f(x)$ is misinterpreted as "f times x" rather than "f of x."

When an equation is used to represent a function, it is convenient to name the function so that it can be referenced easily. For example, you know that the equation $y = 1 - x^2$ describes y as a function of x. Suppose you give this function the name "f." Then you can use the following **function notation.**

Input	*Output*	*Equation*
x	$f(x)$	$f(x) = 1 - x^2$

The symbol $f(x)$ is read as *the value of f at x* or simply f *of x.* The symbol $f(x)$ corresponds to the y-value for a given x. So, you can write $y = f(x)$. Keep in mind that f is the *name* of the function, whereas $f(x)$ is the *value* of the function at x. For instance, the function

$$f(x) = 3 - 2x$$

has *function values* denoted by $f(-1)$, $f(0)$, $f(2)$, and so on. To find these values, substitute the specified input values into the given equation.

For $x = -1$, $\quad f(-1) = 3 - 2(-1) = 3 + 2 = 5.$

For $x = 0$, $\quad f(0) = 3 - 2(0) = 3 - 0 = 3.$

For $x = 2$, $\quad f(2) = 3 - 2(2) = 3 - 4 = -1.$

Although f is often used as a convenient function name and x is often used as the independent variable, you can use other letters. For instance,

$$f(x) = x^2 - 4x + 7, \quad f(t) = t^2 - 4t + 7, \quad \text{and} \quad g(s) = s^2 - 4s + 7$$

all define the same function. In fact, the role of the independent variable is that of a "placeholder." Consequently, the function could be described by

$$f() = ()^2 - 4() + 7.$$

Example 3 ▶   Evaluating a Function

Let $g(x) = -x^2 + 4x + 1$. Find

a. $g(2)$ **b.** $g(t)$ **c.** $g(x + 2)$.

Solution

a. Replacing x with 2 in $g(x) = -x^2 + 4x + 1$ yields the following.

$$g(2) = -(2)^2 + 4(2) + 1 = -4 + 8 + 1 = 5$$

b. Replacing x with t yields the following.

$$g(t) = -(t)^2 + 4(t) + 1 = -t^2 + 4t + 1$$

c. Replacing x with $x + 2$ yields the following.

$$g(x + 2) = -(x + 2)^2 + 4(x + 2) + 1$$
$$= -(x^2 + 4x + 4) + 4x + 8 + 1$$
$$= -x^2 - 4x - 4 + 4x + 8 + 1$$
$$= -x^2 + 5$$

A function defined by two or more equations over a specified domain is called a **piecewise-defined function.**

Example 4 ▶   A Piecewise-Defined Function

Evaluate the function when $x = -1, 0,$ and 1.

$$f(x) = \begin{cases} x^2 + 1, & x < 0 \\ x - 1, & x \geq 0 \end{cases}$$

Solution

Because $x = -1$ is less than 0, use $f(x) = x^2 + 1$ to obtain

$$f(-1) = (-1)^2 + 1 = 2.$$

For $x = 0$, use $f(x) = x - 1$ to obtain

$$f(0) = (0) - 1 = -1.$$

For $x = 1$, use $f(x) = x - 1$ to obtain

$$f(1) = (1) - 1 = 0.$$

Students often have difficulty understanding how to evaluate piecewise-defined functions. You may want to use the following additional examples to demonstrate evaluation of piecewise-defined functions.

a. Evaluate at $x = 0, 1, 3$.

$$f(x) = \begin{cases} \dfrac{x}{2} + 1, & x \leq 1 \\ 3x + 2, & x > 1 \end{cases}$$

Solution

Because $x = 0$ is less than or equal to 1, use $f(x) = (x/2) + 1$ to obtain

$$f(0) = \frac{0}{2} + 1 = 1.$$

For $x = 1$, use $f(x) = (x/2) + 1$ to obtain

$$f(1) = \frac{1}{2} + 1 = 1\frac{1}{2}.$$

For $x = 3$, use $f(x) = 3x + 2$ to obtain

$$f(3) = 3(3) + 2 = 11.$$

b. Evaluate at $x = 0, 3, 5$.

$$f(x) = \begin{cases} x^2 + 3, & x < 2 \\ 7, & 2 \leq x \leq 4 \\ 2x - 1, & x > 4 \end{cases}$$

Solution

Because $x = 0$ is less than 2, use $f(x) = x^2 + 3$ to obtain

$$f(0) = 0^2 + 3 = 3.$$

For $x = 3$, use $f(x) = 7$ to obtain

$$f(3) = 7.$$

For $x = 5$, use $f(x) = 2x - 1$ to obtain

$$f(5) = 2(5) - 1 = 9.$$

The Domain of a Function

The domain of a function can be described explicitly or it can be *implied* by the expression used to define the function. The **implied domain** is the set of all real numbers for which the expression is defined. For instance, the function

$$f(x) = \frac{1}{x^2 - 4}$$ Domain excludes *x*-values that result in division by zero.

has an implied domain that consists of all real *x* other than $x = \pm 2$. These two values are excluded from the domain because division by zero is undefined. Another common type of implied domain is that used to avoid even roots of negative numbers. For example, the function

$$f(x) = \sqrt{x}$$ Domain excludes *x*-values that result in even roots of negative numbers.

is defined only for $x \geq 0$. So, its implied domain is the interval $[0, \infty)$. In general, the domain of a function *excludes* values that would cause division by zero *or that* would result in the even root of a negative number.

Technology

Use a graphing utility to graph $y = \sqrt{4 - x^2}$. What is the domain of this function? Then graph $y = \sqrt{x^2 - 4}$. What is the domain of this function? Do the domains of these two functions overlap? If so, for what values?

Example 5 ▶ **Finding the Domain of a Function**

Find the domain of each function.

a. f: $\{(-3, 0), (-1, 4), (0, 2), (2, 2), (4, -1)\}$ **b.** $g(x) = \dfrac{1}{x + 5}$

c. Volume of a sphere: $V = \frac{4}{3}\pi r^3$ **d.** $h(x) = \sqrt{4 - x^2}$

Solution

a. The domain of f consists of all first coordinates in the set of ordered pairs.

Domain $= \{-3, -1, 0, 2, 4\}$

b. Excluding *x*-values that yield zero in the denominator, the domain of g is the set of all real numbers $x \neq -5$.

c. Because this function represents the volume of a sphere, the values of the radius r must be positive. So, the domain is the set of all real numbers r such that $r > 0$.

d. This function is defined only for *x*-values for which

$$4 - x^2 \geq 0.$$

By solving this inequality, you can conclude that $-2 \leq x \leq 2$. So, the domain is the interval $[-2, 2]$.

In Example 5(c), note that the domain of a function may be implied by the physical context. For instance, from the equation

$$V = \frac{4}{3}\pi r^3$$

you would have no reason to restrict r to positive values, but the physical context implies that a sphere cannot have a negative radius.

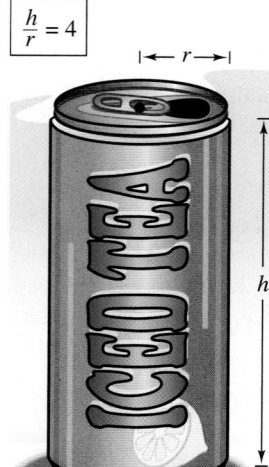

$$\boxed{\dfrac{h}{r} = 4}$$

$\leftarrow r \rightarrow$

h

FIGURE P.48

Applications

Example 6 ▶ **The Dimensions of a Container**

You work in the marketing department of a soft-drink company and are experimenting with a new can for iced tea that is slightly narrower and taller than a standard can. For your experimental can, the ratio of the height to the radius is 4, as shown in Figure P.48.

a. Write the volume of the can as a function of the radius r.

b. Write the volume of the can as a function of the height h.

Solution

a. $V(r) = \pi r^2 h = \pi r^2 (4r) = 4\pi r^3$ Write V as a function of r.

b. $V(h) = \pi \left(\dfrac{h}{4}\right)^2 h = \dfrac{\pi h^3}{16}$ Write V as a function of h.

Example 7 ▶ **The Path of a Baseball**

A baseball is hit at a point 3 feet above ground at a velocity of 100 feet per second and an angle of 45°. The path of the baseball is given by the function

$$f(x) = -0.0032x^2 + x + 3$$

where y and x are measured in feet, as shown in Figure P.49. Will the baseball clear a 10-foot fence located 300 feet from home plate?

Solution

When $x = 300$, the height of the baseball is

$$f(300) = -0.0032(300)^2 + 300 + 3$$

$$= 15 \text{ feet.}$$

So, the ball will clear the fence.

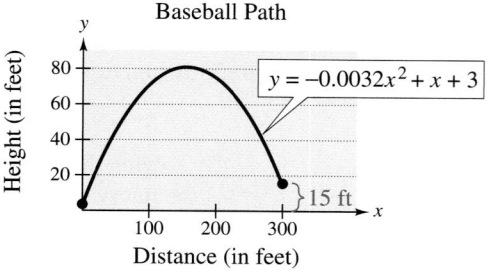

FIGURE P.49

In the equation in Example 7, the height of the baseball is a function of the distance from home plate.

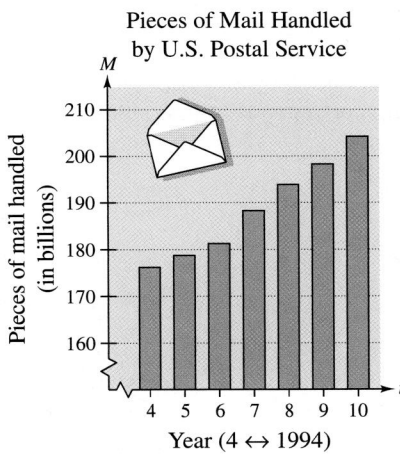

Pieces of Mail Handled
by U.S. Postal Service

FIGURE P.50

Example 8 ▶ **Pieces of Mail Handled**

The number M (in billions) of pieces of mail handled by the U.S. Postal Service increased in a linear pattern from 1994 to 1996, as shown in Figure P.50. Then, in 1997, the number handled took a jump and, until 2000, increased in a *different* linear pattern. These two patterns can be approximated by the function

$$M(t) = \begin{cases} 167.2 + 2.70t, & 4 \le t \le 6 \\ 152.0 + 5.57t, & 7 \le t \le 10 \end{cases}$$

where $t = 4$ represents 1994. Use this function to approximate the total number of pieces of mail handled from 1994 to 2000.   (Source: U.S. Postal Service)

Solution

From 1994 to 1996, use $M(t) = 167.2 + 2.70t$.

$$\underbrace{178.0,}_{1994} \quad \underbrace{180.7,}_{1995} \quad \underbrace{183.4}_{1996}$$

From 1997 to 2000, use $M(t) = 152.0 + 5.57t$.

$$\underbrace{191.0,}_{1997} \quad \underbrace{196.6,}_{1998} \quad \underbrace{202.1,}_{1999} \quad \underbrace{207.7}_{2000}$$

The total of these seven amounts is 1339.5, which implies that the total number of pieces of mail handled was approximately 1.3 trillion.

One of the basic definitions in calculus employs the ratio

$$\frac{f(x + h) - f(x)}{h}, \qquad h \ne 0.$$

This ratio is called a **difference quotient,** as illustrated in Example 9.

Example 9 ▶ **Evaluating a Difference Quotient**

For $f(x) = x^2 - 4x + 7$, find $\dfrac{f(x + h) - f(x)}{h}$.

Solution

$$\frac{f(x + h) - f(x)}{h} = \frac{[(x + h)^2 - 4(x + h) + 7] - (x^2 - 4x + 7)}{h}$$

$$= \frac{x^2 + 2xh + h^2 - 4x - 4h + 7 - x^2 + 4x - 7}{h}$$

$$= \frac{2xh + h^2 - 4h}{h} = \frac{h(2x + h - 4)}{h} = 2x + h - 4, \; h \ne 0$$

The symbol ∫ indicates an example or exercise that highlights algebraic techniques specifically used in calculus.

Activities

1. Evaluate $f(x) = 2 + 3x - x^2$ for
   a. $f(-3)$
   b. $f(x + 1)$
   c. $f(x + \Delta x) - f(x)$.
   Answers: a. -16
           b. $-x^2 + x + 4$
           c. $3\Delta x - 2x\Delta x - (\Delta x)^2$

2. Determine whether y is a function of x.
   $2x^3 + 3x^2y^2 + 1 = 0$
   Answer: No

3. Find the domain: $f(x) = \dfrac{3}{x + 1}$.
   Answer: All real numbers $x \neq -1$

Summary of Function Terminology

Function: A **function** is a relationship between two variables such that to each value of the independent variable there corresponds exactly one value of the dependent variable.

Function Notation: $y = f(x)$
       f is the *name* of the function.
       y is the **dependent variable.**
       x is the **independent variable.**
       $f(x)$ is the *value of the function at x.*

Domain: The **domain** of a function is the set of all values (inputs) of the independent variable for which the function is defined. If x is in the domain of f, f is said to be *defined* at x. If x is not in the domain of f, f is said to be *undefined* at x.

Range: The range of a function is the set of all values (outputs) assumed by the dependent variable (that is, the set of all function values).

Implied Domain: If f is defined by an algebraic expression and the domain is not specified, the **implied domain** consists of all real numbers for which the expression is defined.

Writing ABOUT MATHEMATICS

Everyday Functions In groups of two or three, identify common real-life functions. Consider everyday activities, events, and expenses, such as long distance telephone calls and car insurance. Here are two examples.

a. The statement, "Your happiness is a function of the grade you receive in this course" *is not* a correct mathematical use of the word "function." The word "happiness" is ambiguous.

b. The statement, "Your federal income tax is a function of your adjusted gross income" *is* a correct mathematical use of the word "function." Once you have determined your adjusted gross income, your income tax can be determined.

Describe your functions in words. Avoid using ambiguous words. Can you find an example of a piecewise-defined function?

P.5 Exercises

In Exercises 1–4, is the relationship a function?

1. *Domain*   *Range*   **2.** *Domain*   *Range*

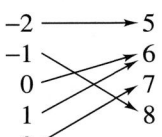   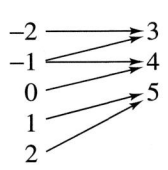

3. *Domain*   *Range*   **4.** *Domain*   *Range*

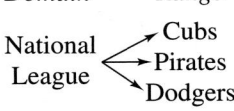   (Year)   (Number of North Atlantic tropical storms and hurricanes)

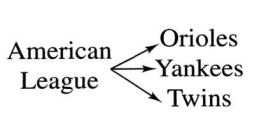   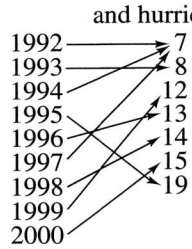

In Exercises 5–8, does the table describe a function? Explain your reasoning.

5.

Input value	-2	-1	0	1	2
Output value	-8	-1	0	1	8

6.

Input value	0	1	2	1	0
Output value	-4	-2	0	2	4

7.

Input value	10	7	4	7	10
Output value	3	6	9	12	15

8.

Input value	0	3	9	12	15
Output value	3	3	3	3	3

In Exercises 9 and 10, which sets of ordered pairs represent functions from *A* to *B*? Explain.

9. $A = \{0, 1, 2, 3\}$ and $B = \{-2, -1, 0, 1, 2\}$
  (a) $\{(0, 1), (1, -2), (2, 0), (3, 2)\}$
  (b) $\{(0, -1), (2, 2), (1, -2), (3, 0), (1, 1)\}$
  (c) $\{(0, 0), (1, 0), (2, 0), (3, 0)\}$
  (d) $\{(0, 2), (3, 0), (1, 1)\}$

10. $A = \{a, b, c\}$ and $B = \{0, 1, 2, 3\}$
  (a) $\{(a, 1), (c, 2), (c, 3), (b, 3)\}$
  (b) $\{(a, 1), (b, 2), (c, 3)\}$
  (c) $\{(1, a), (0, a), (2, c), (3, b)\}$
  (d) $\{(c, 0), (b, 0), (a, 3)\}$

Circulation of Newspapers **In Exercises 11 and 12, use the graph, which shows the circulation (in millions) of daily newspapers in the United States.** (Source: Editor & Publisher Company)

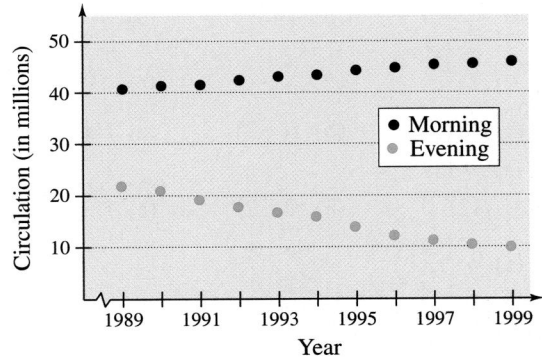

11. Is the circulation of morning newspapers a function of the year? Is the circulation of evening newspapers a function of the year? Explain.

12. Let $f(x)$ represent the circulation of evening newspapers in year x. Find $f(1998)$.

In Exercises 13–22, determine whether the equation represents *y* as a function of *x*.

13. $x^2 + y^2 = 4$ **14.** $x = y^2$

15. $x^2 + y = 4$ **16.** $x + y^2 = 4$

17. $2x + 3y = 4$ **18.** $(x - 2)^2 + y^2 = 4$

19. $y^2 = x^2 - 1$ **20.** $y = \sqrt{x + 5}$

21. $y = |4 - x|$ **22.** $|y| = 4 - x$

In Exercises 23–36, evaluate the function at each specified value of the independent variable and simplify.

23. $f(x) = 2x - 3$
   (a) $f(1)$ 　　(b) $f(-3)$ 　　(c) $f(x - 1)$

24. $g(y) = 7 - 3y$
   (a) $g(0)$ 　　(b) $g\left(\frac{7}{3}\right)$ 　　(c) $g(s + 2)$

25. $V(r) = \frac{4}{3}\pi r^3$
   (a) $V(3)$ 　　(b) $V\left(\frac{3}{2}\right)$ 　　(c) $V(2r)$

26. $h(t) = t^2 - 2t$
   (a) $h(2)$ 　　(b) $h(1.5)$ 　　(c) $h(x + 2)$

27. $f(y) = 3 - \sqrt{y}$
   (a) $f(4)$ 　　(b) $f(0.25)$ 　　(c) $f(4x^2)$

28. $f(x) = \sqrt{x + 8} + 2$
   (a) $f(-8)$ 　　(b) $f(1)$ 　　(c) $f(x - 8)$

29. $q(x) = \dfrac{1}{x^2 - 9}$
   (a) $q(0)$ 　　(b) $q(3)$ 　　(c) $q(y + 3)$

30. $q(t) = \dfrac{2t^2 + 3}{t^2}$
   (a) $q(2)$ 　　(b) $q(0)$ 　　(c) $q(-x)$

31. $f(x) = \dfrac{|x|}{x}$
   (a) $f(2)$ 　　(b) $f(-2)$ 　　(c) $f(x - 1)$

32. $f(x) = |x| + 4$
   (a) $f(2)$ 　　(b) $f(-2)$ 　　(c) $f(x^2)$

33. $f(x) = \begin{cases} 2x + 1, & x < 0 \\ 2x + 2, & x \geq 0 \end{cases}$
   (a) $f(-1)$ 　　(b) $f(0)$ 　　(c) $f(2)$

34. $f(x) = \begin{cases} x^2 + 2, & x \leq 1 \\ 2x^2 + 2, & x > 1 \end{cases}$
   (a) $f(-2)$ 　　(b) $f(1)$ 　　(c) $f(2)$

35. $f(x) = \begin{cases} 3x - 1, & x < -1 \\ 4, & -1 \leq x \leq 1 \\ x^2, & x > 1 \end{cases}$
   (a) $f(-2)$ 　　(b) $f\left(-\frac{1}{2}\right)$ 　　(c) $f(3)$

36. $f(x) = \begin{cases} 4 - 5x, & x \leq -2 \\ 0, & -2 < x \leq 2 \\ x^2 + 1, & x > 2 \end{cases}$
   (a) $f(-3)$ 　　(b) $f(4)$ 　　(c) $f(-1)$

In Exercises 37–42, complete the table.

37. $f(x) = x^2 - 3$

x	$f(x)$
-2	
-1	
0	
1	
2	

38. $g(x) = \sqrt{x - 3}$

x	$g(x)$
3	
4	
5	
6	
7	

39. $h(t) = \frac{1}{2}|t + 3|$

t	$h(t)$
-5	
-4	
-3	
-2	
-1	

40. $f(s) = \dfrac{|s - 2|}{s - 2}$

s	$f(s)$
0	
1	
$\frac{3}{2}$	
$\frac{5}{2}$	
4	

41. $f(x) = \begin{cases} -\frac{1}{2}x + 4, & x \leq 0 \\ (x - 2)^2, & x > 0 \end{cases}$

x	$f(x)$
-2	
-1	
0	
1	
2	

42. $h(x) = \begin{cases} 9 - x^2, & x < 3 \\ x - 3, & x \geq 3 \end{cases}$

x	$h(x)$
1	
2	
3	
4	
5	

In Exercises 43–50, find all real values of x such that f(x) = 0.

43. $f(x) = 15 - 3x$

44. $f(x) = 5x + 1$

45. $f(x) = \dfrac{3x - 4}{5}$

46. $f(x) = \dfrac{12 - x^2}{5}$

47. $f(x) = x^2 - 9$

48. $f(x) = x^2 - 8x + 15$

49. $f(x) = x^3 - x$

50. $f(x) = x^3 - x^2 - 4x + 4$

In Exercises 51–54, find the value(s) of x for which f(x) = g(x).

51. $f(x) = x^2 + 2x + 1, \quad g(x) = 3x + 3$

52. $f(x) = x^4 - 2x^2, \quad g(x) = 2x^2$

53. $f(x) = \sqrt{3x} + 1, \quad g(x) = x + 1$

54. $f(x) = \sqrt{x} - 4, \quad g(x) = 2 - x$

In Exercises 55–68, find the domain of the function.

55. $f(x) = 5x^2 + 2x - 1$

56. $g(x) = 1 - 2x^2$

57. $h(t) = \dfrac{4}{t}$

58. $s(y) = \dfrac{3y}{y + 5}$

59. $g(y) = \sqrt{y - 10}$

60. $f(t) = \sqrt[3]{t + 4}$

61. $f(x) = \sqrt[4]{1 - x^2}$

62. $f(x) = \sqrt[4]{x^2 + 3x}$

63. $g(x) = \dfrac{1}{x} - \dfrac{3}{x + 2}$

64. $h(x) = \dfrac{10}{x^2 - 2x}$

65. $f(s) = \dfrac{\sqrt{s - 1}}{s - 4}$

66. $f(x) = \dfrac{\sqrt{x + 6}}{6 + x}$

67. $f(x) = \dfrac{x - 4}{\sqrt{x}}$

68. $f(x) = \dfrac{x - 5}{\sqrt{x^2 - 9}}$

In Exercises 69–72, assume that the domain of f is the set A = {−2, −1, 0, 1, 2}. Determine the set of ordered pairs that represents the function f.

69. $f(x) = x^2$

70. $f(x) = x^2 - 3$

71. $f(x) = |x| + 2$

72. $f(x) = |x + 1|$

Exploration **In Exercises 73–76, match the data with one of the following functions**

$f(x) = cx, \ g(x) = cx^2, \ h(x) = c\sqrt{|x|}, \ \text{and } r(x) = \dfrac{c}{x}$

and determine the value of the constant c that will make the function fit the data in the table.

73.

x	y
−4	−32
−1	−2
0	0
1	−2
4	−32

74.

x	y
−4	−1
−1	$-\frac{1}{4}$
0	0
1	$\frac{1}{4}$
4	1

75.

x	y
−4	−8
−1	−32
0	Undef.
1	32
4	8

76.

x	y
−4	6
−1	3
0	0
1	3
4	6

In Exercises 77–84, find the difference quotient and simplify your answer.

77. $f(x) = x^2 - x + 1, \quad \dfrac{f(2 + h) - f(2)}{h}, h \neq 0$

78. $f(x) = 5x - x^2, \quad \dfrac{f(5 + h) - f(5)}{h}, h \neq 0$

79. $f(x) = x^3 + 3x, \quad \dfrac{f(x + h) - f(x)}{h}, h \neq 0$

80. $f(x) = 4x^2 - 2x, \quad \dfrac{f(x + h) - f(x)}{h}, h \neq 0$

81. $g(x) = \dfrac{1}{x^2}, \quad \dfrac{g(x) - g(3)}{x - 3}, x \neq 3$

82. $f(t) = \dfrac{1}{t - 2}, \quad \dfrac{f(t) - f(1)}{t - 1}, t \neq 1$

83. $f(x) = \sqrt{5x}, \quad \dfrac{f(x) - f(5)}{x - 5}, x \neq 5$

84. $f(x) = x^{2/3} + 1, \quad \dfrac{f(x) - f(8)}{x - 8}, x \neq 8$

The symbol ∫ indicates an example or exercise that highlights algebraic techniques specifically used in calculus.

85. Geometry   Write the area A of a square as a function of its perimeter P.

86. Geometry   Write the area A of a circle as a function of its circumference C.

87. Maximum Volume   An open box of maximum volume is to be made from a square piece of material 24 centimeters on a side by cutting equal squares from the corners and turning up the sides (see figure).

(a) The table shows the volume V (in cubic centimeters) of the box for various heights x (in centimeters). Use the table to estimate the maximum volume.

Height, x	Volume, V
1	484
2	800
3	972
4	1024
5	980
6	864

(b) Plot the points (x, V). Does the relation defined by the ordered pairs represent V as a function of x?

(c) If V is a function of x, write the function and determine its domain.

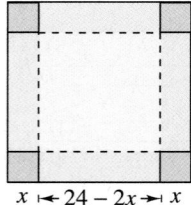
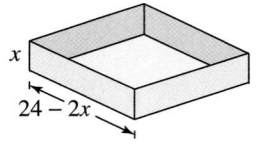

$x \longmapsto 24 - 2x \longrightarrow x$

88. Maximum Profit   The cost per unit in the production of a radio model is \$60. The manufacturer charges \$90 per unit for orders of 100 or less. To encourage large orders, the manufacturer reduces the charge by \$0.15 per radio for each unit ordered in excess of 100 (for example, there would be a charge of \$87 per radio for an order size of 120).

(a) The table shows the profit P (in dollars) for various numbers of units ordered, x. Use the table to estimate the maximum profit.

Units, x	Profit, P
110	3135
120	3240
130	3315
140	3360
150	3375
160	3360
170	3315

(b) Plot the points (x, P). Does the relation defined by the ordered pairs represent P as a function of x?

(c) If P is a function of x, write the function and determine its domain.

89. Geometry   A right triangle is formed in the first quadrant by the x- and y-axes and a line through the point $(2, 1)$ (see figure). Write the area A of the triangle as a function of x, and determine the domain of the function.

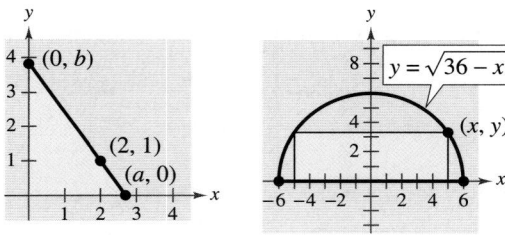

FIGURE FOR 89   FIGURE FOR 90

90. Geometry   A rectangle is bounded by the x-axis and the semicircle $y = \sqrt{36 - x^2}$ (see figure). Write the area A of the rectangle as a function of x, and determine the domain of the function.

91. Average Price   The average price p (in thousands of dollars) of a new mobile home in the United States from 1990 to 1999 (see figure) can be approximated by the model

$$p(t) = \begin{cases} 0.543t^2 - 0.75t + 27.8, & 0 \le t \le 4 \\ 1.89t + 27.1, & 5 \le t \le 9 \end{cases}$$

where $t = 0$ represents 1990. Use this model to find the average prices of a mobile home in 1990, 1994, 1996, and 1999.   (Source: U.S. Census Bureau)

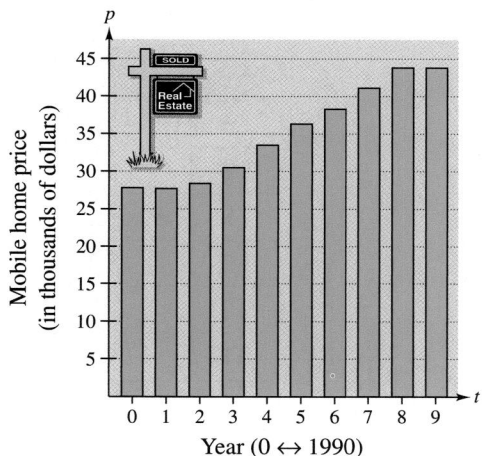

FIGURE FOR 91

92. Postal Regulations A rectangular package to be sent by the U.S. Postal Service can have a maximum combined length and girth (perimeter of a cross section) of 108 inches (see figure).

(a) Write the volume V of the package as a function of x. What is the domain of the function?

 (b) Use a graphing utility to graph your function. Be sure to use the appropriate window setting.

(c) What dimensions will maximize the volume of the package? Explain your answer.

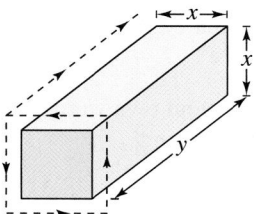

93. Cost, Revenue, and Profit A company produces a product for which the variable cost is $12.30 per unit and the fixed costs are $98,000. The product sells for $17.98. Let x be the number of units produced and sold.

(a) The total cost for a business is the sum of the variable cost and the fixed costs. Write the total cost C as a function of the number of units produced.

(b) Write the revenue R as a function of the number of units sold.

(c) Write the profit P as a function of the number of units sold. (*Note: $P = R - C$.*)

94. Average Cost The inventor of a new game believes that the variable cost for producing the game is $0.95 per unit and the fixed costs are $6000. The inventor sells each game for $1.69. Let x be the number of games sold.

(a) The total cost for a business is the sum of the variable cost and the fixed costs. Write the total cost C as a function of the number of games sold.

(b) Write the average cost per unit $\overline{C} = C/x$ as a function of x.

95. Transportation For groups of 80 or more people, a charter bus company determines the rate per person according to the formula

$$\text{Rate} = 8 - 0.05(n - 80), \qquad n \geq 80$$

where the rate is given in dollars and n is the number of people.

(a) Write the revenue R for the bus company as a function of n.

(b) Use the function in part (a) to complete the table. What can you conclude?

n	90	100	110	120	130	140	150
$R(n)$							

96. Physics The force F (in tons) of water against the face of a dam is estimated by the function $F(y) = 149.76\sqrt{10}\,y^{5/2}$, where y is the depth of the water in feet.

(a) Complete the table. What can you conclude from the table?

y	5	10	20	30	40
$F(y)$					

(b) Use the table to approximate the depth at which the force against the dam is 1,000,000 tons.

(c) Find the depth at which the force against the dam is 1,000,000 tons algebraically.

97. Height of a Balloon A balloon carrying a transmitter ascends vertically from a point 3000 feet from the receiving station.

(a) Draw a diagram that gives a visual representation of the problem. Let h represent the height of the balloon and let d represent the distance between the balloon and the receiving station.

(b) Write the height of the balloon as a function of d. What is the domain of the function?

98. *Path of a Ball* The height y (in feet) of a baseball thrown by a child is

$$y = -\frac{1}{10}x^2 + 3x + 6$$

where x is the horizontal distance (in feet) from where the ball was thrown. Will the ball fly over the head of another child 30 feet away trying to catch the ball? (Assume that the child who is trying to catch the ball holds a baseball glove at a height of 5 feet.)

▶ Model It

99. *Wildlife* The graph shows the number of threatened and endangered fish species in the world from 1996 through 2001. Let $f(t)$ represent the number of threatened and endangered fish species in the year t. (Source: U.S. Fish and Wildlife Service)

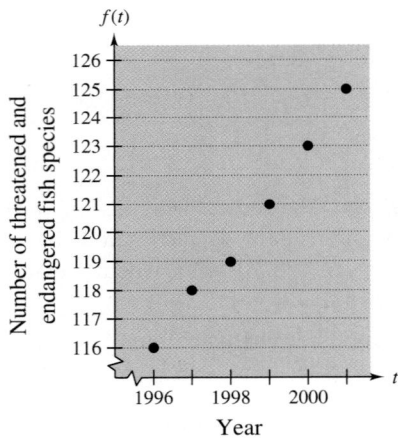

(a) Find

$$\frac{f(2001) - f(1996)}{2001 - 1996}$$

and interpret the result in the context of the problem.

(b) Find a linear model for the data algebraically. Let N represent the number of threatened and endangered fish species and let $x = 6$ correspond to 1996.

(c) Use the model found in part (b) to complete the table.

▶ Model It *(continued)*

x	N
6	
7	
8	
9	
10	
11	

(d) Compare your results from part (c) with the actual data.

 (e) Use a graphing utility to find a linear model for the data. Let $x = 6$ correspond to 1996. How does the model you found in part (b) compare with the model given by the graphing utility?

Synthesis

True or False? **In Exercises 100 and 101, determine whether the statement is true or false. Justify your answer.**

100. The domain of the function $f(x) = x^4 - 1$ is $(-\infty, \infty)$, and the range of $f(x)$ is $(0, \infty)$.

101. The set of ordered pairs $\{(-8, -2), (-6, 0), (-4, 0), (-2, 2), (0, 4), (2, -2)\}$ represents a function.

102. *Writing* In your own words, explain the meanings of *domain* and *range*.

P.6 Analyzing Graphs of Functions

▶ **Why you should learn it**

Graphs of functions can help you visualize relationships between variables in real life. For instance, in Exercise 76 on page 75, you will use the graph of a function to represent visually the merchandise trade balance for the United States.

The Graph of a Function

In Section P.5, you studied functions from an algebraic point of view. In this section, you will study functions from a graphical perspective.

The **graph of a function** f is the collection of ordered pairs $(x, f(x))$ such that x is in the domain of f. As you study this section, remember that

$$x = \text{the directed distance from the } y\text{-axis}$$

$$f(x) = \text{the directed distance from the } x\text{-axis}$$

as shown in Figure P.51.

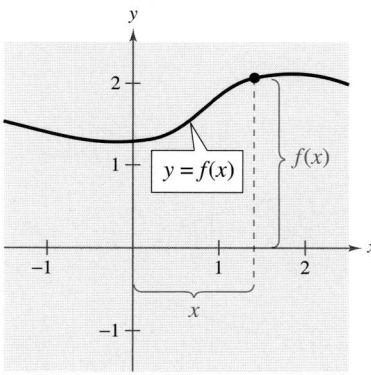

FIGURE P.51

Example 1 ▶ Finding the Domain and Range of a Function

Use the graph of the function f, shown in Figure P.52, to find (a) the domain of f, (b) the function values $f(-1)$ and $f(2)$, and (c) the range of f.

Solution

a. The closed dot at $(-1, 1)$ indicates that $x = -1$ is in the domain of f, whereas the open dot at $(5, 2)$ indicates $x = 5$ is not in the domain. So, the domain of f is all x in the interval $[-1, 5)$.

b. Because $(-1, 1)$ is a point on the graph of f, it follows that $f(-1) = 1$. Similarly, because $(2, -3)$ is a point on the graph of f, it follows that $f(2) = -3$.

c. Because the graph does not extend below $f(2) = -3$ or above $f(0) = 3$, the range of f is the interval $[-3, 3]$.

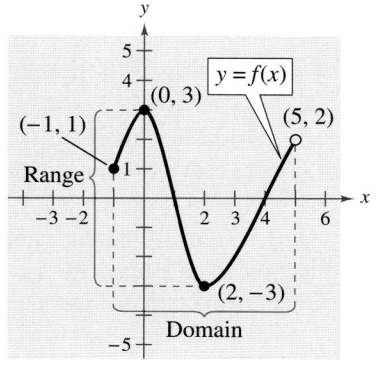

FIGURE P.52

The use of dots (open or closed) at the extreme left and right points of a graph indicates that the graph does not extend beyond these points. If no such dots are shown, assume that the graph extends beyond these points.

By the definition of a function, at most one y-value corresponds to a given x-value. This means that the graph of a function cannot have two or more different points with the same x-coordinate, and no two points on the graph of a function can be vertically above and below each other. It follows, then, that a vertical line can intersect the graph of a function at most once. This observation provides a convenient visual test called the **Vertical Line Test** for functions.

Vertical Line Test for Functions

A set of points in a coordinate plane is the graph of y as a function of x if and only if no *vertical* line intersects the graph at more than one point.

Example 2 ▶ **Vertical Line Test for Functions**

Use the Vertical Line Test to decide whether the graphs in Figure P.53 represent y as a function of x.

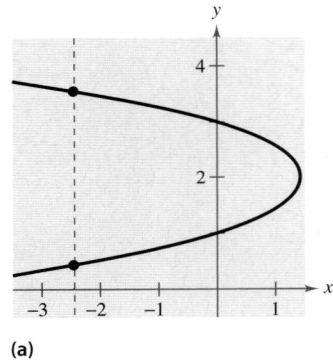

(a)

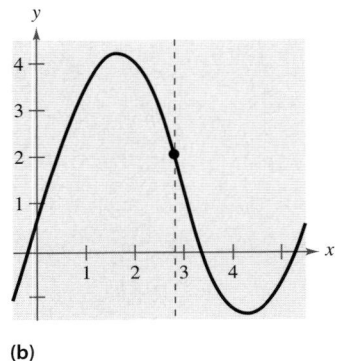

(b)

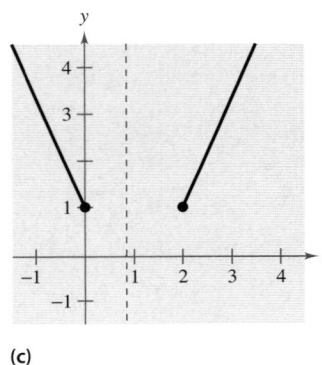

(c)

FIGURE P.53

Solution

a. This *is not* a graph of y as a function of x, because you can find a vertical line that intersects the graph twice. That is, for a particular input x, there is more than one output y.

b. This *is* a graph of y as a function of x, because every vertical line intersects the graph at most once. That is, for a particular input x, there is at most one output y.

c. This *is* a graph of y as a function of x. (Note that if a vertical line does not intersect the graph, it simply means that the function is undefined for that particular value of x.) That is, for a particular input x, there is at most one output y.

Zeros of a Function

If the graph of a function of x has an x-intercept at $(a, 0)$, then a is a **zero** of the function.

> ### Zeros of a Function
>
> The **zeros of a function** f of x are the x-values for which $f(x) = 0$.

Example 3 ▶ **Finding the Zeros of a Function**

Find the zeros of each function.

a. $f(x) = 3x^2 + x - 10$ **b.** $g(x) = \sqrt{10 - x^2}$ **c.** $h(t) = \dfrac{2t - 3}{t + 5}$

Solution

To find the zeros of a function, set the function equal to zero and solve for the independent variable.

a.
$$3x^2 + x - 10 = 0 \quad \text{Set } f(x) \text{ equal to 0.}$$
$$(3x - 5)(x + 2) = 0 \quad \text{Factor.}$$
$$3x - 5 = 0 \implies x = \tfrac{5}{3} \quad \text{Set 1st factor equal to 0.}$$
$$x + 2 = 0 \implies x = -2 \quad \text{Set 2nd factor equal to 0.}$$

The zeros of f are $x = \tfrac{5}{3}$ and $x = -2$. In Figure P.54, note that the graph of f has $\left(\tfrac{5}{3}, 0\right)$ and $(-2, 0)$ as its x-intercepts.

b.
$$\sqrt{10 - x^2} = 0 \quad \text{Set } g(x) \text{ equal to 0.}$$
$$10 - x^2 = 0 \quad \text{Square each side.}$$
$$10 = x^2 \quad \text{Add } x^2 \text{ to each side.}$$
$$\pm\sqrt{10} = x \quad \text{Extract square root.}$$

The zeros of g are $x = -\sqrt{10}$ and $x = \sqrt{10}$. In Figure P.55, note that the graph of g has $\left(-\sqrt{10}, 0\right)$ and $\left(\sqrt{10}, 0\right)$ as its x-intercepts.

c.
$$\frac{2t - 3}{t + 5} = 0 \quad \text{Set } h(t) \text{ equal to 0.}$$
$$2t - 3 = 0 \quad \text{Multiply each side by } t + 5.$$
$$2t = 3 \quad \text{Add 3 to each side.}$$
$$t = \frac{3}{2} \quad \text{Divide each side by 2.}$$

The zero of h is $t = \tfrac{3}{2}$. In Figure P.56, note that the graph of h has $\left(\tfrac{3}{2}, 0\right)$ as its t-intercept.

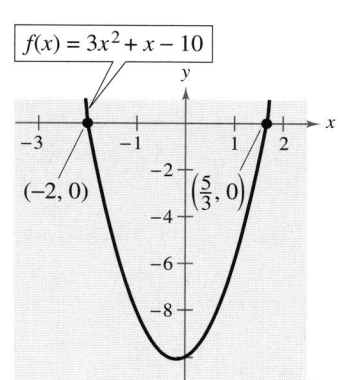

Zeros of f: $x = -2, x = \tfrac{5}{3}$
FIGURE P.54

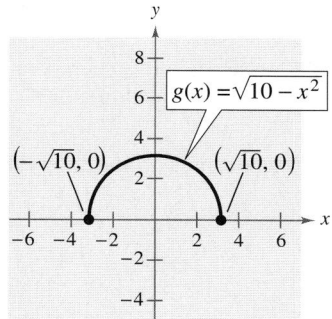

Zeros of g: $x = \pm\sqrt{10}$
FIGURE P.55

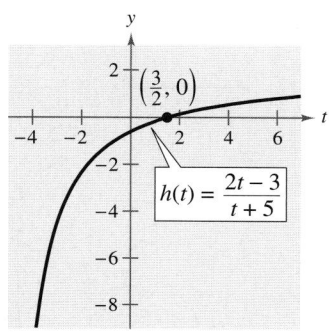

Zero of h: $t = \tfrac{3}{2}$
FIGURE P.56

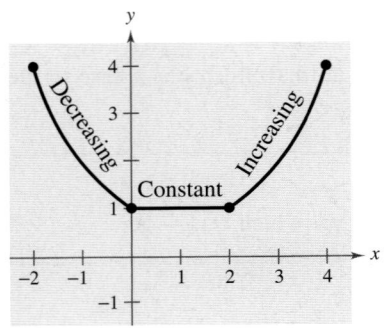

FIGURE P.57

Increasing and Decreasing Functions

The more you know about the graph of a function, the more you know about the function itself. Consider the graph shown in Figure P.57. As you move from *left to right*, this graph decreases, then is constant, and then increases.

Increasing, Decreasing, and Constant Functions

A function f is **increasing** on an interval if, for any x_1 and x_2 in the interval, $x_1 < x_2$ implies $f(x_1) < f(x_2)$.

A function f is **decreasing** on an interval if, for any x_1 and x_2 in the interval, $x_1 < x_2$ implies $f(x_1) > f(x_2)$.

A function f is **constant** on an interval if, for any x_1 and x_2 in the interval, $f(x_1) = f(x_2)$.

Example 4 ▶ Increasing and Decreasing Functions

In Figure P.58, use the graphs to describe the increasing or decreasing behavior of each function.

Solution

a. This function is increasing over the entire real line.

b. This function is increasing on the interval $(-\infty, -1)$, decreasing on the interval $(-1, 1)$, and increasing on the interval $(1, \infty)$.

c. This function is increasing on the interval $(-\infty, 0)$, constant on the interval $(0, 2)$, and decreasing on the interval $(2, \infty)$.

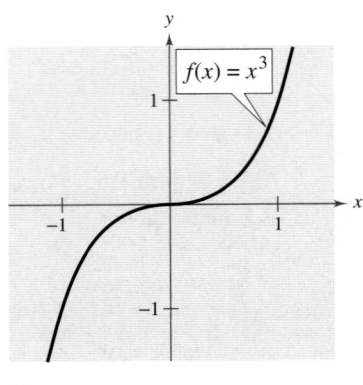

(a)

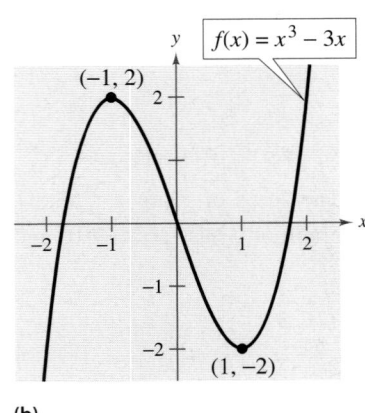

(b)

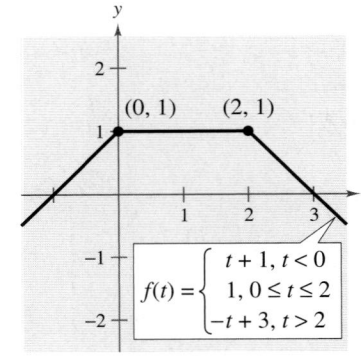

(c)

FIGURE P.58

To help you decide whether a function is increasing, decreasing, or constant on an interval, you can evaluate the function for several values of x. However, calculus is needed to determine, for certain, all intervals on which a function is increasing, decreasing, or constant.

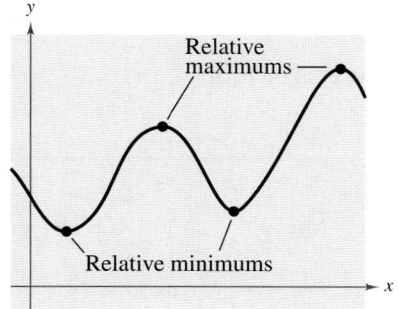

FIGURE P.59

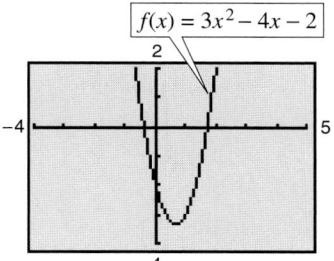

FIGURE P.60

The points at which a function changes its increasing, decreasing, or constant behavior are helpful in determining the **relative minimum** or **relative maximum** values of the function.

Definition of Relative Minimum and Relative Maximum

A function value $f(a)$ is called a **relative minimum** of f if there exists an interval (x_1, x_2) that contains a such that

$$x_1 < x < x_2 \quad \text{implies} \quad f(a) \leq f(x).$$

A function value $f(a)$ is called a **relative maximum** of f if there exists an interval (x_1, x_2) that contains a such that

$$x_1 < x < x_2 \quad \text{implies} \quad f(a) \geq f(x).$$

Figure P.59 shows several different examples of relative minima and relative maxima. By writing a second-degree polynomial equation in standard form, $y = a(x - h)^2 + k$, you can find the *exact point* (h, k) at which it has a relative minimum or relative maximum. For the time being, however, you can use a graphing utility to find reasonable approximations of these points.

Example 5 ► Approximating a Relative Minimum

Use a graphing utility to approximate the relative minimum of the function $f(x) = 3x^2 - 4x - 2$.

Solution

The graph of f is shown in Figure P.60. By using the *zoom* and *trace* features of a graphing utility, you can estimate that the function has a relative minimum at the point

$$(0.67, -3.33). \qquad \text{Relative minimum}$$

By writing this second-degree equation in standard form, $f(x) = 3\left(x - \frac{2}{3}\right)^2 - \frac{10}{3}$, you can determine that the exact point at which the relative minimum occurs is $\left(\frac{2}{3}, -\frac{10}{3}\right)$.

You can also use the *table* feature of a graphing utility to approximate numerically the relative minimum of the function in Example 5. Using a table that begins at 0.6 and increments the value of x by 0.01, you can approximate the minimum of $f(x) = 3x^2 - 4x - 2$ to be $(0.67, -3.33)$.

Technology

If you use a graphing utility to estimate the x- and y-values of a relative minimum or relative maximum, the *automatic zoom* feature will often produce graphs that are nearly flat. To overcome this problem, you can manually change the vertical setting of the viewing window. The graph will stretch vertically if the values of Ymin and Ymax are closer together.

Even and Odd Functions

In Section P.3, you studied different types of symmetry of a graph. In the terminology of functions, a function is said to be **even** if its graph is symmetric with respect to the y-axis and to be **odd** if its graph is symmetric with respect to the origin. The symmetry tests in Section P.3 yield the following tests for even and odd functions.

Tests for Even and Odd Functions

A function $y = f(x)$ is **even** if, for each x in the domain of f,

$$f(-x) = f(x).$$

A function $y = f(x)$ is **odd** if, for each x in the domain of f,

$$f(-x) = -f(x).$$

▶ Exploration ▶

Graph each of the functions with a graphing utility. Determine whether the function is *even*, *odd*, or *neither*.

$$f(x) = x^2 - x^4$$

$$g(x) = 2x^3 + 1$$

$$h(x) = x^5 - 2x^3 + x$$

$$j(x) = 2 - x^6 - x^8$$

$$k(x) = x^5 - 2x^4 + x - 2$$

$$p(x) = x^9 + 3x^5 - x^3 + x$$

What do you notice about the equations of functions that are odd? What do you notice about the equations of functions that are even? Can you describe a way to identify a function as odd or even by inspecting the equation? Can you describe a way to identify a function as neither odd nor even by inspecting the equation?

Example 6 ▶ Even and Odd Functions

a. The function $g(x) = x^3 - x$ is odd because $g(-x) = -g(x)$, as follows.

$$g(-x) = (-x)^3 - (-x) \qquad \text{Substitute } -x \text{ for } x.$$

$$= -x^3 + x \qquad \text{Simplify.}$$

$$= -(x^3 - x) \qquad \text{Distributive Property}$$

$$= -g(x) \qquad \text{Test for odd function}$$

b. The function $h(x) = x^2 + 1$ is even because $h(-x) = h(x)$, as follows.

$$h(-x) = (-x)^2 + 1 \qquad \text{Substitute } -x \text{ for } x.$$

$$= x^2 + 1 \qquad \text{Simplify.}$$

$$= h(x) \qquad \text{Test for even function}$$

The graphs of these two functions are shown in Figure P.61.

Additional Example
Is the function $f(x) = x^3 - 1$ even, odd, or neither?

Solution
Substituting $-x$ for x,
$f(-x) = (-x)^3 - 1 = -x^3 - 1$.
Because $f(x) = x^3 - 1$ and
$f(-x) = -x^3 - 1$, $f(-x) \neq f(x)$ and
$f(-x) \neq -f(x)$. The function is neither even nor odd.

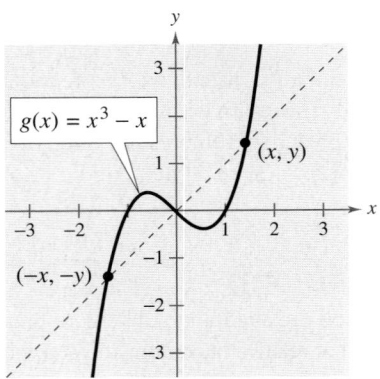

(a) Symmetric to origin: Odd Function

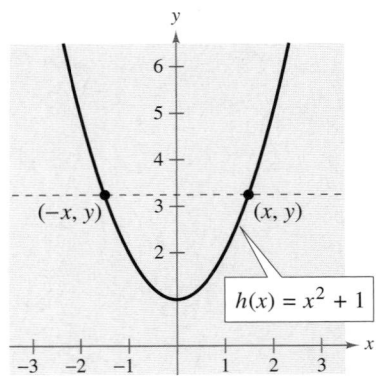

(b) Symmetric to y-axis: Even Function

FIGURE P.61

P.6 **Exercises**

In Exercises 1–4, use the graph of the function to find the domain and range of f.

1.

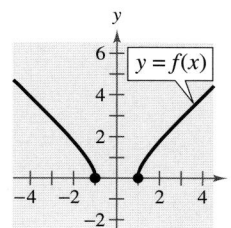

2.

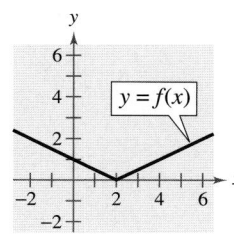

3.

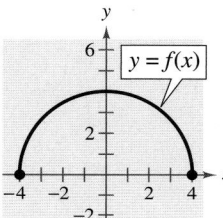

4.

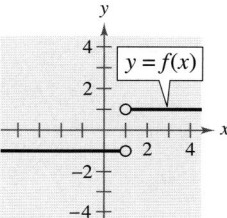

In Exercises 5–8, use the graph of the function to find the indicated function values.

5. (a) $f(-2)$ (b) $f(-1)$
   (c) $f\left(\frac{1}{2}\right)$ (d) $f(1)$

6. (a) $f(-1)$ (b) $f(2)$
   (c) $f(0)$ (d) $f(1)$

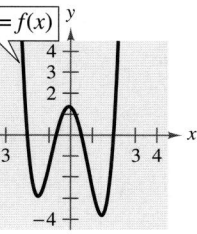

7. (a) $f(-2)$ (b) $f(1)$
   (c) $f(0)$ (d) $f(2)$

8. (a) $f(2)$ (b) $f(1)$
   (c) $f(3)$ (d) $f(-1)$

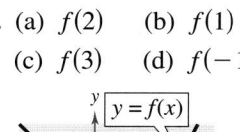
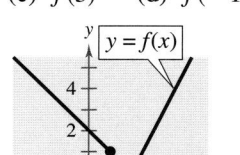
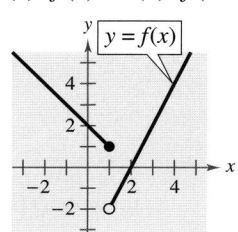

In Exercises 9–14, use the Vertical Line Test to determine whether y is a function of x. To print an enlarged copy of the graph, go to the website *www.mathgraphs.com*.

9. $y = \frac{1}{2}x^2$

10. $y = \frac{1}{4}x^3$

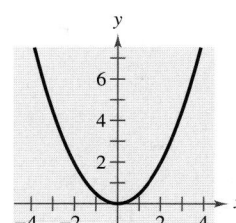
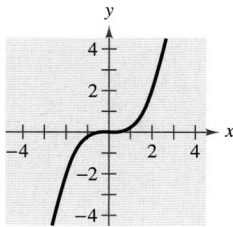

11. $x - y^2 = 1$

12. $x^2 + y^2 = 25$

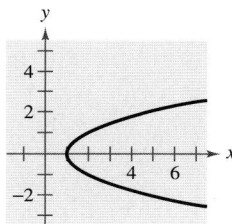
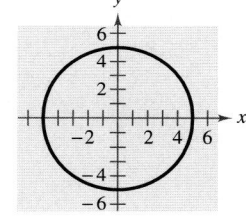

13. $x^2 = 2xy - 1$

14. $x = |y + 2|$

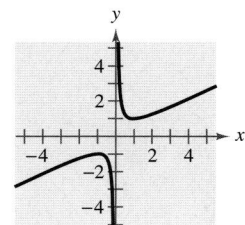
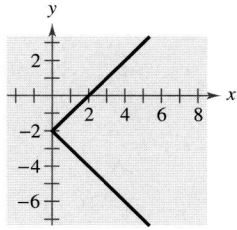

In Exercises 15–24, find the zeros of the function algebraically.

15. $f(x) = 2x^2 - 7x - 30$

16. $f(x) = 3x^2 + 22x - 16$

17. $f(x) = \dfrac{x}{9x^2 - 4}$

18. $f(x) = \dfrac{x^2 - 9x + 14}{4x}$

19. $f(x) = \frac{1}{2}x^3 - x$

20. $f(x) = x^3 - 4x^2 - 9x + 36$

21. $f(x) = 4x^3 - 24x^2 - x + 6$

22. $f(x) = 9x^4 - 25x^2$

23. $f(x) = \sqrt{2x - 1}$

24. $f(x) = \sqrt{3x + 2}$

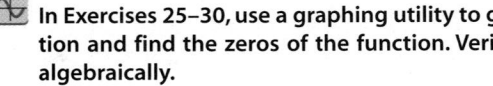 In Exercises 25–30, use a graphing utility to graph the function and find the zeros of the function. Verify your results algebraically.

25. $f(x) = 3 + \dfrac{5}{x}$

26. $f(x) = x(x - 7)$

27. $f(x) = \sqrt{2x + 11}$

28. $f(x) = \sqrt{3x - 14} - 8$

29. $f(x) = \dfrac{3x - 1}{x - 6}$

30. $f(x) = \dfrac{2x^2 - 9}{3 - x}$

In Exercises 31–38, determine the intervals over which the function is increasing, decreasing, or constant.

31. $f(x) = \frac{3}{2}x$

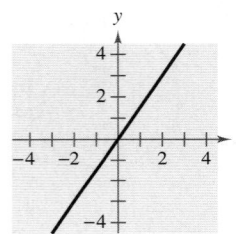

32. $f(x) = x^2 - 4x$

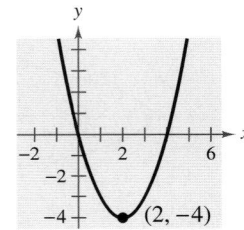

33. $f(x) = x^3 - 3x^2 + 2$

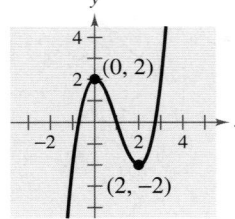

34. $f(x) = \sqrt{x^2 - 1}$

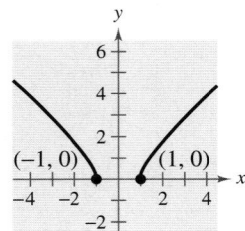

35. $f(x) = \begin{cases} x + 3, & x \le 0 \\ 3, & 0 < x \le 2 \\ 2x + 1, & x > 2 \end{cases}$

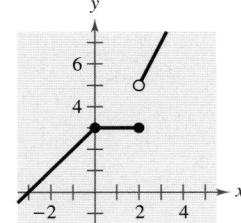

36. $f(x) = \begin{cases} 2x + 1, & x \le -1 \\ x^2 - 2, & x > -1 \end{cases}$

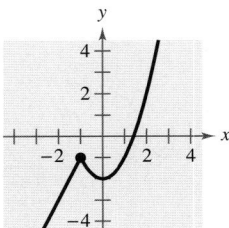

37. $f(x) = |x + 1| + |x - 1|$

38. $f(x) = \dfrac{x^2 + x + 1}{x + 1}$

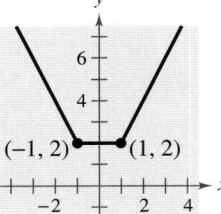
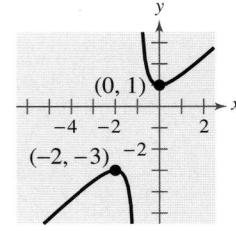

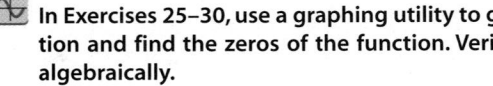 In Exercises 39–48, (a) use a graphing utility to graph the function and visually determine the intervals over which the function is increasing, decreasing, or constant, and (b) make a table of values to verify whether the function is increasing, decreasing, or constant over the intervals you identified in part (a).

39. $f(x) = 3$

40. $g(x) = x$

41. $g(s) = \dfrac{s^2}{4}$

42. $h(x) = x^2 - 4$

43. $f(t) = -t^4$

44. $f(x) = 3x^4 - 6x^2$

45. $f(x) = \sqrt{1 - x}$

46. $f(x) = x\sqrt{x + 3}$

47. $f(x) = x^{3/2}$

48. $f(x) = x^{2/3}$

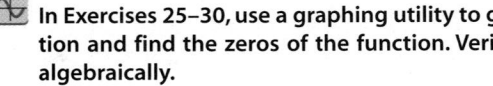 In Exercises 49–52, use a graphing utility to approximate the relative minimum/relative maximum of each function.

49. $f(x) = (x - 4)(x + 2)$

50. $f(x) = 3x^2 - 2x - 5$

51. $f(x) = x(x - 2)(x + 3)$

52. $f(x) = x^3 - 3x^2 - x + 1$

In Exercises 53–60, graph the function and determine the interval(s) for which $f(x) \ge 0$.

53. $f(x) = 4 - x$

54. $f(x) = 4x + 2$

55. $f(x) = x^2 + x$

56. $f(x) = x^2 - 4x$

57. $f(x) = \sqrt{x - 1}$

58. $f(x) = \sqrt{x + 2}$

59. $f(x) = -(1 + |x|)$

60. $f(x) = \frac{1}{2}(2 + |x|)$

In Exercises 61–66, determine whether the function is even, odd, or neither.

61. $f(x) = x^6 - 2x^2 + 3$ **62.** $h(x) = x^3 - 5$

63. $g(x) = x^3 - 5x$ **64.** $f(x) = x\sqrt{1-x^2}$

65. $f(t) = t^2 + 2t - 3$ **66.** $g(s) = 4s^{2/3}$

In Exercises 67–70, write the height h of the rectangle as a function of x.

67.
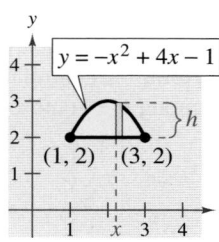

68.
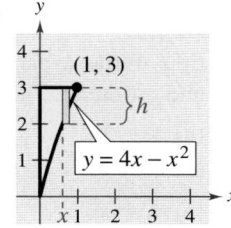

69.
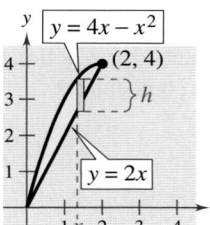

70.
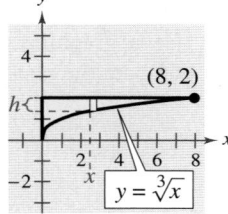

In Exercises 71–74, write the length L of the rectangle as a function of y.

71.
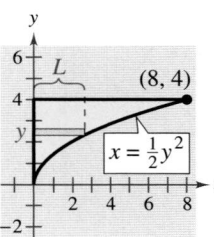

72.
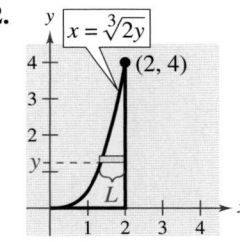

73.
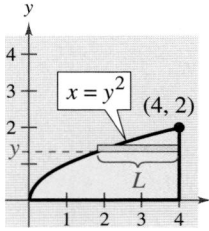

74.
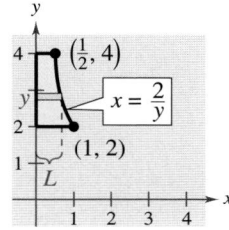

75. *Electronics*   The number of lumens (time rate of flow of light) L from a fluorescent lamp can be approximated by the model

$$L = -0.294x^2 + 97.744x - 664.875, \quad 20 \le x \le 90$$

where x is the wattage of the lamp.

(a) Use a graphing utility to graph the function.

(b) Use the graph from part (a) to estimate the wattage necessary to obtain 2000 lumens.

▶ **Model It**

76. *Data Analysis*   The table shows the amount y (in billions of dollars) of the merchandise trade balance of the United States for the years 1991 through 1999. The merchandise trade balance is the difference between the values of exports and imports. A negative merchandise trade balance indicates that imports exceeded exports. (Source: U.S. International Trade Administration and U.S. Foreign Trade Highlights)

Year, x	Trade balance, y
1991	−66.8
1992	−84.5
1993	−115.6
1994	−150.7
1995	−158.7
1996	−170.2
1997	−181.5
1998	−229.8
1999	−330.0

(a) Use a graphing utility to create a scatter plot of the data.

(b) Use the graph in part (a) to determine whether the data represents y as a function of x.

(c) Use the *regression* feature of a graphing utility to find a cubic model (a model of the form $y = ax^3 + bx^2 + cx + d$) for the data. Let x be the time (in years), with $x = 1$ corresponding to 1991.

(d) What is the domain of the model?

(e) Use a graphing utility to graph the model in the same viewing window you used in part (a).

(f) For which year does the model most accurately estimate the actual data? During which year is it least accurate?

77. *Coordinate Axis Scale* Each function models the specified data for the years 1995 through 2002, with $t = 5$ corresponding to 1995. Estimate a reasonable scale for the vertical axis (e.g., hundreds, thousands, millions, etc.) of the graph and justify your answer. (There are many correct answers.)

(a) $f(t)$ represents the average salary of college professors.

(b) $f(t)$ represents the U.S. population.

(c) $f(t)$ represents the percent of the civilian work force that is unemployed.

78. *Geometry* Corners of equal size are cut from a square with sides of length 8 meters (see figure).

(a) Write the area A of the resulting figure as a function of x. Determine the domain of the function.

 (b) Use a graphing utility to graph the area function over its domain. Use the graph to find the range of the function.

(c) Identify the figure that would result if x were chosen to be the maximum value in the domain of the function. What would be the length of each side of the figure?

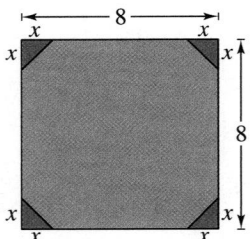

Synthesis

True or False? **In Exercises 79 and 80, determine whether the statement is true or false. Justify your answer.**

79. A function with a square root cannot have a domain that is the set of real numbers.

80. It is possible for an odd function to have the interval $[0, \infty)$ as its domain.

81. If f is an even function, determine whether g is even, odd, or neither. Explain.

(a) $g(x) = -f(x)$ (b) $g(x) = f(-x)$

(c) $g(x) = f(x) - 2$ (d) $g(x) = f(x - 2)$

82. *Think About It* Does the graph in Exercise 11 represent x as a function of y? Explain.

Think About It **In Exercises 83–86, find the coordinates of a second point on the graph of a function f if the given point is on the graph and the function is (a) even and (b) odd.**

83. $\left(-\frac{3}{2}, 4\right)$ **84.** $\left(-\frac{5}{3}, -7\right)$

85. $(4, 9)$ **86.** $(5, -1)$

 87. *Writing* Use a graphing utility to graph each function. Write a paragraph describing any similarities and differences you observe among the graphs.

(a) $y = x$ (b) $y = x^2$

(c) $y = x^3$ (d) $y = x^4$

(e) $y = x^5$ (f) $y = x^6$

88. *Conjecture* Use the results of Exercise 87 to make a conjecture about the graphs of $y = x^7$ and $y = x^8$. Use a graphing utility to graph the functions and compare the results with your conjecture.

P.7 A Library of Functions

▶ What you should learn

- How to identify and graph linear and squaring functions
- How to identify and graph cubic, square root, and reciprocal functions
- How to identify and graph step and other piecewise-defined functions
- How to recognize graphs of common functions

▶ Why you should learn it

Piecewise-defined functions can be used to model real-life situations. For instance, in Exercise 68 on page 84, you will use a piecewise-defined function to model the monthly revenue of a landscaping business.

Michael Newman/PhotoEdit

Additional Example
Write the linear function f for which $f(0) = -1$ and $f(5) = 8$.
Solution
$$m = \frac{y_2 - y_1}{x_2 - x_1} = \frac{8 - (-1)}{5 - 0} = \frac{9}{5}$$
$$y - y_1 = m(x - x_1)$$
$$y - (-1) = \frac{9}{5}(x - 0)$$
$$y + 1 = \frac{9}{5}x$$
$$y = \frac{9}{5}x - 1$$

Linear and Squaring Functions

One of the goals of this text is to enable you to recognize the basic shapes of the graphs of different types of functions. For instance, you know that the graph of the **linear function** $f(x) = ax + b$ is a line with slope $m = a$ and y-intercept at $(0, b)$. The graph of the linear function has the following features.

- The domain of the function is the set of all real numbers.
- The range of the function is the set of all real numbers.
- The graph has one intercept, $(0, b)$.
- The graph is increasing if $m > 0$, decreasing if $m < 0$, and constant if $m = 0$.

Example 1 ▶ **Writing a Linear Function**

Write the linear function f for which $f(1) = 3$ and $f(4) = 0$.

Solution
To find the equation of the line that passes through $(x_1, y_1) = (1, 3)$ and $(x_2, y_2) = (4, 0)$, first find the slope of the line.

$$m = \frac{y_2 - y_1}{x_2 - x_1} = \frac{0 - 3}{4 - 1} = \frac{-3}{3} = -1$$

Next, use the point-slope form of the equation of a line.

$$y - y_1 = m(x - x_1) \qquad \text{Point-slope form}$$
$$y - 3 = -1(x - 1) \qquad \text{Substitute for } x_1, y_1, \text{ and } m.$$
$$y = -x + 4 \qquad \text{Simplify.}$$
$$f(x) = -x + 4 \qquad \text{Function notation}$$

The graph of this function is shown in Figure P.62.

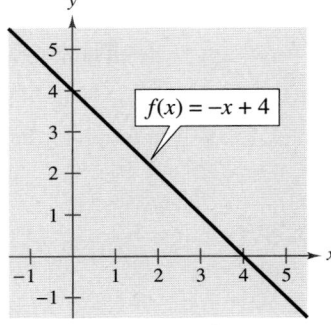

FIGURE **P.62**

There are two special types of linear functions, the **constant function** and the **identity function.** A constant function has the form

$$f(x) = c$$

and has the domain of all real numbers with a range consisting of a single real number c. The graph of a constant function is a horizontal line, as shown in Figure P.63. The identity function has the form

$$f(x) = x.$$

Its domain and range are the set of all real numbers. The identity function has a slope of $m = 1$ and a y-intercept $(0, 0)$. The graph of the identity function is a line for which each x-coordinate equals the corresponding y-coordinate. The graph is always increasing, as shown in Figure P.64.

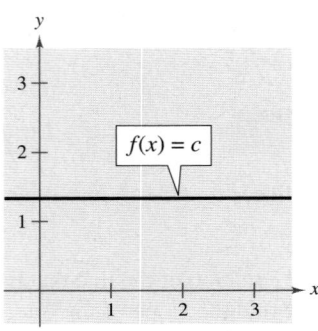

FIGURE P.63

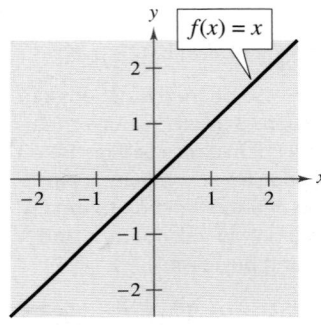

FIGURE P.64

The graph of the **squaring function**

$$f(x) = x^2$$

is a U-shaped curve with the following features.

- The domain of the function is the set of all real numbers.
- The range of the function is the set of all nonnegative real numbers.
- The function is even.
- The graph has an intercept at $(0, 0)$.
- The graph is decreasing on the interval $(-\infty, 0)$ and increasing on the interval $(0, \infty)$.
- The graph is symmetric with respect to the y-axis.
- The graph has a relative minimum at $(0, 0)$.

The graph of the squaring function is shown in Figure P.65.

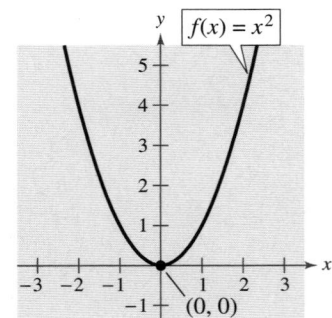

FIGURE P.65

Cubic, Square Root, and Reciprocal Functions

Special features of the graphs of the **cubic, square root,** and **reciprocal functions** are summarized below.

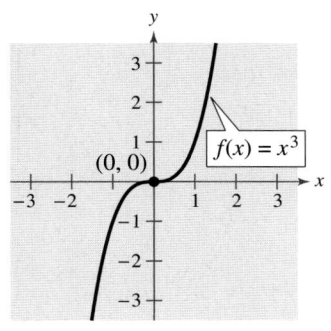

Cubic function
FIGURE P.66

1. The graph of the *cubic* function

$$f(x) = x^3$$

has the following features.

- The domain of the function is the set of all real numbers.
- The range of the function is the set of all real numbers.
- The function is odd.
- The graph has an intercept at $(0, 0)$.
- The graph is increasing on the interval $(-\infty, \infty)$.
- The graph is symmetric with respect to the origin.

The graph of the cubic function is shown in Figure P.66.

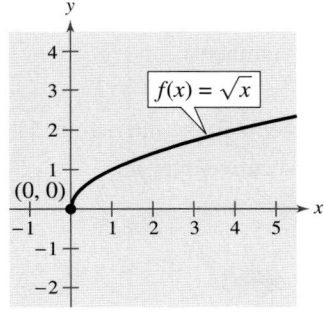

Square root function
FIGURE P.67

2. The graph of the *square root* function

$$f(x) = \sqrt{x}$$

has the following features.

- The domain of the function is the set of all nonnegative real numbers.
- The range of the function is the set of all nonnegative real numbers.
- The graph has an intercept at $(0, 0)$.
- The graph is increasing on the interval $(0, \infty)$.

The graph of the square root function is shown in Figure P.67.

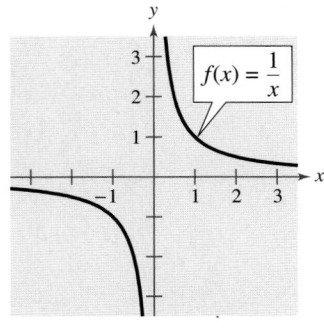

Reciprocal function
FIGURE P.68

3. The graph of the *reciprocal* function

$$f(x) = \frac{1}{x}$$

has the following features.

- The domain of the function is $(-\infty, 0) \cup (0, \infty)$.
- The range of the function is $(-\infty, 0) \cup (0, \infty)$.
- The function is odd.
- The graph does not have any intercepts.
- The graph is decreasing on the intervals $(-\infty, 0)$ and $(0, \infty)$.
- The graph is symmetric with respect to the origin.

The graph of the reciprocal function is shown in Figure P.68.

Step and Piecewise-Defined Functions

Functions whose graphs resemble sets of stairsteps are known as **step functions.** The most famous of the step functions is the **greatest integer function,** which is denoted by $[\![x]\!]$ and defined as

$$f(x) = [\![x]\!] = \textit{the greatest integer less than or equal to x.}$$

Some values of the greatest integer function are as follows.

$$[\![-1]\!] = (\text{greatest integer} \le -1) = -1$$

$$\left[\!\!\left[-\tfrac{1}{2}\right]\!\!\right] = \left(\text{greatest integer} \le -\tfrac{1}{2}\right) = -1$$

$$\left[\!\!\left[\tfrac{1}{10}\right]\!\!\right] = \left(\text{greatest integer} \le \tfrac{1}{10}\right) = 0$$

$$[\![1.5]\!] = (\text{greatest integer} \le 1.5) = 1$$

The graph of the greatest integer function

$$f(x) = [\![x]\!]$$

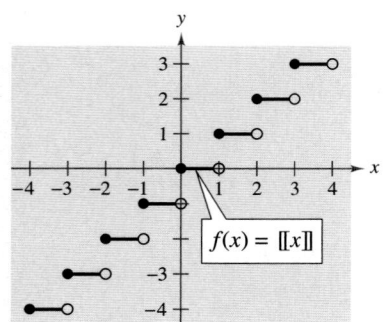

FIGURE P.69

has the following features, as shown in Figure P.69.

- The domain of the function is the set of all real numbers.
- The range of the function is the set of all integers.
- The graph has a y-intercept at $(0, 0)$ and x-intercepts in the interval $[0, 1)$.
- The graph is constant between each pair of consecutive integers.
- The graph jumps vertically one unit at each integer value.

Example 2 ▶   Evaluating a Step Function

Evaluate the function when $x = -1$, 2, and $\tfrac{3}{2}$.

$$f(x) = [\![x]\!] + 1$$

Solution

For $x = -1$, the greatest integer ≤ -1 is -1, so

$$f(-1) = [\![-1]\!] + 1 = -1 + 1 = 0.$$

For $x = 2$, the greatest integer ≤ 2 is 2, so

$$f(2) = [\![2]\!] + 1 = 2 + 1 = 3.$$

For $x = \tfrac{3}{2}$, the greatest integer $\le \tfrac{3}{2}$ is 1, so

$$f\left(\tfrac{3}{2}\right) = \left[\!\!\left[\tfrac{3}{2}\right]\!\!\right] + 1 = 1 + 1 = 2.$$

The graph of $f(x) = [\![x]\!] + 1$ is shown in Figure P.70.

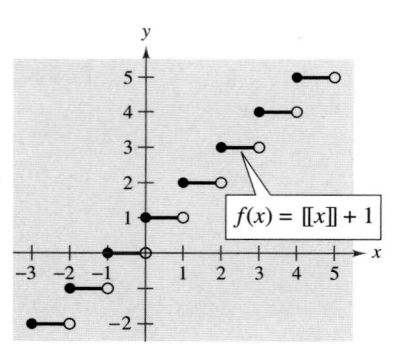

FIGURE P.70

Demonstrate the real-life nature of step functions by discussing Exercises 63 and 64 on page 83. If writing is a part of your course, this section provides a good opportunity for students to find other examples of step functions and write brief essays on their applications.

Recall from Section P.5 that a piecewise-defined function is defined by two or more equations over a specified domain. To graph a piecewise-defined function, graph each equation separately over the specified domain, as shown in Example 3.

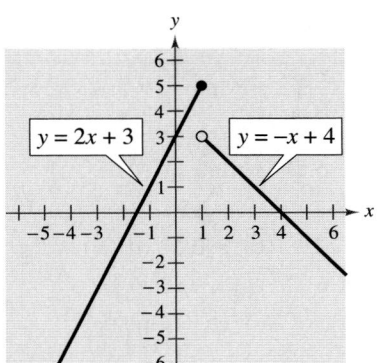

FIGURE P.71

Example 3 ▶ **Graphing a Piecewise-Defined Function**

Sketch the graph of

$$f(x) = \begin{cases} 2x + 3, & x \le 1 \\ -x + 4, & x > 1 \end{cases}.$$

Solution

This piecewise-defined function is composed of two linear functions. At $x = 1$ and to the left of $x = 1$ the graph is the line $y = 2x + 3$, and to the right of $x = 1$ the graph is the line $y = -x + 4$, as shown in Figure P.71.

Common Functions

The eight graphs shown in Figure P.72 represent the most commonly used functions in algebra. Familiarity with the basic characteristics of these simple graphs will help you analyze the shapes of more complicated graphs—in particular, graphs obtained from these graphs by the rigid and nonrigid transformations studied in the next section.

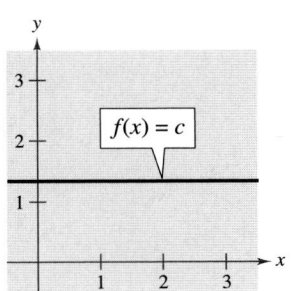

(a) Constant Function

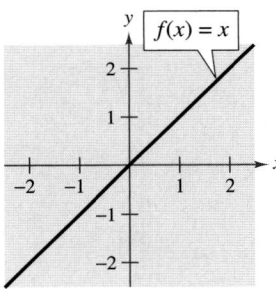

(b) Identity Function

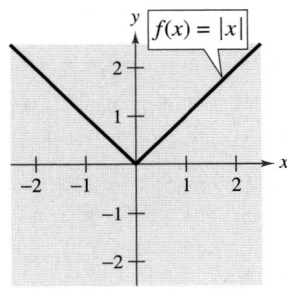

(c) Absolute Value Function

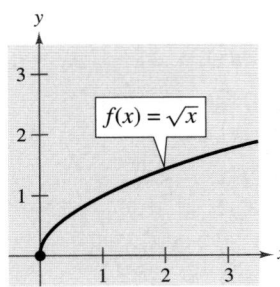

(d) Square Root Function

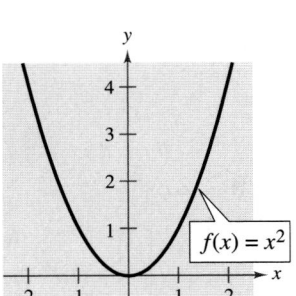

(e) Quadratic Function

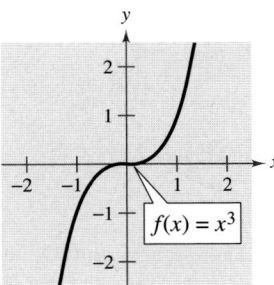

(f) Cubic Function

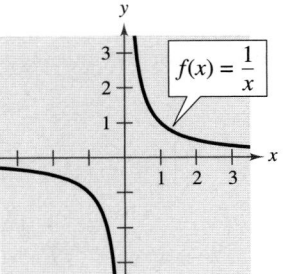

(g) Reciprocal Function

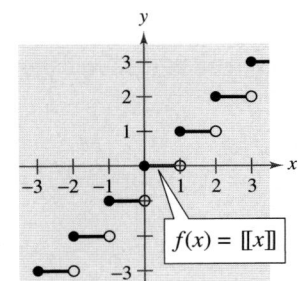

(h) Greatest Integer Function

FIGURE P.72

P.7 Exercises

In Exercises 1–8, write the linear function that has the indicated function values. Then sketch the graph of the function.

1. $f(1) = 4, f(0) = 6$

2. $f(-3) = -8, f(1) = 2$

3. $f(5) = -4, f(-2) = 17$

4. $f(3) = 9, f(-1) = -11$

5. $f(-5) = -1, f(5) = -1$

6. $f(-10) = 12, f(16) = -1$

7. $f(\frac{1}{2}) = -6, f(4) = -3$

8. $f(\frac{2}{3}) = -\frac{15}{2}, f(-4) = -11$

 In Exercises 9–28, use a graphing utility to graph the function. Be sure to choose an appropriate viewing window.

9. $f(x) = -x - \frac{3}{4}$

10. $f(x) = 3x - \frac{5}{2}$

11. $f(x) = -\frac{1}{6}x - \frac{5}{2}$

12. $f(x) = \frac{5}{6} - \frac{2}{3}x$

13. $f(x) = x^2 - 2x$

14. $f(x) = -x^2 + 8x$

15. $h(x) = -x^2 + 4x + 12$

16. $g(x) = x^2 - 6x - 16$

17. $f(x) = x^3 - 1$

18. $f(x) = 8 - x^3$

19. $f(x) = (x - 1)^3 + 2$

20. $g(x) = 2(x + 3)^3 + 1$

21. $f(x) = 4\sqrt{x}$

22. $f(x) = 4 - 2\sqrt{x}$

23. $g(x) = 2 - \sqrt{x + 4}$

24. $h(x) = \sqrt{x + 2} + 3$

25. $f(x) = -\dfrac{1}{x}$

26. $f(x) = 4 + \dfrac{1}{x}$

27. $h(x) = \dfrac{1}{x + 2}$

28. $k(x) = \dfrac{1}{x - 3}$

In Exercise 29–36, evaluate the function for the indicated values.

29. $f(x) = [\![x]\!]$

    (a) $f(2.1)$ (b) $f(2.9)$ (c) $f(-3.1)$ (d) $f(\frac{7}{2})$

30. $g(x) = 2[\![x]\!]$

    (a) $g(-3)$ (b) $g(0.25)$ (c) $g(9.5)$ (d) $g(\frac{11}{3})$

31. $h(x) = [\![x + 3]\!]$

    (a) $h(-2)$ (b) $h(\frac{1}{2})$ (c) $h(4.2)$ (d) $h(-21.6)$

32. $f(x) = 4[\![x]\!] + 7$

    (a) $f(0)$ (b) $f(-1.5)$ (c) $f(6)$ (d) $f(\frac{5}{3})$

33. $h(x) = [\![3x - 1]\!]$

    (a) $h(2.5)$ (b) $h(-3.2)$ (c) $h(\frac{7}{3})$ (d) $h(-\frac{21}{3})$

34. $k(x) = [\![\frac{1}{2}x + 6]\!]$

    (a) $k(5)$ (b) $k(-6.1)$ (c) $k(0.1)$ (d) $k(15)$

35. $g(x) = 3[\![x - 2]\!] + 5$

    (a) $g(-2.7)$ (b) $g(-1)$ (c) $g(0.8)$ (d) $g(14.5)$

36. $g(x) = -7[\![x + 4]\!] + 6$

    (a) $g(\frac{1}{8})$ (b) $g(9)$ (c) $g(-4)$ (d) $g(\frac{3}{2})$

In Exercises 37–42, sketch the graph of the function.

37. $g(x) = -[\![x]\!]$

38. $g(x) = 4[\![x]\!]$

39. $g(x) = [\![x]\!] - 2$

40. $g(x) = [\![x]\!] - 1$

41. $g(x) = [\![x + 1]\!]$

42. $g(x) = [\![x - 3]\!]$

In Exercises 43–50, graph the function.

43. $f(x) = \begin{cases} 2x + 3, & x < 0 \\ 3 - x, & x \geq 0 \end{cases}$

44. $g(x) = \begin{cases} x + 6, & x \leq -4 \\ \frac{1}{2}x - 4, & x > -4 \end{cases}$

45. $f(x) = \begin{cases} \sqrt{4 + x}, & x < 0 \\ \sqrt{4 - x}, & x \geq 0 \end{cases}$

46. $f(x) = \begin{cases} 1 - (x - 1)^2, & x \leq 2 \\ \sqrt{x - 2}, & x > 2 \end{cases}$

47. $f(x) = \begin{cases} x^2 + 5, & x \leq 1 \\ -x^2 + 4x + 3, & x > 1 \end{cases}$

48. $h(x) = \begin{cases} 3 - x^2, & x < 0 \\ x^2 + 2, & x \geq 0 \end{cases}$

49. $h(x) = \begin{cases} 4 - x^2, & x < -2 \\ 3 + x, & -2 \leq x < 0 \\ x^2 + 1, & x \geq 0 \end{cases}$

50. $k(x) = \begin{cases} 2x + 1, & x \leq -1 \\ 2x^2 - 1, & -1 < x \leq 1 \\ 1 - x^2, & x > 1 \end{cases}$

 In Exercises 51 and 52, use a graphing utility to graph the function. State the domain and range of the function. Describe the pattern of the graph.

51. $s(x) = 2(\frac{1}{4}x - [\![\frac{1}{4}x]\!])$

52. $g(x) = 2(\frac{1}{4}x - [\![\frac{1}{4}x]\!])^2$

In Exercises 53–62, identify the common function and the transformed common function shown in the graph. Write an equation for the function shown in the graph. Then use a graphing utility to verify your answer.

53.

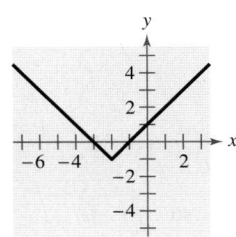

54.

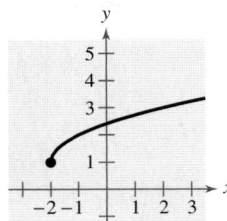

55.

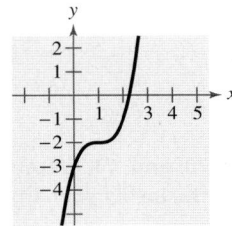

56.

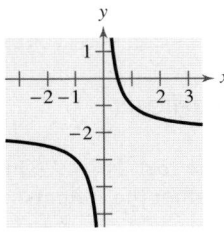

57.

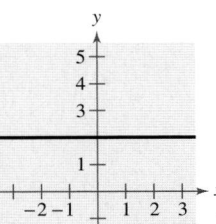

58.

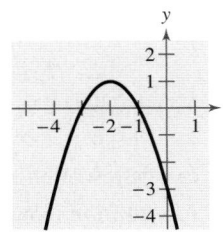

59.

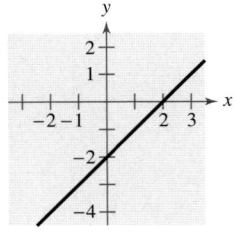

60.

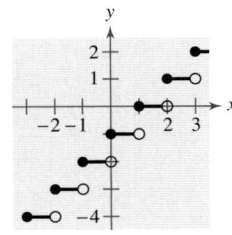

61.

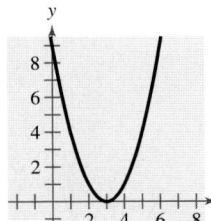

62.

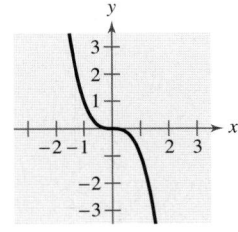

63. ***Communications*** The cost of a telephone call between Denver and Boise is $0.60 for the first minute and $0.42 for each additional minute or portion of a minute. A model for the total cost C (in dollars) of the phone call is

$$C = 0.60 - 0.42[\![1 - t]\!], \quad t > 0$$

where t is the length of the phone call in minutes.

(a) Sketch the graph of the model.

(b) Determine the cost of a call lasting 12 minutes and 30 seconds.

64. ***Communications*** The cost of using a telephone calling card is $1.05 for the first minute and $0.38 for each additional minute or portion of a minute.

(a) A customer needs a model for the cost C of using a calling card for a call lasting t minutes. Which of the following is the appropriate model? Explain.

$$C_1(t) = 1.05 + 0.38[\![t - 1]\!]$$

$$C_2(t) = 1.05 - 0.38[\![-(t - 1)]\!]$$

(b) Graph the appropriate model. Determine the cost of a call lasting 18 minutes and 45 seconds.

65. ***Delivery Charges*** The cost of sending an overnight package from Los Angeles to Miami is $10.75 for a package weighing up to but not including 1 pound and $3.95 for each additional pound or portion of a pound. A model for the total cost C (in dollars) of sending the package is

$$C = 10.75 + 3.95[\![x]\!], \quad x > 0$$

where x is the weight in pounds.

(a) Sketch a graph of the model.

(b) Determine the cost of sending a package that weighs 10.33 pounds.

66. ***Delivery Charges*** The cost of sending an overnight package from New York to Atlanta is $9.80 for a package weighing up to but not including 1 pound and $2.50 for each additional pound or portion of a pound.

(a) Use the greatest integer function to create a model for the cost C of overnight delivery of a package weighing x pounds, $x > 0$.

(b) Sketch the graph of the function.

67. ***Wages*** A mechanic is paid $12.00 per hour for regular time and time-and-a-half for overtime. The weekly wage function is

$$W(h) = \begin{cases} 12h, & 0 < h \le 40 \\ 18(h - 40) + 480, & h > 40 \end{cases}$$

where h is the number of hours worked in a week.

(a) Evaluate $W(30)$, $W(40)$, $W(45)$, and $W(50)$.

(b) The company increased the regular work week to 45 hours. What is the new weekly wage function?

▶ Model It

68. Revenue The table shows the monthly revenue *y* (in thousands of dollars) of a landscaping business for the year 2002, with *x* = 1 representing January.

Month, *x*	Revenue, *y*
1	5.2
2	5.6
3	6.6
4	8.3
5	11.5
6	15.8
7	12.8
8	10.1
9	8.6
10	6.9
11	4.5
12	2.7

A mathematical model that represents this data is

$$f(x) = \begin{cases} -1.97x + 26.3 \\ 0.505x^2 - 1.47x + 6.3 \end{cases}.$$

(a) What is the domain of each part of the piecewise-defined function? How can you tell? Explain your reasoning.

(b) Sketch a graph of the model.

(c) Find $f(5)$ and $f(11)$, and interpret your results in the context of the problem.

(d) How do the values obtained from the model in part (b) compare with the actual data values?

69. Fluid Flow The intake pipe of a 100-gallon tank has a flow rate of 10 gallons per minute, and two drainpipes have flow rates of 5 gallons per minute each. The figure shows the volume *V* of fluid in the tank as a function of time *t*. Determine the combination of the input pipe and drain pipes in which the fluid is flowing in specific subintervals of the 1 hour of time shown on the graph. (There are many correct answers.)

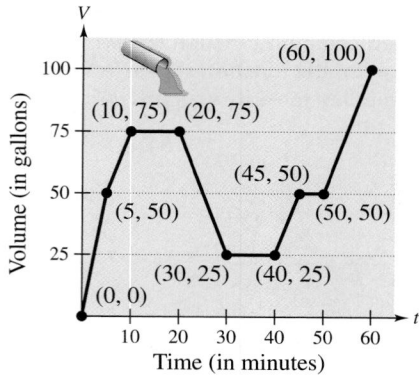

FIGURE FOR **69**

Synthesis

True or False? In Exercises 70 and 71, determine whether the statement is true or false. Justify your answer.

70. A piecewise-defined function will always have at least one *x*-intercept or at least one *y*-intercept.

71. $f(x) = \begin{cases} 2, & 1 \le x < 2 \\ 4, & 2 \le x < 3 \\ 6, & 3 \le x < 4 \end{cases}$

can be rewritten as $f(x) = 2[\![x]\!]$, $1 \le x < 4$.

72. Exploration Write equations for the piecewise-defined function shown in the graph.

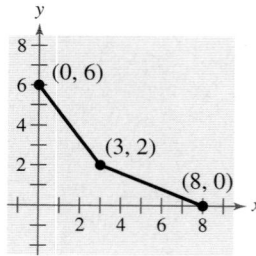

P.8 Shifting, Reflecting, and Stretching Graphs

Chuck Keeler/The Stock Market

Shifting Graphs

Many functions have graphs that are simple transformations of the common graphs summarized in Section P.7. For example, you can obtain the graph of

$$h(x) = x^2 + 2$$

by shifting the graph of $f(x) = x^2$ *up* two units, as shown in Figure P.73. In function notation, h and f are related as follows.

$$h(x) = x^2 + 2 = f(x) + 2 \qquad \text{Upward shift of two units}$$

Similarly, you can obtain the graph of

$$g(x) = (x - 2)^2$$

by shifting the graph of $f(x) = x^2$ to the *right* two units, as shown in Figure P.74. In this case, the functions g and f have the following relationship.

$$g(x) = (x - 2)^2 = f(x - 2) \qquad \text{Right shift of two units}$$

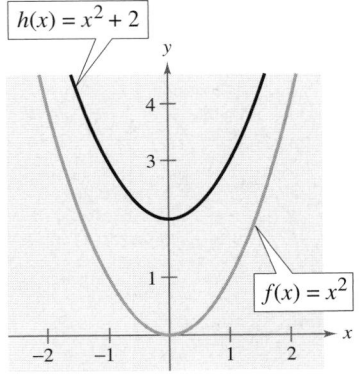

FIGURE P.73

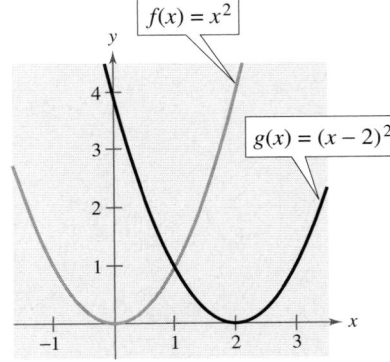

FIGURE P.74

The following list summarizes this discussion about horizontal and vertical shifts.

> **Vertical and Horizontal Shifts**
>
> Let c be a positive real number. **Vertical and horizontal shifts** in the graph of $y = f(x)$ are represented as follows.
>
> 1. Vertical shift c units *upward:* $h(x) = f(x) + c$
>
> 2. Vertical shift c units *downward:* $h(x) = f(x) - c$
>
> 3. Horizontal shift c units to the *right:* $h(x) = f(x - c)$
>
> 4. Horizontal shift c units to the *left:* $h(x) = f(x + c)$

You might also wish to illustrate simple transformations of functions numerically using tables to emphasize what happens to individual ordered pairs. For instance, if you have $f(x) = x^2$, $h(x) = x^2 + 2 = f(x) + 2$, and $g(x) = (x - 2)^2 = f(x - 2)$, you can illustrate these transformations with the following tables.

x	$f(x)$	$h(x) = f(x) + 2$
-2	4	$4 + 2 = 6$
-1	1	$1 + 2 = 3$
0	0	$0 + 2 = 2$
1	1	$1 + 2 = 3$
2	4	$4 + 2 = 6$

x	$x - 2$	$g(x) = f(x - 2)$
0	$0 - 2 = -2$	4
1	$1 - 2 = -1$	1
2	$2 - 2 = 0$	0
3	$3 - 2 = 1$	1
4	$4 - 2 = 2$	4

Some graphs can be obtained from combinations of vertical and horizontal shifts, as demonstrated in Example 1(b). Vertical and horizontal shifts generate a *family of functions*, each with the same shape but at different locations in the plane.

Example 1 ▶   **Shifts in the Graphs of a Function**

Use the graph of $f(x) = x^3$ to sketch the graph of each function.

a. $g(x) = x^3 - 1$

b. $h(x) = (x + 2)^3 + 1$

Solution

a. Relative to the graph of $f(x) = x^3$, the graph of $g(x) = x^3 - 1$ is a downward shift of one unit, as shown in Figure P.75.

b. Relative to the graph of $f(x) = x^3$, the graph of $h(x) = (x + 2)^3 + 1$ involves a left shift of two units and an upward shift of one unit, as shown in Figure P.76.

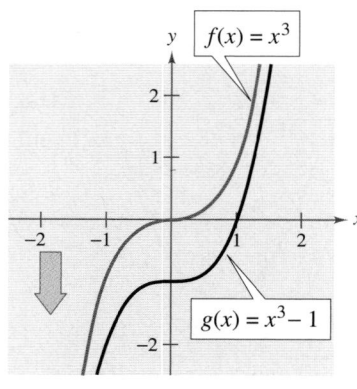

FIGURE P.75

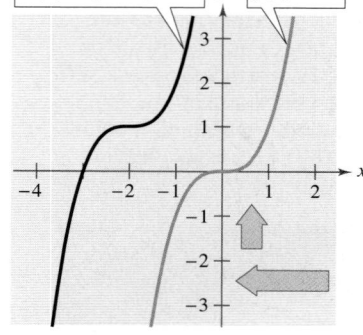

FIGURE P.76

In Figure P.77, notice that the same result is obtained if the vertical shift precedes the horizontal shift *or* if the horizontal shift precedes the vertical shift.

◀ Exploration ▶

Graphing utilities are ideal tools for exploring translations of functions. Graph f, g, and h in same viewing window. Before looking at the graphs, try to predict how the graphs of g and h relate to the graph of f.

a. $f(x) = x^2$, $g(x) = (x - 4)^2$, $h(x) = (x - 4)^2 + 3$

b. $f(x) = x^2$, $g(x) = (x + 1)^2$, $h(x) = (x + 1)^2 - 2$

c. $f(x) = x^2$, $g(x) = (x + 4)^2$, $h(x) = (x + 4)^2 + 2$

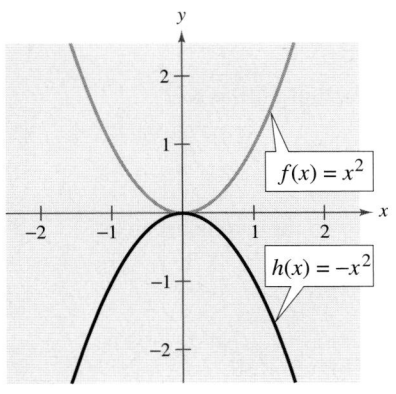

FIGURE P.77

Reflecting Graphs

The second common type of transformation is a **reflection.** For instance, if you consider the x-axis to be a mirror, the graph of

$$h(x) = -x^2$$

is the mirror image (or reflection) of the graph of $f(x) = x^2$, as shown in Figure P.77.

Reflections in the Coordinate Axes

Reflections in the coordinate axes of the graph of $y = f(x)$ are represented as follows.

1. Reflection in the x-axis: $h(x) = -f(x)$

2. Reflection in the y-axis: $h(x) = f(-x)$

Example 2 ▶ **Finding Equations from Graphs**

The graph of the function

$$f(x) = x^4$$

is shown in Figure P.78. Each of the graphs in Figure P.79 is a transformation of the graph of f. Find an equation for each of these functions.

FIGURE P.78

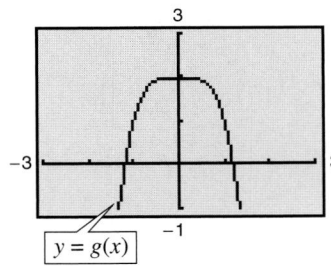

(a)

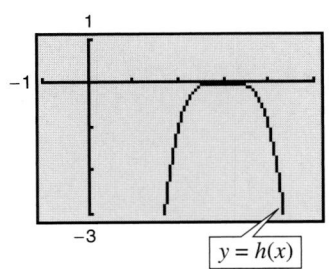

(b)

FIGURE P.79

Solution

a. The graph of g is a reflection in the x-axis *followed by* an upward shift of two units of the graph of $f(x) = x^4$. So, the equation for g is

$$g(x) = -x^4 + 2.$$

b. The graph of h is a horizontal shift of three units to the right *followed by* a reflection in the x-axis of the graph of $f(x) = x^4$. So, the equation for h is

$$h(x) = -(x - 3)^4.$$

◀ **E x p l o r a t i o n** ▶

Reverse the order of transformations in Example 2(a). Do you obtain the same graph? Do the same for Example 2(b). Do you obtain the same graph? Explain.

Activities

1. How are the graphs of $f(x)$ and $g = -f(x)$ related?

   Answer: They are reflections of each other in the x-axis.

2. Compare the graph of $f(x) = |x|$ with the graph of $g(x) = |x - 9|$.

   Answer: $g(x)$ is $f(x)$ shifted to the right nine units.

3. Does the graph of $f(x) = -(x + 1)^3 + 4$ represent a horizontal shift of one unit to the left, followed by a vertical shift of four units upward, followed by a reflection in the x-axis?

   Answer: No, it represents a horizontal shift of one unit to the left, followed by a reflection in the x-axis, followed by a vertical shift of four units upward.

Example 3 ▶ **Reflections and Shifts**

Compare the graph of each function with the graph of $f(x) = \sqrt{x}$.

a. $g(x) = -\sqrt{x}$ **b.** $h(x) = \sqrt{-x}$ **c.** $k(x) = -\sqrt{x + 2}$

Solution

a. The graph of g is a reflection of the graph of f in the x-axis because

$$g(x) = -\sqrt{x}$$
$$= -f(x).$$

The graph of g compared with f is shown in Figure P.80.

b. The graph of h is a reflection of the graph of f in the y-axis because

$$h(x) = \sqrt{-x}$$
$$= f(-x).$$

The graph of h compared with f is shown in Figure P.81.

c. The graph of k is a left shift of two units, followed by a reflection in the x-axis because

$$k(x) = -\sqrt{x + 2}$$
$$= -f(x + 2).$$

The graph of k compared with f is shown in Figure P.82.

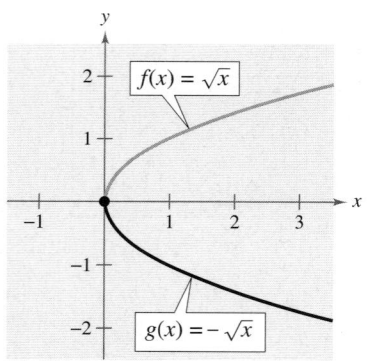

FIGURE P.80

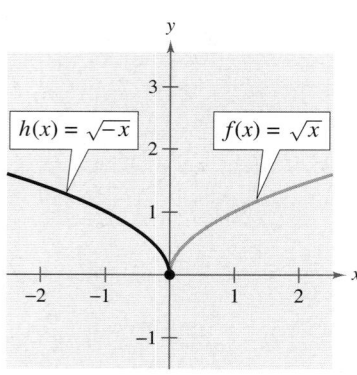

FIGURE P.81

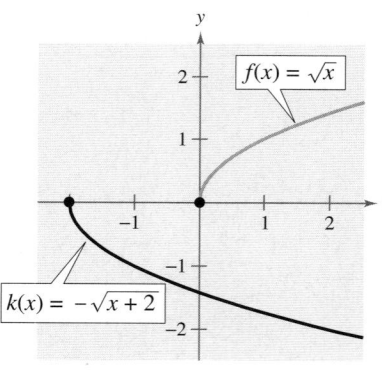

FIGURE P.82

When sketching the graphs of functions involving square roots, remember that the domain must be restricted to exclude negative numbers inside the radical. For instance, here are the domains of the functions in Example 3.

Domain of $g(x) = -\sqrt{x}$: $x \geq 0$

Domain of $h(x) = \sqrt{-x}$: $x \leq 0$

Domain of $k(x) = -\sqrt{x + 2}$: $x \geq -2$

Nonrigid Transformations

Horizontal shifts, vertical shifts, and reflections are **rigid transformations** because the basic shape of the graph is unchanged. These transformations change only the *position* of the graph in the *xy*-plane. **Nonrigid transformations** are those that cause a *distortion*—a change in the shape of the original graph. For instance, a nonrigid transformation of the graph of $y = f(x)$ is represented by $g(x) = cf(x)$, where the transformation is a **vertical stretch** if $c > 1$ and a **vertical shrink** if $0 < c < 1$. Another nonrigid transformation of the graph of $y = f(x)$ is represented by $h(x) = f(cx)$, where the transformation is a **horizontal shrink** if $c > 1$ and a **horizontal stretch** if $0 < c < 1$.

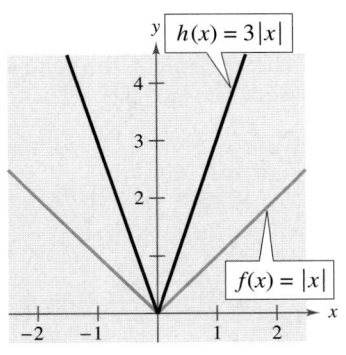

FIGURE P.83

Example 4 ▶ Nonrigid Transformations

Compare the graph of each function with the graph of $f(x) = |x|$.

a. $h(x) = 3|x|$　　**b.** $g(x) = \frac{1}{3}|x|$

Solution

a. Relative to the graph of $f(x) = |x|$, the graph of

$$h(x) = 3|x| = 3f(x)$$

is a vertical stretch (each *y*-value is multiplied by 3) of the graph of *f*. (See Figure P.83.)

b. Similarly, the graph of

$$g(x) = \frac{1}{3}|x| = \frac{1}{3}f(x)$$

is a vertical shrink $\left(\text{each } y\text{-value is multiplied by } \frac{1}{3}\right)$ of the graph of *f*. (See Figure P.84.)

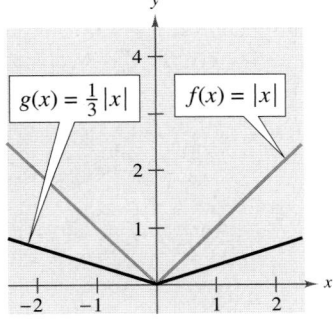

FIGURE P.84

Example 5 ▶ Nonrigid Transformations

Compare the graph of each function with the graph of $f(x) = 2 - x^3$.

a. $g(x) = f(2x)$　　**b.** $h(x) = f\left(\frac{1}{2}x\right)$

Solution

a. Relative to the graph of $f(x) = 2 - x^3$, the graph of

$$g(x) = f(2x) = 2 - (2x)^3 = 2 - 8x^3$$

is a horizontal shrink $\left(\text{each } x\text{-value is multiplied by } \frac{1}{2}\right)$ of the graph of *f*. (See Figure P.85.)

b. Similarly, the graph of

$$h(x) = f\left(\frac{1}{2}x\right) = 2 - \left(\frac{1}{2}x\right)^3 = 2 - \frac{1}{8}x^3$$

is a horizontal stretch (each *x*-value is multiplied by 2) of the graph of *f*. (See Figure P.86.)

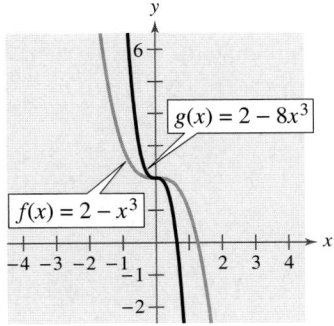

FIGURE P.85

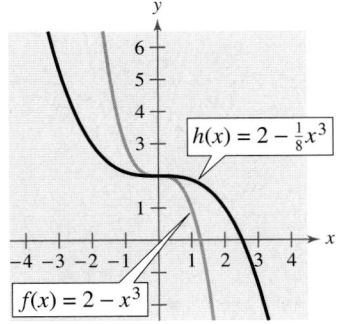

FIGURE P.86

P.8 Exercises

1. For each function, sketch (on the same set of coordinate axes) a graph of each function for $c = -1, 1,$ and 3.

   (a) $f(x) = |x| + c$ (b) $f(x) = |x - c|$

   (c) $f(x) = |x + 4| + c$

2. For each function, sketch (on the same set of coordinate axes) a graph of each function for $c = -3, -1, 1,$ and 3.

   (a) $f(x) = \sqrt{x} + c$ (b) $f(x) = \sqrt{x - c}$

   (c) $f(x) = \sqrt{x - 3} + c$

3. For each function, sketch (on the same set of coordinate axes) a graph of each function for $c = -2, 0,$ and 2.

   (a) $f(x) = [\![x]\!] + c$ (b) $f(x) = [\![x + c]\!]$

   (c) $f(x) = [\![x - 1]\!] + c$

4. For each function, sketch (on the same set of coordinate axes) a graph of each function for $c = -3, -1, 1,$ and 3.

   (a) $f(x) = \begin{cases} x^2 + c, & x < 0 \\ -x^2 + c, & x \ge 0 \end{cases}$

   (b) $f(x) = \begin{cases} (x + c)^2, & x < 0 \\ -(x + c)^2, & x \ge 0 \end{cases}$

5. Use the graph of f to sketch each graph. To print an enlarged copy of the graph, go to the website *www.mathgraphs.com*.

   (a) $y = f(x) + 2$

   (b) $y = f(x - 2)$

   (c) $y = 2f(x)$

   (d) $y = -f(x)$

   (e) $y = f(x + 3)$

   (f) $y = f(-x)$

   (g) $y = f\left(\frac{1}{2}x\right)$

6. Use the graph of f to sketch each graph. To print an enlarged copy of the graph, go to the website *www.mathgraphs.com*.

   (a) $y = f(-x)$

   (b) $y = f(x) + 4$

   (c) $y = 2f(x)$

   (d) $y = -f(x - 4)$

   (e) $y = f(x) - 3$

   (f) $y = -f(x) - 1$

   (g) $y = f(2x)$

7. Use the graph of f to sketch each graph. To print an enlarged copy of the graph, go to the website *www.mathgraphs.com*.

   (a) $y = f(x) - 1$

   (b) $y = f(x - 1)$

   (c) $y = f(-x)$

   (d) $y = f(x + 1)$

   (e) $y = -f(x - 2)$

   (f) $y = \frac{1}{2}f(x)$

   (g) $y = f(2x)$

8. Use the graph of f to sketch each graph. To print an enlarged copy of the graph, go to the website *www.mathgraphs.com*.

   (a) $y = f(x - 5)$

   (b) $y = -f(x) + 3$

   (c) $y = \frac{1}{3}f(x)$

   (d) $y = -f(x + 1)$

   (e) $y = f(-x)$

   (f) $y = f(x) - 10$

   (g) $y = f\left(\frac{1}{3}x\right)$

9. Use the graph of $f(x) = x^2$ to write an equation for each function whose graph is shown.

   (a)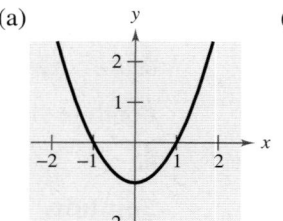

   (b)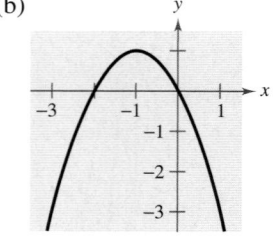

   (c)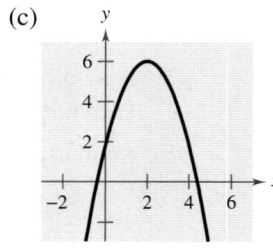

   (d)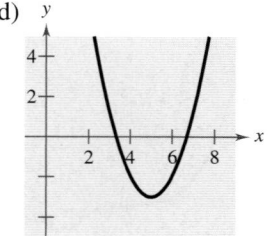

10. Use the graph of $f(x) = x^3$ to write an equation for each function whose graph is shown.

(a)
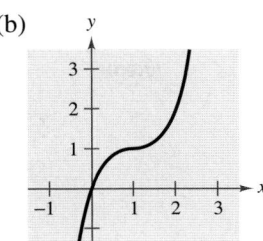

(b)

(c)
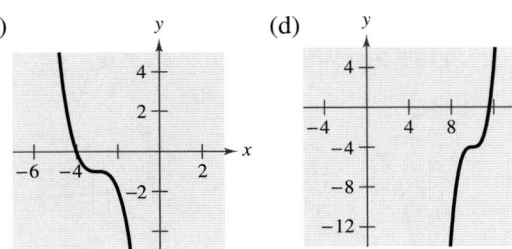

(d)

11. Use the graph of $f(x) = |x|$ to write an equation for each function whose graph is shown.

(a)
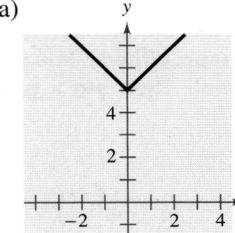

(b)
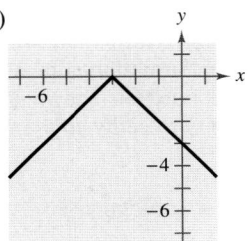

(c)
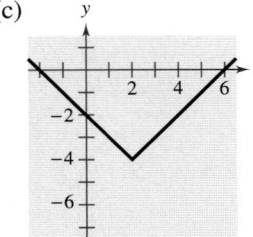

(d)
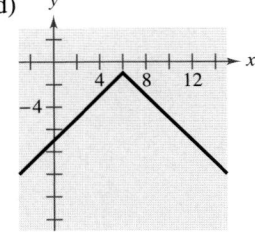

12. Use the graph of $f(x) = \sqrt{x}$ to write an equation for each function whose graph is shown.

(a)
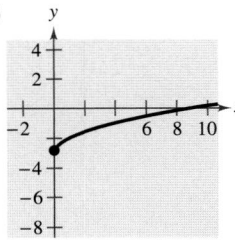

(b)
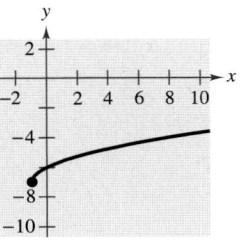

(c)
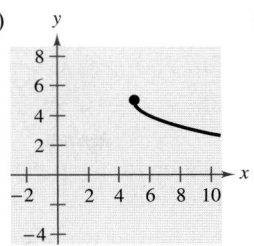

(d)
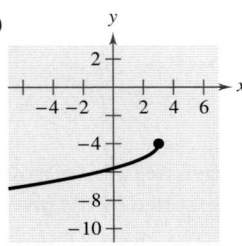

In Exercises 13–18, identify the common function and the transformation shown in the graph. Write an equation for the function shown in the graph.

13.
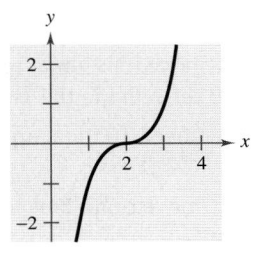

14.
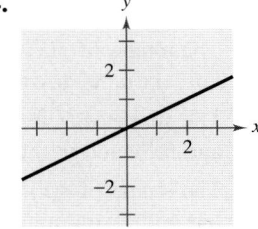

15.
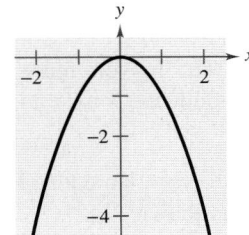

16.
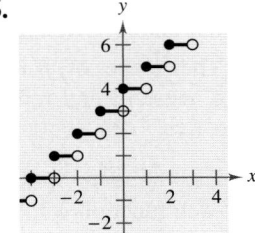

17.
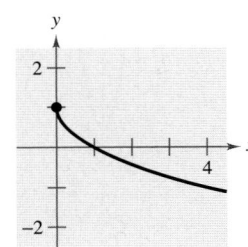

18.
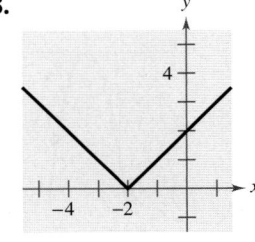

In Exercises 19–38, describe the transformation from a common function that occurs in the function. Then sketch its graph.

19. $f(x) = 12 - x^2$

20. $f(x) = (x - 8)^2$

21. $f(x) = x^3 + 7$

22. $f(x) = -x^3 - 1$

23. $f(x) = 2 - (x + 5)^2$

24. $f(x) = -(x + 10)^2 + 5$

25. $f(x) = (x - 1)^3 + 2$

26. $f(x) = (x + 3)^3 - 10$

27. $f(x) = -|x| - 2$

28. $f(x) = 6 - |x + 5|$

29. $f(x) = -|x + 4| + 8$

30. $f(x) = |-x + 3| + 9$

31. $f(x) = 3 - [\![x]\!]$

32. $f(x) = 2[\![x + 5]\!]$

33. $f(x) = \sqrt{x - 9}$ **34.** $f(x) = \sqrt{x + 4} + 8$

35. $f(x) = \sqrt{7 - x} - 2$ **36.** $f(x) = -\sqrt{x + 1} - 6$

37. $f(x) = \sqrt{\frac{1}{2}x} - 4$ **38.** $f(x) = \sqrt{3x} + 1$

In Exercises 39–46, write an equation for the function that is described by the given characteristics.

39. The shape of $f(x) = x^2$, but moved two units to the right and eight units downward

40. The shape of $f(x) = x^2$, but moved three units to the left, seven units upward, and reflected in the x-axis

41. The shape of $f(x) = x^3$, but moved 13 units to the right

42. The shape of $f(x) = x^3$, but moved six units to the left, six units downward, and reflected in the y-axis

43. The shape of $f(x) = |x|$, but moved 10 units upward and reflected in the x-axis

44. The shape of $f(x) = |x|$, but moved one unit to the left and seven units downward

45. The shape of $f(x) = \sqrt{x}$, but moved six units to the left and reflected in both the x-axis and the y-axis

46. The shape of $f(x) = \sqrt{x}$, but moved nine units downward and reflected in both the x-axis and the y-axis

47. Use the graph of $f(x) = x^2$ to write an equation for each function whose graph is shown.

(a) 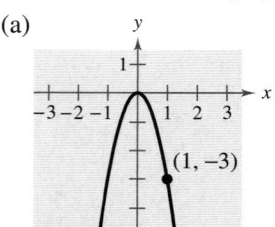   (b)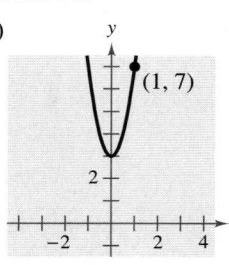

48. Use the graph of $f(x) = x^3$ to write an equation for each function whose graph is shown.

(a) 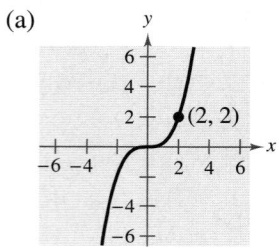   (b)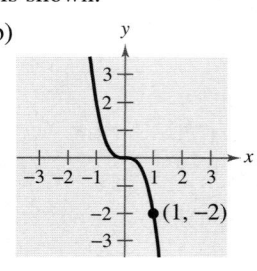

49. Use the graph of $f(x) = |x|$ to write an equation for each function whose graph is shown.

(a) 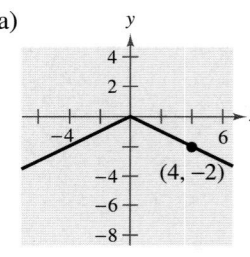   (b)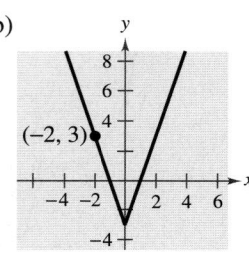

50. Use the graph of $f(x) = \sqrt{x}$ to write an equation for each function whose graph is shown.

(a) 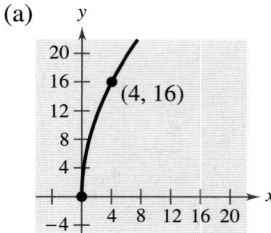   (b)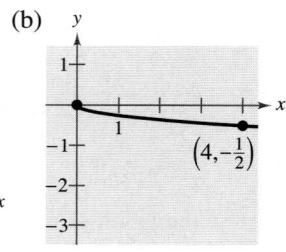

In Exercises 51–56, identify the common function and the transformation shown in the graph. Write an equation for the function shown in the graph. Then use a graphing utility to verify your answer.

51. 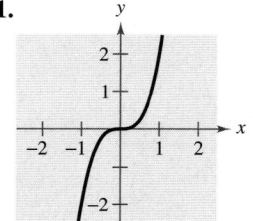   **52.**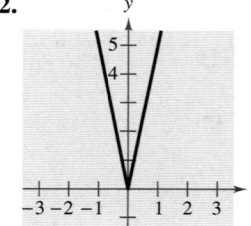

53. 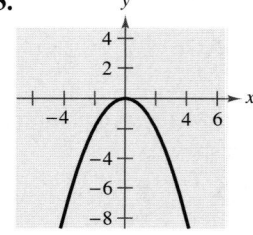   **54.**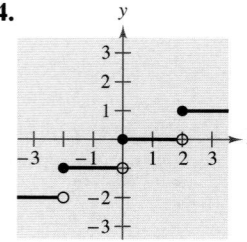

55.

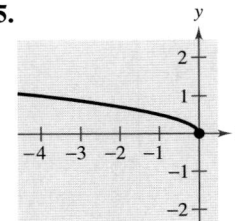

56.

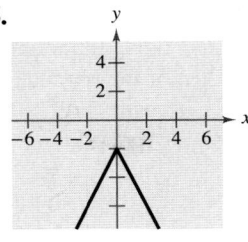

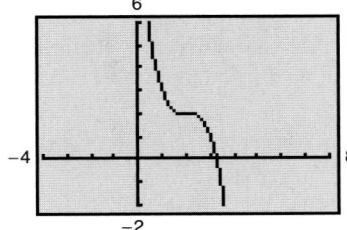
Graphical Analysis In Exercises 57–60, use the viewing window shown to write a possible equation for the transformation of the common function.

57.

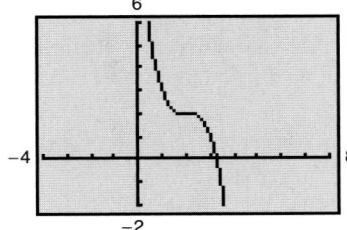

58.

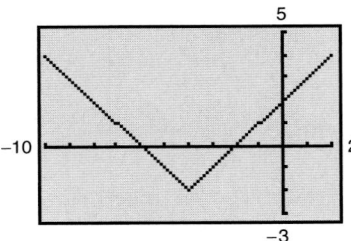

59.

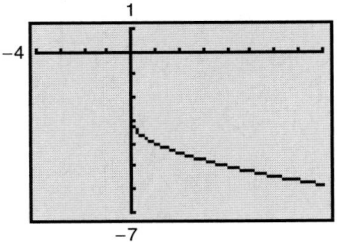

60.

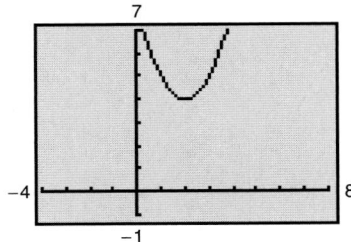

Graphical Reasoning In Exercises 61 and 62, use the graph of *f* to sketch the graph of *g*. To print an enlarged copy of the graph, go to the website *www.mathgraphs.com*.

61.

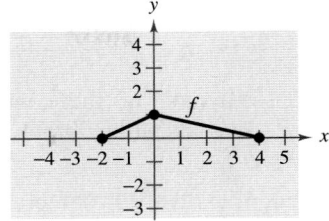

(a) $g(x) = f(x) + 2$ (b) $g(x) = f(x) - 1$
(c) $g(x) = f(-x)$ (d) $g(x) = -2f(x)$
(e) $g(x) = f(4x)$ (f) $g(x) = f\left(\frac{1}{2}x\right)$

62.

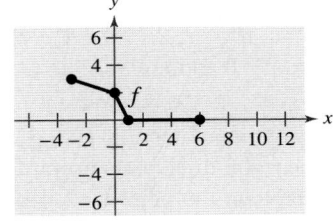

(a) $g(x) = f(x) - 5$ (b) $g(x) = f(x) + \frac{1}{2}$
(c) $g(x) = f(-x)$ (d) $g(x) = -4f(x)$
(e) $g(x) = f(2x) + 1$ (f) $g(x) = f\left(\frac{1}{4}x\right) - 2$

▶ Model It

63. *Fuel Use* The amount of fuel *F* (in billions of gallons) used by trucks from 1980 through 1999 can be approximated by the function

$$F = f(t) = 20.5 + 0.035t^2$$

where $t = 0$ represents 1980. (Source: U.S. Federal Highway Administration)

(a) Describe the transformation of the common function $f(x) = x^2$. Then sketch the graph over the interval $0 \le t \le 19$.

(b) Find and interpret $\dfrac{f(19) - f(0)}{19 - 0}$.

(c) Rewrite the function so that $t = 0$ represents 1990. Explain how you got your answer.

(d) Use the model from part (c) to predict the amount of fuel used by trucks in 2005. Does your answer seem reasonable? Explain.

64. *Finance* The amount M (in trillions of dollars) of mortgage debt outstanding in the United States from 1980 through 1999 can be approximated by the function $M = f(t) = 0.0037(t + 14.979)^2$, where $t = 0$ represents 1980.   (Source: Board of Governors of the Federal Reserve System)

(a) Describe the transformation of the common function $f(x) = x^2$. Then sketch the graph over the interval $0 \le t \le 19$.

(b) Rewrite the function so that $t = 0$ represents 1990. Explain how you got your answer.

Synthesis

True or False? **In Exercises 65 and 66, determine whether the statement is true or false. Justify your answer.**

65. The graphs of $f(x) = |x| + 6$ and $f(x) = |-x| + 6$ are identical.

66. If the graph of the common function $f(x) = x^2$ is moved six units to the right, three units upward, and reflected in the x-axis, then the point $(-2, 19)$ will lie on the graph of the transformation.

67. *Describing Profits* Management originally predicted that the profits from the sales of a new product would be approximated by the graph of the function f shown. The actual profits are shown by the function g along with a verbal description. Use the concepts of transformations of graphs to write g in terms of f.

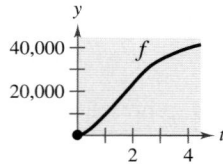

(a) The profits were only three-fourths as large as expected.

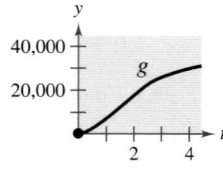

(b) The profits were consistently $10,000 greater than predicted.

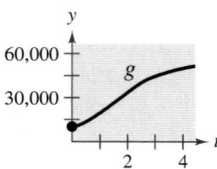

(c) There was a two-year delay in the introduction of the product. After sales began, profits grew as expected.

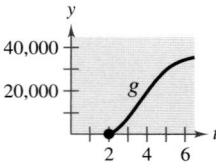

68. Explain why the graph of $y = -f(x)$ is a reflection of the graph of $y = f(x)$ about the x-axis.

69. The graph of $y = f(x)$ passes through the points $(0, 1)$, $(1, 2)$, and $(2, 3)$. Find the corresponding points on the graph of $y = f(x + 2) - 1$.

70. *Think About It* You can use either of two methods to graph a function: plotting points or translating a common function as shown in this section. Which method of graphing do you prefer to use for each function? Explain.

(a) $f(x) = 3x^2 - 4x + 1$

(b) $f(x) = 2(x - 1)^2 - 6$

P.9 Combinations of Functions

What you should learn

- How to add, subtract, multiply, and divide functions
- How to find the composition of one function with another function
- How to use combinations of functions to model and solve real-life problems

Why you should learn it

Combinations of functions can be used to model and solve real-life problems. For instance, in Exercise 33 on page 101, combinations of functions are used to analyze U.S. health expenditures.

Charles Gupton/Tony Stone Images

Arithmetic Combinations of Functions

Just as two real numbers can be combined by the operations of addition, subtraction, multiplication, and division to form other real numbers, two *functions* can be combined to create new functions. For example, the functions $f(x) = 2x - 3$ and $g(x) = x^2 - 1$ can be combined to form the sum, difference, product, and quotient of f and g.

$$f(x) + g(x) = (2x - 3) + (x^2 - 1)$$
$$= x^2 + 2x - 4 \qquad \text{Sum}$$
$$f(x) - g(x) = (2x - 3) - (x^2 - 1)$$
$$= -x^2 + 2x - 2 \qquad \text{Difference}$$
$$f(x)g(x) = (2x - 3)(x^2 - 1)$$
$$= 2x^3 - 3x^2 - 2x + 3 \qquad \text{Product}$$
$$\frac{f(x)}{g(x)} = \frac{2x - 3}{x^2 - 1}, \qquad x \neq \pm 1 \qquad \text{Quotient}$$

The domain of an **arithmetic combination** of functions f and g consists of all real numbers that are common to the domains of f and g. In the case of the quotient $f(x)/g(x)$, there is the further restriction that $g(x) \neq 0$.

Sum, Difference, Product, and Quotient of Functions

Let f and g be two functions with overlapping domains. Then, for all x common to both domains, the *sum*, *difference*, *product*, and *quotient* of f and g are defined as follows.

1. *Sum:* $\qquad (f + g)(x) = f(x) + g(x)$

2. *Difference:* $\quad (f - g)(x) = f(x) - g(x)$

3. *Product:* $\qquad (fg)(x) = f(x) \cdot g(x)$

4. *Quotient:* $\qquad \left(\dfrac{f}{g}\right)(x) = \dfrac{f(x)}{g(x)}, \qquad g(x) \neq 0$

Example 1 ▶ **Finding the Sum of Two Functions**

Given $f(x) = 2x + 1$ and $g(x) = x^2 + 2x - 1$, find $(f + g)(x)$.

Solution

$$(f + g)(x) = f(x) + g(x) = (2x + 1) + (x^2 + 2x - 1) = x^2 + 4x$$

Example 2 ▶ **Finding the Difference of Two Functions**

Given $f(x) = 2x + 1$ and $g(x) = x^2 + 2x - 1$, find $(f - g)(x)$. Then evaluate the difference when $x = 2$.

Solution

The difference of f and g is

$$(f - g)(x) = f(x) - g(x)$$

$$= (2x + 1) - (x^2 + 2x - 1)$$

$$= -x^2 + 2.$$

When $x = 2$, the value of this difference is

$$(f - g)(2) = -(2)^2 + 2$$

$$= -2.$$

In Examples 1 and 2, both f and g have domains that consist of all real numbers. So, the domains of $(f + g)$ and $(f - g)$ are also the set of all real numbers. Remember that any restrictions on the domains of f and g must be considered when forming the sum, difference, product, or quotient of f and g.

Example 3 ▶ **Finding the Domains of Quotients of Functions**

Find the domains of $\left(\dfrac{f}{g}\right)(x)$ and $\left(\dfrac{g}{f}\right)(x)$ for the functions

$$f(x) = \sqrt{x} \quad \text{and} \quad g(x) = \sqrt{4 - x^2}.$$

Solution

The quotient of f and g is

$$\left(\frac{f}{g}\right)(x) = \frac{f(x)}{g(x)} = \frac{\sqrt{x}}{\sqrt{4 - x^2}}$$

and the quotient of g and f is

$$\left(\frac{g}{f}\right)(x) = \frac{g(x)}{f(x)} = \frac{\sqrt{4 - x^2}}{\sqrt{x}}.$$

The domain of f is $[0, \infty)$ and the domain of g is $[-2, 2]$. The intersection of these domains is $[0, 2]$. So, the domains of $\left(\dfrac{f}{g}\right)$ and $\left(\dfrac{g}{f}\right)$ are as follows.

$$\text{Domain of } \left(\frac{f}{g}\right): [0, 2) \qquad \text{Domain of } \left(\frac{g}{f}\right): (0, 2]$$

Can you see why these two domains differ slightly?

Composition of Functions

Another way of combining two functions is to form the **composition** of one with the other. For instance, if $f(x) = x^2$ and $g(x) = x + 1,$ the composition of f with g is

$$f(g(x)) = f(x + 1)$$
$$= (x + 1)^2.$$

This composition is denoted as $(f \circ g).$

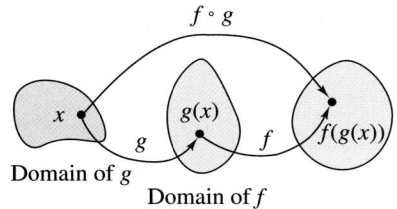

$f \circ g$

Domain of g

Domain of f

FIGURE P.87

Definition of Composition of Two Functions

The **composition** of the function f with the function g is

$$(f \circ g)(x) = f(g(x)).$$

The domain of $(f \circ g)$ is the set of all x in the domain of g such that $g(x)$ is in the domain of $f.$ (See Figure P.87.)

STUDY TIP

The following tables of values help illustrate the composition $(f \circ g)(x)$ given in Example 4.

x	0	1	2	3
$g(x)$	4	3	0	-5

$g(x)$	4	3	0	-5
$f(g(x))$	6	5	2	-3

x	0	1	2	3
$f(g(x))$	6	5	2	-3

Note that the first two tables can be combined (or "composed") to produce the values given in the third table.

Example 4 ▶ **Composition of Functions**

Given $f(x) = x + 2$ and $g(x) = 4 - x^2,$ find the following.

a. $(f \circ g)(x)$ **b.** $(g \circ f)(x)$ **c.** $(g \circ f)(-2)$

Solution

a. The composition of f with g is as follows.

$$(f \circ g)(x) = f(g(x)) \qquad \text{Definition of } f \circ g$$
$$= f(4 - x^2) \qquad \text{Definition of } g(x)$$
$$= (4 - x^2) + 2 \qquad \text{Definition of } f(x)$$
$$= -x^2 + 6 \qquad \text{Simplify.}$$

b. The composition of g with f is as follows.

$$(g \circ f)(x) = g(f(x)) \qquad \text{Definition of } g \circ f$$
$$= g(x + 2) \qquad \text{Definition of } f(x)$$
$$= 4 - (x + 2)^2 \qquad \text{Definition of } g(x)$$
$$= 4 - (x^2 + 4x + 4) \qquad \text{Expand.}$$
$$= -x^2 - 4x \qquad \text{Simplify.}$$

Note that, in this case, $(f \circ g)(x) \ne (g \circ f)(x).$

c. Using the result of part (b), you can write the following.

$$(g \circ f)(-2) = -(-2)^2 - 4(-2) \qquad \text{Substitute.}$$
$$= -4 + 8 \qquad \text{Simplify.}$$
$$= 4 \qquad \text{Simplify.}$$

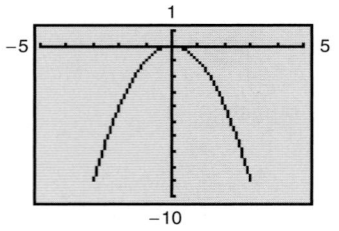
Activities

1. Given $f(x) = 3x^2 + 2$ and $g(x) = 2x$, find $f \circ g$.

   Answer: $(f \circ g)(x) = 12x^2 + 2$

2. Given the functions

   $f(x) = \dfrac{1}{x - 2}$ and $g(x) = \sqrt{x}$,

   find the composition of f with g. Then find the domain of the composition.

   Answer: $(f \circ g)(x) = \dfrac{1}{\sqrt{x} - 2}$. The domain of $(f \circ g)$ is the set of all non-negative real numbers except $x = 4$.

3. Find two functions f and g such that $(f \circ g)(x) = h(x)$. (There are many correct answers.)

   a. $h(x) = \dfrac{1}{\sqrt{3x + 1}}$

   Answer: $f(x) = \dfrac{1}{\sqrt{x}}$ and $g(x) = 3x + 1$

   b. $h(x) = (2x + 3)^4$

   Answer: $f(x) = x^4$ and $g(x) = 2x + 3$

Example 5 ▶ **Finding the Domain of a Composite Function**

Find the composition $(f \circ g)(x)$ for the functions

$$f(x) = x^2 - 9 \quad \text{and} \quad g(x) = \sqrt{9 - x^2}.$$

Then find the domain of $(f \circ g)$.

Solution

$$
\begin{aligned}
(f \circ g)(x) &= f(g(x)) \\
&= f\left(\sqrt{9 - x^2}\right) \\
&= \left(\sqrt{9 - x^2}\right)^2 - 9 \\
&= 9 - x^2 - 9 \\
&= -x^2
\end{aligned}
$$

From this, it might appear that the domain of the composition is the set of all real numbers. Because the domain of f is the set of all real numbers and the domain of g is $-3 \leq x \leq 3$, the domain of $(f \circ g)$ is $-3 \leq x \leq 3$.

In Examples 4 and 5, you formed the composition of two given functions. In calculus, it is also important to be able to identify two functions that make up a given composite function. For instance, the function h given by

$$h(x) = (3x - 5)^3$$

is the composition of f with g, where $f(x) = x^3$ and $g(x) = 3x - 5$. That is,

$$h(x) = (3x - 5)^3 = [g(x)]^3 = f(g(x)).$$

Basically, to "decompose" a composite function, look for an "inner" function and an "outer" function. In the function h above, $g(x) = 3x - 5$ is the inner function and $f(x) = x^3$ is the outer function.

Example 6 ▶ **Finding Components of Composite Functions**

Express the function $h(x) = \dfrac{1}{(x - 2)^2}$ as a composition of two functions.

Solution

One way to write h as a composition of two functions is to take the inner function to be $g(x) = x - 2$ and the outer function to be

$$f(x) = \frac{1}{x^2} = x^{-2}.$$

Then you can write

$$h(x) = \frac{1}{(x - 2)^2} = (x - 2)^{-2} = f(x - 2) = f(g(x)).$$

You are buying an automobile whose price is $18,500. Which of the following options would you choose? Explain.

a. You are given a factory rebate of $2000, followed by a dealer discount of 10%.

b. You are given a dealer discount of 10%, followed by a factory rebate of $2000.

Let $f(x) = x - 2000$ and let $g(x) = 0.9x$. Which option is represented by the composite $f(g(x))$? Which is represented by the composite $g(f(x))$?

Writing About Mathematics

To expand on this activity, you might consider asking your students to use the tables they created in parts (a) and (b), along with a table of values for x and $f(x)$, to demonstrate and explain how the tables can be manipulated to yield tables of values for $h(x)$ and $g(x)$.

Application

Example 7 ▶ **Bacteria Count**

The number N of bacteria in a refrigerated food is

$$N(T) = 20T^2 - 80T + 500, \qquad 2 \le T \le 14$$

where T is the temperature of the food in degrees Celsius. When the food is removed from refrigeration, the temperature is

$$T(t) = 4t + 2, \qquad 0 \le t \le 3$$

where t is the time in hours. (a) Find the composite $N(T(t))$ and interpret its meaning in context. (b) Find the time when the bacterial count reaches 2000.

Solution

a. $N(T(t)) = 20(4t + 2)^2 - 80(4t + 2) + 500$

$\qquad = 20(16t^2 + 16t + 4) - 320t - 160 + 500$

$\qquad = 320t^2 + 320t + 80 - 320t - 160 + 500$

$\qquad = 320t^2 + 420$

The composite function $N(T(t))$ represents the number of bacteria in the food as a function of time.

b. The bacterial count will reach 2000 when $320t^2 + 420 = 2000$. Solve this equation to find that the count will reach 2000 when $t \approx 2.2$ hours. When you solve this equation, note that the negative value is rejected because it is not in the domain of the composite function.

Writing **ABOUT MATHEMATICS**

Analyzing Arithmetic Combinations of Functions

a. Use the graphs of f and $(f + g)$ in Figure P.88 to make a table showing the values of $g(x)$ when $x = 1, 2, 3, 4, 5,$ and 6. Explain your reasoning.

b. Use the graphs of f and $(f - h)$ in Figure P.88 to make a table showing the values of $h(x)$ when $x = 1, 2, 3, 4, 5,$ and 6. Explain your reasoning.

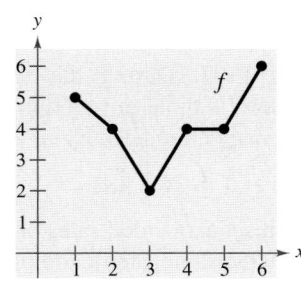
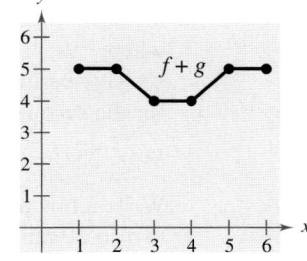
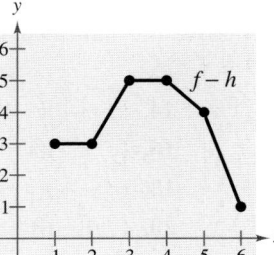

FIGURE P.88

P.9 Exercises

In Exercises 1–4, use the graphs of f and g to graph $h(x) = (f + g)(x)$. To print an enlarged copy of the graph, go to the website **www.mathgraphs.com**.

1.
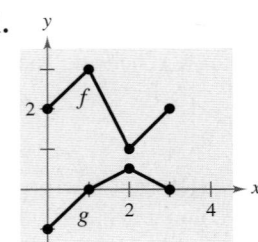

2.
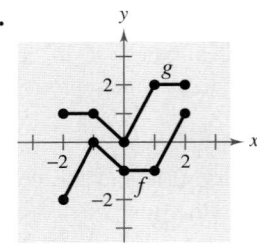

3.
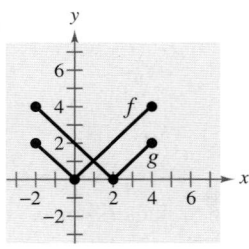

4.
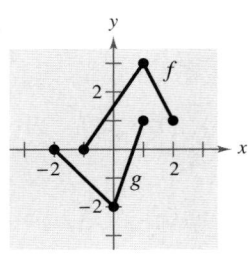

In Exercises 5–12, find (a) $(f + g)(x)$, (b) $(f - g)(x)$, (c) $(fg)(x)$, and (d) $(f/g)(x)$. What is the domain of f/g?

5. $f(x) = x + 2,$ $g(x) = x - 2$

6. $f(x) = 2x - 5,$ $g(x) = 2 - x$

7. $f(x) = x^2,$ $g(x) = 4x - 5$

8. $f(x) = 2x - 5,$ $g(x) = 4$

9. $f(x) = x^2 + 6,$ $g(x) = \sqrt{1 - x}$

10. $f(x) = \sqrt{x^2 - 4},$ $g(x) = \dfrac{x^2}{x^2 + 1}$

11. $f(x) = \dfrac{1}{x},$ $g(x) = \dfrac{1}{x^2}$

12. $f(x) = \dfrac{x}{x + 1},$ $g(x) = x^3$

In Exercises 13–24, evaluate the indicated function for $f(x) = x^2 + 1$ and $g(x) = x - 4$.

13. $(f + g)(2)$

14. $(f - g)(-1)$

15. $(f - g)(0)$

16. $(f + g)(1)$

17. $(f - g)(3t)$

18. $(f + g)(t - 2)$

19. $(fg)(6)$

20. $(fg)(-6)$

21. $\left(\dfrac{f}{g}\right)(5)$

22. $\left(\dfrac{f}{g}\right)(0)$

23. $\left(\dfrac{f}{g}\right)(-1) - g(3)$

24. $(fg)(5) + f(4)$

In Exercises 25–28, graph the functions $f, g,$ and $f + g$ on the same set of coordinate axes.

25. $f(x) = \frac{1}{2}x,$ $g(x) = x - 1$

26. $f(x) = \frac{1}{3}x,$ $g(x) = -x + 4$

27. $f(x) = x^2,$ $g(x) = -2x$

28. $f(x) = 4 - x^2,$ $g(x) = x$

 Graphical Reasoning In Exercises 29 and 30, use a graphing utility to graph $f, g,$ and $f + g$ in the same viewing window. Which function contributes most to the magnitude of the sum when $0 \le x \le 2$? Which function contributes most to the magnitude of the sum when $x > 6$?

29. $f(x) = 3x,$ $g(x) = -\dfrac{x^3}{10}$

30. $f(x) = \dfrac{x}{2},$ $g(x) = \sqrt{x}$

31. *Stopping Distance* The research and development department of an automobile manufacturer has determined that when required to stop quickly to avoid an accident, the distance (in feet) a car travels during the driver's reaction time is given by $R(x) = \frac{3}{4}x,$ where x is the speed of the car in miles per hour. The distance (in feet) traveled while the driver is braking is $B(x) = \frac{1}{15}x^2.$ Find the function that represents the total stopping distance $T.$ Graph the functions $R, B,$ and T on the same set of coordinate axes for $0 \le x \le 60.$

 32. *Sales* From 1997 to 2002, the sales R_1 (in thousands of dollars) for one of two restaurants owned by the same parent company can be modeled by

$$R_1 = 480 - 8t - 0.8t^2, \qquad t = 0, 1, 2, 3, 4, 5$$

where $t = 0$ represents 1997. During the same six-year period, the sales R_2 (in thousands of dollars) for the second restaurant can be modeled by

$$R_2 = 254 + 0.78t, \qquad t = 0, 1, 2, 3, 4, 5.$$

Write a function that represents the total sales of the two restaurants owned by the same parent company. Use a graphing utility to graph the total sales function.

► **Model It**

33. Health Care Costs The table shows the total amount (in billions of dollars) spent on health services and supplies in the United States (including Puerto Rico) for the years 1993 through 1999. The variables y_1, y_2, and y_3 represent out-of-pocket payments, insurance premiums, and other types of payments, respectively. (Source: Centers for Medicare and Medicaid Services)

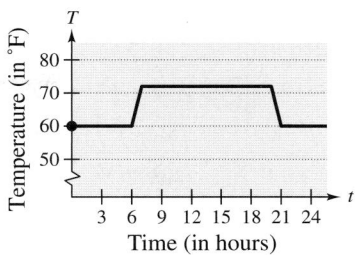

Year	y_1	y_2	y_3
1993	148.9	295.7	39.1
1994	146.2	308.9	40.8
1995	149.2	322.3	44.8
1996	155.0	337.4	47.9
1997	165.5	355.6	52.0
1998	176.1	376.8	54.8
1999	186.5	401.2	58.9

(a) Use the *regression* feature of a graphing utility to find a quadratic model for y_1 and linear models for y_2 and y_3. Let $t = 3$ represent 1993.

(b) Find $y_1 + y_2 + y_3$. What does this sum represent?

(c) Use a graphing utility to graph y_1, y_2, y_3, and $y_1 + y_2 + y_3$ in the same viewing window.

(d) Use the model from part (b) to estimate the total amount spent on health services and supplies in the years 2003 and 2005.

34. Graphical Reasoning An electronically controlled thermostat in a home is programmed to lower the temperature automatically during the night. The temperature in the house T (in degrees Fahrenheit) is given in terms of t, the time in hours on a 24-hour clock (see figure).

(a) Explain why T is a function of t.

(b) Approximate $T(4)$ and $T(15)$.

(c) The thermostat is reprogrammed to produce a temperature H for which $H(t) = T(t - 1)$. How does this change the temperature?

(d) The thermostat is reprogrammed to produce a temperature H for which $H(t) = T(t) - 1$. How does this change the temperature?

(e) Write a piecewise-defined function that represents the graph.

FIGURE FOR **34**

In Exercises 35–38, find (a) $f \circ g$, (b) $g \circ f$, and (c) $f \circ f$.

35. $f(x) = x^2$, $\qquad$ $g(x) = x - 1$

36. $f(x) = 3x + 5$, $\qquad$ $g(x) = 5 - x$

37. $f(x) = \sqrt[3]{x - 1}$, $\qquad$ $g(x) = x^3 + 1$

38. $f(x) = x^3$, $\qquad$ $g(x) = \dfrac{1}{x}$

In Exercises 39–46, find (a) $f \circ g$ and (b) $g \circ f$. Find the domain of each function and each composite function.

39. $f(x) = \sqrt{x + 4}$, $\qquad$ $g(x) = x^2$

40. $f(x) = \sqrt[3]{x - 5}$, $\qquad$ $g(x) = x^3 + 1$

41. $f(x) = x^2 + 1$, $\qquad$ $g(x) = \sqrt{x}$

42. $f(x) = x^{2/3}$, $\qquad$ $g(x) = x^6$

43. $f(x) = |x|$, $\qquad$ $g(x) = x + 6$

44. $f(x) = |x - 4|$, $\qquad$ $g(x) = 3 - x$

45. $f(x) = \dfrac{1}{x}$, $\qquad$ $g(x) = x + 3$

46. $f(x) = \dfrac{3}{x^2 - 1}$, $\qquad$ $g(x) = x + 1$

In Exercises 47–50, use the graphs of f and g to evaluate the functions.

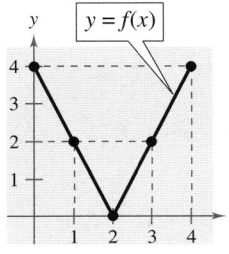

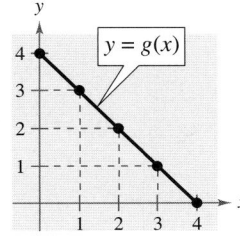

47. (a) $(f + g)(3)$ $\qquad$ (b) $(f/g)(2)$

48. (a) $(f - g)(1)$ $\qquad$ (b) $(fg)(4)$

49. (a) $(f \circ g)(2)$ (b) $(g \circ f)(2)$

50. (a) $(f \circ g)(1)$ (b) $(g \circ f)(3)$

In Exercises 51–58, find two functions f and g such that $(f \circ g)(x) = h(x)$. (There is more than one correct answer.)

51. $h(x) = (2x + 1)^2$ **52.** $h(x) = (1 - x)^3$

53. $h(x) = \sqrt[3]{x^2 - 4}$ **54.** $h(x) = \sqrt{9 - x}$

55. $h(x) = \dfrac{1}{x + 2}$ **56.** $h(x) = \dfrac{4}{(5x + 2)^2}$

57. $h(x) = \dfrac{-x^2 + 3}{4 - x^2}$ **58.** $h(x) = \dfrac{27x^3 + 6x}{10 - 27x^3}$

59. *Geometry* A square concrete foundation is prepared as a base for a cylindrical tank (see figure).

  (a) Write the radius r of the tank as a function of the length x of the sides of the square.

  (b) Write the area A of the circular base of the tank as a function of the radius r.

  (c) Find and interpret $(A \circ r)(x)$.

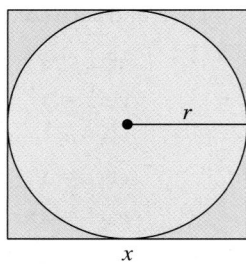

60. *Physics* A pebble is dropped into a calm pond, causing ripples in the form of concentric circles (see figure). The radius (in feet) of the outer ripple is $r(t) = 0.6t$, where t is the time in seconds after the pebble strikes the water. The area of the circle is given by the function $A(r) = \pi r^2$. Find and interpret $(A \circ r)(t)$.

Synthesis

True or False? In Exercises 61 and 62, determine whether the statement is true or false. Justify your answer.

61. If $f(x) = x + 1$ and $g(x) = 6x$, then $(f \circ g)(x) = (g \circ f)(x)$.

62. If you are given two functions $f(x)$ and $g(x)$, you can calculate $(f \circ g)(x)$ if and only if the range of g is a subset of the domain of f.

63. *Think About It* You are a sales representative for an automobile manufacturer. You are paid an annual salary, plus a bonus of 3% of your sales over $500,000. Consider the two functions

$$f(x) = x - 500{,}000 \qquad \text{and} \qquad g(x) = 0.03x.$$

If x is greater than $500,000, which of the following represents your bonus? Explain your reasoning.

  (a) $f(g(x))$ (b) $g(f(x))$

64. *Proof* Prove that the product of two odd functions is an even function, and that the product of two even functions is an even function.

65. *Conjecture* Use examples to hypothesize whether the product of an odd function and an even function is even or odd. Then prove your hypothesis.

P.10 | **Inverse Functions**

▶ **What you should learn**

- How to find inverse functions informally and verify that two functions are inverse functions of each other
- How to use graphs of functions to determine whether functions have inverse functions
- How to use the Horizontal Line Test to determine if functions are one-to-one
- How to find inverse functions algebraically

▶ **Why you should learn it**

Inverse functions can be used to model and solve real-life problems. For instance, in Exercise 79 on page 111, an inverse function can be used to determine the year in which there were a given number of households in the United States.

Inverse Functions

Recall from Section P.5 that a function can be represented by a set of ordered pairs. For instance, the function $f(x) = x + 4$ from the set $A = \{1, 2, 3, 4\}$ to the set $B = \{5, 6, 7, 8\}$ can be written as follows.

$$f(x) = x + 4: \ \{(1, 5), (2, 6), (3, 7), (4, 8)\}$$

In this case, by interchanging the first and second coordinates of each of these ordered pairs, you can form the **inverse function** of f, which is denoted by f^{-1}. It is a function from the set B to the set A, and can be written as follows.

$$f^{-1}(x) = x - 4: \ \{(5, 1), (6, 2), (7, 3), (8, 4)\}$$

Note that the domain of f is equal to the range of f^{-1}, and vice versa, as shown in Figure P.89. Also note that the functions f and f^{-1} have the effect of "undoing" each other. In other words, when you form the composition of f with f^{-1} or the composition of f^{-1} with f, you obtain the identity function.

$$f(f^{-1}(x)) = f(x - 4) = (x - 4) + 4 = x$$
$$f^{-1}(f(x)) = f^{-1}(x + 4) = (x + 4) - 4 = x$$

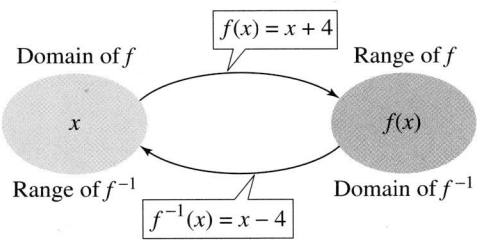

FIGURE P.89

Example 1 ▶ Finding Inverse Functions Informally

Find the inverse function of $f(x) = 4x$. Then verify that both $f(f^{-1}(x))$ and $f^{-1}(f(x))$ are equal to the identity function.

Solution

The function f *multiplies* each input by 4. To "undo" this function, you need to *divide* each input by 4. So, the inverse function of $f(x) = 4x$ is

$$f^{-1}(x) = \frac{x}{4}.$$

You can verify that both $f(f^{-1}(x))$ and $f^{-1}(f(x))$ are equal to the identity function as follows.

$$f(f^{-1}(x)) = f\left(\frac{x}{4}\right) = 4\left(\frac{x}{4}\right) = x \qquad f^{-1}(f(x)) = f^{-1}(4x) = \frac{4x}{4} = x$$

Consider the functions

$$f(x) = x + 2$$

and

$$f^{-1}(x) = x - 2.$$

Evaluate $f(f^{-1}(x))$ and $f^{-1}(f(x))$ for the indicated values of x. What can you conclude about the functions?

x	-10	0	7	45
$f(f^{-1}(x))$				
$f^{-1}(f(x))$				

Definition of Inverse Function

Let f and g be two functions such that

$$f(g(x)) = x \qquad \text{for every } x \text{ in the domain of } g$$

and

$$g(f(x)) = x \qquad \text{for every } x \text{ in the domain of } f.$$

Under these conditions, the function g is the **inverse function** of the function f. The function g is denoted by f^{-1} (read "f-inverse"). So,

$$f(f^{-1}(x)) = x \qquad \text{and} \qquad f^{-1}(f(x)) = x.$$

The domain of f must be equal to the range of f^{-1}, and the range of f must be equal to the domain of f^{-1}.

Don't be confused by the use of -1 to denote the inverse function f^{-1}. In this text, whenever f^{-1} is written, it *always* refers to the inverse function of the function f and *not* to the reciprocal of $f(x)$.

If the function g is the inverse function of the function f, it must also be true that the function f is the inverse function of the function g. For this reason, you can say that the functions f and g are *inverse functions of each other*.

Additional Examples

Determine whether the two functions are inverse functions of each other.

a. $f(x) = 3x - 2, g(x) = \dfrac{1}{3}x + 2$

b. $f(x) = \dfrac{1}{4}x - 3, g(x) = 12 + 4x$

c. $f(x) = 2x + 4, g(x) = \dfrac{1}{2}x - 2$

Solution

The functions in parts (b) and (c) are inverse functions of each other because $f(g(x)) = g(f(x)) = x$.

The functions in part (a) are not inverse functions of each other.

Example 2 ▶ **Verifying Inverse Functions**

Which of the functions is the inverse function of $f(x) = \dfrac{5}{x-2}$?

$$g(x) = \dfrac{x-2}{5} \qquad\qquad h(x) = \dfrac{5}{x} + 2$$

Solution

By forming the composition of f with g, you have

$$f(g(x)) = f\left(\dfrac{x-2}{5}\right)$$

$$= \dfrac{5}{\left(\dfrac{x-2}{5}\right) - 2} \qquad \text{Substitute } \dfrac{x-2}{5} \text{ for } x.$$

$$= \dfrac{25}{x - 12} \neq x.$$

Because this composition is not equal to the identity function x, it follows that g *is not* the inverse function of f. By forming the composition of f with h, you have

$$f(h(x)) = f\left(\dfrac{5}{x} + 2\right) = \dfrac{5}{\left(\dfrac{5}{x} + 2\right) - 2} = \dfrac{5}{\left(\dfrac{5}{x}\right)} = x.$$

So, it appears that h *is* the inverse function of f. You can confirm this by showing that the composition of h with f is also equal to the identity function.

The Graph of an Inverse Function

The graphs of a function f and its inverse function f^{-1} are related to each other in the following way. If the point (a, b) lies on the graph of f, then the point (b, a) must lie on the graph of f^{-1}, and vice versa. This means that the graph of f^{-1} is a *reflection* of the graph of f in the line $y = x$, as shown in Figure P.90.

Example 3 ▶ The Graphs of f and f^{-1}

Sketch the graphs of the inverse functions $f(x) = 2x - 3$ and $f^{-1}(x) = \frac{1}{2}(x + 3)$ on the same rectangular coordinate system and show that the graphs are reflections of each other in the line $y = x$.

Solution

The graphs of f and f^{-1} are shown in Figure P.91. It appears that the graphs are reflections of each other in the line $y = x$. You can further verify this reflective property by testing a few points on each graph. Note in the following list that if the point (a, b) is on the graph of f, the point (b, a) is on the graph of f^{-1}.

Graph of $f(x) = 2x - 3$	*Graph of* $f^{-1}(x) = \frac{1}{2}(x + 3)$
$(-1, -5)$	$(-5, -1)$
$(0, -3)$	$(-3, 0)$
$(1, -1)$	$(-1, 1)$
$(2, 1)$	$(1, 2)$
$(3, 3)$	$(3, 3)$

Example 4 ▶ Finding Inverse Functions Graphically

Sketch the graphs of the inverse functions $f(x) = x^2 (x \geq 0)$ and $f^{-1}(x) = \sqrt{x}$ on the same rectangular coordinate system and show that the graphs are reflections of each other in the line $y = x$.

Solution

The graphs of f and f^{-1} are shown in Figure P.92. It appears that the graphs are reflections of each other in the line $y = x$. You can further verify this reflective property by testing a few points on each graph. Note in the following list that if the point (a, b) is on the graph of f, the point (b, a) is on the graph of f^{-1}.

Graph of $f(x) = x^2, \quad x \geq 0$	*Graph of* $f^{-1}(x) = \sqrt{x}$
$(0, 0)$	$(0, 0)$
$(1, 1)$	$(1, 1)$
$(2, 4)$	$(4, 2)$
$(3, 9)$	$(9, 3)$

Try showing that $f(f^{-1}(x)) = x$ and $f^{-1}(f(x)) = x$.

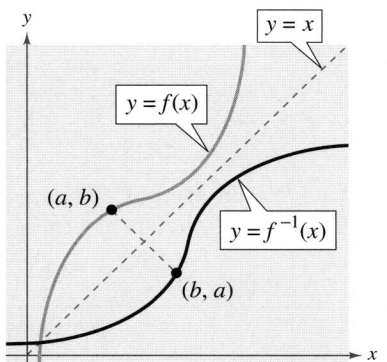

FIGURE P.90

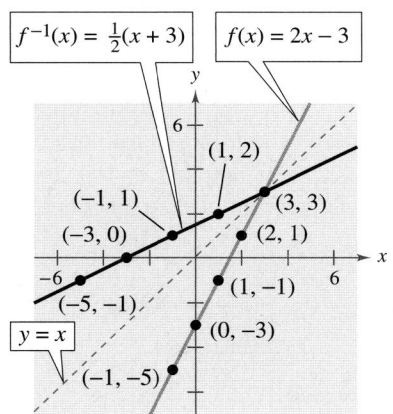

FIGURE P.91

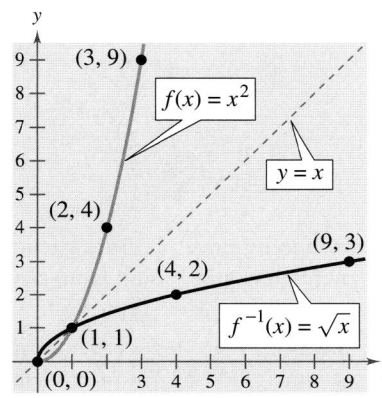

FIGURE P.92

One-to-One Functions

The reflective property of the graphs of inverse functions gives you a nice *geometric* test for determining whether a function has an inverse function. This test is called the **Horizontal Line Test** for inverse functions.

Horizontal Line Test for Inverse Functions

A function f has an inverse function if and only if no *horizontal* line intersects the graph of f at more than one point.

If no horizontal line intersects the graph of f at more than one point, then no x-value is matched with more than one y-value. This is the essential characteristic of what are called **one-to-one** functions.

One-to-One Functions

A function f is **one-to-one** if each value of the dependent variable corresponds to exactly one value of the independent variable. A function f has an inverse function if and only if f is one-to-one.

Consider the function $f(x) = x^2$. The table on the left is a table of values for $f(x) = x^2$. The table of values on the right is made up by interchanging the columns of the first table. The table on the right does not represent a function because the input $x = 4$ is matched with two different outputs: $y = -2$ and $y = 2$. So, $f(x) = x^2$ is not one-to-one and does not have an inverse function.

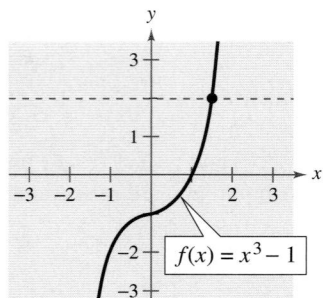

FIGURE P.93

x	$f(x)$
-2	4
-1	1
0	0
1	1
2	4
3	9

x	y
4	-2
1	-1
0	0
1	1
4	2
9	3

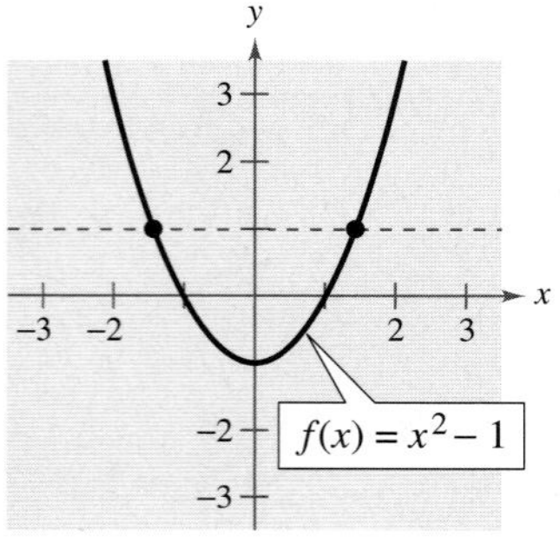

Example 5 ▶ **Applying the Horizontal Line Test**

a. The graph of the function $f(x) = x^3 - 1$ is shown in Figure P.93. Because no horizontal line intersects the graph of f at more than one point, you can conclude that f *is* a one-to-one function and *does* have an inverse function.

b. The graph of the function $f(x) = x^2 - 1$ is shown in Figure P.94. Because it is possible to find a horizontal line that intersects the graph of f at more than one point, you can conclude that f *is not* a one-to-one function and *does not* have an inverse function.

FIGURE P.94

Finding Inverse Functions Algebraically

For simple functions (such as the one in Example 1), you can find inverse functions by inspection. For more complicated functions, however, it is best to use the following guidelines. The key step in these guidelines is Step 3—interchanging the roles of x and y. This step corresponds to the fact that inverse functions have ordered pairs with the coordinates reversed.

> **Finding an Inverse Function**
>
> 1. Use the Horizontal Line Test to decide whether f has an inverse function.
>
> 2. In the equation for $f(x)$, replace $f(x)$ by y.
>
> 3. Interchange the roles of x and y, and solve for y.
>
> 4. Replace y by $f^{-1}(x)$ in the new equation.
>
> 5. Verify that f and f^{-1} are inverse functions of each other by showing that the domain of f is equal to the range of f^{-1}, the range of f is equal to the domain of f^{-1}, and $f(f^{-1}(x)) = x = f^{-1}(f(x))$.

Example 6 ▶ Finding an Inverse Function Algebraically

Find the inverse function of

$$f(x) = \frac{5 - 3x}{2}.$$

Solution

The graph of f is a line, as shown in Figure P.95. This graph passes the Horizontal Line Test. So, you know that f is one-to-one and has an inverse function.

$f(x) = \dfrac{5 - 3x}{2}$	Write original function.
$y = \dfrac{5 - 3x}{2}$	Replace $f(x)$ by y.
$x = \dfrac{5 - 3y}{2}$	Interchange x and y.
$2x = 5 - 3y$	Multiply each side by 2.
$3y = 5 - 2x$	Isolate the y-term.
$y = \dfrac{5 - 2x}{3}$	Solve for y.
$f^{-1}(x) = \dfrac{5 - 2x}{3}$	Replace y by $f^{-1}(x)$.

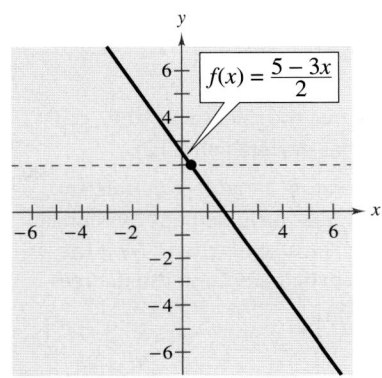

FIGURE P.95

Note that both f and f^{-1} have domains and ranges that consist of the entire set of real numbers. Check that $f(f^{-1}(x)) = x$ and $f^{-1}(f(x)) = x$.

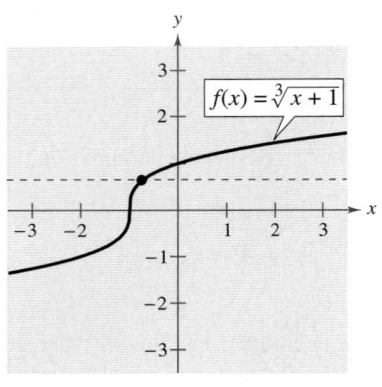

FIGURE P.96

Example 7 ▶ Finding an Inverse Function

Find the inverse function of

$$f(x) = \sqrt[3]{x + 1}.$$

Solution

The graph of f is a curve, as shown in Figure P.96. Because this graph passes the Horizontal Line Test, you know that f is one-to-one and has an inverse function.

$f(x) = \sqrt[3]{x + 1}$	Write original function.
$y = \sqrt[3]{x + 1}$	Replace $f(x)$ by y.
$x = \sqrt[3]{y + 1}$	Interchange x and y.
$x^3 = y + 1$	Cube each side.
$x^3 - 1 = y$	Solve for y.
$x^3 - 1 = f^{-1}(x)$	Replace y by $f^{-1}(x)$.

Both f and f^{-1} have domains and ranges that consist of the entire set of real numbers. You can verify this result numerically as shown in the tables below.

x	$f(x)$
-28	-3
-9	-2
-2	-1
-1	0
0	1
7	2
26	3

x	$f^{-1}(x)$
-3	-28
-2	-9
-1	-2
0	-1
1	0
2	7
3	26

Activities

1. Given $f(x) = 5x - 7$, find $f^{-1}(x)$.

   Answer: $f^{-1}(x) = \dfrac{x + 7}{5}$

2. Show that f and g are inverse functions by showing that $f(g(x)) = x$ and $g(f(x)) = x$.

   $f(x) = 3x^3 + 1$

   $g(x) = \sqrt[3]{\dfrac{x - 1}{3}}$

3. Describe the graphs of functions that have inverse functions and show how the graphs of a function and its inverse function are related.

Writing ABOUT MATHEMATICS

The Existence of an Inverse Function Write a short paragraph describing why the following functions do or do not have inverse functions.

a. Let x represent the retail price of an item (in dollars), and let $f(x)$ represent the sales tax on the item. Assume that the sales tax is 6% of the retail price *and* that the sales tax is rounded to the nearest cent. Does this function have an inverse function? (*Hint:* Can you undo this function? For instance, if you know that the sales tax is $0.12, can you determine exactly what the retail price is?)

b. Let x represent the temperature in degrees Celsius, and let $f(x)$ represent the temperature in degrees Fahrenheit. Does this function have an inverse function? (*Hint:* The formula for converting from degrees Celsius to degrees Fahrenheit is $F = \frac{9}{5}C + 32$.)

P.10 Exercises

In Exercises 1–4, match the graph of the function with the graph of its inverse function. [The graphs of the inverse functions are labeled (a), (b), (c), and (d).]

(a)
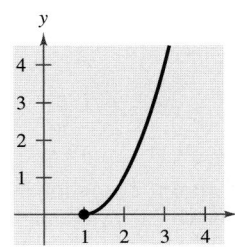

(b)
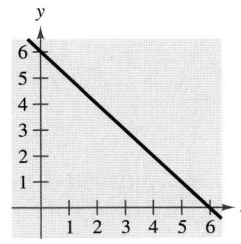

(c)
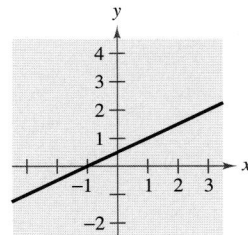

(d)
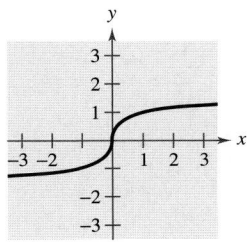

1.
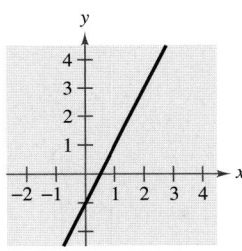

2.
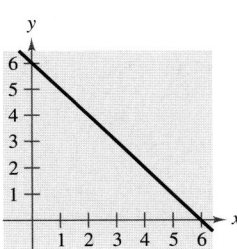

3.
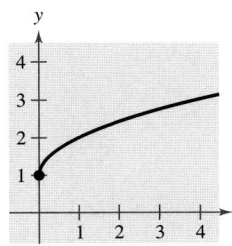

4.
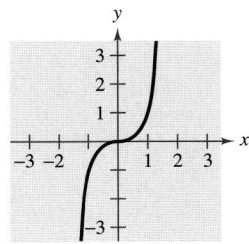

In Exercises 5–12, find the inverse function of f informally. Verify that $f(f^{-1}(x)) = x$ and $f^{-1}(f(x)) = x$.

5. $f(x) = 6x$

6. $f(x) = \frac{1}{3}x$

7. $f(x) = x + 9$

8. $f(x) = x - 4$

9. $f(x) = 3x + 1$

10. $f(x) = \dfrac{x - 1}{5}$

11. $f(x) = \sqrt[3]{x}$

12. $f(x) = x^5$

In Exercises 13–24, show that f and g are inverse functions (a) algebraically and (b) graphically.

13. $f(x) = 2x,$ $\qquad g(x) = \dfrac{x}{2}$

14. $f(x) = x - 5,$ $\qquad g(x) = x + 5$

15. $f(x) = 7x + 1,$ $\qquad g(x) = \dfrac{x - 1}{7}$

16. $f(x) = 3 - 4x,$ $\qquad g(x) = \dfrac{3 - x}{4}$

17. $f(x) = \dfrac{x^3}{8},$ $\qquad g(x) = \sqrt[3]{8x}$

18. $f(x) = \dfrac{1}{x},$ $\qquad g(x) = \dfrac{1}{x}$

19. $f(x) = \sqrt{x - 4},$ $\qquad g(x) = x^2 + 4, \quad x \geq 0$

20. $f(x) = 1 - x^3,$ $\qquad g(x) = \sqrt[3]{1 - x}$

21. $f(x) = 9 - x^2, \quad x \geq 0, \quad g(x) = \sqrt{9 - x}, \quad x \leq 9$

22. $f(x) = \dfrac{1}{1 + x}, \quad x \geq 0$

$\quad g(x) = \dfrac{1 - x}{x}, \quad 0 < x \leq 1$

23. $f(x) = \dfrac{x - 1}{x + 5},$ $\qquad g(x) = -\dfrac{5x + 1}{x - 1}$

24. $f(x) = \dfrac{x + 3}{x - 2},$ $\qquad g(x) = \dfrac{2x + 3}{x - 1}$

In Exercises 25 and 26, does the function have an inverse function?

25.

x	$f(x)$
-1	-2
0	1
1	2
2	1
3	-2
4	-6

26.

x	$f(x)$
-3	10
-2	6
-1	4
0	1
2	-3
3	-10

In Exercises 27 and 28, use the table of values for $y = f(x)$ to complete a table for $y = f^{-1}(x)$.

27.

x	-2	-1	0	1	2	3
$f(x)$	-2	0	2	4	6	8

28.

x	-3	-2	-1	0	1	2
$f(x)$	-10	-7	-4	-1	2	5

In Exercises 29–32, does the function have an inverse function?

29.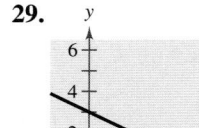

30.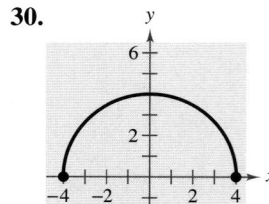

31.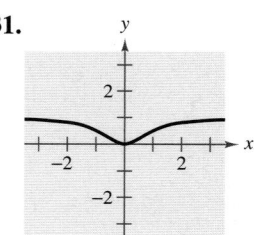

32.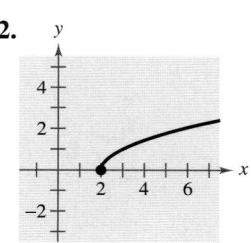

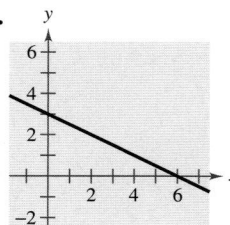 In Exercises 33–38, use a graphing utility to graph the function, and use the Horizontal Line Test to determine whether the function is one-to-one and so has an inverse function.

33. $g(x) = \dfrac{4 - x}{6}$

34. $f(x) = 10$

35. $h(x) = |x + 4| - |x - 4|$

36. $g(x) = (x + 5)^3$

37. $f(x) = -2x\sqrt{16 - x^2}$

38. $f(x) = \frac{1}{8}(x + 2)^2 - 1$

In Exercises 39–54, find the inverse function of f. Then graph both f and f^{-1} on the same set of coordinate axes.

39. $f(x) = 2x - 3$

40. $f(x) = 3x + 1$

41. $f(x) = x^5 - 2$

42. $f(x) = x^3 + 1$

43. $f(x) = \sqrt{x}$

44. $f(x) = x^2, \quad x \geq 0$

45. $f(x) = \sqrt{4 - x^2}, \quad 0 \leq x \leq 2$

46. $f(x) = x^2 - 2, \quad x \leq 0$

47. $f(x) = \dfrac{4}{x}$

48. $f(x) = -\dfrac{2}{x}$

49. $f(x) = \dfrac{x + 1}{x - 2}$

50. $f(x) = \dfrac{x - 3}{x + 2}$

51. $f(x) = \sqrt[3]{x - 1}$

52. $f(x) = x^{3/5}$

53. $f(x) = \dfrac{6x + 4}{4x + 5}$

54. $f(x) = \dfrac{8x - 4}{2x + 6}$

In Exercises 55–68, determine whether the function has an inverse function. If it does, find the inverse function.

55. $f(x) = x^4$

56. $f(x) = \dfrac{1}{x^2}$

57. $g(x) = \dfrac{x}{8}$

58. $f(x) = 3x + 5$

59. $p(x) = -4$

60. $f(x) = \dfrac{3x + 4}{5}$

61. $f(x) = (x + 3)^2, \quad x \geq -3$

62. $q(x) = (x - 5)^2$

63. $f(x) = \begin{cases} x + 3, & x < 0 \\ 6 - x, & x \geq 0 \end{cases}$

64. $f(x) = \begin{cases} -x, & x \leq 0 \\ x^2 - 3x, & x > 0 \end{cases}$

65. $h(x) = -\dfrac{4}{x^2}$

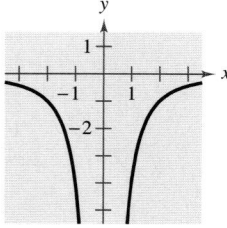

66. $f(x) = |x - 2|, \quad x \leq 2$

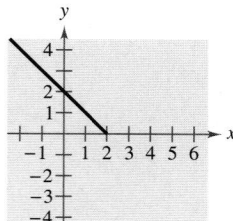
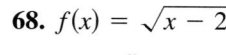

67. $f(x) = \sqrt{2x + 3}$

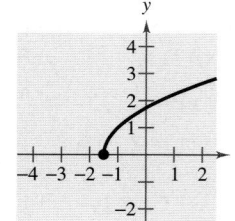

68. $f(x) = \sqrt{x - 2}$

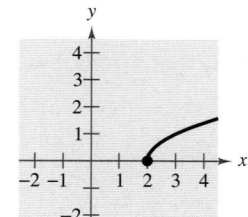

In Exercises 69–74, use the functions $f(x) = \frac{1}{8}x - 3$ and $g(x) = x^3$ to find the indicated value or function.

69. $(f^{-1} \circ g^{-1})(1)$

70. $(g^{-1} \circ f^{-1})(-3)$

71. $(f^{-1} \circ f^{-1})(6)$

72. $(g^{-1} \circ g^{-1})(-4)$

73. $(f \circ g)^{-1}$

74. $g^{-1} \circ f^{-1}$

In Exercises 75–78, use the functions $f(x) = x + 4$ and $g(x) = 2x - 5$ to find the specified function.

75. $g^{-1} \circ f^{-1}$

76. $f^{-1} \circ g^{-1}$

77. $(f \circ g)^{-1}$

78. $(g \circ f)^{-1}$

▶ **Model It**

79. *U.S. Households* The number of households f (in thousands) in the United States from 1994 to 2000 are shown in the table. The time (in years) is given by t, with $t = 4$ corresponding to 1994. (Source: U.S. Census Bureau)

Year, t	Households, $f(t)$
4	97,107
5	98,990
6	99,627
7	101,018
8	102,528
9	103,874
10	104,705

(a) Find $f^{-1}(103,874)$.

(b) What does f^{-1} mean in the context of the problem?

(c) Use the *regression* feature of a graphing utility to find a linear model for the data, $y = mx + b$. (Round m and b to two decimal places.)

(d) Algebraically find the inverse function of the linear model in part (c).

(e) Use the inverse function of the linear model you found in part (d) to approximate $f^{-1}(111, 254)$.

80. *Bottled Water Consumption* The per capita consumption f (in gallons) of bottled water in the United States from 1994 through 1999 is shown in the table. The time (in years) is given by t, with $t = 4$ corresponding to 1994. (Source: U.S. Department of Agriculture)

t	$f(t)$
4	10.7
5	11.6
6	12.5
7	13.1
8	16.0
9	18.1

(a) Does f^{-1} exist?

(b) If f^{-1} exists, what does it represent in the context of the problem?

(c) If f^{-1} exists, find $f^{-1}(16.0)$.

81. *Miles Traveled* The total number f (in billions) of miles traveled by motor vehicles in the United States from 1992 through 1999 is shown in the table below. The time (in years) is given by t, with $t = 2$ corresponding to 1992. (Source: U.S. Federal Highway Administration)

Year, t	Miles traveled, $f(t)$
2	2247
3	2296
4	2358
5	2423
6	2486
7	2562
8	2632
9	2691

(a) Does f^{-1} exist?

(b) If f^{-1} exists, what does it mean in the context of the problem?

(c) If f^{-1} exists, find $f^{-1}(2632)$.

(d) If the table was extended to 2000 and if the total number of miles traveled by motor vehicles for that year was 2423 billion, would f^{-1} exist? Explain.

82. *Hourly Wage* Your wage is $8.00 per hour plus $0.75 for each unit produced per hour. So, your hourly wage y in terms of the number of units produced is $y = 8 + 0.75x$.

(a) Find the inverse function.

(b) What does each variable represent in the inverse function?

(c) Determine the number of units produced when your hourly wage is $22.25.

83. *Diesel Mechanics* The function

$$y = 0.03x^2 + 245.50, \qquad 0 < x < 100$$

approximates the exhaust temperature y in degrees Fahrenheit, where x is the percent load for a diesel engine.

(a) Find the inverse function. What does each variable represent in the inverse function?

(b) Use a graphing utility to graph the inverse function.

(c) The exhaust temperature of the engine must not exceed 500 degrees Fahrenheit. What is the percent load interval?

84. *Cost* You need a total of 50 pounds of two types of ground beef costing $1.25 and $1.60 per pound, respectively. A model for the total cost y of the two types of beef is

$$y = 1.25x + 1.60(50 - x)$$

where x is the number of pounds of the less expensive ground beef.

(a) Find the inverse function of the cost function. What does each variable represent in the inverse function?

(b) Use the context of the problem to determine the domain of the inverse function.

(c) Determine the number of pounds of the less expensive ground beef purchased when the total cost is $73.

Synthesis

True or False? **In Exercises 85 and 86, determine whether the statement is true or false. Justify your answer.**

85. If f is an even function, f^{-1} exists.

86. If the inverse function of f exists and the graph of f has a y-intercept, the y-intercept of f is an x-intercept of f^{-1}.

In Exercises 87–90, use the graph of the function f to create a table of values for the given points. Then create a second table that can be used to find f^{-1}, and sketch the graph of f^{-1} if possible.

87.

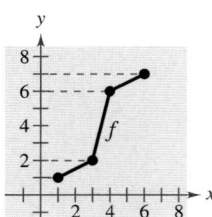

88.

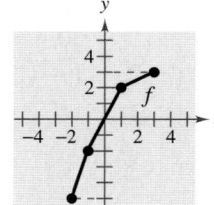

89.

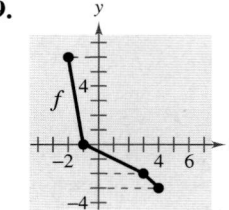

90.

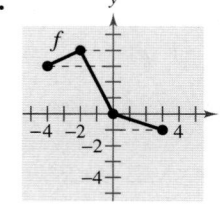

91. *Think About It* The function

$$f(x) = k(2 - x - x^3)$$

has an inverse function, and $f^{-1}(3) = -2$. Find k.

92. *Think About It* The function

$$f(x) = k(x^3 + 3x - 4)$$

has an inverse function, and $f^{-1}(-5) = 2$. Find k.

Chapter Summary

► *What* did you learn?

Review Exercises

P.1 In Exercises 1 and 2, determine which numbers in the set are (a) natural numbers, (b) integers, (c) rational numbers, and (d) irrational numbers.

1. $\left\{ 11, -14, -\frac{8}{9}, \frac{5}{2}, \sqrt{6}, 0.4 \right\}$
2. $\left\{ \sqrt{15}, -22, -\frac{10}{3}, 0, 5.2, \frac{3}{7} \right\}$

In Exercises 3 and 4, use a calculator to find the decimal form of each rational number. If it is a nonterminating decimal, write the repeating pattern. Then plot the numbers on the real number line and place the appropriate inequality sign (< or >) between them.

3. (a) $\frac{5}{6}$ (b) $\frac{7}{8}$ 4. (a) $\frac{9}{25}$ (b) $\frac{5}{7}$

In Exercises 5 and 6, give a verbal description of the subset of real numbers represented by the inequality, and sketch the subset on the real number line.

5. $x \leq 7$ 6. $x > 1$

In Exercises 7 and 8, find the distance between a and b.

7. $a = -92, b = 63$ 8. $a = -112, b = -6$

In Exercises 9 and 10, use absolute value notation to describe the expression.

9. The distance between x and 25 is no more than 10.
10. The distance between z and -16 is greater than 8.

In Exercises 11 and 12, evaluate the expression for each value of x. (If not possible, state the reason.)

	Expression	*Values*	
11.	$-x^2 + x - 1$	(a) $x = 1$	(b) $x = -1$
12.	$\dfrac{x}{x - 3}$	(a) $x = -3$	(b) $x = 3$

In Exercises 13–16, identify the rule of algebra illustrated by the equation.

13. $2x + (3x - 10) = (2x + 3x) - 10$
14. $(t + 4)(2t) = (2t)(t + 4)$
15. $0 + (a - 5) = a - 5$
16. $\dfrac{2}{y + 4} \cdot \dfrac{y + 4}{2} = 1, \quad y \neq -4$

In Exercises 17–22, perform the operations without using a calculator.

17. $|-3| + 4(-2) - 6$ 18. $\dfrac{|-10|}{-10}$

19. $\dfrac{5}{18} \div \dfrac{10}{3}$ 20. $(16 - 8) \div 4$

21. $6[4 - 2(6 + 8)]$ 22. $-4[16 - 3(7 - 10)]$

P.2 In Exercises 23 and 24, determine whether the equation is an identity or a conditional equation.

23. $6 - (x - 2)^2 = 2 + 4x - x^2$
24. $3(x - 2) + 2x = 2(x + 3)$

In Exercises 25–32, solve the equation (if possible) and check your solution.

25. $3x - 2(x + 5) = 10$ 26. $4x + 2(7 - x) = 5$
27. $4(x + 3) - 3 = 2(4 - 3x) - 4$
28. $\frac{1}{2}(x - 3) - 2(x + 1) = 5$

29. $\dfrac{x}{5} - 3 = \dfrac{2x}{3} + 1$ 30. $\dfrac{4x - 3}{6} + \dfrac{x}{4} = x - 2$

31. $\dfrac{18}{x} = \dfrac{10}{x - 4}$ 32. $\dfrac{5}{x - 2} = \dfrac{13}{2x - 3}$

In Exercises 33–42, use any method to solve the quadratic equation.

33. $15 + x - 2x^2 = 0$ 34. $2x^2 - x - 28 = 0$
35. $6 = 3x^2$ 36. $16x^2 = 25$
37. $(x + 4)^2 = 18$ 38. $(x - 8)^2 = 15$
39. $x^2 - 12x + 30 = 0$ 40. $x^2 + 6x - 3 = 0$
41. $-2x^2 - 5x + 27 = 0$ 42. $-20 - 3x + 3x^2 = 0$

In Exercises 43–54, find all solutions of the equation. Check your solutions in the original equation.

43. $5x^4 - 12x^3 = 0$ 44. $4x^3 - 6x^2 = 0$
45. $x^4 - 5x^2 + 6 = 0$
46. $9x^4 + 27x^3 - 4x^2 - 12x = 0$
47. $\sqrt{x + 4} = 3$ 48. $\sqrt{x - 2} - 8 = 0$
49. $\sqrt{2x + 3} + \sqrt{x - 2} = 2$
50. $5\sqrt{x} - \sqrt{x - 1} = 6$
51. $(x - 1)^{2/3} - 25 = 0$ 52. $(x + 2)^{3/4} = 27$
53. $|x - 5| = 10$ 54. $|2x + 3| = 7$

P.3 In Exercises 55–58, determine the quadrant(s) in which (x, y) is located so that the condition(s) is (are) satisfied.

55. $x > 0$ and $y = -2$

56. $y > 0$

57. $(-x, y)$ is in the third quadrant.

58. $xy = 4$

In Exercises 59 and 60, (a) plot the points and (b) find the distance between the points.

59. $(-3, 8), (1, 5)$ **60.** $(5.6, 0), (0, 8.2)$

In Exercises 61 and 62, (a) plot the points and (b) find the midpoint of the line segment joining the points.

61. $(-2, 6), (4, -3)$ **62.** $(0, -1.2), (-3.6, 0)$

In Exercises 63–68, complete a table of values. Use the solution points to sketch the graph of the equation.

63. $y = 3x - 5$ **64.** $y = -\frac{1}{2}x + 2$

65. $y = x^2 - 3x$ **66.** $y = 2x^2 - x - 9$

67. $y = \sqrt{5 - x}$ **68.** $y = \sqrt{x + 2}$

In Exercises 69–72, find the x- and y-intercepts of the graph of the equation.

69. $y = 2x - 9$ **70.** $y = 4 - (x - 4)^2$

71. $y = x\sqrt{9 - x^2}$ **72.** $y = |x + 2| + |3 - x|$

In Exercises 73–78, use the algebraic tests to check for symmetry with respect to both axes and the origin. Then sketch the graph.

73. $y = 5 - x^2$ **74.** $y^2 = 2 - x^2$

75. $y = x^3 + 3$ **76.** $y = -6 - x^3$

77. $x = -|y|$ **78.** $y = |x| + 9$

In Exercises 79–82, find the center and radius of the circle and sketch its graph.

79. $x^2 + y^2 = 9$

80. $x^2 + (y - 8)^2 = 81$

81. $(x + 2)^2 + y^2 = 16$

82. $(x + 4)^2 + \left(y - \frac{3}{2}\right)^2 = 100$

83. Find the standard form of the equation of the circle for which the endpoints of a diameter are $(0, 0)$ and $(4, -6)$.

84. Find the standard form of the equation of the circle for which the endpoints of a diameter are $(-2, -3)$ and $(4, -10)$.

P.4 In Exercises 85–92, sketch the graph of the linear equation.

85. $y = -2x - 7$ **86.** $y = 4x - 3$

87. $y = 6$ **88.** $x = -3$

89. $y = 3x + 13$ **90.** $y = -10x + 9$

91. $y = -\frac{5}{2}x - 1$ **92.** $y = \frac{5}{6}x + 5$

In Exercises 93–98, plot the points and find the slope of the line passing through the points.

93. $(3, -4), (-7, 1)$ **94.** $(-1, 8), (6, 5)$

95. $(-2, 5), (1, 1)$ **96.** $(-3, 2), (8, 2)$

97. $(-4.5, 6), (2.1, 3)$ **98.** $(7, 3.8), (5.2, 0)$

In Exercises 99–104, find an equation of the line that passes through the points.

99. $(0, 0), (0, 10)$

100. $(2, 5), (-2, -1)$

101. $(-1, 4), (2, 0)$

102. $(11, -2), (6, -1)$

103. $(-5, 10), (-1, -6)$

104. $(-3, 4), (-7, 9)$

In Exercises 105–108, find an equation of the line that passes through the given point and has the specified slope. Sketch the line.

	Point	Slope
105.	$(0, -5)$	$m = \frac{3}{2}$
106.	$(-2, 6)$	$m = 0$
107.	$(10, -3)$	$m = -\frac{1}{2}$
108.	$(-8, 5)$	m is undefined.

In Exercises 109–112, write an equation of the line through the point (a) parallel to the given line and (b) perpendicular to the given line.

	Point	Line
109.	$(3, -2)$	$5x - 4y = 8$
110.	$(-8, 3)$	$2x + 3y = 5$
111.	$(4, -1)$	$x = 3$
112.	$(-2, 5)$	$y = -4$

Rate of Change In Exercises 113 and 114, you are given the dollar value of a product in the year 2004 *and* the rate at which the value of the item is expected to change during the next 5 years. Write a linear equation that gives the dollar value V of the product in terms of the year t. (Let $t = 4$ represent 2004.)

2004 Value	Rate
113. $12,500	$850 increase per year
114. $72.95	$5.15 increase per year

115. ***Sales*** During the second and third quarters of the year, a salvage yard had sales of $160,000 and $185,000, respectively. The growth of sales follows a linear pattern. Estimate sales during the fourth quarter.

116. ***Inflation*** The dollar value of a product in 2005 is $85, and the product is expected to increase in value at a rate of $3.75 per year.

(a) Write a linear equation that gives the dollar value V of the product in terms of the year t. (Let $t = 5$ represent 2005.)

(b) Use a graphing utility to graph the equation found in part (a).

(c) Move the cursor along the graph of the sales model to estimate the dollar value of the product in 2010.

P.5 In Exercises 117 and 118, determine which of the sets of ordered pairs represent functions from A to B. Give reasons for your answers.

117. $A = \{10, 20, 30, 40\}$ and $B = \{0, 2, 4, 6\}$

(a) $\{(20, 4), (40, 0), (20, 6), (30, 2)\}$

(b) $\{(10, 4), (20, 4), (30, 4), (40, 4)\}$

(c) $\{(40, 0), (30, 2), (20, 4), (10, 6)\}$

(d) $\{(20, 2), (10, 0), (40, 4)\}$

118. $A = \{u, v, w\}$ and $B = \{-2, -1, 0, 1, 2\}$

(a) $\{(v, -1), (u, 2), (w, 0), (u, -2)\}$

(b) $\{(u, -2), (v, 2), (w, 1)\}$

(c) $\{(u, 2), (v, 2), (w, 1), (w, 1)\}$

(d) $\{(w, -2), (v, 0), (w, 2)\}$

In Exercises 119–122, determine whether the equation represents y as a function of x.

119. $16x - y^4 = 0$

120. $2x - y - 3 = 0$

121. $y = \sqrt{1 - x}$

122. $|y| = x + 2$

In Exercises 123 and 124, evaluate the function as indicated. Simplify your answers.

123. $h(x) = \begin{cases} 2x + 1, & x \le -1 \\ x^2 + 2, & x > -1 \end{cases}$

(a) $h(-2)$ (b) $h(-1)$ (c) $h(0)$ (d) $h(2)$

124. $f(x) = \dfrac{4}{x^2 + 1}$

(a) $f(1)$ (b) $f(-5)$ (c) $f(-t)$ (d) $f(0)$

In Exercises 125–130, determine the domain of the function. Verify your result with a graph.

125. $f(x) = \sqrt{25 - x^2}$ **126.** $f(x) = 3x + 4$

127. $g(s) = \dfrac{5}{3s - 9}$ **128.** $f(x) = \sqrt{x^2 + 8x}$

129. $h(x) = \dfrac{x}{x^2 - x - 6}$ **130.** $h(t) = |t + 1|$

In Exercises 131 and 132, find the difference quotient and simplify your answer.

131. $f(x) = 2x^2 + 3x - 1$, $\dfrac{f(x + h) - f(x)}{h}$, $h \ne 0$

132. $f(x) = x^3 - 5x^2 + x$, $\dfrac{f(x + h) - f(x)}{h}$, $h \ne 0$

133. ***Physics*** The velocity of a ball thrown vertically upward from ground level is $v(t) = -32t + 48$, where t is the time in seconds and v is the velocity in feet per second.

(a) Find the velocity when $t = 1$.

(b) Find the time when the ball reaches its maximum height. [*Hint:* Find the time when $v(t) = 0$.]

(c) Find the velocity when $t = 2$.

134. ***Total Cost*** A hand tool manufacturer produces a product for which the variable cost is $5.35 per unit and the fixed costs are $16,000. The company sells the product for $8.20 and can sell all that it produces.

(a) Find the total cost as a function of x, the number of units produced.

(b) Find the profit as a function of x.

135. ***Geometry*** A wire 24 inches long is to be cut into four pieces to form a rectangle with one side of length x.

(a) Write the area A of the rectangle as a function of x.

(b) Determine the domain of the function.

136. *Mixture Problem* From a full 50-liter container of a 40% concentration of acid, x liters is removed and replaced with 100% acid.

(a) Write the amount of acid in the final mixture as a function of x.

(b) Determine the domain and range of the function.

(c) Determine x if the final mixture is 50% acid.

P.6 In Exercises 137–140, use the Vertical Line Test to determine whether y is a function of x. To print an enlarged copy of the graph, go to the website *www.mathgraphs.com*.

137. $y = (x - 3)^2$

138. $y = -\frac{3}{5}x^3 - 2x + 1$

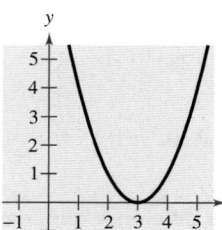
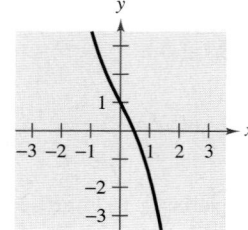

139. $x - 4 = y^2$

140. $x = -|4 - y|$

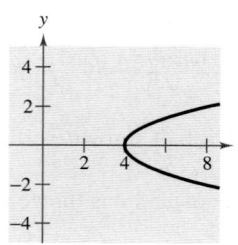
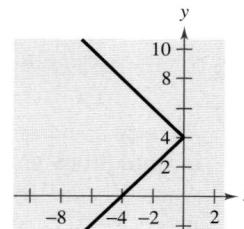

In Exercises 141–144, find the zeros of the function.

141. $f(x) = 5x^2 + 4x - 1$
142. $f(x) = \dfrac{8x + 3}{11 - x}$

143. $f(x) = 3x^2 - 16x + 21$

144. $f(x) = x^3 - x^2 - 25x + 25$

In Exercises 145 and 146, determine the intervals over which the function is increasing, decreasing, or constant.

145. $f(x) = |x| + |x + 1|$ **146.** $f(x) = (x^2 - 4)^2$

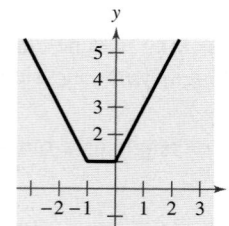
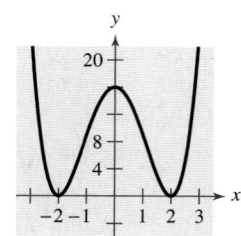

In Exercises 147–150, determine whether the function is even, odd, or neither.

147. $f(x) = x^5 + 4x - 7$ **148.** $f(x) = x^4 - 20x^2$

149. $f(x) = 2x\sqrt{x^2 + 3}$ **150.** $f(x) = \sqrt[5]{6x^2}$

P.7 In Exercises 151–154, write the linear function f such that it has the indicated function values. Sketch a graph of the function.

151. $f(2) = -6,\ f(-1) = 3$

152. $f(0) = -5,\ f(4) = -8$

153. $f\left(-\frac{4}{5}\right) = 2,\ f\left(\frac{11}{5}\right) = 7$

154. $f(3.3) = 5.6,\ f(-4.7) = -1.4$

In Exercises 155–164, graph the function.

155. $f(x) = 3 - x^2$ **156.** $h(x) = x^3 - 2$

157. $f(x) = -\sqrt{x}$ **158.** $f(x) = \sqrt{x + 1}$

159. $g(x) = \dfrac{3}{x}$ **160.** $g(x) = \dfrac{1}{x + 5}$

161. $f(x) = [\![x]\!] - 2$ **162.** $g(x) = [\![x + 4]\!]$

163. $f(x) = \begin{cases} 5x - 3, & x \geq -1 \\ -4x + 5, & x < -1 \end{cases}$

164. $f(x) = \begin{cases} x^2 - 2, & x < -2 \\ 5, & -2 \leq x \leq 0 \\ 8x - 5, & x > 0 \end{cases}$

In Exercises 165 and 166, identify the transformed common function shown in the graph.

165.

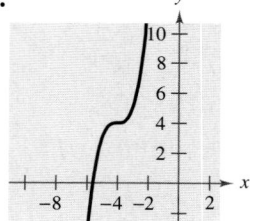

166.

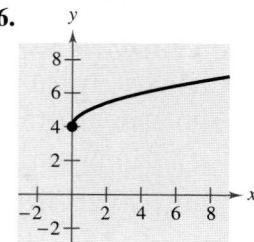

P.8 In Exercises 167–180, identify the transformation of the graph of f and sketch the graph of h.

167. $f(x) = x^2,$ $h(x) = x^2 - 9$

168. $f(x) = x^3,$ $h(x) = (x - 2)^3 + 2$

169. $f(x) = \sqrt{x},$ $h(x) = \sqrt{x - 7}$

170. $f(x) = |x|,$ $h(x) = |x + 3| - 5$

171. $f(x) = x^2,$ $h(x) = -(x + 3)^2 + 1$

172. $f(x) = x^3,$ $h(x) = -(x - 5)^3 - 5$

173. $f(x) = [\![x]\!],$ $h(x) = -[\![x]\!] + 6$

174. $f(x) = \sqrt{x}, \quad h(x) = -\sqrt{x+1} + 9$

175. $f(x) = |x|, \quad h(x) = -|-x + 4| + 6$

176. $f(x) = x^2, \quad h(x) = -(x + 1)^2 - 3$

177. $f(x) = [\![x]\!], \quad h(x) = 5[\![x - 9]\!]$

178. $f(x) = x^3, \quad h(x) = -\frac{1}{3}x^3$

179. $f(x) = \sqrt{x}, \quad h(x) = -2\sqrt{x-4}$

180. $f(x) = |x|, \quad h(x) = \frac{1}{2}|x| - 1$

P.9 In Exercises 181 and 182, find (a) $(f + g)(x)$, (b) $(f - g)(x)$, (c) $(fg)(x)$, and (d) $(f/g)(x)$. What is the domain of f/g?

181. $f(x) = x^2 + 3, \quad g(x) = 2x - 1$

182. $f(x) = x^2 - 4, \quad g(x) = \sqrt{3 - x}$

In Exercises 183 and 184, find (a) $f \circ g$ and (b) $g \circ f$. Find the domain of each function and each composite function.

183. $f(x) = \frac{1}{3}x - 3, \quad g(x) = 3x + 1$

184. $f(x) = x^3 - 4, \quad g(x) = \sqrt[3]{x + 7}$

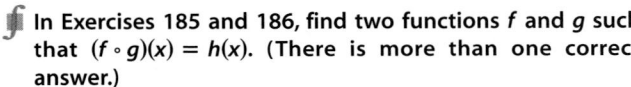 In Exercises 185 and 186, find two functions f and g such that $(f \circ g)(x) = h(x)$. (There is more than one correct answer.)

185. $h(x) = (6x - 5)^3$ **186.** $h(x) = \sqrt[3]{x + 2}$

Data Analysis In Exercises 187 and 188, use the table, which shows the total values (in billions of dollars) of U.S. imports from Mexico and Canada for the years 1995 through 1999. The variables y_1 and y_2 represent the total values of imports from Mexico and Canada, respectively. (Source: U.S. Census Bureau)

Year	y_1	y_2
1995	62.1	144.4
1996	74.3	155.9
1997	85.9	168.2
1998	94.6	173.3
1999	109.7	198.3

187. Use a graphing utility to find quadratic models for y_1 and y_2. Let $t = 5$ represent 1995.

188. Use a graphing utility to graph y_1, y_2, and $y_1 + y_2$ in the same viewing window. Use the model to estimate the total value of U.S. imports from Canada and Mexico in 2005.

P.10 In Exercises 189 and 190, find the inverse function of f informally. Verify that $f(f^{-1}(x)) = x = f^{-1}(f(x))$.

189. $f(x) = x - 7$ **190.** $f(x) = x + 5$

In Exercises 191 and 192, determine whether the function has an inverse function.

191. 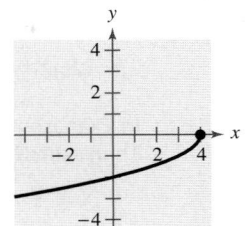   **192.**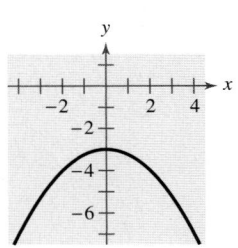

In Exercises 193–196, use the Horizontal Line Test to determine if the function is one-to-one and so has an inverse function.

193. $f(x) = 4 - \frac{1}{3}x$ **194.** $f(x) = (x - 1)^2$

195. $h(t) = \dfrac{2}{t - 3}$ **196.** $g(x) = \sqrt{x + 6}$

In Exercises 197–200, (a) find f^{-1}, (b) sketch the graphs of f and f^{-1} on the same coordinate system, and (c) verify that $f^{-1}(f(x)) = x = f(f^{-1}(x))$.

197. $f(x) = \frac{1}{2}x - 3$ **198.** $f(x) = 5x - 7$

199. $f(x) = \sqrt{x + 1}$ **200.** $f(x) = x^3 + 2$

In Exercises 201 and 202, restrict the domain of the function f to an interval over which the function is increasing, and determine f^{-1} over that interval.

201. $f(x) = 2(x - 4)^2$ **202.** $f(x) = |x - 2|$

Synthesis

True or False? In Exercises 203 and 204, determine whether the statement is true or false. Justify your answer.

203. Relative to the graph of $f(x) = \sqrt{x}$, the function $h(x) = -\sqrt{x + 9} - 13$ is shifted nine units to the left and 13 units downward, then reflected in the x-axis.

204. If f and g are two inverse functions, then the domain of g is equal to the range of f.

205. *Writing* Explain how to tell whether a relation between two variables is a function.

Chapter Test

Take this test as you would take a test in class. When you are finished, check your work against the answers given in the back of the book.

1. Place $<$ or $>$ between the real numbers $-\frac{10}{3}$ and $-|-4|$.

2. Find the distance between the real numbers -5.4 and $3\frac{3}{4}$.

3. Identify the rule of algebra illustrated by $(5 - x) + 0 = 5 - x$.

In Exercises 4 and 5, check for symmetry with respect to both axes and the origin. Then sketch the graph of the equation. Identify any x- and y-intercepts.

4. $y = 4 - |x|$ 5. $y = x^2 - 1$

6. Find the center and radius of the circle given by $(x - 3)^2 + y^2 = 9$. Then sketch its graph.

In Exercises 7–10, solve the equation (if possible).

7. $\frac{2}{3}(x - 1) + \frac{1}{4}x = 10$ 8. $(x - 3)(x + 2) = 14$

9. $2\sqrt{x} - \sqrt{2x + 1} = 1$ 10. $|3x - 1| = 7$

11. Find an equation of the line that passes through the point $(3, 8)$ and is (a) parallel to and (b) perpendicular to the line $-4x + 7y = -5$.

12. Evaluate the function $f(x) = |x + 2| - 15$ at each specified value.
    (a) $f(-8)$ (b) $f(14)$ (c) $f(x - 6)$

13. Determine the domain of $f(x) = \sqrt{100 - x^2}$.

 In Exercises 14–16, (a) use a graphing utility to graph the function, (b) approximate the intervals over which the function is increasing, decreasing, or constant, and (c) determine whether the function is even, odd, or neither.

14. $f(x) = 2x^6 + 5x^4 - x^2$ 15. $f(x) = 4x\sqrt{3 - x}$ 16. $f(x) = |x + 5|$

17. Sketch the graph of $f(x) = \begin{cases} 3x + 7, & x \le -3 \\ 4x^2 - 1, & x > -3 \end{cases}$.

In Exercises 18–20, identify the common function in the transformation. Then sketch a graph of the function.

18. $h(x) = -x^3 - 7$ 19. $h(x) = -\sqrt{x + 5} + 8$ 20. $h(x) = \frac{1}{4}|x + 1| - 3$

In Exercises 21 and 22, find (a) $(f + g)(x)$, (b) $(f - g)(x)$, (c) $(fg)(x)$, (d) $(f/g)(x)$, (e) $(f \circ g)(x)$, and (f) $(g \circ f)(x)$.

21. $f(x) = 3x^2 - 7, g(x) = -x^2 - 4x + 5$ 22. $f(x) = 1/x, g(x) = 2\sqrt{x}$

In Exercises 23–25, determine whether the function has an inverse function, and if so, find the inverse function.

23. $f(x) = x^3 + 8$ 24. $f(x) = |x^2 - 3| + 6$ 25. $f(x) = 3x\sqrt{x}$

26. It costs a company $58 to produce six units of a product and $78 to produce 10 units. How much does it cost to produce 25 units, assuming that the cost function is linear?

Proofs in Mathematics

What does the word *proof* mean to you? In mathematics, the word *proof* is used to mean simply a valid argument. When you are proving a statement or theorem, you must use facts, definitions, and accepted properties in a logical order. You can also use previously proved theorems in your proof. For instance, the Distance Formula is used in the proof of the Midpoint Formula below. There are several different proof methods, which you will see in later chapters.

The Midpoint Formula   *(p. 28)*

The midpoint of the segment joining the points (x_1, y_1) and (x_2, y_2) is given by the Midpoint Formula

$$\text{Midpoint} = \left(\frac{x_1 + x_2}{2}, \frac{y_1 + y_2}{2} \right).$$

The Cartesian Plane

The Cartesian Plane was named after the French mathematician René Descartes (1596–1650). While Descartes was lying in bed sick, he noticed a fly buzzing around on the square ceiling tiles. He discovered that the position of the fly could be described by which ceiling tile the fly landed on. This led to the development of the Cartesian Plane. Descartes felt that a coordinate plane could be used to facilitate description of the positions of objects.

Proof

Using the figure, you must show that $d_1 = d_2$ and $d_1 + d_2 = d_3$.

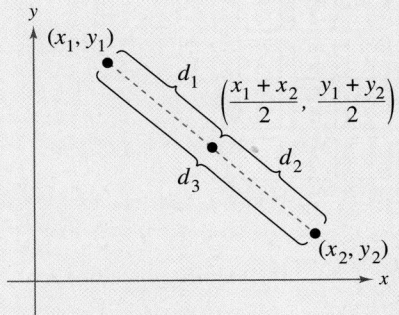

By the Distance Formula, you obtain

$$d_1 = \sqrt{\left(\frac{x_1 + x_2}{2} - x_1 \right)^2 + \left(\frac{y_1 + y_2}{2} - y_1 \right)^2}$$

$$= \frac{1}{2}\sqrt{(x_2 - x_1)^2 + (y_2 - y_1)^2}$$

$$d_2 = \sqrt{\left(x_2 - \frac{x_1 + x_2}{2} \right)^2 + \left(y_2 - \frac{y_1 + y_2}{2} \right)^2}$$

$$= \frac{1}{2}\sqrt{(x_2 - x_1)^2 + (y_2 - y_1)^2}$$

$$d_3 = \sqrt{(x_2 - x_1)^2 + (y_2 - y_1)^2}$$

So, it follows that $d_1 = d_2$ and $d_1 + d_2 = d_3$.

1. As a salesperson, you receive a monthly salary of $2000, plus a commission of 7% of sales. You are offered a new job at $2300 per month, plus a commission of 5% of sales.

   (a) Write a linear equation for your current monthly wage W_1 in terms of your monthly sales S.

   (b) Write a linear equation for the monthly wage W_2 of your new job offer in terms of the monthly sales S.

   (c) Use a graphing utility to graph both equations in the same viewing window. Find the point of intersection. What does it signify?

   (d) You think you can sell $20,000 per month. Should you change jobs? Explain.

2. For the numbers 2 through 9 on a telephone keypad (see figure), create two relations: one mapping numbers onto letters, and the other mapping letters onto numbers. Are both relations functions? Explain.

3. What can be said about the sum and difference of each of the following?

   (a) Two even functions

   (b) Two odd functions

   (c) An odd function and an even function

4. The two functions

   $f(x) = x$ and $g(x) = -x$

   are their own inverse functions. Graph each function and explain why this is true. Graph other linear functions that are their own inverse functions. Find a general formula for a family of linear functions that are their own inverse functions.

5. Prove that a function of the following form is even.

   $$y = a_{2n}x^{2n} + a_{2n-2}x^{2n-2} + \cdots + a_2 x^2 + a_0$$

6. A miniature golf professional is trying to make a hole-in-one on the miniature golf green shown. A coordinate plane is placed over the golf green. The golf ball is at the point $(2.5, 2)$ and the hole is at the point $(9.5, 2)$. The professional wants to bank the ball off the side wall of the green at the point (x, y). Find the coordinates of the point (x, y). Then write an equation for the path of the ball.

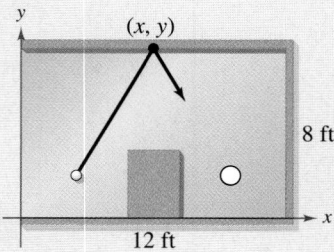

7. At 2:00 P.M. on April 11, 1912, the *Titanic* left Cobh, Ireland, on her voyage to New York City. At 11:40 P.M. on April 14, the *Titanic* struck an iceberg and sank, having covered only about 2100 miles of the approximately 3400-mile trip.

   (a) What was the total length of the *Titanic's* voyage in hours?

   (b) What was the *Titanic's* average speed in miles per hour?

   (c) Write a function relating the *Titanic's* distance from New York City and the number of hours traveled. Find the domain and range of the function.

   (d) Graph the function from part (c).

8. Consider the functions $f(x) = 4x$ and $g(x) = x + 6$.

   (a) Find $(f \circ g)(x)$.

   (b) Find $(f \circ g)^{-1}(x)$.

   (c) Find $f^{-1}(x)$ and $g^{-1}(x)$.

   (d) Find $(g^{-1} \circ f^{-1})(x)$ and compare the result with that of part (b).

   (e) Repeat parts (a) through (d) for $f(x) = x^3 + 1$ and $g(x) = 2x$.

   (f) Write two one-to-one functions f and g, and repeat parts (a) through (d) for these functions.

   (g) Make a conjecture about $(f \circ g)^{-1}(x)$ and $(g^{-1} \circ f^{-1})(x)$.

9. You are in a boat 2 miles from the nearest point on the coast. You are to travel to a point Q, 3 miles down the coast and 1 mile inland (see figure). You can row at 2 miles per hour and walk at 4 miles per hour.

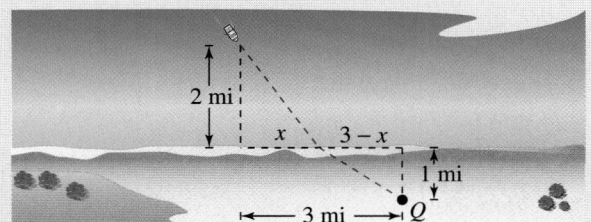

(a) Write the total time T of the trip as a function of x.

(b) Determine the domain of the function.

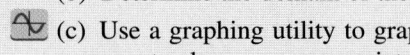 (c) Use a graphing utility to graph the function. Be sure to choose an appropriate viewing window.

(d) Use the *zoom* and *trace* features to find the value of x that minimizes T.

(e) Write a brief paragraph interpreting these values.

10. The Heaviside function $H(x)$ is widely used in engineering applications. (See figure.) To print an enlarged copy of the graph, go to the website *www.mathgraphs.com*.

$$H(x) = \begin{cases} 1, & x \geq 0 \\ 0, & x < 0 \end{cases}$$

Sketch the graph of each function by hand.

(a) $H(x) - 2$ (b) $H(x - 2)$ (c) $-H(x)$

(d) $H(-x)$ (e) $\frac{1}{2}H(x)$ (f) $-H(x - 2) + 2$

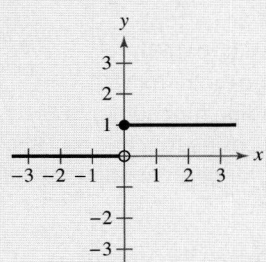

11. Let $f(x) = \dfrac{1}{1 - x}$.

(a) What are the domain and range of f?

(b) Find $f(f(x))$. What is the domain of this function?

(c) Find $f(f(f(x)))$. Is the graph a line? Why or why not?

12. Show that the Associative Property holds for compositions of functions—that is,

$$(f \circ (g \circ h))(x) = ((f \circ g) \circ h)(x).$$

13. Consider the graph of the function f shown in the figure. Use this graph to sketch the graph of each function. To print an enlarged copy of the graph, go to the website *www.mathgraphs.com*.

(a) $f(x + 1)$ (b) $f(x) + 1$ (c) $2f(x)$ (d) $f(-x)$

(e) $-f(x)$ (f) $|f(x)|$ (g) $f(|x|)$

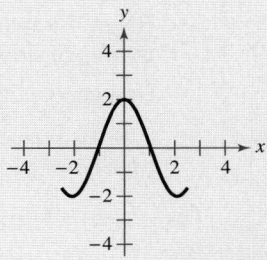

14. Use the graphs of f and f^{-1} to complete each table of function values.

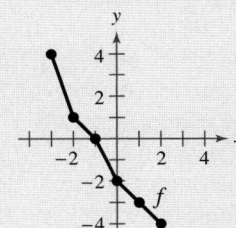

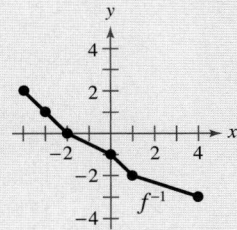

(a)

x	$f(f^{-1}(x))$
-4	
-2	
0	
4	

(b)

x	$(f + f^{-1})(x)$
-3	
-2	
0	
1	

(c)

x	$(f \cdot f^{-1})(x)$
-3	
-2	
0	
1	

(d)

x	$|f^{-1}(x)|$
-4	
-3	
0	
4	

123

How to study Chapter 1

▶ **What you should learn**

In this chapter you will learn the following skills and concepts:

- How to describe an angle and convert between radian and degree measure
- How to identify a unit circle and its relationship to real numbers
- How to evaluate trigonometric functions of any angle
- How to use the fundamental trigonometric identities
- How to sketch the graphs of trigonometric functions and translations of graphs of sine and cosine functions
- How to evaluate the inverse trigonometric functions
- How to evaluate the compositions of trigonometric functions and inverse trigonometric functions

▶ **Important Vocabulary**

As you encounter each new vocabulary term in this chapter, add the term and its definition to your notebook glossary.

Trigonometry (p. 126)
Angle (p. 126)
Initial side (of an angle) (p. 126)
Terminal side (of an angle) (p. 126)
Vertex (of an angle) (p. 126)
Standard position (p. 126)
Coterminal angles (p. 126)
Central angle (p. 127)
Radian (p. 127)
Acute angles (p. 127)
Obtuse angles (p. 127)
Complementary angles (p. 129)
Supplementary angles (p. 129)
Degree (p. 129)
Linear speed (p. 131)
Angular speed (p. 131)
Unit circle (p. 137)

Sine (pp. 138, 144)
Cosecant (pp. 138, 144)
Cosine (pp. 138, 144)
Secant (pp. 138, 144)
Tangent (pp. 138, 144)
Cotangent (pp. 138, 144)
Period (pp. 140, 167)
Angle of elevation (p. 149)
Angle of depression (p. 149)
Reference angles (p. 157)
Amplitude (p. 166)
Phase shift (p. 168)
Damping factor (p. 180)
Inverse sine function (p. 186)
Inverse cosine function (p. 188)
Inverse tangent function (p. 188)

Study Tools

Learning objectives in each section
Chapter Summary (p. 207)
Review Exercises (pp. 208–211)
Chapter Test (p. 212)

Additional Resources

Study and Solutions Guide
Interactive Trigonometry
Videotapes/DVD for Chapter 1
Trigonometry Website
Student Success Organizer

Raymond Forbes/SuperStock

1

Trigonometry

1.1 Radian and Degree Measure

▶ What you should learn
- How to describe angles
- How to use radian measure
- How to use degree measure
- How to use angles to model and solve real-life problems

▶ Why you should learn it

You can use angles to model and solve real-life problems. For instance, in Exercise 101 on page 136, you are asked to use angles to find the speed of a bicycle.

Pablo River/SuperStock

Angles

As derived from the Greek language, the word **trigonometry** means "measurement of triangles." Initially, trigonometry dealt with relationships among the sides and angles of triangles and was used in the development of astronomy, navigation, and surveying. With the development of calculus and the physical sciences in the 17th century, a different perspective arose—one that viewed the classic trigonometric relationships as *functions* with the set of real numbers as their domains. Consequently, the applications of trigonometry expanded to include a vast number of physical phenomena involving rotations and vibrations. These phenomena include sound waves, light rays, planetary orbits, vibrating strings, pendulums, and orbits of atomic particles.

The approach in this text incorporates *both* perspectives, starting with angles and their measure.

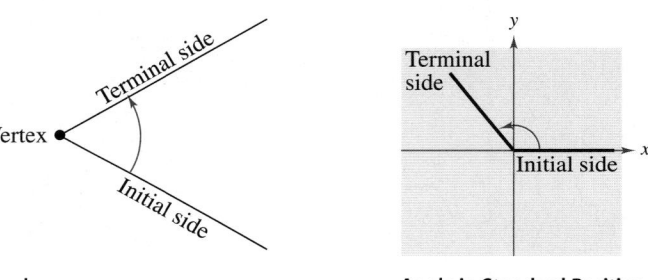

Angle
FIGURE 1.1

Angle in Standard Position
FIGURE 1.2

An **angle** is determined by rotating a ray (half-line) about its endpoint. The starting position of the ray is the **initial side** of the angle, and the position after rotation is the **terminal side,** as shown in Figure 1.1. The endpoint of the ray is the **vertex** of the angle. This perception of an angle fits a coordinate system in which the origin is the vertex and the initial side coincides with the positive *x*-axis. Such an angle is in **standard position,** as shown in Figure 1.2. **Positive angles** are generated by counterclockwise rotation, and **negative angles** by clockwise rotation, as shown in Figure 1.3. Angles are labeled with Greek letters α (alpha), β (beta), and θ (theta), as well as uppercase letters A, B, and C. In Figure 1.4, note that angles α and β have the same initial and terminal sides. Such angles are **coterminal.**

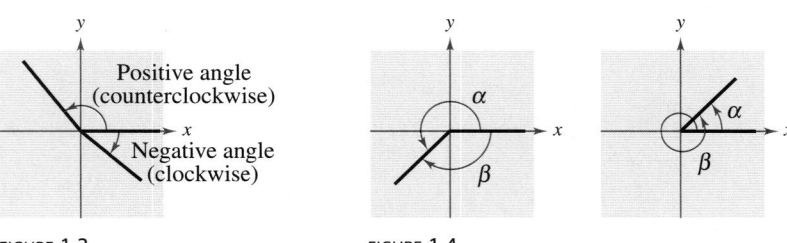

FIGURE 1.3

FIGURE 1.4

Radian Measure

The **measure of an angle** is determined by the amount of rotation from the initial side to the terminal side. One way to measure angles is in *radians*. This type of measure is especially useful in calculus. To define a radian, you can use a **central angle** of a circle, one whose vertex is the center of the circle, as shown in Figure 1.5.

> ### Definition of Radian
>
> One **radian** is the measure of a central angle θ that intercepts an arc s equal in length to the radius r of the circle. See Figure 1.5.

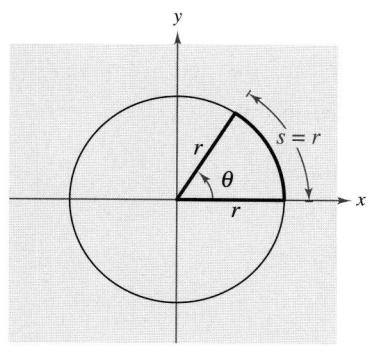

Arc length = radius when θ = 1 radian

FIGURE **1.5**

Because the circumference of a circle is $2\pi r$ units, it follows that a central angle of one full revolution (counterclockwise) corresponds to an arc length of

$$s = 2\pi r.$$

Moreover, because $2\pi \approx 6.28$, there are just over six radius lengths in a full circle, as shown in Figure 1.6. In general, the radian measure of a central angle θ is obtained by dividing the arc length s by r. That is, $s/r = \theta$, where θ is *measured in radians*. Because the units of measure for s and r are the same, this ratio has no units—it is simply a real number.

Because the radian measure of an angle of one full revolution is 2π, you can obtain the following.

$$\frac{1}{2}\text{ revolution} = \frac{2\pi}{2} = \pi \text{ radians}$$

$$\frac{1}{4}\text{ revolution} = \frac{2\pi}{4} = \frac{\pi}{2} \text{ radians}$$

$$\frac{1}{6}\text{ revolution} = \frac{2\pi}{6} = \frac{\pi}{3} \text{ radians}$$

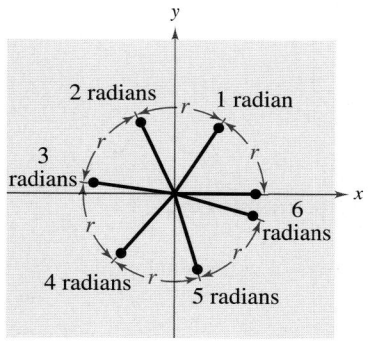

FIGURE **1.6**

These and other common angles are shown in Figure 1.7.

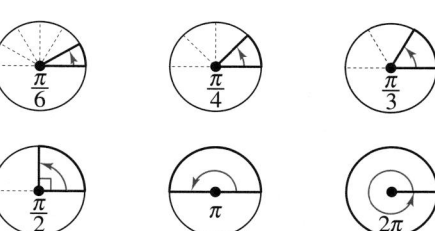

FIGURE **1.7**

Recall that the four quadrants in a coordinate system are numbered I, II, III, and IV. Figure 1.8 on page 128 shows which angles between 0 and 2π lie in each of the four quadrants. Note that angles between 0 and $\pi/2$ are **acute** angles and angles between $\pi/2$ and π are **obtuse** angles.

STUDY TIP

One revolution around a circle of radius r corresponds to an angle of 2π radians because

$$\theta = \frac{s}{r} = \frac{2\pi r}{r} = 2\pi \text{ radians}.$$

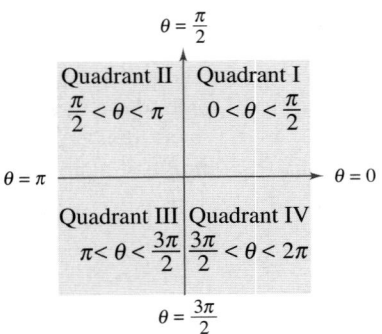

FIGURE 1.8

Two angles are coterminal if they have the same initial and terminal sides. For instance, the angles 0 and 2π are coterminal, as are the angles $\pi/6$ and $13\pi/6$. You can find an angle that is coterminal to a given angle θ by adding or subtracting 2π (one revolution), as demonstrated in Example 1. A given angle θ has infinitely many coterminal angles. For instance, $\theta = \pi/6$ is coterminal with

$$\frac{\pi}{6} + 2n\pi$$

where n is an integer.

Example 1 ▶ **Sketching and Finding Coterminal Angles**

a. For the positive angle $13\pi/6$, subtract 2π to obtain a coterminal angle

$$\frac{13\pi}{6} - 2\pi = \frac{\pi}{6}.$$ See Figure 1.9.

b. For the positive angle $3\pi/4$, subtract 2π to obtain a coterminal angle

$$\frac{3\pi}{4} - 2\pi = -\frac{5\pi}{4}.$$ See Figure 1.10.

Remind your students to work in radians.

c. For the negative angle $-2\pi/3$, add 2π to obtain a coterminal angle

$$-\frac{2\pi}{3} + 2\pi = \frac{4\pi}{3}.$$ See Figure 1.11.

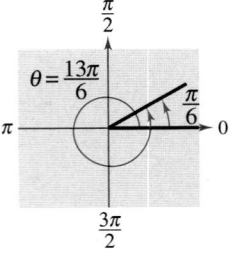

FIGURE 1.9

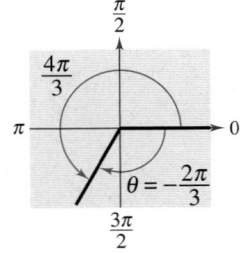

FIGURE 1.10

FIGURE 1.11

The icon identifies examples and concepts related to features of the Learning Tools CD-ROM and the *Interactive* and *Internet* versions of this text. For more details see the chart on pages *xix-xxiii*.

You might point out that complementary and supplementary angles do not necessarily share a common side. For example, the acute angles of a right triangle are complementary because the sum of their measures is $\pi/2$.

Two positive angles α and β are **complementary** (complements of each other) if their sum is $\pi/2$. Two positive angles are **supplementary** (supplements of each other) if their sum is π. See Figure 1.12.

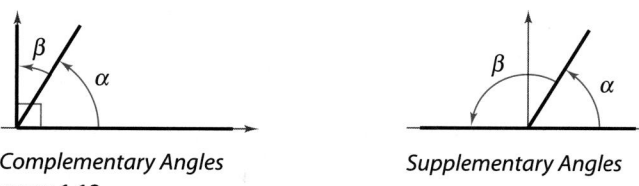

Complementary Angles   *Supplementary Angles*
FIGURE **1.12**

Example 2 ▶ **Complementary and Supplementary Angles**

The *Interactive* CD-ROM and *Internet* versions of this text offer a Try It for each example in the text.

If possible, find the complement and the supplement of (a) $2\pi/5$ and (b) $4\pi/5$.

Solution

a. The complement of $2\pi/5$ is

$$\frac{\pi}{2} - \frac{2\pi}{5} = \frac{5\pi}{10} - \frac{4\pi}{10} = \frac{\pi}{10}.$$

The supplement of $2\pi/5$ is

$$\pi - \frac{2\pi}{5} = \frac{5\pi}{5} - \frac{2\pi}{5} = \frac{3\pi}{5}.$$

b. Because $4\pi/5$ is greater than $\pi/2$, it has no complement. (Remember to use only positive angles for complements.) The supplement is

$$\pi - \frac{4\pi}{5} = \frac{5\pi}{5} - \frac{4\pi}{5} = \frac{\pi}{5}.$$

Degree Measure

A second way to measure angles is in terms of **degrees,** denoted by the symbol °. A measure of one degree (1°) is equivalent to a rotation of $\frac{1}{360}$ of a complete revolution about the vertex. To measure angles, it is convenient to mark degrees on the circumference of a circle, as shown in Figure 1.13. So, a full revolution (counterclockwise) corresponds to 360°, a half revolution to 180°, a quarter revolution to 90°, and so on.

Because 2π radians corresponds to one complete revolution, degrees and radians are related by the equations

$$360° = 2\pi \text{ rad} \qquad \text{and} \qquad 180° = \pi \text{ rad}.$$

From the latter equation, you obtain

$$1° = \frac{\pi}{180} \text{ rad} \qquad \text{and} \qquad 1 \text{ rad} = \left(\frac{180°}{\pi}\right)$$

which lead to the conversion rules at the top of the next page.

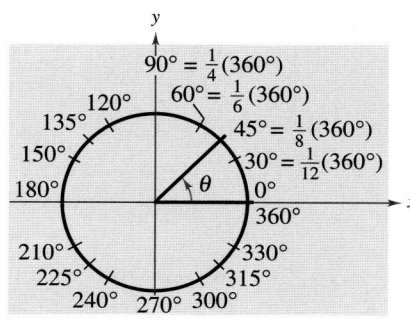

FIGURE **1.13**

Converting from degrees to radians and vice versa should help your students become familiar with radian measure.

Conversions Between Degrees and Radians

1. To convert degrees to radians, multiply degrees by $\dfrac{\pi \text{ rad}}{180°}$.

2. To convert radians to degrees, multiply radians by $\dfrac{180°}{\pi \text{ rad}}$.

To apply these two conversion rules, use the basic relationship $\pi \text{ rad} = 180°$. (See Figure 1.14.)

The *Interactive* CD-ROM and *Internet* versions of this text offer a Quiz for every section of the text.

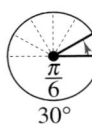
$\dfrac{\pi}{6}$
30°

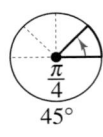
$\dfrac{\pi}{4}$
45°

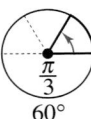
$\dfrac{\pi}{3}$
60°

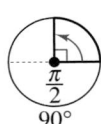
$\dfrac{\pi}{2}$
90°

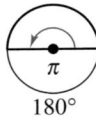
π
180°

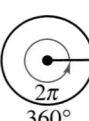
2π
360°

FIGURE **1.14**

When no units of angle measure are specified, *radian measure is implied*. For instance, if you write $\theta = \pi$ or $\theta = 2$, you should mean $\theta = \pi$ radians or $\theta = 2$ radians.

Technology

With calculators it is convenient to use *decimal* degrees to denote fractional parts of degrees. Historically, however, fractional parts of degrees were expressed in *minutes* and *seconds*, using the prime (′) and double prime (″) notations, respectively. That is,

$1' = $ one minute $= \frac{1}{60}(1°)$

$1'' = $ one second $= \frac{1}{3600}(1°)$

Consequently, an angle of 64 degrees, 32 minutes, and 47 seconds is represented by $\theta = 64° \, 32' \, 47''$.

Many calculators have special keys for converting an angle in degrees, minutes, and seconds (D° M′ S″) into decimal degree form, and vice versa.

Example 3 ▶ **Converting from Degrees to Radians**

a. $135° = (135 \text{ deg})\left(\dfrac{\pi \text{ rad}}{180 \text{ deg}}\right) = \dfrac{3\pi}{4}$ radians   Multiply by $\pi/180$.

b. $540° = (540 \text{ deg})\left(\dfrac{\pi \text{ rad}}{180 \text{ deg}}\right) = 3\pi$ radians   Multiply by $\pi/180$.

c. $-270° = (-270 \text{ deg})\left(\dfrac{\pi \text{ rad}}{180 \text{ deg}}\right) = -\dfrac{3\pi}{2}$ radians   Multiply by $\pi/180$.

Example 4 ▶ **Converting from Radians to Degrees**

a. $-\dfrac{\pi}{2} \text{ rad} = \left(-\dfrac{\pi}{2} \text{ rad}\right)\left(\dfrac{180 \text{ deg}}{\pi \text{ rad}}\right) = -90°$ Multiply by $180/\pi$.

b. $\dfrac{9\pi}{2} \text{ rad} = \left(\dfrac{9\pi}{2} \text{ rad}\right)\left(\dfrac{180 \text{ deg}}{\pi \text{ rad}}\right) = 810°$ Multiply by $180/\pi$.

c. $2 \text{ rad} = (2 \text{ rad})\left(\dfrac{180 \text{ deg}}{\pi \text{ rad}}\right) = \dfrac{360°}{\pi} \approx 114.59°$ Multiply by $180/\pi$.

If you have a calculator with a "radian-to-degree" conversion key, try using it to verify the result shown in part (c) of Example 4.

Applications

The *radian measure* formula, $\theta = s/r$, can be used to measure arc length along a circle. Specifically, for a circle of radius r, a central angle θ intercepts an arc of length s given by

$$s = r\theta \qquad \text{Length of circular arc}$$

where θ is measured in radians.

Example 5 ▶ **Finding Arc Length**

A circle has a radius of 4 inches. Find the length of the arc intercepted by a central angle of 240°, as shown in Figure 1.15.

Solution

To use the formula $s = r\theta$, first convert 240° to radian measure.

$$240° = (240 \text{ deg})\left(\frac{\pi \text{ rad}}{180 \text{ deg}}\right) \qquad \text{Convert from degrees to radians.}$$

$$= \frac{4\pi}{3} \text{ radians} \qquad \text{Simplify.}$$

Then, using a radius of $r = 4$ inches, you can find the arc length to be

$$s = r\theta \qquad \text{Length of circular arc}$$

$$= 4\left(\frac{4\pi}{3}\right) \qquad \text{Substitute for } r \text{ and } \theta.$$

$$= \frac{16\pi}{3} \approx 16.76 \text{ inches.} \qquad \text{Simplify.}$$

Note that the units for $r\theta$ are determined by the units for r because θ is given in radian measure and so has no units.

The formula for the length of a circular arc can be used to analyze the motion of a particle moving at a *constant speed* along a circular path.

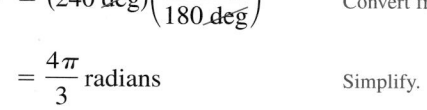
FIGURE 1.15

Because radian measure is so often used, you may want to encourage your students to be as familiar with the radian measure of angles as they are with degree measure. Measuring arc length along a circle is one of many applications in which radian measure is used.

Linear and Angular Speed

Consider a particle moving at a constant speed along a circular arc of radius r. If s is the length of the arc traveled in time t, then the **linear speed** of the particle is

$$\text{Linear speed} = \frac{\text{arc length}}{\text{time}} = \frac{s}{t}.$$

Moreover, if θ is the angle (in radian measure) corresponding to the arc length s, then the **angular speed** of the particle is

$$\text{Angular speed} = \frac{\text{central angle}}{\text{time}} = \frac{\theta}{t}.$$

STUDY TIP

Linear speed measures how fast the particle moves, and angular speed measures how fast the angle changes.

FIGURE **1.16**

 Example 6 ▶ **Finding Linear Speed**

The second hand of a clock is 10.2 centimeters long, as shown in Figure 1.16. Find the linear speed of the tip of this second hand as it passes around the clock face.

Solution

In one revolution, the arc length traveled is

$$s = 2\pi r$$
$$= 2\pi(10.2) \qquad \text{Substitute for } r.$$
$$= 20.4\pi \text{ centimeters.}$$

The time required for the second hand to travel this distance is

$$t = 1 \text{ minute} = 60 \text{ seconds.}$$

So, the linear speed of the tip of the second hand is

$$\text{Linear speed} = \frac{s}{t}$$
$$= \frac{20.4\pi \text{ centimeters}}{60 \text{ seconds}}$$
$$\approx 1.068 \text{ centimeters per second.}$$

Example 7 ▶ **Finding Angular and Linear Speed**

A lawn roller with a 10-inch radius (see Figure 1.17) makes 1.2 revolutions per second.

a. Find the angular speed of the roller in radians per second.

b. Find the speed of the tractor that is pulling the roller.

Solution

a. Because each revolution generates 2π radians, it follows that the roller turns $(1.2)(2\pi) = 2.4\pi$ radians per second. In other words, the angular speed is

$$\text{Angular speed} = \frac{\theta}{t}$$
$$= \frac{2.4\pi \text{ radians}}{1 \text{ second}} = 2.4\pi \text{ radians per second.}$$

b. The linear speed (speed of the tractor) is

$$\text{Linear speed} = \frac{s}{t}$$
$$= \frac{r\theta}{t}$$
$$= \frac{10(2.4\pi) \text{ inches}}{1 \text{ second}} \approx 75.4 \text{ inches per second.}$$

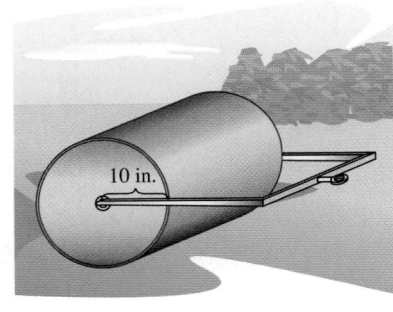

FIGURE **1.17**

Activities

1. Find the supplement of the angle $\theta = \dfrac{5\pi}{7}$.

   Answer: $\dfrac{2\pi}{7}$

2. Convert $60°$ from degrees to radians.

   Answer: $\dfrac{\pi}{3}$

3. On a circle with a radius of 9 inches, find the length of the arc intercepted by a central angle of $140°$.

   Answer: $7\pi \approx 22$ inches

1.1 Exercises

The *Interactive* CD-ROM and *Internet* versions of this text contain step-by-step solutions to all odd-numbered exercises. They also provide Tutorial Exercises for additional help.

In Exercises 1–6, estimate the angle to the nearest one-half radian.

1.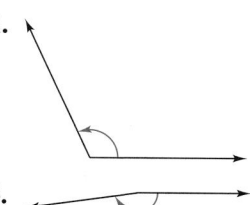

2.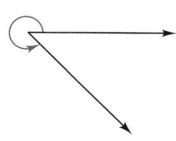

3.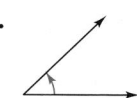

4.

5.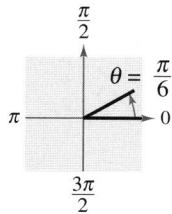

6.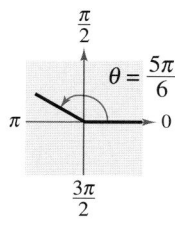

In Exercises 7–12, determine the quadrant in which each angle lies. (The angle measure is given in radians.)

7. (a) $\dfrac{\pi}{5}$ (b) $\dfrac{7\pi}{5}$ **8.** (a) $\dfrac{11\pi}{8}$ (b) $\dfrac{9\pi}{8}$

9. (a) $-\dfrac{\pi}{12}$ (b) -2 **10.** (a) -1 (b) $-\dfrac{11\pi}{9}$

11. (a) 3.5 (b) 2.25 **12.** (a) 6.02 (b) -4.25

In Exercises 13–16, sketch each angle in standard position.

13. (a) $\dfrac{5\pi}{4}$ (b) $-\dfrac{2\pi}{3}$ **14.** (a) $-\dfrac{7\pi}{4}$ (b) $\dfrac{5\pi}{2}$

15. (a) $\dfrac{11\pi}{6}$ (b) -3 **16.** (a) 4 (b) 7π

In Exercises 17–20, determine two coterminal angles (one positive and one negative) for each angle. Give your answers in radians.

17. (a) 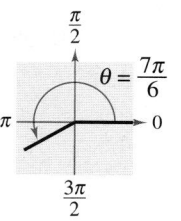 (b)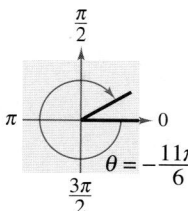

18. (a)

$\theta = \dfrac{7\pi}{6}$

(b)

$\theta = -\dfrac{11\pi}{6}$

19. (a) $\theta = \dfrac{2\pi}{3}$ (b) $\theta = \dfrac{\pi}{12}$

20. (a) $\theta = -\dfrac{9\pi}{4}$ (b) $\theta = -\dfrac{2\pi}{15}$

In Exercises 21–24, find (if possible) the complement and supplement of each angle.

21. (a) $\dfrac{\pi}{3}$ (b) $\dfrac{3\pi}{4}$ **22.** (a) $\dfrac{\pi}{12}$ (b) $\dfrac{11\pi}{12}$

23. (a) 1 (b) 2 **24.** (a) 3 (b) 1.5

In Exercises 25–28, write each angle in radian measure as a multiple of π. (Do not use a calculator.)

25. (a) $30°$ (b) $150°$

26. (a) $315°$ (b) $120°$

27. (a) $-20°$ (b) $-240°$

28. (a) $-270°$ (b) $144°$

In Exercises 29–36, convert the measure from degrees to radians. Round to three decimal places.

29. $115°$ **30.** $87.4°$

31. $-216.35°$ **32.** $-48.27°$

33. $532°$ **34.** $345°$

35. $-0.83°$ **36.** $0.54°$

In Exercises 37–40, write each angle in degree measure. (Do not use a calculator.)

37. (a) $\dfrac{3\pi}{2}$ (b) $\dfrac{7\pi}{6}$

38. (a) $-\dfrac{7\pi}{12}$ (b) $\dfrac{\pi}{9}$

39. (a) $\dfrac{7\pi}{3}$ (b) $-\dfrac{11\pi}{30}$

40. (a) $\dfrac{11\pi}{6}$ (b) $\dfrac{34\pi}{15}$

In Exercises 41–48, convert the measure from radians to degrees. Round to three decimal places.

41. $\dfrac{\pi}{7}$

42. $\dfrac{5\pi}{11}$

43. $\dfrac{15\pi}{8}$

44. $\dfrac{13\pi}{2}$

45. -4.2π

46. 4.8π

47. -2

48. -0.57

In Exercises 49–54, estimate the number of degrees in the angle.

49.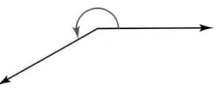

50.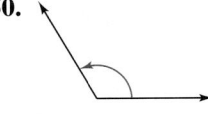

51.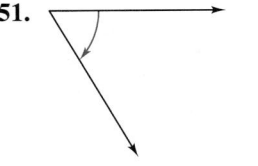

52.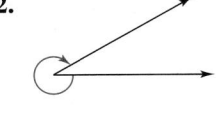

53.

54.

In Exercises 55–58, determine the quadrant in which each angle lies.

55. (a) $130°$ (b) $285°$

56. (a) $8.3°$ (b) $257° \, 30'$

57. (a) $-132° \, 50'$ (b) $-336°$

58. (a) $-260°$ (b) $-3.4°$

In Exercises 59–62, sketch each angle in standard position.

59. (a) $30°$ (b) $150°$

60. (a) $-270°$ (b) $-120°$

61. (a) $405°$ (b) $480°$

62. (a) $-750°$ (b) $-600°$

In Exercises 63–66, determine two coterminal angles (one positive and one negative) for each angle. Give your answers in degrees.

63. (a) 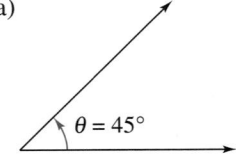   (b)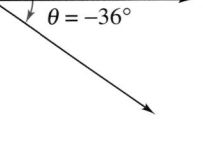

64. (a) 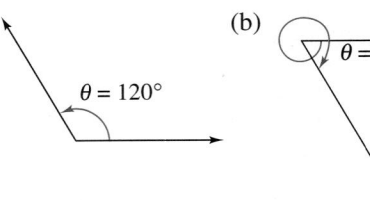 (b)

65. (a) 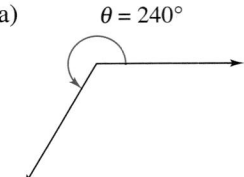 (b)

66. (a) $\theta = -420°$ (b) $\theta = 230°$

In Exercises 67–70, find (if possible) the complement and supplement of each angle.

67. (a) $18°$ (b) $115°$

68. (a) $3°$ (b) $64°$

69. (a) $79°$ (b) $150°$

70. (a) $130°$ (b) $170°$

In Exercises 71–74, convert each angle measure to decimal degree form.

71. (a) $54° \, 45'$ (b) $-128° \, 30'$

72. (a) $245° \, 10'$ (b) $2° \, 12'$

73. (a) $85° \, 18' \, 30''$ (b) $330° \, 25''$

74. (a) $-135° \, 36''$ (b) $-408° \, 16' \, 20''$

In Exercises 75–78, convert each angle measure to D° M′ S″ form.

75. (a) $240.6°$ (b) $-145.8°$

76. (a) $-345.12°$ (b) $0.45°$

77. (a) $2.5°$ (b) $-3.58°$

78. (a) $-0.355°$ (b) $0.7865°$

In Exercises 79–82, find the angle in radians.

79.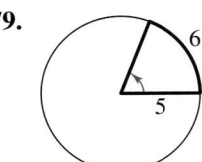

80.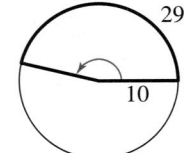

81.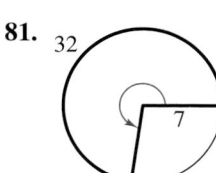

82.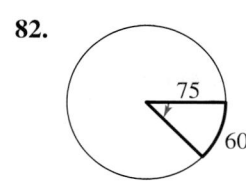

In Exercises 83–86, find the radian measure of the central angle of a circle of radius *r* that intercepts an arc of length *s*.

	Radius r	Arc Length s
83.	27 inches	6 inches
84.	14 feet	8 feet
85.	14.5 centimeters	25 centimeters
86.	80 kilometers	160 kilometers

In Exercises 87–90, find the length of the arc on a circle of radius *r* intercepted by a central angle *θ*.

	Radius r	Central Angle θ
87.	15 inches	180°
88.	9 feet	60°
89.	3 meters	1 radian
90.	20 centimeters	$\pi/4$ radian

Distance Between Cities **In Exercises 91–94, find the distance between the cities. Assume that Earth is a sphere of radius 4000 miles and that the cities are on the same longitude (one city is due north of the other).**

	City	Latitude
91.	Dallas, Texas	32° 47′ 9″ N
	Omaha, Nebraska	41° 15′ 42″ N
92.	San Francisco, California	37° 46′ 39″ N
	Seattle, Washington	47° 36′ 32″ N
93.	Miami, Florida	25° 46′ 37″ N
	Erie, Pennsylvania	42° 7′ 15″ N
94.	Johannesburg, South Africa	26° 10′ S
	Jerusalem, Israel	31° 47′ N

95. ***Difference in Latitudes*** Assuming that Earth is a sphere of radius 6378 kilometers, what is the difference in the latitudes of Syracuse, New York, and Annapolis, Maryland, where Syracuse is 450 kilometers due north of Annapolis?

96. ***Difference in Latitudes*** Assuming that Earth is a sphere of radius 6378 kilometers, what is the difference in the latitudes of Lynchburg, Virginia and Myrtle Beach, South Carolina, where Lynchburg is 400 kilometers due north of Myrtle Beach?

97. ***Instrumentation*** The pointer on a voltmeter is 6 centimeters in length (see figure). Find the angle through which the pointer rotates when it moves 2.5 centimeters on the scale.

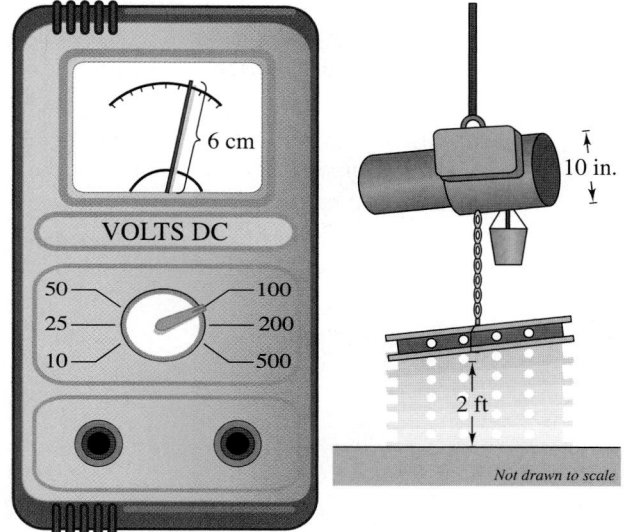

FIGURE FOR 97   FIGURE FOR 98

98. ***Electric Hoist*** An electric hoist is being used to lift a beam (see figure). The diameter of the drum on the hoist is 10 inches, and the beam must be raised 2 feet. Find the number of degrees through which the drum must rotate.

99. ***Angular Speed*** A car is moving at a rate of 65 miles per hour, and the diameter of its wheels is 2.5 feet.

   (a) Find the number of revolutions per minute the wheels are rotating.

   (b) Find the angular speed of the wheels in radians per minute.

100. ***Angular Speed*** A two-inch-diameter pulley on an electric motor that runs at 1700 revolutions per minute is connected by a belt to a four-inch-diameter pulley on a saw arbor.

   (a) Find the angular speed (in radians per minute) of each pulley.

   (b) Find the revolutions per minute of the saw.

▶ Model It

101. *Speed of a Bicycle* The radii of the pedal sprocket, the wheel sprocket, and the wheel of the bicycle in the figure are 4 inches, 2 inches, and 14 inches, respectively. The cyclist is pedaling at a rate of 1 revolution per second.

(a) Find the speed of the bicycle in feet per second and miles per hour.

(b) Use your result from part (a) to write a function for the distance d (in miles) a cyclist travels in terms of the number n of revolutions of the pedal sprocket.

(c) Write a function for the distance d (in miles) a cyclist travels in terms of the time t (in seconds). Compare this function with the function from part (b).

(d) Classify the types of functions you found in parts (b) and (c). Explain your reasoning.

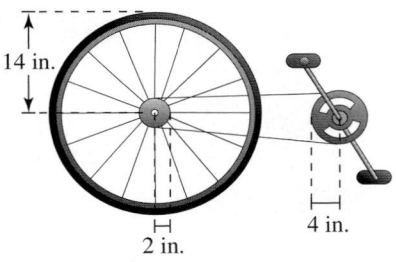

14 in.

2 in.

4 in.

102. *Floppy Disk* The radius of the magnetic disk in a 3.5-inch diskette is 1.68 inches. Find the linear speed of a point on the circumference of the disk if it is rotating at a speed of 360 revolutions per minute.

Synthesis

True or False? In Exercises 103–105, determine whether the statement is true or false. Justify your answer.

103. A measurement of 4 radians corresponds to two complete revolutions from the initial side to the terminal side of an angle.

104. The difference between the measures of two coterminal angles is always a multiple of $360°$ if expressed in degrees and is always a multiple of 2π radians if expressed in radians.

105. An angle that measures $-1260°$ lies in Quadrant III.

106. *Writing* In your own words, explain the meanings of (a) an angle in standard position, (b) a negative angle, (c) coterminal angles, and (d) an obtuse angle.

107. A fan motor turns at a given angular speed. How does the speed of the tips of the blades change if a fan of greater diameter is installed on the motor? Explain.

108. *Think About It* Is a degree or a radian the larger unit of measure? Explain.

109. *Writing* If the radius of a circle is increasing and the magnitude of a central angle is held constant, how is the length of the intercepted arc changing? Explain your reasoning.

110. *Proof* Prove that the area of a circular sector of radius r with central angle θ is $A = \frac{1}{2}\theta r^2$, where θ is measured in radians.

Review

In Exercises 111–114, sketch the graphs of $y = x^5$ and the specified transformation.

111. $f(x) = (x - 2)^5$

112. $f(x) = x^5 - 4$

113. $f(x) = 2 - x^5$

114. $f(x) = -(x + 3)^5$

In Exercises 115–122, simplify the radical expression.

115. $\dfrac{4}{4\sqrt{2}}$

116. $\dfrac{2}{\sqrt{3}}$

117. $\dfrac{2\sqrt{3}}{\sqrt{6}}$

118. $\dfrac{5\sqrt{5}}{2\sqrt{10}}$

119. $\sqrt{2^2 + 6^2}$

120. $\sqrt{18^2 + 12^2}$

121. $\sqrt{18^2 - 6^2}$

122. $\sqrt{17^2 - 9^2}$

▶ **What you should learn**

- How to identify a unit circle and its relationship to real numbers
- How to evaluate trigonometric functions using the unit circle
- How to use the domain and period to evaluate sine and cosine functions
- How to use a calculator to evaluate trigonometric functions

▶ **Why you should learn it**

Trigonometric functions are used to model the movement of an oscillating weight. For instance, in Exercise 57 on page 143, the displacement from equilibrium of an oscillating weight suspended by a spring is modeled as a function of time.

Richard Megna/Fundamental Photographs

The Unit Circle

The two historical perspectives of trigonometry incorporate different methods for introducing the trigonometric functions. Our first introduction to these functions is based on the unit circle.

Consider the **unit circle** given by

$$x^2 + y^2 = 1 \qquad \text{Unit circle}$$

as shown in Figure 1.18.

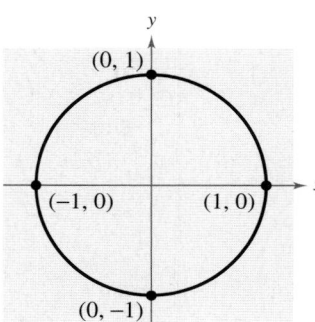

FIGURE 1.18

Imagine that the real number line is wrapped around this circle, with positive numbers corresponding to a counterclockwise wrapping and negative numbers corresponding to a clockwise wrapping, as shown in Figure 1.19.

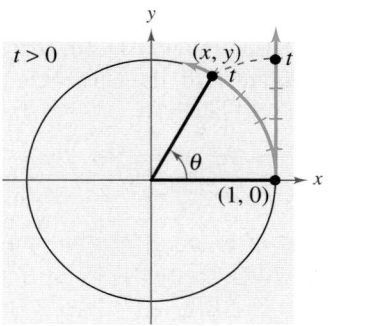
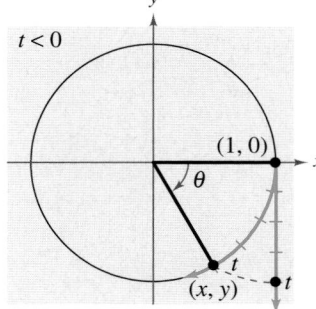

FIGURE 1.19

As the real number line is wrapped around the unit circle, each real number t corresponds to a point (x, y) on the circle. For example, the real number 0 corresponds to the point $(1, 0)$. Moreover, because the unit circle has a circumference of 2π, the real number 2π also corresponds to the point $(1, 0)$.

In general, each real number t also corresponds to a central angle θ (in standard position) whose radian measure is t. With this interpretation of t, the arc length formula $s = r\theta$ (with $r = 1$) indicates that the real number t is the length of the arc intercepted by the angle θ, given in radians.

Try demonstrating the wrapping function by using a spool and thread. Demonstrate the real number t as the length of the thread. Then wrap the thread around the spool to show the correspondence between t and the point (x, y) on the circle. For example, if $t = \pi/2$, the point on the unit circle to which it corresponds is $(0, 1)$.

The Trigonometric Functions

From the preceding discussion, it follows that the coordinates x and y are two functions of the real variable t. You can use these coordinates to define the six trigonometric functions of t.

sine	cosecant
cosine	secant
tangent	cotangent

These six functions are normally abbreviated sin, csc, cos, sec, tan, and cot, respectively.

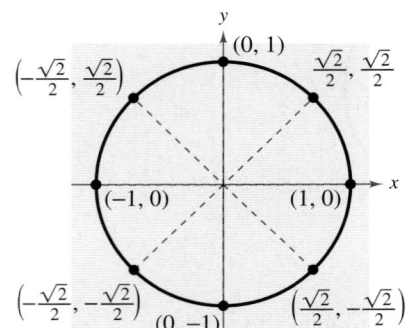

FIGURE **1.20**

Definitions of Trigonometric Functions

Let t be a real number and let (x, y) be the point on the unit circle corresponding to t.

$$\sin t = y \qquad\qquad\qquad \csc t = \frac{1}{y}, \qquad y \neq 0$$

$$\cos t = x \qquad\qquad\qquad \sec t = \frac{1}{x}, \qquad x \neq 0$$

$$\tan t = \frac{y}{x}, \qquad x \neq 0 \qquad\qquad \cot t = \frac{x}{y}, \qquad y \neq 0$$

Note that the functions in the second column are the *reciprocals* of the corresponding functions in the first column.

In the definitions of the trigonometric functions, note that the tangent and secant are not defined when $x = 0$. For instance, because $t = \pi/2$ corresponds to $(x, y) = (0, 1)$, it follows that $\tan(\pi/2)$ and $\sec(\pi/2)$ are *undefined*. Similarly, the cotangent and cosecant are not defined when $y = 0$. For instance, because $t = 0$ corresponds to $(x, y) = (1, 0)$, cot 0 and csc 0 are *undefined*.

In Figure 1.20, the unit circle has been divided into eight equal arcs, corresponding to t-values of

$$0, \frac{\pi}{4}, \frac{\pi}{2}, \frac{3\pi}{4}, \pi, \frac{5\pi}{4}, \frac{3\pi}{2}, \frac{7\pi}{4}, \text{ and } 2\pi.$$

Similarly, in Figure 1.21, the unit circle has been divided into 12 equal arcs, corresponding to t-values of

$$0, \frac{\pi}{6}, \frac{\pi}{3}, \frac{\pi}{2}, \frac{2\pi}{3}, \frac{5\pi}{6}, \pi, \frac{7\pi}{6}, \frac{4\pi}{3}, \frac{3\pi}{2}, \frac{5\pi}{3}, \frac{11\pi}{6}, \text{ and } 2\pi.$$

Using the (x, y) coordinates in Figures 1.20 and 1.21, you can easily evaluate the trigonometric functions for common t-values. This procedure is demonstrated in Examples 1 and 2.

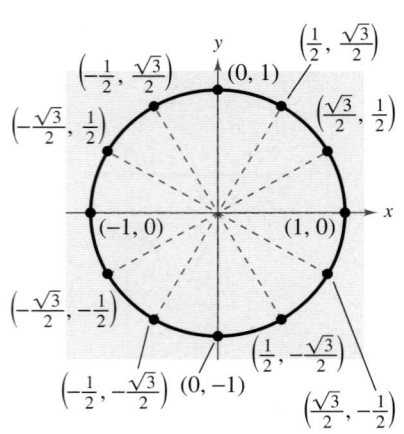

FIGURE **1.21**

| Example 1 ▶ | Evaluating Trigonometric Functions | |

Evaluate the six trigonometric functions at each real number.

a. $t = \dfrac{\pi}{6}$ **b.** $t = \dfrac{5\pi}{4}$ **c.** $t = 0$ **d.** $t = \pi$

Solution

For each t-value, begin by finding the corresponding point (x, y) on the unit circle. Then use the definitions of trigonometric functions listed on page 138.

a. $t = \pi/6$ corresponds to the point $(x, y) = \left(\sqrt{3}/2, 1/2\right)$.

$$\sin \frac{\pi}{6} = y = \frac{1}{2} \qquad\qquad \csc \frac{\pi}{6} = \frac{1}{y} = \frac{1}{1/2} = 2$$

$$\cos \frac{\pi}{6} = x = \frac{\sqrt{3}}{2} \qquad\qquad \sec \frac{\pi}{6} = \frac{1}{x} = \frac{2}{\sqrt{3}} = \frac{2\sqrt{3}}{3}$$

$$\tan \frac{\pi}{6} = \frac{y}{x} = \frac{1/2}{\sqrt{3}/2} = \frac{1}{\sqrt{3}} \qquad\qquad \cot \frac{\pi}{6} = \frac{x}{y} = \frac{\sqrt{3}/2}{1/2} = \sqrt{3}$$

b. $t = 5\pi/4$ corresponds to the point $(x, y) = \left(-\sqrt{2}/2, -\sqrt{2}/2\right)$.

$$\sin \frac{5\pi}{4} = y = -\frac{\sqrt{2}}{2} \qquad\qquad \csc \frac{5\pi}{4} = \frac{1}{y} = -\frac{2}{\sqrt{2}} = -\sqrt{2}$$

$$\cos \frac{5\pi}{4} = x = -\frac{\sqrt{2}}{2} \qquad\qquad \sec \frac{5\pi}{4} = \frac{1}{x} = -\frac{2}{\sqrt{2}} = -\sqrt{2}$$

$$\tan \frac{5\pi}{4} = \frac{y}{x} = \frac{-\sqrt{2}/2}{-\sqrt{2}/2} = 1 \qquad\qquad \cot \frac{5\pi}{4} = \frac{x}{y} = \frac{-\sqrt{2}/2}{-\sqrt{2}/2} = 1$$

c. $t = 0$ corresponds to the point $(x, y) = (1, 0)$.

$$\sin 0 = y = 0 \qquad\qquad \csc 0 = \frac{1}{y} \text{ is undefined.}$$

$$\cos 0 = x = 1 \qquad\qquad \sec 0 = \frac{1}{x} = \frac{1}{1} = 1$$

$$\tan 0 = \frac{y}{x} = \frac{0}{1} = 0 \qquad\qquad \cot 0 = \frac{x}{y} \text{ is undefined.}$$

d. $t = \pi$ corresponds to the point $(x, y) = (-1, 0)$.

$$\sin \pi = y = 0 \qquad\qquad \csc \pi = \frac{1}{y} \text{ is undefined.}$$

$$\cos \pi = x = -1 \qquad\qquad \sec \pi = \frac{1}{x} = \frac{1}{-1} = -1$$

$$\tan \pi = \frac{y}{x} = \frac{0}{-1} = 0 \qquad\qquad \cot \pi = \frac{x}{y} \text{ is undefined.}$$

Additional Example

Evaluate the six trigonometric functions at $t = \dfrac{5\pi}{2}$.

Solution

Moving counterclockwise around the unit circle one and a quarter revolutions, you find that $t = 5\pi/2$ corresponds to the point $(x, y) = (0, 1)$.

$$\sin \frac{5\pi}{2} = y = 1 \qquad \csc \frac{5\pi}{2} = \frac{1}{y} = 1$$

$$\cos \frac{5\pi}{2} = x = 0$$

$$\sec \frac{5\pi}{2} = \frac{1}{x} \text{ is undefined.}$$

$$\tan \frac{5\pi}{2} = \frac{y}{x} \text{ is undefined.}$$

$$\cot \frac{5\pi}{2} = \frac{x}{y} = 0$$

◀ **Exploration** ▶

With your graphing utility in *radian* and *parametric* modes, enter the equations

X1T = cos T and Y1T = sin T

and use the following settings.

Tmin = 0, Tmax = 6.3,
Tstep = 0.1
Xmin = -1.5, Xmax = 1.5,
Xscl = 1
Ymin = -1, Ymax = 1,
Yscl = 1

1. Graph the entered equations and describe the graph.

2. Use the *trace* feature to move the cursor around the graph. What do the *t*-values represent? What do the *x*- and *y*-values represent?

3. What are the least and greatest values of *x* and *y*?

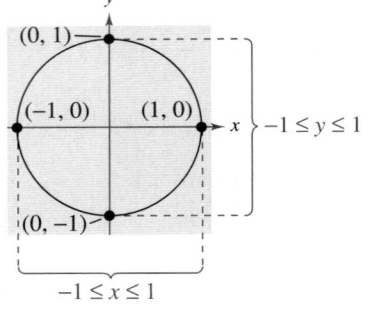

FIGURE 1.22

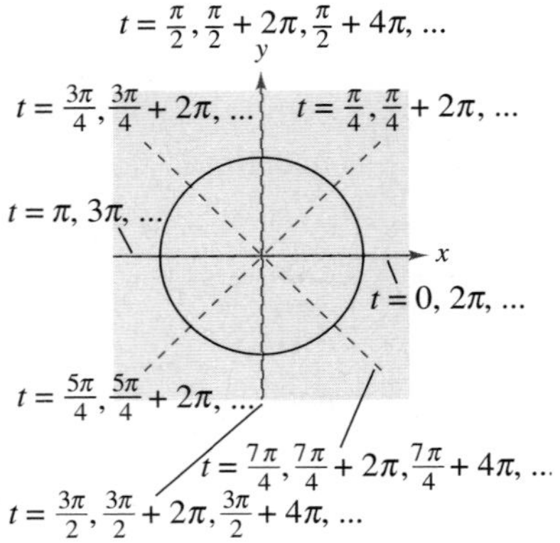

Example 2 ▶   Evaluating Trigonometric Functions

Evaluate the six trigonometric functions at $t = -\dfrac{\pi}{3}$.

Solution

Moving *clockwise* around the unit circle, it follows that $t = -\pi/3$ corresponds to the point $(x, y) = \left(1/2, -\sqrt{3}/2\right)$.

$$\sin\left(-\frac{\pi}{3}\right) = -\frac{\sqrt{3}}{2} \qquad\qquad \csc\left(-\frac{\pi}{3}\right) = -\frac{2}{\sqrt{3}} = -\frac{2\sqrt{3}}{3}$$

$$\cos\left(-\frac{\pi}{3}\right) = \frac{1}{2} \qquad\qquad \sec\left(-\frac{\pi}{3}\right) = 2$$

$$\tan\left(-\frac{\pi}{3}\right) = \frac{-\sqrt{3}/2}{1/2} = -\sqrt{3} \quad \cot\left(-\frac{\pi}{3}\right) = \frac{1/2}{-\sqrt{3}/2} = -\frac{1}{\sqrt{3}} = -\frac{\sqrt{3}}{3}$$

Domain and Period of Sine and Cosine

The *domain* of the sine and cosine functions is the set of all real numbers. To determine the *range* of these two functions, consider the unit circle shown in Figure 1.22. Because $r = 1$, it follows that $\sin t = y$ and $\cos t = x$. Moreover, because (x, y) is on the unit circle, you know that $-1 \le y \le 1$ and $-1 \le x \le 1$. So, the values of sine and cosine also range between -1 and 1.

$$-1 \le \quad y \quad \le 1 \qquad\qquad -1 \le \quad x \quad \le 1$$
$$\text{and}$$
$$-1 \le \sin t \le 1 \qquad\qquad -1 \le \cos t \le 1$$

Adding 2π to each value of t in the interval $[0, 2\pi]$ completes a second revolution around the unit circle, as shown in Figure 1.23. The values of $\sin(t + 2\pi)$ and $\cos(t + 2\pi)$ correspond to those of $\sin t$ and $\cos t$. Similar results can be obtained for repeated revolutions (positive or negative) on the unit circle. This leads to the general result

$$\sin(t + 2\pi n) = \sin t$$

and

$$\cos(t + 2\pi n) = \cos t$$

for any integer n and real number t. Functions that behave in such a repetitive (or cyclic) manner are called **periodic.**

Definition of Periodic Function

A function f is **periodic** if there exists a positive real number c such that

$$f(t + c) = f(t)$$

for all t in the domain of f. The smallest number c for which f is periodic is called the **period** of f.

FIGURE 1.23

STUDY TIP

From the definition of periodic function, it follows that the sine and cosine functions are periodic and have a period of 2π. The other four trigonometric functions are also periodic, and more will be said about this in Section 1.6.

Example 3 ▶ **Using the Period to Evaluate the Sine and Cosine**

a. Because $\dfrac{13\pi}{6} = 2\pi + \dfrac{\pi}{6}$, you have

$$\sin \frac{13\pi}{6} = \sin\left(2\pi + \frac{\pi}{6}\right) = \sin \frac{\pi}{6} = \frac{1}{2}.$$

b. Because $-\dfrac{7\pi}{2} = -4\pi + \dfrac{\pi}{2}$, you have

$$\cos\left(-\frac{7\pi}{2}\right) = \cos\left(-4\pi + \frac{\pi}{2}\right) = \cos \frac{\pi}{2} = 0.$$

Recall from Section P.6 that a function f is *even* if $f(-t) = f(t)$, and is *odd* if $f(-t) = -f(t)$.

Even and Odd Trigonometric Functions

The cosine and secant functions are *even*.

$$\cos(-t) = \cos t \qquad \sec(-t) = \sec t$$

The sine, cosecant, tangent, and cotangent functions are *odd*.

$$\sin(-t) = -\sin t \qquad \csc(-t) = -\csc t$$
$$\tan(-t) = -\tan t \qquad \cot(-t) = -\cot t$$

Students may have difficulty evaluating cosecant, secant, and/or cotangent functions using a calculator. Try having them rewrite the expression in terms of sine, cosine, or tangent before evaluating.

Example: $\csc 1.3 = \dfrac{1}{\sin 1.3} \approx 1.0378$

Evaluating Trigonometric Functions with a Calculator

When evaluating a trigonometric function with a calculator, you need to set the calculator to the desired *mode* of measurement (degrees or radians).

Most calculators do not have keys for the cosecant, secant, and cotangent functions. To evaluate these functions, you can use the $\boxed{x^{-1}}$ key with their respective reciprocal functions sine, cosine, and tangent. For example, to evaluate $\csc(\pi/8)$, use the fact that

$$\csc \frac{\pi}{8} = \frac{1}{\sin(\pi/8)}$$

and enter the following keystroke sequence in radian mode.

$\boxed{(}\ \boxed{\text{SIN}}\ \boxed{(}\boxed{\pi}\ \boxed{\div}\ 8\ \boxed{)}\ \boxed{)}\ \boxed{x^{-1}}\ \boxed{\text{ENTER}}$ Display 2.6131259

Technology

When evaluating trigonometric functions with a calculator, remember to enclose all fractional angle measures in parentheses. For instance, if you want to evaluate $\sin \theta$ for $\theta = \pi/6$, you should enter

$\boxed{\text{SIN}}\ \boxed{(}\boxed{\pi}\ \boxed{\div}\ 6\ \boxed{)}\ \boxed{\text{ENTER}}$

rather than

$\boxed{\text{SIN}}\ \boxed{\pi}\ \boxed{\div}\ 6\ \boxed{\text{ENTER}}$.

The first set of keystrokes yields the correct value of 0.5. The second set yields the incorrect value of 0.

Example 4 ▶ **Using a Calculator**

Function	*Mode*	*Calculator Keystrokes*	*Display*
a. $\sin 2\pi/3$	Radian	$\boxed{\text{SIN}}\ \boxed{(}\ 2\ \boxed{\pi}\ \boxed{\div}\ 3\ \boxed{)}\ \boxed{\text{ENTER}}$	0.8660254
b. $\cot 1.5$	Radian	$\boxed{(}\ \boxed{\text{TAN}}\ 1.5\ \boxed{)}\ \boxed{x^{-1}}\ \boxed{\text{ENTER}}$	0.0709148

1.2 Exercises

In Exercises 1–4, determine the exact values of the six trigonometric functions of the angle θ.

1.

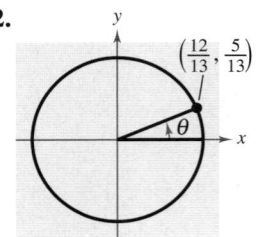

$\left(-\frac{8}{17}, \frac{15}{17}\right)$

2.

$\left(\frac{12}{13}, \frac{5}{13}\right)$

3.

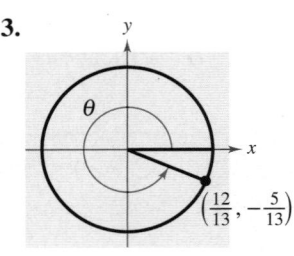

$\left(\frac{12}{13}, -\frac{5}{13}\right)$

4.

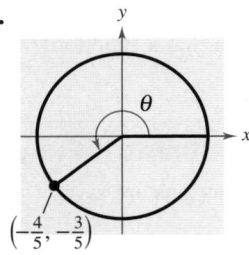

$\left(-\frac{4}{5}, -\frac{3}{5}\right)$

In Exercises 5–12, find the point (x, y) on the unit circle that corresponds to the real number t.

5. $t = \dfrac{\pi}{4}$ 　　　　　　**6.** $t = \dfrac{\pi}{3}$

7. $t = \dfrac{7\pi}{6}$ 　　　　　　**8.** $t = \dfrac{5\pi}{4}$

9. $t = \dfrac{4\pi}{3}$ 　　　　　　**10.** $t = \dfrac{5\pi}{3}$

11. $t = \dfrac{3\pi}{2}$ 　　　　　　**12.** $t = \pi$

In Exercises 13–22, evaluate (if possible) the sine, cosine, and tangent of the real number.

13. $t = \dfrac{\pi}{4}$ 　　　　　　**14.** $t = \dfrac{\pi}{3}$

15. $t = -\dfrac{\pi}{6}$ 　　　　　**16.** $t = -\dfrac{\pi}{4}$

17. $t = -\dfrac{7\pi}{4}$ 　　　　　**18.** $t = -\dfrac{4\pi}{3}$

19. $t = \dfrac{11\pi}{6}$ 　　　　　**20.** $t = \dfrac{5\pi}{3}$

21. $t = -\dfrac{3\pi}{2}$ 　　　　　**22.** $t = -2\pi$

In Exercises 23–28, evaluate (if possible) the six trigonometric functions of the real number.

23. $t = \dfrac{3\pi}{4}$ 　　　　　　**24.** $t = \dfrac{5\pi}{6}$

25. $t = -\dfrac{\pi}{2}$ 　　　　　**26.** $t = \dfrac{3\pi}{2}$

27. $t = \dfrac{4\pi}{3}$ 　　　　　　**28.** $t = \dfrac{7\pi}{4}$

In Exercises 29–36, evaluate the trigonometric function using its period as an aid.

29. $\sin 5\pi$ 　　　　　　**30.** $\cos 5\pi$

31. $\cos \dfrac{8\pi}{3}$ 　　　　　**32.** $\sin \dfrac{9\pi}{4}$

33. $\cos\left(-\dfrac{15\pi}{2}\right)$ 　　　**34.** $\sin \dfrac{19\pi}{6}$

35. $\sin\left(-\dfrac{9\pi}{4}\right)$ 　　　**36.** $\cos\left(-\dfrac{8\pi}{3}\right)$

In Exercises 37–42, use the value of the trigonometric function to evaluate the indicated functions.

37. $\sin t = \frac{1}{3}$ 　　　　**38.** $\sin(-t) = \frac{3}{8}$

(a) $\sin(-t)$ 　　　　　(a) $\sin t$

(b) $\csc(-t)$ 　　　　　(b) $\csc t$

39. $\cos(-t) = -\frac{1}{5}$ 　　**40.** $\cos t = -\frac{3}{4}$

(a) $\cos t$ 　　　　　　(a) $\cos(-t)$

(b) $\sec(-t)$ 　　　　　(b) $\sec(-t)$

41. $\sin t = \frac{4}{5}$ 　　　　**42.** $\cos t = \frac{4}{5}$

(a) $\sin(\pi - t)$ 　　　　(a) $\cos(\pi - t)$

(b) $\sin(t + \pi)$ 　　　　(b) $\cos(t + \pi)$

In Exercises 43–52, use a calculator to evaluate the expression. Round to four decimal places.

43. $\sin \dfrac{\pi}{4}$ 　　　　　　**44.** $\tan \dfrac{\pi}{3}$

45. $\csc 1.3$ 　　　　　　**46.** $\cot 1$

47. $\cos(-1.7)$ 　　　　　**48.** $\cos(-2.5)$

49. $\csc 0.8$ 　　　　　　**50.** $\sec 1.8$

51. $\sec 22.8$ 　　　　　**52.** $\sin(-0.9)$

Estimation In Exercises 53 and 54, use the figure and a straightedge to approximate the value of each trigonometric function. To print an enlarged copy of the graph, go to the website *www.mathgraphs.com*.

53. (a) sin 5 (b) cos 2   **54.** (a) sin 0.75 (b) cos 2.5

Estimation In Exercises 55 and 56, use the figure and a straightedge to approximate the solution of each equation, where $0 \le t < 2\pi$. To print an enlarged copy of the graph, go to the website *www.mathgraphs.com*.

55. (a) $\sin t = 0.25$ (b) $\cos t = -0.25$
56. (a) $\sin t = -0.75$ (b) $\cos t = 0.75$

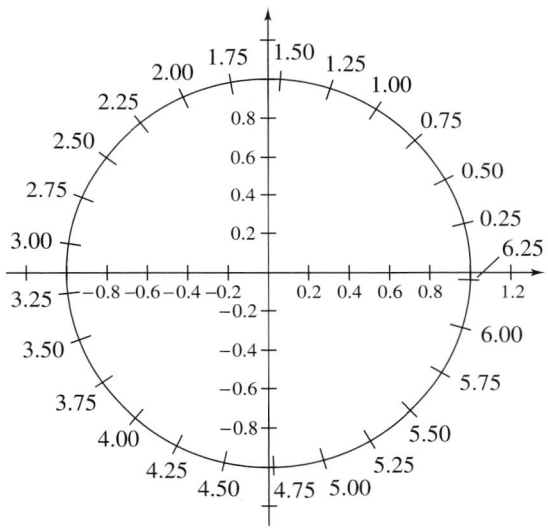

FIGURE FOR 53–56

▶ Model It

57. *Harmonic Motion* The displacement from equilibrium of an oscillating weight suspended by a spring and subject to the damping effect of friction is

$$y(t) = \frac{1}{4}e^{-t}\cos 6t$$

where y is the displacement in feet and t is the time in seconds.

(a) Complete the table.

t	0	$\frac{1}{4}$	$\frac{1}{2}$	$\frac{3}{4}$	1
y					

▶ Model It *(continued)*

(b) Use the *table* feature of a graphing utility to approximate the time when the weight reaches equilibrium.

(c) What appears to happen to the displacement as t increases?

58. *Harmonic Motion* The displacement from equilibrium of an oscillating weight suspended by a spring is $y(t) = \frac{1}{4}\cos 6t$, where y is the displacement in feet and t is the time in seconds. Find the displacement when (a) $t = 0$, (b) $t = \frac{1}{4}$, and (c) $t = \frac{1}{2}$.

Synthesis

True or False? In Exercises 59 and 60, determine whether the statement is true or false. Justify your answer.

59. Because $\sin(-t) = -\sin t$, it can be said that the sine of a negative angle is a negative number.

60. $\tan a = \tan(a - 6\pi)$

61. *Exploration* Let (x_1, y_1) and (x_2, y_2) be points on the unit circle corresponding to $t = t_1$ and $t = \pi - t_1$, respectively.

(a) Identify the symmetry of the points (x_1, y_1) and (x_2, y_2).

(b) Make a conjecture about any relationship between $\sin t_1$ and $\sin(\pi - t_1)$.

(c) Make a conjecture about any relationship between $\cos t_1$ and $\cos(\pi - t_1)$.

62. Use the unit circle to verify that the cosine and secant functions are even and that the sine, cosecant, tangent, and cotangent functions are odd.

63. Verify that $\sin(t_1 + t_2) \ne \sin t_1 + \sin t_2$ by approximating $\sin 0.25$, $\sin 0.75$, and $\sin 1$.

64. *Think About It* Because $f(t) = \sin t$ is an odd function and $g(t) = \cos t$ is an even function, what can be said about the function $h(t) = f(t)g(t)$?

Review

In Exercises 65–68, find the inverse function of the one-to-one function f.

65. $f(x) = \frac{1}{2}(3x - 2)$ **66.** $f(x) = \frac{1}{4}x^3 + 1$

67. $f(x) = \sqrt{x^2 - 4}$, $x \ge 2$ **68.** $f(x) = \dfrac{x + 2}{x - 4}$

1.3   Right Triangle Trigonometry

▶ **What you should learn**

- How to evaluate trigonometric functions of acute angles
- How to use the fundamental trigonometric identities
- How to use a calculator to evaluate trigonometric functions
- How to use trigonometric functions to model and solve real-life problems

▶ **Why you should learn it**

Trigonometric functions are often used to analyze real-life situations. For instance, in Exercise 63 on page 153, you are asked to use trigonometric functions to find the height of a helium-filled balloon.

Joseph Sohm; ChromoSohm

The Six Trigonometric Functions

Our second look at the trigonometric functions is from a *right triangle* perspective. Consider a right triangle, with one acute angle labeled θ, as shown in Figure 1.24. Relative to the angle θ, the three sides of the triangle are the **hypotenuse,** the **opposite side** (the side opposite the angle θ), and the **adjacent side** (the side adjacent to the angle θ).

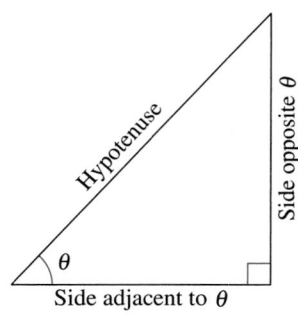

Side adjacent to θ

FIGURE **1.24**

Using the lengths of these three sides, you can form six ratios that define the six trigonometric functions of the acute angle θ.

    sine   cosecant   cosine   secant   tangent   cotangent

In the following definition, it is important to see that $0° < \theta < 90°$ and that for such angles the value of each trigonometric function is *positive*.

Right Triangle Definitions of Trigonometric Functions

Let θ be an *acute* angle of a right triangle. The six trigonometric functions of the angle θ are defined as follows. (Note that the functions in the second row are the *reciprocals* of the corresponding functions in the first row.)

$$\sin \theta = \frac{\text{opp}}{\text{hyp}} \qquad \cos \theta = \frac{\text{adj}}{\text{hyp}} \qquad \tan \theta = \frac{\text{opp}}{\text{adj}}$$

$$\csc \theta = \frac{\text{hyp}}{\text{opp}} \qquad \sec \theta = \frac{\text{hyp}}{\text{adj}} \qquad \cot \theta = \frac{\text{adj}}{\text{opp}}$$

The abbreviations opp, adj, and hyp represent the lengths of the three sides of a right triangle.

    opp = the length of the side *opposite* θ

    adj = the length of the side *adjacent to* θ

    hyp = the length of the *hypotenuse*

You may wish to review the Pythagorean Theorem before presenting the examples in this section.

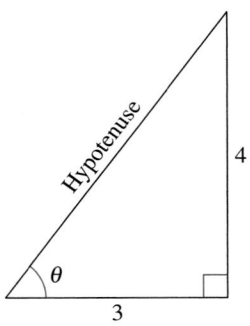

FIGURE **1.25**

Historical Note

Georg Joachim Rhaeticus (1514–1576) was the leading Teutonic mathematical astronomer of the 16th century. He was the first to define the trigonometric functions as ratios of the sides of a right triangle.

Example 1 ► Evaluating Trigonometric Functions

Use the triangle in Figure 1.25 to find the values of the six trigonometric functions of θ.

Solution

By the Pythagorean Theorem, $(\text{hyp})^2 = (\text{opp})^2 + (\text{adj})^2$, it follows that

$$\begin{aligned} \text{hyp} &= \sqrt{4^2 + 3^2} \\ &= \sqrt{25} \\ &= 5. \end{aligned}$$

So, the six trigonometric functions of θ are

$$\sin \theta = \frac{\text{opp}}{\text{hyp}} = \frac{4}{5} \qquad \csc \theta = \frac{\text{hyp}}{\text{opp}} = \frac{5}{4}$$

$$\cos \theta = \frac{\text{adj}}{\text{hyp}} = \frac{3}{5} \qquad \sec \theta = \frac{\text{hyp}}{\text{adj}} = \frac{5}{3}$$

$$\tan \theta = \frac{\text{opp}}{\text{adj}} = \frac{4}{3} \qquad \cot \theta = \frac{\text{adj}}{\text{opp}} = \frac{3}{4}.$$

In Example 1, you were given the lengths of two sides of the right triangle, but not the angle θ. Often, you will be asked to find the trigonometric functions of a *given* acute angle θ. To do this, construct a right triangle having θ as one of its angles.

Example 2 ► Evaluating Trigonometric Functions of 45°

Find the values of sin 45°, cos 45°, and tan 45°.

Solution

Construct a right triangle having 45° as one of its acute angles, as shown in Figure 1.26. Choose the length of the adjacent side to be 1. From geometry, you know that the other acute angle is also 45°. So, the triangle is isosceles and the length of the opposite side is also 1. Using the Pythagorean Theorem, you find the length of the hypotenuse to be $\sqrt{2}$.

$$\sin 45° = \frac{\text{opp}}{\text{hyp}} = \frac{1}{\sqrt{2}} = \frac{\sqrt{2}}{2}$$

$$\cos 45° = \frac{\text{adj}}{\text{hyp}} = \frac{1}{\sqrt{2}} = \frac{\sqrt{2}}{2}$$

$$\tan 45° = \frac{\text{opp}}{\text{adj}} = \frac{1}{1} = 1$$

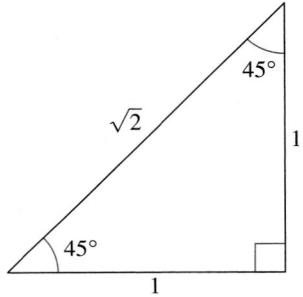

FIGURE **1.26**

Example 3 ▶ Evaluating Trigonometric Functions of 30° and 60°

Use the equilateral triangle shown in Figure 1.27 to find the values of sin 60°, cos 60°, sin 30°, and cos 30°.

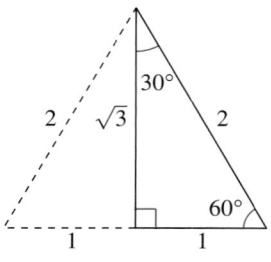

FIGURE **1.27**

Consider having your students construct the triangle in Figure 1.27 with angles in the corresponding radian measures, then find the six trigonometric functions for each of the acute angles.

Solution

Use the Pythagorean Theorem and the equilateral triangle in Figure 1.27 to verify the lengths of the sides shown in Figure 1.27. For $\theta = 60°$, you have adj = 1, opp = $\sqrt{3}$, and hyp = 2. So,

$$\sin 60° = \frac{\text{opp}}{\text{hyp}} = \frac{\sqrt{3}}{2} \qquad \text{and} \qquad \cos 60° = \frac{\text{adj}}{\text{hyp}} = \frac{1}{2}.$$

For $\theta = 30°$, adj = $\sqrt{3}$, opp = 1, and hyp = 2. So,

$$\sin 30° = \frac{\text{opp}}{\text{hyp}} = \frac{1}{2} \qquad \text{and} \qquad \cos 30° = \frac{\text{adj}}{\text{hyp}} = \frac{\sqrt{3}}{2}.$$

> ## STUDY TIP
>
> Because the angles 30°, 45°, and 60° ($\pi/6$, $\pi/4$, and $\pi/3$) occur frequently in trigonometry, you should learn to construct the triangles shown in Figures 1.26 and 1.27.

The triangles in Figures 1.25, 1.26, and 1.27 are useful problem-solving aids. Encourage your students to draw diagrams when they solve problems similar to Examples 1, 2, and 3.

> ### Sines, Cosines, and Tangents of Special Angles
>
> $$\sin 30° = \sin \frac{\pi}{6} = \frac{1}{2} \qquad \cos 30° = \cos \frac{\pi}{6} = \frac{\sqrt{3}}{2} \qquad \tan 30° = \tan \frac{\pi}{6} = \frac{\sqrt{3}}{3}$$
>
> $$\sin 45° = \sin \frac{\pi}{4} = \frac{\sqrt{2}}{2} \qquad \cos 45° = \cos \frac{\pi}{4} = \frac{\sqrt{2}}{2} \qquad \tan 45° = \tan \frac{\pi}{4} = 1$$
>
> $$\sin 60° = \sin \frac{\pi}{3} = \frac{\sqrt{3}}{2} \qquad \cos 60° = \cos \frac{\pi}{3} = \frac{1}{2} \qquad \tan 60° = \tan \frac{\pi}{3} = \sqrt{3}$$

In the box, note that $\sin 30° = \frac{1}{2} = \cos 60°$. This occurs because 30° and 60° are complementary angles. In general, it can be shown from the right triangle definitions that *cofunctions of complementary angles are equal*. That is, if θ is an acute angle, the following relationships are true.

$$\sin(90° - \theta) = \cos \theta \qquad\qquad \cos(90° - \theta) = \sin \theta$$

$$\tan(90° - \theta) = \cot \theta \qquad\qquad \cot(90° - \theta) = \tan \theta$$

$$\sec(90° - \theta) = \csc \theta \qquad\qquad \csc(90° - \theta) = \sec \theta$$

Trigonometric Identities

In trigonometry, a great deal of time is spent studying relationships between trigonometric functions (identities).

These identities will be used many times in trigonometry and later in calculus. Encourage your students to learn them well.

Fundamental Trigonometric Identities

Reciprocal Identities

$$\sin \theta = \frac{1}{\csc \theta} \qquad \cos \theta = \frac{1}{\sec \theta} \qquad \tan \theta = \frac{1}{\cot \theta}$$

$$\csc \theta = \frac{1}{\sin \theta} \qquad \sec \theta = \frac{1}{\cos \theta} \qquad \cot \theta = \frac{1}{\tan \theta}$$

Quotient Identities

$$\tan \theta = \frac{\sin \theta}{\cos \theta} \qquad \cot \theta = \frac{\cos \theta}{\sin \theta}$$

Pythagorean Identities

$$\sin^2 \theta + \cos^2 \theta = 1 \qquad 1 + \tan^2 \theta = \sec^2 \theta$$

$$1 + \cot^2 \theta = \csc^2 \theta$$

Note that $\sin^2 \theta$ represents $(\sin \theta)^2$, $\cos^2 \theta$ represents $(\cos \theta)^2$, and so on.

Example 4 ▶ **Applying Trigonometric Identities**

Let θ be an acute angle such that $\sin \theta = 0.6$. Find the values of (a) $\cos \theta$ and (b) $\tan \theta$ using trigonometric identities.

Solution

a. To find the value of $\cos \theta$, use the Pythagorean identity

$$\sin^2 \theta + \cos^2 \theta = 1.$$

So, you have

$$(0.6)^2 + \cos^2 \theta = 1 \qquad \text{Substitute 0.6 for } \sin \theta.$$

$$\cos^2 \theta = 1 - (0.6)^2 = 0.64 \qquad \text{Subtract } (0.6)^2 \text{ from each side.}$$

$$\cos \theta = \sqrt{0.64} = 0.8. \qquad \text{Extract the positive square root.}$$

b. Now, knowing the sine and cosine of θ, you can find the tangent of θ to be

$$\tan \theta = \frac{\sin \theta}{\cos \theta}$$

$$= \frac{0.6}{0.8}$$

$$= 0.75.$$

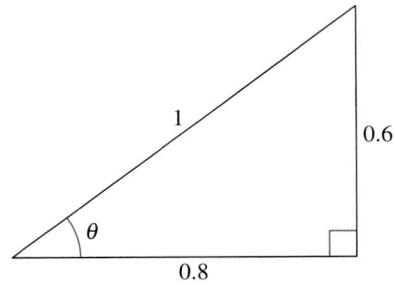

FIGURE **1.28**

Use the definitions of $\cos \theta$ and $\tan \theta$, and the triangle shown in Figure 1.28, to check these results.

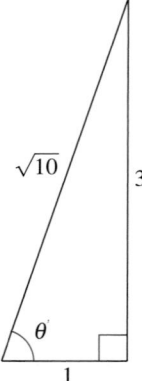

FIGURE **1.29**

Example 5 ▶   Applying Trigonometric Identities

Let θ be an acute angle such that $\tan \theta = 3$. Find the value of each trigonometric function using trigonometric identities.

a. $\cot \theta$

b. $\sec \theta$

Solution

a. $\cot \theta = \dfrac{1}{\tan \theta} = \dfrac{1}{3}$ Reciprocal identity

b. $\sec^2 \theta = 1 + \tan^2 \theta$ Pythagorean identity

$\sec^2 \theta = 1 + 3^2$

$\sec^2 \theta = 10$

$\sec \theta = \sqrt{10}$

Use the definitions of $\cot \theta$ and $\sec \theta$, and the triangle shown in Figure 1.29, to check these results.

STUDY TIP

You can also use the reciprocal identities for sine, cosine, and tangent to evaluate the cosecant, secant, and cotangent functions with a calculator. For instance, you could use the following keystroke sequence to evaluate sec 28°.

1 ÷ COS 28 ENTER

The calculator should display 1.1325701.

One of the most common errors students make when they evaluate trigonometric functions with a calculator is not having their calculators set to the correct mode (radian vs. degree).

Evaluating Trigonometric Functions with a Calculator

To use a calculator to evaluate trigonometric functions of angles measured in degrees, first set the calculator to *degree* mode and then proceed as demonstrated in Section 1.2. For instance, you can find values of cos 28° and sec 28° as follows.

Function	Calculator Keystrokes	Display
$\cos 28°$	COS 28 ENTER	0.8829476
$\sec 28°$	(COS 28) x^{-1} ENTER	1.1325701

 Throughout this text, angles are assumed to be measured in radians unless noted otherwise. For example, sin 1 means the sine of 1 radian and sin 1° means the sine of 1 degree.

Example 6 ▶   Using a Calculator

Use a calculator to evaluate $\sec(5°\,40'\,12'')$.

Solution

Begin by converting to decimal form.

$$5°\,40'\,12'' = 5° + \left(\frac{40}{60}\right)° + \left(\frac{12}{3600}\right)° = 5.67°$$

Then, use a calculator to evaluate sec 5.67°.

$$\sec(5°\,40'\,12'') = \sec 5.67° = \frac{1}{\cos 5.67°} \approx 1.00492$$

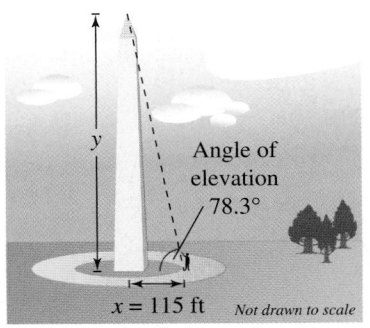

FIGURE **1.30**

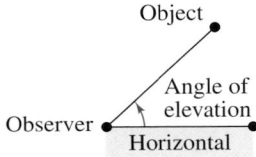

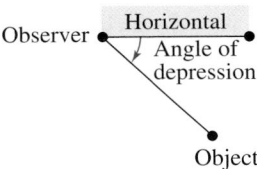

FIGURE **1.31**

Activity

Use the right triangle shown, to find each of the six trigonometric functions of the angle θ.

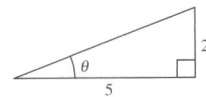

Answer:

$\sin \theta = \dfrac{2\sqrt{29}}{29}, \csc \theta = \dfrac{\sqrt{29}}{2},$

$\cos \theta = \dfrac{5\sqrt{29}}{29}, \sec \theta = \dfrac{\sqrt{29}}{5},$

$\tan \theta = \dfrac{2}{5}, \cot \theta = \dfrac{5}{2}$

Applications Involving Right Triangles

Many applications of trigonometry involve a process called **solving right triangles.** In this type of application, you are usually given one side of a right triangle and one of the acute angles and asked to find one of the other sides, or you are given two sides and asked to find one of the acute angles.

Example 7 ▶ Using Trigonometry to Solve a Right Triangle

A surveyor is standing 115 feet from the base of the Washington Monument, as shown in Figure 1.30. The surveyor measures the angle of elevation to the top of the monument as 78.3°. How tall is the Washington Monument?

Solution

From Figure 1.30, you can see that

$$\tan 78.3° = \frac{\text{opp}}{\text{adj}} = \frac{y}{x}$$

where $x = 115$ and y is the height of the monument. So, the height of the Washington Monument is

$$y = x \tan 78.3° \approx 115(4.82882) \approx 555 \text{ feet.}$$

The term **angle of elevation** represents the angle from the horizontal upward to an object. For objects that lie below the horizontal, it is common to use the term **angle of depression,** as shown in Figure 1.31.

Example 8 ▶ Using Trigonometry to Solve a Right Triangle

You are 200 yards from a river. Rather than walking directly to the river, you walk 400 yards along a straight path to the river's edge. Find the acute angle θ between this path and the river's edge, as illustrated in Figure 1.32.

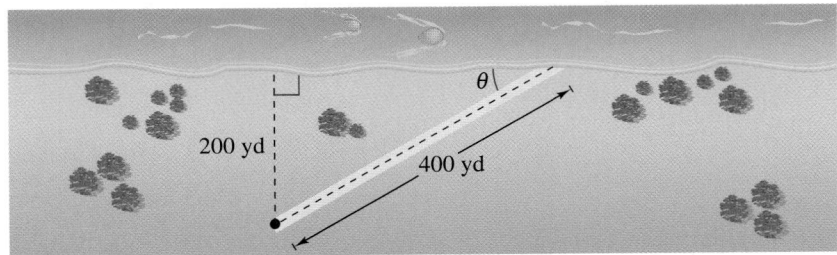

FIGURE **1.32**

Solution

From Figure 1.32, you can see that the sine of the angle θ is

$$\sin \theta = \frac{\text{opp}}{\text{hyp}} = \frac{200}{400} = \frac{1}{2}.$$

So, $\theta = 30°$.

By now you are able to recognize that the acute angle that satisfies the equation $\sin\theta = \frac{1}{2}$ is $\theta = 30°$. Suppose, however, that you were given the equation $\sin\theta = 0.6$ and were asked to find the acute angle θ. Because

$$\sin 30° = \frac{1}{2}$$

$$= 0.5000$$

and

$$\sin 45° = \frac{1}{\sqrt{2}}$$

$$\approx 0.7071$$

you might guess that θ lies somewhere between $30°$ and $45°$. A more precise value of θ can be found using the *inverse sine* key on a calculator. To do this, you can use the following keystroke sequence in *degree* mode.

[sin⁻¹] .6 [ENTER]   Display 36.8699

So, you can conclude that if $\sin\theta = 0.6$ and $0° < \theta < 90°$, then $\theta \approx 36.87°$.

Example 9 ▶ Solving a Right Triangle

Specifications for a loading dock ramp require a rise of 1 foot for each 3 feet of horizontal length. In Figure 1.33, find the lengths of sides b and c and find the measure of θ.

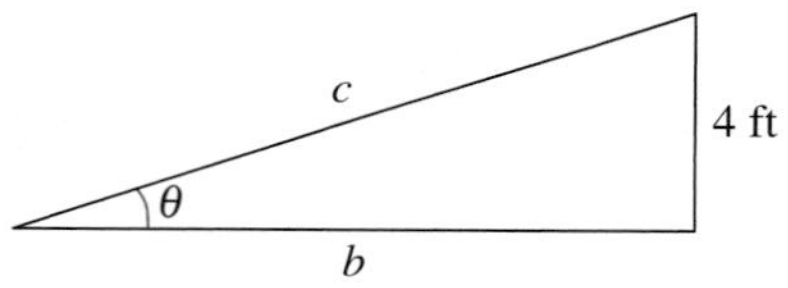

Solution

From the given specifications, you can write

$$\frac{\text{rise}}{\text{run}} = \frac{1}{3} = \frac{4 \text{ ft}}{b \text{ ft}}$$

which implies that $b = 12$ feet. Using the Pythagorean Theorem, you can write

$c^2 = a^2 + b^2$	Pythagorean Theorem
$c^2 = 4^2 + 12^2$	Substitute for a and b.
$c^2 = 160$	Simplify.
$c = 4\sqrt{10}.$	Extract positive square root.

So, $c = 4\sqrt{10} \approx 12.65$ feet. To solve for θ, you can write

$$\tan\theta = \frac{4}{12}$$

$$= \frac{1}{3}.$$

Then, using the calculator keystrokes in *degree* mode

[tan⁻¹] [(] 1 [÷] 3 [)] [ENTER]

you obtain $\theta \approx 18.43°$.

FIGURE 1.33

Triangle with side c (hypotenuse), side b (base), height 4 ft, and angle θ.

Activities

1. A 10-foot ladder leans against the side of a house. The ladder makes an angle of 60° with the ground. How far up the side of the house does the ladder reach?

   Answer: $5\sqrt{3} \approx 8.66$ feet

2. Using your calculator, find the value of θ, $0 \leq \theta \leq \frac{\pi}{2}$, given that $\sin\theta = 0.2962$.

   Answer: $\theta \approx 0.3007$

1.3 Exercises

In Exercises 1–4, find the exact values of the six trigono-metric functions of the angle θ shown in the figure. (Use the Pythagorean Theorem to find the third side of the triangle.)

1.

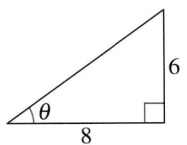

2.

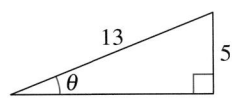

3.

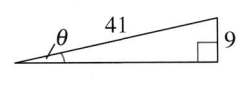

4.

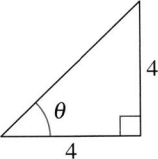

In Exercises 5–8, find the exact values of the six trigono-metric functions of the angle θ for each of the two triangles. Explain why the function values are the same.

5.

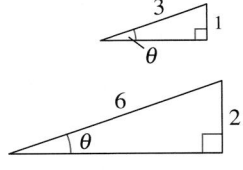

6.

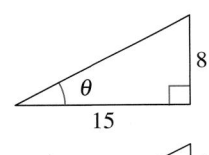

7.

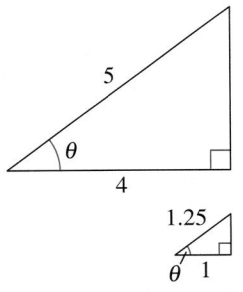

8.

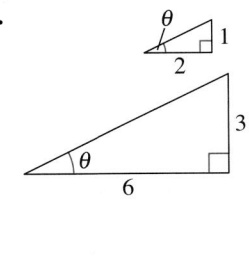

In Exercises 9–16, sketch a right triangle corresponding to the trigonometric function of the acute angle θ. Use the Pythagorean Theorem to determine the third side and then find the other five trigonometric functions of θ.

9. $\sin \theta = \frac{3}{4}$ **10.** $\cos \theta = \frac{5}{7}$

11. $\sec \theta = 2$ **12.** $\cot \theta = 5$

13. $\tan \theta = 3$ **14.** $\sec \theta = 6$

15. $\cot \theta = \frac{3}{2}$ **16.** $\csc \theta = \frac{17}{4}$

In Exercises 17–22, use the given function value(s), and trigonometric identities (including the cofunction identi-ties), to find the indicated trigonometric functions.

17. $\sin 60° = \dfrac{\sqrt{3}}{2}, \quad \cos 60° = \dfrac{1}{2}$

    (a) $\tan 60°$ (b) $\sin 30°$

    (c) $\cos 30°$ (d) $\cot 60°$

18. $\sin 30° = \dfrac{1}{2}, \quad \tan 30° = \dfrac{\sqrt{3}}{3}$

    (a) $\csc 30°$ (b) $\cot 60°$

    (c) $\cos 30°$ (d) $\cot 30°$

19. $\csc \theta = \dfrac{\sqrt{13}}{2}, \quad \sec \theta = \dfrac{\sqrt{13}}{3}$

    (a) $\sin \theta$ (b) $\cos \theta$

    (c) $\tan \theta$ (d) $\sec(90° - \theta)$

20. $\sec \theta = 5, \quad \tan \theta = 2\sqrt{6}$

    (a) $\cos \theta$ (b) $\cot \theta$

    (c) $\cot(90° - \theta)$ (d) $\sin \theta$

21. $\cos \alpha = \frac{1}{3}$

    (a) $\sec \alpha$ (b) $\sin \alpha$

    (c) $\cot \alpha$ (d) $\sin(90° - \alpha)$

22. $\tan \beta = 5$

    (a) $\cot \beta$ (b) $\cos \beta$

    (c) $\tan(90° - \beta)$ (d) $\csc \beta$

In Exercises 23–26, evaluate the trigonometric function by memory or by constructing an appropriate triangle for the given special angle.

23. (a) $\cos 60°$ (b) $\csc 30°$ (c) $\tan 60°$

24. (a) $\cot 45°$ (b) $\cos 45°$ (c) $\csc 45°$

25. (a) $\sin 45°$ (b) $\cos 30°$ (c) $\tan 30°$

26. (a) $\sin 60°$ (b) $\tan 45°$ (c) $\sec 30°$

In Exercises 27–36, use a calculator to evaluate each function. Round your answers to four decimal places. (Be sure the calculator is in the correct angle mode.)

27. (a) $\sin 10°$ (b) $\cos 80°$

28. (a) $\tan 23.5°$ (b) $\cot 66.5°$

29. (a) $\sin 16.35°$ (b) $\csc 16.35°$

30. (a) $\cos 16° 18'$ (b) $\sin 73° 56'$

31. (a) $\sec 42° 12'$ (b) $\csc 48° 7'$

32. (a) $\cos 4° 50' 15''$ (b) $\sec 4° 50' 15''$

33. (a) $\cot 11° 15'$ (b) $\tan 11° 15'$

34. (a) $\sec 56° 8' 10''$ (b) $\cos 56° 8' 10''$

35. (a) $\csc 32° 40' 3''$ (b) $\tan 44° 28' 16''$

36. (a) $\sec\left(\frac{9}{5} \cdot 20 + 32\right)°$ (b) $\cot\left(\frac{9}{5} \cdot 30 + 32\right)°$

In Exercises 37–42, find the values of θ in degrees $(0° < \theta < 90°)$ and radians $(0 < \theta < \pi/2)$ without the aid of a calculator.

37. (a) $\sin \theta = \dfrac{1}{2}$ (b) $\csc \theta = 2$

38. (a) $\cos \theta = \dfrac{\sqrt{2}}{2}$ (b) $\tan \theta = 1$

39. (a) $\sec \theta = 2$ (b) $\cot \theta = 1$

40. (a) $\tan \theta = \sqrt{3}$ (b) $\cos \theta = \dfrac{1}{2}$

41. (a) $\csc \theta = \dfrac{2\sqrt{3}}{3}$ (b) $\sin \theta = \dfrac{\sqrt{2}}{2}$

42. (a) $\cot \theta = \dfrac{\sqrt{3}}{3}$ (b) $\sec \theta = \sqrt{2}$

In Exercises 43–46, find the values of θ in degrees $(0° < \theta < 90°)$ and radians $(0 < \theta < \pi/2)$ by using a calculator.

43. (a) $\sin \theta = 0.0145$ (b) $\sin \theta = 0.4565$

44. (a) $\cos \theta = 0.9848$ (b) $\cos \theta = 0.8746$

45. (a) $\tan \theta = 0.0125$ (b) $\tan \theta = 2.3545$

46. (a) $\sin \theta = 0.3746$ (b) $\cos \theta = 0.3746$

In Exercises 47–56, use trigonometric identities to transform the left side of the equation into the right side.

47. $\tan \theta \cot \theta = 1$

48. $\cos \theta \sec \theta = 1$

49. $\tan \alpha \cos \alpha = \sin \alpha$

50. $\cot \alpha \sin \alpha = \cos \alpha$

51. $(1 + \cos \theta)(1 - \cos \theta) = \sin^2 \theta$

52. $(1 + \sin \theta)(1 - \sin \theta) = \cos^2 \theta$

53. $(\sec \theta + \tan \theta)(\sec \theta - \tan \theta) = 1$

54. $\sin^2 \theta - \cos^2 \theta = 2 \sin^2 \theta - 1$

55. $\dfrac{\sin \theta}{\cos \theta} + \dfrac{\cos \theta}{\sin \theta} = \csc \theta \sec \theta$

56. $\dfrac{\tan \beta + \cot \beta}{\tan \beta} = \csc^2 \beta$

In Exercises 57–60, solve for x, y, or r, as indicated.

57. Solve for x.   **58.** Solve for y.

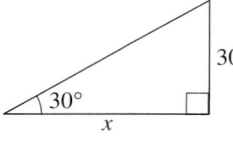   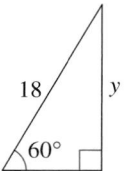

59. Solve for x.   **60.** Solve for r.

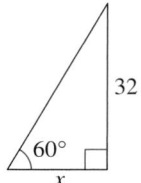   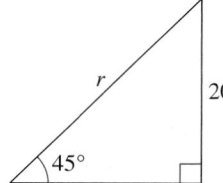

61. *Height*   A six-foot person walks from the base of a broadcasting tower directly toward the tip of the shadow cast by the tower. When the person is 132 feet from the tower and 3 feet from the tip of the shadow, the person's shadow starts to appear beyond the tower's shadow.

  (a) Draw a right triangle that gives a visual representation of the problem. Show the known quantities of the triangle and use a variable to indicate the height of the tower.

  (b) Use a trigonometric function to write an equation involving the unknown quantity.

  (c) What is the height of the tower?

62. *Height*   A six-foot person standing 20 feet from a streetlight casts a 10-foot shadow (see figure). What is the height of the streetlight?

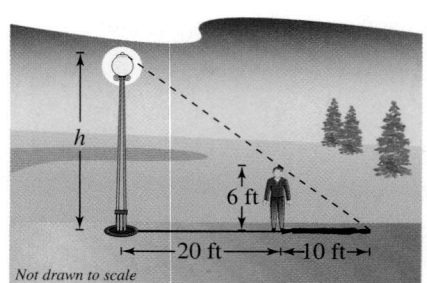

Not drawn to scale

▶ **Model It**

63. *Height* A 20-meter line is used to tether a helium-filled balloon. Because of a breeze, the line makes an angle of approximately 85° with the ground.

  (a) Draw a right triangle that gives a visual representation of the problem. Show the known quantities of the triangle and use a variable to indicate the height of the balloon.

  (b) Use a trigonometric function to write an equation involving the unknown quantity.

  (c) What is the height of the balloon?

  (d) The breeze becomes stronger and the angle the balloon makes with the ground decreases. How does this affect the triangle you drew in part (a)?

  (e) Complete the table, which shows the height (in meters) of the balloon for decreasing angle measures θ.

Angle, θ	80°	70°	60°	50°
Height				

Angle, θ	40°	30°	20°	10°
Height				

  (f) As the angle the balloon makes with the ground approaches 0°, how does this affect the height of the balloon? Draw a right triangle to explain your reasoning.

64. *Angle of Elevation* A ramp 20 feet in length rises to a loading platform that is $3\frac{1}{3}$ feet off the ground.

  (a) Draw a right triangle that gives a visual representation of the problem. Show the known quantities of the triangle and use a variable to indicate the angle of elevation of the ramp.

  (b) Use a trigonometric function to write an equation involving the unknown quantity.

  (c) What is the angle of elevation of the ramp?

65. *Width of a River* A biologist wants to know the width w of a river in order to set instruments for studying the pollutants in the water. From point A, the biologist walks downstream 100 feet and sights to point C (see figure). From this sighting, it is determined that $\theta = 54°$. How wide is the river?

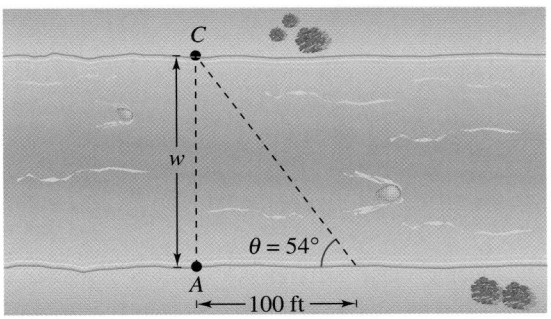

66. *Height of a Mountain* In traveling across flat land, you notice a mountain directly in front of you. Its angle of elevation (to the peak) is 3.5°. After you drive 13 miles closer to the mountain, the angle of elevation is 9°. Approximate the height of the mountain.

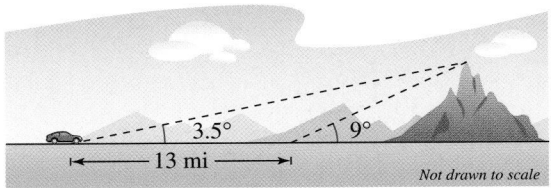

67. *Machine Shop Calculations* A steel plate has the form of one-fourth of a circle with a radius of 60 centimeters. Two two-centimeter holes are to be drilled in the plate positioned as shown in the figure. Find the coordinates of the center of each hole.

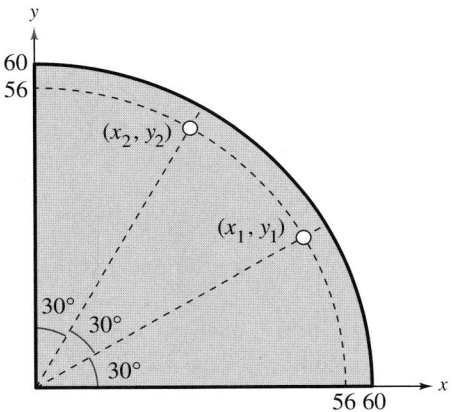

68. *Machine Shop Calculations* A tapered shaft has a diameter of 5 centimeters at the small end and is 15 centimeters long (see figure). The taper is 3°. Find the diameter d of the large end of the shaft.

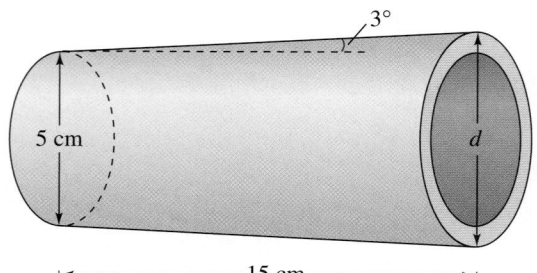

69. *Geometry* Use a compass to sketch a quarter of a circle of radius 10 centimeters. Using a protractor, construct an angle of 20° in standard position (see figure). Drop a perpendicular from the point of intersection of the terminal side of the angle and the arc of the circle. By actual measurement, calculate the coordinates (x, y) of the point of intersection and use these measurements to approximate the six trigonometric functions of a 20° angle.

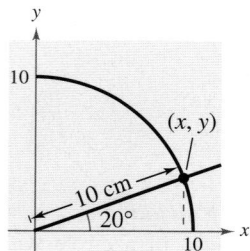

70. *Geometry* Repeat Exercise 69 using a 75° angle.

Synthesis

True or False? In Exercises 71–76, determine whether the statement is true or false. Justify your answer.

71. $\sin 60° \csc 60° = 1$ **72.** $\sec 30° = \csc 60°$

73. $\sin 45° + \cos 45° = 1$

74. $\cot^2 10° - \csc^2 10° = -1$

75. $\dfrac{\sin 60°}{\sin 30°} = \sin 2°$ **76.** $\tan[(5°)^2] = \tan^2(5°)$

77. *Writing* In right triangle trigonometry, explain why $\sin 30° = \frac{1}{2}$ regardless of the size of the triangle.

78. *Think About It* You are given only the value $\tan \theta$. Is it possible to find the value of $\sec \theta$ without finding the measure of θ? Explain.

79. *Exploration*

(a) Complete the table.

θ	0.1	0.2	0.3	0.4	0.5
$\sin \theta$					

(b) As θ approaches 0, how do θ and $\sin \theta$ compare? Explain.

80. *Exploration*

(a) Complete the table.

θ	0°	18°	36°	54°	72°	90°
$\sin \theta$						
$\cos \theta$						

(b) Discuss the behavior of the sine function for θ in the range from 0° to 90°.

(c) Discuss the behavior of the cosine function for θ in the range from 0° to 90°.

(d) Use the definitions of the sine and cosine functions to explain the results of parts (b) and (c).

Review

In Exercises 81–84, perform the operations and simplify.

81. $\dfrac{x^2 - 6x}{x^2 + 4x - 12} \cdot \dfrac{x^2 + 12x + 36}{x^2 - 36}$

82. $\dfrac{2t^2 + 5t - 12}{9 - 4t^2} \div \dfrac{t^2 - 16}{4t^2 + 12t + 9}$

83. $\dfrac{3}{x + 2} - \dfrac{2}{x - 2} + \dfrac{x}{x^2 + 4x + 4}$

84. $\dfrac{\left(\dfrac{3}{x} - \dfrac{1}{4}\right)}{\left(\dfrac{12}{x} - 1\right)}$

In Exercises 85 and 86, solve for x.

85. $\dfrac{2}{x + 3} + \dfrac{4}{x - 2} = \dfrac{12}{x^2 + x - 6}$

86. $\dfrac{3x + 2}{x^2 + x - 2} = \dfrac{4}{x + 2} - \dfrac{2}{1 - x}$

1.4 Trigonometric Functions of Any Angle

▶ **What you should learn**

- How to evaluate trigonometric functions of any angle
- How to use reference angles to evaluate trigonometric functions
- How to evaluate trigonometric functions of real numbers

▶ **Why you should learn it**

You can use trigonometric functions to model and solve real-life problems. For instance, in Exercise 91 on page 162, you are asked to use trigonometric functions to model the monthly normal temperatures in New York City and Fairbanks, Alaska.

James Urbach/SuperStock

Introduction

In Section 1.3, the definitions of trigonometric functions were restricted to acute angles. In this section, the definitions are extended to cover *any* angle. If θ is an *acute* angle, these definitions coincide with those given in the preceding section.

Definitions of Trigonometric Functions of Any Angle

Let θ be an angle in standard position with (x, y) a point on the terminal side of θ and $r = \sqrt{x^2 + y^2} \neq 0$.

$$\sin \theta = \frac{y}{r} \qquad\qquad \cos \theta = \frac{x}{r}$$

$$\tan \theta = \frac{y}{x}, \quad x \neq 0 \qquad \cot \theta = \frac{x}{y}, \quad y \neq 0$$

$$\sec \theta = \frac{r}{x}, \quad x \neq 0 \qquad \csc \theta = \frac{r}{y}, \quad y \neq 0$$

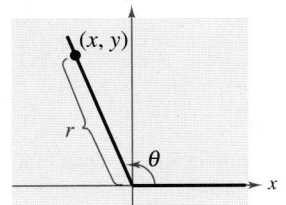

Because $r = \sqrt{x^2 + y^2}$ *cannot* be zero, it follows that the sine and cosine functions are defined for any real value of θ. However, if $x = 0$, the tangent and secant of θ are undefined. For example, the tangent of $90°$ is undefined. Similarly, if $y = 0$, the cotangent and cosecant of θ are undefined.

Example 1 ▶ **Evaluating Trigonometric Functions**

Let $(-3, 4)$ be a point on the terminal side of θ. Find the sine, cosine, and tangent of θ.

Solution

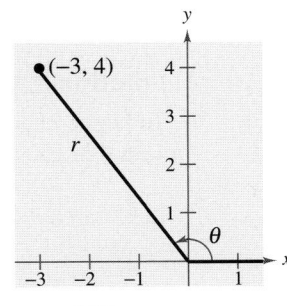

FIGURE 1.34

Referring to Figure 1.34, you can see that $x = -3$, $y = 4$, and

$$r = \sqrt{x^2 + y^2} = \sqrt{(-3)^2 + 4^2} = \sqrt{25} = 5.$$

So, you have the following.

$$\sin \theta = y/r = 4/5, \qquad \cos \theta = x/r = -3/5, \qquad \tan \theta = y/x = -4/3$$

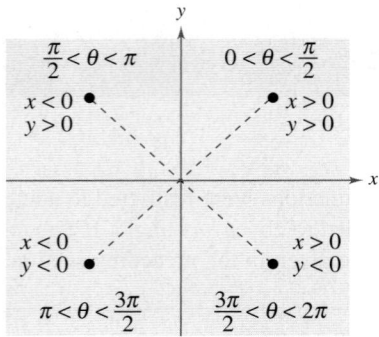

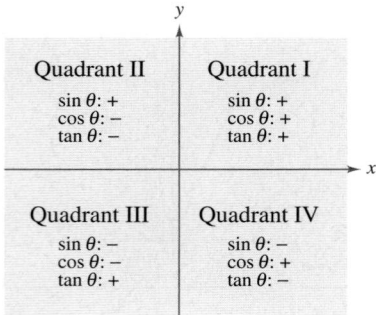

FIGURE **1.35**

The *signs* of the trigonometric functions in the four quadrants can be determined easily from the definitions of the functions. For instance, because $\cos \theta = x/r$, it follows that $\cos \theta$ is positive wherever $x > 0$, which is in Quadrants I and IV. (Remember, r is always positive.) In a similar manner, you can verify the results shown in Figure 1.35.

Example 2 ▶ **Evaluating Trigonometric Functions**

Given $\tan \theta = -\frac{5}{4}$ and $\cos \theta > 0$, find $\sin \theta$ and $\sec \theta$.

Solution

Note that θ lies in Quadrant IV because that is the only quadrant in which the tangent is negative and the cosine is positive. Moreover, using

$$\tan \theta = \frac{y}{x}$$

$$= -\frac{5}{4}$$

and the fact that y is negative in Quadrant IV, you can let $y = -5$ and $x = 4$. So, $r = \sqrt{16 + 25} = \sqrt{41}$ and you have

$$\sin \theta = \frac{y}{r} = \frac{-5}{\sqrt{41}}$$

$$\approx -0.7809$$

$$\sec \theta = \frac{r}{x} = \frac{\sqrt{41}}{4}$$

$$\approx 1.6008.$$

Example 3 ▶ **Trigonometric Functions of Quadrant Angles**

Evaluate the sine function at the four quadrant angles 0, $\dfrac{\pi}{2}$, π, and $\dfrac{3\pi}{2}$.

Solution

To begin, choose a point on the terminal side of each angle, as shown in Figure 1.36. For each of the four points, $r = 1$, and you have

$$\sin 0 = \frac{y}{r} = \frac{0}{1} = 0 \qquad\qquad (x, y) = (1, 0)$$

$$\sin \frac{\pi}{2} = \frac{y}{r} = \frac{1}{1} = 1 \qquad\qquad (x, y) = (0, 1)$$

$$\sin \pi = \frac{y}{r} = \frac{0}{1} = 0 \qquad\qquad (x, y) = (-1, 0)$$

$$\sin \frac{3\pi}{2} = \frac{y}{r} = \frac{-1}{1} = -1. \qquad (x, y) = (0, -1)$$

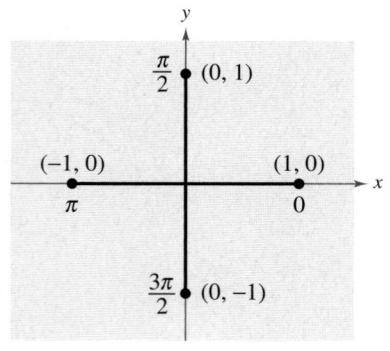

FIGURE **1.36**

Sketching several angles with their reference angles helps reinforce the fact that the reference angle is the acute angle formed with the horizontal.

Reference Angles

The values of the trigonometric functions of angles greater than 90° (or less than 0°) can be determined from their values at corresponding acute angles called **reference angles.**

> ## Definition of Reference Angle
>
> Let θ be an angle in standard position. Its **reference angle** is the acute angle θ' formed by the terminal side of θ and the horizontal axis.

Figure 1.37 shows the reference angles for θ in Quadrants II, III, and IV.

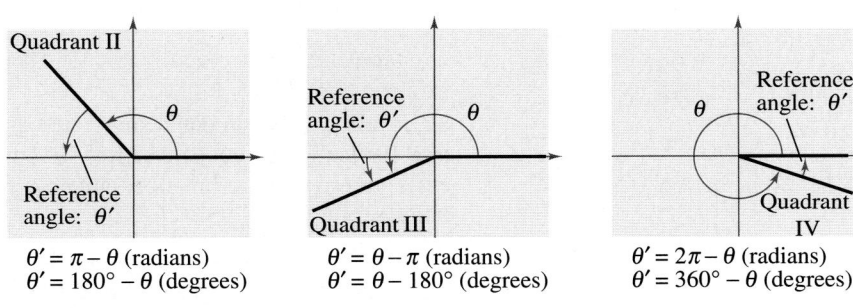

$\theta' = \pi - \theta$ (radians)
$\theta' = 180° - \theta$ (degrees)

$\theta' = \theta - \pi$ (radians)
$\theta' = \theta - 180°$ (degrees)

$\theta' = 2\pi - \theta$ (radians)
$\theta' = 360° - \theta$ (degrees)

FIGURE **1.37**

Example 4 ▶ **Finding Reference Angles**

Find the reference angle θ'.

a. $\theta = 300°$ **b.** $\theta = 2.3$ **c.** $\theta = -135°$

Solution

a. Because 300° lies in Quadrant IV, the angle it makes with the x-axis is

$$\theta' = 360° - 300°$$

$$= 60°. \qquad \text{Degrees}$$

Figure 1.38 shows the angle $\theta = 300°$ and its reference angle $\theta' = 60°$.

b. Because 2.3 lies between $\pi/2 \approx 1.5708$ and $\pi \approx 3.1416$, it follows that it is in Quadrant II and its reference angle is

$$\theta' = \pi - 2.3$$

$$\approx 0.8416. \qquad \text{Radians}$$

Figure 1.39 shows the angle $\theta = 2.3$ and its reference angle $\theta' = \pi - 2.3$.

c. First, determine that $-135°$ is coterminal with 225°, which lies in Quadrant III. So, the reference angle is

$$\theta' = 225° - 180°$$

$$= 45°. \qquad \text{Degrees}$$

Figure 1.40 shows the angle $\theta = -135°$ and its reference angle $\theta' = 45°$.

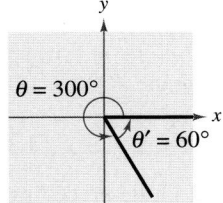

FIGURE **1.38**

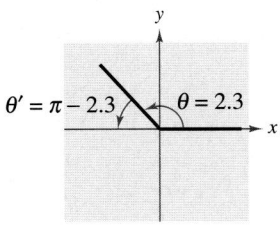

FIGURE **1.39**

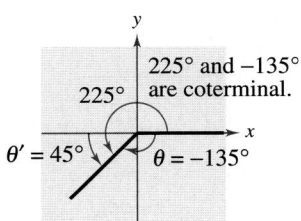

FIGURE **1.40**

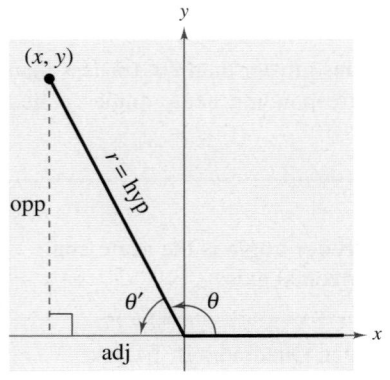

opp = $|y|$, adj = $|x|$

FIGURE 1.41

Trigonometric Functions of Real Numbers

To see how a reference angle is used to evaluate a trigonometric function, consider the point (x, y) on the terminal side of θ, as shown in Figure 1.41. By definition, you know that

$$\sin \theta = \frac{y}{r} \quad \text{and} \quad \tan \theta = \frac{y}{x}.$$

For the right triangle with acute angle θ' and sides of lengths $|x|$ and $|y|$, you have

$$\sin \theta' = \frac{\text{opp}}{\text{hyp}} = \frac{|y|}{r}$$

and

$$\tan \theta' = \frac{\text{opp}}{\text{adj}} = \frac{|y|}{|x|}.$$

So, it follows that $\sin \theta$ and $\sin \theta'$ are equal, *except possibly in sign*. The same is true for $\tan \theta$ and $\tan \theta'$ *and* for the other four trigonometric functions. In all cases, the sign of the function value can be determined by the quadrant in which θ lies.

Evaluating Trigonometric Functions of Any Angle

To find the value of a trigonometric function of any angle θ,

1. determine the function value for the associated reference angle θ';

2. depending on the quadrant in which θ lies, affix the appropriate sign to the function value.

By using reference angles and the special angles discussed in the preceding section, you can greatly extend the scope of *exact* trigonometric values. For instance, knowing the function values of 30° means that you know the function values of all angles for which 30° is a reference angle. For convenience, the table below shows the exact values of the trigonometric functions of special angles and quadrant angles.

Trigonometric Values of Common Angles

θ (degrees)	0°	30°	45°	60°	90°	180°	270°
θ (radians)	0	$\frac{\pi}{6}$	$\frac{\pi}{4}$	$\frac{\pi}{3}$	$\frac{\pi}{2}$	π	$\frac{3\pi}{2}$
$\sin \theta$	0	$\frac{1}{2}$	$\frac{\sqrt{2}}{2}$	$\frac{\sqrt{3}}{2}$	1	0	-1
$\cos \theta$	1	$\frac{\sqrt{3}}{2}$	$\frac{\sqrt{2}}{2}$	$\frac{1}{2}$	0	-1	0
$\tan \theta$	0	$\frac{\sqrt{3}}{3}$	1	$\sqrt{3}$	Undef.	0	Undef.

Trigonometric Functions of Nonacute Angles

Evaluate each trigonometric function.

a. $\cos \dfrac{4\pi}{3}$ **b.** $\tan(-210°)$ **c.** $\csc \dfrac{11\pi}{4}$

Solution

a. Because $\theta = 4\pi/3$ lies in Quadrant III, the reference angle is $\theta' = (4\pi/3) - \pi = \pi/3$, as shown in Figure 1.42. Moreover, the cosine is negative in Quadrant III, so

$$\cos \frac{4\pi}{3} = (-)\cos \frac{\pi}{3}$$

$$= -\frac{1}{2}.$$

Emphasize the importance of reference angles in evaluating trigonometric functions of angles greater than 90°.

b. Because $-210° + 360° = 150°$, it follows that $-210°$ is coterminal with the second-quadrant angle $150°$. So, the reference angle is $\theta' = 180° - 150° = 30°$, as shown in Figure 1.43. Finally, because the tangent is negative in Quadrant II, you have

$$\tan(-210°) = (-)\tan 30°$$

$$= -\frac{\sqrt{3}}{3}.$$

c. Because $(11\pi/4) - 2\pi = 3\pi/4$, it follows that $11\pi/4$ is coterminal with the second-quadrant angle $3\pi/4$. So, the reference angle is $\theta' = \pi - (3\pi/4) = \pi/4$, as shown in Figure 1.44. Because the cosecant is positive in Quadrant II, you have

$$\csc \frac{11\pi}{4} = (+)\csc \frac{\pi}{4}$$

$$= \frac{1}{\sin(\pi/4)}$$

$$= \sqrt{2}.$$

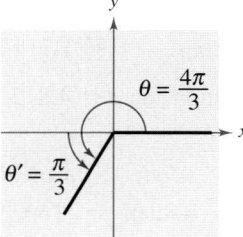

FIGURE 1.42

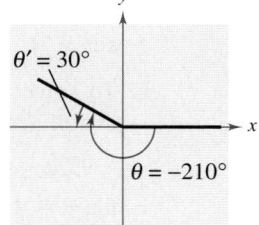

FIGURE 1.43

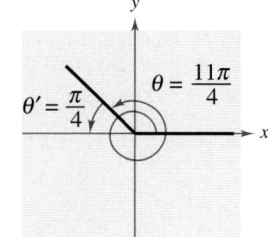

FIGURE 1.44

Example 6 ▶ **Using Trigonometric Identities**

Let θ be an angle in Quadrant II such that $\sin \theta = \frac{1}{3}$. Find (a) $\cos \theta$ and (b) $\tan \theta$ by using trigonometric identities.

Solution

a. Using the Pythagorean identity $\sin^2 \theta + \cos^2 \theta = 1$, you obtain

$$\left(\frac{1}{3}\right)^2 + \cos^2 \theta = 1 \qquad \text{Substitute } \tfrac{1}{3} \text{ for } \sin \theta.$$

$$\cos^2 \theta = 1 - \frac{1}{9} = \frac{8}{9}.$$

Because $\cos \theta < 0$ in Quadrant II, you can use the negative root to obtain

$$\cos \theta = -\frac{\sqrt{8}}{\sqrt{9}}$$

$$= -\frac{2\sqrt{2}}{3}.$$

Students often have difficulty determining angles, especially when the functions given are csc, sec, and/or cot. Have your students rewrite the expression in terms of sin, cos, or tan, whichever is applicable, before evaluating.

b. Using the trigonometric identity $\tan \theta = \sin \theta / \cos \theta$, you obtain

$$\tan \theta = \frac{1/3}{-2\sqrt{2}/3} \qquad \text{Substitute for } \sin \theta \text{ and } \cos \theta.$$

$$= -\frac{1}{2\sqrt{2}}$$

$$= -\frac{\sqrt{2}}{4}.$$

Example 7 ▶ **Using a Calculator**

a. Use a calculator to evaluate $\cot 410°$ and $\sin(-7)$.

b. Use a calculator to solve $\tan \theta = 4.812$, $0 \le \theta < 2\pi$.

Additional Examples

a. Solve the equation
   $\cot \theta = -4.3315, 0° \le \theta < 360°$.
b. Solve the equation $\csc \theta = 1.1034$,
   $0 \le \theta < 2\pi$.
c. Solve the equation
   $\sec \theta = -1.3054, 0 \le \theta < 2\pi$.

Solutions

a. $\tan \theta = \dfrac{1}{\cot \theta} = -0.230866905$;

   $\theta \approx 167°, 347°$

b. $\sin \theta = \dfrac{1}{\csc \theta} = 0.906289650$;

   $\theta \approx 1.13, 2.01$

c. $\cos \theta = \dfrac{1}{\sec \theta} = -0.766048721$;

   $\theta \approx 2.44, 3.84$

Solution

Function	Mode	Calculator Keystrokes	Display
a. $\cot 410°$	Degree	(TAN 410) x^{-1} ENTER	0.8390996
$\sin(-7)$	Radian	SIN ((−) 7) ENTER	−0.6569866

b. To solve the equation $\tan \theta = 4.812$, you can use the inverse tangent key, as follows.

Equation	Mode	Calculator Keystrokes	Display
$\tan \theta = 4.812$	Radian	TAN⁻¹ 4.812 ENTER	1.365898912

The angle $\theta \approx 1.3659$ lies in Quadrant I. A second value of θ lies in Quadrant III (tangent is positive) and is

$$\theta = \pi + 1.3659 \approx 4.5075.$$

1.4 Exercises

In Exercises 1–4, determine the exact values of the six trigonometric functions of the angle θ.

1. (a)
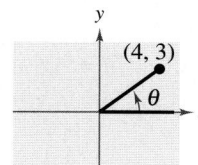

(b)
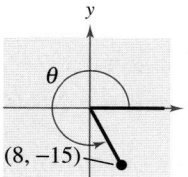

2. (a)
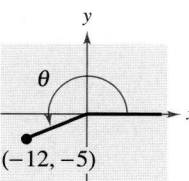

(b)
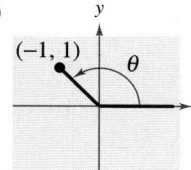

3. (a)
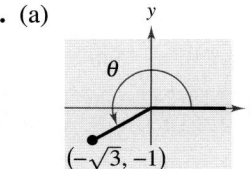

(b)
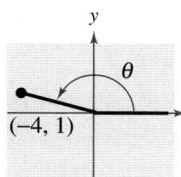

4. (a)
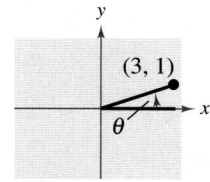

(b)
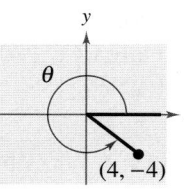

In Exercises 5–10, the point is on the terminal side of an angle in standard position. Determine the exact values of the six trigonometric functions of the angle.

5. $(7, 24)$

6. $(8, 15)$

7. $(-4, 10)$

8. $(-5, -2)$

9. $(-3.5, 6.8)$

10. $\left(3\frac{1}{2}, -7\frac{3}{4}\right)$

In Exercises 11–14, state the quadrant in which θ lies.

11. $\sin \theta < 0$ and $\cos \theta < 0$

12. $\sin \theta > 0$ and $\cos \theta > 0$

13. $\sin \theta > 0$ and $\tan \theta < 0$

14. $\sec \theta > 0$ and $\cot \theta < 0$

In Exercises 15–24, find the values of the six trigonometric functions of θ.

Function Value	Constraint
15. $\sin \theta = \frac{3}{5}$	θ lies in Quadrant II.
16. $\cos \theta = -\frac{4}{5}$	θ lies in Quadrant III.
17. $\tan \theta = -\frac{15}{8}$	$\sin \theta < 0$
18. $\cos \theta = \frac{8}{17}$	$\tan \theta < 0$
19. $\cot \theta = -3$	$\cos \theta > 0$
20. $\csc \theta = 4$	$\cot \theta < 0$
21. $\sec \theta = -2$	$\sin \theta > 0$
22. $\sin \theta = 0$	$\sec \theta = -1$
23. $\cot \theta$ is undefined.	$\pi/2 \leq \theta \leq 3\pi/2$
24. $\tan \theta$ is undefined.	$\pi \leq \theta \leq 2\pi$

In Exercises 25–28, the terminal side of θ lies on the given line in the specified quadrant. Find the values of the six trigonometric functions of θ by finding a point on the line.

Line	Quadrant
25. $y = -x$	II
26. $y = \frac{1}{3}x$	III
27. $2x - y = 0$	III
28. $4x + 3y = 0$	IV

In Exercises 29–36, evaluate the trigonometric function of the quadrant angle.

29. $\cos \pi$

30. $\cos \dfrac{3\pi}{2}$

31. $\sec \dfrac{3\pi}{2}$

32. $\sec \pi$

33. $\tan \dfrac{\pi}{2}$

34. $\tan \pi$

35. $\csc \pi$

36. $\cot \dfrac{\pi}{2}$

In Exercises 37–44, find the reference angle θ′, and sketch θ and θ′ in standard position.

37. $\theta = 203°$

38. $\theta = 309°$

39. $\theta = -245°$

40. $\theta = -145°$

41. $\theta = \dfrac{2\pi}{3}$

42. $\theta = \dfrac{7\pi}{4}$

43. $\theta = 3.5$

44. $\theta = \dfrac{11\pi}{3}$

In Exercises 45–58, evaluate the sine, cosine, and tangent of the angle without using a calculator.

45. $225°$

46. $300°$

47. $750°$

48. $-405°$

49. $-150°$

50. $-840°$

51. $\dfrac{4\pi}{3}$

52. $\dfrac{\pi}{4}$

53. $-\dfrac{\pi}{6}$

54. $-\dfrac{\pi}{2}$

55. $\dfrac{11\pi}{4}$

56. $\dfrac{10\pi}{3}$

57. $-\dfrac{3\pi}{2}$

58. $-\dfrac{25\pi}{4}$

In Exercises 59–68, use a calculator to evaluate the trigonometric function. Round your answer to four decimal places. (Be sure the calculator is set in the correct angle mode.)

59. $\sin 10°$

60. $\sec 225°$

61. $\cos(-110°)$

62. $\csc(-330°)$

63. $\tan 4.5$

64. $\cot 1.35$

65. $\tan \dfrac{\pi}{9}$

66. $\tan\left(-\dfrac{\pi}{9}\right)$

67. $\sin(-0.65)$

68. $\sin 0.65$

In Exercises 69–74, find two solutions of the equation. Give your answers in degrees ($0° \le \theta < 360°$) and radians ($0 \le \theta < 2\pi$). Do not use a calculator.

69. (a) $\sin \theta = \dfrac{1}{2}$ (b) $\sin \theta = -\dfrac{1}{2}$

70. (a) $\cos \theta = \dfrac{\sqrt{2}}{2}$ (b) $\cos \theta = -\dfrac{\sqrt{2}}{2}$

71. (a) $\csc \theta = \dfrac{2\sqrt{3}}{3}$ (b) $\cot \theta = -1$

72. (a) $\sec \theta = 2$ (b) $\sec \theta = -2$

73. (a) $\tan \theta = 1$ (b) $\cot \theta = -\sqrt{3}$

74. (a) $\sin \theta = \dfrac{\sqrt{3}}{2}$ (b) $\sin \theta = -\dfrac{\sqrt{3}}{2}$

In Exercises 75–78, use a calculator to approximate two values of θ ($0° \le \theta < 360°$) that satisfy the equation. Round the values to two decimal places.

75. $\sin \theta = 0.8191$

76. $\cos \theta = 0.8746$

77. $\cos \theta = -0.4367$

78. $\sin \theta = -0.6514$

In Exercises 79–84, use a calculator to approximate two values of θ ($0 \le \theta < 2\pi$) that satisfy the equation. Round the values to three decimal places.

79. $\cos \theta = 0.9848$

80. $\sin \theta = 0.0175$

81. $\tan \theta = 1.192$

82. $\cot \theta = 5.671$

83. $\sec \theta = -2.6667$

84. $\cos \theta = -0.3214$

In Exercises 85–90, find the indicated trigonometric value in the specified quadrant.

	Function	Quadrant	Trigonometric Value
85.	$\sin \theta = -\frac{3}{5}$	IV	$\cos \theta$
86.	$\cot \theta = -3$	II	$\sin \theta$
87.	$\tan \theta = \frac{3}{2}$	III	$\sec \theta$
88.	$\csc \theta = -2$	IV	$\cot \theta$
89.	$\cos \theta = \frac{5}{8}$	I	$\sec \theta$
90.	$\sec \theta = -\frac{9}{4}$	III	$\tan \theta$

▶ Model It

91. *Data Analysis* The table shows the monthly normal temperatures (in degrees Fahrenheit) for selected months for New York City (*N*) and Fairbanks, Alaska (*F*). (Source: National Climatic Data Center)

Month	N	F
January	32	-10
April	53	31
July	77	63
October	58	25
December	37	-7

(a) Use the *regression* feature of a graphing utility to find a model of the form

$$y = a \sin(bt + c) + d$$

for each city. Let t represent the month, with $t = 1$ corresponding to January.

(b) Use the models from part (a) to find the monthly normal temperatures for the two cities in February, March, May, June, August, September, and November.

(c) Compare the models for the two cities.

92. *Sales* A company that produces snowboards, which is a seasonal product, forecasts monthly sales over the next 2 years to be

$$S = 23.1 + 0.442t + 4.3 \cos \frac{\pi t}{6}$$

where S is measured in thousands of units and t is the time in months, with $t = 1$ representing January 2003. Predict sales for each of the following months.

(a) February 2003   (b) February 2004

(c) June 2003   (d) June 2004

93. *Harmonic Motion* The displacement from equilibrium of an oscillating weight suspended by a spring is

$$y(t) = 2 \cos 6t$$

where y is the displacement in centimeters and t is the time in seconds. Find the displacement when (a) $t = 0$, (b) $t = \frac{1}{4}$, and (c) $t = \frac{1}{2}$.

94. *Harmonic Motion* The displacement from equilibrium of an oscillating weight suspended by a spring and subject to the damping effect of friction is

$$y(t) = 2e^{-t} \cos 6t$$

where y is the displacement in centimeters and t is the time in seconds. Find the displacement when (a) $t = 0$, (b) $t = \frac{1}{4}$, and (c) $t = \frac{1}{2}$.

95. *Electric Circuits* The current I (in amperes) when 100 volts is applied to a circuit is

$$I = 5e^{-2t} \sin t$$

where t is the time in seconds after the voltage is applied. Approximate the current $t = 0.7$ second after the voltage is applied.

96. *Distance* An airplane, flying at an altitude of 6 miles, is on a flight path that passes directly over an observer (see figure). If θ is the angle of elevation from the observer to the plane, find the distance d from the observer to the plane when (a) $\theta = 30°$, (b) $\theta = 90°$, and (c) $\theta = 120°$.

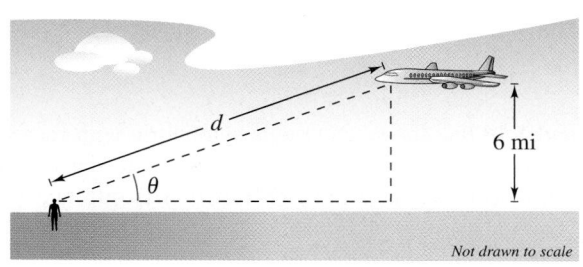

Not drawn to scale

97. *Writing* Consider an angle in standard position with $r = 12$ centimeters, as shown in the figure. Write a short paragraph describing the changes in the magnitudes of x, y, $\sin \theta$, $\cos \theta$, and $\tan \theta$ as θ increases continuously from $0°$ to $90°$.

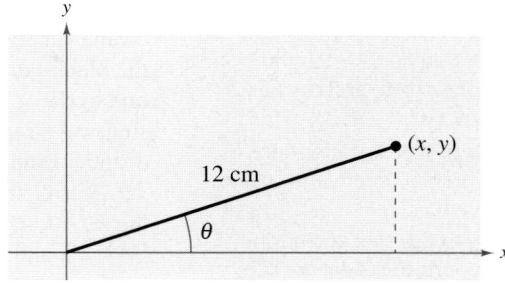

98. *Writing* Explain how reference angles are used to find the trigonometric functions of obtuse angles.

Review

In Exercises 99–106, graph the function. Identify the intercepts, asymptotes, and domain of the function.

99. $y = x - 8$

100. $y = 6 - 7x$

101. $y = x^2 + 3x - 4$

102. $y = 2x^2 - 5x$

103. $f(x) = x^3 - 8$

104. $g(x) = x^4 + 2x^2 - 3$

105. $g(x) = \sqrt{x + 5}$

106. $f(x) = \sqrt{4x - 1}$

1.5 Graphs of Sine and Cosine Functions

What you should learn

- How to sketch the graphs of basic sine and cosine functions
- How to use amplitude and period to help sketch the graphs of sine and cosine functions
- How to sketch translations of the graphs of sine and cosine functions
- How to use sine and cosine functions to model real-life data

Why you should learn it

Sine and cosine functions are often used in scientific calculations. For instance, in Exercise 74 on page 173, you can use a trigonometric function to model the percent of the moon's face that is illuminated for any given day in 2005.

Basic Sine and Cosine Curves

In this section you will study techniques for sketching the graphs of the sine and cosine functions. The graph of the sine function is a **sine curve.** In Figure 1.45, the black portion of the graph represents one period of the function and is called **one cycle** of the sine curve. The gray portion of the graph indicates that the basic sine wave repeats indefinitely in the positive and negative directions. The graph of the cosine function is shown in Figure 1.46.

Recall from Section 1.2 that the domain of the sine and cosine functions is the set of all real numbers. Moreover, the range of each function is the interval $[-1, 1]$, and each function has a period of 2π. Do you see how this information is consistent with the basic graphs shown in Figures 1.45 and 1.46?

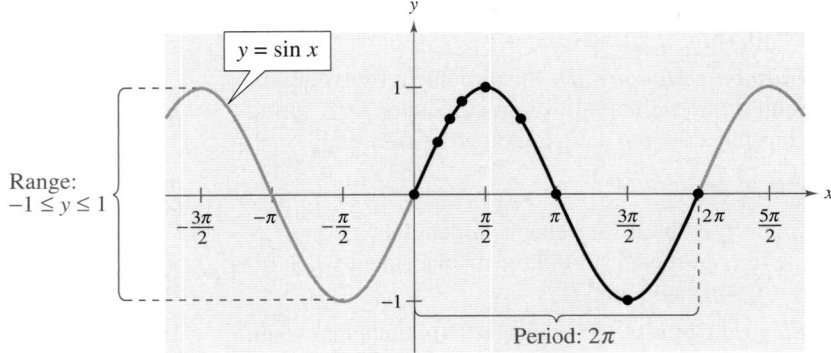

FIGURE 1.45

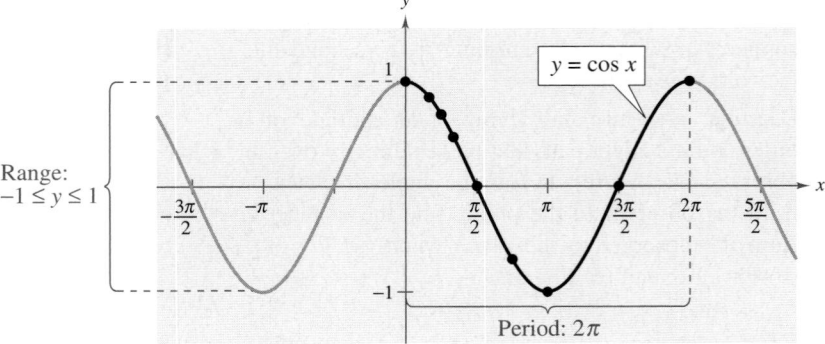

FIGURE 1.46

Note in Figures 1.45 and 1.46 that the sine curve is symmetric with respect to the *origin*, whereas the cosine curve is symmetric with respect to the *y-axis*. These properties of symmetry occur because the sine function is odd and the cosine function is even.

To sketch the graphs of the basic sine and cosine functions by hand, it helps to note five **key points** in one period of each graph: the *intercepts*, *maximum points*, and *minimum points* (see Figure 1.47).

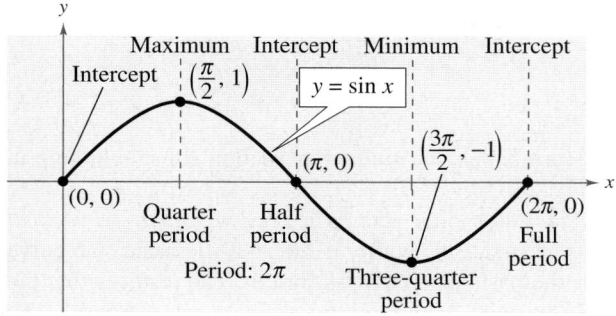
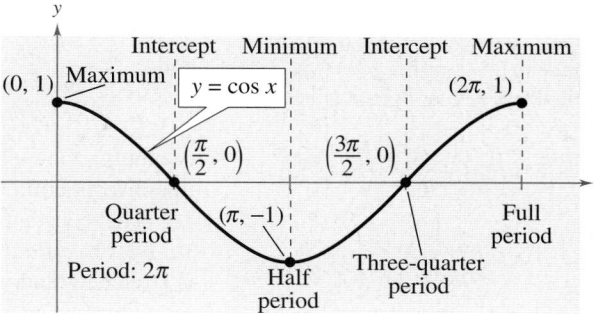

FIGURE **1.47**

Example 1 ▶ **Using Key Points to Sketch a Sine Curve**

Sketch the graph of $y = 2 \sin x$ on the interval $[-\pi, 4\pi]$.

Solution

Note that

$$y = 2 \sin x = 2(\sin x)$$

indicates that the y-values for the key points will have twice the magnitude of those on the graph of $y = \sin x$. Divide the period 2π into four equal parts to get the key points for $y = 2 \sin x$.

Intercept	*Maximum*	*Intercept*	*Minimum*		*Intercept*
$(0, 0)$,	$\left(\dfrac{\pi}{2}, 2\right)$,	$(\pi, 0)$,	$\left(\dfrac{3\pi}{2}, -2\right)$,	and	$(2\pi, 0)$

By connecting these key points with a smooth curve and extending the curve in both directions over the interval $[-\pi, 4\pi]$, you obtain the graph shown in Figure 1.48.

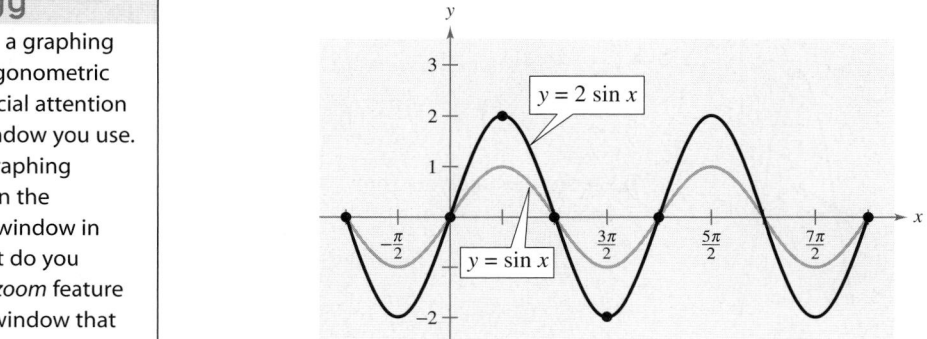

FIGURE **1.48**

Amplitude and Period

To graph the examples in this section, your students must know the basic graphs of $y = \sin x$ and $y = \cos x$. For example, to sketch the graph of $y = 3 \sin x$, your students must be able to identify that because $a = 3$, the amplitude is 3 times the amplitude of $y = \sin x$.

In the remainder of this section you will study the graphic effect of each of the constants a, b, c, and d in equations of the forms

$$y = d + a \sin(bx - c)$$

and

$$y = d + a \cos(bx - c).$$

A quick review of the transformations you studied in Section P.8 should help in this investigation.

The constant factor a in $y = a \sin x$ acts as a *scaling factor*—a *vertical stretch* or *vertical shrink* of the basic sine curve. If $|a| > 1$, the basic sine curve is stretched, and if $|a| < 1$, the basic sine curve is shrunk. The result is that the graph of $y = a \sin x$ ranges between $-a$ and a instead of between -1 and 1. The absolute value of a is the **amplitude** of the function $y = a \sin x$. The range of the function $y = a \sin x$ for $a > 0$ is $-a \leq y \leq a$.

Definition of Amplitude of Sine and Cosine Curves

The **amplitude** of $y = a \sin x$ and $y = a \cos x$ represents half the distance between the maximum and minimum values of the function and is given by

$$\text{Amplitude} = |a|.$$

To help students learn how to determine and locate key points (intercepts, minimums, maximums), have them mark each of the points on their graphs and then check their graphs using a graphing utility.

Example 2 ▶   **Scaling: Vertical Shrinking and Stretching**

On the same coordinate axes, sketch the graph of each function.

a. $y = \dfrac{1}{2} \cos x$

b. $y = 3 \cos x$

Solution

a. Because the amplitude of $y = \frac{1}{2} \cos x$ is $\frac{1}{2}$, the maximum value is $\frac{1}{2}$ and the minimum value is $-\frac{1}{2}$. Divide one cycle, $0 \leq x \leq 2\pi$, into four equal parts to get the key points

Maximum	Intercept	Minimum	Intercept		Maximum
$\left(0, \dfrac{1}{2}\right)$,	$\left(\dfrac{\pi}{2}, 0\right)$,	$\left(\pi, -\dfrac{1}{2}\right)$,	$\left(\dfrac{3\pi}{2}, 0\right)$,	and	$\left(2\pi, \dfrac{1}{2}\right)$.

b. A similar analysis shows that the amplitude of $y = 3 \cos x$ is 3, and the key points are

Maximum	Intercept	Minimum	Intercept		Maximum
$(0, 3)$,	$\left(\dfrac{\pi}{2}, 0\right)$,	$(\pi, -3)$,	$\left(\dfrac{3\pi}{2}, 0\right)$,	and	$(2\pi, 3)$.

The graphs of these two functions are shown in Figure 1.49.

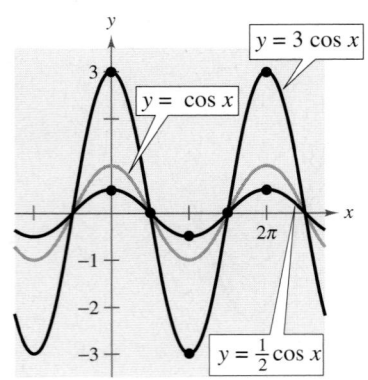

FIGURE **1.49**

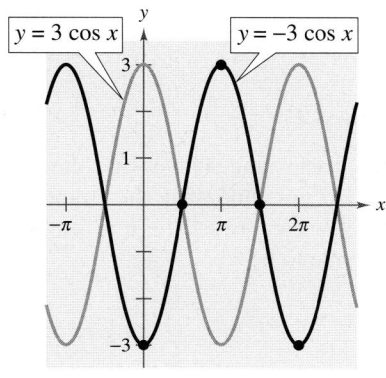

FIGURE **1.50**

You know from Section P.8 that the graph of $y = -f(x)$ is a **reflection** in the x-axis of the graph of $y = f(x)$. For instance, the graph of

$$y = -3 \cos x$$

is a reflection of the graph of

$$y = 3 \cos x$$

as shown in Figure 1.50.

Because $y = a \sin x$ completes one cycle from $x = 0$ to $x = 2\pi$, it follows that $y = a \sin bx$ completes one cycle from $x = 0$ to $x = 2\pi/b$.

> ### Period of Sine and Cosine Functions
>
> Let b be a positive real number. The **period** of $y = a \sin bx$ and $y = a \cos bx$ is $2\pi/b$.

Note that if $0 < b < 1$, the period of $y = a \sin bx$ is greater than 2π and represents a *horizontal stretching* of the graph of $y = a \sin x$. Similarly, if $b > 1$, the period of $y = a \sin bx$ is less than 2π and represents a *horizontal shrinking* of the graph of $y = a \sin x$. If b is negative, the identities $\sin(-x) = -\sin x$ and $\cos(-x) = \cos x$ are used to rewrite the function.

◀ Exploration ▶

Sketch the graph of $y = \cos bx$ for $b = \frac{1}{2}$, 2, and 3. How does the value of b affect the graph? How many complete cycles occur between 0 and 2π for each value of b?

Example 3 ▶ Scaling: Horizontal Stretching

Sketch the graph of $y = \sin \dfrac{x}{2}$.

Solution

The amplitude is 1. Moreover, because $b = \frac{1}{2}$, the period is

$$\frac{2\pi}{b} = \frac{2\pi}{\frac{1}{2}} = 4\pi. \qquad \text{Substitute for } b.$$

Now, divide the period-interval $[0, 4\pi]$ into four equal parts with the values π, 2π, and 3π to obtain the key points on the graph.

Intercept	Maximum	Intercept	Minimum		Intercept
$(0, 0)$,	$(\pi, 1)$,	$(2\pi, 0)$,	$(3\pi, -1)$,	and	$(4\pi, 0)$

The graph is shown in Figure 1.51.

STUDY TIP

In general, to divide a period-interval into four equal parts, successively add "period/4," starting with the left endpoint of the interval. For instance, for the period-interval $[-\pi/6, \pi/2]$ of length $2\pi/3$, you would successively add

$$\frac{2\pi/3}{4} = \frac{\pi}{6}$$

to get $-\pi/6$, 0, $\pi/6$, $\pi/3$, and $\pi/2$.

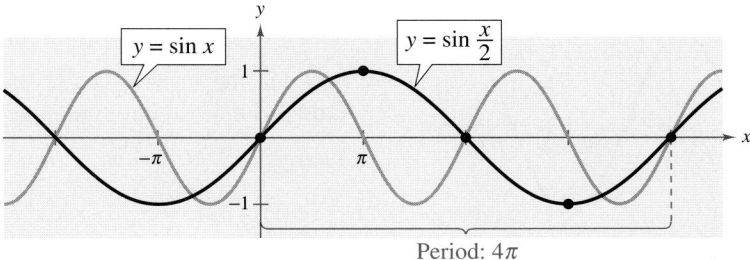

FIGURE **1.51**

Translations of Sine and Cosine Curves

The constant c in the general equations

$$y = a \sin(bx - c) \quad \text{and} \quad y = a \cos(bx - c)$$

creates a *horizontal translation* (shift) of the basic sine and cosine curves. Comparing $y = a \sin bx$ with $y = a \sin(bx - c)$, you find that the graph of $y = a \sin(bx - c)$ completes one cycle from $bx - c = 0$ to $bx - c = 2\pi$. By solving for x, you can find the interval for one cycle to be

Left endpoint   Right endpoint

$$\overbrace{\frac{c}{b}}^{} \le x \le \underbrace{\frac{c}{b} + \frac{2\pi}{b}}_{\text{Period}}.$$

This implies that the period of $y = a \sin(bx - c)$ is $2\pi/b$, and the graph of $y = a \sin bx$ is shifted by an amount c/b. The number c/b is the **phase shift.**

◀ **E x p l o r a t i o n** ▶

Sketch the graph of

$$y = \sin(x - c)$$

where $c = -\pi/4$, 0, and $\pi/4$. How does the value of c affect the graph?

Graphs of Sine and Cosine Functions

The graphs of $y = a \sin(bx - c)$ and $y = a \cos(bx - c)$ have the following characteristics. (Assume $b > 0$.)

$$\text{Amplitude} = |a| \qquad \text{Period} = \frac{2\pi}{b}$$

The left and right endpoints of a one-cycle interval can be determined by solving the equations $bx - c = 0$ and $bx - c = 2\pi$.

Example 4 ▶   **Horizontal Translation**

Sketch the graph of $y = \frac{1}{2} \sin(x - \pi/3)$.

Solution

The amplitude is $\frac{1}{2}$ and the period is 2π. By solving the equations

$$x - \frac{\pi}{3} = 0 \quad \Longrightarrow \quad x = \frac{\pi}{3}$$

and

$$x - \frac{\pi}{3} = 2\pi \quad \Longrightarrow \quad x = \frac{7\pi}{3}$$

you see that the interval $[\pi/3, 7\pi/3]$ corresponds to one cycle of the graph. Dividing this interval into four equal parts produces the key points

Intercept	*Maximum*	*Intercept*	*Minimum*		*Intercept*
$\left(\frac{\pi}{3}, 0\right)$,	$\left(\frac{5\pi}{6}, \frac{1}{2}\right)$,	$\left(\frac{4\pi}{3}, 0\right)$,	$\left(\frac{11\pi}{6}, -\frac{1}{2}\right)$,	and	$\left(\frac{7\pi}{3}, 0\right)$.

The graph is shown in Figure 1.52.

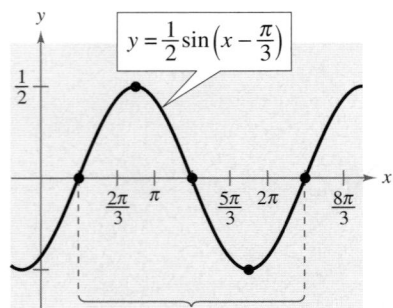

FIGURE **1.52**

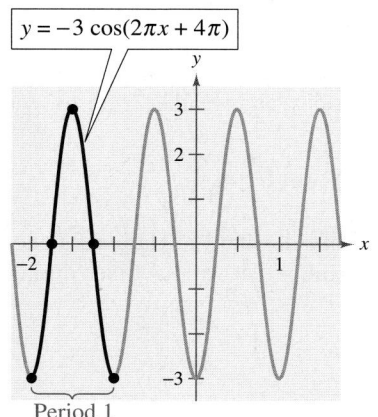

$y = -3 \cos(2\pi x + 4\pi)$

Period 1

FIGURE **1.53**

Activities

1. Describe the relationship between the graphs of $f(x) = \sin x$ and $g(x) = 3 \sin(2x + 1)$.

   Answer: The amplitude of the basic sine curve is 1, whereas the amplitude of $g(x)$ is 3. The period of the basic sine curve is 2π, whereas the period of $g(x)$ is π. Lastly, the graph of $g(x)$ has a phase shift $\frac{1}{2}$ unit to the left of the graph of $f(x) = \sin x$.

2. Determine the amplitude and period.

   $y = \frac{1}{2} \cos(\pi x - 1)$

   Answer: Amplitude, $\frac{1}{2}$; period, 2

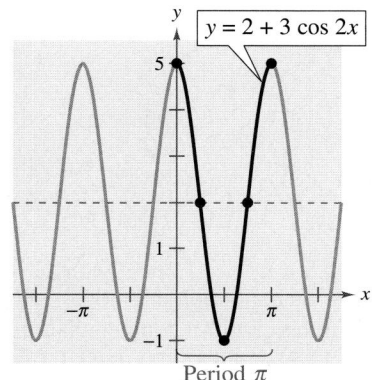

$y = 2 + 3 \cos 2x$

Period π

FIGURE **1.54**

Example 5 ▶   **Horizontal Translation**

Sketch the graph of

$$y = -3 \cos(2\pi x + 4\pi).$$

Solution

The amplitude is 3 and the period is $2\pi/2\pi = 1$. By solving the equations

$$2\pi x + 4\pi = 0$$
$$2\pi x = -4\pi$$
$$x = -2$$

and

$$2\pi x + 4\pi = 2\pi$$
$$2\pi x = -2\pi$$
$$x = -1$$

you see that the interval $[-2, -1]$ corresponds to one cycle of the graph. Dividing this interval into four equal parts produces the key points

Minimum	*Intercept*	*Maximum*	*Intercept*		*Minimum*
$(-2, -3)$,	$\left(-\frac{7}{4}, 0\right)$,	$\left(-\frac{3}{2}, 3\right)$,	$\left(-\frac{5}{4}, 0\right)$,	and	$(-1, -3)$.

The graph is shown in Figure 1.53.

The final type of transformation is the *vertical translation* caused by the constant d in the equations

$$y = d + a \sin(bx - c)$$

and

$$y = d + a \cos(bx - c).$$

The shift is d units upward for $d > 0$ and downward for $d < 0$. In other words, the graph oscillates about the horizontal line $y = d$ instead of the x-axis.

Example 6 ▶   **Vertical Translation**

Sketch the graph of

$$y = 2 + 3 \cos 2x.$$

Solution

The amplitude is 3 and the period is π. The key points over the interval $[0, \pi]$ are

$$(0, 5), \quad \left(\frac{\pi}{4}, 2\right), \quad \left(\frac{\pi}{2}, -1\right), \quad \left(\frac{3\pi}{4}, 2\right), \quad \text{and} \quad (\pi, 5).$$

The graph is shown in Figure 1.54.

Mathematical Modeling

Sine and cosine functions can be used to model many real-life situations, including electric currents, musical tones, radio waves, tides, and weather patterns.

Example 7 ▶ Finding a Trigonometric Model

Throughout the day, the depth of water at the end of a dock in Bar Harbor, Maine, varies with the tides. The table shows the depths (in feet) at various times during the morning. (Source: Nautical Software, Inc.)

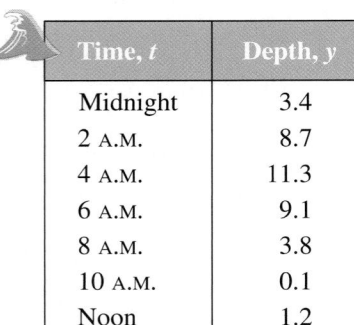

Time, t	Depth, y
Midnight	3.4
2 A.M.	8.7
4 A.M.	11.3
6 A.M.	9.1
8 A.M.	3.8
10 A.M.	0.1
Noon	1.2

a. Use a trigonometric function to model this data.

b. Find the depths at 9 A.M. and 3 P.M.

c. A boat needs at least 10 feet of water to moor at the dock. During what times in the afternoon can it safely dock?

Solution

a. Begin by graphing the data, as shown in Figure 1.55. You can use either a sine or cosine model. Suppose you use a cosine model of the form

$$y = a \cos(bt - c) + d.$$

The amplitude is given by

$$a = \frac{1}{2}[(\text{high}) - (\text{low})] = \frac{1}{2}(11.3 - 0.1) = 5.6.$$

The period is

$$p = 2[(\text{low time}) - (\text{high time})] = 2(10 - 4) = 12$$

which implies that

$$b = \frac{2\pi}{p} \approx 0.524.$$

Because high tide occurs 4 hours after midnight, you can conclude that $c/b = 4$, so $c \approx 2.094$. Moreover, because the average depth is $\frac{1}{2}(11.3 + 0.1) = 5.7$, it follows that $d = 5.7$. So, you can model the depth with the function

$$y = 5.6 \cos(0.524t - 2.094) + 5.7.$$

b. The depths at 9 A.M. and 3 P.M. are as follows.

$$y = 5.6 \cos(0.524 \cdot 9 - 2.094) + 5.7$$

$$\approx 0.84 \text{ foot} \qquad\qquad\qquad\qquad 9 \text{ A.M.}$$

$$y = 5.6 \cos(0.524 \cdot 15 - 2.094) + 5.7$$

$$\approx 10.57 \text{ feet} \qquad\qquad\qquad\qquad 3 \text{ P.M.}$$

c. To find out when the depth y is at least 10 feet, you can graph the model with the line $y = 10$, as shown in Figure 1.56. From the graph, it follows that the depth is at least 10 feet between 2:42 P.M. ($t \approx 14.7$) and 5:18 P.M. ($t \approx 17.3$).

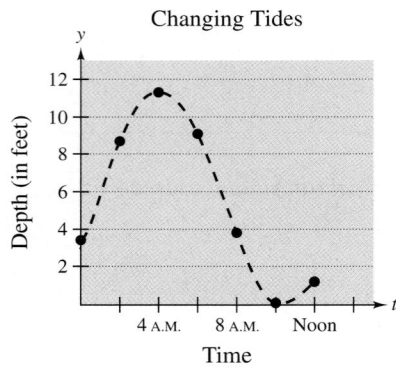

FIGURE **1.55**

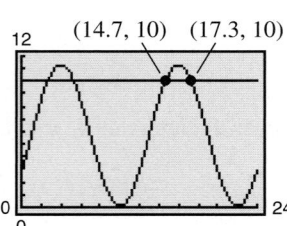

FIGURE **1.56**

1.5 Exercises

In Exercises 1–14, find the period and amplitude.

1. $y = 3 \sin 2x$

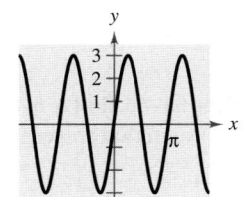

2. $y = 2 \cos 3x$

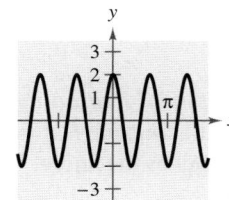

3. $y = \frac{5}{2} \cos \frac{x}{2}$

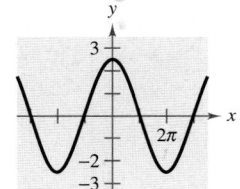

4. $y = -3 \sin \frac{x}{3}$

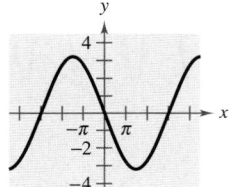

5. $y = \frac{1}{2} \sin \frac{\pi x}{3}$

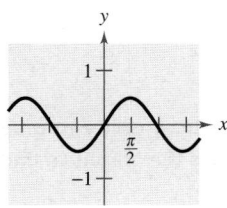

6. $y = \frac{3}{2} \cos \frac{\pi x}{2}$

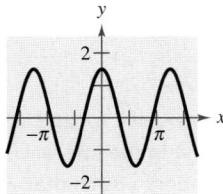

7. $y = -2 \sin x$

8. $y = -\cos \frac{2x}{3}$

9. $y = 3 \sin 10x$

10. $y = \frac{1}{3} \sin 8x$

11. $y = \frac{1}{2} \cos \frac{2x}{3}$

12. $y = \frac{5}{2} \cos \frac{x}{4}$

13. $y = \frac{1}{4} \sin 2\pi x$

14. $y = \frac{2}{3} \cos \frac{\pi x}{10}$

In Exercises 15–22, describe the relationship between the graphs of f and g. Consider amplitude, period, and shifts.

15. $f(x) = \sin x$
$g(x) = \sin(x - \pi)$

16. $f(x) = \cos x$
$g(x) = \cos(x + \pi)$

17. $f(x) = \cos 2x$
$g(x) = -\cos 2x$

18. $f(x) = \sin 3x$
$g(x) = \sin(-3x)$

19. $f(x) = \cos x$
$g(x) = \cos 2x$

20. $f(x) = \sin x$
$g(x) = \sin 3x$

21. $f(x) = \sin 2x$
$g(x) = 3 + \sin 2x$

22. $f(x) = \cos 4x$
$g(x) = -2 + \cos 4x$

In Exercises 23–26, describe the relationship between the graphs of f and g. Consider amplitude, period, and shifts.

23.

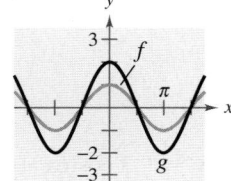

24.

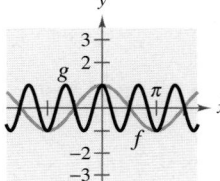

25.

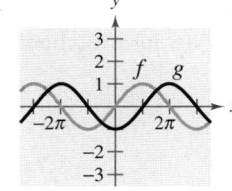

26.

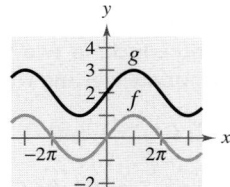

In Exercises 27–34, graph f and g on the same set of coordinate axes. (Include two full periods.)

27. $f(x) = -2 \sin x$
$g(x) = 4 \sin x$

28. $f(x) = \sin x$
$g(x) = \sin \frac{x}{3}$

29. $f(x) = \cos x$
$g(x) = 1 + \cos x$

30. $f(x) = 2 \cos 2x$
$g(x) = -\cos 4x$

31. $f(x) = -\frac{1}{2} \sin \frac{x}{2}$
$g(x) = 3 - \frac{1}{2} \sin \frac{x}{2}$

32. $f(x) = 4 \sin \pi x$
$g(x) = 4 \sin \pi x - 3$

33. $f(x) = 2 \cos x$
$g(x) = 2 \cos(x + \pi)$

34. $f(x) = -\cos x$
$g(x) = -\cos(x - \pi)$

In Exercises 35–52, sketch the graph of the function. (Include two full periods.)

35. $y = -2 \sin 6x$

36. $y = -3 \cos 4x$

37. $y = \cos 2\pi x$

38. $y = \sin \dfrac{\pi x}{4}$

39. $y = -\sin \dfrac{2\pi x}{3}$

40. $y = -10 \cos \dfrac{\pi x}{6}$

41. $y = \sin\left(x - \dfrac{\pi}{4}\right)$

42. $y = \sin(x - \pi)$

43. $y = 3 \cos(x + \pi)$

44. $y = 4 \cos\left(x + \dfrac{\pi}{4}\right)$

45. $y = 2 - \sin \dfrac{2\pi x}{3}$

46. $y = -3 + 5 \cos \dfrac{\pi t}{12}$

47. $y = 2 + \frac{1}{10} \cos 60\pi x$

48. $y = 2 \cos x - 3$

49. $y = 3 \cos(x + \pi) - 3$

50. $y = 4 \cos\left(x + \dfrac{\pi}{4}\right) + 4$

51. $y = \dfrac{2}{3} \cos\left(\dfrac{x}{2} - \dfrac{\pi}{4}\right)$

52. $y = -3 \cos(6x + \pi)$

 In Exercises 53–58, use a graphing utility to graph the function. Include two full periods. Be sure to choose an appropriate viewing window.

53. $y = -2 \sin(4x + \pi)$

54. $y = -4 \sin\left(\dfrac{2}{3}x - \dfrac{\pi}{3}\right)$

55. $y = \cos\left(2\pi x - \dfrac{\pi}{2}\right) + 1$

56. $y = 3 \cos\left(\dfrac{\pi x}{2} + \dfrac{\pi}{2}\right) - 2$

57. $y = -0.1 \sin\left(\dfrac{\pi x}{10} + \pi\right)$

58. $y = \frac{1}{100} \sin 120\pi t$

Graphical Reasoning In Exercises 59–62, find *a* and *d* for the function $f(x) = a \cos x + d$ such that the graph of *f* matches the figure.

59.

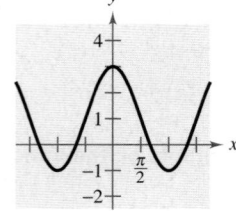

60.

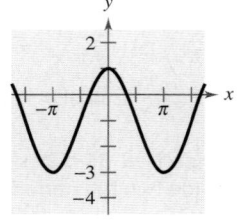

61.

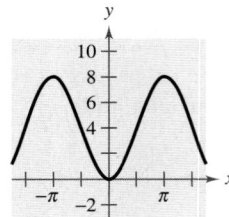

62.

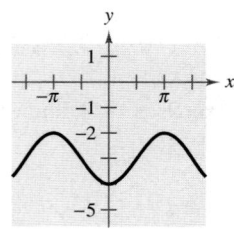

Graphical Reasoning In Exercises 63–66, find *a*, *b*, and *c* for the function $f(x) = a \sin(bx - c)$ such that the graph of *f* matches the figure.

63.

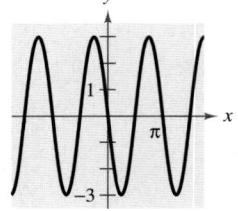

64.

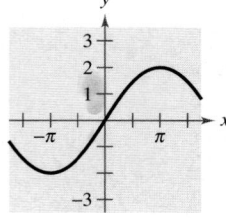

65.

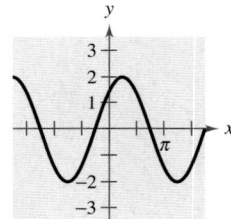

66.

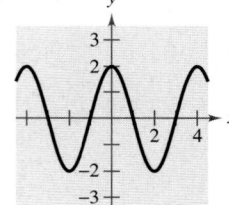

 In Exercises 67 and 68, use a graphing utility to graph y_1 and y_2 in the interval $[-2\pi, 2\pi]$. Use the graphs to find real numbers *x* such that $y_1 = y_2$.

67. $y_1 = \sin x$

$y_2 = -\frac{1}{2}$

68. $y_1 = \cos x$

$y_2 = -1$

69. ***Respiratory Cycle*** For a person at rest, the velocity *v* (in liters per second) of air flow during a respiratory cycle (the time from the beginning of one breath to the beginning of the next) is

$$v = 0.85 \sin \dfrac{\pi t}{3}$$

where *t* is the time in seconds. (Inhalation occurs when $v > 0$, and exhalation occurs when $v < 0$.)

(a) Find the time for one full respiratory cycle.

(b) Find the number of cycles per minute.

(c) Sketch the graph of the velocity function.

70. *Respiratory Cycle* After exercising for a few minutes, a person has a respiratory cycle for which the velocity of air flow is approximated by

$$v = 1.75 \sin \frac{\pi t}{2}$$

where t is the time in seconds. (Inhalation occurs when $v > 0$, and exhalation occurs when $v < 0$.)

(a) Find the time for one full respiratory cycle.

(b) Find the number of cycles per minute.

(c) Sketch the graph of the velocity function.

71. *Piano Tuning* When tuning a piano, a technician strikes a tuning fork for the A above middle C and sets up a wave motion that can be approximated by $y = 0.001 \sin 880\pi t$, where t is the time in seconds.

(a) What is the period of the function?

(b) The frequency f is given by $f = 1/p$. What is the frequency of the note?

72. *Health* The function

$$P = 100 - 20 \cos \frac{5\pi t}{3}$$

approximates the blood pressure P in millimeters of mercury at time t in seconds for a person at rest.

(a) Find the period of the function.

(b) Find the number of heartbeats per minute.

73. *Data Analysis* The table shows the maximum daily high temperatures for Tallahassee T and Chicago C (in degrees Fahrenheit) for month t, with $t = 1$ corresponding to January. (Source: Southeast Regional Climate Center and National Climatic Data Center)

t	T	C	t	T	C
1	63.9	29.0	7	91.3	83.7
2	67.1	33.5	8	91.0	81.8
3	73.1	45.8	9	88.1	74.8
4	79.9	58.6	10	80.8	63.3
5	86.5	70.1	11	72.2	48.4
6	90.5	79.6	12	65.5	34.0

(a) A model for the temperature in Tallahassee is

$$T(t) = 77.60 + 13.70 \cos\left(\frac{\pi t}{6} - 3.67\right).$$

Find a trigonometric model for Chicago.

 (b) Use a graphing utility to graph the data points and the model for the temperatures in Tallahassee. How well does the model fit the data?

 (c) Use a graphing utility to graph the data points and the model for the temperatures in Chicago. How well does the model fit the data?

(d) Use the models to estimate the average maximum temperature in each city. Which term of the models did you use? Explain.

(e) What is the period of each model? Are the periods what you expected? Explain.

(f) Which city has the greater variability in temperature throughout the year? Which factor of the models determines this variability? Explain.

▶ **Model It**

74. *Data Analysis* The percent y of the moon's face that is illuminated on day x of the year 2005, where $x = 70$ represents March 11, is shown in the table. (Source: U.S. Naval Observatory)

x	y
76	0.5
84	1.0
91	0.5
98	0.0
106	0.5
114	1.0

(a) Create a scatter plot of the data.

(b) Find a trigonometric model that fits the data.

(c) Add the graph of your model in part (b) to the scatter plot. How well does the model fit the data?

(d) What is the period of the model?

(e) Estimate the moon's percent illumination for May 8, 2005.

75. *Fuel Consumption* The daily consumption C (in gallons) of diesel fuel on a farm is modeled by

$$C = 30.3 + 21.6 \sin\left(\frac{2\pi t}{365} + 10.9\right)$$

where t is the time in days, with $t = 1$ corresponding to January 1.

(a) What is the period of the model? Is it what you expected? Explain.

(b) What is the average daily fuel consumption? Which term of the model did you use? Explain.

 (c) Use a graphing utility to graph the model. Use the graph to approximate the time of the year when consumption exceeds 40 gallons per day.

Synthesis

True or False? In Exercises 76–78, determine whether the statement is true or false. Justify your answer.

76. The graph of the function $f(x) = \sin(x + 2\pi)$ translates the graph of $f(x) = \sin x$ exactly one period to the right so that the two graphs look identical.

77. The function $y = \frac{1}{2}\cos 2x$ has an amplitude that is twice that of the function $y = \cos x$.

78. The graph of $y = -\cos x$ is a reflection of the graph of $y = \sin(x + \pi/2)$ in the x-axis.

Conjecture In Exercises 79 and 80, graph f and g on the same set of coordinate axes. Include two full periods. Make a conjecture about the functions.

79. $f(x) = \sin x$, $\quad g(x) = \cos\left(x - \frac{\pi}{2}\right)$

80. $f(x) = \sin x$, $\quad g(x) = -\cos\left(x + \frac{\pi}{2}\right)$

 81. *Writing* Use a graphing utility to graph the function $y = a \sin x$ for $a = \frac{1}{2}$, $a = \frac{3}{2}$, and $a = -3$. Write a paragraph describing the changes in the graph corresponding to the specified changes in a.

82. *Writing* Use a graphing utility to graph the function $y = d + \sin x$ for $d = 2$, $d = 3.5$, and $d = -2$. Write a paragraph describing the changes in the graph corresponding to the specified changes in d.

83. *Writing* Use a graphing utility to graph the function $y = \sin bx$ for $b = \frac{1}{2}$, $b = \frac{3}{2}$, and $b = 4$. Write a paragraph describing the changes in the graph corresponding to the specified changes in b.

84. *Writing* Use a graphing utility to graph the function $y = \sin(x - c)$ for $c = 1$, $c = 3$, and $c = -2$. Write a paragraph describing the changes in the graph corresponding to the specified changes in c.

85. *Exploration* Using calculus, it can be shown that the sine and cosine functions can be approximated by the polynomials

$$\sin x \approx x - \frac{x^3}{3!} + \frac{x^5}{5!} \quad \text{and} \quad \cos x \approx 1 - \frac{x^2}{2!} + \frac{x^4}{4!}$$

where x is in radians.

(a) Use a graphing utility to graph the sine function and its polynomial approximation in the same viewing window. How do the graphs compare?

(b) Use a graphing utility to graph the cosine function and its polynomial approximation in the same viewing window. How do the graphs compare?

(c) Study the patterns in the polynomial approximations of the sine and cosine functions and guess the next term in each. Then repeat parts (a) and (b). How did the accuracy of the approximations change when additional terms were added?

86. *Exploration* Use the polynomial approximations for the sine and cosine functions from Exercise 85 to approximate the following functional values. Compare the results with those given by a calculator. Is the error in the approximation the same in each case? Explain.

(a) $\sin\dfrac{1}{2}$ (b) $\sin 1$ (c) $\sin\dfrac{\pi}{6}$

(d) $\cos(-0.5)$ (e) $\cos 1$ (f) $\cos\dfrac{\pi}{4}$

Review

In Exercises 87–90, perform the operations and simplify.

87. $\dfrac{4}{x} + \dfrac{4}{1 - x}$

88. $\dfrac{2}{x + 5} - \dfrac{2}{x - 5}$

89. $\dfrac{3}{x - 1} - \dfrac{2}{x(x - 1)}$

90. $\dfrac{x}{x - 5} + \dfrac{1}{2}$

In Exercises 91–94, find the domain of the function.

91. $f(x) = \dfrac{2}{11 - x}$

92. $f(x) = \dfrac{\sqrt{x - 3}}{x - 8}$

93. $f(x) = \sqrt{81 - x^2}$

94. $f(x) = \sqrt[3]{4 - x^2}$

1.6 Graphs of Other Trigonometric Functions

▶ **Why you should learn it**

Trigonometric functions can be used to model biological patterns. For instance, Exercise 75 on page 184 shows how trigonometric functions can model predator-prey cycles.

Tom Brakefield/Corbis

Graph of the Tangent Function

Recall that the tangent function is odd. That is, $\tan(-x) = -\tan x$. Consequently, the graph of $y = \tan x$ is symmetric with respect to the origin. You also know from the identity $\tan x = \sin x/\cos x$ that the tangent is undefined for values at which $\cos x = 0$. Two such values are $x = \pm\pi/2 \approx \pm 1.5708$.

x	$-\dfrac{\pi}{2}$	-1.57	-1.5	$-\dfrac{\pi}{4}$	0	$\dfrac{\pi}{4}$	1.5	1.57	$\dfrac{\pi}{2}$
$\tan x$	Undef.	-1255.8	-14.1	-1	0	1	14.1	1255.8	Undef.

As indicated in the table, $\tan x$ increases without bound as x approaches $\pi/2$ from the left, and decreases without bound as x approaches $-\pi/2$ from the right. So, the graph of $y = \tan x$ has *vertical asymptotes* at $x = \pi/2$ and $x = -\pi/2$, as shown in Figure 1.57. Moreover, because the period of the tangent function is π, vertical asymptotes also occur when $x = \pi/2 + n\pi$, where n is an integer. The domain of the tangent function is the set of all real numbers other than $x = \pi/2 + n\pi$, and the range is the set of all real numbers.

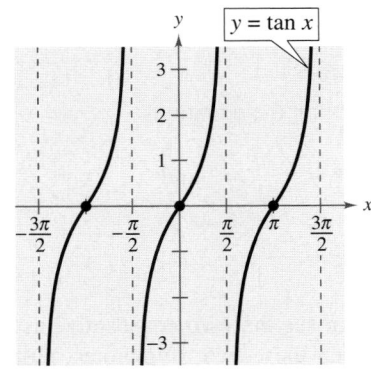

Period: π
Domain: all $x \neq \frac{\pi}{2} + n\pi$
Range: $(-\infty, \infty)$
Vertical asymptotes: $x = \frac{\pi}{2} + n\pi$

FIGURE 1.57

Sketching the graph of $y = a \tan(bx - c)$ is similar to sketching the graph of $y = a \sin(bx - c)$ in that you locate key points that identify the intercepts and asymptotes. Two consecutive asymptotes can be found by solving the equations

$$bx - c = -\frac{\pi}{2} \qquad \text{and} \qquad bx - c = \frac{\pi}{2}.$$

The midpoint between two consecutive asymptotes is an x-intercept of the graph. The period of the function $y = a \tan(bx - c)$ is the distance between two consecutive asymptotes. The amplitude of a tangent function is not defined. After plotting the asymptotes and the x-intercept, plot a few additional points between the two asymptotes and sketch one cycle. Finally, sketch one or two additional cycles to the left and right.

Consider reviewing period, range, and domain for all six trigonometric functions, especially emphasizing the period difference in the tangent and cotangent functions.

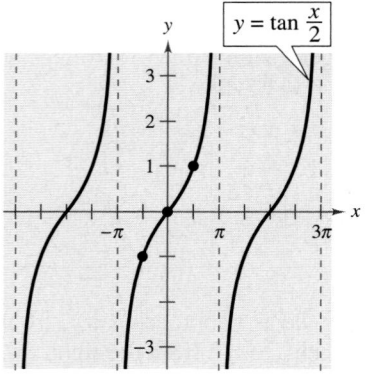

FIGURE **1.58**

Example 1 ▶ **Sketching the Graph of a Tangent Function**

Sketch the graph of $y = \tan \dfrac{x}{2}$.

Solution

By solving the equations

$$\frac{x}{2} = -\frac{\pi}{2} \quad \text{and} \quad \frac{x}{2} = \frac{\pi}{2}$$

$$x = -\pi \qquad\qquad x = \pi$$

you can see that two consecutive asymptotes occur at $x = -\pi$ and $x = \pi$. Between these two asymptotes, plot a few points, including the x-intercept, as shown in the table. Three cycles of the graph are shown in Figure 1.58.

x	$-\dfrac{\pi}{2}$	0	$\dfrac{\pi}{2}$
$\tan \dfrac{x}{2}$	-1	0	1

Example 2 ▶ **Sketching the Graph of a Tangent Function**

Sketch the graph of $y = -3 \tan 2x$.

Solution

By solving the equations

$$2x = -\frac{\pi}{2} \quad \text{and} \quad 2x = \frac{\pi}{2}$$

$$x = -\frac{\pi}{4} \qquad\qquad x = \frac{\pi}{4}$$

you can see that two consecutive asymptotes occur at $x = -\pi/4$ and $x = \pi/4$. Between these two asymptotes, plot a few points, including the x-intercept, as shown in the table. Three cycles of the graph are shown in Figure 1.59.

x	$-\dfrac{\pi}{8}$	0	$\dfrac{\pi}{8}$
$-3 \tan 2x$	3	0	-3

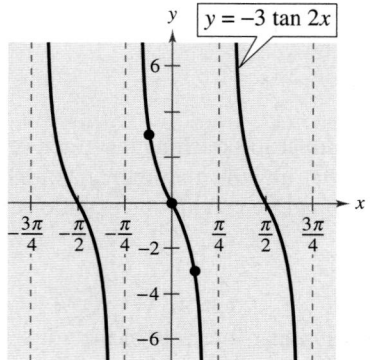

FIGURE **1.59**

By comparing the graphs in Examples 1 and 2, you can see that the graph of $y = a \tan(bx - c)$ increases between consecutive vertical asymptotes when $a > 0$, and decreases between consecutive vertical asymptotes when $a < 0$. In other words, the graph for $a < 0$ is a reflection in the x-axis of the graph for $a > 0$.

Graph of the Cotangent Function

The graph of the cotangent function is similar to the graph of the tangent function. It also has a period of π. However, from the identity

$$y = \cot x = \frac{\cos x}{\sin x}$$

you can see that the cotangent function has vertical asymptotes when $\sin x$ is zero, which occurs at $x = n\pi$, where n is an integer. The graph of the cotangent function is shown in Figure 1.60.

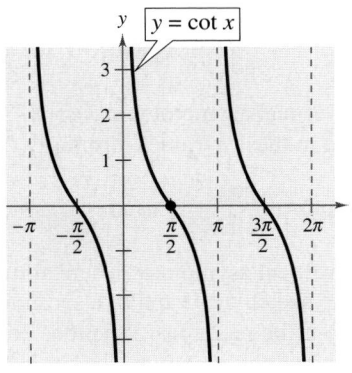

Period: π
Domain: all $x \neq n\pi$
Range: $(-\infty, \infty)$
Vertical asymptotes: $x = n\pi$

FIGURE **1.60**

Example 3 ▶ **Sketching the Graph of a Cotangent Function**

Sketch the graph of $y = 2 \cot \dfrac{x}{3}$.

Solution

To locate two consecutive vertical asymptotes of the graph, solve the equations $x/3 = 0$ and $x/3 = \pi$, as follows.

$$\frac{x}{3} = 0 \qquad \text{and} \qquad \frac{x}{3} = \pi$$

$$x = 0 \qquad\qquad\qquad x = 3\pi$$

Then, between these two asymptotes, plot a few points, including the x-intercept, as shown in the table. Three cycles of the graph are shown in Figure 1.61. (Note that the period is 3π, the distance between consecutive asymptotes.)

x	$\dfrac{3\pi}{4}$	$\dfrac{3\pi}{2}$	$\dfrac{9\pi}{4}$
$2 \cot \dfrac{x}{3}$	2	0	-2

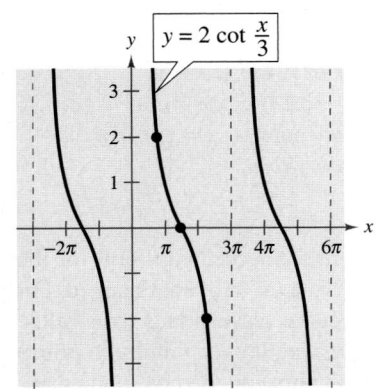

FIGURE **1.61**

Graphs of the Reciprocal Functions

Using a graphing utility to graph cosecant and secant functions can help reinforce the need to use the reciprocal functions.

The graphs of the two remaining trigonometric functions can be obtained from the graphs of the sine and cosine functions using the reciprocal identities

$$\csc x = \frac{1}{\sin x} \qquad \text{and} \qquad \sec x = \frac{1}{\cos x}.$$

For instance, at a given value of x, the y-coordinate of $\sec x$ is the reciprocal of the y-coordinate of $\cos x$. Of course, when $\cos x = 0$, the reciprocal does not exist. Near such values of x, the behavior of the secant function is similar to that of the tangent function. In other words, the graphs of

$$\tan x = \frac{\sin x}{\cos x} \qquad \text{and} \qquad \sec x = \frac{1}{\cos x}$$

have vertical asymptotes at $x = \pi/2 + n\pi$, where n is an integer, and the cosine is zero at these x-values. Similarly,

$$\cot x = \frac{\cos x}{\sin x} \qquad \text{and} \qquad \csc x = \frac{1}{\sin x}$$

have vertical asymptotes where $\sin x = 0$—that is, at $x = n\pi$.

To sketch the graph of a secant or cosecant function, you should first make a sketch of its reciprocal function. For instance, to sketch the graph of $y = \csc x$, first sketch the graph of $y = \sin x$. Then take reciprocals of the y-coordinates to obtain points on the graph of $y = \csc x$. This procedure is used to obtain the graphs shown in Figure 1.62.

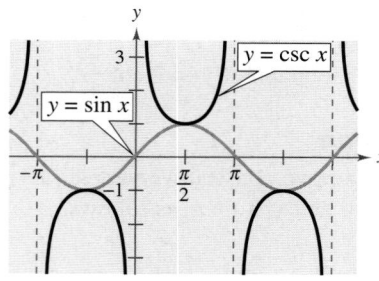

Period: 2π
Domain: all $x \ne n\pi$
Range: $(-\infty, -1]$ and $[1, \infty)$
Vertical asymptotes: $x = n\pi$
Symmetry: origin

FIGURE 1.62

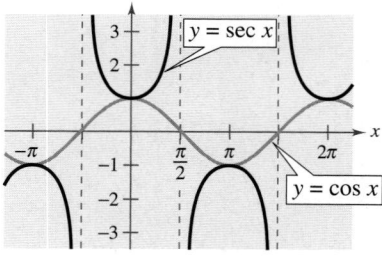

Period: 2π
Domain: all $x \ne \frac{\pi}{2} + n\pi$
Range: $(-\infty, -1]$ and $[1, \infty)$
Vertical asymptotes: $x = \frac{\pi}{2} + n\pi$
Symmetry: y-axis

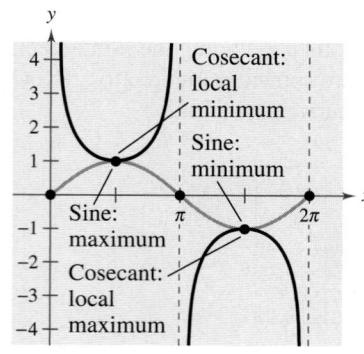

FIGURE 1.63

In comparing the graphs of the secant and cosecant functions with those of the sine and cosine functions, note that the "hills" and "valleys" are interchanged. For example, a hill (or maximum point) on the sine curve corresponds to a valley (a local minimum) on the cosecant curve. Similarly, a valley (or minimum point) on the sine curve corresponds to a hill (a local maximum) on the cosecant curve. Additionally, x-intercepts of the sine and cosine functions become vertical asymptotes of the cosecant and secant functions, as shown in Figure 1.63.

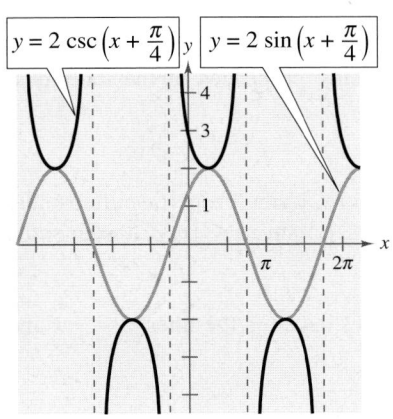

FIGURE **1.64**

Example 4 ▶ **Sketching the Graph of a Cosecant Function**

Sketch the graph of $y = 2 \csc\left(x + \dfrac{\pi}{4}\right)$.

Solution

Begin by sketching the graph of

$$y = 2 \sin\left(x + \frac{\pi}{4}\right).$$

For this function, the amplitude is 2 and the period is 2π. By solving the equations

$$x + \frac{\pi}{4} = 0 \qquad \text{and} \qquad x + \frac{\pi}{4} = 2\pi$$

$$x = -\frac{\pi}{4} \qquad\qquad\qquad x = \frac{7\pi}{4}$$

you can see that one cycle of the sine function corresponds to the interval from $x = -\pi/4$ to $x = 7\pi/4$. The graph of this sine function is represented by the gray curve in Figure 1.64. Because the sine function is zero at the midpoint and endpoints of this interval, the corresponding cosecant function

$$y = 2 \csc\left(x + \frac{\pi}{4}\right)$$

$$= 2\left(\frac{1}{\sin[x + (\pi/4)]}\right)$$

has vertical asymptotes at $x = -\pi/4$, $x = 3\pi/4$, $x = 7\pi/4$, etc. The graph of the cosecant function is represented by the black curve in Figure 1.64.

Example 5 ▶ **Sketching the Graph of a Secant Function**

Sketch the graph of $y = \sec 2x$.

Solution

Begin by sketching the graph of $y = \cos 2x$, as indicated by the gray curve in Figure 1.65. Then form the graph of $y = \sec 2x$ as the black curve in the figure. Note that the x-intercepts of $y = \cos 2x$

$$\left(-\frac{\pi}{4}, 0\right), \qquad \left(\frac{\pi}{4}, 0\right), \qquad \left(\frac{3\pi}{4}, 0\right), \ldots$$

correspond to the vertical asymptotes

$$x = -\frac{\pi}{4}, \qquad x = \frac{\pi}{4}, \qquad x = \frac{3\pi}{4}, \ldots$$

of the graph of $y = \sec 2x$. Moreover, notice that the period of $y = \cos 2x$ and $y = \sec 2x$ is π.

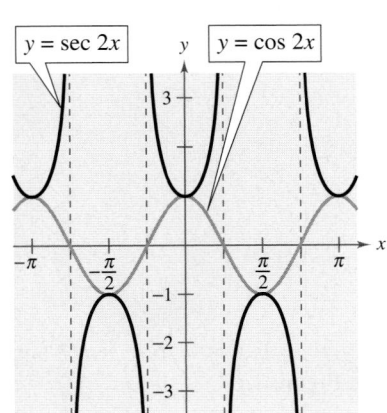

FIGURE **1.65**

Damped Trigonometric Graphs

A *product* of two functions can be graphed using properties of the individual functions. For instance, consider the function

$$f(x) = x \sin x$$

as the product of the functions $y = x$ and $y = \sin x$. Using properties of absolute value and the fact that $|\sin x| \leq 1$, you have $0 \leq |x||\sin x| \leq |x|$. Consequently,

$$-|x| \leq x \sin x \leq |x|$$

which means that the graph of $f(x) = x \sin x$ lies between the lines $y = -x$ and $y = x$. Furthermore, because

$$f(x) = x \sin x = \pm x \qquad \text{at} \qquad x = \frac{\pi}{2} + n\pi$$

and

$$f(x) = x \sin x = 0 \qquad \text{at} \qquad x = n\pi$$

the graph of f touches the line $y = -x$ or the line $y = x$ at $x = \pi/2 + n\pi$ and has x-intercepts at $x = n\pi$. A sketch of f is shown in Figure 1.66. In the function $f(x) = x \sin x$, the factor x is called the **damping factor.**

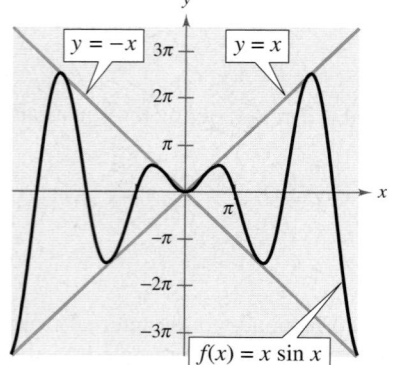

FIGURE 1.66

Example 6 ▶ **Damped Sine Wave**

Sketch the graph of

$$f(x) = x^2 \sin 3x.$$

Solution

Consider $f(x)$ as the product of the two functions

$$y = x^2 \qquad \text{and} \qquad y = \sin 3x$$

each of which has the set of real numbers as its domain. For any real number x, $x^2 \geq 0$ and $|\sin 3x| \leq 1$. So, $x^2 |\sin 3x| \leq x^2$, which means that

$$-x^2 \leq x^2 \sin 3x \leq x^2.$$

Furthermore, because

$$f(x) = x^2 \sin 3x = \pm x^2 \qquad \text{at} \qquad x = \frac{\pi}{6} + \frac{n\pi}{3}$$

and

$$f(x) = x^2 \sin 3x = 0 \qquad \text{at} \qquad x = \frac{n\pi}{3}$$

the graph of f touches the curves $y = -x^2$ and $y = x^2$ at $x = \pi/6 + n\pi/3$ and has intercepts at $x = n\pi/3$. A sketch is shown in Figure 1.67.

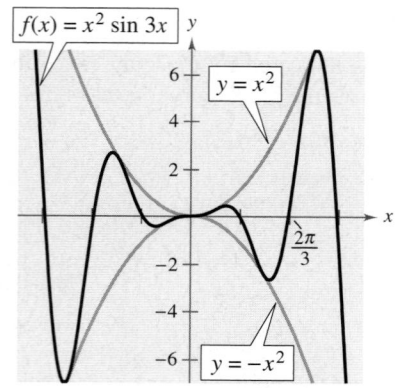

FIGURE 1.67

Figure 1.68 summarizes the six basic trigonometric functions.

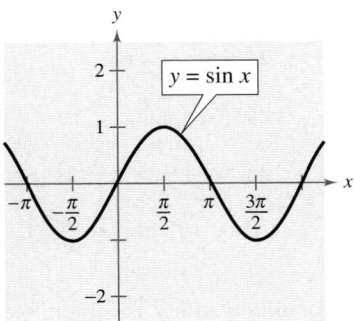

Domain: *all reals*
Range: $[-1, 1]$
Period: 2π

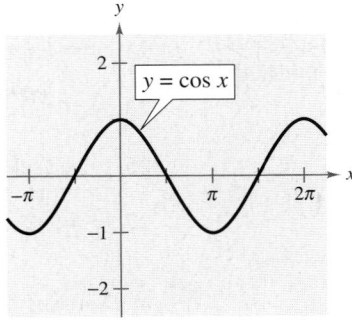

Domain: *all reals*
Range: $[-1, 1]$
Period: 2π

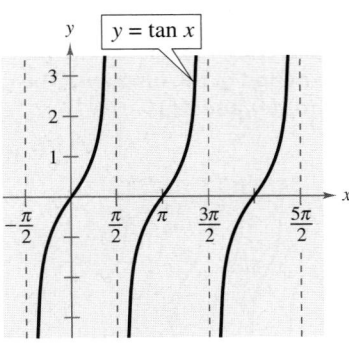

Domain: *all $x \neq \frac{\pi}{2} + n\pi$*
Range: $(-\infty, \infty)$
Period: π

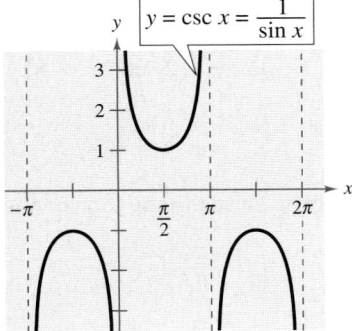

Domain: *all $x \neq n\pi$*
Range: $(-\infty, -1]$ *and* $[1, \infty)$
Period: 2π
FIGURE 1.68

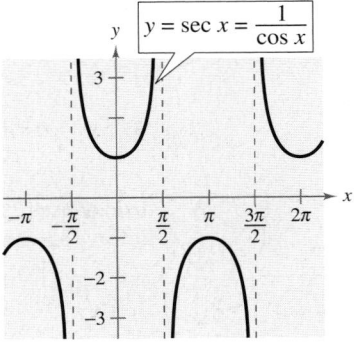

Domain: *all $x \neq \frac{\pi}{2} + n\pi$*
Range: $(-\infty, -1]$ *and* $[1, \infty)$
Period: 2π

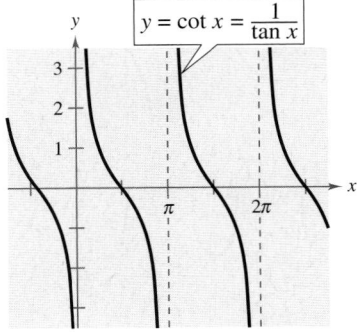

Domain: *all $x \neq n\pi$*
Range: $(-\infty, \infty)$
Period: π

Writing ABOUT MATHEMATICS

Combining Trigonometric Functions Recall from Section P.9 that functions can be combined arithmetically. This also applies to trigonometric functions. For each of the functions

$$h(x) = x + \sin x \quad \text{and} \quad h(x) = \cos x - \sin 3x$$

(a) identify two simpler functions f and g that comprise the combination, (b) use a table to show how to obtain the numerical values of $h(x)$ from the numerical values of $f(x)$ and $g(x)$, and (c) use a graph of f and g to show how h may be formed.

Can you find functions

$$f(x) = d + a\sin(bx + c) \quad \text{and} \quad g(x) = d + a\cos(bx + c)$$

such that $f(x) + g(x) = 0$ for all x?

1.6 Exercises

In Exercises 1–6, match the function with its graph. State the period of the function. [The graphs are labeled (a), (b), (c), (d), (e), and (f).]

(a)
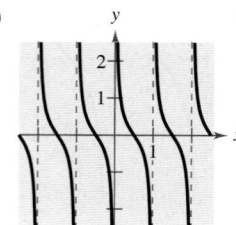

(b)
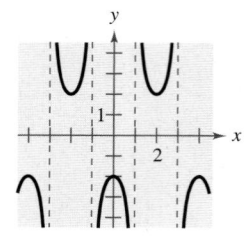

(c)
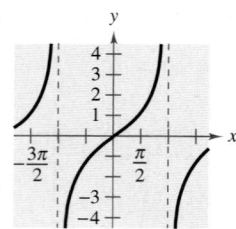

(d)
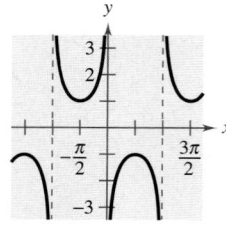

(e)
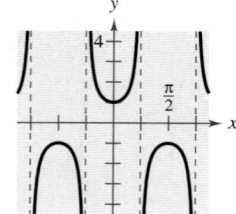

(f)
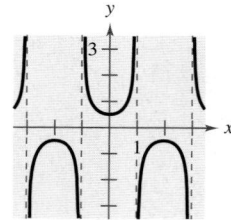

1. $y = \sec 2x$

2. $y = \tan \dfrac{x}{2}$

3. $y = \dfrac{1}{2} \cot \pi x$

4. $y = -\csc x$

5. $y = \dfrac{1}{2} \sec \dfrac{\pi x}{2}$

6. $y = -2 \sec \dfrac{\pi x}{2}$

In Exercises 7–28, sketch the graph of the function. Include two full periods.

7. $y = \frac{1}{3} \tan x$

8. $y = \frac{1}{4} \tan x$

9. $y = \tan 3x$

10. $y = -3 \tan \pi x$

11. $y = -\frac{1}{2} \sec x$

12. $y = \frac{1}{4} \sec x$

13. $y = \csc \pi x$

14. $y = 3 \csc 4x$

15. $y = \sec \pi x - 1$

16. $y = -2 \sec 4x + 2$

17. $y = \csc \dfrac{x}{2}$

18. $y = \csc \dfrac{x}{3}$

19. $y = \cot \dfrac{x}{2}$

20. $y = 3 \cot \dfrac{\pi x}{2}$

21. $y = \frac{1}{2} \sec 2x$

22. $y = -\frac{1}{2} \tan x$

23. $y = \tan \dfrac{\pi x}{4}$

24. $y = \tan(x + \pi)$

25. $y = \csc(\pi - x)$

26. $y = \sec(\pi - x)$

27. $y = \dfrac{1}{4} \csc\left(x + \dfrac{\pi}{4}\right)$

28. $y = 2 \cot\left(x + \dfrac{\pi}{2}\right)$

In Exercises 29–38, use a graphing utility to graph the function. Include two full periods.

29. $y = \tan \dfrac{x}{3}$

30. $y = -\tan 2x$

31. $y = -2 \sec 4x$

32. $y = \sec \pi x$

33. $y = \tan\left(x - \dfrac{\pi}{4}\right)$

34. $y = \dfrac{1}{4} \cot\left(x - \dfrac{\pi}{2}\right)$

35. $y = -\csc(4x - \pi)$

36. $y = 2 \sec(2x - \pi)$

37. $y = 0.1 \tan\left(\dfrac{\pi x}{4} + \dfrac{\pi}{4}\right)$

38. $y = \dfrac{1}{3} \sec\left(\dfrac{\pi x}{2} + \dfrac{\pi}{2}\right)$

In Exercises 39–46, use a graph to solve the equation on the interval $[-2\pi, 2\pi]$.

39. $\tan x = 1$

40. $\tan x = \sqrt{3}$

41. $\cot x = -\dfrac{\sqrt{3}}{3}$

42. $\cot x = 1$

43. $\sec x = -2$

44. $\sec x = 2$

45. $\csc x = \sqrt{2}$

46. $\csc x = -\dfrac{2\sqrt{3}}{3}$

In Exercises 47 and 48, use the graph of the function to determine whether the function is even, odd, or neither.

47. $f(x) = \sec x$

48. $f(x) = \tan x$

49. *Graphical Reasoning* Consider the functions

$$f(x) = 2 \sin x \quad \text{and} \quad g(x) = \dfrac{1}{2} \csc x$$

on the interval $(0, \pi)$.

(a) Graph f and g in the same coordinate plane.

(b) Approximate the interval on which $f > g$.

(c) Describe the behavior of each of the functions as x approaches π. How is the behavior of g related to the behavior of f as x approaches π?

50. *Graphical Reasoning* Consider the functions

$$f(x) = \tan \frac{\pi x}{2} \quad \text{and} \quad g(x) = \frac{1}{2} \sec \frac{\pi x}{2}$$

on the interval $(-1, 1)$.

(a) Use a graphing utility to graph f and g in the same viewing window.

(b) Approximate the interval on which $f < g$.

(c) Approximate the interval on which $2f < 2g$. How does the result compare with that of part (b)? Explain.

In Exercises 51–54, use a graphing utility to graph the two equations in the same viewing window. Determine analytically whether the expressions are equivalent.

51. $y_1 = \sin x \csc x, \quad y_2 = 1$

52. $y_1 = \sin x \sec x, \quad y_2 = \tan x$

53. $y_1 = \dfrac{\cos x}{\sin x}, \quad y_2 = \cot x$

54. $y_1 = \sec^2 x - 1, \quad y_2 = \tan^2 x$

In Exercises 55–58, match the function with its graph. Describe the behavior of the function as x approaches zero. [The graphs are labeled (a), (b), (c), and (d).]

(a)

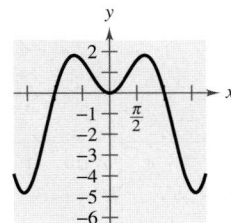

(b)

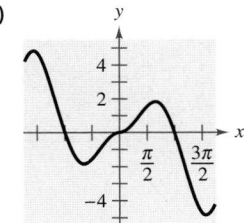

(c)

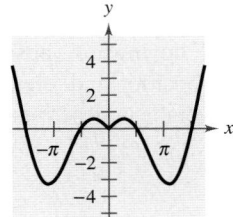

(d)

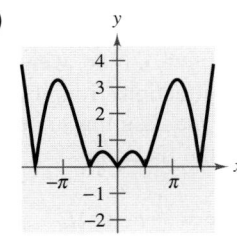

55. $f(x) = |x \cos x|$

56. $f(x) = x \sin x$

57. $g(x) = |x| \sin x$

58. $g(x) = |x| \cos x$

Conjecture In Exercises 59–62, graph the functions f and g. Use the graphs to make a conjecture about the relationship between the functions.

59. $f(x) = \sin x + \cos\left(x + \dfrac{\pi}{2}\right), \quad g(x) = 0$

60. $f(x) = \sin x - \cos\left(x + \dfrac{\pi}{2}\right), \quad g(x) = 2 \sin x$

61. $f(x) = \sin^2 x, \quad g(x) = \frac{1}{2}(1 - \cos 2x)$

62. $f(x) = \cos^2 \dfrac{\pi x}{2}, \quad g(x) = \dfrac{1}{2}(1 + \cos \pi x)$

In Exercises 63–66, use a graphing utility to graph the function and the damping factor of the function in the same viewing window. Describe the behavior of the function as x increases without bound.

63. $f(x) = x \cos \pi x$

64. $f(x) = x^2 \cos x$

65. $g(x) = x^3 \sin x$

66. $h(x) = x^3 \cos x$

Exploration In Exercises 67–72, use a graphing utility to graph the function. Describe the behavior of the function as x approaches zero.

67. $y = \dfrac{6}{x} + \cos x, \quad x > 0$

68. $y = \dfrac{4}{x} + \sin 2x, \quad x > 0$

69. $g(x) = \dfrac{\sin x}{x}$

70. $f(x) = \dfrac{1 - \cos x}{x}$

71. $f(x) = \sin \dfrac{1}{x}$

72. $h(x) = x \sin \dfrac{1}{x}$

73. *Distance* A plane flying at an altitude of 7 miles above a radar antenna will pass directly over the radar antenna (see figure). Let d be the ground distance from the antenna to the point directly under the plane and let x be the angle of elevation to the plane from the antenna. (d is positive as the plane approaches the antenna.) Write d as a function of x and graph the function over the interval $0 < x < \pi$.

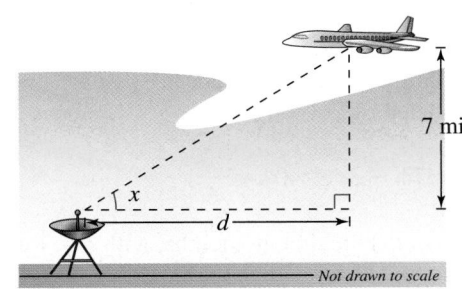

7 mi

Not drawn to scale

74. Television Coverage A television camera is on a reviewing platform 27 meters from the street on which a parade will be passing from left to right (see figure). Express the distance d from the camera to a particular unit in the parade as a function of the angle x, and graph the function over the interval $-\pi/2 < x < \pi/2$. (Consider x as negative when a unit in the parade approaches from the left.)

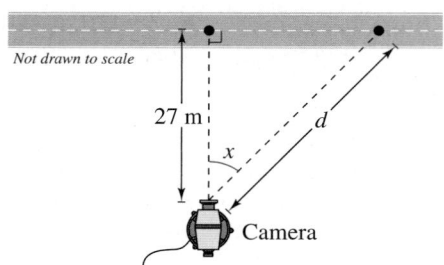

Not drawn to scale

27 m

Camera

▶ **Model It**

75. Predator-Prey Model The population of coyotes (a predator) at time t (in months) in a region is estimated to be

$$C = 5000 + 2000 \sin \frac{\pi t}{12}$$

and the population of rabbits (its prey) is estimated to be

$$R = 25{,}000 + 15{,}000 \cos \frac{\pi t}{12}.$$

(a) Use a graphing utility to graph both models in the same viewing window. Use the window setting $0 \le t \le 100$.

(b) Use the graphs of the models in part (a) to explain the oscillations in the size of each population.

(c) The cycles of each population follow a periodic pattern. Find the period of each model and describe several factors that could be contributing to the cyclical patterns.

---

76. Sales The projected monthly sales S (in thousands of units) of lawn mowers (a seasonal product) are modeled by

$$S = 74 + 3t - 40 \cos \frac{\pi t}{6}$$

where t is the time in months, with $t = 1$ corresponding to January. Graph the sales function over 1 year.

77. Meterology The normal monthly high temperatures H in degrees Fahrenheit for Erie, Pennsylvania are approximated by

$$H(t) = 54.33 - 20.38 \cos \frac{\pi t}{6} - 15.69 \sin \frac{\pi t}{6}$$

and the normal monthly low temperatures L are approximated by

$$L(t) = 39.36 - 15.70 \cos \frac{\pi t}{6} - 14.16 \sin \frac{\pi t}{6}$$

where t is the time in months, with $t = 1$ corresponding to January (see figure). (Source: National Oceanic and Atmospheric Administration)

(a) What is the period of each function?

(b) During what part of the year is the difference between the normal high and low temperatures greatest? When is it smallest?

(c) The sun is northernmost in the sky around June 21, but the graph shows the warmest temperatures at a later date. Approximate the lag time of the temperatures relative to the position of the sun.

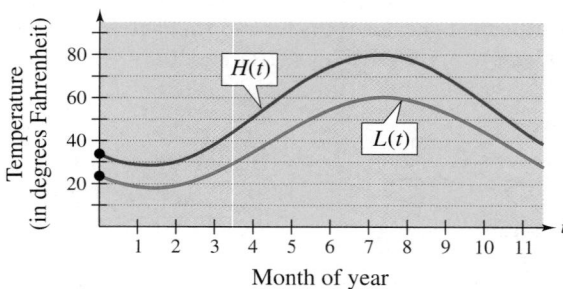

78. Harmonic Motion An object weighing W pounds is suspended from the ceiling by a steel spring (see figure). The weight is pulled downward (positive direction) from its equilibrium position and released. The resulting motion of the weight is described by the function

$$y = \frac{1}{2} t \cos 4t, \qquad t > 0$$

where y is the distance in feet and t is the time in seconds.

(a) Use a graphing utility to graph the function.

(b) Describe the behavior of the displacement function for increasing values of time t.

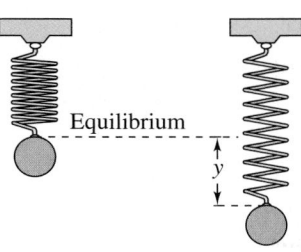

FIGURE FOR 78

Synthesis

True or False? In Exercises 79 and 80, determine whether the statement is true or false. Justify your answer.

79. The graph of $y = \csc x$ can be obtained on a calculator by graphing the reciprocal of $y = \sin x$.

80. The graph of $y = \sec x$ can be obtained on a calculator by graphing a translation of the reciprocal of $y = \sin x$.

81. **Writing** Describe the behavior of $f(x) = \tan x$ as x approaches $\pi/2$ from the left and from the right.

82. **Writing** Describe the behavior of $f(x) = \csc x$ as x approaches π from the left and from the right.

83. **Exploration** Consider the function

$$f(x) = x - \cos x.$$

 (a) Use a graphing utility to graph the function and verify that there exists a zero between 0 and 1. Use the graph to approximate the zero.

(b) Starting with $x_0 = 1$, generate a sequence $x_1, x_2, x_3, \ldots$, where $x_n = \cos(x_{n-1})$. For example,

$$x_0 = 1$$
$$x_1 = \cos(x_0)$$
$$x_2 = \cos(x_1)$$
$$x_3 = \cos(x_2)$$
$$\vdots$$

What value does the sequence approach?

84. **Approximation** Using calculus, it can be shown that the tangent function can be approximated by the polynomial

$$\tan x \approx x + \frac{2x^3}{3!} + \frac{16x^5}{5!}$$

where x is in radians. Use a graphing utility to graph the tangent function and its polynomial approximation in the same viewing window. How do the graphs compare?

85. **Approximation** Using calculus, it can be shown that the secant function can be approximated by the polynomial

$$\sec x \approx 1 + \frac{x^2}{2!} + \frac{5x^4}{4!}$$

where x is in radians. Use a graphing utility to graph the secant function and its polynomial approximation in the same viewing window. How do the graphs compare?

86. **Pattern Recognition**

(a) Use a graphing utility to graph each function.

$$y_1 = \frac{4}{\pi}\left(\sin \pi x + \frac{1}{3}\sin 3\pi x\right)$$

$$y_2 = \frac{4}{\pi}\left(\sin \pi x + \frac{1}{3}\sin 3\pi x + \frac{1}{5}\sin 5\pi x\right)$$

(b) Identify the pattern started in part (a) and find a function y_3 that continues the pattern one more term. Use a graphing utility to graph y_3.

(c) The graphs in parts (a) and (b) approximate the periodic function in the figure. Find a function y_4 that is a better approximation.

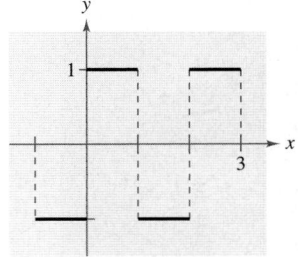

Review

In Exercises 87–94, solve the equation by any convenient method.

87. $x^2 = 64$

88. $(x - 5)^2 = 8$

89. $4x^2 - 12x + 9 = 0$

90. $9x^2 + 12x + 3 = 0$

91. $x^2 - 6x + 4 = 0$

92. $2x^2 - 4x - 6 = 0$

93. $50 + 5x = 3x^2$

94. $2x^2 + 4x - 9 = 2(x - 1)^2$

1.7 Inverse Trigonometric Functions

John Foxx/Imagestate

Inverse Sine Function

Recall from Section P.10 that, for a function to have an inverse function, it must be one-to-one—that is, it must pass the Horizontal Line Test. From Figure 1.69 you can see that $y = \sin x$ does not pass the test because different values of x yield the same y-value.

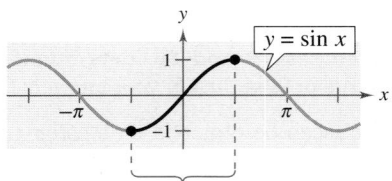

Sin x has an inverse function on this interval.

FIGURE **1.69**

However, if you restrict the domain to the interval $-\pi/2 \le x \le \pi/2$ (corresponding to the black portion of the graph in Figure 1.69), the following properties hold.

1. On the interval $[-\pi/2, \pi/2]$, the function $y = \sin x$ is increasing.
2. On the interval $[-\pi/2, \pi/2]$, $y = \sin x$ takes on its full range of values, $-1 \le \sin x \le 1$.
3. On the interval $[-\pi/2, \pi/2]$, $y = \sin x$ is one-to-one.

So, on the restricted domain $-\pi/2 \le x \le \pi/2$, $y = \sin x$ has a unique inverse function called the **inverse sine function.** It is denoted by

$$y = \arcsin x \qquad \text{or} \qquad y = \sin^{-1} x.$$

The notation $\sin^{-1} x$ is consistent with the inverse function notation $f^{-1}(x)$. The arcsin x notation (read as "the arcsine of x") comes from the association of a central angle with its subtended *arc length* on a unit circle. So, arcsin x means the angle (or arc) whose sine is x. Both notations, arcsin x and $\sin^{-1} x$, are commonly used in mathematics, so remember that $\sin^{-1} x$ denotes the *inverse* sine function rather than $1/\sin x$. The values of arcsin x lie in the interval $-\pi/2 \le \arcsin x \le \pi/2$. The graph of $y = \arcsin x$ is shown in Figure 1.70 on page 187.

STUDY TIP

When evaluating the inverse sine function, it helps to remember the phrase "the arcsine of x is the angle (or number) whose sine is x."

Definition of Inverse Sine Function

The **inverse sine function** is defined by

$$y = \arcsin x \qquad \text{if and only if} \qquad \sin y = x$$

where $-1 \le x \le 1$ and $-\pi/2 \le y \le \pi/2$. The domain of $y = \arcsin x$ is $[-1, 1]$, and the range is $[-\pi/2, \pi/2]$.

You may need to remind your students that the values of the inverse sine function are always in radians.

You may wish to illustrate the reflections of $y = \sin x$ and $y = \arcsin x$ about the line $y = x$. Consider using a graphing utility to do this.

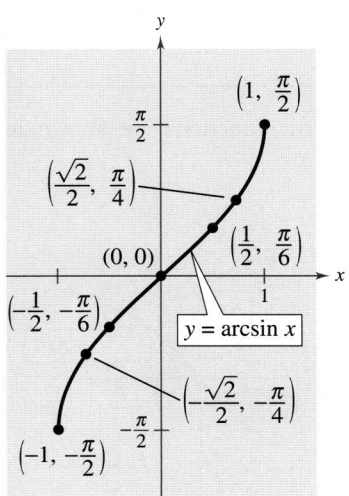

FIGURE 1.70

Example 1 ▶ **Evaluating the Inverse Sine Function**

If possible, find the exact value.

a. $\arcsin\left(-\dfrac{1}{2}\right)$

b. $\sin^{-1}\dfrac{\sqrt{3}}{2}$

c. $\sin^{-1}2$

Solution

a. Because $\sin\left(-\dfrac{\pi}{6}\right) = -\dfrac{1}{2}$ for $-\dfrac{\pi}{2} \le y \le \dfrac{\pi}{2}$, it follows that

$$\arcsin\left(-\dfrac{1}{2}\right) = -\dfrac{\pi}{6}.$$ Angle whose sine is $-\frac{1}{2}$

b. Because $\sin(\pi/3) = \sqrt{3}/2$ for $-\pi/2 \le y \le \pi/2$, it follows that

$$\sin^{-1}\dfrac{\sqrt{3}}{2} = \dfrac{\pi}{3}.$$ Angle whose sine is $\sqrt{3}/2$

c. It is not possible to evaluate $y = \sin^{-1}x$ when $x = 2$ because there is no angle whose sine is 2. Remember that the domain of the inverse sine function is $[-1, 1]$.

Example 2 ▶ **Graphing the Arcsine Function**

Sketch a graph of

$$y = \arcsin x.$$

Solution

By definition, the equations $y = \arcsin x$ and $\sin y = x$ are equivalent for $-\pi/2 \le y \le \pi/2$. So, their graphs are the same. From the interval $[-\pi/2, \pi/2]$, you can assign values to y in the second equation to make a table of values. Then plot the points and draw a smooth curve through the points.

y	$-\dfrac{\pi}{2}$	$-\dfrac{\pi}{4}$	$-\dfrac{\pi}{6}$	0	$\dfrac{\pi}{6}$	$\dfrac{\pi}{4}$	$\dfrac{\pi}{2}$
$x = \sin y$	-1	$-\dfrac{\sqrt{2}}{2}$	$-\dfrac{1}{2}$	0	$\dfrac{1}{2}$	$\dfrac{\sqrt{2}}{2}$	1

The resulting graph for $y = \arcsin x$ is shown in Figure 1.70. Note that it is the reflection (in the line $y = x$) of the black portion of the graph in Figure 1.69. Be sure you see that Figure 1.70 shows the *entire* graph of the inverse sine function. Remember that the range of $y = \arcsin x$ is the closed interval $[-\pi/2, \pi/2]$.

Other Inverse Trigonometric Functions

The cosine function is decreasing and one-to-one on the interval $0 \leq x \leq \pi$, as shown in Figure 1.71.

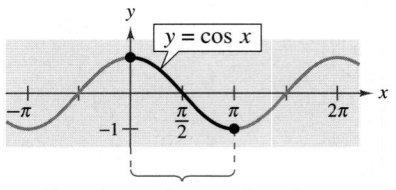

Cos x has an inverse function
on this interval.

FIGURE **1.71**

Consequently, on this interval the cosine function has an inverse function—the **inverse cosine function**—denoted by

$$y = \arccos x \quad \text{or} \quad y = \cos^{-1} x.$$

Similarly, you can define an **inverse tangent function** by restricting the domain of $y = \tan x$ to the interval $(-\pi/2, \pi/2)$. The following list summarizes the definitions of the three most common inverse trigonometric functions. The remaining three are defined in Exercises 100–102.

You may need to point out to your students that the range for each of these functions is different. Students should know these ranges well to ensure that their answers are within the correct range.

Definitions of the Inverse Trigonometric Functions

Function	Domain	Range
$y = \arcsin x$ if and only if $\sin y = x$	$-1 \leq x \leq 1$	$-\dfrac{\pi}{2} \leq y \leq \dfrac{\pi}{2}$
$y = \arccos x$ if and only if $\cos y = x$	$-1 \leq x \leq 1$	$0 \leq y \leq \pi$
$y = \arctan x$ if and only if $\tan y = x$	$-\infty < x < \infty$	$-\dfrac{\pi}{2} < y < \dfrac{\pi}{2}$

The graphs of these three inverse trigonometric functions are shown in Figure 1.72.

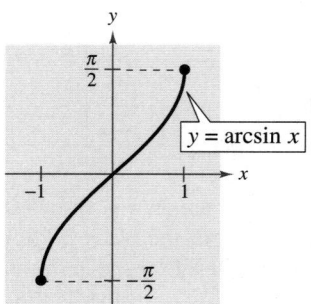

DOMAIN: $[-1, 1]$
RANGE: $\left[-\dfrac{\pi}{2}, \dfrac{\pi}{2}\right]$

FIGURE **1.72**

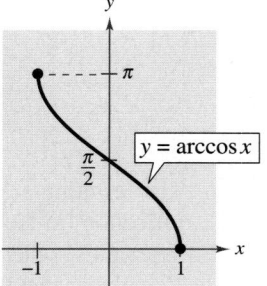

DOMAIN: $[-1, 1]$
RANGE: $[0, \pi]$

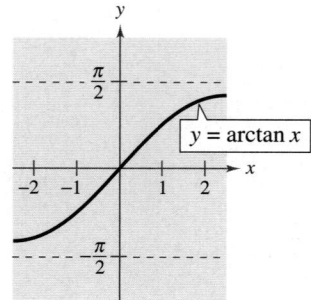

DOMAIN: $(-\infty, \infty)$
RANGE: $\left(-\dfrac{\pi}{2}, \dfrac{\pi}{2}\right)$

Example 3 ▶ **Evaluating Inverse Trigonometric Functions**

Find the exact value.

a. $\arccos \dfrac{\sqrt{2}}{2}$ **b.** $\cos^{-1}(-1)$

c. $\arctan 0$ **d.** $\tan^{-1}(-1)$

Solution

a. Because $\cos(\pi/4) = \sqrt{2}/2$, and $\pi/4$ lies in $[0, \pi]$, it follows that

$$\arccos \frac{\sqrt{2}}{2} = \frac{\pi}{4}.$$ Angle whose cosine is $\sqrt{2}/2$

b. Because $\cos \pi = -1$, and π lies in $[0, \pi]$, it follows that

$$\cos^{-1}(-1) = \pi.$$ Angle whose cosine is -1

c. Because $\tan 0 = 0$, and 0 lies in $(-\pi/2, \pi/2)$, it follows that

$$\arctan 0 = 0.$$ Angle whose tangent is 0

d. Because $\tan(-\pi/4) = -1$, and $-\pi/4$ lies in $(-\pi/2, \pi/2)$, it follows that

$$\tan^{-1}(-1) = -\frac{\pi}{4}.$$ Angle whose tangent is -1

Example 4 ▶ **Calculators and Inverse Trigonometric Functions**

Use a calculator to approximate the value (if possible).

a. $\arctan(-8.45)$

b. $\sin^{-1} 0.2447$

c. $\arccos 2$

Solution

Function	*Mode*	*Calculator Keystrokes*
a. $\arctan(-8.45)$	Radian	[TAN⁻¹] [(] [(−)] 8.45 [)] [ENTER]

From the display, it follows that $\arctan(-8.45) \approx -1.453001$.

| **b.** $\sin^{-1} 0.2447$ | Radian | [SIN⁻¹] 0.2447 [ENTER] |

From the display, it follows that $\arcsin 0.2447 \approx 0.2472103$.

| **c.** $\arccos 2$ | Radian | [COS⁻¹] 2 [ENTER] |

In real number mode, the calculator should display an *error message* because the domain of the inverse cosine function is $[-1, 1]$.

STUDY TIP

It is important to remember that the domain of the inverse sine function and the inverse cosine function is $[-1, 1]$, as indicated in Example 4(c).

In Example 4, if you had set the calculator to degree mode, the displays would have been in degrees rather than radians. This convention is peculiar to calculators. By definition, the values of inverse trigonometric functions are *always in radians*.

Compositions of Functions

Recall from Section P.10 that for all x in the domains of f and f^{-1}, inverse functions have the properties

$$f(f^{-1}(x)) = x \qquad \text{and} \qquad f^{-1}(f(x)) = x.$$

Inverse Properties of Trigonometric Functions

If $-1 \le x \le 1$ and $-\pi/2 \le y \le \pi/2$, then

$$\sin(\arcsin x) = x \qquad \text{and} \qquad \arcsin(\sin y) = y.$$

If $-1 \le x \le 1$ and $0 \le y \le \pi$, then

$$\cos(\arccos x) = x \qquad \text{and} \qquad \arccos(\cos y) = y.$$

If x is a real number and $-\pi/2 < y < \pi/2$, then

$$\tan(\arctan x) = x \qquad \text{and} \qquad \arctan(\tan y) = y.$$

Keep in mind that these inverse properties do not apply for arbitrary values of x and y. For instance,

$$\arcsin\left(\sin\frac{3\pi}{2}\right) = \arcsin(-1) = -\frac{\pi}{2} \ne \frac{3\pi}{2}.$$

In other words, the property

$$\arcsin(\sin y) = y$$

is not valid for values of y outside the interval $[-\pi/2, \pi/2]$.

Example 5 ▶ Using Inverse Properties

If possible, find the exact value.

a. $\tan[\arctan(-5)]$ **b.** $\arcsin\left(\sin\dfrac{5\pi}{3}\right)$ **c.** $\cos(\cos^{-1}\pi)$

Solution

a. Because -5 lies in the domain of the arctan function, the inverse property applies, and you have

$$\tan[\arctan(-5)] = -5.$$

b. In this case, $5\pi/3$ does not lie within the range of the arcsine function, $-\pi/2 \le y \le \pi/2$. However, $5\pi/3$ is coterminal with

$$\frac{5\pi}{3} - 2\pi = -\frac{\pi}{3}$$

which does lie in the range of the arcsine function, and you have

$$\arcsin\left(\sin\frac{5\pi}{3}\right) = \arcsin\left[\sin\left(-\frac{\pi}{3}\right)\right] = -\frac{\pi}{3}.$$

c. The expression $\cos(\cos^{-1}\pi)$ is not defined because $\cos^{-1}\pi$ is not defined. Remember that the domain of the inverse cosine function is $[-1, 1]$.

Example 6 shows how to use right triangles to find exact values of compositions of inverse functions. Then, Example 7 shows how to use triangles to convert a trigonometric expression into an algebraic expression. This conversion technique is used frequently in calculus.

Example 6 ▶ **Evaluating Compositions of Functions**

Find the exact value.

a. $\tan\left(\arccos\dfrac{2}{3}\right)$ **b.** $\cos\left[\arcsin\left(-\dfrac{3}{5}\right)\right]$

Solution

a. If you let $u = \arccos\frac{2}{3}$, then $\cos u = \frac{2}{3}$. Because $\cos u$ is positive, u is a *first*-quadrant angle. You can sketch and label angle u as shown in Figure 1.73. Consequently,

$$\tan\left(\arccos\frac{2}{3}\right) = \tan u = \frac{\text{opp}}{\text{adj}} = \frac{\sqrt{5}}{2}.$$

b. If you let $u = \arcsin\left(-\frac{3}{5}\right)$, then $\sin u = -\frac{3}{5}$. Because $\sin u$ is negative, u is a *fourth*-quadrant angle. You can sketch and label angle u as shown in Figure 1.74. Consequently,

$$\cos\left[\arcsin\left(-\frac{3}{5}\right)\right] = \cos u = \frac{\text{adj}}{\text{hyp}} = \frac{4}{5}.$$

Example 7 ▶ **Some Problems from Calculus**

Write each of the following as an algebraic expression in x.

a. $\sin(\arccos 3x),\quad 0 \le x \le \dfrac{1}{3}$ **b.** $\cot(\arccos 3x),\quad 0 \le x < \dfrac{1}{3}$

Solution

If you let $u = \arccos 3x$, then $\cos u = 3x$. Because

$$\cos u = \frac{\text{adj}}{\text{hyp}} = \frac{3x}{1}$$

you can sketch a right triangle with acute angle u, as shown in Figure 1.75. From this triangle, you can easily convert each expression to algebraic form.

a. $\sin(\arccos 3x) = \sin u = \dfrac{\text{opp}}{\text{hyp}} = \sqrt{1 - 9x^2},\quad 0 \le x \le \dfrac{1}{3}$

b. $\cot(\arccos 3x) = \cot u = \dfrac{\text{adj}}{\text{opp}} = \dfrac{3x}{\sqrt{1 - 9x^2}},\quad 0 \le x < \dfrac{1}{3}$

In Example 7, similar arguments can be made for x-values lying in the interval $\left[-\frac{1}{3}, 0\right]$.

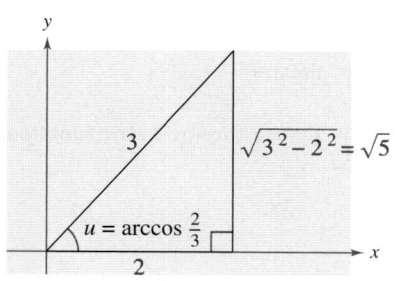

Angle whose cosine is $\frac{2}{3}$
FIGURE **1.73**

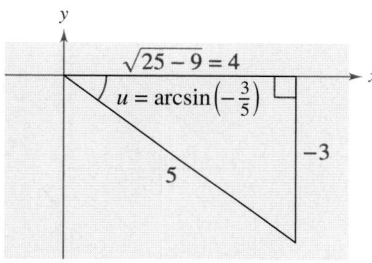

Angle whose sine is $-\frac{3}{5}$
FIGURE **1.74**

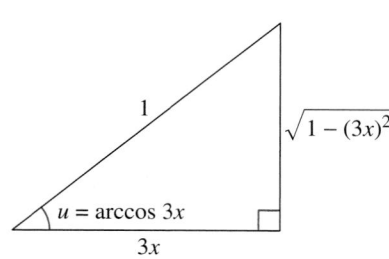

Angle whose cosine is $3x$
FIGURE **1.75**

You may want to review with students how to rationalize the denominator of a fractional expression.

1.7 Exercises

In Exercises 1–16, evaluate the expression without the aid of a calculator.

1. $\arcsin \frac{1}{2}$ 2. $\arcsin 0$

3. $\arccos \frac{1}{2}$ 4. $\arccos 0$

5. $\arctan \dfrac{\sqrt{3}}{3}$ 6. $\arctan(-1)$

7. $\cos^{-1}\left(-\dfrac{\sqrt{3}}{2}\right)$ 8. $\sin^{-1}\left(-\dfrac{\sqrt{2}}{2}\right)$

9. $\arctan\left(-\sqrt{3}\right)$ 10. $\arctan \sqrt{3}$

11. $\arccos\left(-\dfrac{1}{2}\right)$ 12. $\arcsin \dfrac{\sqrt{2}}{2}$

13. $\sin^{-1}\dfrac{\sqrt{3}}{2}$ 14. $\tan^{-1}\left(-\dfrac{\sqrt{3}}{3}\right)$

15. $\tan^{-1} 0$ 16. $\cos^{-1} 1$

In Exercises 17–32, use a calculator to approximate the expression. Round your result to two decimal places.

17. $\arccos 0.28$ 18. $\arcsin 0.45$

19. $\arcsin(-0.75)$ 20. $\arccos(-0.7)$

21. $\arctan(-3)$ 22. $\arctan 15$

23. $\sin^{-1} 0.31$ 24. $\cos^{-1} 0.26$

25. $\arccos(-0.41)$ 26. $\arcsin(-0.125)$

27. $\arctan 0.92$ 28. $\arctan 2.8$

29. $\arcsin \frac{3}{4}$ 30. $\arccos\left(-\frac{1}{3}\right)$

31. $\tan^{-1} \frac{7}{2}$ 32. $\tan^{-1}\left(-\frac{95}{7}\right)$

In Exercises 33 and 34, determine the missing coordinates of the points on the graph of the function.

33.
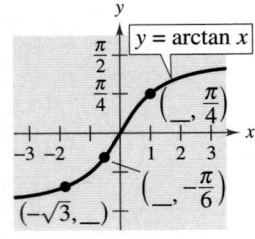

34.
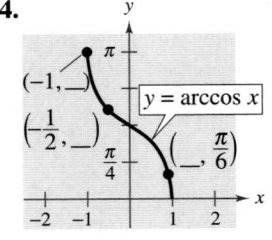

 In Exercises 35 and 36, use a graphing utility to graph f, g, and $y = x$ in the same viewing window to verify geometrically that g is the inverse function of f. (Be sure to restrict the domain of f properly.)

35. $f(x) = \tan x$, $g(x) = \arctan x$

36. $f(x) = \sin x$, $g(x) = \arcsin x$

In Exercises 37–42, use an inverse trigonometric function to write θ as a function of x.

37.
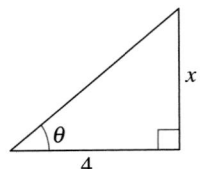

38.
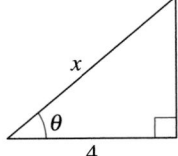

39.
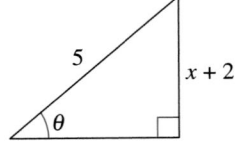

40.
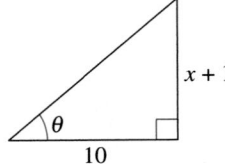

41.
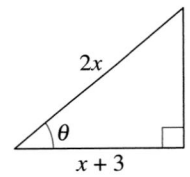

42.
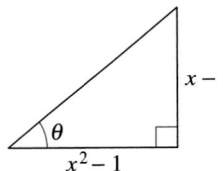

In Exercises 43–48, use the properties of inverse trigonometric functions to evaluate the expression.

43. $\sin(\arcsin 0.3)$

44. $\tan(\arctan 25)$

45. $\cos[\arccos(-0.1)]$

46. $\sin[\arcsin(-0.2)]$

47. $\arcsin(\sin 3\pi)$

48. $\arccos\left(\cos \dfrac{7\pi}{2}\right)$

In Exercises 49–58, find the exact value of the expression. (Hint: Make a sketch of a right triangle.)

49. $\sin\left(\arctan \frac{3}{4}\right)$ 50. $\sec\left(\arcsin \frac{4}{5}\right)$

51. $\cos(\tan^{-1} 2)$ 52. $\sin\left(\cos^{-1} \dfrac{\sqrt{5}}{5}\right)$

53. $\cos\left(\arcsin \frac{5}{13}\right)$ 54. $\csc\left[\arctan\left(-\frac{5}{12}\right)\right]$

55. $\sec\left[\arctan\left(-\frac{3}{5}\right)\right]$ 56. $\tan\left[\arcsin\left(-\frac{3}{4}\right)\right]$

57. $\sin\left[\arccos\left(-\frac{2}{3}\right)\right]$ 58. $\cot\left(\arctan \frac{5}{8}\right)$

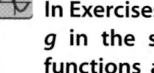 In Exercises 59–68, write an algebraic expression that is equivalent to the expression. (*Hint:* Sketch a right triangle, as demonstrated in Example 7.)

59. $\cot(\arctan x)$

60. $\sin(\arctan x)$

61. $\cos(\arcsin 2x)$

62. $\sec(\arctan 3x)$

63. $\sin(\arccos x)$

64. $\sec[\arcsin(x - 1)]$

65. $\tan\left(\arccos \dfrac{x}{3}\right)$

66. $\cot\left(\arctan \dfrac{1}{x}\right)$

67. $\csc\left(\arctan \dfrac{x}{\sqrt{2}}\right)$

68. $\cos\left(\arcsin \dfrac{x - h}{r}\right)$

 In Exercises 69 and 70, use a graphing utility to graph *f* and *g* in the same viewing window to verify that the two functions are equal. Explain why they are equal. Identify any asymptotes of the graphs.

69. $f(x) = \sin(\arctan 2x)$, $g(x) = \dfrac{2x}{\sqrt{1 + 4x^2}}$

70. $f(x) = \tan\left(\arccos \dfrac{x}{2}\right)$, $g(x) = \dfrac{\sqrt{4 - x^2}}{x}$

In Exercises 71–74, fill in the blank.

71. $\arctan \dfrac{9}{x} = \arcsin(\quad)$, $x \neq 0$

72. $\arcsin \dfrac{\sqrt{36 - x^2}}{6} = \arccos(\quad)$, $0 \leq x \leq 6$

73. $\arccos \dfrac{3}{\sqrt{x^2 - 2x + 10}} = \arcsin(\quad)$

74. $\arccos \dfrac{x - 2}{2} = \arctan(\quad)$, $|x - 2| \leq 2$

In Exercises 75–82, sketch a graph of the function.

75. $y = 2 \arccos x$

76. $y = \arcsin \dfrac{x}{2}$

77. $f(x) = \arcsin(x - 1)$

78. $g(t) = \arccos(t + 2)$

79. $f(x) = \arctan 2x$

80. $f(x) = \dfrac{\pi}{2} + \arctan x$

81. $h(v) = \tan(\arccos v)$

82. $f(x) = \arccos \dfrac{x}{4}$

 In Exercises 83–88, use a graphing utility to graph the function.

83. $f(x) = 2 \arccos(2x)$

84. $f(x) = \pi \arcsin(4x)$

85. $f(x) = \arctan(2x - 3)$

86. $f(x) = -3 + \arctan(\pi x)$

87. $f(x) = \pi - \sin^{-1}\left(\dfrac{2}{3}\right)$ **88.** $f(x) = \dfrac{\pi}{2} + \cos^{-1}\left(\dfrac{1}{\pi}\right)$

 In Exercises 89 and 90, write the function in terms of the sine function by using the identity

$$A \cos \omega t + B \sin \omega t = \sqrt{A^2 + B^2} \sin\left(\omega t + \arctan \dfrac{A}{B}\right).$$

Use a graphing utility to graph both forms of the function. What does the graph imply?

89. $f(t) = 3 \cos 2t + 3 \sin 2t$

90. $f(t) = 4 \cos \pi t + 3 \sin \pi t$

91. ***Docking a Boat***   A boat is pulled in by means of a winch located on a dock 5 feet above the deck of the boat (see figure). Let θ be the angle of elevation from the boat to the winch and let *s* be the length of the rope from the winch to the boat.

(a) Write θ as a function of *s*.

(b) Find θ when $s = 40$ feet and $s = 20$ feet.

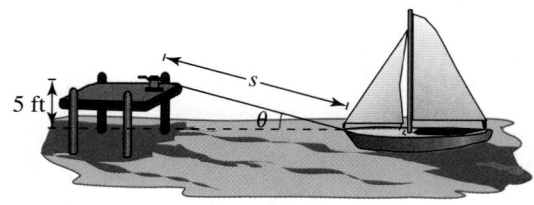

92. ***Photography***   A television camera at ground level is filming the lift-off of a space shuttle at a point 750 meters from the launch pad (see figure). Let θ be the angle of elevation to the shuttle and let *s* be the height of the shuttle.

(a) Write θ as a function of *s*.

(b) Find θ when $s = 300$ meters and $s = 1200$ meters.

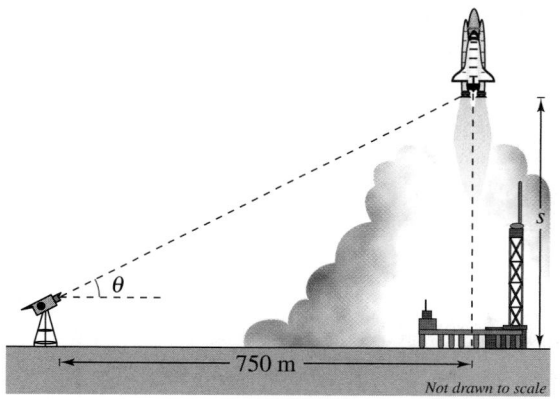

750 m

Not drawn to scale

▶ **Model It**

93. *Photography* A photographer is taking a picture of a three-foot-tall painting hung in an art gallery. The camera lens is 1 foot below the lower edge of the painting (see figure). The angle β subtended by the camera lens x feet from the painting is

$$\beta = \arctan \frac{3x}{x^2 + 4}, \qquad x > 0.$$

(a) Use a graphing utility to graph β as a function of x.

(b) Move the cursor along the graph to approximate the distance from the picture when β is maximum.

(c) Identify the asymptote of the graph and discuss its meaning in the context of the problem.

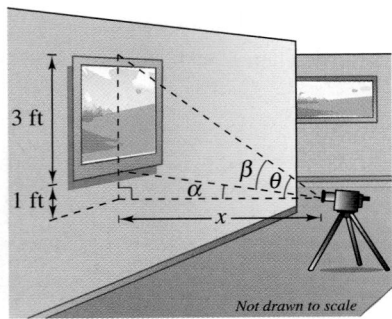

Not drawn to scale

94. *Granular Angle of Repose* Different types of granular substances naturally settle at different angles when stored in cone-shaped piles. This angle θ is called the *angle of repose* (see figure). When rock salt is stored in a cone-shaped pile 11 feet high, the diameter of the pile's base is about 34 feet. (Source: Bulk-Store Structures, Inc.)

(a) Find the angle of repose for rock salt.

(b) How tall is a pile of rock salt that has a base diameter of 40 feet?

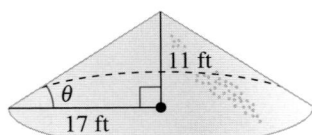

95. *Granular Angle of Repose* When whole corn is stored in a cone-shaped pile 20 feet high, the diameter of the pile's base is about 82 feet.

(a) Find the angle of repose for whole corn.

(b) How tall is a pile of corn that has a base diameter of 100 feet?

96. *Angle of Elevation* An airplane flies at an altitude of 6 miles toward a point directly over an observer. Consider θ and x as shown in the figure.

(a) Write θ as a function of x.

(b) Find θ when $x = 7$ miles and $x = 1$ mile.

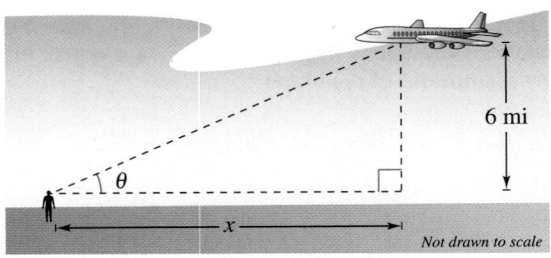

Not drawn to scale

97. *Security Patrol* A security car with its spotlight on is parked 20 meters from a warehouse. Consider θ and x as shown in the figure.

(a) Write θ as a function of x.

(b) Find θ when $x = 5$ meters and $x = 12$ meters.

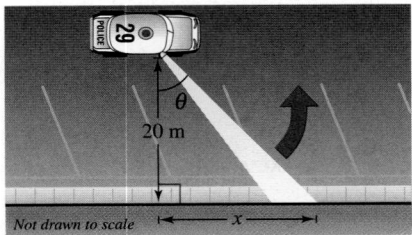

Not drawn to scale

Synthesis

True or False? In Exercises 98 and 99, determine whether the statement is true or false. Justify your answer.

98. $\sin \dfrac{5\pi}{6} = \dfrac{1}{2}$ ⟹ $\arcsin \dfrac{1}{2} = \dfrac{5\pi}{6}$

99. $\tan \dfrac{5\pi}{4} = 1$ ⟹ $\arctan 1 = \dfrac{5\pi}{4}$

100. Define the inverse cotangent function by restricting the domain of the cotangent function to the interval $(0, \pi)$, and sketch its graph.

101. Define the inverse secant function by restricting the domain of the secant function to the intervals $[0, \pi/2)$ and $(\pi/2, \pi]$, and sketch its graph.

102. Define the inverse cosecant function by restricting the domain of the cosecant function to the intervals $[-\pi/2, 0)$ and $(0, \pi/2]$, and sketch its graph.

103. Use the results of Exercises 100–102 to evaluate each expression without using a calculator.

(a) $\operatorname{arcsec} \sqrt{2}$ (b) $\operatorname{arcsec} 1$

(c) $\operatorname{arccot}\left(-\sqrt{3}\right)$ (d) $\operatorname{arccsc} 2$

104. *Area* In calculus, it is shown that the area of the region bounded by the graphs of $y = 0$, $y = 1/(x^2 + 1)$, $x = a$, and $x = b$ is given by

$$\text{Area} = \arctan b - \arctan a$$

(see figure). Find the area for the following values of a and b.

(a) $a = 0, b = 1$ (b) $a = -1, b = 1$

(c) $a = 0, b = 3$ (d) $a = -1, b = 3$

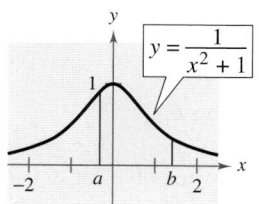

105. *Think About It* Use a graphing utility to graph the functions

$$f(x) = \sqrt{x} \text{ and } g(x) = 6 \arctan x.$$

For $x > 0$, it appears that $g > f$. Explain why you know that there exists a positive real number a such that $g < f$ for $x > a$. Approximate the number a.

106. *Think About It* Consider the functions

$$f(x) = \sin x \quad \text{and} \quad f^{-1}(x) = \arcsin x.$$

(a) Use a graphing utility to graph the composite functions $f \circ f^{-1}$ and $f^{-1} \circ f$.

(b) Explain why the graphs in part (a) are not the graph of the line $y = x$. Why do the graphs of $f \circ f^{-1}$ and $f^{-1} \circ f$ differ?

Proof **In Exercises 107–112, prove the identity.**

107. $\arcsin(-x) = -\arcsin x$

108. $\arctan(-x) = -\arctan x$

109. $\arccos(-x) = \pi - \arccos x$

110. $\arctan x + \arctan \dfrac{1}{x} = \dfrac{\pi}{2}, \quad x > 0$

111. $\arcsin x + \arccos x = \dfrac{\pi}{2}$

112. $\arcsin x = \arctan \dfrac{x}{\sqrt{1 - x^2}}$

Review

In Exercises 113–116, sketch a right triangle corresponding to the trigonometric function of the acute angle θ. Use the Pythagorean Theorem to determine the third side.

113. $\sin \theta = \frac{3}{4}$

114. $\tan \theta = 2$

115. $\cos \theta = \frac{5}{6}$

116. $\sec \theta = 3$

117. *Partnership Costs* A group of people agree to share equally in the cost of a \$250,000 endowment to a college. If they could find two more people to join the group, each person's share of the cost would decrease by \$6250. How many people are presently in the group?

118. *Speed* A boat travels at a speed of 18 miles per hour in still water. It travels 35 miles upstream and then returns to the starting point in a total of 4 hours. Find the speed of the current.

1.8 Applications and Models

▶ What you should learn

- How to solve real-life problems involving right triangles
- How to solve real-life problems involving directional bearings
- How to solve real-life problems involving harmonic motion

▶ Why you should learn it

Trigonometric functions frequently model real-life problems involving cyclical patterns in business. For instance, in Exercise 63 on page 206, you can find a trigonometric model for the sales of an outerwear manufacturer.

Applications Involving Right Triangles

In this section the three angles of a right triangle are denoted by the letters A, B, and C (where C is the right angle), and the lengths of the sides opposite these angles by the letters a, b, and c (where c is the hypotenuse).

Example 1 ▶ Solving a Right Triangle

Solve the right triangle shown in Figure 1.76 for all unknown sides and angles.

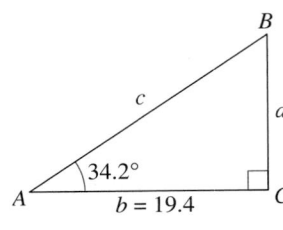

FIGURE **1.76**

Solution

Because $C = 90°$, it follows that $A + B = 90°$ and $B = 90° - 34.2° = 55.8°$. To solve for a, use the fact that

$$\tan A = \frac{\text{opp}}{\text{adj}} = \frac{a}{b} \implies a = b \tan A.$$

So, $a = 19.4 \tan 34.2° \approx 13.18$. Similarly, to solve for c, use the fact that

$$\cos A = \frac{\text{adj}}{\text{hyp}} = \frac{b}{c} \implies c = \frac{b}{\cos A}.$$

So, $c = \dfrac{19.4}{\cos 34.2°} \approx 23.46$.

Example 2 ▶ Finding a Side of a Right Triangle

A safety regulation states that the maximum angle of elevation for a rescue ladder is 72°. A fire department's longest ladder is 110 feet. What is the maximum safe rescue height?

Solution

A sketch is shown in Figure 1.77. From the equation $\sin A = a/c$, it follows that

$$a = c \sin A = 110 \sin 72° \approx 104.6.$$

So, the maximum safe rescue height is about 104.6 feet above the height of the fire truck.

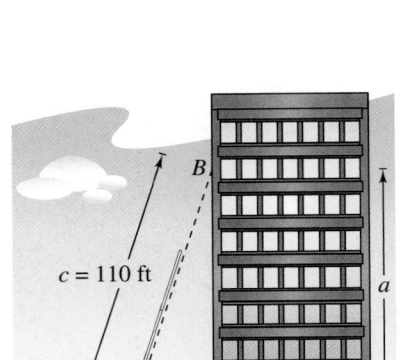

FIGURE **1.77**

Example 3 ▶ Finding a Side of a Right Triangle

At a point 200 feet from the base of a building, the angle of elevation to the *bottom* of a smokestack is 35°, whereas the angle of elevation to the *top* is 53°, as shown in Figure 1.78. Find the height *s* of the smokestack alone.

Solution

Note from Figure 1.78 that this problem involves two right triangles. For the smaller right triangle, use the fact that

$$\tan 35° = \frac{a}{200}$$

to conclude that the height of the building is

$$a = 200 \tan 35°.$$

For the larger right triangle, use the equation

$$\tan 53° = \frac{a + s}{200}$$

to conclude that $a + s = 200 \tan 53°$. So, the height of the smokestack is

$$s = 200 \tan 53° - a$$

$$= 200 \tan 53° - 200 \tan 35°$$

$$\approx 125.4 \text{ feet.}$$

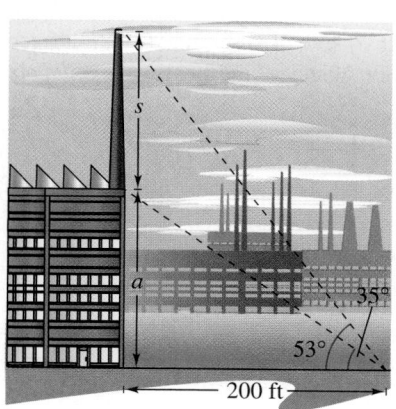

FIGURE **1.78**

Example 4 ▶ Finding an Acute Angle of a Right Triangle

A swimming pool is 20 meters long and 12 meters wide. The bottom of the pool is slanted so that the water depth is 1.3 meters at the shallow end and 4 meters at the deep end, as shown in Figure 1.79. Find the angle of depression of the bottom of the pool.

Solution

Using the tangent function, you can see that

$$\tan A = \frac{\text{opp}}{\text{adj}}$$

$$= \frac{2.7}{20}$$

$$= 0.135.$$

So, the angle of depression is

$$A = \arctan 0.135$$

$$\approx 0.13419 \text{ radian}$$

$$\approx 7.69°.$$

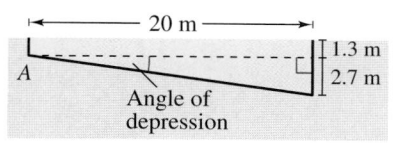

FIGURE **1.79**

Trigonometry and Bearings

In surveying and navigation, directions are generally given in terms of **bearings.** A bearing measures the acute angle that a path or line of sight makes with a fixed north-south line, as shown in Figure 1.80. For instance, the bearing S 35° E in Figure 1.80 means 35 degrees east of south.

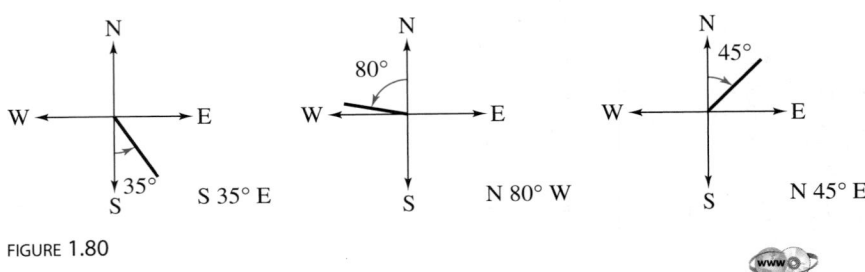

FIGURE **1.80**

Example 5 ▶ **Finding Directions in Terms of Bearings**

A ship leaves port at noon and heads due west at 20 knots, or 20 nautical miles (nm) per hour. At 2 P.M. the ship changes course to N 54° W, as shown in Figure 1.81. Find the ship's bearing and distance from the port of departure at 3 P.M.

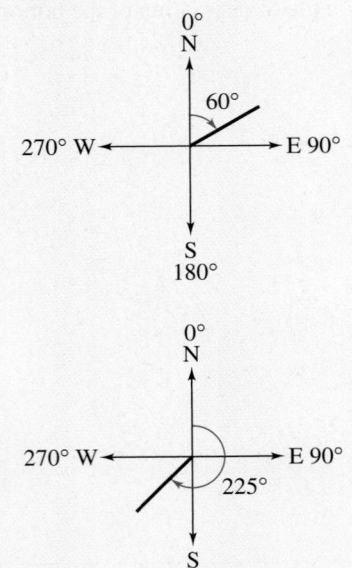

FIGURE **1.81**

Solution

For triangle BCD, you have $B = 90° - 54° = 36°$. The two sides of this triangle can be determined to be

$$b = 20 \sin 36° \qquad \text{and} \qquad d = 20 \cos 36°.$$

For triangle ACD, you can find angle A as follows.

$$\tan A = \frac{b}{d + 40} = \frac{20 \sin 36°}{20 \cos 36° + 40} \approx 0.2092494$$

$$A \approx \arctan 0.2092494 \approx 0.2062732 \text{ radian} \approx 11.82°$$

The angle with the north-south line is $90° - 11.82° = 78.18°$. So, the bearing of the ship is N 78.18° W. Finally, from triangle ACD, you have $\sin A = b/c$, which yields

$$c = \frac{b}{\sin A} = \frac{20 \sin 36°}{\sin 11.82°}$$

$$\approx 57.4 \text{ nautical miles.} \qquad \text{Distance from port}$$

Harmonic Motion

The periodic nature of the trigonometric functions is useful for describing the motion of a point on an object that vibrates, oscillates, rotates, or is moved by wave motion.

For example, consider a ball that is bobbing up and down on the end of a spring, as shown in Figure 1.82. Suppose that 10 centimeters is the maximum distance the ball moves vertically upward or downward from its equilibrium (at rest) position. Suppose further that the time it takes for the ball to move from its maximum displacement above zero to its maximum displacement below zero and back again is $t = 4$ seconds. Assuming the ideal conditions of perfect elasticity and no friction or air resistance, the ball would continue to move up and down in a uniform and regular manner.

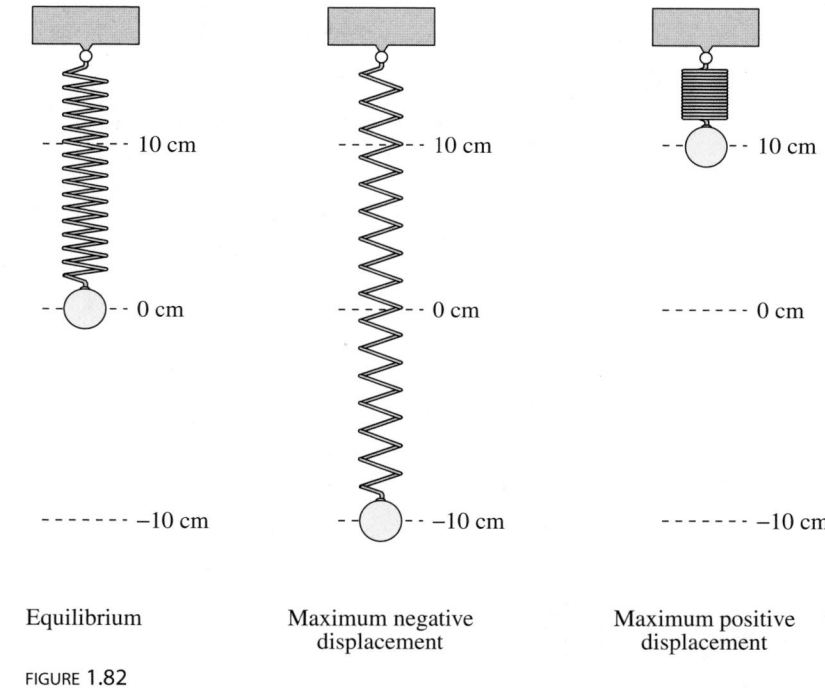

Equilibrium Maximum negative Maximum positive
displacement displacement

FIGURE 1.82

From this spring you can conclude that the period (time for one complete cycle) of the motion is

Period = 4 seconds

and that its amplitude (maximum displacement from equilibrium) is

Amplitude = 10 centimeters.

Motion of this nature can be described by a sine or cosine function, and is called **simple harmonic motion.**

Definition of Simple Harmonic Motion

A point that moves on a coordinate line is said to be in **simple harmonic motion** if its distance d from the origin at time t is given by either

$$d = a \sin \omega t \qquad \text{or} \qquad d = a \cos \omega t$$

where a and ω are real numbers such that $\omega > 0$. The motion has amplitude $|a|$, period $2\pi/\omega$, and frequency $\omega/(2\pi)$.

Example 6 ▶   **Simple Harmonic Motion**

Write the equation for the simple harmonic motion of the ball described in Figure 1.82, where the period is 4 seconds. What is the frequency of this harmonic motion?

Solution

Because the spring is at equilibrium ($d = 0$) when $t = 0$, you use the equation

$$d = a \sin \omega t.$$

Moreover, because the maximum displacement from zero is 10 and the period is 4, you have

$$\text{Amplitude} = |a| = 10$$

$$\text{Period} = \frac{2\pi}{\omega} = 4 \quad \Longrightarrow \quad \omega = \frac{\pi}{2}.$$

Consequently, the equation of motion is

$$d = 10 \sin \frac{\pi}{2} t.$$

Note that the choice of $a = 10$ or $a = -10$ depends on whether the ball initially moves up or down. The frequency is

$$\text{Frequency} = \frac{\omega}{2\pi}$$

$$= \frac{\pi/2}{2\pi}$$

$$= \frac{1}{4} \text{ cycle per second.}$$

FIGURE **1.83**

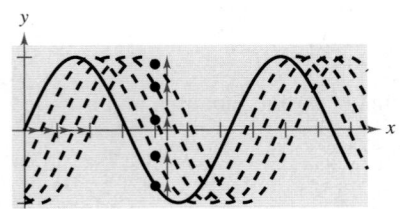

FIGURE **1.84**

One illustration of the relationship between sine waves and harmonic motion can be seen in the wave motion resulting when a stone is dropped into a calm pool of water. The waves move outward in roughly the shape of sine (or cosine) waves, as shown in Figure 1.83. As an example, suppose you are fishing and your fishing bob is attached so that it does not move horizontally. As the waves move outward from the dropped stone, your fishing bob will move up and down in simple harmonic motion, as shown in Figure 1.84.

Example 7 ▶   **Simple Harmonic Motion**

Given the equation for simple harmonic motion

$$d = 6 \cos \frac{3\pi}{4} t$$

find (a) the maximum displacement, (b) the frequency, (c) the value of d when $t = 4$, and (d) the least positive value of t for which $d = 0$.

Solution

The given equation has the form $d = a \cos \omega t$, with $a = 6$ and $\omega = 3\pi/4$.

a. The maximum displacement (from the point of equilibrium) is given by the amplitude. So, the maximum displacement is 6.

b. Frequency $= \dfrac{\omega}{2\pi}$

$$= \frac{3\pi/4}{2\pi} = \frac{3}{8} \text{ cycle per unit of time}$$

c. $d = 6 \cos \left[\dfrac{3\pi}{4} (4) \right]$

$$= 6 \cos 3\pi$$

$$= 6(-1)$$

$$= -6$$

d. To find the least positive value of t for which $d = 0$, solve the equation

$$d = 6 \cos \frac{3\pi}{4} t = 0$$

to obtain

$$\frac{3\pi}{4} t = \frac{\pi}{2}, \frac{3\pi}{2}, \frac{5\pi}{2}, \dots \qquad \Longrightarrow \qquad t = \frac{2}{3}, 2, \frac{10}{3}, \dots$$

So, the least positive value of t is $t = \frac{2}{3}$.

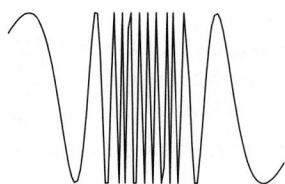

(a)

(b)

FIGURE **1.85**

Writing ABOUT MATHEMATICS

Radio Waves   Many different physical phenomena can be characterized by wave motion. These phenomena include electromagnetic waves such as radio waves, television waves, and microwaves. Radio waves transmit sound in two different ways. For an AM station, the *amplitude* of the wave is modified to carry sound. The letters AM stand for "amplitude modulation." An FM radio signal has its *frequency* modified in order to carry sound, hence the term "frequency modulation." The FM radio signal is preferred by listeners because of its low-noise and wide-bandwidth qualities. Of the two graphs in Figure 1.85, one shows an AM wave and the other shows an FM wave. Which is which? Explain your reasoning.

1.8 Exercises

In Exercises 1–10, solve the right triangle shown in the figure. Round your answer to two decimal places.

1. $A = 20°$, $b = 10$ **2.** $B = 54°$, $c = 15$

3. $B = 71°$, $b = 24$ **4.** $A = 8.4°$, $a = 40.5$

5. $a = 6$, $b = 10$ **6.** $a = 25$, $c = 35$

7. $b = 16$, $c = 52$ **8.** $b = 1.32$, $c = 9.45$

9. $A = 12°15'$, $c = 430.5$

10. $B = 65°12'$, $a = 14.2$

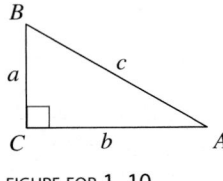

FIGURE FOR 1–10

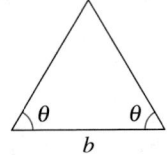

FIGURE FOR 11–14

In Exercises 11–14, find the altitude of the isosceles triangle shown in the figure. Round your answer to two decimal places.

11. $\theta = 52°$, $b = 4$ inches

12. $\theta = 18°$, $b = 10$ meters

13. $\theta = 41°$, $b = 46$ inches

14. $\theta = 27°$, $b = 11$ feet

15. *Length*   The sun is 25° above the horizon. Find the length of a shadow cast by a silo that is 50 feet tall (see figure).

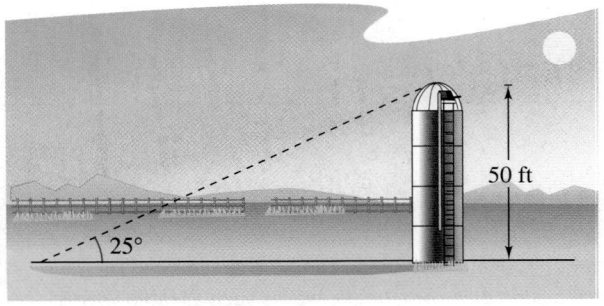

16. *Length*   The sun is 20° above the horizon. Find the length of a shadow cast by a building that is 600 feet tall.

17. *Height*   A ladder 20 feet long leans against the side of a house. Find the height h from the top of the ladder to the ground if the angle of elevation of the ladder is 80°.

18. *Height*   The length of a shadow of a tree is 125 feet when the angle of elevation of the sun is 33°. Approximate the height h of the tree.

19. *Height*   From a point 50 feet in front of a church, the angles of elevation to the base of the steeple and the top of the steeple are 35° and 47° 40′, respectively.

(a) Draw right triangles that represent the problem. Label the known and unknown quantities.

(b) Use a trigonometric function to write an equation involving the unknown height of the steeple.

(c) Find the height of the steeple.

20. *Height*   You are standing 100 feet from the base of a platform from which people are bungee jumping. The angle of elevation from your position to the top of the platform from which they jump is 51°. From what height are the people jumping?

21. *Depth*   The sonar of a navy cruiser detects a submarine that is 4000 feet from the cruiser. The angle between the water line and the submarine is 34° (see figure). How deep is the submarine?

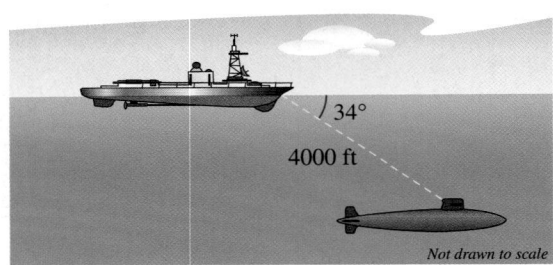

34°

4000 ft

Not drawn to scale

22. *Angle of Elevation*   An amateur radio operator erects a 75-foot vertical tower for an antenna. Find the angle of elevation to the top of the tower at a point on level ground 50 feet from its base.

23. *Angle of Elevation*   The height of an outdoor basketball backboard is $12\frac{1}{2}$ feet, and the backboard casts a shadow $17\frac{1}{3}$ feet long.

(a) Draw a right triangle that represents the problem. Label the known and unknown quantities.

(b) Use a trigonometric function to write an equation involving the unknown angle of elevation of the sun.

(c) Find the angle of elevation of the sun.

24. *Angle of Depression* A Global Positioning System satellite orbits 12,500 miles above Earth's surface. Find the angle of depression from the satellite to the horizon. Assume the radius of Earth is 4000 miles.

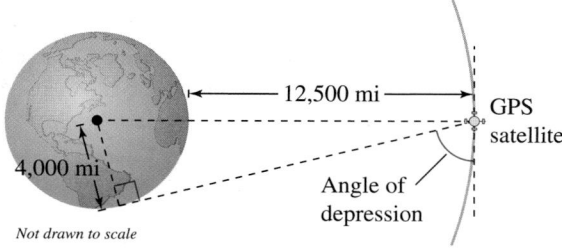

Not drawn to scale

25. *Angle of Depression* A cellular telephone tower that is 150 feet tall is placed on top of a mountain that is 1200 feet above sea level. What is the angle of depression from the top of the tower to a cell phone user who is 5 horizontal miles away and 400 feet above sea level?

26. *Airplane Ascent* During takeoff, an airplane's angle of climb is 18° and its speed is 275 feet per second.

   (a) Find the plane's altitude after 1 minute.

   (b) How long will it take the plane to climb to an altitude of 10,000 feet?

27. *Mountain Descent* A sign on a roadway at the top of a mountain indicates that for the next 4 miles the grade is 10.5° (see figure). Find the change in elevation for a car descending the mountain.

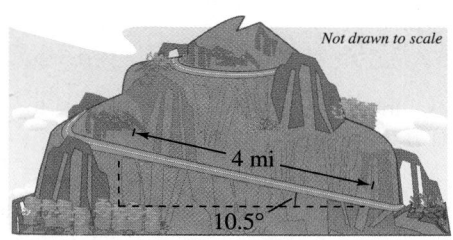

Not drawn to scale

28. *Mountain Descent* A roadway sign at the top of a mountain indicates that for the next 4 miles the grade is 12%. Find the angle of the grade and the change in elevation for a car descending the mountain.

29. *Navigation* An airplane flying at 600 miles per hour has a bearing of 52°. After flying for 1.5 hours, how far north and how far east will the plane have traveled from its point of departure?

30. *Surveying* A surveyor wishes to find the distance across a swamp (see figure). The bearing from A to B is N 32° W. The surveyor walks 50 meters from A, and at the point C the bearing to B is N 68° W. Find (a) the bearing from A to C and (b) the distance from A to B.

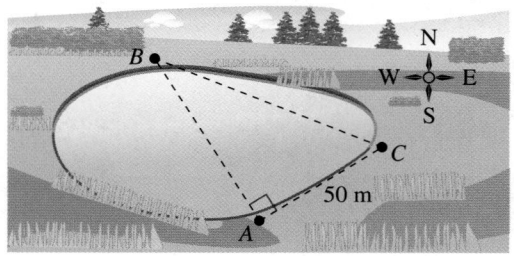

31. *Location of a Fire* Two fire towers are 30 kilometers apart, where tower A is due west of tower B. A fire is spotted from the towers, and the bearings from A and B are E 14° N and W 34° N, respectively (see figure). Find the distance d of the fire from the line segment AB.

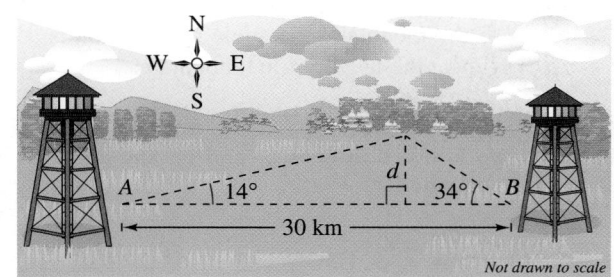

Not drawn to scale

32. *Navigation* A ship is 45 miles east and 30 miles south of port. The captain wants to sail directly to port. What bearing should be taken?

33. *Distance* An observer in a lighthouse 350 feet above sea level observes two ships directly offshore. The angles of depression to the ships are 4° and 6.5° (see figure). How far apart are the ships?

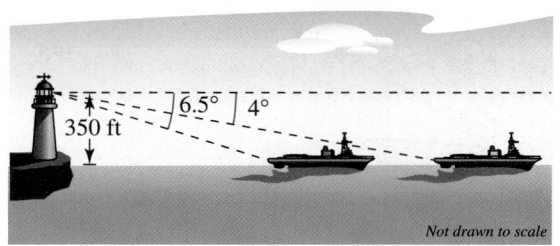

Not drawn to scale

34. Distance A passenger in an airplane at an altitude of 10 kilometers sees two towns directly to the east of the plane. The angles of depression to the towns are 28° and 55° (see figure). How far apart are the towns?

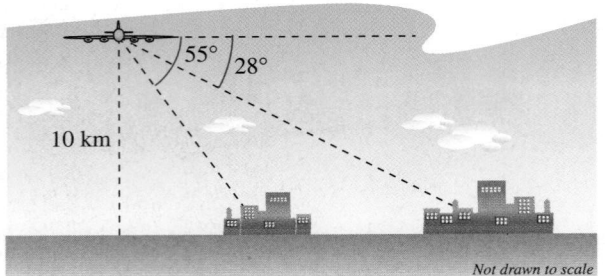

Not drawn to scale

35. Altitude A plane is observed approaching your home and you assume that its speed is 550 miles per hour. The angle of elevation of the plane is 16° at one time and 57° one minute later. Approximate the altitude of the plane.

36. Height While traveling across flat land, you notice a mountain directly in front of you. The angle of elevation to the peak is 2.5°. After you drive 17 miles closer to the mountain, the angle of elevation is 9°. Approximate the height of the mountain.

Geometry In Exercises 37 and 38, find the angle α between two nonvertical lines L_1 and L_2. The angle α satisfies the equation

$$\tan \alpha = \left| \frac{m_2 - m_1}{1 + m_2 m_1} \right|$$

where m_1 and m_2 are the slopes of L_1 and L_2, respectively. (Assume that $m_1 m_2 \neq -1$.)

37. L_1: $3x - 2y = 5$
L_2: $x + y = 1$

38. L_1: $2x - y = 8$
L_2: $x - 5y = -4$

39. Geometry Determine the angle between the diagonal of a cube and the diagonal of its base, as shown in the figure.

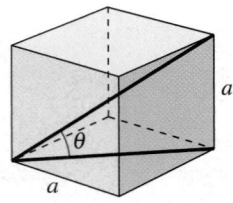

FIGURE FOR 39

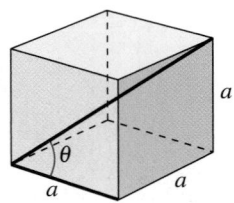

FIGURE FOR 40

40. Geometry Determine the angle between the diagonal of a cube and its edge, as shown in the figure.

41. Geometry Find the length of the sides of a regular pentagon inscribed in a circle of radius 25 inches.

42. Geometry Find the length of the sides of a regular hexagon inscribed in a circle of radius 25 inches.

43. Hardware Express the distance y across the flat sides of a hexagonal nut as a function of r, as shown in the figure.

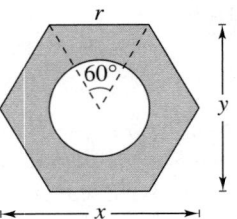

44. Bolt Holes The figure shows a circular piece of sheet metal that has a diameter of 40 centimeters and contains 12 equally spaced bolt holes. Determine the straight-line distance between the centers of consecutive bolt holes.

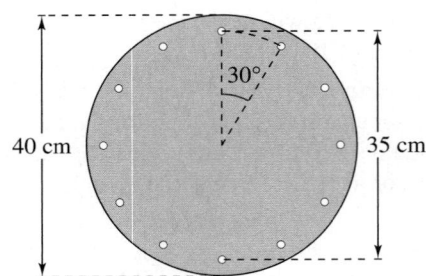

Trusses In Exercises 45 and 46, find all the unknown lengths of the members of the truss.

45.

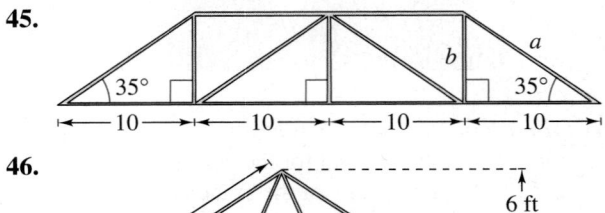

46.

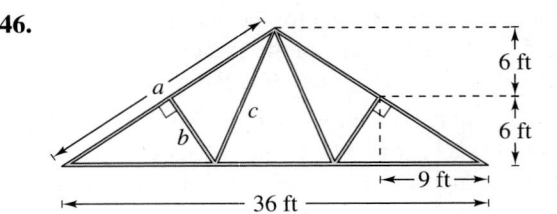

Harmonic Motion In Exercises 47–50, for the simple harmonic motion described by the trigonometric function, find (a) the maximum displacement, (b) the frequency, and (c) the least positive value of t for which $d = 0$.

47. $d = 4 \cos 8\pi t$

48. $d = \frac{1}{2} \cos 20\pi t$

49. $d = \frac{1}{16} \sin 120\pi t$

50. $d = \frac{1}{64} \sin 792\pi t$

Harmonic Motion In Exercises 51–54, find a model for simple harmonic motion satisfying the specified conditions.

	Displacement ($t = 0$)	Amplitude	Period
51.	0	4 centimeters	2 seconds
52.	0	3 meters	6 seconds
53.	3 inches	3 inches	1.5 seconds
54.	2 feet	2 feet	10 seconds

55. **Tuning Fork** A point on the end of a tuning fork moves in simple harmonic motion described by $d = a \sin \omega t$. Find ω given that the tuning fork for middle C has a frequency of 264 vibrations per second.

56. **Wave Motion** A buoy oscillates in simple harmonic motion as waves go past. It is noted that the buoy moves a total of 3.5 feet from its low point to its high point (see figure), and that it returns to its high point every 10 seconds. Write an equation that describes the motion of the buoy if its high point is at $t = 0$.

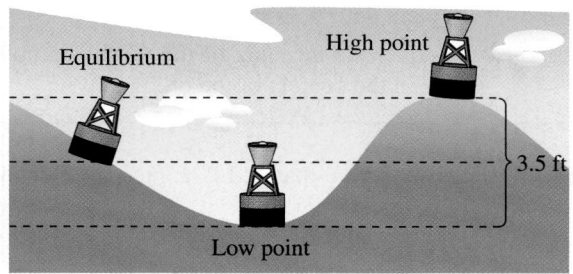

57. **Oscillation of a Spring** A ball that is bobbing up and down on the end of a spring has a maximum displacement of 3 inches. Its motion (in ideal conditions) is modeled by $y = \frac{1}{4} \cos 16t$ ($t > 0$), where y is in feet and t is in seconds.

   (a) Graph the function.

   (b) What is the period of the oscillations?

   (c) Determine the first time the weight passes the point of equilibrium ($y = 0$).

Synthesis

True or False? In Exercises 58 and 59, determine whether the statement is true or false. Justify your answer.

58. The Leaning Tower of Pisa is not vertical, but if you know the exact angle of elevation θ to the 191-foot tower when you stand near it, then you can determine the exact distance to the tower d by using the formula

$$\tan \theta = \frac{191}{d}.$$

59. For the harmonic motion of a ball bobbing up and down on the end of a spring, one period can be described as the length of one coil of the spring.

60. **Numerical and Graphical Analysis** A two-meter-high fence is 3 meters from the side of a grain storage bin. A grain elevator must reach from ground level outside the fence to the storage bin (see figure). The objective is to determine the shortest elevator that meets the constraints.

   (a) Complete four rows of the table.

θ	L_1	L_2	$L_1 + L_2$
0.1	$\dfrac{2}{\sin 0.1}$	$\dfrac{3}{\cos 0.1}$	23.0
0.2	$\dfrac{2}{\sin 0.2}$	$\dfrac{3}{\cos 0.2}$	13.1

   (b) Use a graphing utility to generate additional rows of the table. Use the table to estimate the minimum length of the elevator.

   (c) Write the length $L_1 + L_2$ as a function of θ.

   (d) Use a graphing utility to graph the function. Use the graph to estimate the minimum length. How does your estimate compare with that of part (b)?

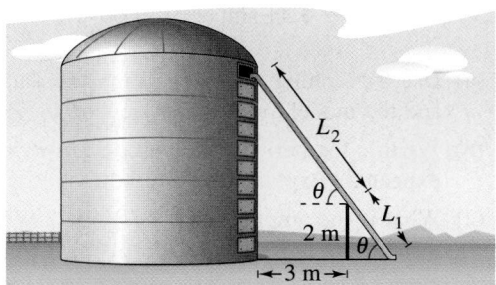

61. *Numerical and Graphical Analysis* The cross section of an irrigation canal is an isosceles trapezoid of which three of the sides are 8 feet long (see figure). The objective is to find the angle θ that maximizes the area of the cross section. [*Hint:* The area of a trapezoid is $(h/2)(b_1 + b_2)$.]

(a) Complete seven rows of the table.

Base 1	Base 2	Altitude	Area
8	$8 + 16 \cos 10°$	$8 \sin 10°$	22.1
8	$8 + 16 \cos 20°$	$8 \sin 20°$	42.5

(b) Use a graphing utility to generate additional rows of the table. Use the table to estimate the maximum cross-sectional area.

(c) Write the area A as a function of θ.

(d) Use a graphing utility to graph the function. Use the graph to estimate the maximum cross-sectional area. How does your estimate compare with that of part (b)?

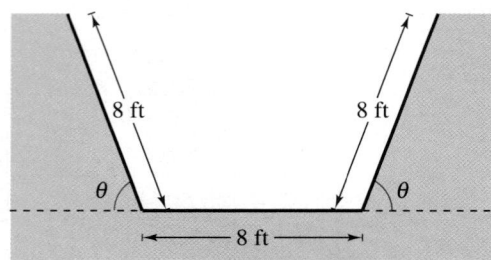

8 ft   8 ft

θ θ

8 ft

62. *Data Analysis* The times S of sunset (Greenwich Mean Time) at 40° north latitude on the 15th of each month are: 1(16:59), 2(17:35), 3(18:06), 4(18:38), 5(19:08), 6(19:30), 7(19:28), 8(18:57), 9(18:09), 10(17:21), 11(16:44), 12(16:36). The month is represented by t, with $t = 1$ corresponding to January. A model (in which minutes have been converted to the decimal parts of an hour) for this data is

$$S(t) = 18.09 + 1.41 \sin\left(\frac{\pi t}{6} + 4.60\right).$$

(a) Use a graphing utility to graph the data points and the model in the same viewing window.

(b) What is the period of the model? Is it what you expected? Explain.

(c) What is the amplitude of the model? What does it represent in the model? Explain.

▶ **Model It**

63. *Data Analysis* The table shows the average sales S (in millions of dollars) of an outerwear manufacturer for each month t, where $t = 1$ represents January.

Time, t	Sales, S
1	13.46
2	11.15
3	8.00
4	4.85
5	2.54
6	1.70
7	2.54
8	4.85
9	8.00
10	11.15
11	13.46
12	14.3

(a) Create a scatter plot of the data.

(b) Find a trigonometric model that fits the data. Graph the model on your scatter plot. How well does the model fit the data?

(c) What is the period of the model? Do you think it is reasonable given the context? Explain your reasoning.

(d) Interpret the meaning of the model's amplitude in the context of the problem.

---

64. *Writing* Is it true that N 24° E means 24 degrees north of east? Explain.

Review

In Exercises 65–72, sketch a graph of the equation.

65. $3x - 2y = 4$

66. $5y - 3x = 12$

67. $(y - 2)^2 = 8(x + 2)$

68. $(x + 3)^2 = 5y - 8$

69. $\dfrac{x^2}{4} + y^2 = 1$

70. $2x^2 + y^2 - 4 = 0$

71. $\dfrac{x^2}{4} + \dfrac{y^2}{4} = 1$

72. $(x - 2)^2 + y^2 = 25$

Chapter Summary

▶ *What* did you learn?

Review Exercises

1.1 **In Exercises 1–4, estimate the angle to the nearest one-half radian.**

1.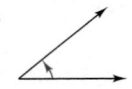

2.

3.

4.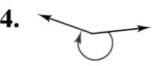

In Exercises 5–12, sketch the angle in standard position. List one positive and one negative coterminal angle.

5. $\dfrac{11\pi}{4}$ **6.** $\dfrac{2\pi}{9}$

7. $-\dfrac{4\pi}{3}$ **8.** $-\dfrac{23\pi}{3}$

9. $70°$ **10.** $280°$

11. $-110°$ **12.** $-405°$

In Exercises 13–16, convert the measure from radians to degrees. Round your answer to two decimal places.

13. $\dfrac{5\pi}{7}$ **14.** $-\dfrac{11\pi}{6}$

15. -3.5 **16.** 5.7

In Exercises 17–20, convert the measure from degrees to radians. Round your answer to four decimal places.

17. $480°$ **18.** $-127.5°$

19. $-33°\,45'$ **20.** $196°\,77'$

21. *Phonograph* Compact discs have all but replaced phonograph records. Phonograph records are vinyl discs that rotate on a turntable. A typical record album is 12 inches in diameter and plays at $33\frac{1}{3}$ revolutions per minute.

(a) What is the angular speed of a record album?

(b) What is the linear speed of the outer edge of a record album?

22. *Bicycle* At what speed is a bicyclist traveling when his 27-inch-diameter tires are rotating at an angular speed of 5π radians per second?

1.2 **In Exercises 23–26, find the point (x, y) on the unit circle that corresponds to the real number t.**

23. $t = \dfrac{2\pi}{3}$ **24.** $t = \dfrac{3\pi}{4}$

25. $t = \dfrac{5\pi}{6}$ **26.** $t = -\dfrac{4\pi}{3}$

In Exercises 27–30, evaluate (if possible) the six trigonometric functions of the real number.

27. $t = \dfrac{7\pi}{6}$ **28.** $t = \dfrac{\pi}{4}$

29. $t = -\dfrac{2\pi}{3}$ **30.** $t = 2\pi$

In Exercises 31–34, evaluate the trigonometric function using its period as an aid.

31. $\sin \dfrac{11\pi}{4}$ **32.** $\cos 4\pi$

33. $\sin\left(-\dfrac{17\pi}{6}\right)$ **34.** $\cos\left(-\dfrac{13\pi}{3}\right)$

In Exercises 35–38, use a calculator to evaluate the trigonometric function. Round your answer to two decimal places.

35. $\tan 33$ **36.** $\csc 10.5$

37. $\sec \dfrac{12\pi}{5}$ **38.** $\sin\left(-\dfrac{\pi}{9}\right)$

1.3 **In Exercises 39–42, find the exact values of the six trigonometric functions of the angle θ shown in the figure.**

39.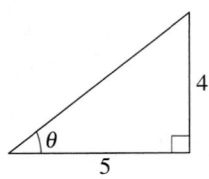

40.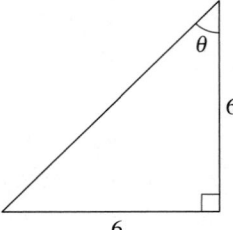

41.

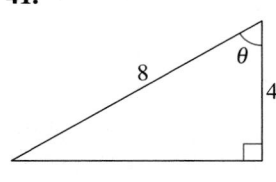

42.

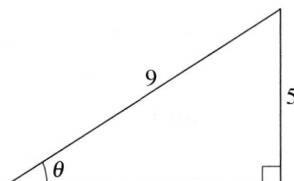

In Exercises 43–46, use the given function value and trigonometric identities (including the cofunction identities) to find the indicated trigonometric functions.

43. $\sin \theta = \frac{1}{3}$

    (a) $\csc \theta$ (b) $\cos \theta$

    (c) $\sec \theta$ (d) $\tan \theta$

44. $\tan \theta = 4$

    (a) $\cot \theta$ (b) $\sec \theta$

    (c) $\cos \theta$ (d) $\csc \theta$

45. $\csc \theta = 4$

    (a) $\sin \theta$ (b) $\cos \theta$

    (c) $\sec \theta$ (d) $\tan \theta$

46. $\csc \theta = 5$

    (a) $\sin \theta$ (b) $\cot \theta$

    (c) $\tan \theta$ (d) $\sec(90° - \theta)$

In Exercises 47–52, use a calculator to evaluate the trigonometric function. Round your answer to two decimal places.

47. $\tan 33°$ **48.** $\csc 11°$

49. $\sin 34.2°$ **50.** $\sec 79.3°$

51. $\cot 15° \, 14'$ **52.** $\cos 78° \, 11' 58''$

53. *Railroad Grade*   A train travels 3.5 kilometers on a straight track with a grade of $1° \, 10'$ (see figure). What is the vertical rise of the train in that distance?

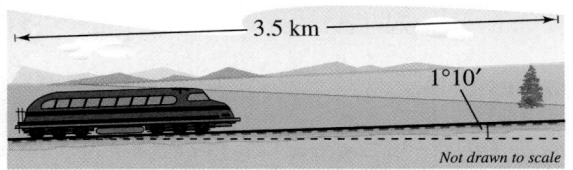

Not drawn to scale

54. *Guy Wire*   A guy wire runs from the ground to the top of a 25-foot telephone pole. The angle formed between the wire and the ground is 52°. How far from the base of the pole is the wire attached to the ground?

1.4 In Exercises 55–62, find the exact values of the six trigonometric functions of the angle θ (in standard position) whose terminal side passes through the point.

55. $(12, 16)$ **56.** $(3, -4)$

57. $\left(\frac{2}{3}, \frac{5}{2}\right)$ **58.** $\left(-\frac{10}{3}, -\frac{2}{3}\right)$

59. $(-0.5, 4.5)$ **60.** $(0.3, 0.4)$

61. $(x, 4x), \ x > 0$ **62.** $(-2x, -3x), \ x > 0$

In Exercises 63–68, find the remaining five trigonometric functions of θ satisfying the condition.

	Function Value	*Constraint*
63.	$\sec \theta = \frac{6}{5}$	$\tan \theta < 0$
64.	$\csc \theta = \frac{3}{2}$	$\cos \theta < 0$
65.	$\sin \theta = \frac{3}{8}$	$\cos \theta < 0$
66.	$\tan \theta = \frac{5}{4}$	$\cos \theta < 0$
67.	$\cos \theta = -\frac{2}{5}$	$\sin \theta > 0$
68.	$\sin \theta = -\frac{2}{4}$	$\cos \theta > 0$

In Exercises 69–76, evaluate the sine, cosine, and tangent of the angle without using a calculator.

69. $\dfrac{\pi}{3}$ **70.** $\dfrac{\pi}{4}$

71. $-\dfrac{7\pi}{3}$ **72.** $-\dfrac{5\pi}{4}$

73. $495°$ **74.** $-150°$

75. $-240°$ **76.** $315°$

In Exercises 77–82, use a calculator to evaluate the trigonometric function of the real number. Round your answer to two decimal places.

77. $\sin 4$ **78.** $\tan 3$

79. $\sin(-3.2)$ **80.** $\cot(-4.8)$

81. $\sec 12\pi/5$ **82.** $\tan(-25\pi/7)$

1.5 In Exercises 83–90, sketch a graph of the function. Include two full periods.

83. $y = \sin x$ **84.** $y = \cos x$

85. $f(x) = 5 \sin \dfrac{2x}{5}$ **86.** $f(x) = 8 \cos\left(-\dfrac{x}{4}\right)$

87. $y = 2 + \sin x$ **88.** $y = -4 - \cos \pi x$

89. $g(t) = \frac{5}{2} \sin(t - \pi)$ **90.** $g(t) = 3 \cos(t + \pi)$

91. *Sound Waves* Sound waves can be modeled by sine functions of the form $y = a \sin bx$, where x is measured in seconds.

(a) Write an equation of a sound wave whose amplitude is 2 and whose period is $\frac{1}{264}$ second.

(b) What is the frequency of the sound wave described in part (a)?

92. *Sound Waves* Use the cosine function $y = a \cos bx$ to model the sound wave described in Exercise 91.

`1.6` In Exercises 93–102, sketch a graph of the function. Include two full periods.

93. $f(x) = \tan x$

94. $f(t) = \tan\left(t - \frac{\pi}{4}\right)$

95. $f(x) = \cot x$

96. $g(t) = 2 \cot 2t$

97. $f(x) = \sec x$

98. $h(t) = \sec\left(t - \frac{\pi}{4}\right)$

99. $f(x) = \csc x$

100. $f(t) = 3 \csc\left(2t + \frac{\pi}{4}\right)$

101. $f(x) = x \cos x$

102. $g(x) = x^4 \cos x$

`1.7` In Exercises 103–108, evaluate the expression. If necessary, round your answer to two decimal places.

103. $\arcsin\left(-\frac{1}{2}\right)$

104. $\arcsin(-1)$

105. $\arcsin 0.4$

106. $\arcsin 0.213$

107. $\sin^{-1}(-0.44)$

108. $\sin^{-1} 0.89$

In Exercises 109–112, evaluate the expression without the aid of a calculator.

109. $\arccos \dfrac{\sqrt{3}}{2}$

110. $\arccos \dfrac{\sqrt{2}}{2}$

111. $\cos^{-1}(-1)$

112. $\cos^{-1} \dfrac{\sqrt{3}}{2}$

In Exercises 113–120, use a calculator to approximate the value of the expression. Round your answer to two decimal places.

113. $\arccos 0.324$

114. $\arccos(-0.888)$

115. $\arctan 0.123$

116. $\arctan 2.34$

117. $\arctan 5.783$

118. $\arctan 99.1$

119. $\tan^{-1}(-1.5)$

120. $\tan^{-1} 8.2$

In Exercises 121–128, find the exact value of the expression.

121. $\sin(\arcsin 0.72)$

122. $\cos(\arccos 0.25)$

---

123. $\arctan(\tan \pi)$

124. $\arccos[\cos(-5\pi)]$

125. $\cos\left(\arctan \frac{3}{4}\right)$

126. $\tan\left(\arccos \frac{3}{5}\right)$

127. $\sec\left(\arctan \frac{12}{5}\right)$

128. $\cot\left[\arcsin\left(-\frac{12}{13}\right)\right]$

`1.8` **129.** *Angle of Elevation* The height of a radio transmission tower is 70 meters, and it casts a shadow of length 30 meters (see figure). Find the angle of elevation of the sun.

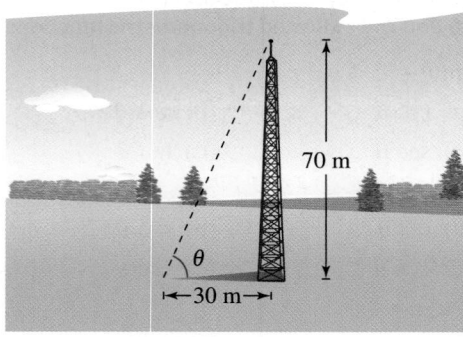

130. *Height* Your football has landed at the edge of the roof of your school building. When you are 25 feet from the base of the building, the angle of elevation to your football is 21°. How high off the ground is your football?

131. *Distance* From city A to city B, a plane flies 650 miles at a bearing of 48°. From city B to city C, the plane flies 810 miles at a bearing of 115°. Find the distance from A to C and the bearing from A to C.

132. *Wave Motion* Your fishing bobber oscillates in simple harmonic motion from the waves in the lake where you fish. Your bobber moves a total of 1.5 inches from its high point to its low point and returns to its high point every 3 seconds. Write an equation modeling the motion of your bobber if it is at its high point at time $t = 0$.

Synthesis

True or False? In Exercises 133–136, determine whether the statement is true or false. Justify your answer.

133. The tangent function is often useful for modeling simple harmonic motion.

134. The inverse sine function $y = \arcsin x$ cannot be defined as a function over any interval that is greater than the interval defined as $-\pi/2 \le y \le \pi/2$.

135. $y = \sin \theta$ is not a function because $\sin 30° = \sin 150°$.

136. Because $\tan 3\pi/4 = -1$, $\arctan(-1) = 3\pi/4$.

In Exercises 137–140, match the function $y = a \sin bx$ with its graph. Base your selection solely on your interpretation of the constants a and b. Explain your reasoning. [The graphs are labeled (a), (b), (c), and (d).]

(a)
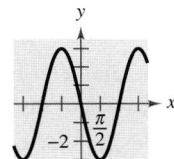

(b)
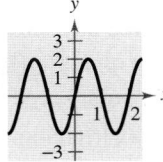

(c)
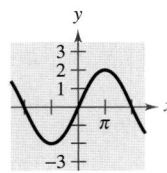

(d)
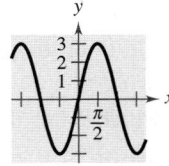

137. $y = 3 \sin x$

138. $y = -3 \sin x$

139. $y = 2 \sin \pi x$

140. $y = 2 \sin \dfrac{x}{2}$

141. *Writing* Describe the behavior of $f(\theta) = \sec \theta$ at the zeros of $g(\theta) = \cos \theta$. Explain your reasoning.

142. *Conjecture*

 (a) Use a graphing utility to complete the table.

θ	0.1	0.4	0.7	1.0	1.3
$\tan\left(\theta - \dfrac{\pi}{2}\right)$					
$-\cot \theta$					

(b) Make a conjecture about the relationship between $\tan\left(\theta - \dfrac{\pi}{2}\right)$ and $-\cot \theta$.

143. *Writing* When graphing the sine and cosine functions, determining the amplitude is part of the analysis. Explain why this is not true for the other four trigonometric functions.

144. *Graphical Reasoning* The formulas for the area of a circular sector and arc length are $A = \frac{1}{2}r^2 \theta$ and $s = 2\theta$, respectively. (r is the radius and θ is the angle measured in radians.)

(a) If $\theta = 0.8$, write the area and arc length as functions of r. What is the domain of each function? Use a graphing utility to graph the functions. Use the graphs to determine which function changes more rapidly as r increases. Explain.

(b) If $r = 10$ centimeters, express the area and arc length of the functions of θ. What is the domain of each function? Use a graphing utility to graph and identify the functions.

145. *Writing* Describe a real-life application that can be represented by a simple harmonic motion model and is different from any that you've seen in this chapter. Explain which function you would use to model your application, and why. Explain how you would determine the amplitude, period, and frequency of the model for your application.

Chapter Test

Take this test as you would take a test in class. When you are finished, check your work against the answers given in the back of the book.

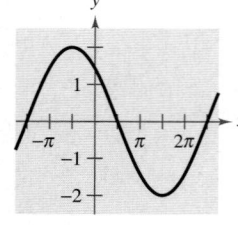

(−2, 6)

FIGURE FOR 3

The *Interactive* CD-ROM and *Internet* versions of this text offer Chapter Pre-Tests and Chapter Post-Tests, both of which have randomly generated exercises with diagnostic capabilities.

1. Consider an angle that measures $5\pi/4$ radians.
   (a) Sketch the angle in standard position.
   (b) Determine two coterminal angles (one positive and one negative).
   (c) Convert the angle to degree measure.

2. A truck is moving at a rate of 90 kilometers per hour, and the diameter of its wheels is 1 meter. Find the angular speed of the wheels in radians per minute.

3. Find the exact values of the six trigonometric functions of the angle θ shown in the figure.

4. Given that $\tan \theta = \frac{3}{2}$, find the other five trigonometric functions of θ.

5. Determine the reference angle θ' of the angle $\theta = 290°$ and sketch θ and θ' in standard position.

6. Determine the quadrant in which θ lies if $\sec \theta < 0$ and $\tan \theta > 0$.

7. Find two values of θ in degrees ($0 \le \theta < 360°$) if $\cos \theta = -\sqrt{3}/2$. (Do not use a calculator.)

8. Use a calculator to approximate two values of θ in radians ($0 \le \theta < 2\pi$) if $\csc \theta = 1.030$. Round the result to two decimal places.

In Exercises 9 and 10, find the remaining five trigonometric functions of θ satisfying the conditions.

9. $\cos \theta = \frac{3}{5},\ \tan \theta < 0$ 10. $\sec \theta = -\frac{17}{8},\ \sin \theta > 0$

In Exercises 11 and 12, graph the function through two full periods without the aid of a graphing utility.

11. $g(x) = -2 \sin\left(x - \dfrac{\pi}{4}\right)$ 12. $f(\alpha) = \dfrac{1}{2} \tan 2\alpha$

In Exercises 13 and 14, use a graphing utility to graph the function. If the function is periodic, find its period.

13. $y = \sin 2\pi x + 2 \cos \pi x$ 14. $y = 6t \cos(0.25t),\quad 0 \le t \le 32$

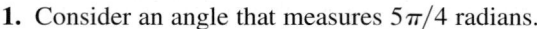

y

1

−π   π   2π

−1

−2

FIGURE FOR 15

15. Find a, b, and c for the function $f(x) = a \sin(bx + c)$ such that the graph of f matches the figure.

16. Find the exact value of $\tan\left(\arccos \frac{2}{3}\right)$ without the aid of a calculator.

17. Graph the function $f(x) = 2 \arcsin\left(\frac{1}{2}x\right)$.

18. A plane is 80 miles south and 95 miles east of Cleveland Hopkins International Airport. What bearing should be taken to fly directly to the airport?

19. Write the equation for the simple harmonic motion of a ball on a spring that starts at its lowest point of 6 inches below equilibrium, bounces to its maximum height of 6 inches above equilibrium, and returns to its lowest point in a total of 2 seconds.

Proofs in Mathematics

The Pythagorean Theorem

The Pythagorean Theorem is one of the most famous theorems in mathematics. More than 100 different proofs now exist. James A. Garfield, the twentieth president of the United States, developed a proof of the Pythagorean Theorem in 1876. His proof, shown below, involved the fact that a trapezoid can be formed from two congruent right triangles and an isosceles right triangle.

> ### The Pythagorean Theorem
>
> In a right triangle, the sum of the squares of the lengths of the legs is equal to the square of the length of the hypotenuse, where a and b are the legs and c is the hypotenuse.
>
> $$a^2 + b^2 = c^2$$
>
>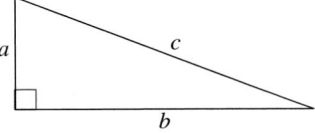

Proof

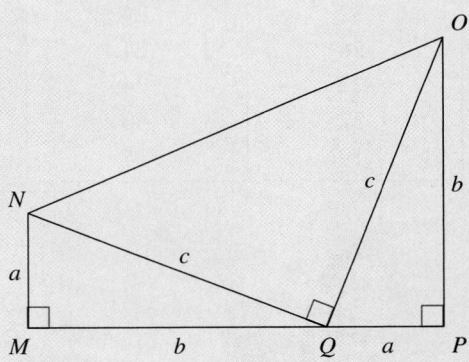

$$\text{Area of trapezoid } MNOP = \text{Area of } \triangle MNQ + \text{Area of } \triangle PQO + \text{Area of } \triangle NOQ$$

$$\frac{1}{2}(a + b)(a + b) = \frac{1}{2}ab + \frac{1}{2}ab + \frac{1}{2}c^2$$

$$\frac{1}{2}(a + b)(a + b) = ab + \frac{1}{2}c^2$$

$$(a + b)(a + b) = 2ab + c^2$$

$$a^2 + 2ab + b^2 = 2ab + c^2$$

$$a^2 + b^2 = c^2$$

P.S. Problem Solving

1. The restaurant at the top of the Space Needle in Seattle, Washington is circular and has a radius of 47.25 feet. The dining part of the restaurant revolves, making about one complete revolution every 48 minutes. A dinner party was seated at the edge of the revolving restaurant at 6:45 P.M. and was finished at 8:57 P.M.

   (a) Find the angle through which the dinner party rotated.

   (b) Find the distance the party traveled during dinner.

2. A bicycle's gear ratio is the number of times the freewheel turns for every one turn of the chainwheel (see figure). The table shows the numbers of teeth in the freewheel and chainwheel for the first five gears of an 18-speed touring bicycle. The chainwheel completes one rotation for each gear. Find the angle through which the freewheel turns for each gear. Give your answers in both degrees and radians.

Gear number	Number of teeth in freewheel	Number of teeth in chainwheel
1	32	24
2	26	24
3	22	24
4	32	40
5	19	24

Freewheel

Chainwheel

3. A surveyor in a helicopter is trying to determine the width of an island, as shown in the figure.

   (a) What is the shortest distance d the helicopter would have to travel to land on the island?

   (b) What is the horizontal distance x that the helicopter would have to travel before it would be directly over the nearer end of the island?

   (c) Find the width w of the island. Explain how you obtained your answer.

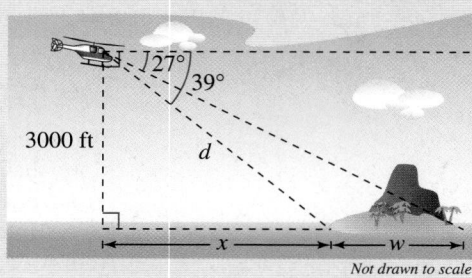

Not drawn to scale

FIGURE FOR 3

4. Use the figure below.

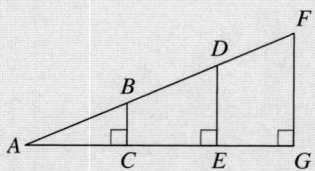

   (a) Explain why $\triangle ABC$, $\triangle ADE$, and $\triangle AFG$ are similar triangles.

   (b) What does similarity imply about the ratios
   $$\frac{BC}{AB}, \frac{DE}{AD}, \text{ and } \frac{FG}{AF}?$$

   (c) Does the value of sin A depend on which triangle from part (a) is used to calculate it? Would the value of sin A change if it were found using a different right triangle that was similar to the three given triangles?

   (d) Do your conclusions from part (c) apply to the other five trigonometric functions? Explain.

5. Use a graphing utility to graph h, and use the graph to decide whether h is even, odd, or neither.

   (a) $h(x) = \cos^2 x$ (b) $h(x) = \sin^2 x$

6. If f is an even function and g is an odd function, use the results of Exercise 5 to make a conjecture about h where

   (a) $h(x) = [f(x)]^2$ (b) $h(x) = [g(x)]^2$.

7. The model for the height h of a Ferris wheel car is

   $$h = 50 + 50 \sin 8\pi t$$

   where t is the time in minutes. (The Ferris wheel has a radius of 50 feet.) This model yields a height of 50 feet when $t = 0$. Alter the model so that the height of the car is 1 foot when $t = 0$.

8. A popular theory that attempts to explain the ups and downs of everyday life states that each of us has three cycles, called biorhythms, which begin at birth. These three cycles can be modeled by sine waves.

Physical (23 days): $P = \sin \dfrac{2\pi t}{23}, \quad t \geq 0$

Emotional (28 days): $E = \sin \dfrac{2\pi t}{28}, \quad t \geq 0$

Intellectual (33 days): $I = \sin \dfrac{2\pi t}{33}, \quad t \geq 0$

where t is the number of days since birth. Consider a person who was born on July 20, 1984.

(a) Use a graphing utility to graph the three models in the same viewing window for $7300 \leq t \leq 7380$

(b) Describe the person's biorhythms during the month of September 2004.

(c) Calculate the person's three energy levels on September 22, 2004.

9. (a) Use a graphing utility to graph the functions

$f(x) = 2 \cos 2x + 3 \sin 3x$

and

$g(x) = 2 \cos 2x + 3 \sin 4x.$

(b) Use the graphs from part (a) to find the period of each function.

(c) If α and β are positive integers, is the function

$h(x) = A \cos \alpha x + B \sin \beta x$

periodic? Explain your reasoning.

10. Two trigonometric functions f and g have periods of 2, and their graphs intersect at $x = 5.35$.

(a) Give one smaller and one larger positive value of x at which the functions have the same value.

(b) Determine one negative value of x at which the graphs intersect.

(c) Is it true that $f(13.35) = g(-4.65)$? Explain your reasoning.

11. The function f is periodic, with period c. So, $f(t + c) = f(t)$. Are the following equal? Explain.

(a) $f(t - 2c) = f(t)$ (b) $f\left(t + \tfrac{1}{2}c\right) = f\left(\tfrac{1}{2}t\right)$

(c) $f\left(\tfrac{1}{2}(t + c)\right) = f\left(\tfrac{1}{2}t\right)$

12. If you stand in shallow water and look at an object below the surface of the water, the object will look farther away from you than it really is. This is because when light rays pass between air and water, the water refracts, or bends, the light rays. The index of refraction for water is 1.333. This is the ratio of the sine of θ_1 and the sine of θ_2 (see figure).

(a) You are standing in water that is 2 feet deep and are looking at a rock at angle $\theta_1 = 60°$ (measured from a line perpendicular to the surface of the water). Find θ_2.

(b) Find the distances x and y.

(c) Find the distance d between where the rock is and where it appears to be.

(d) What happens to d as you move closer to the rock? Explain your reasoning.

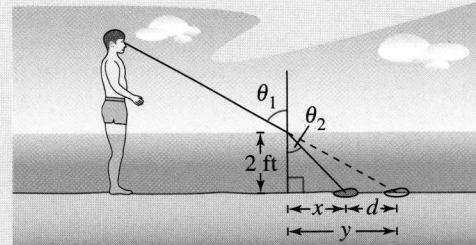

13. In calculus it can be shown that the arctangent function can be approximated by the polynomial

$$\arctan x \approx x - \dfrac{x^3}{3} + \dfrac{x^5}{5} - \dfrac{x^7}{7}$$

where x is in radians.

(a) Use a graphing utility to graph the arctangent function and its polynomial approximation in the same viewing window. How do the graphs compare?

(b) Study the pattern in the polynomial approximation of the arctangent function and guess the next term. Then repeat part (a). How does the accuracy of the approximation change when additional terms are added?

215

▶ **What you should learn**

In this chapter you will learn the following skills and concepts:

- How to use fundamental trigonometric identities to evaluate trigonometric functions and simplify trigonometric expressions
- How to verify trigonometric identities
- How to use standard algebraic techniques and inverse trigonometric functions to solve trigonometric equations
- How to use sum and difference formulas, multiple-angle formulas, power-reducing formulas, half-angle formulas, and product-to-sum formulas to rewrite and evaluate trigonometric functions

▶ **Important Vocabulary**

As you encounter each new vocabulary term in this chapter, add the term and its definition to your notebook glossary.

Sum and difference formulas
  (p. 244)
Reduction formulas (p. 246)
Double-angle formulas (p. 251)
Power-reducing formulas (p. 253)
Half-angle formulas (p. 254)
Product-to-sum formulas (p. 255)
Sum-to-product formulas (p. 256)

Study Tools

Learning objectives in each section
Chapter Summary (p. 262)
Review Exercises (pp. 263–265)
Chapter Test (p. 266)

Additional Resources

Study and Solutions Guide
Interactive Trigonometry
Videotapes/DVD for Chapter 2
Trigonometry Website
Student Success Organizer

Mike Powell/Allsport

2

Analytic Trigonometry

2.1 Using Fundamental Identities

▶ **What you should learn**

- How to recognize and write the fundamental trigonometric identities
- How to use the fundamental trigonometric identities to evaluate trigonometric functions, simplify trigonometric expressions, and rewrite trigonometric expressions

▶ **Why you should learn it**

Fundamental trigonometric identities can be used to simplify trigonometric expressions. For instance, in Exercise 99 on page 225, you can use trigonometric identities to simplify an expression for the coefficient of friction.

Introduction

In Chapter 1, you studied the basic definitions, properties, graphs, and applications of the individual trigonometric functions. In this chapter, you will learn how to use the fundamental identities to do the following.

1. Evaluate trigonometric functions.
2. Simplify trigonometric expressions.
3. Develop additional trigonometric identities.
4. Solve trigonometric equations.

Fundamental Trigonometric Identities

Reciprocal Identities

$$\sin u = \frac{1}{\csc u} \qquad \cos u = \frac{1}{\sec u} \qquad \tan u = \frac{1}{\cot u}$$

$$\csc u = \frac{1}{\sin u} \qquad \sec u = \frac{1}{\cos u} \qquad \cot u = \frac{1}{\tan u}$$

Quotient Identities

$$\tan u = \frac{\sin u}{\cos u} \qquad \cot u = \frac{\cos u}{\sin u}$$

Pythagorean Identities

$$\sin^2 u + \cos^2 u = 1 \qquad 1 + \tan^2 u = \sec^2 u \qquad 1 + \cot^2 u = \csc^2 u$$

Cofunction Identities

$$\sin\left(\frac{\pi}{2} - u\right) = \cos u \qquad \cos\left(\frac{\pi}{2} - u\right) = \sin u$$

$$\tan\left(\frac{\pi}{2} - u\right) = \cot u \qquad \cot\left(\frac{\pi}{2} - u\right) = \tan u$$

$$\sec\left(\frac{\pi}{2} - u\right) = \csc u \qquad \csc\left(\frac{\pi}{2} - u\right) = \sec u$$

Even/Odd Identities

$$\sin(-u) = -\sin u \qquad \cos(-u) = \cos u \qquad \tan(-u) = -\tan u$$

$$\csc(-u) = -\csc u \qquad \sec(-u) = \sec u \qquad \cot(-u) = -\cot u$$

Pythagorean identities are sometimes used in radical form such as

$$\sin u = \pm\sqrt{1 - \cos^2 u}$$

or

$$\tan u = \pm\sqrt{\sec^2 u - 1}$$

where the sign depends on the choice of u.

Using the Fundamental Identities

One common use of trigonometric identities is to use given values of trigonometric functions to evaluate other trigonometric functions.

Technology

You can use a graphing utility to check the result of Example 2. To do this, graph

$$y_1 = \sin x \cos^2 x - \sin x$$

and

$$y_2 = -\sin^3 x$$

in the same viewing window, as shown below. Because Example 2 shows the equivalence algebraically and the two graphs appear to coincide, you can conclude that the expressions are equivalent.

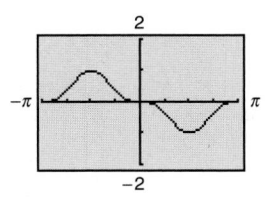

Remind students that they must use an algebraic approach to prove that two expressions are equivalent. A graphical approach can only confirm that the simplification found using algebraic techniques is correct.

Example 1 ▶ **Using Identities to Evaluate a Function**

Use the values $\sec u = -\frac{3}{2}$ and $\tan u > 0$ to find the values of all six trigonometric functions.

Solution

Using a reciprocal identity, you have

$$\cos u = \frac{1}{\sec u} = \frac{1}{-3/2} = -\frac{2}{3}.$$

Using a Pythagorean identity, you have

$$\sin^2 u = 1 - \cos^2 u$$

$$= 1 - \left(-\frac{2}{3}\right)^2$$

$$= 1 - \frac{4}{9} = \frac{5}{9}.$$

Because $\sec u < 0$ and $\tan u > 0$, it follows that u lies in Quadrant III. Moreover, because $\sin u$ is negative when u is in Quadrant III, you can choose the negative root and obtain $\sin u = -\sqrt{5}/3$. Now, knowing the values of the sine and cosine, you can find the values of all six trigonometric functions.

$$\sin u = -\frac{\sqrt{5}}{3} \qquad\qquad \csc u = \frac{1}{\sin u} = -\frac{3}{\sqrt{5}}$$

$$\cos u = -\frac{2}{3} \qquad\qquad \sec u = \frac{1}{\cos u} = -\frac{3}{2}$$

$$\tan u = \frac{\sin u}{\cos u} = \frac{-\sqrt{5}/3}{-2/3} = \frac{\sqrt{5}}{2} \qquad \cot u = \frac{1}{\tan u} = \frac{2}{\sqrt{5}}$$

Example 2 ▶ **Simplifying a Trigonometric Expression**

Simplify $\sin x \cos^2 x - \sin x$.

Solution

Factor the expression and then use a fundamental identity.

$$\sin x \cos^2 x - \sin x = \sin x(\cos^2 x - 1) \qquad \text{Monomial factor}$$

$$= -\sin x(1 - \cos^2 x) \qquad \text{Factor out } -1.$$

$$= -\sin x(\sin^2 x) \qquad \text{Pythagorean identity}$$

$$= -\sin^3 x \qquad \text{Multiply.}$$

The icon identifies examples and concepts related to features of the Learning Tools CD-ROM and the *Interactive* and *Internet* versions of this text. For more details see the chart on pages *xix-xxiii*.

Example 3 ▶ **Factoring Trigonometric Expressions**

Factor each expression.

a. $\sec^2 \theta - 1$

b. $4 \tan^2 \theta + \tan \theta - 3$

Solution

a. Here you have the difference of two squares, which factors as

$$\sec^2 \theta - 1 = (\sec \theta - 1)(\sec \theta + 1).$$

b. This expression has the polynomial form $ax^2 + bx + c$, and it factors as

$$4 \tan^2 \theta + \tan \theta - 3 = (4 \tan \theta - 3)(\tan \theta + 1).$$

On occasion, factoring or simplifying can best be done by first rewriting the expression in terms of just *one* trigonometric function or in terms of *sine and cosine only*. These strategies are illustrated in Examples 4 and 5, respectively.

Example 4 ▶ **Factoring a Trigonometric Expression**

Factor $\csc^2 x - \cot x - 3$.

Solution

You can use the identity $\csc^2 x = 1 + \cot^2 x$ to rewrite the expression in terms of the cotangent.

$$
\begin{aligned}
\csc^2 x - \cot x - 3 &= (1 + \cot^2 x) - \cot x - 3 && \text{Pythagorean identity} \\
&= \cot^2 x - \cot x - 2 && \text{Combine like terms.} \\
&= (\cot x - 2)(\cot x + 1) && \text{Factor.}
\end{aligned}
$$

Example 5 ▶ **Simplifying a Trigonometric Expression**

Simplify $\sin t + \cot t \cos t$.

Solution

Begin by rewriting $\cot t$ in terms of sine and cosine.

$$
\begin{aligned}
\sin t + \cot t \cos t &= \sin t + \left(\frac{\cos t}{\sin t} \right) \cos t && \text{Quotient identity} \\
&= \frac{\sin^2 t + \cos^2 t}{\sin t} && \text{Add fractions.} \\
&= \frac{1}{\sin t} && \text{Pythagorean identity} \\
&= \csc t && \text{Reciprocal identity}
\end{aligned}
$$

Example 6 ▶ **Adding Trigonometric Expressions**

Perform the addition and simplify.

$$\frac{\sin \theta}{1 + \cos \theta} + \frac{\cos \theta}{\sin \theta}$$

Solution

$$\frac{\sin \theta}{1 + \cos \theta} + \frac{\cos \theta}{\sin \theta} = \frac{(\sin \theta)(\sin \theta) + (\cos \theta)(1 + \cos \theta)}{(1 + \cos \theta)(\sin \theta)}$$

$$= \frac{\sin^2 \theta + \cos^2 \theta + \cos \theta}{(1 + \cos \theta)(\sin \theta)} \qquad \text{Multiply.}$$

$$= \frac{1 + \cos \theta}{(1 + \cos \theta)(\sin \theta)} \qquad \text{Pythagorean identity}$$

$$= \frac{1}{\sin \theta} \qquad \text{Divide out common factor.}$$

$$= \csc \theta \qquad \text{Reciprocal identity}$$

The last two examples in this section involve techniques for rewriting expressions in forms that are used in calculus.

Example 7 ▶ **Rewriting a Trigonometric Expression**

Rewrite $\dfrac{1}{1 + \sin x}$ so that it is *not* in fractional form.

Solution

From the Pythagorean identity $\cos^2 x = 1 - \sin^2 x = (1 - \sin x)(1 + \sin x)$, you can see that by multiplying both the numerator and the denominator by $(1 - \sin x)$ you produce a monomial denominator.

$$\frac{1}{1 + \sin x} = \frac{1}{1 + \sin x} \cdot \frac{1 - \sin x}{1 - \sin x} \qquad \begin{array}{l}\text{Multiply numerator and} \\ \text{denominator by } (1 - \sin x).\end{array}$$

$$= \frac{1 - \sin x}{1 - \sin^2 x} \qquad \text{Multiply.}$$

$$= \frac{1 - \sin x}{\cos^2 x} \qquad \text{Pythagorean identity}$$

$$= \frac{1}{\cos^2 x} - \frac{\sin x}{\cos^2 x} \qquad \text{Separate fractions.}$$

$$= \frac{1}{\cos^2 x} - \frac{\sin x}{\cos x} \cdot \frac{1}{\cos x} \qquad \text{Product of fractions}$$

$$= \sec^2 x - \tan x \sec x \qquad \text{Identities}$$

Example 8 ▶   **Trigonometric Substitution**

Use the substitution $x = 2 \tan \theta$, $0 < \theta < \pi/2$, to express

$$\sqrt{4 + x^2}$$

as a trigonometric function of θ.

Solution

Begin by letting $x = 2 \tan \theta$. Then, you can obtain

$$
\begin{aligned}
\sqrt{4 + x^2} &= \sqrt{4 + (2 \tan \theta)^2} && \text{Substitute } 2 \tan \theta \text{ for } x.\\
&= \sqrt{4 + 4 \tan^2 \theta} && \text{Rule of exponents}\\
&= \sqrt{4(1 + \tan^2 \theta)} && \text{Factor.}\\
&= \sqrt{4 \sec^2 \theta} && \text{Pythagorean identity}\\
&= 2 \sec \theta. && \sec \theta > 0 \text{ for } 0 < \theta < \pi/2
\end{aligned}
$$

Figure 2.1 shows the right triangle illustration of the trigonometric substitution in Example 8. You can use this triangle to check the solution of Example 8. For $0 < \theta < \pi/2$, you have

$$\text{opp} = x, \quad \text{adj} = 2, \quad \text{and} \quad \text{hyp} = \sqrt{4 + x^2}.$$

With these expressions, you can write the following.

$$\sec \theta = \frac{\text{hyp}}{\text{adj}}$$

$$\sec \theta = \frac{\sqrt{4 + x^2}}{2}$$

$$2 \sec \theta = \sqrt{4 + x^2}$$

So, the solution checks.

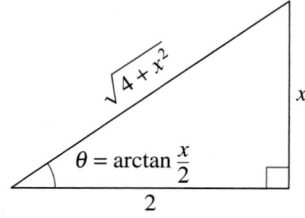

Angle whose tangent is $\frac{x}{2}$.

FIGURE **2.1**

2.1 Exercises

The *Interactive* CD-ROM and *Internet* versions of this text contain step-by-step solutions to all odd-numbered exercises. They also provide Tutorial Exercises for additional help.

In Exercises 1–14, use the given values to evaluate (if possible) the remaining trigonometric functions.

1. $\sin x = \dfrac{\sqrt{3}}{2}, \quad \cos x = -\dfrac{1}{2}$

2. $\tan x = \dfrac{\sqrt{3}}{3}, \quad \cos x = -\dfrac{\sqrt{3}}{2}$

3. $\sec \theta = \sqrt{2}, \quad \sin \theta = -\dfrac{\sqrt{2}}{2}$

4. $\csc \theta = \frac{5}{3}, \quad \tan \theta = \frac{3}{4}$

5. $\tan x = \frac{5}{12}, \quad \sec x = -\frac{13}{12}$

6. $\cot \phi = -3, \quad \sin \phi = \dfrac{\sqrt{10}}{10}$

7. $\sec \phi = \dfrac{3}{2}, \quad \csc \phi = -\dfrac{3\sqrt{5}}{5}$

8. $\cos\left(\dfrac{\pi}{2} - x\right) = \dfrac{3}{5}, \quad \cos x = \dfrac{4}{5}$

9. $\sin(-x) = -\dfrac{1}{3}, \quad \tan x = -\dfrac{\sqrt{2}}{4}$

10. $\sec x = 4, \quad \sin x > 0$

11. $\tan \theta = 2, \quad \sin \theta < 0$

12. $\csc \theta = -5, \quad \cos \theta < 0$

13. $\sin \theta = -1, \quad \cot \theta = 0$

14. $\tan \theta$ is undefined, $\quad \sin \theta > 0$

In Exercises 15–20, match the trigonometric expression with one of the following.

(a) $\sec x$ (b) -1 (c) $\cot x$

(d) 1 (e) $-\tan x$ (f) $\sin x$

15. $\sec x \cos x$ **16.** $\tan x \csc x$

17. $\cot^2 x - \csc^2 x$ **18.** $(1 - \cos^2 x)(\csc x)$

19. $\dfrac{\sin(-x)}{\cos(-x)}$ **20.** $\dfrac{\sin[(\pi/2) - x]}{\cos[(\pi/2) - x]}$

In Exercises 21–26, match the trigonometric expression with one of the following.

(a) $\csc x$ (b) $\tan x$ (c) $\sin^2 x$

(d) $\sin x \tan x$ (e) $\sec^2 x$ (f) $\sec^2 x + \tan^2 x$

21. $\sin x \sec x$ **22.** $\cos^2 x(\sec^2 x - 1)$

23. $\sec^4 x - \tan^4 x$ **24.** $\cot x \sec x$

25. $\dfrac{\sec^2 x - 1}{\sin^2 x}$ **26.** $\dfrac{\cos^2[(\pi/2) - x]}{\cos x}$

In Exercises 27–44, use the fundamental identities to simplify the expression. There is more than one correct form of each answer.

27. $\cot \theta \sec \theta$ **28.** $\cos \beta \tan \beta$

29. $\sin \phi(\csc \phi - \sin \phi)$ **30.** $\sec^2 x(1 - \sin^2 x)$

31. $\dfrac{\cot x}{\csc x}$ **32.** $\dfrac{\csc \theta}{\sec \theta}$

33. $\dfrac{1 - \sin^2 x}{\csc^2 x - 1}$ **34.** $\dfrac{1}{\tan^2 x + 1}$

35. $\sec \alpha \cdot \dfrac{\sin \alpha}{\tan \alpha}$ **36.** $\dfrac{\tan^2 \theta}{\sec^2 \theta}$

37. $\cos\left(\dfrac{\pi}{2} - x\right)\sec x$ **38.** $\cot\left(\dfrac{\pi}{2} - x\right)\cos x$

39. $\dfrac{\cos^2 y}{1 - \sin y}$ **40.** $\cos t(1 + \tan^2 t)$

41. $\sin \beta \tan \beta + \cos \beta$ **42.** $\csc \phi \tan \phi + \sec \phi$

43. $\cot u \sin u + \tan u \cos u$

44. $\sin \theta \sec \theta + \cos \theta \csc \theta$

In Exercises 45–56, factor the expression and use the fundamental identities to simplify. There is more than one correct form of each answer.

45. $\tan^2 x - \tan^2 x \sin^2 x$

46. $\sin^2 x \csc^2 x - \sin^2 x$

47. $\sin^2 x \sec^2 x - \sin^2 x$

48. $\cos^2 x + \cos^2 x \tan^2 x$

49. $\dfrac{\sec^2 x - 1}{\sec x - 1}$

50. $\dfrac{\cos^2 x - 4}{\cos x - 2}$

51. $\tan^4 x + 2 \tan^2 x + 1$

52. $1 - 2 \cos^2 x + \cos^4 x$

53. $\sin^4 x - \cos^4 x$

54. $\sec^4 x - \tan^4 x$

55. $\csc^3 x - \csc^2 x - \csc x + 1$

56. $\sec^3 x - \sec^2 x - \sec x + 1$

In Exercises 57–60, perform the multiplication and use the fundamental identities to simplify. There is more than one correct form of each answer.

57. $(\sin x + \cos x)^2$

58. $(\cot x + \csc x)(\cot x - \csc x)$

59. $(2 \csc x + 2)(2 \csc x - 2)$

60. $(3 - 3 \sin x)(3 + 3 \sin x)$

In Exercises 61–64, perform the addition or subtraction and use the fundamental identities to simplify. There is more than one correct form of each answer.

61. $\dfrac{1}{1 + \cos x} + \dfrac{1}{1 - \cos x}$

62. $\dfrac{1}{\sec x + 1} - \dfrac{1}{\sec x - 1}$

63. $\dfrac{\cos x}{1 + \sin x} + \dfrac{1 + \sin x}{\cos x}$

64. $\tan x - \dfrac{\sec^2 x}{\tan x}$

In Exercises 65–68, rewrite the expression so that it is not in fractional form. There is more than one correct form of each answer.

65. $\dfrac{\sin^2 y}{1 - \cos y}$

66. $\dfrac{5}{\tan x + \sec x}$

67. $\dfrac{3}{\sec x - \tan x}$

68. $\dfrac{\tan^2 x}{\csc x + 1}$

 Numerical and Graphical Analysis In Exercises 69–72, use a graphing utility to complete the table and graph the functions. Make a conjecture about y_1 and y_2.

x	0.2	0.4	0.6	0.8	1.0	1.2	1.4
y_1							
y_2							

69. $y_1 = \cos\left(\dfrac{\pi}{2} - x\right)$, $y_2 = \sin x$

70. $y_1 = \sec x - \cos x$, $y_2 = \sin x \tan x$

71. $y_1 = \dfrac{\cos x}{1 - \sin x}$, $y_2 = \dfrac{1 + \sin x}{\cos x}$

72. $y_1 = \sec^4 x - \sec^2 x$, $y_2 = \tan^2 x + \tan^4 x$

 In Exercises 73–76, use a graphing utility to determine which of the six trigonometric functions is equal to the expression. Verify your answer algebraically.

73. $\cos x \cot x + \sin x$

74. $\sec x \csc x - \tan x$

75. $\dfrac{1}{\sin x}\left(\dfrac{1}{\cos x} - \cos x\right)$

76. $\dfrac{1}{2}\left(\dfrac{1 + \sin \theta}{\cos \theta} + \dfrac{\cos \theta}{1 + \sin \theta}\right)$

In Exercises 77–82, use the trigonometric substitution to write the algebraic expression as a trigonometric function of θ, where $0 < \theta < \pi/2$.

77. $\sqrt{9 - x^2}$, $x = 3 \cos \theta$

78. $\sqrt{64 - 16x^2}$, $x = 2 \cos \theta$

79. $\sqrt{x^2 - 9}$, $x = 3 \sec \theta$

80. $\sqrt{x^2 - 4}$, $x = 2 \sec \theta$

81. $\sqrt{x^2 + 25}$, $x = 5 \tan \theta$

82. $\sqrt{x^2 + 100}$, $x = 10 \tan \theta$

In Exercises 83–86, use the trigonometric substitution to write the algebraic equation as a trigonometric function of θ, where $-\pi/2 < \theta < \pi/2$. Then find $\sin \theta$ and $\cos \theta$.

83. $3 = \sqrt{9 - x^2}$, $x = 3 \sin \theta$

84. $3 = \sqrt{36 - x^2}$, $x = 6 \sin \theta$

85. $2\sqrt{2} = \sqrt{16 - 4x^2}$, $x = 2 \cos \theta$

86. $-5\sqrt{3} = \sqrt{100 - x^2}$, $x = 10 \cos \theta$

 In Exercises 87–90, use a graphing utility to solve the equation for θ, where $0 \le \theta < 2\pi$.

87. $\sin \theta = \sqrt{1 - \cos^2 \theta}$ **88.** $\cos \theta = -\sqrt{1 - \sin^2 \theta}$

89. $\sec \theta = \sqrt{1 + \tan^2 \theta}$ **90.** $\csc \theta = \sqrt{1 + \cot^2 \theta}$

In Exercises 91–98, use a calculator to demonstrate the identity for each value of θ.

91. $\csc^2 \theta - \cot^2 \theta = 1$

(a) $\theta = 132°$, (b) $\theta = \dfrac{2\pi}{7}$

92. $\tan^2 \theta + 1 = \sec^2 \theta$

(a) $\theta = 346°$, (b) $\theta = 3.1$

93. $\cos\left(\dfrac{\pi}{2} - \theta\right) = \sin \theta$

(a) $\theta = 80°$, (b) $\theta = 0.8$

94. $\sin(-\theta) = -\sin \theta$

(a) $\theta = 250°$, (b) $\theta = \dfrac{1}{2}$

95. $\cot \theta = \dfrac{\cos \theta}{\sin \theta}$

(a) $\theta = 25°$, (b) $\theta = \dfrac{\pi}{8}$

96. $1 + \cot^2 \theta = \csc^2 \theta$

   (a) $\theta = 240°$, (b) $\theta = 2.2$

97. $\tan\left(\dfrac{\pi}{2} - \theta\right) = \cot \theta$

   (a) $\theta = 5°$, (b) $\theta = \dfrac{11\pi}{12}$

98. $\cos(-\theta) = \cos \theta$

   (a) $\theta = 125°$, (b) $\theta = \dfrac{5\pi}{6}$

99. *Friction* The forces acting on an object weighing W units on an inclined plane positioned at an angle of θ with the horizontal (see figure) are modeled by

$$\mu W \cos \theta = W \sin \theta$$

where μ is the coefficient of friction. Solve the equation for μ and simplify the result.

100. *Rate of Change* The rate of change of the function

$$f(x) = -\csc x - \sin x$$

is given by the expression

$$\csc x \cot x - \cos x.$$

Show that this expression can also be written as

$$\cos x \cot^2 x.$$

Synthesis

True or False? In Exercises 101 and 102, determine whether the statement is true or false. Justify your answer.

101. The even and odd trigonometric identities are helpful for determining whether the value of a trigonometric function is positive or negative.

102. A cofunction identity can be used to transform a tangent function so that it can be represented by a cosecant function.

Calculus In Exercises 103–106, fill in the blanks. (*Note:* The notation $x \to c^+$ indicates that x approaches c from the right and $x \to c^-$ indicates that x approaches c from the left.)

103. As $x \to \dfrac{\pi^-}{2}$, $\sin x \to$ ⬚ and $\csc x \to$ ⬚ .

104. As $x \to 0^+$, $\cos x \to$ ⬚ and $\sec x \to$ ⬚ .

105. As $x \to \dfrac{\pi^-}{2}$, $\tan x \to$ ⬚ and $\cot x \to$ ⬚ .

106. As $x \to \pi^+$, $\sin x \to$ ⬚ and $\csc x \to$ ⬚ .

In Exercises 107–112, determine whether or not the equation is an identity, and give a reason for your answer.

107. $\cos \theta = \sqrt{1 - \sin^2 \theta}$

108. $\cot \theta = \sqrt{\csc^2 \theta + 1}$

109. $\dfrac{(\sin k\theta)}{(\cos k\theta)} = \tan \theta$, k is a constant.

110. $\dfrac{1}{(5 \cos \theta)} = 5 \sec \theta$

111. $\sin \theta \csc \theta = 1$

112. $\sin \theta \csc \phi = 1$

113. Use the definitions of sine and cosine to derive the Pythagorean identity $\sin^2 \theta + \cos^2 \theta = 1$.

114. *Writing* Use the Pythagorean identity

$$\sin^2 \theta + \cos^2 \theta = 1$$

to derive the other Pythagorean identities,

$$1 + \tan^2 \theta = \sec^2 \theta$$

and

$$1 + \cot^2 \theta = \csc^2 \theta.$$

Discuss how to remember these identities and other fundamental identities.

Review

In Exercises 115 and 116, perform the operation and simplify.

115. $\left(\sqrt{x} + 5\right)\left(\sqrt{x} - 5\right)$ **116.** $\left(2\sqrt{z} + 3\right)^2$

In Exercises 117–120, perform the addition or subtraction and simplify.

117. $\dfrac{1}{x + 5} + \dfrac{x}{x - 8}$ **118.** $\dfrac{6x}{x - 4} - \dfrac{3}{4 - x}$

119. $\dfrac{2x}{x^2 - 4} - \dfrac{7}{x + 4}$ **120.** $\dfrac{x}{x^2 - 25} + \dfrac{x^2}{x - 5}$

2.2 Verifying Trigonometric Identities

▶ **What you should learn**

• How to plan a strategy for verifying trigonometric identities
• How to verify trigonometric identities

▶ **Why you should learn it**

You can use trigonometric identities to rewrite trigonometric equations that model real-life situations. For instance, in Exercise 58 on page 232, you can use trigonometric identities to simplify the equation that models the length of a shadow cast by a gnomon.

Robert Ginn/PhotoEdit

Introduction

In this section, you will study techniques for verifying trigonometric identities. In the next section, you will study techniques for solving trigonometric equations. The key to verifying identities *and* solving equations is the ability to use the fundamental identities and the rules of algebra to rewrite trigonometric expressions.

Remember that a *conditional equation* is an equation that is true for only some of the values in its domain. For example, the conditional equation

$$\sin x = 0 \qquad \text{Conditional equation}$$

is true only for $x = n\pi$, where n is an integer. When you find these values, you are *solving* the equation.

On the other hand, an equation that is true for all real values in the domain of the variable is an *identity*. For example, the familiar equation

$$\sin^2 x = 1 - \cos^2 x \qquad \text{Identity}$$

is true for all real numbers x. So, it is an identity.

Although there are similarities, verifying that a trigonometric equation is an identity is quite different from solving an equation. There is no well-defined set of rules to follow in verifying trigonometric identities, and the process is best learned by practice.

For instance, to verify that the trigonometric equation $\tan\theta\cos\theta = \sin\theta$ is an identity, begin by working with the more complicated, left side of the equation.

$$\tan\theta\cos\theta = \left(\frac{\sin\theta}{\cos\theta}\right)\cos\theta \qquad \text{Rewrite } \tan\theta \text{ as } \frac{\sin\theta}{\cos\theta}.$$

$$= \left(\frac{\sin\theta}{\cos\theta}\right)\cos\theta \qquad \text{Divide out } \cos\theta.$$

$$= \sin\theta \qquad \text{Simplify.}$$

The result shows that the left side of the equation is equal to the right side. So, the identity has been verified.

Guidelines for Verifying Trigonometric Identities

1. Work with one side of the equation at a time. It is often better to work with the more complicated side first.

2. Look for opportunities to factor an expression, add fractions, square a binomial, or create a monomial denominator.

3. Look for opportunities to use the fundamental identities. Note which functions are in the final expression you want. Sines and cosines pair up well, as do secants and tangents, and cosecants and cotangents.

4. If the preceding guidelines do not help, try converting all terms to sines and cosines.

5. Always try *something*. Even paths that lead to dead ends provide insights.

You may want to review the distinctions among expressions, equations, and identities. Have your students look at some algebraic identities and conditional equations before starting this section. It is important for them to understand what it means to verify an identity and not try to solve it as an equation.

Verifying Trigonometric Identities

Example 1 ▶ **Verifying a Trigonometric Identity**

Verify the identity $\dfrac{\sec^2 \theta - 1}{\sec^2 \theta} = \sin^2 \theta$.

Solution

Because the left side is more complicated, start with it.

$$\dfrac{\sec^2 \theta - 1}{\sec^2 \theta} = \dfrac{(\tan^2 \theta + 1) - 1}{\sec^2 \theta} \qquad \text{Pythagorean identity}$$

$$= \dfrac{\tan^2 \theta}{\sec^2 \theta} \qquad \text{Simplify.}$$

$$= \tan^2 \theta(\cos^2 \theta) \qquad \text{Reciprocal identity}$$

$$= \dfrac{\sin^2 \theta}{(\cos^2 \theta)} \, (\cos^2 \theta) \qquad \text{Quotient identity}$$

$$= \sin^2 \theta \qquad \text{Simplify.}$$

Here is another way to verify the identity in Example 1.

$$\dfrac{\sec^2 \theta - 1}{\sec^2 \theta} = \dfrac{\sec^2 \theta}{\sec^2 \theta} - \dfrac{1}{\sec^2 \theta} \qquad \text{Rewrite as the difference of fractions.}$$

$$= 1 - \cos^2 \theta \qquad \text{Reciprocal identity}$$

$$= \sin^2 \theta \qquad \text{Pythagorean identity}$$

As you can see, there can be more than one way to verify an identity. Your method may differ from that used by your instructor or fellow students. Here is a good chance to be creative and establish your own style, but try to be as efficient as possible.

Example 2 ▶ **Combining Fractions Before Using Identities**

Verify the identity $\dfrac{1}{1 - \sin \alpha} + \dfrac{1}{1 + \sin \alpha} = 2 \sec^2 \alpha$.

Solution

$$\dfrac{1}{1 - \sin \alpha} + \dfrac{1}{1 + \sin \alpha} = \dfrac{1 + \sin \alpha + 1 - \sin \alpha}{(1 - \sin \alpha)(1 + \sin \alpha)} \qquad \text{Add fractions.}$$

$$= \dfrac{2}{1 - \sin^2 \alpha} \qquad \text{Simplify.}$$

$$= \dfrac{2}{\cos^2 \alpha} \qquad \text{Pythagorean identity}$$

$$= 2 \sec^2 \alpha \qquad \text{Reciprocal identity}$$

Example 3 ▶   **Verifying a Trigonometric Identity**

Verify the identity

$$(\tan^2 x + 1)(\cos^2 x - 1) = -\tan^2 x.$$

Solution

By applying identities before multiplying, you obtain the following.

$$(\tan^2 x + 1)(\cos^2 x - 1) = (\sec^2 x)(-\sin^2 x) \qquad \text{Pythagorean identities}$$

$$= -\frac{\sin^2 x}{\cos^2 x} \qquad \text{Reciprocal identity}$$

$$= -\left(\frac{\sin x}{\cos x}\right)^2 \qquad \text{Rule of exponents}$$

$$= -\tan^2 x \qquad \text{Quotient identity}$$

Example 4 ▶   **Converting to Sines and Cosines**

Verify the identity

$$\tan x + \cot x = \sec x \csc x.$$

Solution

In this case there appear to be no fractions to add, no products to find, and no opportunities to use the Pythagorean identities. So, try converting the left side into sines and cosines to see what happens.

$$\tan x + \cot x = \frac{\sin x}{\cos x} + \frac{\cos x}{\sin x} \qquad \text{Quotient identities}$$

$$= \frac{\sin^2 x + \cos^2 x}{\cos x \sin x} \qquad \text{Add fractions.}$$

$$= \frac{1}{\cos x \sin x} \qquad \text{Pythagorean identity}$$

$$= \frac{1}{\cos x} \cdot \frac{1}{\sin x} \qquad \text{Product of fractions}$$

$$= \sec x \csc x \qquad \text{Reciprocal identities}$$

Recall from algebra that *rationalizing the denominator* using conjugates is, on occasion, a powerful simplification technique. A related form of this technique works for simplifying trigonometric expressions as well. For instance, to simplify $1/(1 - \cos x)$, multiply the numerator and the denominator by $1 + \cos x$.

$$\frac{1}{1 - \cos x} = \frac{1}{1 - \cos x}\left(\frac{1 + \cos x}{1 + \cos x}\right) = \frac{1 + \cos x}{1 - \cos^2 x} = \frac{1 + \cos x}{\sin^2 x}$$

$$= \csc^2 x(1 + \cos x)$$

This technique is demonstrated in the next example.

Example 5 ▶ **Verifying Trigonometric Identities**

Verify the identity $\sec y + \tan y = \dfrac{\cos y}{1 - \sin y}$.

Solution

Begin with the *right* side. Note that you can create a monomial denominator by multiplying the numerator and denominator by $1 + \sin y$.

$$\frac{\cos y}{1 - \sin y} = \frac{\cos y}{1 - \sin y}\left(\frac{1 + \sin y}{1 + \sin y}\right) \qquad \text{Multiply numerator and denominator by } 1 + \sin y.$$

$$= \frac{\cos y + \cos y \sin y}{1 - \sin^2 y} \qquad \text{Multiply.}$$

$$= \frac{\cos y + \cos y \sin y}{\cos^2 y} \qquad \text{Pythagorean identity}$$

$$= \frac{\cos y}{\cos^2 y} + \frac{\cos y \sin y}{\cos^2 y} \qquad \text{Separate fractions.}$$

$$= \frac{1}{\cos y} + \frac{\sin y}{\cos y} \qquad \text{Simplify.}$$

$$= \sec y + \tan y \qquad \text{Identities}$$

In Examples 1 through 5, you have been verifying trigonometric identities by working with one side of the equation and converting to the form given on the other side. On occasion, it is practical to work with each side *separately*, to obtain one common form equivalent to both sides. This is illustrated in Example 6.

Example 6 ▶ **Working with Each Side Separately**

Verify the identity $\dfrac{\cot^2 \theta}{1 + \csc \theta} = \dfrac{1 - \sin \theta}{\sin \theta}$.

Solution

Working with the left side, you have

$$\frac{\cot^2 \theta}{1 + \csc \theta} = \frac{\csc^2 \theta - 1}{1 + \csc \theta} \qquad \text{Pythagorean identity}$$

$$= \frac{(\csc \theta - 1)(\cancel{\csc \theta + 1})}{\cancel{1 + \csc \theta}} \qquad \text{Factor.}$$

$$= \csc \theta - 1. \qquad \text{Simplify.}$$

Now, simplifying the right side, you have

$$\frac{1 - \sin \theta}{\sin \theta} = \frac{1}{\sin \theta} - \frac{\sin \theta}{\sin \theta} \qquad \text{Separate fractions.}$$

$$= \csc \theta - 1. \qquad \text{Reciprocal identity}$$

The identity is verified because both sides are equal to $\csc \theta - 1$.

STUDY TIP

The technique of cross multiplication is not used in verifying trigonometric identities, because you do not know that the expressions are equal. Cross multiplication is used in solving an equation, when you know that the left side equals the right side.

In Example 7, powers of trigonometric functions are rewritten as more complicated sums of products of trigonometric functions. This is a common procedure used in calculus.

Example 7 ▶   **Three Examples from Calculus**

Verify each identity.

a. $\tan^4 x = \tan^2 x \sec^2 x - \tan^2 x$

b. $\sin^3 x \cos^4 x = (\cos^4 x - \cos^6 x) \sin x$

c. $\csc^4 x \cot x = \csc^2 x(\cot x + \cot^3 x)$

Solution

a. $\tan^4 x = (\tan^2 x)(\tan^2 x)$ Separate factors.

$\qquad\quad = \tan^2 x(\sec^2 x - 1)$ Pythagorean identity

$\qquad\quad = \tan^2 x \sec^2 x - \tan^2 x$ Multiply.

b. $\sin^3 x \cos^4 x = \sin^2 x \cos^4 x \sin x$ Separate factors.

$\qquad\qquad\quad = (1 - \cos^2 x)\cos^4 x \sin x$ Pythagorean identity

$\qquad\qquad\quad = (\cos^4 x - \cos^6 x) \sin x$ Multiply.

c. $\csc^4 x \cot x = \csc^2 x \csc^2 x \cot x$ Separate factors.

$\qquad\qquad\quad = \csc^2 x(1 + \cot^2 x) \cot x$ Pythagorean identity

$\qquad\qquad\quad = \csc^2 x(\cot x + \cot^3 x)$ Multiply.

Writing ABOUT MATHEMATICS

Alternative Writing About Mathematics

a. Ask students to assemble a list of techniques and strategies for rewriting trigonometric expressions such as those demonstrated in the examples of this section.

b. Ask students to work in pairs. Each student should create an identity equation from the fundamental trigonometric identities. Partners then trade and verify one another's identities. Then have students write a brief explanation of the techniques they used to create the identities.

Error Analysis   You are tutoring a student in trigonometry. One of the homework problems your student encounters asks whether the following statement is an identity.

$$\tan^2 x \sin^2 x \overset{?}{=} \frac{5}{6}\tan^2 x$$

Your student does not attempt to verify the equivalence algebraically, but mistakenly uses only a graphical approach. Using range settings of

Xmin = -3π	Ymin = -20
Xmax = 3π	Ymax = 20
Xscl = $\pi/2$	Yscl = 1

your student graphs both sides of the expression on a graphing utility and concludes that the statement is an identity.

What is wrong with your student's reasoning? Explain. Discuss the limitations of verifying identities graphically.

2.2 Exercises

In Exercises 1–40, verify the identity.

1. $\sin t \csc t = 1$

2. $\sec y \cos y = 1$

3. $(1 + \sin \alpha)(1 - \sin \alpha) = \cos^2 \alpha$

4. $\cot^2 y(\sec^2 y - 1) = 1$

5. $\cos^2 \beta - \sin^2 \beta = 1 - 2\sin^2 \beta$

6. $\cos^2 \beta - \sin^2 \beta = 2\cos^2 \beta - 1$

7. $\tan^2 \theta + 4 = \sec^2 \theta + 3$

8. $2 - \sec^2 z = 1 - \tan^2 z$

9. $\sin^2 \alpha - \sin^4 \alpha = \cos^2 \alpha - \cos^4 \alpha$

10. $\cos x + \sin x \tan x = \sec x$

11. $\dfrac{\csc^2 \theta}{\cot \theta} = \csc \theta \sec \theta$

12. $\dfrac{\cot^3 t}{\csc t} = \cos t(\csc^2 t - 1)$

13. $\dfrac{\cot^2 t}{\csc t} = \csc t - \sin t$

14. $\dfrac{1}{\tan \beta} + \tan \beta = \dfrac{\sec^2 \beta}{\tan \beta}$

15. $\sin^{1/2} x \cos x - \sin^{5/2} x \cos x = \cos^3 x \sqrt{\sin x}$

16. $\sec^6 x(\sec x \tan x) - \sec^4 x(\sec x \tan x) = \sec^5 x \tan^3 x$

17. $\dfrac{1}{\sec x \tan x} = \csc x - \sin x$

18. $\dfrac{\sec \theta - 1}{1 - \cos \theta} = \sec \theta$

19. $\cot \alpha + \tan \alpha = \csc \alpha \sec \alpha$

20. $\sec x - \cos x = \sin x \tan x$

21. $\dfrac{1}{\tan x} + \dfrac{1}{\cot x} = \tan x + \cot x$

22. $\dfrac{1}{\sin x} - \dfrac{1}{\csc x} = \csc x - \sin x$

23. $\dfrac{\cos \theta \cot \theta}{1 - \sin \theta} - 1 = \csc \theta$

24. $\dfrac{1 + \sin \theta}{\cos \theta} + \dfrac{\cos \theta}{1 + \sin \theta} = 2\sec \theta$

25. $\dfrac{1}{\sin x + 1} + \dfrac{1}{\csc x + 1} = 1$

26. $\cos x - \dfrac{\cos x}{1 - \tan x} = \dfrac{\sin x \cos x}{\sin x - \cos x}$

27. $\tan\left(\dfrac{\pi}{2} - \theta\right)\tan \theta = 1$

28. $\dfrac{\cos[(\pi/2) - x]}{\sin[(\pi/2) - x]} = \tan x$

29. $\dfrac{\csc(-x)}{\sec(-x)} = -\cot x$

30. $(1 + \sin y)[1 + \sin(-y)] = \cos^2 y$

31. $\dfrac{\sin x \cos y + \cos x \sin y}{\cos x \cos y - \sin x \sin y} = \dfrac{\tan x + \tan y}{1 - \tan x \tan y}$

32. $\dfrac{\tan x + \tan y}{1 - \tan x \tan y} = \dfrac{\cot x + \cot y}{\cot x \cot y - 1}$

33. $\dfrac{\tan x + \cot y}{\tan x \cot y} = \tan y + \cot x$

34. $\dfrac{\cos x - \cos y}{\sin x + \sin y} + \dfrac{\sin x - \sin y}{\cos x + \cos y} = 0$

35. $\sqrt{\dfrac{1 + \sin \theta}{1 - \sin \theta}} = \dfrac{1 + \sin \theta}{|\cos \theta|}$

36. $\sqrt{\dfrac{1 - \cos \theta}{1 + \cos \theta}} = \dfrac{1 - \cos \theta}{|\sin \theta|}$

37. $\cos^2 \beta + \cos^2\left(\dfrac{\pi}{2} - \beta\right) = 1$

38. $\sec^2 y - \cot^2\left(\dfrac{\pi}{2} - y\right) = 1$

39. $\sin t \csc\left(\dfrac{\pi}{2} - t\right) = \tan t$

40. $\sec^2\left(\dfrac{\pi}{2} - x\right) - 1 = \cot^2 x$

In Exercises 41–52, use a graphing utility to determine whether the equation is an identity. If it is, confirm it algebraically.

41. $2\sec^2 x - 2\sec^2 x \sin^2 x - \sin^2 x - \cos^2 x = 1$

42. $\csc x(\csc x - \sin x) + \dfrac{\sin x - \cos x}{\sin x} + \cot x = \csc^2 x$

43. $2 + \cos^2 x - 3\cos^4 x = \sin^2 x(3 + 2\cos^2 x)$

44. $\tan^4 x + \tan^2 x - 3 = \sec^2 x(4\tan^2 x - 3)$

45. $\csc^4 x - 2\csc^2 x + 1 = \cot^4 x$

46. $(\sin^4 \beta - 2\sin^2 \beta + 1)\cos \beta = \cos^5 \beta$

47. $\sec^4 \theta - \tan^4 \theta = 1 + 2\tan^2 \theta$

48. $\csc^4 \theta - \cot^4 \theta = 2\csc^2 \theta - 1$

49. $\dfrac{\cos x}{1 - \sin x} = \dfrac{1 - \sin x}{\cos x}$

50. $\dfrac{\cot \alpha}{\csc \alpha + 1} = \dfrac{\csc \alpha + 1}{\cot \alpha}$

51. $\dfrac{\tan^3 \alpha - 1}{\tan \alpha - 1} = \tan^2 \alpha + \tan \alpha + 1$

52. $\dfrac{\sin^3 \beta + \cos^3 \beta}{\sin \beta + \cos \beta} = 1 - \sin \beta \cos \beta$

In Exercises 53–56, use the cofunction identities to evaluate the expression without the aid of a calculator.

53. $\sin^2 25° + \sin^2 65°$

54. $\cos^2 55° + \cos^2 35°$

55. $\cos^2 20° + \cos^2 52° + \cos^2 38° + \cos^2 70°$

56. $\sin^2 12° + \sin^2 40° + \sin^2 50° + \sin^2 78°$

57. *Rate of Change*   The rate of change of the function

$$f(x) = \sin x + \csc x$$

with respect to change in the variable x is given by the expression $\cos x - \csc x \cot x$. Show that the expression for the rate of change can also be $-\cos x \cot^2 x$.

▶ Model It

58. *Shadow Length*   The length s of a shadow cast by a vertical gnomon (a device used to tell time) of height h when the angle of the sun above the horizon is θ (see figure) can be modeled by the equation

$$s = \frac{h \sin(90° - \theta)}{\sin \theta}.$$

(a) Verify that the equation above is equal to $h \cot \theta$.

(b) Use a graphing utility to complete the table. Let $h = 5$ feet.

θ	10°	20°	30°	40°	50°
s					

θ	60°	70°	80°	90°
s				

▶ Model It (continued)

(c) Use your table from part (b) to determine the angles of the sun for which the length of the shadow is the greatest and the least.

(d) Based on your results from part (c), what time of day do you think it is when the angle of the sun above the horizon is 90°?

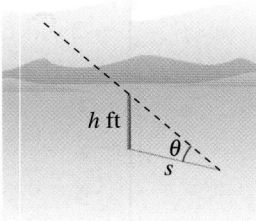

Synthesis

True or False?   **In Exercises 59 and 60, determine whether the statement is true or false. Justify your answer.**

59. The equation

$$\sin^2 \theta + \cos^2 \theta = 1 + \tan^2 \theta$$

is an identity, because $\sin^2(0) + \cos^2(0) = 1$ and $1 + \tan^2(0) = 1$.

60. The equation $1 + \tan^2 \theta = 1 + \cot^2 \theta$ is *not* an identity, because it is true that $1 + \tan^2(\pi/6) = 1\frac{1}{3}$, and $1 + \cot^2(\pi/6) = 4$.

Think About It   **In Exercises 61 and 62, explain why the equation is *not* an identity and find one value of the variable for which the equation is not true.**

61. $\sin \theta = \sqrt{1 - \cos^2 \theta}$

62. $\tan \theta = \sqrt{\sec^2 \theta - 1}$

In Exercises 63–66, use the Quadratic Formula to solve the quadratic equation.

63. $x^2 - 6x - 12 = 0$

64. $x^2 + 5x - 7 = 0$

65. $3x^2 + 6x - 12 = 0$

66. $8x^2 - 4x - 3 = 0$

2.3 Solving Trigonometric Equations

▶ **What you should learn**

- How to use standard algebraic techniques to solve trigonometric equations
- How to solve trigonometric equations of quadratic type
- How to solve trigonometric equations involving multiple angles
- How to use inverse trigonometric functions to solve trigonometric equations

▶ **Why you should learn it**

You can use trigonometric equations to solve a variety of real-life problems. For instance, in Exercise 75 on page 243, you can solve a trigonometric equation to help answer questions about the unemployment rate in the United States.

Introduction

To solve a trigonometric equation, use standard algebraic techniques such as collecting like terms and factoring. Your preliminary goal in solving a trigonometric equation is to isolate the trigonometric function involved in the equation. For example, to solve the equation $2 \sin x = 1$, divide each side by 2 to obtain

$$\sin x = \frac{1}{2}.$$

To solve for x, note in Figure 2.2 that the equation $\sin x = \frac{1}{2}$ has solutions $x = \pi/6$ and $x = 5\pi/6$ in the interval $[0, 2\pi)$. Moreover, because $\sin x$ has a period of 2π, there are infinitely many other solutions, which can be written as

$$x = \frac{\pi}{6} + 2n\pi \qquad \text{and} \qquad x = \frac{5\pi}{6} + 2n\pi \qquad \text{General solution}$$

where n is an integer, as shown in Figure 2.2.

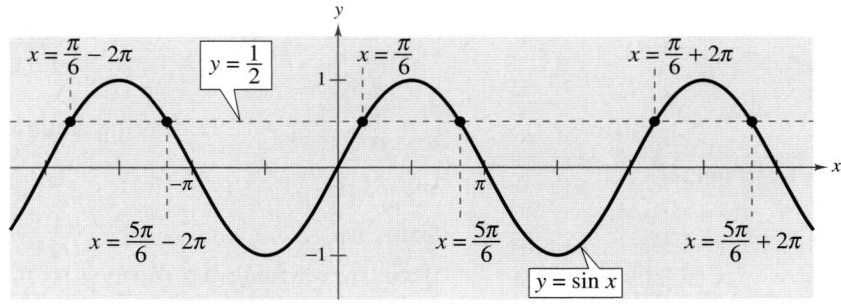

FIGURE 2.2

Another way to show that the equation $\sin x = \frac{1}{2}$ has infinitely many solutions is indicated in Figure 2.3. Any angles that are coterminal with $\pi/6$ or $5\pi/6$ will also be solutions of the equation.

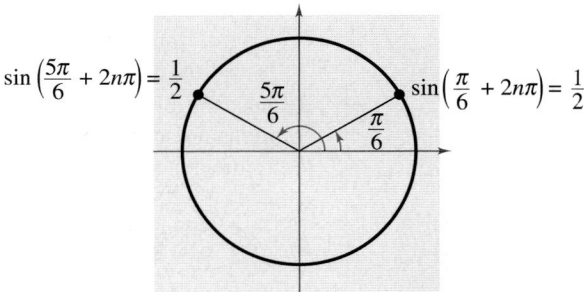

FIGURE 2.3

Encourage your students to give exact answers (rather than decimal approximations using a calculator) when solving these trigonometric equations.

Example 1 ▶ **Collecting Like Terms**

Solve $\sin x + \sqrt{2} = -\sin x$.

Solution

Begin by rewriting the equation so that $\sin x$ is isolated on one side of the equation.

$$\sin x + \sqrt{2} = -\sin x \qquad \text{Write original equation.}$$

$$\sin x + \sin x + \sqrt{2} = 0 \qquad \text{Add } \sin x \text{ to each side.}$$

$$\sin x + \sin x = -\sqrt{2} \qquad \text{Subtract } \sqrt{2} \text{ from each side.}$$

$$2\sin x = -\sqrt{2} \qquad \text{Combine like terms.}$$

$$\sin x = -\frac{\sqrt{2}}{2} \qquad \text{Divide each side by 2.}$$

Because $\sin x$ has a period of 2π, first find all solutions in the interval $[0, 2\pi)$. These solutions are $x = 5\pi/4$ and $x = 7\pi/4$. Finally, add $2n\pi$ to each of these solutions to get the general form

$$x = \frac{5\pi}{4} + 2n\pi \qquad \text{and} \qquad x = \frac{7\pi}{4} + 2n\pi \qquad \text{General solution}$$

where n is an integer.

Example 2 ▶ **Extracting Square Roots**

Solve $3\tan^2 x - 1 = 0$.

Solution

Begin by rewriting the equation so that $\tan x$ is isolated on one side of the equation.

$$3\tan^2 x - 1 = 0 \qquad \text{Write original equation.}$$

$$3\tan^2 x = 1 \qquad \text{Add 1 to each side.}$$

$$\tan^2 x = \frac{1}{3} \qquad \text{Divide each side by 3.}$$

$$\tan x = \pm\frac{1}{\sqrt{3}} \qquad \text{Extract square roots.}$$

Because $\tan x$ has a period of π, first find all solutions in the interval $[0, \pi)$. These solutions are $x = \pi/6$ and $x = 5\pi/6$. Finally, add $n\pi$ to each of these solutions to get the general form

$$x = \frac{\pi}{6} + n\pi \qquad \text{and} \qquad x = \frac{5\pi}{6} + n\pi \qquad \text{General solution}$$

where n is an integer.

Technology

The solutions in Examples 1 and 2 are obtained analytically. You can use a graphing utility to confirm the solutions graphically. For instance, to confirm the solutions found in Example 2, graph

$$y = 3\tan^2 x - 1$$

as shown below. Then use the *zero* or *root* feature or the *zoom* and *trace* features to approximate the x-intercepts.

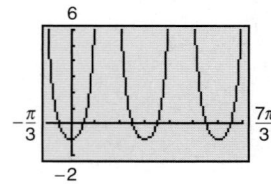

The equations in Examples 1 and 2 involved only one trigonometric function. When two or more functions occur in the same equation, collect all terms on one side and try to separate the functions by factoring or by using appropriate identities. This may produce factors that yield no solutions, as illustrated in Example 3.

Example 3 ▶ **Factoring**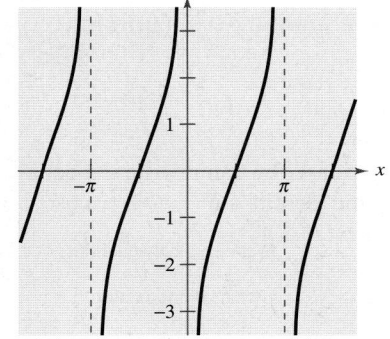

Solve $\cot x \cos^2 x = 2 \cot x$.

Solution

Begin by rewriting the equation so that all terms are collected on one side of the equation.

$$\cot x \cos^2 x = 2 \cot x \qquad \text{Write original equation.}$$

$$\cot x \cos^2 x - 2 \cot x = 0 \qquad \text{Subtract 2 cot } x \text{ from each side.}$$

$$\cot x(\cos^2 x - 2) = 0 \qquad \text{Factor.}$$

By setting each of these factors equal to zero, you obtain

$$\cot x = 0 \qquad \text{and} \qquad \cos^2 x - 2 = 0$$

$$x = \frac{\pi}{2} \qquad\qquad \cos^2 x = 2$$

$$\cos x = \pm\sqrt{2}.$$

The equation $\cot x = 0$ has the solution $x = \pi/2$. No solution is obtained for $\cos x = \pm\sqrt{2}$ because $\pm\sqrt{2}$ are outside the range of the cosine function. Because $\cot x$ has a period of π, the general form of the solution is obtained by adding multiples of π to $x = \pi/2$, to get

$$x = \frac{\pi}{2} + n\pi \qquad \text{General solution}$$

where n is an integer. You can confirm this graphically by sketching the graph of $y = \cot x \cos^2 x - 2 \cot x$, as shown in Figure 2.4.

FIGURE **2.4**

In Example 3, do not make the mistake of dividing each side of the equation by $\cot x$. If you do this, you lose the solutions. Can you see why?

Equations of Quadratic Type

Many trigonometric equations are of quadratic type. Here are a couple of examples.

Quadratic in sin x	*Quadratic in sec x*
$2\sin^2 x - \sin x - 1 = 0$	$\sec^2 x - 3\sec x - 2 = 0$
$2(\sin x)^2 - \sin x - 1 = 0$	$(\sec x)^2 - 3(\sec x) - 2 = 0$

To solve equations of this type, factor the quadratic or, if this is not possible, use the Quadratic Formula.

Example 4 ▶ **Factoring an Equation of Quadratic Type**

Find all solutions of $2 \sin^2 x - \sin x - 1 = 0$ in the interval $[0, 2\pi)$.

Solution

Begin by treating the equation as a quadratic in $\sin x$ and factoring.

$$(2 \sin x + 1)(\sin x - 1) = 0 \qquad \text{Factor.}$$

Setting each factor equal to zero, you can find the solutions in the interval $[0, 2\pi)$.

$$2 \sin x + 1 = 0 \qquad \text{and} \qquad \sin x - 1 = 0$$

$$\sin x = -\frac{1}{2} \qquad\qquad\qquad \sin x = 1$$

$$x = \frac{7\pi}{6}, \frac{11\pi}{6} \qquad\qquad\qquad x = \frac{\pi}{2}$$

When working with an equation of quadratic type, be sure that the equation involves a *single* trigonometric function, as shown in the next example.

Example 5 ▶ **Rewriting with a Single Trigonometric Function**

Solve $2 \sin^2 x + 3 \cos x - 3 = 0$.

Solution

This equation contains both sine and cosine functions. You can rewrite the equation so that it has only cosine functions by using the identity $\sin^2 x = 1 - \cos^2 x$.

$2 \sin^2 x + 3 \cos x - 3 = 0$	Write original equation.
$2(1 - \cos^2 x) + 3 \cos x - 3 = 0$	Pythagorean identity
$2 \cos^2 x - 3 \cos x + 1 = 0$	Multiply each side by -1.
$(2 \cos x - 1)(\cos x - 1) = 0$	Factor.

By setting each factor equal to zero, you can find the solutions in the interval $[0, 2\pi)$.

$$2 \cos x - 1 = 0 \qquad \text{and} \qquad \cos x - 1 = 0$$

$$\cos x = \frac{1}{2} \qquad\qquad\qquad \cos x = 1$$

$$\qquad\qquad\qquad\qquad\qquad x = 0$$

$$x = \frac{\pi}{3}, \frac{5\pi}{3}$$

Because $\cos x$ has a period of 2π, the general form of the solution is obtained by adding multiples of 2π to get

$$x = 2n\pi, \quad x = \frac{\pi}{3} + 2n\pi, \quad x = \frac{5\pi}{3} + 2n\pi \qquad \text{General solution}$$

where n is an integer.

The algebraic model can be difficult for students to visualize. Encourage them to write or visualize the algebraic model, rewrite the corresponding trigonometric equation in terms of a single trigonometric function, and then solve for x. (See Exercises 43 and 44.)

Sometimes you must square each side of an equation to obtain a quadratic, as demonstrated in the next example. Because this procedure can introduce extraneous solutions, you should check any solutions in the original equation to see whether they are valid or extraneous.

Example 6 ▶ Squaring and Converting to Quadratic Type

Find all solutions of $\cos x + 1 = \sin x$ in the interval $[0, 2\pi)$.

Solution

It is not clear how to rewrite this equation in terms of a single trigonometric function. See what happens when you square each side of the equation.

$\cos x + 1 = \sin x$	Write original equation.
$\cos^2 x + 2\cos x + 1 = \sin^2 x$	Square each side.
$\cos^2 x + 2\cos x + 1 = 1 - \cos^2 x$	Pythagorean identity
$\cos^2 x + \cos^2 x + 2\cos x + 1 - 1 = 0$	Rewrite equation.
$2\cos^2 x + 2\cos x = 0$	Combine like terms.
$2\cos x(\cos x + 1) = 0$	Factor.

Setting each factor equal to zero produces

$$2\cos x = 0 \qquad \text{and} \qquad \cos x + 1 = 0$$

$$\cos x = 0 \qquad\qquad\qquad \cos x = -1$$

$$x = \frac{\pi}{2}, \frac{3\pi}{2} \qquad\qquad\qquad x = \pi.$$

Because you squared the original equation, check for extraneous solutions.

Check for $x = \pi/2$

$$\cos \frac{\pi}{2} + 1 \overset{?}{=} \sin \frac{\pi}{2} \qquad\qquad \text{Substitute } \pi/2 \text{ for } x.$$

$$0 + 1 = 1 \qquad\qquad \text{Solution checks. } ✓$$

Check for $x = 3\pi/2$

$$\cos \frac{3\pi}{2} + 1 \overset{?}{=} \sin \frac{3\pi}{2} \qquad\qquad \text{Substitute } 3\pi/2 \text{ for } x.$$

$$0 + 1 \neq -1 \qquad\qquad \text{Solution does not check.}$$

Check for $x = \pi$

$$\cos \pi + 1 \overset{?}{=} \sin \pi \qquad\qquad \text{Substitute } \pi \text{ for } x.$$

$$-1 + 1 = 0 \qquad\qquad \text{Solution checks. } ✓$$

Of the three possible solutions, $x = 3\pi/2$ is extraneous. So, in the interval $[0, 2\pi)$, the only two solutions are $x = \pi/2$ and $x = \pi$.

STUDY TIP

You square each side of the equation in Example 6 because the squares of the sine and cosine functions are related by a Pythagorean identity. The same is true for the squares of the secant and tangent functions and the cosecant and cotangent functions.

◀ **Exploration** ▶

Use a graphing utility to confirm the solutions found in Example 6 in two different ways. Do both methods produce the same x-values? Which method do you prefer? Why?

1. Graph both sides of the equation and find the x-coordinates of the points at which the graphs intersect.

    Left side: $y = \cos x + 1$

    Right side: $y = \sin x$

2. Graph the equation

    $y = \cos x + 1 - \sin x$

    and find the x-intercepts of the graph.

Functions Involving Multiple Angles

The next two examples involve trigonometric functions of multiple angles of the forms $\sin ku$ and $\cos ku$. To solve equations of these forms, first solve the equation for ku, then divide your result by k.

Example 7 ► Functions of Multiple Angles

Find all solutions of $2 \cos 3t - 1 = 0$.

Solution

$2 \cos 3t - 1 = 0$	Write original equation.
$2 \cos 3t = 1$	Add 1 to each side.
$\cos 3t = \dfrac{1}{2}$	Divide each side by 2.

In the interval $[0, 2\pi)$, you know that $3t = \pi/3$ and $3t = 5\pi/3$ are the only solutions, so, in general, you have

$$3t = \frac{\pi}{3} + 2n\pi \qquad \text{and} \qquad 3t = \frac{5\pi}{3} + 2n\pi.$$

Dividing these results by 3, you obtain the general solution

$$t = \frac{\pi}{9} + \frac{2n\pi}{3} \qquad \text{and} \qquad t = \frac{5\pi}{9} + \frac{2n\pi}{3} \qquad \text{General solution}$$

where n is an integer.

Example 8 ► Functions of Multiple Angles

Find all solutions of $3 \tan(x/2) + 3 = 0$.

Solution

$3 \tan \dfrac{x}{2} + 3 = 0$	Write original equation.
$3 \tan \dfrac{x}{2} = -3$	Subtract 3 from each side.
$\tan \dfrac{x}{2} = -1$	Divide each side by 3.

In the interval $[0, \pi)$, you know that $x/2 = 3\pi/4$ is the only solution, so, in general, you have

$$\frac{x}{2} = \frac{3\pi}{4} + n\pi.$$

Multiplying this result by 2, you obtain the general solution

$$x = \frac{3\pi}{2} + 2n\pi \qquad \text{General solution}$$

where n is an integer.

Using Inverse Functions

In the next example, you will see how inverse trigonometric functions can be used to solve an equation.

> ### Example 9 ▶ Using Inverse Functions
>
> Find all solutions of $\sec^2 x - 2\tan x = 4$.

Solution

$$\sec^2 x - 2\tan x = 4 \qquad\qquad \text{Write original equation.}$$

$$1 + \tan^2 x - 2\tan x - 4 = 0 \qquad\qquad \text{Pythagorean identity}$$

$$\tan^2 x - 2\tan x - 3 = 0 \qquad\qquad \text{Combine like terms.}$$

$$(\tan x - 3)(\tan x + 1) = 0 \qquad\qquad \text{Factor.}$$

Setting each factor equal to zero, you obtain two solutions in the interval $(-\pi/2, \pi/2)$. [Recall that the range of the inverse tangent function is $(-\pi/2, \pi/2)$.]

$$\tan x - 3 = 0 \qquad \text{and} \qquad \tan x + 1 = 0$$

$$\tan x = 3 \qquad\qquad\qquad \tan x = -1$$

$$x = \arctan 3 \qquad\qquad\qquad x = -\frac{\pi}{4}$$

Finally, because $\tan x$ has a period of π, you obtain the general solution by adding multiples of π

$$x = \arctan 3 + n\pi \qquad \text{and} \qquad x = -\frac{\pi}{4} + n\pi \qquad \text{General solution}$$

where n is an integer. You can use a calculator to approximate the value of $\arctan 3$.

Writing ABOUT MATHEMATICS

Equations with No Solutions   One of the following equations has solutions and the other two do not. Which two equations do not have solutions?

a. $\sin^2 x - 5\sin x + 6 = 0$

b. $\sin^2 x - 4\sin x + 6 = 0$

c. $\sin^2 x - 5\sin x - 6 = 0$

Find conditions involving the constants b and c that will guarantee that the equation

$$\sin^2 x + b\sin x + c = 0$$

has at least one solution on some interval of length 2π.

2.3 Exercises

In Exercises 1–6, verify that the x-values are solutions of the equation.

1. $2 \cos x - 1 = 0$

   (a) $x = \dfrac{\pi}{3}$ (b) $x = \dfrac{5\pi}{3}$

2. $\sec x - 2 = 0$

   (a) $x = \dfrac{\pi}{3}$ (b) $x = \dfrac{5\pi}{3}$

3. $3 \tan^2 2x - 1 = 0$

   (a) $x = \dfrac{\pi}{12}$ (b) $x = \dfrac{5\pi}{12}$

4. $2 \cos^2 4x - 1 = 0$

   (a) $x = \dfrac{\pi}{16}$ (b) $x = \dfrac{3\pi}{16}$

5. $2 \sin^2 x - \sin x - 1 = 0$

   (a) $x = \dfrac{\pi}{2}$ (b) $x = \dfrac{7\pi}{6}$

6. $\csc^4 x - 4 \csc^2 x = 0$

   (a) $x = \dfrac{\pi}{6}$ (b) $x = \dfrac{5\pi}{6}$

In Exercises 7–20, solve the equation.

7. $2 \cos x + 1 = 0$ **8.** $2 \sin x + 1 = 0$

9. $\sqrt{3} \csc x - 2 = 0$ **10.** $\tan x + \sqrt{3} = 0$

11. $3 \sec^2 x - 4 = 0$ **12.** $3 \cot^2 x - 1 = 0$

13. $\sin x(\sin x + 1) = 0$

14. $(3 \tan^2 x - 1)(\tan^2 x - 3) = 0$

15. $4 \cos^2 x - 1 = 0$ **16.** $\sin^2 x = 3 \cos^2 x$

17. $2 \sin^2 2x = 1$ **18.** $\tan^2 3x = 3$

19. $\tan 3x(\tan x - 1) = 0$

20. $\cos 2x(2 \cos x + 1) = 0$

In Exercises 21–32, find all solutions of the equation in the interval [0, 2π).

21. $\cos^3 x = \cos x$ **22.** $\sec^2 x - 1 = 0$

23. $3 \tan^3 x = \tan x$ **24.** $2 \sin^2 x = 2 + \cos x$

25. $\sec^2 x - \sec x = 2$ **26.** $\sec x \csc x = 2 \csc x$

27. $2 \sin x + \csc x = 0$ **28.** $\sec x + \tan x = 1$

29. $2 \cos^2 x + \cos x - 1 = 0$

30. $2 \sin^2 x + 3 \sin x + 1 = 0$

31. $2 \sec^2 x + \tan^2 x - 3 = 0$

32. $\cos x + \sin x \tan x = 2$

In Exercises 33–38, find all solutions of the equation.

33. $\cos 2x = \dfrac{1}{2}$ **34.** $\sin 2x = -\dfrac{\sqrt{3}}{2}$

35. $\tan 3x = 1$ **36.** $\sec 4x = 2$

37. $\cos \dfrac{x}{2} = \dfrac{\sqrt{2}}{2}$ **38.** $\sin \dfrac{x}{2} = -\dfrac{\sqrt{3}}{2}$

In Exercises 39–42, find the x-intercepts of the graph.

39. $y = \sin \dfrac{\pi x}{2} + 1$ **40.** $y = \sin \pi x + \cos \pi x$

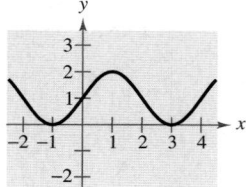   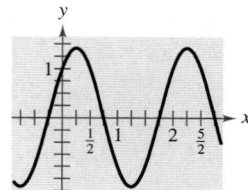

41. $y = \tan^2\left(\dfrac{\pi x}{6}\right) - 3$ **42.** $y = \sec^4\left(\dfrac{\pi x}{8}\right) - 4$

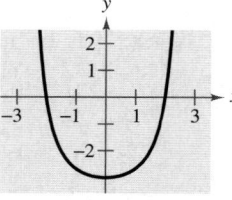   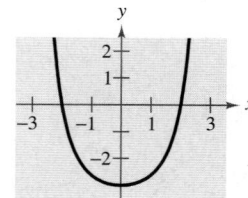

In Exercises 43 and 44, solve both equations. How do the solutions of the algebraic equation compare with the solutions of the trigonometric equation?

43. $6y^2 - 13y + 6 = 0$

   $6 \cos^2 x - 13 \cos x + 6 = 0$

44. $y^2 + y - 20 = 0$

   $\sin^2 x + \sin x - 20 = 0$

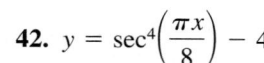 **In Exercises 45–54, use a graphing utility to approximate the solutions of the equation in the interval [0, 2π).**

45. $2 \sin x + \cos x = 0$

46. $4 \sin^3 x + 2 \sin^2 x - 2 \sin x - 1 = 0$

47. $\dfrac{1 + \sin x}{\cos x} + \dfrac{\cos x}{1 + \sin x} = 4$ 48. $\dfrac{\cos x \cot x}{1 - \sin x} = 3$

49. $x \tan x - 1 = 0$

50. $x \cos x - 1 = 0$

51. $\sec^2 x + 0.5 \tan x - 1 = 0$

52. $\csc^2 x + 0.5 \cot x - 5 = 0$

53. $2 \tan^2 x + 7 \tan x - 15 = 0$

54. $6 \sin^2 x - 7 \sin x + 2 = 0$

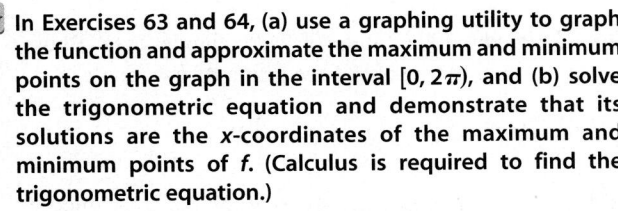 In Exercises 55–58, use the Quadratic Formula to solve the equation in the interval $[0, 2\pi)$. Then use a graphing utility to approximate the angle x.

55. $12 \sin^2 x - 13 \sin x + 3 = 0$

56. $3 \tan^2 x + 4 \tan x - 4 = 0$

57. $\tan^2 x + 3 \tan x + 1 = 0$

58. $4 \cos^2 x - 4 \cos x - 1 = 0$

In Exercises 59–62, use inverse functions where needed to find all solutions of the equation in the interval $[0, 2\pi)$.

59. $\tan^2 x - 6 \tan x + 5 = 0$

60. $\sec^2 x + \tan x - 3 = 0$

61. $2 \cos^2 x - 5 \cos x + 2 = 0$

62. $2 \sin^2 x - 7 \sin x + 3 = 0$

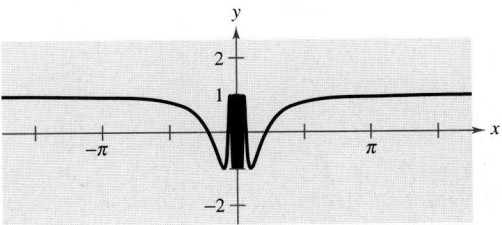 In Exercises 63 and 64, (a) use a graphing utility to graph the function and approximate the maximum and minimum points on the graph in the interval $[0, 2\pi)$, and (b) solve the trigonometric equation and demonstrate that its solutions are the x-coordinates of the maximum and minimum points of f. (Calculus is required to find the trigonometric equation.)

Function	Trigonometric Equation
63. $f(x) = \sin x + \cos x$	$\cos x - \sin x = 0$
64. $f(x) = 2 \sin x + \cos 2x$	$2 \cos x - 4 \sin x \cos x = 0$

Fixed Point In Exercises 65 and 66, find the smallest positive fixed point of the function f. [A fixed point of a function f is a real number c such that $f(c) = c$.]

65. $f(x) = \tan \dfrac{\pi x}{4}$ 66. $f(x) = \cos x$

67. *Graphical Reasoning* Consider the function

$$f(x) = \cos \dfrac{1}{x}$$

and its graph shown in the figure.

(a) What is the domain of the function?

(b) Identify any symmetry or asymptotes of the graph.

(c) Describe the behavior of the function as $x \to 0$.

(d) How many solutions does the equation

$$\cos \dfrac{1}{x} = 0$$

have in the interval $[-1, 1]$?

(e) Does the equation $\cos(1/x) = 0$ have a greatest solution? If so, approximate the solution. If not, explain why.

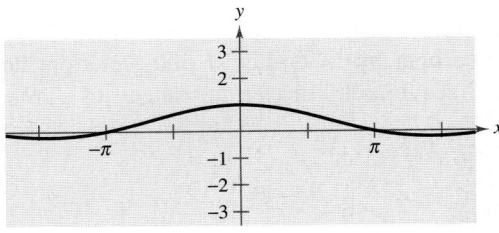

FIGURE FOR 67

68. *Graphical Reasoning* Consider the function

$$f(x) = \dfrac{\sin x}{x}$$

and its graph shown in the figure.

(a) What is the domain of the function?

(b) Identify any symmetry or asymptotes of the graph.

(c) Describe the behavior of the function as $x \to 0$.

(d) How many solutions does the equation

$$\dfrac{\sin x}{x} = 0$$

have in the interval $[-8, 8]$? Find the solutions.

69. *Harmonic Motion* A weight is oscillating on the end of a spring (see figure). The position of the weight relative to the point of equilibrium is

$$y = \frac{1}{12}(\cos 8t - 3 \sin 8t)$$

where y is the displacement in meters and t is the time in seconds. Find the times when the weight is at the point of equilibrium $(y = 0)$ for $0 \le t \le 1$.

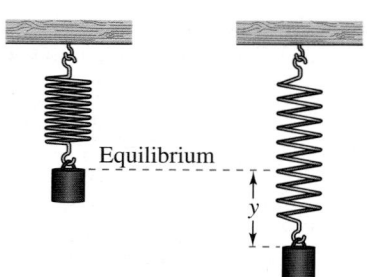

70. *Damped Harmonic Motion* The displacement from equilibrium of a weight oscillating on the end of a spring is

$$y = 1.56t^{-1/2} \cos 1.9t$$

where y is the displacement in feet and t is the time in seconds. Use a graphing utility to graph the displacement function for $0 \le t \le 10$. Find the time beyond which the displacement does not exceed 1 foot from equilibrium.

71. *Sales* The monthly sales (in thousands of units) of a seasonal product are approximated by

$$S = 74.50 + 43.75 \sin \frac{\pi t}{6}$$

where t is the time in months, with $t = 1$ corresponding to January. Determine the months when sales exceed 100,000 units.

72. *Projectile Motion* A batted baseball leaves the bat at an angle of θ with the horizontal and an initial velocity of $v_0 = 100$ feet per second. The ball is caught by an outfielder 300 feet from home plate (see figure). Find θ if the range r of a projectile is

$$r = \frac{1}{32}v_0^2 \sin 2\theta.$$

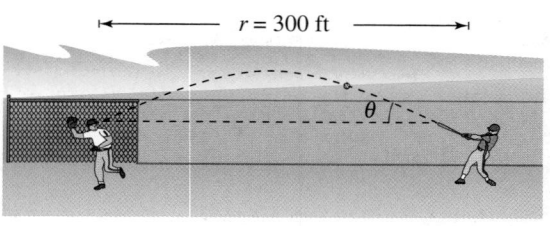

Not drawn to scale

FIGURE FOR **72**

73. *Projectile Motion* A sharpshooter intends to hit a target at a distance of 1000 yards with a gun that has a muzzle velocity of 1200 feet per second (see figure). Neglecting air resistance, determine the gun's minimum angle of elevation θ if the range r is

$$r = \frac{1}{32}v_0^2 \sin 2\theta.$$

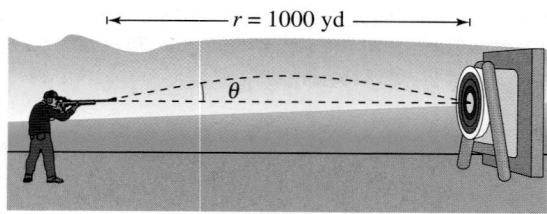

Not drawn to scale

74. *Geometry* The area of a rectangle (see figure) inscribed in one arc of the graph of $y = \cos x$ is

$$A = 2x \cos x, \qquad 0 < x < \frac{\pi}{2}.$$

(a) Use a graphing utility to graph the area function, and approximate the area of the largest inscribed rectangle.

(b) Determine the values of x for which $A \ge 1$.

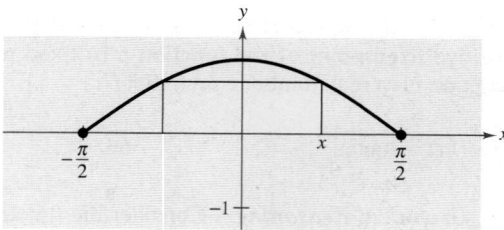

▶ Model It

75. *Data Analysis* The table shows the unemployment rates r for the years 1990 through 2001 in the United States. The time t is measured in years, with $t = 0$ corresponding to 1990. (Source: U.S. Bureau of Labor Statistics)

Time, t	Rate, r	Time, t	Rate, r
0	5.6	6	5.4
1	6.8	7	4.9
2	7.5	8	4.5
3	6.9	9	4.2
4	6.1	10	4.0
5	5.6	11	4.8

(a) Create a scatter plot of the data.

(b) Which of the following models best represents the data? Explain your reasoning.

 (1) $r = 1.39 \sin(0.48t + 0.42) + 5.51$

 (2) $r = 1.39 \sin(0.48t - 0.01) + 5.51$

 (3) $r = \sin(0.10t + 5.61) + 4.80$

 (4) $r = 896 \sin(0.57t - 2.05) + 6.48$

(c) What term in the model gives the average unemployment rate? What is the rate?

(d) Economists study the lengths of business cycles such as unemployment rates. Based on this short span of time, use the model to give the length of this cycle.

(e) Use the model to estimate the next time the unemployment rate will be 6.5% or more.

76. *Quadratic Approximation* Consider the function

$f(x) = 3 \sin(0.6x - 2).$

(a) Approximate the zero of the function in the interval $[0, 6]$.

(b) A quadratic approximation agreeing with f at $x = 5$ is $g(x) = -0.45x^2 + 5.52x - 13.70$. Use a graphing utility to graph f and g in the same viewing window. Describe the result.

(c) Use the Quadratic Formula to find the zeros of g. Compare the zero in the interval $[0, 6]$ with the result of part (a).

Synthesis

True or False? In Exercises 77 and 78, determine whether the statement is true or false. Justify your answer.

77. The equation $2 \sin 4t - 1 = 0$ has four times the number of solutions in the interval $[0, 2\pi)$ as the equation $2 \sin t - 1 = 0$.

78. If you correctly solve a trigonometric equation down to the statement $\sin x = 3.4$, then you can finish solving the equation by using an inverse function.

In Exercises 79 and 80, use the graph to approximate the number of points of intersection of the graphs of y_1 and y_2.

79. $y_1 = 2 \sin x$

 $y_2 = 3x + 1$

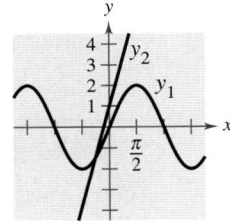

80. $y_1 = 2 \sin x$

 $y_2 = \frac{1}{2}x + 1$

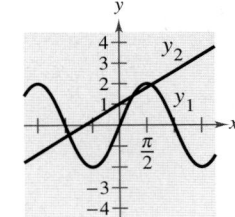

Review

In Exercises 81 and 82, solve triangle ABC by finding all missing angle measures and side lengths.

81.

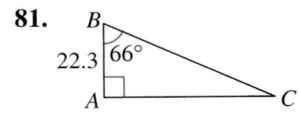

82.

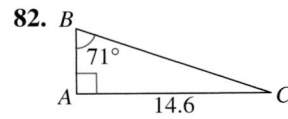

In Exercises 83–86, use reference angles to find the sine, cosine, and tangent of an angle with the given measure.

83. $390°$

84. $570°$

85. $-1845°$

86. $-1410°$

87. *Angle of Depression* Find the angle of depression from the top of a lighthouse 250 feet above water level to the water line of a ship 2 miles offshore.

88. *Height* From a point 100 feet in front of the public library, the angles of elevation to the base of the flagpole and the top of the pole are 28° and 39° 45′, respectively. The flagpole is mounted on the front of the library's roof. Find the height of the flagpole.

2.4 Sum and Difference Formulas

▶ **What you should learn**

- How to use sum and difference formulas to evaluate trigonometric functions, verify identities, and solve trigonometric equations

▶ **Why you should learn it**

You can use identities to rewrite trigonometric expressions. For instance, in Exercise 75 on page 249, you can use an identity to rewrite a trigonometric expression in a form that helps you analyze a harmonic motion equation.

Using Sum and Difference Formulas

In this and the following section, you will study the uses of several trigonometric identities and formulas.

Sum and Difference Formulas

$$\sin(u + v) = \sin u \cos v + \cos u \sin v$$
$$\sin(u - v) = \sin u \cos v - \cos u \sin v$$
$$\cos(u + v) = \cos u \cos v - \sin u \sin v$$
$$\cos(u - v) = \cos u \cos v + \sin u \sin v$$

$$\tan(u + v) = \frac{\tan u + \tan v}{1 - \tan u \tan v}$$
$$\tan(u - v) = \frac{\tan u - \tan v}{1 + \tan u \tan v}$$

For a proof of the sum and difference formulas, see Proofs in Mathematics on page 267.

◀ **E x p l o r a t i o n** ▶

Use a graphing utility to graph $y_1 = \cos(x + 2)$ and $y_2 = \cos x + \cos 2$ in the same viewing window. What can you conclude about the graphs? Is it true that $\cos(x + 2) = \cos x + \cos 2$?

Use a graphing utility to graph $y_1 = \sin(x + 4)$ and $y_2 = \sin x + \sin 4$ in the same viewing window. What can you conclude about the graphs? Is it true that $\sin(x + 4) = \sin x + \sin 4$?

Examples 1 and 2 show how **sum and difference formulas** can be used to find exact values of trigonometric functions involving sums or differences of special angles.

Example 1 ▶ Evaluating a Trigonometric Function

Find the exact value of $\cos 75°$.

Solution

To find the *exact* value of $\cos 75°$, use the fact that $75° = 30° + 45°$. Consequently, the formula for $\cos(u + v)$ yields

$$\cos 75° = \cos(30° + 45°)$$
$$= \cos 30° \cos 45° - \sin 30° \sin 45°$$
$$= \frac{\sqrt{3}}{2}\left(\frac{\sqrt{2}}{2}\right) - \frac{1}{2}\left(\frac{\sqrt{2}}{2}\right) = \frac{\sqrt{6} - \sqrt{2}}{4}.$$

Try checking this result on your calculator. You will find that $\cos 75° \approx 0.259$.

Historical Note
Hipparchus, considered the most eminent of Greek astronomers, was born about 160 B.C. in Nicaea. He was credited with the invention of trigonometry. He also derived the sum and difference formulas for $\sin(A \pm B)$ and $\cos(A \pm B)$.

Example 2 ▶ **Evaluating a Trigonometric Expression**

Find the exact value of $\sin \dfrac{\pi}{12}$.

Solution
Using the fact that

$$\frac{\pi}{12} = \frac{\pi}{3} - \frac{\pi}{4}$$

together with the formula for $\sin(u - v)$, you obtain

$$\sin \frac{\pi}{12} = \sin\left(\frac{\pi}{3} - \frac{\pi}{4}\right)$$

$$= \sin \frac{\pi}{3} \cos \frac{\pi}{4} - \cos \frac{\pi}{3} \sin \frac{\pi}{4}$$

$$= \frac{\sqrt{3}}{2}\left(\frac{\sqrt{2}}{2}\right) - \frac{1}{2}\left(\frac{\sqrt{2}}{2}\right)$$

$$= \frac{\sqrt{6} - \sqrt{2}}{4}.$$

Example 3 ▶ **Evaluating a Trigonometric Expression**

Find the exact value of $\sin 42° \cos 12° - \cos 42° \sin 12°$.

Solution
Recognizing that this expression fits the formula for $\sin(u - v)$, you can write

$$\sin 42° \cos 12° - \cos 42° \sin 12° = \sin(42° - 12°)$$

$$= \sin 30°$$

$$= \frac{1}{2}.$$

Example 4 ▶ **An Application of a Sum Formula**

Write $\cos(\arctan 1 + \arccos x)$ as an algebraic expression.

Solution
This expression fits the formula for $\cos(u + v)$. Angles $u = \arctan 1$ and $v = \arccos x$ are shown in Figure 2.5. So

$$\cos(u + v) = \cos(\arctan 1)\cos(\arccos x) - \sin(\arctan 1)\sin(\arccos x)$$

$$= \frac{1}{\sqrt{2}} \cdot x - \frac{1}{\sqrt{2}} \cdot \sqrt{1 - x^2}$$

$$= \frac{x - \sqrt{1 - x^2}}{\sqrt{2}}.$$

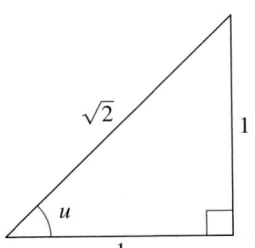

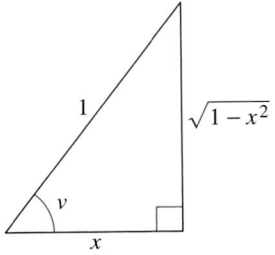

FIGURE 2.5

Example 5 shows how to use a difference formula to prove the cofunction identity

$$\cos\left(\frac{\pi}{2} - x\right) = \sin x.$$

Example 5 ▶   **Proving a Cofunction Identity**

Prove the cofunction identity $\cos\left(\dfrac{\pi}{2} - x\right) = \sin x$.

Solution

Using the formula for $\cos(u - v)$, you have

$$\cos\left(\frac{\pi}{2} - x\right) = \cos\frac{\pi}{2}\cos x + \sin\frac{\pi}{2}\sin x$$

$$= (0)(\cos x) + (1)(\sin x) = \sin x.$$

Sum and difference formulas can be used to derive **reduction formulas** involving expressions such as

$$\sin\left(\theta + \frac{n\pi}{2}\right) \quad \text{and} \quad \cos\left(\theta + \frac{n\pi}{2}\right), \quad \text{where } n \text{ is an integer.}$$

Example 6 ▶   **Deriving Reduction Formulas**

Simplify each expression.

a. $\cos\left(\theta - \dfrac{3\pi}{2}\right)$ **b.** $\tan(\theta + 3\pi)$

Solution

a. Using the formula for $\cos(u - v)$, you have

$$\cos\left(\theta - \frac{3\pi}{2}\right) = \cos\theta\cos\frac{3\pi}{2} + \sin\theta\sin\frac{3\pi}{2}$$

$$= (\cos\theta)(0) + (\sin\theta)(-1)$$

$$= -\sin\theta.$$

b. Using the formula for $\tan(u + v)$, you have

$$\tan(\theta + 3\pi) = \frac{\tan\theta + \tan 3\pi}{1 - \tan\theta\tan 3\pi}$$

$$= \frac{\tan\theta + 0}{1 - (\tan\theta)(0)}$$

$$= \tan\theta.$$

The next example was taken from calculus. It is used to derive the derivative of the sine function.

Activities

1. Use the sum and difference formulas to find the exact value of cos 50°.

   Answer: $\dfrac{\sqrt{6} + \sqrt{2}}{4}$

2. Rewrite the expression using the sum and difference formulas.

   $\dfrac{\tan 40° + \tan 10°}{1 - \tan 40° \tan 10°}$

   Answer: $\tan(40° + 10°) = \tan 50°$

3. Verify the identity

   $\sin\left(\dfrac{\pi}{2} - \theta\right) = \cos\theta.$

   Answer:

   $\sin\left(\dfrac{\pi}{2} - \theta\right) = \sin\dfrac{\pi}{2}\cos\theta - \cos\dfrac{\pi}{2}\sin\theta$

   $= (1)\cos\theta - (0)\sin\theta$

   $= \cos\theta$

Example 7 ▶ An Application from Calculus

Verify that

$$\frac{\sin(x + h) - \sin x}{h} = (\cos x)\left(\frac{\sin h}{h}\right) - (\sin x)\left(\frac{1 - \cos h}{h}\right)$$

where $h \neq 0$.

Solution

Using the formula for $\sin(u + v)$, you have

$$\frac{\sin(x + h) - \sin x}{h} = \frac{\sin x \cos h + \cos x \sin h - \sin x}{h}$$

$$= \frac{\cos x \sin h - \sin x(1 - \cos h)}{h}$$

$$= (\cos x)\left(\frac{\sin h}{h}\right) - (\sin x)\left(\frac{1 - \cos h}{h}\right).$$

Example 8 ▶ Solving a Trigonometric Equation

Find all solutions of

$$\sin\left(x + \frac{\pi}{4}\right) + \sin\left(x - \frac{\pi}{4}\right) = -1$$

in the interval $[0, 2\pi)$.

Solution

Using sum and difference formulas, rewrite the equation as

$$\sin x \cos \frac{\pi}{4} + \cos x \sin \frac{\pi}{4} + \sin x \cos \frac{\pi}{4} - \cos x \sin \frac{\pi}{4} = -1$$

$$2 \sin x \cos \frac{\pi}{4} = -1$$

$$2(\sin x)\left(\frac{\sqrt{2}}{2}\right) = -1$$

$$\sin x = -\frac{1}{\sqrt{2}}$$

$$\sin x = -\frac{\sqrt{2}}{2}.$$

So, the only solutions in the interval $[0, 2\pi)$ are

$$x = \frac{5\pi}{4} \quad \text{and} \quad x = \frac{7\pi}{4}.$$

These solutions are checked graphically in Figure 2.6.

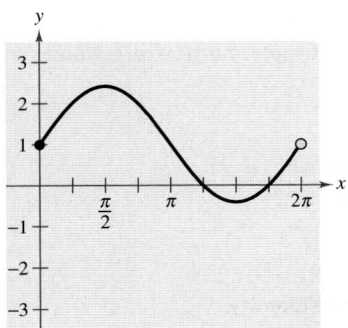

FIGURE 2.6

2.4 **Exercises**

In Exercises 1–6, find the exact value of each expression.

1. (a) $\cos\left(\dfrac{\pi}{4} + \dfrac{\pi}{3}\right)$ (b) $\cos\dfrac{\pi}{4} + \cos\dfrac{\pi}{3}$

2. (a) $\sin\left(\dfrac{3\pi}{4} + \dfrac{5\pi}{6}\right)$ (b) $\sin\dfrac{3\pi}{4} + \sin\dfrac{5\pi}{6}$

3. (a) $\sin\left(\dfrac{7\pi}{6} - \dfrac{\pi}{3}\right)$ (b) $\sin\dfrac{7\pi}{6} - \sin\dfrac{\pi}{3}$

4. (a) $\cos\left(\dfrac{2\pi}{3} - \dfrac{\pi}{6}\right)$ (b) $\cos\dfrac{2\pi}{3} + \cos\dfrac{\pi}{6}$

5. (a) $\cos(120° + 45°)$ (b) $\cos 120° + \cos 45°$

6. (a) $\sin(135° - 30°)$ (b) $\sin 135° - \cos 30°$

In Exercises 7–14, find the exact values of the sine, cosine, and tangent of the angle.

7. $105° = 60° + 45°$ **8.** $165° = 135° + 30°$

9. $195° = 225° - 30°$ **10.** $255° = 300° - 45°$

11. $\dfrac{11\pi}{12} = \dfrac{3\pi}{4} + \dfrac{\pi}{6}$ **12.** $\dfrac{7\pi}{12} = \dfrac{\pi}{3} + \dfrac{\pi}{4}$

13. $\dfrac{17\pi}{12} = \dfrac{9\pi}{4} - \dfrac{5\pi}{6}$ **14.** $-\dfrac{\pi}{12} = \dfrac{\pi}{6} - \dfrac{\pi}{4}$

In Exercises 15–22, find the exact values of the sine, cosine, and tangent of the angle.

15. $285°$ **16.** $-105°$

17. $-165°$ **18.** $15°$

19. $\dfrac{13\pi}{12}$ **20.** $-\dfrac{7\pi}{12}$

21. $-\dfrac{13\pi}{12}$ **22.** $\dfrac{5\pi}{12}$

In Exercises 23–30, write the expression as the sine, cosine, or tangent of an angle.

23. $\cos 25° \cos 15° - \sin 25° \sin 15°$

24. $\sin 140° \cos 50° + \cos 140° \sin 50°$

25. $\dfrac{\tan 325° - \tan 86°}{1 + \tan 325° \tan 86°}$ **26.** $\dfrac{\tan 140° - \tan 60°}{1 + \tan 140° \tan 60°}$

27. $\sin 3 \cos 1.2 - \cos 3 \sin 1.2$

28. $\cos\dfrac{\pi}{7} \cos\dfrac{\pi}{5} - \sin\dfrac{\pi}{7} \sin\dfrac{\pi}{5}$

29. $\dfrac{\tan 2x + \tan x}{1 - \tan 2x \tan x}$

30. $\cos 3x \cos 2y + \sin 3x \sin 2y$

In Exercises 31–36, find the exact value of the expression.

31. $\sin 330° \cos 30° - \cos 330° \sin 30°$

32. $\cos 15° \cos 60° + \sin 15° \sin 60°$

33. $\sin\dfrac{\pi}{12} \cos\dfrac{\pi}{4} + \cos\dfrac{\pi}{12} \sin\dfrac{\pi}{4}$

34. $\cos\dfrac{\pi}{16} \cos\dfrac{3\pi}{16} - \sin\dfrac{\pi}{16} \sin\dfrac{3\pi}{16}$

35. $\dfrac{\tan 25° + \tan 110°}{1 - \tan 25° \tan 110°}$

36. $\dfrac{\tan(5\pi/4) - \tan(\pi/12)}{1 + \tan(5\pi/4) \tan(\pi/12)}$

In Exercises 37–44, find the exact value of the trigonometric function given that $\sin u = \frac{5}{13}$ and $\cos v = -\frac{3}{5}$. (Both u and v are in Quadrant II.)

37. $\sin(u + v)$ **38.** $\cos(u - v)$

39. $\cos(u + v)$ **40.** $\sin(v - u)$

41. $\tan(u + v)$ **42.** $\csc(u - v)$

43. $\sec(v - u)$ **44.** $\cot(u + v)$

In Exercises 45–50, find the exact value of the trigonometric function given that $\sin u = -\frac{7}{25}$ and $\cos v = -\frac{4}{5}$. (Both u and v are in Quadrant III.)

45. $\cos(u + v)$ **46.** $\sin(u + v)$

47. $\tan(u - v)$ **48.** $\cot(v - u)$

49. $\sec(u + v)$ **50.** $\cos(u - v)$

In Exercises 51–54, write the trigonometric expression as an algebraic expression.

51. $\sin(\arcsin x + \arccos x)$

52. $\sin(\arctan 2x - \arccos x)$

53. $\cos(\arccos x + \arcsin x)$

54. $\cos(\arccos x - \arctan x)$

In Exercises 55–64, verify the identity.

55. $\sin(3\pi - x) = \sin x$ **56.** $\sin\left(\dfrac{\pi}{2} + x\right) = \cos x$

57. $\sin\left(\dfrac{\pi}{6} + x\right) = \dfrac{1}{2}(\cos x + \sqrt{3}\sin x)$

58. $\cos\left(\dfrac{5\pi}{4} - x\right) = -\dfrac{\sqrt{2}}{2}(\cos x + \sin x)$

59. $\cos(\pi - \theta) + \sin\left(\dfrac{\pi}{2} + \theta\right) = 0$

60. $\tan\left(\dfrac{\pi}{4} - \theta\right) = \dfrac{1 - \tan\theta}{1 + \tan\theta}$

61. $\cos(x + y)\cos(x - y) = \cos^2 x - \sin^2 y$

62. $\sin(x + y)\sin(x - y) = \sin^2 x - \sin^2 y$

63. $\sin(x + y) + \sin(x - y) = 2\sin x \cos y$

64. $\cos(x + y) + \cos(x - y) = 2\cos x \cos y$

In Exercises 65–68, simplify the expression algebraically and use a graphing utility to confirm your answer graphically.

65. $\cos\left(\dfrac{3\pi}{2} - x\right)$ **66.** $\cos(\pi + x)$

67. $\sin\left(\dfrac{3\pi}{2} + \theta\right)$ **68.** $\tan(\pi + \theta)$

In Exercises 69–72, find all solutions of the equation in the interval $[0, 2\pi)$.

69. $\sin\left(x + \dfrac{\pi}{3}\right) + \sin\left(x - \dfrac{\pi}{3}\right) = 1$

70. $\sin\left(x + \dfrac{\pi}{6}\right) - \sin\left(x - \dfrac{\pi}{6}\right) = \dfrac{1}{2}$

71. $\cos\left(x + \dfrac{\pi}{4}\right) - \cos\left(x - \dfrac{\pi}{4}\right) = 1$

72. $\tan(x + \pi) + 2\sin(x + \pi) = 0$

In Exercises 73 and 74, use a graphing utility to approximate the solutions in the interval $[0, 2\pi)$.

73. $\cos\left(x + \dfrac{\pi}{4}\right) + \cos\left(x - \dfrac{\pi}{4}\right) = 1$

74. $\tan(x + \pi) - \cos\left(x + \dfrac{\pi}{2}\right) = 0$

▶ **Model It**

75. *Harmonic Motion*   A weight is attached to a spring suspended vertically from a ceiling. When a driving force is applied to the system, the weight moves vertically from its equilibrium position, and this motion is modeled by

▶ **Model It** *(continued)*

$$y = \dfrac{1}{3}\sin 2t + \dfrac{1}{4}\cos 2t$$

where y is the distance from equilibrium measured in feet and t is the time in seconds.

(a) Use the identity

$$a\sin B\theta + b\cos B\theta = \sqrt{a^2 + b^2}\,\sin(B\theta + C)$$

where $C = \arctan(b/a)$, $a > 0$, to write the model in the form

$$y = \sqrt{a^2 + b^2}\,\sin(Bt + C).$$

(b) Find the amplitude of the oscillations of the weight.

(c) Find the frequency of the oscillations of the weight.

76. *Standing Waves*   The equation of a standing wave is obtained by adding the displacements of two waves traveling in opposite directions (see figure). Assume that each of the waves has amplitude A, period T, and wavelength λ. If the models for these waves are

$$y_1 = A\cos 2\pi\left(\dfrac{t}{T} - \dfrac{x}{\lambda}\right) \quad \text{and}$$

$$y_2 = A\cos 2\pi\left(\dfrac{t}{T} + \dfrac{x}{\lambda}\right)$$

show that

$$y_1 + y_2 = 2A\cos\dfrac{2\pi t}{T}\cos\dfrac{2\pi x}{\lambda}.$$

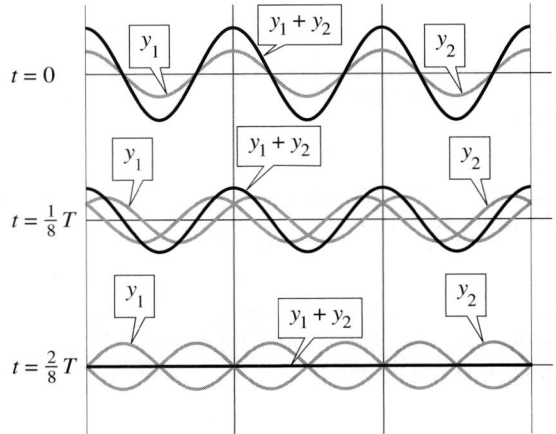

Synthesis

True or False? **In Exercises 77–80, determine whether the statement is true or false. Justify your answer.**

77. $\sin(u \pm v) = \sin u \pm \sin v$

78. $\cos(u \pm v) = \cos u \pm \cos v$

79. $\cos\left(x - \dfrac{\pi}{2}\right) = -\sin x$

80. $\sin\left(x - \dfrac{\pi}{2}\right) = -\cos x$

In Exercises 81–84, verify the identity.

81. $\cos(n\pi + \theta) = (-1)^n \cos \theta$, n is an integer

82. $\sin(n\pi + \theta) = (-1)^n \sin \theta$, n is an integer

83. $a \sin B\theta + b \cos B\theta = \sqrt{a^2 + b^2} \sin(B\theta + C)$,
where $C = \arctan(b/a)$ and $a > 0$

84. $a \sin B\theta + b \cos B\theta = \sqrt{a^2 + b^2} \cos(B\theta - C)$,
where $C = \arctan(a/b)$ and $b > 0$

In Exercises 85–88, use the formulas given in Exercises 83 and 84 to write the trigonometric expression in the following forms.

(a) $\sqrt{a^2 + b^2} \sin(B\theta + C)$

(b) $\sqrt{a^2 + b^2} \cos(B\theta - C)$

85. $\sin \theta + \cos \theta$

86. $3 \sin 2\theta + 4 \cos 2\theta$

87. $12 \sin 3\theta + 5 \cos 3\theta$

88. $\sin 2\theta - \cos 2\theta$

In Exercises 89 and 90, use the formulas given in Exercises 83 and 84 to write the trigonometric expression in the form $a \sin B\theta + b \cos B\theta$.

89. $2 \sin\left(\theta + \dfrac{\pi}{2}\right)$

90. $5 \cos\left(\theta + \dfrac{3\pi}{4}\right)$

In Exercises 91 and 92, use the figure, which shows two lines whose equations are

$$y_1 = m_1 x + b_1 \qquad \text{and} \qquad y_2 = m_2 x + b_2.$$

Assume that both lines have positive slopes. Derive a formula for the angle between the two lines. Then use your formula to find the angle between the given pair of lines.

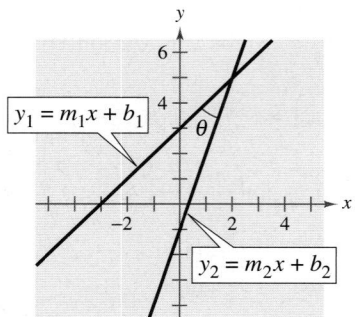

91. $y = x$ and $y = \sqrt{3}x$ **92.** $y = x$ and $y = \dfrac{1}{\sqrt{3}}x$

93. ***Conjecture***   Consider the function

$$f(\theta) = \sin^2\left(\theta + \dfrac{\pi}{4}\right) + \sin^2\left(\theta - \dfrac{\pi}{4}\right).$$

Use a graphing utility to graph the function and use the graph to create an identity. Prove your conjecture.

94. ***Conjecture***   Three squares of side s are placed side by side (see figure). Make a conjecture about the relationship between the sum $u + v$ and w. Prove your conjecture by using the identity for the tangent of the sum of two angles.

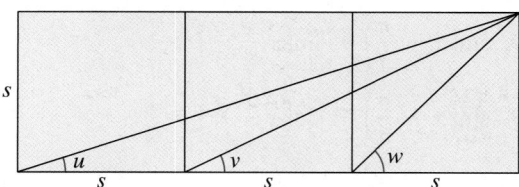

95. ***Proof***   Write a proof for the formula for $\sin(u + v)$.

96. ***Proof***   Write a proof for the formula for $\sin(u - v)$.

Review

In Exercises 97–100, find the inverse function of f. **Verify that** $f(f^{-1}(x)) = x$ **and** $f^{-1}(f(x)) = x$.

97. $f(x) = 5(x - 3)$ **98.** $f(x) = \dfrac{7 - x}{8}$

99. $f(x) = x^2 - 8$ **100.** $f(x) = \sqrt{x - 16}$

Multiple-Angle and Product-to-Sum Formulas

▶ What you should learn

- How to use multiple-angle formulas to rewrite and evaluate trigonometric functions
- How to use power-reducing formulas to rewrite and evaluate trigonometric functions
- How to use half-angle formulas to rewrite and evaluate trigonometric functions
- How to use product-to-sum and sum-to-product formulas to rewrite and evaluate trigonometric functions

▶ Why you should learn it

You can use a variety of trigonometric formulas to rewrite trigonometric functions in more convenient forms. For instance, in Exercise 121 on page 260, you can use a half-angle formula to determine the apex angle of a sound wave cone from the speed of an airplane.

NASA-Liaison Agency

Consider having your students prove the double-angle formulas. Double-angle formulas are often used in calculus.

Multiple-Angle Formulas

In this section you will study four other categories of trigonometric identities.

1. The first category involves *functions of multiple angles* such as $\sin ku$ and $\cos ku$.
2. The second category involves *squares of trigonometric functions* such as $\sin^2 u$.
3. The third category involves *functions of half-angles* such as $\sin(u/2)$.
4. The fourth category involves *products of trigonometric functions* such as $\sin u \cos v$.

You should learn the **double-angle formulas** because they are used most often. For proofs of the formulas, see Proofs in Mathematics on page 268.

Double-Angle Formulas

$$\sin 2u = 2 \sin u \cos u \qquad\qquad \cos 2u = \cos^2 u - \sin^2 u$$

$$\tan 2u = \frac{2 \tan u}{1 - \tan^2 u} \qquad\qquad\qquad = 2 \cos^2 u - 1$$

$$= 1 - 2 \sin^2 u$$

Example 1 ▶ Solving a Multiple-Angle Equation

Find all solutions of $2 \cos x + \sin 2x = 0$.

Solution

Begin by rewriting the equation so that it involves functions of x (rather than $2x$). Then factor and solve as usual.

$2 \cos x + \sin 2x = 0$	Write original equation.
$2 \cos x + 2 \sin x \cos x = 0$	Double-angle formula
$2 \cos x(1 + \sin x) = 0$	Factor.
$2 \cos x = 0 \quad\text{and}\quad 1 + \sin x = 0$	Set factors equal to zero.
$x = \dfrac{\pi}{2}, \dfrac{3\pi}{2} \qquad\qquad x = \dfrac{3\pi}{2}$	Solutions in $[0, 2\pi)$

So, the general solution is

$$x = \frac{\pi}{2} + 2n\pi \qquad\text{and}\qquad x = \frac{3\pi}{2} + 2n\pi$$

where n is an integer. Try verifying these solutions graphically.

Example 2 ▶   **Using Double-Angle Formulas to Analyze Graphs**

Use a double-angle formula to rewrite the equation

$$y = 4 \cos^2 x - 2.$$

Then sketch the graph of the equation over the interval $[0, 2\pi]$.

Solution

Using a double-angle formula, you can rewrite the original function as

$$y = 4 \cos^2 x - 2$$
$$= 2(2 \cos^2 x - 1)$$
$$= 2 \cos 2x.$$

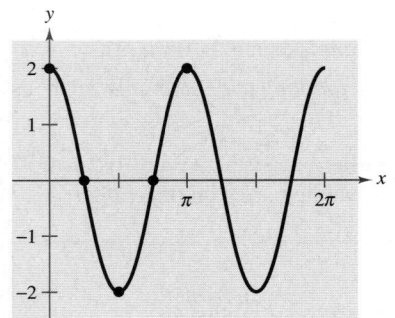

FIGURE **2.7**

Using the techniques discussed in Section 1.5, you can recognize that the graph of this function has an amplitude of 2 and a period of π. The key points in the interval $[0, \pi]$ are as follows.

Maximum	Intercept	Minimum	Intercept	Maximum
$(0, 2)$	$\left(\dfrac{\pi}{4}, 0\right)$	$\left(\dfrac{\pi}{2}, -2\right)$	$\left(\dfrac{3\pi}{4}, 0\right)$	$(\pi, 2)$

Two cycles of the graph are shown in Figure 2.7.

Example 3 ▶   **Evaluating Functions Involving Double Angles**

Use the following to find $\sin 2\theta$, $\cos 2\theta$, and $\tan 2\theta$.

$$\cos \theta = \frac{5}{13}, \qquad \frac{3\pi}{2} < \theta < 2\pi$$

Solution

From Figure 2.8, you can see that $\sin \theta = y/r = -12/13$. Consequently, using each of the double-angle formulas, you can write

$$\sin 2\theta = 2 \sin \theta \cos \theta = 2\left(-\frac{12}{13}\right)\left(\frac{5}{13}\right) = -\frac{120}{169}$$

$$\cos 2\theta = 2 \cos^2 \theta - 1 = 2\left(\frac{25}{169}\right) - 1 = -\frac{119}{169}$$

$$\tan 2\theta = \frac{\sin 2\theta}{\cos 2\theta} = \frac{120}{119}.$$

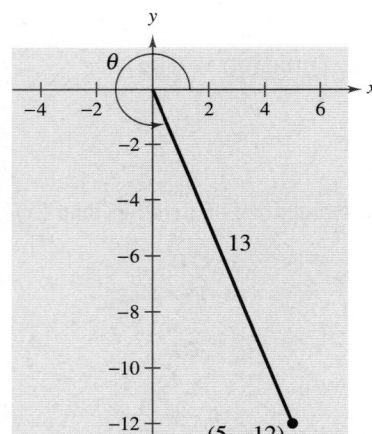

FIGURE **2.8**

The double-angle formulas are not restricted to angles 2θ and θ. Other *double* combinations, such as 4θ and 2θ or 6θ and 3θ, are also valid. Here are two examples.

$$\sin 4\theta = 2 \sin 2\theta \cos 2\theta \qquad \text{and} \qquad \cos 6\theta = \cos^2 3\theta - \sin^2 3\theta$$

By using double-angle formulas together with the sum formulas given in the preceding section, you can form other multiple-angle formulas.

Example 4 ▶ Deriving a Triple-Angle Formula

$$\sin 3x = \sin(2x + x)$$
$$= \sin 2x \cos x + \cos 2x \sin x$$
$$= 2 \sin x \cos x \cos x + (1 - 2 \sin^2 x)\sin x$$
$$= 2 \sin x \cos^2 x + \sin x - 2 \sin^3 x$$
$$= 2 \sin x(1 - \sin^2 x) + \sin x - 2 \sin^3 x$$
$$= 2 \sin x - 2 \sin^3 x + \sin x - 2 \sin^3 x$$
$$= 3 \sin x - 4 \sin^3 x$$

Power-Reducing Formulas

The double-angle formulas can be used to obtain the following **power-reducing formulas.** Example 5 shows a typical power reduction that is used in calculus.

Power-reducing formulas are often used in calculus.

Power-Reducing Formulas

$$\sin^2 u = \frac{1 - \cos 2u}{2} \qquad \cos^2 u = \frac{1 + \cos 2u}{2} \qquad \tan^2 u = \frac{1 - \cos 2u}{1 + \cos 2u}$$

For a proof of the power-reducing formulas, see Proofs in Mathematics on page 268.

Example 5 ▶ Reducing a Power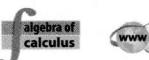

Rewrite $\sin^4 x$ as a sum of first powers of the cosines of multiple angles.

Solution

Note the repeated use of power-reducing formulas.

$$\sin^4 x = (\sin^2 x)^2 \qquad\qquad \text{Rewrite.}$$

$$= \left(\frac{1 - \cos 2x}{2}\right)^2 \qquad\qquad \text{Power-reducing formula}$$

$$= \frac{1}{4}(1 - 2\cos 2x + \cos^2 2x) \qquad\qquad \text{Expand.}$$

$$= \frac{1}{4}\left(1 - 2\cos 2x + \frac{1 + \cos 4x}{2}\right) \qquad\qquad \text{Power-reducing formula}$$

$$= \frac{1}{4} - \frac{1}{2}\cos 2x + \frac{1}{8} + \frac{1}{8}\cos 4x \qquad\qquad \text{Distributive Property}$$

$$= \frac{1}{8}(3 - 4\cos 2x + \cos 4x) \qquad\qquad \text{Factor out common factor.}$$

Half-Angle Formulas

You can derive some useful alternative forms of the power-reducing formulas by replacing u with $u/2$. The results are called **half-angle formulas.**

Half-Angle Formulas

$$\sin \frac{u}{2} = \pm\sqrt{\frac{1 - \cos u}{2}}$$

$$\cos \frac{u}{2} = \pm\sqrt{\frac{1 + \cos u}{2}}$$

$$\tan \frac{u}{2} = \frac{1 - \cos u}{\sin u} = \frac{\sin u}{1 + \cos u}$$

The signs of $\sin \dfrac{u}{2}$ and $\cos \dfrac{u}{2}$ depend on the quadrant in which $\dfrac{u}{2}$ lies.

Example 6 ▶ **Using a Half-Angle Formula**

Find the exact value of $\sin 105°$.

Solution

Begin by noting that $105°$ is half of $210°$. Then, using the half-angle formula for $\sin(u/2)$ and the fact that $105°$ lies in Quadrant II, you have

$$\sin 105° = \sqrt{\frac{1 - \cos 210°}{2}}$$

$$= \sqrt{\frac{1 - (-\cos 30°)}{2}}$$

$$= \sqrt{\frac{1 + \left(\sqrt{3}/2\right)}{2}}$$

$$= \frac{\sqrt{2 + \sqrt{3}}}{2}.$$

The positive square root is chosen because $\sin \theta$ is positive in Quadrant II.

Use your calculator to verify the result obtained in Example 6. That is, evaluate $\sin 105°$ and $\left(\sqrt{2 + \sqrt{3}}\right)/2$.

$$\sin 105° \approx 0.9659258$$

$$\frac{\sqrt{2 + \sqrt{3}}}{2} \approx 0.9659258$$

You can see that both values are approximately 0.9659258.

Example 7 ▶ **Solving a Trigonometric Equation**

Find all solutions of

$$2 - \sin^2 x = 2 \cos^2 \frac{x}{2}$$

in the interval $[0, 2\pi)$.

Solution

$2 - \sin^2 x = 2 \cos^2 \dfrac{x}{2}$	Write original equation.
$2 - \sin^2 x = 2\left(\pm\sqrt{\dfrac{1 + \cos x}{2}}\right)^2$	Half-angle formula
$2 - \sin^2 x = 2\left(\dfrac{1 + \cos x}{2}\right)$	Simplify.
$2 - \sin^2 x = 1 + \cos x$	Simplify.
$2 - (1 - \cos^2 x) = 1 + \cos x$	Pythagorean identity
$\cos^2 x - \cos x = 0$	Simplify.
$\cos x(\cos x - 1) = 0$	Factor.

By setting the factors $\cos x$ and $\cos x - 1$ equal to zero, you find that the solutions in the interval $[0, 2\pi)$ are

$$x = \frac{\pi}{2}, \quad x = \frac{3\pi}{2}, \quad \text{and} \quad x = 0.$$

STUDY TIP

A common error is to write

$$2 \cos^2 \frac{x}{2}$$

as $\cos^2 x$, rather than to use the correct identity

$$2 \cos^2 \frac{x}{2} = 2\left(\frac{1 + \cos x}{2}\right).$$

Product-to-Sum Formulas

Each of the following **product-to-sum formulas** is easily verified using the sum and difference formulas discussed in the preceding section.

Product-to-Sum Formulas

$$\sin u \sin v = \frac{1}{2}[\cos(u - v) - \cos(u + v)]$$

$$\cos u \cos v = \frac{1}{2}[\cos(u - v) + \cos(u + v)]$$

$$\sin u \cos v = \frac{1}{2}[\sin(u + v) + \sin(u - v)]$$

$$\cos u \sin v = \frac{1}{2}[\sin(u + v) - \sin(u - v)]$$

> **Example 8** ▶ **Writing Products as Sums**

Rewrite the product $\cos 5x \sin 4x$ as a sum or difference.

Solution

Using the appropriate product-to-sum formula, you obtain

$$\cos 5x \sin 4x = \frac{1}{2}[\sin(5x + 4x) - \sin(5x - 4x)]$$

$$= \frac{1}{2}\sin 9x - \frac{1}{2}\sin x.$$

Occasionally, it is useful to reverse the procedure and write a sum of trigonometric functions as a product. This can be accomplished with the following **sum-to-product formulas.**

Sum-to-Product Formulas

$$\sin x + \sin y = 2 \sin\left(\frac{x + y}{2}\right) \cos\left(\frac{x - y}{2}\right)$$

$$\sin x - \sin y = 2 \cos\left(\frac{x + y}{2}\right) \sin\left(\frac{x - y}{2}\right)$$

$$\cos x + \cos y = 2 \cos\left(\frac{x + y}{2}\right) \cos\left(\frac{x - y}{2}\right)$$

$$\cos x - \cos y = -2 \sin\left(\frac{x + y}{2}\right) \sin\left(\frac{x - y}{2}\right)$$

For a proof of the sum-to-product formulas, see Proofs in Mathematics on page 269.

> **Example 9** ▶ **Using a Sum-to-Product Formula**

Find the exact value of $\cos 195° + \cos 105°$.

Solution

Using the appropriate sum-to-product formula, you obtain

$$\cos 195° + \cos 105° = 2 \cos\left(\frac{195° + 105°}{2}\right) \cos\left(\frac{195° - 105°}{2}\right)$$

$$= 2 \cos 150° \cos 45°$$

$$= 2\left(-\frac{\sqrt{3}}{2}\right)\left(\frac{\sqrt{2}}{2}\right)$$

$$= -\frac{\sqrt{6}}{2}.$$

Example 10 ▶ **Solving a Trigonometric Equation**

Find all solutions of $\sin 5x + \sin 3x = 0$.

Solution

$$\sin 5x + \sin 3x = 0 \qquad \text{Write original equation.}$$

$$2 \sin\left(\frac{5x + 3x}{2}\right) \cos\left(\frac{5x - 3x}{2}\right) = 0 \qquad \text{Sum-to-product formula}$$

$$2 \sin 4x \cos x = 0 \qquad \text{Simplify.}$$

By setting the factor $2 \sin 4x$ equal to zero, you can find that the solutions in the interval $[0, 2\pi)$ are

$$x = 0, \frac{\pi}{4}, \frac{\pi}{2}, \frac{3\pi}{4}, \pi, \frac{5\pi}{4}, \frac{3\pi}{2}, \frac{7\pi}{4}.$$

The equation $\cos x = 0$ yields no additional solutions, and you can conclude that the solutions are of the form

$$x = \frac{n\pi}{4}$$

where n is an integer. These solutions are verified graphically in Figure 2.9.

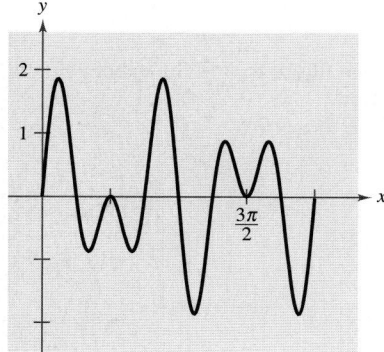

FIGURE **2.9**

Example 11 ▶ **Verifying a Trigonometric Identity**

Verify the identity

$$\frac{\sin t + \sin 3t}{\cos t + \cos 3t} = \tan 2t.$$

Solution

Using appropriate sum-to-product formulas, you have

$$\frac{\sin t + \sin 3t}{\cos t + \cos 3t} = \frac{2 \sin 2t \cos(-t)}{2 \cos 2t \cos(-t)}$$

$$= \frac{\sin 2t}{\cos 2t}$$

$$= \tan 2t.$$

Writing ABOUT MATHEMATICS

Deriving an Area Formula   Describe how you can use a double-angle formula or a half-angle formula to derive a formula for the area of an isosceles triangle. Use a labeled sketch to illustrate your derivation. Then write two examples that show how your formula can be used.

2.5 **Exercises**

In Exercises 1–8, use the figure to find the exact value of the trigonometric function.

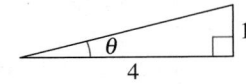

1. $\sin \theta$ **2.** $\tan \theta$

3. $\cos 2\theta$ **4.** $\sin 2\theta$

5. $\tan 2\theta$ **6.** $\sec 2\theta$

7. $\csc 2\theta$ **8.** $\cot 2\theta$

In Exercises 9–18, find the exact solutions of the equation in the interval $[0, 2\pi)$.

9. $\sin 2x - \sin x = 0$ **10.** $\sin 2x + \cos x = 0$

11. $4 \sin x \cos x = 1$ **12.** $\sin 2x \sin x = \cos x$

13. $\cos 2x - \cos x = 0$ **14.** $\cos 2x + \sin x = 0$

15. $\tan 2x - \cot x = 0$ **16.** $\tan 2x - 2 \cos x = 0$

17. $\sin 4x = -2 \sin 2x$ **18.** $(\sin 2x + \cos 2x)^2 = 1$

In Exercises 19–22, use a double-angle formula to rewrite the expression.

19. $6 \sin x \cos x$

20. $6 \cos^2 x - 3$

21. $4 - 8 \sin^2 x$

22. $(\cos x + \sin x)(\cos x - \sin x)$

In Exercises 23–28, find the exact values of sin 2u, cos 2u, and tan 2u using the double-angle formulas.

23. $\sin u = -\dfrac{4}{5}, \quad \pi < u < \dfrac{3\pi}{2}$

24. $\cos u = -\dfrac{2}{3}, \quad \dfrac{\pi}{2} < u < \pi$

25. $\tan u = \dfrac{3}{4}, \quad 0 < u < \dfrac{\pi}{2}$

26. $\cot u = -4, \quad \dfrac{3\pi}{2} < u < 2\pi$

27. $\sec u = -\dfrac{5}{2}, \quad \dfrac{\pi}{2} < u < \pi$

28. $\csc u = 3, \quad \dfrac{\pi}{2} < u < \pi$

In Exercises 29–34, use the power-reducing formulas to rewrite the expression in terms of the first power of the cosine.

29. $\cos^4 x$ **30.** $\sin^8 x$

31. $\sin^2 x \cos^2 x$ **32.** $\sin^4 x \cos^4 x$

33. $\sin^2 x \cos^4 x$ **34.** $\sin^4 x \cos^2 x$

In Exercises 35–40, use the figure to find the exact value of the trigonometric function.

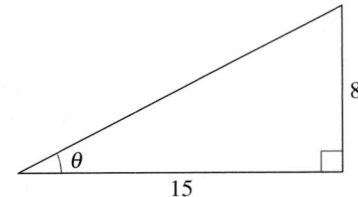

35. $\cos \dfrac{\theta}{2}$ **36.** $\sin \dfrac{\theta}{2}$

37. $\tan \dfrac{\theta}{2}$ **38.** $\sec \dfrac{\theta}{2}$

39. $\csc \dfrac{\theta}{2}$ **40.** $\cot \dfrac{\theta}{2}$

In Exercises 41–48, use the half-angle formulas to determine the exact values of the sine, cosine, and tangent of the angle.

41. $75°$ **42.** $165°$

43. $112° \, 30'$ **44.** $67° \, 30'$

45. $\dfrac{\pi}{8}$ **46.** $\dfrac{\pi}{12}$

47. $\dfrac{3\pi}{8}$ **48.** $\dfrac{7\pi}{12}$

In Exercises 49–54, find the exact values of sin(u/2), cos(u/2), and tan(u/2) using the half-angle formulas.

49. $\sin u = \dfrac{5}{13}, \quad \dfrac{\pi}{2} < u < \pi$

50. $\cos u = \dfrac{3}{5}, \quad 0 < u < \dfrac{\pi}{2}$

51. $\tan u = -\dfrac{5}{8}, \quad \dfrac{3\pi}{2} < u < 2\pi$

52. $\cot u = 3, \quad \pi < u < \dfrac{3\pi}{2}$

53. $\csc u = -\dfrac{5}{3}, \quad \pi < u < \dfrac{3\pi}{2}$

54. $\sec u = -\dfrac{7}{2}, \quad \dfrac{\pi}{2} < u < \pi$

In Exercises 55–58, use the half-angle formulas to simplify the expression.

55. $\sqrt{\dfrac{1 - \cos 6x}{2}}$ **56.** $\sqrt{\dfrac{1 + \cos 4x}{2}}$

57. $-\sqrt{\dfrac{1 - \cos 8x}{1 + \cos 8x}}$ **58.** $-\sqrt{\dfrac{1 - \cos(x - 1)}{2}}$

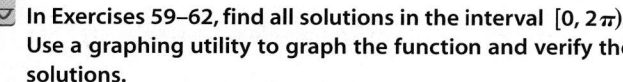 In Exercises 59–62, find all solutions in the interval $[0, 2\pi)$. Use a graphing utility to graph the function and verify the solutions.

59. $\sin \dfrac{x}{2} + \cos x = 0$

60. $\sin \dfrac{x}{2} + \cos x - 1 = 0$

61. $\cos \dfrac{x}{2} - \sin x = 0$

62. $\tan \dfrac{x}{2} - \sin x = 0$

In Exercises 63–74, use the product-to-sum formulas to write the product as a sum or difference.

63. $6 \sin \dfrac{\pi}{4} \cos \dfrac{\pi}{4}$ **64.** $4 \cos \dfrac{\pi}{3} \sin \dfrac{5\pi}{6}$

65. $\cos 4\theta \sin 6\theta$ **66.** $3 \sin 2\alpha \sin 3\alpha$

67. $5 \cos(-5\beta) \cos 3\beta$ **68.** $\cos 2\theta \cos 4\theta$

69. $\sin(x + y) \sin(x - y)$

70. $\sin(x + y) \cos(x - y)$

71. $\cos(\theta - \pi) \sin(\theta + \pi)$

72. $\sin(\theta + \pi) \sin(\theta - \pi)$

73. $10 \cos 75° \cos 15°$ **74.** $6 \sin 45° \cos 15°$

In Exercises 75–86, use the sum-to-product formulas to write the sum or difference as a product.

75. $\sin 60° + \sin 30°$ **76.** $\cos 120° + \cos 30°$

77. $\cos \dfrac{3\pi}{4} - \cos \dfrac{\pi}{4}$ **78.** $\sin \dfrac{5\pi}{4} - \sin \dfrac{3\pi}{4}$

79. $\sin 5\theta - \sin 3\theta$ **80.** $\sin 3\theta + \sin \theta$

81. $\cos 6x + \cos 2x$ **82.** $\sin x + \sin 5x$

83. $\sin(\alpha + \beta) - \sin(\alpha - \beta)$

84. $\cos(\phi + 2\pi) + \cos \phi$

85. $\cos\left(\theta + \dfrac{\pi}{2}\right) - \cos\left(\theta - \dfrac{\pi}{2}\right)$

86. $\sin\left(x + \dfrac{\pi}{2}\right) + \sin\left(x - \dfrac{\pi}{2}\right)$

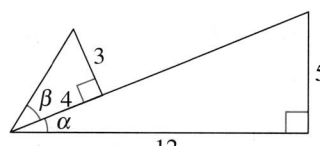 In Exercises 87–90, find all solutions in the interval $[0, 2\pi)$. Use a graphing utility to graph the function and verify the solutions.

87. $\sin 6x + \sin 2x = 0$

88. $\cos 2x - \cos 6x = 0$

89. $\dfrac{\cos 2x}{\sin 3x - \sin x} - 1 = 0$

90. $\sin^2 3x - \sin^2 x = 0$

In Exercises 91–94, use the figure and trigonometric identities to find the exact value of the trigonometric function in two ways.

91. $\sin^2 \alpha$ **92.** $\cos^2 \alpha$

93. $\sin \alpha \cos \beta$ **94.** $\cos \alpha \sin \beta$

In Exercises 95–110, verify the identity.

95. $\csc 2\theta = \dfrac{\csc \theta}{2 \cos \theta}$

96. $\sec 2\theta = \dfrac{\sec^2 \theta}{2 - \sec^2 \theta}$

97. $\cos^2 2\alpha - \sin^2 2\alpha = \cos 4\alpha$

98. $\cos^4 x - \sin^4 x = \cos 2x$

99. $(\sin x + \cos x)^2 = 1 + \sin 2x$

100. $\sin \dfrac{\alpha}{3} \cos \dfrac{\alpha}{3} = \dfrac{1}{2} \sin \dfrac{2\alpha}{3}$

101. $1 + \cos 10y = 2 \cos^2 5y$

102. $\dfrac{\cos 3\beta}{\cos \beta} = 1 - 4\sin^2 \beta$

103. $\sec \dfrac{u}{2} = \pm \sqrt{\dfrac{2\tan u}{\tan u + \sin u}}$

104. $\tan \dfrac{u}{2} = \csc u - \cot u$

105. $\dfrac{\sin x \pm \sin y}{\cos x + \cos y} = \tan \dfrac{x \pm y}{2}$

106. $\dfrac{\sin x + \sin y}{\cos x - \cos y} = -\cot \dfrac{x - y}{2}$

107. $\dfrac{\cos 4x + \cos 2x}{\sin 4x + \sin 2x} = \cot 3x$

108. $\dfrac{\cos t + \cos 3t}{\sin 3t - \sin t} = \cot t$

109. $\sin\left(\dfrac{\pi}{6} + x\right) + \sin\left(\dfrac{\pi}{6} - x\right) = \cos x$

110. $\cos\left(\dfrac{\pi}{3} + x\right) + \cos\left(\dfrac{\pi}{3} - x\right) = \cos x$

 In Exercises 111–114, use a graphing utility to verify the identity. Confirm that it is an identity algebraically.

111. $\cos 3\beta = \cos^3 \beta - 3\sin^2 \beta \cos \beta$

112. $\sin 4\beta = 4\sin \beta \cos \beta (1 - 2\sin^2 \beta)$

113. $(\cos 4x - \cos 2x)/(2\sin 3x) = -\sin x$

114. $(\cos 3x - \cos x)/(\sin 3x - \sin x) = -\tan 2x$

In Exercises 115 and 116, graph the function by hand in the interval $[0, 2\pi)$ by using the power-reducing formulas.

115. $f(x) = \sin^2 x$

116. $f(x) = \cos^2 x$

In Exercises 117 and 118, write the trigonometric expression as an algebraic expression.

117. $\sin(2 \arcsin x)$

118. $\cos(2 \arccos x)$

119. *Geometry* The length of each of the two equal sides of an isosceles triangle is 10 meters (see figure). The angle between the two sides is θ.

   (a) Write the area of the triangle as a function of $\theta/2$.

   (b) Write the area of the triangle as a function of θ. Determine the value of θ such that the area is a maximum.

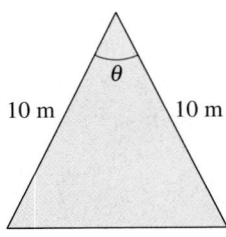

FIGURE FOR **119**

120. *Projectile Motion* The range of a projectile fired at an angle θ with the horizontal and with an initial velocity of v_0 feet per second is

$$r = \dfrac{1}{32} v_0^2 \sin 2\theta$$

where r is measured in feet. Determine the expression for the range in terms of θ.

▶ **Model It**

121. *Mach Number* The mach number M of an airplane is the ratio of its speed to the speed of sound. When an airplane travels faster than the speed of sound, the sound waves form a cone behind the airplane. The mach number is related to the apex angle θ of the cone by

$$\sin \dfrac{\theta}{2} = \dfrac{1}{M}.$$

   (a) Find the angle θ that corresponds to a mach number of 1.

   (b) Find the angle θ that corresponds to a mach number of 4.5.

   (c) The speed of sound is about 760 miles per hour. Determine the speed of an object with the mach numbers from parts (a) and (b).

   (d) Rewrite the equation in terms of θ.

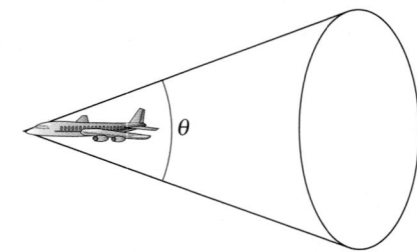

122. *Railroad Track*   When two railroad tracks merge, the overlapping portions of the tracks are in the shapes of circular arcs (see figure). The radius of each arc r (in feet) and the angle θ are related by

$$\frac{x}{2} = 2r \sin^2 \frac{\theta}{2}.$$

Write a formula for x in terms of $\cos \theta$.

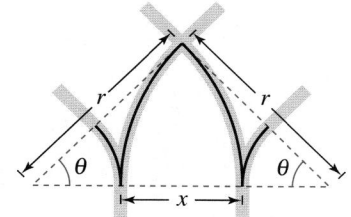

Synthesis

True or False?   **In Exercises 123 and 124, determine whether the statement is true or false. Justify your answer.**

123. Because the sine function is an odd function, for a negative number u, $\sin 2u = -2 \sin u \cos u$.

124. $\sin \dfrac{u}{2} = -\sqrt{\dfrac{1 - \cos u}{2}}$ when u is in the second quadrant.

 In Exercises 125 and 126, (a) use a graphing utility to graph the function and approximate the maximum and minimum points on the graph in the interval $[0, 2\pi)$ and (b) solve the trigonometric equation and verify that its solutions are the x-coordinates of the maximum and minimum points of f. (Calculus is required to find the trigonometric equation.)

Function	*Trigonometric Equation*
125. $f(x) = 4 \sin \dfrac{x}{2} + \cos x$	$2 \cos \dfrac{x}{2} - \sin x = 0$
126. $f(x) = \cos 2x - 2 \sin x$	$-2 \cos x(2 \sin x + 1) = 0$

127. *Exploration*   Consider the function

$$f(x) = \sin^4 x + \cos^4 x.$$

(a) Use the power-reducing formulas to write the function in terms of cosine to the first power.

 (b) Determine another way of rewriting the function. Use a graphing utility to rule out incorrectly rewritten functions.

 (c) Add a trigonometric term to the function so that it becomes a perfect square trinomial. Rewrite the function as a perfect square trinomial minus the term that you added. Use a graphing utility to rule out incorrectly rewritten functions.

 (d) Rewrite the result of part (c) in terms of the sine of a double angle. Use a graphing utility to rule out incorrectly rewritten functions.

(e) When you rewrite a trigonometric expression, the result may not be the same as a friend's. Does this mean that one of you is wrong? Explain.

128. *Conjecture*   Consider the function

$$f(x) = 2 \sin x \left(2 \cos^2 \frac{x}{2} - 1 \right).$$

 (a) Use a graphing utility to graph the function.

(b) Make a conjecture about the function that is an identity with f.

(c) Verify your conjecture analytically.

Review

129. *Profit*   The total profit for a car manufacturer in October was 16% higher than it was in September. The total profit for the 2 months was \$507,600. Find the profit for each month.

130. *Mixture Problem*   A 55-gallon barrel contains a mixture with a concentration of 30%. How much of this mixture must be withdrawn and replaced by 100% concentrate to bring the mixture up to 50% concentration?

131. *Distance*   A baseball diamond has the shape of a square in which the distance between each of the consecutive bases is 90 feet. Approximate the distance from home plate to second base.

Chapter Summary

▶ *What* did you learn?

Review Exercises

In Exercises 1–6, name the trigonometric function that is equivalent to the expression.

1. $\dfrac{1}{\cos x}$

2. $\dfrac{1}{\sin x}$

3. $\dfrac{1}{\sec x}$

4. $\dfrac{1}{\tan x}$

5. $\dfrac{\cos x}{\sin x}$

6. $\sqrt{1 + \tan^2 x}$

In Exercises 7–10, use the given values and trigonometric identities to evaluate (if possible) the other trigonometric functions of the angle.

7. $\sin x = \frac{3}{5}, \quad \cos x = \frac{4}{5}$

8. $\tan \theta = \dfrac{2}{3}, \quad \sec \theta = \dfrac{\sqrt{13}}{3}$

9. $\sin\left(\dfrac{\pi}{2} - x\right) = \dfrac{\sqrt{2}}{2}, \quad \sin x = -\dfrac{\sqrt{2}}{2}$

10. $\csc\left(\dfrac{\pi}{2} - \theta\right) = 9, \quad \sin \theta = \dfrac{4\sqrt{5}}{9}$

In Exercises 11–22, use the fundamental trigonometric identities to simplify the trigonometric expression.

11. $\dfrac{1}{\cot^2 x + 1}$

12. $\dfrac{\tan \theta}{1 - \cos^2 \theta}$

13. $\tan^2 x(\csc^2 x - 1)$

14. $\cot^2 x(\sin^2 x)$

15. $\dfrac{\sin\left(\dfrac{\pi}{2} - \theta\right)}{\sin \theta}$

16. $\dfrac{\cot\left(\dfrac{\pi}{2} - u\right)}{\cos u}$

17. $\cos^2 x + \cos^2 x \cot^2 x$

18. $\tan^2 \theta \csc^2 \theta - \tan^2 \theta$

19. $(\tan x + 1)^2 \cos x$

20. $(\sec x - \tan x)^2$

21. $\dfrac{1}{\csc \theta + 1} - \dfrac{1}{\csc \theta - 1}$

22. $\dfrac{\cos^2 x}{1 - \sin x}$

23. *Rate of Change* The rate of change of the function $f(x) = \csc x - \cot x$ is the expression $\csc^2 x - \csc x \cot x$. Show that this expression can also be written as

$$\dfrac{(1 - \cos x)}{\sin^2 x}.$$

24. *Rate of Change* The rate of change of the function $f(x) = 2\sqrt{\sin x}$ is the expression $\sin^{-1/2} x \cos x$. Show that this expression can also be written as $\cot x\sqrt{\sin x}$.

In Exercises 25–32, verify the identity.

25. $\cos x(\tan^2 x + 1) = \sec x$

26. $\sec^2 x \cot x - \cot x = \tan x$

27. $\cos\left(x + \dfrac{\pi}{2}\right) = -\sin x$

28. $\cot\left(\dfrac{\pi}{2} - x\right) = \tan x$

29. $\dfrac{1}{\tan \theta \csc \theta} = \cos \theta$

30. $\dfrac{1}{\tan x \csc x \sin x} = \cot x$

31. $\sin^5 x \cos^2 x = (\cos^2 x - 2\cos^4 x + \cos^6 x) \sin x$

32. $\cos^3 x \sin^2 x = (\sin^2 x - \sin^4 x) \cos x$

In Exercises 33–38, solve the equation.

33. $\sin x = \sqrt{3} - \sin x$

34. $4 \cos \theta = 1 + 2 \cos \theta$

35. $3\sqrt{3} \tan u = 3$

36. $\frac{1}{2} \sec x - 1 = 0$

37. $3 \csc^2 x = 4$

38. $4 \tan^2 u - 1 = \tan^2 u$

In Exercises 39–46, find all solutions of the equation in the interval $[0, 2\pi)$.

39. $2 \cos^2 x - \cos x = 1$

40. $2 \sin^2 x - 3 \sin x = -1$

41. $\cos^2 x + \sin x = 1$

42. $\sin^2 x + 2 \cos x = 2$

43. $2 \sin 2x - \sqrt{2} = 0$

44. $\sqrt{3} \tan 3x = 0$

45. $\cos 4x(\cos x - 1) = 0$

46. $3 \csc^2 5x = -4$

In Exercises 47–50, use inverse functions where needed to find all solutions of the equation in the interval $[0, 2\pi)$.

47. $\sin^2 x - 2 \sin x = 0$

48. $2 \cos^2 x + 3 \cos x = 0$

49. $\tan^2 \theta + \tan \theta - 12 = 0$

50. $\sec^2 x + 6 \tan x + 4 = 0$

2.4 In Exercises 51–54, find the exact values of the sine, cosine, and tangent of the angle by using a sum or difference formula.

51. $285° = 315° - 30°$

52. $345° = 300° + 45°$

53. $\dfrac{25\pi}{12} = \dfrac{11\pi}{6} + \dfrac{\pi}{4}$

54. $\dfrac{19\pi}{12} = \dfrac{11\pi}{6} - \dfrac{\pi}{4}$

In Exercises 55–58, write the expression as the sine, cosine, or tangent of an angle.

55. $\sin 60° \cos 45° - \cos 60° \sin 45°$

56. $\cos 45° \cos 120° - \sin 45° \sin 120°$

57. $\dfrac{\tan 25° + \tan 10°}{1 - \tan 25° \tan 10°}$

58. $\dfrac{\tan 68° - \tan 115°}{1 + \tan 68° \tan 115°}$

In Exercises 59–64, find the exact value of the trigonometric function given that $\sin u = \frac{3}{4}$, $\cos v = -\frac{5}{13}$, and u and v are in Quadrant II.

59. $\sin(u + v)$

60. $\tan(u + v)$

61. $\cos(u - v)$

62. $\sin(u - v)$

63. $\cos(u + v)$

64. $\tan(u - v)$

In Exercises 65–68, find all solutions of the equation in the interval $[0, 2\pi)$.

65. $\sin\left(x + \dfrac{\pi}{4}\right) - \sin\left(x - \dfrac{\pi}{4}\right) = 1$

66. $\cos\left(x + \dfrac{\pi}{6}\right) - \cos\left(x - \dfrac{\pi}{6}\right) = 1$

67. $\sin\left(x + \dfrac{\pi}{2}\right) - \sin\left(x - \dfrac{\pi}{2}\right) = \sqrt{3}$

68. $\cos\left(x + \dfrac{3\pi}{4}\right) - \cos\left(x - \dfrac{3\pi}{4}\right) = 0$

2.5 In Exercises 69 and 70, use double-angle formulas to verify the identity algebraically and use a graphing utility to confirm it graphically.

69. $\sin 4x = 8 \cos^3 x \sin x - 4 \cos x \sin x$

70. $\tan^2 x = \dfrac{1 - \cos 2x}{1 + \cos 2x}$

In Exercises 71 and 72, find the exact values of $\sin 2u$, $\cos 2u$, and $\tan 2u$ using the double-angle formulas.

71. $\sin u = -\dfrac{4}{5}, \quad \pi < u < \dfrac{3\pi}{2}$

72. $\cos u = -\dfrac{2}{\sqrt{5}}, \quad \dfrac{\pi}{2} < u < \pi$

73. *Projectile Motion* A baseball leaves the hand of the person at first base at an angle of θ with the horizontal and at an initial velocity of $v_0 = 80$ feet per second. The ball is caught by the person at second base 100 feet away. Find θ if the range r of a projectile is

$$r = \dfrac{1}{32} v_0^2 \sin 2\theta.$$

74. *Projectile Motion* Use the equation in Exercise 73 to find θ when a golf ball is hit at an initial velocity of 50 feet per second and lands 77 feet away.

In Exercises 75–78, use the power-reducing formulas to rewrite the expression in terms of the first power of the cosine.

75. $\tan^2 2x$

76. $\cos^2 3x$

77. $\sin^2 x \tan^2 x$

78. $\cos^2 x \tan^2 x$

In Exercises 79–82, use the half-angle formulas to determine the exact values of the sine, cosine, and tangent of the angle.

79. $-75°$

80. $15°$

81. $\dfrac{19\pi}{12}$

82. $-\dfrac{17\pi}{12}$

In Exercises 83 and 84, use the half-angle formulas to simplify the expression.

83. $-\sqrt{\dfrac{1 + \cos 10x}{2}}$

84. $\dfrac{\sin 6x}{1 + \cos 6x}$

85. Find the exact values of $\sin(u/2)$, $\cos(u/2)$, and $\tan(u/2)$ for $\sin u = \frac{3}{5}$, $0 < u < \frac{\pi}{2}$.

86. *Geometry* A trough for feeding cattle is 4 meters long and its cross sections are isosceles triangles with the two equal sides being $\frac{1}{2}$ meter (see figure on page 589). The angle between the two sides is θ.

(a) Write the trough's volume as a function of $\dfrac{\theta}{2}$.

(b) Write the volume of the trough as a function of θ and determine the value of θ such that the volume is maximum.

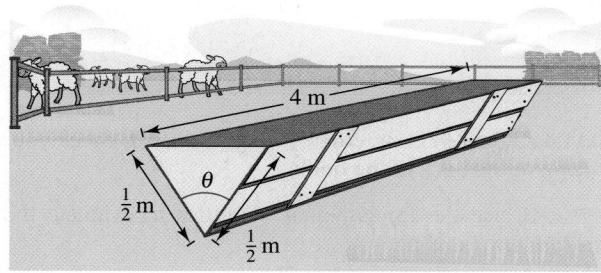

FIGURE FOR 86

In Exercises 87–90, use the product-to-sum formulas to write the product as a sum or difference.

87. $\cos \dfrac{\pi}{6} \sin \dfrac{\pi}{6}$

88. $6 \sin 15° \sin 45°$

89. $\cos 5\theta \cos 3\theta$

90. $4 \sin 3\alpha \cos 2\alpha$

In Exercises 91–94, use the sum-to-product formulas to write the sum or difference as a product.

91. $\sin 60° + \sin 90°$

92. $\cos 3\theta + \cos 2\theta$

93. $\cos\left(x + \dfrac{\pi}{6}\right) - \cos\left(x - \dfrac{\pi}{6}\right)$

94. $\sin\left(x + \dfrac{\pi}{4}\right) - \sin\left(x - \dfrac{\pi}{4}\right)$

95. ***Harmonic Motion*** A weight is attached to a spring suspended vertically from a ceiling. When a driving force is applied to the system, the weight moves vertically from its equilibrium position, and this motion is described by the model

$$y = 1.5 \sin 8t - 0.5 \cos 8t$$

where y is the distance from equilibrium measured in feet and t is the time in seconds.

(a) Write the model in the form

$$y = \sqrt{a^2 + b^2} \sin(Bt + C).$$

(b) Find the amplitude of the oscillations of the weight.

(c) Find the frequency of the oscillations of the weight.

Synthesis

True or False? **In Exercises 96–99, determine whether the statement is true or false. Justify your answer.**

96. If $\dfrac{\pi}{2} < \theta < \pi$, then $\cos \dfrac{\theta}{2} < 0$.

97. $\sin(x + y) = \sin x + \sin y$

98. $4 \sin(-x) \cos(-x) = -2 \sin 2x$

99. $4 \sin 45° \cos 15° = 1 + \sqrt{3}$

100. List the reciprocal identities, quotient identities, and Pythagorean identities from memory.

101. ***Think About It*** If a trigonometric equation has an infinite number of solutions, is it true that the equation is an identity? Explain.

102. ***Think About It*** Explain why you know from observation that the equation $a \sin x - b = 0$ has no solution if $|a| < |b|$.

In Exercises 103 and 104, use the graphs of y_1 and y_2 to determine how to change one function to form the identity $y_1 = y_2$.

103. $y_1 = \sec^2\left(\dfrac{\pi}{2} - x\right)$

$y_2 = \cot^2 x$

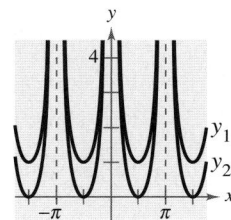

104. $y_1 = \dfrac{\cos 3x}{\cos x}$

$y_2 = (2 \sin x)^2$

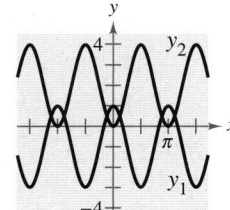

In Exercises 105 and 106, use the *zero* or *root* feature of a graphing utility to approximate the zeros of the function.

105. $y = \sqrt{x + 3} + 4 \cos x$

106. $y = 2 - \dfrac{1}{2}x^2 + 3 \sin \dfrac{\pi x}{2}$

Chapter Test

Take this test as you would take a test in class. When you are finished, check your work against the answers given in the back of the book.

1. If $\tan \theta = \frac{3}{2}$ and $\cos \theta < 0$, use the fundamental identities to evaluate the other five trigonometric functions of θ.

2. Use the fundamental identities to simplify $\csc^2 \beta (1 - \cos^2 \beta)$.

3. Factor and simplify $\dfrac{\sec^4 x - \tan^4 x}{\sec^2 x + \tan^2 x}$.   4. Add and simplify $\dfrac{\cos \theta}{\sin \theta} + \dfrac{\sin \theta}{\cos \theta}$.

5. Determine the values of θ, $0 \le \theta < 2\pi$, for which $\tan \theta = -\sqrt{\sec^2 \theta - 1}$ is true.

 6. Use a graphing utility to graph the functions $y_1 = \cos x + \sin x \tan x$ and $y_2 = \sec x$. Make a conjecture about y_1 and y_2. Verify the result analytically.

In Exercises 7–12, verify the identity.

7. $\sin \theta \sec \theta = \tan \theta$ 8. $\sec^2 x \tan^2 x + \sec^2 x = \sec^4 x$

9. $\dfrac{\csc \alpha + \sec \alpha}{\sin \alpha + \cos \alpha} = \cot \alpha + \tan \alpha$ 10. $\cos\left(x + \dfrac{\pi}{2}\right) = -\sin x$

11. $\sin(n\pi + \theta) = (-1)^n \sin \theta$, n is an integer.

12. $(\sin x + \cos x)^2 = 1 + \sin 2x$

13. Rewrite $\sin^4 x \tan^2 x$ in terms of the first power of the cosine.

14. Use a half-angle formula to simplify the expression $\dfrac{\sin 4\theta}{1 + \cos 4\theta}$.

15. Write $4 \cos 2\theta \sin 4\theta$ as a sum or difference.

16. Write $\sin 3\theta - \sin 4\theta$ as a product.

In Exercises 17–20, find all solutions of the equation in the interval $[0, 2\pi)$.

17. $\tan^2 x + \tan x = 0$ 18. $\sin 2\alpha - \cos \alpha = 0$

19. $4 \cos^2 x - 3 = 0$ 20. $\csc^2 x - \csc x - 2 = 0$

 21. Use a graphing utility to approximate the solutions of the equation $3 \cos x - x = 0$ accurate to three decimal places.

22. Find the exact value of $\cos 105°$ using the fact that $105° = 135° - 30°$.

23. Use the figure to find the exact values of $\sin 2u$ and $\tan 2u$.

 24. Cheyenne, Wyoming has a latitude of 41°N. At this latitude, the position of the sun at sunrise can be modeled by

$$D = 31 \sin\left(\dfrac{2\pi}{365}t - 1.4\right)$$

where t is the time (in days) and $t = 1$ represents January 1. In this model, D represents the number of degrees north or south of due east that the sun rises. Use a graphing utility to determine the days on which the sun is more than 20° north of due east at sunrise.

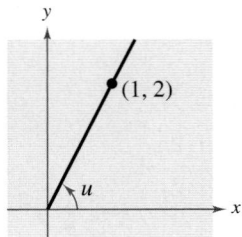

FIGURE FOR **23**

Proofs in Mathematics

Sum and Difference Formulas *(p. 244)*

$$\sin(u + v) = \sin u \cos v + \cos u \sin v$$

$$\sin(u - v) = \sin u \cos v - \cos u \sin v$$

$$\tan(u + v) = \frac{\tan u + \tan v}{1 - \tan u \tan v}$$

$$\cos(u + v) = \cos u \cos v - \sin u \sin v$$

$$\cos(u - v) = \cos u \cos v + \sin u \sin v$$

$$\tan(u - v) = \frac{\tan u - \tan v}{1 + \tan u \tan v}$$

Proof

You can use the figures at the left for the proofs of the formulas for $\cos(u \pm v)$. In the top figure, let A be the point $(1, 0)$ and then use u and v to locate the points $B = (x_1, y_1)$, $C = (x_2, y_2)$, and $D = (x_3, y_3)$ on the unit circle. So, $x_i^2 + y_i^2 = 1$ for $i = 1, 2,$ and 3. For convenience, assume that $0 < v < u < 2\pi$. In the bottom figure, note that arcs AC and BD have the same length. So, line segments AC and BD are also equal in length, which implies that

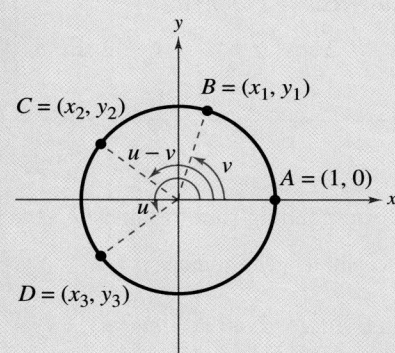

$$\sqrt{(x_2 - 1)^2 + (y_2 - 0)^2} = \sqrt{(x_3 - x_1)^2 + (y_3 - y_1)^2}$$

$$x_2^2 - 2x_2 + 1 + y_2^2 = x_3^2 - 2x_1x_3 + x_1^2 + y_3^2 - 2y_1y_3 + y_1^2$$

$$(x_2^2 + y_2^2) + 1 - 2x_2 = (x_3^2 + y_3^2) + (x_1^2 + y_1^2) - 2x_1x_3 - 2y_1y_3$$

$$1 + 1 - 2x_2 = 1 + 1 - 2x_1x_3 - 2y_1y_3$$

$$x_2 = x_3x_1 + y_3y_1.$$

Finally, by substituting the values $x_2 = \cos(u - v)$, $x_3 = \cos u$, $x_1 = \cos v$, $y_3 = \sin u$, and $y_1 = \sin v$, you obtain $\cos(u - v) = \cos u \cos v + \sin u \sin v$. The formula for $\cos(u + v)$ can be established by considering $u + v = u - (-v)$ and using the formula just derived to obtain

$$\cos(u + v) = \cos[u - (-v)] = \cos u \cos (-v) + \sin u \sin(-v)$$

$$= \cos u \cos v - \sin u \sin v.$$

You can use the sum and difference formulas for sine and cosine to prove the formulas for $\tan(u \pm v)$.

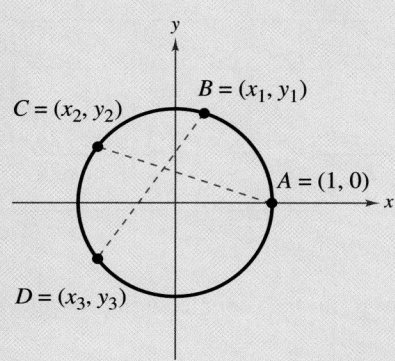

$$\tan(u \pm v) = \frac{\sin(u \pm v)}{\cos(u \pm v)} \qquad \text{Quotient identity}$$

$$= \frac{\sin u \cos v \pm \cos u \sin v}{\cos u \cos v \mp \sin u \sin v} \qquad \text{Sum and difference formulas}$$

$$= \frac{\dfrac{\sin u \cos v \pm \cos u \sin v}{\cos u \cos v}}{\dfrac{\cos u \cos v \mp \sin u \sin v}{\cos u \cos v}} \qquad \text{Divide numerator and denominator by } \cos u \cos v.$$

$$= \dfrac{\dfrac{\sin u \cos v}{\cos u \cos v} \pm \dfrac{\cos u \sin v}{\cos u \cos v}}{\dfrac{\cos u \cos v}{\cos u \cos v} \mp \dfrac{\sin u \sin v}{\cos u \cos v}}$$ Separate fractions.

$$= \dfrac{\dfrac{\sin u}{\cos u} \pm \dfrac{\sin v}{\cos v}}{1 \mp \dfrac{\sin u}{\cos u} \cdot \dfrac{\sin v}{\cos v}}$$ Product of fractions.

$$= \dfrac{\tan u \pm \tan v}{1 \mp \tan u \tan v}$$ Quotient identity

Trigonometry and Astronomy

Trigonometry was used by early astronomers to calculate measurements in the universe. Trigonometry was used to calculate the circumference of Earth and the distance from Earth to the moon. Another major accomplishment in astronomy using trigonometry was computing distances to stars.

Double-Angle Formulas   *(p. 251)*

$$\sin 2u = 2 \sin u \cos u \qquad \cos 2u = \cos^2 u - \sin^2 u$$

$$\tan 2u = \dfrac{2 \tan u}{1 - \tan^2 u} \qquad\qquad = 2\cos^2 u - 1 = 1 - 2\sin^2 u$$

Proof

To prove all three formulas, let $v = u$ in the corresponding sum formulas.

$$\sin 2u = \sin(u + u) = \sin u \cos u + \cos u \sin u = 2 \sin u \cos u$$

$$\cos 2u = \cos(u + u) = \cos u \cos u - \sin u \sin u = \cos^2 u - \sin^2 u$$

$$\tan 2u = \tan(u + u) = \dfrac{\tan u + \tan u}{1 - \tan u \tan u} = \dfrac{2 \tan u}{1 - \tan^2 u}$$

Power-Reducing Formulas   *(p. 253)*

$$\sin^2 u = \dfrac{1 - \cos 2u}{2} \qquad \cos^2 u = \dfrac{1 + \cos 2u}{2} \qquad \tan^2 u = \dfrac{1 - \cos 2u}{1 + \cos 2u}$$

Proof

To prove the first formula, solve for $\sin^2 u$ in the double-angle formula $\cos 2u = 1 - 2\sin^2 u$, as follows.

$$\cos 2u = 1 - 2\sin^2 u$$ Write double-angle formula.

$$2\sin^2 u = 1 - \cos 2u$$ Subtract cos $2u$ from and add 2 sin² u to each side.

$$\sin^2 u = \dfrac{1 - \cos 2u}{2}$$ Divide each side by 2.

In a similar way you can prove the second formula, by solving for $\cos^2 u$ in the double-angle formula

$$\cos 2u = 2\cos^2 u - 1.$$

To prove the third formula, use a quotient identity, as follows.

$$\tan^2 u = \frac{\sin^2 u}{\cos^2 u}$$

$$= \frac{\dfrac{1 - \cos 2u}{2}}{\dfrac{1 + \cos 2u}{2}}$$

$$= \frac{1 - \cos 2u}{1 + \cos 2u}$$

Sum-to-Product Formulas   *(p. 256)*

$$\sin x + \sin y = 2 \sin\left(\frac{x + y}{2}\right)\cos\left(\frac{x - y}{2}\right)$$

$$\sin x - \sin y = 2 \cos\left(\frac{x + y}{2}\right)\sin\left(\frac{x - y}{2}\right)$$

$$\cos x + \cos y = 2 \cos\left(\frac{x + y}{2}\right)\cos\left(\frac{x - y}{2}\right)$$

$$\cos x - \cos y = -2 \sin\left(\frac{x + y}{2}\right)\sin\left(\frac{x - y}{2}\right)$$

Proof

To prove the first formula, let $x = u + v$ and $y = u - v$. Then substitute $u = (x + y)/2$ and $v = (x - y)/2$ in the product-to-sum formula.

$$\sin u \cos v = \frac{1}{2}\left[\sin(u + v) + \sin(u - v)\right]$$

$$\sin\left(\frac{x + y}{2}\right)\cos\left(\frac{x - y}{2}\right) = \frac{1}{2}(\sin x + \sin y)$$

$$2 \sin\left(\frac{x + y}{2}\right)\cos\left(\frac{x - y}{2}\right) = \sin x + \sin y$$

The other sum-to-product formulas can be proved in a similar manner.

---

1. (a) Write each of the other trigonometric functions of θ in terms of $\sin \theta$.

   (b) Write each of the other trigonometric functions of θ in terms of $\cos \theta$.

2. Verify that for all integers n,

$$\cos\left[\frac{(2n + 1)\pi}{2}\right] = 0.$$

3. Verify that for all integers n,

$$\sin\left[\frac{(12n + 1)\pi}{6}\right] = \frac{1}{2}.$$

 4. A particular sound wave is modeled by

$$p(t) = \frac{1}{4\pi}(p_1(t) + 30p_2(t) + p_3(t) + p_5(t) + 30p_6(t))$$

where $p_n(t) = \frac{1}{n}\sin(524n\pi t)$, and t is the time in seconds.

   (a) Find the sine components $p_n(t)$ and use a graphing utility to graph each component. Then verify the graph of p that is shown.

   (b) Find the period of each sine component of p. Is p periodic? If so, what is its period?

   (c) Use the *zero* or *root* feature or the *zoom* and *trace* features of a graphing utility to find the t-intercepts of the graph of p over one cycle.

   (d) Use the *maximum* and *minimum* features of a graphing utility to approximate the absolute maximum and absolute minimum values of p over one cycle.

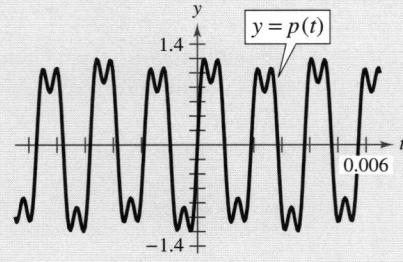

5. The path traveled by an object (neglecting air resistance) that is projected at an initial height of h_0 feet, an initial velocity of v feet per second, and an initial angle θ is given by

$$y = -\frac{16}{v^2 \cos^2 \theta}x^2 + (\tan \theta)x + h_0$$

where x and y are measured in feet. Find a formula for the maximum height of an object projected from ground level at velocity v and angle θ. To do this, find half of the horizontal distance

$$\frac{1}{32}v^2 \sin 2\theta$$

and then substitute it for x in the general model for the path of a projectile (where $h_0 = 0$).

6. Verify the following identity used in calculus.

$$\frac{\cos(x + h) - \cos x}{h} = \cos x\left(\frac{\cos h - 1}{h}\right) - \sin x\left(\frac{\sin h}{h}\right)$$

7. Use the figure to derive the formulas for

$$\sin \frac{\theta}{2}, \cos \frac{\theta}{2}, \text{ and } \tan \frac{\theta}{2}$$

when θ is an acute angle.

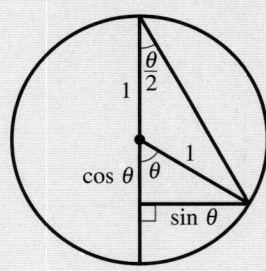

8. The force F (in pounds) on a person's back when he or she bends over at an angle θ is modeled by

$$F = \frac{0.6W \sin(\theta + 90°)}{\sin 12°}$$

where W is the person's weight (in pounds).

   (a) Simplify the model.

   (b) Use a graphing utility to graph the model, where $W = 185$ and $0° < \theta < 90°$.

   (c) At what angle is the force a maximum? At what angle is the force a minimum?

9. The number of hours of daylight that occur at any location on Earth depends on the time of year and the latitude of the location. The following equations model the number of hours of daylight in Seward, Alaska (60° latitude) and New Orleans, Louisiana (30° latitude).

$$D = 12.2 - 6.4 \cos\left[\frac{\pi(t + 0.2)}{182.6}\right] \qquad \text{Seward}$$

$$D = 12.2 - 1.9 \cos\left[\frac{\pi(t + 0.2)}{182.6}\right] \qquad \text{New Orleans}$$

In these models, D represents the number of hours of daylight and t represents the day, with $t = 0$ corresponding to January 1.

(a) Use a graphing utility to graph both models in the same viewing window. Use a viewing window of $0 \le t \le 365$.

(b) Find the days of the year on which both cities receive the same amount of daylight. What are these days called?

(c) Which city has the greater variation in the number of daylight hours? Which constant in each model would you use to determine the difference between the greatest and least numbers of hours of daylight?

(d) Determine the period of each model.

10. The tide, or depth of the ocean near the shore, changes throughout the day. The water depth d (in feet) of a bay can be modeled by

$$d = 35 - 28 \cos\frac{\pi}{6.2}t$$

where t is the time in hours, with $t = 0$ corresponding to 12:00 A.M.

(a) Algebraically find the times at which the high and low tides occur.

(b) Algebraically find the time(s) at which the water depth is 3.5 feet.

(c) Use a graphing utility to verify your results from parts (a) and (b).

11. Find the solution of each inequality in the interval $[0, 2\pi]$.

(a) $\sin x \ge 0.5$ (b) $\cos x \le -0.5$

(c) $\tan x < \sin x$ (d) $\cos x \ge \sin x$

12. The index of refraction n of a transparent material is the ratio of the speed of light in a vacuum to the speed of light in the material. Some common materials and their indices are air (1.00), water (1.33), and glass (1.50). Triangular prisms are often used to measure the index of refraction based on the formula

$$n = \frac{\sin\left(\frac{\theta}{2} + \frac{\alpha}{2}\right)}{\sin\frac{\theta}{2}}.$$

For the prism shown in the figure, $\alpha = 60°$.

(a) Write the index of refraction as a function of $\cot(\theta/2)$.

(b) Find θ for a prism made of glass.

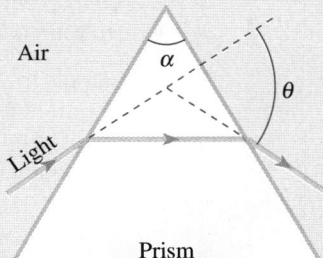

13. (a) Write a sum formula for $\sin (u + v + w)$.

(b) Write a sum formula for $\tan (u + v + w)$.

14. (a) Derive a formula for $\cos 3\theta$.

(b) Derive a formula for $\cos 4\theta$.

15. The heights h (in inches) of pistons 1 and 2 in an automobile engine can be modeled by

$$h_1 = 3.75 \sin 733t + 7.5$$

and

$$h_2 = 3.75 \sin 733\left(t + \frac{4\pi}{3}\right) + 7.5$$

where t is measured in seconds.

(a) Use a graphing utility to graph the heights of these two pistons in the same viewing window for $0 \le t \le 1$.

(b) How often are the pistons at the same height?

How to study Chapter 3

► **What you should learn**

In this chapter you will learn the following skills and concepts:

- How to use the Law of Sines and the Law of Cosines to solve oblique triangles
- How to find the areas of oblique triangles
- How to write the component forms of vectors and perform basic vector operations
- How to find the direction angles of vectors and the angle between two vectors

► **Important Vocabulary**

As you encounter each new vocabulary term in this chapter, add the term and its definition to your notebook glossary.

Oblique triangle (p. 274)
Law of Sines (p. 274)
Law of Cosines (p.283)
Directed line segment (p. 291)
Initial point (p. 291)
Terminal point (p. 291)
Magnitude of directed line segment (p. 291)
Vector **v** in the plane (p. 291)
Standard position (p. 292)
Component form of a vector **v** (p. 292)
Zero vector (p. 292)
Magnitude of **v** (p.292)
Unit vector (p. 292)
Parallelogram law (p. 293)
Resultant (p. 293)

Standard unit vectors (p. 296)
Linear combination of vectors (p. 296)
Direction angle (p. 297)
Dot product (p. 304)
Angle between two nonzero vectors (p. 305)
Orthogonal vectors (p. 306)
Work (p. 310)

Study Tools

Learning objectives in each section
Chapter Summary (p. 313)
Review Exercises (pp. 314–316)
Chapter Test (p. 317)
Cumulative Test for Chapters 1–3
(pp. 318, 319)

Additional Resources

Study and Solutions Guide
Interactive Trigonometry
Videotapes/DVD for Chapter 3
Trigonometry Website
Student Success Organizer

Michael Rutherford/SuperStock

3

Additional Topics in Trigonometry

3.1 Law of Sines

▶ **What you should learn**

- How to use the Law of Sines to solve oblique triangles (AAS or ASA)
- How to use the Law of Sines to solve oblique triangles (SSA)
- How to find the areas of oblique triangles
- How to use the Law of Sines to model and solve real-life problems

▶ **Why you should learn it**

You can use the Law of Sines to solve real-life problems involving oblique triangles. For instance, in Exercise 44 on page 282, you can use the Law of Sines to determine the length of the shadow of the Leaning Tower of Pisa.

John Lawrence/Imagestate

You may want to illustrate each of the four cases (AAS or ASA, SSA, SSS, SAS) as a quick geometry review for your students.

Introduction

In Chapter 1, you studied techniques for solving right triangles. In this section and the next, you will solve **oblique triangles**—triangles that have no right angles. As standard notation, the angles of a triangle are labeled A, B, and C, and their opposite sides are labeled a, b, and c, as shown in Figure 3.1.

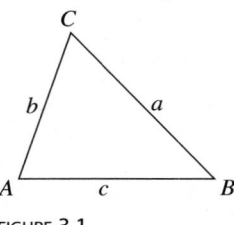

FIGURE **3.1**

To solve an oblique triangle, you need to know the measure of at least one side and any two other parts of the triangle—either two sides, two angles, or one angle and one side. This breaks down into the following four cases.

1. Two angles and any side (AAS or ASA)
2. Two sides and an angle opposite one of them (SSA)
3. Three sides (SSS)
4. Two sides and their included angle (SAS)

The first two cases can be solved using the **Law of Sines,** whereas the last two cases require the Law of Cosines (see Section 3.2).

Law of Sines

If ABC is a triangle with sides a, b, and c, then

$$\frac{a}{\sin A} = \frac{b}{\sin B} = \frac{c}{\sin C}.$$

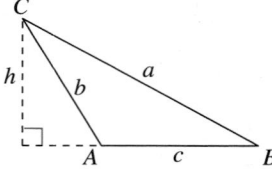

| A is acute. | A is obtuse. |

The Law of Sines can also be written in the reciprocal form

$$\frac{\sin A}{a} = \frac{\sin B}{b} = \frac{\sin C}{c}.$$

For a proof of the Law of Sines, see Proofs in Mathematics on page 320.

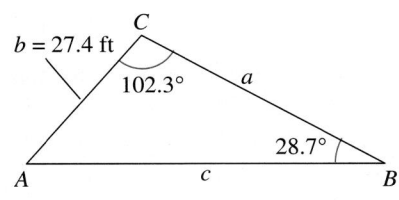

$b = 27.4$ ft

$102.3°$

a

$28.7°$

c

A B C

FIGURE **3.2**

STUDY TIP

When solving triangles, a careful sketch is useful as a quick test for the feasibility of an answer. Remember that the longest side lies opposite the largest angle, and the shortest side lies opposite the smallest angle.

Example 1 ▶ **Given Two Angles and One Side—AAS**

For the triangle in Figure 3.2, $C = 102.3°$, $B = 28.7°$, and $b = 27.4$ feet. Find the remaining angle and sides.

Solution

The third angle of the triangle is

$$A = 180° - B - C$$
$$= 180° - 28.7° - 102.3°$$
$$= 49.0°.$$

By the Law of Sines, you have

$$\frac{a}{\sin A} = \frac{b}{\sin B} = \frac{c}{\sin C}.$$

Using $b = 27.4$ produces

$$a = \frac{b}{\sin B}(\sin A) = \frac{27.4}{\sin 28.7°}(\sin 49.0°) \approx 43.06 \text{ feet}$$

and

$$c = \frac{b}{\sin B}(\sin C) = \frac{27.4}{\sin 28.7°}(\sin 102.3°) \approx 55.75 \text{ feet}.$$

Example 2 ▶ **Given Two Angles and One Side—ASA**

A pole tilts *toward* the sun at an 8° angle from the vertical, and it casts a 22-foot shadow. The angle of elevation from the tip of the shadow to the top of the pole is 43°. How tall is the pole?

Solution

From Figure 3.3, note that $A = 43°$ and $B = 90° + 8° = 98°$. So, the third angle is

$$C = 180° - A - B$$
$$= 180° - 43° - 98°$$
$$= 39°.$$

By the Law of Sines, you have

$$\frac{a}{\sin A} = \frac{c}{\sin C}.$$

Because $c = 22$ feet, the length of the pole is

$$a = \frac{c}{\sin C}(\sin A) = \frac{22}{\sin 39°}(\sin 43°) \approx 23.84 \text{ feet}.$$

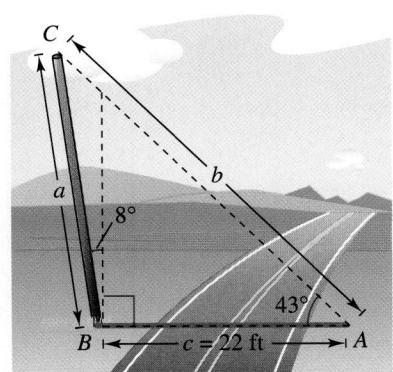

C

a

$8°$

b

$43°$

B $c = 22$ ft A

FIGURE **3.3**

For practice, try reworking Example 2 for a pole that tilts *away from* the sun under the same conditions.

The icon identifies examples and concepts related to features of the Learning Tools CD-ROM and the *Interactive* and *Internet* versions of this text. For more details see the chart on pages *xix-xxiii*.

The Ambiguous Case (SSA)

In Examples 1 and 2 you saw that two angles and one side determine a unique triangle. However, if two sides and one opposite angle are given, three possible situations can occur: (1) no such triangle exists, (2) one such triangle exists, or (3) two distinct triangles may satisfy the conditions.

The Ambiguous Case (SSA)

Consider a triangle in which you are given a, b, and A. $(h = b \sin A)$

	A is acute.	A is acute.	A is acute.	A is acute.	A is obtuse.	A is obtuse.
Sketch	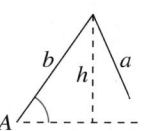	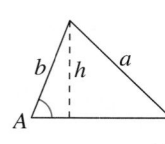	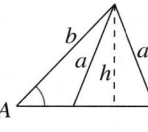	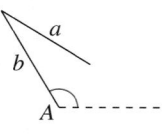		
Necessary condition	$a < h$	$a = h$	$a > b$	$h < a < b$	$a \leq b$	$a > b$
Triangles possible	None	One	One	Two	None	One

Example 3 ▶ Single-Solution Case—SSA

For the triangle in Figure 3.4, $a = 22$ inches, $b = 12$ inches, and $A = 42°$. Find the remaining side and angles.

Solution

By the Law of Sines, you have

$$\frac{\sin B}{b} = \frac{\sin A}{a} \qquad \text{Reciprocal form}$$

$$\sin B = b\left(\frac{\sin A}{a}\right) \qquad \text{Multiply each side by } b.$$

$$\sin B = 12\left(\frac{\sin 42°}{22}\right) \qquad \text{Substitute for } A, a, \text{ and } b.$$

$$B \approx 21.41°. \qquad B \text{ is acute.}$$

Now, you can determine that

$$C \approx 180° - 42° - 21.41° = 116.59°.$$

Then, the remaining side is

$$\frac{c}{\sin C} = \frac{a}{\sin A}$$

$$c = \frac{a}{\sin A}(\sin C) = \frac{22}{\sin 42°}(\sin 116.59°) \approx 29.40 \text{ inches.}$$

$b = 12$ in. C $a = 22$ in.

$42°$

A c B

One solution: a > b

FIGURE **3.4**

The *Interactive* CD-ROM and *Internet* versions of this text offer a Try It for each example in the text.

Encourage your students to sketch the triangle, keeping in mind that the longest side lies opposite the largest angle of the triangle. For practice, suggest that students also find h.

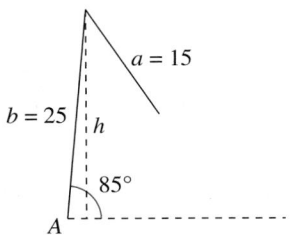

No solution: a < h

FIGURE **3.5**

Example 4 ▶ **No-Solution Case—SSA**

Show that there is no triangle for which $a = 15$, $b = 25$, and $A = 85°$.

Solution

Begin by making the sketch shown in Figure 3.5. From this figure it appears that no triangle is formed. You can verify this using the Law of Sines.

$$\frac{\sin B}{b} = \frac{\sin A}{a} \qquad \text{Reciprocal form}$$

$$\sin B = b\left(\frac{\sin A}{a}\right) \qquad \text{Multiply each side by } b.$$

$$\sin B = 25\left(\frac{\sin 85°}{15}\right) \approx 1.660 > 1$$

This contradicts the fact that $|\sin B| \leq 1$. So, no triangle can be formed having sides $a = 15$ and $b = 25$ and an angle of $A = 85°$.

Example 5 ▶ **Two-Solution Case—SSA**

Find two triangles for which $a = 12$ meters, $b = 31$ meters, and $A = 20.5°$.

Solution

By the Law of Sines, you have

$$\frac{\sin B}{b} = \frac{\sin A}{a} \qquad \text{Reciprocal form}$$

$$\sin B = b\left(\frac{\sin A}{a}\right) = 31\left(\frac{\sin 20.5°}{12}\right) \approx 0.9047.$$

There are two angles $B_1 \approx 64.8°$ and $B_2 \approx 180° - 64.8° \approx 115.2°$ between 0° and 180° whose sine is 0.9047. For $B_1 \approx 64.8°$, you obtain

$$C \approx 180° - 20.5° - 64.8° = 94.7°$$

$$c = \frac{a}{\sin A}(\sin C) = \frac{12}{\sin 20.5°}(\sin 94.7°) \approx 34.15 \text{ meters.}$$

For $B_2 \approx 115.2°$, you obtain

$$C \approx 180° - 20.5° - 115.2° = 44.3°$$

$$c = \frac{a}{\sin A}(\sin C) = \frac{12}{\sin 20.5°}(\sin 44.3°) \approx 23.93 \text{ meters.}$$

The resulting triangles are shown in Figure 3.6.

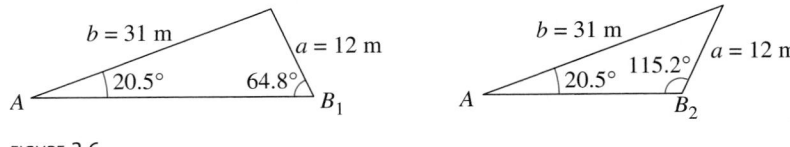

FIGURE **3.6**

STUDY TIP

To see how to obtain the height of the obtuse triangle in Figure 3.7, notice the use of the reference angle $180° - A$ and the difference formula for sine, as follows.

$h = b \sin(180° - A)$

$\quad = b(\sin 180° \cos A$

$\qquad - \cos 180° \sin A)$

$\quad = b[0 \cdot \cos A - (-1) \cdot \sin A]$

$\quad = b \sin A$

Area of an Oblique Triangle

The procedure used to prove the Law of Sines leads to a simple formula for the area of an oblique triangle. Referring to Figure 3.7, note that each triangle has a height of $h = b \sin A$. Consequently, the area of each triangle is

$$\text{Area} = \frac{1}{2}(\text{base})(\text{height}) = \frac{1}{2}(c)(b \sin A) = \frac{1}{2}bc \sin A.$$

By similar arguments, you can develop the formulas

$$\text{Area} = \frac{1}{2}ab \sin C = \frac{1}{2}ac \sin B.$$

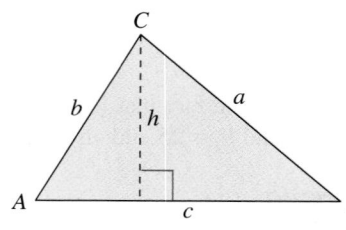
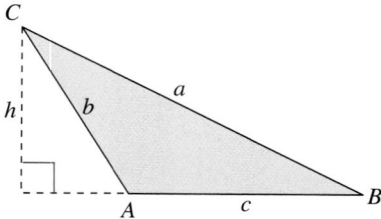

A is acute   A is obtuse

FIGURE **3.7**

Activities

1. Use the given information to find (if possible) the remaining side and angles of the oblique triangle. If two solutions exist, find both.

   $A = 58°,\ a = 20,\ c = 10$

   Answer: $B = 97°,\ C = 25°,$
   $\qquad\qquad b = 23.4$

2. Use the given information to find (if possible) the remaining side and angles of the oblique triangle. If two solutions exist, find both.

   $B = 78°,\ b = 207,\ c = 210$

   Answer: Two solutions
   $A = 19.1°,\ a = 69.2,\ C = 82.9°$
   $A = 4.9°,\ a = 18.1,\ C = 97.1°$

3. Find the area of the triangle with $B = 120°,\ a = 32,\ c = 50$.

   Answer: Area = 692.8 square units

Area of an Oblique Triangle

The area of any triangle is one-half the product of the lengths of two sides times the sine of their included angle. That is,

$$\text{Area} = \frac{1}{2}bc \sin A = \frac{1}{2}ab \sin C = \frac{1}{2}ac \sin B.$$

Note that if angle A is 90°, the formula gives the area for a right triangle:

$$\text{Area} = \frac{1}{2}bc(\sin 90°) = \frac{1}{2}bc = \frac{1}{2}(\text{base})(\text{height}). \qquad \text{sin } 90° = 1$$

Similar results are obtained for angles C and B equal to 90°.

Example 6 ▶ Finding the Area of a Triangular Lot

Find the area of a triangular lot having two sides of lengths 90 meters and 52 meters and an included angle of 102°.

Solution

Consider $a = 90$ meters, $b = 52$ meters, and angle $C = 102°$, as shown in Figure 3.8. Then, the area of the triangle is

$$\text{Area} = \frac{1}{2}ab \sin C = \frac{1}{2}(90)(52)(\sin 102°) \approx 2289 \text{ square meters.}$$

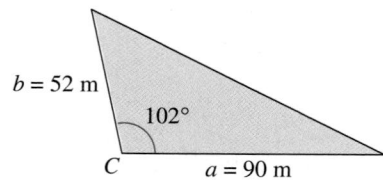

$b = 52$ m
102°
C $a = 90$ m

FIGURE **3.8**

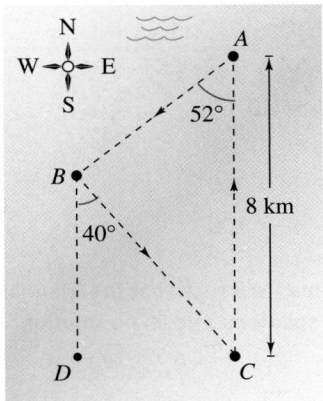

FIGURE **3.9**

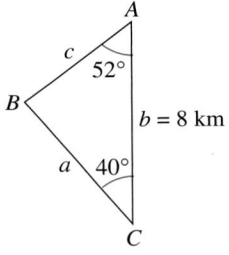

FIGURE **3.10**

Application

Example 7 ▶   An Application of the Law of Sines

The course for a boat race starts at point A and proceeds in the direction S 52° W to point B, then in the direction S 40° E to point C, and finally back to A, as shown in Figure 3.9. Point C lies 8 kilometers directly south of point A. Approximate the total distance of the race course.

Solution

Because lines BD and AC are parallel, it follows that $\angle BCA \cong \angle DBC$. Consequently, triangle ABC has the measures shown in Figure 3.10. For angle B, you have $B = 180° - 52° - 40° = 88°$. Using the Law of Sines

$$\frac{a}{\sin 52°} = \frac{b}{\sin 88°} = \frac{c}{\sin 40°}$$

you can let $b = 8$ and obtain

$$a = \frac{8}{\sin 88°}(\sin 52°) \approx 6.308$$

and

$$c = \frac{8}{\sin 88°}(\sin 40°) \approx 5.145.$$

The total length of the course is approximately

$$\text{Length} \approx 8 + 6.308 + 5.145$$

$$= 19.453 \text{ kilometers.}$$

Alternative Writing About Mathematics: Error Analysis

You are a math instructor, and one of your students hands in the following solution. Discuss what is wrong with your student's solution. How could you help the student avoid making a similar mistake in the future? Use a diagram to illustrate your explanation.

Find side c in a triangle that has $a = 5.8$, $b = 7$, and $C = 82°$.

Student's Solution

Because this is an SSA situation, I can use the Law of Sines.

$$\frac{5.8}{7} = \frac{c}{\sin 82°}$$

$$(\sin 82°)\frac{5.8}{7} = 0.821$$

Writing ABOUT MATHEMATICS

Using the Law of Sines   In this section, you have been using the Law of Sines to solve *oblique* triangles. Can the Law of Sines also be used to solve a right triangle? If so, write a short paragraph explaining how to use the Law of Sines to solve each triangle. Is there an easier way to solve these triangles?

a. (AAS)

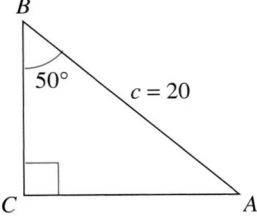

b. (ASA)

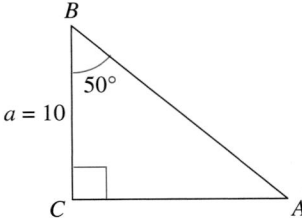

3.1 Exercises

The *Interactive* CD-ROM and *Internet* versions of this text contain step-by-step solutions to all odd-numbered exercises. They also provide Tutorial Exercises for additional help.

In Exercises 1–18, use the information to solve the triangle.

1.

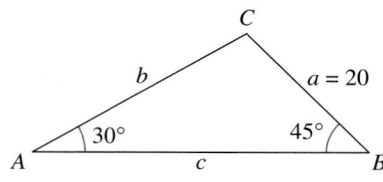

2.

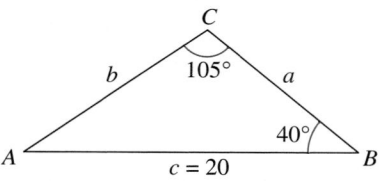

3.

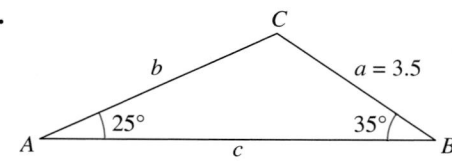

4.

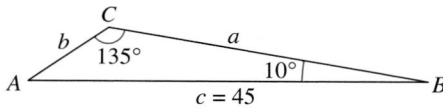

5. $A = 36°$, $a = 8$, $b = 5$

6. $A = 60°$, $a = 9$, $c = 10$

7. $A = 102.4°$, $C = 16.7°$, $a = 21.6$

8. $A = 24.3°$, $C = 54.6°$, $c = 2.68$

9. $A = 83° 20'$, $C = 54.6°$, $c = 18.1$

10. $A = 5° 40'$, $B = 8° 15'$, $b = 4.8$

11. $B = 15° 30'$, $a = 4.5$, $b = 6.8$

12. $B = 2° 45'$, $b = 6.2$, $c = 5.8$

13. $C = 145°$, $b = 4$, $c = 14$

14. $A = 100°$, $a = 125$, $c = 10$

15. $A = 110° 15'$, $a = 48$, $b = 16$

16. $C = 85° 20'$, $a = 35$, $c = 50$

17. $A = 55°$, $B = 42°$, $c = \frac{3}{4}$

18. $B = 28°$, $C = 104°$, $a = 3\frac{5}{8}$

In Exercises 19–24, use the information to solve (if possible) the triangle. If two solutions exist, find both.

19. $A = 58°$, $a = 4.5$, $b = 12.8$

20. $A = 58°$, $a = 11.4$, $b = 12.8$

21. $A = 76°$, $a = 18$, $b = 20$

22. $A = 76°$, $a = 34$, $b = 21$

23. $A = 110°$, $a = 125$, $b = 200$

24. $A = 110°$, $a = 125$, $b = 100$

In Exercises 25–28, find a value for b such that the triangle has (a) one solution, (b) two solutions, and (c) no solution.

25. $A = 36°$, $a = 5$

26. $A = 60°$, $a = 10$

27. $A = 10°$, $a = 10.8$

28. $A = 88°$, $a = 315.6$

In Exercises 29–34, find the area of the triangle having the indicated sides and angle.

29. $C = 120°$, $a = 4$, $b = 6$

30. $B = 130°$, $a = 62$, $c = 20$

31. $A = 43° 45'$, $b = 57$, $c = 85$

32. $A = 5° 15'$, $b = 4.5$, $c = 22$

33. $B = 72° 30'$, $a = 105$, $c = 64$

34. $C = 84° 30'$, $a = 16$, $b = 20$

35. *Height*   Because of prevailing winds, a tree grew so that it was leaning 4° from the vertical. At a point 35 meters from the tree, the angle of elevation to the top of the tree is 23° (see figure). Find the height h of the tree.

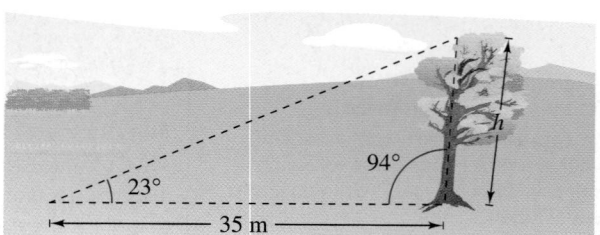

36. *Height*   A flagpole at a right angle to the horizontal is located on a slope that makes an angle of 12° with the horizontal. The flagpole's shadow is 16 meters long and points directly up the slope. The angle of elevation from the tip of the shadow to the sun is 20°.

(a) Draw a triangle that represents the problem. Show the known quantities on the triangle and use a variable to indicate the height of the flagpole.

(b) Write an equation involving the unknown quantity.

(c) Find the height of the flagpole.

37. *Angle of Elevation*   A 10-meter telephone pole casts a 17-meter shadow directly down a slope when the angle of elevation of the sun is 42° (see figure). Find θ, the angle of elevation of the ground.

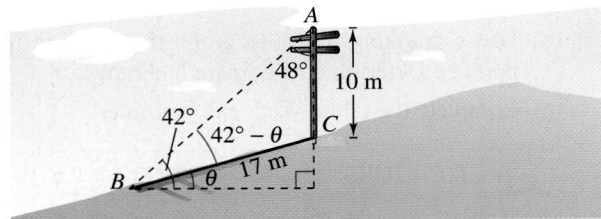

38. *Flight Path*   A plane flies 500 kilometers with a bearing of 316° from Naples to Elgin (see figure). The plane then flies 720 kilometers from Elgin to Canton. Find the bearing of the flight from Elgin to Canton.

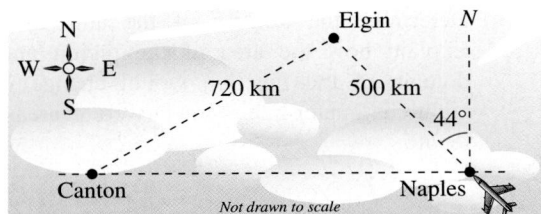

Not drawn to scale

39. *Bridge Design*   A bridge is to be built across a small lake from a gazebo to a dock (see figure). The bearing from the gazebo to the dock is S 41° W. From a tree 100 meters from the gazebo, the bearings to the gazebo and the dock are S 74° E and S 28° E, respectively. Find the distance from the gazebo to the dock.

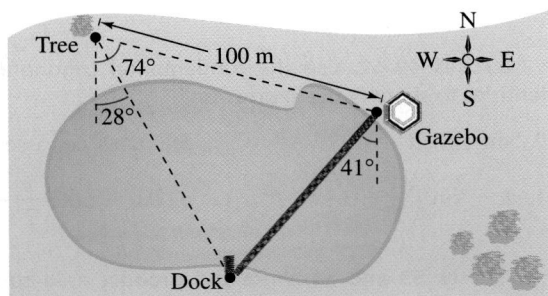

40. *Railroad Track Design*   The circular arc of a railroad curve has a chord of length 3000 feet and a central angle of 40°.

(a) Draw a figure that visually represents the problem. Show the known quantities on the figure and use variables r and s to represent the radius of the arc and the length of the arc, respectively.

(b) Find the radius r of the circular arc.

(c) Find the length s of the circular arc.

41. *Glide Path*   A pilot has just started on the glide path for landing at an airport where the length of the runway is 9000 feet. The angles of depression from the plane to the ends of the runway are 17.5° and 18.8°.

(a) Draw a figure that visually represents the problem.

(b) Find the air distance the plane must travel until touching down on the near end of the runway.

(c) Find the ground distance the plane must travel until touching down.

(d) Find the altitude of the plane when the pilot begins the descent.

42. *Locating a Fire*   The bearing from the Pine Knob fire tower to the Colt Station fire tower is N 65° E, and the two towers are 30 kilometers apart. A fire spotted by rangers in each tower has a bearing of N 80° E from Pine Knob and S 70° E from Colt Station. Find the distance of the fire from each tower.

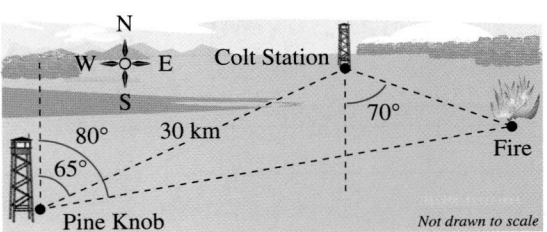

Not drawn to scale

43. *Distance*   A boat is sailing due east parallel to the shoreline at a speed of 10 miles per hour. At a given time, the bearing to the lighthouse is S 70° E, and 15 minutes later the bearing is S 63° E (see figure). The lighthouse is located at the shoreline. What is the distance from the boat to the shoreline?

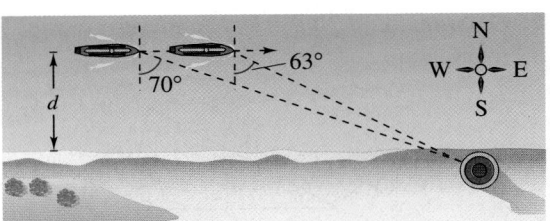

▶ Model It

44. Shadow Length   The Leaning Tower of Pisa in Italy is characterized by its tilt. The tower leans because it was built on a layer of unstable soil—clay, sand, and water. The tower is approximately 58.36 meters tall from its foundation, and 55 meters tall from the ground (see figure). The top of the tower leans about 4.47 meters off center.

(a) Find the angle α of lean of the tower.

(b) Write β as a function of θ, where θ is the angle of elevation to the sun.

(c) Use the Law of Sines to write an equation for the length d of the shadow cast by the tower.

(d) Use a graphing utility to complete the table.

θ	10°	20°	30°	40°	50°	60°
d						

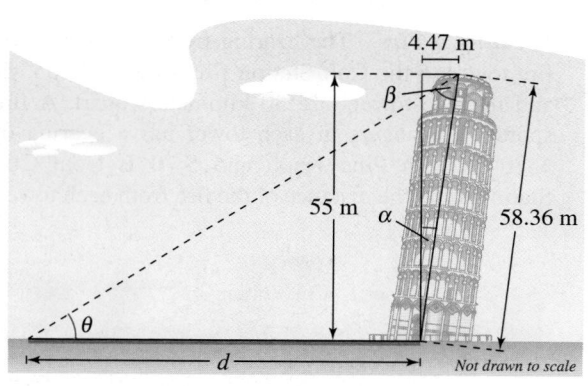

4.47 m

β

55 m α 58.36 m

θ

d

Not drawn to scale

Synthesis

True or False?   **In Exercises 45 and 46, determine whether the statement is true or false. Justify your answer.**

45. If a triangle contains an obtuse angle, then it must be oblique.

46. Two angles and one side of a triangle do not necessarily determine a unique triangle.

47. Graphical and Numerical Analysis   In the figure, α and β are positive angles.

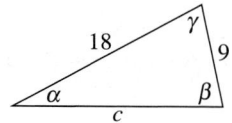

γ

18   9

α β

c

(a) Write α as a function of β.

(b) Use a graphing utility to graph the function. Determine its domain and range.

(c) Use the result of part (a) to write c as a function of β.

(d) Use a graphing utility to graph the function in part (c). Determine its domain and range.

(e) Complete the table. What can you infer?

β	0.4	0.8	1.2	1.6	2.0	2.4	2.8
α							
c							

48. Graphical Analysis

(a) Write the area A of the shaded region in the figure as a function of θ.

(b) Use a graphing utility to graph the area function.

(c) Determine the domain of the area function. Explain how the area of the region and the domain of the function would change if the eight-centimeter line segment were decreased in length.

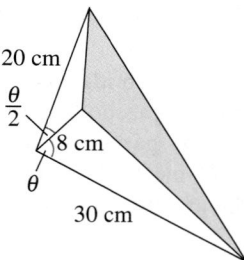

20 cm

$\dfrac{\theta}{2}$

8 cm

θ

30 cm

Review

In Exercises 49–52, use the fundamental trigonometric identities to simplify the expression.

49. $\sin x \cot x$ **50.** $\tan x \cos x \sec x$

51. $1 - \sin^2\left(\dfrac{\pi}{2} - x\right)$ **52.** $1 + \cot^2\left(\dfrac{\pi}{2} - x\right)$

In Exercises 53 and 54, write the product as a sum or difference.

53. $6 \sin 8\theta \cos 3\theta$

54. $2 \cos 5\theta \sin 2\theta$

3.2 Law of Cosines

▶ **What you should learn**

- How to use the Law of Cosines to solve oblique triangles (SSS or SAS)
- How to use the Law of Cosines to model and solve real-life problems
- How to use Heron's Area Formula to find the area of a triangle

▶ **Why you should learn it**

You can use the Law of Cosines to solve real-life problems involving oblique triangles. For instance, in Exercise 44 on page 289, you can use the Law of Cosines to determine the maximum distance a piston moves in one cycle.

Rosenthal/SuperStock

Introduction

Two cases remain in the list of conditions needed to solve an oblique triangle—SSS and SAS. If you are given three sides (SSS), or two sides and their included angle (SAS), none of the ratios in the Law of Sines would be complete. In such cases, you can use the **Law of Cosines.**

Law of Cosines

Standard Form	*Alternative Form*
$a^2 = b^2 + c^2 - 2bc \cos A$	$\cos A = \dfrac{b^2 + c^2 - a^2}{2bc}$
$b^2 = a^2 + c^2 - 2ac \cos B$	$\cos B = \dfrac{a^2 + c^2 - b^2}{2ac}$
$c^2 = a^2 + b^2 - 2ab \cos C$	$\cos C = \dfrac{a^2 + b^2 - c^2}{2ab}$

For a proof of the Law of Cosines, see Proofs in Mathematics on page 321.

Example 1 ▶ **Three Sides of a Triangle—SSS**

Find the three angles of the triangle in Figure 3.11.

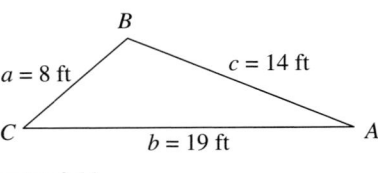

FIGURE **3.11**

Solution

It is a good idea first to find the angle opposite the longest side—side b in this case (see Figure 3.11). Using the Law of Cosines, you find that

$$\cos B = \frac{a^2 + c^2 - b^2}{2ac} = \frac{8^2 + 14^2 - 19^2}{2(8)(14)} \approx -0.45089.$$

Because $\cos B$ is negative, you know that B is an *obtuse* angle given by $B \approx 116.80°$. At this point, it is simpler to use the Law of Sines to determine A.

$$\sin A = a\left(\frac{\sin B}{b}\right) \approx 8\left(\frac{\sin 116.80°}{19}\right) \approx 0.37583$$

Because B is obtuse, you know that A must be acute, because a triangle can have, at most, one obtuse angle. So, $A \approx 22.08°$ and $C \approx 180° - 22.08° - 116.80° = 41.12°$.

In cases where the Law of Cosines must be used, encourage your students to solve for the largest angle first, then finish the problem using either the Law of Sines or the Law of Cosines.

◀ **Exploration** ▶

What familiar formula do you obtain when you use the third form of the Law of Cosines

$$c^2 = a^2 + b^2 - 2ab \cos C$$

and you let $C = 90°$? What is the relationship between the Law of Cosines and this formula?

Do you see why it was wise to find the largest angle *first* in Example 1? Knowing the cosine of an angle, you can determine whether the angle is acute or obtuse. That is,

$\cos \theta > 0$ for $0° < \theta < 90°$ Acute

$\cos \theta < 0$ for $90° < \theta < 180°$.   Obtuse

So, in Example 1, once you found that angle B was obtuse, you knew that angles A and C were both acute. If the largest angle is acute, the remaining two angles are acute also.

Example 2 ▶ **Two Sides and the Included Angle—SAS**

Find the remaining angles and side of the triangle in Figure 3.12.

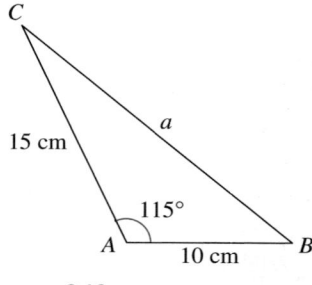

FIGURE **3.12**

Solution

Use the Law of Cosines to find the unknown side a in the figure.

$$a^2 = b^2 + c^2 - 2bc \cos A$$

$$a^2 = 15^2 + 10^2 - 2(15)(10) \cos 115°$$

$$a^2 \approx 451.79$$

$$a \approx 21.26$$

Because $a \approx 21.26$ centimeters, you now know the ratio $\sin A / a$ and you can use the Law of Sines

$$\frac{\sin B}{b} = \frac{\sin A}{a}$$

to solve for B.

$$\sin B = b\left(\frac{\sin A}{a}\right)$$

$$= 15\left(\frac{\sin 115°}{21.26}\right)$$

$$\approx 0.63945$$

So, $B = \arcsin 0.63945 \approx 39.75°$ and $C \approx 180° - 115° - 39.75° = 25.25°$.

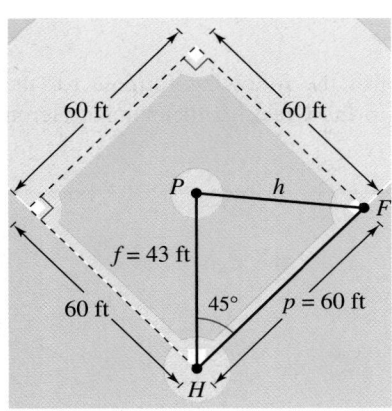

FIGURE **3.13**

Applications

Example 3 ▶ An Application of the Law of Cosines

The pitcher's mound on a women's softball field is 43 feet from home plate and the distance between the bases is 60 feet, as shown in Figure 3.13. (The pitcher's mound is not halfway between home plate and second base.) How far is the pitcher's mound from first base?

Solution

In triangle *HPF*, $H = 45°$ (line *HP* bisects the right angle at *H*), $f = 43$, and $p = 60$. Using the Law of Cosines for this SAS case, you have

$$h^2 = f^2 + p^2 - 2fp \cos H$$

$$= 43^2 + 60^2 - 2(43)(60) \cos 45°$$

$$\approx 1800.3$$

So, the approximate distance from the pitcher's mound to first base is

$$h \approx \sqrt{1800.3} \approx 42.43 \text{ feet.}$$

Example 4 ▶ An Application of the Law of Cosines

A ship travels 60 miles due east, then adjusts its course northward, as shown in Figure 3.14. After traveling 80 miles in that direction, the ship is 139 miles from its point of departure. Describe the bearing from point *B* to point *C*.

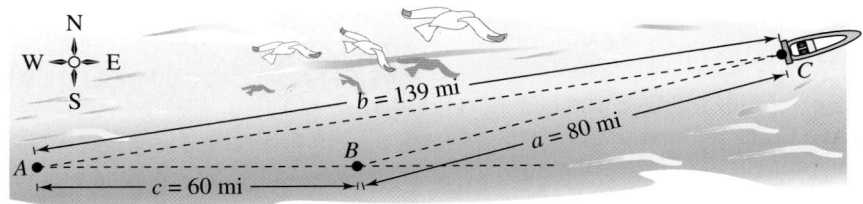

FIGURE **3.14**

Solution

You have $a = 80$, $b = 139$, and $c = 60$; so, using the alternative form of the Law of Cosines, you have

$$\cos B = \frac{a^2 + c^2 - b^2}{2ac}$$

$$= \frac{80^2 + 60^2 - 139^2}{2(80)(60)}$$

$$\approx -0.97094.$$

So, $B \approx \arccos(-0.97094) \approx 166.15°$, and thus the bearing measured from due north from point *B* to point *C* is $166.15° - 90° = 76.15°$, or N 76.15° E.

Heron's Area Formula

The Law of Cosines can be used to establish the following formula for the area of a triangle. This formula is credited to the Greek mathematician Heron (c. 100 B.C.).

Heron's Area Formula

Given any triangle with sides of lengths a, b, and c, the area of the triangle is

$$\text{Area} = \sqrt{s(s - a)(s - b)(s - c)}$$

where $s = (a + b + c)/2$.

For a proof of Heron's Area Formula, see Proofs in Mathematics on page 322.

Example 5 ▶ **Using Heron's Area Formula**

Find the area of a triangle having sides of lengths $a = 43$ meters, $b = 53$ meters, and $c = 72$ meters.

Solution

Because $s = (a + b + c)/2 = 168/2 = 84$, Heron's Area Formula yields

$$\text{Area} = \sqrt{s(s - a)(s - b)(s - c)}$$
$$= \sqrt{84(41)(31)(12)} \approx 1131.89 \text{ square meters.}$$

 **ABOUT MATHEMATICS**

The Area of a Triangle You have now studied three different formulas for the area of a triangle. Use the most appropriate formula to find the area of each triangle below. Show your work and give your reasons for choosing each formula.

Standard Formula   Area $= \frac{1}{2}bh$
Oblique Triangle   Area $= \frac{1}{2}bc \sin A = \frac{1}{2}ab \sin C = \frac{1}{2}ac \sin B$
Heron's Area Formula   Area $= \sqrt{s(s - a)(s - b)(s - c)}$

a.
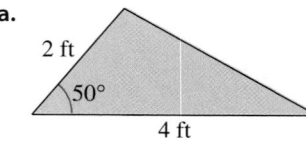

b.
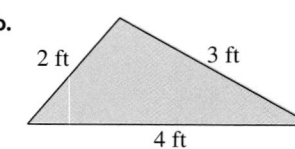

c.
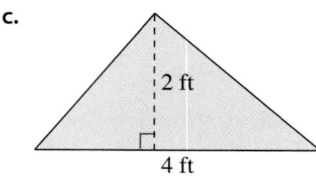

d.
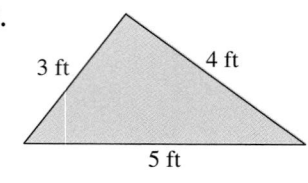

3.2 Exercises

In Exercises 1–16, use the Law of Cosines to solve the triangle.

1.

C, $b = 10$, $a = 7$, A, B, $c = 15$

2.

C, $b = 3$, $a = 8$, A, B, $c = 9$

3.

C, $b = 15$, a, $30°$, A, B, $c = 30$

4.

C, $b = 4.5$, $a = 10$, $105°$, A, B, c

5. $a = 11$, $b = 14$, $c = 20$

6. $a = 55$, $b = 25$, $c = 72$

7. $a = 75.4$, $b = 52$, $c = 52$

8. $a = 1.42$, $b = 0.75$, $c = 1.25$

9. $A = 135°$, $b = 4$, $c = 9$

10. $A = 55°$, $b = 3$, $c = 10$

11. $B = 10° 35'$, $a = 40$, $c = 30$

12. $B = 75° 20'$, $a = 6.2$, $c = 9.5$

13. $B = 125° 40'$, $a = 32$, $c = 32$

14. $C = 15° 15'$, $a = 6.25$, $b = 2.15$

15. $C = 43°$, $a = \frac{4}{9}$, $b = \frac{7}{9}$

16. $C = 103°$, $a = \frac{3}{8}$, $b = \frac{3}{4}$

In Exercises 17–22, complete the table by solving the parallelogram shown in the figure. (The lengths of the diagonals are given by c and d.)

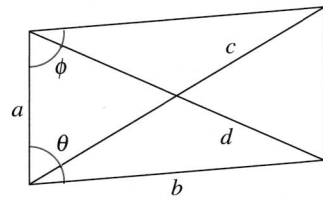

	a	b	c	d	θ	ϕ
17.	5	8			$45°$	
18.	25	35				$120°$
19.	10	14	20			
20.	40	60		80		
21.	15		25	20		
22.		25	50	35		

In Exercises 23–28, use Heron's Area Formula to find the area of the triangle.

23. $a = 5$, $b = 7$, $c = 10$

24. $a = 12$, $b = 15$, $c = 9$

25. $a = 2.5$, $b = 10.2$, $c = 9$

26. $a = 75.4$, $b = 52$, $c = 52$

27. $a = 12.32$, $b = 8.46$, $c = 15.05$

28. $a = 3.05$, $b = 0.75$, $c = 2.45$

29. *Navigation* A boat race runs along a triangular course marked by buoys A, B, and C. The race starts with the boats headed west for 3700 meters. The other two sides of the course lie to the north of the first side, and their lengths are 1700 meters and 3000 meters. Draw a figure that gives a visual representation of the problem, and find the bearings for the last two legs of the race.

30. *Navigation* A plane flies 810 miles from Niagara to Cuyahoga with a bearing of $75°$. Then it flies 648 miles from Cuyahoga to Rosemount with a bearing of $32°$. Draw a figure that visually represents the problem, and find the straight-line distance and bearing from Niagara to Rosemount.

31. *Surveying* To approximate the length of a marsh, a surveyor walks 250 meters from point A to point B, then turns $75°$ and walks 220 meters to point C (see figure). Approximate the length AC of the marsh.

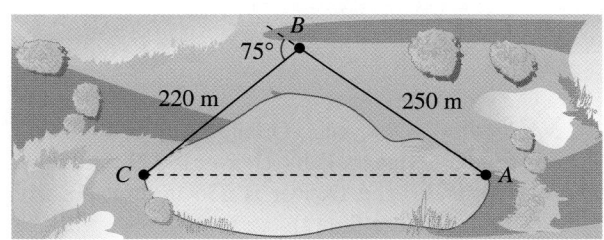

32. *Surveying* A triangular parcel of land has 115 meters of frontage, and the other boundaries have lengths of 76 meters and 92 meters. What angles does the frontage make with the two other boundaries?

33. *Surveying* A triangular parcel of ground has sides of lengths 725 feet, 650 feet, and 575 feet. Find the measure of the largest angle.

34. *Streetlight Design* Determine the angle θ in the design of the streetlight shown in the figure.

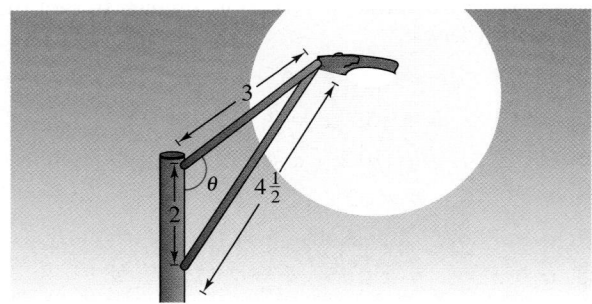

35. *Distance* Two ships leave a port at 9 A.M. One travels at a bearing of N 53° W at 12 miles per hour, and the other travels at a bearing of S 67° W at 16 miles per hour. Approximate how far apart they are at noon that day.

36. *Length* A 100-foot vertical tower is to be erected on the side of a hill that makes a 6° angle with the horizontal (see figure). Find the length of each of the two guy wires that will be anchored 75 feet uphill and downhill from the base of the tower.

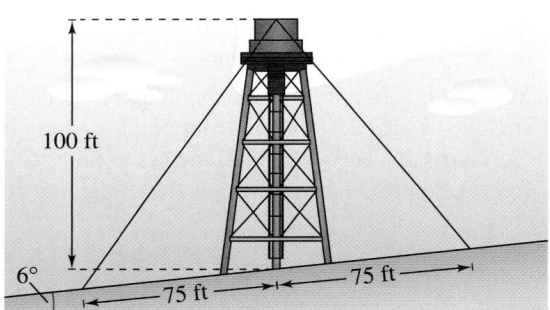

37. *Navigation* On a map, Orlando is 178 millimeters due south of Niagara Falls, Denver is 273 millimeters from Orlando, and Denver is 235 millimeters from Niagara Falls (see figure).

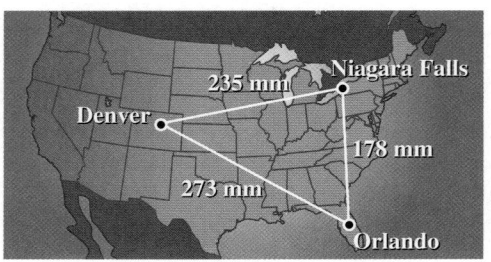

(a) Find the bearing of Denver from Orlando.

(b) Find the bearing of Denver from Niagara Falls.

38. *Navigation* On a map, Minneapolis is 165 millimeters due west of Albany, Phoenix is 216 millimeters from Minneapolis, and Phoenix is 368 millimeters from Albany (see figure).

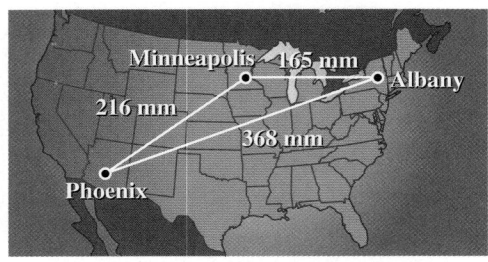

(a) Find the bearing of Minneapolis from Phoenix.

(b) Find the bearing of Albany from Phoenix.

39. *Baseball* On a baseball diamond with 90-foot sides, the pitcher's mound is 60.5 feet from home plate. How far is it from the pitcher's mound to third base?

40. *Baseball* The baseball player in center field is playing approximately 330 feet from the television camera that is behind home plate. A batter hits a fly ball that goes to the wall 420 feet from the camera (see figure). The camera turns 8° to follow the play. Approximately how far does the center fielder have to run to make the catch?

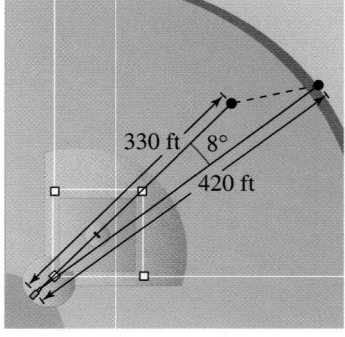

41. *Aircraft Tracking* To determine the distance between two aircraft, a tracking station continuously determines the distance to each aircraft and the angle A between them (see figure). Determine the distance a between the planes when $A = 42°$, $b = 35$ miles, and $c = 20$ miles.

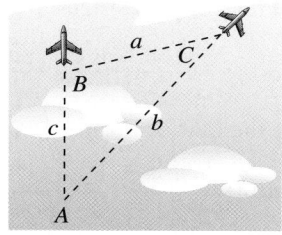

FIGURE FOR 41

42. *Aircraft Tracking* Use the figure for Exercise 41 to determine the distance a between the planes when $A = 11°$, $b = 20$ miles, and $c = 20$ miles.

43. *Trusses* Q is the midpoint of the line segment $\overline{PR}$ in the truss rafter shown in the figure. What are the lengths of the line segments $\overline{PQ}$, $\overline{QS}$, and $\overline{RS}$?

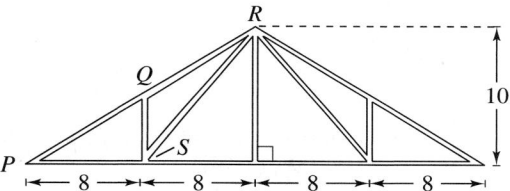

▶ **Model It**

44. *Engine Design* An engine has a seven-inch connecting rod fastened to a crank (see figure).

(a) Use the Law of Cosines to write an equation giving the relationship between x and θ.

(b) Write x as a function of θ. (Select the sign that yields positive values of x.)

(c) Use a graphing utility to graph the function in part (b).

(d) Use the graph in part (c) to determine the maximum distance the piston moves in one cycle.

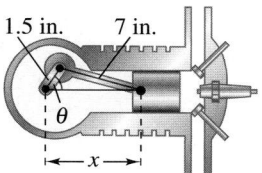

---

45. *Paper Manufacturing* In a process with continuous paper, the paper passes across three rollers of radii 3 inches, 4 inches, and 6 inches (see figure). The centers of the three-inch and six-inch rollers are d inches apart, and the length of the arc in contact with the paper on the four-inch roller is s inches. Complete the table.

d (inches)	9	10	12	13	14	15	16
θ (degrees)							
s (inches)							

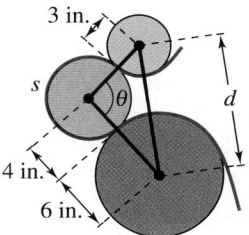

FIGURE FOR 45

46. *Awning Design* A retractable awning above a patio door lowers at an angle of 50° from the exterior wall at a height of 10 feet above the ground (see figure). No direct sunlight is to enter the door when the angle of elevation of the sun is greater than 70°. What is the length x of the awning?

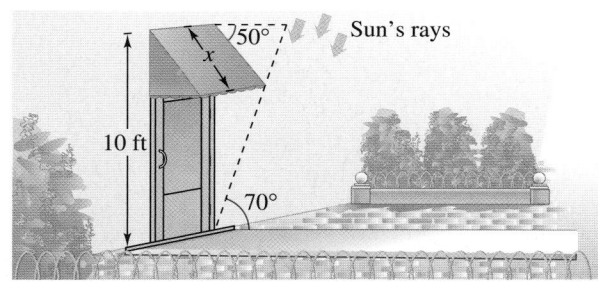

47. *Geometry* The lengths of the sides of a triangular parcel of land are approximately 200 feet, 500 feet, and 600 feet. Approximate the area of the parcel.

48. *Geometry* A parking lot has the shape of a parallelogram (see figure). The lengths of two adjacent sides are 70 meters and 100 meters. The angle between the two sides is 70°. What is the area of the parking lot?

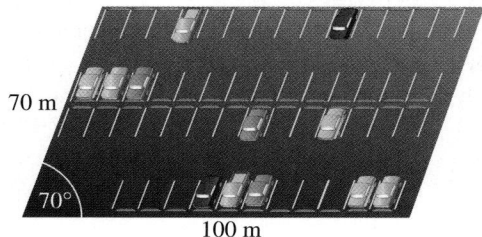

Synthesis

True or False?   In Exercises 49–51, determine whether the statement is true or false. Justify your answer.

49. In Heron's Area Formula

$$\text{Area} = \sqrt{s(s-a)(s-b)(s-c)}$$

s is the average of the lengths of the three sides of the triangle.

50. In addition to SSS and SAS, the Law of Cosines can be used to solve triangles with SSA conditions.

51. A triangle with side lengths of 10 centimeters, 16 centimeters, and 5 centimeters can be solved using the Law of Cosines.

52. ***Circumscribed and Inscribed Circles***   Let R and r be the radii of the circumscribed and inscribed circles of a triangle ABC, respectively (see figure), and let

$$s = \frac{a+b+c}{2}.$$

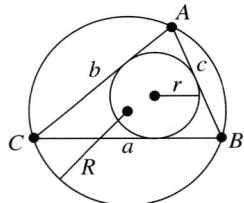

(a) Prove that $2R = \dfrac{a}{\sin A} = \dfrac{b}{\sin B} = \dfrac{c}{\sin C}.$

(b) Prove that $r = \sqrt{\dfrac{(s-a)(s-b)(s-c)}{s}}.$

Circumscribed and Inscribed Circles   In Exercises 53 and 54, use the results of Exercise 52.

53. Given a triangle with

$a = 25, b = 55,$ and $c = 72$

find the areas of (a) the triangle, (b) the circumscribed circle, and (c) the inscribed circle.

54. Find the length of the largest circular running track that can be built on a triangular piece of property with sides of lengths 200 feet, 250 feet, and 325 feet.

55. ***Proof***   Use the Law of Cosines to prove that

$$\frac{1}{2}bc(1+\cos A) = \frac{a+b+c}{2} \cdot \frac{-a+b+c}{2}.$$

56. ***Proof***   Use the Law of Cosines to prove that

$$\frac{1}{2}bc(1-\cos A) = \frac{a-b+c}{2} \cdot \frac{a+b-c}{2}.$$

Review

In Exercises 57–62, evaluate the expression without the aid of a calculator.

57. $\arcsin(-1)$

58. $\arccos 0$

59. $\arctan \sqrt{3}$

60. $\arctan -\sqrt{3}$

61. $\arcsin\left(-\dfrac{\sqrt{3}}{2}\right)$

62. $\arccos\left(-\dfrac{\sqrt{3}}{2}\right)$

In Exercises 63–66, write an algebraic expression that is equivalent to the given expression.

63. $\sec(\arcsin 2x)$

64. $\tan(\arccos 3x)$

65. $\cot[\arctan(x-2)]$

66. $\cos\left(\arcsin \dfrac{x-1}{2}\right)$

In Exercises 67–70, use trigonometric substitution to write the algebraic equation as a trigonometric function of θ, where $-\pi/2 < \theta < \pi/2$. Then find $\sec \theta$ and $\csc \theta$.

67. $5 = \sqrt{25-x^2}, \quad x = 5\sin\theta$

68. $-\sqrt{2} = \sqrt{4-x^2}, \quad x = 2\cos\theta$

69. $-\sqrt{3} = \sqrt{x^2-9}, \quad x = 3\sec\theta$

70. $12 = \sqrt{36+x^2}, \quad x = 6\tan\theta$

In Exercises 71 and 72, write the sum or difference as a product.

71. $\cos\dfrac{5\pi}{6} - \cos\dfrac{\pi}{3}$

72. $\sin\left(x - \dfrac{\pi}{2}\right) - \sin\left(x + \dfrac{\pi}{2}\right)$

3.3 Vectors in the Plane

▶ **Why you should learn it**

You can use vectors to model and solve real-life problems involving magnitude and direction. For instance, in Exercise 80 on page 302, you can use vectors to determine the true direction of a commercial jet.

Bill Bachman/Photo Researchers, Inc.

Introduction

Quantities such as force and velocity involve both *magnitude* and *direction* and cannot be completely characterized by a single real number. To represent such a quantity, you can use a **directed line segment,** as shown in Figure 3.15. The directed line segment $\overrightarrow{PQ}$ has **initial point** P and **terminal point** Q. Its **magnitude** (or length) is denoted by $\|PQ\|$ and can be found using the Distance Formula.

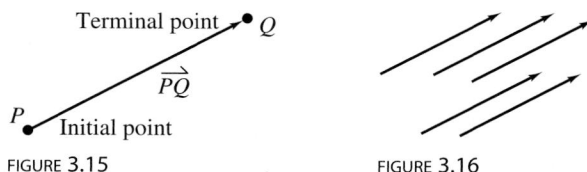

FIGURE 3.15   FIGURE 3.16

Two directed line segments that have the same magnitude and direction are equivalent. For example, the directed line segments in Figure 3.16 are all equivalent. The set of all directed line segments that are equivalent to the directed line segment $\overrightarrow{PQ}$ is a **vector v in the plane,** written $\mathbf{v} = \overrightarrow{PQ}$. Vectors are denoted by lowercase, boldface letters such as $\mathbf{u}$, $\mathbf{v}$, and $\mathbf{w}$.

Example 1 ▶   **Vector Representation by Directed Line Segments**

Let $\mathbf{u}$ be represented by the directed line segment from $P = (0, 0)$ to $Q = (3, 2)$, and let $\mathbf{v}$ be represented by the directed line segment from $R = (1, 2)$ to $S = (4, 4)$, as shown in Figure 3.17. Show that $\mathbf{u} = \mathbf{v}$.

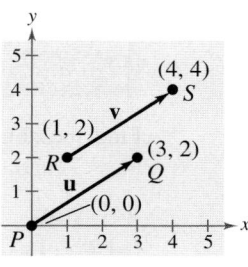

FIGURE 3.17

Solution

From the Distance Formula, it follows that $\overrightarrow{PQ}$ and $\overrightarrow{RS}$ have the *same magnitude*.

$$\|\overrightarrow{PQ}\| = \sqrt{(3 - 0)^2 + (2 - 0)^2} = \sqrt{13}$$
$$\|\overrightarrow{RS}\| = \sqrt{(4 - 1)^2 + (4 - 2)^2} = \sqrt{13}$$

Moreover, both line segments have the *same direction* because they are both directed toward the upper right on lines having a slope of $\frac{2}{3}$. So, $\overrightarrow{PQ}$ and $\overrightarrow{RS}$ have the same magnitude and direction, and it follows that $\mathbf{u} = \mathbf{v}$.

Component Form of a Vector

The directed line segment whose initial point is the origin is often the most convenient representative of a set of equivalent directed line segments. This representative of the vector **v** is in **standard position.**

A vector whose initial point is at the origin $(0, 0)$ can be uniquely represented by the coordinates of its terminal point (v_1, v_2). This is the **component form of a vector v,** written

$$\mathbf{v} = \langle v_1, v_2 \rangle.$$

The coordinates v_1 and v_2 are the *components* of **v.** If both the initial point and the terminal point lie at the origin, **v** is the **zero vector** and is denoted by $\mathbf{0} = \langle 0, 0 \rangle$.

> ### Component Form of a Vector
>
> The component form of the vector with initial point $P = (p_1, p_2)$ and terminal point $Q = (q_1, q_2)$ is
>
> $$\overrightarrow{PQ} = \langle q_1 - p_1, q_2 - p_2 \rangle = \langle v_1, v_2 \rangle = \mathbf{v}.$$
>
> The **magnitude** (or length) of **v** is
>
> $$\|\mathbf{v}\| = \sqrt{(q_1 - p_1)^2 + (q_2 - p_2)^2} = \sqrt{v_1^2 + v_2^2}.$$
>
> If $\|\mathbf{v}\| = 1$, **v** is a **unit vector.** Moreover, $\|\mathbf{v}\| = 0$ if and only if **v** is the zero vector **0**.

Two vectors $\mathbf{u} = \langle u_1, u_2 \rangle$ and $\mathbf{v} = \langle v_1, v_2 \rangle$ are *equal* if and only if $u_1 = v_1$ and $u_2 = v_2$. For instance, in Example 1, the vector **u** from $P = (0, 0)$ to $Q = (3, 2)$ is

$$\mathbf{u} = \overrightarrow{PQ} = \langle 3 - 0, 2 - 0 \rangle = \langle 3, 2 \rangle$$

and the vector **v** from $R = (1, 2)$ to $S = (4, 4)$ is

$$\mathbf{v} = \overrightarrow{RS} = \langle 4 - 1, 4 - 2 \rangle = \langle 3, 2 \rangle.$$

Example 2 ▶   Finding the Component Form of a Vector

Find the component form and magnitude of the vector **v** that has initial point $(4, -7)$ and terminal point $(-1, 5)$.

Solution

Let $P = (4, -7) = (p_1, p_2)$ and let $Q = (-1, 5) = (q_1, q_2)$, as shown in Figure 3.18. Then, the components of $\mathbf{v} = \langle v_1, v_2 \rangle$ are

$$v_1 = q_1 - p_1 = -1 - 4 = -5$$
$$v_2 = q_2 - p_2 = 5 - (-7) = 12.$$

So, $\mathbf{v} = \langle -5, 12 \rangle$ and the magnitude of **v** is

$$\|\mathbf{v}\| = \sqrt{(-5)^2 + 12^2}$$
$$= \sqrt{169} = 13.$$

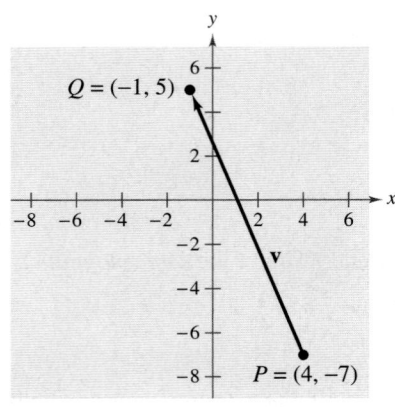

FIGURE 3.18

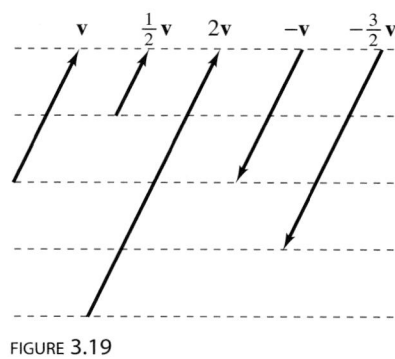

FIGURE **3.19**

Vector Operations ⓦⓦⓦ

The two basic vector operations are **scalar multiplication** and **vector addition.** In operations with vectors, numbers are usually referred to as **scalars.** In this text, scalars will always be real numbers. Geometrically, the product of a vector **v** and a scalar k is the vector that is $|k|$ times as long as **v.** If k is positive, k**v** has the same direction as **v,** and if k is negative, k**v** has the direction opposite that of **v,** as shown in Figure 3.19.

To add two vectors geometrically, position them (without changing length or direction) so that the initial point of one coincides with the terminal point of the other. The sum **u** + **v** is formed by joining the initial point of the second vector **v** with the terminal point of the first vector **u,** as shown in Figure 3.20. This technique is called the **parallelogram law** for vector addition because the vector **u** + **v,** often called the **resultant** of vector addition, is the diagonal of a parallelogram having **u** and **v** as its adjacent sides.

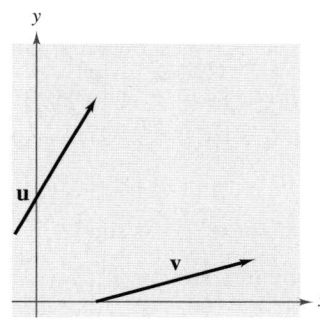

FIGURE **3.20**

The graphical representation of the difference of two vectors may not be obvious to your students. You may want to go over this carefully.

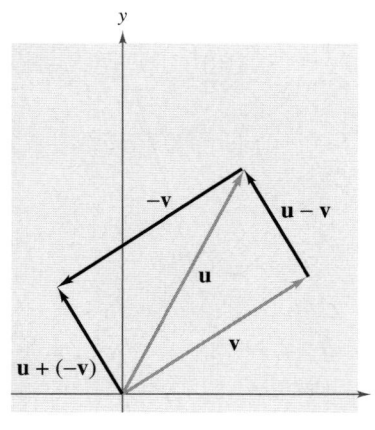

$$\mathbf{u} - \mathbf{v} = \mathbf{u} + (-\mathbf{v})$$

FIGURE **3.21**

Definitions of Vector Addition and Scalar Multiplication

Let $\mathbf{u} = \langle u_1, u_2 \rangle$ and $\mathbf{v} = \langle v_1, v_2 \rangle$ be vectors and let k be a scalar (a real number). Then the *sum* of **u** and **v** is the vector

$$\mathbf{u} + \mathbf{v} = \langle u_1 + v_1, u_2 + v_2 \rangle \qquad \text{Sum}$$

and the *scalar multiple* of k times **u** is the vector

$$k\mathbf{u} = k\langle u_1, u_2 \rangle = \langle ku_1, ku_2 \rangle. \qquad \text{Scalar multiple}$$

The *negative* of $\mathbf{v} = \langle v_1, v_2 \rangle$ is

$$-\mathbf{v} = (-1)\mathbf{v}$$

$$= \langle -v_1, -v_2 \rangle \qquad \text{Negative}$$

and the *difference* of **u** and **v** is

$$\mathbf{u} - \mathbf{v} = \mathbf{u} + (-\mathbf{v}) \qquad \text{Add } (-\mathbf{v}). \text{ See Figure 3.21.}$$

$$= \langle u_1 - v_1, u_2 - v_2 \rangle. \qquad \text{Difference}$$

To represent $\mathbf{u} - \mathbf{v}$ geometrically, you can use directed line segments with the *same* initial point. The difference $\mathbf{u} - \mathbf{v}$ is the vector from the terminal point of **v** to the terminal point of **u,** which is equal to $\mathbf{u} + (-\mathbf{v})$, as shown in Figure 3.21.

The component definitions of vector addition and scalar multiplication are illustrated in Example 3. In this example, notice that each of the vector operations can be interpreted geometrically.

Example 3 ▶ **Vector Operations**

Let $\mathbf{v} = \langle -2, 5 \rangle$ and $\mathbf{w} = \langle 3, 4 \rangle$, and find each of the following vectors.

a. $2\mathbf{v}$　　**b.** $\mathbf{w} - \mathbf{v}$　　**c.** $\mathbf{v} + 2\mathbf{w}$

Solution

a. Because $\mathbf{v} = \langle -2, 5 \rangle$, you have

$$2\mathbf{v} = 2\langle -2, 5 \rangle$$
$$= \langle 2(-2), 2(5) \rangle$$
$$= \langle -4, 10 \rangle.$$

A sketch of $2\mathbf{v}$ is shown in Figure 3.22.

b. The difference of $\mathbf{w}$ and $\mathbf{v}$ is

$$\mathbf{w} - \mathbf{v} = \langle 3 - (-2), 4 - 5 \rangle$$
$$= \langle 5, -1 \rangle.$$

A sketch of $\mathbf{w} - \mathbf{v}$ is shown in Figure 3.23.

c. The sum of $\mathbf{v}$ and $2\mathbf{w}$ is

$$\mathbf{v} + 2\mathbf{w} = \langle -2, 5 \rangle + 2\langle 3, 4 \rangle$$
$$= \langle -2, 5 \rangle + \langle 2(3), 2(4) \rangle$$
$$= \langle -2, 5 \rangle + \langle 6, 8 \rangle$$
$$= \langle -2 + 6, 5 + 8 \rangle$$
$$= \langle 4, 13 \rangle.$$

A sketch of $\mathbf{v} + 2\mathbf{w}$ is shown in Figure 3.24.

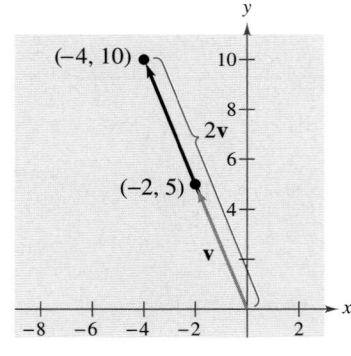

FIGURE **3.22**

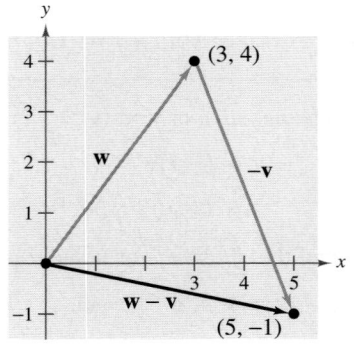

FIGURE **3.23**

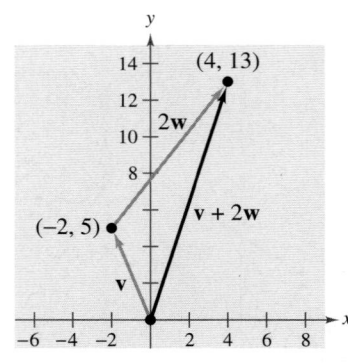

FIGURE **3.24**

Vector addition and scalar multiplication share many of the properties of ordinary arithmetic.

Properties of Vector Addition and Scalar Multiplication

Let **u**, **v**, and **w** be vectors and let c and d be scalars. Then the following properties are true.

1. $\mathbf{u} + \mathbf{v} = \mathbf{v} + \mathbf{u}$
2. $(\mathbf{u} + \mathbf{v}) + \mathbf{w} = \mathbf{u} + (\mathbf{v} + \mathbf{w})$
3. $\mathbf{u} + \mathbf{0} = \mathbf{u}$
4. $\mathbf{u} + (-\mathbf{u}) = \mathbf{0}$
5. $c(d\mathbf{u}) = (cd)\mathbf{u}$
6. $(c + d)\mathbf{u} = c\mathbf{u} + d\mathbf{u}$
7. $c(\mathbf{u} + \mathbf{v}) = c\mathbf{u} + c\mathbf{v}$
8. $1(\mathbf{u}) = \mathbf{u}, \, 0(\mathbf{u}) = \mathbf{0}$
9. $\|c\mathbf{v}\| = |c| \, \|\mathbf{v}\|$

Property 9 can be stated as follows: the magnitude of the vector $c\mathbf{v}$ is the absolute value of c times the magnitude of **v**.

Unit Vectors

In many applications of vectors, it is useful to find a unit vector that has the same direction as a given nonzero vector **v**. To do this, you can divide **v** by its magnitude to obtain

$$\mathbf{u} = \text{unit vector} = \frac{\mathbf{v}}{\|\mathbf{v}\|} = \left(\frac{1}{\|\mathbf{v}\|}\right)\mathbf{v}. \qquad \text{Unit vector in direction of } \mathbf{v}$$

Note that **u** is a scalar multiple of **v**. The vector **u** has magnitude of 1 and the same direction as **v**. The vector **u** is called a **unit vector in the direction of v**.

Example 4 ▶ **Finding a Unit Vector**

Find a unit vector in the direction of $\mathbf{v} = \langle -2, 5\rangle$ and verify that the result has a magnitude of 1.

Solution

The unit vector in the direction of **v** is

$$\frac{\mathbf{v}}{\|\mathbf{v}\|} = \frac{\langle -2, 5\rangle}{\sqrt{(-2)^2 + (5)^2}}$$

$$= \frac{1}{\sqrt{29}}\langle -2, 5\rangle$$

$$= \left\langle \frac{-2}{\sqrt{29}}, \frac{5}{\sqrt{29}}\right\rangle.$$

This vector has a magnitude of 1 because

$$\sqrt{\left(\frac{-2}{\sqrt{29}}\right)^2 + \left(\frac{5}{\sqrt{29}}\right)^2} = \sqrt{\frac{4}{29} + \frac{25}{29}} = \sqrt{\frac{29}{29}} = 1.$$

Historical Note
William Rowan Hamilton (1805–1865), an Irish mathematician, did some of the earliest work with vectors. Hamilton spent many years developing a system of vector-like quantities called quaternions. Although Hamilton was convinced of the benefits of quaternions, the operations he defined did not produce good models for physical phenomena. It wasn't until the latter half of the nineteenth century that the Scottish physicist James Maxwell (1831–1879) restructured Hamilton's quaternions in a form useful for representing physical quantities such as force, velocity, and acceleration.

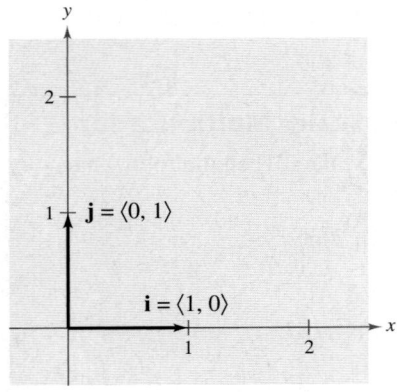

FIGURE 3.25

The unit vectors $\langle 1, 0 \rangle$ and $\langle 0, 1 \rangle$ are called the **standard unit vectors** and are denoted by

$$\mathbf{i} = \langle 1, 0 \rangle \quad \text{and} \quad \mathbf{j} = \langle 0, 1 \rangle$$

as shown in Figure 3.25. (Note that the lowercase letter $\mathbf{i}$ is written in boldface to distinguish it from the imaginary number $i = \sqrt{-1}$.) These vectors can be used to represent any vector $\mathbf{v} = \langle v_1, v_2 \rangle$, as follows.

$$\mathbf{v} = \langle v_1, v_2 \rangle$$
$$= v_1 \langle 1, 0 \rangle + v_2 \langle 0, 1 \rangle$$
$$= v_1 \mathbf{i} + v_2 \mathbf{j}$$

The scalars v_1 and v_2 are called the **horizontal** and **vertical components of v,** respectively. The vector sum

$$v_1 \mathbf{i} + v_2 \mathbf{j}$$

is called a **linear combination** of the vectors $\mathbf{i}$ and $\mathbf{j}$. Any vector in the plane can be expressed as a linear combination of the standard unit vectors $\mathbf{i}$ and $\mathbf{j}$.

Example 5 ▶ Writing a Linear Combination of Unit Vectors

Let $\mathbf{u}$ be the vector with initial point $(2, -5)$ and terminal point $(-1, 3)$. Write $\mathbf{u}$ as a linear combination of the standard unit vectors $\mathbf{i}$ and $\mathbf{j}$.

Solution

Begin by writing the component form of the vector $\mathbf{u}$.

$$\mathbf{u} = \langle -1 - 2, 3 + 5 \rangle$$
$$= \langle -3, 8 \rangle$$
$$= -3\mathbf{i} + 8\mathbf{j}$$

This result is shown graphically in Figure 3.26.

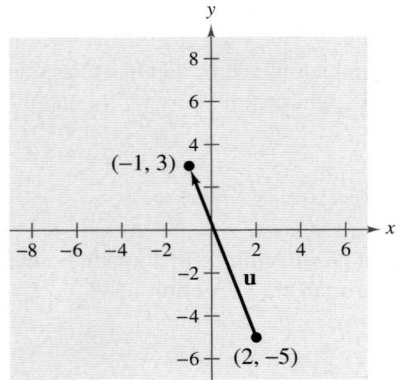

FIGURE 3.26

Example 6 ▶ Vector Operations

Let $\mathbf{u} = -3\mathbf{i} + 8\mathbf{j}$ and let $\mathbf{v} = 2\mathbf{i} - \mathbf{j}$. Find $2\mathbf{u} - 3\mathbf{v}$.

Solution

You could solve this problem by converting $\mathbf{u}$ and $\mathbf{v}$ to component form. This, however, is not necessary. It is just as easy to perform the operations in unit vector form.

$$2\mathbf{u} - 3\mathbf{v} = 2(-3\mathbf{i} + 8\mathbf{j}) - 3(2\mathbf{i} - \mathbf{j})$$
$$= -6\mathbf{i} + 16\mathbf{j} - 6\mathbf{i} + 3\mathbf{j}$$
$$= -12\mathbf{i} + 19\mathbf{j}$$

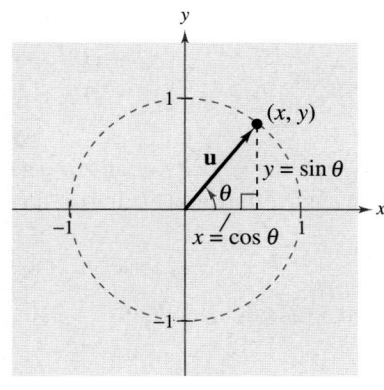

FIGURE 3.27 $\|\mathbf{u}\| = 1$

Direction Angles

If **u** is a *unit vector* such that θ is the angle (measured counterclockwise) from the positive x-axis to **u**, the terminal point of **u** lies on the unit circle and you have

$$\mathbf{u} = \langle x, y \rangle = \langle \cos \theta, \sin \theta \rangle = (\cos \theta)\mathbf{i} + (\sin \theta)\mathbf{j}$$

as shown in Figure 3.27. The angle θ is the **direction angle** of the vector **u**.

Suppose that **u** is a unit vector with direction angle θ. If **v** is any vector that makes an angle θ with the positive x-axis, it has the same direction as **u** and you can write

$$\mathbf{v} = \|\mathbf{v}\| \langle \cos \theta, \sin \theta \rangle$$
$$= \|\mathbf{v}\| (\cos \theta)\mathbf{i} + \|\mathbf{v}\| (\sin \theta)\mathbf{j}.$$

Because $\mathbf{v} = a\mathbf{i} + b\mathbf{j} = \|\mathbf{v}\| (\cos \theta) \mathbf{i} + \|\mathbf{v}\| (\sin \theta) \mathbf{j}$, it follows that the direction angle θ for **v** is determined from

$$\tan \theta = \frac{\sin \theta}{\cos \theta} \qquad \text{Quotient identity}$$

$$= \frac{\|\mathbf{v}\| \sin \theta}{\|\mathbf{v}\| \cos \theta} \qquad \text{Multiply numerator and denominator by } \|\mathbf{v}\|.$$

$$= \frac{b}{a}. \qquad \text{Simplify.}$$

Example 7 ▶ Finding Direction Angles of Vectors

Find the direction angle of each vector.

a. $\mathbf{u} = 3\mathbf{i} + 3\mathbf{j}$

b. $\mathbf{v} = 3\mathbf{i} - 4\mathbf{j}$

Solution

a. The direction angle is

$$\tan \theta = \frac{b}{a} = \frac{3}{3} = 1.$$

So, $\theta = 45°$, as shown in Figure 3.28.

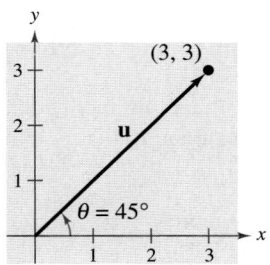

FIGURE 3.28

b. The direction angle is

$$\tan \theta = \frac{b}{a} = \frac{-4}{3}.$$

Moreover, because $\mathbf{v} = 3\mathbf{i} - 4\mathbf{j}$ lies in Quadrant IV, θ lies in Quadrant IV and its reference angle is

$$\theta = \left| \arctan\left(-\frac{4}{3} \right) \right| \approx |-53.13°| = 53.13°.$$

So, it follows that $\theta \approx 360° - 53.13° = 306.87°$, as shown in Figure 3.29.

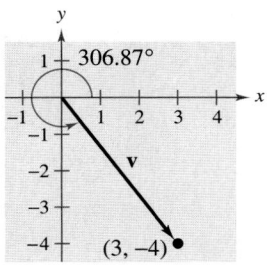

FIGURE 3.29

Applications of Vectors

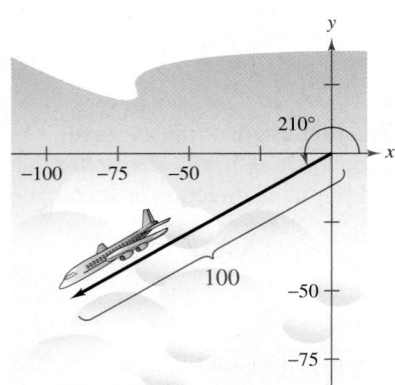

FIGURE **3.30**

Example 8 ▶ Finding the Component Form of a Vector

Find the component form of the vector that represents the velocity of an airplane descending at a speed of 100 miles per hour at an angle 30° below the horizontal, as shown in Figure 3.30.

Solution

The velocity vector $\mathbf{v}$ has a magnitude of 100 and a direction angle of $\theta = 210°$.

$$\mathbf{v} = \|\mathbf{v}\| (\cos \theta)\mathbf{i} + \|\mathbf{v}\| (\sin \theta)\mathbf{j}$$
$$= 100(\cos 210°)\mathbf{i} + 100(\sin 210°)\mathbf{j}$$
$$= 100\left(\frac{-\sqrt{3}}{2}\right)\mathbf{i} + 100\left(\frac{-1}{2}\right)\mathbf{j}$$
$$= -50\sqrt{3}\,\mathbf{i} - 50\mathbf{j}$$
$$= \langle -50\sqrt{3}, -50 \rangle$$

You can check that $\mathbf{v}$ has a magnitude of 100, as follows.

$$\|\mathbf{v}\| = \sqrt{\left(-50\sqrt{3}\right)^2 + (50)^2}$$
$$= \sqrt{7500 + 2500}$$
$$= \sqrt{10,000}$$
$$= 100$$

Example 9 ▶ Using Vectors to Determine Weight

A force of 600 pounds is required to pull a boat and trailer up a ramp inclined at 15° from the horizontal. Find the combined weight of the boat and trailer.

Solution

Based on Figure 3.31, you can make the following observations.

$\|\overrightarrow{BA}\| =$ force of gravity $=$ combined weight of boat and trailer

$\|\overrightarrow{BC}\| =$ force against ramp

$\|\overrightarrow{AC}\| =$ force required to move boat up ramp $=$ 600 pounds

By construction, triangles BWD and ABC are similar. So, angle ABC is 15°, and so in triangle ABC you have

$$\sin 15° = \frac{\|\overrightarrow{AC}\|}{\|\overrightarrow{BA}\|} = \frac{600}{\|\overrightarrow{BA}\|}$$

$$\|BA\| = \frac{600}{\sin 15°} \approx 2318.$$

Consequently, the combined weight is approximately 2318 pounds. (In Figure 3.31, note that $\overrightarrow{AC}$ is parallel to the ramp.)

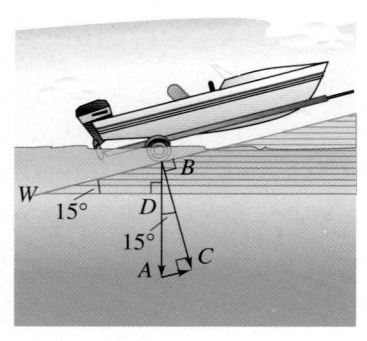

FIGURE **3.31**

Example 10 ▶ Using Vectors to Find Speed and Direction

An airplane is traveling at a speed of 500 miles per hour on a bearing of 330° at a fixed altitude with a negligible wind velocity as shown in Figure 3.32(a). When the airplane reaches a certain point, it encounters a wind with a velocity of 70 miles per hour in the direction N 45° E, as shown in Figure 3.32(b). What are the resultant speed and direction of the airplane?

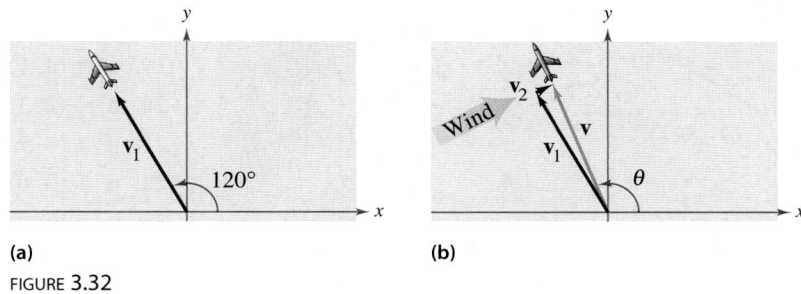

(a)   (b)

FIGURE 3.32

Solution

Using Figure 3.32, the velocity of the airplane (alone) is

$$\mathbf{v}_1 = 500\langle \cos 120°, \sin 120° \rangle$$
$$= \langle -250, 250\sqrt{3} \rangle$$

and the velocity of the wind is

$$\mathbf{v}_2 = 70\langle \cos 45°, \sin 45° \rangle$$
$$= \langle 35\sqrt{2}, 35\sqrt{2} \rangle.$$

So, the velocity of the airplane (in the wind) is

$$\mathbf{v} = \mathbf{v}_1 + \mathbf{v}_2$$
$$= \langle -250 + 35\sqrt{2}, 250\sqrt{3} + 35\sqrt{2} \rangle$$
$$\approx \langle -200.5, 482.5 \rangle$$

and the resultant speed of the airplane is

$$\|\mathbf{v}\| = \sqrt{(-200.5)^2 + (482.5)^2}$$
$$\approx 522.5 \text{ miles per hour.}$$

Finally, if θ is the direction angle of the flight path, you have

$$\tan \theta = \frac{482.5}{-200.5}$$
$$\approx -2.4065$$

which implies that

$$\theta \approx 180° + \arctan(-2.4065) \approx 180° - 67.4° = 112.6°.$$

So, the true direction of the airplane is 337.4°.

3.3 Exercises

In Exercises 1–12, find the component form and the magnitude of the vector **v**.

1.
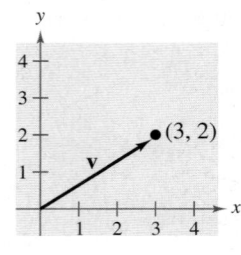

2.
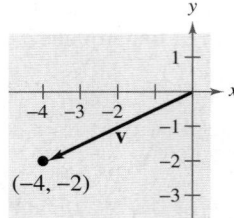

3.
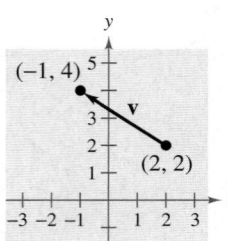

4.
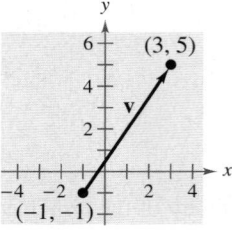

5.
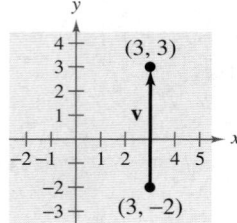

6.
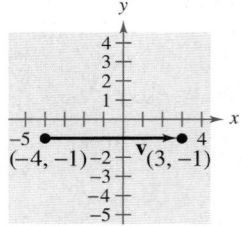

Initial Point	*Terminal Point*
7. $(-1, 5)$	$(15, 12)$
8. $(1, 11)$	$(9, 3)$
9. $(-3, -5)$	$(5, 1)$
10. $(-3, 11)$	$(9, 40)$
11. $(1, 3)$	$(-8, -9)$
12. $(-2, 7)$	$(5, -17)$

In Exercises 13–18, use the figure to sketch a graph of the specified vector. To print an enlarged copy of the graph, go to the website, *www.mathgraphs.com*.

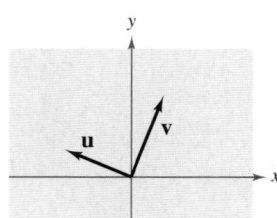

13. $-\mathbf{v}$

14. $5\mathbf{v}$

15. $\mathbf{u} + \mathbf{v}$

16. $\mathbf{u} - \mathbf{v}$

17. $\mathbf{u} + 2\mathbf{v}$

18. $\mathbf{v} - \frac{1}{2}\mathbf{u}$

In Exercises 19–26, find (a) $\mathbf{u} + \mathbf{v}$, (b) $\mathbf{u} - \mathbf{v}$, and (c) $2\mathbf{u} - 3\mathbf{v}$. Then sketch the resultant vector.

19. $\mathbf{u} = \langle 2, 1 \rangle$, $\mathbf{v} = \langle 1, 3 \rangle$

20. $\mathbf{u} = \langle 2, 3 \rangle$, $\mathbf{v} = \langle 4, 0 \rangle$

21. $\mathbf{u} = \langle -5, 3 \rangle$, $\mathbf{v} = \langle 0, 0 \rangle$

22. $\mathbf{u} = \langle 0, 0 \rangle$, $\mathbf{v} = \langle 2, 1 \rangle$

23. $\mathbf{u} = \mathbf{i} + \mathbf{j}$, $\mathbf{v} = 2\mathbf{i} - 3\mathbf{j}$

24. $\mathbf{u} = -2\mathbf{i} + \mathbf{j}$, $\mathbf{v} = -\mathbf{i} + 2\mathbf{j}$

25. $\mathbf{u} = 2\mathbf{i}$, $\mathbf{v} = \mathbf{j}$

26. $\mathbf{u} = 3\mathbf{j}$, $\mathbf{v} = 2\mathbf{i}$

In Exercises 27–36, find a unit vector in the direction of the given vector.

27. $\mathbf{u} = \langle 3, 0 \rangle$

28. $\mathbf{u} = \langle 0, -2 \rangle$

29. $\mathbf{v} = \langle -2, 2 \rangle$

30. $\mathbf{v} = \langle 5, -12 \rangle$

31. $\mathbf{v} = 6\mathbf{i} - 2\mathbf{j}$

32. $\mathbf{v} = \mathbf{i} + \mathbf{j}$

33. $\mathbf{w} = 4\mathbf{j}$

34. $\mathbf{w} = -6\mathbf{i}$

35. $\mathbf{w} = \mathbf{i} - 2\mathbf{j}$

36. $\mathbf{w} = 7\mathbf{j} - 3\mathbf{i}$

In Exercises 37–40, find the vector **v** with the given magnitude and the same direction as **u**.

	Magnitude	*Direction*
37.	$\|\mathbf{v}\| = 5$	$\mathbf{u} = \langle 3, 3 \rangle$
38.	$\|\mathbf{v}\| = 6$	$\mathbf{u} = \langle -3, 3 \rangle$
39.	$\|\mathbf{v}\| = 9$	$\mathbf{u} = \langle 2, 5 \rangle$
40.	$\|\mathbf{v}\| = 10$	$\mathbf{u} = \langle -10, 0 \rangle$

In Exercises 41–46, find the component form of **v** and sketch the specified vector operations geometrically, where $\mathbf{u} = 2\mathbf{i} - \mathbf{j}$ and $\mathbf{w} = \mathbf{i} + 2\mathbf{j}$.

41. $\mathbf{v} = \frac{3}{2}\mathbf{u}$

42. $\mathbf{v} = \frac{3}{4}\mathbf{w}$

43. $\mathbf{v} = \mathbf{u} + 2\mathbf{w}$

44. $\mathbf{v} = -\mathbf{u} + \mathbf{w}$

45. $\mathbf{v} = \frac{1}{2}(3\mathbf{u} + \mathbf{w})$

46. $\mathbf{v} = \mathbf{u} - 2\mathbf{w}$

In Exercises 47–50, find the magnitude and direction angle of the vector **v**.

47. $\mathbf{v} = 3(\cos 60°\mathbf{i} + \sin 60°\mathbf{j})$

48. $\mathbf{v} = 8(\cos 135°\mathbf{i} + \sin 135°\mathbf{j})$

49. $\mathbf{v} = 6\mathbf{i} - 6\mathbf{j}$ **50.** $\mathbf{v} = -5\mathbf{i} + 4\mathbf{j}$

In Exercises 51–58, find the component form of v given its magnitude and the angle it makes with the positive x-axis. Sketch v.

Magnitude	Angle
51. $\|\mathbf{v}\| = 3$	$\theta = 0°$
52. $\|\mathbf{v}\| = 1$	$\theta = 45°$
53. $\|\mathbf{v}\| = \frac{7}{2}$	$\theta = 150°$
54. $\|\mathbf{v}\| = \frac{5}{2}$	$\theta = 45°$
55. $\|\mathbf{v}\| = 3\sqrt{2}$	$\theta = 150°$
56. $\|\mathbf{v}\| = 4\sqrt{3}$	$\theta = 90°$
57. $\|\mathbf{v}\| = 2$	$\mathbf{v}$ in the direction $\mathbf{i} + 3\mathbf{j}$
58. $\|\mathbf{v}\| = 3$	$\mathbf{v}$ in the direction $3\mathbf{i} + 4\mathbf{j}$

In Exercises 59–62, find the component form of the sum of u and v with direction angles θ_u and θ_v.

Magnitude	Angle
59. $\|\mathbf{u}\| = 5$	$\theta_u = 0°$
$\|\mathbf{v}\| = 5$	$\theta_v = 90°$
60. $\|\mathbf{u}\| = 4$	$\theta_u = 60°$
$\|\mathbf{v}\| = 4$	$\theta_v = 90°$
61. $\|\mathbf{u}\| = 20$	$\theta_u = 45°$
$\|\mathbf{v}\| = 50$	$\theta_v = 180°$
62. $\|\mathbf{u}\| = 50$	$\theta_u = 30°$
$\|\mathbf{v}\| = 30$	$\theta_v = 110°$

In Exercises 63–66, use the Law of Cosines to find the angle α between the vectors. (Assume $0° \leq \alpha \leq 180°$.)

63. $\mathbf{v} = \mathbf{i} + \mathbf{j}, \quad \mathbf{w} = 2\mathbf{i} - 2\mathbf{j}$

64. $\mathbf{v} = 3\mathbf{i} - 2\mathbf{j}, \quad \mathbf{w} = 2\mathbf{i} + 2\mathbf{j}$

65. $\mathbf{v} = \mathbf{i} + \mathbf{j}, \quad \mathbf{w} = 3\mathbf{i} - \mathbf{j}$

66. $\mathbf{v} = \mathbf{i} + 2\mathbf{j}, \quad \mathbf{w} = 2\mathbf{i} - \mathbf{j}$

Resultant Force In Exercises 67 and 68, find the angle between the forces given the magnitude of their resultant. (*Hint:* Write force 1 as a vector in the direction of the positive x-axis and force 2 as a vector at an angle θ with the positive x-axis.)

Force 1	Force 2	Resultant Force
67. 45 pounds	60 pounds	90 pounds
68. 3000 pounds	1000 pounds	3750 pounds

69. ***Resultant Force*** Forces with magnitudes of 125 newtons and 300 newtons act on a hook (see figure). The angle between the two forces is 45°. Find the direction and magnitude of the resultant of these forces.

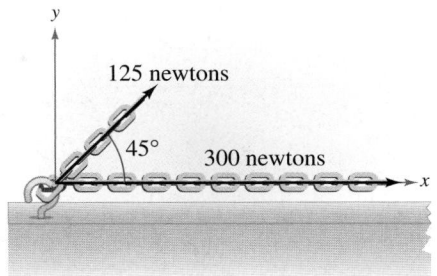

70. ***Resultant Force*** Forces with magnitudes of 2000 newtons and 900 newtons act on a machine part at angles of 30° and −45°, respectively, with the x-axis (see figure). Find the direction and magnitude of the resultant of these forces.

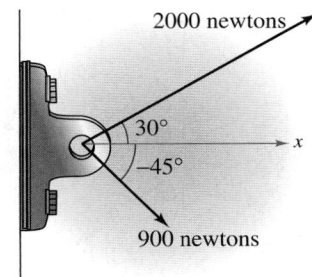

71. ***Resultant Force*** Three forces with magnitudes of 75 pounds, 100 pounds, and 125 pounds act on an object at angles of 30°, 45°, and 120°, respectively, with the positive x-axis. Find the direction and magnitude of the resultant of these forces.

72. ***Resultant Force*** Three forces with magnitudes of 70 pounds, 40 pounds, and 60 pounds act on an object at angles of −30°, 45°, and 135°, respectively, with the positive x-axis. Find the direction and magnitude of the resultant of these forces.

73. *Velocity*   A ball is thrown with an initial velocity of 70 feet per second, at an angle of 35° with the horizontal (see figure). Find the vertical and horizontal components of the velocity.

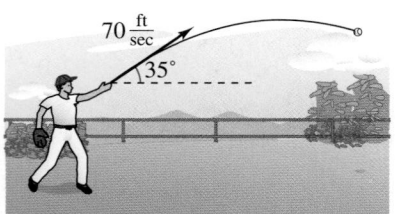

74. *Velocity*   A gun with a muzzle velocity of 1200 feet per second is fired at an angle of 6° with the horizontal. Find the vertical and horizontal components of the velocity.

Cable Tension   **In Exercises 75 and 76, use the figure to determine the tension in each cable supporting the load.**

75.   **76.**

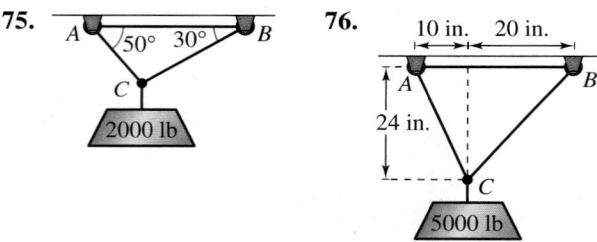

77. *Tow Line Tension*   A loaded barge is being towed by two tugboats, and the magnitude of the resultant is 6000 pounds directed along the axis of the barge (see figure). Find the tension in the tow lines if they each make an 18° angle with the axis of the barge.

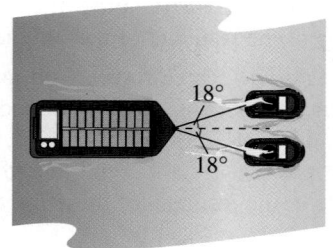

78. *Rope Tension*   To carry a 100-pound cylindrical weight, two people lift on the ends of short ropes that are tied to an eyelet on the top center of the cylinder. Each rope makes a 20° angle with the vertical. Draw a figure that gives a visual representation of the problem, and find the tension in the ropes.

79. *Navigation*   An airplane is flying in the direction of 148°, with an airspeed of 875 kilometers per hour. Because of the wind, its groundspeed and direction are 800 kilometers per hour and 140°, respectively (see figure). Find the direction and speed of the wind.

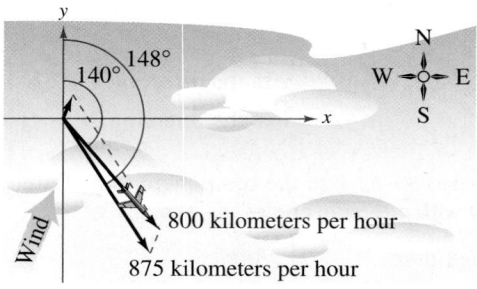

▶ **Model It**

80. *Navigation*   A commercial jet is flying from Miami to Seattle. The jet's velocity with respect to the air is 580 miles per hour, and its bearing is 332°. The wind, at the altitude of the plane, is blowing from the southwest with a velocity of 60 miles per hour.

  (a) Draw a figure that gives a visual representation of the problem.

  (b) Write the velocity of the wind as a vector in component form.

  (c) Write the velocity of the jet relative to the air in component form.

  (d) What is the speed of the jet with respect to the ground?

  (e) What is the true direction of the jet?

81. *Work*   A heavy implement is pulled 30 feet across a floor, using a force of 100 pounds. The force is 50° above the horizontal (see figure). Find the work done. (Use the formula for work, $W = FD$, where F is the component of the force in the direction of motion and D is the distance.)

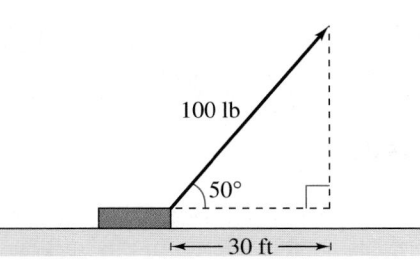

100 lb

50°

← 30 ft →

FIGURE FOR 81

82. ***Rope Tension*** A tetherball weighing 1 pound is pulled outward from the pole by a horizontal force **u** until the rope makes a 45° angle with the pole (see figure). Determine the resulting tension in the rope and the magnitude of **u**.

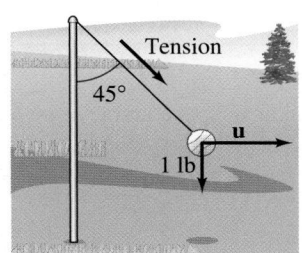

Tension

45°

u

1 lb

Synthesis

True or False? **In Exercises 83 and 84, decide whether the statement is true or false. Justify your answer.**

83. If **u** and **v** have the same magnitude and direction, then **u** = **v**.

84. If **u** = $a\mathbf{i} + b\mathbf{j}$ is a unit vector, then $a^2 + b^2 = 1$.

85. ***Think About It*** Consider two forces of equal magnitude acting on a point.

(a) If the magnitude of the resultant is the sum of the magnitudes of the two forces, make a conjecture about the angle between the forces.

(b) If the resultant of the forces is **0**, make a conjecture about the angle between the forces.

(c) Can the magnitude of the resultant be greater than the sum of the magnitudes of the two forces? Explain.

86. ***Graphical Reasoning*** Consider two forces

$\mathbf{F}_1 = \langle 10, 0 \rangle$ and $\mathbf{F}_2 = 5\langle \cos \theta, \sin \theta \rangle$.

(a) Find $\|\mathbf{F}_1 + \mathbf{F}_2\|$ as a function of θ.

(b) Use a graphing utility to graph the function in part (a) for $0 \le \theta < 2\pi$.

(c) Use the graph in part (b) to determine the range of the function. What is its maximum, and for what value of θ does it occur? What is its minimum, and for what value of θ does it occur?

(d) Explain why the magnitude of the resultant is never 0.

87. ***Proof*** Prove that $(\cos \theta)\mathbf{i} + (\sin \theta)\mathbf{j}$ is a unit vector for any value of θ.

88. ***Technology*** Write a program for your graphing utility that graphs two vectors and their difference given the vectors in component form.

In Exercises 89 and 90, use the program in Exercise 88 to find the difference of the vectors shown in the figure.

89.

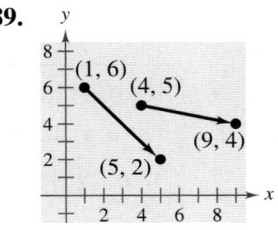

(1, 6)
(4, 5)
(5, 2)
(9, 4)

90.

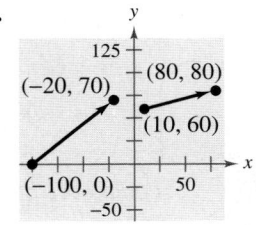

(−20, 70)
(80, 80)
(10, 60)
(−100, 0)

Review

In Exercises 91–94, use the specified trigonometric substitution to write the algebraic expression as a trigonometric function of θ, where $0 < \theta < \pi/2$.

91. $\sqrt{x^2 - 64}, \quad x = 8 \sec \theta$

92. $\sqrt{64 - x^2}, \quad x = 8 \sin \theta$

93. $\sqrt{x^2 + 36}, \quad x = 6 \tan \theta$

94. $\sqrt{(x^2 - 25)^3}, \quad x = 5 \sec \theta$

In Exercises 95–98, solve the equation.

95. $\cos x(\cos x + 1) = 0$

96. $\sin x(2 \sin x + \sqrt{2}) = 0$

97. $3 \sec x \sin x - 2\sqrt{3} \sin x = 0$

98. $\cos x \csc x + \cos x \sqrt{2} = 0$

3.4 Vectors and Dot Products

▶ **What you should learn**

- How to find the dot product of two vectors and use the Properties of the Dot Product
- How to find the angle between two vectors and determine whether two vectors are orthogonal
- How to write a vector as the sum of two vector components
- How to use vectors to find the work done by a force

▶ **Why you should learn it**

You can use the dot product of two vectors to solve real-life problems involving two vector quantities. For instance, in Exercise 48 on page 312, you can use the dot product to find the force necessary to keep a truck from rolling down a hill.

Alan Thornton/Tony Stone Images

The Dot Product of Two Vectors

So far you have studied two vector operations—vector addition and multiplication by a scalar—each of which yields another vector. In this section you will study a third vector operation, the **dot product.** This product yields a scalar, rather than a vector.

Definition of Dot Product

The **dot product** of $\mathbf{u} = \langle u_1, u_2 \rangle$ and $\mathbf{v} = \langle v_1, v_2 \rangle$ is

$$\mathbf{u} \cdot \mathbf{v} = u_1 v_1 + u_2 v_2.$$

Properties of the Dot Product

Let $\mathbf{u}$, $\mathbf{v}$, and $\mathbf{w}$ be vectors in the plane or in space and let c be a scalar.

1. $\mathbf{u} \cdot \mathbf{v} = \mathbf{v} \cdot \mathbf{u}$

2. $\mathbf{0} \cdot \mathbf{v} = 0$

3. $\mathbf{u} \cdot (\mathbf{v} + \mathbf{w}) = \mathbf{u} \cdot \mathbf{v} + \mathbf{u} \cdot \mathbf{w}$

4. $\mathbf{v} \cdot \mathbf{v} = \|\mathbf{v}\|^2$

5. $c(\mathbf{u} \cdot \mathbf{v}) = c\mathbf{u} \cdot \mathbf{v} = \mathbf{u} \cdot c\mathbf{v}$

For proofs of the Properties of the Dot Product, see Proofs in Mathematics on page 323.

Example 1 ▶ Finding Dot Products

Find each dot product.

a. $\langle 4, 5 \rangle \cdot \langle 2, 3 \rangle$ **b.** $\langle 2, -1 \rangle \cdot \langle 1, 2 \rangle$ **c.** $\langle 0, 3 \rangle \cdot \langle 4, -2 \rangle$

Solution

a. $\langle 4, 5 \rangle \cdot \langle 2, 3 \rangle = 4(2) + 5(3)$
$$= 8 + 15$$
$$= 23$$

b. $\langle 2, -1 \rangle \cdot \langle 1, 2 \rangle = 2(1) + (-1)(2) = 2 - 2 = 0$

c. $\langle 0, 3 \rangle \cdot \langle 4, -2 \rangle = 0(4) + 3(-2) = 0 - 6 = -6$

In Example 1, be sure you see that the dot product of two vectors is a scalar (a real number), not a vector. Moreover, notice that the dot product can be positive, zero, or negative.

Example 2 ▶ **Using Properties of Dot Products**

Let $\mathbf{u} = \langle -1, 3 \rangle$, $\mathbf{v} = \langle 2, -4 \rangle$, and $\mathbf{w} = \langle 1, -2 \rangle$. Find each dot product.

a. $(\mathbf{u} \cdot \mathbf{v})\mathbf{w}$ **b.** $\mathbf{u} \cdot 2\mathbf{v}$

Solution

Begin by finding the dot product of $\mathbf{u}$ and $\mathbf{v}$.

$$\begin{aligned}
\mathbf{u} \cdot \mathbf{v} &= \langle -1, 3 \rangle \cdot \langle 2, -4 \rangle \\
&= (-1)(2) + 3(-4) \\
&= -14
\end{aligned}$$

a. $\begin{aligned}[t] (\mathbf{u} \cdot \mathbf{v})\mathbf{w} &= -14\langle 1, -2 \rangle \\ &= \langle -14, 28 \rangle \end{aligned}$

b. $\begin{aligned}[t] \mathbf{u} \cdot 2\mathbf{v} &= 2(\mathbf{u} \cdot \mathbf{v}) \\ &= 2(-14) \\ &= -28 \end{aligned}$

Notice that the first product is a vector, whereas the second is a scalar. Can you see why?

Example 3 ▶ **Dot Product and Magnitude**

The dot product of $\mathbf{u}$ with itself is 5. What is the magnitude of $\mathbf{u}$?

Solution

Because $\|\mathbf{u}\|^2 = \mathbf{u} \cdot \mathbf{u}$ and $\mathbf{u} \cdot \mathbf{u} = 5$, it follows that

$$\begin{aligned}
\|\mathbf{u}\| &= \sqrt{\mathbf{u} \cdot \mathbf{u}} \\
&= \sqrt{5}.
\end{aligned}$$

The Angle Between Two Vectors

The **angle between two nonzero vectors** is the angle θ, $0 \le \theta \le \pi$, between their respective standard position vectors, as shown in Figure 3.33 on page 306. This angle can be found using the dot product. (Note that the angle between the zero vector and another vector is not defined.)

Angle Between Two Vectors

If θ is the angle between two nonzero vectors $\mathbf{u}$ and $\mathbf{v}$, then

$$\cos \theta = \frac{\mathbf{u} \cdot \mathbf{v}}{\|\mathbf{u}\|\,\|\mathbf{v}\|}.$$

For a proof of the angle between two vectors, see Proofs in Mathematics on page 323.

Example 4 ▶ **Finding the Angle Between Two Vectors**

Find the angle between $\mathbf{u} = \langle 4, 3 \rangle$ and $\mathbf{v} = \langle 3, 5 \rangle$.

Solution

$$\cos \theta = \frac{\mathbf{u} \cdot \mathbf{v}}{\|\mathbf{u}\| \, \|\mathbf{v}\|}$$

$$= \frac{\langle 4, 3 \rangle \cdot \langle 3, 5 \rangle}{\|\langle 4, 3 \rangle\| \, \|\langle 3, 5 \rangle\|}$$

$$= \frac{27}{5\sqrt{34}}$$

This implies that the angle between the two vectors is

$$\theta = \arccos \frac{27}{5\sqrt{34}} \approx 22.2°$$

as shown in Figure 3.33.

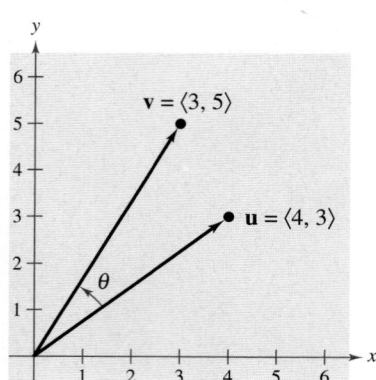

FIGURE **3.33**

Rewriting the expression for the angle between two vectors in the form

$$\mathbf{u} \cdot \mathbf{v} = \|\mathbf{u}\| \, \|\mathbf{v}\| \cos \theta \qquad \text{Alternative form of dot product}$$

produces an alternative way to calculate the dot product. From this form, you can see that because $\|\mathbf{u}\|$ and $\|\mathbf{v}\|$ are always positive, $\mathbf{u} \cdot \mathbf{v}$ and $\cos \theta$ will always have the same sign. Figure 3.34 shows the five possible orientations of two vectors.

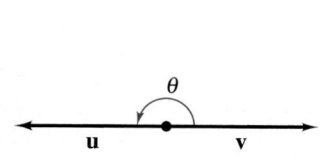

$\theta = \pi$
$\cos \theta = -1$
Opposite Direction

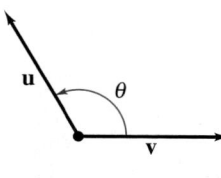

$\dfrac{\pi}{2} < \theta < \pi$
$-1 < \cos \theta < 0$
Obtuse Angle

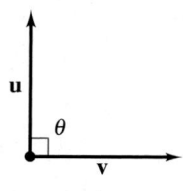

$\theta = \dfrac{\pi}{2}$
$\cos \theta = 0$
90° Angle

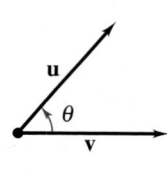

$0 < \theta < \dfrac{\pi}{2}$
$0 < \cos \theta < 1$
Acute Angle

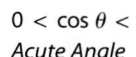
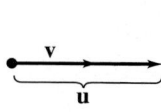

$\theta = 0$
$\cos \theta = 1$
Same Direction

FIGURE **3.34**

Definition of Orthogonal Vectors

The vectors $\mathbf{u}$ and $\mathbf{v}$ are **orthogonal** if $\mathbf{u} \cdot \mathbf{v} = 0$.

The terms "orthogonal" and "perpendicular" mean essentially the same thing—meeting at right angles. Even though the angle between the zero vector and another vector is not defined, it is convenient to extend the definition of orthogonality to include the zero vector. In other words, the zero vector is orthogonal to every vector $\mathbf{u}$, because $\mathbf{0} \cdot \mathbf{u} = 0$.

Example 5 ▶ Determining Orthogonal Vectors

Are the vectors $\mathbf{u} = \langle 2, -3 \rangle$ and $\mathbf{v} = \langle 6, 4 \rangle$ orthogonal?

Solution

Begin by finding the dot product of the two vectors.

$$\mathbf{u} \cdot \mathbf{v} = \langle 2, -3 \rangle \cdot \langle 6, 4 \rangle = 2(6) + (-3)(4) = 0$$

Because the dot product is 0, the two vectors are orthogonal (see Figure 3.35).

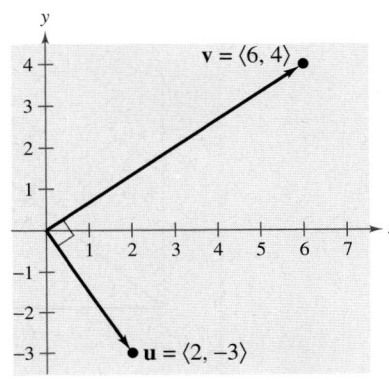

FIGURE **3.35**

Finding Vector Components

You have already seen applications in which two vectors are added to produce a resultant vector. Many applications in physics and engineering pose the reverse problem—decomposing a given vector into the sum of two **vector components.**

Consider a boat on an inclined ramp, as shown in Figure 3.36. The force **F** due to gravity pulls the boat *down* the ramp and *against* the ramp. These two orthogonal forces, $\mathbf{w}_1$ and $\mathbf{w}_2$, are vector components of **F**. That is,

$$\mathbf{F} = \mathbf{w}_1 + \mathbf{w}_2. \qquad \text{Vector components of } \mathbf{F}$$

The negative of component $\mathbf{w}_1$ represents the force needed to keep the boat from rolling down the ramp, whereas $\mathbf{w}_2$ represents the force that the tires must withstand against the ramp. A procedure for finding $\mathbf{w}_1$ and $\mathbf{w}_2$ is shown on the following page.

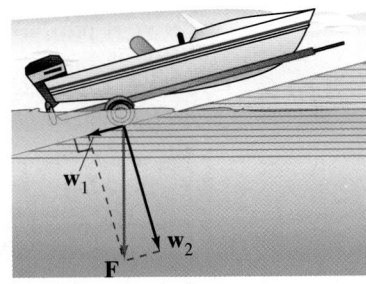

FIGURE **3.36**

Definition of Vector Components

Let $\mathbf{u}$ and $\mathbf{v}$ be nonzero vectors such that

$$\mathbf{u} = \mathbf{w}_1 + \mathbf{w}_2$$

where $\mathbf{w}_1$ and $\mathbf{w}_2$ are orthogonal and $\mathbf{w}_1$ is parallel to (or a scalar multiple of) $\mathbf{v}$, as shown in Figure 3.37. The vectors $\mathbf{w}_1$ and $\mathbf{w}_2$ are called vector components of $\mathbf{u}$. The vector $\mathbf{w}_1$ is the **projection** of $\mathbf{u}$ onto $\mathbf{v}$ and is denoted by

$$\mathbf{w}_1 = \text{proj}_{\mathbf{v}}\mathbf{u}.$$

The vector $\mathbf{w}_2$ is given by $\mathbf{w}_2 = \mathbf{u} - \mathbf{w}_1$.

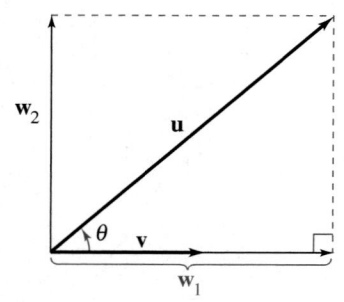   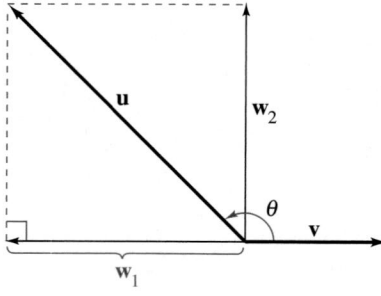

θ is acute.   *θ is obtuse.*

FIGURE 3.37

From the definition of vector components, you can see that it is easy to find the component $\mathbf{w}_2$ once you have found the projection of $\mathbf{u}$ onto $\mathbf{v}$. To find the projection, you can use the dot product, as follows.

$$\mathbf{u} = \mathbf{w}_1 + \mathbf{w}_2 = c\mathbf{v} + \mathbf{w}_2 \qquad \text{$\mathbf{w}_1$ is a scalar multiple of $\mathbf{v}$.}$$

$$\mathbf{u} \cdot \mathbf{v} = (c\mathbf{v} + \mathbf{w}_2) \cdot \mathbf{v} \qquad \text{Take dot product of each side with $\mathbf{v}$.}$$

$$= c\mathbf{v} \cdot \mathbf{v} + \mathbf{w}_2 \cdot \mathbf{v}$$

$$= c\|\mathbf{v}\|^2 + 0 \qquad \text{$\mathbf{w}_2$ and $\mathbf{v}$ are orthogonal.}$$

So,

$$c = \frac{\mathbf{u} \cdot \mathbf{v}}{\|\mathbf{v}\|^2}$$

and

$$\mathbf{w}_1 = \text{proj}_{\mathbf{v}}\,\mathbf{u} = c\mathbf{v} = \frac{\mathbf{u} \cdot \mathbf{v}}{\|\mathbf{v}\|^2}\,\mathbf{v}.$$

Projection of u onto v

Let $\mathbf{u}$ and $\mathbf{v}$ be nonzero vectors. The projection of $\mathbf{u}$ onto $\mathbf{v}$ is

$$\text{proj}_{\mathbf{v}}\mathbf{u} = \left(\frac{\mathbf{u} \cdot \mathbf{v}}{\|\mathbf{v}\|^2}\right)\mathbf{v}.$$

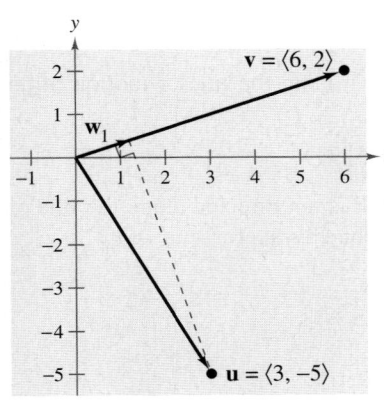

FIGURE **3.38**

Example 6 ► Decomposing a Vector into Components

Find the projection of $\mathbf{u} = \langle 3, -5 \rangle$ onto $\mathbf{v} = \langle 6, 2 \rangle$. Then write $\mathbf{u}$ as the sum of two orthogonal vectors, one of which is $\text{proj}_\mathbf{v}\mathbf{u}$.

Solution

The projection of $\mathbf{u}$ onto $\mathbf{v}$ is

$$\mathbf{w}_1 = \text{proj}_\mathbf{v}\mathbf{u} = \left(\frac{\mathbf{u} \cdot \mathbf{v}}{\|\mathbf{v}\|^2}\right)\mathbf{v} = \left(\frac{8}{40}\right)\langle 6, 2 \rangle = \left\langle \frac{6}{5}, \frac{2}{5} \right\rangle$$

as shown in Figure 3.38. The other component, $\mathbf{w}_2$, is

$$\mathbf{w}_2 = \mathbf{u} - \mathbf{w}_1 = \langle 3, -5 \rangle - \left\langle \frac{6}{5}, \frac{2}{5} \right\rangle = \left\langle \frac{9}{5}, -\frac{27}{5} \right\rangle.$$

So,

$$\mathbf{u} = \mathbf{w}_1 + \mathbf{w}_2 = \left\langle \frac{6}{5}, \frac{2}{5} \right\rangle + \left\langle \frac{9}{5}, -\frac{27}{5} \right\rangle = \langle 3, -5 \rangle.$$

Example 7 ► Finding a Force

A 200-pound cart sits on a ramp inclined at 30°, as shown in Figure 3.39. What force is required to keep the cart from rolling down the ramp?

Solution

Because the force due to gravity is vertical and downward, you can represent the gravitational force by the vector

$$\mathbf{F} = -200\mathbf{j}. \qquad \text{Force due to gravity}$$

To find the force required to keep the cart from rolling down the ramp, project $\mathbf{F}$ onto a unit vector $\mathbf{v}$ in the direction of the ramp, as follows.

$$\mathbf{v} = (\cos 30°)\mathbf{i} + (\sin 30°)\mathbf{j} = \frac{\sqrt{3}}{2}\mathbf{i} + \frac{1}{2}\mathbf{j} \qquad \text{Unit vector along ramp}$$

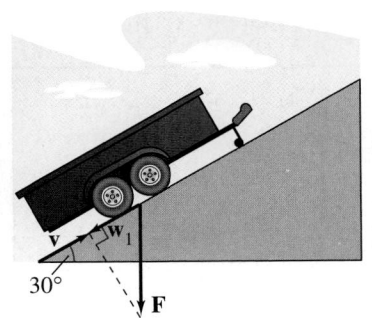

FIGURE **3.39**

Therefore, the projection of $\mathbf{F}$ onto $\mathbf{v}$ is

$$\begin{aligned}
\mathbf{w}_1 &= \text{proj}_\mathbf{v}\mathbf{F} \\
&= \left(\frac{\mathbf{F} \cdot \mathbf{v}}{\|\mathbf{v}\|^2}\right)\mathbf{v} \\
&= (\mathbf{F} \cdot \mathbf{v})\mathbf{v} \\
&= (-200)\left(\frac{1}{2}\right)\mathbf{v} \\
&= -100\left(\frac{\sqrt{3}}{2}\mathbf{i} + \frac{1}{2}\mathbf{j}\right).
\end{aligned}$$

The magnitude of this force is 100, and so a force of 100 pounds is required to keep the cart from rolling down the ramp.

Work

The work W done by a *constant* force $\mathbf{F}$ acting along the line of motion of an object is

$$W = (\text{magnitude of force})(\text{distance}) = \|\mathbf{F}\| \, \|\overrightarrow{PQ}\|$$

as shown in Figure 3.40. If the constant force $\mathbf{F}$ is not directed along the line of motion, as shown in Figure 3.41, the work W done by the force is

$$W = \|\text{proj}_{\overrightarrow{PQ}}\mathbf{F}\| \, \|\overrightarrow{PQ}\|$$
$$= (\cos \theta)\|\mathbf{F}\| \, \|\overrightarrow{PQ}\| \qquad \|\text{proj}_{\overrightarrow{PQ}}\mathbf{F}\| = (\cos \theta)\|\mathbf{F}\|$$
$$= \mathbf{F} \cdot \overrightarrow{PQ}. \qquad \cos \theta(\mathbf{F} \cdot \overrightarrow{PQ})/(\|\mathbf{F}\| \, \|\overrightarrow{PQ}\|)$$

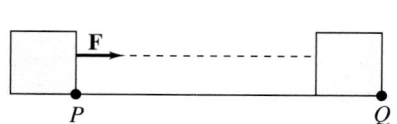

Force acts along the line of motion.
FIGURE **3.40**

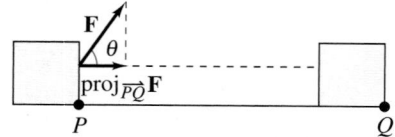

Force acts at angle θ with the line of motion.
FIGURE **3.41**

This notion of work is summarized in the following definition.

Definition of Work

The **work** W done by a constant force $\mathbf{F}$ as its point of application moves along the vector $\overrightarrow{PQ}$ is given by either of the following.

1. $W = \|\text{proj}_{\overrightarrow{PQ}}\mathbf{F}\| \, \|\overrightarrow{PQ}\|$ Projection form

2. $W = \mathbf{F} \cdot \overrightarrow{PQ}$ Dot product form

---

Example 8 ▶ **Finding Work**

To close a sliding door, a person pulls on a rope with a constant force of 50 pounds at a constant angle of 60°, as shown in Figure 3.42. Find the work done in moving the door 12 feet to its closed position.

Solution

Using a projection, you can calculate the work as follows.

$$W = \|\text{proj}_{\overrightarrow{PQ}}\mathbf{F}\| \, \|\overrightarrow{PQ}\| \qquad \text{Projection form for work}$$
$$= (\cos 60°)\|\mathbf{F}\| \, \|\overrightarrow{PQ}\|$$
$$= \frac{1}{2}(50)(12) = 300 \text{ foot-pounds}$$

So, the work done is 300 foot-pounds. You can verify this result by finding the vectors $\mathbf{F}$ and $\overrightarrow{PQ}$ and calculating their dot product.

FIGURE **3.42**

3.4 Exercises

In Exercises 1–4, find the dot product of u and v.

1. $u = \langle 6, 1 \rangle$

$v = \langle -2, 3 \rangle$

2. $u = \langle 5, 12 \rangle$

$v = \langle -3, 2 \rangle$

3. $u = 4i - 2j$

$v = i - j$

4. $u = 3i + 4j$

$v = 7i - 2j$

In Exercises 5–8, use the vectors $u = \langle 2, 2 \rangle$ and $v = \langle -3, 4 \rangle$ to find the indicated quantity. State whether the result is a vector or a scalar.

5. $u \cdot u$

6. $\|v\| + 3$

7. $(u \cdot v)v$

8. $3u \cdot v$

In Exercises 9–14, use the dot product to find the magnitude of u.

9. $u = \langle -5, 12 \rangle$

10. $u = \langle 2, -4 \rangle$

11. $u = 20i + 25j$

12. $u = 12i - 16j$

13. $u = 6j$

14. $u = -21i$

In Exercises 15–24, find the angle θ between the vectors.

15. $u = \langle 1, 0 \rangle$

$v = \langle 0, -2 \rangle$

16. $u = \langle 3, 2 \rangle$

$v = \langle 4, 0 \rangle$

17. $u = 3i + 4j$

$v = -2j$

18. $u = 2i - 3j$

$v = i - 2j$

19. $u = 2i - j$

$v = 6i + 4j$

20. $u = -6i - 3j$

$v = -8i + 4j$

21. $u = 5i + 5j$

$v = -6i + 6j$

22. $u = 2i - 3j$

$v = 4i + 3j$

23. $u = \cos\left(\dfrac{\pi}{3}\right)i + \sin\left(\dfrac{\pi}{3}\right)j$

$v = \cos\left(\dfrac{3\pi}{4}\right)i + \sin\left(\dfrac{3\pi}{4}\right)j$

24. $u = \cos\left(\dfrac{\pi}{4}\right)i + \sin\left(\dfrac{\pi}{4}\right)j$

$v = \cos\left(\dfrac{\pi}{2}\right)i + \sin\left(\dfrac{\pi}{2}\right)j$

In Exercises 25–28, use vectors to find the interior angles of the triangle with the given vertices.

25. $(1, 2), (3, 4), (2, 5)$

26. $(-3, -4), (1, 7), (8, 2)$

27. $(-3, 0), (2, 2), (0, 6)$

28. $(-3, 5), (-1, 9), (7, 9)$

In Exercises 29 and 30, find $u \cdot v$, where θ is the angle between u and v.

29. $\|u\| = 4, \|v\| = 10, \theta = \dfrac{2\pi}{3}$

30. $\|u\| = 100, \|v\| = 250, \theta = \dfrac{\pi}{6}$

In Exercises 31–36, determine whether u and v are orthogonal, parallel, or neither.

31. $u = \langle -12, 30 \rangle$

$v = \langle \frac{1}{2}, -\frac{5}{4} \rangle$

32. $u = \langle 3, 15 \rangle$

$v = \langle -1, 5 \rangle$

33. $u = \frac{1}{4}(3i - j)$

$v = 5i + 6j$

34. $u = i$

$v = -2i + 2j$

35. $u = 2i - 2j$

$v = -i - j$

36. $u = \langle \cos\theta, \sin\theta \rangle$

$v = \langle \sin\theta, -\cos\theta \rangle$

In Exercises 37–40, find the projection of u onto v and the vector component of u orthogonal to v.

37. $u = \langle 2, 2 \rangle$

$v = \langle 6, 1 \rangle$

38. $u = \langle 4, 2 \rangle$

$v = \langle 1, -2 \rangle$

39. $u = \langle 0, 3 \rangle$

$v = \langle 2, 15 \rangle$

40. $u = \langle -3, -2 \rangle$

$v = \langle -4, -1 \rangle$

In Exercises 41–44, find two vectors in opposite directions that are orthogonal to the vector u. (The answers are not unique.)

41. $u = \langle 3, 5 \rangle$

42. $u = \langle -8, 3 \rangle$

43. $u = \frac{1}{2}i - \frac{2}{3}j$

44. $u = -\frac{5}{2}i - 3j$

Work **In Exercises 45 and 46, find the work done in moving a particle from P to Q if the magnitude and direction of the force are given by v.**

45. $P = (0, 0), \quad Q = (4, 7), \quad v = \langle 1, 4 \rangle$

46. $P = (1, 3), \quad Q = (-3, 5), \quad v = -2i + 3j$

47. ***Revenue*** The vector $u = \langle 1650, 3200 \rangle$ gives the numbers of units of two types of baking pans produced by a company. The vector $v = \langle 15.25, 10.50 \rangle$ gives the price (in dollars) of each pan, respectively.

(a) Find the dot product $u \cdot v$ and explain what information it gives.

(b) Identify the vector operation used to increase the prices by 5%.

▶ Model It

48. Braking Load   A truck with a gross weight of 30,000 pounds is parked on a slope of $d°$ (see figure). Assume that the only force to overcome is the force of gravity.

(a) Find the force required to keep the truck from rolling down the hill in terms of the slope d.

(b) Use a graphing utility to complete the table.

d	0°	1°	2°	3°	4°	5°
Force						

d	6°	7°	8°	9°	10°
Force					

(c) Find the force perpendicular to the hill when $d = 5°$.

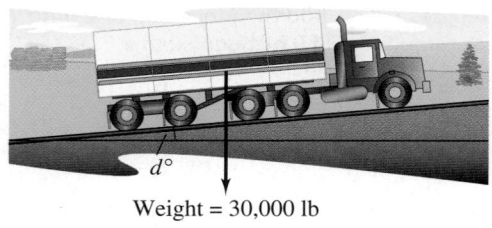

Weight = 30,000 lb

49. Work   A 25-kilogram (245-newton) bag of sugar is lifted 3 meters. Determine the work done.

50. Work   Determine the work done by a crane lifting a 2400-pound car 5 feet.

51. Work   A force of 45 pounds in the direction of 30° above the horizontal is required to slide a table across a floor (see figure). The table is dragged 20 feet. Determine the work done.

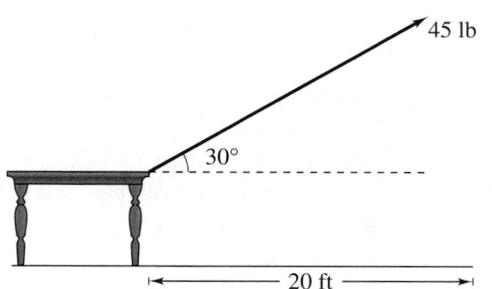

45 lb

30°

20 ft

52. Work   A tractor pulls a log 800 meters, and the tension in the cable connecting the tractor and log is approximately 1600 kilograms (15,691 newtons). The direction of the force is 35° above the horizontal. Approximate the work done.

Synthesis

True or False?   In Exercises 53 and 54, determine whether the statement is true or false. Justify your answer.

53. The work W done by a constant force **F** acting along the line of motion of an object is represented by a vector.

54. A sliding door moves along the line of vector $\overrightarrow{PQ}$. If a force is applied to the door along a vector that is orthogonal to $\overrightarrow{PQ}$, then no work is done.

55. Think About It   What is known about θ, the angle between two nonzero vectors **u** and **v**, under each condition?

(a) $\mathbf{u} \cdot \mathbf{v} = 0$ (b) $\mathbf{u} \cdot \mathbf{v} > 0$ (c) $\mathbf{u} \cdot \mathbf{v} < 0$

56. Think About It   What can be said about the vectors **u** and **v** under each condition?

(a) The projection of **u** onto **v** equals **u**.

(b) The projection of **u** onto **v** equals **0**.

57. Proof   Use vectors to prove that the diagonals of a rhombus are perpendicular.

58. Proof   Prove the following.

$$\|\mathbf{u} - \mathbf{v}\|^2 = \|\mathbf{u}\|^2 + \|\mathbf{v}\|^2 - 2\mathbf{u} \cdot \mathbf{v}$$

Review

In Exercises 59 and 60, perform the operation.

59. $\sqrt{42} \cdot \sqrt{24}$ **60.** $\sqrt{18} \cdot \sqrt{112}$

In Exercises 61–64, find the exact solutions of the equation in the interval $[0, 2\pi)$.

61. $\sin 2x - \sqrt{3} \sin x = 0$ **62.** $\sin 2x + \sqrt{2} \cos x = 0$

63. $2 \tan x = \tan 2x$ **64.** $\cos 2x - 3 \sin x = 2$

In Exercises 65–68, find the exact value of the trigonometric function given that $\sin u = -\frac{12}{13}$ and $\cos v = \frac{24}{25}$. (Both u and v are in Quadrant IV.)

65. $\sin(u - v)$ **66.** $\sin(u + v)$

67. $\cos(v - u)$ **68.** $\tan(u - v)$

Chapter Summary

▶ *What* did you learn?

Section 3.1	Review Exercises
☐ How to use the Law of Sines to solve oblique triangles (AAS, ASA, or SSA)	1–12
☐ How to find the areas of oblique triangles	13–16
☐ How to use the Law of Sines to model and solve real-life problems	17–20

Section 3.2	
☐ How to use the Law of Cosines to solve oblique triangles (SSS or SAS)	21–28
☐ How to use the Law of Cosines to model and solve real-life problems	29, 30
☐ How to use Heron's Area Formula to find the area of a triangle	31–34

Section 3.3	
☐ How to represent vectors as directed line segments	35–38
☐ How to write the component forms of vectors	39–44
☐ How to perform basic vector operations and represent them graphically	45–48
☐ How to write vectors as linear combinations of unit vectors	49–54
☐ How to find the direction angles of vectors	55–60
☐ How to use vectors to model and solve real-life problems	61–64

Section 3.4	
☐ How to find the dot product of two vectors and use the Properties of the Dot Product	65–72
☐ How to find the angle between two vectors and determine whether two vectors are orthogonal	73–80
☐ How to write a vector as the sum of two vector components	81–84
☐ How to use vectors to find the work done by a force	85, 86

Review Exercises

3.1 **In Exercises 1–12, use the Law of Sines to solve (if possible) the triangle. If two solutions exist, list both. Round your answers to two decimal places.**

1.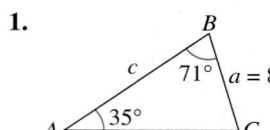

2.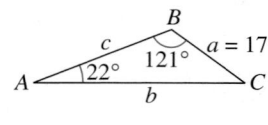

3. $B = 72°$, $C = 82°$, $b = 54$

4. $B = 10°$, $C = 20°$, $c = 33$

5. $A = 16°$, $B = 98°$, $c = 8.4$

6. $A = 95°$, $B = 45°$, $c = 104.8$

7. $A = 24°$, $C = 48°$, $b = 27.5$

8. $B = 64°$, $C = 36°$, $a = 367$

9. $B = 150°$, $b = 30$, $c = 10$

10. $B = 150°$, $a = 10$, $b = 3$

11. $A = 75°$, $a = 51.2$, $b = 33.7$

12. $B = 25°$, $a = 6.2$, $b = 4$

In Exercises 13–16, use the information to find the area of the triangle.

13. $A = 27°$, $b = 5$, $c = 7$

14. $B = 80°$, $a = 4$, $c = 8$

15. $C = 123°$, $a = 16$, $b = 5$

16. $A = 11°$, $b = 22$, $c = 21$

17. *Height* From a certain distance, the angle of elevation to the top of a building is 17°. At a point 50 meters closer to the building, the angle of elevation is 31°. Approximate the height of the building.

18. *Geometry* Find the length of the side w of the parallelogram.

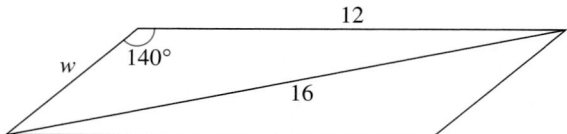

19. *Height* A tree stands on a hillside of slope 28°(from the horizontal). From a point 75 feet down the hill, the angle of elevation to the top of the tree is 45° (see figure). What is the height of the tree?

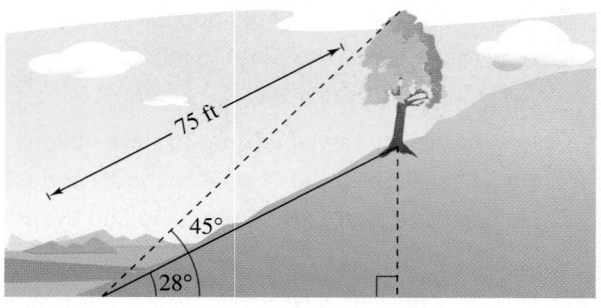

20. *River Width* A surveyor finds that a tree on the opposite bank of a river flowing due east has a bearing of N 22° 30′ E from a certain point and a bearing of N 15° W from a point 400 feet downstream. What is the width of the river?

3.2 **In Exercises 21–28, use the Law of Cosines to solve the triangle. Round your answers to two decimal places.**

21. $a = 5$, $b = 8$, $c = 10$

22. $a = 80$, $b = 60$, $c = 100$

23. $a = 2.5$, $b = 5.0$, $c = 4.5$

24. $a = 16.4$, $b = 8.8$, $c = 12.2$

25. $B = 110°$, $a = 4$, $c = 4$

26. $B = 150°$, $a = 10$, $c = 20$

27. $C = 43°$, $a = 22.5$, $b = 31.4$

28. $A = 62°$, $b = 11.34$, $c = 19.52$

29. *Surveying* To approximate the length of a marsh, a surveyor walks 425 meters from point A to point B. Then the surveyor turns 65° and walks 300 meters to point C (see figure). Approximate the length AC of the marsh.

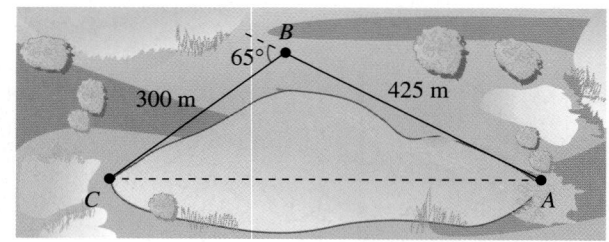

30. *Navigation* Two planes leave Raleigh-Durham Airport at approximately the same time. One is flying 425 miles per hour at a bearing of 355°, and the other is flying 530 miles per hour at a bearing of 67°. Draw a figure that gives a visual representation of the problem and determine the distance between the planes after they have flown for 2 hours.

In Exercises 31–34, use Heron's Area Formula to find the area of the triangle.

31. $a = 4$, $b = 5$, $c = 7$

32. $a = 15$, $b = 8$, $c = 10$

33. $a = 12.3$, $b = 15.8$, $c = 3.7$

34. $a = 38.1$, $b = 26.7$, $c = 19.4$

3.3 **In Exercises 35–38, graph the vector with the specified initial point and terminal point.**

Initial Point	*Terminal Point*
35. $(0, 0)$	$(8, 7)$
36. $(3, 4)$	$(-5, -7)$
37. $(-3, 9)$	$(8, -4)$
38. $(-6, -8)$	$(8, 3)$

In Exercises 39–44, find the component form of the vector v satisfying the conditions.

39.

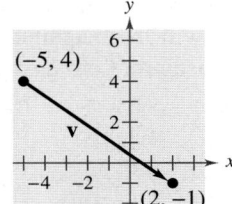

40.

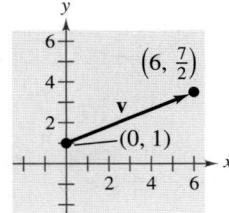

41. Initial point: $(0, 10)$; Terminal point: $(7, 3)$

42. Initial point: $(1, 5)$; Terminal point: $(15, 9)$

43. $\|v\| = 8$, $\quad \theta = 120°$

44. $\|v\| = \frac{1}{2}$, $\quad \theta = 225°$

In Exercises 45–48, find the component form of the vector given that u = 6i − 5j and v = 10i + 3j. Then sketch your result.

45. $2u + v$

46. $4u - 5v$

47. $3v$

48. $\frac{1}{2}v$

In Exercises 49–52, write vector u as a linear combination of the standard unit vectors i and j.

49. $u = \langle -3, 4 \rangle$

50. $u = \langle -6, -8 \rangle$

51. u has initial point $(3, 4)$ and terminal point $(9, 8)$.

52. u has initial point $(-2, 7)$ and terminal point $(5, -9)$.

In Exercises 53 and 54, write the vector v in the form $\|v\|(\cos \theta i + \sin \theta j)$.

53. $v = -10i + 10j$

54. $v = 4i - j$

In Exercises 55–60, find the magnitude and the direction angle of the vector v.

55. $v = 7(\cos 60°i + \sin 60°j)$

56. $v = 3(\cos 150°i + \sin 150°j)$

57. $v = 5i + 4j$

58. $v = -4i + 7j$

59. $v = -3i - 3j$

60. $v = 8i - j$

61. *Resultant Force* Forces of 85 pounds and 50 pounds act on a single point. The angle between the forces is 15°. Describe the resultant force.

62. *Rope Tension* A 180-pound weight is supported by two ropes, as shown in the figure. Find the tension exerted on each rope.

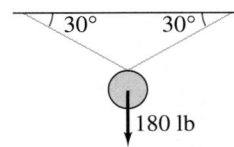

180 lb

63. *Navigation* An airplane has an airspeed of 430 miles per hour at a bearing of 135°. The wind velocity is 35 miles per hour in the direction of N 30° E. What are the groundspeed and the direction of the plane?

64. *Navigation* An airplane has an airspeed of 724 kilometers per hour at a bearing of 30°. The wind velocity is 32 kilometers per hour from the west. What are the groundspeed and the direction of the plane?

| 3.4 | **In Exercises 65–68, find the dot product of u and v.**

65. $\mathbf{u} = \langle 6, 7 \rangle$

$\quad \mathbf{v} = \langle -3, 9 \rangle$

66. $\mathbf{u} = \langle -7, 12 \rangle$

$\quad \mathbf{v} = \langle -4, -14 \rangle$

67. $\mathbf{u} = 3\mathbf{i} + 7\mathbf{j}$

$\quad \mathbf{v} = 11\mathbf{i} - 5\mathbf{j}$

68. $\mathbf{u} = -7\mathbf{i} + 2\mathbf{j}$

$\quad \mathbf{v} = 16\mathbf{i} - 12\mathbf{j}$

In Exercises 69–72, use the vectors $\mathbf{u} = \langle -3, 4 \rangle$ and $\mathbf{v} = \langle 2, 1 \rangle$ to find the quantity. State whether the result is a vector or a scalar.

69. $2\mathbf{u} \cdot \mathbf{u}$

70. $\|\mathbf{v}\|^2$

71. $\mathbf{u}(\mathbf{u} \cdot \mathbf{v})$

72. $3\mathbf{u} \cdot \mathbf{v}$

In Exercises 73–76, find the angle between u and v.

73. $\mathbf{u} = \cos \dfrac{7\pi}{4}\mathbf{i} + \sin \dfrac{7\pi}{4}\mathbf{j}$

$\quad \mathbf{v} = \cos \dfrac{5\pi}{6}\mathbf{i} + \sin \dfrac{5\pi}{6}\mathbf{j}$

74. $\mathbf{u} = \cos 45°\mathbf{i} + \sin 45°\mathbf{j}$

$\quad \mathbf{v} = \cos 300°\mathbf{i} + \sin 300°\mathbf{j}$

75. $\mathbf{u} = \langle 2\sqrt{2}, -4 \rangle, \quad \mathbf{v} = \langle -\sqrt{2}, 1 \rangle$

76. $\mathbf{u} = \langle 3, \sqrt{3} \rangle, \quad \mathbf{v} = \langle 4, 3\sqrt{3} \rangle$

In Exercises 77–80, determine whether u and v are orthogonal, parallel, or neither.

77. $\mathbf{u} = \langle -3, 8 \rangle$

$\quad \mathbf{v} = \langle 8, 3 \rangle$

78. $\mathbf{u} = \langle \frac{1}{4}, -\frac{1}{2} \rangle$

$\quad \mathbf{v} = \langle -2, 4 \rangle$

79. $\mathbf{u} = -\mathbf{i}$

$\quad \mathbf{v} = \mathbf{i} + 2\mathbf{j}$

80. $\mathbf{u} = -2\mathbf{i} + \mathbf{j}$

$\quad \mathbf{v} = 3\mathbf{i} + 6\mathbf{j}$

In Exercises 81–84, find proj$_\mathbf{v}\mathbf{u}$ and the vector component of u orthogonal to v.

81. $\mathbf{u} = \langle -4, 3 \rangle, \ \mathbf{v} = \langle -8, -2 \rangle$

82. $\mathbf{u} = \langle 5, 6 \rangle, \ \mathbf{v} = \langle 10, 0 \rangle$

83. $\mathbf{u} = \langle 2, 7 \rangle, \ \mathbf{v} = \langle 1, -1 \rangle$

84. $\mathbf{u} = \langle -3, 5 \rangle, \ \mathbf{v} = \langle -5, 2 \rangle$

Work **In Exercises 85 and 86, find the work done in moving a particle from P to Q if the magnitude and direction of the force are given by v.**

85. $P = (5, 3), Q = (8, 9), \mathbf{v} = \langle 2, 7 \rangle$

86. $P = (-2, -9), Q = (-12, 8), \mathbf{v} = 3\mathbf{i} - 6\mathbf{j}$

Synthesis

True or False? **In Exercises 87–90, determine whether the statement is true or false. Justify your answer.**

87. The Law of Sines is true if one of the angles in the triangle is a right angle.

88. When the Law of Sines is used, the solution is always unique.

89. If $\mathbf{u}$ is a unit vector in the direction of $\mathbf{v}$, then $\mathbf{v} = \|\mathbf{v}\|\, \mathbf{u}$.

90. If $\mathbf{v} = a\mathbf{i} + b\mathbf{j} = \mathbf{0}$, then $a = -b$.

91. State the Law of Sines from memory.

92. State the Law of Cosines from memory.

93. What characterizes a vector in the plane?

94. Which vectors in the figure appear to be equivalent?

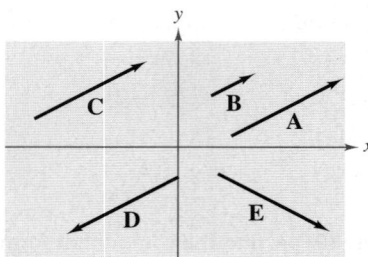

95. The vectors $\mathbf{u}$ and $\mathbf{v}$ have the same magnitudes in the two figures. In which figure will the magnitude of the sum be greater? Give a reason for your answer.

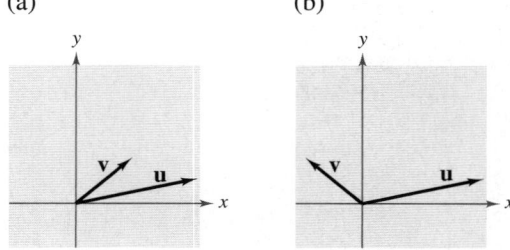

   (a)   (b)

96. Give a geometric description of the scalar multiple $k\mathbf{u}$ of the vector $\mathbf{u}$, for $k > 0$ and for $k < 0$.

97. Give a geometric description of the sum of the vectors $\mathbf{u}$ and $\mathbf{v}$.

Chapter Test

Take this test as you would take a test in class. When you are finished, check your work against the answers given in the back of the book.

In Exercises 1–6, use the information to solve the triangle. If two solutions exist, find both.

1. $A = 24°$, $B = 68°$, $a = 12.2$
2. $B = 104°$, $C = 33°$, $a = 18.1$
3. $A = 24°$, $a = 11.2$, $b = 13.4$
4. $a = 4.0$, $b = 7.3$, $c = 12.4$
5. $B = 100°$, $a = 15$, $b = 23$
6. $C = 123°$, $a = 41$, $b = 57$

7. A triangular parcel of land has borders of lengths 60 meters, 70 meters, and 82 meters. Find the area of the parcel of land.

8. An airplane flies 370 miles from point A to point B with a bearing of 24°. It then flies 240 miles from point B to point C with a bearing of 37° (see figure). Find the distance and bearing from point A to point C.

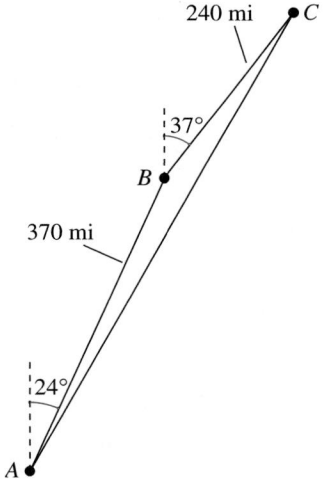

240 mi • C

37°

B •

370 mi

24°

A •

FIGURE FOR **8**

In Exercises 9 and 10, find the component form of the vector v with the given components.

9. Initial point of **v**: $(-3, 7)$; Terminal point of **v**: $(11, -16)$
10. Magnitude of **v**: $\|\mathbf{v}\| = 12$; Direction of **v**: $\mathbf{u} = \langle 3, -5 \rangle$

In Exercises 11–13, u $= \langle 3, 5 \rangle$ and v $= \langle -7, 1 \rangle$. Find the resultant vector and sketch its graph.

11. $\mathbf{u} + \mathbf{v}$
12. $\mathbf{u} - \mathbf{v}$
13. $5\mathbf{u} - 3\mathbf{v}$

14. Find a unit vector in the direction of $\mathbf{u} = \langle 4, -3 \rangle$.

15. Forces with magnitudes of 250 pounds and 130 pounds act on an object at angles of 45° and −60°, respectively, with the x-axis. Find the direction and magnitude of the resultant of these forces.

16. Find the angle between the vectors $\mathbf{u} = \langle -1, 5 \rangle$ and $\mathbf{v} = \langle 3, -2 \rangle$.

17. Are the vectors $\mathbf{u} = \langle 6, 10 \rangle$ and $\mathbf{v} = \langle 2, 3 \rangle$ orthogonal?

18. Find the projection of $\mathbf{u} = \langle 6, 7 \rangle$ onto $\mathbf{v} = \langle -5, -1 \rangle$ and the vector component of $\mathbf{u}$ orthogonal to $\mathbf{v}$.

Cumulative Test for Chapters 1–3

Take this test to review the material from earlier chapters. When you are finished, check your work against the answers given in the back of the book.

1. Consider the angle $\theta = -120°$.
   (a) Sketch the angle in standard position.
   (b) Determine a coterminal angle in the interval $[0°, 360°)$.
   (c) Convert the angle to radian measure.
   (d) Find the reference angle θ'.
   (e) Find the exact values of the six trigonometric functions of θ.

2. Convert the angle of measure 2.35 radians to degrees. Round the answer to one decimal place.

3. Find $\cos \theta$ if $\tan \theta = -\frac{4}{3}$ and $\sin \theta < 0$.

In Exercises 4 and 5, find the period and amplitude, and sketch the graph of the trigonometric function.

4. $f(x) = 3 - 2 \sin \pi x$

5. $g(x) = \frac{1}{2} \tan\left(x - \frac{\pi}{2}\right)$

6. Find a, b, and c such that the graph of the function $h(x) = a \cos(bx + c)$ matches the graph in the figure.

7. Sketch the graph of the function $f(x) = \frac{1}{2}x \sin x$ over the interval $-3\pi \le x \le 3\pi$.

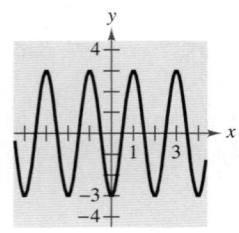

FIGURE FOR 6

In Exercises 8 and 9, find the exact value of the expression without the aid of a calculator.

8. $\tan(\arctan 6.7)$

9. $\tan\left(\arcsin \frac{3}{5}\right)$

10. Write an algebraic expression equivalent to $\sin(\arccos 2x)$.

11. Use the fundamental identities to simplify: $\cos\left(\frac{\pi}{2} - x\right) \csc x$.

12. Subtract and simplify: $\dfrac{\sin \theta - 1}{\cos \theta} - \dfrac{\cos \theta}{\sin \theta - 1}$.

In Exercises 13–15, prove the identity.

13. $\cot^2 \alpha(\sec^2 \alpha - 1) = 1$

14. $\sin(x + y) \sin(x - y) = \sin^2 x - \sin^2 y$

15. $\sin^2 x \cos^2 x = \frac{1}{8}(1 - \cos 4x)$

In Exercises 16 and 17, find all solutions of the equation in the interval $[0, 2\pi)$.

16. $2 \cos^2 \beta - \cos \beta = 0$

17. $3 \tan \theta - \cot \theta = 0$

18. Use the Quadratic Formula to solve the equation in the interval $[0, 2\pi)$: $\sin^2 x + 2 \sin x + 1 = 0$.

19. Given that $\sin u = \frac{12}{13}$, $\cos v = \frac{3}{5}$, and angles u and v are both in Quadrant I, find $\tan(u - v)$.

20. If $\tan \theta = \frac{1}{2}$, find the exact value of $\tan(2\theta)$.

21. If $\tan \theta = \frac{4}{3}$, find the exact value of $\sin \frac{\theta}{2}$.

22. Write the product $5 \sin \frac{3\pi}{4} \cdot \cos \frac{7\pi}{4}$ as a sum or difference.

In Exercises 23–26, use the information to solve the triangle shown in the figure.

23. $A = 30°$, $a = 9$, $b = 8$

24. $A = 30°$, $b = 8$, $c = 10$

25. $A = 30°$, $C = 90°$, $b = 10$

26. $a = 4$, $b = 8$, $c = 9$

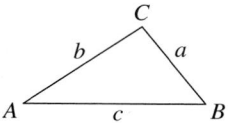

FIGURE FOR 23–26

27. Two sides of a triangle have lengths 7 inches and 12 inches. Their included angle measures 60°. Find the area of the triangle.

28. Find the area of a triangle with sides of lengths 11 inches, 16 inches, and 17 inches.

29. Write vector $\mathbf{u} = \langle 3, 5 \rangle$ as a linear combination of the standard unit vectors $\mathbf{i}$ and $\mathbf{j}$.

30. Find $\mathbf{u} \cdot \mathbf{v}$ for $\mathbf{u} = 3\mathbf{i} + 4\mathbf{j}$ and $\mathbf{v} = \mathbf{i} - 2\mathbf{j}$.

31. Find the projection of $\mathbf{u} = \langle 8, -2 \rangle$ onto $\mathbf{v} = \langle 1, 5 \rangle$ and the vector component of $\mathbf{u}$ orthogonal to $\mathbf{v}$.

32. From a point 200 feet from a flagpole, the angles of elevation to the bottom and top of the flag are $16° \, 45'$ and $18°$, respectively. Approximate the height of the flag to the nearest foot.

33. A compact disc can have an angular speed up to 3142 radians per minute. At this angular speed, how many revolutions per minute would the CD make? How long would it take the CD to make 10,000 revolutions?

34. To determine the angle of elevation of a star in the sky, you get the star in your line of vision with the backboard of a basketball hoop that is 5 feet higher than your eyes (see figure). Your horizontal distance from the backboard is 12 feet. What is the angle of elevation of the star?

35. Write a model for a particle in simple harmonic motion with a displacement of 4 inches and a period of 8 seconds.

36. An airplane's velocity with respect to the air is 500 kilometers per hour, with a bearing of 30°. The wind at the altitude of the plane has a velocity of 50 kilometers per hour with a bearing of N 60° E. What is the true direction of the plane, and what is its speed relative to the ground?

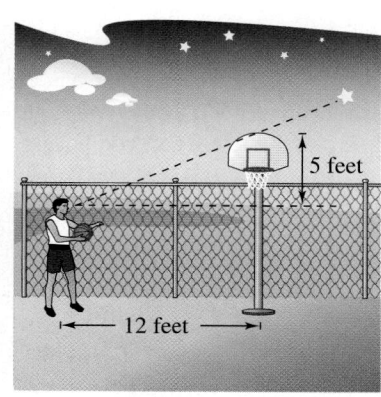

FIGURE FOR 34

Proofs in Mathematics

Law of Tangents

Besides the Law of Sines and the Law of Cosines in trigonometry there is also a Law of Tangents developed by Francois Viète (1540–1603). The Law of Tangents follows from the Law of Sines and the sum-to-product formulas for sine and is defined as follows.

$$\frac{a + b}{a - b} = \frac{\tan(A + B)/2}{\tan(A - B)/2}$$

The Law of Tangents can be used to solve a triangle when two sides and the included angle are given (SAS). Before calculators were invented, the Law of Tangents was used to solve the SAS case instead of the Law of Cosines, because computation with a table of tangent values was easier.

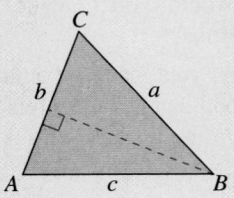

A is acute

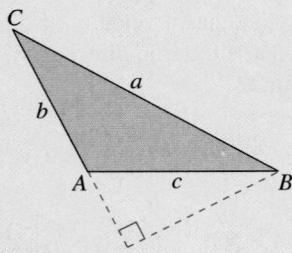

A is obtuse

Law of Sines   (p. 274)

If ABC is a triangle with sides a, b, and c, then

$$\frac{a}{\sin A} = \frac{b}{\sin B} = \frac{c}{\sin C}.$$

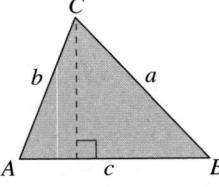 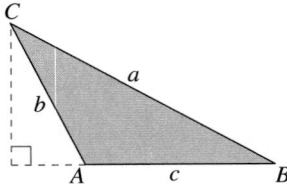

Proof

Let h be the altitude of either triangle found in the figure above. Then you have

$$\sin A = \frac{h}{b} \qquad \text{or} \qquad h = b \sin A$$

$$\sin B = \frac{h}{a} \qquad \text{or} \qquad h = a \sin B.$$

Equating these two values of h, you have

$$a \sin B = b \sin A \qquad \text{or} \qquad \frac{a}{\sin A} = \frac{b}{\sin B}.$$

Note that $\sin A \neq 0$ and $\sin B \neq 0$ because no angle of a triangle can have a measure of 0° or 180°. In a similar manner, construct an altitude from vertex B to side AC (extended in the obtuse triangle), as shown at the left. Then you have

$$\sin A = \frac{h}{c} \qquad \text{or} \qquad h = c \sin A$$

$$\sin C = \frac{h}{a} \qquad \text{or} \qquad h = a \sin C.$$

Equating these two values of h, you have

$$a \sin C = c \sin A \qquad \text{or} \qquad \frac{a}{\sin A} = \frac{c}{\sin C}.$$

By the Transitive Property of Equality you know that

$$\frac{a}{\sin A} = \frac{b}{\sin B} = \frac{c}{\sin C}.$$

So, the Law of Sines is established.

Law of Cosines *(p. 283)*

Standard Form	*Alternative Form*
$a^2 = b^2 + c^2 - 2bc \cos A$	$\cos A = \dfrac{b^2 + c^2 - a^2}{2bc}$
$b^2 = a^2 + c^2 - 2ac \cos B$	$\cos B = \dfrac{a^2 + c^2 - b^2}{2ac}$
$c^2 = a^2 + b^2 - 2ab \cos C$	$\cos C = \dfrac{a^2 + b^2 - c^2}{2ab}$

Proof

To prove the first formula, consider the top triangle at the left, which has three acute angles. Note that vertex B has coordinates $(c, 0)$. Furthermore, C has coordinates (x, y), where $x = b \cos A$ and $y = b \sin A$. Because a is the distance from vertex C to vertex B, it follows that

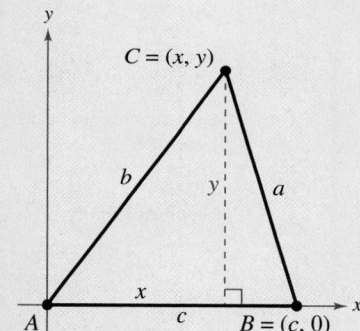

$$a = \sqrt{(x - c)^2 + (y - 0)^2} \qquad \text{Distance Formula}$$

$$a^2 = (x - c)^2 + (y - 0)^2 \qquad \text{Square each side.}$$

$$a^2 = (b \cos A - c)^2 + (b \sin A)^2 \qquad \text{Substitute for } x \text{ and } y.$$

$$a^2 = b^2 \cos^2 A - 2bc \cos A + c^2 + b^2 \sin^2 A \qquad \text{Expand.}$$

$$a^2 = b^2(\sin^2 A + \cos^2 A) + c^2 - 2ab \cos A \qquad \text{Factor out } b^2.$$

$$a^2 = b^2 + c^2 - 2bc \cos A. \qquad \sin^2 A + \cos^2 A = 1$$

To prove the second formula, consider the bottom triangle at the left, which also has three acute angles. Note that vertex A has coordinates $(c, 0)$. Furthermore, C has coordinates (x, y), where $x = a \cos B$ and $y = a \sin B$. Because b is the distance from vertex C to vertex A, it follows that

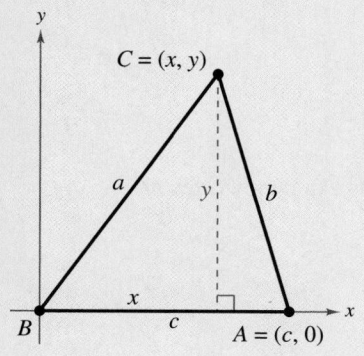

$$b = \sqrt{(x - c)^2 + (y - 0)^2} \qquad \text{Distance Formula}$$

$$b^2 = (x - c)^2 + (y - 0)^2 \qquad \text{Square each side.}$$

$$b^2 = (a \cos B - c)^2 + (a \sin B)^2 \qquad \text{Substitute for } x \text{ and } y.$$

$$b^2 = a^2 \cos^2 B - 2ac \cos B + c^2 + a^2 + \sin^2 B \qquad \text{Expand.}$$

$$b^2 = a^2(\sin^2 B + \cos^2 B) + c^2 - 2ac \cos B \qquad \text{Factor out } a^2.$$

$$b^2 = a^2 + c^2 - 2ac \cos B. \qquad \sin^2 B + \cos^2 B = 1$$

Try using a similar argument to establish the third formula.

Heron's Area Formula *(p. 286)*

Given any triangle with sides of lengths a, b, and c, the area of the triangle is

$$\text{Area} = \sqrt{s(s-a)(s-b)(s-c)}$$

where $s = \dfrac{(a+b+c)}{2}$.

Proof

From Section 3.1, you know that

$$\text{Area} = \frac{1}{2}bc\sin A \qquad\qquad \text{Formula for the area of an oblique triangle.}$$

$$= \frac{1}{4}b^2c^2\sin^2 A \qquad\qquad \text{Square the right side.}$$

$$= \sqrt{\frac{1}{4}b^2c^2\sin^2 A} \qquad\qquad \text{Take the square root of the right side.}$$

$$= \sqrt{\frac{1}{4}b^2c^2(1-\cos^2 A)} \qquad \text{Pythagorean Identity}$$

$$= \sqrt{\left[\frac{1}{2}bc(1+\cos A)\right]\left[\frac{1}{2}bc(1-\cos A)\right]}. \qquad \text{Factor.}$$

Using the Law of Cosines, you can show that

$$\frac{1}{2}bc(1+\cos A) = \frac{a+b+c}{2}\cdot\frac{-a+b+c}{2}$$

and

$$\frac{1}{2}bc(1-\cos A) = \frac{a-b+c}{2}\cdot\frac{a+b-c}{2}.$$

Letting $s = (a+b+c)/2$, these two equations can be rewritten as

$$\frac{1}{2}bc(1+\cos A) = s(s-a)$$

and

$$\frac{1}{2}bc(1-\cos A) = (s-b)(s-c).$$

By substituting into the last formula for area, you can conclude that

$$\text{Area} = \sqrt{s(s-a)(s-b)(s-c)}.$$

Properties of the Dot Product *(p. 304)*

Let **u**, **v**, and **w** be vectors in the place or in space and let c be a scalar.

1. $\mathbf{u} \cdot \mathbf{v} = \mathbf{v} \cdot \mathbf{u}$
2. $\mathbf{0} \cdot \mathbf{v} = 0$
3. $\mathbf{u} \cdot (\mathbf{v} + \mathbf{w}) = \mathbf{u} \cdot \mathbf{v} + \mathbf{u} \cdot \mathbf{w}$
4. $\mathbf{v} \cdot \mathbf{v} = \|\mathbf{v}\|^2$
5. $c(\mathbf{u} \cdot \mathbf{v}) = c\mathbf{u} \cdot \mathbf{v} = \mathbf{u} \cdot c\mathbf{v}$

Proof

Let $\mathbf{u} = \langle u_1, u_2 \rangle$, $\mathbf{v} = \langle v_1, v_2 \rangle$, $\mathbf{w} = \langle w_1, w_2 \rangle$, $\mathbf{0} = \langle 0, 0 \rangle$, and let c be a scalar.

1. $\mathbf{u} \cdot \mathbf{v} = u_1 v_1 + u_2 v_2 = v_1 u_1 + v_2 u_2 = \mathbf{v} \cdot \mathbf{u}$
2. $\mathbf{0} \cdot \mathbf{v} = 0 \cdot v_1 + 0 \cdot v_2 = 0$
3. $\mathbf{u} \cdot (\mathbf{v} + \mathbf{w}) = \mathbf{u} \cdot \langle v_1 + w_1, v_2 + w_2 \rangle$
$$= u_1(v_1 + w_1) + u_2(v_2 + w_2)$$
$$= u_1 v_1 + u_1 w_1 + u_2 v_2 + u_2 w_2$$
$$= (u_1 v_1 + u_2 v_2) + (u_1 w_1 + u_2 w_2) = \mathbf{u} \cdot \mathbf{v} + \mathbf{u} \cdot \mathbf{w}$$
4. $\mathbf{v} \cdot \mathbf{v} = v_1^2 + v_2^2 = \left(\sqrt{v_1^2 + v_2^2} \right)^2 = \|\mathbf{v}\|^2$
5. $c(\mathbf{u} \cdot \mathbf{v}) = c(\langle u_1, u_2 \rangle \cdot \langle v_1, v_2 \rangle)$
$$= c(u_1 v_1 + u_2 v_2) = (cu_1)v_1 + (cu_2)v_2$$
$$= \langle cu_1, cu_2 \rangle \cdot \langle v_1, v_2 \rangle = c\mathbf{u} \cdot \mathbf{v}$$

Angle Between Two Vectors *(p. 305)*

If θ is the angle between two nonzero vectors **u** and **v**, then $\cos \theta = \dfrac{\mathbf{u} \cdot \mathbf{v}}{\|\mathbf{u}\| \, \|\mathbf{v}\|}$.

Proof

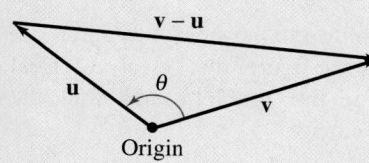

Consider the triangle determined by vectors **u**, **v**, and **v** − **u**, as shown in the figure. By the Law of Cosines, you can write

$$\|\mathbf{v} - \mathbf{u}\|^2 = \|\mathbf{u}\|^2 + \|\mathbf{v}\|^2 - 2\|\mathbf{u}\| \, \|\mathbf{v}\| \cos \theta$$
$$(\mathbf{v} - \mathbf{u}) \cdot (\mathbf{v} - \mathbf{u}) = \|\mathbf{u}\|^2 + \|\mathbf{v}\|^2 - 2\|\mathbf{u}\| \, \|\mathbf{v}\| \cos \theta$$
$$(\mathbf{v} - \mathbf{u}) \cdot \mathbf{v} - (\mathbf{v} - \mathbf{u}) \cdot \mathbf{u} = \|\mathbf{u}\|^2 + \|\mathbf{v}\|^2 - 2\|\mathbf{u}\| \, \|\mathbf{v}\| \cos \theta$$
$$\mathbf{v} \cdot \mathbf{v} - \mathbf{u} \cdot \mathbf{v} - \mathbf{v} \cdot \mathbf{u} + \mathbf{u} \cdot \mathbf{u} = \|\mathbf{u}\|^2 + \|\mathbf{v}\|^2 - 2\|\mathbf{u}\| \, \|\mathbf{v}\| \cos \theta$$
$$\|\mathbf{v}\|^2 - 2\mathbf{u} \cdot \mathbf{v} + \|\mathbf{u}\|^2 = \|\mathbf{u}\|^2 + \|\mathbf{v}\|^2 - 2\|\mathbf{u}\| \, \|\mathbf{v}\| \cos \theta$$

$$\cos \theta = \frac{\mathbf{u} \cdot \mathbf{v}}{\|\mathbf{u}\| \, \|\mathbf{v}\|}.$$

1. In the figure, a beam of light is directed at the blue mirror, reflected to the red mirror, and then reflected back to the blue mirror. Find the distance PT that the light travels from the red mirror back to the blue mirror.

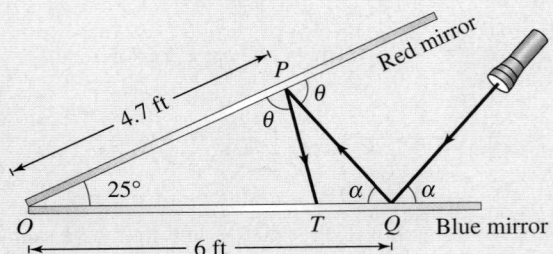

2. A triathlete sets a course to swim S 25° E from a point on shore to a buoy $\frac{3}{4}$ mile away. After swimming 300 yards through a strong current, the triathlete is off course at a bearing of S 35° E. Find the bearing and distance the triathlete needs to swim to correct her course.

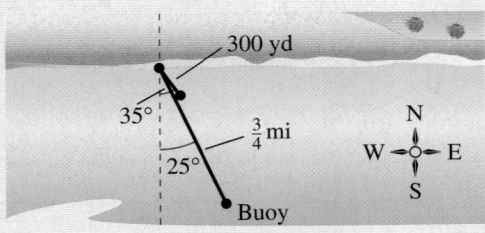

3. A hiking party is lost in a national park. Two ranger stations have received an emergency SOS signal from the party. Station B is 75 miles due east of station A. The bearing from station A to the signal is S 60° E and the bearing from station B to the signal is S 75° W.

(a) Find the distance from each station to the SOS signal.

(b) A rescue party is in the park 20 miles from station A at a bearing of S 80° E. Find the distance and the bearing the rescue party must travel to reach the lost hiking party.

4. You are seeding a triangular courtyard. One side of the courtyard is 52 feet long and the other side is 46 feet long. The angle opposite the 52-foot side is 65°.

(a) Draw a diagram that gives a visual representation of the problem.

(b) How long is the third side of the courtyard?

(c) One bag of grass covers an area of 50 square feet. How many bags of grass will you need to cover the courtyard?

5. For each pair of vectors, find the following.

(i) $\|\mathbf{u}\|$ (ii) $\|\mathbf{v}\|$ (iii) $\|\mathbf{u} + \mathbf{v}\|$

(iv) $\left\|\dfrac{\mathbf{u}}{\|\mathbf{u}\|}\right\|$ (v) $\left\|\dfrac{\mathbf{v}}{\|\mathbf{v}\|}\right\|$ (vi) $\left\|\dfrac{\mathbf{u} + \mathbf{v}}{\|\mathbf{u} + \mathbf{v}\|}\right\|$

(a) $\mathbf{u} = \langle 1, -1 \rangle$
$\mathbf{v} = \langle -1, 2 \rangle$

(b) $\mathbf{u} = \langle 0, 1 \rangle$
$\mathbf{v} = \langle 3, -3 \rangle$

(c) $\mathbf{u} = \langle 1, \frac{1}{2} \rangle$
$\mathbf{v} = \langle 2, 3 \rangle$

(d) $\mathbf{u} = \langle 2, -4 \rangle$
$\mathbf{v} = \langle 5, 5 \rangle$

6. A skydiver is falling at a constant downward velocity of 120 miles per hour. In the figure, vector $\mathbf{u}$ represents the skydiver's velocity. A steady breeze pushes the skydiver to the east at 40 miles per hour. Vector $\mathbf{v}$ represents the wind velocity.

(a) Write the vectors $\mathbf{u}$ and $\mathbf{v}$ in component form.

(b) Let $\mathbf{s} = \mathbf{u} + \mathbf{v}$. Use the figure to sketch $\mathbf{s}$. To print an enlarged copy of the graph, go to the website, *www.mathgraphs.com*.

(c) Find the magnitude of $\mathbf{s}$. What information does the magnitude give you about the skydiver's fall?

(d) If there were no wind, the skydiver would fall in a path perpendicular to Earth. At what angle to the ground is the path of the skydiver when the skydiver is affected by the 40 mile per hour wind from due west?

(e) The skydiver is blown to the west at 30 miles per hour. Draw a new figure that gives a visual representation of the problem and find the skydiver's new velocity.

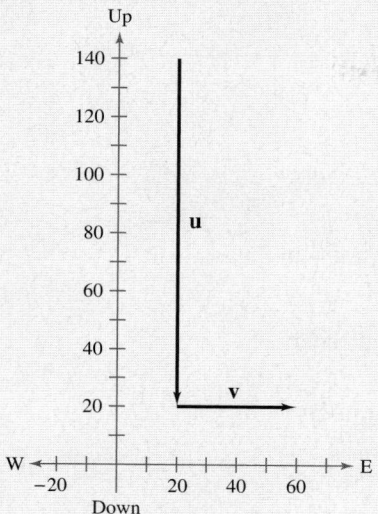

FIGURE FOR 6

7. Write the vector **w** in terms of **u** and **v**, given that the terminal point of **w** bisects the line segment.

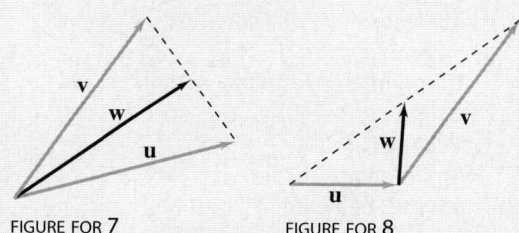

FIGURE FOR 7   FIGURE FOR 8

8. Prove that if **u** is orthogonal to **v** and **w**, then **u** is orthogonal to

$$cv + dw$$

for any scalars c and d.

9. Two forces of the same magnitude F_1 and F_2 act at angles θ_1 and θ_2, respectively. Use a diagram to compare the work done by F_1 with the work done by F_2 in moving along the vector PQ if

(a) $\theta_1 = -\theta_2$

(b) $\theta_1 = 60°$ and $\theta_2 = 30°$.

10. Four basic forces are in action during flight: weight, lift, thrust, and drag. To fly through the air, an object must overcome its own *weight*. To do this, it must create an upward force called *lift*. To generate lift, a forward motion called *thrust* is needed. The thrust must be great enough to overcome air resistance, which is called *drag*.

For a commercial jet aircraft, a quick climb is important to maximize efficiency, because the performance of an aircraft at high altitudes is enhanced. In addition, it is necessary to clear obstacles such as buildings and mountains and reduce noise in residential areas. In the diagram, the angle θ is called the climb angle. The velocity of the plane can be represented by a vector **v** with a vertical component $\|\mathbf{v}\| \sin \theta$ (called climb speed) and a horizontal component $\|\mathbf{v}\| \cos \theta$, where $\|\mathbf{v}\|$ is the speed of the plane. When taking off, a pilot must decide how much of the thrust to apply to each component. The more the thrust is applied to the horizontal component, the faster the airplane will gain speed. The more the thrust is applied to the vertical component, the quicker the airplane will climb.

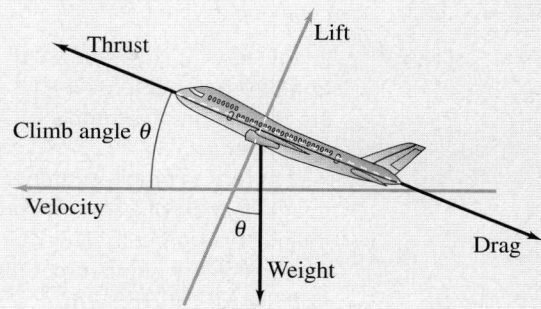

(a) Complete the table for an airplane that has a speed of $\|\mathbf{v}\| = 100$ miles per hour.

θ	0.5°	1.0°	1.5°	2.0°	2.5°	3.0°
$\|\mathbf{v}\| \sin \theta$						
$\|\mathbf{v}\| \cos \theta$						

(b) Does an airplane's speed equal the sum of the vertical and horizontal components of its velocity? If not, how could you find the speed of an airplane whose velocity components were known?

(c) Use the result of part (b) to find the speed of an airplane with the given velocity components.

(i) $\|\mathbf{v}\| \sin \theta = 5.235$ miles per hour
$\|\mathbf{v}\| \cos \theta = 149.909$ miles per hour

(ii) $\|\mathbf{v}\| \sin \theta = 10.463$ miles per hour
$\|\mathbf{v}\| \cos \theta = 149.634$ miles per hour

325

How to study Chapter 4

▶ **What you should learn**

In this chapter you will learn the following skills and concepts:

- How to perform operations with complex numbers
- How to determine the number of zeros of polynomial functions
- How to find the zeros of polynomial functions
- How to multiply and divide complex numbers written in trigonometric form
- How to find powers and nth roots of complex numbers

▶ **Important Vocabulary**

As you encounter each new vocabulary term in this chapter, add the term and its definition to your notebook glossary.

Imaginary unit i (p. 328)
Complex number (p. 328)
Standard form of a complex
  number (p. 328)
Real part of a complex number (p. 328)
Imaginary part of a complex number (p. 328)
Imaginary number (p. 328)
Pure imaginary number (p. 328)
Additive identity (p. 329)
Additive inverse (p. 329)
Complex conjugates (p. 331)
Discriminant (p. 336)

Complex plane (p. 342)
Real axis (p. 342)
Imaginary axis (p. 342)
Absolute value of a complex
  number (p. 342)
Trigonometric form of a complex
  number (p. 343)
Modulus (p. 343)
Argument (p. 343)
nth root of a complex number
  (p. 350)
nth roots of unity (p. 352)

Study Tools

Learning objectives in each section
Chapter Summary (p. 355)
Review Exercises (pp. 356–357)
Chapter Test (p. 358)

Additional Resources

Study and Solutions Guide
Interactive Trigonometry
Videotapes/DVD for Chapter 4
Trigonometry Website
Student Success Organizer

Gregory Sams/SPL/Photo Researchers, Inc.

4

Complex Numbers

4.1 Complex Numbers

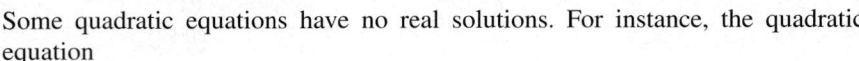

▶ **What you should learn**

- How to use the imaginary unit i to write complex numbers
- How to add, subtract, and multiply complex numbers
- How to use complex conjugates to write the quotient of two complex numbers in standard form
- How to find complex solutions of quadratic equations

▶ **Why you should learn it**

You can use complex numbers to model and solve real-life problems in electronics. For instance, in Exercise 84 on page 334, you will learn how to use complex numbers to find the impedance of an electrical circuit.

Chris Hamilton/Corbis

The Imaginary Unit i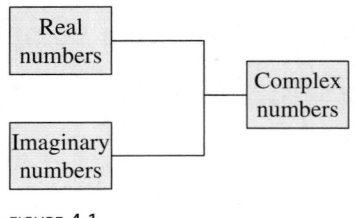

Some quadratic equations have no real solutions. For instance, the quadratic equation

$$x^2 + 1 = 0 \qquad \text{Equation with no real solution}$$

has no real solution because there is no real number x that can be squared to produce -1. To overcome this deficiency, mathematicians created an expanded system of numbers using the **imaginary unit i,** defined as

$$i = \sqrt{-1} \qquad \text{Imaginary unit}$$

where $i^2 = -1$. By adding real numbers to real multiples of this imaginary unit, the set of **complex numbers** is obtained. Each complex number can be written in the **standard form $a + bi$.** The real number a is called the **real part** of the complex number $a + bi$, and the number bi (where b is a real number) is called the **imaginary part** of the complex number.

Definition of a Complex Number

If a and b are real numbers, the number $a + bi$ is a **complex number,** and it is said to be written in **standard form.** If $b = 0$, the number $a + bi = a$ is a real number. If $b \neq 0$, the number $a + bi$ is called an **imaginary number.** A number of the form bi, where $b \neq 0$, is called a **pure imaginary number.**

The set of real numbers is a subset of the set of complex numbers, as shown in Figure 4.1. This is true because every real number a can be written as a complex number using $b = 0$. That is, for every real number a, you can write $a = a + 0i$.

```
┌─────────┐
│  Real   │──────────┐
│ numbers │          │   ┌──────────┐
└─────────┘          ├───│ Complex  │
┌─────────┐          │   │ numbers  │
│Imaginary│──────────┘   └──────────┘
│ numbers │
└─────────┘
```

FIGURE 4.1

Equality of Complex Numbers

Two complex numbers $a + bi$ and $c + di$, written in standard form, are equal to each other

$$a + bi = c + di \qquad \text{Equality of two complex numbers}$$

if and only if $a = c$ and $b = d$.

The icon 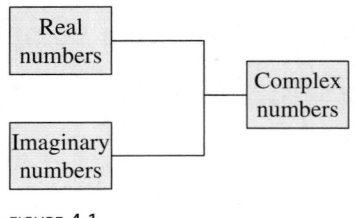 identifies examples and concepts related to features of the Learning Tools CD-ROM and the *Interactive* and *Internet* versions of this text. For more details see the chart on pages *xix-xxiii*.

Operations with Complex Numbers

To add (or subtract) two complex numbers, you add (or subtract) the real and imaginary parts of the numbers separately.

For each operation on complex numbers, you can show the parallel operation on polynomials.

Addition and Subtraction of Complex Numbers

If $a + bi$ and $c + di$ are two complex numbers written in standard form, their sum and difference are defined as follows.

Sum: $(a + bi) + (c + di) = (a + c) + (b + d)i$

Difference: $(a + bi) - (c + di) = (a - c) + (b - d)i$

The **additive identity** in the complex number system is zero (the same as in the real number system). Furthermore, the **additive inverse** of the complex number $a + bi$ is

$$-(a + bi) = -a - bi. \qquad \text{Additive inverse}$$

So, you have

$$(a + bi) + (-a - bi) = 0 + 0i = 0.$$

The *Interactive* CD-ROM and *Internet* versions of this text offer a Try It for each example in the text.

Example 1 ► Adding and Subtracting Complex Numbers

a. $(4 + 7i) + (1 - 6i) = 4 + 7i + 1 - 6i$ Remove parentheses.

$$= (4 + 1) + (7i - 6i) \qquad \text{Group like terms.}$$

$$= 5 + i \qquad \text{Write in standard form.}$$

b. $(1 + 2i) - (4 + 2i) = 1 + 2i - 4 - 2i$ Remove parentheses.

$$= (1 - 4) + (2i - 2i) \qquad \text{Group like terms.}$$

$$= -3 + 0 \qquad \text{Simplify.}$$

$$= -3 \qquad \text{Write in standard form.}$$

c. $3i - (-2 + 3i) - (2 + 5i) = 3i + 2 - 3i - 2 - 5i$

$$= (2 - 2) + (3i - 3i - 5i)$$

$$= 0 - 5i$$

$$= -5i$$

d. $(3 + 2i) + (4 - i) - (7 + i) = 3 + 2i + 4 - i - 7 - i$

$$= (3 + 4 - 7) + (2i - i - i)$$

$$= 0 + 0i$$

$$= 0$$

Note in Examples 1(b) and 1(d) that the sum of two complex numbers can be a real number.

◀ **E x p l o r a t i o n** ▶

Complete the following.

$i^1 = i$ $i^7 = $ ⬜

$i^2 = -1$ $i^8 = $ ⬜

$i^3 = -i$ $i^9 = $ ⬜

$i^4 = 1$ $i^{10} = $ ⬜

$i^5 = $ ⬜ $i^{11} = $ ⬜

$i^6 = $ ⬜ $i^{12} = $ ⬜

What pattern do you see? Write a brief description of how you would find i raised to any positive integer power.

Many of the properties of real numbers are valid for complex numbers as well. Here are some examples.

Associative Properties of Addition and Multiplication

Commutative Properties of Addition and Multiplication

Distributive Property of Multiplication Over Addition

Notice below how these properties are used when two complex numbers are multiplied.

$$(a + bi)(c + di) = a(c + di) + bi(c + di) \qquad \text{Distributive Property}$$

$$= ac + (ad)i + (bc)i + (bd)i^2 \qquad \text{Distributive Property}$$

$$= ac + (ad)i + (bc)i + (bd)(-1) \qquad i^2 = -1$$

$$= ac - bd + (ad)i + (bc)i \qquad \text{Commutative Property}$$

$$= (ac - bd) + (ad + bc)i \qquad \text{Associative Property}$$

Rather than trying to memorize this multiplication rule, you should simply remember how the Distributive Property is used to multiply two complex numbers.

Example 2 ▶   **Multiplying Complex Numbers**

a. $4(-2 + 3i) = 4(-2) + 4(3i)$ Distributive Property

$\qquad\qquad = -8 + 12i$ Simplify.

b. $(2 - i)(4 + 3i) = 2(4 + 3i) - i(4 + 3i)$ Distributive Property

$\qquad\qquad = 8 + 6i - 4i - 3i^2$ Distributive Property

$\qquad\qquad = 8 + 6i - 4i - 3(-1)$ $i^2 = -1$

$\qquad\qquad = (8 + 3) + (6i - 4i)$ Group like terms.

$\qquad\qquad = 11 + 2i$ Write in standard form.

c. $(3 + 2i)(3 - 2i) = 3(3 - 2i) + 2i(3 - 2i)$ Distributive Property

$\qquad\qquad = 9 - 6i + 6i - 4i^2$ Distributive Property

$\qquad\qquad = 9 - 6i + 6i - 4(-1)$ $i^2 = -1$

$\qquad\qquad = 9 + 4$ Simplify.

$\qquad\qquad = 13$ Write in standard form.

d. $(3 + 2i)^2 = (3 + 2i)(3 + 2i)$ Square of a binomial

$\qquad\qquad = 3(3 + 2i) + 2i(3 + 2i)$ Distributive Property

$\qquad\qquad = 9 + 6i + 6i + 4i^2$ Distributive Property

$\qquad\qquad = 9 + 6i + 6i + 4(-1)$ $i^2 = -1$

$\qquad\qquad = 9 + 12i - 4$ Simplify.

$\qquad\qquad = 5 + 12i$ Write in standard form.

Complex Conjugates

Notice in Example 2(c) that the product of two complex numbers can be a real number. This occurs with pairs of complex numbers of the form $a + bi$ and $a - bi$, called **complex conjugates.**

$$(a + bi)(a - bi) = a^2 - abi + abi - b^2i^2$$
$$= a^2 - b^2(-1)$$
$$= a^2 + b^2$$

Example 3 ▶ **Multiplying Conjugates**

Multiply each complex number by its complex conjugate.

a. $1 + i$ **b.** $4 - 3i$

Solution

a. The complex conjugate of $1 + i$ is $1 - i$.

$$(1 + i)(1 - i) = 1^2 - i^2 = 1 - (-1) = 2$$

b. The complex conjugate of $4 - 3i$ is $4 + 3i$.

$$(4 - 3i)(4 + 3i) = 4^2 - (3i)^2 = 16 - 9i^2 = 16 - 9(-1) = 25$$

To write the quotient of $a + bi$ and $c + di$ in standard form, where c and d are not both zero, multiply the numerator and denominator by the complex conjugate of the *denominator* to obtain

$$\frac{a + bi}{c + di} = \frac{a + bi}{c + di}\left(\frac{c - di}{c - di}\right)$$

$$= \frac{(ac + bd) + (bc - ad)i}{c^2 + d^2}. \qquad \text{Standard form}$$

Example 4 ▶ **Writing a Quotient of Complex Numbers in Standard Form**

$$\frac{2 + 3i}{4 - 2i} = \frac{2 + 3i}{4 - 2i}\left(\frac{4 + 2i}{4 + 2i}\right) \qquad \text{Multiply numerator and denominator by complex conjugate of denominator.}$$

$$= \frac{8 + 4i + 12i + 6i^2}{16 - 4i^2} \qquad \text{Expand.}$$

$$= \frac{8 - 6 + 16i}{16 + 4} \qquad i^2 = -1$$

$$= \frac{2 + 16i}{20} \qquad \text{Simplify.}$$

$$= \frac{1}{10} + \frac{4}{5}i \qquad \text{Write in standard form.}$$

Complex Solutions of Quadratic Equations

When using the Quadratic Formula to solve a quadratic equation, you often obtain a result such as $\sqrt{-3}$, which you know is not a real number. By factoring out $i = \sqrt{-1}$, you can write this number in standard form.

$$\sqrt{-3} = \sqrt{3(-1)} = \sqrt{3}\sqrt{-1} = \sqrt{3}\,i$$

The number $\sqrt{3}\,i$ is called the *principal square root* of -3.

STUDY TIP

The definition of principal square root uses the rule

$$\sqrt{ab} = \sqrt{a}\sqrt{b}$$

for $a > 0$ and $b < 0$. This rule is not valid if *both* a and b are negative. For example,

$$\sqrt{-5}\sqrt{-5} = \sqrt{5(-1)}\sqrt{5(-1)}$$
$$= \sqrt{5}\,i\sqrt{5}\,i$$
$$= \sqrt{25}\,i^2$$
$$= 5i^2 = -5$$

whereas

$$\sqrt{(-5)(-5)} = \sqrt{25} = 5.$$

To avoid problems with square roots of negative numbers, be sure to convert to standard form *before* multiplying.

> **Principal Square Root of a Negative Number**
>
> If a is a positive number, the **principal square root** of the negative number $-a$ is defined as
>
> $$\sqrt{-a} = \sqrt{a}\,i.$$

Example 5 ▶   Writing Complex Numbers in Standard Form

a. $\sqrt{-3}\sqrt{-12} = \sqrt{3}\,i\sqrt{12}\,i = \sqrt{36}\,i^2 = 6(-1) = -6$

b. $\sqrt{-48} - \sqrt{-27} = \sqrt{48}\,i - \sqrt{27}\,i = 4\sqrt{3}\,i - 3\sqrt{3}\,i = \sqrt{3}\,i$

c. $\left(-1 + \sqrt{-3}\right)^2 = \left(-1 + \sqrt{3}\,i\right)^2$
$$= (-1)^2 - 2\sqrt{3}\,i + \left(\sqrt{3}\right)^2(i^2)$$
$$= 1 - 2\sqrt{3}\,i + 3(-1)$$
$$= -2 - 2\sqrt{3}\,i$$

Example 6 ▶   Complex Solutions of a Quadratic Equation

Solve (a) $x^2 + 4 = 0$ and (b) $3x^2 - 2x + 5 = 0$.

Solution

a. $x^2 + 4 = 0$ Write original equation.

$\quad\quad x^2 = -4$ Subtract 4 from each side.

$\quad\quad x = \pm 2i$ Extract square roots.

b. $3x^2 - 2x + 5 = 0$ Write original equation.

$$x = \frac{-(-2) \pm \sqrt{(-2)^2 - 4(3)(5)}}{2(3)}$$ Quadratic Formula

$$= \frac{2 \pm \sqrt{-56}}{6}$$ Simplify.

$$= \frac{2 \pm 2\sqrt{14}\,i}{6}$$ Write $\sqrt{-56}$ in standard form.

$$= \frac{1}{3} \pm \frac{\sqrt{14}}{3}\,i$$ Write in standard form.

Activities

1. Perform the indicated operations and write the result in standard form.
$$\left(4 - \sqrt{-9}\right)\left(2 + \sqrt{-9}\right)$$
Answer: $17 + 6i$

2. Write $\dfrac{3 + i}{i}$ in standard form.

Answer: $1 - 3i$

3. Use the Quadratic Formula to solve $2x^2 - 5x + 7 = 0$.

Answer: $x = \dfrac{5}{4} \pm \dfrac{\sqrt{31}}{4}\,i$

4.1 Exercises

The *Interactive* CD-ROM and *Internet* versions of this text contain step-by-step solutions to all odd-numbered exercises. They also provide Tutorial Exercises for additional help.

In Exercises 1–4, find real numbers a and b such that the equation is true.

1. $a + bi = -10 + 6i$

2. $a + bi = 13 + 4i$

3. $(a - 1) + (b + 3)i = 5 + 8i$

4. $(a + 6) + 2bi = 6 - 5i$

In Exercises 5–16, write the complex number in standard form.

5. $4 + \sqrt{-9}$

6. $3 + \sqrt{-16}$

7. $2 - \sqrt{-27}$

8. $1 + \sqrt{-8}$

9. $\sqrt{-75}$

10. $\sqrt{-4}$

11. 8

12. 45

13. $-6i + i^2$

14. $-4i^2 + 2i$

15. $\sqrt{-0.09}$

16. $\sqrt{-0.0004}$

In Exercises 17–26, perform the addition or subtraction and write the result in standard form.

17. $(5 + i) + (6 - 2i)$

18. $(13 - 2i) + (-5 + 6i)$

19. $(8 - i) - (4 - i)$

20. $(3 + 2i) - (6 + 13i)$

21. $\left(-2 + \sqrt{-8}\right) + \left(5 - \sqrt{-50}\right)$

22. $\left(8 + \sqrt{-18}\right) - \left(4 + 3\sqrt{2}i\right)$

23. $13i - (14 - 7i)$

24. $22 + (-5 + 8i) + 10i$

25. $-\left(\frac{3}{2} + \frac{5}{2}i\right) + \left(\frac{5}{3} + \frac{11}{3}i\right)$

26. $(1.6 + 3.2i) + (-5.8 + 4.3i)$

In Exercises 27–40, perform the operation and write the result in standard form.

27. $\sqrt{-6} \cdot \sqrt{-2}$

28. $\sqrt{-5} \cdot \sqrt{-10}$

29. $\left(\sqrt{-10}\right)^2$

30. $\left(\sqrt{-75}\right)^2$

31. $(1 + i)(3 - 2i)$

32. $(6 - 2i)(2 - 3i)$

33. $6i(5 - 2i)$

34. $-8i(9 + 4i)$

35. $\left(\sqrt{14} + \sqrt{10}i\right)\left(\sqrt{14} - \sqrt{10}i\right)$

36. $\left(3 + \sqrt{-5}\right)\left(7 - \sqrt{-10}\right)$

37. $(4 + 5i)^2$

38. $(2 - 3i)^2$

39. $(2 + 3i)^2 + (2 - 3i)^2$

40. $(1 - 2i)^2 - (1 + 2i)^2$

In Exercises 41–48, write the complex conjugate of the complex number. Then multiply the number by its complex conjugate.

41. $6 + 3i$

42. $7 - 12i$

43. $-1 - \sqrt{5}i$

44. $-3 + \sqrt{2}i$

45. $\sqrt{-20}$

46. $\sqrt{-15}$

47. $\sqrt{8}$

48. $1 + \sqrt{8}$

In Exercises 49–58, write the quotient in standard form.

49. $\dfrac{5}{i}$

50. $-\dfrac{14}{2i}$

51. $\dfrac{2}{4 - 5i}$

52. $\dfrac{5}{1 - i}$

53. $\dfrac{3 + i}{3 - i}$

54. $\dfrac{6 - 7i}{1 - 2i}$

55. $\dfrac{6 - 5i}{i}$

56. $\dfrac{8 + 16i}{2i}$

57. $\dfrac{3i}{(4 - 5i)^2}$

58. $\dfrac{5i}{(2 + 3i)^2}$

In Exercises 59–62, perform the operation and write the result in standard form.

59. $\dfrac{2}{1 + i} - \dfrac{3}{1 - i}$

60. $\dfrac{2i}{2 + i} + \dfrac{5}{2 - i}$

61. $\dfrac{i}{3 - 2i} + \dfrac{2i}{3 + 8i}$

62. $\dfrac{1 + i}{i} - \dfrac{3}{4 - i}$

In Exercises 63–72, use the Quadratic Formula to solve the quadratic equation.

63. $x^2 - 2x + 2 = 0$

64. $x^2 + 6x + 10 = 0$

65. $4x^2 + 16x + 17 = 0$

66. $9x^2 - 6x + 37 = 0$

67. $4x^2 + 16x + 15 = 0$

68. $16t^2 - 4t + 3 = 0$

69. $\frac{3}{2}x^2 - 6x + 9 = 0$

70. $\frac{7}{8}x^2 - \frac{3}{4}x + \frac{5}{16} = 0$

71. $1.4x^2 - 2x - 10 = 0$

72. $4.5x^2 - 3x + 12 = 0$

In Exercises 73–80, simplify the complex number and write it in standard form.

73. $-6i^3 + i^2$

74. $4i^2 - 2i^3$

75. $-5i^5$

76. $(-i)^3$

77. $(\sqrt{-75})^3$

78. $(\sqrt{-2})^6$

79. $\dfrac{1}{i^3}$

80. $\dfrac{1}{(2i)^3}$

81. Cube each complex number.

(a) 2 (b) $-1 + \sqrt{3}i$ (c) $-1 - \sqrt{3}i$

82. Raise each complex number to the fourth power.

(a) 2 (b) -2 (c) $2i$ (d) $-2i$

83. Express each of the powers of i as i, $-i$, 1, or -1.

(a) i^{40} (b) i^{25} (c) i^{50} (d) i^{67}

▶ Model It

84. *Impedance* The opposition to current in an electrical circuit is called its impedance. The impedance z in a parallel circuit with two pathways satisfies the equation

$$\frac{1}{z} = \frac{1}{z_1} + \frac{1}{z_2}$$

where z_1 is the impedance (in ohms) of pathway 1 and z_2 is the impedance of pathway 2.

(a) The impedance of each pathway in a parallel circuit is found by adding the impedances of all components in the pathway. Use the table to find z_1 and z_2.

(b) Find the impedance z.

	Resistor	Inductor	Capacitor
Symbol	$-\!\!\bigwedge\!\!\bigwedge\!\!-$ $a\Omega$	$-\!\!\text{000}\!\!-$ $b\Omega$	$-\!\!\mid\!\!\vdash$ $c\Omega$
Impedance	a	bi	$-ci$

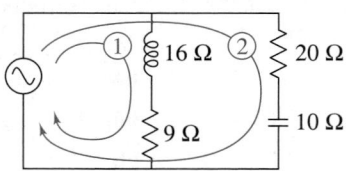

Synthesis

True or False? **In Exercises 85–87, determine whether the statement is true or false. Justify your answer.**

85. There is no complex number that is equal to its complex conjugate.

86. $-i\sqrt{6}$ is a solution of $x^4 - x^2 + 14 = 56$.

87. $i^{44} + i^{150} - i^{74} - i^{109} + i^{61} = -1$

88. *Error Analysis*   Describe the error.

$$\sqrt{-6}\sqrt{-6} = \sqrt{(-6)(-6)} = \sqrt{36} = 6$$

89. *Proof* Prove that the complex conjugate of the product of two complex numbers $a_1 + b_1 i$ and $a_2 + b_2 i$ is the product of their complex conjugates.

90. *Proof* Prove that the complex conjugate of the sum of two complex numbers $a_1 + b_1 i$ and $a_2 + b_2 i$ is the sum of their complex conjugates.

Review

In Exercises 91–94, perform the operation and write the result in standard form.

91. $(4 + 3x) + (8 - 6x - x^2)$

92. $(x^3 - 3x^2) - (6 - 2x - 4x^2)$

93. $\left(3x - \frac{1}{2}\right)(x + 4)$

94. $(2x - 5)^2$

In Exercises 95–98, solve the equation and check your solution.

95. $-x - 12 = 19$

96. $8 - 3x = -34$

97. $4(5x - 6) - 3(6x + 1) = 0$

98. $5[x - (3x + 11)] = 20x - 15$

99. *Volume of an Oblate Spheroid*

Solve for a: $V = \frac{4}{3}\pi a^2 b$

100. *Newton's Law of Universal Gravitation*

Solve for r: $F = \alpha \dfrac{m_1 m_2}{r^2}$

101. *Mixture Problem*   A five-liter container contains a mixture with a concentration of 50%. How much of this mixture must be withdrawn and replaced by 100% concentrate to bring the mixture up to 60% concentration?

4.2 Complex Solutions of Equations

▶ **Why you should learn it**

Finding zeros of polynomial functions is an important part of solving real-life problems. For instance, in Exercise 64 on page 341, the zeros of a polynomial function can help you analyze the revenue function for Manpower, Inc.

The Number of Solutions of a Polynomial Equation

The Fundamental Theorem of Algebra implies that a polynomial equation of degree n has precisely n solutions in the complex number system. These solutions can be real or complex and may be repeated. The Fundamental Theorem of Algebra and the Linear Factorization Theorem are listed below for your review. For a proof of the Linear Factorization Theorem, see Proofs in Mathematics on page 359.

The Fundamental Theorem of Algebra

If $f(x)$ is a polynomial of degree n, where $n > 0$, then f has at least one zero in the complex number system.

Linear Factorization Theorem

If $f(x)$ is a polynomial of degree n, where $n > 0$, then f has precisely n linear factors $f(x) = a_n(x - c_1)(x - c_2) \cdots (x - c_n)$, where $c_1, c_2, \ldots, c_n$ are complex numbers.

Example 1 ▶ **Solutions of Polynomial Equations**

a. The first-degree equation $x - 2 = 0$ has exactly *one* solution: $x = 2$.

b. The second-degree equation

$$x^2 - 6x + 9 = 0 \qquad \text{Second-degree equation}$$

$$(x - 3)(x - 3) = 0 \qquad \text{Factor.}$$

has exactly *two* solutions: $x = 3$ and $x = 3$. (This is called a *repeated solution*.)

c. The third-degree equation

$$x^3 + 4x = 0 \qquad \text{Third-degree equation}$$

$$x(x - 2i)(x + 2i) = 0 \qquad \text{Factor.}$$

has exactly *three* solutions: $x = 0$, $x = 2i$, and $x = -2i$.

d. The fourth-degree equation

$$x^4 - 1 = 0 \qquad \text{Fourth-degree equation}$$

$$(x - 1)(x + 1)(x - i)(x + i) = 0 \qquad \text{Factor.}$$

has exactly *four* solutions: $x = 1$, $x = -1$, $x = i$, and $x = -i$.

You can use a graph to check the number of *real* solutions of an equation. For instance, to check the number of real solutions of $x^4 - 1 = 0$, sketch the graph of $f(x) = x^4 - 1$. As shown in Figure 4.2, the graph has two x-intercepts, which implies that the equation has two real solutions.

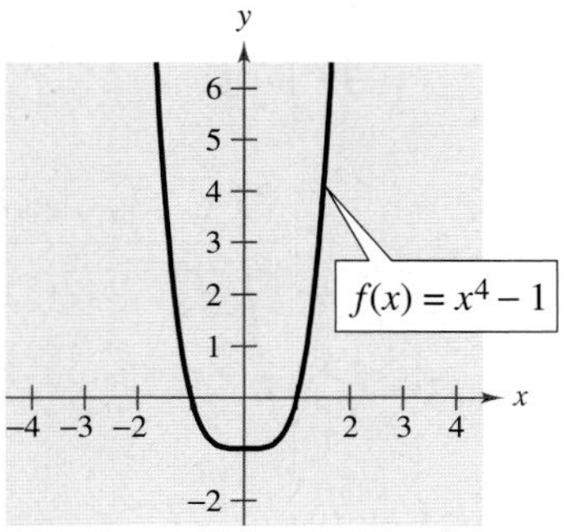

$f(x) = x^4 - 1$

FIGURE **4.2**

Every second-degree equation, $ax^2 + bx + c = 0$, has precisely two solutions given by the Quadratic Formula.

$$x = \frac{-b \pm \sqrt{b^2 - 4ac}}{2a}$$

The expression inside the radical, $b^2 - 4ac$, is called the **discriminant,** and can be used to determine whether the solutions are real, repeated, or complex.

1. If $b^2 - 4ac < 0$, the equation has two complex solutions.
2. If $b^2 - 4ac = 0$, the equation has one repeated real solution.
3. If $b^2 - 4ac > 0$, the equation has two distinct real solutions.

Example 2 ▶ Using the Discriminant

Use the discriminant to find the number of real solutions of each equation.

a. $4x^2 - 20x + 25 = 0$ **b.** $13x^2 + 7x + 2 = 0$ **c.** $5x^2 - 8x = 0$

Solution

a. For this equation, $a = 4$, $b = -20$, and $c = 25$. So, the discriminant is

$$b^2 - 4ac = (-20)^2 - 4(4)(25) = 400 - 400 = 0.$$

Because the discriminant is zero, there is one repeated real solution.

b. For this equation, $a = 13$, $b = 7$, and $c = 2$. So, the discriminant is

$$b^2 - 4ac = 7^2 - 4(13)(2) = 49 - 104 = -55.$$

Because the discriminant is negative, there are two complex solutions.

c. For this equation, $a = 5$, $b = -8$, and $c = 0$. So, the discriminant is

$$b^2 - 4ac = (-8)^2 - 4(5)(0) = 64 - 0 = 64.$$

Because the discriminant is positive, there are two real solutions.

Figure 4.3 shows the graphs of the functions corresponding to the equations in Example 2. Notice that with one repeated solution, the graph touches the x-axis at its x-intercept. With two complex solutions, the graph has no x-intercepts. With two real solutions, the graph crosses the x-axis at its x-intercepts.

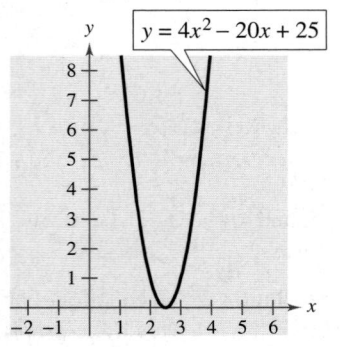

(a)

FIGURE **4.3**

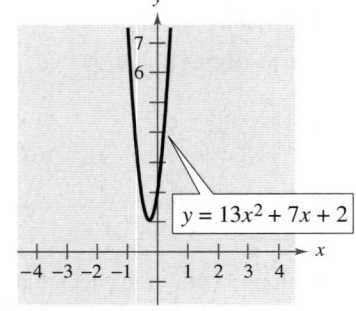

(b)

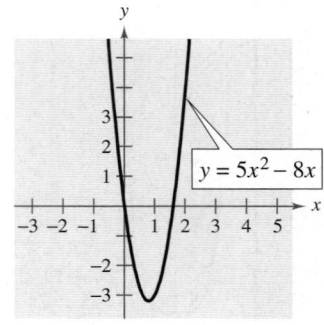

(c)

Finding Solutions of Polynomial Equations

Example 3 ▶ Solving a Quadratic Equation

Solve $x^2 + 2x + 2 = 0$. Write complex solutions in standard form.

Solution

Using $a = 1$, $b = 2$, and $c = 2$, you can apply the Quadratic Formula as follows.

$$x = \frac{-b \pm \sqrt{b^2 - 4ac}}{2a} \qquad \text{Quadratic Formula}$$

$$= \frac{-2 \pm \sqrt{2^2 - 4(1)(2)}}{2(1)} \qquad \text{Substitute 1 for } a, \text{ 2 for } b, \text{ and 2 for } c.$$

$$= \frac{-2 \pm \sqrt{-4}}{2} \qquad \text{Simplify.}$$

$$= \frac{-2 \pm 2i}{2} \qquad \text{Simplify.}$$

$$= -1 \pm i \qquad \text{Standard form}$$

In Example 3, the two complex solutions are **conjugates.** That is, they are of the form $a \pm bi$. This is not a coincidence, as indicated by the following theorem.

Complex Solutions Occur in Conjugate Pairs

If $a + bi$, $b \neq 0$, is a solution of a polynomial equation with real coefficients, the conjugate $a - bi$ is also a solution of the equation.

Be sure you see that this result is true only if the polynomial has *real* coefficients. For instance, the result applies to the equation $x^2 + 1 = 0$, but not to the equation $x - i = 0$.

Example 4 ▶ Solving a Polynomial Equation

Solve $x^4 - x^2 - 20 = 0$.

Solution

$$x^4 - x^2 - 20 = 0 \qquad \text{Write original equation.}$$

$$(x^2 - 5)(x^2 + 4) = 0 \qquad \text{Partially factor.}$$

$$\left(x + \sqrt{5}\right)\left(x - \sqrt{5}\right)(x + 2i)(x - 2i) = 0 \qquad \text{Factor completely.}$$

Setting each factor equal to zero yields the solutions $x = -\sqrt{5}$, $x = \sqrt{5}$, $x = -2i$, and $x = 2i$.

Finding Zeros of Polynomial Functions

The problem of finding the zeros of a polynomial function is essentially the same problem as finding the solutions of a polynomial equation. For instance, the zeros of the polynomial function

$$f(x) = 3x^2 - 4x + 5$$

are simply the solutions of the polynomial equation

$$3x^2 - 4x + 5 = 0.$$

You may want to point out to students that a graphing utility is helpful in determining real zeros, which in turn are useful in finding complex zeros. Try demonstrating this technique with the function $f(x) = x^5 + x^3 + 2x^2 - 12x + 8$, which has zeros $x = 1$, $x = -2$, $x = 2i$, and $x = -2i$. (Note that $x = 1$ is a repeated zero.)

Example 5 ▶   Finding the Zeros of a Polynomial Function

Find all the zeros of

$$f(x) = x^4 - 3x^3 + 6x^2 + 2x - 60$$

given that $1 + 3i$ is a zero of f.

Solution

Because complex zeros occur in conjugate pairs, you know that $1 - 3i$ is also a zero of f. This means that both

$$[x - (1 + 3i)] \quad \text{and} \quad [x - (1 - 3i)]$$

are factors of f. Multiplying these two factors produces

$$[x - (1 + 3i)][x - (1 - 3i)] = [(x - 1) - 3i][(x - 1) + 3i]$$
$$= (x - 1)^2 - 9i^2$$
$$= x^2 - 2x + 1 - 9(-1)$$
$$= x^2 - 2x + 10.$$

Using long division, you can divide $x^2 - 2x + 10$ into f to obtain the following.

$$
\begin{array}{r}
x^2 - x - 6 \\
x^2 - 2x + 10 \overline{)\ x^4 - 3x^3 + 6x^2 + 2x - 60} \\
\underline{x^4 - 2x^3 + 10x^2} \\
-x^3 - 4x^2 + 2x \\
\underline{-x^3 + 2x^2 - 10x} \\
-6x^2 + 12x - 60 \\
\underline{-6x^2 + 12x - 60} \\
0
\end{array}
$$

So, you have

$$f(x) = (x^2 - 2x + 10)(x^2 - x - 6)$$
$$= (x^2 - 2x + 10)(x - 3)(x + 2)$$

and you can conclude that the zeros of f are $x = 1 + 3i$, $x = 1 - 3i$, $x = 3$, and $x = -2$.

Example 6 ▶ **Finding a Polynomial with Given Zeros**

Find a fourth-degree polynomial function with real coefficients that has $-1, -1$, and $3i$ as zeros.

Solution

Because $3i$ is a zero *and* the polynomial is stated to have real coefficients, you know that the conjugate $-3i$ must also be a zero. So, $f(x)$ can be written as

$$f(x) = a(x + 1)(x + 1)(x - 3i)(x + 3i).$$

For simplicity, let $a = 1$ to obtain

$$f(x) = (x^2 + 2x + 1)(x^2 + 9)$$
$$= x^4 + 2x^3 + 10x^2 + 18x + 9.$$

Example 7 ▶ **Finding a Polynomial with Given Zeros**

Find a cubic polynomial function f with real coefficients that has 2 and $1 - i$ as zeros, such that $f(1) = 3$.

Solution

Because $1 - i$ is a zero of f, so is $1 + i$. So,

$$f(x) = a(x - 2)[x - (1 - i)][x - (1 + i)]$$
$$= a(x - 2)[(x - 1) + i][(x - 1) - i]$$
$$= a(x - 2)[(x - 1)^2 - i^2]$$
$$= a(x - 2)(x^2 - 2x + 2)$$
$$= a(x^3 - 4x^2 + 6x - 4).$$

To find the value of a, use the fact that $f(1) = 3$ and obtain

$$f(1) = a[1^3 - 4(1)^2 + 6(1) - 4]$$
$$3 = -a$$
$$-3 = a.$$

So, $a = -3$ and it follows that

$$f(x) = -3(x^3 - 4x^2 + 6x - 4)$$
$$= -3x^3 + 12x^2 - 18x + 12.$$

Activities

1. Write $f(x) = x^4 - 16$ as a product of linear factors.

   Answer:
   $f(x) = (x - 2)(x + 2)(x - 2i)(x + 2i)$

2. Find a third-degree polynomial function with integer coefficients that has $2, 3 + i$, and $3 - i$ as zeros.

   Answer: $f(x) = x^3 - 8x^2 + 22x - 20$

3. Use the zero $x = 4i$ to find the zeros of $f(x) = x^4 + 4x^3 + 11x^2 + 64x - 80$.

   Answer: $-5, 1, 4i, -4i$

Writing ABOUT MATHEMATICS

Solutions, Zeros, and Intercepts Write a paragraph explaining the relationships among the solutions of a polynomial equation, the zeros of a polynomial function, and the x-intercepts of the graph of a polynomial function. Include examples in your paragraph.

4.2 Exercises

In Exercises 1–4, determine the number of solutions of the equation in the complex number system.

1. $2x^3 + 3x + 1 = 0$

2. $x^6 + 4x^2 + 12 = 0$

3. $50 - 2x^4 = 0$

4. $14 - x + 4x^2 - 7x^5 = 0$

In Exercises 5–12, use the discriminant to determine the number of real solutions of the quadratic equation.

5. $2x^2 - 5x + 5 = 0$ **6.** $2x^2 - x - 1 = 0$

7. $\frac{1}{5}x^2 + \frac{6}{5}x - 8 = 0$ **8.** $\frac{1}{3}x^2 - 5x + 25 = 0$

9. $2x^2 - x - 15 = 0$ **10.** $-2x^2 + 11x - 2 = 0$

11. $x^2 + 2x + 10 = 0$ **12.** $x^2 - 4x + 53 = 0$

In Exercises 13–26, solve the equation. Write complex solutions in standard form.

13. $x^2 - 5 = 0$ **14.** $3x^2 - 1 = 0$

15. $(x + 5)^2 - 6 = 0$ **16.** $16 - (x - 1)^2 = 0$

17. $x^2 - 8x + 16 = 0$ **18.** $4x^2 + 4x + 1 = 0$

19. $x^2 + 2x + 5 = 0$ **20.** $54 + 16x - x^2 = 0$

21. $4x^2 - 4x + 5 = 0$

22. $4x^2 - 4x + 21 = 0$

23. $230 + 20x - 0.5x^2 = 0$

24. $125 - 30x + 0.4x^2 = 0$

25. $8 + (x + 3)^2 = 0$

26. $6 - (x - 1)^2 = 0$

Graphical and Analytical Analysis In Exercises 27–30, use a graphing utility to graph the function. Find all the zeros of the function. Is there a relationship between the number of real zeros and the number of x-intercepts of the graph? Explain.

27. $f(x) = x^3 - 4x^2 + x - 4$

28. $f(x) = x^3 - 4x^2 - 4x + 16$

29. $f(x) = x^4 + 4x^2 + 4$

30. $f(x) = x^4 - 3x^2 - 4$

In Exercises 31–48, find all the zeros of the function and write the polynomial as a product of linear factors.

31. $f(x) = x^2 + 25$ **32.** $f(x) = x^2 - x + 56$

33. $h(x) = x^2 - 4x + 1$ **34.** $g(x) = x^2 + 10x + 23$

35. $f(x) = x^4 - 81$

36. $f(y) = y^4 - 625$

37. $f(z) = z^2 - 2z + 2$

38. $h(x) = x^2 - 6x - 10$

39. $g(x) = x^3 + 3x^2 - 3x - 9$

40. $f(x) = x^3 - 8x^2 - 12x + 96$

41. $h(x) = x^3 - 4x^2 + 16x - 64$

42. $h(x) = x^3 + 5x^2 + 2x + 10$

43. $f(x) = 2x^3 - x^2 + 36x - 18$

44. $g(x) = 4x^3 + 3x^2 + 96x + 72$

45. $g(x) = x^4 - 4x^3 + 36x^2 - 144x$

46. $h(x) = x^4 + x^3 + 100x^2 + 100x$

47. $f(x) = x^4 + 10x^2 + 9$

48. $f(x) = x^4 + 29x^2 + 100$

In Exercises 49–56, use the given zero to find all the zeros of the function.

Function	Zero
49. $f(x) = 2x^3 + 3x^2 + 50x + 75$	$5i$
50. $f(x) = x^3 + x^2 + 9x + 9$	$3i$
51. $f(x) = 2x^4 - x^3 + 7x^2 - 4x - 4$	$2i$
52. $g(x) = x^3 - 7x^2 - x + 87$	$5 + 2i$
53. $g(x) = 4x^3 + 23x^2 + 34x - 10$	$-3 + i$
54. $h(x) = 3x^3 - 4x^2 + 8x + 8$	$1 - \sqrt{3}i$
55. $f(x) = x^4 + 3x^3 - 5x^2 - 21x + 22$	$-3 + \sqrt{2}i$
56. $f(x) = x^3 + 4x^2 + 14x + 20$	$-1 - 3i$

In Exercises 57–62, find a polynomial function with integer coefficients that has the given zeros. (There are many correct answers.)

57. $1, 5i, -5i$

58. $4, 3i, -3i$

59. $6, -5 + 2i, -5 - 2i$

60. $2, 4 + i, 4 - i$

61. $\frac{2}{3}, -1, 3 + \sqrt{2}i$

62. $-5, -5, 1 + \sqrt{3}i$

63. Height of a Baseball A baseball is thrown upward from ground level with an initial velocity of 48 feet per second, and its height h (in feet) is

$$h(t) = -16t^2 + 48t, \quad 0 \le t \le 3$$

where t is the time (in seconds). You are told that the ball reaches a height of 64 feet. Is this possible?

▶ **Model It**

64. Data Analysis The revenues R (in millions of dollars) for Manpower, Inc. for the years 1997 through 2001 are shown in the table. (Source: Manpower, Inc.)

Year	Revenue, R
1997	7,258.5
1998	8,814.3
1999	9,770.1
2000	10,842.8
2001	10,483.8

(a) Use the *regression* feature of a graphing utility to find a quadratic model for the data. Let t represent the year, with $t = 7$ corresponding to 1997.

(b) Use a graphing utility to graph the model you found in part (a).

(c) Use your graph from part (b) to determine the year in which revenue reached $11 billion. Is this possible?

(d) Determine algebraically the year in which revenue reached $11 billion. Is this possible? Explain.

Synthesis

True or False? In Exercises 65 and 66, decide whether the statement is true or false. Justify your answer.

65. It is possible for a third-degree polynomial function with integer coefficients to have no real zeros.

66. If $x = -i$ is a zero of the function $f(x) = x^3 + ix^2 + ix - 1$, then $x = i$ must also be a zero of f.

Think About It In Exercises 67–72, determine (if possible) the zeros of the function g if the function f has zeros at $x = r_1, x = r_2,$ and $x = r_3$.

67. $g(x) = -f(x)$

68. $g(x) = 3f(x)$

69. $g(x) = f(x - 5)$

70. $g(x) = f(2x)$

71. $g(x) = 3 + f(x)$

72. $g(x) = f(-x)$

73. (a) Find a quadratic function f (with integer coefficients) that has $\pm\sqrt{b}\,i$ as zeros. Assume that b is a positive integer.

(b) Find a quadratic function f (with integer coefficients) that has $a \pm bi$ as zeros. Assume that b is a positive integer.

74. Graphical Reasoning The graph of one of the following functions is shown below. Identify the function shown in the graph. Explain why each of the others is not the correct function. Use a graphing utility to verify your result.

(a) $f(x) = x^2(x + 2)(x - 3.5)$

(b) $g(x) = (x + 2)(x - 3.5)$

(c) $h(x) = (x + 2)(x - 3.5)(x^2 + 1)$

(d) $k(x) = (x + 1)(x + 2)(x - 3.5)$

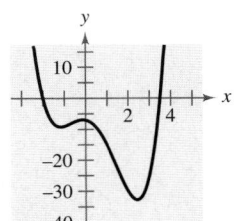

Review

In Exercises 75–78, perform the operation and simplify.

75. $(-3 + 6i) - (8 - 3i)$

76. $(12 - 5i) + 16i$

77. $(6 - 2i)(1 + 7i)$

78. $(9 - 5i)(9 + 5i)$

In Exercises 79–84, use the graph of f to sketch the graph of g. To print an enlarged copy of the graph, go to the website www.mathgraphs.com.

79. $g(x) = f(x - 2)$

80. $g(x) = f(x) - 2$

81. $g(x) = 2f(x)$

82. $g(x) = f(-x)$

83. $g(x) = f(2x)$

84. $g(x) = f\left(\tfrac{1}{2}x\right)$

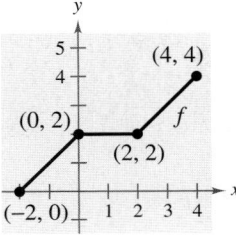

4.3 Trigonometric Form of a Complex Number

▶ **What you should learn**
- How to plot complex numbers in the complex plane
- How to write the trigonometric forms of complex numbers
- How to multiply and divide complex numbers written in trigonometric form

▶ **Why you should learn it**

You can perform the operations of multiplication and division on complex numbers by learning to write complex numbers in trigonometric form. For instance, in Exercises 61–68 on page 348, you can multiply and divide complex numbers in trigonometric form and standard form.

The Complex Plane

Just as real numbers can be represented by points on the real number line, you can represent a complex number

$$z = a + bi$$

as the point (a, b) in a coordinate plane (the **complex plane**). The horizontal axis is called the **real axis** and the vertical axis is called the **imaginary axis,** as shown in Figure 4.4.

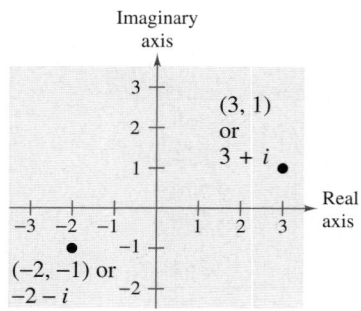

FIGURE 4.4

The **absolute value** of the complex number $a + bi$ is defined as the distance between the origin $(0, 0)$ and the point (a, b).

> **Definition of the Absolute Value of a Complex Number**
> The **absolute value** of the complex number $z = a + bi$ is
> $$|a + bi| = \sqrt{a^2 + b^2}.$$

If the complex number $a + bi$ is a real number (that is, if $b = 0$), then this definition agrees with that given for the absolute value of a real number

$$|a + 0i| = \sqrt{a^2 + 0^2} = |a|.$$

Example 1 ▶ Finding the Absolute Value of a Complex Number

Plot $z = -2 + 5i$ and find its absolute value.

Solution

The number is plotted in Figure 4.5. It has an absolute value of

$$|z| = \sqrt{(-2)^2 + 5^2}$$
$$= \sqrt{29}.$$

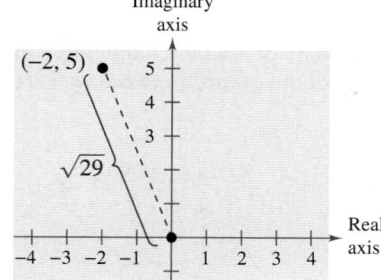

FIGURE 4.5

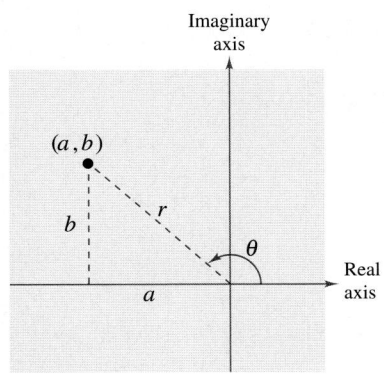

FIGURE **4.6**

Trigonometric Form of a Complex Number

In Section 4.1, you learned how to add, subtract, multiply, and divide complex numbers. To work effectively with *powers* and *roots* of complex numbers, it is helpful to write complex numbers in trigonometric form. In Figure 4.6, consider the nonzero complex number $a + bi$. By letting θ be the angle from the positive real axis (measured counterclockwise) to the line segment connecting the origin and the point (a, b), you can write

$$a = r \cos \theta \qquad \text{and} \qquad b = r \sin \theta$$

where $r = \sqrt{a^2 + b^2}$. Consequently, you have

$$a + bi = (r \cos \theta) + (r \sin \theta)i$$

from which you can obtain the **trigonometric form of a complex number.**

Trigonometric Form of a Complex Number

The **trigonometric form** of the complex number $z = a + bi$ is

$$z = r(\cos \theta + i \sin \theta)$$

where $a = r \cos \theta$, $b = r \sin \theta$, $r = \sqrt{a^2 + b^2}$, and $\tan \theta = b/a$. The number r is the **modulus** of z, and θ is called an **argument** of z.

The trigonometric form of a complex number is also called the *polar form.* Because there are infinitely many choices for θ, the trigonometric form of a complex number is not unique. Normally, θ is restricted to the interval $0 \leq \theta < 2\pi$, although on occasion it is convenient to use $\theta < 0$.

Example 2 ▶ Writing a Complex Number in Trigonometric Form

Write the complex number $z = -2 - 2\sqrt{3}\,i$ in trigonometric form.

Solution

The absolute value of z is

$$r = \left| -2 - 2\sqrt{3}\,i \right| = \sqrt{(-2)^2 + \left(-2\sqrt{3}\right)^2} = \sqrt{16} = 4$$

and the angle θ is

$$\tan \theta = \frac{b}{a} = \frac{-2\sqrt{3}}{-2} = \sqrt{3}.$$

Because $\tan(\pi/3) = \sqrt{3}$ and $z = -2 - 2\sqrt{3}\,i$ lies in Quadrant III, you choose θ to be $\theta = \pi + \pi/3 = 4\pi/3$. So, the trigonometric form is

$$z = r(\cos \theta + i \sin \theta)$$

$$= 4\left(\cos \frac{4\pi}{3} + i \sin \frac{4\pi}{3} \right).$$

See Figure 4.7.

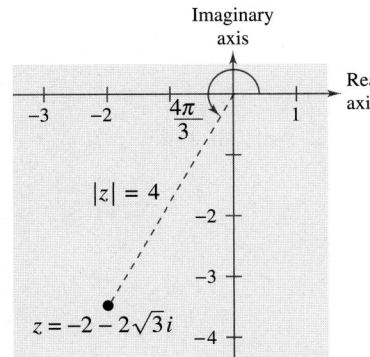

FIGURE **4.7**

Reasons for using the trigonometric forms of complex numbers are to multiply efficiently, raise to powers, and find roots of complex numbers.

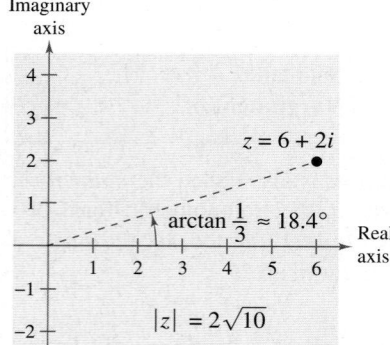

Imaginary
axis

$z = 6 + 2i$

$\arctan \dfrac{1}{3} \approx 18.4°$

Real
axis

$|z| = 2\sqrt{10}$

FIGURE **4.8**

Example 3 ▶ **Writing a Complex Number in Trigonometric Form**

Write the complex number in trigonometric form.

$$z = 6 + 2i$$

Solution

The absolute value of z is

$$r = |6 + 2i|$$
$$= \sqrt{6^2 + 2^2}$$
$$= \sqrt{40}$$
$$= 2\sqrt{10}$$

and the angle θ is

$$\tan \theta = \frac{b}{a} = \frac{2}{6} = \frac{1}{3}.$$

Because θ is in Quadrant I, you can conclude that

$$\theta = \arctan \frac{1}{3} \approx 0.32175 \text{ radian} \approx 18.4°.$$

So, the trigonometric form of z is

$$z = r(\cos \theta + i \sin \theta)$$
$$= 2\sqrt{10}\left[\cos\left(\arctan \frac{1}{3}\right) + i \sin\left(\arctan \frac{1}{3}\right)\right]$$
$$\approx 2\sqrt{10}(\cos 18.4° + i \sin 18.4°).$$

This result is illustrated graphically in Figure 4.8.

Example 4 ▶ **Writing a Complex Number in Standard Form**

Write the complex number in standard form $a + bi$.

$$z = \sqrt{8}\left[\cos\left(-\frac{\pi}{3}\right) + i \sin\left(-\frac{\pi}{3}\right)\right]$$

Solution

Because $\cos(-\pi/3) = \frac{1}{2}$ and $\sin(-\pi/3) = -\sqrt{3}/2$, you can write

$$z = \sqrt{8}\left[\cos\left(-\frac{\pi}{3}\right) + i \sin\left(-\frac{\pi}{3}\right)\right]$$
$$= 2\sqrt{2}\left(\frac{1}{2} - \frac{\sqrt{3}}{2}i\right)$$
$$= \sqrt{2} - \sqrt{6}i.$$

Multiplication and Division of Complex Numbers

The trigonometric form adapts nicely to multiplication and division of complex numbers. Suppose you are given two complex numbers

$$z_1 = r_1(\cos \theta_1 + i \sin \theta_1) \quad \text{and} \quad z_2 = r_2(\cos \theta_2 + i \sin \theta_2).$$

The product of z_1 and z_2 is

$$z_1 z_2 = r_1 r_2 (\cos \theta_1 + i \sin \theta_1)(\cos \theta_2 + i \sin \theta_2)$$

$$= r_1 r_2 [(\cos \theta_1 \cos \theta_2 - \sin \theta_1 \sin \theta_2) + i(\sin \theta_1 \cos \theta_2 + \cos \theta_1 \sin \theta_2)].$$

Using the sum and difference formulas for cosine and sine, you can rewrite this equation as

$$z_1 z_2 = r_1 r_2 [\cos(\theta_1 + \theta_2) + i \sin(\theta_1 + \theta_2)].$$

This establishes the first part of the following rule. The second part is left to you (see Exercise 75).

Product and Quotient of Two Complex Numbers

Let $z_1 = r_1(\cos \theta_1 + i \sin \theta_1)$ and $z_2 = r_2(\cos \theta_2 + i \sin \theta_2)$ be complex numbers.

$$z_1 z_2 = r_1 r_2 [\cos(\theta_1 + \theta_2) + i \sin(\theta_1 + \theta_2)] \qquad \text{Product}$$

$$\frac{z_1}{z_2} = \frac{r_1}{r_2} [\cos(\theta_1 - \theta_2) + i \sin(\theta_1 - \theta_2)], \quad z_2 \neq 0 \qquad \text{Quotient}$$

Note that this rule says that to *multiply* two complex numbers you multiply moduli and add arguments, whereas to *divide* two complex numbers you divide moduli and subtract arguments.

Example 5 ▶ **Dividing Complex Numbers**

Find the quotient, z_1/z_2, of the complex numbers.

$$z_1 = 24(\cos 300° + i \sin 300°) \qquad z_2 = 8(\cos 75° + i \sin 75°)$$

Solution

$$\frac{z_1}{z_2} = \frac{24(\cos 300° + i \sin 300°)}{8(\cos 75° + i \sin 75°)}$$

$$= \frac{24}{8}[\cos(300° - 75°) + i \sin(300° - 75°)]$$

$$= 3(\cos 225° + i \sin 225°)$$

$$= 3\left[\left(-\frac{\sqrt{2}}{2}\right) + i\left(-\frac{\sqrt{2}}{2}\right)\right]$$

$$= -\frac{3\sqrt{2}}{2} - \frac{3\sqrt{2}}{2}i$$

Example 6 ▶ **Multiplying Complex Numbers**

Find the product of the complex numbers.

$$z_1 = 2\left(\cos\frac{2\pi}{3} + i\sin\frac{2\pi}{3}\right) \qquad z_2 = 8\left(\cos\frac{11\pi}{6} + i\sin\frac{11\pi}{6}\right)$$

Solution

$$z_1 z_2 = 2\left(\cos\frac{2\pi}{3} + i\sin\frac{2\pi}{3}\right) \times 8\left(\cos\frac{11\pi}{6} + i\sin\frac{11\pi}{6}\right)$$

$$= 16\left[\cos\left(\frac{2\pi}{3} + \frac{11\pi}{6}\right) + i\sin\left(\frac{2\pi}{3} + \frac{11\pi}{6}\right)\right]$$

$$= 16\left(\cos\frac{5\pi}{2} + i\sin\frac{5\pi}{2}\right)$$

$$= 16\left(\cos\frac{\pi}{2} + i\sin\frac{\pi}{2}\right)$$

$$= 16[0 + i(1)]$$

$$= 16i$$

You can check the result in Example 6 by first converting the complex numbers to the standard forms $z_1 = -1 + \sqrt{3}i$ and $z_2 = 4\sqrt{3} - 4i$ and then multiplying algebraically, as in Section 4.1.

$$z_1 z_2 = \left(-1 + \sqrt{3}i\right)\left(4\sqrt{3} - 4i\right)$$

$$= -4\sqrt{3} + 4i + 12i + 4\sqrt{3}$$

$$= 16i$$

Activities

1. Find the absolute value and the trigonometric form of the complex number $z = 4 - 4i$.

   Answer: $r = 4\sqrt{2}$

   $$z = 4\sqrt{2}\left(\cos\frac{7\pi}{4} + i\sin\frac{7\pi}{4}\right)$$

2. Multiply, using the trigonometric form, then rewrite your answer in standard form.

   $$\left[3\left(\cos\frac{\pi}{3} + i\sin\frac{\pi}{3}\right)\right] \times$$
   $$\left[\frac{2}{3}\left(\cos\frac{5\pi}{3} + i\sin\frac{5\pi}{3}\right)\right]$$

   Answer: $2(\cos 2\pi + i\sin 2\pi) = 2$

3. Represent the complex number $-4 + 2i$ graphically.

   Answer:

   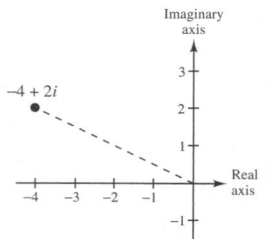

Writing ABOUT MATHEMATICS

Multiplying Complex Numbers Graphically Discuss how you can graphically approximate the product of the complex numbers. Then, approximate the values of the products and check your answers analytically.

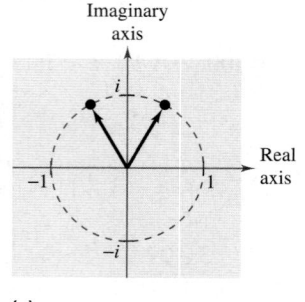

(a)

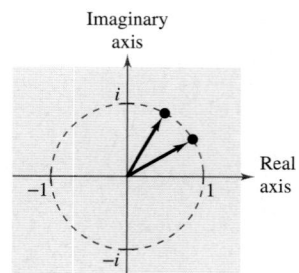

(b)

4.3 Exercises

In Exercises 1–6, plot the complex number and find its absolute value.

1. $-7i$

2. -7

3. $-4 + 4i$

4. $5 - 12i$

5. $6 - 7i$

6. $-8 + 3i$

In Exercises 7–10, write the complex number in trigonometric form.

7.
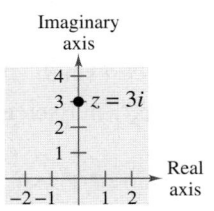

8.
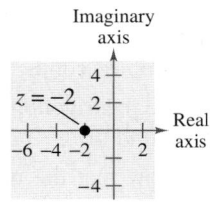

9.
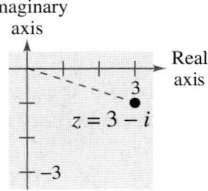

10.
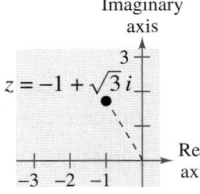

In Exercises 11–26, represent the complex number graphically, and find the trigonometric form of the number.

11. $3 - 3i$

12. $2 + 2i$

13. $\sqrt{3} + i$

14. $4 - 4\sqrt{3}i$

15. $-2(1 + \sqrt{3}i)$

16. $\frac{5}{2}(\sqrt{3} - i)$

17. $-5i$

18. $4i$

19. $-7 + 4i$

20. $3 - i$

21. 7

22. 4

23. $3 + \sqrt{3}i$

24. $2\sqrt{2} - i$

25. $-3 - i$

26. $1 + 3i$

 In Exercises 27–34, use a graphing utility to represent the complex number in trigonometric form.

27. $5 + 2i$

28. $8 + 3i$

29. $-3 + i$

30. $-5 - i$

31. $3\sqrt{2} - 7i$

32. $4\sqrt{5} - 4i$

33. $-8 - 5\sqrt{3}i$

34. $-9 - 2\sqrt{10}i$

In Exercises 35–44, represent the complex number graphically, and find the standard form of the number.

35. $3(\cos 120° + i \sin 120°)$

36. $5(\cos 135° + i \sin 135°)$

37. $\frac{3}{2}(\cos 300° + i \sin 300°)$

38. $\frac{1}{4}(\cos 225° + i \sin 225°)$

39. $3.75\left(\cos \dfrac{3\pi}{4} + i \sin \dfrac{3\pi}{4}\right)$

40. $6\left(\cos \dfrac{5\pi}{12} + i \sin \dfrac{5\pi}{12}\right)$

41. $8\left(\cos \dfrac{\pi}{2} + i \sin \dfrac{\pi}{2}\right)$

42. $7(\cos 0 + i \sin 0)$

43. $3[\cos(18° \, 45') + i \sin(18° \, 45')]$

44. $6[\cos(230° \, 30') + i \sin(230° \, 30')]$

 In Exercises 45–48, use a graphing utility to represent the complex number in standard form.

45. $5\left(\cos \dfrac{\pi}{9} + i \sin \dfrac{\pi}{9}\right)$

46. $10\left(\cos \dfrac{2\pi}{5} + i \sin \dfrac{2\pi}{5}\right)$

47. $3(\cos 165.5° + i \sin 165.5°)$

48. $9(\cos 58° + i \sin 58°)$

In Exercises 49–60, perform the operation and leave the result in trigonometric form.

49. $\left[2\left(\cos \dfrac{\pi}{4} + i \sin \dfrac{\pi}{4}\right)\right]\left[6\left(\cos \dfrac{\pi}{12} + i \sin \dfrac{\pi}{12}\right)\right]$

50. $\left[\dfrac{3}{4}\left(\cos \dfrac{\pi}{3} + i \sin \dfrac{\pi}{3}\right)\right]\left[4\left(\cos \dfrac{3\pi}{4} + i \sin \dfrac{3\pi}{4}\right)\right]$

51. $\left[\frac{5}{3}(\cos 140° + i \sin 140°)\right]\left[\frac{2}{3}(\cos 60° + i \sin 60°)\right]$

52. $[0.5(\cos 100° + i \sin 100°)] \times$
$\qquad [0.8(\cos 300° + i \sin 300°)]$

53. $[0.45(\cos 310° + i \sin 310°)] \times$
$\qquad [0.60(\cos 200° + i \sin 200°)]$

54. $(\cos 5° + i \sin 5°)(\cos 20° + i \sin 20°)$

55. $\dfrac{\cos 50° + i \sin 50°}{\cos 20° + i \sin 20°}$

56. $\dfrac{2(\cos 120° + i \sin 120°)}{4(\cos 40° + i \sin 40°)}$

57. $\dfrac{\cos(5\pi/3) + i \sin(5\pi/3)}{\cos \pi + i \sin \pi}$

58. $\dfrac{5(\cos 4.3 + i \sin 4.3)}{4(\cos 2.1 + i \sin 2.1)}$

59. $\dfrac{12(\cos 52° + i \sin 52°)}{3(\cos 110° + i \sin 110°)}$

60. $\dfrac{6(\cos 40° + i \sin 40°)}{7(\cos 100° + i \sin 100°)}$

In Exercises 61–68, (a) give the trigonometric forms of the complex numbers, (b) perform the indicated operation using the trigonometric forms, and (c) perform the indicated operation using the standard forms, and check your result with that of part (b).

61. $(2 + 2i)(1 - i)$

62. $(\sqrt{3} + i)(1 + i)$

63. $-2i(1 + i)$

64. $4(1 - \sqrt{3}i)$

65. $\dfrac{3 + 4i}{1 - \sqrt{3}i}$

66. $\dfrac{1 + \sqrt{3}i}{6 - 3i}$

67. $\dfrac{5}{2 + 3i}$

68. $\dfrac{4i}{-4 + 2i}$

In Exercises 69–72, sketch the graphs of all complex numbers z satisfying the given condition.

69. $|z| = 2$

70. $|z| = 3$

71. $\theta = \dfrac{\pi}{6}$

72. $\theta = \dfrac{5\pi}{4}$

Synthesis

True or False? In Exercises 73 and 74, determine whether the statement is true or false. Justify your answer.

73. Although the square of the complex number bi is given by $(bi)^2 = -b^2$, the absolute value of the complex number $z = a + bi$ is defined as

$$|a + bi| = \sqrt{a^2 + b^2}.$$

74. The product of two complex numbers

$$z_1 = r_1(\cos \theta_1 + i \sin \theta_1)$$

and

$$z_2 = r_2(\cos \theta_2 + i \sin \theta_2),$$

is zero only when $r_1 = 0$ and/or $r_2 = 0$.

75. Given two complex numbers $z_1 = r_1(\cos \theta_1 + i \sin \theta_1)$ and $z_2 = r_2(\cos \theta_2 + i \sin \theta_2)$, $z_2 \neq 0$, show that

$$\dfrac{z_1}{z_2} = \dfrac{r_1}{r_2}[\cos(\theta_1 - \theta_2) + i \sin(\theta_1 - \theta_2)].$$

76. Show that $\bar{z} = r[\cos(-\theta) + i \sin(-\theta)]$ is the complex conjugate of $z = r(\cos \theta + i \sin \theta)$.

77. Use the trigonometric forms of z and $\bar{z}$ in Exercise 76 to find (a) $z\bar{z}$ and (b) $z/\bar{z}$, $\bar{z} \neq 0$.

78. Show that the negative of $z = r(\cos \theta + i \sin \theta)$ is $-z = r[\cos(\theta + \pi) + i \sin(\theta + \pi)]$.

Review

In Exercises 79–84, solve the right triangle shown in the figure. Approximate the result to two decimal places.

79. $A = 22°$, $a = 8$

80. $B = 66°$, $a = 33.5$

81. $A = 30°$, $b = 112.6$

82. $B = 6°$, $b = 211.2$

83. $A = 42° \, 15'$, $c = 11.2$

84. $B = 81° \, 30'$, $c = 6.8$

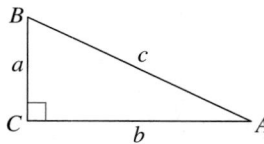

Harmonic Motion In Exercises 85–88, for the simple harmonic motion described by the trigonometric function, find the maximum displacement and the lowest possible value of t for which $d = 0$.

85. $d = 16 \cos \dfrac{\pi}{4}t$

86. $d = \dfrac{1}{8} \cos 12\pi t$

87. $d = \dfrac{1}{16} \sin \dfrac{5}{4}\pi t$

88. $d = \dfrac{1}{12} \sin 60\pi t$

4.4 DeMoivre's Theorem

▶ **What you should learn**

- How to use DeMoivre's Theorem to find powers of complex numbers
- How to find *n*th roots of complex numbers

▶ **Why you should learn it**

You can use the trigonometric form of a complex number to perform operations with complex numbers. For instance, in Exercises 45–60 on page 353, you can use the trigonometric forms of complex numbers to help you solve polynomial equations.

Powers of Complex Numbers

To raise a complex number to a power, consider repeated use of the multiplication rule.

$$z = r(\cos \theta + i \sin \theta)$$

$$z^2 = r(\cos \theta + i \sin \theta)r(\cos \theta + i \sin \theta) = r^2(\cos 2\theta + i \sin 2\theta)$$

$$z^3 = r^2(\cos 2\theta + i \sin 2\theta)r(\cos \theta + i \sin \theta) = r^3(\cos 3\theta + i \sin 3\theta)$$

$$z^4 = r^4(\cos 4\theta + i \sin 4\theta)$$

$$z^5 = r^5(\cos 5\theta + i \sin 5\theta)$$

$$\vdots$$

This pattern leads to the following important theorem, which is named after the French mathematician Abraham DeMoivre (1667–1754).

DeMoivre's Theorem

If $z = r(\cos \theta + i \sin \theta)$ is a complex number and n is a positive integer, then

$$z^n = [r(\cos \theta + i \sin \theta)]^n = r^n(\cos n\theta + i \sin n\theta).$$

Example 1 ▶ Finding a Power of a Complex Number

Use DeMoivre's Theorem to find $\left(-1 + \sqrt{3}i\right)^{12}$.

Solution

First convert the complex number to trigonometric form using

$$r = \sqrt{(-1)^2 + \left(\sqrt{3}\right)^2} = 2 \text{ and } \theta = \arctan \frac{\sqrt{3}}{-1} = \frac{2\pi}{3}.$$

$$-1 + \sqrt{3}i = 2\left(\cos \frac{2\pi}{3} + i \sin \frac{2\pi}{3}\right)$$

Then, by DeMoivre's Theorem, you have

$$(-1 + \sqrt{3}i)^{12} = \left[2\left(\cos \frac{2\pi}{3} + i \sin \frac{2\pi}{3}\right)\right]^{12}$$

$$= 2^{12}\left[\cos(12)\frac{2\pi}{3} + i \sin(12)\frac{2\pi}{3}\right]$$

$$= 4096(\cos 8\pi + i \sin 8\pi)$$

$$= 4096(1 + 0)$$

$$= 4096.$$

Roots of Complex Numbers

Recall that a consequence of the Fundamental Theorem of Algebra is that a polynomial equation of degree n has n solutions in the complex number system. So, the equation $x^6 = 1$ has six solutions, and in this particular case you can find the six solutions by factoring and using the Quadratic Formula.

$$x^6 - 1 = (x^3 - 1)(x^3 + 1)$$
$$= (x - 1)(x^2 + x + 1)(x + 1)(x^2 - x + 1) = 0$$

Consequently, the solutions are

$$x = \pm 1, \qquad x = \frac{-1 \pm \sqrt{3}i}{2}, \qquad \text{and} \qquad x = \frac{1 \pm \sqrt{3}i}{2}.$$

Each of these numbers is a sixth root of 1. In general, the **nth root** of a complex number is defined as follows.

Definition of *n*th Root of a Complex Number

The complex number $u = a + bi$ is an **nth root** of the complex number z if

$$z = u^n = (a + bi)^n.$$

To find a formula for an nth root of a complex number, let u be an nth root of z, where

$$u = s(\cos \beta + i \sin \beta)$$

and

$$z = r(\cos \theta + i \sin \theta).$$

By DeMoivre's Theorem and the fact that $u^n = z$, you have

$$s^n (\cos n\beta + i \sin n\beta) = r(\cos \theta + i \sin \theta).$$

Taking the absolute value of each side of this equation, it follows that $s^n = r$. Substituting back into the previous equation and dividing by r, you get

$$\cos n\beta + i \sin n\beta = \cos \theta + i \sin \theta.$$

So, it follows that

$$\cos n\beta = \cos \theta$$

and

$$\sin n\beta = \sin \theta.$$

Because both sine and cosine have a period of 2π, these last two equations have solutions if and only if the angles differ by a multiple of 2π. Consequently, there must exist an integer k such that

$$n\beta = \theta + 2\pi k$$

$$\beta = \frac{\theta + 2\pi k}{n}.$$

By substituting this value of β into the trigonometric form of u, you get the result stated on the following page.

Historical Note

Abraham DeMoivre (1667–1754) is remembered for his work in probability theory and DeMoivre's Theorem. His *The Doctrine of Chances* (published in 1718) includes the theory of recurring series and the theory of partial fractions.

◀ **E x p l o r a t i o n** ▶

The nth roots of a complex number are useful for solving some polynomial equations. For instance, explain how you can use DeMoivre's Theorem to solve the polynomial equation

$$x^4 + 16 = 0.$$

[*Hint*: Write -16 as $16(\cos \pi + i \sin \pi)$.]

> ### nth Roots of a Complex Number
>
> For a positive integer n, the complex number $z = r(\cos \theta + i \sin \theta)$ has exactly n distinct nth roots given by
>
> $$\sqrt[n]{r}\left(\cos \frac{\theta + 2\pi k}{n} + i \sin \frac{\theta + 2\pi k}{n}\right)$$
>
> where $k = 0, 1, 2, \ldots, n - 1$.

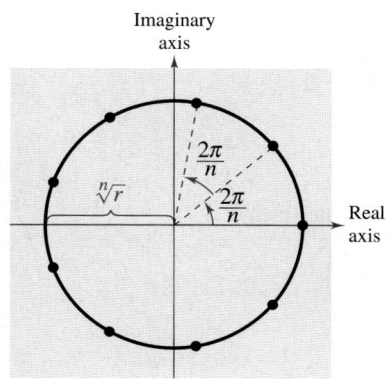

FIGURE **4.9**

When k exceeds $n - 1$, the roots begin to repeat. For instance, if $k = n$, the angle

$$\frac{\theta + 2\pi n}{n} = \frac{\theta}{n} + 2\pi$$

is coterminal with θ/n, which is also obtained when $k = 0$.

The formula for the nth roots of a complex number z has a nice geometrical interpretation, as shown in Figure 4.9. Note that because the nth roots of z all have the same magnitude $\sqrt[n]{r}$, they all lie on a circle of radius $\sqrt[n]{r}$ with center at the origin. Furthermore, because successive nth roots have arguments that differ by $2\pi/n$, the n roots are equally spaced around the circle.

You have already found the sixth roots of 1 by factoring and by using the Quadratic Formula. Example 2 shows how you can solve the same problem with the formula for nth roots.

Example 2 ▶ Finding the nth Roots of a Real Number

Find all the sixth roots of 1.

Solution

First write 1 in the trigonometric form $1 = 1(\cos 0 + i \sin 0)$. Then, by the nth root formula, with $n = 6$ and $r = 1$, the roots have the form

$$\sqrt[6]{1}\left(\cos \frac{0 + 2\pi k}{6} + i \sin \frac{0 + 2\pi k}{6}\right)$$

or simply $\cos(\pi k/3) + i \sin(\pi k/3)$. So, for $k = 0, 1, 2, 3, 4,$ and 5, the sixth roots are as follows. (See Figure 4.10.)

$$\cos 0 + i \sin 0 = 1$$

$$\cos \frac{\pi}{3} + i \sin \frac{\pi}{3} = \frac{1}{2} + \frac{\sqrt{3}}{2}i \qquad \text{Increment by } \frac{2\pi}{n} = \frac{2\pi}{6} = \frac{\pi}{3}$$

$$\cos \frac{2\pi}{3} + i \sin \frac{2\pi}{3} = -\frac{1}{2} + \frac{\sqrt{3}}{2}i$$

$$\cos \pi + i \sin \pi = -1$$

$$\cos \frac{4\pi}{3} + i \sin \frac{4\pi}{3} = -\frac{1}{2} - \frac{\sqrt{3}}{2}i$$

$$\cos \frac{5\pi}{3} + i \sin \frac{5\pi}{3} = \frac{1}{2} - \frac{\sqrt{3}}{2}i$$

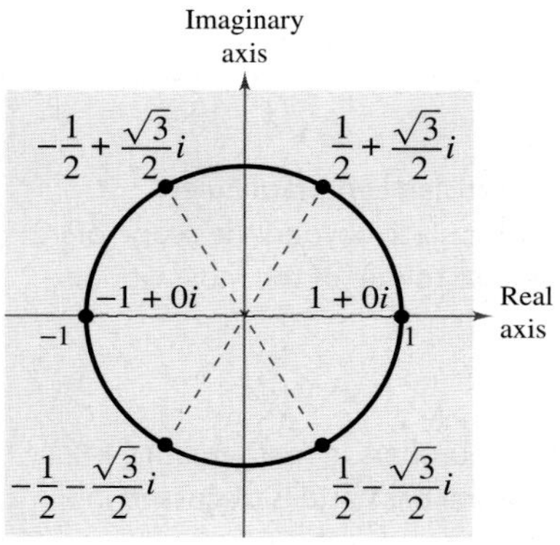

FIGURE **4.10**

In Figure 4.10, notice that the roots obtained in Example 2 all have a magnitude of 1 and are equally spaced around the unit circle. Also notice that the complex roots occur in conjugate pairs, as discussed in Section 4.2. The n distinct nth roots of 1 are called the **nth roots of unity.**

Example 3 ▶ Finding the nth Roots of a Complex Number

Find the three cube roots of

$$z = -2 + 2i.$$

Solution

Because z lies in Quadrant II, the trigonometric form of z is

$$z = -2 + 2i$$

$$= \sqrt{8}\,(\cos 135° + i \sin 135°). \qquad \theta = \arctan(2/-2) = 135°$$

By the formula for nth roots, the cube roots have the form

$$\sqrt[6]{8}\left(\cos\frac{135° + 360°k}{3} + i \sin\frac{135° + 360°k}{3}\right).$$

Finally, for $k = 0$, 1, and 2, you obtain the roots

$$\sqrt[6]{8}\left[\left(\cos\frac{135° + 360°(0)}{3} + i \sin\frac{135° + 360°(0)}{3}\right)\right] = \sqrt{2}(\cos 45° + i \sin 45°)$$

$$= 1 + i$$

$$\sqrt[6]{8}\left[\left(\cos\frac{135° + 360°(1)}{3} + i \sin\frac{135° + 360°(1)}{3}\right)\right] = \sqrt{2}(\cos 165° + i \sin 165°)$$

$$\approx -1.3660 + 0.3660i$$

$$\sqrt[6]{8}\left[\left(\cos\frac{135° + 360°(2)}{3} + i \sin\frac{135° + 360°(2)}{3}\right)\right] = \sqrt{2}(\cos 285° + i \sin 285°)$$

$$\approx 0.3660 - 1.3660i.$$

Activities

1. Use DeMoivre's Theorem to find $\left(-2 - 2\sqrt{3}i\right)^3$.

   Answer: $64(\cos 4\pi + i \sin 4\pi) = 64$

2. Find the cube roots of

   $$8\left(-\frac{1}{2} + \frac{\sqrt{3}}{2}i\right) =$$

   $$8\left(\cos\frac{2\pi}{3} + i \sin\frac{2\pi}{3}\right).$$

   Answer: $2\left(\cos\frac{2\pi}{9} + i \sin\frac{2\pi}{9}\right)$

   $$2\left(\cos\frac{8\pi}{9} + i \sin\frac{8\pi}{9}\right)$$

   $$2\left(\cos\frac{14\pi}{9} + i \sin\frac{14\pi}{9}\right)$$

3. Find all of the solutions of the equation $x^4 + 1 = 0$.

   Answer:

   $$\cos\frac{\pi}{4} + i \sin\frac{\pi}{4} = \frac{\sqrt{2}}{2} + \frac{\sqrt{2}}{2}i$$

   $$\cos\frac{3\pi}{4} + i \sin\frac{3\pi}{4} = -\frac{\sqrt{2}}{2} + \frac{\sqrt{2}}{2}i$$

   $$\cos\frac{5\pi}{4} + i \sin\frac{5\pi}{4} = -\frac{\sqrt{2}}{2} - \frac{\sqrt{2}}{2}i$$

   $$\cos\frac{7\pi}{4} + i \sin\frac{7\pi}{4} = \frac{\sqrt{2}}{2} - \frac{\sqrt{2}}{2}i$$

Writing ABOUT MATHEMATICS

A Famous Mathematical Formula The famous formula

$$e^{a + bi} = e^a(\cos b + i \sin b)$$

is called Euler's Formula, after the German mathematician Leonhard Euler (1707–1783). Although the interpretation of this formula is beyond the scope of this text, we decided to include it because it gives rise to one of the most wonderful equations in mathematics.

$$e^{\pi i} + 1 = 0$$

This elegant equation relates the five most famous numbers in mathematics—0, 1, π, e, and i—in a single equation (e is called the natural base and is discussed in Section 5.1). Show how Euler's Formula can be used to derive this equation.

4.4 Exercises

In Exercises 1–24, use DeMoivre's Theorem to find the indicated power of the complex number. Write the result in standard form.

1. $(1 + i)^5$

2. $(2 + 2i)^6$

3. $(-1 + i)^{10}$

4. $(3 - 2i)^8$

5. $2(\sqrt{3} + i)^7$

6. $4(1 - \sqrt{3}i)^3$

7. $[5(\cos 20° + i \sin 20°)]^3$

8. $[3(\cos 150° + i \sin 150°)]^4$

9. $\left(\cos \dfrac{\pi}{4} + i \sin \dfrac{\pi}{4}\right)^{12}$

10. $\left[2\left(\cos \dfrac{\pi}{2} + i \sin \dfrac{\pi}{2}\right)\right]^8$

11. $[5(\cos 3.2 + i \sin 3.2)]^4$

12. $(\cos 0 + i \sin 0)^{20}$

13. $(3 - 2i)^5$

14. $(2 + 5i)^6$

15. $(\sqrt{5} - 4i)^3$

16. $(\sqrt{3} + 2i)^4$

17. $[3(\cos 15° + i \sin 15°)]^4$

18. $[2(\cos 10° + i \sin 10°)]^8$

19. $[5(\cos 95° + i \sin 95°)]^3$

20. $[4(\cos 110° + i \sin 110°)]^4$

21. $\left[2\left(\cos \dfrac{\pi}{10} + i \sin \dfrac{\pi}{10}\right)\right]^5$

22. $\left[2\left(\cos \dfrac{\pi}{8} + i \sin \dfrac{\pi}{8}\right)\right]^6$

23. $\left[3\left(\cos \dfrac{2\pi}{3} + i \sin \dfrac{2\pi}{3}\right)\right]^3$

24. $\left[3\left(\cos \dfrac{\pi}{12} + i \sin \dfrac{\pi}{12}\right)\right]^5$

In Exercises 25–44, (a) use the formula on page 351 to find the indicated roots of the complex number, (b) represent each of the roots graphically, and (c) express each of the roots in standard form.

25. Square roots of $5(\cos 120° + i \sin 120°)$

26. Square roots of $16(\cos 60° + i \sin 60°)$

27. Cube roots of $8\left(\cos \dfrac{2\pi}{3} + i \sin \dfrac{2\pi}{3}\right)$

28. Cube roots of $64\left(\cos \dfrac{\pi}{3} + i \sin \dfrac{\pi}{3}\right)$

29. Fifth roots of $243\left(\cos \dfrac{\pi}{6} + i \sin \dfrac{\pi}{6}\right)$

30. Fifth roots of $32\left(\cos \dfrac{5\pi}{6} + i \sin \dfrac{5\pi}{6}\right)$

31. Square roots of $-25i$

32. Square roots of $-36i$

33. Fourth roots of $81i$

34. Fourth roots of $625i$

35. Cube roots of $-\dfrac{125}{2}(1 + \sqrt{3}i)$

36. Cube roots of $-4\sqrt{2}(1 - i)$

37. Fourth roots of 16

38. Fourth roots of i

39. Fifth roots of 1

40. Cube roots of 1000

41. Cube roots of -125

42. Fourth roots of -4

43. Fifth roots of $128(-1 + i)$

44. Sixth roots of $64i$

In Exercises 45–60, use the formula on page 351 to find all the solutions of the equation and represent the solutions graphically.

45. $x^4 + i = 0$

46. $x^3 - i = 0$

47. $x^6 + 1 = 0$

48. $x^3 + 1 = 0$

49. $x^5 + 243 = 0$

50. $x^3 + 125 = 0$

51. $x^5 - 32 = 0$

52. $x^3 - 27 = 0$

53. $x^4 + 16i = 0$

54. $x^3 + 27i = 0$

55. $x^4 - 16i = 0$

56. $x^6 + 64i = 0$

57. $x^3 - (1 - i) = 0$

58. $x^5 - (1 - i) = 0$

59. $x^6 + (1 + i) = 0$

60. $x^4 + (1 + i) = 0$

Synthesis

True or False? **In Exercises 61 and 62, determine whether the statement is true or false. Justify your answer.**

61. Geometrically, the nth roots of any complex number z are all equally spaced around the unit circle centered at the origin.

62. By DeMoivre's Theorem,
$$\left(4 + \sqrt{6}i\right)^8 = \cos(32) + i\sin\left(8\sqrt{6}\right).$$

63. Show that $-\frac{1}{2}\left(1 + \sqrt{3}i\right)$ is a sixth root of 1.

64. Show that $2^{-1/4}(1 - i)$ is a fourth root of -2.

Graphical Reasoning **In Exercises 65 and 66, use the graph of the roots of a complex number.**

(a) Write each of the roots in trigonometric form.

(b) Identify the complex number whose roots are given.

 (c) Use a graphing utility to verify the results of part (b).

65.

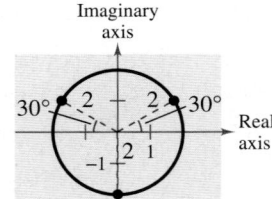

66.

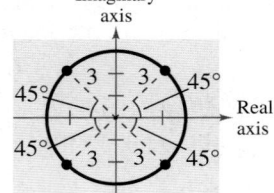

Review

In Exercises 67–76, use the figure and trigonometric identities to find the exact value of the trigonometric function.

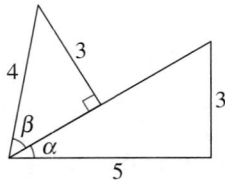

67. $\cos(\alpha + \beta)$ **68.** $\sin(\alpha + \beta)$

69. $\sin(\alpha - \beta)$ **70.** $\cos(\beta - \alpha)$

71. $\tan(\alpha + \beta)$ **72.** $\tan(\beta - \alpha)$

73. $\tan 2\alpha$ **74.** $\sin 2\beta$

75. $\sin \dfrac{\beta}{2}$ **76.** $\cos \dfrac{\alpha}{2}$

In Exercises 77–80, find a unit vector in the direction of the given vector.

77. $\mathbf{u} = \langle 10, 0 \rangle$ **78.** $\mathbf{v} = \langle -3, 7 \rangle$

79. $\mathbf{v} = 12\mathbf{i} - 5\mathbf{j}$ **80.** $\mathbf{w} = 8\mathbf{j}$

In Exercises 81–88, use the dot product to find the length of u.

81. $\mathbf{u} = \langle -3, 4 \rangle$

82. $\mathbf{u} = \langle -5, 7 \rangle$

83. $\mathbf{u} = \langle -9, 40 \rangle$

84. $\mathbf{u} = \langle -5, -12 \rangle$

85. $\mathbf{u} = 22\mathbf{i} + 3\mathbf{j}$

86. $\mathbf{u} = 16\mathbf{i} + 4\mathbf{j}$

87. $\mathbf{u} = 13\mathbf{i} + 6\mathbf{j}$

88. $\mathbf{u} = 24\mathbf{i} + 16\mathbf{j}$

Chapter Summary

▶ *What* did you learn?

Review Exercises

In Exercises 1–4, write the complex number in standard form.

1. $6 + \sqrt{-4}$
2. $3 - \sqrt{-25}$
3. $i^2 + 3i$
4. $-5i + i^2$

In Exercises 5–10, perform the operation and write the result in standard form.

5. $(7 + 5i) + (-4 + 2i)$
6. $\left(\dfrac{\sqrt{2}}{2} - \dfrac{\sqrt{2}}{2}i \right) - \left(\dfrac{\sqrt{2}}{2} + \dfrac{\sqrt{2}}{2}i \right)$
7. $5i(13 - 8i)$
8. $(1 + 6i)(5 - 2i)$
9. $(10 - 8i)(2 - 3i)$
10. $i(6 + i)(3 - 2i)$

In Exercises 11 and 12, write the quotient in standard form.

11. $\dfrac{6 + i}{4 - i}$
12. $\dfrac{3 + 2i}{5 + i}$

In Exercises 13 and 14, perform the operation and write the result in standard form.

13. $\dfrac{4}{2 - 3i} + \dfrac{2}{1 + i}$
14. $\dfrac{1}{2 + i} - \dfrac{5}{1 + 4i}$

In Exercises 15–18, find all solutions of the equation.

15. $3x^2 + 1 = 0$
16. $2 + 8x^2 = 0$
17. $x^2 - 2x + 10 = 0$
18. $6x^2 + 3x + 27 = 0$

In Exercises 19–22, determine the number of solutions of the equation in the complex number system.

19. $x^5 - 2x^4 + 3x^2 - 5 = 0$
20. $-2x^6 + 7x^3 + x^2 + 4x - 19 = 0$
21. $\frac{1}{2}x^4 + \frac{2}{3}x^3 - x^2 + \frac{3}{10} = 0$
22. $\frac{3}{4}x^3 + \frac{1}{2}x^2 + \frac{3}{2}x + 2 = 0$

In Exercises 23–26, use the discriminant to determine the number of real solutions of the equation.

23. $6x^2 + x - 2 = 0$
24. $9x^2 - 12x + 4 = 0$
25. $0.13x^2 - 0.45x + 0.65 = 0$
26. $4x^2 + \frac{4}{3}x + \frac{1}{9} = 0$

In Exercises 27–30, solve the equation.

27. $x^2 - 2x = 0$
28. $6x - x^2 = 0$
29. $x^2 + 8x + 10 = 0$
30. $3 + 4x - x^2 = 0$

In Exercises 31–36, find all the zeros of the function.

31. $r(x) = 2x^2 + 2x + 3$
32. $s(x) = 2x^2 + 5x + 4$
33. $f(x) = 2x^3 - 3x^2 + 50x - 75$
34. $f(x) = 4x^3 - x^2 + 128x - 32$
35. $f(x) = 4x^4 + 3x^2 - 10$
36. $f(x) = 5x^4 + 126x^2 + 25$

In Exercises 37–44, use the given zero to find all the zeros of the function. Write the polynomial as a product of linear factors.

	Function	Zero
37.	$f(x) = x^3 + 3x^2 - 24x + 28$	2
38.	$f(x) = 10x^3 + 21x^2 - x - 6$	-2
39.	$f(x) = x^3 + 3x^2 - 5x + 25$	-5
40.	$g(x) = x^3 - 8x^2 + 29x - 52$	4
41.	$h(x) = 2x^3 - 19x^2 + 58x + 34$	$5 + 3i$
42.	$f(x) = 5x^3 - 4x^2 + 20x - 16$	$2i$
43.	$f(x) = x^4 + 5x^3 + 2x^2 - 50x - 84$	$-3 + \sqrt{5}i$
44.	$g(x) = x^4 - 6x^3 + 18x^2 - 26x + 21$	$2 + \sqrt{3}i$

In Exercises 45–52, find a polynomial function with integer coefficients that has the given zeros. (There are many correct answers.)

45. $1, 1, \frac{1}{4}, -\frac{2}{3}$
46. $-2, 2, 3, 3$
47. $3, 2 - \sqrt{3}, 2 + \sqrt{3}$
48. $5, 1 - \sqrt{2}, 1 + \sqrt{2}$
49. $\frac{2}{3}, 4, \sqrt{3}i, -\sqrt{3}i$
50. $2, -3, 1 - 2i, 1 + 2i$
51. $-\sqrt{2}i, \sqrt{2}i, -5i, 5i$
52. $-2i, 2i, -4i, 4i$

53. *Profit* The demand equation for a product is $p = 140 - 0.0001x$, where p is the unit price (in dollars) of the product and x is the number of units produced and sold. The cost equation for the product is $C = 75x + 100{,}000$, where C is the total cost (in dollars) and x is the number of units produced. The total profit obtained by producing and selling x units is

$$P = xp - C.$$

You work in the marketing department of the company that produces this product and are asked to determine a price p that would yield a profit of 9 million dollars. Is this possible? Explain.

54. *Profit* Rework Exercise 53 for a cost equation of $C = 80x + 150{,}000$.

4.3 In Exercises 55–58, plot the complex number and find its absolute value.

55. $8i$

56. $-6i$

57. $5 + 3i$

58. $-10 - 4i$

In Exercises 59–62, write the trigonometric form of the complex number.

59. $5 - 5i$

60. $5 + 12i$

61. $-3\sqrt{3} + 3i$

62. -9

In Exercises 63 and 64, (a) express the two complex numbers in trigonometric form, and (b) use the trigonometric form to find $z_1 z_2$ and z_1/z_2.

63. $z_1 = 2\sqrt{3} - 2i$, $\quad z_2 = -10i$

64. $z_1 = -3(1 + i)$, $\quad z_2 = 2(\sqrt{3} + i)$

4.4 In Exercises 65–68, use DeMoivre's Theorem to find the indicated power of the complex number. Express the result in standard form.

65. $\left[5\left(\cos \dfrac{\pi}{12} + i \sin \dfrac{\pi}{12}\right)\right]^4$

66. $\left[2\left(\cos \dfrac{4\pi}{15} + i \sin \dfrac{4\pi}{15}\right)\right]^5$

67. $(2 + 3i)^6$

68. $(1 - i)^8$

In Exercises 69 and 70, use the formula on page 351 to find the roots of the complex number.

69. Sixth roots of $-729i$

70. Fourth roots of 256

In Exercises 71–74, find all solutions of the equation and represent the solutions graphically.

71. $x^4 + 81 = 0$

72. $x^5 - 32 = 0$

73. $x^3 + 8i = 0$

74. $(x^3 - 1)(x^2 + 1) = 0$

Synthesis

True or False? In Exercises 75–77, determine whether the statement is true or false. Justify your answer.

75. $\sqrt{-18}\,\sqrt{-2} = \sqrt{(-18)(-2)}$

76. The equation $325x^2 - 717x + 398 = 0$ has no solution.

77. A fourth-degree polynomial can have -5, $128i$, $4i$, and 5 as its zeros.

78. Write quadratic equations that have (a) two distinct real solutions, (b) two complex solutions, and (c) no real solution.

Graphical Reasoning In Exercises 79 and 80, use the graph of the roots of a complex number.

(a) Write each of the roots in trigonometric form.

(b) Identify the complex number whose roots are given.

(c) Use a graphing utility to verify the results of part (b).

79. 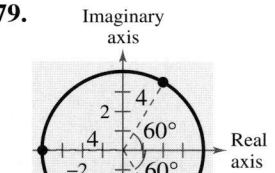 **80.**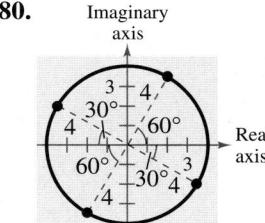

81. The figure shows z_1 and z_2. Describe $z_1 z_2$ and z_1/z_2.

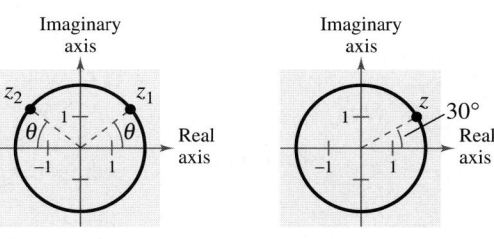

FIGURE FOR 81   FIGURE FOR 82

82. One of the fourth roots of a complex number z is shown in the figure.

(a) How many roots are not shown?

(b) Describe the other roots.

Chapter Test

The *Interactive* CD-ROM and *Internet* versions of this text offer Chapter Pre-Tests and Chapter Post-Tests, both of which have randomly generated exercises with diagnostic capabilities.

Take this test as you would take a test in class. When you are finished, check your work against the answers given in the back of the book.

1. Write the complex number $-3 + \sqrt{-81}$ in standard form.

In Exercises 2–4, perform the operations and write the result in standard form.

2. $10i - \left(3 + \sqrt{-25}\right)$ **3.** $(2 + 6i)^2$ **4.** $\left(2 + \sqrt{3}i\right)\left(2 - \sqrt{3}i\right)$

5. Write the quotient in standard form: $\dfrac{5}{2 + i}$.

6. Use the Quadratic Formula to solve the equation $2x^2 - 2x + 3 = 0$.

In Exercises 7 and 8, determine the number of solutions of the equation in the complex number system.

7. $x^5 + x^3 - x + 1 = 0$ **8.** $x^4 - 3x^3 + 2x^2 - 4x - 5 = 0$

In Exercises 9 and 10, find all the zeros of the function.

9. $f(x) = x^3 - 6x^2 + 5x - 30$ **10.** $f(x) = x^4 - 2x^2 - 24$

In Exercises 11 and 12, use the given zero(s) to find all the zeros of the function. Write the polynomial as a product of linear factors.

Function	Zero(s)
11. $h(x) = x^4 - 2x^2 - 8$	$-2, 2$
12. $g(v) = 2v^3 - 11v^2 + 22v - 15$	$\dfrac{3}{2}$

In Exercises 13 and 14, find a polynomial function with integer coefficients that has the given zeros. (There are many correct answers.)

13. $0, 3, 3 + i, 3 - i$ **14.** $1 + \sqrt{6}i, 1 - \sqrt{6}i, 3, 3$

15. Is it possible for a polynomial function with integer coefficients to have exactly one complex zero? Explain.

16. Write the complex number $z = 5 - 5i$ in trigonometric form.

17. Write the complex number $z = 6(\cos 120° + i \sin 120°)$ in standard form.

In Exercises 18 and 19, use DeMoivre's Theorem to find the indicated power of the complex number.

18. $\left[3\left(\cos \dfrac{7\pi}{6} + i \sin \dfrac{7\pi}{6}\right)\right]^8$ **19.** $(3 - 3i)^6$

20. Find the fourth roots of $256\left(1 + \sqrt{3}i\right)$.

21. Find all solutions of the equation $x^3 - 27i = 0$ and represent the solutions graphically.

Proofs in Mathematics

The Linear Factorization Theorem is closely related to the Fundamental Theorem of Algebra. The Fundamental Theorem of Algebra has a long and interesting history. In the early work with polynomial equations, The Fundamental Theorem of Algebra was thought to have been not true, because imaginary solutions were not considered. In fact, in the very early work by mathematicians such as Abu al-Khwarizmi (c. 800 A.D.), negative solutions were also not considered.

Once imaginary numbers were accepted, several mathematicians attempted to give a general proof of the Fundamental Theorem of Algebra. These mathematicians included Gottfried von Leibniz (1702), Jean D`Alembert (1740), Leonhard Euler (1749), Joseph-Louis Lagrange (1772), and Pierre Simon Laplace (1795). The mathematician usually credited with the first correct proof of the Fundamental Theorem of Algebra is Carl Friedrich Gauss, who published the proof in his doctoral thesis in 1799.

Linear Factorization Theorem   *(p. 335)*

If $f(x)$ is a polynomial of degree n, where $n > 0$, then f has precisely n linear factors

$$f(x) = a_n(x - c_1)(x - c_2) \cdots (x - c_n)$$

where $c_1, c_2, \ldots, c_n$ are complex numbers.

Proof

Using the Fundamental Theorem of Algebra, you know that f must have at least one zero, c_1. Consequently, $(x - c_1)$ is a factor of $f(x)$, and you have

$$f(x) = (x - c_1)f_1(x).$$

If the degree of $f_1(x)$ is greater than zero, you again apply the Fundamental Theorem to conclude that f_1 must have a zero c_2, which implies that

$$f(x) = (x - c_1)(x - c_2)f_2(x).$$

It is clear that the degree of $f_1(x)$ is $n - 1$, that the degree of $f_2(x)$ is $n - 2$, and that you can repeatedly apply the Fundamental Theorem n times until you obtain

$$f(x) = a_n(x - c_1)(x - c_2) \cdots (x - c_n)$$

where a_n is the leading coefficient of the polynomial $f(x)$.

P.S. Problem Solving

1. (a) The complex numbers

   $$z = 2, z = \frac{-2 + 2\sqrt{3}i}{2}, \text{ and } z = \frac{-2 - 2\sqrt{3}i}{2}$$

   are represented graphically (see figure). Evaluate the expression z^3 for each complex number. What do you observe?

   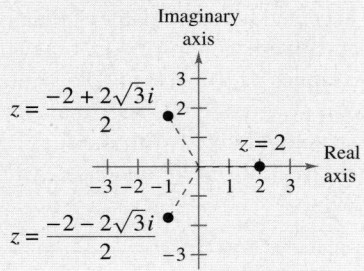

2. (b) The complex numbers

   $$z = 3, z = \frac{-3 + 3\sqrt{3}i}{2}, \text{ and } z = \frac{-3 - 3\sqrt{3}i}{2}$$

   are represented graphically (see figure). Evaluate the expression z^3 for each complex number. What do you observe?

   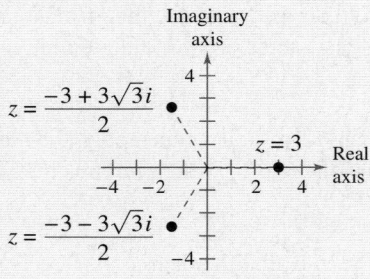

   (c) Use your results from parts (a) and (b) to generalize your findings.

2. The multiplicative inverse of z is a complex number z_m such that $z \cdot z_m = 1$. Find the multiplicative inverse of each complex number.

   (a) $z = 1 + i$

   (b) $z = 3 - i$

   (c) $z = -2 + 8i$

3. Show that the product of a complex number $a + bi$ and its conjugate is a real number.

4. Let

   $z = a + bi, \bar{z} = a - bi, w = c + di$, and $\bar{w} = c - di$.

   Prove each statement.

   (a) $\overline{z + w} = \bar{z} + \bar{w}$

   (b) $\overline{z - w} = \bar{z} - \bar{w}$

   (c) $\overline{zw} = \bar{z} \cdot \bar{w}$

   (d) $\overline{z/w} = \bar{z}/\bar{w}$

   (e) $(\bar{z})^2 = \overline{z^2}$

   (f) $\bar{\bar{z}} = z$

   (g) $\bar{z} = z$ if z is real.

5. Find the values of k such that the equation

   $$x^2 - 2kx + k = 0$$

   has (a) two real solutions and (b) two complex solutions.

6. Use a graphing utility to graph the function

   $$f(x) = x^4 - 4x^2 + k$$

   for different values of k. Find values of k such that the zeros of f satisfy the specified characteristics. (Some parts do not have unique answers.)

   (a) Four real zeros

   (b) Two real zeros and two complex zeros

   (c) Four complex zeros

7. Will the answers to Exercise 6 change for the function g?

   (a) $g(x) = f(x - 2)$

   (b) $g(x) = f(2x)$

8. A third-degree polynomial function f has real zeros $-2, \frac{1}{2}$, and 3, and its leading coefficient is negative. Write an equation for f. Sketch the graph of f. How many different polynomial functions are possible for f?

9. Use the information in the table.

Interval	Value of $f(x)$
$(-\infty, -2)$	Positive
$(-2, 1)$	Negative
$(1, 4)$	Negative
$(4, \infty)$	Positive

(a) What are the three real zeros of the polynomial function f?

(b) What can be said about the behavior of the graph of f at $x = 1$?

(c) What is the least possible degree of f? Explain. Can the degree of f ever be odd? Explain.

(d) Is the leading coefficient of f positive or negative? Explain.

(e) Write an equation for f.

(f) Sketch a graph of the function you wrote in part (e).

10. Use the information in the table.

Interval	Value of $f(x)$
$(-\infty, -2)$	Negative
$(-2, 0)$	Positive
$(0, 2)$	Positive
$(2, \infty)$	Negative

(a) What are the three real zeros of the polynomial function f?

(b) What can be said about the behavior of the graph of f at $x = 0$?

(c) What is the least possible degree of f? Explain. Can the degree of f ever be odd? Explain.

(d) Is the leading coefficient of f positive or negative? Explain.

(e) Write an equation for f.

(f) Sketch a graph of the function you wrote in part (e).

11. (a) Complete the table.

Function	Zeros	Sum of zeros	Product of zeros
$f_1(x) = x^2 - 5x + 6$			
$f_2(x) = x^3 - 7x + 6$			
$f_3(x) = x^4 + 2x^3 + x^2 + 8x - 12$			
$f_4(x) = x^5 - 3x^4 - 9x^3 + 25x^2 - 6x$			

(b) Use the table to make a conjecture relating the sum of the zeros of a polynomial function to the coefficients of the polynomial function.

(c) Use the table to make a conjecture relating the product of the zeros of a polynomial function to the coefficients of the polynomial function.

12. Use the Quadratic Formula and, if necessary, DeMoivre's Theorem to solve each equation with complex coefficients.

(a) $x^2 - (4 + 2i)x + 2 + 4i = 0$

(b) $x^2 - (3 + 2i)x + 5 + i = 0$

(c) $2x^2 + (5 - 8i)x - 13 - i = 0$

(d) $3x^2 - (11 + 14i)x + 1 - 9i = 0$

13. Show that the solutions to

$$|z - 1| \cdot |\bar{z} - 1| = 1$$

are the points (x, y) in the complex plane such that $(x - 1)^2 + y^2 = 1$. Identify the graph of the solution set. $\bar{z}$ is the conjugate of z. (*Hint:* Let $z = x + yi$.)

14. Let $z = a + bi$ and $\bar{z} = a - bi$. Show that the equation

$$z^2 - \bar{z}^2 = 0$$

has only real solutions, whereas the equation

$$z^2 + \bar{z}^2 = 0$$

has complex solutions.

How to study Chapter 5

► **What you should learn**

In this chapter you will learn the following skills and concepts:

- How to recognize and evaluate exponential and logarithmic functions
- How to graph exponential and logarithmic functions
- How to use the change-of-base formula to rewrite and evaluate logarithmic expressions
- How to use properties of logarithms to evaluate, rewrite, expand, or condense logarithmic expressions
- How to solve exponential and logarithmic equations
- How to use exponential growth models, exponential decay models, Gaussian models, logistic growth models, and logarithmic models to solve real-life problems

► **Important Vocabulary**

As you encounter each new vocabulary term in this chapter, add the term and its definition to your notebook glossary.

Algebraic functions (p. 364)
Transcendental functions (p. 364)
Exponential function *f* with base *a* (p. 364)
Natural base *e* (p. 368)
Natural exponential function (p. 368)
Continuous compounding (p. 369)
Logarithmic function with base *a* (p. 375)
Common logarithmic function (p. 376)
Natural logarithmic function (p. 379)

Exponential growth model (p. 402)
Exponential decay model (p. 402)
Gaussian model (p. 402)
Logistic growth model (p. 402)
Logarithmic models (p. 402)
Bell-shaped curve (p. 406)
Logistic curve (p. 407)
Sigmoidal curve (p. 407)

Study Tools

Learning objectives in each section
Chapter Summary (p. 415)
Review Exercises (pp. 416–419)
Chapter Test (p. 420)

Additional Resources

Study and Solutions Guide
Interactive Trigonometry
Videotapes/DVD for Chapter 5
Trigonometry Website
Student Success Organizer

Bill Ross/Corbis

5

Exponential and Logarithmic Functions

5.1 Exponential Functions and Their Graphs

▶ **What you should learn**

- How to recognize and evaluate exponential functions with base a
- How to graph exponential functions
- How to recognize and evaluate exponential functions with base e
- How to use exponential functions to model and solve real-life applications

▶ **Why you should learn it**

Exponential functions can be used to model and solve real-life problems. For instance, in Exercise 59 on page 373, an exponential function is used to model the amount of defoliation caused by the gypsy moth.

Jenny Hager/The Image Works

Exponential Functions

So far, this book has dealt only with **algebraic functions,** which include polynomial functions and rational functions. In this chapter you will study two types of nonalgebraic functions—*exponential* functions and *logarithmic* functions. These functions are examples of **transcendental functions.**

Definition of Exponential Function

The **exponential function f with base a** is denoted by

$$f(x) = a^x$$

where $a > 0$, $a \neq 1$, and x is any real number.

The base $a = 1$ is excluded because it yields $f(x) = 1^x = 1$. This is a constant function, not an exponential function.

You already know how to evaluate a^x for integer and rational values of x. For example, you know that $4^3 = 64$ and $4^{1/2} = 2$. However, to evaluate 4^x for any real number x, you need to interpret forms with *irrational* exponents. For the purposes of this book, it is sufficient to think of

$$a^{\sqrt{2}} \quad (\text{where } \sqrt{2} \approx 1.41421356)$$

as the number that has the successively closer approximations

$$a^{1.4}, a^{1.41}, a^{1.414}, a^{1.4142}, a^{1.41421}, \ldots.$$

Example 1 shows how to use a calculator to evaluate exponential expressions.

Example 1 ▶ **Evaluating Exponential Functions**

Use a calculator to evaluate each function at the indicated value of x.

Function	Value
a. $f(x) = 2^x$	$x = -3.1$
b. $f(x) = 2^{-x}$	$x = \pi$
c. $f(x) = 12^x$	$x = \frac{5}{7}$
d. $f(x) = 0.6^x$	$x = \frac{3}{2}$

Solution

Function Value	Graphing Calculator Keystrokes	Display
a. $f(-3.1) = 2^{-3.1}$	2 $\boxed{\wedge}$ $\boxed{(-)}$ 3.1 $\boxed{\text{ENTER}}$	0.1166291
b. $f(\pi) = 2^{-\pi}$	2 $\boxed{\wedge}$ $\boxed{(-)}$ π $\boxed{\text{ENTER}}$	0.1133147
c. $f\left(\frac{5}{7}\right) = 12^{5/7}$	12 $\boxed{\wedge}$ $\boxed{(}$ 5 $\boxed{\div}$ 7 $\boxed{)}$ $\boxed{\text{ENTER}}$	5.8998877
d. $f\left(\frac{3}{2}\right) = (0.6)^{3/2}$	.6 $\boxed{\wedge}$ $\boxed{(}$ 3 $\boxed{\div}$ 2 $\boxed{)}$ $\boxed{\text{ENTER}}$	0.4647580

Graphs of Exponential Functions

The graphs of all exponential functions have similar characteristics, as shown in Examples 2, 3, and 4.

Example 2 ▶ Graphs of $y = a^x$

In the same coordinate plane, sketch the graph of each function.

a. $f(x) = 2^x$ **b.** $g(x) = 4^x$

Solution

The table below lists some values for each function, and Figure 5.1 shows the graphs of the two functions. Note that both graphs are increasing. Moreover, the graph of $g(x) = 4^x$ is increasing more rapidly than the graph of $f(x) = 2^x$.

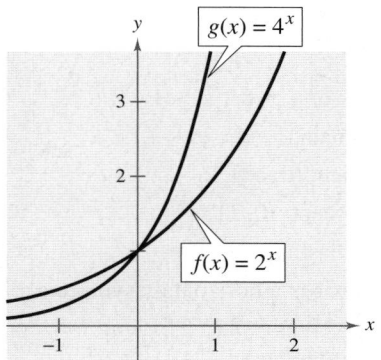

FIGURE **5.1**

x	2^x	4^x
-2	$\frac{1}{4}$	$\frac{1}{16}$
-1	$\frac{1}{2}$	$\frac{1}{4}$
0	1	1
1	2	4
2	4	16
3	8	64

STUDY TIP

The table in Example 2 was evaluated by hand. You could, of course, use a graphing utility to construct tables with even more values.

Example 3 ▶ Graphs of $y = a^{-x}$

In the same coordinate plane, sketch the graph of each function.

a. $F(x) = 2^{-x}$ **b.** $G(x) = 4^{-x}$

Solution

The table below lists some values for each function, and Figure 5.2 shows the graphs of the two functions. Note that both graphs are decreasing. Moreover, the graph of $G(x) = 4^{-x}$ is decreasing more rapidly than the graph of $F(x) = 2^{-x}$.

x	-3	-2	-1	0	1	2
2^{-x}	8	4	2	1	$\frac{1}{2}$	$\frac{1}{4}$
4^{-x}	64	16	4	1	$\frac{1}{4}$	$\frac{1}{16}$

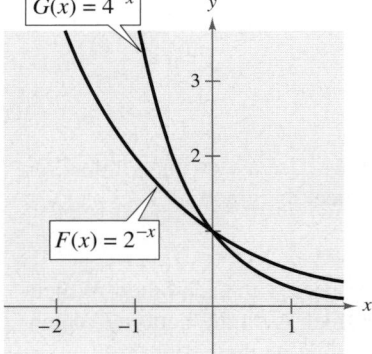

FIGURE **5.2**

In Example 3, note that the functions $F(x) = 2^{-x}$ and $G(x) = 4^{-x}$ can be rewritten with positive exponents.

$$F(x) = 2^{-x} = \left(\frac{1}{2}\right)^x \quad \text{and} \quad G(x) = 4^{-x} = \left(\frac{1}{4}\right)^x$$

The icon  identifies examples and concepts related to features of the Learning Tools CD-ROM and the *Interactive* and *Internet* versions of this text. For more details see the chart on pages *xix–xxiii*.

Comparing the functions in Examples 2 and 3, observe that

$$F(x) = 2^{-x} = f(-x) \qquad \text{and} \qquad G(x) = 4^{-x} = g(-x).$$

Consequently, the graph of F is a reflection (in the y-axis) of the graph of f. The graphs of G and g have the same relationship. The graphs in Figures 5.1 and 5.2 are typical of the exponential functions $y = a^x$ and $y = a^{-x}$. They have one y-intercept and one horizontal asymptote (the x-axis), and they are continuous. The basic characteristics of these exponential functions are summarized in Figures 5.3 and 5.4.

STUDY TIP

Notice that the range of an exponential function is $(0, \infty)$, which means that $a^x > 0$ for all values of x.

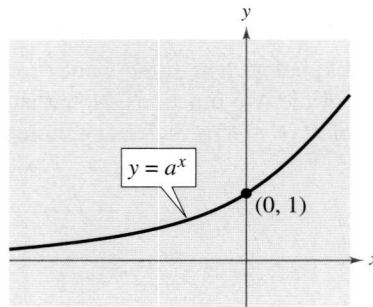

FIGURE **5.3**

Graph of $y = a^x$, $a > 1$
- Domain: $(-\infty, \infty)$
- Range: $(0, \infty)$
- Intercept: $(0, 1)$
- Increasing
- x-Axis is a horizontal asymptote ($a^x \to 0$ as $x \to -\infty$)
- Continuous

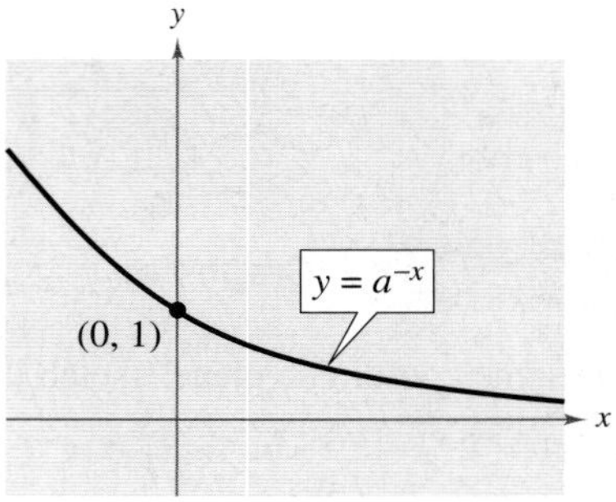

FIGURE **5.4**

Graph of $y = a^{-x}$, $a > 1$
- Domain: $(-\infty, \infty)$
- Range: $(0, \infty)$
- Intercept: $(0, 1)$
- Decreasing
- x-Axis is a horizontal asymptote ($a^{-x} \to 0$ as $x \to \infty$)
- Continuous

◀ **Exploration** ▶

Use a graphing utility to graph

$$y = a^x$$

for $a = 3$, 5, and 7 in the same viewing window. (Use a viewing window in which $-2 \le x \le 1$ and $0 \le y \le 2$.) For instance, the graph of

$$y = 3^x$$

is shown in Figure 5.5. How do the graphs compare with each other? Which graph is on the top in the interval $(-\infty, 0)$? Which is on the bottom? Which graph is on the top in the interval $(0, \infty)$? Which is on the bottom?

Repeat this experiment with the graphs of $y = b^x$ for $b = \frac{1}{3}, \frac{1}{5}$, and $\frac{1}{7}$. (Use a viewing window in which $-1 \le x \le 2$ and $0 \le y \le 2$.) What can you conclude about the shape of the graph of $y = b^x$ and the value of b?

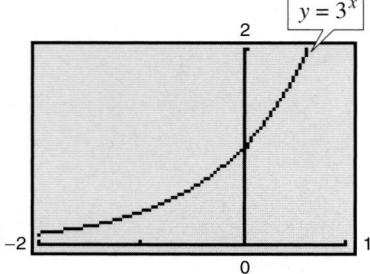

FIGURE **5.5**

In the following example, notice how the graph of $y = a^x$ can be used to sketch the graphs of functions of the form $f(x) = b \pm a^{x+c}$.

Example 4 ▶ **Transformations of Graphs of Exponential Functions**

Each of the following graphs is a transformation of the graph of $f(x) = 3^x$.

a. Because $g(x) = 3^{x+1} = f(x + 1)$, the graph of g can be obtained by shifting the graph of f one unit to the *left*, as shown in Figure 5.6.

b. Because $h(x) = 3^x - 2 = f(x) - 2$, the graph of h can be obtained by shifting the graph of f *downward* two units, as shown in Figure 5.7.

c. Because $k(x) = -3^x = -f(x)$, the graph of k can be obtained by *reflecting* the graph of f in the *x*-axis, as shown in Figure 5.8.

d. Because $j(x) = 3^{-x} = f(-x)$, the graph of j can be obtained by *reflecting* the graph of f in the *y*-axis, as shown in Figure 5.9.

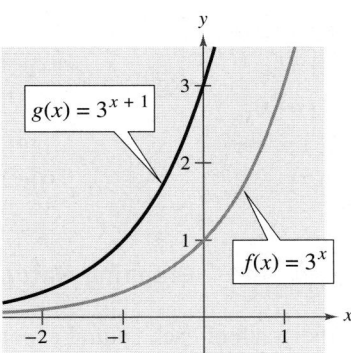

FIGURE 5.6   *Horizontal shift*

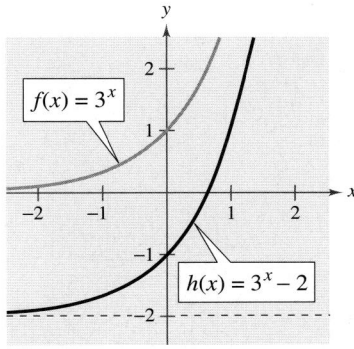

FIGURE 5.7   *Vertical shift*

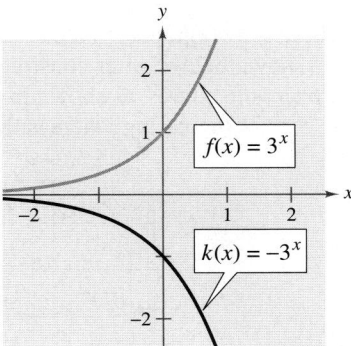

FIGURE 5.8   *Reflection in x-axis*

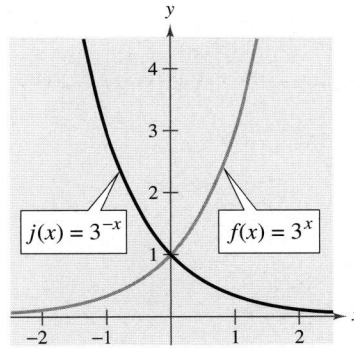

FIGURE 5.9   *Reflection in y-axis*

Notice that the transformations in Figures 5.6, 5.8, and 5.9 keep the *x*-axis as a horizontal asymptote, but the transformation in Figure 5.7 yields a new horizontal asymptote of $y = -2$. Also, be sure to note how the *y*-intercept is affected by each transformation.

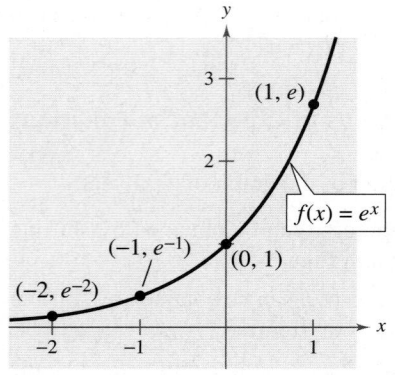

FIGURE **5.10**

The Natural Base *e*

In many applications, the most convenient choice for a base is the irrational number

$$e \approx 2.718281828 \ldots .$$

This number is called the **natural base.** The function $f(x) = e^x$ is called the **natural exponential function.** Its graph is shown in Figure 5.10. Be sure you see that for the exponential function $f(x) = e^x$, e is the constant 2.718281828 . . . , whereas x is the variable.

Example 5 ▶ Evaluating the Natural Exponential Function

Use a calculator to evaluate the function $f(x) = e^x$ at each indicated value of x.
a. $x = -2$ **b.** $x = -1$ **c.** $x = 0.25$ **d.** $x = -0.3$

Solution

Function Value	Graphing Calculator Keystrokes	Display
a. $f(-2) = e^{-2}$	$\boxed{e^x}$ $\boxed{(-)}$ 2 $\boxed{\text{ENTER}}$	0.1353353
b. $f(-1) = e^{-1}$	$\boxed{e^x}$ $\boxed{(-)}$ 1 $\boxed{\text{ENTER}}$	0.3678794
c. $f(0.25) = e^{0.25}$	$\boxed{e^x}$ 0.25 $\boxed{\text{ENTER}}$	1.2840254
d. $f(-0.3) = e^{-0.3}$	$\boxed{e^x}$ $\boxed{(-)}$ 0.3 $\boxed{\text{ENTER}}$	0.7408182

Example 6 ▶ Graphing Natural Exponential Functions

Sketch the graph of each natural exponential function.

a. $f(x) = 2e^{0.24x}$ **b.** $g(x) = \frac{1}{2}e^{-0.58x}$

Solution

To sketch these two graphs, you can use a graphing utility to construct a table of values, as shown below. After constructing the table, plot the points and connect them with smooth curves, as shown in Figures 5.11 and 5.12. Note that the graph in Figure 5.11 is increasing whereas the graph in Figure 5.12 is decreasing.

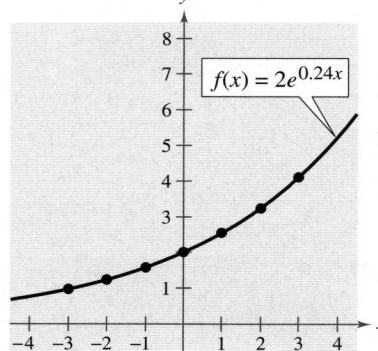

FIGURE **5.11**

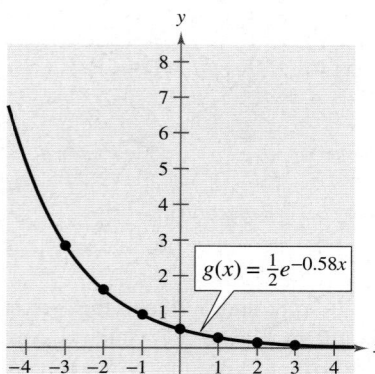

FIGURE **5.12**

x	$f(x)$	$g(x)$
-3	0.974	2.849
-2	1.238	1.595
-1	1.573	0.893
0	2.000	0.500
1	2.542	0.280
2	3.232	0.157
3	4.109	0.088

Use the formula

$$A = P\left(1 + \frac{r}{n}\right)^{nt}$$

to calculate the amount in an account when $P = \$3000$, $r = 6\%, t = 10$ years, and compounding is done (1) by the day, (2) by the hour, (3) by the minute, and (4) by the second. Does increasing the number of compoundings per year result in unlimited growth of the amount in the account? Explain.

Activities

1. Sketch the graphs of the functions $f(x) = e^x$ and $g(x) = 1 + e^x$ on the same coordinate system.

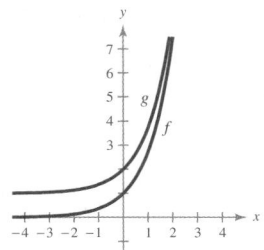

2. Determine the balance A at the end of 20 years if $1500 is invested at 6.5% interest and the interest is compounded (a) quarterly and (b) continuously.

 Answer: (a) $5446.73 (b) $5503.95

3. Determine the amount of money that should be invested at 9% interest, compounded monthly, to produce a final balance of $30,000 in 15 years.

 Answer: $7816.48

Applications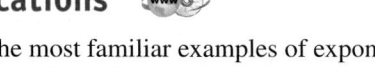

One of the most familiar examples of exponential growth is that of an investment earning *continuously compounded interest.* Using exponential functions, you can now *develop* a formula for the balance in an account that pays compound interest, and show how it leads to continuous compounding.

   Suppose a principal P is invested at an annual interest rate r, compounded once a year. If the interest is added to the principal at the end of the year, the new balance P_1 is

$$P_1 = P + Pr$$
$$= P(1 + r).$$

This pattern of multiplying the previous principal by $1 + r$ is then repeated each successive year, as shown below.

Year	Balance After Each Compounding
0	$P = P$
1	$P_1 = P(1 + r)$
2	$P_2 = P_1(1 + r) = P(1 + r)(1 + r) = P(1 + r)^2$
3	$P_3 = P_2(1 + r) = P(1 + r)^2(1 + r) = P(1 + r)^3$
⋮	
t	$P_t = P(1 + r)^t$

   To accommodate more frequent (quarterly, monthly, or daily) compounding of interest, let n be the number of compoundings per year and let t be the number of years. Then the rate per compounding is r/n and the account balance after t years is

$$A = P\left(1 + \frac{r}{n}\right)^{nt}.$$ Amount (balance) with n compoundings per year

If you let the number of compoundings n increase without bound, the process approaches what is called **continuous compounding.** In the formula for n compoundings per year, let $m = n/r$. This produces

$$A = P\left(1 + \frac{r}{n}\right)^{nt}$$ Amount with n compoundings per year

$$= P\left(1 + \frac{r}{mr}\right)^{mrt}$$ Substitute mr for n.

$$= P\left(1 + \frac{1}{m}\right)^{mrt}$$ Simplify.

$$= P\left[\left(1 + \frac{1}{m}\right)^m\right]^{rt}.$$ Property of exponents

As m increases without bound, it can be shown that $[1 + (1/m)]^m$ approaches e. (Try the values $m = 10, 10,000,$ and $10,000,000.$) From this, you can conclude that the formula for continuous compounding is

$$A = Pe^{rt}.$$ Substitute e for $(1 + 1/m)^m$.

Formulas for Compound Interest

After t years, the balance A in an account with principal P and annual interest rate r (in decimal form) is given by the following formulas.

1. For n compoundings per year: $A = P\left(1 + \dfrac{r}{n}\right)^{nt}$

2. For continuous compounding: $A = Pe^{rt}$

Example 7 ▶ **Compound Interest**

A total of $12,000 is invested at an annual interest rate of 9%. Find the balance after 5 years if it is compounded

a. quarterly.

b. monthly.

c. continuously.

Solution

a. For quarterly compoundings, you have $n = 4$. So, in 5 years at 9%, the balance is

$$A = P\left(1 + \frac{r}{n}\right)^{nt} \qquad \text{Formula for compound interest}$$

$$= 12{,}000\left(1 + \frac{0.09}{4}\right)^{4(5)} \qquad \text{Substitute for } P, r, n, \text{ and } t.$$

$$\approx \$18{,}726.11. \qquad \text{Use a calculator.}$$

b. For monthly compoundings, you have $n = 12$. So, in 5 years at 9%, the balance is

$$A = P\left(1 + \frac{r}{n}\right)^{nt} \qquad \text{Formula for compound interest}$$

$$= 12{,}000\left(1 + \frac{0.09}{12}\right)^{12(5)} \qquad \text{Substitute for } P, r, n, \text{ and } t.$$

$$\approx \$18{,}788.17. \qquad \text{Use a calculator.}$$

c. For continuous compounding, the balance is

$$A = Pe^{rt} \qquad \text{Formula for continuous compounding}$$

$$= 12{,}000e^{0.09(5)} \qquad \text{Substitute for } P, r, \text{ and } t.$$

$$\approx \$18{,}819.75. \qquad \text{Use a calculator.}$$

In Example 7, note that continuous compounding yields more than quarterly or monthly compounding. This is typical of the two types of compounding. That is, for a given principal, interest rate, and time, continuous compounding will always yield a larger balance than compounding n times a year.

Example 8 ▶   Radioactive Decay

In 1986, a nuclear reactor accident occurred in Chernobyl in what was then the Soviet Union. The explosion spread highly toxic radioactive chemicals, such as plutonium, over hundreds of square miles, and the government evacuated the city and the surrounding area. To see why the city is now uninhabited, consider the model

$$P = 10\left(\frac{1}{2}\right)^{t/24,360}$$

which represents the amount of plutonium P that remains (from an initial amount of 10 pounds) after t years. Sketch the graph of this function over the interval from $t = 0$ to $t = 100,000$, where $t = 0$ represents 1986. How much of the 10 pounds will remain in the year 2005? How much of the 10 pounds will remain after 100,000 years?

Solution

The graph of this function is shown in Figure 5.13. Note from this graph that plutonium has a *half-life* of about 24,360 years. That is, after 24,360 years, *half* of the original amount will remain. After another 24,360 years, one-quarter of the original amount will remain, and so on. In the year 2005 ($t = 19$), there will still be

$$P = 10\left(\frac{1}{2}\right)^{19/24,360} \approx 10\left(\frac{1}{2}\right)^{0.0007800} \approx 9.995 \text{ pounds}$$

of plutonium remaining. After 100,000 years, there will still be

$$P = 10\left(\frac{1}{2}\right)^{100,000/24,360} \approx 10\left(\frac{1}{2}\right)^{4.105} \approx 0.581 \text{ pound}$$

of plutonium remaining.

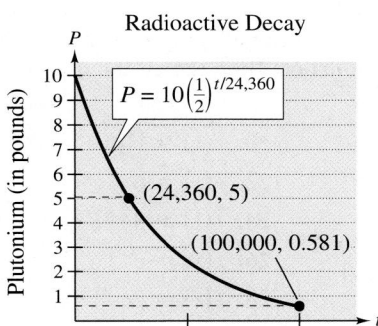

Radioactive Decay

$P = 10\left(\frac{1}{2}\right)^{t/24,360}$

(24,360, 5)

(100,000, 0.581)

Years of decay

FIGURE **5.13**

The *Interactive* CD-ROM and *Internet* versions of this text offer a Quiz for every section of the text.

One way your students might approach this problem is to create a table, covering x-values from -2 through 3, for each of the functions and compare this table with the given tables. If this method is used, you might consider dividing your class into groups of three or six and having the groups assign one or two functions to each member. They should then pool their results and work cooperatively to determine that each function has a y-intercept of $(0, 8)$.

Another approach is a graphical one: the groups can create scatter plots of the data shown in the table and compare them with sketches of the graphs of the given functions. Consider assigning students to groups of four and giving the responsibility for sketching three graphs to each group member.

Writing ABOUT MATHEMATICS

Identifying Exponential Functions Which of the following functions generated the two tables below? Discuss how you were able to decide. What do these functions have in common? Are any of them the same? If so, explain why.

a. $f_1(x) = 2^{(x+3)}$ **b.** $f_2(x) = 8\left(\frac{1}{2}\right)^x$ **c.** $f_3(x) = \left(\frac{1}{2}\right)^{(x-3)}$

d. $f_4(x) = \left(\frac{1}{2}\right)^x + 7$ **e.** $f_5(x) = 7 + 2^x$ **f.** $f_6(x) = (8)2^x$

x	-1	0	1	2	3
$g(x)$	7.5	8	9	11	15

x	-2	-1	0	1	2
$h(x)$	32	16	8	4	2

Create two different exponential functions of the forms $y = a(b)^x$ and $y = c^x + d$ with y-intercepts of $(0, -3)$.

5.1 Exercises

The *Interactive* CD-ROM and *Internet* versions of this text contain step-by-step solutions to all odd-numbered exercises. They also provide Tutorial Exercises for additional help.

In Exercises 1–6, evaluate the function at the indicated value of *x*. Round your result to three decimal places.

Function	Value
1. $f(x) = 3.4^x$	$x = 5.6$
2. $f(x) = 2.3^x$	$x = \frac{3}{2}$
3. $f(x) = 5^x$	$x = -\pi$
4. $g(x) = 5000(2^x)$	$x = -1.5$
5. $h(x) = e^{-x}$	$x = \frac{3}{4}$
6. $f(x) = e^x$	$x = 3.2$

In Exercises 7–10, match the exponential function with its graph. [The graphs are labeled (a), (b), (c), and (d).]

(a)
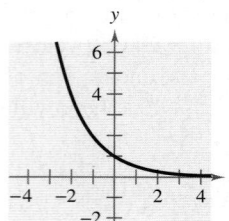

(b)
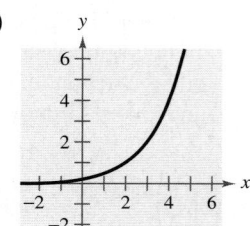

(c)
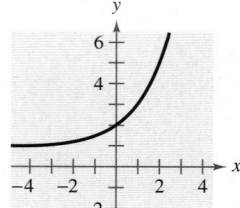

(d)
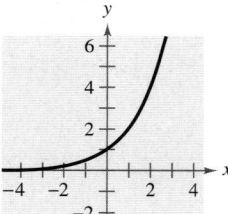

7. $f(x) = 2^x$

8. $f(x) = 2^x + 1$

9. $f(x) = 2^{-x}$

10. $f(x) = 2^{x-2}$

In Exercises 11–18, use the graph of *f* to describe the transformation that yields the graph of *g*.

11. $f(x) = 3^x,$ $g(x) = 3^{x-4}$

12. $f(x) = 4^x,$ $g(x) = 4^x + 1$

13. $f(x) = -2^x,$ $g(x) = 5 - 2^x$

14. $f(x) = 10^x,$ $g(x) = 10^{-x+3}$

15. $f(x) = \left(\frac{3}{5}\right)^x,$ $g(x) = -\left(\frac{3}{5}\right)^{x+4}$

16. $f(x) = \left(\frac{7}{2}\right)^x,$ $g(x) = -\left(\frac{7}{2}\right)^{-x+6}$

17. $f(x) = 0.3^x,$ $g(x) = -0.3^x + 5$

18. $f(x) = 3.6^x,$ $g(x) = -3.6^{-x} + 8$

 In Exercises 19–32, use a graphing utility to construct a table of values for the function. Then sketch the graph of the function.

19. $f(x) = \left(\frac{1}{2}\right)^x$ **20.** $f(x) = \left(\frac{1}{2}\right)^{-x}$

21. $f(x) = 6^{-x}$ **22.** $f(x) = 6^x$

23. $f(x) = 2^{x-1}$ **24.** $f(x) = 3^{x+2}$

25. $f(x) = e^x$ **26.** $f(x) = e^{-x}$

27. $f(x) = 3e^{x+4}$ **28.** $f(x) = 2e^{-0.5x}$

29. $f(x) = 2e^{x-2} + 4$ **30.** $f(x) = 2 + e^{x-5}$

31. $f(x) = 4^{x-3} + 3$ **32.** $f(x) = -4^{x-3} - 3$

In Exercises 33–42, use a graphing utility to graph the exponential function.

33. $y = 2^{-x^2}$ **34.** $y = 3^{-|x|}$

35. $y = 3^{x-2} + 1$ **36.** $y = 4^{x+1} - 2$

37. $y = 1.08^{-5x}$ **38.** $y = 1.08^{5x}$

39. $s(t) = 2e^{0.12t}$ **40.** $s(t) = 3e^{-0.2t}$

41. $g(x) = 1 + e^{-x}$ **42.** $h(x) = e^{x-2}$

Compound Interest In Exercises 43–46, complete the table to determine the balance *A* for *P* dollars invested at rate *r* for *t* years and compounded *n* times per year.

n	1	2	4	12	365	Continuous
A						

43. $P = \$2500, r = 8\%, t = 10$ years

44. $P = \$1000, r = 6\%, t = 10$ years

45. $P = \$2500, r = 8\%, t = 20$ years

46. $P = \$1000, r = 6\%, t = 40$ years

Compound Interest In Exercises 47–50, complete the table to determine the balance *A* for $12,000 invested at rate *r* for *t* years, compounded continuously.

t	10	20	30	40	50
A					

47. $r = 8\%$ **48.** $r = 6\%$

49. $r = 6.5\%$ **50.** $r = 7.5\%$

51. Trust Fund On the day of a child's birth, a deposit of $25,000 is made in a trust fund that pays 8.75% interest, compounded continuously. Determine the balance in this account on the child's 25th birthday.

52. Trust Fund A deposit of $5000 is made in a trust fund that pays 7.5% interest, compounded continuously. It is specified that the balance will be given to the college from which the donor graduated after the money has earned interest for 50 years. How much will the college receive?

53. Inflation If the annual rate of inflation averages 4% over the next 10 years, the approximate cost C of goods or services during any year in that decade will be modeled by $C(t) = P(1.04)^t$, where t is the time in years and P is the present cost. The price of an oil change for your car is presently $23.95. Estimate the price 10 years from now.

54. Demand The demand equation for a product is

$$p = 5000\left(1 - \frac{4}{4 + e^{-0.002x}}\right).$$

 (a) Use a graphing utility to graph the demand function for $x > 0$ and $p > 0$.

(b) Find the price p for a demand of $x = 500$ units.

 (c) Use the graph in part (a) to approximate the greatest price that will still yield a demand of at least 600 units.

55. Population Growth A certain type of bacterium increases according to the model $P(t) = 100e^{0.2197t}$, where t is the time in hours. Find (a) $P(0)$, (b) $P(5)$, and (c) $P(10)$.

56. Population Growth The population of a town increases according to the model $P(t) = 2500e^{0.0293t}$, where t is the time in years, with $t = 0$ corresponding to 2000. Use the model to estimate the population in (a) 2010 and (b) 2020.

57. Radioactive Decay Let Q represent a mass of radioactive radium (^{226}Ra) (in grams), whose half-life is 1620 years. The quantity of radium present after t years is

$$Q = 25\left(\frac{1}{2}\right)^{t/1620}.$$

(a) Determine the initial quantity (when $t = 0$).

(b) Determine the quantity present after 1000 years.

 (c) Use a graphing utility to graph the function over the interval $t = 0$ to $t = 5000$.

58. Radioactive Decay Let Q represent a mass of carbon 14 (^{14}C) (in grams), whose half-life is 5730 years. The quantity of carbon 14 present after t years is

$$Q = 10\left(\frac{1}{2}\right)^{t/5730}.$$

(a) Determine the initial quantity (when $t = 0$).

(b) Determine the quantity present after 2000 years.

(c) Sketch the graph of this function over the interval $t = 0$ to $t = 10,000$.

▶ **Model It**

59. Data Analysis To estimate the amount of defoliation caused by the gypsy moth during a given year, a forester counts the number x of egg masses on $\frac{1}{40}$ of an acre (circle of radius 18.6 feet) in the fall. The percent of defoliation y the next spring is shown in the table. (Source: USDA, Forest Service)

Egg masses, x	Percent of defoliation, y
0	12
25	44
50	81
75	96
100	99

A model for the data is

$$y = \frac{100}{1 + 7e^{-0.069x}}.$$

 (a) Use a graphing utility to create a scatter plot of the data and graph the model in the same viewing window.

(b) Create a table that compares the model with the sample data.

(c) Estimate the percent of defoliation if 36 egg masses are counted on $\frac{1}{40}$ acre.

 (d) You observe that $\frac{2}{3}$ of a forest is defoliated the following spring. Use the graph in part (a) to estimate the number of egg masses per $\frac{1}{40}$ acre.

60. Data Analysis A meteorologist measures the atmospheric pressure P (in pascals) at altitude h (in kilometers). The data is shown in the table.

Altitude, h	Pressure, P
0	101,293
5	54,735
10	23,294
15	12,157
20	5,069

A model for the data is given by

$$P = 102{,}303e^{-0.137h}.$$

(a) Sketch a scatter plot of the data and graph the model on the same set of axes.

(b) Estimate the atmospheric pressure at a height of 8 kilometers.

Synthesis

True or False? In Exercises 61 and 62, determine whether the statement is true or false. Justify your answer.

61. The line $y = -2$ is an asymptote for the graph of $f(x) = 10^x - 2$.

62. $e = \dfrac{271{,}801}{99{,}990}$.

Think About It In Exercises 63–66, use properties of exponents to determine which functions (if any) are the same.

63. $f(x) = 3^{x-2}$
$g(x) = 3^x - 9$
$h(x) = \frac{1}{9}(3^x)$

64. $f(x) = 4^x + 12$
$g(x) = 2^{2x+6}$
$h(x) = 64(4^x)$

65. $f(x) = 16(4^{-x})$
$g(x) = \left(\frac{1}{4}\right)^{x-2}$
$h(x) = 16(2^{-2x})$

66. $f(x) = 5^{-x} + 3$
$g(x) = 5^{3-x}$
$h(x) = -5^{x-3}$

67. Graph the functions $y = 3^x$ and $y = 4^x$ and use the graphs to solve the inequalities.

(a) $4^x < 3^x$ (b) $4^x > 3^x$

68. Graph the functions $y = \left(\frac{1}{2}\right)^x$ and $y = \left(\frac{1}{4}\right)^x$ and use the graphs to solve the inequalities.

(a) $\left(\frac{1}{4}\right)^x < \left(\frac{1}{2}\right)^x$ (b) $\left(\frac{1}{4}\right)^x > \left(\frac{1}{2}\right)^x$

69. Use a graphing utility to graph each function. Use the graph to find any asymptotes of the function.

(a) $f(x) = \dfrac{8}{1 + e^{-0.5x}}$ (b) $g(x) = \dfrac{8}{1 + e^{-0.5/x}}$

70. Use a graphing utility to graph each function. Use the graph to find where the function is increasing and decreasing, and approximate any relative maximum or minimum values.

(a) $f(x) = x^2 e^{-x}$ (b) $g(x) = x2^{3-x}$

71. Graphical Analysis Use a graphing utility to graph

$$f(x) = \left(1 + \frac{0.5}{x}\right)^x \quad \text{and} \quad g(x) = e^{0.5}$$

in the same viewing window. What is the relationship between f and g as x increases and decreases without bound?

72. Conjecture Use the result of Exercise 71 to make a conjecture about the value of $[1 + (r/x)]^x$ as x increases without bound. Create a table that illustrates your conjecture for $r = 1$.

73. Think About It Which functions are exponential?

(a) $3x$ (b) $3x^2$
(c) 3^x (d) 2^{-x}

74. Writing Explain why $2^{\sqrt{2}}$ is greater than 2, but less than 4.

Review

In Exercises 75–78, solve for y.

75. $2x - 7y + 14 = 0$ **76.** $x^2 + 3y = 4$

77. $x^2 + y^2 = 25$ **78.** $x - |y| = 2$

▶ **What you should learn**

- How to recognize and evaluate logarithmic functions with base a
- How to graph logarithmic functions
- How to recognize and evaluate natural logarithmic functions
- How to use logarithmic functions to model and solve real-life applications

▶ **Why you should learn it**

You can use logarithmic functions to model and solve real-life problems. For instance, in Exercise 57 on page 383, you can use a logarithmic function to approximate the length of a home mortgage.

Ryan McVay/Getty Images

Logarithmic Functions

In Section P.10, you studied the concept of an inverse function. There, you learned that if a function is one-to-one—that is, if the function has the property that no horizontal line intersects the graph of the function more than once—the function must have an inverse function. By looking back at the graphs of the exponential functions introduced in Section 5.1, you will see that every function of the form

$$f(x) = a^x$$

passes the Horizontal Line Test and therefore must have an inverse function. This inverse function is called the **logarithmic function with base a.**

Definition of Logarithmic Function with Base a

For $x > 0$ and $0 < a \neq 1$,

$$y = \log_a x \text{ if and only if } x = a^y.$$

The function given by

$$f(x) = \log_a x$$

is called the **logarithmic function with base a.**

The equations

$$y = \log_a x \quad \text{and} \quad x = a^y$$

are equivalent. The first equation is in logarithmic form and the second is in exponential form.

When evaluating logarithms, remember that *a logarithm is an exponent.* This means that $\log_a x$ is the exponent to which a must be raised to obtain x. For instance, $\log_2 8 = 3$ because 2 must be raised to the third power to get 8.

Example 1 ▶ **Evaluating Logarithms**

Use the definition of logarithmic function to evaluate each logarithm at the indicated value of x.

a. $f(x) = \log_2 x, \quad x = 32$ **b.** $f(x) = \log_3 x, \quad x = 1$

c. $f(x) = \log_4 x, \quad x = 2$ **d.** $f(x) = \log_{10} x, \quad x = \frac{1}{100}$

Solution

a. $f(32) = \log_2 32 = 5$ because $2^5 = 32.$

b. $f(1) = \log_3 1 = 0$ because $3^0 = 1.$

c. $f(2) = \log_4 2 = \frac{1}{2}$ because $4^{1/2} = \sqrt{4} = 2.$

d. $f\left(\frac{1}{100}\right) = \log_{10} \frac{1}{100} = -2$ because $10^{-2} = \frac{1}{10^2} = \frac{1}{100}.$

Complete the table for
$f(x) = 10^x$.

x	-2	-1	0	1	2
$f(x)$					

Complete the table for
$f(x) = \log_{10} x$.

x	$\frac{1}{100}$	$\frac{1}{10}$	1	10	100
$f(x)$					

Compare the two tables. What is the relationship between $f(x) = 10^x$ and $f(x) = \log_{10} x$?

The logarithmic function with base 10 is called the **common logarithmic function.** On most calculators, this function is denoted by ⌊LOG⌋. Example 2 shows how to use a calculator to evaluate common logarithmic functions. You will learn how to use a calculator to calculate logarithms to any base in the next section.

Example 2 ▶ **Evaluating Common Logarithms on a Calculator**

Use a calculator to evaluate the function $f(x) = \log_{10} x$ at each value of x.

a. $x = 10$ **b.** $x = \frac{1}{3}$
c. $x = 2.5$ **d.** $x = -2$

Solution

Function Value	Graphing Calculator Keystrokes	Display
a. $f(10) = \log_{10} 10$	⌊LOG⌋ 10 ⌊ENTER⌋	1
b. $f\left(\frac{1}{3}\right) = \log_{10} \frac{1}{3}$	⌊LOG⌋ ⌊(⌋ 1 ⌊÷⌋ 3 ⌊)⌋ ⌊ENTER⌋	-0.4771213
c. $f(2.5) = \log_{10} 2.5$	⌊LOG⌋ 2.5 ⌊ENTER⌋	0.3979400
d. $f(-2) = \log_{10}(-2)$	⌊LOG⌋ ⌊(−)⌋ 2 ⌊ENTER⌋	ERROR

Note that the calculator displays an error message (or a complex number) when you try to evaluate $\log_{10}(-2)$. The reason for this is that there is no real number power to which 10 can be raised to obtain -2.

The following properties follow directly from the definition of the logarithmic function with base a.

The logarithmic function can be one of the most difficult concepts for students to understand. Remind students that a logarithm is an exponent. Converting back and forth from logarithmic form to exponential form supports this concept.

Properties of Logarithms

1. $\log_a 1 = 0$ because $a^0 = 1$.

2. $\log_a a = 1$ because $a^1 = a$.

3. $\log_a a^x = x$ and $a^{\log_a x} = x$ Inverse Properties

4. If $\log_a x = \log_a y$, then $x = y$.   One-to-One Property

---

Example 3 ▶ **Using Properties of Logarithms**

a. Solve for x: $\log_2 x = \log_2 3$ **b.** Solve for x: $\log_4 4 = x$
c. Simplify: $\log_5 5^x$ **d.** Simplify: $6^{\log_6 20}$

Solution

a. Using the One-to-One Property (Property 4), you can conclude that $x = 3$.
b. Using Property 2, you can conclude that $x = 1$.
c. Using the Inverse Property (Property 3), it follows that $\log_5 5^x = x$.
d. Using the Inverse Property (Property 3), it follows that $6^{\log_6 20} = 20$.

Graphs of Logarithmic Functions

To sketch the graph of $y = \log_a x$, you can use the fact that the graphs of inverse functions are reflections of each other in the line $y = x$.

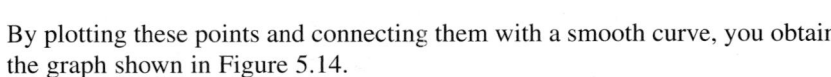

Example 4 ► Graphs of Exponential and Logarithmic Functions

In the same coordinate plane, sketch the graph of each function.

a. $f(x) = 2^x$ **b.** $g(x) = \log_2 x$

Solution

a. For $f(x) = 2^x$, construct a table of values.

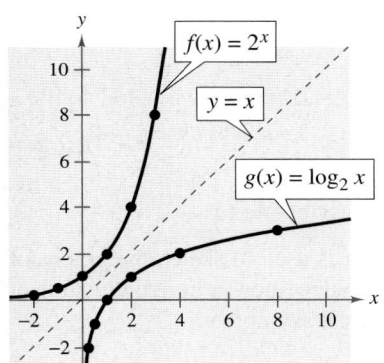

FIGURE **5.14**

x	$f(x) = 2^x$
-2	$\frac{1}{4}$
-1	$\frac{1}{2}$
0	1
1	2
2	4
3	8

By plotting these points and connecting them with a smooth curve, you obtain the graph shown in Figure 5.14.

b. Because $g(x) = \log_2 x$ is the inverse of $f(x) = 2^x$, the graph of g is obtained by plotting the points $(f(x), x)$ and connecting them with a smooth curve. The graph of g is a reflection of the graph of f in the line $y = x$, as shown in Figure 5.14.

Example 5 ► Sketching the Graph of a Logarithmic Function

Sketch the graph of the common logarithmic function $f(x) = \log_{10} x$. Identify the x-intercept and the vertical asymptote.

Solution

Begin by constructing a table of values. Note that some of the values can be obtained without a calculator by using the Inverse Property of Logarithms. Others require a calculator. Next, plot the points and connect them with a smooth curve, as shown in Figure 5.15. The x-intercept of the graph is $(1, 0)$ and the vertical asymptote is $x = 0$ (y-axis).

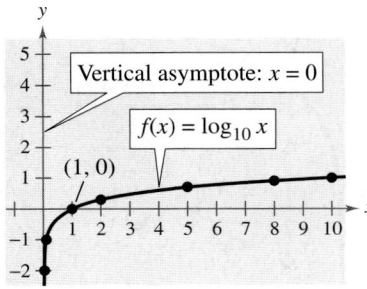

FIGURE **5.15**

	Without calculator				With calculator		
x	$\frac{1}{100}$	$\frac{1}{10}$	1	10	2	5	8
$\log_{10} x$	-2	-1	0	1	0.301	0.699	0.903

The nature of the graph in Figure 5.15 is typical of functions of the form $f(x) = \log_a x, a > 1$. They have one x-intercept and one vertical asymptote. Notice how slowly the graph rises for $x > 1$. The basic characteristics of logarithmic graphs are summarized in Figure 5.16.

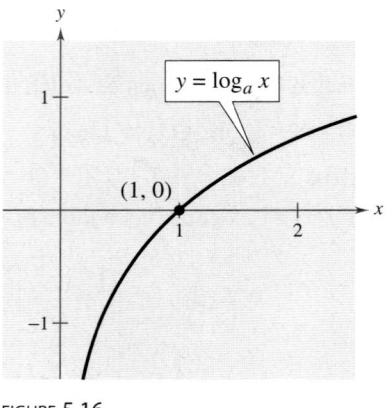

FIGURE **5.16**

Graph of $y = \log_a x, a > 1$
- Domain: $(0, \infty)$
- Range: $(-\infty, \infty)$
- x-Intercept: $(1, 0)$
- Increasing
- One-to-one, therefore has an inverse function
- y-Axis is a vertical asymptote $(\log_a x \to -\infty \text{ as } x \to 0^+)$.
- Continuous
- Reflection of graph of $y = a^x$ about the line $y = x$

The basic characteristics of the graph of $f(x) = a^x$ are shown below to illustrate the inverse relation between the functions $f(x) = a^x$ and $g(x) = \log_a x$.

- Domain: $(-\infty, \infty)$ • Range: $(0, \infty)$
- y-Intercept: $(0,1)$ • x-Axis is a horizontal asymptote $(a^x \to 0 \text{ as } x \to -\infty)$.

In the next example, the graph of $y = \log_a x$ is used to sketch the graphs of functions of the form $f(x) = b \pm \log_a(x + c)$. Notice how a horizontal shift of the graph results in a horizontal shift of the vertical asymptote.

STUDY TIP

You can use your understanding of transformations to identify vertical asymptotes of logarithmic functions. For instance, in Example 6(a) the graph of $g(x) = f(x - 1)$ shifts the graph of $f(x)$ one unit to the right. So the vertical asymptote of $g(x)$ is $x = 1$, one unit to the right of the asymptote of the graph of $f(x)$.

Example 6 ▶ **Shifting Graphs of Logarithmic Functions**

The graph of each of the functions is similar to the graph of $f(x) = \log_{10} x$.
a. Because $g(x) = \log_{10}(x - 1) = f(x - 1)$, the graph of g can be obtained by shifting the graph of f one unit to the right, as shown in Figure 5.17.
b. Because $h(x) = 2 + \log_{10} x = 2 + f(x)$, the graph of h can be obtained by shifting the graph of f two units upward, as shown in Figure 5.18.

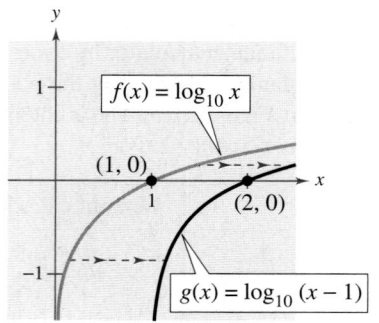

FIGURE **5.17**

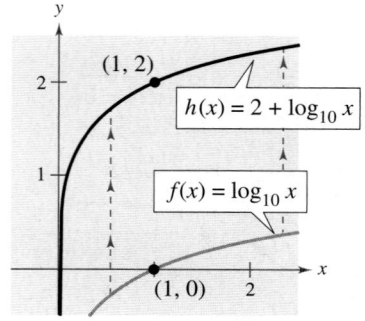

FIGURE **5.18**

The Natural Logarithmic Function

By looking back at the graph of the natural exponential function introduced in Section 5.1, you will see that $f(x) = e^x$ is one-to-one and so has an inverse function. This inverse function is called the **natural logarithmic function** and is denoted by the special symbol ln x, read as "the natural log of x" or "el en of x."

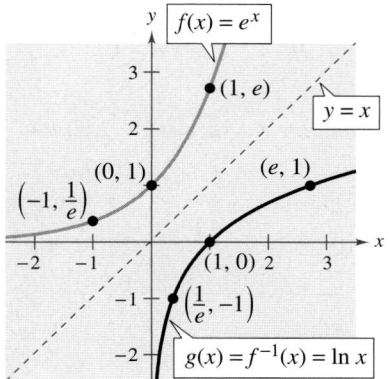

The Natural Logarithmic Function

The function defined by

$$f(x) = \log_e x = \ln x, \quad x > 0$$

is called the **natural logarithmic function.**

The definition above implies that the natural logarithmic function and the natural exponential function are inverse functions of each other. So, every logarithmic equation can be written in an equivalent exponential form and every exponential equation can be written in logarithmic form. That is, $y = \ln x$ and $x = e^y$ are equivalent equations.

Because the functions $f(x) = e^x$ and $g(x) = \ln x$ are inverse functions of each other, their graphs are reflections of each other in the line $y = x$. This reflective property is illustrated in Figure 5.19.

The four properties of logarithms listed on page 376 are also valid for natural logarithms.

Reflection of graph of $f(x) = e^x$ about the line $y = x$

FIGURE 5.19

STUDY TIP

Notice that as with every other logarithmic function, the domain of the natural logarithmic function is the set of *positive real numbers*—be sure you see that ln x is not defined for zero or for negative numbers.

Properties of Natural Logarithms

1. $\ln 1 = 0$ because $e^0 = 1$.

2. $\ln e = 1$ because $e^1 = e$.

3. $\ln e^x = x$ and $e^{\ln x} = x$ Inverse Properties

4. If $\ln x = \ln y$, then $x = y$.   One-to-One Property

Example 7 ▶ **Using Properties of Natural Logarithms**

Use the properties of natural logarithms to simplify each expression.

a. $\ln \dfrac{1}{e}$ **b.** $e^{\ln 5}$ **c.** $\dfrac{\ln 1}{3}$ **d.** $2 \ln e$

Solution

a. $\ln \dfrac{1}{e} = \ln e^{-1} = -1$ Inverse Property

b. $e^{\ln 5} = 5$ Inverse Property

c. $\dfrac{\ln 1}{3} = \dfrac{0}{3} = 0$ Property 1

d. $2 \ln e = 2(1) = 2$ Property 2

On most calculators, the natural logarithm is denoted by $\boxed{\text{LN}}$, as illustrated in Example 8.

Example 8 ▶ **Evaluating the Natural Logarithmic Function**

Use a calculator to evaluate the function $f(x) = \ln x$ for each value of x.

a. $x = 2$ **b.** $x = 0.3$ **c.** $x = -1$ **d.** $x = 1 + \sqrt{2}$

Solution

	Function Value	Graphing Calculator Keystrokes	Display
a.	$f(2) = \ln 2$	$\boxed{\text{LN}}$ 2 $\boxed{\text{ENTER}}$	0.6931472
b.	$f(0.3) = \ln 0.3$	$\boxed{\text{LN}}$ .3 $\boxed{\text{ENTER}}$	-1.2039728
c.	$f(-1) = \ln(-1)$	$\boxed{\text{LN}}$ $\boxed{(-)}$ 1 $\boxed{\text{ENTER}}$	ERROR
d.	$f(1 + \sqrt{2}) = \ln(1 + \sqrt{2})$	$\boxed{\text{LN}}$ $\boxed{(}$ 1 $\boxed{+}$ $\boxed{\sqrt{}}$ 2 $\boxed{)}$ $\boxed{\text{ENTER}}$	0.8813736

In Example 8, be sure you see that $\ln(-1)$ gives an error message on most calculators. This occurs because the domain of $\ln x$ is the set of positive real numbers (see Figure 5.19). So, $\ln(-1)$ is undefined.

Example 9 ▶ **Finding the Domains of Logarithmic Functions**

Find the domain of each function.

a. $f(x) = \ln(x - 2)$
b. $g(x) = \ln(2 - x)$
c. $h(x) = \ln x^2$

Solution

a. Because $\ln(x - 2)$ is defined only if $x - 2 > 0$, it follows that the domain of f is $(2, \infty)$. The graph of f is shown in Figure 5.20.

b. Because $\ln(2 - x)$ is defined only if $2 - x > 0$, it follows that the domain of g is $(-\infty, 2)$. The graph of g is shown in Figure 5.21.

c. Because $\ln x^2$ is defined only if $x^2 > 0$, it follows that the domain of h is all real numbers except $x = 0$. The graph of h is shown in Figure 5.22.

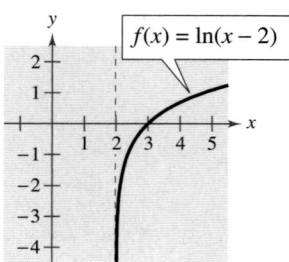

FIGURE **5.20**

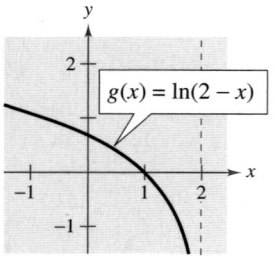

FIGURE **5.21**

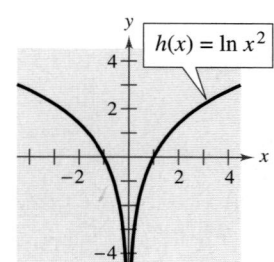

FIGURE **5.22**

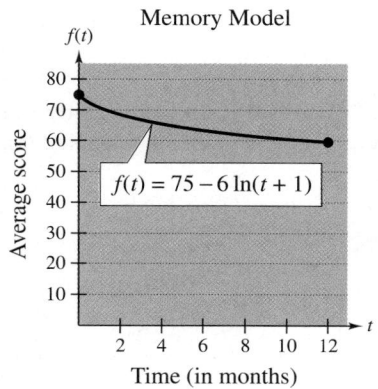

Memory Model

f(t)

$f(t) = 75 - 6\ln(t + 1)$

FIGURE 5.23

Application

Human Memory Model

Students participating in a psychology experiment attended several lectures on a subject and were given an exam. Every month for a year after the exam, the students were retested to see how much of the material they remembered. The average scores for the group are given by the *human memory model*

$$f(t) = 75 - 6\ln(t + 1), \quad 0 \le t \le 12$$

where t is the time in months. The graph of f is shown in Figure 5.23.

a. What was the average score on the original ($t = 0$) exam?

b. What was the average score at the end of $t = 2$ months?

c. What was the average score at the end of $t = 6$ months?

Solution

a. The original average score was

$$f(0) = 75 - 6\ln(0 + 1) \qquad \text{Substitute 0 for } t.$$
$$= 75 - 6\ln 1 \qquad \text{Simplify.}$$
$$= 75 - 6(0) \qquad \text{Property of natural logarithms}$$
$$= 75. \qquad \text{Solution}$$

b. After 2 months, the average score was

$$f(2) = 75 - 6\ln(2 + 1) \qquad \text{Substitute 2 for } t.$$
$$= 75 - 6\ln 3 \qquad \text{Simplify.}$$
$$\approx 75 - 6(1.0986) \qquad \text{Use a calculator.}$$
$$\approx 68.4. \qquad \text{Solution}$$

c. After 6 months, the average score was

$$f(6) = 75 - 6\ln(6 + 1) \qquad \text{Substitute 6 for } t.$$
$$= 75 - 6\ln 7 \qquad \text{Simplify.}$$
$$\approx 75 - 6(1.9459) \qquad \text{Use a calculator.}$$
$$\approx 63.3. \qquad \text{Solution}$$

Alternative Writing About Mathematics

Use a graphing utility to graph $f(x) = \ln x$. How will the graphs of $h(x) = \ln x + 5$, $j(x) = \ln(x - 3)$, and $l(x) = \ln x - 4$ differ from the graph of f?

How will the basic graph of f be affected when a constant c is introduced: $g(x) = c \ln x$? Use a graphing utility to graph g with several different positive values of c, and summarize the effect of c.

Writing ABOUT MATHEMATICS

Analyzing a Human Memory Model Use a graphing utility to determine the time in months when the average score in Example 10 was 60. Explain your method of solving the problem. Describe another way that you can use a graphing utility to determine the answer.

5.2 **Exercises**

In Exercises 1–8, write the logarithmic equation in exponential form. For example, the exponential form of $\log_5 25 = 2$ is $5^2 = 25$.

1. $\log_4 64 = 3$ **2.** $\log_3 81 = 4$

3. $\log_7 \frac{1}{49} = -2$ **4.** $\log_{10} \frac{1}{1000} = -3$

5. $\log_{32} 4 = \frac{2}{5}$ **6.** $\log_{16} 8 = \frac{3}{4}$

7. $\ln \frac{1}{2} = -0.693 \ldots$ **8.** $\ln 4 = 1.386 \ldots$

In Exercises 9–18, write the exponential equation in logarithmic form. For example, the logarithmic form of $2^3 = 8$ is $\log_2 8 = 3$.

9. $5^3 = 125$ **10.** $8^2 = 64$

11. $81^{1/4} = 3$ **12.** $9^{3/2} = 27$

13. $6^{-2} = \frac{1}{36}$ **14.** $10^{-3} = 0.001$

15. $e^3 = 20.0855 \ldots$ **16.** $e^{1/2} = 1.6487 \ldots$

17. $e^x = 4$ **18.** $u^v = w$

In Exercises 19–26, evaluate the function at the indicated value of x without using a calculator.

	Function	Value
19.	$f(x) = \log_2 x$	$x = 16$
20.	$f(x) = \log_{16} x$	$x = 4$
21.	$f(x) = \log_7 x$	$x = 1$
22.	$f(x) = \log_{10} x$	$x = 10$
23.	$g(x) = \ln x$	$x = e^3$
24.	$g(x) = \ln x$	$x = e^{-2}$
25.	$g(x) = \log_a x$	$x = a^2$
26.	$g(x) = \log_b x$	$x = b^{-3}$

In Exercises 27–32, use a calculator to evaluate the function at the indicated value of x. Round your result to three decimal places.

	Function	Value
27.	$f(x) = \log_{10} x$	$x = \frac{4}{5}$
28.	$f(x) = \log_{10} x$	$x = 12.5$
29.	$f(x) = \ln x$	$x = 18.42$
30.	$f(x) = 3 \ln x$	$x = 0.32$
31.	$g(x) = 2 \ln x$	$x = 0.75$
32.	$g(x) = -\ln x$	$x = \frac{1}{2}$

In Exercises 33–38, use the graph of $y = \log_3 x$ to match the given function with its graph. [The graphs are labeled (a), (b), (c), (d), (e), and (f).]

(a) 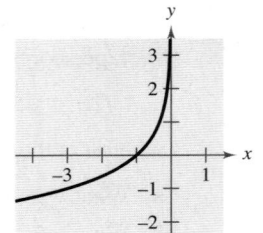   (b)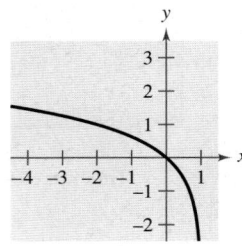

(c) 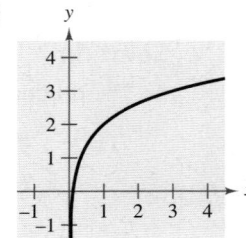   (d)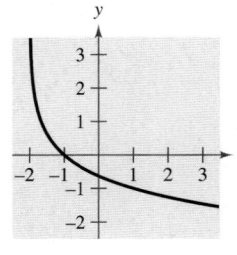

(e) 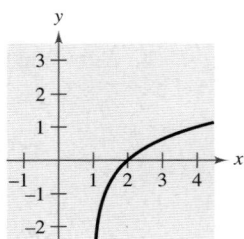   (f)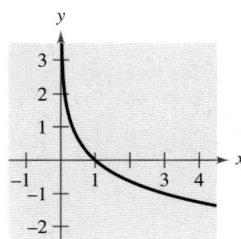

33. $f(x) = \log_3 x + 2$ **34.** $f(x) = -\log_3 x$

35. $f(x) = -\log_3(x + 2)$ **36.** $f(x) = \log_3(x - 1)$

37. $f(x) = \log_3(1 - x)$ **38.** $f(x) = -\log_3(-x)$

In Exercises 39–50, find the domain, x-intercept, and vertical asymptote of the logarithmic function and sketch its graph.

39. $f(x) = \log_4 x$ **40.** $g(x) = \log_6 x$

41. $y = -\log_3 x + 2$ **42.** $h(x) = \log_4(x - 3)$

43. $f(x) = -\log_6(x + 2)$ **44.** $y = \log_5(x - 1) + 4$

45. $y = \log_{10}\left(\frac{x}{5}\right)$ **46.** $y = \log_{10}(-x)$

47. $f(x) = \ln(x - 2)$ **48.** $h(x) = \ln(x + 1)$

49. $g(x) = \ln(-x)$ **50.** $f(x) = \ln(3 - x)$

In Exercises 51–56, use a graphing utility to graph the function. Be sure to use an appropriate viewing window.

51. $f(x) = \log_{10}(x + 1)$

52. $f(x) = \log_{10}(x - 1)$

53. $f(x) = \ln(x - 1)$

54. $f(x) = \ln(x + 2)$

55. $f(x) = \ln x + 2$

56. $f(x) = 3 \ln x - 1$

▶ Model It

57. *Monthly Payment* The model

$$t = 12.542 \ln\left(\frac{x}{x - 1000}\right), \qquad x > 1000$$

approximates the length of a home mortgage of $150,000 at 8% in terms of the monthly payment. In the model, t is the length of the mortgage in years and x is the monthly payment in dollars (see figure).

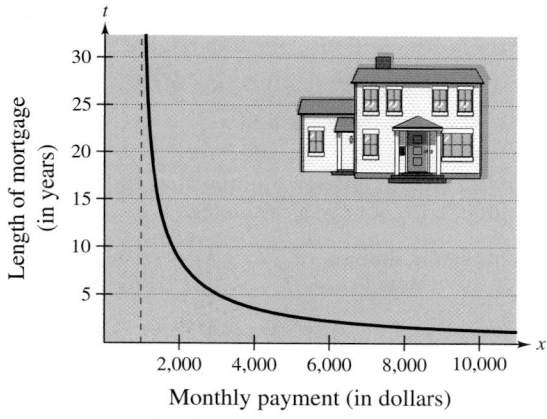

(a) Use the model to approximate the length of a $150,000 mortgage at 8% when the monthly payment is $1100.65 and when the monthly payment is $1254.68.

(b) Approximate the total amount paid over the term of the mortgage with a monthly payment of $1100.65 and with a monthly payment of $1254.68.

(c) Approximate the total interest charge for a monthly payment of $1100.65 and for a monthly payment of $1254.68.

(d) What is the vertical asymptote for the model? Interpret its meaning in the context of the problem.

58. *Compound Interest* A principal P, invested at $9\frac{1}{2}\%$ and compounded continuously, increases to an amount K times the original principal after t years, where t is given by

$$t = \frac{\ln K}{0.095}.$$

(a) Complete the table and interpret your results.

K	1	2	4	6	8	10	12
t							

(b) Sketch a graph of the function.

59. *Population* The time t in years for the world population to double if it is increasing at a continuous rate of r is given by

$$t = \frac{\ln 2}{r}.$$

(a) Complete the table.

r	0.005	0.01	0.015	0.02	0.025	0.03
t						

(b) Sketch a graph of the function.

(c) Use a reference source to decide which value of r best approximates the actual rate of growth for the world population.

60. *Sound Intensity* The relationship between the number of decibels β and the intensity of a sound I in watts per square meter is

$$\beta = 10 \log_{10}\left(\frac{I}{10^{-12}}\right).$$

(a) Determine the number of decibels of a sound with an intensity of 1 watt per square meter.

(b) Determine the number of decibels of a sound with an intensity of 10^{-2} watt per square meter.

(c) The intensity of the sound in part (a) is 100 times as great as that in part (b). Is the number of decibels 100 times as great? Explain.

61. *Human Memory Model*   Students in a mathematics class were given an exam and then retested monthly with an equivalent exam. The average scores for the class are given by the human memory model

$$f(t) = 80 - 17 \log_{10}(t + 1), \quad 0 \le t \le 12$$

where t is the time in months.

 (a) Use a graphing utility to graph the model over the specified domain.

(b) What was the average score on the original exam ($t = 0$)?

(c) What was the average score after 4 months?

(d) What was the average score after 10 months?

62. (a) Complete the table for the function

$$f(x) = \frac{\ln x}{x}.$$

x	1	5	10	10^2	10^4	10^6
$f(x)$						

(b) Use the table in part (a) to determine what value $f(x)$ approaches as x increases without bound.

(c) Use a graphing utility to confirm the result of part (b).

Synthesis

True or False?   **In Exercises 63 and 64, determine whether the statement is true or false. Justify your answer.**

63. You can determine the graph of $f(x) = \log_6 x$ by graphing $g(x) = 6^x$ and reflecting it about the x-axis.

64. The graph of $f(x) = \log_3 x$ contains the point $(27, 3)$.

In Exercises 65–68, describe the relationship between the graphs of f and g. What is the relationship between the functions f and g?

65. $f(x) = 3^x$

$g(x) = \log_3 x$

66. $f(x) = 5^x$

$g(x) = \log_5 x$

67. $f(x) = e^x$

$g(x) = \ln x$

68. $f(x) = 10^x$

$g(x) = \log_{10} x$

69. *Graphical Analysis*   Use a graphing utility to graph f and g in the same viewing window and determine which is increasing at the greater rate as x approaches $+\infty$. What can you conclude about the rate of growth of the natural logarithmic function?

(a) $f(x) = \ln x, \quad g(x) = \sqrt{x}$

(b) $f(x) = \ln x, \quad g(x) = \sqrt[4]{x}$

70. *Think About It*   The table of values was obtained by evaluating a function. Determine which of the statements may be true and which must be false.

x	y
1	0
2	1
8	3

(a) y is an exponential function of x.

(b) y is a logarithmic function of x.

(c) x is an exponential function of y.

(d) y is a linear function of x.

In Exercises 71–73, answer the question for the function $f(x) = \log_{10} x$. Do not use a calculator.

71. What is the domain of f?

72. What is f^{-1}?

73. If x is a real number between 1000 and 10,000, in which interval will $f(x)$ be found?

74. *Writing*   Explain why $\log_a x$ is defined only for $0 < a < 1$ and $a > 1$.

 In Exercises 75 and 76, (a) use a graphing utility to graph the function, (b) use the graph to determine the intervals in which the function is increasing and decreasing, and (c) approximate any relative maximum or minimum values of the function.

75. $f(x) = |\ln x|$ **76.** $h(x) = \ln(x^2 + 1)$

Review

In Exercises 77 and 78, translate the statement into an algebraic expression.

77. The total cost for auto repairs if the cost of parts was $83.95 and there were t hours of labor at $37.50 per hour

78. The area of a rectangle if the length is 10 units more than the width w

5.3 Properties of Logarithms

▶ What you should learn

- How to use the change-of-base formula to rewrite and evaluate logarithmic expressions
- How to use properties of logarithms to evaluate or rewrite logarithmic expressions
- How to use properties of logarithms to expand or condense logarithmic expressions
- How to use logarithmic functions to model and solve real-life applications

▶ Why you should learn it

Logarithmic functions are often used to model scientific observations. For instance, in Exercise 81 on page 390, a logarithmic function is used to model human memory.

Gary Conner/PhotoEdit

Change of Base

Most calculators have only two types of log keys, one for common logarithms (base 10) and one for natural logarithms (base e). Although common logs and natural logs are the most frequently used, you may occasionally need to evaluate logarithms to other bases. To do this, you can use the following **change-of-base formula.**

Change-of-Base Formula

Let a, b, and x be positive real numbers such that $a \neq 1$ and $b \neq 1$. Then $\log_a x$ can be converted to a different base as follows.

Base b	*Base 10*	*Base e*
$\log_a x = \dfrac{\log_b x}{\log_b a}$	$\log_a x = \dfrac{\log_{10} x}{\log_{10} a}$	$\log_a x = \dfrac{\ln x}{\ln a}$

One way to look at the change-of-base formula is that logarithms to base a are simply *constant multiples* of logarithms to base b. The constant multiplier is $1/(\log_b a)$.

Example 1 ▶ **Changing Bases Using Common Logarithms**

a. $\log_4 30 = \dfrac{\log_{10} 30}{\log_{10} 4}$ $\log_a x = \dfrac{\log_{10} x}{\log_{10} a}$

$\approx \dfrac{1.47712}{0.60206}$ Use a calculator.

≈ 2.4534 Simplify.

b. $\log_2 14 = \dfrac{\log_{10} 14}{\log_{10} 2} \approx \dfrac{1.14613}{0.30103} \approx 3.8074$

Example 2 ▶ **Changing Bases Using Natural Logarithms**

a. $\log_4 30 = \dfrac{\ln 30}{\ln 4}$ $\log_a x = \dfrac{\ln x}{\ln a}$

$\approx \dfrac{3.40120}{1.38629}$ Use a calculator.

≈ 2.4535 Simplify.

b. $\log_2 14 = \dfrac{\ln 14}{\ln 2} \approx \dfrac{2.63906}{0.69315} \approx 3.8073$

Encourage your students to know these properties well. They will be used for solving logarithmic and exponential equations, as well as in calculus.

Properties of Logarithms

You know from the preceding section that the logarithmic function with base a is the *inverse function* of the exponential function with base a. So, it makes sense that the properties of exponents should have corresponding properties involving logarithms. For instance, the exponential property $a^0 = 1$ has the corresponding logarithmic property $\log_a 1 = 0$.

Properties of Logarithms

Let a be a positive number such that $a \neq 1$, and let n be a real number. If u and v are positive real numbers, the following properties are true.

Logarithm with Base a	*Natural Logarithm*
1. $\log_a(uv) = \log_a u + \log_a v$	**1.** $\ln(uv) = \ln u + \ln v$
2. $\log_a \dfrac{u}{v} = \log_a u - \log_a v$	**2.** $\ln \dfrac{u}{v} = \ln u - \ln v$
3. $\log_a u^n = n \log_a u$	**3.** $\ln u^n = n \ln u$

Remind your students to note the domain when applying properties of logarithms to a logarithmic function. For example, the domain of $f(x) = \ln x^2$ is all real $x \neq 0$, whereas the domain of $g(x) = 2 \ln x$ is all real $x > 0$.

For a proof of the properties listed above, see Proofs in Mathematics on page 421.

Example 3 ▶ Using Properties of Logarithms

Write each logarithm in terms of $\ln 2$ and $\ln 3$.

a. $\ln 6$ **b.** $\ln \dfrac{2}{27}$

Solution

a. $\ln 6 = \ln(2 \cdot 3)$ Rewrite 6 as $2 \cdot 3$.

$\qquad = \ln 2 + \ln 3$ Property 1

b. $\ln \dfrac{2}{27} = \ln 2 - \ln 27$ Property 2

$\qquad = \ln 2 - \ln 3^3$ Rewrite 27 as 3^3.

$\qquad = \ln 2 - 3 \ln 3$ Property 3

Example 4 ▶ Using Properties of Logarithms

Use the properties of logarithms to verify that $-\log_{10} \frac{1}{100} = \log_{10} 100$.

Solution

$-\log_{10} \frac{1}{100} = -\log_{10}(100^{-1})$ Rewrite $\frac{1}{100}$ as 100^{-1}.

$\qquad = -(-1)\log_{10} 100$ Property 3

$\qquad = \log_{10} 100$ Simplify.

Try checking this result on your calculator.

The Granger Collection

Historical Note
John Napier, a Scottish mathematician, developed logarithms as a way to simplify some of the tedious calculations of his day. Beginning in 1594, Napier worked about 20 years on the invention of logarithms. Napier was only partially successful in his quest to simplify tedious calculations. Nonetheless, the development of logarithms was a step forward and received immediate recognition.

Rewriting Logarithmic Expressions

The properties of logarithms are useful for rewriting logarithmic expressions in forms that simplify the operations of algebra. This is true because these properties convert complicated products, quotients, and exponential forms into simpler sums, differences, and products, respectively.

A common error made in expanding logarithmic expressions is to rewrite $\log ax^n$ as $n \log ax$ instead of as $\log a + n \log x$.

Example 5 ▶ **Expanding Logarithmic Expressions**

Expand each logarithmic expression.

a. $\log_4 5x^3y$ **b.** $\ln \dfrac{\sqrt{3x-5}}{7}$

Solution

a. $\log_4 5x^3y = \log_4 5 + \log_4 x^3 + \log_4 y$ Property 1

$= \log_4 5 + 3 \log_4 x + \log_4 y$ Property 3

b. $\ln \dfrac{\sqrt{3x-5}}{7} = \ln \dfrac{(3x-5)^{1/2}}{7}$ Rewrite using rational exponent.

$= \ln(3x-5)^{1/2} - \ln 7$ Property 2

$= \dfrac{1}{2} \ln(3x-5) - \ln 7$ Property 3

In Example 5, the properties of logarithms were used to *expand* logarithmic expressions. In Example 6, this procedure is reversed and the properties of logarithms are used to *condense* logarithmic expressions.

A common error made in condensing logarithmic expressions is to rewrite $\log x - \log y$ as $\dfrac{\log x}{\log y}$ instead of as $\log \dfrac{x}{y}$.

Example 6 ▶ **Condensing Logarithmic Expressions**

Condense each logarithmic expression.

a. $\dfrac{1}{2} \log_{10} x + 3 \log_{10}(x+1)$ **b.** $2 \ln(x+2) - \ln x$

c. $\dfrac{1}{3}[\log_2 x + \log_2(x+1)]$

Solution

a. $\dfrac{1}{2} \log_{10} x + 3 \log_{10}(x+1) = \log_{10} x^{1/2} + \log_{10}(x+1)^3$ Property 3

$= \log_{10}\left[\sqrt{x}(x+1)^3\right]$ Property 1

b. $2 \ln(x+2) - \ln x = \ln(x+2)^2 - \ln x$ Property 3

$= \ln \dfrac{(x+2)^2}{x}$ Property 2

c. $\dfrac{1}{3}[\log_2 x + \log_2(x+1)] = \dfrac{1}{3}\{\log_2[x(x+1)]\}$ Property 1

$= \log_2[x(x+1)]^{1/3}$ Property 3

$= \log_2 \sqrt[3]{x(x+1)}$ Rewrite with a radical.

Application

One method of determining how the x- and y-values for a set of nonlinear data are related begins by taking the natural log of each of the x- and y-values. If the points are graphed and fall on a straight line, then you can determine that the x- and y-values are related by the equation

$$\ln y = m \ln x$$

where m is the slope of the straight line.

Example 7 ▶ Finding a Mathematical Model

The table shows the mean distance x and the period (the time it takes a planet to orbit the sun) y for each of the six planets that are closest to the sun. In the table, the mean distance is given in terms of astronomical units (where Earth's mean distance is defined as 1.0), and the period is given in terms of years. Find an equation that expresses y as a function of x.

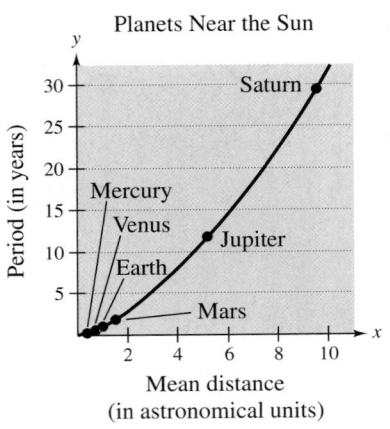

Planets Near the Sun

FIGURE 5.24

Planet	Mean distance, x	Period, y
Mercury	0.387	0.241
Venus	0.723	0.615
Earth	1.000	1.000
Mars	1.524	1.881
Jupiter	5.203	11.862
Saturn	9.555	29.458

Solution

The points in the table are plotted in Figure 5.24. From this figure it is not clear how to find an equation that relates y and x. To solve this problem, take the natural log of each of the x- and y-values in the table. This produces the following results.

Planet	$\ln x$	$\ln y$
Mercury	-0.949	-1.423
Venus	-0.324	-0.486
Earth	0.000	0.000
Mars	0.421	0.632
Jupiter	1.649	2.473
Saturn	2.257	3.383

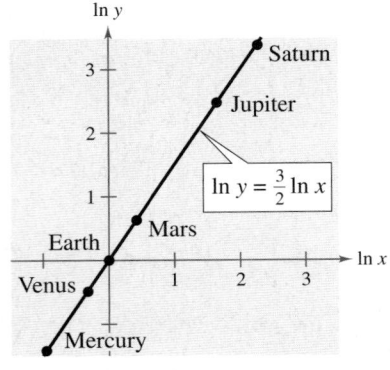

FIGURE 5.25

Now, by plotting the points in the second table, you can see that all six of the points appear to lie in a line (see Figure 5.25). You can use a graphical approach or an algebraic approach to find that the slope of this line is $\frac{3}{2}$. You can therefore conclude that $\ln y = \frac{3}{2}\ln x$.

5.3 Exercises

In Exercises 1–8, rewrite the logarithm as a ratio of (a) common logarithms and (b) natural logarithms.

1. $\log_5 x$

2. $\log_3 x$

3. $\log_{1/5} x$

4. $\log_{1/3} x$

5. $\log_x \frac{3}{10}$

6. $\log_x \frac{3}{4}$

7. $\log_{2.6} x$

8. $\log_{7.1} x$

In Exercises 9–16, evaluate the logarithm using the change-of-base formula. Round your result to three decimal places.

9. $\log_3 7$

10. $\log_7 4$

11. $\log_{1/2} 4$

12. $\log_{1/4} 5$

13. $\log_9 0.4$

14. $\log_{20} 0.125$

15. $\log_{15} 1250$

16. $\log_3 0.015$

In Exercises 17–38, use the properties of logarithms to expand the expression as a sum, difference, and/or constant multiple of logarithms. (Assume all variables are positive.)

17. $\log_4 5x$

18. $\log_3 10z$

19. $\log_8 x^4$

20. $\log_{10} \frac{y}{2}$

21. $\log_5 \frac{5}{x}$

22. $\log_6 \frac{1}{z^3}$

23. $\ln \sqrt{z}$

24. $\ln \sqrt[3]{t}$

25. $\ln xyz^2$

26. $\log_{10} 4x^2 y$

27. $\ln z(z-1)^2, \ z > 1$

28. $\ln\left(\frac{x^2-1}{x^3}\right), \ x > 1$

29. $\log_2 \frac{\sqrt{a-1}}{9}, \ a > 1$

30. $\ln \frac{6}{\sqrt{x^2+1}}$

31. $\ln \sqrt[3]{\frac{x}{y}}$

32. $\ln \sqrt{\frac{x^2}{y^3}}$

33. $\ln \frac{x^4\sqrt{y}}{z^5}$

34. $\log_2 \frac{\sqrt{x}\,y^4}{z^4}$

35. $\log_5 \frac{x^2}{y^2 z^3}$

36. $\log_{10} \frac{xy^4}{z^5}$

37. $\ln \sqrt[4]{x^3(x^2+3)}$

38. $\ln \sqrt{x^2(x+2)}$

In Exercises 39–56, condense the expression to the logarithm of a single quantity.

39. $\ln x + \ln 3$

40. $\ln y + \ln t$

41. $\log_4 z - \log_4 y$

42. $\log_5 8 - \log_5 t$

43. $2 \log_2(x+4)$

44. $\frac{2}{3} \log_7(z-2)$

45. $\frac{1}{4} \log_3 5x$

46. $-4 \log_6 2x$

47. $\ln x - 3 \ln(x+1)$

48. $2 \ln 8 + 5 \ln (z-4)$

49. $\log_{10} x - 2 \log_{10} y + 3 \log_{10} z$

50. $3 \log_3 x + 4 \log_3 y - 4 \log_3 z$

51. $\ln x - 4[\ln(x+2) + \ln(x-2)]$

52. $4[\ln z + \ln(z+5)] - 2 \ln(z-5)$

53. $\frac{1}{3}[2 \ln(x+3) + \ln x - \ln(x^2-1)]$

54. $2[3 \ln x - \ln(x+1) - \ln(x-1)]$

55. $\frac{1}{3}[\log_8 y + 2 \log_8(y+4)] - \log_8(y-1)$

56. $\frac{1}{2}[\log_4(x+1) + 2 \log_4(x-1)] + 6 \log_4 x$

In Exercises 57 and 58, compare the logarithmic quantities. If two are equal, explain why.

57. $\dfrac{\log_2 32}{\log_2 4}$, $\quad \log_2 \dfrac{32}{4}$, $\quad \log_2 32 - \log_2 4$

58. $\log_7 \sqrt{70}$, $\quad \log_7 35$, $\quad \frac{1}{2} + \log_7 \sqrt{10}$

In Exercises 59–74, find the exact value of the logarithm without using a calculator. (If this is not possible, state the reason.)

59. $\log_3 9$

60. $\log_5 \frac{1}{125}$

61. $\log_2 \sqrt[4]{8}$

62. $\log_6 \sqrt[3]{6}$

63. $\log_4 16^{1.2}$

64. $\log_3 81^{-0.2}$

65. $\log_3(-9)$

66. $\log_2(-16)$

67. $\ln e^{4.5}$

68. $3 \ln e^4$

69. $\ln \frac{1}{\sqrt{e}}$

70. $\ln \sqrt[4]{e^3}$

71. $\ln e^2 + \ln e^5$

72. $2 \ln e^6 - \ln e^5$

73. $\log_5 75 - \log_5 3$

74. $\log_4 2 + \log_4 32$

In Exercises 75–80, use the properties of logarithms to rewrite and simplify the logarithmic expression.

75. $\log_4 8$

76. $\log_2(4^2 \cdot 3^4)$

77. $\log_5 \frac{1}{250}$

78. $\log_{10} \frac{9}{300}$

79. $\ln(5e^6)$

80. $\ln \dfrac{6}{e^2}$

▶ Model It

81. *Human Memory Model* Students participating in a psychology experiment attended several lectures and were given an exam. Every month for a year after the exam, the students were retested to see how much of the material they remembered. The average scores for the group can be modeled by the human memory model

$$f(t) = 90 - 15 \log_{10}(t + 1), \qquad 0 \le t \le 12$$

where t is the time in months.

(a) What was the average score on the original exam ($t = 0$)?

(b) What was the average score after 6 months?

(c) What was the average score after 12 months?

(d) When will the average score decrease to 75?

(e) Use the properties of logarithms to write the function in another form.

(f) Sketch the graph of the function over the specified domain.

---

82. *Sound Intensity* The relationship between the number of decibels β and the intensity of a sound I in watts per square meter is

$$\beta = 10 \log_{10}\left(\frac{I}{10^{-12}}\right).$$

Use the properties of logarithms to write the formula in simpler form, and determine the number of decibels of a sound with an intensity of 10^{-6} watt per square meter.

Synthesis

True or False? **In Exercises 83–88, determine whether the statement is true or false given that $f(x) = \ln x$. Justify your answer.**

83. $f(0) = 0$

84. $f(ax) = f(a) + f(x), \qquad a > 0, x > 0$

85. $f(x - 2) = f(x) - f(2), \qquad x > 2$

86. $\sqrt{f(x)} = \frac{1}{2}f(x)$

87. If $f(u) = 2f(v)$, then $v = u^2$.

88. If $f(x) < 0$, then $0 < x < 1$.

89. *Proof* Prove that $\log_b \dfrac{u}{v} = \log_b u - \log_b v$.

90. *Proof* Prove that $\log_b u^n = n \log_b u$.

In Exercises 91–96, use the change-of-base formula to rewrite the logarithm as a ratio of logarithms. Then use a graphing utility to graph both functions in the same viewing window to verify that the functions are equivalent.

91. $f(x) = \log_2 x$

92. $f(x) = \log_4 x$

93. $f(x) = \log_{1/2} x$

94. $f(x) = \log_{1/4} x$

95. $f(x) = \log_{11.8} x$

96. $f(x) = \log_{12.4} x$

97. *Think About It* Consider the functions below.

$$f(x) = \ln \frac{x}{2}, \quad g(x) = \frac{\ln x}{\ln 2}, \quad h(x) = \ln x - \ln 2$$

Which two functions should have identical graphs? Verify your answer by sketching the graphs of all three functions on the same set of coordinate axes.

98. *Exploration* For how many integers between 1 and 20 can the natural logarithms be approximated given that $\ln 2 \approx 0.6931$, $\ln 3 \approx 1.0986$, and $\ln 5 \approx 1.6094$? Approximate these logarithms (do not use a calculator).

Review

In Exercises 99–102, simplify the expression.

99. $\dfrac{24xy^{-2}}{16x^{-3}y}$

100. $\left(\dfrac{2x^2}{3y}\right)^{-3}$

101. $(18x^3y^4)^{-3}(18x^3y^4)^3$

102. $xy(x^{-1} + y^{-1})^{-1}$

In Exercises 103–106, solve the equation.

103. $3x^2 + 2x - 1 = 0$

104. $4x^2 - 5x + 1 = 0$

105. $\dfrac{2}{3x + 1} = \dfrac{x}{4}$

106. $\dfrac{5}{x - 1} = \dfrac{2x}{3}$

▶ What you should learn

- How to solve simple exponential and logarithmic equations
- How to solve more complicated exponential equations
- How to solve more complicated logarithmic equations
- How to use exponential and logarithmic equations to model and solve real-life applications

▶ Why you should learn it

Applications of exponential and logarithmic equations are found in consumer safety testing. For instance, in Exercise 119, on page 400, a logarithmic function is used to model crumple zones for automobile crash tests.

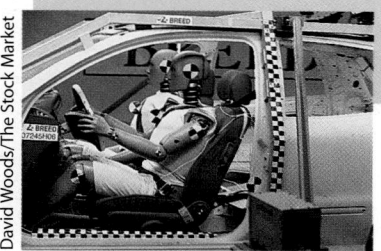

David Woods/The Stock Market

Introduction

So far in this chapter, you have studied the definitions, graphs, and properties of exponential and logarithmic functions. In this section, you will study procedures for *solving equations* involving these exponential and logarithmic functions.

There are two basic strategies for solving exponential or logarithmic equations. The first is based on the One-to-One Properties and the second is based on the Inverse Properties. For $a > 0$ and $a \neq 1$, the following properties are true for all x and y for which $\log_a x$ and $\log_a y$ are defined.

One-to-One Properties

$a^x = a^y$ if and only if $x = y$.

$\log_a x = \log_a y$ if and only if $x = y$.

Inverse Properties

$a^{\log_a x} = x$

$\log_a a^x = x$

Example 1 ▶ **Solving Simple Equations**

	Original Equation	Rewritten Equation	Solution	Property
a.	$2^x = 32$	$2^x = 2^5$	$x = 5$	One-to-One
b.	$\ln x - \ln 3 = 0$	$\ln x = \ln 3$	$x = 3$	One-to-One
c.	$\left(\frac{1}{3}\right)^x = 9$	$3^{-x} = 3^2$	$x = -2$	One-to-One
d.	$e^x = 7$	$\ln e^x = \ln 7$	$x = \ln 7$	Inverse
e.	$\ln x = -3$	$e^{\ln x} = e^{-3}$	$x = e^{-3}$	Inverse
f.	$\log_{10} x = -1$	$10^{\log_{10} x} = 10^{-1}$	$x = 10^{-1} = \frac{1}{10}$	Inverse

---

The strategies used in Example 1 are summarized as follows.

Strategies for Solving Exponential and Logarithmic Equations

1. Rewrite the original equation in a form that allows the use of the One-to-One Properties of exponential or logarithmic functions.

2. Rewrite an *exponential* equation in logarithmic form and apply the Inverse Property of logarithmic functions.

3. Rewrite a *logarithmic* equation in exponential form and apply the Inverse Property of exponential functions.

Solving Exponential Equations

Example 2 ▶ Solving Exponential Equations

Solve each equation and approximate the result to three decimal places.

a. $4^x = 72$ **b.** $3(2^x) = 42$

Solution

a.

$4^x = 72$	Write original equation.
$\log_4 4^x = \log_4 72$	Take logarithm (base 4) of each side.
$x = \log_4 72$	Inverse Property
$x = \dfrac{\ln 72}{\ln 4}$	Change-of-base formula
$x \approx 3.085$	Use a calculator.

The solution is $x = \log_4 72 \approx 3.085$. Check this in the original equation.

b.

$3(2^x) = 42$	Write original equation.
$2^x = 14$	Divide each side by 3.
$\log_2 2^x = \log_2 14$	Take log (base 2) of each side.
$x = \log_2 14$	Inverse Property
$x = \dfrac{\ln 14}{\ln 2}$	Change-of-base formula
$x \approx 3.807$	Use a calculator.

The solution is $x = \log_2 14 \approx 3.807$. Check this in the original equation.

In Example 2(a), the exact solution is $x = \log_4 72$ and the approximate solution is $x \approx 3.085$. An exact answer is preferred when the solution is an intermediate step in a larger problem. For a final answer, an approximate solution is easier to comprehend.

Example 3 ▶ Solving an Exponential Equation

Solve $e^x + 5 = 60$ and approximate the result to three decimal places.

Solution

$e^x + 5 = 60$	Write original equation.
$e^x = 55$	Subtract 5 from each side.
$\ln e^x = \ln 55$	Take natural log of each side.
$x = \ln 55$	Inverse Property
$x \approx 4.007$	Use a calculator.

The solution is $x = \ln 55 \approx 4.007$. Check this in the original equation.

Technology

When solving an exponential or logarithmic equation, remember that you can check your solution graphically by "graphing the left and right sides separately" and using the *intersect* feature of your graphing utility to determine the point of intersection. For instance, to check the solution of the equation in Example 2(a), you can graph

$$y_1 = 4^x \quad \text{and} \quad y_2 = 72$$

in the same viewing window, as shown below. Using the *intersect* feature of your graphing utility, you can determine that the graphs intersect when $x \approx 3.085$, which confirms the solution found in Example 2(a).

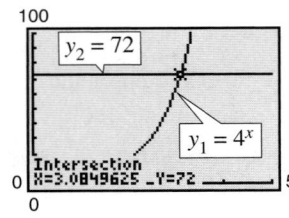

Additional Example

Additional Example

$2^x = 10$

$\ln 2^x = \ln 10$

$x \ln 2 = \ln 10$

$x = \dfrac{\ln 10}{\ln 2} \approx 3.322$

Note: Using the change-of-base formula or the definition of a logarithmic function, you could write this solution as $x = \log_2 10$.

Example 4 ▶ Solving an Exponential Equation

Solve $2(3^{2t-5}) - 4 = 11$ and approximate the result to three decimal places.

Solution

$2(3^{2t-5}) - 4 = 11$	Write original equation.
$2(3^{2t-5}) = 15$	Add 4 to each side.
$3^{2t-5} = \dfrac{15}{2}$	Divide each side by 2.
$\log_3 3^{2t-5} = \log_3 \dfrac{15}{2}$	Take log (base 3) of each side.
$2t - 5 = \log_3 \dfrac{15}{2}$	Inverse Property
$2t = 5 + \log_3 7.5$	Add 5 to each side.
$t = \dfrac{5}{2} + \dfrac{1}{2} \log_3 7.5$	Divide each side by 2.
$t \approx 3.417$	Use a calculator.

The solution is $t = \frac{5}{2} + \frac{1}{2} \log_3 7.5 \approx 3.417$. Check this in the original equation.

STUDY TIP

Remember that to evaluate a logarithm such as $\log_3 7.5$, you need to use the change-of-base formula.

$$\log_3 7.5 = \dfrac{\ln 7.5}{\ln 3} \approx 1.834$$

When an equation involves two or more exponential expressions, you can still use a procedure similar to that demonstrated in Examples 2, 3, and 4. However, the algebra is a bit more complicated.

Example 5 ▶ Solving an Exponential Equation of Quadratic Type

Solve $e^{2x} - 3e^x + 2 = 0$.

Solution

$e^{2x} - 3e^x + 2 = 0$	Write original equation.
$(e^x)^2 - 3e^x + 2 = 0$	Write in quadratic form.
$(e^x - 2)(e^x - 1) = 0$	Factor.
$e^x - 2 = 0$	Set 1st factor equal to 0.
$x = \ln 2$	Solution
$e^x - 1 = 0$	Set 2nd factor equal to 0.
$x = 0$	Solution

The solutions are $x = \ln 2$ and $x = 0$. Check these in the original equation.

To ensure that students first solve for the unknown variable algebraically and then use their calculators, you can require both exact algebraic solutions and approximate numerical answers.

In Example 5, use a graphing utility to graph $y = e^{2x} - 3e^x + 2$. The graph should have two x-intercepts: one at $x = \ln 2 \approx 0.693$ and one at $x = 0$.

Solving Logarithmic Equations

To solve a logarithmic equation such as

$$\ln x = 3 \qquad \text{Logarithmic form}$$

write the equation in exponential form as follows.

$$e^{\ln x} = e^3 \qquad \text{Exponentiate each side.}$$

$$x = e^3 \qquad \text{Exponential form}$$

This procedure is called *exponentiating* each side of an equation.

Example 6 ▶ Solving a Logarithmic Equation

a. Solve $\ln x = 2$.

b. Solve $\log_3(5x - 1) = \log_3(x + 7)$.

Solution

a.

$\ln x = 2$	Write original equation.
$e^{\ln x} = e^2$	Exponentiate each side.
$x = e^2$	Inverse Property

The solution is $x = e^2$. Check this in the original equation.

b.

$\log_3(5x - 1) = \log_3(x + 7)$	Write original equation.
$5x - 1 = x + 7$	One-to-One Property
$4x = 8$	Add $-x$ and 1 to each side.
$x = 2$	Divide each side by 4.

The solution is $x = 2$. Check this in the original equation.

Example 7 ▶ Solving a Logarithmic Equation

Solve $5 + 2 \ln x = 4$ and approximate the result to three decimal places.

Solution

$5 + 2 \ln x = 4$	Write original equation.
$2 \ln x = -1$	Subtract 5 from each side.
$\ln x = -\dfrac{1}{2}$	Divide each side by 2.
$e^{\ln x} = e^{-1/2}$	Exponentiate each side.
$x = e^{-1/2}$	Inverse Property
$x \approx 0.607$	Use a calculator.

The solution is $x = e^{-1/2} \approx 0.607$. Check this in the original equation.

Example 8 ▶ **Solving a Logarithmic Equation**

Solve $2 \log_5 3x = 4$.

Solution

$2 \log_5 3x = 4$	Write original equation.
$\log_5 3x = 2$	Divide each side by 2.
$5^{\log_5 3x} = 5^2$	Exponentiate each side (base 5).
$3x = 25$	Inverse Property
$x = \dfrac{25}{3}$	Divide each side by 3.

The solution is $x = \frac{25}{3}$. Check this in the original equation.

STUDY TIP

Notice in Example 9 that the logarithmic part of the equation is condensed into a single logarithm before exponentiating each side of the equation.

Because the domain of a logarithmic function generally does not include all real numbers, you should be sure to check for extraneous solutions of logarithmic equations.

Example 9 ▶ **Checking for Extraneous Solutions**

Solve $\log_{10} 5x + \log_{10}(x - 1) = 2$.

Solution

$\log_{10} 5x + \log_{10}(x - 1) = 2$	Write original equation.
$\log_{10}[5x(x - 1)] = 2$	Product Property of Logarithms
$10^{\log_{10}(5x^2 - 5x)} = 10^2$	Exponentiate each side (base 10).
$5x^2 - 5x = 100$	Inverse Property
$x^2 - x - 20 = 0$	Write in general form.
$(x - 5)(x + 4) = 0$	Factor.
$x - 5 = 0$	Set 1st factor equal to 0.
$x = 5$	Solution
$x + 4 = 0$	Set 2nd factor equal to 0.
$x = -4$	Solution

The solutions appear to be $x = 5$ and $x = -4$. However, when you check these in the original equation, you can see that $x = 5$ is the only solution.

Technology

You can use a graphing utility to verify that the equation in Example 9 has $x = 5$ as its only solution. Graph

$$y_1 = \log_{10} 5x + \log_{10}(x - 1)$$

and

$$y_2 = 2$$

in the same viewing window. From the graph shown below, it appears that the graphs of the two equations intersect at one point. Use the *intersect* feature or the *zoom* and *trace* features to determine that $x = 5$ is an approximate solution. You can verify this algebraically by substituting $x = 5$ into the original equation.

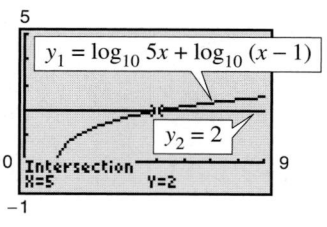

In Example 9, the domain of $\log_{10} 5x$ is $x > 0$ and the domain of $\log_{10}(x - 1)$ is $x > 1$, so the domain of the original equation is $x > 1$. Because the domain is all real numbers greater than 1, the solution $x = -4$ is extraneous.

Applications

Example 10 ▶	Doubling an Investment	

You have deposited $500 in an account that pays 6.75% interest, compounded continuously. How long will it take your money to double?

Solution

Using the formula for continuous compounding, you can find that the balance in the account is

$$A = Pe^{rt}$$

$$A = 500e^{0.0675t}.$$

To find the time required for the balance to double, let $A = 1000$ and solve the resulting equation for t.

$500e^{0.0675t} = 1000$	Let $A = 1000$.
$e^{0.0675t} = 2$	Divide each side by 500.
$\ln e^{0.0675t} = \ln 2$	Take natural log of each side.
$0.0675t = \ln 2$	Inverse Property
$t = \dfrac{\ln 2}{0.0675}$	Divide each side by 0.0675.
$t \approx 10.27$	Use a calculator.

The balance in the account will double after approximately 10.27 years. This result is demonstrated graphically in Figure 5.26.

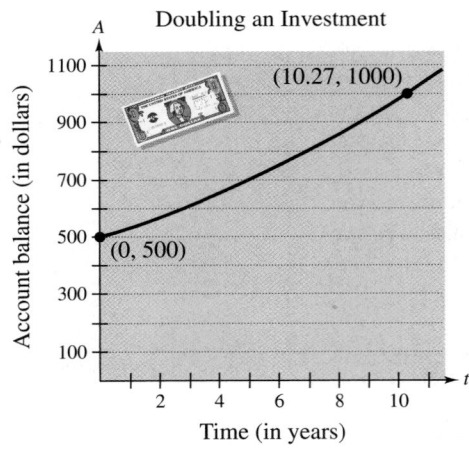

FIGURE **5.26**

In Example 10, an approximate answer of 10.27 years is given. Within the context of the problem, the exact solution, $(\ln 2)/0.0675$ years, does not make sense as an answer.

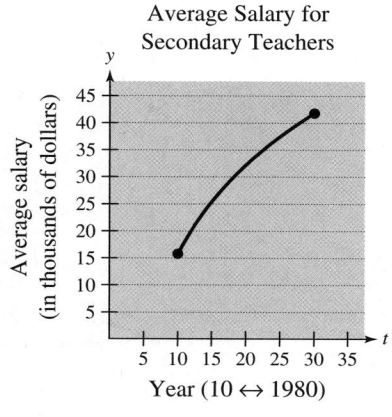

Average Salary for Secondary Teachers

FIGURE 5.27

Example 11 ▶ **Average Salary for Secondary Teachers**

For selected years from 1980 to 2000, the average salary for secondary teachers y (in thousands of dollars) for the year t can be modeled by the equation

$$y = -38.8 + 23.7 \ln t, \qquad 10 \le t \le 30$$

where $t = 10$ represents 1980 (see Figure 5.27). During which year did the average salary for secondary teachers reach 2.5 times its 1980 level of $16.5 thousand? (Source: National Education Association)

Solution

$-38.8 + 23.7 \ln t = y$	Write original equation.
$-38.8 + 23.7 \ln t = 41.25$	Let $y = (2.5)(16.5) = 41.25$.
$23.7 \ln t = 80.05$	Add 38.8 to each side.
$\ln t \approx 3.378$	Divide each side by 23.7.
$e^{\ln t} \approx e^{3.378}$	Exponentiate each side.
$t \approx e^{3.378}$	Inverse Property
$t \approx 29$	Use a calculator.

The solution is $t \approx 29$ years. Because $t = 10$ represents 1980, it follows that the average salary for secondary teachers reached 2.5 times its 1980 level in 1999.

Writing ABOUT MATHEMATICS

Comparing Mathematical Models The table shows the U.S. Postal Service rates y for sending an express mail package for selected years from 1985 through 2001, where $x = 5$ represents 1985. (Source: U.S. Postal Service)

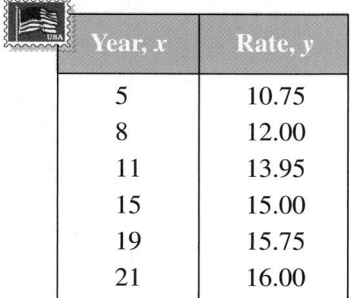

Year, x	Rate, y
5	10.75
8	12.00
11	13.95
15	15.00
19	15.75
21	16.00

a. Create a scatter plot of the data. Find a linear model for the data, and add its graph to your scatter plot. According to this model, when will the rate for sending an express mail package reach $17.50?

b. Create a new table showing values for $\ln x$ and $\ln y$ and create a scatter plot of this transformed data. Use the method illustrated in Example 7 in Section 5.3 to find a model for the transformed data, and add its graph to your scatter plot. According to this model, when will the rate for sending an express mail package reach $17.50?

c. Solve the model in part (b) for y, and add its graph to your scatter plot in part (a). Which model better fits the original data? Which model will better predict future shipments? Explain.

5.4 Exercises

In Exercises 1–6, determine whether each x-value is a solution (or an approximate solution) of the equation.

1. $4^{2x-7} = 64$

   (a) $x = 5$

   (b) $x = 2$

2. $2^{3x+1} = 32$

   (a) $x = -1$

   (b) $x = 2$

3. $3e^{x+2} = 75$

   (a) $x = -2 + e^{25}$

   (b) $x = -2 + \ln 25$

   (c) $x \approx 1.219$

4. $5^{2x+3} = 812$

   (a) $x = -\frac{3}{2} + \log_5 \sqrt{812}$

   (b) $x \approx 0.581$

   (c) $x = -1.5 + \dfrac{\ln 812}{\ln 5}$

5. $\log_4(3x) = 3$

   (a) $x \approx 20.356$

   (b) $x = -4$

   (c) $x = \frac{64}{3}$

6. $\ln(x - 1) = 3.8$

   (a) $x = 1 + e^{3.8}$

   (b) $x \approx 45.701$

   (c) $x = 1 + \ln 3.8$

In Exercises 7–26, solve for x.

7. $4^x = 16$

8. $3^x = 243$

9. $5^x = 625$

10. $3^x = 729$

11. $7^x = \frac{1}{49}$

12. $8^x = 4$

13. $\left(\frac{1}{2}\right)^x = 32$

14. $\left(\frac{1}{4}\right)^x = 64$

15. $\left(\frac{3}{4}\right)^x = \frac{27}{64}$

16. $\left(\frac{2}{3}\right)^x = \frac{4}{9}$

17. $\ln x - \ln 2 = 0$

18. $\ln x - \ln 5 = 0$

19. $e^x = 2$

20. $e^x = 4$

21. $\ln x = -1$

22. $\ln x = -7$

23. $\log_4 x = 3$

24. $\log_5 x = -3$

25. $\log_{10} x = -1$

26. $\log_{10} x - 2 = 0$

In Exercises 27–30, approximate the point of intersection of the graphs of f and g. Then solve the equation $f(x) = g(x)$ algebraically.

27. $f(x) = 2^x$

   $g(x) = 8$

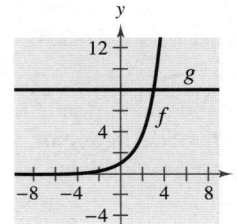

28. $f(x) = 27^x$

   $g(x) = 9$

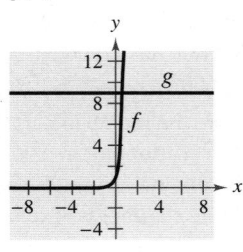

29. $f(x) = \log_3 x$

   $g(x) = 2$

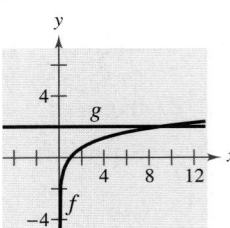

30. $f(x) = \ln(x - 4)$

   $g(x) = 0$

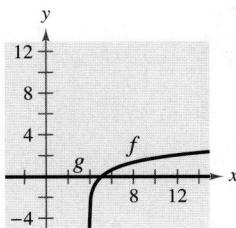

In Exercises 31–68, solve the exponential equation algebraically. Approximate the result to three decimal places.

31. $4(3^x) = 20$

32. $2(5^x) = 32$

33. $2e^x = 10$

34. $4e^x = 91$

35. $e^x - 9 = 19$

36. $6^x + 10 = 47$

37. $3^{2x} = 80$

38. $6^{5x} = 3000$

39. $5^{-t/2} = 0.20$

40. $4^{-3t} = 0.10$

41. $3^{x-1} = 27$

42. $2^{x-3} = 32$

43. $2^{3-x} = 565$

44. $8^{-2-x} = 431$

45. $8(10^{3x}) = 12$

46. $5(10^{x-6}) = 7$

47. $3(5^{x-1}) = 21$

48. $8(3^{6-x}) = 40$

49. $e^{3x} = 12$

50. $e^{2x} = 50$

51. $500e^{-x} = 300$

52. $1000e^{-4x} = 75$

53. $7 - 2e^x = 5$

54. $-14 + 3e^x = 11$

55. $6(2^{3x-1}) - 7 = 9$

56. $8(4^{6-2x}) + 13 = 41$

57. $e^{2x} - 4e^x - 5 = 0$

58. $e^{2x} - 5e^x + 6 = 0$

59. $e^{2x} - 3e^x - 4 = 0$

60. $e^{2x} + 9e^x + 36 = 0$

61. $\dfrac{500}{100 - e^{x/2}} = 20$

62. $\dfrac{400}{1 + e^{-x}} = 350$

63. $\dfrac{3000}{2 + e^{2x}} = 2$

64. $\dfrac{119}{e^{6x} - 14} = 7$

65. $\left(1 + \dfrac{0.065}{365}\right)^{365t} = 4$

66. $\left(4 - \dfrac{2.471}{40}\right)^{9t} = 21$

67. $\left(1 + \dfrac{0.10}{12}\right)^{12t} = 2$

68. $\left(16 - \dfrac{0.878}{26}\right)^{3t} = 30$

In Exercises 69–76, use a graphing utility to solve the equation. Approximate the result to three decimal places. Verify your result algebraically.

69. $6e^{1-x} = 25$ **70.** $-4e^{-x-1} + 15 = 0$

71. $3e^{3x/2} = 962$ **72.** $8e^{-2x/3} = 11$

73. $e^{0.09t} = 3$ **74.** $-e^{1.8x} + 7 = 0$

75. $e^{0.125t} - 8 = 0$ **76.** $e^{2.724x} = 29$

In Exercises 77–104, solve the logarithmic equation algebraically. Approximate the result to three decimal places.

77. $\ln x = -3$ **78.** $\ln x = 2$

79. $\ln 2x = 2.4$ **80.** $\ln 4x = 1$

81. $\log_{10} x = 6$ **82.** $\log_{10} 3z = 2$

83. $6 \log_3(0.5x) = 11$

84. $5 \log_{10}(x - 2) = 11$

85. $3 \ln 5x = 10$

86. $2 \ln x = 7$

87. $\ln \sqrt{x + 2} = 1$

88. $\ln \sqrt{x - 8} = 5$

89. $7 + 3 \ln x = 5$

90. $2 - 6 \ln x = 10$

91. $\ln x - \ln(x + 1) = 2$

92. $\ln x + \ln(x + 1) = 1$

93. $\ln x + \ln(x - 2) = 1$

94. $\ln x + \ln(x + 3) = 1$

95. $\ln(x + 5) = \ln(x - 1) - \ln(x + 1)$

96. $\ln(x + 1) - \ln(x - 2) = \ln x$

97. $\log_2(2x - 3) = \log_2(x + 4)$

98. $\log_{10}(x - 6) = \log_{10}(2x + 1)$

99. $\log_{10}(x + 4) - \log_{10} x = \log_{10}(x + 2)$

100. $\log_2 x + \log_2(x + 2) = \log_2(x + 6)$

101. $\log_4 x - \log_4(x - 1) = \frac{1}{2}$

102. $\log_3 x + \log_3(x - 8) = 2$

103. $\log_{10} 8x - \log_{10}\left(1 + \sqrt{x}\right) = 2$

104. $\log_{10} 4x - \log_{10}\left(12 + \sqrt{x}\right) = 2$

 In Exercises 105–108, use a graphing utility to solve the equation. Approximate the result to three decimal places. Verify your result algebraically.

105. $7 = 2^x$ **106.** $500 = 1500e^{-x/2}$

107. $3 - \ln x = 0$ **108.** $10 - 4 \ln(x - 2) = 0$

Compound Interest In Exercises 109 and 110, find the time required for a $1000 investment to double at interest rate r, compounded continuously.

109. $r = 0.085$

110. $r = 0.12$

Compound Interest In Exercises 111 and 112, find the time required for a $1000 investment to triple at interest rate r, compounded continuously.

111. $r = 0.085$

112. $r = 0.12$

113. *Demand* The demand equation for a microwave oven is

$$p = 500 - 0.5(e^{0.004x}).$$

Find the demand x for a price of (a) $p = \$350$ and (b) $p = \$300$.

114. *Demand* The demand equation for a hand-held electronic organizer is

$$p = 5000\left(1 - \frac{4}{4 + e^{-0.002x}}\right).$$

Find the demand x for a price of (a) $p = \$600$ and (b) $p = \$400$.

115. *Forest Yield* The yield V (in millions of cubic feet per acre) for a forest at age t years is

$$V = 6.7e^{-48.1/t}.$$

 (a) Use a graphing utility to graph the function.

(b) Determine the horizontal asymptote of the function. Interpret its meaning in the context of the problem.

(c) Find the time necessary to obtain a yield of 1.3 million cubic feet.

116. *Trees per Acre* The number of trees per acre N of a species is approximated by the model

$$N = 68(10^{-0.04x}), \qquad 5 \le x \le 40$$

where x is the average diameter of the trees 3 feet above the ground. Use the model to approximate the average diameter of the trees in a test plot when $N = 21$.

117. *Average Heights* The percent of American males between the ages of 18 and 24 who are no more than x inches tall is

$$m(x) = \frac{100}{1 + e^{-0.6114(x - 69.71)}}$$

and the percent of American females between the ages of 18 and 24 who are no more than x inches tall is

$$f(x) = \frac{100}{1 + e^{-0.66607(x - 64.51)}}$$

where m and f are the percents and x is the height in inches. (Source: U.S. National Center for Health Statistics)

(a) Use the graph to determine any horizontal asymptotes of the functions. Interpret the meaning in the context of the problem.

(b) What is the average height of each sex?

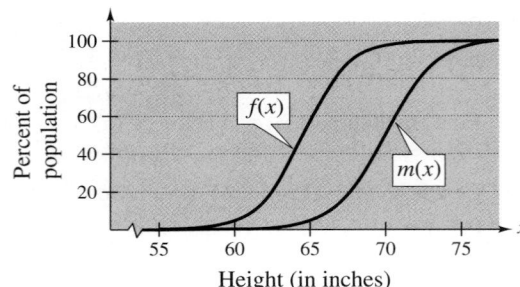

Height (in inches)

118. *Learning Curve* In a group project in learning theory, a mathematical model for the proportion P of correct responses after n trials was found to be

$$P = \frac{0.83}{1 + e^{-0.2n}}.$$

(a) Use a graphing utility to graph the function.

(b) Use the graph to determine any horizontal asymptotes of the function. Interpret the meaning of the upper asymptote in the context of this problem.

(c) After how many trials will 60% of the responses be correct?

▶ **Model It**

119. *Automobiles* Automobiles are designed with crumple zones that help protect their occupants in crashes. The crumple zones allow the occupants to move short distances when the automobiles come to abrupt stops. The greater the distance moved, the fewer g's the crash victims experience. (One g is equal to the acceleration due to gravity. For very short periods of time, humans have withstood as much as 40 g's.) In crash tests with vehicles moving at 90 kilometers per hour, analysts measured the numbers of g's experienced during deceleration by crash dummies that were permitted to move x meters during impact. The data is shown in the table.

x	g's
0.2	158
0.4	80
0.6	53
0.8	40
1.0	32

A model for this data is

$$y = -3.00 + 11.88 \ln x + \frac{36.94}{x}$$

where y is the number of g's.

(a) Complete the table using the model.

x	0.2	0.4	0.6	0.8	1.0
y					

 (b) Use a graphing utility to graph the data points and the model in the same viewing window. How do they compare?

(c) Use the model to estimate the distance traveled during impact if the passenger deceleration must not exceed 30 g's.

(d) Do you think it is practical to lower the number of g's experienced during impact to fewer than 23? Explain your reasoning.

120. *Data Analysis* An object at a temperature of 160°C was removed from a furnace and placed in a room at 20°C. The temperature T of the object was measured each hour h and recorded in the table. A model for this data is $T = 20[1 + 7(2^{-h})]$. The graph of this model is shown in the figure.

Hour, h	Temperature, T
0	160°
1	90°
2	56°
3	38°
4	29°
5	24°

(a) Use the graph to identify the horizontal asymptote of the model and interpret the asymptote in the context of the problem.

(b) Use the model to approximate the time when the temperature of the object was 100°C.

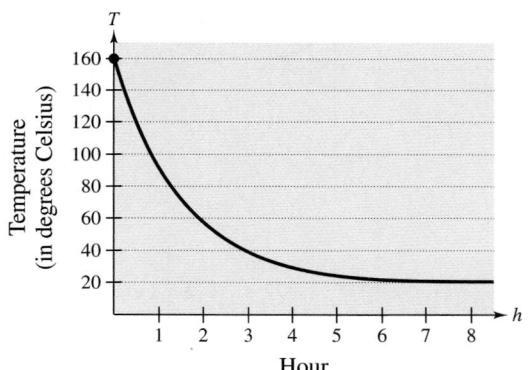

Hour

Synthesis

True or False? **In Exercises 121–124, rewrite each verbal statement as an equation. Then decide whether the statement is true or false. Justify your answer.**

121. The logarithm of the product of two numbers is equal to the sum of the logarithms of the numbers.

122. The logarithm of the sum of two numbers is equal to the product of the logarithms of the numbers.

123. The logarithm of the difference of two numbers is equal to the difference of the logarithms of the numbers.

124. The logarithm of the quotient of two numbers is equal to the difference of the logarithms of the numbers.

125. *Think About It* Is it possible for a logarithmic equation to have more than one extraneous solution? Explain.

126. *Finance* You are investing P dollars at an annual interest rate of r, compounded continuously, for t years. Which of the following would result in the highest value of the investment? Explain your reasoning.

(a) Double the amount you invest.

(b) Double your interest rate.

(c) Double the number of years.

127. *Think About It* Are the times required for the investments in Exercises 109 and 110 to quadruple twice as long as the times for them to double? Give a reason for your answer and verify your answer algebraically.

128. *Writing* Write two or three sentences stating the general guidelines that you follow when solving (a) exponential equations and (b) logarithmic equations.

Review

In Exercises 129–132, simplify the expression.

129. $\sqrt{48x^2y^5}$

130. $\sqrt{32} - 2\sqrt{25}$

131. $\sqrt[3]{25} \cdot \sqrt[3]{15}$

132. $\dfrac{3}{\sqrt{10} - 2}$

In Exercises 133–136, find a mathematical model for the verbal statement.

133. M varies directly as the cube of p.

134. t varies inversely as the cube of s.

135. d varies jointly as a and b.

136. x is inversely proportional to $b - 3$.

In Exercises 137–140, evaluate the logarithm using the change-of-base formula. Approximate your result to three decimal places.

137. $\log_6 9$

138. $\log_3 4$

139. $\log_{3/4} 5$

140. $\log_8 22$

5.5 Exponential and Logarithmic Models

Vittoriano Rastelli/Corbis

▶ What you should learn

- How to recognize the five most common types of models involving exponential and logarithmic functions
- How to use exponential growth and decay functions to model and solve real-life problems
- How to use Gaussian functions to model and solve real-life problems
- How to use logistic growth functions to model and solve real-life problems
- How to use logarithmic functions to model and solve real-life problems

▶ Why you should learn it

Exponential growth and decay models are often used to model the population of a country. For instance, in Exercise 36 on page 410, you will use exponential growth and decay models to compare the populations of several countries.

Introduction

The five most common types of mathematical models involving exponential functions and logarithmic functions are as follows.

1. **Exponential growth model:** $y = ae^{bx}, \quad b > 0$
2. **Exponential decay model:** $y = ae^{-bx}, \quad b > 0$
3. **Gaussian model:** $y = ae^{-(x-b)^2/c}$
4. **Logistic growth model:** $y = \dfrac{a}{1 + be^{-rx}}$
5. **Logarithmic models:** $y = a + b \ln x, \quad y = a + b \log_{10} x$

The graphs of the basic forms of these functions are shown in Figure 5.28.

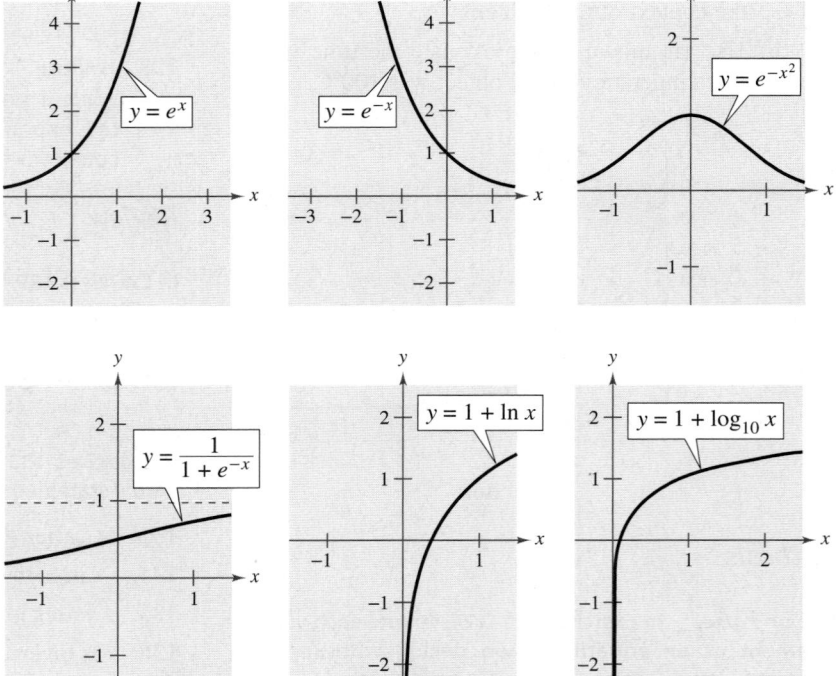

FIGURE 5.28

You can often gain quite a bit of insight into a situation modeled by an exponential or logarithmic function by identifying and interpreting the function's asymptotes. Use the graphs in Figure 5.28 to identify the asymptotes of each function.

This section shows students real-world applications for logarithmic and exponential functions.

Exponential Growth and Decay

Example 1 ▶ Population Increase

Estimates of the world population (in millions) from 1995 through 2003 are shown in the table. The scatter plot of the data is shown in Figure 5.29. (Source: U.S. Census Bureau)

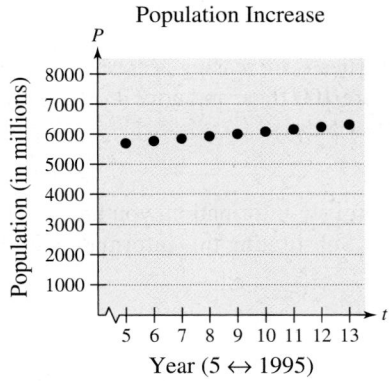

Population Increase

FIGURE 5.29

Year	Population
1995	5691
1996	5769
1997	5847
1998	5925
1999	6003

Year	Population
2000	6080
2001	6157
2002	6234
2003	6311

An exponential growth model that approximates this data is

$$P = 5340e^{0.012922t}, \quad 5 \le t \le 13$$

where P is the population (in millions) and $t = 5$ represents 1995. Compare the values given by the model with the estimates given by the U.S. Census Bureau. According to this model, when will the world population reach 6.8 billion?

Solution

The following table compares the two sets of population figures. The graph of the model is shown in Figure 5.30.

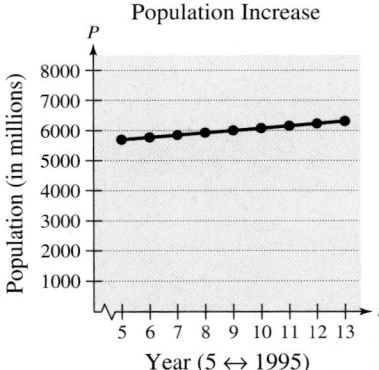

Population Increase

FIGURE 5.30

Year	1995	1996	1997	1998	1999	2000	2001	2002	2003
Population	5691	5769	5847	5925	6003	6080	6157	6234	6311
Model	5696	5771	5846	5922	5999	6077	6156	6236	6317

To find when the world population will reach 6.8 billion, let $P = 6800$ in the model and solve for t.

$5340e^{0.012922t} = P$	Write original model.
$5340e^{0.012922t} = 6800$	Let $P = 6800$.
$e^{0.012922t} \approx 1.27341$	Divide each side by 5340.
$\ln e^{0.012922t} \approx \ln 1.27341$	Take natural log of each side.
$0.012922t \approx 0.241698$	Inverse Property
$t \approx 18.7$	Divide each side by 0.012922.

According to the model, the world population will reach 6.8 billion in 2008.

Technology

Some graphing utilities have curve-fitting capabilities that can be used to find models that represent data. If you have such a graphing utility, try using it to find a model for the data given in Example 1. How does your model compare with the model given in Example 1?

Additional Example

Radioactive iodine is a by-product of some types of nuclear reactors. Its half-life is 60 days. That is, after 60 days, a given amount of radioactive iodine will have decayed to half the original amount. Suppose a contained nuclear accident occurs and gives off an initial amount C of radioactive iodine.

a. Write an equation for the amount of radioactive iodine present at any time t following the accident.

b. How long will it take for the radioactive iodine to decay to a level of 20% of the original amount?

Solution

a. Knowing that half the original amount remains after 60 days, you can use the exponential decay model $y = ae^{-bt}$ to obtain

$$\frac{1}{2}C = Ce^{-b(60)}$$

$$\frac{1}{2} = e^{-60b}$$

$$-\ln 2 = -60b$$

$$b = \frac{\ln 2}{60} \approx 0.0116.$$

So, $y = Ce^{-0.0116t}$.

b. The time required for the radioactive iodine to decay to 20% of the original amount is

$$Ce^{-0.0116t} = (0.2)C$$

$$e^{-0.0116t} = 0.2$$

$$-0.0116t = \ln 0.2$$

$$t = \frac{\ln 0.2}{-0.0116} \approx 139 \text{ days.}$$

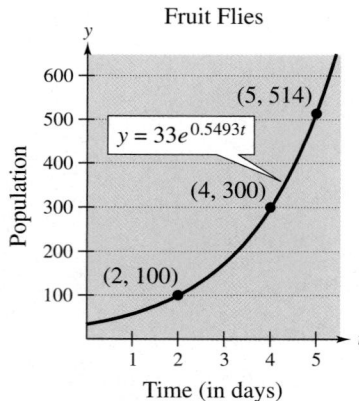

FIGURE **5.31**

In Example 1, you were given the exponential growth model. But suppose this model were not given; how could you find such a model? One technique for doing this is demonstrated in Example 2.

Example 2 ▶   **Modeling Population Growth**

In a research experiment, a population of fruit flies is increasing according to the law of exponential growth. After 2 days there are 100 flies, and after 4 days there are 300 flies. How many flies will there be after 5 days?

Solution

Let y be the number of flies at time t. From the given information, you know that $y = 100$ when $t = 2$ and $y = 300$ when $t = 4$. Substituting this information into the model $y = ae^{bt}$ produces

$$100 = ae^{2b} \quad \text{and} \quad 300 = ae^{4b}.$$

To solve for b, solve for a in the first equation.

$$100 = ae^{2b} \quad\Longrightarrow\quad a = \frac{100}{e^{2b}} \qquad \text{Solve for } a \text{ in the first equation.}$$

Then substitute the result into the second equation.

$$300 = ae^{4b} \qquad\qquad\qquad \text{Write second equation.}$$

$$300 = \left(\frac{100}{e^{2b}}\right)e^{4b} \qquad\qquad \text{Substitute } 100/e^{2b} \text{ for } a.$$

$$\frac{300}{100} = e^{2b} \qquad\qquad\qquad \text{Divide each side by 100.}$$

$$\ln 3 = 2b \qquad\qquad\qquad \text{Take natural log of each side.}$$

$$\frac{1}{2}\ln 3 = b \qquad\qquad\qquad \text{Solve for } b.$$

Using $b = \frac{1}{2}\ln 3$ and the equation you found for a, you can determine that

$$a = \frac{100}{e^{2[(1/2)\ln 3]}} \qquad\qquad \text{Substitute } (1/2)\ln 3 \text{ for } b.$$

$$= \frac{100}{e^{\ln 3}} \qquad\qquad\qquad \text{Simplify.}$$

$$= \frac{100}{3} \qquad\qquad\qquad \text{Inverse Property}$$

$$\approx 33. \qquad\qquad\qquad \text{Simplify.}$$

So, with $a \approx 33$ and $b = \frac{1}{2}\ln 3 \approx 0.5493$, the exponential growth model is

$$y = 33e^{0.5493t}$$

as shown in Figure 5.31. This implies that, after 5 days, the population will be

$$y = 33e^{0.5493(5)} \approx 514 \text{ flies.}$$

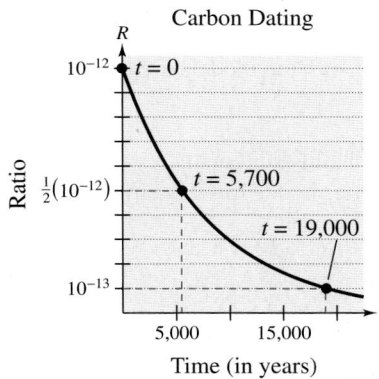

FIGURE **5.32**

In living organic material, the ratio of the number of radioactive carbon isotopes (carbon 14) to the number of nonradioactive carbon isotopes (carbon 12) is about 1 to 10^{12}. When organic material dies, its carbon 12 content remains fixed, whereas its radioactive carbon 14 begins to decay with a half-life of about 5700 years. To estimate the age of dead organic material, scientists use the following formula, which denotes the ratio of carbon 14 to carbon 12 present at any time t (in years).

$$R = \frac{1}{10^{12}} e^{-t/8223}$$ Carbon dating model

The graph of R is shown in Figure 5.32. Note that R decreases as t increases.

Example 3 ▶ **Carbon Dating**

The ratio of carbon 14 to carbon 12 in a newly discovered fossil is

$$R = \frac{1}{10^{13}}.$$

Estimate the age of the fossil.

Solution

In the carbon dating model, substitute the given value of R to obtain the following.

$$\frac{1}{10^{12}} e^{-t/8223} = R$$ Write original model.

$$\frac{e^{-t/8223}}{10^{12}} = \frac{1}{10^{13}}$$ Let $R = \frac{1}{10^{13}}$.

$$e^{-t/8223} = \frac{1}{10}$$ Multiply each side by 10^{12}.

$$\ln e^{-t/8223} = \ln \frac{1}{10}$$ Take natural log of each side.

$$-\frac{t}{8223} \approx -2.3026$$ Inverse Property

$$t \approx 18{,}934$$ Multiply each side by -8223.

So, to the nearest thousand years, you can estimate the age of the fossil to be 19,000 years.

The carbon dating model in Example 3 assumed that the carbon 14/carbon 12 ratio was one part in 10,000,000,000,000. Suppose an error in measurement occurred and the actual ratio was only one part in 8,000,000,000,000. The fossil age corresponding to the actual ratio would then be approximately 17,000 years. Try checking this result.

STUDY TIP

An exponential model can be used to determine the *decay* of radioactive isotopes. For instance, to find how much of an initial 10 grams of ^{226}Ra isotope with a half-life of 1620 years is left after 500 years, you would use the exponential decay model.

$$y = ae^{-bt}$$

$$\frac{1}{2}(10) = 10e^{-b(1620)}$$

$$b = -\ln\left(\frac{1}{2}\right)/1620$$

Using the value of b found above and $a = 10$, the amount left is

$$y = 10e^{-[-\ln(1/2)/1620](500)}$$

$$y \approx 8.07 \text{ grams.}$$

Gaussian Models

As mentioned at the beginning of this section, Gaussian models are of the form

$$y = ae^{-(x-b)^2/c}.$$

This type of model is commonly used in probability and statistics to represent populations that are **normally distributed.** One model for this situation takes the form

$$y = \frac{1}{\sigma\sqrt{2\pi}} e^{-x^2/(2\sigma^2)}$$

where σ is the standard deviation (σ is the lowercase Greek letter sigma). The graph of a Gaussian model is called a **bell-shaped curve.**

The average value for a population can be found from the bell-shaped curve by observing where the maximum y-value of the function occurs. The x-value corresponding to the maximum y-value of the function represents the average value of the independent variable—in this case, x.

Example 4 ▶ **SAT Scores**

In 2001, the Scholastic Aptitude Test (SAT) math scores for college-bound seniors roughly followed a normal distribution

$$y = 0.0035e^{-(x-514)^2/25,538}, \quad 200 \le x \le 800$$

where x is the SAT score for mathematics. Sketch the graph of this function. From the graph, estimate the average SAT score. (Source: College Board)

Solution

The graph of the function is shown in Figure 5.33. From the graph, you can see that the average mathematics score for college-bound seniors in 2001 was 514.

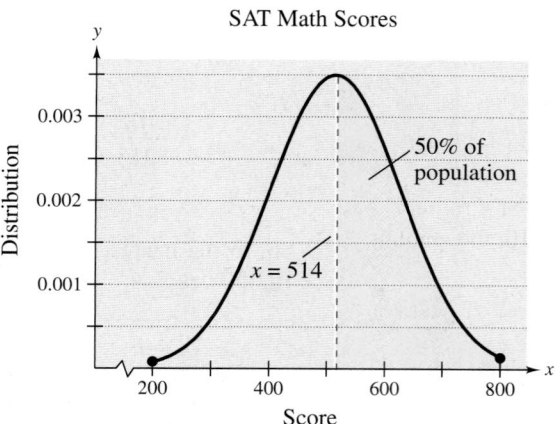

FIGURE 5.33

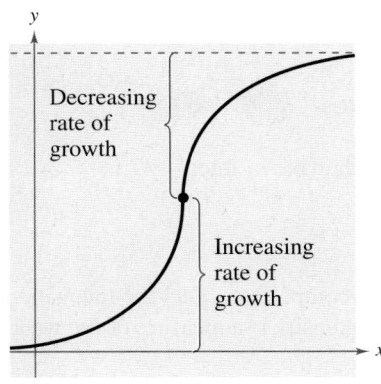

FIGURE **5.34**

Logistic Growth Models

Some populations initially have rapid growth, followed by a declining rate of growth, as indicated by the graph in Figure 5.34. One model for describing this type of growth pattern is the **logistic curve** given by the function

$$y = \frac{a}{1 + be^{-rx}}$$

where y is the population size and x is the time. An example is a bacteria culture that is initially allowed to grow under ideal conditions, and then under less favorable conditions that inhibit growth. A logistic growth curve is also called a **sigmoidal curve.**

Example 5 ▶ **Spread of a Virus**

On a college campus of 5000 students, one student returns from vacation with a contagious and long-lasting flu virus. The spread of the virus is modeled by

$$y = \frac{5000}{1 + 4999e^{-0.8t}}, \quad t \geq 0$$

where y is the total number of students infected after t days. The college will cancel classes when 40% or more of the students are infected.

a. How many students are infected after 5 days?

b. After how many days will the college cancel classes?

Solution

a. After 5 days, the number of students infected is

$$y = \frac{5000}{1 + 4999e^{-0.8(5)}} = \frac{5000}{1 + 4999e^{-4}} \approx 54.$$

b. Classes are canceled when the number infected is $(0.40)(5000) = 2000$.

$$2000 = \frac{5000}{1 + 4999e^{-0.8t}}$$

$$1 + 4999e^{-0.8t} = 2.5$$

$$e^{-0.8t} \approx \frac{1.5}{4999}$$

$$\ln e^{-0.8t} \approx \ln \frac{1.5}{4999}$$

$$-0.8t \approx \ln \frac{1.5}{4999}$$

$$t = -\frac{1}{0.8} \ln \frac{1.5}{4999}$$

$$t \approx 10.1$$

So, after 10 days, at least 40% of the students will be infected, and classes will be canceled. The graph of the function is shown in Figure 5.35.

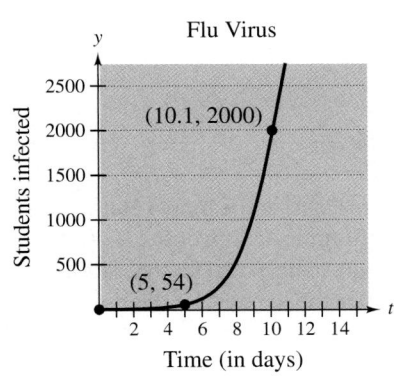

FIGURE **5.35**

AFP/Corbis

On January 13, 2001 an earthquake of magnitude 7.7 in El Salvador caused 185 landslides and killed over 800 people.

Alternative Writing About Mathematics

Use your school's library, the Internet, or some other reference source to find an application that fits one of the five models discussed in this section. After you have collected data for the model, plot the corresponding points and find an equation that describes the points you have plotted.

t	Year	Population
1	1910	91.97
2	1920	105.71
3	1930	122.78
4	1940	131.67
5	1950	151.33
6	1960	179.32
7	1970	203.30
8	1980	226.54
9	1990	248.72
10	2000	281.42

Logarithmic Models

Example 6 ▶ **Magnitude of Earthquakes**

On the Richter scale, the magnitude R of an earthquake of intensity I is

$$R = \log_{10} \frac{I}{I_0}$$

where $I_0 = 1$ is the minimum intensity used for comparison. Find the intensities per unit of area for the following earthquakes. (Intensity is a measure of the wave energy of an earthquake.)

a. Tokyo and Yokohama, Japan in 1923: $R = 8.3$.

b. El Salvador in 2001: $R = 7.7$.

Solution

a. Because $I_0 = 1$ and $R = 8.3$, you have

$$8.3 = \log_{10} \frac{I}{1} \qquad \text{Substitute 1 for } I_0 \text{ and 8.3 for } R.$$

$$10^{8.3} = 10^{\log_{10} I} \qquad \text{Exponentiate each side.}$$

$$I = 10^{8.3} \approx 199{,}526{,}000. \qquad \text{Inverse property of exponents and logs}$$

b. For $R = 7.7$, you have

$$7.7 = \log_{10} \frac{I}{1} \qquad \text{Substitute 1 for } I_0 \text{ and 7.7 for } R.$$

$$10^{7.7} = 10^{\log_{10} I} \qquad \text{Exponentiate each side.}$$

$$I = 10^{7.7} \approx 50{,}119{,}000. \qquad \text{Inverse property of exponents and logs}$$

Note that an increase of 0.6 unit on the Richter scale (from 7.7 to 8.3) represents an increase in intensity by a factor of

$$\frac{199{,}526{,}000}{50{,}119{,}000} \approx 4.$$

In other words, the earthquake in 1923 had an intensity about 4 times greater than that of the 2001 earthquake.

Writing ABOUT MATHEMATICS

Comparing Population Models The population (in millions) of the United States from 1910 to 2000 is shown in the table at the left. (Source: U.S. Census Bureau) Least squares regression analysis gives the best quadratic model for this data as $P = 1.0317t^2 + 9.668t + 81.38$ and the best exponential model for this data as $P = 82.367e^{0.125t}$. Which model better fits the data? Describe the method you used to reach your conclusion.

5.5 Exercises

In Exercises 1–6, match the function with its graph. [The graphs are labeled (a), (b), (c), (d), (e), and (f).]

(a)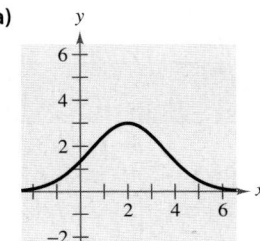

(b)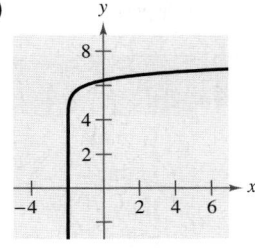

(c)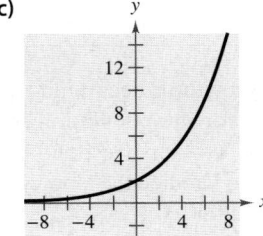

(d)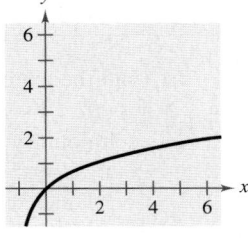

(e)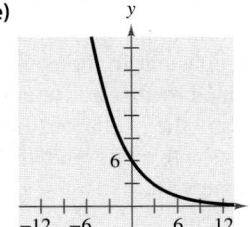

(f)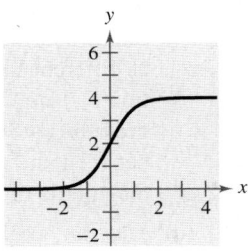

1. $y = 2e^{x/4}$ **2.** $y = 6e^{-x/4}$

3. $y = 6 + \log_{10}(x + 2)$ **4.** $y = 3e^{-(x-2)^2/5}$

5. $y = \ln(x + 1)$ **6.** $y = \dfrac{4}{1 + e^{-2x}}$

Compound Interest **In Exercises 7–14, complete the table for a savings account in which interest is compounded continuously.**

Initial Investment	Annual % Rate	Time to Double	Amount After 10 Years
7. $1000	12%		
8. $20,000	$10\frac{1}{2}$%		
9. $750		$7\frac{3}{4}$ yr	
10. $10,000		12 yr	
11. $500			$1505.00
12. $600			$19,205.00
13.	4.5%		$10,000.00
14.	8%		$20,000.00

Compound Interest **In Exercises 15 and 16, determine the principal P that must be invested at rate r, compounded monthly, so that $500,000 will be available for retirement in t years.**

15. $r = 7\frac{1}{2}\%, t = 20$

16. $r = 12\%, t = 40$

Compound Interest **In Exercises 17 and 18, determine the time necessary for $1000 to double if it is invested at interest rate r compounded (a) annually, (b) monthly, (c) daily, and (d) continuously.**

17. $r = 11\%$

18. $r = 10\frac{1}{2}\%$

19. ***Compound Interest*** Complete the table for the time t necessary for P dollars to triple if interest is compounded continuously at rate r.

r	2%	4%	6%	8%	10%	12%
t						

20. ***Modeling Data*** Draw a scatter plot of the data in Exercise 19. Use the *regression* feature of a graphing utility to find a model for the data.

21. ***Compound Interest*** Complete the table for the time t necessary for P dollars to triple if interest is compounded annually at rate r.

r	2%	4%	6%	8%	10%	12%
t						

22. ***Modeling Data*** Draw a scatter plot of the data in Exercise 21. Use the *regression* feature of a graphing utility to find a model for the data.

23. ***Comparing Models*** If $1 is invested in an account over a 10-year period, the amount in the account, where t represents the time in years, is

$$A = 1 + 0.075[\![t]\!] \quad \text{or} \quad A = e^{0.07t}$$

depending on whether the account pays simple interest at $7\frac{1}{2}\%$ or continuous compound interest at 7%. Graph each function on the same set of axes. Which grows at the faster rate? (Remember that $[\![t]\!]$ is the greatest integer function discussed in Section P.7.)

24. *Comparing Models* If \$1 is invested in an account over a 10-year period, the amount in the account, where t represents the time in years, is

$$A = 1 + 0.06 [\![t]\!] \quad \text{or} \quad A = \left(1 + \frac{0.055}{365}\right)^{[\![365t]\!]}$$

depending on whether the account pays simple interest at 6% or compound interest at $5\frac{1}{2}\%$ compounded daily. Use a graphing utility to graph each function in the same viewing window. Which grows at the faster rate?

Radioactive Decay **In Exercises 25–30, complete the table for the radioactive isotope.**

Isotope	Half-life (years)	Initial Quantity	Amount After 1000 Years
25. ^{226}Ra	1620	10 g	
26. ^{226}Ra	1620		1.5 g
27. ^{14}C	5730		2 g
28. ^{14}C	5730	3 g	
29. ^{239}Pu	24,360		2.1 g
30. ^{239}Pu	24,360		0.4 g

In Exercises 31–34, find the exponential model $y = ae^{bx}$ that fits the points in the graph or table.

31.

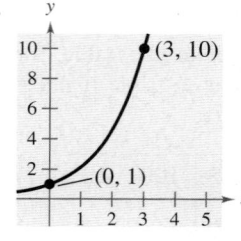

32.

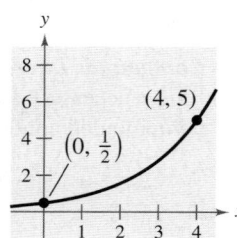

33.

x	0	4
y	5	1

34.

x	0	3
y	1	$\frac{1}{4}$

35. *Population* The population P of Texas (in thousands) from 1991 through 2000 can be modeled by

$$P = 16,968e^{0.019t}$$

where $t = 1$ represents the year 1991. According to this model, when will the population reach 22 million? (Source: U.S. Census Bureau)

▶ **Model It**

36. *Population* The table shows the populations (in millions) of five countries in 2000 and the projected populations (in millions) for the year 2010. (Source: U.S. Census Bureau)

Country	2000	2010
Canada	31.3	34.3
China	1261.8	1359.1
Italy	57.6	57.4
United Kingdom	59.5	60.6
United States	275.6	300.1

(a) Find the exponential growth or decay model $y = ae^{bt}$ or $y = ae^{-bt}$ for the population in each country by letting $t = 0$ correspond to 2000. Use the model to predict the population of each country in 2030.

(b) You can see that the populations of the United States and the United Kingdom are growing at different rates. What constant in the equation $y = ae^{bt}$ is determined by these different growth rates? Discuss the relationship between the different growth rates and the magnitude of the constant.

(c) You can see that the population of China is increasing while the population of Italy is decreasing. What constant in the equation $y = ae^{bt}$ reflects this difference? Explain.

37. *Population* The population P of Charlotte, North Carolina (in thousands) is

$$P = 548e^{kt}$$

where $t = 0$ represents the year 2000. In 1970, the population was 241,000. Find the value of k, and use this result to predict the population in the year 2010. (Source: U.S. Census Bureau)

38. *Population* The population P of Lincoln, Nebraska (in thousands) is

$$P = 224e^{kt}$$

where $t = 0$ represents the year 2000. In 1980, the population was 172,000. Find the value of k, and use this result to predict the population in the year 2020. (Source: U.S. Census Bureau)

39. Bacteria Growth The number N of bacteria in a culture is modeled by

$$N = 100e^{kt}$$

where t is the time in hours. If $N = 300$ when $t = 5$, estimate the time required for the population to double in size.

40. Bacteria Growth The number N of bacteria in a culture is modeled by $N = 250e^{kt}$, where t is the time in hours. If $N = 280$ when $t = 10$, estimate the time required for the population to double in size.

41. Radioactive Decay The half-life of radioactive radium (^{226}Ra) is 1620 years. What percent of a present amount of radioactive radium will remain after 100 years?

42. Radioactive Decay Carbon 14 dating assumes that the carbon dioxide on Earth today has the same radioactive content as it did centuries ago. If this is true, the amount of ^{14}C absorbed by a tree that grew several centuries ago should be the same as the amount of ^{14}C absorbed by a tree growing today. A piece of ancient charcoal contains only 15% as much radioactive carbon as a piece of modern charcoal. How long ago was the tree burned to make the ancient charcoal if the half-life of ^{14}C is 5730 years?

43. Depreciation A car that cost $22,000 new has a book value of $13,000 after 2 years.

(a) Find the straight-line model $V = mt + b$.

(b) Find the exponential model $V = ae^{kt}$.

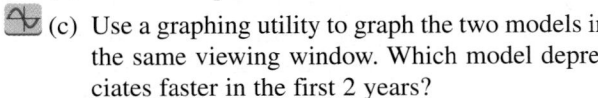 (c) Use a graphing utility to graph the two models in the same viewing window. Which model depreciates faster in the first 2 years?

(d) Find the book values of the car after 1 year and after 3 years using each model.

(e) Interpret the slope of the straight-line model.

44. Depreciation A computer that costs $2000 new has a book value of $500 after 2 years.

(a) Find the straight-line model $V = mt + b$.

(b) Find the exponential model $V = ae^{kt}$.

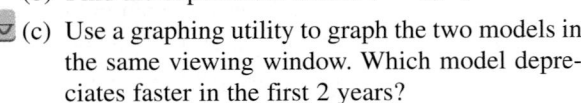 (c) Use a graphing utility to graph the two models in the same viewing window. Which model depreciates faster in the first 2 years?

(d) Find the book values of the computer after 1 year and after 3 years using each model.

(e) Interpret the slope of the straight-line model.

45. Sales The sales S (in thousands of units) of a new CD burner after it has been on the market t years are modeled by $S(t) = 100(1 - e^{kt})$. Fifteen thousand units of the new product were sold the first year.

(a) Complete the model by solving for k.

(b) Sketch the graph of the model.

(c) Use the model to estimate the number of units sold after 5 years.

46. Sales After discontinuing all advertising for a tool kit in 1998, the manufacturer noted that sales began to drop according to the model

$$S = \frac{500,000}{1 + 0.6e^{kt}}$$

where S represents the number of units sold and $t = 0$ represents 1998. In 2000, the company sold 300,000 units.

(a) Complete the model by solving for k.

(b) Estimate sales in 2005.

47. Sales The sales S (in thousands of units) of a cleaning solution after x hundred dollars is spent on advertising are modeled by $S = 10(1 - e^{kx})$. When $500 is spent on advertising, 2500 units are sold.

(a) Complete the model by solving for k.

(b) Estimate the number of units that will be sold if advertising expenditures are raised to $700.

48. Profit Because of a slump in the economy, a department store finds that its annual profits have dropped from $742,000 in 2000 to $632,000 in 2002. The profit follows an exponential pattern of decline. What is the expected profit for 2005? (Let $t = 0$ represent 2000.)

49. Learning Curve The management at a plastics factory has found that the maximum number of units a worker can produce in a day is 30. The learning curve for the number N of units produced per day after a new employee has worked t days is

$$N = 30(1 - e^{kt}).$$

After 20 days on the job, a new employee produces 19 units.

(a) Find the learning curve for this employee (first, find the value of k).

(b) How many days should pass before this employee is producing 25 units per day?

50. *Population Growth*   A conservation organization releases 100 animals of an endangered species into a game preserve. The organization believes that the preserve has a carrying capacity of 1000 animals and that the growth of the herd will be modeled by the logistic curve

$$p(t) = \frac{1000}{1 + 9e^{-0.1656t}}$$

where t is measured in months (see figure).

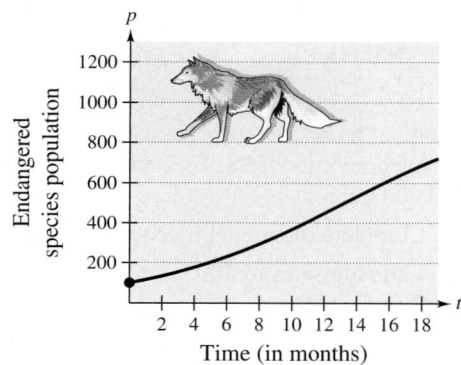

(a) Estimate the population after 5 months.

(b) After how many months will the population be 500?

(c) Use a graphing utility to graph the function. Use the graph to determine the horizontal asymptotes, and interpret the meaning of the larger p-value in the context of the problem.

Geology   In Exercises 51 and 52, use the Richter scale for measuring the magnitudes of earthquakes.

51. Find the magnitude R of an earthquake of intensity I (let $I_0 = 1$).

(a) $I = 80,500,000$

(b) $I = 48,275,000$

(c) $I = 251,200$

52. Find the intensity I of an earthquake measuring R on the Richter scale (let $I_0 = 1$).

(a) Chile in 1906, $R = 8.2$

(b) Los Angeles in 1971, $R = 6.7$

(c) India in 2001, $R = 7.7$

Intensity of Sound   In Exercises 53–56, use the following information for determining sound intensity. The level of sound β, in decibels, with an intensity of I is

$$\beta = 10 \log_{10} \frac{I}{I_0}$$

where I_0 is an intensity of 10^{-12} watt per square meter, corresponding roughly to the faintest sound that can be heard by the human ear. In Exercises 53 and 54, find the level of sound, β.

53. (a) $I = 10^{-10}$ watt per m² (faint whisper)

     (b) $I = 10^{-5}$ watt per m² (busy street corner)

     (c) $I = 10^{-2.5}$ watt per m² (air hammer)

     (d) $I = 10^0$ watt per m² (threshold of pain)

54. (a) $I = 10^{-9}$ watt per m² (whisper)

     (b) $I = 10^{-3.5}$ watt per m² (jet 4 miles from takeoff)

     (c) $I = 10^{-3}$ watt per m² (diesel truck at 25 feet)

     (d) $I = 10^{-0.5}$ watt per m² (auto horn at 3 feet)

55. Due to the installation of noise suppression materials, the noise level in an auditorium was reduced from 93 to 80 decibels. Find the percent decrease in the intensity level of the noise as a result of the installation of these materials.

56. Due to the installation of a muffler, the noise level of an engine was reduced from 88 to 72 decibels. Find the percent decrease in the intensity level of the noise as a result of the installation of the muffler.

pH Levels   In Exercises 57–62, use the acidity model given by pH $= -\log_{10}[H^+]$, where acidity (pH) is a measure of the hydrogen ion concentration $[H^+]$ (measured in moles of hydrogen per liter) of a solution.

57. Find the pH if $[H^+] = 2.3 \times 10^{-5}$.

58. Find the pH if $[H^+] = 11.3 \times 10^{-6}$.

59. Compute $[H^+]$ for a solution in which pH $= 5.8$.

60. Compute $[H^+]$ for a solution in which pH $= 3.2$.

61. A fruit has a pH of 2.5 and an antacid tablet has a pH of 9.5. The hydrogen ion concentration of the fruit is how many times the concentration of the tablet?

62. The pH of a solution is decreased by one unit. The hydrogen ion concentration is increased by what factor?

63. Forensics At 8:30 A.M., a coroner was called to the home of a person who had died during the night. In order to estimate the time of death, the coroner took the person's temperature twice. At 9:00 A.M. the temperature was 85.7°F, and at 11:00 A.M. the temperature was 82.8°F. From these two temperatures, the coroner was able to determine that the time elapsed since death and the body temperature were related by the formula

$$t = -10 \ln \frac{T - 70}{98.6 - 70}$$

where t is the time in hours elapsed since the person died and T is the temperature (in degrees Fahrenheit) of the person's body. Assume that the person had a normal body temperature of 98.6°F at death, and that the room temperature was a constant 70°F. (This formula is derived from a general cooling principle called Newton's Law of Cooling.) Use the formula to estimate the time of death of the person.

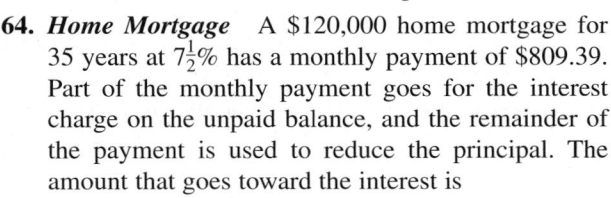

64. Home Mortgage A $120,000 home mortgage for 35 years at $7\frac{1}{2}\%$ has a monthly payment of $809.39. Part of the monthly payment goes for the interest charge on the unpaid balance, and the remainder of the payment is used to reduce the principal. The amount that goes toward the interest is

$$u = M - \left(M - \frac{Pr}{12}\right)\left(1 + \frac{r}{12}\right)^{12t}$$

and the amount that goes toward the reduction of the principal is

$$v = \left(M - \frac{Pr}{12}\right)\left(1 + \frac{r}{12}\right)^{12t}.$$

In these formulas, P is the size of the mortgage, r is the interest rate, M is the monthly payment, and t is the time in years.

(a) Use a graphing utility to graph each function in the same viewing window. (The viewing window should show all 35 years of mortgage payments.)

(b) In the early years of the mortgage, the larger part of the monthly payment goes for what purpose? Approximate the time when the monthly payment is evenly divided between interest and principal reduction.

(c) Repeat parts (a) and (b) for a repayment period of 20 years ($M = \$966.71$). What can you conclude?

65. Home Mortgage The total interest u paid on a home mortgage of P dollars at interest rate r for t years is

$$u = P\left[\frac{rt}{1 - \left(\dfrac{1}{1 + r/12}\right)^{12t}} - 1\right].$$

Consider a $120,000 home mortgage at $7\frac{1}{2}\%$.

(a) Use a graphing utility to graph the total interest function.

(b) Approximate the length of the mortgage for which the total interest paid is the same as the size of the mortgage. Is it possible that some people are paying twice as much in interest charges as the size of the mortgage?

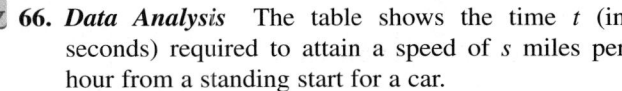

66. Data Analysis The table shows the time t (in seconds) required to attain a speed of s miles per hour from a standing start for a car.

Speed, s	Time, t
30	3.4
40	5.0
50	7.0
60	9.3
70	12.0
80	15.8
90	20.0

Two models for this data are as follows.

$$t_1 = 40.757 + 0.556s - 15.817 \ln s$$

$$t_2 = 1.2259 + 0.0023s^2$$

(a) Use a graphing utility to fit a linear model t_3 and an exponential model t_4 to the data.

(b) Use a graphing utility to graph the data points and each model in the same viewing window.

(c) Create a table comparing the data with estimates obtained from each model.

(d) Use the results of part (c) to find the sum of the absolute values of the differences between the data and estimated values given by each model. Based on the four sums, which model do you think better fits the data? Explain.

Synthesis

True or False? **In Exercises 67–70, determine whether the statement is true or false. Justify your answer.**

67. The domain of a logistic growth function cannot be the set of real numbers.

68. A logistic growth function will always have an x-intercept.

69. The graph of

$$f(x) = \frac{4}{1 + 6e^{-2x}} + 5$$

is the graph of

$$g(x) = \frac{4}{1 + 6e^{-2x}}$$

shifted to the right five units.

70. The graph of a Gaussian model will never have an x-intercept.

71. Identify each model as linear, logarithmic, exponential, logistic, or none of the above. Explain your reasoning.

(a)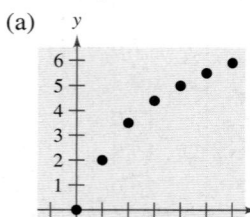

(b)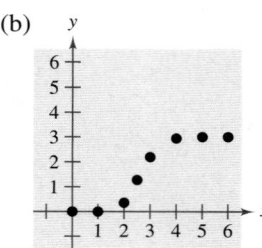

(c)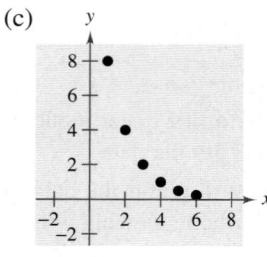

(d)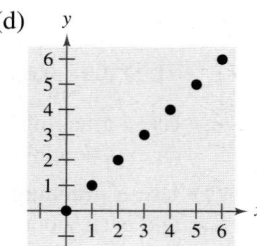

(e)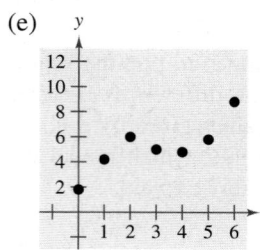

(f)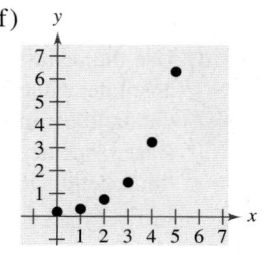

72. ***Writing*** Use your school's library, the Internet, or some other reference source to write a paper describing John Napier's work with logarithms.

Review

In Exercises 73–82, sketch the graph of the equation.

73. $y = 10 - 3x$

74. $y = -4x - 1$

75. $y = -2x^2 - 3$

76. $y = 2x^2 - 7x - 30$

77. $3x^2 - 4y = 0$

78. $-x^2 - 8y = 0$

79. $y = \dfrac{4}{1 - 3x}$

80. $y = \dfrac{x^2}{-x - 2}$

81. $x^2 + (y - 8)^2 = 25$

82. $(x - 4)^2 + (y + 7) = 4$

In Exercises 83–86, graph the exponential function.

83. $f(x) = 2^{x-1} + 5$

84. $f(x) = -2^{-x-1} - 1$

85. $f(x) = 3^x - 4$

86. $f(x) = -3^x + 4$

Chapter Summary

▶ *What* did you learn?

Review Exercises

 5.1 In Exercises 1–6, evaluate the expression. Approximate your result to three decimal places.

1. $(6.1)^{2.4}$ **2.** $-14(5^{-0.8})$

3. $2^{-0.5\pi}$ **4.** $\sqrt[5]{1278}$

5. $60^{\sqrt{3}}$ **6.** $7^{-\sqrt{11}}$

In Exercises 7–10, match the function with its graph. [The graphs are labeled (a), (b), (c), and (d).]

(a)
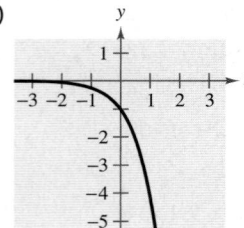

(b)
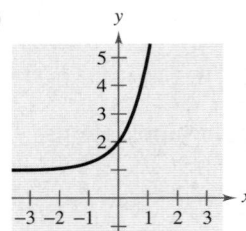

(c)
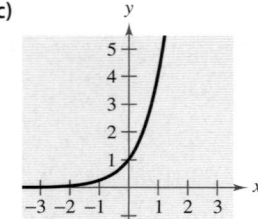

(d)
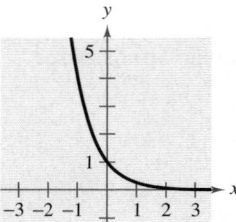

7. $f(x) = 4^x$ **8.** $f(x) = 4^{-x}$

9. $f(x) = -4^x$ **10.** $f(x) = 4^x + 1$

In Exercises 11–14, use the graph of f to describe the transformation that yields the graph of g.

11. $f(x) = 5^x$, $\quad g(x) = 5^{x-1}$

12. $f(x) = 4^x$, $\quad g(x) = 4^x - 3$

13. $f(x) = \left(\frac{1}{2}\right)^x$, $\quad g(x) = -\left(\frac{1}{2}\right)^{x+2}$

14. $f(x) = \left(\frac{2}{3}\right)^x$, $\quad g(x) = 8 - \left(\frac{2}{3}\right)^x$

 In Exercises 15–22, use a graphing utility to construct a table of values. Then sketch the graph of the function.

15. $f(x) = 4^{-x} + 4$ **16.** $f(x) = -4^x - 3$

17. $f(x) = -2.65^{x+1}$ **18.** $f(x) = 2.65^{x-1}$

19. $f(x) = 5^{x-2} + 4$ **20.** $f(x) = 2^{x-6} - 5$

21. $f(x) = \left(\frac{1}{2}\right)^{-x} + 3$ **22.** $f(x) = \left(\frac{1}{8}\right)^{x+2} - 5$

In Exercises 23–26, evaluate the function $f(x) = e^x$ for the indicated value of x. Approximate your result to three decimal places.

23. $x = 8$ **24.** $x = \dfrac{5}{8}$

25. $x = -1.7$ **26.** $x = 0.278$

In Exercises 27–30, use a graphing utility to construct a table of values. Then sketch the graph of the function.

27. $h(x) = e^{-x/2}$ **28.** $h(x) = 2 - e^{-x/2}$

29. $f(x) = e^{x+2}$ **30.** $s(t) = 4e^{-2/t}$, $\quad t > 0$

Compound Interest In Exercises 31 and 32, complete the table to determine the balance A for P dollars invested at rate r for t years and compounded n times per year.

n	1	2	4	12	365	Continuous
A						

31. $P = \$3500$, $r = 6.5\%$, $t = 10$ years

32. $P = \$2000$, $r = 5\%$, $t = 30$ years

33. ***Waiting Times*** The average time between incoming calls at a switchboard is 3 minutes. The probability F of waiting less than t minutes until the next incoming call is approximated by the model $F(t) = 1 - e^{-t/3}$. A call has just come in. Find the probability that the next call will be within

(a) $\frac{1}{2}$ minute.   (b) 2 minutes.   (c) 5 minutes.

34. ***Depreciation*** After t years, the value V of a car that cost $\$14,000$ is $V(t) = 14,000\left(\frac{3}{4}\right)^t$.

(a) Use a graphing utility to graph the function.

(b) Find the value of the car 2 years after it was purchased.

(c) According to the model, when does the car depreciate most rapidly? Is this realistic? Explain.

35. ***Trust Fund*** On the day a person was born, a deposit of $\$50,000$ was made in a trust fund that pays 8.75% interest, compounded continuously.

(a) Find the balance on the person's 35th birthday.

(b) How much longer would the person have to wait to get twice as much?

36. ***Radioactive Decay*** Let Q represent a mass of plutonium 241 (^{241}Pu) (in grams), whose half-life is 13 years. The quantity of plutonium 241 present after t years is

$$Q = 100\left(\frac{1}{2}\right)^{t/13}.$$

(a) Determine the initial quantity (when $t = 0$).

(b) Determine the quantity present after 10 years.

(c) Sketch the graph of this function over the interval $t = 0$ to $t = 100$.

5.2 In Exercises 37 and 38, write the exponential equation in logarithmic form.

37. $4^3 = 64$ **38.** $25^{3/2} = 125$

In Exercises 39–42, evaluate the function at the indicated value of x without using a calculator.

Function	Value
39. $f(x) = \log_{10} x$	$x = 1000$
40. $g(x) = \log_9 x$	$x = 3$
41. $g(x) = \log_2 x$	$x = \frac{1}{8}$
42. $f(x) = \log_4 x$	$x = \frac{1}{4}$

In Exercises 43–48, find the domain, x-intercept, and vertical asymptote of the logarithmic function and sketch its graph.

43. $g(x) = \log_7 x$ **44.** $g(x) = \log_5 x$

45. $f(x) = \log_{10}\left(\frac{x}{3}\right)$ **46.** $f(x) = 6 + \log_{10} x$

47. $f(x) = 4 - \log_{10}(x + 5)$

48. $f(x) = \log_{10}(x - 3) + 1$

In Exercises 49–54, use your calculator to evaluate the function $f(x) = \ln x$ for the indicated value of x. Approximate your result to three decimal places if necessary.

49. $x = 22.6$ **50.** $x = 0.98$

51. $x = e^{-12}$ **52.** $x = e^7$

53. $x = \sqrt{7} + 5$ **54.** $x = \frac{\sqrt{3}}{8}$

In Exercises 55–58, find the domain, x-intercept, and vertical asymptote of the logarithmic function and sketch its graph.

55. $f(x) = \ln x + 3$ **56.** $f(x) = \ln(x - 3)$

57. $h(x) = \ln(x^2)$ **58.** $f(x) = \frac{1}{4} \ln x$

59. ***Antler Spread*** The antler spread a (in inches) and shoulder height h (in inches) of an adult male American elk are related by the model

$$h = 116 \log_{10}(a + 40) - 176.$$

Approximate the shoulder height of a male American elk with an antler spread of 55 inches.

60. ***Snow Removal*** The number of miles s of roads cleared of snow is approximated by the model

$$s = 25 - \frac{13 \ln(h/12)}{\ln 3}, \quad 2 \le h \le 15$$

where h is the depth of the snow in inches. Use this model to find s when $h = 10$ inches.

5.3 In Exercises 61–64, evaluate the logarithm using the change-of-base formula. Do each problem twice, once with common logarithms and once with natural logarithms. Approximate the results to three decimal places.

61. $\log_4 9$ **62.** $\log_{12} 200$

63. $\log_{1/2} 5$ **64.** $\log_3 0.28$

In Exercises 65–72, use the properties of logarithms to expand the expression as a sum, difference, and/or multiple of logarithms.

65. $\log_5 5x^2$ **66.** $\log_{10} 7x^4$

67. $\log_3 \dfrac{6}{\sqrt[3]{x}}$ **68.** $\log_7 \dfrac{\sqrt{x}}{4}$

69. $\ln x^2 y^2 z$ **70.** $\ln 3xy^2$

71. $\ln\left(\dfrac{x + 3}{xy}\right)$ **72.** $\ln\left(\dfrac{y - 1}{4}\right)^2, \quad y > 1$

In Exercises 73–80, condense the expression to the logarithm of a single quantity.

73. $\log_2 5 + \log_2 x$

74. $\log_6 y - 2 \log_6 z$

75. $\ln x - \frac{1}{4} \ln y$

76. $3 \ln x + 2 \ln(x + 1)$

77. $\frac{1}{3} \log_8(x + 4) + 7 \log_8 y$

78. $-2 \log_{10} x - 5 \log_{10}(x + 6)$

79. $\frac{1}{2} \ln(2x - 1) - 2 \ln(x + 1)$

80. $5 \ln(x - 2) - \ln(x + 2) - 3 \ln x$

81. *Climb Rate* The time t (in minutes) for a small plane to climb to an altitude of h feet is modeled by

$$t = 50 \log_{10} \frac{18,000}{18,000 - h}$$

where 18,000 feet is the plane's absolute ceiling.

(a) Determine the domain of the function appropriate for the context of the problem.

 (b) Use a graphing utility to graph the time function and identify any asymptotes.

(c) As the plane approaches its absolute ceiling, what can be said about the time required to increase its altitude further?

(d) Find the time for the plane to climb to an altitude of 4000 feet.

82. *Human Memory Model* Students in a sociology class were given an exam and then were retested monthly with an equivalent exam. The average score for the class was given by the human memory model

$$f(t) = 85 - 14 \log_{10}(t + 1), \quad 0 \le t \le 4$$

where t is the time in months. How did the average score change over the four-month period?

5.4 In Exercises 83–92, solve for x.

83. $8^x = 512$ **84.** $3^x = 729$

85. $6^x = \frac{1}{216}$ **86.** $5^x = \frac{1}{25}$

87. $e^x = 3$ **88.** $e^x = 6$

89. $\log_4 x = 2$ **90.** $\log_6 x = -1$

91. $\ln x = 4$ **92.** $\ln x = -3$

In Exercises 93–102, solve the exponential equation. Approximate your result to three decimal places.

93. $e^x = 12$ **94.** $e^{3x} = 25$

95. $3e^{-5x} = 132$ **96.** $14e^{3x+2} = 560$

97. $2^x + 13 = 35$ **98.** $6^x - 28 = -8$

99. $-4(5^x) = -68$ **100.** $2(12^x) = 190$

101. $e^{2x} - 7e^x + 10 = 0$ **102.** $e^{2x} - 6e^x + 8 = 0$

 In Exercises 103–106, use a graphing utility to graph and solve the equation. Approximate the result to two decimal places.

103. $2^{0.6x} - 3x = 0$ **104.** $4^{-0.2x} + x = 0$

105. $25e^{-0.3x} = 12$ **106.** $4e^{1.2x} = 9$

In Exercises 107–118, solve the logarithmic equation. Approximate the result to three decimal places.

107. $\ln 3x = 8.2$ **108.** $\ln 5x = 7.2$

109. $2 \ln 4x = 15$ **110.** $4 \ln 3x = 15$

111. $\ln x - \ln 3 = 2$ **112.** $\ln \sqrt{x + 8} = 3$

113. $\ln \sqrt{x + 1} = 2$ **114.** $\ln x - \ln 5 = 4$

115. $\log_{10}(x - 1) = \log_{10}(x - 2) - \log_{10}(x + 2)$

116. $\log_{10}(x + 2) - \log_{10} x = \log_{10}(x + 5)$

117. $\log_{10}(1 - x) = -1$

118. $\log_{10}(-x - 4) = 2$

 In Exercises 119–122, use a graphing utility to graph and solve the equation. Approximate the result to two decimal places.

119. $2 \ln(x + 3) + 3x = 8$

120. $6 \log_{10}(x^2 + 1) - x = 0$

121. $4 \ln(x + 5) - x = 10$

122. $x - 2 \log_{10}(x + 4) = 0$

123. *Compound Interest* $7550 is deposited in an account that pays 7.25% interest, compounded continuously. How long will it take the money to triple?

124. *Demand* The demand equation for a 32-inch television is modeled by $p = 500 - 0.5e^{0.004x}$. Find the demand x for a price of (a) $p = \$450$ and (b) $p = \$400$.

5.5 In Exercises 125–130, match the function with its graph. [The graphs are labeled (a), (b), (c), (d), (e), and (f).]

(a)

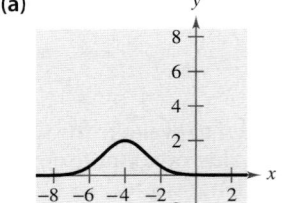

(b)

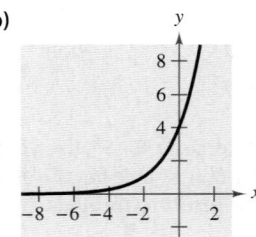

(c)

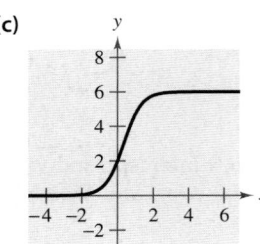

(d)

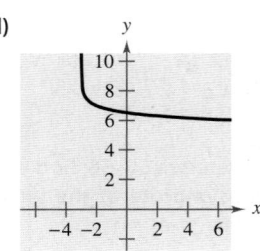

(e)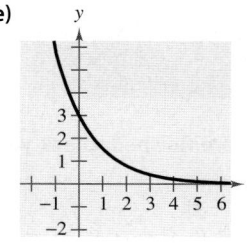

(f)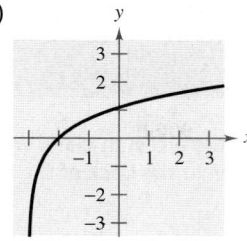

125. $y = 3e^{-2x/3}$

126. $y = 4e^{2x/3}$

127. $y = \ln(x + 3)$

128. $y = 7 - \log_{10}(x + 3)$

129. $y = 2e^{-(x+4)^2/3}$

130. $y = \dfrac{6}{1 + 2e^{-2x}}$

131. *Population* The population P of Phoenix, Arizona (in thousands) from 1970 through 2000 can be modeled by $P = 590e^{0.027t}$, where t represents the year, with $t = 0$ corresponding to 1970. According to this model, when will the population reach 1.5 million? (Source: U.S. Census Bureau)

132. *Radioactive Decay* The half-life of radioactive uranium II (^{234}U) is 250,000 years. What percent of a present amount of radioactive uranium II will remain after 5000 years?

133. *Compound Interest* A deposit of $10,000 is made in a savings account for which the interest is compounded continuously. The balance will double in 5 years.

(a) What is the annual interest rate for this account?

(b) Find the balance after 1 year.

134. *Bacteria Growth* The number N of bacteria in a culture is given by the model $N = 200e^{kt}$, where t is the time in hours. If $N = 350$ when $t = 5$, estimate the time required for the population to triple in size.

In Exercises 135 and 136, find the exponential function $y = ae^{bx}$ that passes through the points.

135. $(0, 2), (4, 3)$

136. $\left(0, \frac{1}{2}\right), (5, 5)$

 137. *Test Scores* The test scores for a biology test follow a normal distribution modeled by

$$y = 0.0499e^{-(x - 71)^2/128}, \quad 40 \le x \le 100$$

where x is the test score.

(a) Use a graphing utility to graph the equation.

(b) From the graph, estimate the average test score.

138. *Typing Speed* In a typing class, the average number of words per minute typed after t weeks of lessons was found to be

$$N = \frac{157}{1 + 5.4e^{-0.12t}}.$$

Find the time necessary to type (a) 50 words per minute and (b) 75 words per minute.

139. *Sound Intensity* The relationship between the number of decibels β and the intensity of a sound I in watts per square centimeter is

$$\beta = 10 \log_{10}\left(\frac{I}{10^{-16}}\right).$$

Determine the intensity of a sound in watts per square centimeter if the decibel level is 125.

140. *Geology* On the Richter scale, the magnitude R of an earthquake of intensity I is

$$R = \log_{10}\frac{I}{I_0}$$

where $I_0 = 1$ is the minimum intensity used for comparison. Find the intensity per unit of area for each value of R.

(a) $R = 8.4$ (b) $R = 6.85$ (c) $R = 9.1$

Synthesis

True or False? **In Exercises 141 and 142, determine whether the equation or statement is true or false. Justify your answer.**

141. $\log_b b^{2x} = 2x$

142. $\ln(x + y) = \ln x + \ln y$

143. The graphs of $y = e^{kt}$ are shown for $k = a, b, c,$ and d. Use the graphs to order $a, b, c,$ and d. Which of the four values are negative? Which are positive?

(a)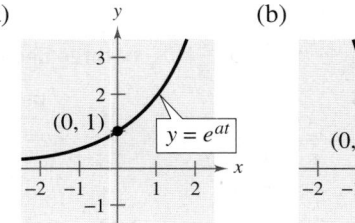

$(0, 1)$ $y = e^{at}$

(b)

$y = e^{bt}$ $(0, 1)$

(c)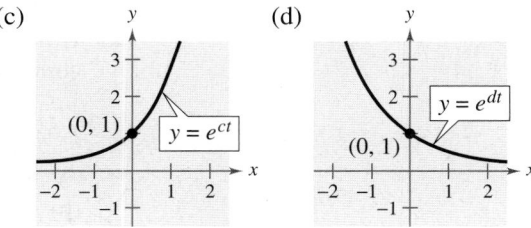

$(0, 1)$ $y = e^{ct}$

(d)

$y = e^{dt}$ $(0, 1)$

Chapter Test

Take this test as you would take a test in class. When you are finished, check your work against the answers given in the back of the book.

In Exercises 1–4, evaluate the expression. Approximate your result to three decimal places.

1. $12.4^{2.79}$ **2.** $4^{3\pi/2}$ **3.** $e^{-7/10}$ **4.** $e^{3.1}$

In Exercises 5–7, construct a table of values. Then sketch the graph of the function.

5. $f(x) = 10^{-x}$ **6.** $f(x) = -6^{x-2}$ **7.** $f(x) = 1 - e^{2x}$

8. Evaluate (a) $\log_7 7^{-0.89}$ and (b) $4.6 \ln e^2$.

In Exercises 9–11, construct a table of values. Then sketch the graph of the function. Identify any asymptotes.

9. $f(x) = -\log_{10} x - 6$ **10.** $f(x) = \ln(x - 4)$ **11.** $f(x) = 1 + \ln(x + 6)$

In Exercises 12–14, evaluate the expression. Approximate your result to three decimal places.

12. $\log_7 44$ **13.** $\log_{2/5} 0.9$ **14.** $\log_{24} 68$

In Exercises 15 and 16, use the properties of logarithms to expand the expression as a sum, difference, and/or multiple of logarithms.

15. $\log_2 3a^4$ **16.** $\ln \dfrac{5\sqrt{x}}{6}$

In Exercises 17 and 18, condense the expression to the logarithm of a single quantity.

17. $\log_3 13 + \log_3 y$ **18.** $4 \ln x - 4 \ln y$

In Exercises 19 and 20, solve the equation algebraically. Approximate your result to three decimal places.

19. $\dfrac{1025}{8 + e^{4x}} = 5$ **20.** $\log_{10} x - \log_{10}(8 - 5x) = 2$

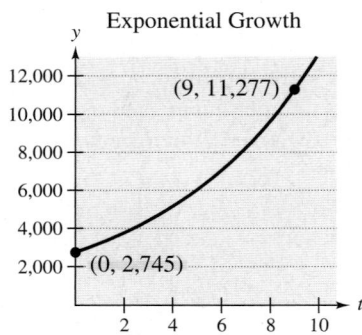

FIGURE FOR **21**

21. Find an exponential growth model for the graph shown in the figure.

22. The half-life of radioactive actinium (^{227}Ac) is 22 years. What percent of a present amount of radioactive actinium will remain after 19 years?

23. A model that can be used for predicting the height H (in centimeters) of a child based on his or her age is $H = 70.228 + 5.104x + 9.222 \ln x$, $\frac{1}{4} \le x \le 6$, where x is the age of the child in years. (Source: Snapshots of Applications in Mathematics)

(a) Construct a table of values. Then sketch the graph of the model.

(b) Use the graph from part (a) to estimate the height of a four-year-old child. Then calculate the actual height using the model.

Proofs in Mathematics

Each of the following three properties of logarithms can be proved by using properties of exponential functions.

Slide Rules

The slide rule was invented by William Oughtred (1574–1660) in 1625. The slide rule is a computational device with a sliding portion and a fixed portion. A slide rule enables you to perform multiplication by using Property 1 of the Properties of Logarithms. There are other slide rules that allow for the calculation of roots and trigonometric functions. Slide rules were used by mathematicians and engineers until the invention of the handheld calculator in 1972.

Properties of Logarithms *(p. 386)*

Let a be a positive number such that $a \neq 1$, and let n be a real number. If u and v are positive real numbers, the following properties are true.

Logarithm with Base a

1. $\log_a(uv) = \log_a u + \log_a v$

2. $\log_a \dfrac{u}{v} = \log_a u - \log_a v$

3. $\log_a u^n = n \log_a u$

Natural Logarithm

1. $\ln(uv) = \ln u + \ln v$

2. $\ln \dfrac{u}{v} = \ln u - \ln v$

3. $\ln u^n = n \ln u$

Proof

Let

$$x = \log_a u \quad \text{and} \quad y = \log_a v.$$

The corresponding exponential forms of these two equations are

$$a^x = u \quad \text{and} \quad a^y = v.$$

To prove Property 1, multiply u and v to obtain

$$uv = a^x a^y = a^{x+y}.$$

The corresponding logarithmic form of $uv = a^{x+y}$ is $\log_a(uv) = x + y$. So,

$$\log_a(uv) = \log_a u + \log_a v.$$

To prove Property 2, divide u by v to obtain

$$\frac{u}{v} = \frac{a^x}{a^y} = a^{x-y}.$$

The corresponding logarithmic form of $u/v = a^{x-y}$ is $\log_a(u/v) = x - y$. So,

$$\log_a(u/v) = \log_a u - \log_a v.$$

To prove Property 3, substitute a^x for u in the expression $\log_a u^n$, as follows.

$$\log_a u^n = \log_a(a^x)^n \qquad \text{Substitute } a^x \text{ for } u.$$
$$= \log_a a^{nx} \qquad \text{Property of exponents}$$
$$= nx \qquad \text{Inverse Property of Logarithms}$$
$$= n \log_a u \qquad \text{Substitute } \log_a u \text{ for } x.$$

So, $\log_a u^n = n \log_a u.$

1. Graph the exponential function $y = a^x$ for $a = 0.5$, 1.2, and 2.0. Which of these curves intersects the line $y = x$? Determine all positive numbers a for which the curve $y = a^x$ intersects the line $y = x$.

2. Use a graphing utility to graph $y_1 = e^x$ and each of the functions $y_2 = x^2$, $y_3 = x^3$, $y_4 = \sqrt{x}$, and $y_5 = |x|$. Which function increases at the fastest rate as x approaches $+\infty$?

3. Use the result of Exercise 2 to make a conjecture about the rate of growth of $y_1 = e^x$ and $y = x^n$, where n is a natural number and x approaches $+\infty$.

4. Use the results of Exercises 2 and 3 to describe what is implied when it is stated that a quantity is growing exponentially.

5. Given the exponential function

   $$f(x) = a^x$$

   show that

   (a) $f(u + v) = f(u) \cdot f(v)$.

   (b) $f(2x) = [f(x)]^2$.

6. Given that

   $$f(x) = \frac{e^x + e^{-x}}{2} \text{ and } g(x) = \frac{e^x - e^{-x}}{2}$$

   show that

   $$[f(x)]^2 - [g(x)]^2 = 1.$$

7. Use a graphing utility to compare the graph of the function $y = e^x$ with the graph of each given function. [$n!$ (read "n factorial") is defined as $n! = 1 \cdot 2 \cdot 3 \cdot \cdot \cdot (n - 1) \cdot n$.]

   (a) $y_1 = 1 + \dfrac{x}{1!}$

   (b) $y_2 = 1 + \dfrac{x}{1!} + \dfrac{x^2}{2!}$

   (c) $y_3 = 1 + \dfrac{x}{1!} + \dfrac{x^2}{2!} + \dfrac{x^3}{3!}$

8. Identify the pattern of successive polynomials given in Exercise 7. Extend the pattern one more term and compare the graph of the resulting polynomial function with the graph of $y = e^x$. What do you think this pattern implies?

9. Graph the function

   $$f(x) = e^x - e^{-x}.$$

   From the graph, the function appears to be one-to-one. Assuming that the function has an inverse function, find $f^{-1}(x)$.

10. Find a pattern for $f^{-1}(x)$ if

    $$f(x) = \frac{a^x + 1}{a^x - 1}$$

    where $a > 0$, $a \neq 1$.

11. By observation, identify the equation that corresponds to the graph. Explain your reasoning.

    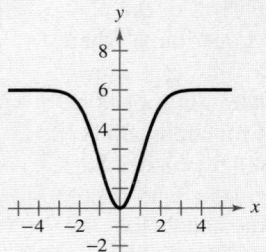

    (a) $y = 6e^{-x^2/2}$

    (b) $y = \dfrac{6}{1 + e^{-x/2}}$

    (c) $y = 6(1 - e^{-x^2/2})$

12. There are two options for investing $500. The first earns 7% compounded annually and the second earns 7% simple interest. The figure shows the growth of each investment over a 30-year period.

    (a) Identify which graph represents each type of investment. Explain your reasoning.

    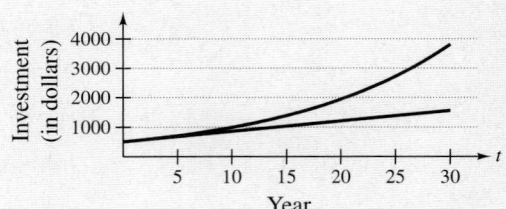

    (b) Verify your answer in part (a) by finding the equations that model the investment growth and graphing the models.

13. Two different samples of radioactive isotopes are decaying. The isotopes have initial amounts of c_1 and c_2, as well as half-lives of k_1 and k_2, respectively. Find the time required for the samples to decay to equal amounts.

14. A lab culture initially contains 500 bacteria. Two hours later, the number of bacteria has decreased to 200. Find the exponential decay model of the form

$$B = B_0 a^{kt}$$

that can be used to approximate the number of bacteria after t hours.

15. The table shows the colonial population estimates of the American colonies from 1700 to 1780. (Source: U.S. Census Bureau)

Year	Population
1700	250,900
1710	331,700
1720	466,200
1730	629,400
1740	905,600
1750	1,170,800
1760	1,593,600
1770	2,148,100
1780	2,780,400

In each of the following, let y represent the population in the year t, with $t = 0$ corresponding to 1700.

(a) Use the *regression* feature of a graphing utility to find an exponential model for the data.

(b) Use the *regression* feature of the graphing utility to find a quadratic model for the data.

(c) Use the graphing utility to plot the data and the models from parts (a) and (b) in the same viewing window.

(d) Which model is a better fit for the data? Would you use this model to predict the population of the United States in 2010? Explain your reasoning.

16. Show that $\dfrac{\log_a x}{\log_{a/b} x} = 1 + \log_a \dfrac{1}{b}$.

17. Solve $(\ln x)^2 = \ln x^2$.

 18. Use a graphing utility to compare the graph of the function $y = \ln x$ with the graph of each given function.

(a) $y_1 = x - 1$

(b) $y_2 = (x - 1) - \frac{1}{2}(x - 1)^2$

(c) $y_3 = (x - 1) - \frac{1}{2}(x - 1)^2 + \frac{1}{3}(x - 1)^3$

19. Identify the pattern of successive polynomials given in Exercise 18. Extend the pattern one more term and compare the graph of the resulting polynomial function with the graph of $y = \ln x$. What do you think the pattern implies?

20. Using

$$y = ab^x \text{ and } y = ax^b$$

take the natural logarithm of each side of each equation. What are the slope and y-intercept of the line relating x and $\ln y$ for $y = ab^x$? What are the slope and y-intercept of the line relating $\ln x$ and $\ln y$ for $y = ax^b$?

In Exercises 21 and 22, use the model

$$y = 80.4 - 11 \ln x, \quad 100 \le x \le 1500$$

which approximates the minimum required ventilation rate in terms of the air space per child in a public school classroom. In the model, x is the air space per child in cubic feet and y is the ventilation rate in cubic feet per minute.

21. Use a graphing utility to graph the function and approximate the required ventilation rate if there is 300 cubic feet of air space per child.

22. A classroom is designed for 30 students. The air conditioning system in the room has the capacity of moving 450 cubic feet of air per minute.

(a) Determine the ventilation rate per child, assuming that the room is filled to capacity.

(b) Estimate the air space required per child.

(c) Determine the minimum number of square feet of floor space required for the room if the ceiling height is 30 feet.

How to study Chapter 6

▶ **What you should learn**

In this chapter you will learn the following skills and concepts:

• How to find the inclination of a line, the angle between two lines, and the distance between a point and a line

• How to write the standard form of the equation of a parabola, an ellipse, and a hyperbola

• How to eliminate the *xy*-term in the equation of a conic and use the discriminant to identify a conic

• How to rewrite a set of parametric equations as a rectangular equation and find a set of parametric equations for a graph

• How to write equations in polar form and graph polar equations

▶ **Important Vocabulary**

As you encounter each new vocabulary term in this chapter, add the term and its definition to your notebook glossary.

Inclination (p. 426)
Angle between two lines (p. 427)
Conic section or conic (pp. 433, 490)
Parabola (pp. 434, 490)
Directrix (p. 434)
Focus or foci (pp. 434, 442)
Focal chord (p. 436)
Latus rectum (p. 436)
Tangent (p. 436)
Ellipse (pp. 442, 490)
Vertices (pp. 442, 451)
Major axis (p. 442)
Minor axis (p. 442)
Center (pp. 442, 451)
Eccentricity (pp. 446, 455, 490)

Hyperbola (pp. 451, 490)
Branches (p. 451)
Transverse axis (p. 451)
Asymptotes (p. 453)
Conjugate axis (p. 453)
Invariant under rotation (p. 464)
Discriminant (p. 464)
Parameter (p. 468)
Parametric equations (p. 468)
Plane curve (p. 468)
Orientation (p. 469)
Polar coordinate system (p. 476)
Pole or origin (p. 476)
Polar axis (p. 476)
Polar coordinates (p. 476)

Study Tools

Learning objectives in each section
Chapter Summary (p. 497)
Review Exercises (pp. 498–501)
Chapter Test (p. 502)
Cumulative Test for Chapters 4–6
  (pp. 503–504)

Additional Resources

Study and Solutions Guide
Interactive Trigonometry
Videotapes/DVD for Chapter 6
Trigonometry Website
Student Success Organizer

www.PerfectPhoto.CA/© Rob vanNostrand

6

Topics in Analytic Geometry

6.1 | Lines

▶ **What you should learn**
- How to find the inclination of a line
- How to find the angle between two lines
- How to find the distance between a point and a line

▶ **Why you should learn it**

The inclination of a line can be used to measure heights indirectly. For instance, in Exercise 56 on page 432, the inclination of a line can be used to determine the change in elevation from the top of the Johnstown Inclined Plane.

AP/Wide World Photos

Inclination of a Line

In a previous course, you should have learned that the graph of the linear equation

$$y = mx + b$$

is a nonvertical line with slope m and y-intercept $(0, b)$. There, the slope of a line was described as the rate of change in y with respect to x. In this section, you will look at the slope of a line in terms of the angle of inclination of the line.

Every nonhorizontal line must intersect the x-axis. The angle formed by such an intersection determines the **inclination** of the line, as specified in the following definition.

Definition of Inclination

The **inclination** of a nonhorizontal line is the positive angle θ (less than π) measured counterclockwise from the x-axis to the line. (See Figure 6.1.)

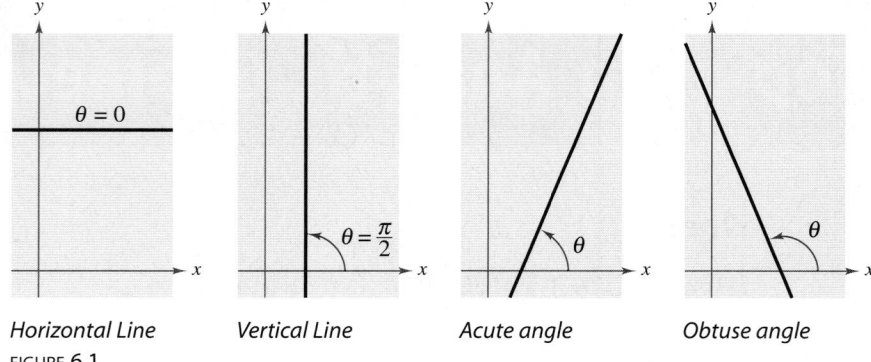

Horizontal Line   Vertical Line   Acute angle   Obtuse angle
FIGURE **6.1**

The inclination of a line is related to its slope in the following manner.

Inclination and Slope

If a nonvertical line has inclination θ and slope m, then

$$m = \tan \theta.$$

For a proof of the relation between inclination and slope, see Proofs in Mathematics on page 505.

The icon (www ○) identifies examples and concepts related to features of the Learning Tools CD-ROM and the *Interactive* and *Internet* versions of this text. For more details see the chart on pages *xix–xxiii*.

| Example 1 ▶ | **Finding the Inclination of a Line** | |

Find the inclination of the line $2x + 3y = 6$.

Solution

The slope of this line is $m = -\frac{2}{3}$. So, its inclination is determined from the equation

$$\tan \theta = -\frac{2}{3}.$$

From Figure 6.2, it follows that $\frac{\pi}{2} < \theta < \pi$. This means that

$$\theta = \pi + \arctan\left(-\frac{2}{3}\right)$$

$$\approx \pi + (-0.588)$$

$$= \pi - 0.588$$

$$\approx 2.554.$$

The angle of inclination is about 2.554 radians or about 146.3°.

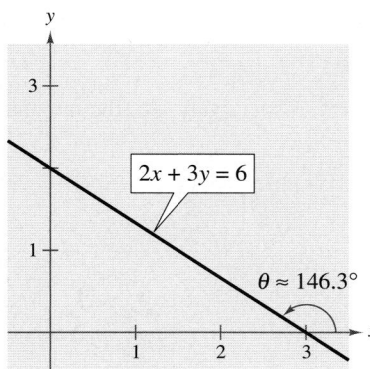

FIGURE 6.2

The *Interactive* CD-ROM and *Internet* versions of this text offer a Try It for each example in the text.

The Angle Between Two Lines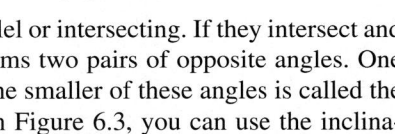

Two distinct lines in a plane are either parallel or intersecting. If they intersect and are nonperpendicular, their intersection forms two pairs of opposite angles. One pair is acute and the other pair is obtuse. The smaller of these angles is called the **angle between the two lines.** As shown in Figure 6.3, you can use the inclinations of the two lines to find the angle between the two lines. Specifically, if two lines have inclinations θ_1 and θ_2, the angle between the two lines is

$$\theta = \theta_2 - \theta_1$$

where $\theta_1 < \theta_2$. You can use the formula for the tangent of the difference of two angles

$$\tan \theta = \tan(\theta_2 - \theta_1)$$

$$= \frac{\tan \theta_2 - \tan \theta_1}{1 + \tan \theta_1 \tan \theta_2}$$

to obtain the formula for the angle between two lines.

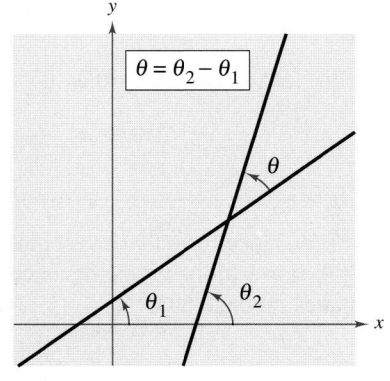

FIGURE 6.3

Angle Between Two Lines

If two nonperpendicular lines have slopes m_1 and m_2, the angle between the two lines is

$$\tan \theta = \left| \frac{m_2 - m_1}{1 + m_1 m_2} \right|.$$

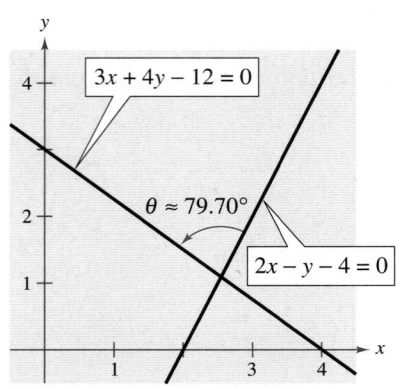

FIGURE 6.4

Example 2 ▶ Finding the Angle Between Two Lines

Find the angle between the two lines.

Line 1: $2x - y - 4 = 0$ *Line 2:* $3x + 4y - 12 = 0$

Solution

The two lines have slopes of $m_1 = 2$ and $m_2 = -\frac{3}{4}$, respectively. So, the tangent of the angle between the two lines is

$$\tan \theta = \left| \frac{m_2 - m_1}{1 + m_1 m_2} \right| = \left| \frac{(-3/4) - 2}{1 + (2)(-3/4)} \right| = \left| \frac{-11/4}{-2/4} \right| = \frac{11}{2}.$$

Finally, you can conclude that the angle is

$$\theta = \arctan \frac{11}{2} \approx 1.391 \text{ radians} \approx 79.70°$$

as shown in Figure 6.4.

The Distance Between a Point and a Line

Finding the distance between a line and a point not on the line is an application of perpendicular lines. This distance is defined as the length of the perpendicular line segment joining the point to the line, as shown in Figure 6.5.

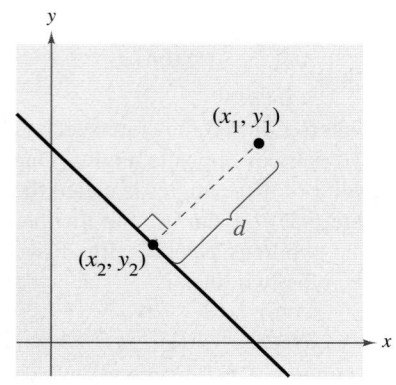

FIGURE 6.5

> **Distance Between a Point and a Line**
>
> The distance between the point (x_1, y_1) and the line $Ax + By + C = 0$ is
>
> $$d = \frac{|Ax_1 + By_1 + C|}{\sqrt{A^2 + B^2}}.$$

Remember that the values of A, B, and C in this distance formula correspond to the general equation of a line, $Ax + By + C = 0$. For a proof of the distance between a point and a line, see Proofs in Mathematics on page 505.

Example 3 ▶ Finding the Distance Between a Point and a Line

Find the distance between the point $(4, 1)$ and the line $y = 2x + 1$.

Solution

The general form of the equation is

$$-2x + y - 1 = 0.$$

So, the distance between the point and the line is

$$d = \frac{|-2(4) + 1(1) - 1|}{\sqrt{(-2)^2 + 1^2}} = \frac{8}{\sqrt{5}} \approx 3.58 \text{ units.}$$

The line and the point are shown in Figure 6.6.

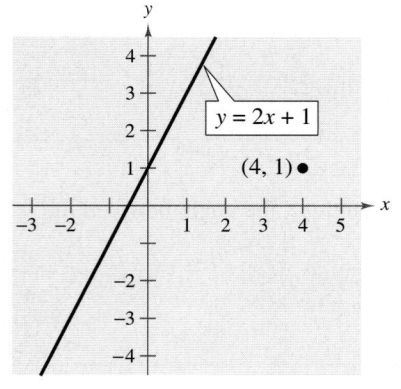

FIGURE 6.6

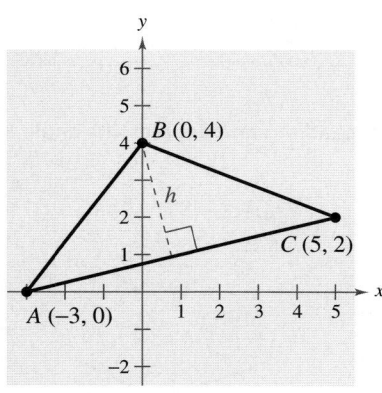

FIGURE **6.7**

Activities

1. Find the inclination of the line $5x - 4y = 20$.
   Answer: $\theta = 0.896$ radian or $51.34°$
2. Find the angle θ between the lines $x + 2y = 5$ and $3x - y = 6$.
   Answer: $\theta = 1.429$ radians or $81.87°$
3. Find the distance between the point $(3, -1)$ and the line $4x - 3y - 12 = 0$.
   Answer: $d = \frac{3}{5}$ unit

The *Interactive* CD-ROM and *Internet* versions of this text offer a Quiz for every section of the text.

Example 4 ▶   An Application of Two Distance Formulas

Figure 6.7 shows a triangle with vertices $A(-3, 0)$, $B(0, 4)$, and $C(5, 2)$.

a. Find the altitude from vertex B to side AC.

b. Find the area of the triangle.

Solution

a. To find the altitude, use the formula for the distance between line AC and the point $(0, 4)$. The equation of line AC is obtained as follows.

$$\text{Slope: } m = \frac{2 - 0}{5 + 3} = \frac{2}{8} = \frac{1}{4}$$

Equation:	$y - 0 = \dfrac{1}{4}(x + 3)$	Point-slope form
	$4y = x + 3$	Multiply each side by 4.
	$x - 4y + 3 = 0$	General form

So, the distance between this line and the point $(0, 4)$ is

$$\text{Altitude} = h = \frac{|1(0) + (-4)(4) + 3|}{\sqrt{1^2 + (-4)^2}} = \frac{13}{\sqrt{17}} \text{ units.}$$

b. Using the formula for the distance between two points, you can find the length of the base AC to be

$b = \sqrt{(5 + 3)^2 + (2 - 0)^2}$	Distance Formula
$= \sqrt{8^2 + 2^2}$	Simplify.
$= \sqrt{68}$	Simplify.
$= 2\sqrt{17}$ units.	Simplify.

Finally, the area of the triangle in Figure 6.7 is

$A = \dfrac{1}{2}bh$	Formula for the area of a triangle
$= \dfrac{1}{2}\left(2\sqrt{17}\right)\left(\dfrac{13}{\sqrt{17}}\right)$	Substitute for b and h.
$= 13$ square units.	Simplify.

Group Activity: Graphing Utility

Put your students in groups of two. Ask each group to write a graphing calculator program that finds the angle between two lines based on user input. Ask each group to demonstrate the program on an example or exercise in this section.

Writing ABOUT MATHEMATICS

Inclination and the Angle Between Two Lines   Discuss why the inclination of a line can be an angle that is larger than $\pi/2$, but the angle between two lines cannot be larger than $\pi/2$. Decide whether the following statement is true or false: "The inclination of a line is the angle between the line and the x-axis." Explain.

6.1 **Exercises**

In Exercises 1–8, find the slope of the line with inclination θ.

1.

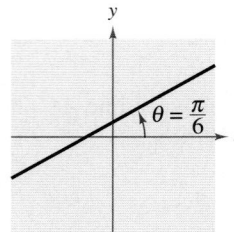

2.

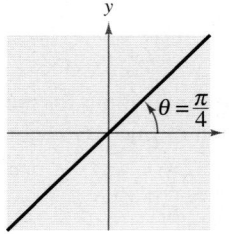

3.

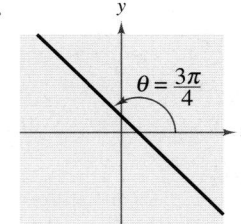

4.

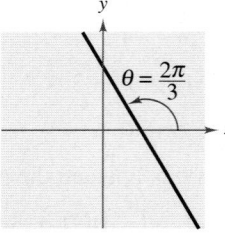

5. $\theta = \dfrac{\pi}{3}$ radians

6. $\theta = \dfrac{5\pi}{6}$ radians

7. $\theta = 1.27$ radians

8. $\theta = 2.88$ radians

In Exercises 9–14, find, in radians and degrees, the inclination θ of the line with a slope of m.

9. $m = -1$

10. $m = -2$

11. $m = 1$

12. $m = 2$

13. $m = \dfrac{3}{4}$

14. $m = -\dfrac{5}{2}$

In Exercises 15–18, find, in radians and degrees, the inclination θ of the line passing through the points.

15. $(6, 1), (10, 8)$

16. $(12, 8), (-4, -3)$

17. $(-2, 20), (10, 0)$

18. $(0, 100), (50, 0)$

In Exercises 19–22, find, in radians and degrees, the inclination θ of the line.

19. $6x - 2y + 8 = 0$

20. $4x + 5y - 9 = 0$

21. $5x + 3y = 0$

22. $x - y - 10 = 0$

In Exercises 23–32, find, in radians and degrees, the angle θ between the lines.

23. $3x + y = 3$
$x - y = 2$

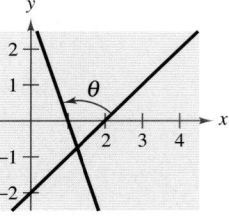

24. $x + 3y = 2$
$x - 2y = -3$

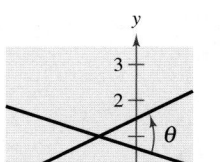

25. $x - y = 0$
$3x - 2y = -1$

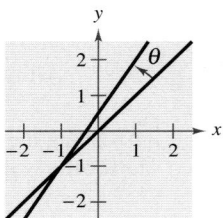

26. $2x - y = 2$
$4x + 3y = 24$

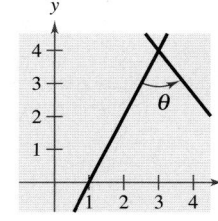

27. $x - 2y = 7$
$6x + 2y = 5$

28. $5x + 2y = 16$
$3x - 5y = -1$

29. $x + 2y = 8$
$x - 2y = 2$

30. $3x - 5y = 3$
$3x + 5y = 12$

31. $0.05x - 0.03y = 0.21$
$0.07x + 0.02y = 0.16$

32. $0.02x - 0.05y = -0.19$
$0.03x + 0.04y = 0.52$

Angle Measurement **In Exercises 33–36, find the slope of each side of the triangle and use the slopes to find the measures of the interior angles.**

33.

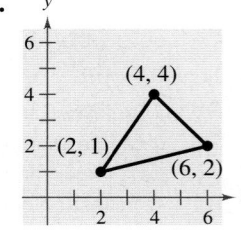

34.

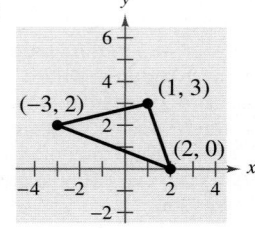

35. 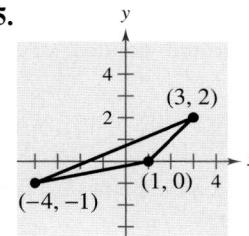 **36.**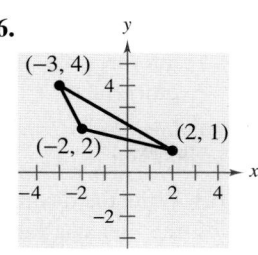

In Exercises 37–44, find the distance between the point and the line.

Point	Line
37. $(0, 0)$	$4x + 3y = 0$
38. $(0, 0)$	$2x - y = 4$
39. $(2, 3)$	$4x + 3y = 10$
40. $(-2, 1)$	$x - y = 2$
41. $(6, 2)$	$x + 1 = 0$
42. $(10, 8)$	$y - 4 = 0$
43. $(0, 8)$	$6x - y = 0$
44. $(4, 2)$	$x - y = 20$

Area **In Exercises 45–48, (a) find the altitude from vertex B of the triangle to side AC, and (b) find the area of the triangle.**

45. $A = (0, 0), B = (1, 4), C = (4, 0)$

46. $A = (0, 0), B = (4, 5), C = (5, -2)$

47. $A = \left(-\frac{1}{2}, \frac{1}{2}\right), B = (2, 3), C = \left(\frac{5}{2}, 0\right)$

48. $A = (-4, -5), B = (3, 10), C = (6, 12)$

In Exercises 49 and 50, find the distance between the parallel lines.

49. $x + y = 1$
$x + y = 5$

50. $3x - 4y = 1$
$3x - 4y = 10$

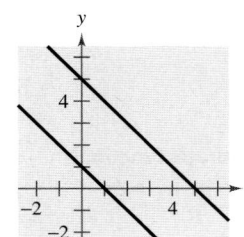 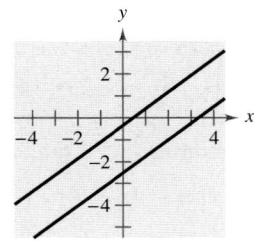

51. Grade of a Road A straight road rises with an inclination of 0.10 radian from the horizontal (see figure). Find the slope of the road and the change in elevation over a two-mile stretch of the road.

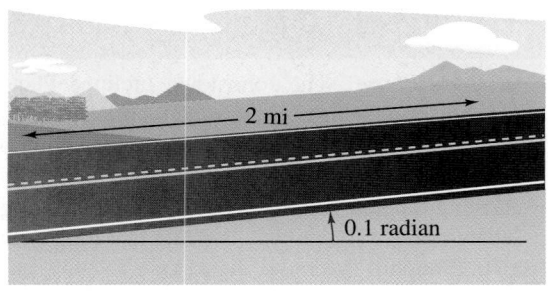

FIGURE FOR 51

52. Grade of a Road A straight road rises with an inclination of 0.20 radian from the horizontal. Find the slope of the road and the change in elevation over a one-mile stretch of the road.

53. Conveyor Design A moving conveyor is built so that it rises 1 meter for each 3 meters of horizontal travel.

(a) Find the inclination of the conveyor.

(b) The conveyor runs between two floors in a factory. The distance between the floors is 5 meters. Find the length of the conveyor.

54. Pitch of a Roof A roof has a rise of 3 feet for every horizontal change of 5 feet (see figure). Find the inclination of the roof.

55. Truss Find the angles α and β shown in the drawing of the roof truss.

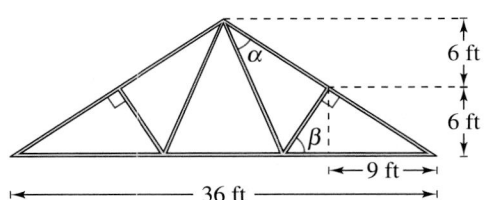

▶ Model It

56. *Inclined Plane* The Johnstown Inclined Plane in Johnstown, Pennsylvania is an inclined railway that was designed to carry people to the hilltop community of Westmont. It also proved useful in carrying people and vehicles to safety during severe floods. The railway is 896.5 feet long with a 70.9% uphill grade (see figure).

(a) Find the inclination θ of the railway.

(b) Find the change in elevation from the top of the railway.

(c) Using the origin as the base of the inclined plane, find the equation of the line that models the railway track.

(d) Sketch a graph of the equation you found in part (c).

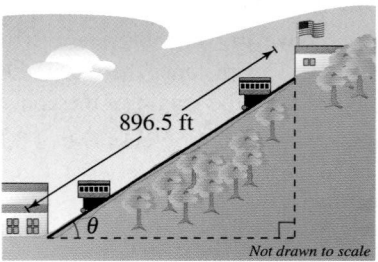

896.5 ft

θ

Not drawn to scale

Synthesis

True or False? In Exercises 57 and 58, determine whether the statement is true or false. Justify your answer.

57. A line that has an inclination greater than $\pi/2$ radians has a negative slope.

58. To find the angle between two lines whose angles of inclination θ_1 and θ_2 are known, substitute θ_1 and θ_2 for m_1 and m_2, respectively, in the formula for the angle between two lines.

59. *Exploration* Consider a line with slope m and y-intercept $(0, 4)$.

(a) Write the distance d between the origin and the line as a function of m.

(b) Graph the function in part (a).

(c) Find the slope that yields the maximum distance between the origin and the line.

(d) Find the asymptote of the graph in part (b) and interpret its meaning in the context of the problem.

60. *Exploration* Consider a line with slope m and y-intercept $(0, 4)$.

(a) Write the distance d between the point $(3, 1)$ and the line as a function of m.

(b) Graph the function in part (a).

(c) Find the slope that yields the maximum distance between the point and the line.

(d) Is it possible for the distance to be 0? If so, what is the slope of the line that yields a distance of 0?

(e) Find the asymptote of the graph in part (b) and interpret its meaning in the context of the problem.

Review

In Exercises 61–66, find all x-intercepts and y-intercepts of the graph of the quadratic function.

61. $f(x) = (x - 7)^2$

62. $f(x) = (x + 9)^2$

63. $f(x) = (x - 5)^2 - 5$

64. $f(x) = (x + 11)^2 + 12$

65. $f(x) = x^2 - 7x - 1$

66. $f(x) = x^2 + 9x - 22$

In Exercises 67–72, write the quadratic function in standard form by completing the square. Identify the vertex of the function.

67. $f(x) = 3x^2 + 2x - 16$

68. $f(x) = 2x^2 - x - 21$

69. $f(x) = 5x^2 + 34x - 7$

70. $f(x) = -x^2 - 8x - 15$

71. $f(x) = 6x^2 - x - 12$

72. $f(x) = -8x^2 - 34x - 21$

In Exercises 73–76, graph the quadratic function.

73. $f(x) = (x - 4)^2 + 3$

74. $f(x) = 6 - (x + 1)^2$

75. $g(x) = 2x^2 - 3x + 1$

76. $g(x) = -x^2 + 6x - 8$

6.2 Introduction to Conics: Parabolas

▶ **What you should learn**

- How to recognize a conic as the intersection of a plane and a double-napped cone
- How to write the standard form of the equation of a parabola
- How to use the reflective property of parabolas to solve real-life problems

▶ **Why you should learn it**

Parabolas can be used to model and solve many types of real-life problems. For instance, in Exercise 62 on page 439, a parabola is used to model the cables of the Golden Gate Bridge.

Conics

Conic sections were discovered during the classical Greek period, 600 to 300 B.C. The early Greeks were concerned largely with the geometric properties of conics. It was not until the 17th century that the broad applicability of conics became apparent and played a prominent role in the early development of calculus.

Each **conic section** (or simply **conic**) is the intersection of a plane and a double-napped cone. Notice in Figure 6.8 that in the formation of the four basic conics, the intersecting plane does not pass through the vertex of the cone. When the plane does pass through the vertex, the resulting figure is a **degenerate conic,** as shown in Figure 6.9.

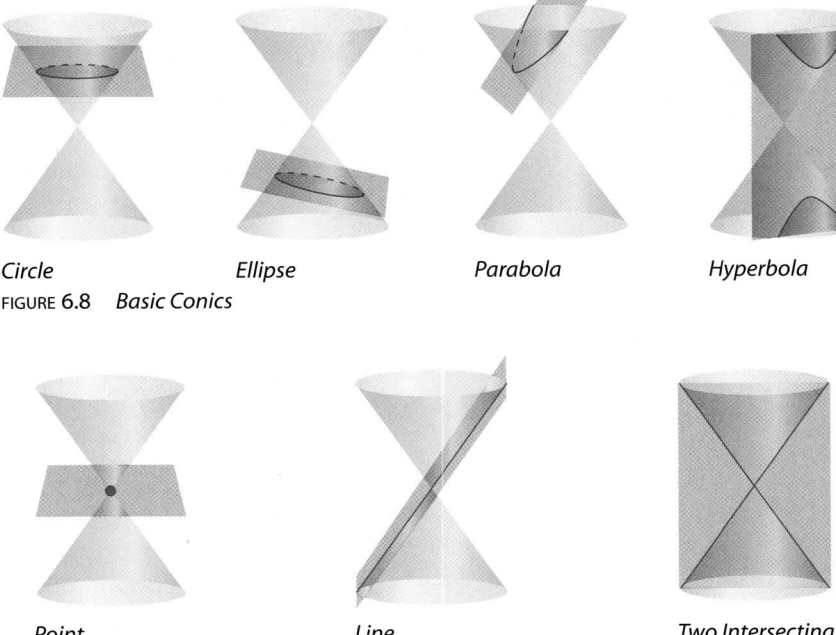

Circle   Ellipse   Parabola   Hyperbola

FIGURE **6.8** *Basic Conics*

Point   Line   Two Intersecting Lines

FIGURE **6.9** *Degenerate Conics*

There are several ways to approach the study of conics. You could begin by defining conics in terms of the intersections of planes and cones, as the Greeks did, or you could define them algebraically, in terms of the general second-degree equation

$$Ax^2 + Bxy + Cy^2 + Dx + Ey + F = 0.$$

However, you will study a third approach, in which each of the conics is defined as a **locus** (collection) of points satisfying a geometric property. For example, a circle is defined as the collection of all points (x, y) that are equidistant from a fixed point (h, k). This leads to the standard equation of a circle

$$(x - h)^2 + (y - k)^2 = r^2. \qquad \text{Equation of circle}$$

This study of conics is from a locus-of-points approach, which leads to the development of the standard equation for each conic. Your students should know the standard equations of all conics well. Make sure they understand the relationship of h and k to the horizontal and vertical shifts.

Parabolas

In Section P.3, you learned that the graph of the quadratic equation

$$y = ax^2 + bx + c$$

is a parabola that opens upward or downward. The following definition of a parabola is more general in the sense that it is independent of the orientation of the parabola.

Definition of Parabola

A **parabola** is the set of all points (x, y) in a plane that are equidistant from a fixed line (**directrix**) and a fixed point (**focus**) not on the line.

The midpoint between the focus and the directrix is called the **vertex,** and the line passing through the focus and the vertex is called the **axis** of the parabola. Note in Figure 6.10 that a parabola is symmetric with respect to its axis. Using the definition of a parabola, you can derive the following **standard form** of the equation of a parabola whose directrix is parallel to the x-axis or to the y-axis.

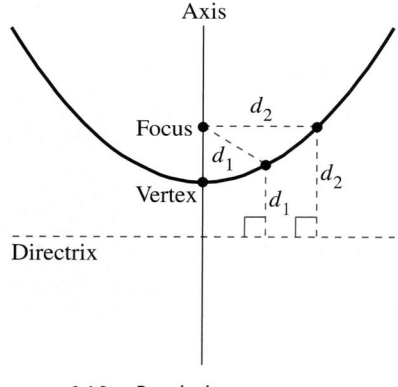

FIGURE 6.10 *Parabola*

Standard Equation of a Parabola

The **standard form of the equation of a parabola** with vertex at (h, k) is as follows.

$$(x - h)^2 = 4p(y - k), \ p \neq 0 \qquad \text{Vertical axis, directrix: } y = k - p$$

$$(y - k)^2 = 4p(x - h), \ p \neq 0 \qquad \text{Horizontal axis, directrix: } x = h - p$$

The focus lies on the axis p units (*directed distance*) from the vertex. If the vertex is at the origin $(0, 0)$, the equation takes one of the following forms.

$$x^2 = 4py \qquad \text{Vertical axis}$$

$$y^2 = 4px \qquad \text{Horizontal axis}$$

See Figure 6.11.

For a proof of the standard form of the equation of a parabola, see Proofs in Mathematics on page 506.

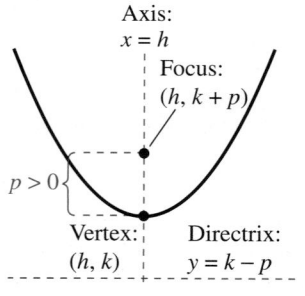

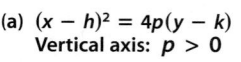

(a) $(x - h)^2 = 4p(y - k)$
   Vertical axis: $p > 0$

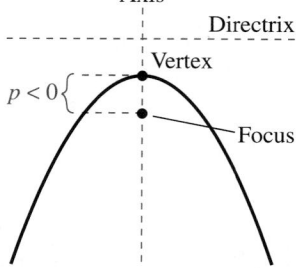

(b) $(x - h)^2 = 4p(y - k)$
   Vertical axis: $p < 0$

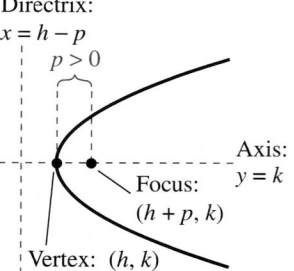

(c) $(y - k)^2 = 4p(x - h)$
   Horizontal axis: $p > 0$

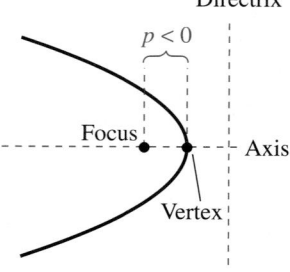

(d) $(y - k)^2 = 4p(x - h)$
   Horizontal axis: $p < 0$

FIGURE 6.11

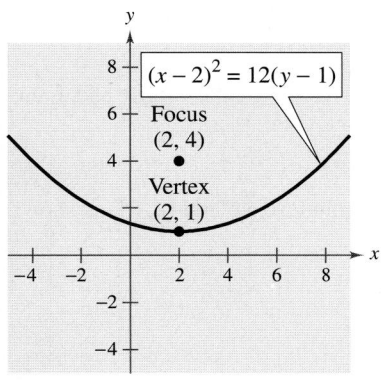

FIGURE **6.12**

Example 1 ▶ **Finding the Standard Equation of a Parabola**

Find the standard form of the equation of the parabola with vertex $(2, 1)$ and focus $(2, 4)$.

Solution

Because the axis of the parabola is vertical, consider the equation

$$(x - h)^2 = 4p(y - k)$$

where $h = 2$, $k = 1$, and $p = 4 - 1 = 3$. So, the standard form is

$$(x - 2)^2 = 12(y - 1).$$

You can obtain the more common quadratic form as follows.

$$x^2 - 4x + 4 = 12y - 12 \qquad \text{Multiply.}$$

$$x^2 - 4x + 16 = 12y \qquad \text{Add 12 to each side.}$$

$$\frac{1}{12}(x^2 - 4x + 16) = y \qquad \text{Divide each side by 12.}$$

The graph of this parabola is shown in Figure 6.12.

Example 2 ▶ **Finding the Focus of a Parabola**

Find the focus of the parabola

$$y = -\frac{1}{2}x^2 - x + \frac{1}{2}.$$

Solution

To find the focus, convert to standard form by completing the square.

$$y = -\frac{1}{2}x^2 - x + \frac{1}{2} \qquad \text{Write original equation.}$$

$$-2y = x^2 + 2x - 1 \qquad \text{Multiply each side by } -2.$$

$$1 - 2y = x^2 + 2x \qquad \text{Add 1 to each side.}$$

$$1 + 1 - 2y = x^2 + 2x + 1 \qquad \text{Complete the square.}$$

$$2 - 2y = x^2 + 2x + 1 \qquad \text{Combine like terms.}$$

$$-2(y - 1) = (x + 1)^2 \qquad \text{Standard form}$$

Comparing this equation with

$$(x - h)^2 = 4p(y - k)$$

you can conclude that $h = -1$, $k = 1$, and $p = -\frac{1}{2}$. Because p is negative, the parabola opens downward, as shown in Figure 6.13. So, the focus of the parabola is

$$(h, k + p) = \left(-1, \frac{1}{2}\right). \qquad \text{Focus}$$

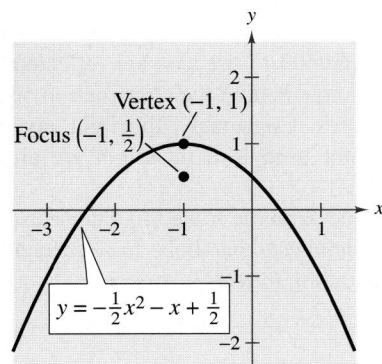

FIGURE **6.13**

Technology

Try using a graphing utility to confirm the equation found in Example 3. To do this, it helps to graph the equation using two separate equations:

$$y_1 = \sqrt{8x} \qquad \text{Upper part}$$

and

$$y_2 = -\sqrt{8x}. \qquad \text{Lower part}$$

Example 3 ▶   Vertex at the Origin

Find the standard equation of the parabola with vertex at the origin and focus $(2, 0)$.

Solution

The axis of the parabola is horizontal, passing through $(0, 0)$ and $(2, 0)$, as shown in Figure 6.14.

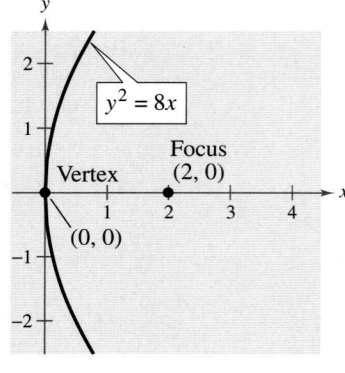

FIGURE **6.14**

So, the standard form is $y^2 = 4px$, where $h = 0$, $k = 0$, and $p = 2$. So, the equation is $y^2 = 8x$.

Application

A line segment that passes through the focus of a parabola and has endpoints on the parabola is called a **focal chord.** The specific focal chord perpendicular to the axis of the parabola is called the **latus rectum.**

    Parabolas occur in a wide variety of applications. For instance, a parabolic reflector can be formed by revolving a parabola around its axis. The resulting surface has the property that all incoming rays parallel to the axis are reflected through the focus of the parabola. This is the principle behind the construction of the parabolic mirrors used in reflecting telescopes. Conversely, the light rays emanating from the focus of a parabolic reflector used in a flashlight are all parallel to one another, as shown in Figure 6.15.

    A line is **tangent** to a parabola at a point on the parabola if the line intersects, but does not cross, the parabola at the point. Tangent lines to parabolas have special properties related to the use of parabolas in constructing reflective surfaces.

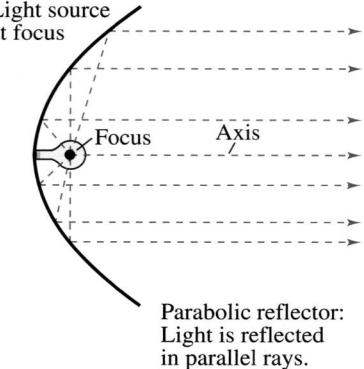

Parabolic reflector: Light is reflected in parallel rays.

FIGURE **6.15**

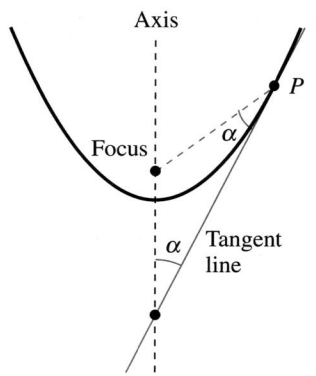

FIGURE **6.16**

Reflective Property of a Parabola

The tangent line to a parabola at a point P makes equal angles with the following two lines (see Figure 6.16).

1. The line passing through P and the focus

2. The axis of the parabola

Your students may question why $d_1 = \frac{1}{4} - b$ rather than $b - \frac{1}{4}$. You may need to explain that the distance must be positive, so the order of subtraction is important.

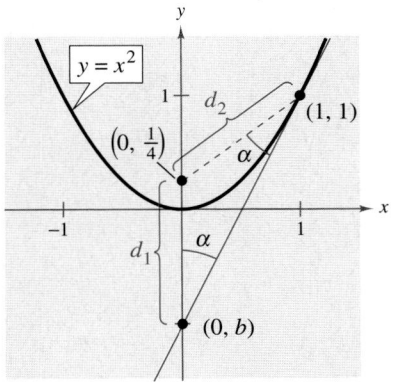

$y = x^2$

FIGURE **6.17**

Activities

1. Find the vertex, focus, and directrix of the parabola $x^2 - 6x - 4y + 5 = 0$.
   Answer: Vertex $(3, -1)$; Focus $(3, 0)$; Directrix $y = -2$
2. Find an equation of the parabola with vertex $(4, 0)$ and directrix $x = 5$.
   Answer: $y^2 + 4x - 16 = 0$
3. Find an equation of the tangent line to the parabola $y = 2x^2$ at the point $(1, 2)$.
   Answer: $y = 4x - 2$

Example 4 ▶ Finding the Tangent Line at a Point on a Parabola

Find the equation of the tangent line to the parabola given by $y = x^2$ at the point $(1, 1)$.

Solution

For this parabola, $p = \frac{1}{4}$ and the focus is $\left(0, \frac{1}{4}\right)$, as shown in Figure 6.17. You can find the y-intercept $(0, b)$ of the tangent line by equating the lengths of the two sides of the isosceles triangle shown in Figure 6.17:

$$d_1 = \frac{1}{4} - b$$

and

$$d_2 = \sqrt{(1 - 0)^2 + [1 - (1/4)]^2} = \frac{5}{4}.$$

Setting $d_1 = d_2$ produces

$$\frac{1}{4} - b = \frac{5}{4}$$

$$b = -1.$$

So, the slope of the tangent line is

$$m = \frac{1 - (-1)}{1 - 0} = 2$$

and the equation of the tangent line in slope-intercept form is

$$y = 2x - 1.$$

Writing ABOUT MATHEMATICS

Television Antenna Dishes Cross sections of television antenna dishes are parabolic in shape. Write a paragraph explaining why these dishes are parabolic.

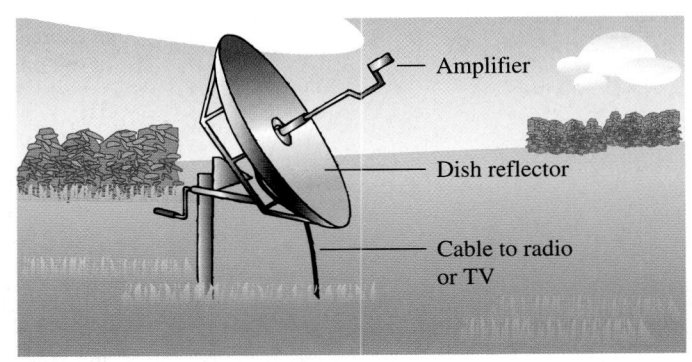

6.2 Exercises

In Exercises 1–4, describe in words how a plane could intersect with the double-napped cone shown to form the conic section.

1. Circle   **2.** Ellipse

3. Parabola   **4.** Hyperbola

In Exercises 5–10, match the equation with its graph. [The graphs are labeled (a), (b), (c), (d), (e), and (f).]

(a)
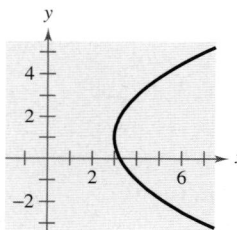

(b)
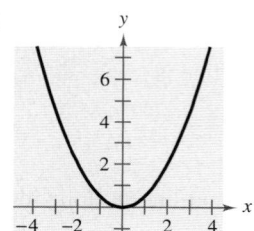

(c)
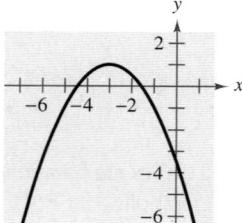

(d)
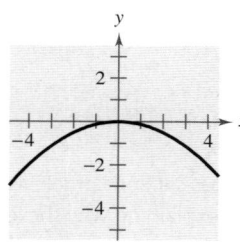

(e)
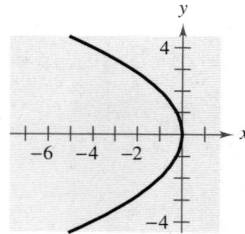

(f)
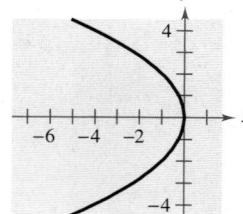

5. $y^2 = -4x$ **6.** $x^2 = 2y$

7. $x^2 = -8y$ **8.** $y^2 = -12x$

9. $(y - 1)^2 = 4(x - 3)$

10. $(x + 3)^2 = -2(y - 1)$

In Exercises 11–24, find the vertex, focus, and directrix of the parabola and sketch its graph.

11. $y = \frac{1}{2}x^2$ **12.** $y = -2x^2$

13. $y^2 = -6x$ **14.** $y^2 = 3x$

15. $x^2 + 6y = 0$ **16.** $x + y^2 = 0$

17. $(x - 1)^2 + 8(y + 2) = 0$

18. $(x + 5) + (y - 1)^2 = 0$

19. $\left(x + \frac{3}{2}\right)^2 = 4(y - 2)$

20. $\left(x + \frac{1}{2}\right)^2 = 4(y - 1)$

21. $y = \frac{1}{4}(x^2 - 2x + 5)$

22. $x = \frac{1}{4}(y^2 + 2y + 33)$

23. $y^2 + 6y + 8x + 25 = 0$

24. $y^2 - 4y - 4x = 0$

 In Exercises 25–28, find the vertex, focus, and directrix of the parabola. Use a graphing utility to graph the parabola.

25. $x^2 + 4x + 6y - 2 = 0$

26. $x^2 - 2x + 8y + 9 = 0$

27. $y^2 + x + y = 0$

28. $y^2 - 4x - 4 = 0$

 In Exercises 29 and 30, the equations of a parabola and a tangent line to the parabola are given. Use a graphing utility to graph both equations in the same viewing window. Determine the coordinates of the point of tangency.

	Parabola	*Tangent Line*
29.	$y^2 - 8x = 0$	$x - y + 2 = 0$
30.	$x^2 + 12y = 0$	$x + y - 3 = 0$

In Exercises 31–42, find the standard form of the equation of the parabola with its vertex at the origin.

31.
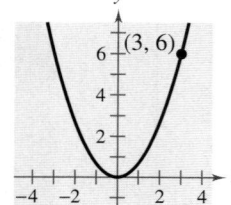

32.
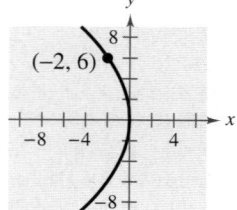

33. Focus: $\left(0, -\frac{3}{2}\right)$ **34.** Focus: $(2, 0)$

35. Focus: $(-2, 0)$ **36.** Focus: $(0, -2)$

37. Directrix: $y = -1$ **38.** Directrix: $y = 3$

39. Directrix: $x = 2$ **40.** Directrix: $x = -3$

41. Horizontal axis and passes through the point $(4, 6)$

42. Vertical axis and passes through the point $(-3, -3)$

In Exercises 43–52, find the standard form of the equation of the parabola.

43.
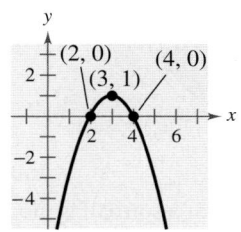

44.
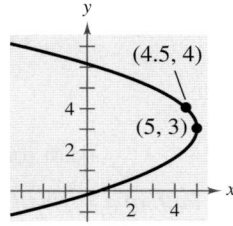

45.
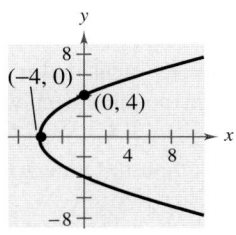

46.
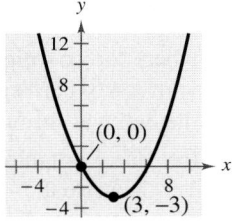

47. Vertex: $(5, 2)$; Focus: $(3, 2)$

48. Vertex: $(-1, 2)$; Focus: $(-1, 0)$

49. Vertex: $(0, 4)$; Directrix: $y = 2$

50. Vertex: $(-2, 1)$; Directrix: $x = 1$

51. Focus: $(2, 2)$; Directrix: $x = -2$

52. Focus: $(0, 0)$; Directrix: $y = 8$

In Exercises 53 and 54, change the equation of the parabola so that its graph matches the description.

53. $(y - 3)^2 = 6(x + 1)$; upper half of parabola

54. $(y + 1)^2 = 2(x - 4)$; lower half of parabola

In Exercises 55–58, find an equation of the tangent line to the parabola at the given point, and find the x-intercept of the line.

55. $x^2 = 2y$, $(4, 8)$ **56.** $x^2 = 2y$, $\left(-3, \frac{9}{2}\right)$

57. $y = -2x^2$, $(-1, -2)$ **58.** $y = -2x^2$, $(2, -8)$

59. ***Revenue*** The revenue R generated by the sale of x units of a patio furniture set is $(x - 106)^2 = -\frac{4}{5}(R - 14{,}045)$. Use a graphing utility to graph the function and approximate the number of sales that will maximize revenue.

 60. ***Revenue*** The revenue R generated by the sale of x units of a digital camera is

$$(x - 135)^2 = -\frac{5}{7}(R - 25{,}515).$$

Use a graphing utility to graph the function and approximate the number of sales that will maximize revenue.

61. ***Satellite Antenna*** The receiver in a parabolic television dish antenna is 4.5 feet from the vertex and is located at the focus (see figure). Find an equation of a cross section of the reflector. (Assume that the dish is directed upward and the vertex is at the origin.)

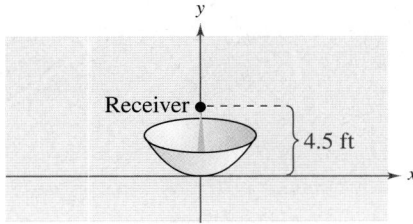

▶ Model It

62. ***Suspension Bridge*** Each cable of the Golden Gate Bridge is suspended (in the shape of a parabola) between two towers that are 1280 meters apart. The top of each tower is 152 meters above the roadway. The cables touch the roadway midway between the towers.

(a) Draw a sketch of the bridge. Locate the origin of a rectangular coordinate system at the center of the roadway. Label the coordinates of the known points.

(b) Write an equation that models the cables.

(c) Complete the table by finding the height y of the suspension cables over the roadway at a distance of x meters from the center of the bridge.

Distance, x	Height, y
0	
250	
400	
500	
1000	

63. *Road Design* Roads are often designed with parabolic surfaces to allow rain to drain off. A particular road that is 32 feet wide is 0.4 foot higher in the center than it is on the sides (see figure).

(a) Find an equation of the parabola that models the road surface. (Assume that the origin is at the center of the road.)

(b) How far from the center of the road is the road surface 0.1 foot lower than in the middle?

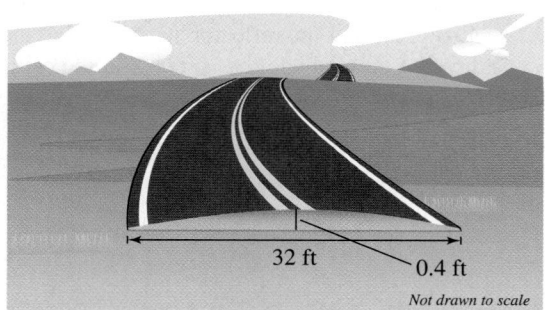

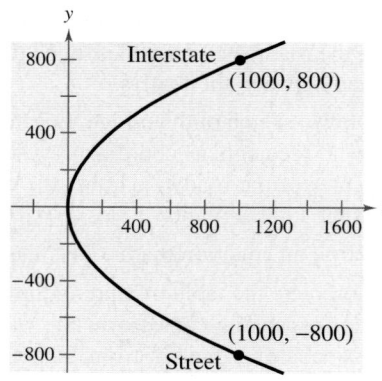

Cross section of road surface

64. *Highway Design* Highway engineers design a parabolic curve for an entrance ramp from a straight street to an interstate highway (see figure). Find an equation of the parabola.

65. *Satellite Orbit* An Earth satellite in a 100-mile-high circular orbit around Earth has a velocity of approximately 17,500 miles per hour. If this velocity is multiplied by $\sqrt{2}$, the satellite will have the minimum velocity necessary to escape Earth's gravity and it will follow a parabolic path with the center of Earth as the focus (see figure).

(a) Find the escape velocity of the satellite.

(b) Find an equation of the parabolic path of the satellite (assume that the radius of Earth is 4000 miles).

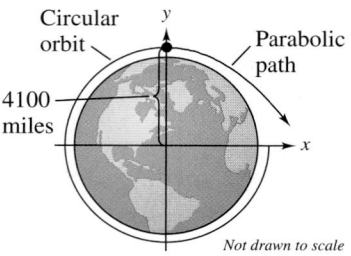

FIGURE FOR **65**

66. *Path of a Softball* The path of a softball is modeled by

$$-12.5(y - 7.125) = (x - 6.25)^2$$

where the coordinates x and y are measured in feet, with $x = 0$ corresponding to the position from which the ball was thrown.

(a) Use a graphing utility to graph the trajectory of the softball.

(b) Use the *trace* feature of the graphing utility to approximate the highest point and the range of the trajectory.

Projectile Motion In Exercises 67 and 68, consider the path of a projectile projected horizontally with a velocity of v feet per second at a height of s feet, where the model for the path is

$$x^2 = -\frac{v^2}{16}(y - s).$$

In this model, air resistance is disregarded and y is the height (in feet) of the projectile and x is the horizontal distance (in feet) the projectile travels.

67. A ball is thrown from the top of a 75-foot tower with a velocity of 32 feet per second.

(a) Find the equation of the parabolic path.

(b) How far does the ball travel horizontally before striking the ground?

68. A bomber flying due east at 550 miles per hour at an altitude of 42,000 feet releases a bomb. Determine the distance the bomb travels horizontally before striking the ground.

Synthesis

True or False? **In Exercises 69 and 70, determine whether the statement is true or false. Justify your answer.**

69. It is possible for a parabola to intersect its directrix.

70. If the vertex and focus of a parabola are on a horizontal line, then the directrix of the parabola is vertical.

71. ***Exploration*** Consider the parabola $x^2 = 4py$.

 (a) Use a graphing utility to graph the parabola for $p = 1$, $p = 2$, $p = 3$, and $p = 4$. Describe the effect on the graph when p increases.

(b) Locate the focus for each parabola in part (a).

(c) For each parabola in part (a), find the length of the chord passing through the focus and parallel to the directrix (see figure). How can the length of this chord be determined directly from the standard form of the equation of the parabola?

(d) Explain how the result of part (c) can be used as a sketching aid when graphing parabolas.

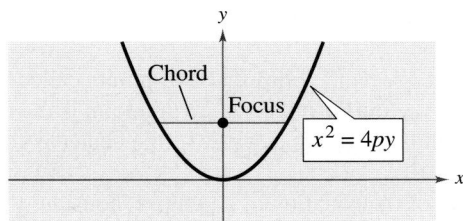

72. ***Geometry*** The area of the shaded region in the figure is

$$A = \frac{8}{3}p^{1/2}b^{3/2}.$$

(a) Find the area when $p = 2$ and $b = 4$.

(b) Give a geometric explanation of why the area approaches 0 as p approaches 0.

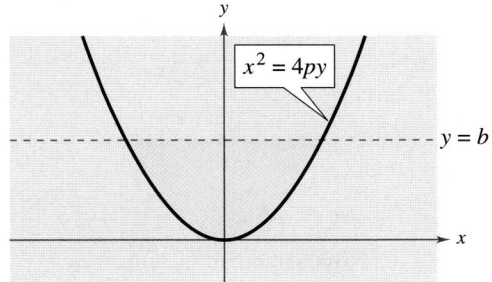

73. ***Exploration*** Let (x_1, y_1) be the coordinates of a point on the parabola $x^2 = 4py$. The equation of the line tangent to the parabola at the point is

$$y - y_1 = \frac{x_1}{2p}(x - x_1).$$

What is the slope of the tangent line?

74. ***Writing*** In your own words, state the reflective property of a parabola.

Review

75. Find a polynomial with integer coefficients that has the zeros 3, $2 + i$, and $2 - i$.

76. Find all the zeros of

$$f(x) = 2x^3 - 3x^2 + 50x - 75$$

if one of the zeros is $x = \frac{3}{2}$.

77. Find all the zeros of the function

$$g(x) = 6x^4 + 7x^3 - 29x^2 - 28x + 20$$

if two of the zeros are $x = \pm 2$.

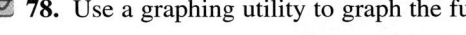 **78.** Use a graphing utility to graph the function

$$h(x) = 2x^4 + x^3 - 19x^2 - 9x + 9.$$

Use the graph to approximate the zeros of h.

In Exercises 79–86, use the information to solve the triangle.

79. $A = 35°, a = 10, b = 7$

80. $B = 54°, b = 18, c = 11$

81. $A = 40°, B = 51°, c = 3$

82. $B = 26°, C = 104°, a = 19$

83. $a = 7, b = 10, c = 16$

84. $a = 58, b = 28, c = 75$

85. $A = 65°, b = 5, c = 12$

86. $B = 71°, a = 21, c = 29$

6.3 Ellipses

▶ **What you should learn**

- How to write the standard form of the equation of an ellipse
- How to use properties of ellipses to model and solve real-life problems
- How to find the eccentricity of an ellipse

▶ **Why you should learn it**

Ellipses can be used to model and solve many types of real-life problems. For instance, in Exercise 52 on page 449, an ellipse is used to model the orbit of Halley's comet.

Introduction

The second type of conic is called an **ellipse,** and is defined as follows.

> **Definition of Ellipse**
>
> An **ellipse** is the set of all points (x, y) in a plane, the sum of whose distances from two distinct fixed points **(foci)** is constant. See Figure 6.18.

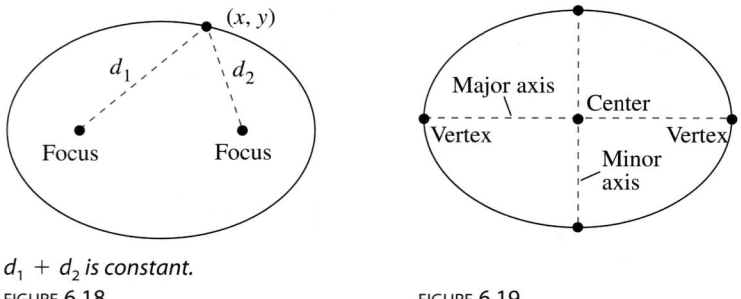

$d_1 + d_2$ is constant.
FIGURE **6.18**

FIGURE **6.19**

The line through the foci intersects the ellipse at two points called **vertices.** The chord joining the vertices is the **major axis,** and its midpoint is the **center** of the ellipse. The chord perpendicular to the major axis at the center is the **minor axis** of the ellipse. See Figure 6.19.

To derive the standard form of the equation of an ellipse, consider the ellipse in Figure 6.20 with the following points: center, (h, k); vertices, $(h \pm a, k)$; foci, $(h \pm c, k)$. Note that the center is the midpoint of the segment joining the foci.

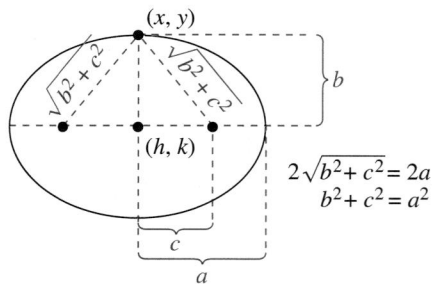

FIGURE **6.20**

The sum of the distances from any point on the ellipse to the two foci is constant. Using a vertex point, this constant sum is

$$(a + c) + (a - c) = 2a \qquad \text{Length of major axis}$$

or simply the length of the major axis. Now, if you let (x, y) be *any* point on the ellipse, the sum of the distances between (x, y) and the two foci must also be $2a$.

When discussing ellipses, you might also choose to discuss the latera recta as background for Exercises 54–58.

That is,

$$\sqrt{[x - (h - c)]^2 + (y - k)^2} + \sqrt{[x - (h + c)]^2 + (y - k)^2} = 2a.$$

Finally, in Figure 6.20, you can see that $b^2 = a^2 - c^2$, which implies that the equation of the ellipse is

$$b^2(x - h)^2 + a^2(y - k)^2 = a^2 b^2$$

$$\frac{(x - h)^2}{a^2} + \frac{(y - k)^2}{b^2} = 1.$$

You would obtain a similar equation in the derivation by starting with a vertical major axis. Both results are summarized as follows.

Standard Equation of an Ellipse

The standard form of the equation of an ellipse, with center (h, k) and major and minor axes of lengths $2a$ and $2b$, respectively, where $0 < b < a$, is

$$\frac{(x - h)^2}{a^2} + \frac{(y - k)^2}{b^2} = 1 \qquad \text{Major axis is horizontal.}$$

$$\frac{(x - h)^2}{b^2} + \frac{(y - k)^2}{a^2} = 1. \qquad \text{Major axis is vertical.}$$

The foci lie on the major axis, c units from the center, with $c^2 = a^2 - b^2$. If the center is at the origin $(0, 0)$, the equation takes one of the following forms.

$$\frac{x^2}{a^2} + \frac{y^2}{b^2} = 1 \quad \begin{array}{l}\text{Major axis is}\\\text{horizontal.}\end{array} \qquad \frac{x^2}{b^2} + \frac{y^2}{a^2} = 1 \quad \begin{array}{l}\text{Major axis is}\\\text{vertical.}\end{array}$$

Figure 6.21 shows both the horizontal and vertical orientations for an ellipse.

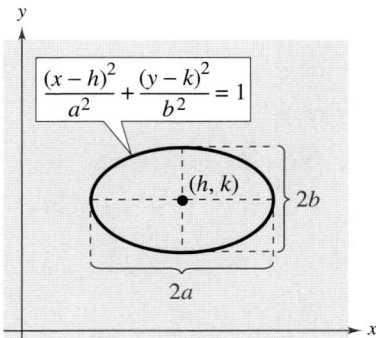

Major axis is horizontal.
FIGURE **6.21**

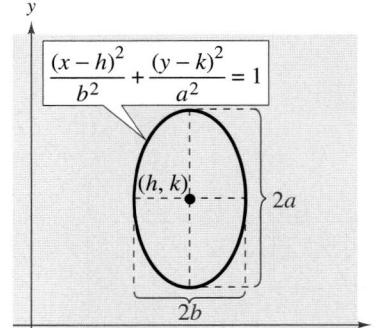

Major axis is vertical.

You can visualize the definition of an ellipse by imagining two thumbtacks placed at the foci, as shown in Figure 6.22. If the ends of a fixed length of string are fastened to the thumbtacks and the string is drawn taut with a pencil, the path traced by the pencil will be an ellipse.

FIGURE **6.22**

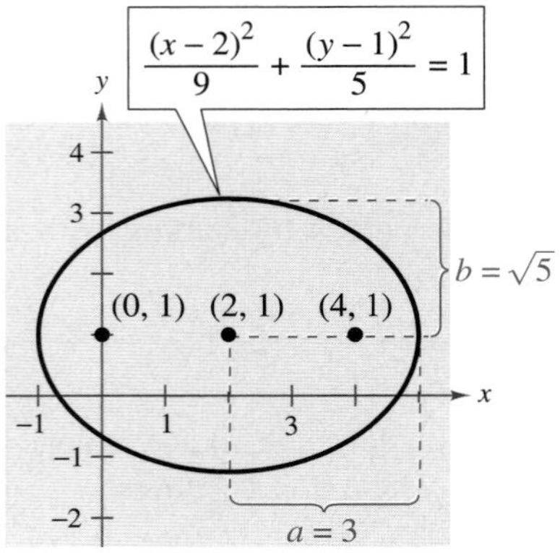

Example 1 ▶   Finding the Standard Equation of an Ellipse

Find the standard form of the equation of the ellipse having foci at $(0, 1)$ and $(4, 1)$ and a major axis of length 6.

Solution

Because the foci occur at $(0, 1)$ and $(4, 1)$, the center of the ellipse is $(2, 1)$ and the distance from the center to one of the foci is $c = 2$. Because $2a = 6$, you know that $a = 3$. Now, from $c^2 = a^2 - b^2$, you have

$$b = \sqrt{a^2 - c^2} = \sqrt{3^2 - 2^2} = \sqrt{5}.$$

Because the major axis is horizontal, the standard equation is

$$\frac{(x - 2)^2}{3^2} + \frac{(y - 1)^2}{(\sqrt{5})^2} = 1.$$

This equation simplifies to

$$\frac{(x - 2)^2}{9} + \frac{(y - 1)^2}{5} = 1. \qquad \text{See Figure 6.23.}$$

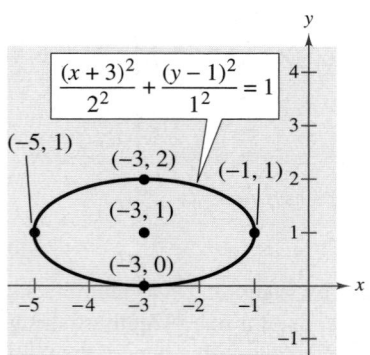

$$\frac{(x - 2)^2}{9} + \frac{(y - 1)^2}{5} = 1$$

$b = \sqrt{5}$

$(0, 1) \quad (2, 1) \quad (4, 1)$

$a = 3$

FIGURE **6.23**

In Example 1, note the use of the equation $c^2 = a^2 - b^2$. Do not confuse this equation with the Pythagorean Theorem—there is a difference in sign.

Example 2 ▶   Writing an Equation in Standard Form

Sketch the graph of the ellipse $x^2 + 4y^2 + 6x - 8y + 9 = 0$.

Solution

Begin by writing the original equation in standard form. In the fourth step, note that 9 and 4 are added to *both* sides of the equation when completing the squares.

$$x^2 + 4y^2 + 6x - 8y + 9 = 0 \qquad \text{Write original equation.}$$

$$\left(x^2 + 6x + \quad\right) + \left(4y^2 - 8y + \quad\right) = -9 \qquad \text{Group terms.}$$

$$\left(x^2 + 6x + \quad\right) + 4\left(y^2 - 2y + \quad\right) = -9 \qquad \text{Factor 4 out of } y\text{-terms.}$$

$$(x^2 + 6x + 9) + 4(y^2 - 2y + 1) = -9 + 9 + 4(1)$$

$$(x + 3)^2 + 4(y - 1)^2 = 4 \qquad \text{Write in completed square form.}$$

$$\frac{(x + 3)^2}{4} + \frac{(y - 1)^2}{1} = 1 \qquad \text{Divide each side by 4.}$$

$$\frac{(x + 3)^2}{2^2} + \frac{(y - 1)^2}{1^2} = 1 \qquad \text{Write in standard form.}$$

From this standard form, it follows that the center is $(h, k) = (-3, 1)$. Because the denominator of the x-term is $a^2 = 2^2$, the endpoints of the major axis lie two units to the right and left of the center. Similarly, because the denominator of the y-term is $b^2 = 1^2$, the endpoints of the minor axis lie one unit up and down from the center. The ellipse is shown in Figure 6.24.

Remind your students that completing the square must be performed twice to write the equation of the ellipse in standard form in Example 2.

$$\frac{(x + 3)^2}{2^2} + \frac{(y - 1)^2}{1^2} = 1$$

$(-5, 1)$

$(-3, 2)$

$(-1, 1)$

$(-3, 1)$

$(-3, 0)$

FIGURE **6.24**

Example 3 ▶ Analyzing an Ellipse

Find the center, vertices, and foci of the ellipse $4x^2 + y^2 - 8x + 4y - 8 = 0$.

Solution

By completing the square, you can write the original equation in standard form.

$$4x^2 + y^2 - 8x + 4y - 8 = 0$$

$$\left(4x^2 - 8x + \right) + \left(y^2 + 4y + \right) = 8$$

$$4\left(x^2 - 2x + \right) + \left(y^2 + 4y + \right) = 8$$

$$4(x^2 - 2x + 1) + (y^2 + 4y + 4) = 8 + 4(1) + 4$$

$$4(x - 1)^2 + (y + 2)^2 = 16$$

$$\frac{(x - 1)^2}{4} + \frac{(y + 2)^2}{16} = 1$$

$$\frac{(x - 1)^2}{2^2} + \frac{(y + 2)^2}{4^2} = 1$$

The major axis is vertical, where $h = 1$, $k = -2$, $a = 4$, $b = 2$, and

$$c = \sqrt{a^2 - b^2} = \sqrt{16 - 4} = \sqrt{12} = 2\sqrt{3}.$$

So, you have the following.

Center: $(1, -2)$	Vertices: $(1, -6)$	Foci: $\left(1, -2 - 2\sqrt{3}\right)$
	$(1, 2)$	$\left(1, -2 + 2\sqrt{3}\right)$

The graph of the ellipse is shown in Figure 6.25.

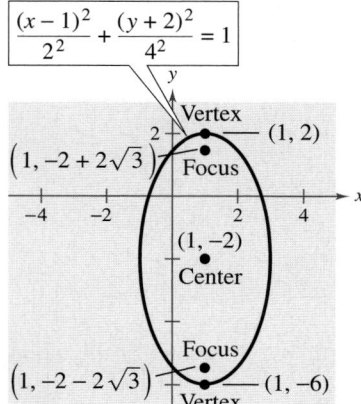

$$\frac{(x - 1)^2}{2^2} + \frac{(y + 2)^2}{4^2} = 1$$

FIGURE **6.25**

Technology

You can use a graphing utility to graph an ellipse by graphing the upper and lower portions in the same viewing window. For instance, to graph the ellipse in Example 3, first solve for y to get

$$y_1 = -2 + 4\sqrt{1 - \frac{(x - 1)^2}{4}} \quad \text{and} \quad y_2 = -2 - 4\sqrt{1 - \frac{(x - 1)^2}{4}}.$$

Use a viewing window in which $-6 \le x \le 9$ and $-7 \le y \le 3$. You should obtain the graph shown below.

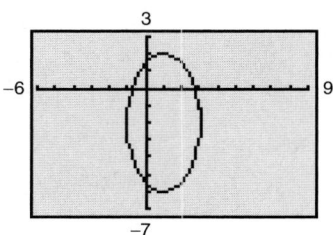

Application

Ellipses have many practical and aesthetic uses. For instance, machine gears, supporting arches, and acoustic designs often involve elliptical shapes. The orbits of satellites and planets are also ellipses. Example 4 investigates the elliptical orbit of the moon about Earth.

Example 4 ▶	An Application Involving an Elliptical Orbit

The moon travels about Earth in an elliptical orbit with Earth at one focus, as shown in Figure 6.26. The major and minor axes of the orbit have lengths of 768,806 kilometers and 767,746 kilometers, respectively. Find the greatest and smallest distances (the *apogee* and *perigee*) from Earth's center to the moon's center.

Solution

Because $2a = 768,806$ and $2b = 767,746$, you have

$$a = 384,403 \text{ and } b = 383,873$$

which implies that

$$c = \sqrt{a^2 - b^2}$$
$$= \sqrt{384,403^2 - 383,873^2}$$
$$\approx 20,179.$$

So, the greatest distance between the center of Earth and the center of the moon is

$$a + c \approx 404,582 \text{ kilometers}$$

and the smallest distance is

$$a - c \approx 364,224 \text{ kilometers}.$$

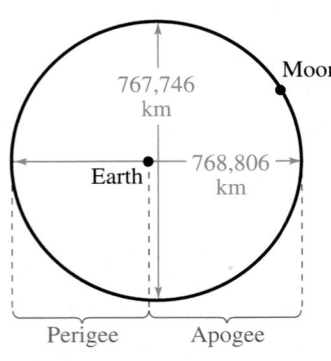

FIGURE **6.26**

Eccentricity

One of the reasons it was difficult for early astronomers to detect that the orbits of the planets are ellipses is that the foci of the planetary orbits are relatively close to their centers, and so the orbits are nearly circular. To measure the ovalness of an ellipse, you can use the concept of **eccentricity.**

Definition of Eccentricity

The **eccentricity** e of an ellipse is given by the ratio

$$e = \frac{c}{a}.$$

Note that $0 < e < 1$ for *every* ellipse.

Activities

1. Find the center, foci, vertices, and eccentricity of the ellipse
$$\frac{(x - 2)^2}{25} + \frac{(y + 1)^2}{9} = 1.$$
Answer: Center $(2, -1)$;
Foci $(-2, -1)$, $(6, -1)$;
Vertices $(-3, -1)$, $(7, -1)$; $e = \frac{4}{5}$

2. Rewrite the equation of the ellipse in standard form:
$$9x^2 + 5y^2 + 36x - 30y + 36 = 0.$$
Answer: $\dfrac{(x + 2)^2}{5} + \dfrac{(y - 3)^2}{9} = 1$

To see how this ratio is used to describe the shape of an ellipse, note that because the foci of an ellipse are located along the major axis between the vertices and the center, it follows that

$$0 < c < a.$$

For an ellipse that is nearly circular, the foci are close to the center and the ratio c/a is small, as shown in Figure 6.27. On the other hand, for an elongated ellipse, the foci are close to the vertices, and the ratio c/a is close to 1, as shown in Figure 6.28.

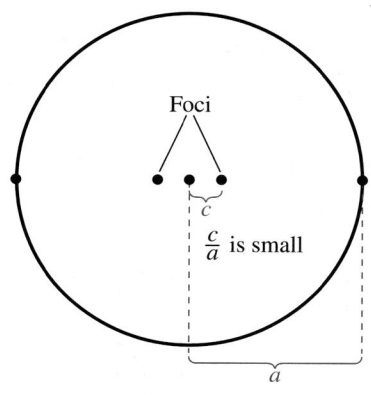

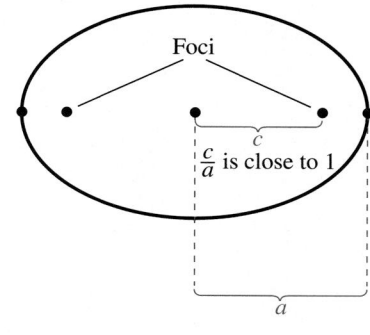

FIGURE **6.27**

FIGURE **6.28**

The orbit of the moon has an eccentricity of $e \approx 0.0525$, and the eccentricities of the nine planetary orbits are as follows.

Mercury:	$e \approx 0.2056$	Saturn:	$e \approx 0.0543$
Venus:	$e \approx 0.0068$	Uranus:	$e \approx 0.0460$
Earth:	$e \approx 0.0167$	Neptune:	$e \approx 0.0082$
Mars:	$e \approx 0.0934$	Pluto:	$e \approx 0.2481$
Jupiter:	$e \approx 0.0484$		

The time it takes Saturn to orbit the sun is equal to 29.5 Earth years.

Writing ABOUT MATHEMATICS

Ellipses and Circles

a. Show that the equation of an ellipse can be written as
$$\frac{(x - h)^2}{a^2} + \frac{(y - k)^2}{a^2(1 - e^2)} = 1.$$

b. For the equation in part (a), let $a = 4$, $h = 1$, and $k = 2$, and use a graphing utility to graph the ellipse for $e = 0.95$, $e = 0.75$, $e = 0.5$, $e = 0.25$, and $e = 0.1$. Discuss the changes in the shape of the ellipse as e approaches 0.

c. Make a conjecture about the shape of the graph in part (b) when $e = 0$. What is the equation of this ellipse? What is another name for an ellipse with an eccentricity of 0?

6.3 Exercises

In Exercises 1–6, match the equation with its graph. [The graphs are labeled (a), (b), (c), (d), (e), and (f).]

(a)
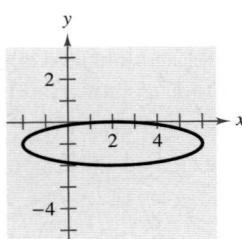

(b)
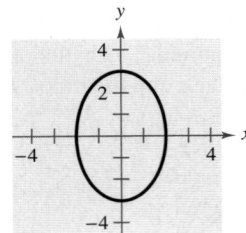

(c)
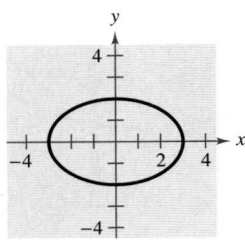

(d)
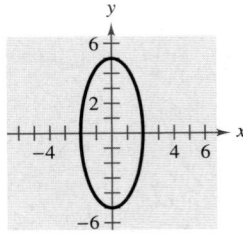

(e)
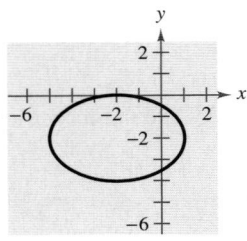

(f)
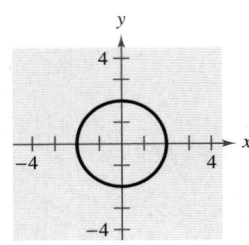

1. $\dfrac{x^2}{4} + \dfrac{y^2}{9} = 1$

2. $\dfrac{x^2}{9} + \dfrac{y^2}{4} = 1$

3. $\dfrac{x^2}{4} + \dfrac{y^2}{25} = 1$

4. $\dfrac{x^2}{4} + \dfrac{y^2}{4} = 1$

5. $\dfrac{(x-2)^2}{16} + (y+1)^2 = 1$

6. $\dfrac{(x+2)^2}{9} + \dfrac{(y+2)^2}{4} = 1$

In Exercises 7–22, find the center, vertices, foci, and eccentricity of the ellipse, and sketch its graph.

7. $\dfrac{x^2}{25} + \dfrac{y^2}{16} = 1$

8. $\dfrac{x^2}{81} + \dfrac{y^2}{144} = 1$

9. $\dfrac{x^2}{5} + \dfrac{y^2}{9} = 1$

10. $\dfrac{x^2}{64} + \dfrac{y^2}{28} = 1$

11. $\dfrac{(x+3)^2}{16} + \dfrac{(y-5)^2}{25} = 1$

12. $\dfrac{(x-4)^2}{12} + \dfrac{(y+3)^2}{16} = 1$

13. $\dfrac{(x+5)^2}{9/4} + (y-1)^2 = 1$

14. $(x+2)^2 + \dfrac{(y+4)^2}{1/4} = 1$

15. $9x^2 + 4y^2 + 36x - 24y + 36 = 0$

16. $9x^2 + 4y^2 - 54x + 40y + 37 = 0$

17. $x^2 + 5y^2 - 8x - 30y - 39 = 0$

18. $3x^2 + y^2 + 18x - 2y - 8 = 0$

19. $6x^2 + 2y^2 + 18x - 10y + 2 = 0$

20. $x^2 + 4y^2 - 6x + 20y - 2 = 0$

21. $16x^2 + 25y^2 - 32x + 50y + 16 = 0$

22. $9x^2 + 25y^2 - 36x - 50y + 60 = 0$

In Exercises 23–26, use a graphing utility to graph the ellipse. Find the center, foci, and vertices. (Recall that it may be necessary to solve the equation for y and obtain two equations.)

23. $5x^2 + 3y^2 = 15$

24. $3x^2 + 4y^2 = 12$

25. $12x^2 + 20y^2 - 12x + 40y - 37 = 0$

26. $36x^2 + 9y^2 + 48x - 36y - 72 = 0$

In Exercises 27–34, find the standard form of the equation of the ellipse with center at the origin.

27.
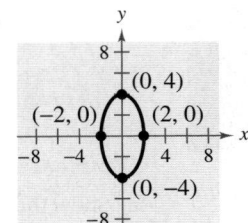

28.
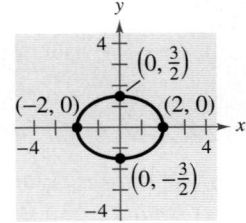

29. Vertices: $(\pm 6, 0)$; Foci: $(\pm 2, 0)$

30. Vertices: $(0, \pm 8)$; Foci: $(0, \pm 4)$

31. Foci: $(\pm 5, 0)$; Major axis of length 12

32. Foci: $(\pm 2, 0)$; Major axis of length 8

33. Vertices: $(0, \pm 5)$; Passes through the point $(4, 2)$

34. Major axis vertical; Passes through the points $(0, 4)$ and $(2, 0)$

In Exercises 35–46, find the standard form of the equation of the ellipse.

35.

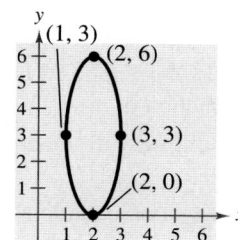

36.

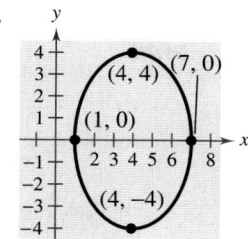

37.

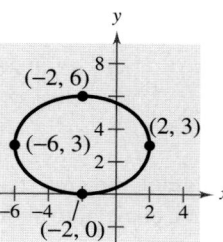

38.

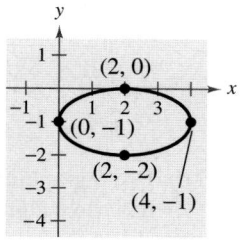

39. Vertices: $(0, 4)$, $(4, 4)$; Minor axis of length 2

40. Foci: $(0, 0)$, $(4, 0)$; Major axis of length 8

41. Foci: $(0, 0)$, $(0, 8)$; Major axis of length 16

42. Center: $(2, -1)$; Vertex: $\left(2, \frac{1}{2}\right)$; Minor axis of length 2

43. Center: $(0, 4)$; $a = 2c$; Vertices: $(-4, 4)$, $(4, 4)$

44. Center: $(3, 2)$; $a = 3c$; Foci: $(1, 2)$, $(5, 2)$

45. Vertices: $(0, 2)$, $(4, 2)$; Endpoints of the minor axis: $(2, 3)$, $(2, 1)$

46. Vertices: $(5, 0)$, $(5, 12)$; Endpoints of the minor axis: $(1, 6)$, $(9, 6)$

47. Find an equation of the ellipse with vertices $(\pm 5, 0)$ and eccentricity $e = \frac{3}{5}$.

48. Find an equation of the ellipse with vertices $(0, \pm 8)$ and eccentricity $e = \frac{1}{2}$.

49. *Architecture* A semielliptical arch over a tunnel for a road through a mountain has a major axis of 80 feet and a height at the center of 30 feet.

(a) Draw a rectangular coordinate system on a sketch of the tunnel with the center of the road entering the tunnel at the origin. Identify the coordinates of the known points.

(b) Find an equation of the semielliptical arch over the tunnel.

(c) Determine the height of the arch 5 feet from each edge of the tunnel.

50. *Architecture* A fireplace arch is to be constructed in the shape of a semiellipse. The opening is to have a height of 2 feet at the center and a width of 6 feet along the base (see figure). The contractor draws the outline of the ellipse using tacks as described near the beginning of this section. Give the required positions of the tacks and the length of the string.

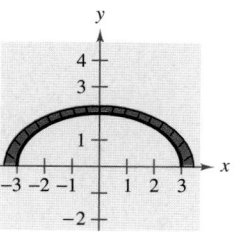

51. *Geometry* The area of the ellipse in the figure is twice the area of the circle. What is the length of the major axis? (*Hint:* The area of an ellipse is $A = \pi ab$.)

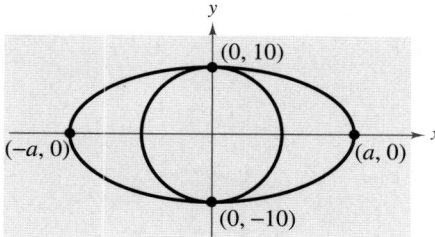

▶ **Model It**

52. *Comet Orbit* Halley's comet has an elliptical orbit, with the sun at one focus. The eccentricity of the orbit is approximately 0.97. The length of the major axis of the orbit is approximately 36.18 astronomical units. (An astronomical unit is about 93 million miles.)

(a) Find an equation of the orbit. Place the center of the orbit at the origin, and place the major axis on the x-axis.

(b) Use a graphing utility to graph the equation of the orbit.

(c) Find the greatest (apogee) and smallest (perigee) distances from the sun's center to the comet's center.

53. *Satellite Orbit* The first artificial satellite to orbit Earth was Sputnik I (launched by Russia in 1957). Its highest point above Earth's surface was 938 kilometers, and its lowest point was 212 kilometers (see figure). The radius of Earth is 6378 kilometers. Find the eccentricity of the orbit.

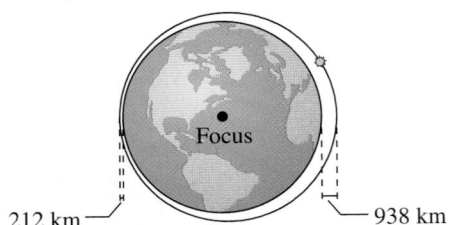

212 km   Focus   938 km

54. *Geometry* A line segment through a focus of an ellipse with endpoints on the ellipse and perpendicular to the major axis is called a **latus rectum** of the ellipse. Therefore, an ellipse has two latera recta. Knowing the length of the latera recta is helpful in sketching an ellipse because it yields other points on the curve (see figure). Show that the length of each latus rectum is $2b^2/a$.

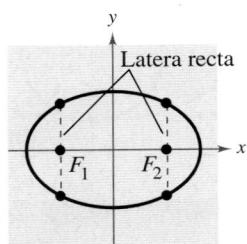

Latera recta

F_1 F_2

In Exercises 55–58, sketch the graph of the ellipse, using latera recta (see Exercise 54).

55. $\dfrac{x^2}{9} + \dfrac{y^2}{16} = 1$ **56.** $\dfrac{x^2}{4} + \dfrac{y^2}{1} = 1$

57. $9x^2 + 4y^2 = 36$ **58.** $5x^2 + 3y^2 = 15$

Synthesis

True or False? **In Exercises 59 and 60, determine whether the statement is true or false. Justify your answer.**

59. The graph of $x^2 + 4y^4 - 4 = 0$ is an ellipse.

60. It is easier to distinguish the graph of an ellipse from the graph of a circle if the eccentricity of the ellipse is large (close to 1).

61. *Exploration* Consider the ellipse

$$\frac{x^2}{a^2} + \frac{y^2}{b^2} = 1, \quad a + b = 20.$$

(a) The area of the ellipse is $A = \pi ab$. Write the area of the ellipse as a function of a.

(b) Find the equation of an ellipse with an area of 264 square centimeters.

(c) Complete the table using your equation from part (a), and make a conjecture about the shape of the ellipse with maximum area.

a	8	9	10	11	12	13
A						

(d) Use a graphing utility to graph the area function and use the graph to make a conjecture about the shape of the ellipse that yields a maximum area.

62. *Think About It* Near the beginning of this section it was noted that an ellipse can be drawn using two thumbtacks, a string of fixed length (greater than the distance between the two tacks), and a pencil. If the ends of the string are fastened at the tacks and the string is drawn taut with a pencil, the path traced by the pencil is an ellipse.

(a) What is the length of the string in terms of a?

(b) Explain why the path is an ellipse.

Review

In Exercises 63–66, solve the equation.

63. $x^2 - 10x = 0$ **64.** $100 - (x - 5)^2 = 0$

65. $x^3 - x = 0$ **66.** $16x^2 - 40x + 25 = 0$

In Exercises 67–70, evaluate the function at each specified value of the independent variable and simplify.

67. $f(x) = 5x - 8$

   (a) $f(9)$ (b) $f(-4)$ (c) $f(x - 7)$

68. $f(x) = x^2 - 10x$

   (a) $f(4)$ (b) $f(-8)$ (c) $f(x - 4)$

69. $f(x) = \sqrt{x - 12} - 9$

   (a) $f(12)$ (b) $f(40)$ (c) $f\left(-\sqrt{36}\right)$

70. $f(x) = \begin{cases} -3x, & x < -1 \\ 2x + 1, & x \geq -1 \end{cases}$

   (a) $f(-1)$ (b) $f(7)$ (c) $f(-6)$

6.4 Hyperbolas

▶ What you should learn

- How to write the standard form of the equation of a hyperbola
- How to find the asymptotes of a hyperbola
- How to use properties of hyperbolas to solve real-life problems
- How to classify a conic from its general equation

▶ Why you should learn it

Hyperbolas can be used to model and solve many types of real-life problems. For instance, in Exercise 39 on page 459, hyperbolas are used in long distance radio navigation for aircraft and ships.

Introduction

The third type of conic is called a **hyperbola.** The definition of a hyperbola is similar to that of an ellipse. The difference is that for an ellipse the *sum* of the distances between the foci and a point on the ellipse is fixed, whereas for a hyperbola the *difference* of the distances between the foci and a point on the hyperbola is fixed.

Definition of Hyperbola

A **hyperbola** is the set of all points (x, y) in a plane, the difference of whose distances from two distinct fixed points **(foci)** is a positive constant. See Figure 6.29.

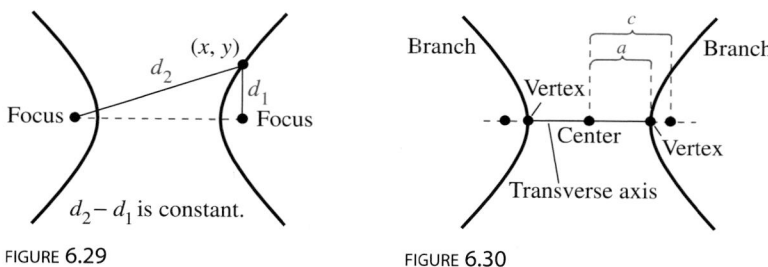

$d_2 - d_1$ is constant.

FIGURE **6.29**   FIGURE **6.30**

The graph of a hyperbola has two disconnected **branches.** The line through the two foci intersects the hyperbola at its two **vertices.** The line segment connecting the vertices is the **transverse axis,** and the midpoint of the transverse axis is the **center** of the hyperbola. See Figure 6.30. The development of the standard form of the equation of a hyperbola is similar to that of an ellipse.

Standard Equation of a Hyperbola

The **standard form of the equation of a hyperbola** with center (h, k) is

$$\frac{(x - h)^2}{a^2} - \frac{(y - k)^2}{b^2} = 1 \qquad \text{Transverse axis is horizontal.}$$

$$\frac{(y - k)^2}{a^2} - \frac{(x - h)^2}{b^2} = 1. \qquad \text{Transverse axis is vertical.}$$

The vertices are a units from the center, and the foci are c units from the center. Moreover, $c^2 = a^2 + b^2$. If the center of the hyperbola is at the origin $(0, 0)$, the equation takes one of the following forms.

$$\frac{x^2}{a^2} - \frac{y^2}{b^2} = 1 \qquad \text{Transverse axis is horizontal.} \qquad \frac{y^2}{a^2} - \frac{x^2}{b^2} = 1 \qquad \text{Transverse axis is vertical.}$$

Note that a, b, and c are related differently for hyperbolas than for ellipses.

Figure 6.31 shows both the horizontal and vertical orientations for a hyperbola.

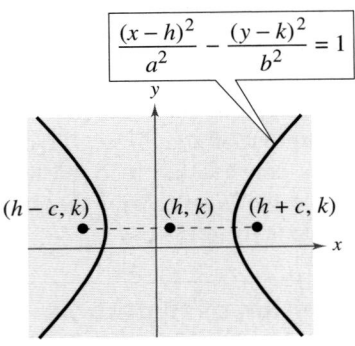

$$\frac{(x-h)^2}{a^2} - \frac{(y-k)^2}{b^2} = 1$$

Transverse axis is horizontal.

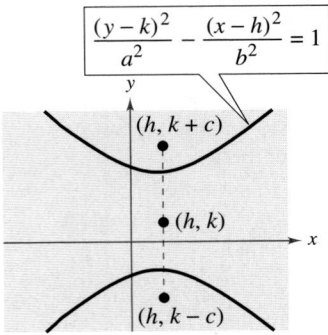

$$\frac{(y-k)^2}{a^2} - \frac{(x-h)^2}{b^2} = 1$$

Transverse axis is vertical.

FIGURE **6.31**

Example 1 ▶ **Finding the Standard Equation of a Hyperbola**

Find the standard form of the equation of the hyperbola with foci $(-1, 2)$ and $(5, 2)$ and vertices $(0, 2)$ and $(4, 2)$.

Solution

Because the vertices occur at $(0, 2)$ and $(4, 2)$, the center of the hyperbola occurs at the point $(2, 2)$. Furthermore, $c = 3$ and $a = 2$, and it follows that

$$b = \sqrt{c^2 - a^2} = \sqrt{3^2 - 2^2} = \sqrt{9 - 4} = \sqrt{5}.$$

So, the equation of the hyperbola is

$$\frac{(x-2)^2}{2^2} - \frac{(y-2)^2}{(\sqrt{5})^2} = 1. \qquad \text{See Figure 6.32.}$$

This equation simplifies to

$$\frac{(x-2)^2}{4} - \frac{(y-2)^2}{5} = 1.$$

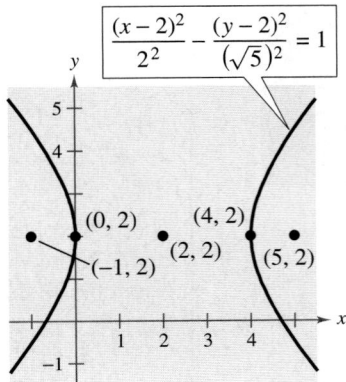

$$\frac{(x-2)^2}{2^2} - \frac{(y-2)^2}{(\sqrt{5})^2} = 1$$

FIGURE **6.32**

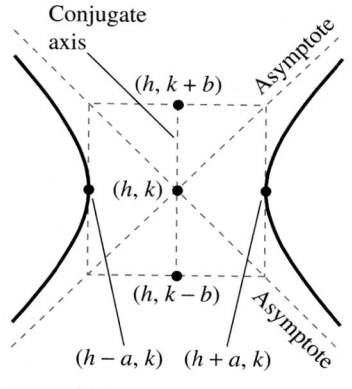

Conjugate
axis

$(h, k + b)$

(h, k)

$(h, k - b)$

$(h - a, k)$ $(h + a, k)$

Asymptote

Asymptote

FIGURE **6.33**

Asymptotes of a Hyperbola

Each hyperbola has two **asymptotes** that intersect at the center of the hyperbola, as shown in Figure 6.33. The asymptotes pass through the vertices of a rectangle of dimensions $2a$ by $2b$, with its center at (h, k). The line segment of length $2b$ joining $(h, k + b)$ and $(h, k - b)$ [or $(h + b, k)$ and $(h - b, k)$] is the **conjugate axis** of the hyperbola.

Asymptotes of a Hyperbola

The equations for the asymptotes of a hyperbola are

$$y = k \pm \frac{b}{a}(x - h) \qquad \text{Transverse axis is horizontal.}$$

$$y = k \pm \frac{a}{b}(x - h). \qquad \text{Transverse axis is vertical.}$$

Example 2 ▶   **Using Asymptotes to Sketch a Hyperbola**

Sketch the hyperbola whose equation is $4x^2 - y^2 = 16$.

Solution

Divide each side of the original equation by 16, and rewrite the equation in standard form.

$$\frac{x^2}{2^2} - \frac{y^2}{4^2} = 1 \qquad \text{Write in standard form.}$$

From this, you can conclude that $a = 2$, $b = 4$, and the transverse axis is horizontal. So, the vertices occur at $(-2, 0)$ and $(2, 0)$, and the ends of the conjugate axis occur at $(0, -4)$ and $(0, 4)$. Using these four points, you are able to sketch the rectangle shown in Figure 6.34. Finally, by drawing the asymptotes through the corners of this rectangle, you can complete the sketch shown in Figure 6.35.

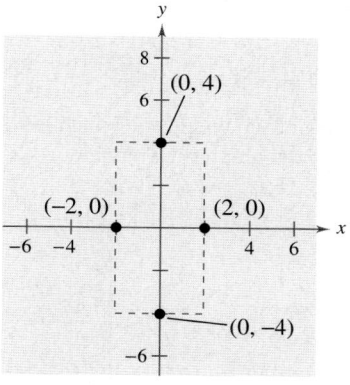

FIGURE **6.34**

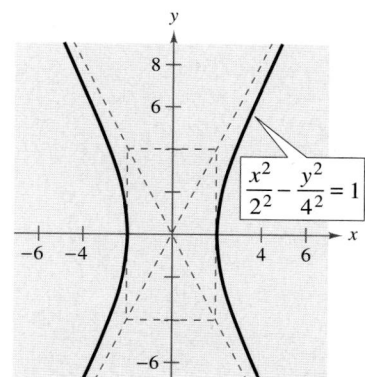

FIGURE **6.35**

   Finding the Asymptotes of a Hyperbola

Sketch the hyperbola given by $4x^2 - 3y^2 + 8x + 16 = 0$ and find the equations of its asymptotes.

Solution

$$4x^2 - 3y^2 + 8x + 16 = 0 \qquad \text{Write original equation.}$$

$$(4x^2 + 8x) - 3y^2 = -16 \qquad \text{Group terms.}$$

$$4(x^2 + 2x) - 3y^2 = -16 \qquad \text{Factor.}$$

$$4(x^2 + 2x + 1) - 3y^2 = -16 + 4 \qquad \text{Add 4 to each side.}$$

$$4(x + 1)^2 - 3y^2 = -12 \qquad \text{Complete the square.}$$

$$-\frac{(x + 1)^2}{3} + \frac{y^2}{4} = 1 \qquad \text{Divide each side by } -12.$$

$$\frac{y^2}{2^2} - \frac{(x + 1)^2}{(\sqrt{3})^2} = 1 \qquad \text{Write in standard form.}$$

From this equation you can conclude that the hyperbola is centered at $(-1, 0)$, has vertices $(-1, 2)$ and $(-1, -2)$, and has a conjugate axis with ends $\left(-1 - \sqrt{3}, 0\right)$ and $\left(-1 + \sqrt{3}, 0\right)$. To sketch the hyperbola, draw a rectangle through these four points. The asymptotes are the lines passing through the corners of the rectangle, as shown in Figure 6.36. Finally, using $a = 2$ and $b = \sqrt{3}$, you can conclude that the equations of the asymptotes are

$$y = \frac{2}{\sqrt{3}}(x + 1) \qquad \text{and} \qquad y = -\frac{2}{\sqrt{3}}(x + 1).$$

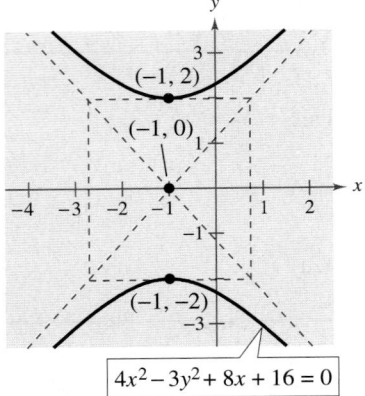

FIGURE 6.36

Technology

You can use a graphing utility to graph a hyperbola by graphing the upper and lower portions in the same viewing window. For instance, to graph the hyperbola in Example 3, first solve for *y* to get

$$y_1 = 2\sqrt{1 + \frac{(x + 1)^2}{3}} \qquad \text{and} \qquad y_2 = -2\sqrt{1 + \frac{(x + 1)^2}{3}}.$$

Use a viewing window in which $-9 \le x \le 9$ and $-6 \le y \le 6$. You should obtain the graph shown below.

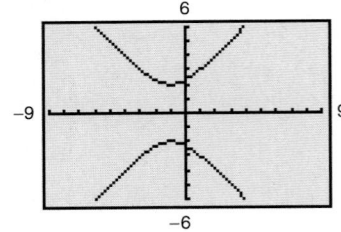

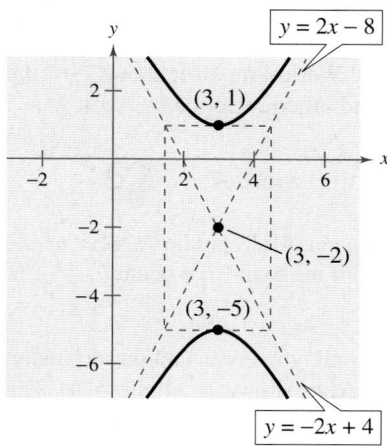

FIGURE **6.37**

www

Example 4 ▶ **Using Asymptotes to Find the Standard Equation**

Find the standard form of the equation of the hyperbola having vertices $(3, -5)$ and $(3, 1)$ and having asymptotes

$$y = 2x - 8 \qquad \text{and} \qquad y = -2x + 4$$

as shown in Figure 6.37.

Solution

Because the vertices occur at $(3, -5)$ and $(3, 1)$, the center of the hyperbola is $(3, -2)$. Furthermore, the hyperbola has a vertical transverse axis with $a = 3$. From the original equations, you can determine the slopes of the asymptotes to be

$$m_1 = 2 = \frac{a}{b} \qquad \text{and} \qquad m_2 = -2 = -\frac{a}{b}$$

and, because $a = 3$

$$2 = \frac{a}{b} \quad \Longrightarrow \quad 2 = \frac{3}{b} \quad \Longrightarrow \quad b = \frac{3}{2}.$$

So, the standard form of the equation is

$$\frac{(y + 2)^2}{3^2} - \frac{(x - 3)^2}{\left(\dfrac{3}{2}\right)^2} = 1.$$

As with ellipses, the **eccentricity** of a hyperbola is

$$e = \frac{c}{a} \qquad \text{Eccentricity}$$

and because $c > a$, it follows that $e > 1$. If the eccentricity is large, the branches of the hyperbola are nearly flat, as shown in Figure 6.38. If the eccentricity is close to 1, the branches of the hyperbola are more pointed, as shown in Figure 6.39.

Additional Example

Find the standard form of the equation of the hyperbola having vertices $(-4, 2)$ and $(1, 2)$ and having foci $(-7, 2)$ and $(4, 2)$.

Answer: $\dfrac{\left(x + \dfrac{3}{2}\right)^2}{\left(\dfrac{5}{2}\right)^2} - \dfrac{(y - 2)^2}{\left(\sqrt{24}\right)^2} = 1$

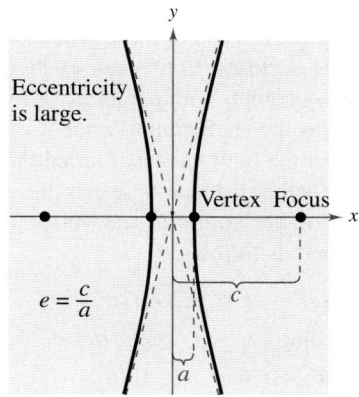

FIGURE **6.38**

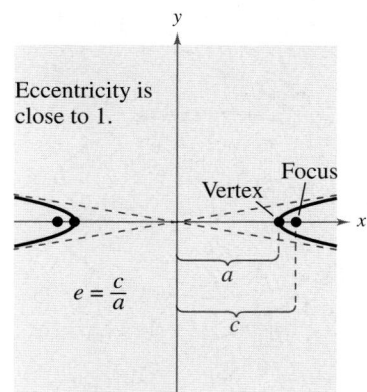

FIGURE **6.39**

Applications

The following application was developed during World War II. It shows how the properties of hyperbolas can be used in radar and other detection systems.

Example 5 ▶   An Application Involving Hyperbolas

Two microphones, 1 mile apart, record an explosion. Microphone A receives the sound 2 seconds before microphone B. Where did the explosion occur?

Solution

Assuming that sound travels at 1100 feet per second, you know that the explosion took place 2200 feet farther from B than from A, as shown in Figure 6.40. The locus of all points that are 2200 feet closer to A than to B is one branch of the hyperbola

$$\frac{x^2}{a^2} - \frac{y^2}{b^2} = 1$$

where

$$c = \frac{5280}{2} = 2640$$

and

$$a = \frac{2200}{2} = 1100.$$

So, $b^2 = c^2 - a^2 = 2640^2 - 1100^2 = 5,759,600$, and you can conclude that the explosion occurred somewhere on the right branch of the hyperbola

$$\frac{x^2}{1,210,000} - \frac{y^2}{5,759,600} = 1.$$

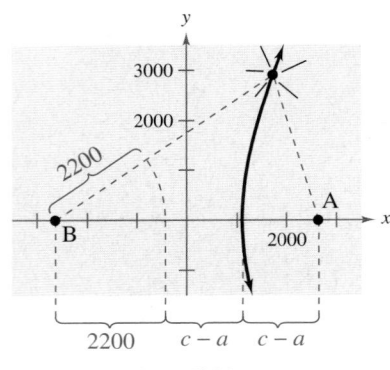

2c = 5280

2200 + 2(c − a) = 5280

FIGURE 6.40

Another interesting application of conic sections involves the orbits of comets in our solar system. Of the 610 comets identified prior to 1970, 245 have elliptical orbits, 295 have parabolic orbits, and 70 have hyperbolic orbits. The center of the sun is a focus of each of these orbits, and each orbit has a vertex at the point where the comet is closest to the sun, as shown in Figure 6.41. Undoubtedly, there have been many comets with parabolic or hyperbolic orbits that were not identified. We only get to see such comets *once*. Comets with elliptical orbits, such as Halley's comet, are the only ones that remain in our solar system.

If p is the distance between the vertex and the focus in meters, and v is the velocity of the comet at the vertex in meters per second, the type of orbit is determined as follows.

1. Ellipse: $v < \sqrt{2GM/p}$
2. Parabola: $v = \sqrt{2GM/p}$
3. Hyperbola: $v > \sqrt{2GM/p}$

In each of these equations, $M \approx 1.991 \times 10^{30}$ kilograms (the mass of the sun) and $G \approx 6.67 \times 10^{-11}$ cubic meters per kilogram-second squared.

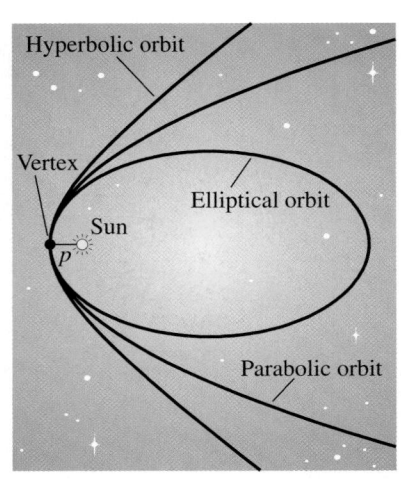

FIGURE 6.41

General Equations of Conics

> ## Classifying a Conic from Its General Equation
>
> The graph of $Ax^2 + Cy^2 + Dx + Ey + F = 0$ is one of the following.
>
> **1.** *Circle:* $A = C$
>
> **2.** *Parabola:* $AC = 0$ $A = 0$ or $C = 0$, but not both.
>
> **3.** *Ellipse:* $AC > 0$ A and C have like signs.
>
> **4.** *Hyperbola:* $AC < 0$ A and C have unlike signs.

The test above is valid if the graph is a conic. The test does not apply to equations such as $x^2 + y^2 = -1$, whose graph is not a conic.

Example 6 ▶ **Classifying Conics from General Equations**

Classify each graph.

a. $4x^2 - 9x + y - 5 = 0$

b. $4x^2 - y^2 + 8x - 6y + 4 = 0$

c. $2x^2 + 4y^2 - 4x + 12y = 0$

d. $2x^2 + 2y^2 - 8x + 12y + 2 = 0$

Solution

a. For the equation $4x^2 - 9x + y - 5 = 0$, you have

$$AC = 4(0) = 0. \qquad \text{Parabola}$$

So, the graph is a parabola.

b. For the equation $4x^2 - y^2 + 8x - 6y + 4 = 0$, you have

$$AC = 4(-1) < 0. \qquad \text{Hyperbola}$$

So, the graph is a hyperbola.

c. For the equation $2x^2 + 4y^2 - 4x + 12y = 0$, you have

$$AC = 2(4) > 0. \qquad \text{Ellipse}$$

So, the graph is an ellipse.

d. For the equation $2x^2 + 2y^2 - 8x + 12y + 2 = 0$, you have

$$A = C = 2. \qquad \text{Circle}$$

So, the graph is a circle.

Historical Note
Caroline Herschel (1750–1848) was the first woman to be credited with detecting a new comet. During her long life, this English astronomer discovered a total of eight new comets.

The Granger Collection

Writing **ABOUT MATHEMATICS**

Sketching Conics Sketch each of the conics described in Example 6. Write a paragraph describing the procedures that allow you to sketch the conics efficiently.

6.4 **Exercises**

In Exercises 1–4, match the equation with its graph. [The graphs are labeled (a), (b), (c), and (d).]

(a)
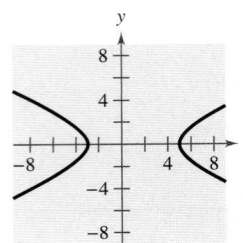

(b)
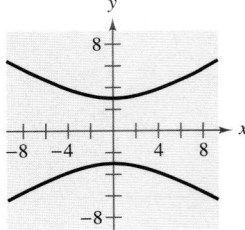

(c)
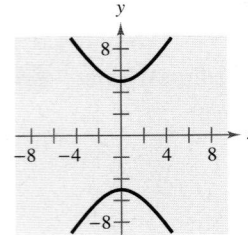

(d)
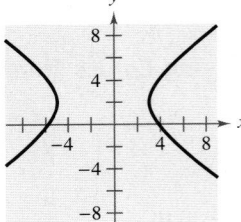

1. $\dfrac{y^2}{9} - \dfrac{x^2}{25} = 1$ **2.** $\dfrac{y^2}{25} - \dfrac{x^2}{9} = 1$

3. $\dfrac{(x-1)^2}{16} - \dfrac{y^2}{4} = 1$ **4.** $\dfrac{(x+1)^2}{16} - \dfrac{(y-2)^2}{9} = 1$

In Exercises 5–16, find the center, vertices, foci, and the equations of the asymptotes of the hyperbola, and sketch its graph.

5. $x^2 - y^2 = 1$ **6.** $\dfrac{x^2}{9} - \dfrac{y^2}{25} = 1$

7. $\dfrac{y^2}{25} - \dfrac{x^2}{81} = 1$ **8.** $\dfrac{x^2}{36} - \dfrac{y^2}{4} = 1$

9. $\dfrac{(x-1)^2}{4} - \dfrac{(y+2)^2}{1} = 1$

10. $\dfrac{(x+3)^2}{144} - \dfrac{(y-2)^2}{25} = 1$

11. $\dfrac{(y+6)^2}{1/9} - \dfrac{(x-2)^2}{1/4} = 1$

12. $\dfrac{(y-1)^2}{1/4} - \dfrac{(x+3)^2}{1/16} = 1$

13. $9x^2 - y^2 - 36x - 6y + 18 = 0$

14. $x^2 - 9y^2 + 36y - 72 = 0$

15. $x^2 - 9y^2 + 2x - 54y - 80 = 0$

16. $16y^2 - x^2 + 2x + 64y + 63 = 0$

In Exercises 17–20, find the center, vertices, foci, and the equations of the asymptotes of the hyperbola. Use a graphing utility to graph the hyperbola and its asymptotes.

17. $2x^2 - 3y^2 = 6$ **18.** $6y^2 - 3x^2 = 18$

19. $9y^2 - x^2 + 2x + 54y + 62 = 0$

20. $9x^2 - y^2 + 54x + 10y + 55 = 0$

In Exercises 21–26, find the standard form of the equation of the hyperbola with center at the origin.

21. Vertices: $(0, \pm 2)$; Foci: $(0, \pm 4)$

22. Vertices: $(\pm 4, 0)$; Foci: $(\pm 6, 0)$

23. Vertices: $(\pm 1, 0)$; Asymptotes: $y = \pm 5x$

24. Vertices: $(0, \pm 3)$; Asymptotes: $y = \pm 3x$

25. Foci: $(0, \pm 8)$; Asymptotes: $y = \pm 4x$

26. Foci: $(\pm 10, 0)$; Asymptotes: $y = \pm \frac{3}{4}x$

In Exercises 27–38, find the standard form of the equation of the hyperbola.

27. Vertices: $(2, 0)$, $(6, 0)$; Foci: $(0, 0)$, $(8, 0)$

28. Vertices: $(2, 3)$, $(2, -3)$; Foci: $(2, 6)$, $(2, -6)$

29. Vertices: $(4, 1)$, $(4, 9)$; Foci: $(4, 0)$, $(4, 10)$

30. Vertices: $(-2, 1)$, $(2, 1)$; Foci: $(-3, 1)$, $(3, 1)$

31. Vertices: $(2, 3)$, $(2, -3)$;
Passes through the point $(0, 5)$

32. Vertices: $(-2, 1)$, $(2, 1)$;
Passes through the point $(5, 4)$

33. Vertices: $(0, 4)$, $(0, 0)$;
Passes through the point $\left(\sqrt{5}, -1\right)$

34. Vertices: $(1, 2)$, $(1, -2)$;
Passes through the point $\left(0, \sqrt{5}\right)$

35. Vertices: $(1, 2)$, $(3, 2)$;
Asymptotes: $y = x$, $y = 4 - x$

36. Vertices: $(3, 0)$, $(3, 6)$;
Asymptotes: $y = 6 - x$, $y = x$

37. Vertices: $(0, 2)$, $(6, 2)$;
Asymptotes: $y = \frac{2}{3}x$, $y = 4 - \frac{2}{3}x$

38. Vertices: $(3, 0)$, $(3, 4)$;
Asymptotes: $y = \frac{2}{3}x$, $y = 4 - \frac{2}{3}x$

▶ Model It

39. **LORAN** Long distance radio navigation for aircraft and ships uses synchronized pulses transmitted by widely separated transmitting stations. These pulses travel at the speed of light (186,000 miles per second). The difference in the times of arrival of these pulses at an aircraft or ship is constant on a hyperbola having the transmitting stations as foci. Assume that two stations, 300 miles apart, are positioned on the rectangular coordinate system at points with coordinates $(-150, 0)$ and $(150, 0)$, and that a ship is traveling on a hyperbolic path with coordinates $(x, 75)$ (see figure).

(a) Find the x-coordinate of the position of the ship if the time difference between the pulses from the transmitting stations is 1000 microseconds (0.001 second).

(b) Determine the distance between the ship and Station 1 when the ship reaches the shore.

(c) The ship wants to enter a bay located between the two stations. The bay is 30 miles from Station 1. What should the time difference between the pulses be?

(d) The ship is 60 miles offshore when the time difference in part (c) is obtained. What is the position of the ship?

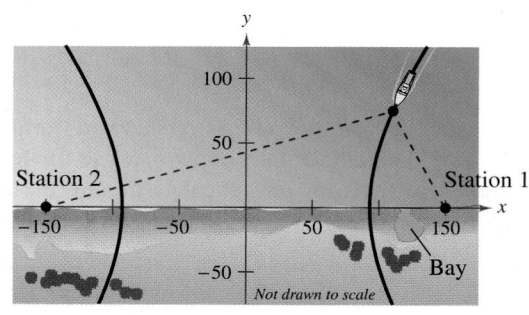

Not drawn to scale

40. **Hyperbolic Mirror** A hyperbolic mirror (used in some telescopes) has the property that a light ray directed at a focus will be reflected to the other focus. The focus of a hyperbolic mirror (see figure) has coordinates $(24, 0)$. Find the vertex of the mirror if the mount at the top edge of the mirror has coordinates $(24, 24)$.

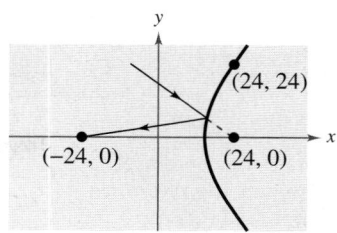

FIGURE FOR **40**

In Exercises 41–48, classify the graph of the equation as a circle, a parabola, an ellipse, or a hyperbola.

41. $x^2 + y^2 - 6x + 4y + 9 = 0$

42. $x^2 + 4y^2 - 6x + 16y + 21 = 0$

43. $4x^2 - y^2 - 4x - 3 = 0$

44. $y^2 - 6y - 4x + 21 = 0$

45. $4x^2 + 3y^2 + 8x - 24y + 51 = 0$

46. $4y^2 - 2x^2 - 4y - 8x - 15 = 0$

47. $25x^2 - 10x - 200y - 119 = 0$

48. $4y^2 + 4x^2 - 24x + 35 = 0$

Synthesis

True or False? In Exercises 49 and 50, determine whether the statement is true or false. Justify your answer.

49. In the standard form of the equation of a hyperbola, the larger the ratio of b to a, the larger the eccentricity of the hyperbola.

50. In the standard form of the equation of a hyperbola, the trivial solution of two intersecting lines occurs when $b = 0$.

51. Consider a hyperbola centered at the origin with a horizontal transverse axis. Use the definition of a hyperbola to derive its standard form.

52. *Writing* Explain how the central rectangle of a hyperbola can be used to sketch its asymptotes.

Review

In Exercises 53–56, factor the polynomial completely.

53. $x^3 - 16x$

54. $x^2 + 14x + 49$

55. $2x^3 - 24x^2 + 72x$

56. $6x^3 - 11x^2 - 10x$

In Exercises 57–60, sketch a graph of the function. Include two full periods.

57. $y = 2 \cos x + 1$

58. $y = \sin \pi x$

59. $y = \tan 2x$

60. $y = -\frac{1}{2} \sec x$

6.5 Rotation of Conics

▶ **What you should learn**

• How to rotate the coordinate axes to eliminate the *xy*-term in the equation of a conic
• How to use the discriminant to classify a conic

▶ **Why you should learn it**

As illustrated in Exercises 7–18 on page 466, rotation of the coordinate axes can help you identify the graph of a general second-degree equation.

Rotation

In the preceding section, you learned that the equation of a conic with axes parallel to one of the coordinate axes has a standard form that can be written in the general form

$$Ax^2 + Cy^2 + Dx + Ey + F = 0. \qquad \text{Horizontal or vertical axis}$$

In this section, you will study the equations of conics whose axes are rotated so that they are not parallel to either the *x*-axis or the *y*-axis. The general equation for such conics contains an *xy*-term.

$$Ax^2 + Bxy + Cy^2 + Dx + Ey + F = 0 \qquad \text{Equation in } xy\text{-plane}$$

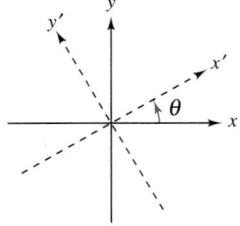

FIGURE 6.42

To eliminate this *xy*-term, you can use a procedure called **rotation of axes.** The objective is to rotate the *x*- and *y*-axes until they are parallel to the axes of the conic. The rotated axes are denoted as the *x'*-axis and the *y'*-axis, as shown in Figure 6.42. After the rotation, the equation of the conic in the new *x'y'*-plane will have the form

$$A'(x')^2 + C'(y')^2 + D'x' + E'y' + F' = 0. \qquad \text{Equation in } x'y'\text{-plane}$$

Because this equation has no *xy*-term, you can obtain a standard form by completing the square. The following theorem identifies how much to rotate the axes to eliminate the *xy*-term and also the equations for determining the new coefficients A', C', D', E', and F'.

Rotation of Axes to Eliminate an *xy*-Term

The general second-degree equation $Ax^2 + Bxy + Cy^2 + Dx + Ey + F = 0$ can be rewritten as

$$A'(x')^2 + C'(y')^2 + D'x' + E'y' + F' = 0$$

by rotating the coordinate axes through an angle θ, where

$$\cot 2\theta = \frac{A - C}{B}.$$

The coefficients of the new equation are obtained by making the substitutions $x = x' \cos \theta - y' \sin \theta$ and $y = x' \sin \theta + y' \cos \theta$.

Example 1 ▶ **Rotation of Axes for a Hyperbola**

Write the equation $xy - 1 = 0$ in standard form.

Solution

Because $A = 0$, $B = 1$, and $C = 0$, you have

$$\cot 2\theta = \frac{A - C}{B} = 0 \quad \Longrightarrow \quad 2\theta = \frac{\pi}{2} \quad \Longrightarrow \quad \theta = \frac{\pi}{4}$$

which implies that

$$x = x' \cos \frac{\pi}{4} - y' \sin \frac{\pi}{4}$$

$$= x' \left(\frac{1}{\sqrt{2}} \right) - y' \left(\frac{1}{\sqrt{2}} \right)$$

$$= \frac{x' - y'}{\sqrt{2}}$$

and

$$y = x' \sin \frac{\pi}{4} + y' \cos \frac{\pi}{4}$$

$$= x' \left(\frac{1}{\sqrt{2}} \right) + y' \left(\frac{1}{\sqrt{2}} \right)$$

$$= \frac{x' + y'}{\sqrt{2}}.$$

The equation in the $x'y'$-system is obtained by substituting these expressions in the equation $xy - 1 = 0$.

$$\left(\frac{x' - y'}{\sqrt{2}} \right)\left(\frac{x' + y'}{\sqrt{2}} \right) - 1 = 0$$

$$\frac{(x')^2 - (y')^2}{2} - 1 = 0$$

$$\frac{(x')^2}{(\sqrt{2})^2} - \frac{(y')^2}{(\sqrt{2})^2} = 1 \qquad \text{Write in standard form.}$$

In the $x'y'$-system, this is a hyperbola centered at the origin with vertices at $(\pm \sqrt{2}, 0)$, as shown in Figure 6.43. To find the coordinates of the vertices in the xy-system, substitute the coordinates $(\pm \sqrt{2}, 0)$ in the equations

$$x = \frac{x' - y'}{\sqrt{2}} \qquad \text{and} \qquad y = \frac{x' + y'}{\sqrt{2}}.$$

This substitution yields the vertices $(1, 1)$ and $(-1, -1)$ in the xy-system. Note also that the asymptotes of the hyperbola have equations $y' = \pm x'$, which correspond to the original x- and y-axes.

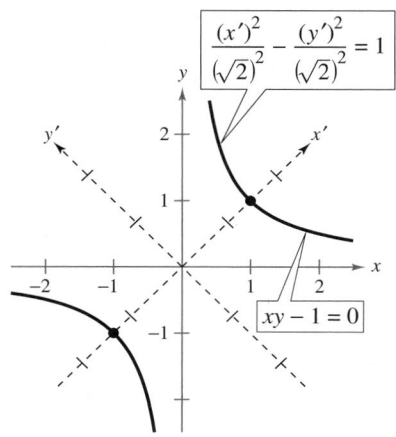

$$\frac{(x')^2}{(\sqrt{2})^2} - \frac{(y')^2}{(\sqrt{2})^2} = 1$$

$$xy - 1 = 0$$

Vertices:
In $x'y'$-system: $(\sqrt{2}, 0), (-\sqrt{2}, 0)$
In xy-system: $(1, 1), (-1, -1)$
FIGURE **6.43**

Example 2 ▶   **Rotation of Axes for an Ellipse**

Sketch the graph of $7x^2 - 6\sqrt{3}\,xy + 13y^2 - 16 = 0$.

Solution

Because $A = 7$, $B = -6\sqrt{3}$, and $C = 13$, you have

$$\cot 2\theta = \frac{A - C}{B} = \frac{7 - 13}{-6\sqrt{3}} = \frac{1}{\sqrt{3}}$$

which implies that $\theta = \pi/6$. The equation in the $x'y'$-system is obtained by making the substitutions

$$x = x'\cos\frac{\pi}{6} - y'\sin\frac{\pi}{6}$$

$$= x'\left(\frac{\sqrt{3}}{2}\right) - y'\left(\frac{1}{2}\right)$$

$$= \frac{\sqrt{3}\,x' - y'}{2}$$

and

$$y = x'\sin\frac{\pi}{6} + y'\cos\frac{\pi}{6}$$

$$= x'\left(\frac{1}{2}\right) + y'\left(\frac{\sqrt{3}}{2}\right)$$

$$= \frac{x' + \sqrt{3}\,y'}{2}$$

in the original equation. So, you have

$$7x^2 - 6\sqrt{3}\,xy + 13y^2 - 16 = 0$$

$$7\left(\frac{\sqrt{3}\,x' - y'}{2}\right)^2 - 6\sqrt{3}\left(\frac{\sqrt{3}\,x' - y'}{2}\right)\left(\frac{x' + \sqrt{3}\,y'}{2}\right)$$

$$+ 13\left(\frac{x' + \sqrt{3}\,y'}{2}\right)^2 - 16 = 0$$

which simplifies to

$$4(x')^2 + 16(y')^2 - 16 = 0$$

$$4(x')^2 + 16(y')^2 = 16$$

$$\frac{(x')^2}{4} + \frac{(y')^2}{1} = 1$$

$$\frac{(x')^2}{2^2} + \frac{(y')^2}{1^2} = 1. \qquad \text{Write in standard form.}$$

This is the equation of an ellipse centered at the origin with vertices $(\pm 2, 0)$ in the $x'y'$-system, as shown in Figure 6.44.

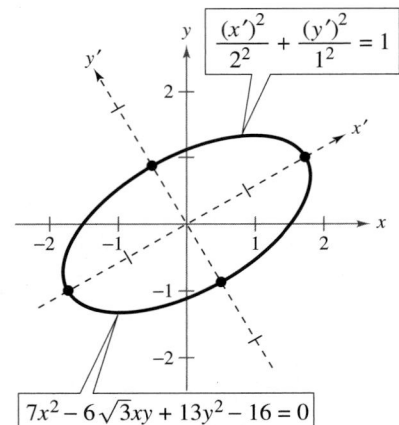

$$\frac{(x')^2}{2^2} + \frac{(y')^2}{1^2} = 1$$

$$7x^2 - 6\sqrt{3}xy + 13y^2 - 16 = 0$$

Vertices:
In $x'y'$-system: $(\pm 2, 0), (0, \pm 1)$
In xy-system: $\left(\sqrt{3}, 1\right), \left(-\sqrt{3}, -1\right),$
$$\left(\frac{1}{2}, -\frac{\sqrt{3}}{2}\right), \left(-\frac{1}{2}, \frac{\sqrt{3}}{2}\right)$$

FIGURE **6.44**

Example 3 ▶ **Rotation of Axes for a Parabola**

Sketch the graph of $x^2 - 4xy + 4y^2 + 5\sqrt{5}y + 1 = 0$.

Solution

Because $A = 1$, $B = -4$, and $C = 4$, you have

$$\cot 2\theta = \frac{A - C}{B} = \frac{1 - 4}{-4} = \frac{3}{4}.$$

Using the identity $\cot 2\theta = (\cot^2\theta - 1)/(2 \cot \theta)$ produces

$$\cot 2\theta = \frac{\cot^2 \theta - 1}{2 \cot \theta} = \frac{3}{4}$$

from which you obtain the equation

$$4 \cot^2 \theta - 4 = 6 \cot \theta$$

$$4 \cot^2 \theta - 6 \cot \theta - 4 = 0$$

$$(2 \cot \theta - 4)(2 \cot \theta + 1) = 0.$$

Considering $0 < \theta < \pi/2$, you have $2 \cot \theta = 4$. So,

$$\cot \theta = 2 \qquad \Longrightarrow \qquad \theta \approx 26.6°.$$

From the triangle in Figure 6.45, you obtain $\sin \theta = 1/\sqrt{5}$ and $\cos \theta = 2/\sqrt{5}$. Consequently, you use the substitutions

$$x = x' \cos \theta - y' \sin \theta$$

$$= x'\left(\frac{2}{\sqrt{5}}\right) - y'\left(\frac{1}{\sqrt{5}}\right) = \frac{2x' - y'}{\sqrt{5}}$$

$$y = x' \sin \theta + y' \cos \theta$$

$$= x'\left(\frac{1}{\sqrt{5}}\right) + y'\left(\frac{2}{\sqrt{5}}\right) = \frac{x' + 2y'}{\sqrt{5}}.$$

Substituting these expressions in the original equation, you have

$$x^2 - 4xy + 4y^2 + 5\sqrt{5}y + 1 = 0$$

$$\left(\frac{2x' - y'}{\sqrt{5}}\right)^2 - 4\left(\frac{2x' - y'}{\sqrt{5}}\right)\left(\frac{x' + 2y'}{\sqrt{5}}\right) + 4\left(\frac{x' + 2y'}{\sqrt{5}}\right)^2 + 5\sqrt{5}\left(\frac{x' + 2y'}{\sqrt{5}}\right) + 1 = 0$$

which simplifies as follows.

$$5(y')^2 + 5x' + 10y' + 1 = 0$$

$$5(y' + 1)^2 = -5x' + 4 \qquad \text{Complete the square.}$$

$$(y' + 1)^2 = (-1)\left(x' - \frac{4}{5}\right) \qquad \text{Write in standard form.}$$

The graph of this equation is a parabola with vertex $\left(\frac{4}{5}, -1\right)$. Its axis is parallel to the x'-axis in the $x'y'$-system, as shown in Figure 6.46.

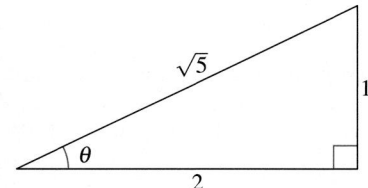

FIGURE 6.45

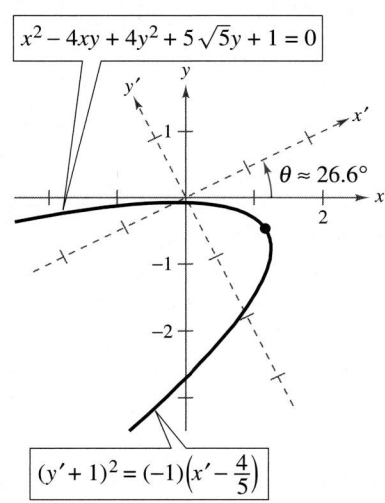

$x^2 - 4xy + 4y^2 + 5\sqrt{5}y + 1 = 0$

$\theta \approx 26.6°$

$(y' + 1)^2 = (-1)\left(x' - \frac{4}{5}\right)$

Vertex:

In $x'y'$-system: $\left(\frac{4}{5}, -1\right)$

In xy-system: $\left(\frac{13}{5\sqrt{5}}, -\frac{6}{5\sqrt{5}}\right)$

FIGURE 6.46

Invariants Under Rotation

In the rotation of axes theorem listed at the beginning of this section, note that the constant term is the same in both equations, $F' = F$. Such quantities are **invariant under rotation.** The next theorem lists some other rotation invariants.

Rotation Invariants

The rotation of the coordinate axes through an angle θ that transforms the equation $Ax^2 + Bxy + Cy^2 + Dx + Ey + F = 0$ into the form

$$A'(x')^2 + C'(y')^2 + D'x' + E'y' + F' = 0$$

has the following rotation invariants.

1. $F = F'$

2. $A + C = A' + C'$

3. $B^2 - 4AC = (B')^2 - 4A'C'$

You can use the results of this theorem to classify the graph of a second-degree equation *with* an xy-term in much the same way you do for a second-degree equation *without* an xy-term. Note that because $B' = 0$, the invariant $B^2 - 4AC$ reduces to

$$B^2 - 4AC = -4A'C'. \qquad \text{Discriminant}$$

This quantity is called the **discriminant** of the equation

$$Ax^2 + Bxy + Cy^2 + Dx + Ey + F = 0.$$

Now, from the classification procedure given in Section 6.4, you know that the sign of $A'C'$ determines the type of graph for the equation

$$A'(x')^2 + C'(y')^2 + D'x' + E'y' + F' = 0.$$

Consequently, the sign of $B^2 - 4AC$ will determine the type of graph for the original equation, as given in the following classification.

Classification of Conics by the Discriminant

The graph of the equation $Ax^2 + Bxy + Cy^2 + Dx + Ey + F = 0$ is, except in degenerate cases, determined by its discriminant as follows.

1. *Ellipse or circle:* $B^2 - 4AC < 0$

2. *Parabola:* $\qquad B^2 - 4AC = 0$

3. *Hyperbola:* $\qquad B^2 - 4AC > 0$

If the *xy*-term exists, your students should be able to recognize immediately the fact that rotation occurs. They should then use the discriminant to classify the conic before rotating the axes.

For example, in the general equation

$$3x^2 + 7xy + 5y^2 - 6x - 7y + 15 = 0$$

you have $A = 3$, $B = 7$, and $C = 5$. So the discriminant is

$$B^2 - 4AC = 7^2 - 4(3)(5) = 49 - 60 = -11.$$

Because $-11 < 0$, the graph of the equation is an ellipse or a circle.

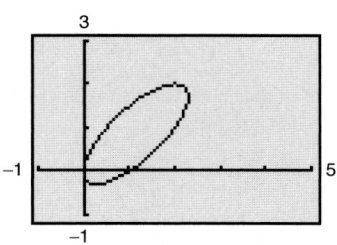

FIGURE **6.47**

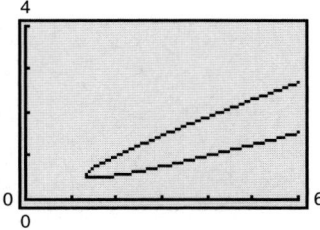

FIGURE **6.48**

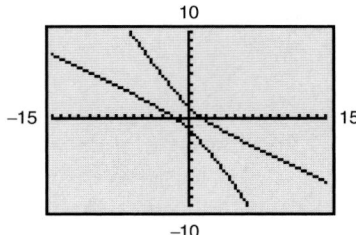

FIGURE **6.49**

Activities

1. Use the discriminant to classify each equation.

a. $x^2 - 4xy + y^2 + 2x - 3y + 1 = 0$
b. $4x^2 - xy + 4y^2 + 2x - 3y + 1 = 0$
c. $x^2 - xy + 4y^2 + 2x - 3y + 1 = 0$
d. $x^2 - 4xy + 4y^2 + 2x - 3y + 1 = 0$

Answer: (a) Hyperbola, (b) Ellipse or circle, (c) Ellipse or circle, (d) Parabola

2. Find the angle of rotation necessary to eliminate the *xy*-term in the equation $3x^2 + 2xy - y^2 + x - 1 = 0$.

Answer: $\theta \approx 13.28°$

3. Rotate the axes to eliminate the *xy*-term in the equation $x^2 + 2xy + y^2 - \sqrt{2}y = 0$.

Answer: $2(x')^2 - x' - y' = 0,\ \theta = \dfrac{\pi}{4}$

Example 4 ▶ **Rotation and Graphing Utilities**

For each of the following, classify the graph, use the Quadratic Formula to solve for *y*, and then use a graphing utility to graph the equation.

a. $2x^2 - 3xy + 2y^2 - 2x = 0$ **b.** $x^2 - 6xy + 9y^2 - 2y + 1 = 0$
c. $3x^2 + 8xy + 4y^2 - 7 = 0$

Solution

a. Because $B^2 - 4AC = 9 - 16 < 0$, the graph is a circle or an ellipse. Solve for *y* as follows.

$$2x^2 - 3xy + 2y^2 - 2x = 0 \qquad \text{Write original equation.}$$

$$2y^2 - 3xy + (2x^2 - 2x) = 0 \qquad \text{Quadratic form } ay^2 + by + c = 0$$

$$y = \frac{-(-3x) \pm \sqrt{(-3x)^2 - 4(2)(2x^2 - 2x)}}{2(2)}$$

Graph both of the equations to obtain the ellipse shown in Figure 6.47.

$$y = \frac{3x + \sqrt{9x^2 - 16(x^2 - x)}}{4} \qquad \text{Top half of ellipse}$$

$$y = \frac{3x - \sqrt{9x^2 - 16(x^2 - x)}}{4} \qquad \text{Bottom half of ellipse}$$

b. Because $B^2 - 4AC = 36 - 36 = 0$, the graph is a parabola.

$$x^2 - 6xy + 9y^2 - 2y + 1 = 0 \qquad \text{Write original equation.}$$

$$9y^2 - (6x + 2)y + (x^2 + 1) = 0 \qquad \text{Quadratic form } ay^2 + by + c = 0$$

$$y = \frac{(6x + 2) \pm \sqrt{(6x + 2)^2 - 4(9)(x^2 + 1)}}{18}$$

Graphing the resulting two equations gives the parabola shown in Figure 6.48.

c. Because $B^2 - 4AC = 64 - 48 > 0$, the graph is a hyperbola.

$$3x^2 + 8xy + 4y^2 - 7 = 0 \qquad \text{Write original equation.}$$

$$4y^2 + 8xy + (3x^2 - 7) = 0 \qquad \text{Quadratic form } ay^2 + by + c = 0$$

$$y = \frac{-8x \pm \sqrt{(8x)^2 - 4(4)(3x^2 - 7)}}{8}$$

The graphs of the resulting two equations yield the hyperbola shown in Figure 6.49.

Writing **ABOUT MATHEMATICS**

Classifying a Graph as a Hyperbola The graph of $f(x) = 1/x$ is a hyperbola. Discuss how you could use the techniques in this section to verify this, and then do so. Compare your statement with that of another student.

6.5 Exercises

In Exercises 1–6, the $x'y'$-coordinate system has been rotated θ degrees from the xy-coordinate system. The coordinates of a point in the xy-coordinate system are given. Find the coordinates of the point in the rotated coordinate system.

1. $\theta = 90°$, $(0, 3)$

2. $\theta = 45°$, $(3, 3)$

3. $\theta = 30°$, $(1, 3)$

4. $\theta = 60°$, $(3, 1)$

5. $\theta = 45°$, $(2, 1)$

6. $\theta = 30°$, $(2, 4)$

In Exercises 7–18, rotate the axes to eliminate the xy-term. Sketch the graph of the resulting equation, showing both sets of axes.

7. $xy + 1 = 0$

8. $xy - 2 = 0$

9. $x^2 - 2xy + y^2 - 1 = 0$

10. $xy + x - 2y + 3 = 0$

11. $xy - 2y - 4x = 0$

12. $2x^2 - 3xy - 2y^2 + 10 = 0$

13. $5x^2 - 6xy + 5y^2 - 12 = 0$

14. $13x^2 + 6\sqrt{3}xy + 7y^2 - 16 = 0$

15. $3x^2 - 2\sqrt{3}xy + y^2 + 2x + 2\sqrt{3}y = 0$

16. $16x^2 - 24xy + 9y^2 - 60x - 80y + 100 = 0$

17. $9x^2 + 24xy + 16y^2 + 90x - 130y = 0$

18. $9x^2 + 24xy + 16y^2 + 80x - 60y = 0$

In Exercises 19–26, use a graphing utility to graph the conic. Determine the angle θ through which the axes are rotated. Explain how you used the graphing utility to obtain the graph.

19. $x^2 + 2xy + y^2 = 20$

20. $x^2 - 4xy + 2y^2 = 6$

21. $17x^2 + 32xy - 7y^2 = 75$

22. $40x^2 + 36xy + 25y^2 = 52$

23. $32x^2 + 48xy + 8y^2 = 50$

24. $24x^2 + 18xy + 12y^2 = 34$

25. $4x^2 - 12xy + 9y^2 + (4\sqrt{13} - 12)x$
$\qquad - (6\sqrt{13} + 8)y = 91$

26. $6x^2 - 4xy + 8y^2 + (5\sqrt{5} - 10)x$
$\qquad - (7\sqrt{5} + 5)y = 80$

In Exercises 27–32, match the graph with its equation. [The graphs are labeled (a), (b), (c), (d), (e), and (f).]

(a)
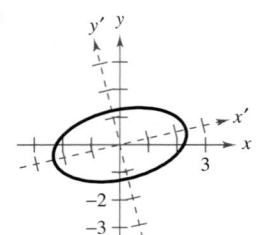

(b)
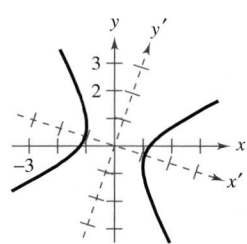

(c)
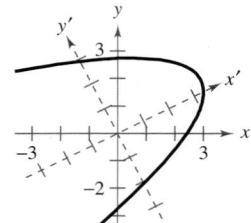

(d)
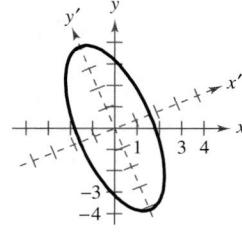

(e)
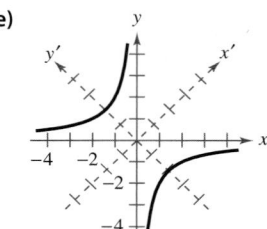

(f)
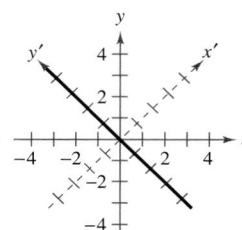

27. $xy + 2 = 0$

28. $x^2 + 2xy + y^2 = 0$

29. $-2x^2 + 3xy + 2y^2 + 3 = 0$

30. $x^2 - xy + 3y^2 - 5 = 0$

31. $3x^2 + 2xy + y^2 - 10 = 0$

32. $x^2 - 4xy + 4y^2 + 10x - 30 = 0$

In Exercises 33–40, use the discriminant to classify the graph. Then use the Quadratic Formula to solve for y and use a graphing utility to graph the equation.

33. $16x^2 - 8xy + y^2 - 10x + 5y = 0$

34. $x^2 - 4xy - 2y^2 - 6 = 0$

35. $12x^2 - 6xy + 7y^2 - 45 = 0$

36. $2x^2 + 4xy + 5y^2 + 3x - 4y - 20 = 0$

37. $x^2 - 6xy - 5y^2 + 4x - 22 = 0$

38. $36x^2 - 60xy + 25y^2 + 9y = 0$

39. $x^2 + 4xy + 4y^2 - 5x - y - 3 = 0$

40. $x^2 + xy + 4y^2 + x + y - 4 = 0$

In Exercises 41–44, sketch (if possible) the graph of the degenerate conic.

41. $y^2 - 9x^2 = 0$

42. $x^2 + y^2 - 2x + 6y + 10 = 0$

43. $x^2 + 2xy + y^2 - 1 = 0$

44. $x^2 - 10xy + y^2 = 0$

In Exercises 45–58, find any points of intersection of the graphs algebraically and then verify using a graphing utility.

45. $-x^2 + y^2 + 4x - 6y + 4 = 0$
$x^2 + y^2 - 4x - 6y + 12 = 0$

46. $-x^2 - y^2 - 8x + 20y - 7 = 0$
$x^2 + 9y^2 + 8x + 4y + 7 = 0$

47. $-4x^2 - y^2 - 16x + 24y - 16 = 0$
$4x^2 + y^2 + 40x - 24y + 208 = 0$

48. $x^2 - 4y^2 - 20x - 64y - 172 = 0$
$16x^2 + 4y^2 - 320x + 64y + 1600 = 0$

49. $x^2 - y^2 - 12x + 16y - 64 = 0$
$x^2 + y^2 - 12x - 16y + 64 = 0$

50. $x^2 + 4y^2 - 2x - 8y + 1 = 0$
$-x^2 + 2x - 4y - 1 = 0$

51. $-16x^2 - y^2 + 24y - 80 = 0$
$16x^2 + 25y^2 - 400 = 0$

52. $16x^2 - y^2 + 16y - 128 = 0$
$y^2 - 48x - 16y - 32 = 0$

53. $x^2 + y^2 - 4 = 0$
$3x - y^2 = 0$

54. $4x^2 + 9y^2 - 36y = 0$
$x^2 + 9y - 27 = 0$

55. $x^2 + 2y^2 - 4x + 6y - 5 = 0$
$-x + y - 4 = 0$

56. $x^2 + 2y^2 - 4x + 6y - 5 = 0$
$x^2 - 4x - y + 4 = 0$

57. $xy + x - 2y + 3 = 0$
$x^2 + 4y^2 - 9 = 0$

58. $5x^2 - 2xy + 5y^2 - 12 = 0$
$x + y - 1 = 0$

Synthesis

True or False? **In Exercises 59 and 60, determine whether the statement is true or false. Justify your answer.**

59. The graph of the equation

$$x^2 + xy + ky^2 + 6x + 10 = 0$$

where k is any constant less than $\frac{1}{4}$, is a hyperbola.

60. After a rotation of axes is used to eliminate the xy-term from an equation of the form

$$Ax^2 + Bxy + Cy^2 + Dx + Ey + F = 0$$

the coefficients of the x^2- and y^2-terms remain A and C, respectively.

61. Show that the equation $x^2 + y^2 = r^2$ is invariant under rotation of axes.

62. Find the lengths of the major and minor axes of the ellipse graphed in Exercise 14.

Review

In Exercises 63–70, graph the function.

63. $f(x) = |x + 3|$

64. $f(x) = |x - 4| + 1$

65. $g(x) = \sqrt{4 - x^2}$

66. $g(x) = \sqrt{3x - 2}$

67. $h(t) = -(t - 2)^3 + 3$

68. $h(t) = \frac{1}{2}(t + 4)^3$

69. $f(t) = [\![t - 5]\!] + 1$

70. $f(t) = -2[\![t]\!] + 3$

In Exercises 71–74, find the area of the triangle.

71. $C = 110°$, $a = 8$, $b = 12$

72. $B = 70°$, $a = 25$, $c = 16$

73. $a = 11$, $b = 18$, $c = 10$

74. $a = 23$, $b = 35$, $c = 27$

75. ***Height*** A 75-foot line is attached to a kite. When the kite has pulled the line taut, the angle of elevation to the kite is approximately 60°. Approximate the height of the kite.

76. ***Navigation*** A ship leaves port at noon and has a bearing of S 27° W. Its speed is 20 knots. How many nautical miles south and how many nautical miles west will the ship have traveled by 6:00 P.M.?

6.6 Parametric Equations

▶ **What you should learn**

- How to evaluate a set of parametric equations for a given value of the parameter
- How to sketch the curve that is represented by a set of parametric equations
- How to rewrite a set of parametric equations as a single rectangular equation
- How to find a set of parametric equations for a graph

▶ **Why you should learn it**

Parametric equations are useful for modeling the path of an object. For instance, in Exercise 59 on page 474, you will use a set of parametric equations to model the path of a baseball.

Plane Curves

Up to this point you have been representing a graph by a single equation involving the *two* variables x and y. In this section, you will study situations in which it is useful to introduce a *third* variable to represent a curve in the plane.

To see the usefulness of this procedure, consider the path followed by an object that is propelled into the air at an angle of 45°. If the initial velocity of the object is 48 feet per second, it can be shown that the object follows the parabolic path

$$y = -\frac{x^2}{72} + x. \qquad \text{Rectangular equation}$$

However, this equation does not tell the whole story. Although it does tell you *where* the object has been, it doesn't tell you *when* the object was at a given point (x, y) on the path. To determine this time, you can introduce a third variable t, called a **parameter**. It is possible to write both x and y as functions of t to obtain the **parametric equations**

$$x = 24\sqrt{2}\,t \qquad \text{Parametric equation for } x$$

$$y = -16t^2 + 24\sqrt{2}\,t. \qquad \text{Parametric equation for } y$$

From this set of equations you can determine that at time $t = 0$, the object is at the point $(0, 0)$. Similarly, at time $t = 1$, the object is at the point $\left(24\sqrt{2}, 24\sqrt{2} - 16\right)$, and so on, as shown in Figure 6.50.

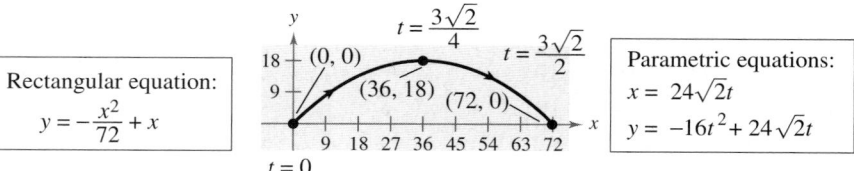

FIGURE 6.50 *Curvilinear Motion: Two Variables for Position, One Variable for Time*

For this particular motion problem, x and y are continuous functions of t, and the resulting path is a **plane curve.** (For this text, it is sufficient to think of a *continuous function* as one whose graph can be traced without lifting the pencil from the paper.)

Definition of Plane Curve

If f and g are continuous functions of t on an interval I, the set of ordered pairs $(f(t), g(t))$ is a **plane curve** C. The equations

$$x = f(t) \qquad \text{and} \qquad y = g(t)$$

are **parametric equations** for C, and t is the **parameter.**

Sketching a Plane Curve

When sketching a curve represented by a pair of parametric equations, you still plot points in the xy-plane. Each set of coordinates (x, y) is determined from a value chosen for the parameter t. Plotting the resulting points in the order of *increasing* values of t traces the curve in a specific direction. This is called the **orientation** of the curve.

Example 1 ▶ Sketching a Curve

Sketch the curve described by the parametric equations

$$x = t^2 - 4 \quad \text{and} \quad y = \frac{t}{2}, \quad -2 \le t \le 3.$$

Solution

Using values of t in the interval, the parametric equations yield the points (x, y) shown in the table.

t	x	y
-2	0	-1
-1	-3	$-1/2$
0	-4	0
1	-3	$1/2$
2	0	1
3	5	$3/2$

By plotting these points in the order of increasing t, you obtain the curve C shown in Figure 6.51. Note that the arrows on the curve indicate its orientation as t increases from -2 to 3.

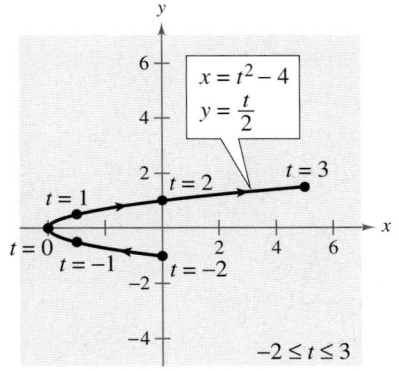

FIGURE **6.51**

Note that the graph shown in Figure 6.51 does not define y as a function of x. This points out one benefit of parametric equations—they can be used to represent graphs that are more general than graphs of functions.

It often happens that two different sets of parametric equations have the same graph. For example, the set of parametric equations

$$x = 4t^2 - 4 \quad \text{and} \quad y = t, \quad -1 \le t \le \frac{3}{2}$$

has the same graph as the set given in Example 1. However, by comparing the values of t in Figures 6.51 and 6.52, you see that this second graph is traced out more *rapidly* (considering t as time) than the first graph. So, in applications, different parametric representations can be used to represent various *speeds* at which objects travel along a given path.

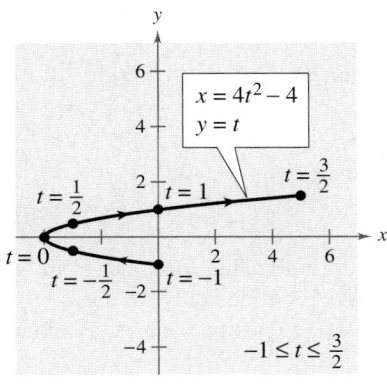

FIGURE **6.52**

Point out to your students the importance of knowing the orientation of a curve, and thus the usefulness of parametric equations.

Eliminating the Parameter

Example 1 uses simple point plotting to sketch the curve. This tedious process can sometimes be simplified by finding a rectangular equation (in x and y) that has the same graph. This process is called **eliminating the parameter.**

Parametric equations	⟹	Solve for t in one equation.	⟹	Substitute in other equation.	⟹	Rectangular equation
$x = t^2 - 4$ $y = t/2$		$t = 2y$		$x = (2y)^2 - 4$		$x = 4y^2 - 4$

Now you can recognize that the equation $x = 4y^2 - 4$ represents a parabola with a horizontal axis and vertex $(-4, 0)$.

When converting equations from parametric to rectangular form, you may need to alter the domain of the rectangular equation so that its graph matches the graph of the parametric equations. Such a situation is demonstrated in Example 2.

Emphasize that converting equations from parametric to rectangular form is primarily an aid in graphing.

Example 2 ▶ **Eliminating the Parameter**

Sketch the curve represented by the equations

$$x = \frac{1}{\sqrt{t+1}} \qquad \text{and} \qquad y = \frac{t}{t+1}$$

by eliminating the parameter and adjusting the domain of the resulting rectangular equation.

Solution

Solving for t in the equation for x, you have

$$x = \frac{1}{\sqrt{t+1}} \quad ⟹ \quad x^2 = \frac{1}{t+1}$$

which implies that

$$t = \frac{1 - x^2}{x^2}.$$

Now, substituting in the equation for y, you obtain

$$y = \frac{t}{t+1} = \frac{\dfrac{(1-x^2)}{x^2}}{\left[\dfrac{(1-x^2)}{x^2}\right] + 1} = \frac{\dfrac{1-x^2}{x^2}}{\dfrac{1-x^2}{x^2} + 1} \cdot \frac{x^2}{x^2} = 1 - x^2.$$

The rectangular equation, $y = 1 - x^2$, is defined for all values of x, but from the parametric equation for x you can see that the curve is defined only when $t > -1$. This implies that you should restrict the domain of x to positive values, as shown in Figure 6.53.

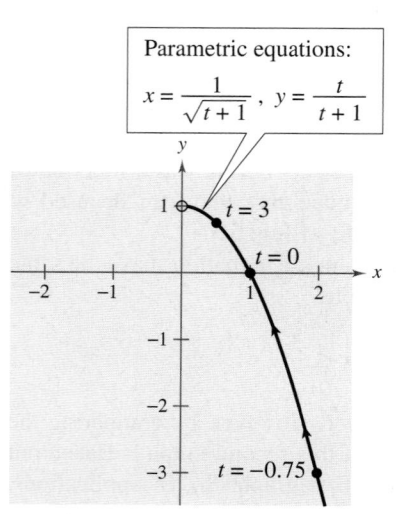

Parametric equations:
$$x = \frac{1}{\sqrt{t+1}}, \quad y = \frac{t}{t+1}$$

$t = 3$

$t = 0$

$t = -0.75$

FIGURE 6.53

It is not necessary for the parameter in a set of parametric equations to represent time. The next example uses an *angle* as the parameter.

Example 3 ▶ **Eliminating the Parameter**

Sketch the curve represented by

$$x = 3 \cos \theta \quad \text{and} \quad y = 4 \sin \theta, \quad 0 \le \theta \le 2\pi$$

by eliminating the parameter.

Solution

Begin by solving for $\cos \theta$ and $\sin \theta$ in the equations.

$$\cos \theta = \frac{x}{3} \quad \text{and} \quad \sin \theta = \frac{y}{4} \qquad \text{Solve for } \cos \theta \text{ and } \sin \theta.$$

Make use of the identity $\sin^2 \theta + \cos^2 \theta = 1$ to form an equation involving only x and y.

$$\cos^2 \theta + \sin^2 \theta = 1 \qquad \text{Trigonometric identity}$$

$$\left(\frac{x}{3}\right)^2 + \left(\frac{y}{4}\right)^2 = 1 \qquad \text{Substitute } \frac{x}{3} \text{ for } \cos \theta \text{ and } \frac{y}{4} \text{ for } \sin \theta.$$

$$\frac{x^2}{9} + \frac{y^2}{16} = 1 \qquad \text{Rectangular equation}$$

From this rectangular equation, you can see that the graph is an ellipse centered at $(0, 0)$, with vertices $(0, 4)$ and $(0, -4)$ and minor axis of length $2b = 6$, as shown in Figure 6.54. Note that the elliptic curve is traced out *counterclockwise* as θ varies from 0 to 2π.

In Examples 2 and 3, it is important to realize that eliminating the parameter is primarily an *aid to curve sketching*. If the parametric equations represent the path of a moving object, the graph alone is not sufficient to describe the object's motion. You still need the parametric equations to tell you the *position*, *direction*, and *speed* at a given time.

Finding Parametric Equations for a Graph

You have been studying techniques for sketching the graph represented by a set of parametric equations. Now consider the reverse problem—that is, how can you find a set of parametric equations for a given graph or a given physical description? From the discussion following Example 1, you know that such a representation is not unique. That is, the equations

$$x = 4t^2 - 4 \quad \text{and} \quad y = t, \ -1 \le t \le \frac{3}{2}$$

produced the same graph as the equations

$$x = t^2 - 4 \quad \text{and} \quad y = \frac{t}{2}, \ -2 \le t \le 3.$$

This is further demonstrated in Example 4.

STUDY TIP

To eliminate the parameter in equations involving trigonometric functions, try using the identities

$$\sin^2 \theta + \cos^2 \theta = 1$$

$$\sec^2 \theta - \tan^2 \theta = 1$$

as shown in Example 3.

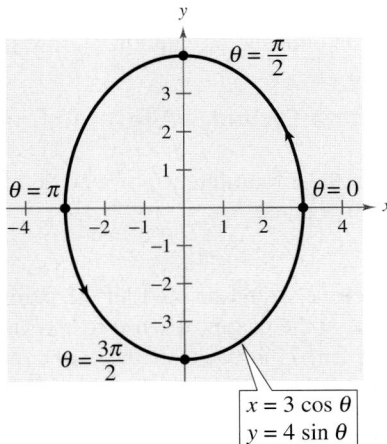

$$x = 3 \cos \theta$$
$$y = 4 \sin \theta$$

FIGURE **6.54**

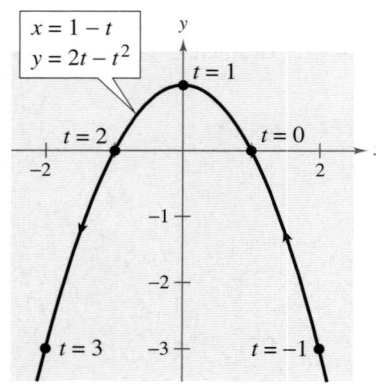

FIGURE 6.55

STUDY TIP

In Example 5, $\widehat{PD}$ represents the arc of the circle between points P and D.

Technology

Use a graphing utility in *parametric* mode to obtain a graph similar to Figure 6.56 by graphing the following equations.

$$X_{1T} = T - \sin T$$

$$Y_{1T} = 1 - \cos T$$

Example 4 ▶   **Finding Parametric Equations for a Graph**

Find a set of parametric equations to represent the graph of $y = 1 - x^2$, using the following parameters.

a. $t = x$ **b.** $t = 1 - x$

Solution

a. Letting $t = x$, you obtain the parametric equations

$$x = t \quad \text{and} \quad y = 1 - x^2 = 1 - t^2.$$

b. Letting $t = 1 - x$, you obtain the parametric equations

$$x = 1 - t \quad \text{and} \quad y = 1 - x^2 = 1 - (1 - t)^2 = 2t - t^2.$$

In Figure 6.55, note how the resulting curve is oriented by the increasing values of t. For part (a), the curve would have the opposite orientation.

Example 5 ▶   **Parametric Equations for a Cycloid**

Describe the **cycloid** traced out by a point P on the circumference of a circle of radius a as the circle rolls along a straight line in a plane.

Solution

As the parameter, let θ be the measure of the circle's rotation, and let the point $P = (x, y)$ begin at the origin. When $\theta = 0$, P is at the origin; when $\theta = \pi$, P is at a maximum point $(\pi a, 2a)$; and when $\theta = 2\pi$, P is back on the x-axis at $(2\pi a, 0)$. From Figure 6.56, you can see that $\angle APC = 180° - \theta$. So, you have

$$\sin \theta = \sin(180° - \theta) = \sin(\angle APC) = \frac{AC}{a} = \frac{BD}{a}$$

$$\cos \theta = -\cos(180° - \theta) = -\cos(\angle APC) = \frac{AP}{-a}$$

which implies that $AP = -a \cos \theta$ and $BD = a \sin \theta$. Because the circle rolls along the x-axis, you know that $OD = \widehat{PD} = a\theta$. Furthermore, because $BA = DC = a$, you have

$$x = OD - BD = a\theta - a \sin \theta \quad \text{and} \quad y = BA + AP = a - a \cos \theta.$$

So, the parametric equations are $x = a(\theta - \sin \theta)$ and $y = a(1 - \cos \theta)$.

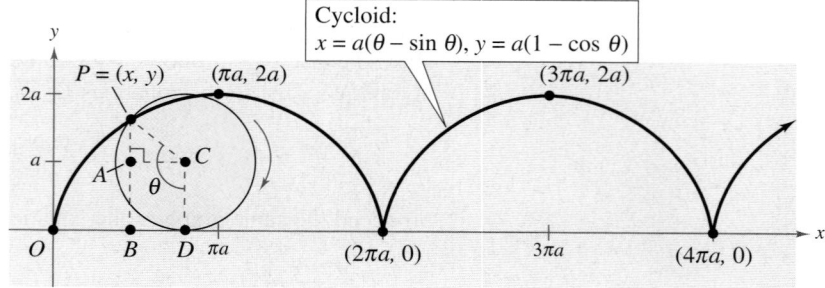

FIGURE 6.56

6.6 Exercises

1. Consider the parametric equations $x = \sqrt{t}$ and $y = 3 - t$.

(a) Create a table of x- and y-values using $t = 0, 1, 2, 3,$ and 4.

(b) Plot the points (x, y) generated in part (a), and sketch a graph of the parametric equations.

(c) Find the rectangular equation by eliminating the parameter. Sketch its graph. How do the graphs differ?

2. Consider the parametric equations $x = 4 \cos^2 \theta$ and $y = 2 \sin \theta$.

(a) Create a table of x- and y-values using $\theta = -\pi/2, -\pi/4, 0, \pi/4,$ and $\pi/2$.

(b) Plot the points (x, y) generated in part (a), and sketch a graph of the parametric equations.

(c) Find the rectangular equation by eliminating the parameter. Sketch its graph. How do the graphs differ?

In Exercises 3–22, sketch the curve represented by the parametric equations (indicate the orientation of the curve) by eliminating the parameter and adjusting the domain of the resulting rectangular equation.

3. $x = 3t - 3$
$y = 2t + 1$

4. $x = 3 - 2t$
$y = 2 + 3t$

5. $x = \frac{1}{4}t$
$y = t^2$

6. $x = t$
$y = t^3$

7. $x = t + 2$
$y = t^2$

8. $x = \sqrt{t}$
$y = 1 - t$

9. $x = t + 1$
$y = t/(t + 1)$

10. $x = t - 1$
$y = t/(t - 1)$

11. $x = 2(t + 1)$
$y = |t - 2|$

12. $x = |t - 1|$
$y = t + 2$

13. $x = 3 \cos \theta$
$y = 3 \sin \theta$

14. $x = 2 \cos \theta$
$y = 3 \sin \theta$

15. $x = 4 \sin 2\theta$
$y = 2 \cos 2\theta$

16. $x = \cos \theta$
$y = 2 \sin 2\theta$

17. $x = 4 + 2 \cos \theta$
$y = -1 + \sin \theta$

18. $x = 4 + 2 \cos \theta$
$y = 2 + 3 \sin \theta$

19. $x = e^{-t}$
$y = e^{3t}$

20. $x = e^{2t}$
$y = e^t$

21. $x = t^3$
$y = 3 \ln t$

22. $x = \ln 2t$
$y = 2t^2$

In Exercises 23 and 24, determine how the plane curves differ from each other.

23. (a) $x = t$
$y = 2t + 1$

(b) $x = \cos \theta$
$y = 2 \cos \theta + 1$

(c) $x = e^{-t}$
$y = 2e^{-t} + 1$

(d) $x = e^t$
$y = 2e^t + 1$

24. (a) $x = t$
$y = t^2 - 1$

(b) $x = t^2$
$y = t^4 - 1$

(c) $x = \sin t$
$y = \sin^2 t - 1$

(d) $x = e^t$
$y = e^{2t} - 1$

In Exercises 25–28, eliminate the parameter and obtain the standard form of the rectangular equation.

25. Line through (x_1, y_1) and (x_2, y_2):
$x = x_1 + t(x_2 - x_1), \; y = y_1 + t(y_2 - y_1)$

26. Circle:
$x = h + r \cos \theta, \; y = k + r \sin \theta$

27. Ellipse:
$x = h + a \cos \theta, \; y = k + b \sin \theta$

28. Hyperbola:
$x = h + a \sec \theta, \; y = k + b \tan \theta$

In Exercises 29–36, use the results of Exercises 25–28 to find a set of parametric equations for the line or conic.

29. Line: Passes through $(0, 0)$ and $(6, -3)$

30. Line: Passes through $(2, 3)$ and $(6, -3)$

31. Circle: Center: $(3, 2)$; Radius: 4

32. Circle: Center: $(-3, 2)$; Radius: 5

33. Ellipse: Vertices: $(\pm 4, 0)$; Foci: $(\pm 3, 0)$

34. Ellipse: Vertices: $(4, 7), (4, -3)$;
Foci: $(4, 5), (4, -1)$

35. Hyperbola: Vertices: $(\pm 4, 0)$; Foci: $(\pm 5, 0)$

36. Hyperbola: Vertices: $(\pm 2, 0)$; Foci: $(\pm 4, 0)$

In Exercises 37–44, find a set of parametric equations for the rectangular equation using (a) $t = x$ and (b) $t = 2 - x$.

37. $y = 3x - 2$

38. $x = 3y - 2$

39. $y = x^2$

40. $y = x^3$

41. $y = x^2 + 1$

42. $y = 2 - x$

43. $y = \dfrac{1}{x}$

44. $y = \dfrac{1}{2x}$

 In Exercises 45–52, use a graphing utility to graph the curve represented by the parametric equations.

45. Cycloid: $x = 4(\theta - \sin \theta)$, $y = 4(1 - \cos \theta)$

46. Cycloid: $x = \theta + \sin \theta$, $y = 1 - \cos \theta$

47. Prolate cycloid: $x = \theta - \frac{3}{2} \sin \theta$, $y = 1 - \frac{3}{2} \cos \theta$

48. Prolate cycloid: $x = 2\theta - 4 \sin \theta$, $y = 2 - 4 \cos \theta$

49. Hypocycloid: $x = 3 \cos^3 \theta$, $y = 3 \sin^3 \theta$

50. Curtate cycloid: $x = 8\theta - 4 \sin \theta$, $y = 8 - 4 \cos \theta$

51. Witch of Agnesi: $x = 2 \cot \theta$, $y = 2 \sin^2 \theta$

52. Folium of Descartes: $x = \dfrac{3t}{1 + t^3}$, $y = \dfrac{3t^2}{1 + t^3}$

In Exercises 53–56, match the parametric equations with the correct graph and describe the domain and range. [The graphs are labeled (a), (b), (c), and (d).]

(a)

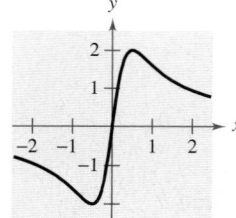

(b)

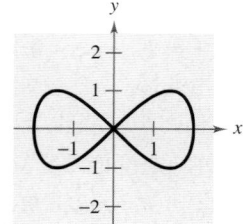

(c)

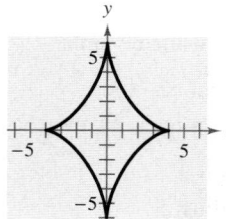

(d)

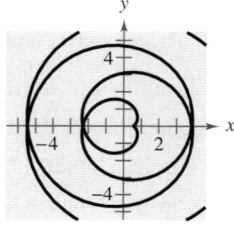

53. Lissajous curve: $x = 2 \cos \theta$

$y = \sin 2\theta$

54. Evolute of ellipse: $x = 4 \cos^3 \theta$

$y = 6 \sin^3 \theta$

55. Involute of circle: $x = \frac{1}{2}(\cos \theta + \theta \sin \theta)$

$y = \frac{1}{2}(\sin \theta - \theta \cos \theta)$

56. Serpentine curve: $x = \frac{1}{2} \cot \theta$

$y = 4 \sin \theta \cos \theta$

 Projectile Motion **A projectile is launched at a height of h feet above the ground and at an angle θ with the horizontal. The initial velocity is v_0 feet per second and the path of the projectile is modeled by the parametric equations**

$$x = (v_0 \cos \theta)t \quad \text{and} \quad y = h + (v_0 \sin \theta)t - 16t^2.$$

In Exercises 57 and 58, use a graphing utility to graph the paths of a projectile launched from ground level at each value of θ and v_0. For each case, use the graph to approximate the maximum height and the range of the projectile.

57. (a) $\theta = 60°$, $v_0 = 88$ feet per second

(b) $\theta = 60°$, $v_0 = 132$ feet per second

(c) $\theta = 45°$, $v_0 = 88$ feet per second

(d) $\theta = 45°$, $v_0 = 132$ feet per second

58. (a) $\theta = 15°$, $v_0 = 60$ feet per second

(b) $\theta = 15°$, $v_0 = 100$ feet per second

(c) $\theta = 30°$, $v_0 = 60$ feet per second

(d) $\theta = 30°$, $v_0 = 100$ feet per second

▶ **Model It**

59. *Sports* The center field fence in Yankee Stadium is 7 feet high and 408 feet from home plate. A baseball is hit 3 feet above the ground. It leaves the bat at an angle of θ degrees with the horizontal at a speed of 100 miles per hour (see figure).

(a) Write a set of parametric equations that model the path of the baseball.

 (b) Use a graphing utility to graph the path of the baseball when $\theta = 15°$. Is the hit a home run?

 (c) Use a graphing utility to graph the path of the baseball when $\theta = 23°$. Is the hit a home run?

(d) Find the minimum angle required for the hit to be a home run.

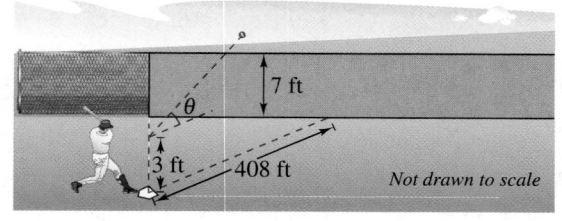

Not drawn to scale

60. Sports An archer releases an arrow from a bow 5 feet above the ground. The arrow leaves the bow at an angle of $10°$ with the horizontal and at an initial speed of 240 feet per second.

(a) Write a set of parametric equations that model the path of the arrow.

(b) Assuming the ground is level, find the distance the arrow travels before it hits the ground. (Ignore air resistance.)

(c) Use a graphing utility to graph the path of the arrow and approximate its maximum height.

(d) Find the total time the arrow is in the air.

61. Projectile Motion Eliminate the parameter t from the parametric equations $x = (v_0 \cos \theta)t$ and $y = h + (v_0 \sin \theta)t - 16t^2$ for the motion of a projectile to show that the rectangular equation is

$$y = -\frac{16 \sec^2 \theta}{v_0^2}x^2 + (\tan \theta)x + h.$$

62. Path of a Projectile The path of a projectile is given by the rectangular equation

$$y = 7 + x - 0.02x^2.$$

(a) Use the result of Exercise 61 to find h, v_0, and θ. Find the parametric equations of the path.

(b) Use a graphing utility to graph the rectangular equation for the path of the projectile. Confirm your answer in part (a) by sketching the curve represented by the parametric equations.

(c) Use a graphing utility to approximate the maximum height of the projectile and its range.

63. Curtate Cycloid A wheel of radius a units rolls along a straight line without slipping. The curve traced by a point P that is b units from the center $(b < a)$ is called a **curtate cycloid** (see figure). Use the angle θ shown in the figure to find a set of parametric equations for the curve.

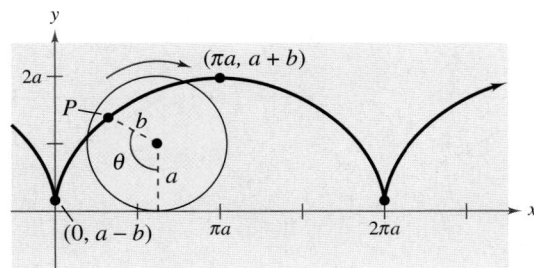

64. Epicycloid A circle of radius one unit rolls around the outside of a circle of radius two units without slipping. The curve traced by a point on the circumference of the smaller circle is called an **epicycloid** (see figure). Use the angle θ shown in the figure to find a set of parametric equations for the curve.

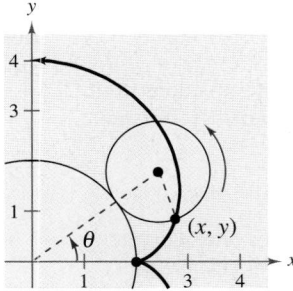

Synthesis

True or False? In Exercises 65 and 66, determine whether the statement is true or false. Justify your answer.

65. The two sets of parametric equations $x = t$, $y = t^2 + 1$ and $x = 3t$, $y = 9t^2 + 1$ have the same rectangular equation.

66. The graph of the parametric equations $x = t^2$ and $y = t^2$ is the line $y = x$.

Review

In Exercises 67–70, find the reference angle θ', and sketch θ and θ' in standard position.

67. $\theta = 105°$

68. $\theta = 230°$

69. $\theta = -\dfrac{2\pi}{3}$

70. $\theta = \dfrac{5\pi}{6}$

In Exercises 71–74, sketch a graph of the function.

71. $y = \arcsin(x + 1)$

72. $y = \arccos(x - 1)$

73. $f(x) = 2 \arctan x$

74. $f(x) = \arctan \dfrac{x}{2}$

6.7 Polar Coordinates

▶ **What you should learn**

- How to plot points on the polar coordinate system
- How to convert points from rectangular to polar form and vice versa
- How to convert equations from rectangular to polar form and vice versa

▶ **Why you should learn it**

Polar coordinates offer a different mathematical perspective on graphing. For instance, in Exercises 1–8 on page 480, you are asked to find multiple representations of polar coordinates.

Introduction

So far, you have been representing graphs of equations as collections of points (x, y) on the rectangular coordinate system, where x and y represent the directed distances from the coordinate axes to the point (x, y). In this section you will study a different system called the **polar coordinate system.**

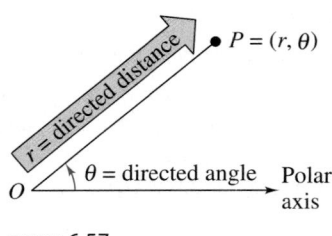

FIGURE 6.57

To form the polar coordinate system in the plane, fix a point O, called the **pole** (or **origin**), and construct from O an initial ray called the **polar axis,** as shown in Figure 6.57. Then each point P in the plane can be assigned **polar coordinates** (r, θ) as follows.

1. $r = directed\ distance$ from O to P
2. $\theta = directed\ angle$, counterclockwise from polar axis to segment $\overline{OP}$

Example 1 ▶ Plotting Points on the Polar Coordinate System

a. The point $(r, \theta) = (2, \pi/3)$ lies two units from the pole on the terminal side of the angle $\theta = \pi/3$, as shown in Figure 6.58.

b. The point $(r, \theta) = (3, -\pi/6)$ lies three units from the pole on the terminal side of the angle $\theta = -\pi/6$, as shown in Figure 6.59.

c. The point $(r, \theta) = (3, 11\pi/6)$ coincides with the point $(3, -\pi/6)$, as shown in Figure 6.60.

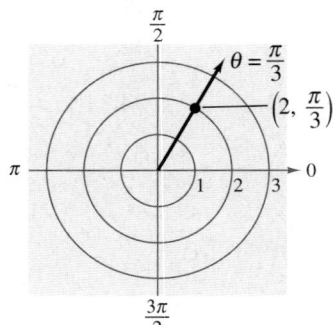

FIGURE 6.58

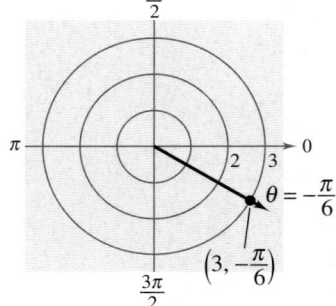

FIGURE 6.59

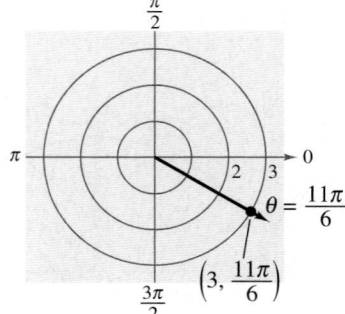

FIGURE 6.60

Most graphing calculators have a *polar* graphing mode. If yours does, try graphing the equation $r = 3$. (Use a setting of $-6 \le x \le 6$ and $-4 \le y \le 4$.) You should obtain a circle of radius 3.

a. Use the *trace* feature to cursor around the circle. Can you locate the point $(3, 5\pi/4)$?

b. Can you find other polar representations of the point $(3, 5\pi/4)$? If so, explain how you did it.

In rectangular coordinates, each point (x, y) has a unique representation. This is not true for polar coordinates. For instance, the coordinates (r, θ) and $(r, \theta + 2\pi)$ represent the same point, as illustrated in Example 1. Another way to obtain multiple representations of a point is to use negative values for r. Because r is a *directed distance*, the coordinates (r, θ) and $(-r, \theta + \pi)$ represent the same point. In general, the point (r, θ) can be represented as

$$(r, \theta) = (r, \theta \pm 2n\pi) \qquad \text{or} \qquad (r, \theta) = (-r, \theta \pm (2n + 1)\pi)$$

where n is any integer. Moreover, the pole is represented by $(0, \theta)$, where θ is any angle.

Example 2 ▶ **Multiple Representations of Points**

Plot the point $(3, -3\pi/4)$ and find three additional polar representations of this point, using $-2\pi < \theta < 2\pi$.

Solution

The point is shown in Figure 6.61. Three other representations are as follows.

$$\left(3, -\frac{3\pi}{4} + 2\pi\right) = \left(3, \frac{5\pi}{4}\right) \qquad \text{Add } 2\pi \text{ to } \theta.$$

$$\left(-3, -\frac{3\pi}{4} - \pi\right) = \left(-3, -\frac{7\pi}{4}\right) \qquad \text{Replace } r \text{ by } -r; \text{ subtract } \pi \text{ from } \theta.$$

$$\left(-3, -\frac{3\pi}{4} + \pi\right) = \left(-3, \frac{\pi}{4}\right) \qquad \text{Replace } r \text{ by } -r; \text{ add } \pi \text{ to } \theta.$$

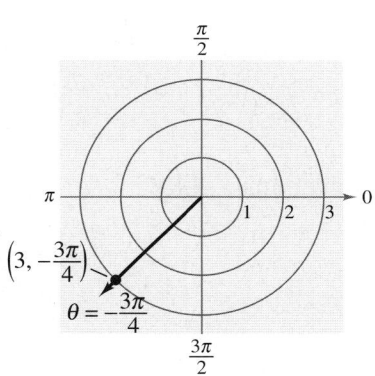

$$\left(3, -\frac{3\pi}{4}\right) = \left(3, \frac{5\pi}{4}\right) = \left(-3, -\frac{7\pi}{4}\right) = \left(-3, \frac{\pi}{4}\right) = \dots$$

FIGURE **6.61**

Coordinate Conversion

To establish the relationship between polar and rectangular coordinates, let the polar axis coincide with the positive *x*-axis and the pole with the origin, as shown in Figure 6.62. Because (x, y) lies on a circle of radius r, it follows that $r^2 = x^2 + y^2$. Moreover, for $r > 0$, the definitions of the trigonometric functions imply that

$$\tan \theta = \frac{y}{x}, \qquad \cos \theta = \frac{x}{r}, \qquad \text{and} \qquad \sin \theta = \frac{y}{r}.$$

If $r < 0$, you can show that the same relationships hold.

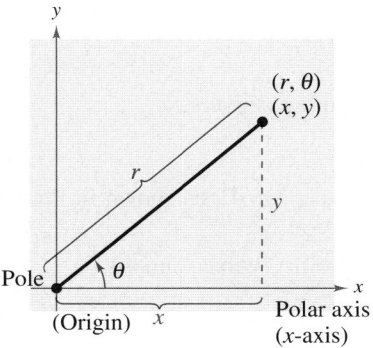

FIGURE **6.62**

Coordinate Conversion

The polar coordinates (r, θ) are related to the rectangular coordinates (x, y) as follows.

$$x = r \cos \theta \qquad \text{and} \qquad \tan \theta = \frac{y}{x}$$

$$y = r \sin \theta \qquad \qquad r^2 = x^2 + y^2$$

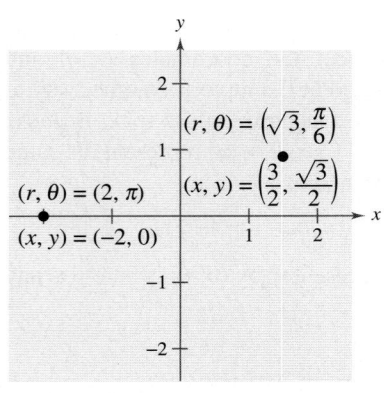

FIGURE **6.63**

Activities

1. Find three additional polar representations of the point $\left(-2, \frac{\pi}{3}\right)$.

   Answer: $\left(2, \frac{4\pi}{3}\right), \left(2, -\frac{2\pi}{3}\right)$, and $\left(-2, -\frac{5\pi}{3}\right)$

2. Convert the point $(-4, 2)$ from rectangular to polar form.

   Answer: $\left(2\sqrt{5}, -0.4636\right)$ or $\left(-2\sqrt{5}, 2.6779\right)$

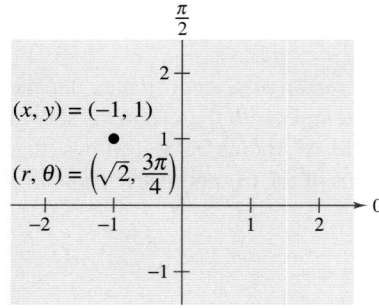

FIGURE **6.64**

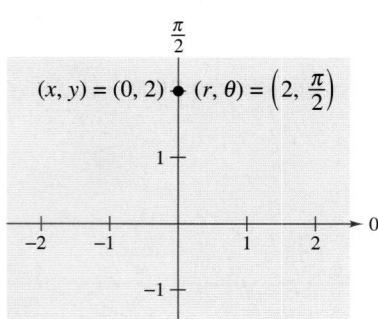

FIGURE **6.65**

Example 3 ▶ **Polar-to-Rectangular Conversion**

Convert the points to rectangular coordinates. (See Figure 6.63.)

a. $(2, \pi)$ **b.** $\left(\sqrt{3}, \pi/6\right)$

Solution

a. For the point $(r, \theta) = (2, \pi)$, you have

$$x = r \cos \theta = 2 \cos \pi = -2$$

and

$$y = r \sin \theta = 2 \sin \pi = 0.$$

The rectangular coordinates are $(x, y) = (-2, 0)$.

b. For the point $(r, \theta) = \left(\sqrt{3}, \pi/6\right)$, you have

$$x = \sqrt{3} \cos \frac{\pi}{6} = \sqrt{3}\left(\frac{\sqrt{3}}{2}\right) = \frac{3}{2}$$

and

$$y = \sqrt{3} \sin \frac{\pi}{6} = \sqrt{3}\left(\frac{1}{2}\right) = \frac{\sqrt{3}}{2}.$$

The rectangular coordinates are $(x, y) = \left(3/2, \sqrt{3}/2\right)$.

Example 4 ▶ **Rectangular-to-Polar Conversion**

Convert the points to polar coordinates.

a. $(-1, 1)$ **b.** $(0, 2)$

Solution

a. For the second-quadrant point $(x, y) = (-1, 1)$, you have

$$\tan \theta = \frac{y}{x}$$

$$\tan \theta = -1$$

$$\theta = \frac{3\pi}{4}.$$

Because θ lies in the same quadrant as (x, y), use positive r.

$$r = \sqrt{x^2 + y^2} = \sqrt{(-1)^2 + (1)^2} = \sqrt{2}$$

So, *one* set of polar coordinates is $(r, \theta) = \left(\sqrt{2}, 3\pi/4\right)$, as shown in Figure 6.64.

b. Because the point $(x, y) = (0, 2)$ lies on the positive y-axis, choose

$$\theta = \frac{\pi}{2} \quad \text{and} \quad r = 2.$$

This implies that *one* set of polar coordinates is $(r, \theta) = (2, \pi/2)$, as shown in Figure 6.65.

Multiple representations of points and equations in the polar system can cause confusion. You may want to discuss several examples.

Activities

1. Convert the polar equation $r = 3 \cos \theta$ to rectangular form.

   Answer: $x^2 + y^2 - 3x = 0$

2. Convert the rectangular equation $x = 4$ to polar form.

   Answer: $r = 4 \sec \theta$

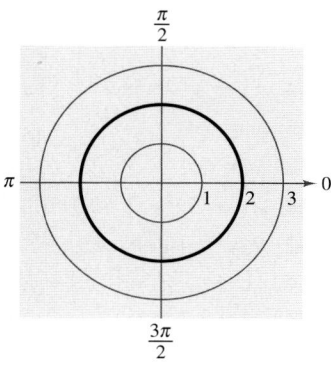

FIGURE **6.66**

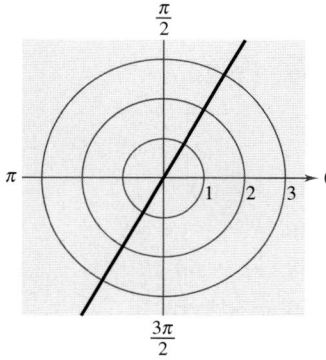

FIGURE **6.67**

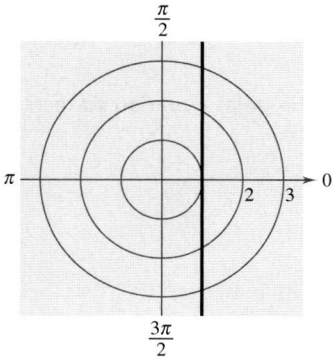

FIGURE **6.68**

Equation Conversion

By comparing Examples 3 and 4, you can see that point conversion from the polar to the rectangular system is straightforward, whereas point conversion from the rectangular to the polar system is more involved. For equations, the opposite is true. To convert a rectangular equation to polar form, you simply replace x by $r \cos \theta$ and y by $r \sin \theta$. For instance, the rectangular equation $y = x^2$ can be written in polar form as follows.

$$y = x^2 \qquad \text{Rectangular equation}$$

$$r \sin \theta = (r \cos \theta)^2 \qquad \text{Polar equation}$$

$$r = \sec \theta \tan \theta \qquad \text{Simplest form}$$

On the other hand, converting a polar equation to rectangular form requires considerable ingenuity.

Example 5 demonstrates several polar-to-rectangular conversions that enable you to sketch the graphs of some polar equations.

Example 5 ▶ **Converting Polar Equations to Rectangular Form**

Describe the graph of each polar equation and find the corresponding rectangular equation.

a. $r = 2$ **b.** $\theta = \dfrac{\pi}{3}$ **c.** $r = \sec \theta$

Solution

a. The graph of the polar equation $r = 2$ consists of all points that are two units from the pole. In other words, this graph is a circle centered at the origin with a radius of 2, as shown in Figure 6.66. You can confirm this by converting to rectangular form, using the relationship $r^2 = x^2 + y^2$.

$$r = 2 \qquad \Longrightarrow \qquad r^2 = 2^2 \qquad \Longrightarrow \qquad x^2 + y^2 = 2^2$$

Polar equation Rectangular equation

b. The graph of the polar equation $\theta = \pi/3$ consists of all points on the line that makes an angle of $\pi/3$ with the positive polar axis, as shown in Figure 6.67. To convert to rectangular form, make use of the relationship $\tan \theta = y/x$.

$$\theta = \frac{\pi}{3} \qquad \Longrightarrow \qquad \tan \theta = \sqrt{3} \qquad \Longrightarrow \qquad y = \sqrt{3}x$$

Polar equation Rectangular equation

c. The graph of the polar equation $r = \sec \theta$ is not evident by simple inspection, so convert to rectangular form by using the relationship $r \cos \theta = x$.

$$r = \sec \theta \qquad \Longrightarrow \qquad r \cos \theta = 1 \qquad \Longrightarrow \qquad x = 1$$

Polar equation Rectangular equation

Now you see that the graph is a vertical line, as shown in Figure 6.68.

6.7 Exercises

In Exercises 1–8, plot the point given in polar coordinates 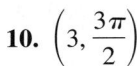 and find two additional polar representations.

1. $\left(4, -\dfrac{\pi}{3}\right)$

2. $\left(-1, -\dfrac{3\pi}{4}\right)$

3. $\left(0, -\dfrac{7\pi}{6}\right)$

4. $\left(16, \dfrac{5\pi}{2}\right)$

5. $\left(\sqrt{2}, 2.36\right)$

6. $(-3, -1.57)$

7. $\left(2\sqrt{2}, 4.71\right)$

8. $(-5, -2.36)$

In Exercises 9–16, a point in polar coordinates is given. Convert the point to rectangular coordinates.

9. $\left(3, \dfrac{\pi}{2}\right)$

10. $\left(3, \dfrac{3\pi}{2}\right)$

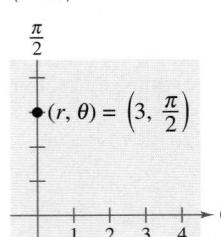

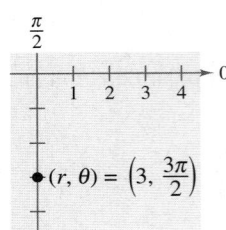

11. $\left(-1, \dfrac{5\pi}{4}\right)$

12. $(0, -\pi)$

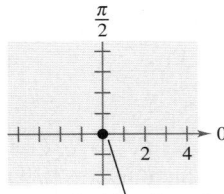

13. $\left(2, \dfrac{3\pi}{4}\right)$

14. $\left(-2, \dfrac{7\pi}{6}\right)$

15. $(-2.5, 1.1)$

16. $(8.25, 3.5)$

In Exercises 17–26, a point in rectangular coordinates is given. Convert the point to polar coordinates.

17. $(1, 1)$

18. $(-3, -3)$

19. $(-6, 0)$

20. $(0, -5)$

21. $(-3, 4)$

22. $(3, -1)$

23. $\left(-\sqrt{3}, -\sqrt{3}\right)$

24. $\left(\sqrt{3}, -1\right)$

25. $(6, 9)$

26. $(5, 12)$

In Exercises 27–32, use a graphing utility to find one set of polar coordinates of the point given in rectangular coordinates.

27. $(3, -2)$

28. $(-5, 2)$

29. $\left(\sqrt{3}, 2\right)$

30. $\left(3\sqrt{2}, 3\sqrt{2}\right)$

31. $\left(\dfrac{5}{2}, \dfrac{4}{3}\right)$

32. $\left(\dfrac{7}{4}, \dfrac{3}{2}\right)$

In Exercises 33–48, convert the rectangular equation to polar form.

33. $x^2 + y^2 = 9$

34. $x^2 + y^2 = 16$

35. $y = 4$

36. $y = x$

37. $x = 10$

38. $x = 4a$

39. $3x - y + 2 = 0$

40. $3x + 5y - 2 = 0$

41. $xy = 16$

42. $2xy = 1$

43. $y^2 - 8x - 16 = 0$

44. $(x^2 + y^2)^2 = 9(x^2 - y^2)$

45. $x^2 + y^2 = a^2$

46. $x^2 + y^2 = 9a^2$

47. $x^2 + y^2 - 2ax = 0$

48. $x^2 + y^2 - 2ay = 0$

In Exercises 49–64, convert the polar equation to rectangular form.

49. $r = 4 \sin \theta$

50. $r = 2 \cos \theta$

51. $\theta = \dfrac{2\pi}{3}$

52. $\theta = \dfrac{5\pi}{3}$

53. $r = 4$

54. $r = 10$

55. $r = 4 \csc \theta$

56. $r = -3 \sec \theta$

57. $r^2 = \cos \theta$

58. $r^2 = \sin 2\theta$

59. $r = 2 \sin 3\theta$

60. $r = 3 \cos 2\theta$

61. $r = \dfrac{2}{1 + \sin \theta}$

62. $r = \dfrac{1}{1 - \cos \theta}$

63. $r = \dfrac{6}{2 - 3 \sin \theta}$

64. $r = \dfrac{6}{2 \cos \theta - 3 \sin \theta}$

In Exercises 65–70, convert the polar equation to rectangular form and sketch its graph.

65. $r = 6$

66. $r = 8$

67. $\theta = \dfrac{\pi}{6}$

68. $\theta = \dfrac{3\pi}{4}$

69. $r = 3 \sec \theta$

70. $r = 2 \csc \theta$

Synthesis

True or False? **In Exercises 71 and 72, determine whether the statement is true or false. Justify your answer.**

71. If $\theta_1 = \theta_2 + 2\pi n$ for some integer n, then (r, θ_1) and (r, θ_2) represent the same point on the polar coordinate system.

72. If $|r_1| = |r_2|$, then (r_1, θ) and (r_2, θ) represent the same point on the polar coordinate system.

73. Convert the polar equation $r = 2(h \cos \theta + k \sin \theta)$ to rectangular form and verify that it is the equation of a circle. Find the radius and the rectangular coordinates of the center of the circle.

74. Convert the polar equation $r = \cos \theta + 3 \sin \theta$ to rectangular form and identify the graph.

75. ***Think About It***

(a) Show that the distance between the points (r_1, θ_1) and (r_2, θ_2) is $\sqrt{r_1^2 + r_2^2 - 2r_1 r_2 \cos(\theta_1 - \theta_2)}$.

(b) Describe the positions of the points relative to each other for $\theta_1 = \theta_2$. Simplify the Distance Formula for this case. Is the simplification what you expected? Explain.

(c) Simplify the Distance Formula for $\theta_1 - \theta_2 = 90°$. Is the simplification what you expected? Explain.

(d) Choose two points on the polar coordinate system and find the distance between them. Then choose different polar representations of the same two points and apply the Distance Formula again. Discuss the result.

76. ***Exploration***

(a) Set the window format of your graphing utility on rectangular coordinates and locate the cursor at any position off the coordinate axes. Move the cursor horizontally and observe any changes in the displayed coordinates of the points. Explain the changes. Now repeat the process moving the cursor vertically.

(b) Set the window format of your graphing utility on polar coordinates and locate the cursor at any position off the coordinate axes. Move the cursor horizontally and observe any changes in the displayed coordinates of the points. Explain the changes. Now repeat the process moving the cursor vertically.

(c) Explain why the results of parts (a) and (b) are not the same.

Review

In Exercises 77–80, use the properties of logarithms to expand the expression as a sum, difference, and/or constant multiple of logarithms. (Assume that all variables are positive.)

77. $\log_6 \dfrac{x^2 z}{3y}$

78. $\log_4 \dfrac{\sqrt{2x}}{y}$

79. $\ln x(x + 4)^2$

80. $\ln 5x^2(x^2 + 1)$

In Exercises 81–84, condense the expression to the logarithm of a single quantity.

81. $\log_7 x - \log_7 3y$

82. $\log_5 a + 8 \log_5(x + 1)$

83. $\dfrac{1}{2}\ln x + \ln(x - 2)$

84. $\ln 6 + \ln y - \ln(x - 3)$

6.8   Graphs of Polar Equations

▶ **Why you should learn it**

Equations of several common figures are simpler in polar form than in rectangular form. For instance, Exercise 6 on page 488 shows the graph of a circle and its polar equation.

Introduction

In previous chapters you spent a lot of time learning how to sketch graphs on rectangular coordinate systems. You began with the basic point-plotting method, which was then enhanced by sketching aids such as symmetry, intercepts, asymptotes, periods, and shifts. This section approaches curve sketching on the polar coordinate system similarly, beginning with a demonstration of point plotting.

Example 1 ▶   **Graphing a Polar Equation by Point Plotting**

Sketch the graph of the polar equation $r = 4 \sin \theta$.

Solution

The sine function is periodic, so you can get a full range of r-values by considering values of θ in the interval $0 \leq \theta \leq 2\pi$, as shown in the following table.

θ	r
0	0
$\dfrac{\pi}{6}$	2
$\dfrac{\pi}{3}$	$2\sqrt{3}$
$\dfrac{\pi}{2}$	4
$\dfrac{2\pi}{3}$	$2\sqrt{3}$
$\dfrac{5\pi}{6}$	2
π	0
$\dfrac{7\pi}{6}$	-2
$\dfrac{3\pi}{2}$	-4
$\dfrac{11\pi}{6}$	-2
2π	0

Emphasize setting up the table of θ values. Your students will benefit from labeling the points as they plot them.

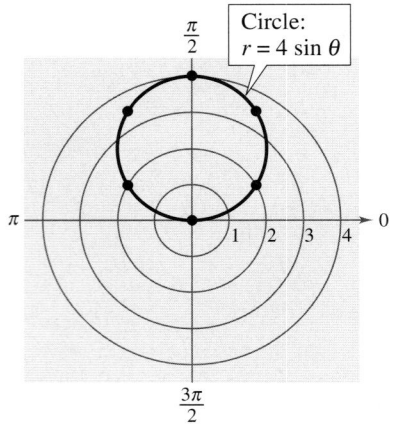

FIGURE **6.69**

If you plot these points as shown in Figure 6.69, it appears that the graph is a circle of radius 2 whose center is at the point $(x, y) = (0, 2)$.

Symmetry

In Figure 6.69, note that as θ increases from 0 to 2π the graph is traced out twice. Moreover, note that the graph is *symmetric with respect to the line* $\theta = \pi/2$. Had you known about this symmetry and retracing ahead of time, you could have used fewer points.

Symmetry with respect to the line $\theta = \pi/2$ is one of three important types of symmetry to consider in polar curve sketching. (See Figure 6.70.)

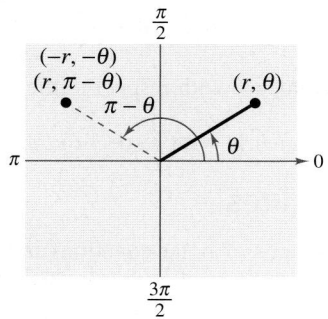

Symmetry with Respect to the Line $\theta = \dfrac{\pi}{2}$

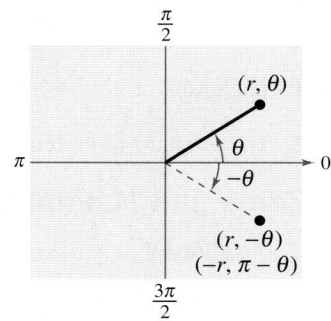

Symmetry with Respect to the Polar Axis

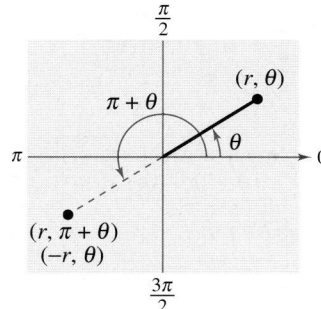

Symmetry with Respect to the Pole

FIGURE 6.70

Point out to your students that these tests are sufficient for showing symmetry; however, they are not necessary. A polar graph can exhibit symmetry even when the tests fail to indicate symmetry.

Tests for Symmetry in Polar Coordinates

The graph of a polar equation is symmetric with respect to the following if the given substitution yields an equivalent equation.

1. *The line* $\theta = \pi/2$:   Replace (r, θ) by $(r, \pi - \theta)$ or $(-r, -\theta)$.

2. *The polar axis:*   Replace (r, θ) by $(r, -\theta)$ or $(-r, \pi - \theta)$.

3. *The pole:*   Replace (r, θ) by $(r, \pi + \theta)$ or $(-r, \theta)$.

Example 2 ▶ Using Symmetry to Sketch a Polar Graph

Use symmetry to sketch the graph of $r = 3 + 2 \cos \theta$.

Solution

Replacing (r, θ) by $(r, -\theta)$ produces $r = 3 + 2 \cos(-\theta) = 3 + 2 \cos \theta$. So, you can conclude that the curve is symmetric with respect to the polar axis. Plotting the points in the table and using polar axis symmetry, you obtain the graph of a **limaçon**, as shown in Figure 6.71.

θ	0	$\dfrac{\pi}{3}$	$\dfrac{\pi}{2}$	$\dfrac{2\pi}{3}$	π
r	5	4	3	2	1

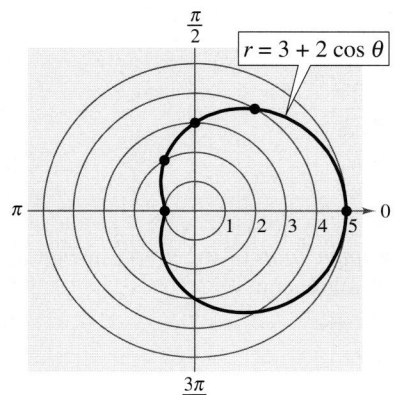

$r = 3 + 2 \cos \theta$

FIGURE 6.71

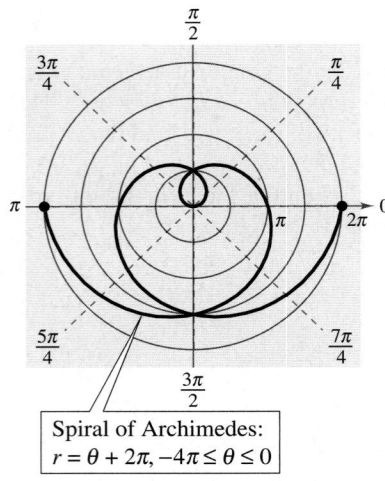

Spiral of Archimedes:
$r = \theta + 2\pi, -4\pi \le \theta \le 0$

FIGURE 6.72

The three tests for symmetry in polar coordinates listed on page 483 are sufficient to guarantee symmetry, but they are not necessary. For instance, Figure 6.72 shows the graph of $r = \theta + 2\pi$ to be symmetric with respect to the line $\theta = \pi/2$, and yet the tests on page 483 fail to indicate symmetry.

The equations discussed in Examples 1 and 2 are of the form

$$r = 4 \sin \theta = f(\sin \theta) \qquad \text{and} \qquad r = 3 + 2 \cos \theta = g(\cos \theta).$$

The graph of the first equation is symmetric with respect to the line $\theta = \pi/2$, and the graph of the second equation is symmetric with respect to the polar axis. This observation can be generalized to yield the following *quick tests for symmetry*.

1. The graph of $r = f(\sin \theta)$ is symmetric with respect to the line $\theta = \dfrac{\pi}{2}$.

2. The graph of $r = g(\cos \theta)$ is symmetric with respect to the polar axis.

Zeros and Maximum *r*-Values

Two additional aids to sketching graphs of polar equations involve knowing the θ-values for which $|r|$ is maximum and knowing the θ-values for which $r = 0$. For instance, in Example 1, the maximum value of $|r|$ for $r = 4 \sin \theta$ is $|r| = 4$, and this occurs when $\theta = \pi/2$, as shown in Figure 6.69. Moreover, $r = 0$ when $\theta = 0$.

Example 3 ▶ **Sketching a Polar Graph**

Sketch the graph of

$$r = 1 - 2 \cos \theta.$$

Solution

From the equation $r = 1 - 2 \cos \theta$, you can obtain the following.

Symmetry: With respect to the polar axis

Maximum value of $|r|$: $r = 3$ when $\theta = \pi$

Zero of r: $r = 0$ when $\theta = \dfrac{\pi}{3}$

The table shows several θ-values in the interval $[0, \pi]$. By plotting the corresponding points, you can sketch the graph shown in Figure 6.73.

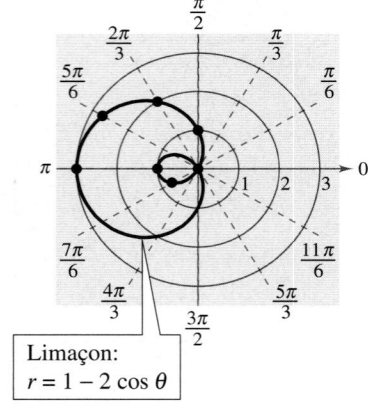

Limaçon:
$r = 1 - 2 \cos \theta$

FIGURE 6.73

θ	0	$\dfrac{\pi}{6}$	$\dfrac{\pi}{3}$	$\dfrac{\pi}{2}$	$\dfrac{2\pi}{3}$	$\dfrac{5\pi}{6}$	π
r	-1	-0.73	0	1	2	2.73	3

Note how the negative r-values determine the *inner loop* of the graph in Figure 6.73. This graph, like the one in Figure 6.71, is a limaçon.

Some curves reach their zeros and maximum r-values at more than one point, as shown in Example 4.

Example 4 ▶ **Sketching a Polar Graph**

Sketch the graph of $r = 2 \cos 3\theta$.

Solution

Symmetry:	With respect to the polar axis
Maximum value of $|r|$:	$|r| = 2$ when $3\theta = 0, \pi, 2\pi, 3\pi$ or $\theta = 0, \pi/3, 2\pi/3, \pi$
Zeros of r:	$r = 0$ when $3\theta = \pi/2, 3\pi/2, 5\pi/2$ or $\theta = \pi/6, \pi/2, 5\pi/6$

θ	0	$\dfrac{\pi}{12}$	$\dfrac{\pi}{6}$	$\dfrac{\pi}{4}$	$\dfrac{\pi}{3}$	$\dfrac{5\pi}{12}$	$\dfrac{\pi}{2}$
r	2	$\sqrt{2}$	0	$-\sqrt{2}$	-2	$-\sqrt{2}$	0

By plotting these points and using the specified symmetry, zeros, and maximum values, you can obtain the graph shown in Figure 6.74. This graph is called a **rose curve,** and each of the loops on the graph is called a *petal* of the rose curve. Note how the entire curve is generated as θ increases from 0 to π.

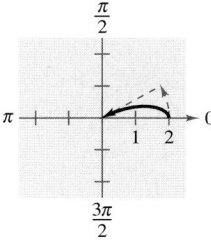

$0 \le \theta \le \dfrac{\pi}{6}$

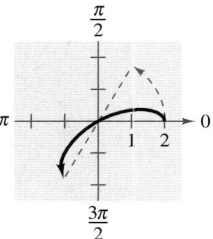

$0 \le \theta \le \dfrac{\pi}{3}$

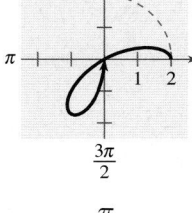

$0 \le \theta \le \dfrac{\pi}{2}$

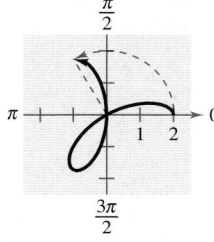

$0 \le \theta \le \dfrac{2\pi}{3}$

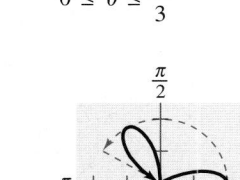

$0 \le \theta \le \dfrac{5\pi}{6}$

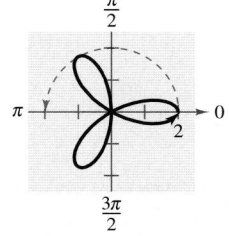

$0 \le \theta \le \pi$

FIGURE **6.74**

Technology

Use a graphing utility in *polar* mode to verify the graph of $r = 2 \cos 3\theta$ shown in Figure 6.74.

Special Polar Graphs

Several important types of graphs have equations that are simpler in polar form than in rectangular form. For example, the circle

$$r = 4 \sin \theta$$

in Example 1 has the more complicated rectangular equation

$$x^2 + (y - 2)^2 = 4.$$

Several other types of graphs that have simple polar equations are shown below.

Limaçons

$r = a \pm b \cos \theta$

$r = a \pm b \sin \theta$

$(a > 0, b > 0)$

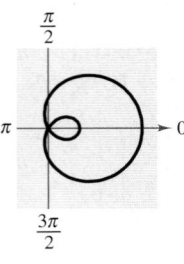

$\dfrac{a}{b} < 1$

Limaçon with inner loop

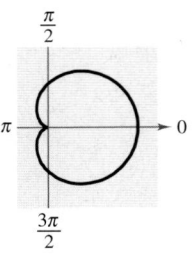

$\dfrac{a}{b} = 1$

Cardioid (heart-shaped)

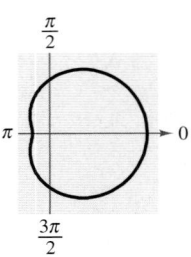

$1 < \dfrac{a}{b} < 2$

Dimpled limaçon

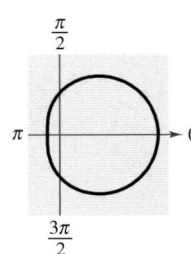

$\dfrac{a}{b} \geq 2$

Convex limaçon

Rose Curves

n petals if n is odd,

$2n$ petals if n is even

$(n \geq 2)$

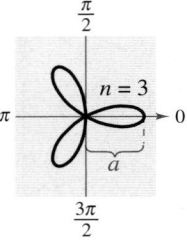

$r = a \cos n\theta$

Rose curve

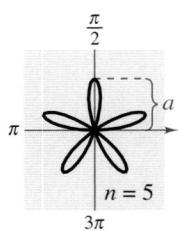

$r = a \cos n\theta$

Rose curve

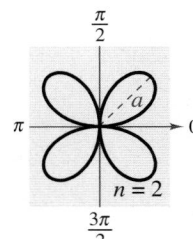

$r = a \sin n\theta$

Rose curve

$r = a \sin n\theta$

Rose curve

Circles and Lemniscates

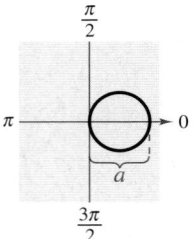

$r = a \cos \theta$

Circle

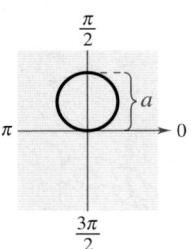

$r = a \sin \theta$

Circle

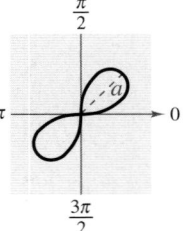

$r^2 = a^2 \sin 2\theta$

Lemniscate

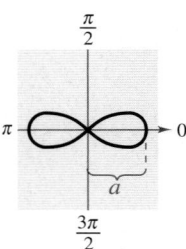

$r^2 = a^2 \cos 2\theta$

Lemniscate

The quick tests for symmetry presented in this section can be especially useful when graphing these special polar graphs.

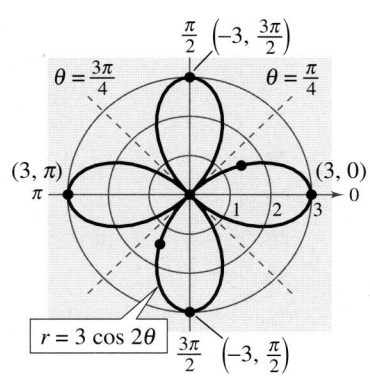

$$\left(-3, \frac{3\pi}{2}\right)$$

$\theta = \frac{3\pi}{4}$ $\theta = \frac{\pi}{4}$

$(3, \pi)$ $(3, 0)$

$r = 3\cos 2\theta$ $\left(-3, \frac{\pi}{2}\right)$

FIGURE **6.75**

Activities

1. Test $r^2 = 3 \sin \theta$ for symmetry.

   Answer: Symmetric with respect to the pole

2. Find the maximum value of $|r|$ and any zeros of r for the polar equation $r = 2 + 2 \cos \theta$.

   Answer: Maximum value of $|r| = 4$ when $\theta = 0$, and $r = 0$ when $\theta = \pi$.

3. Identify the shape of the graph of the polar equation $r = 2 \sin 3\theta$.

   Answer: Rose curve with 3 petals

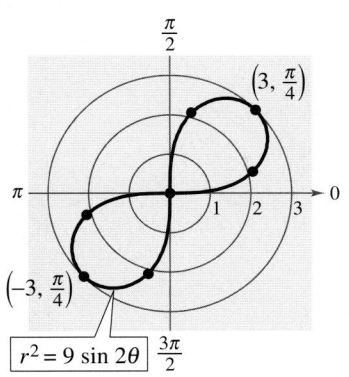

$$\left(3, \frac{\pi}{4}\right)$$

$$\left(-3, \frac{\pi}{4}\right)$$

$r^2 = 9 \sin 2\theta$

FIGURE **6.76**

Example 5 ▶ **Sketching a Rose Curve**

Sketch the graph of $r = 3 \cos 2\theta$.

Solution

Type of curve:	Rose curve with $2n = 4$ petals
Symmetry:	With respect to polar axis, the line $\theta = \pi/2$, and the pole
Maximum value of $|r|$:	$|r| = 3$ when $\theta = 0, \pi/2, \pi, 3\pi/2$
Zeros of r:	$r = 0$ when $\theta = \pi/4, 3\pi/4$

Using this information together with the additional points shown in the following table, you obtain the graph shown in Figure 6.75.

θ	0	$\dfrac{\pi}{6}$	$\dfrac{\pi}{4}$	$\dfrac{\pi}{3}$
r	3	$\dfrac{3}{2}$	0	$-\dfrac{3}{2}$

Example 6 ▶ **Sketching a Lemniscate**

Sketch the graph of $r^2 = 9 \sin 2\theta$.

Solution

Type of curve:	Lemniscate
Symmetry:	With respect to the pole
Maximum value of $|r|$:	$|r| = 3$ when $\theta = \dfrac{\pi}{4}$
Zeros of r:	$r = 0$ when $\theta = 0, \dfrac{\pi}{2}$

If $\sin 2\theta < 0$, this equation has no solution points. So, you restrict the values of θ to those for which $\sin 2\theta \geq 0$.

$$0 \leq \theta \leq \frac{\pi}{2} \quad \text{or} \quad \pi \leq \theta \leq \frac{3\pi}{2}$$

Moreover, using symmetry, you need to consider only the first of these two intervals. By finding a few additional points (see table below), you can obtain the graph shown in Figure 6.76.

θ	0	$\dfrac{\pi}{12}$	$\dfrac{\pi}{4}$	$\dfrac{5\pi}{12}$	$\dfrac{\pi}{2}$
$r = \pm 3\sqrt{\sin 2\theta}$	0	$\dfrac{\pm 3}{\sqrt{2}}$	± 3	$\dfrac{\pm 3}{\sqrt{2}}$	0

6.8 Exercises

In Exercises 1–6, identify the type of polar graph.

1.

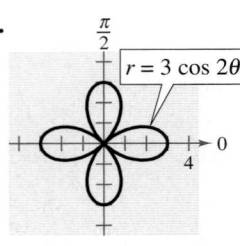

$r = 3\cos 2\theta$

2.

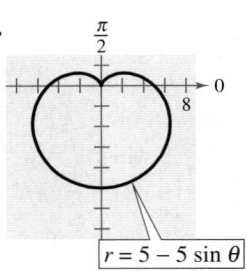

$r = 5 - 5\sin\theta$

3.

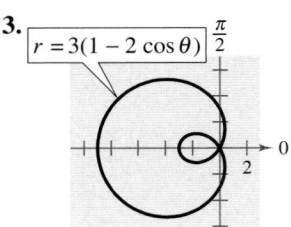

$r = 3(1 - 2\cos\theta)$

4.

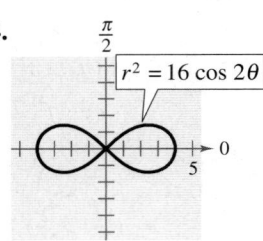

$r^2 = 16\cos 2\theta$

5.

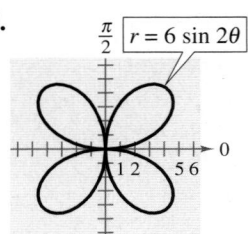

$r = 6\sin 2\theta$

6.

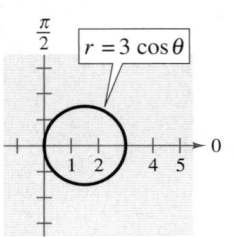

$r = 3\cos\theta$

In Exercises 7–12, test for symmetry with respect to $\theta = \pi/2$, the polar axis, and the pole.

7. $r = 5 + 4\cos\theta$

8. $r = 16\cos 3\theta$

9. $r = \dfrac{2}{1 + \sin\theta}$

10. $r = \dfrac{3}{2 + \cos\theta}$

11. $r^2 = 16\cos 2\theta$

12. $r^2 = 36\sin 2\theta$

In Exercises 13–16, find the maximum value of $|r|$ and any zeros of r.

13. $r = 10(1 - \sin\theta)$

14. $r = 6 + 12\cos\theta$

15. $r = 4\cos 3\theta$

16. $r = 3\sin 2\theta$

In Exercises 17–40, sketch the graph of the polar equation.

17. $r = 5$

18. $r = 2$

19. $r = \dfrac{\pi}{6}$

20. $r = -\dfrac{3\pi}{4}$

21. $r = 3\sin\theta$

22. $r = 4\cos\theta$

23. $r = 3(1 - \cos\theta)$

24. $r = 4(1 - \sin\theta)$

25. $r = 4(1 + \sin\theta)$

26. $r = 2(1 + \cos\theta)$

27. $r = 3 + 6\sin\theta$

28. $r = 4 - 3\sin\theta$

29. $r = 1 - 2\sin\theta$

30. $r = 1 - 2\cos\theta$

31. $r = 3 - 4\cos\theta$

32. $r = 4 + 3\cos\theta$

33. $r = 5\sin 2\theta$

34. $r = 3\cos 2\theta$

35. $r = 2\sec\theta$

36. $r = 5\csc\theta$

37. $r = \dfrac{3}{\sin\theta - 2\cos\theta}$

38. $r = \dfrac{6}{2\sin\theta - 3\cos\theta}$

39. $r^2 = 9\cos 2\theta$

40. $r^2 = 4\sin\theta$

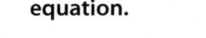 In Exercises 41–46, use a graphing utility to graph the polar equation.

41. $r = 8\cos\theta$

42. $r = \cos 2\theta$

43. $r = 3(2 - \sin\theta)$

44. $r = 2\cos(3\theta - 2)$

45. $r = 8\sin\theta\cos^2\theta$

46. $r = 2\csc\theta + 5$

 In Exercises 47–52, use a graphing utility to graph the polar equation. Find an interval for θ for which the graph is traced only once.

47. $r = 3 - 4\cos\theta$

48. $r = 5 + 4\cos\theta$

49. $r = 2\cos\left(\dfrac{3\theta}{2}\right)$

50. $r = 3\sin\left(\dfrac{5\theta}{2}\right)$

51. $r^2 = 9\sin 2\theta$

52. $r^2 = \dfrac{1}{\theta}$

 In Exercises 53–56, use a graphing utility to graph the polar equation and show that the indicated line is an asymptote of the graph.

	Name of Graph	Polar Equation	Asymptote
53.	Conchoid	$r = 2 - \sec\theta$	$x = -1$
54.	Conchoid	$r = 2 + \csc\theta$	$y = 1$
55.	Hyperbolic spiral	$r = \dfrac{3}{\theta}$	$y = 3$
56.	Strophoid	$r = 2\cos 2\theta\sec\theta$	$x = -2$

Synthesis

True or False? In Exercises 57 and 58, determine whether the statement is true or false. Justify your answer.

57. In the polar coordinate system, if a graph that has symmetry with respect to the polar axis were folded on the line $\theta = 0$, the portion of the graph above the polar axis would coincide with the portion of the graph below the polar axis.

58. In the polar coordinate system, if a graph that has symmetry with respect to the pole were folded on the line $\theta = 3\pi/4$, the portion of the graph on one side of the fold would coincide with the portion of the graph on the other side of the fold.

59. *Exploration*   Sketch the graph of $r = 6 \cos \theta$ over each interval. Describe the part of the graph obtained in each case.

(a) $0 \leq \theta \leq \dfrac{\pi}{2}$

(b) $\dfrac{\pi}{2} \leq \theta \leq \pi$

(c) $-\dfrac{\pi}{2} \leq \theta \leq \dfrac{\pi}{2}$

(d) $\dfrac{\pi}{4} \leq \theta \leq \dfrac{3\pi}{4}$

 60. *Graphical Reasoning*   Use a graphing utility to graph the polar equation

$$r = 6[1 + \cos(\theta - \phi)]$$

for (a) $\phi = 0$, (b) $\phi = \pi/4$, and (c) $\phi = \pi/2$. Use the graphs to describe the effect of the angle ϕ. Write the equation as a function of $\sin \theta$ for part (c).

61. The graph of $r = f(\theta)$ is rotated about the pole through an angle ϕ. Show that the equation of the rotated graph is $r = f(\theta - \phi)$.

62. Consider the graph of $r = f(\sin \theta)$.

(a) Show that if the graph is rotated counterclockwise $\pi/2$ radians about the pole, the equation of the rotated graph is $r = f(-\cos \theta)$.

(b) Show that if the graph is rotated counterclockwise π radians about the pole, the equation of the rotated graph is $r = f(-\sin \theta)$.

(c) Show that if the graph is rotated counterclockwise $3\pi/2$ radians about the pole, the equation of the rotated graph is $r = f(\cos \theta)$.

In Exercises 63–66, use the results of Exercises 61 and 62.

63. Write an equation for the limaçon $r = 2 - \sin \theta$ after it has been rotated by the given amount.

(a) $\dfrac{\pi}{4}$ (b) $\dfrac{\pi}{2}$ (c) π (d) $\dfrac{3\pi}{2}$

64. Write an equation for the rose curve $r = 2 \sin 2\theta$ after it has been rotated by the given amount.

(a) $\dfrac{\pi}{6}$ (b) $\dfrac{\pi}{2}$ (c) $\dfrac{2\pi}{3}$ (d) π

65. Sketch the graph of each equation.

(a) $r = 1 - \sin \theta$ (b) $r = 1 - \sin\left(\theta - \dfrac{\pi}{4}\right)$

66. Sketch the graph of each equation.

(a) $r = 3 \sec \theta$ (b) $r = 3 \sec\left(\theta - \dfrac{\pi}{4}\right)$

(c) $r = 3 \sec\left(\theta + \dfrac{\pi}{3}\right)$ (d) $r = 3 \sec\left(\theta - \dfrac{\pi}{2}\right)$

 67. *Exploration*   Use a graphing utility to graph and identify $r = 2 + k \sin \theta$ for $k = 0$, 1, 2, and 3.

 68. *Exploration*   Consider the equation $r = 3 \sin k\theta$.

(a) Use a graphing utility to graph the equation for $k = 1.5$. Find the interval for θ over which the graph is traced only once.

(b) Use a graphing utility to graph the equation for $k = 2.5$. Find the interval for θ over which the graph is traced only once.

(c) Is it possible to find an interval for θ over which the graph is traced only once for any rational number k? Explain.

Review

In Exercises 69 and 70, find the standard form of the equation of the ellipse. Then sketch the ellipse.

69. Vertices: $(-4, 2)$, $(2, 2)$; Minor axis of length 4

70. Foci: $(3, 2)$, $(3, -4)$; Major axis of length 8

6.9 Polar Equations of Conics

Digital Image © 1996 Corbis; Original image courtesy of NASA/Corbis

▶ **What you should learn**

- How to define conics in terms of eccentricity
- How to write equations of conics in polar form
- How to use equations of conics in polar form to model real-life problems

▶ **Why you should learn it**

The orbits of planets and satellites can be modeled with polar equations. For instance, in Exercise 57 on page 495, a polar equation is used to model the orbit of a satellite.

Alternative Definition of Conic

In Sections 6.3 and 6.4, you learned that the rectangular equations of ellipses and hyperbolas take simple forms when the origin lies at their *centers*. As it happens, there are many important applications of conics in which it is more convenient to use one of the *foci* as the origin. In this section you will learn that polar equations of conics take simple forms if one of the foci lies at the pole.

To begin, consider the following alternative definition of conic that uses the concept of eccentricity.

Alternative Definition of Conic

The locus of a point in the plane that moves so that its distance from a fixed point (focus) is in a constant ratio to its distance from a fixed line (directrix) is a **conic.** The constant ratio is the **eccentricity** of the conic and is denoted by e. Moreover, the conic is an **ellipse** if $e < 1$, a **parabola** if $e = 1$, and a **hyperbola** if $e > 1$. (See Figure 6.77.)

In Figure 6.77, note that for each type of conic, the focus is at the pole.

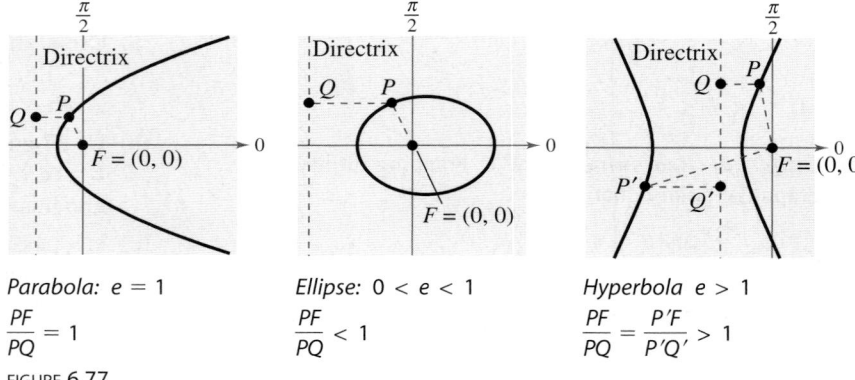

Parabola: $e = 1$

$\dfrac{PF}{PQ} = 1$

Ellipse: $0 < e < 1$

$\dfrac{PF}{PQ} < 1$

Hyperbola $e > 1$

$\dfrac{PF}{PQ} = \dfrac{P'F}{P'Q'} > 1$

FIGURE 6.77

Polar Equations of Conics

The benefit of locating a focus of a conic at the pole is that the equation of the conic takes on a simpler form. For a proof of the polar equations of conics, see Proofs in Mathematics on page 507.

Polar Equations of Conics

The graph of a polar equation of the form

1. $r = \dfrac{ep}{1 \pm e \cos \theta}$ or **2.** $r = \dfrac{ep}{1 \pm e \sin \theta}$

is a conic, where $e > 0$ is the eccentricity and $|p|$ is the distance between the focus (pole) and the directrix.

The equations

$$r = \frac{ep}{1 \pm e \cos \theta} \qquad \text{Vertical directrix}$$

correspond to conics with vertical directrices, and the equations

$$r = \frac{ep}{1 \pm e \sin \theta} \qquad \text{Horizontal directrix}$$

correspond to conics with horizontal directrices. Moreover, the converse is also true—that is, any conic with a focus at the pole and having a horizontal or vertical directrix can be represented by one of the given equations.

Example 1 ▶ **Sketching a Conic from Its Polar Equation**

Identify the conic $r = \dfrac{15}{3 - 2 \cos \theta}$ and sketch its graph.

Solution

To identify the type of conic, rewrite the equation as

$$r = \frac{15}{3 - 2 \cos \theta} = \frac{5}{1 - (2/3) \cos \theta}. \qquad \begin{array}{l} \text{Divide numerator and} \\ \text{denominator by 3.} \end{array}$$

Because $e = \frac{2}{3} < 1$, you can conclude that the graph is an ellipse. You can sketch the upper half of the ellipse by plotting points from $\theta = 0$ to $\theta = \pi$, as shown in Figure 6.78. Using symmetry with respect to the polar axis, you can sketch the lower half.

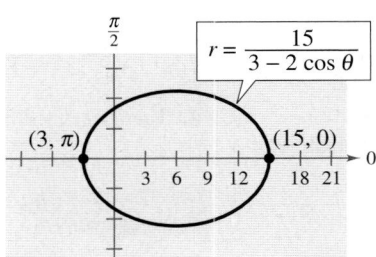

FIGURE **6.78**

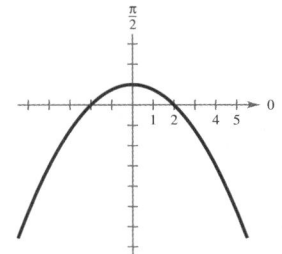
For the ellipse in Figure 6.78, the major axis is horizontal and the vertices lie at $(15, 0)$ and $(3, \pi)$. So, the length of the *major* axis is $2a = 18$. To find the length of the *minor* axis, you can use the equations $e = c/a$ and $b^2 = a^2 - c^2$ to conclude that

$$b^2 = a^2 - c^2 = a^2 - (ea)^2 = a^2(1 - e^2). \qquad \text{Ellipse}$$

Because $e = \frac{2}{3}$, you have $b^2 = 9^2\left[1 - \left(\frac{2}{3}\right)^2\right] = 45$, which implies that $b = \sqrt{45} = 3\sqrt{5}$. So, the length of the minor axis is $2b = 6\sqrt{5}$. A similar analysis for hyperbolas yields

$$b^2 = c^2 - a^2 = (ea)^2 - a^2 = a^2(e^2 - 1). \qquad \text{Hyperbola}$$

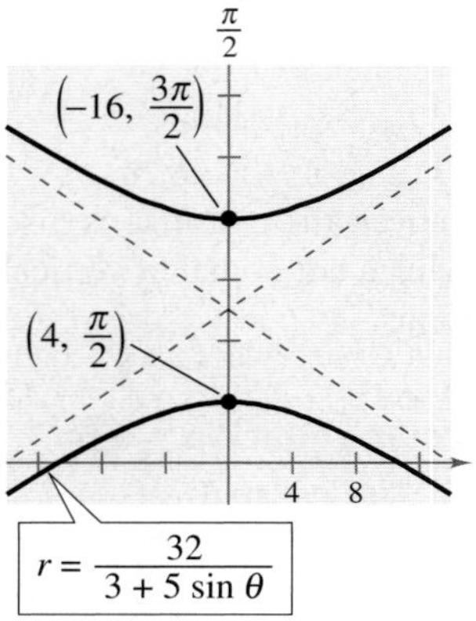

Example 2 ▶ **Sketching a Conic from Its Polar Equation**

Identify the conic $r = 32/(3 + 5 \sin \theta)$ and sketch its graph.

Solution

Dividing the numerator and denominator by 3, you have

$$r = \frac{32/3}{1 + (5/3) \sin \theta}.$$

Because $e = \frac{5}{3} > 1$, the graph is a hyperbola. The transverse axis of the hyperbola lies on the line $\theta = \pi/2$, and the vertices occur at $(4, \pi/2)$ and $(-16, 3\pi/2)$. Because the length of the transverse axis is 12, you can see that $a = 6$. To find b, write

$$b^2 = a^2(e^2 - 1) = 6^2 \left[\left(\frac{5}{3} \right)^2 - 1 \right] = 64.$$

So, $b = 8$. Finally, you can use a and b to determine the asymptotes of the hyperbola and obtain the sketch shown in Figure 6.79.

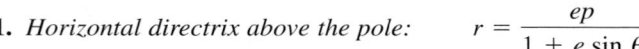

$$r = \frac{32}{3 + 5 \sin \theta}$$

FIGURE **6.79**

In the next example, you are asked to find a polar equation of a specified conic. To do this, let p be the distance between the pole and the directrix.

1. *Horizontal directrix above the pole:* $r = \dfrac{ep}{1 + e \sin \theta}$

2. *Horizontal directrix below the pole:* $r = \dfrac{ep}{1 - e \sin \theta}$

3. *Vertical directrix to the right of the pole:* $r = \dfrac{ep}{1 + e \cos \theta}$

4. *Vertical directrix to the left of the pole:* $r = \dfrac{ep}{1 - e \cos \theta}$

Technology

Most graphing utilities have a *polar* mode. Try using a graphing utility set in *polar* mode to verify the four orientations shown at the right. Remember that e must be positive, but p can be positive or negative.

Example 3 ▶ **Finding the Polar Equation of a Conic**

Find the polar equation of the parabola whose focus is the pole and whose directrix is the line $y = 3$.

Solution

From Figure 6.80, you can see that the directrix is horizontal and above the pole, so you can choose an equation of the form

$$r = \frac{ep}{1 + e \sin \theta}.$$

Moreover, because the eccentricity of a parabola is $e = 1$ and the distance between the pole and the directrix is $p = 3$, you have the equation

$$r = \frac{3}{1 + \sin \theta}.$$

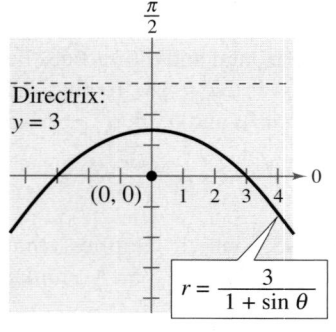

FIGURE **6.80**

Activities

1. Find a polar equation of the parabola with focus at the pole and directrix $y = 2$.

   Answer: $r = \dfrac{2}{1 + \sin \theta}$

2. Identify the conic and sketch its graph.

   $r = \dfrac{4}{3 - 2 \cos \theta}$

   Answer: Ellipse

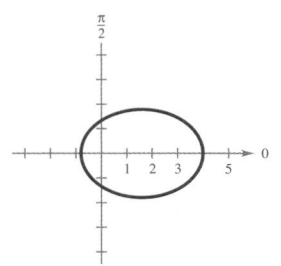

Applications

Kepler's Laws (listed below), named after the German astronomer Johannes Kepler (1571–1630), can be used to describe the orbits of the planets about the sun.

1. Each planet moves in an elliptical orbit with the sun at one focus.

2. A ray from the sun to the planet sweeps out equal areas of the ellipse in equal times.

3. The square of the period (the time it takes for a planet to orbit the sun) is proportional to the cube of the mean distance between the planet and the sun.

Although Kepler simply stated these laws on the basis of observation, they were later validated by Isaac Newton (1642–1727). In fact, Newton was able to show that each law can be deduced from a set of universal laws of motion and gravitation that govern the movement of all heavenly bodies, including comets and satellites. This is illustrated in the next example, which involves the comet named after the English mathematician and physicist Edmund Halley (1656–1742).

   If you use Earth as a reference with a period of 1 year and a distance of 1 astronomical unit (an *astronomical unit* is defined as the mean distance between Earth and the sun, or about 93 million miles), the proportionality constant in Kepler's third law is 1. For example, because Mars has a mean distance to the sun of $d = 1.523$ astronomical units, its period P is given by $d^3 = P^2$. So, the period of Mars is $P = 1.88$ years.

Example 4 ▶ **Halley's Comet**

Halley's comet has an elliptical orbit with an eccentricity of $e \approx 0.97$. The length of the major axis of the orbit is approximately 36.18 astronomical units. Find a polar equation for the orbit. How close does Halley's comet come to the sun?

Solution

Using a vertical axis, as shown in Figure 6.81, choose an equation of the form $r = ep/(1 + e \sin\theta)$. Because the vertices of the ellipse occur when $\theta = \pi/2$ and $\theta = 3\pi/2$, you can determine the length of the major axis to be the sum of the r-values of the vertices. That is,

$$2a = \frac{0.97p}{1 + 0.97} + \frac{0.97p}{1 - 0.97} \approx 32.83p \approx 36.18.$$

So, $p \approx 1.102$ and $ep \approx (0.97)(1.102) \approx 1.069$. Using this value of ep in the equation, you have

$$r = \frac{1.069}{1 + 0.97 \sin \theta}$$

where r is measured in astronomical units. To find the closest point to the sun (the focus), substitute $\theta = \pi/2$ in this equation to obtain

$$r = \frac{1.069}{1 + 0.97 \sin(\pi/2)} \approx 0.54 \text{ astronomical unit} \approx 50{,}000{,}000 \text{ miles.}$$

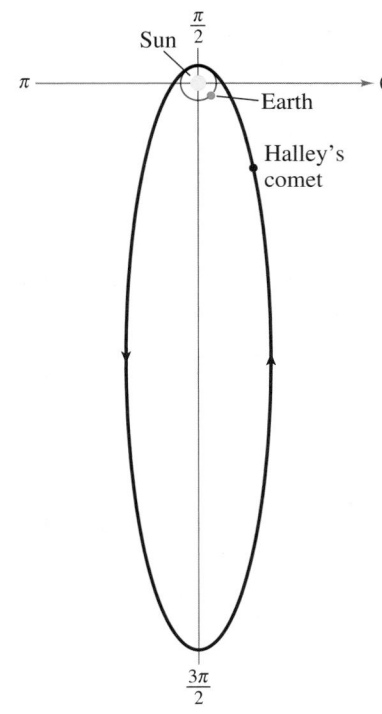

FIGURE **6.81**

6.9 Exercises

In Exercises 1–4, write the polar equation of the conic for $e = 1$, $e = 0.5$, and $e = 1.5$. Identify the conic for each equation. Verify your answers with a graphing utility.

1. $r = \dfrac{4e}{1 + e \cos \theta}$

2. $r = \dfrac{4e}{1 - e \cos \theta}$

3. $r = \dfrac{4e}{1 - e \sin \theta}$

4. $r = \dfrac{4e}{1 + e \sin \theta}$

In Exercises 5–10, match the polar equation with its graph. [The graphs are labeled (a), (b), (c), (d), (e), and (f).]

(a)
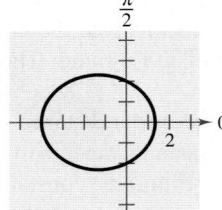

(b)
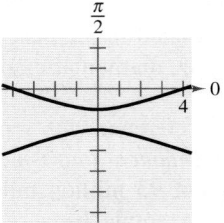

(c)
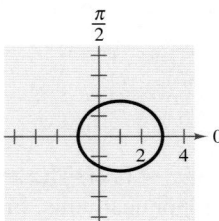

(d)
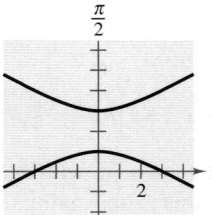

(e)
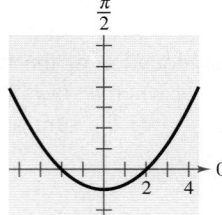

(f)
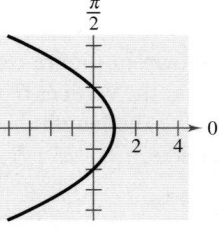

5. $r = \dfrac{2}{1 + \cos \theta}$

6. $r = \dfrac{3}{2 - \cos \theta}$

7. $r = \dfrac{3}{1 + 2 \sin \theta}$

8. $r = \dfrac{2}{1 - \sin \theta}$

9. $r = \dfrac{4}{2 + \cos \theta}$

10. $r = \dfrac{4}{1 - 3 \sin \theta}$

In Exercises 11–24, identify the conic and sketch its graph.

11. $r = \dfrac{2}{1 - \cos \theta}$

12. $r = \dfrac{3}{1 + \sin \theta}$

13. $r = \dfrac{5}{1 + \sin \theta}$

14. $r = \dfrac{6}{1 + \cos \theta}$

15. $r = \dfrac{2}{2 - \cos \theta}$

16. $r = \dfrac{3}{3 + \sin \theta}$

17. $r = \dfrac{6}{2 + \sin \theta}$

18. $r = \dfrac{9}{3 - 2 \cos \theta}$

19. $r = \dfrac{3}{2 + 4 \sin \theta}$

20. $r = \dfrac{5}{-1 + 2 \cos \theta}$

21. $r = \dfrac{3}{2 - 6 \cos \theta}$

22. $r = \dfrac{3}{2 + 6 \sin \theta}$

23. $r = \dfrac{4}{2 - \cos \theta}$

24. $r = \dfrac{2}{2 + 3 \sin \theta}$

 In Exercises 25–28, use a graphing utility to graph the polar equation. Identify the graph.

25. $r = \dfrac{-1}{1 - \sin \theta}$

26. $r = \dfrac{-5}{2 + 4 \sin \theta}$

27. $r = \dfrac{3}{-4 + 2 \cos \theta}$

28. $r = \dfrac{4}{1 - 2 \cos \theta}$

 In Exercises 29–32, use a graphing utility to graph the rotated conic.

29. $r = \dfrac{2}{1 - \cos(\theta - \pi/4)}$ (See Exercise 11.)

30. $r = \dfrac{3}{3 + \sin(\theta - \pi/3)}$ (See Exercise 16.)

31. $r = \dfrac{6}{2 + \sin(\theta + \pi/6)}$ (See Exercise 17.)

32. $r = \dfrac{5}{-1 + 2 \cos(\theta + 2\pi/3)}$ (See Exercise 20.)

In Exercises 33–48, find a polar equation of the conic with its focus at the pole.

Conic	Eccentricity	Directrix
33. Parabola	$e = 1$	$x = -1$
34. Parabola	$e = 1$	$y = -2$
35. Ellipse	$e = \frac{1}{2}$	$y = 1$
36. Ellipse	$e = \frac{3}{4}$	$y = -3$
37. Hyperbola	$e = 2$	$x = 1$
38. Hyperbola	$e = \frac{3}{2}$	$x = -1$

Conic	Vertex or Vertices
39. Parabola	$(1, -\pi/2)$
40. Parabola	$(6, 0)$
41. Parabola	$(5, \pi)$
42. Parabola	$(10, \pi/2)$
43. Ellipse	$(2, 0), (10, \pi)$
44. Ellipse	$(2, \pi/2), (4, 3\pi/2)$
45. Ellipse	$(20, 0), (4, \pi)$
46. Hyperbola	$(2, 0), (8, 0)$
47. Hyperbola	$(1, 3\pi/2), (9, 3\pi/2)$
48. Hyperbola	$(4, \pi/2), (1, \pi/2)$

49. *Planetary Motion* The planets travel in elliptical orbits with the sun at one focus. Assume that the focus is at the pole, the major axis lies on the polar axis, and the length of the major axis is $2a$ (see figure). Show that the polar equation of the orbit is

$$r = \frac{(1 - e^2)a}{1 - e\cos\theta}$$

where e is the eccentricity.

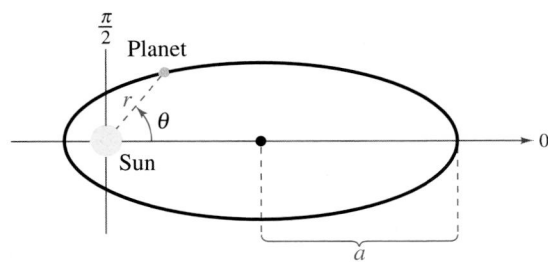

50. *Planetary Motion* Use the result of Exercise 49 to show that the minimum distance (*perihelion distance*) from the sun to the planet is $r = a(1 - e)$ and the maximum distance (*aphelion distance*) is $r = a(1 + e)$.

In Exercises 51–56, use the results of Exercises 49 and 50 to find the polar equation of the planet's orbit and the perihelion and aphelion distances.

51. Earth $a = 92.960 \times 10^6$ miles
  $e = 0.0167$

52. Saturn $a = 1.429 \times 10^9$ kilometers
  $e = 0.0543$

53. Pluto $a = 5.900 \times 10^9$ kilometers
  $e = 0.2481$

54. Mercury $a = 35.98 \times 10^6$ miles
  $e = 0.2056$

55. Mars $a = 141.00 \times 10^6$ miles
  $e = 0.0934$

56. Jupiter $a = 778.40 \times 10^6$ kilometers
  $e = 0.0484$

▶ Model It

57. *Satellite Tracking* A satellite in a 100-mile-high circular orbit around Earth has a velocity of approximately 17,500 miles per hour. If this velocity is multiplied by $\sqrt{2}$, the satellite will have the minimum velocity necessary to escape Earth's gravity and it will follow a parabolic path with the center of Earth as the focus (see figure).

(a) Find a polar equation of the parabolic path of the satellite (assume the radius of Earth is 4000 miles).

(b) Use a graphing utility to graph the equation you found in part (a).

(c) Find the distance between the surface of the Earth and the satellite when $\theta = 30°$.

(d) Find the distance between the surface of Earth and the satellite when $\theta = 60°$.

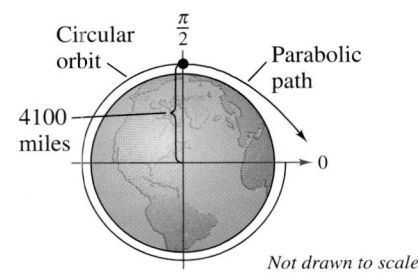

Not drawn to scale

Synthesis

58. *Exploration* The equation

$$r = \frac{ep}{1 \pm e \sin \theta}$$

is the equation of an ellipse with $e < 1$. What happens to the lengths of both the major axis and the minor axis when the value of e remains fixed and the value of p changes? Use an example to explain your reasoning.

True or False? **In Exercises 59 and 60, determine whether the statement is true or false. Justify your answer.**

59. For a given value of $e > 1$ over the interval $\theta = 0$ to $\theta = 2\pi$, the graph of

$$r = \frac{ex}{1 - e \cos \theta}$$

is the same as the graph of

$$r = \frac{e(-x)}{1 + e \cos \theta}.$$

60. The graph of

$$r = \frac{4}{-3 - 3 \sin \theta}$$

has a horizontal directrix above the pole.

61. Show that the polar equation of the ellipse

$$\frac{x^2}{a^2} + \frac{y^2}{b^2} = 1 \quad \text{is} \quad r^2 = \frac{b^2}{1 - e^2 \cos^2 \theta}.$$

62. Show that the polar equation of the hyperbola

$$\frac{x^2}{a^2} - \frac{y^2}{b^2} = 1 \quad \text{is} \quad r^2 = \frac{-b^2}{1 - e^2 \cos^2 \theta}.$$

In Exercises 63–68, use the results of Exercises 61 and 62 to write the polar form of the equation of the conic.

63. $\dfrac{x^2}{169} + \dfrac{y^2}{144} = 1$

64. $\dfrac{x^2}{25} + \dfrac{y^2}{16} = 1$

65. $\dfrac{x^2}{9} - \dfrac{y^2}{16} = 1$

66. $\dfrac{x^2}{36} - \dfrac{y^2}{4} = 1$

67. Hyperbola   One focus: $(5, \pi/2)$
      Vertices: $(4, \pi/2), (4, -\pi/2)$

68. Ellipse   One focus: $(4, 0)$
      Vertices: $(5, 0), (5, \pi)$

Review

In Exercises 69–74, solve the trigonometric equation.

69. $4\sqrt{3} \tan \theta - 3 = 1$

70. $6 \cos x - 2 = 1$

71. $12 \sin^2 \theta = 9$

72. $9 \csc^2 x - 10 = 2$

73. $2 \cot x = 5 \cos \dfrac{\pi}{2}$

74. $\sqrt{2} \sec \theta = 2 \csc \dfrac{\pi}{4}$

In Exercises 75–78, find the value of the trigonometric function given that u and v are in Quadrant IV and $\sin u = -\frac{3}{5}$ and $\cos v = 1/\sqrt{2}$.

75. $\cos(u + v)$

76. $\sin(u + v)$

77. $\cos(u - v)$

78. $\sin(u - v)$

In Exercises 79 and 80, find the exact values of $\sin 2u$, $\cos 2u$, and $\tan 2u$ using the double-angle formulas.

79. $\sin u = \dfrac{4}{5}, \ \dfrac{\pi}{2} < u < \pi$

80. $\tan u = -\sqrt{3}, \ \dfrac{3\pi}{2} < u < 2\pi$

Chapter Summary

▶ *What* did you learn?

Review Exercises

6.1 **In Exercises 1–4, find, in radians and degrees, the inclination θ of the line with the given characteristics.**

1. Passes through the points $(-1, 2)$ and $(2, 5)$
2. Passes through the points $(3, 4)$ and $(-2, 7)$
3. Equation: $y = 2x + 4$
4. Equation: $6x - 7y - 5 = 0$

In Exercises 5–8, find, in radians and degrees, the angle θ between the lines.

5. $4x + y = 2$
   $-5x + y = -1$

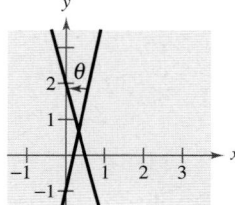

6. $-5x + 3y = 3$
   $-2x + 3y = 1$

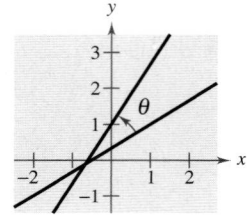

7. $2x - 7y = 8$
   $0.4x + y = 0$

8. $0.02x + 0.07y = 0.18$
   $0.09x - 0.04y = 0.17$

In Exercises 9 and 10, find the distance between the point and the line.

	Point	*Line*
9.	$(1, 2)$	$x - y - 3 = 0$
10.	$(0, 4)$	$x + 2y - 2 = 0$

6.2 **In Exercises 11 and 12, state what type of conic is formed by the intersection of the plane and the double-napped cone.**

11.

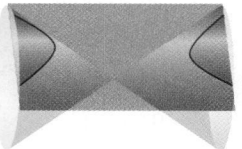

12.

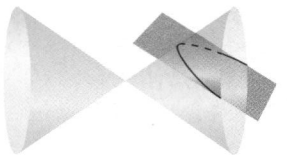

In Exercises 13–16, find the standard form of the equation of the parabola.

13. Vertex: $(4, 2)$
   Focus: $(4, 0)$

14. Vertex: $(2, 0)$
   Focus: $(0, 0)$

15. Vertex: $(0, 2)$
   Directrix: $x = -3$

16. Vertex: $(2, 2)$
   Directrix: $y = 0$

In Exercises 17 and 18, find an equation of a tangent line to the parabola at the given point, and find the x-intercept of the line.

17. $x^2 = -2y$, $(2, -2)$
18. $x^2 = -2y$, $(-4, -8)$

19. **Architecture**   A parabolic archway is 12 meters high at the vertex. At a height of 10 meters, the width of the archway is 8 meters (see figure). How wide is the archway at ground level?

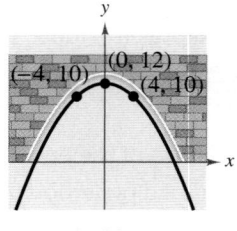

FIGURE FOR 19

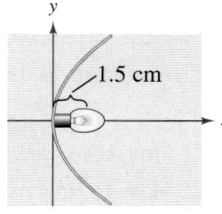

FIGURE FOR 20

20. **Flashlight**   The light bulb in a flashlight is at the focus of its parabolic reflector, 1.5 centimeters from the vertex of the reflector (see figure). Write an equation of a cross section of the flashlight's reflector with its focus on the positive x-axis and its vertex at the origin.

6.3 **In Exercises 21–24, find the standard form of the equation of the ellipse.**

21. Vertices: $(-3, 0), (7, 0)$; Foci: $(0, 0), (4, 0)$
22. Vertices: $(2, 0), (2, 4)$; Foci: $(2, 1), (2, 3)$
23. Vertices: $(0, 1), (4, 1)$;
   Endpoints of the minor axis: $(2, 0), (2, 2)$
24. Vertices: $(-4, -1), (-4, 11)$;
   Endpoints of the minor axis: $(-6, 5), (-2, 5)$

25. **Architecture**   A semielliptical archway is set on pillars that are 10 feet apart. Its height (atop the pillars) is 4 feet. Where should the foci be placed in order to sketch the semielliptical arch?

26. *Wading Pool* You are building a wading pool that is in the shape of an ellipse. Your plans include an equation for the elliptical shape of the pool measured in feet as

$$\frac{x^2}{324} + \frac{y^2}{196} = 1.$$

Find the longest distance across the pool, the shortest distance, and the distance between the foci.

In Exercises 27–30, find the center, vertices, foci, and eccentricity of the ellipse.

27. $\dfrac{(x + 2)^2}{81} + \dfrac{(y - 1)^2}{100} = 1$

28. $\dfrac{(x - 5)^2}{1} + \dfrac{(y + 3)^2}{36} = 1$

29. $16x^2 + 9y^2 - 32x + 72y + 16 = 0$

30. $4x^2 + 25y^2 + 16x - 150y + 141 = 0$

6.4 **In Exercises 31–34, find the standard form of the equation of the hyperbola.**

31. Vertices: $(0, \pm 1)$; Foci: $(0, \pm 3)$

32. Vertices: $(2, 2), (-2, 2)$; Foci: $(4, 2), (-4, 2)$

33. Foci: $(0, 0), (8, 0)$; Asymptotes: $y = \pm 2(x - 4)$

34. Foci: $(3, \pm 2)$; Asymptotes: $y = \pm 2(x - 3)$

In Exercises 35–38, find the center, vertices, foci, and the equations of the asymptotes of the hyperbola, and sketch its graph.

35. $\dfrac{(x - 3)^2}{16} - \dfrac{(y + 5)^2}{4} = 1$

36. $\dfrac{(y - 1)^2}{4} - x^2 = 1$

37. $9x^2 - 16y^2 - 18x - 32y - 151 = 0$

38. $-4x^2 + 25y^2 - 8x + 150y + 121 = 0$

39. *LORAN* Radio transmitting station A is located 200 miles east of transmitting station B. A ship is in an area to the north and 40 miles west of station A. Synchronized radio pulses transmitted at 186,000 miles per second by the two stations are received 0.0005 second sooner from station A than from station B. How far north is the ship?

40. *Locating an Explosion* Two of your friends live 4 miles apart and on the same "east-west" street, and you live halfway between them. You are talking on a three-way phone call when you hear an explosion. Six seconds later, your friend to the east hears the explosion, and your friend to the west hears it 8 seconds after you do. Find equations of two hyperbolas that would locate the explosion. (Sound travels at a rate of 1100 feet per second.)

In Exercises 41 and 42, classify the graph of the equation as a circle, a parabola, an ellipse, or a hyperbola.

41. $5x^2 - 2y^2 + 10x - 4y + 17 = 0$

42. $-4y^2 + 5x + 3y + 7 = 0$

6.5 **In Exercises 43–46, rotate the axes to eliminate the xy-term. Sketch the graph of the resulting equation, showing both sets of axes.**

43. $xy - 4 = 0$

44. $x^2 - 10xy + y^2 + 1 = 0$

45. $5x^2 - 2xy + 5y^2 - 12 = 0$

46. $4x^2 + 8xy + 4y^2 + 7\sqrt{2}x + 9\sqrt{2}y = 0$

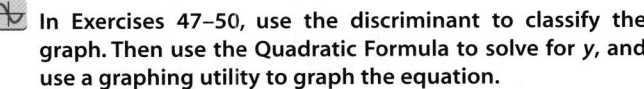 **In Exercises 47–50, use the discriminant to classify the graph. Then use the Quadratic Formula to solve for y, and use a graphing utility to graph the equation.**

47. $16x^2 - 24xy + 9y^2 - 30x - 40y = 0$

48. $13x^2 - 8xy + 7y^2 - 45 = 0$

49. $x^2 + y^2 + 2xy + 2\sqrt{2}x - 2\sqrt{2}y + 2 = 0$

50. $x^2 - 10xy + y^2 + 1 = 0$

6.6 **In Exercises 51–54, evaluate the parametric equations $x = 3 \cos \theta$ and $y = 2 \sin^2 \theta$ for the given value of θ.**

51. $\theta = 0$

52. $\theta = \dfrac{\pi}{3}$

53. $\theta = \dfrac{\pi}{6}$

54. $\theta = -\dfrac{\pi}{4}$

In Exercises 55–60, sketch the curve represented by the parametric equations (indicate the orientation of the curve) and, where possible, write the corresponding rectangular equation by eliminating the parameter. Verify your result with a graphing utility.

55. $x = 2t$
$y = 4t$

56. $x = 1 + 4t$
$y = 2 - 3t$

57. $x = t^2$
$y = \sqrt{t}$

58. $x = t + 4$
$y = t^2$

59. $x = 6 \cos \theta$
$y = 6 \sin \theta$

60. $x = 3 + 3 \cos \theta$
$y = 2 + 5 \sin \theta$

61. Find a parametric representation of the circle with the center $(5, 4)$ and radius 6.

62. Find a parametric representation of the ellipse with center $(-3, 4)$, major axis horizontal and eight units in length, and minor axis six units in length.

63. Find a parametric representation of the hyperbola with vertices $(0, \pm 4)$ and foci $(0, \pm 5)$.

64. *Involute of a Circle* The *involute* of a circle is described by the endpoint P of a string that is held taut as it is unwound from a spool (see figure). The spool does not rotate. Show that a parametric representation of the involute of a circle is

$$x = r(\cos \theta + \theta \sin \theta)$$

$$y = r(\sin \theta - \theta \cos \theta).$$

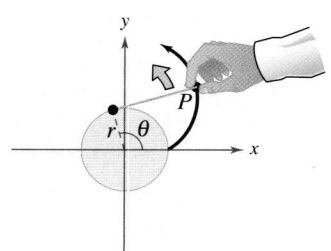

6.7 In Exercises 65–68, plot the point given in polar coordinates and find two additional polar representations.

65. $\left(2, \dfrac{\pi}{4}\right)$

66. $\left(-5, -\dfrac{\pi}{3}\right)$

67. $(-7, 4.19)$

68. $\left(\sqrt{3}, 2.62\right)$

In Exercises 69–72, a point in polar coordinates is given. Convert the point to rectangular coordinates.

69. $\left(-1, \dfrac{\pi}{3}\right)$

70. $\left(2, \dfrac{5\pi}{4}\right)$

71. $\left(3, \dfrac{3\pi}{4}\right)$

72. $\left(0, \dfrac{\pi}{2}\right)$

In Exercises 73–76, a point in rectangular coordinates is given. Convert the point to polar coordinates.

73. $(0, 2)$

74. $\left(-\sqrt{5}, \sqrt{5}\right)$

75. $(4, 6)$

76. $(3, -4)$

In Exercises 77–80, convert the rectangular equation to polar form.

77. $x^2 + y^2 - 6y = 0$

78. $x^2 + y^2 - 4x = 0$

79. $xy = 5$

80. $xy = -2$

In Exercises 81–84, convert the polar equation to rectangular form.

81. $r = 3 \cos \theta$

82. $r = 8 \sin \theta$

83. $r^2 = \sin \theta$

84. $r^2 = \cos 2\theta$

6.8 In Exercises 85–94, determine the symmetry of r, the maximum value of $|r|$, and any zeros of r. Then sketch the graph of the polar equation.

85. $r = 4$

86. $r = 11$

87. $r = 4 \sin 2\theta$

88. $r = \cos 5\theta$

89. $r = -2(1 + \cos \theta)$

90. $r = 3 - 4 \cos \theta$

91. $r = 2 + 6 \sin \theta$

92. $r = 5 - 5 \cos \theta$

93. $r = -3 \cos 2\theta$

94. $r = \cos 2\theta$

In Exercises 95–98, identify the type of polar graph.

95. $r = 3(2 - \cos \theta)$

96. $r = 3(1 - 2 \cos \theta)$

97. $r = 4 \cos 3\theta$

98. $r^2 = 9 \cos 2\theta$

6.9 In Exercises 99–102, identify the conic and sketch its graph.

99. $r = \dfrac{1}{1 + 2 \sin \theta}$

100. $r = \dfrac{2}{1 + \sin \theta}$

101. $r = \dfrac{4}{5 - 3 \cos \theta}$

102. $r = \dfrac{16}{4 + 5 \cos \theta}$

In Exercises 103–106, find a polar equation of the conic with its focus at the pole.

103. Parabola   Vertex: $(2, \pi)$

104. Parabola   Vertex: $(2, \pi/2)$

105. Ellipse   Vertices: $(5, 0), (1, \pi)$

106. Hyperbola   Vertices: $(1, 0), (7, 0)$

107. *Explorer 18* On November 26, 1963, the United States launched Explorer 18. Its low and high points above the surface of Earth were 119 miles and 122,000 miles, respectively (see figure). The center of Earth is at one focus of the orbit. Find the polar equation of the orbit and find the distance between the surface of Earth (assume a radius of 4000 miles) and the satellite when $\theta = \pi/3$ radians.

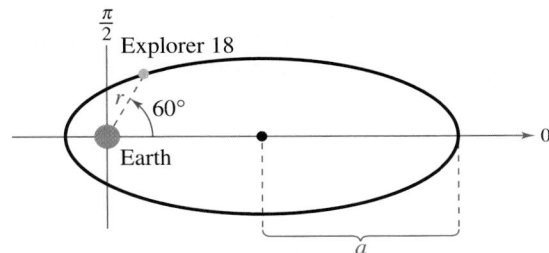

108. *Asteroid* An asteroid takes a parabolic path with Earth as its focus. It is about 6,000,000 miles from Earth at its closest approach. Write the polar equation of the path of the asteroid with its vertex at $\theta = \pi/2$. Find the distance between the asteroid and Earth when $\theta = -\pi/3$.

Synthesis

True or False? In Exercises 109–112, determine whether the statement is true or false. Justify your answer.

109. When $B = 0$ in an equation of the form $Ax^2 + Bxy + Cy^2 + Dx + Ey + F = 0$, the graph of the equation can be a parabola only if $C = 0$ also.

110. The graph of $(x^2/4) - y^4 = 1$ is a hyperbola.

111. Only one set of parametric equations can represent the line $y = 3 - 2x$.

112. There is a unique polar coordinate representation of each point in the plane.

113. Consider an ellipse with the major axis horizontal and 10 units in length. The number b in the standard form of the equation of the ellipse must be less than what real number? Explain the change in the shape of the ellipse as b approaches this number.

114. The graph of the parametric equations $x = 2 \sec t$ and $y = 3 \tan t$ is shown in the figure. Would the graph change for the equations $x = 2 \sec(-t)$ and $y = 3 \tan(-t)$? If so, how would it change?

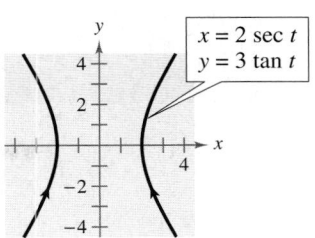

FIGURE FOR **114**

115. A moving object is modeled by the parametric equations $x = 4 \cos t$ and $y = 3 \sin t$, where t is time (see figure). How would the orbit change for the following?

(a) $x = 4 \cos 2t$, $y = 3 \sin 2t$

(b) $x = 5 \cos t$, $y = 3 \sin t$

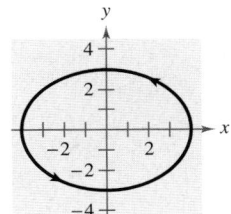

116. Identify the type of symmetry each of the following polar points has with the point in the figure.

(a) $\left(-4, \dfrac{\pi}{6}\right)$

(b) $\left(4, -\dfrac{\pi}{6}\right)$

(c) $\left(-4, -\dfrac{\pi}{6}\right)$

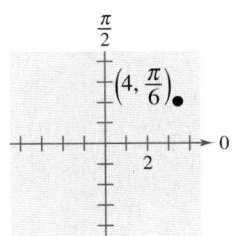

117. What is the relationship between the graphs of the rectangular and polar equations?

(a) $x^2 + y^2 = 25$, $r = 5$

(b) $x - y = 0$, $\theta = \dfrac{\pi}{4}$

Chapter Test

Take this test as you would take a test in class. When you are finished, check your work against the answers given in the back of the book.

The *Interactive* CD-ROM and *Internet* versions of this text offer Chapter Pre-Tests and Chapter Post-Tests, both of which have randomly generated exercises with diagnostic capabilities.

1. Find the inclination of the line $2x - 7y + 3 = 0$.

2. Find the angle between the lines $3x + 2y - 4 = 0$ and $4x - y + 6 = 0$.

3. Find the distance between the point $(7, 5)$ and the line $y = 5 - x$.

In Exercises 4–7, classify the conic and write the equation in standard form. Identify the center, vertices, foci, and asymptotes (if any). Then sketch the graph of the conic.

4. $y^2 - 4x + 4 = 0$ 5. $x^2 - 4y^2 - 4x = 0$

6. $9x^2 + 16y^2 + 54x - 32y - 47 = 0$

7. $2x^2 + 2y^2 - 8x - 4y + 9 = 0$

8. Find the standard form of the equation of the parabola with vertex $(3, -2)$, with vertical axis, and passing through the point $(0, 4)$.

9. Find the standard form of the equation of the hyperbola with foci $(0, 0)$ and $(0, 4)$ and asymptotes $y = \pm\frac{1}{2}x + 2$.

10. (a) Determine the number of degrees the axis must be rotated to eliminate the xy-term of the conic $x^2 + 6xy + y^2 - 6 = 0$.

    (b) Graph the conic and use a graphing utility to confirm your result.

11. Sketch the curve represented by the parametric equations $x = 2 + 3\cos\theta$ and $y = 2\sin\theta$. Eliminate the parameter and write the corresponding rectangular equation.

12. Find a set of parametric equations of the line passing through the points $(2, -3)$ and $(6, 4)$. (The answer is not unique.)

13. Convert the polar coordinate $(-2, 5\pi/6)$ to rectangular form.

14. Convert the rectangular coordinate $(2, -2)$ to polar form and find two additional representations of this point.

15. Convert the rectangular equation $x^2 + y^2 - 4y = 0$ to polar form.

In Exercises 16–19, sketch the graph of the polar equation. Identify the type of graph.

16. $r = \dfrac{4}{1 + \cos\theta}$ 17. $r = \dfrac{4}{2 + \cos\theta}$

18. $r = 2 + 3\sin\theta$ 19. $r = 3\sin 2\theta$

20. A straight road rises with an inclination of 0.15 radian from the horizontal. Find the slope of the road and the change in elevation over a one-mile stretch of the road.

21. A baseball is hit 3 feet above the ground toward the left field fence. The fence is 10 feet high and 375 feet from home plate. The path of the baseball can be modeled by the parametric equations $x = (115\cos\theta)t$ and $y = 3 + (115\sin\theta)t - 16t^2$. Does the baseball go over the fence when it is hit at an angle of $\theta = 30°$? Does the baseball go over the fence when $\theta = 35°$?

Cumulative Test for Chapters 4–6

Take this test to review material from earlier chapters. When you are finished, check your work against the answers given in the back of the book.

1. Write the complex number $3 - \sqrt{-25}$ in standard form.

In Exercises 2–4, perform the operations and write the result in standard form.

2. $6i - \left(2 + \sqrt{-81}\right)$ **3.** $(2i - 3)^2$ **4.** $\left(\sqrt{3} + i\right)\left(\sqrt{3} - i\right)$

5. Write the quotient in standard form: $\dfrac{4i}{1 + 2i}$.

In Exercises 6 and 7, find all the zeros of the function.

6. $f(x) = x^3 + 2x^2 + 4x + 8$ **7.** $f(x) = x^4 + 4x^3 - 21x^2$

8. Find a polynomial with integer coefficients that has -5, -2, and $2 + \sqrt{3}i$ as its zeros.

9. Find the trigonometric form of the complex number $-2 + 2i$.

10. Find the product of $[4(\cos 30° + i \sin 30°)][6(\cos 120° + i \sin 120°)]$. Write the answer in standard form.

In Exercises 11 and 12, use DeMoivre's Theorem to find the indicated power of the complex number.

11. $\left[2\left(\cos \dfrac{2\pi}{3} + i \sin \dfrac{2\pi}{3}\right)\right]^4$ **12.** $\left(-\sqrt{3} - i\right)^6$

13. Find the three cube roots of 1.

14. Write all the solutions of the equation $x^4 - 81i = 0$.

 In Exercises 15 and 16, use the graph of f to describe the transformation that yields the graph of g. Use a graphing utility to graph both equations in the same viewing window.

15. $f(x) = \left(\frac{2}{5}\right)^x$, $g(x) = -\left(\frac{2}{5}\right)^{-x+3}$

16. $f(x) = 2.2^x$, $g(x) = -2.2^x + 4$

In Exercises 17–20, use a calculator to evaluate each expression. Approximate your result to three decimal places.

17. $\log_{10} 98$ **18.** $\log_{10}\left(\frac{6}{7}\right)$ **19.** $\ln \sqrt{31}$ **20.** $\ln\left(\sqrt{40} - 5\right)$

In Exercises 21–23, evaluate the logarithm using the change-of-base formula. Approximate your answer to three decimal places.

21. $\log_7 1.8$ **22.** $\log_3 0.149$ **23.** $\log_{1/2} 17$

24. Use the logarithmic properties to expand $\ln\left(\dfrac{x^2 - 16}{x^4}\right)$, where $x > 4$.

25. Write $2 \ln x - \frac{1}{2} \ln(x + 5)$ as a logarithm of a single quantity.

In Exercises 26–29, solve the equation.

26. $6e^{2x} = 72$

27. $4^{x-5} + 21 = 30$

28. $\log_2 x + \log_2 5 = 6$

29. $\ln 4x - \ln 2 = 8$

30. Use a graphing utility to graph

$$f(x) = \frac{1000}{1 + 4e^{-0.2x}}$$

and determine the horizontal asymptotes.

31. The number of bacteria N in a culture is given by the model $N = 175e^{kt}$, where t is the time in hours. If $N = 420$ when $t = 8$, estimate the time required for the population to double in size.

32. Find the angle between the lines $2x + y - 3 = 0$ and $x - 3y + 6 = 0$.

33. Find the distance between the point $(6, -3)$ and the line $y = 2x - 4$.

In Exercises 34–37, classify the conic and write the equation in standard form. Identify the center, vertices, foci, and asymptotes (if any). Then sketch the graph.

34. $9x^2 + 4y^2 - 36x + 8y + 4 = 0$

35. $4x^2 - y^2 - 4 = 0$

36. $x^2 + y^2 + 2x - 6y - 12 = 0$

37. $y^2 + 2x + 2 = 0$

38. Find an equation in rectangular coordinates of the circle with center $(2, -4)$ and passing through the point $(0, 4)$.

39. Find an equation in rectangular coordinates of the hyperbola with foci $(0, 0)$ and $(0, 6)$ and asymptotes $y = \pm\dfrac{2\sqrt{5}}{5}x + 3$.

40. (a) Determine the number of degrees the axes must be rotated to eliminate the xy-term of the conic $x^2 + xy + y^2 + 2x - 3y - 30 = 0$.

(b) Graph the conic and use a graphing utility to confirm your result.

41. Sketch the curve represented by the parametric equations $x = 3 + 4\cos\theta$ and $y = \sin\theta$. Eliminate the parameter and write the corresponding rectangular equation.

42. Find a set of parametric equations of the line passing through the points $(3, -2)$ and $(-3, 4)$. (The answer is not unique.)

43. Convert the rectangular equation $x^2 + y^2 - 6y = 0$ to polar form.

In Exercises 44 and 45, sketch the graph of the polar equation.

44. $r = \dfrac{3}{2 + \cos\theta}$

45. $r = \dfrac{4}{1 + \sin\theta}$

46. Match the polar equation with its graph at the left.

(a) $r = 2 + 3\sin\theta$

(b) $r = 3\sin\theta$

(c) $r = 3\sin 2\theta$

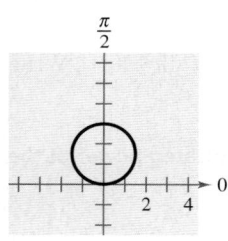

(i)

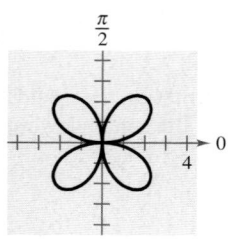

(ii)

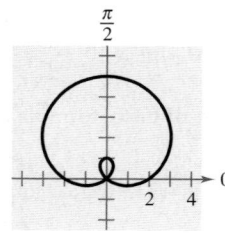

(iii)

FIGURE FOR **46**

Proofs in Mathematics

Inclination and Slope   *(p. 426)*

If a nonvertical line has inclination θ and slope m, then $m = \tan \theta$.

Proof

If $m = 0$, the line is horizontal and $\theta = 0$. So, the result is true for horizontal lines because $m = 0 = \tan 0$.

If the line has a positive slope, it will intersect the x-axis. Label this point $(x_1, 0)$, as shown in the figure. If (x_2, y_2) is a second point on the line, the slope is

$$m = \frac{y_2 - 0}{x_2 - x_1} = \frac{y_2}{x_2 - x_1} = \tan \theta.$$

The case in which the line has a negative slope is left for you to prove.

Distance Between a Point and a Line   *(p. 428)*

The distance between the point (x_1, y_1) and the line $Ax + By + C = 0$ is

$$d = \frac{|Ax_1 + By_1 + C|}{\sqrt{A^2 + B^2}}.$$

Proof

For simplicity's sake, assume that the given line is neither horizontal nor vertical (see figure). By writing the equation $Ax + By + C = 0$ in slope-intercept form

$$y = -\frac{A}{B}x - \frac{C}{B}$$

you can see that the line has a slope of $m = -A/B$. So, the slope of the line passing through (x_1, y_1) and perpendicular to the given line is B/A, and its equation is $y - y_1 = (B/A)(x - x_1)$. These two lines intersect at the point (x_2, y_2), where

$$x_2 = \frac{B(Bx_1 - Ay_1) - AC}{A^2 + B^2} \quad \text{and} \quad y_2 = \frac{A(-Bx_1 + Ay_1) - BC}{A^2 + B^2}.$$

Finally, the distance between (x_1, y_1) and (x_2, y_2) is

$$d = \sqrt{(x_2 - x_1)^2 + (y_2 - y_1)^2}$$

$$= \sqrt{\left(\frac{B^2x_1 - ABy_1 - AC}{A^2 + B^2} - x_1\right)^2 + \left(\frac{-ABx_1 + A^2y_1 - BC}{A^2 + B^2} - y_1\right)^2}$$

$$= \sqrt{\frac{A^2(Ax_1 + By_1 + C)^2 + B^2(Ax_1 + By_1 + C)^2}{(A^2 + B^2)^2}}$$

$$= \frac{|Ax_1 + By_1 + C|}{\sqrt{A^2 + B^2}}.$$

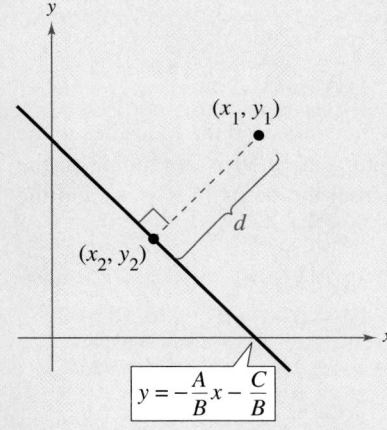

Parabolic Paths

There are many natural occurrences of parabolas in real life. For instance, the famous astronomer Galileo discovered in the 17th century that an object that is projected upward and obliquely to the pull of gravity travels in a parabolic path. Examples of this are the center of gravity of a jumping dolphin and the path of water molecules of a drinking fountain.

Standard Equation of a Parabola   *(p. 434)*

The standard form of the equation of a parabola with vertex at (h, k) is as follows.

$$(x - h)^2 = 4p(y - k), \quad p \neq 0 \qquad \text{Vertical axis, directrix: } y = k - p$$

$$(y - k)^2 = 4p(x - h), \quad p \neq 0 \qquad \text{Horizontal axis, directrix: } x = h - p$$

The focus lies on the axis p units (*directed distance*) from the vertex. If the vertex is at the origin $(0, 0)$, the equation takes one of the following forms.

$$x^2 = 4py \qquad \text{Vertical axis}$$

$$y^2 = 4py \qquad \text{Horizontal axis}$$

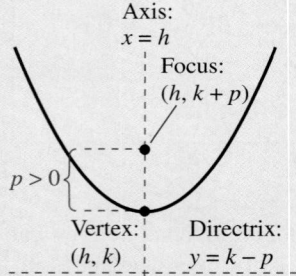

Axis: $x = h$
Focus: $(h, k + p)$
$p > 0$
Vertex: (h, k)
Directrix: $y = k - p$

Parabola with vertical axis

Proof

For the case in which the directrix is parallel to the x-axis and the focus lies above the vertex, as shown in the top figure, if (x, y) is any point on the parabola, then, by definition, it is equidistant from the focus $(h, k + p)$ and the directrix $y = k - p$. So, you have

$$\sqrt{(x - h)^2 + [y - (k + p)]^2} = y - (k - p)$$

$$(x - h)^2 + [y - (k + p)]^2 = [y - (k - p)]^2$$

$$(x - h)^2 + y^2 - 2y(k + p) + (k + p)^2 = y^2 - 2y(k - p) + (k - p)^2$$

$$(x - h)^2 + y^2 - 2ky - 2py + k^2 + 2pk + p^2 = y^2 - 2ky + 2py + k^2 - 2pk + p^2$$

$$(x - h)^2 - 2py + 2pk = 2py - 2pk$$

$$(x - h)^2 = 4p(y - k).$$

For the case in which the directrix is parallel to the y-axis and the focus lies to the right of the vertex, as shown in the bottom figure, if (x, y) is any point on the parabola, then, by definition, it is equidistant from the focus $(h + p, k)$ and the directrix $x = h - p$. So, you have

$$\sqrt{[x - (h + p)]^2 + (y - k)^2} = x - (h - p)$$

$$[x - (h + p)]^2 + (y - k)^2 = [x - (h - p)]^2$$

$$x^2 - 2x(h + p) + (h + p)^2 + (y - k)^2 = x^2 - 2x(h - p) + (h - p)^2$$

$$x^2 - 2hx - 2px + h^2 + 2ph + p^2 + (y - k)^2 = x^2 - 2hx + 2px + h^2 - 2ph + p^2$$

$$-2px + 2ph + (y - k)^2 = 2px - 2ph$$

$$(y - k)^2 = 4p(x - h).$$

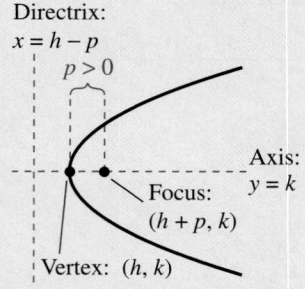

Directrix: $x = h - p$
$p > 0$
Axis: $y = k$
Focus: $(h + p, k)$
Vertex: (h, k)

Parabola with horizontal axis

Polar Equations of Conics   *(p. 490)*

The graph of a polar equation of the form

1. $r = \dfrac{ep}{1 \pm e \cos \theta}$

or

2. $r = \dfrac{ep}{1 \pm e \sin \theta}$

is a conic, where $e > 0$ is the eccentricity and $|p|$ is the distance between the focus (pole) and the directrix.

Proof

A proof for $r = ep/(1 + e \cos \theta)$ with $p > 0$ is listed here. The proofs of the other cases are similar. In the figure, consider a vertical directrix, p units to the right of the focus $F = (0, 0)$. If $P = (r, \theta)$ is a point on the graph of

$$r = \frac{ep}{1 + e \cos \theta}$$

the distance between P and the directrix is

$$
\begin{aligned}
PQ &= |p - x| \\
&= |p - r \cos \theta| \\
&= \left| p - \left(\frac{ep}{1 + e \cos \theta} \right) \cos \theta \right| \\
&= \left| p \left(1 - \frac{e \cos \theta}{1 + e \cos \theta} \right) \right| \\
&= \left| \frac{p}{1 + e \cos \theta} \right| \\
&= \left| \frac{r}{e} \right|.
\end{aligned}
$$

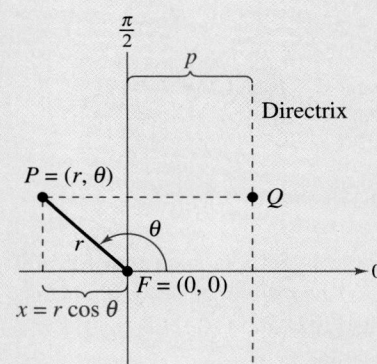

Moreover, because the distance between P and the pole is simply $PF = |r|$, the ratio of PF to PQ is

$$
\begin{aligned}
\frac{PF}{PQ} &= \frac{|r|}{|r/e|} \\
&= |e| \\
&= e
\end{aligned}
$$

and, by definition, the graph of the equation must be a conic.

1. Several mountain climbers are located in a mountain pass between two peaks. The angles of elevation to the two peaks are 0.84 radian and 1.10 radians. A range finder shows that the distances to the peaks are 3250 feet and 6700 feet, respectively (see figure).

   (a) Find the angle between the two lines of sight to the peaks.

   (b) Approximate the amount of vertical climb that is necessary to reach the summit of each peak.

   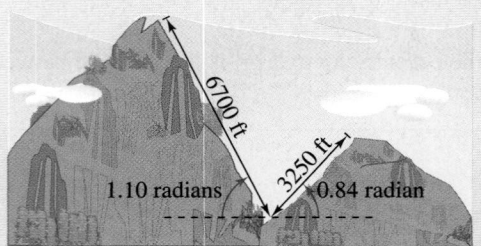

2. Statuary Hall is an elliptical room in the United States Capitol in Washington D.C. The room is also called the Whispering Gallery because a person standing at one focus of the room can hear even a whisper spoken by a person standing at the other focus. This occurs because any sound that is emitted from one focus of an ellipse will reflect off the side of the ellipse to the other focus. Statuary Hall is 46 feet wide and 97 feet long.

   (a) Find an equation that models the shape of the room.

   (b) How far apart are the two foci?

   (c) What is the area of the floor of the room? (The area of an ellipse is $A = \pi ab$.)

3. Find the equation(s) of all parabolas that have the x-axis as the axis of symmetry and focus at the origin.

4. Find the area of the square inscribed in the ellipse below.

   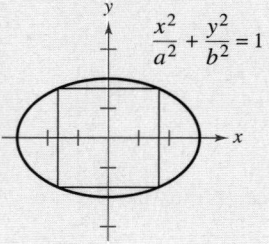

   $$\frac{x^2}{a^2} + \frac{y^2}{b^2} = 1$$

5. A tour boat travels between two islands that are 12 miles apart (see figure). For a trip between the islands, there is enough fuel for a 20-mile trip.

   (a) Explain why the region in which the boat can travel is bounded by an ellipse.

   (b) Let $(0, 0)$ represent the center of the ellipse. Find the coordinates of each island.

   (c) The boat travels from one island, straight past the other island to the vertex of the ellipse, and back to the second island. How many miles does the boat travel? Use your answer to find the coordinates of the vertex.

   (d) Use the results from parts (b) and (c) to write an equation for the ellipse that bounds the region in which the boat can travel.

   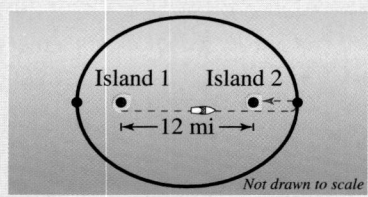

6. Find an equation of the hyperbola such that for any point on the hyperbola, the difference between its distances from the points $(2, 2)$ and $(10, 2)$ is 6.

7. Prove that the graph of the equation

   $$Ax^2 + Cy^2 + Dx + Ey + F = 0$$

   is one of the following (except in degenerate cases).

	Conic	Condition
(a)	Circle	$A = C$
(b)	Parabola	$A = 0$ or $C = 0$ (but not both)
(c)	Ellipse	$AC > 0$
(d)	Hyperbola	$AC < 0$

8. The following sets of parametric equations model projectile motion.

$$x = (v_0 \cos \theta)t \qquad x = (v_0 \cos \theta)t$$
$$y = (v_0 \sin \theta)t \qquad y = h + (v_0 \sin \theta)t - 16t^2$$

(a) Under what circumstances would you use each model?

(b) Eliminate the parameter for each set of equations.

(c) In which case is the path of the moving object not affected by a change in the velocity v? Explain.

9. As t increases, the ellipse given by the parametric equations

$$x = \cos t \text{ and } y = 2 \sin t$$

is traced out *counterclockwise*. Find a parametric representation for which the same ellipse is traced out *clockwise*.

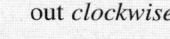 **10.** A **hypocycloid** has the parametric equations

$$x = (a - b) \cos t + b \cos \left(\frac{a - b}{b} t \right)$$

and

$$y = (a - b) \sin t - b \sin \left(\frac{a - b}{b} t \right).$$

Use a graphing utility to graph the hypocycloid for each value of a and b. Describe each graph.

(a) $a = 2, b = 1$ (b) $a = 3, b = 1$
(c) $a = 4, b = 1$ (d) $a = 10, b = 1$
(e) $a = 3, b = 2$ (f) $a = 4, b = 3$

11. The curve given by the parametric equations

$$x = \frac{1 - t^2}{1 + t^2}$$

and

$$y = \frac{t(1 - t^2)}{1 + t^2}$$

is called a **strophoid**.

(a) Find a rectangular equation of the strophoid.

(b) Find a polar equation of the strophoid.

 (c) Use a graphing utility to graph the strophoid.

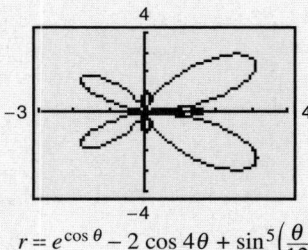 **12.** The rose curves described in this chapter are of the form

$$r = a \cos n\theta \qquad \text{or} \qquad r = a \sin n\theta$$

where n is a positive integer that is greater than or equal to 2. Use a graphing utility to graph $r = a \cos n\theta$ and $r = a \sin n\theta$ for some noninteger values of n. Describe the graphs.

13. What conic section does the polar equation

$$r = a \sin \theta + b \cos \theta$$

represent?

14. The graph of the polar equation

$$r = e^{\cos \theta} - 2 \cos 4\theta + \sin^5 \left(\frac{\theta}{12} \right)$$

is called *the butterfly curve*, as shown in the figure.

(a) The graph below was produced using $0 \le \theta \le 2\pi$. Does this show the entire graph? Explain your reasoning.

(b) Approximate the maximum r-value of the graph. Does this value change if you use $0 \le \theta \le 4\pi$ instead of $0 \le \theta \le 2\pi$? Explain.

$$r = e^{\cos \theta} - 2 \cos 4\theta + \sin^5 \left(\frac{\theta}{12} \right)$$

15. Use a graphing utility to graph the polar equation

$$r = \cos 5\theta + n \cos \theta$$

for $0 \le \theta \le \pi$ for the integers $n = -5$ to $n = 5$. As you graph these equations, you should see the graph change shape from a heart to a bell. Write a short paragraph explaining what values of n produce the heart portion of the curve and what values of n produce the bell portion.

Answers to Odd-Numbered Exercises and Tests

Chapter P

Section P.1 *(page 9)*

1. (a) $5, 1, 2$ (b) $-9, 5, 0, 1, -4, 2, -11$

   (c) $-9, -\frac{7}{2}, 5, \frac{2}{3}, 0, 1, -4, 2, -11$ (d) $\sqrt{2}$

3. (a) 1 (b) $-13, 1, -6$

   (c) $2.01, 0.666\ldots, -13, 1, -6$ (d) $0.010110111\ldots$

5. (a) $\frac{6}{3}, 8$ (b) $\frac{6}{3}, -1, 8, -22$

   (c) $-\frac{1}{3}, \frac{6}{3}, -7.5, -1, 8, -22$ (d) $-\pi, \frac{1}{2}\sqrt{2}$

7. 0.625 **9.** $0.\overline{123}$ **11.** $-1 < 2.5$

13. 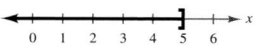 $-4 > -8$

15. 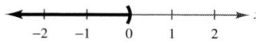 $\frac{3}{2} < 7$

17. 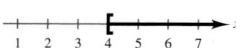 $\frac{5}{6} > \frac{2}{3}$

19. $x \le 5$ denotes the set of all real numbers less than or equal to 5. Unbounded

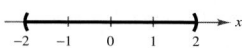

21. $x < 0$ denotes the set of all negative real numbers. Unbounded

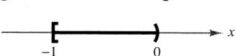

23. $x \ge 4$ denotes the set of all real numbers greater than or equal to 4. Unbounded

25. $-2 < x < 2$ denotes the set of all real numbers greater than -2 and less than 2. Bounded

27. $-1 \le x < 0$ denotes the set of all negative real numbers greater than or equal to -1. Bounded

29. $-2 < x \le 4$ **31.** $y \ge 0$

33. $10 \le t \le 22$ **35.** $W > 65$

37. This interval consists of all real numbers greater than or equal to 0 and less than 8.

39. This interval consists of all real numbers greater than -6.

41. 10 **43.** 5 **45.** -1 **47.** -1 **49.** -1

51. $|-3| > -|-3|$ **53.** $-5 = -|5|$

55. $-|-2| = -|2|$ **57.** 4 **59.** 51 **61.** $\frac{5}{2}$ **63.** $\frac{128}{75}$

65. $|\$113,356 - \$112,700| = \$656 > \$500$

   $0.05(\$112,700) = \$5635$

   Because the actual expenses differ from the budget by more than $500, there is failure to meet the "budget variance test."

67. $|\$37,335 - \$37,640| = \$305 < \$500$

   $0.05(\$37,640) = \$1882$

   Because the difference between the actual expenses and the budget is less than $500 and less than 5% of the budgeted amount, there is compliance with the "budget variance test."

69. (a)

Year	Expenditures (in billions)	Surplus or deficit (in billions)
1960	$92.2	$0.3 (s)
1970	$195.6	$2.8 (d)
1980	$590.9	$73.8 (d)
1990	$1253.2	$221.2 (d)
2000	$1788.8	$236.4 (s)

(b)

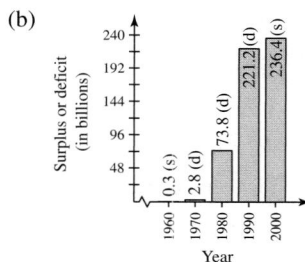

71. $|57 - 236| = 179$ miles   **73.** $|60 - 23| = 37°$

75. $|x - 5| \le 3$ **77.** $|y| \ge 6$

79. $7x$ and 4 are the terms; 7 is the coefficient.

81. $\sqrt{3}x^2$, $-8x$, and -11 are the terms; $\sqrt{3}$ and -8 are the coefficients.

83. $4x^3$, $x/2$, and -5 are the terms; 4 and $\frac{1}{2}$ are the coefficients.

85. (a) -10 (b) -6 **87.** (a) 14 (b) 2

89. (a) Division by 0 is undefined.   (b) 0

91. Commutative Property of Addition

93. Multiplicative Inverse Property

95. Distributive Property

97. Multiplicative Identity Property

99. Associative and Commutative Properties of Multiplication

101. $\dfrac{1}{2}$ **103.** $\dfrac{3}{8}$ **105.** 48 **107.** $\dfrac{5x}{12}$

109. (a)

n	1	0.5	0.01	0.0001	0.000001
$5/n$	5	10	500	50,000	5,000,000

(b) The value of $5/n$ approaches infinity as n approaches 0.

111. False. If $a < b$, then $\dfrac{1}{a} > \dfrac{1}{b}$, where $a \neq b \neq 0$.

113. (a) No. If one variable is negative and the other is positive, the expressions are unequal.

(b) $|u + v| \leq |u| + |v|$

The expressions are equal when u and v have the same sign. If u and v differ in sign, $|u + v|$ is less than $|u| + |v|$.

115. The only even prime number is 2, because its only factors are itself and 1.

117. (a) Negative   (b) Negative

119. Yes. $|a| = -a$ if $a < 0$.

Section P.2   *(page 21)*

1. Identity   **3.** Conditional equation   **5.** Identity

7. Identity   **9.** Conditional equation

11. 4   **13.** -9 **15.** 5   **17.** 9   **19.** No solution

21. -4 **23.** $-\dfrac{6}{5}$ **25.** 9

27. No solution. The x-terms sum to zero.

29. 10   **31.** 4   **33.** 3   **35.** 0

37. No solution. The variable is divided out.

39. No solution. The solution is extraneous.

41. 5   **43.** No solution. The solution is extraneous.

45. 0   **47.** All real numbers

49. $2x^2 + 8x - 3 = 0$ **51.** $x^2 - 6x + 6 = 0$

53. $3x^2 - 90x - 10 = 0$ **55.** $0, -\dfrac{1}{2}$ **57.** $4, -2$

59. -5 **61.** $3, -\dfrac{1}{2}$ **63.** $2, -6$ **65.** $-\dfrac{20}{3}, -4$

67. $-a$ **69.** $\pm 7; \pm 7.00$ **71.** $\pm\sqrt{11}; \pm 3.32$

73. $\pm 3\sqrt{3}; \pm 5.20$ **75.** 8, 16; 8.00, 16.00

77. $-2 \pm \sqrt{14}; 1.74, -5.74$ **79.** $\dfrac{1 \pm 3\sqrt{2}}{2}; 2.62, -1.62$

81. 2; 2.00   **83.** 0, 2   **85.** $4, -8$ **87.** $-3 \pm \sqrt{7}$

89. $1 \pm \dfrac{\sqrt{6}}{3}$ **91.** $2 \pm 2\sqrt{3}$ **93.** $\dfrac{1}{2}, -1$

95. $\dfrac{1}{4}, -\dfrac{3}{4}$ **97.** $1 \pm \sqrt{3}$ **99.** $-7 \pm \sqrt{5}$

101. $-4 \pm 2\sqrt{5}$ **103.** $\dfrac{2}{3} \pm \dfrac{\sqrt{7}}{3}$ **105.** $-\dfrac{4}{3}$

107. $-\dfrac{1}{2} \pm \sqrt{2}$ **109.** $\dfrac{2}{7}$ **111.** $2 \pm \dfrac{\sqrt{6}}{2}$

113. $6 \pm \sqrt{11}$ **115.** $-\dfrac{3}{8} \pm \dfrac{\sqrt{265}}{8}$

117. 0.976, -0.643 **119.** 1.355, -14.071

121. 1.687, -0.488 **123.** $-0.290, -2.200$

125. $1 \pm \sqrt{2}$ **127.** $6, -12$ **129.** $\dfrac{1}{2} \pm \sqrt{3}$

131. $-\dfrac{1}{2}$ **133.** $\dfrac{3}{4} \pm \dfrac{\sqrt{97}}{4}$ **135.** $0, \pm\dfrac{3\sqrt{2}}{2}$

137. ± 3 **139.** -6 **141.** $-3, 0$ **143.** $3, \pm 1$

145. ± 1 **147.** $\pm\sqrt{3}, \pm 1$ **149.** $\pm\dfrac{1}{2}, \pm 4$

151. $1, -2$ **153.** 50   **155.** 26   **157.** -16

159. $2, -5$ **161.** 0   **163.** 9   **165.** $-3 \pm 16\sqrt{2}$

167. $\pm\sqrt{14}$ **169.** 1   **171.** $2, -\dfrac{3}{2}$ **173.** $\dfrac{-3 \pm \sqrt{21}}{6}$

175. $4, -5$ **177.** $\dfrac{1 \pm \sqrt{31}}{3}$ **179.** $3, -2$

181. $\sqrt{3}, -3$ **183.** $3, \dfrac{-1 - \sqrt{17}}{2}$

185. (a) 61.2 inches

(b) Yes. The estimated height of a male with a 19-inch femur is 69.4 inches.

(c)

Height, x	Female femur length	Male femur length
60	15.48	14.79
70	19.80	19.28
80	24.12	23.77
90	28.44	28.26
100	32.76	32.75
110	37.08	37.24

100 inches

(d) $x \approx 100.59$; There would not be a problem because it is not likely for either a male or a female to be 100 inches tall (or 8 feet 4 inches tall).

187. $y = -0.25t + 8$; after about 28 hours

189. $\dfrac{5\sqrt{2}}{2} \approx 3.54$ centimeters

191. ≈ 550 miles per hour and 600 miles per hour

193. 500 units

195. False. $x(3 - x) = 10$
$$3x - x^2 = 10$$
The equation cannot be written in the form $ax + b = 0$.

197. False. $|x| = 0$ has only one solution to check, 0.

199. Equivalent equations have the same solution set, and one is derived from the other by steps for generating equivalent equations.
$$2x = 5, \quad 2x + 3 = 8$$

201. (a) and (b) $x = -5, -\frac{10}{3}$
(c) The method used in part (a) reduces the number of algebraic steps.

203. $a = 9, b = 9$ **205.** $a = 4, b = 24$

Section P.3   *(page 34)*

1. $A: (2, 6),\ B: (-6, -2),\ C: (4, -4),\ D: (-3, 2)$

3. $(-3, 4)$

5. Quadrant IV   **7.** Quadrant II

9. Quadrant III or IV   **11.** Quadrant III

13. Quadrant I or III

15.

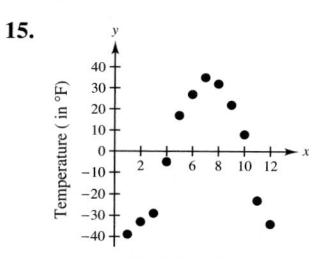

17. (a) 4, 3, 5   (b) $4^2 + 3^2 = 5^2$

19. (a) $10, 3, \sqrt{109}$ (b) $10^2 + 3^2 = \left(\sqrt{109}\right)^2$

21. (a)   (b) 17
(c) $\left(0, \frac{5}{2}\right)$

23. (a)   (b) $2\sqrt{10}$
(c) $(2, 3)$

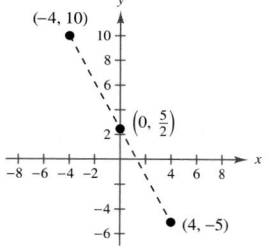

25. (a)   (b) $\dfrac{\sqrt{82}}{3}$
(c) $\left(-1, \frac{7}{6}\right)$

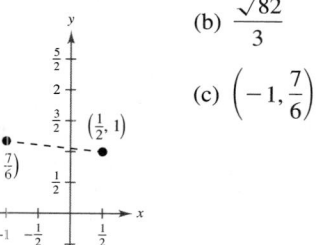

27. (a)   (b) $\sqrt{110.97}$
(c) $(1.25, 3.6)$

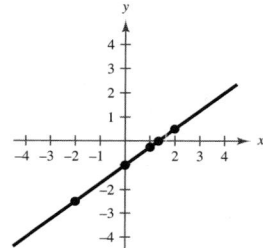

29. (a) The number of artists elected each year seems to be nearly steady except for the first few years. From six to eight artists will be elected in 2004.
(b) The Rock an Roll Hall of Fame was opened in 1986.

31. $5\sqrt{74} \approx 43$ yards   **33.** \$1969.45 million

35. (a) Yes   (b) Yes   **37.** (a) No   (b) Yes

39.

x	-2	0	1	$\frac{4}{3}$	2
y	$-\frac{5}{2}$	-1	$-\frac{1}{4}$	0	$\frac{1}{2}$
(x, y)	$\left(-2, -\frac{5}{2}\right)$	$(0, -1)$	$\left(1, -\frac{1}{4}\right)$	$\left(\frac{4}{3}, 0\right)$	$\left(2, \frac{1}{2}\right)$

41. x-intercepts: $(\pm 2, 0)$ **43.** x-intercept: $\left(\frac{6}{5}, 0\right)$
y-intercept: $(0, 16)$ y-intercept: $(0, -6)$

45. x-intercept: $(-4, 0)$ **47.** x-intercept: $\left(\frac{7}{3}, 0\right)$
y-intercept: $(0, 2)$ y-intercept: $(0, 7)$

49. x-intercepts: $(0, 0), (2, 0)$ **51.** x-intercept: $(6, 0)$
y-intercept: $(0, 0)$ y-intercepts: $\left(0, \pm\sqrt{6}\right)$

53. y-axis symmetry   **55.** Origin symmetry

57. Origin symmetry   **59.** x-axis symmetry

61.

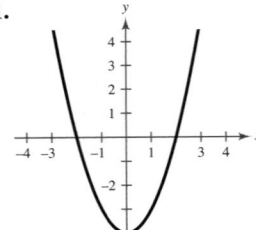

63.

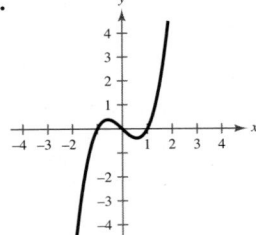

65.

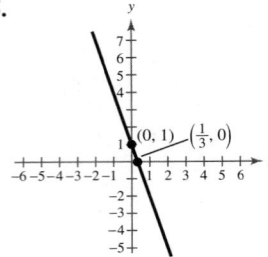

67.

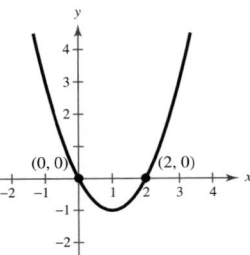

69.

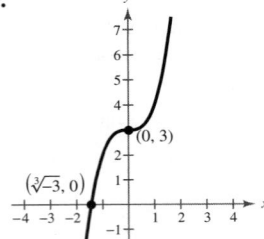

71.

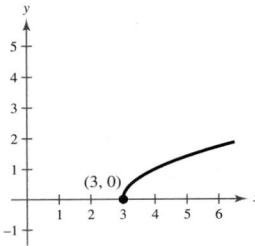

73.

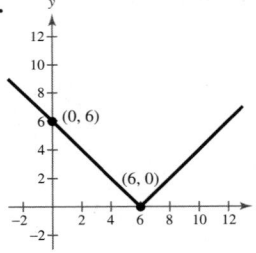

75.

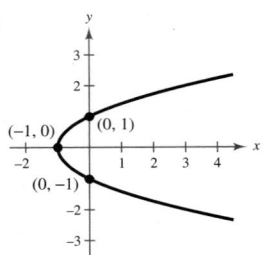

77. $(x - 2)^2 + (y + 1)^2 = 16$

79. $(x + 1)^2 + (y - 2)^2 = 5$

81. $(x - 3)^2 + (y - 4)^2 = 25$

83. Center: $(0, 0)$; Radius: 5

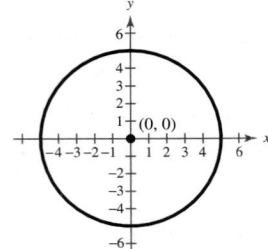

85. Center: $(1, -3)$; Radius: 3

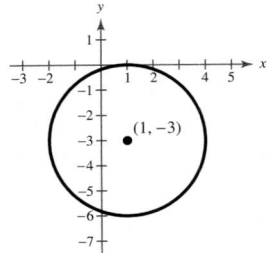

87. Center: $\left(\frac{1}{2}, \frac{1}{2}\right)$; Radius: $\frac{3}{2}$ **89.**

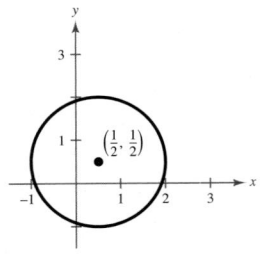

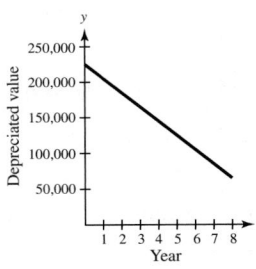

91. (a)

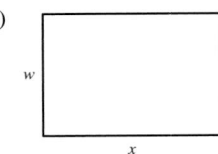

(b) Answers will vary.

(c)

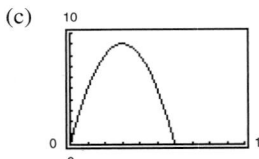

(d) $x = 3, w = 3$

93. (a) and (b)

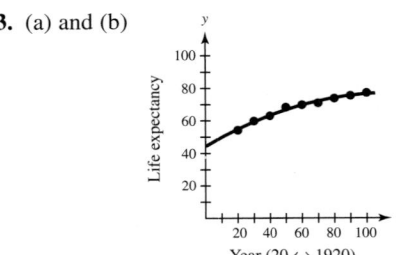

The curve seems to be a good fit for the data.

(c) 2005: 76.8 years; 2010: 77.0 years

(d) No. The model will eventually start to decrease.

95. False. The Midpoint Formula would be used 15 times.

97. False. To find y-intercepts, let $x = 0$ and solve the equation for y.

99. Point on x-axis: $y = 0$; Point on y-axis: $x = 0$

101. Use the Midpoint Formula to prove that the diagonals of the parallelogram bisect each other.
$$\left(\frac{b+a}{2}, \frac{c+0}{2}\right) = \left(\frac{a+b}{2}, \frac{c}{2}\right)$$
$$\left(\frac{a+b+0}{2}, \frac{c+0}{2}\right) = \left(\frac{a+b}{2}, \frac{c}{2}\right)$$

103. (a) $a = 1, b = 0$ (b) $a = 0, b = 1$

Section P.4   *(page 47)*

1. (a) L_2 (b) L_3 (c) L_1

3.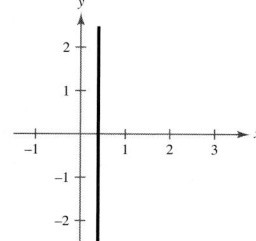

5. $\frac{8}{5}$ **7.** -4

9. $m = 5$;
y-intercept: $(0, 3)$

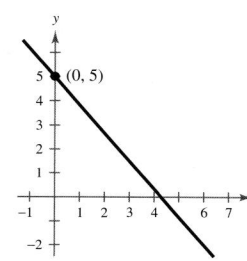

11. $m = -\frac{1}{2}$;
y-intercept: $(0, 4)$

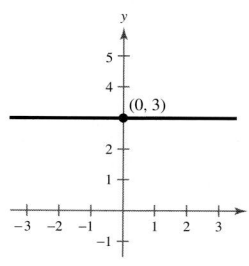

13. m is undefined. There is no y-intercept.

15. $m = -\frac{7}{6}$;
y-intercept: $(0, 5)$

17. $m = 0$;
y-intercept: $(0, 3)$

19. m is undefined. There is no y-intercept.

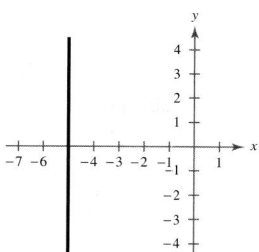

21.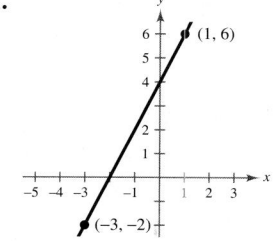

$m = 2$

23.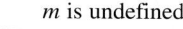

m is undefined.

25.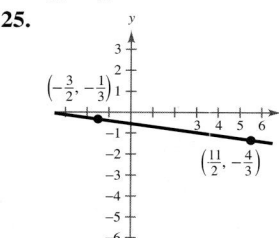

$m = -\frac{1}{7}$

27.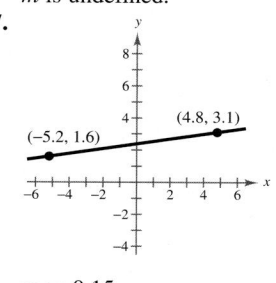

$m = 0.15$

29. Answers will vary. Sample answer:
$(0, 1), (3, 1), (-1, 1)$

31. Answers will vary. Sample answer:
$(6, -5), (7, -4), (8, -3)$

33. Answers will vary. Sample answer:
$(-8, 0), (-8, 2), (-8, 3)$

35. Answers will vary. Sample answer:
$(-4, 6), (-3, 8), (-2, 10)$

37. Answers will vary. Sample answer:
$(9, -1), (11, 0), (13, 1)$

39. Perpendicular   **41.** Parallel

43. (a) Sales increasing 135 units per year
   (b) No change in sales
   (c) Sales decreasing 40 units per year

45. (a) Greatest increase: 2000-2001
      Smallest increase: 1991-1992
   (b) 0.205
   (c) Each year, the earnings per share increase by $0.205.

47. (a) and (b)

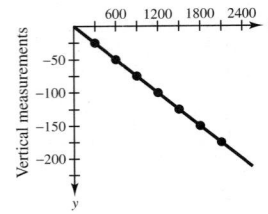

Horizontal measurements

(c) $y = -\frac{1}{12}x$

(d) For every 12 horizontal measurements, the vertical measurement decreases by 1.

(e) "8.3% grade"

49. $V = 125t + 2165$

51. b; The slope is -20, which represents the decrease in the amount of the loan each week.

52. c; The slope is 2, which represents the hourly wage per unit produced.

53. a; The slope is 0.32, which represents the increase in travel cost for each mile driven.

54. d; The slope is -100, which represents the decrease in the value of the word processor each year.

55. $y = 3x - 2$

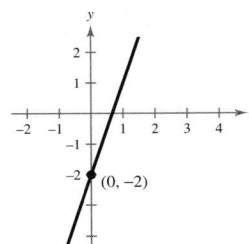

57. $y = -2x$

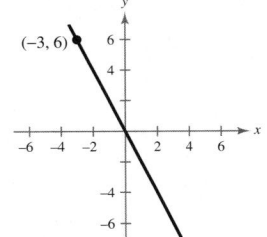

59. $y = -\frac{1}{3}x + \frac{4}{3}$

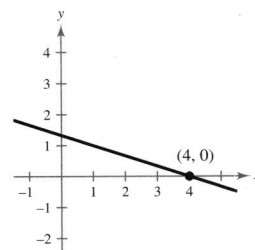

61. $x = 6$

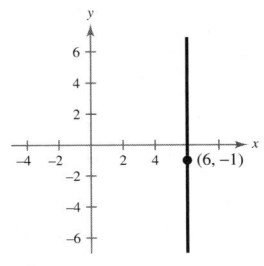

63. $y = \frac{5}{2}$

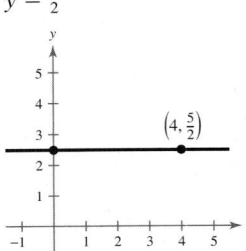

65. $y = 5x + 27.3$

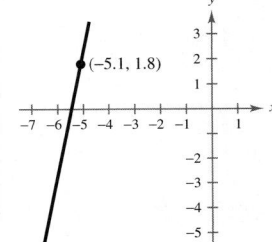

67. $y = -\frac{3}{5}x + 2$

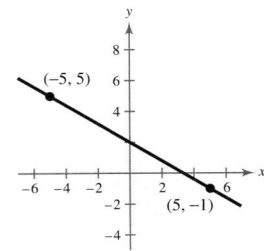

69. $x = -8$

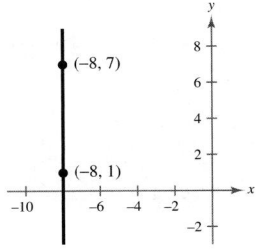

71. $y = -\frac{1}{2}x + \frac{3}{2}$

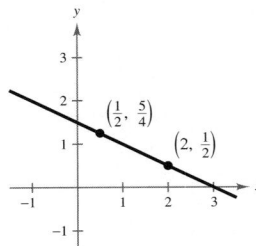

73. $y = -\frac{6}{5}x - \frac{18}{25}$

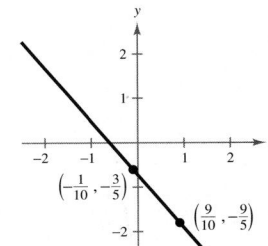

75. $y = 0.4x + 0.2$

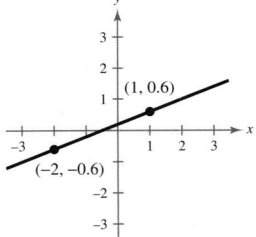

77. $y = -1$

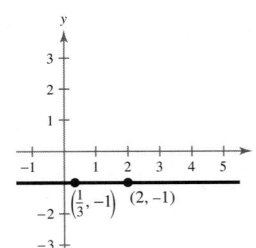

79. $x = \frac{7}{3}$

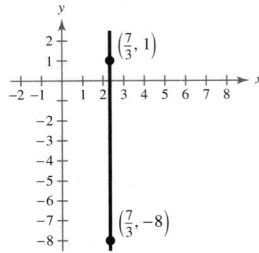

81. $3x + 2y - 6 = 0$ **83.** $12x + 3y + 2 = 0$

85. $x + y - 3 = 0$

87. (a) $y = 2x - 3$ (b) $y = -\frac{1}{2}x + 2$

89. (a) $y = -\frac{3}{4}x + \frac{3}{8}$ (b) $y = \frac{4}{3}x + \frac{127}{72}$

91. (a) $y = 0$ (b) $x = -1$

93. (a) $x = 2$ (b) $y = 5$

95. (a) $y = x + 4.3$ (b) $y = -x + 9.3$

97. Line (b) is perpendicular to line (c).

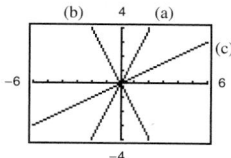

99. Line (a) is parallel to line (b).

Line (c) is perpendicular to line (a) and line (b).

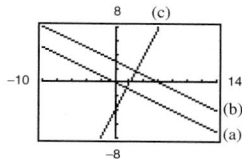

101. $3x - 2y - 1 = 0$ **103.** $80x + 12y + 139 = 0$

105. $y = 0.694t + 0.18$; 2005: $7.12, 2010: $10.59

107. $43,900

109. $V = -175t + 875$ **111.** $S = 0.85L$

113. (a) $C = 16.75t + 36,500$

(b) $R = 27t$

(c) $P = 10.25t - 36,500$

(d) $t \approx 3561$ hours

115. (a) 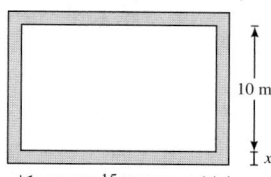   (b) $y = 8x + 50$

(c) 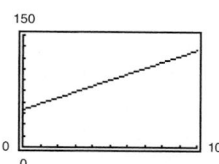   (d) $m = 8$, 8 meters

117. $C = 0.35x + 120$

119. (a) and (b)

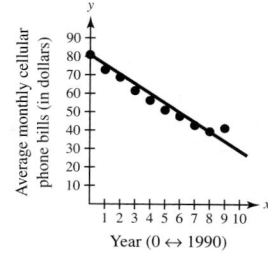

(c) Answers will vary. Sample answer:
$y = -5.2x + 80.9$

(d) Answers will vary. Sample answer: The initial cost is $80.90. Each year, the cellular phone bill decreases an additional $5.20.

(e) The model is accurate.

(f) Answers will vary. Sample answer: $2.90

121. False. The slope with the greatest magnitude corresponds to the steepest line.

123. Find the distance between each two points and use the Pythagorean Theorem.

125. No. The slope cannot be determined without knowing the scale on the y-axis. The slopes could be the same.

127. V-intercept: initial cost; Slope: annual depreciation

Section P.5   *(page 61)*

1. Yes   **3.** No

5. Yes, each input value has exactly one output value.

7. No, the input values of 7 and 10 each have two different output values.

9. (a) Function

(b) Not a function, because the element 1 in A corresponds to two elements, -2 and 1, in B.

(c) Function

(d) Not a function, because not every element in A is matched with an element in B.

11. Each is a function. For each year there corresponds one and only one circulation.

13. Not a function   **15.** Function   **17.** Function

19. Not a function   **21.** Function

23. (a) -1 (b) -9 (c) $2x - 5$

25. (a) 36π (b) $\frac{9}{2}\pi$ (c) $\frac{32}{3}\pi r^3$

27. (a) 1 (b) 2.5 (c) $3 - 2|x|$

29. (a) $-\dfrac{1}{9}$ (b) Undefined   (c) $\dfrac{1}{y^2 + 6y}$

31. (a) 1 (b) -1 (c) $\dfrac{|x - 1|}{x - 1}$

33. (a) -1 (b) 2 (c) 6

35. (a) -7 (b) 4 (c) 9

37.

x	-2	-1	0	1	2
$f(x)$	1	-2	-3	-2	1

39.

t	-5	-4	-3	-2	-1
$h(t)$	1	$\frac{1}{2}$	0	$\frac{1}{2}$	1

41.

x	-2	-1	0	1	2
$f(x)$	5	$\frac{9}{2}$	4	1	0

43. 5 **45.** $\frac{4}{3}$ **47.** ± 3 **49.** $0, \pm 1$

51. $2, -1$ **53.** $3, 0$ **55.** All real numbers

57. All real numbers $t \neq 0$

59. $y \geq 10$ **61.** $-1 \leq x \leq 1$

63. All real numbers $x \neq 0, -2$ **65.** $s \geq 1, s \neq 4$

67. $x > 0$

69. $\{(-2, 4), (-1, 1), (0, 0), (1, 1), (2, 4)\}$

71. $\{(-2, 4), (-1, 3), (0, 2), (1, 3), (2, 4)\}$

73. $g(x) = cx^2; c = -2$ **75.** $r(x) = \dfrac{c}{x}; c = 32$

77. $3 + h, h \neq 0$

79. $3x^2 + 3xh + h^2 + 3, h \neq 0$

81. $-\dfrac{x + 3}{9x^2}, x \neq 3$ **83.** $\dfrac{\sqrt{5x} - 5}{x - 5}$

85. $A = \dfrac{P^2}{16}$

87. (a) The maximum volume is 1024 cubic centimeters.

(b)

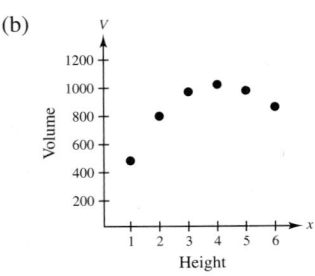

Yes, V is a function of x.

(c) $V = x(24 - 2x)^2, 0 < x < 12$

89. $A = \dfrac{x^2}{2(x - 2)}, x > 2$

91. 1990: \$27,800; 1994: \$33,488;
1996: \$38,440; 1999: \$44,110

93. (a) $C = 12.30x + 98,000$ (b) $R = 17.98x$

(c) $P = 5.68x - 98,000$

95. (a) $R = \dfrac{240n - n^2}{20}, n \geq 80$

(b)

n	90	100	110	120	130	140	150
$R(n)$	\$675	\$700	\$715	\$720	\$715	\$700	\$675

The revenue is maximum when 120 people take the trip.

97. (a)

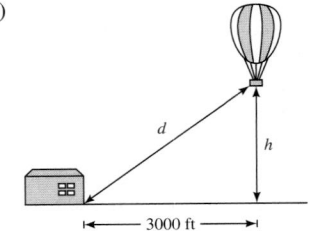

(b) $h = \sqrt{d^2 - 3000^2}, d \geq 3000$

99. (a) The result is 1.8 which is the average increase per year of the number of threatened and endangered fish.

(b) $\begin{cases} 2x + 104 & 6 \leq x \leq 7 \\ 2x + 103 & 8 \leq x \leq 11 \end{cases}$

(c)

x	6	7	8	9	10	11
N	116	118	119	121	123	125

(d) The results are the same.

(e) $y = 1.8x + 105$

The answer in part (b) is more accurate.

101. True. Each x-value corresponds to one y-value.

Section P.6 *(page 73)*

1. Domain: $(-\infty, -1], [1, \infty)$ **3.** Domain: $[-4, 4]$
Range: $[0, \infty)$ Range: $[0, 4]$

5. (a) 0 (b) -1 (c) 0 (d) -2

7. (a) -3 (b) 0 (c) 1 (d) -3

9. Function **11.** Not a function **13.** Function

15. $-\frac{5}{2}, 6$ **17.** 0 **19.** $0, \pm\sqrt{2}$ **21.** $\pm\frac{1}{2}, 6$ **23.** $\frac{1}{2}$

25.

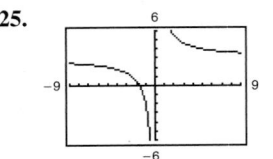

$-\frac{5}{3}$

27.

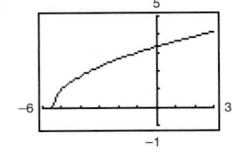

$-\frac{11}{2}$

29.

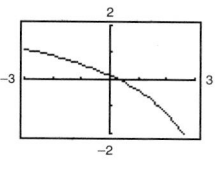

$\frac{1}{3}$

31. Increasing on $(-\infty, \infty)$

33. Increasing on $(-\infty, 0)$ and $(2, \infty)$
Decreasing on $(0, 2)$

35. Increasing on $(-\infty, 0)$ and $(2, \infty)$
Constant on $(0, 2)$

37. Increasing on $(1, \infty)$
Decreasing on $(-\infty, -1)$
Constant on $(-1, 1)$

39. (a)

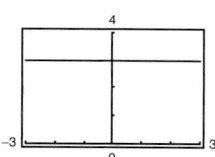

Constant on $(-\infty, \infty)$

(b)

x	-2	-1	0	1	2
$f(x)$	3	3	3	3	3

41. (a)

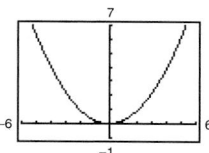

Decreasing on $(-\infty, 0)$; Increasing on $(0, \infty)$

(b)

s	-4	-2	0	2	4
$g(s)$	4	1	0	1	4

43. (a)

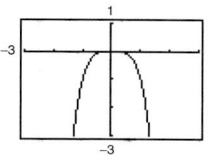

Increasing on $(-\infty, 0)$; Decreasing on $(0, \infty)$

(b)

t	-2	-1	0	1	2
$f(t)$	-16	-1	0	-1	-16

45. (a)

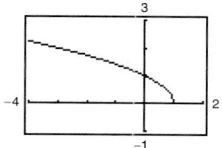

Decreasing on $(-\infty, 1)$

(b)

x	-3	-2	-1	0	1
$f(x)$	2	$\sqrt{3}$	$\sqrt{2}$	1	0

47. (a)

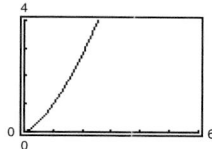

Increasing on $(0, \infty)$

(b)

x	0	1	2	3	4
$f(x)$	0	1	2.8	5.2	8

49.

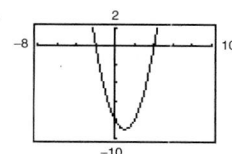

Relative minimum: $(1, -9)$

51.

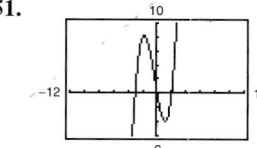

Relative maximum: $(-1.79, 8.21)$
Relative minimum: $(1.12, -4.06)$

53.

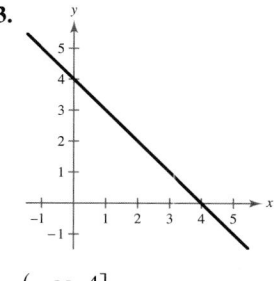

$(-\infty, 4]$

55.

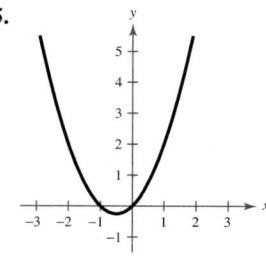

$(-\infty, -1], [0, \infty)$

57.

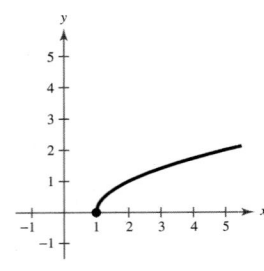

59.

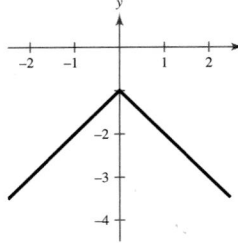

$[1, \infty)$

$f(x) < 0$ for all x

61. Even   **63.** Odd   **65.** Neither even nor odd

67. $h = -x^2 + 4x - 3$ **69.** $h = 2x - x^2$

71. $L = \frac{1}{2}y^2$ **73.** $L = 4 - y^2$

75. (a)

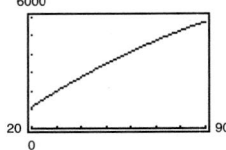

(b) 30 watts

77. (a) ten thousands   (b) ten millions   (c) percents

79. False. The function $f(x) = \sqrt{x^2 + 1}$ has a domain of all real numbers.

81. (a) Even. The graph is a reflection in the x-axis.

(b) Even. The graph is a reflection in the y-axis.

(c) Even. The graph is a vertical translation of f.

(d) Neither. The graph is a horizontal translation of f.

83. (a) $\left(\frac{3}{2}, 4\right)$ (b) $\left(\frac{3}{2}, -4\right)$

85. (a) $(-4, 9)$ (b) $(-4, -9)$

87. (a)

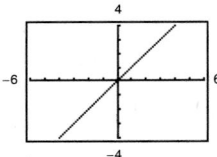

(b)

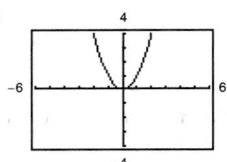

(c)

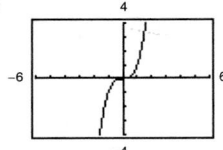

(d)

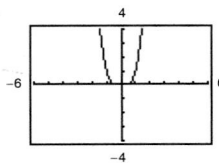

(e)

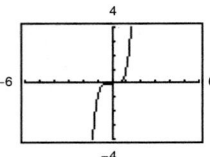

(f)

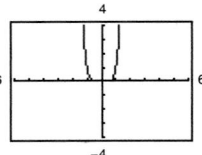

All the graphs pass through the origin. The graphs of the odd powers of x are symmetric with respect to the origin, and the graphs of the even powers are symmetric with respect to the y-axis. As the powers increase, the graphs become flatter in the interval $-1 < x < 1$.

Section P.7 *(page 82)*

1. $f(x) = -2x + 6$ **3.** $f(x) = -3x + 11$

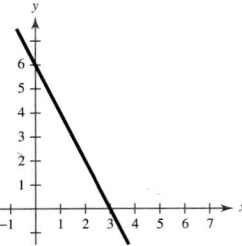

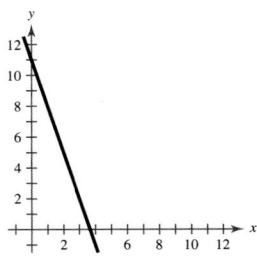

5. $f(x) = -1$ **7.** $f(x) = \frac{6}{7}x - \frac{45}{7}$

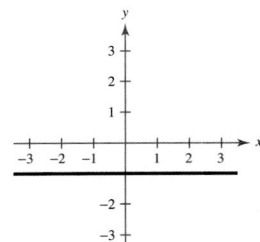

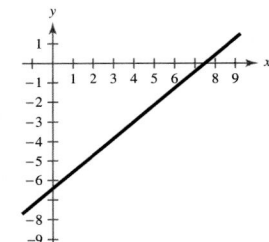

9.

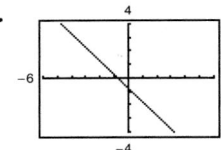

11.

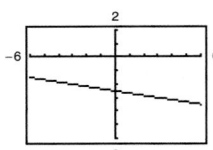

13.

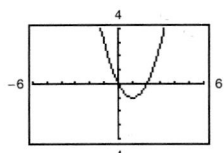

15.

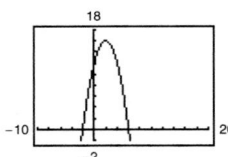

17.

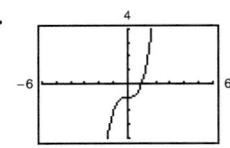

19.

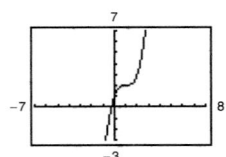

21.

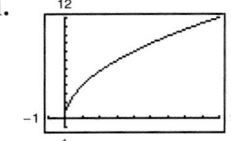

23.

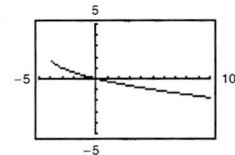

25. 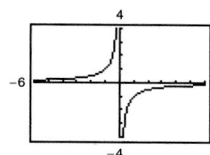   **27.**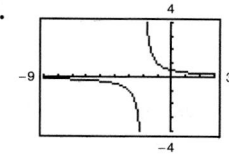

51. 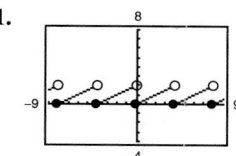   Domain: $(-\infty, \infty)$
Range: $[0, 2)$
Sawtooth pattern

29. (a) 2   (b) 2   (c) -4 (d) 3
31. (a) 1   (b) 3   (c) 7   (d) -19
33. (a) 6   (b) -11 (c) 6   (d) -22
35. (a) -10 (b) -4 (c) -1 (d) 41

53. $f(x) = |x|$; $g(x) = |x + 2| - 1$
55. $f(x) = x^3$; $g(x) = (x - 1)^3 - 2$
57. $f(x) = 2$; $g(x) = 2$
59. $f(x) = x$; $g(x) = x - 2$
61. $f(x) = x^2$; $g(x) = (x - 3)^2$

37. 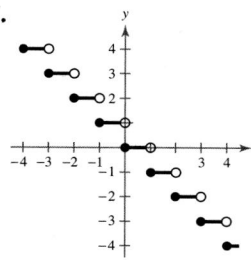   **39.**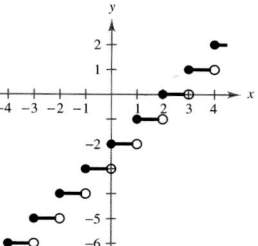

63. (a)   (b) $5.64

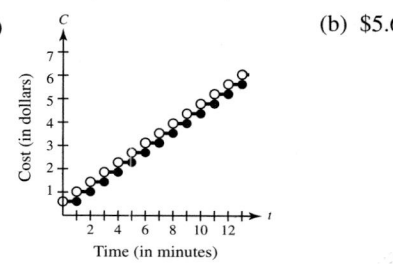

41.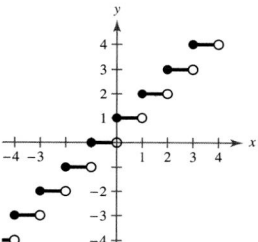

65. (a)   (b) $50.25

43. 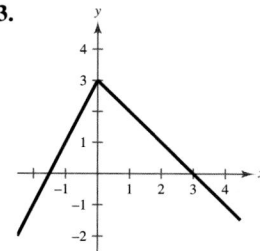   **45.**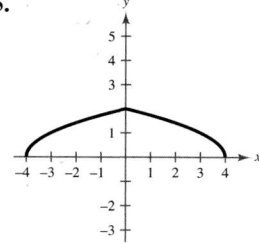

67. (a) $W(30) = 360$; $W(40) = 480$;
$W(45) = 570$; $W(50) = 660$

(b) $W(h) = \begin{cases} 12h, & 0 < h \le 45 \\ 18(h - 45) + 540, & h > 45 \end{cases}$

47. 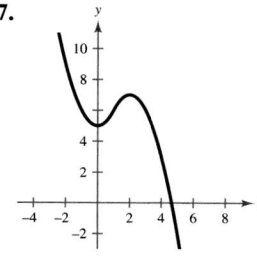   **49.**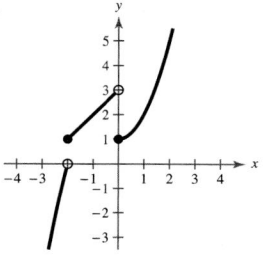

69.

Interval	Input Pipe	Drainpipe 1	Drainpipe 2
$[0, 5]$	Open	Closed	Closed
$[5, 10]$	Open	Open	Closed
$[10, 20]$	Closed	Closed	Closed
$[20, 30]$	Closed	Closed	Open
$[30, 40]$	Open	Open	Open
$[40, 45]$	Open	Closed	Open
$[45, 50]$	Open	Open	Open
$[50, 60]$	Open	Open	Closed

71. True. The solution sets are the same.

Section P.8   *(page 90)*

1. (a)

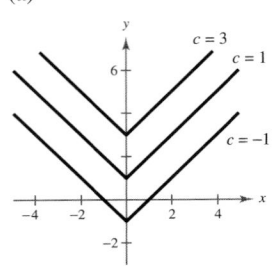

(b)

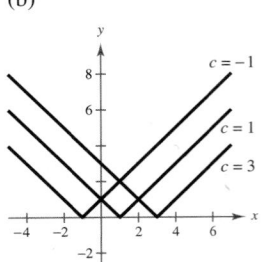

(c)

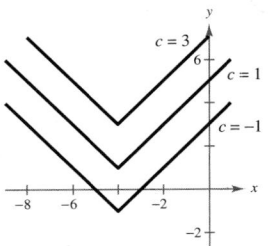

(c)

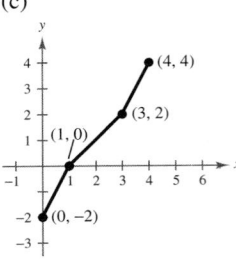

(d)

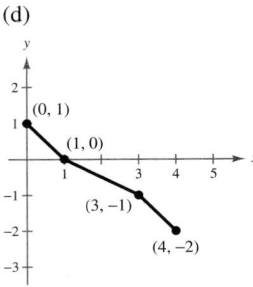

(e)

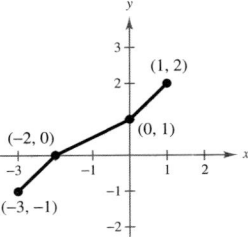

(f)

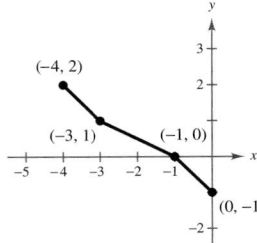

(g)

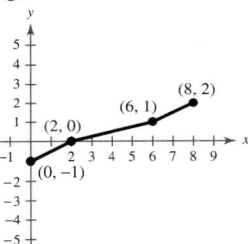

3. (a)

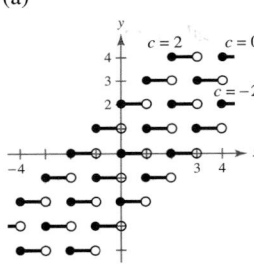

(b)

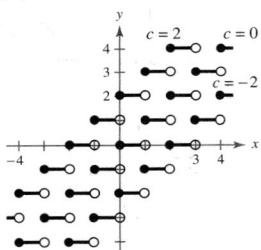

(c)

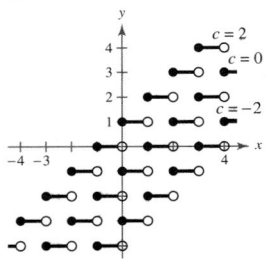

7. (a)

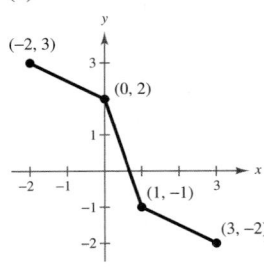

(b)

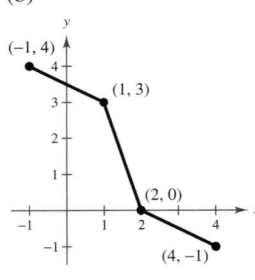

(c)

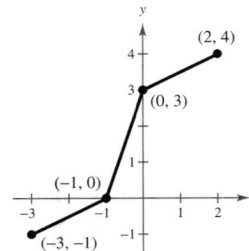

(d)

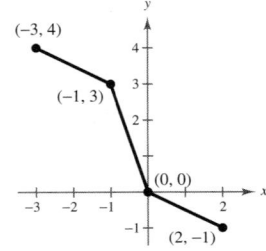

5. (a)

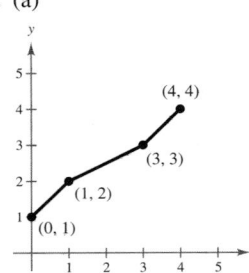

(b)

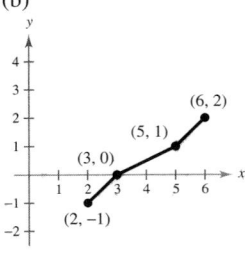

(e)

(f)

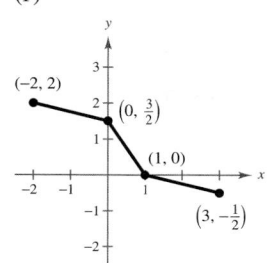

(g)

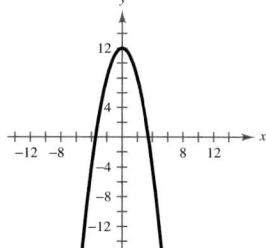

9. (a) $y = x^2 - 1$ (b) $y = 1 - (x + 1)^2$

(c) $y = -(x - 2)^2 + 6$ (d) $y = (x - 5)^2 - 3$

11. (a) $y = |x| + 5$ (b) $y = -|x + 3|$

(c) $y = |x - 2| - 4$ (d) $y = -|x - 6| - 1$

13. Horizontal shift of $y = x^3$; $y = (x - 2)^3$

15. Reflection in the x-axis of $y = x^2$; $y = -x^2$

17. Reflection in the x-axis and vertical shift of $y = \sqrt{x}$; $y = 1 - \sqrt{x}$

19. Reflection in the x-axis, and vertical shift 12 units upward, of $f(x) = x^2$

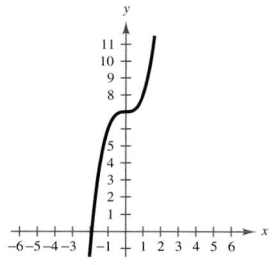

21. Vertical shift seven units upward of $f(x) = x^3$

23. Reflection in the x-axis, vertical shift two units upward, and horizontal shift five units to the left, of $f(x) = x^2$

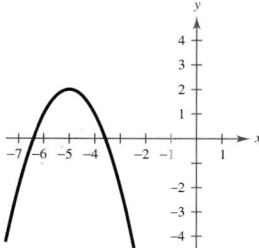

25. Vertical shift two units upward, and horizontal shift one unit to the right, of $f(x) = x^3$

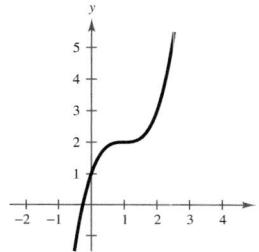

27. Reflection in the x-axis, and vertical shift two units downward, of $f(x) = |x|$

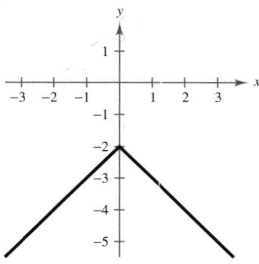

29. Reflection in the x-axis, vertical shift eight units upward, and horizontal shift four units to the left, of $f(x) = |x|$

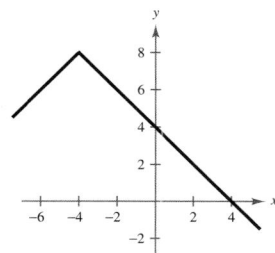

31. Reflection in the x-axis, and vertical shift three units upward, of $f(x) = [\![x]\!]$

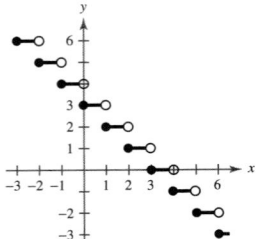

33. Horizontal shift nine units to the right of $f(x) = \sqrt{x}$

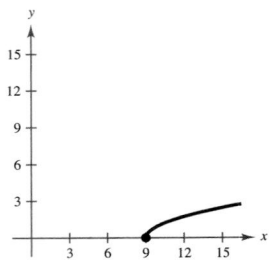

35. Reflection in the y-axis, vertical shift two units downward, and horizontal shift seven units to the right, of $f(x) = \sqrt{x}$

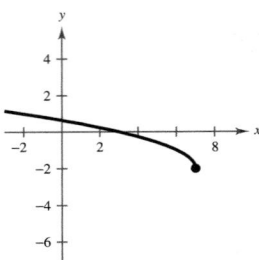

37. Vertical shift four units downward, and horizontal stretch of two, of $f(x) = \sqrt{x}$

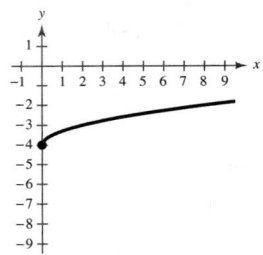

39. $f(x) = (x - 2)^2 - 8$ **41.** $f(x) = (x - 13)^3$

43. $f(x) = -|x| - 10$ **45.** $f(x) = -\sqrt{-x} + 6$

47. (a) $y = -3x^2$ (b) $y = 4x^2 + 3$

49. (a) $y = -\frac{1}{2}|x|$ (b) $y = 3|x| - 3$

51. Vertical stretch of $y = x^3$; $y = 2x^3$

53. Reflection in the x-axis and vertical shrink of $y = x^2$; $y = -\frac{1}{2}x^2$

55. Reflection in the y-axis and vertical shrink of $y = \sqrt{x}$; $y = \frac{1}{2}\sqrt{-x}$

57. $y = -(x - 2)^3 + 2$ **59.** $y = -\sqrt{x} - 3$

61. (a)

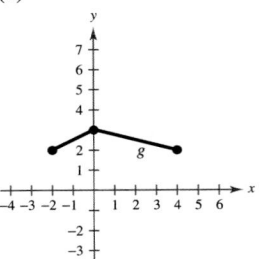

(b)

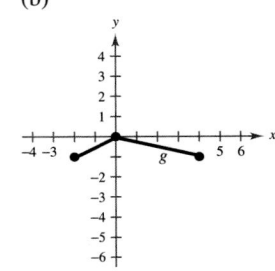

(c)

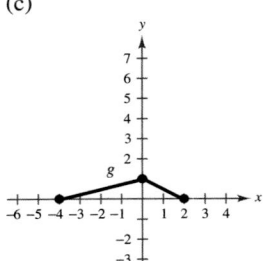

(d)

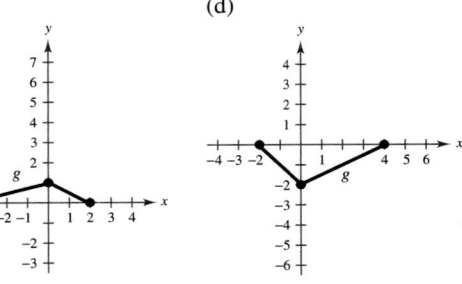

(e)

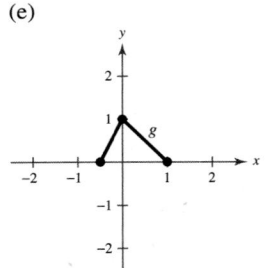

(f)

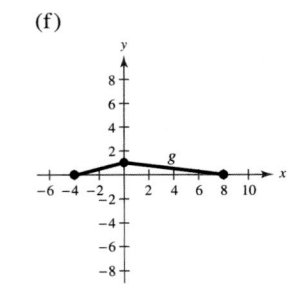

63. (a) Vertical shift of 20.5 units upward and vertical shrink of 0.035

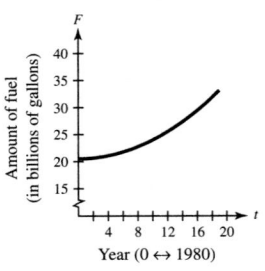

(b) 0.665-billion-gallon increase in fuel usage by trucks each year

(c) $f(t) = 20.5 + 0.035(t + 10)^2$. The graph was shifted 10 units to the left.

(d) 42.375 billion gallons. Yes.

65. True. $|-x| = |x|$

67. (a) $g(t) = \frac{3}{4} f(t)$ (b) $g(t) = f(t) + 10,000$
  (c) $g(t) = f(t - 2)$

69. $(-2, 0), (-1, 1), (0, 2)$

Section P.9 *(page 100)*

1. 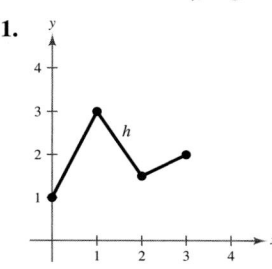   **3.**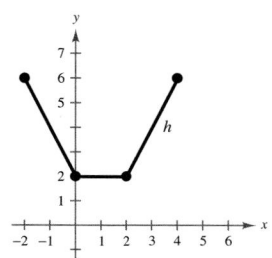

5. (a) $2x$ (b) 4 (c) $x^2 - 4$ (d) $\dfrac{x + 2}{x - 2};\ x \neq 2$

7. (a) $x^2 + 4x - 5$ (b) $x^2 - 4x + 5$
  (c) $4x^3 - 5x^2$ (d) $\dfrac{x^2}{4x - 5};\ x \neq \dfrac{5}{4}$

9. (a) $x^2 + 6 + \sqrt{1 - x}$ (b) $x^2 + 6 - \sqrt{1 - x}$
  (c) $(x^2 + 6)\sqrt{1 - x}$ (d) $\dfrac{(x^2 + 6)\sqrt{1 - x}}{1 - x};\ x < 1$

11. (a) $\dfrac{x + 1}{x^2}$ (b) $\dfrac{x - 1}{x^2}$ (c) $\dfrac{1}{x^3}$ (d) $x;\ x \neq 0$

13. 3 **15.** 5 **17.** $9t^2 - 3t + 5$ **19.** 74

21. 26 **23.** $\frac{3}{5}$

25. 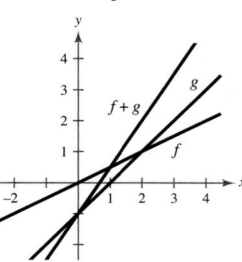   **27.**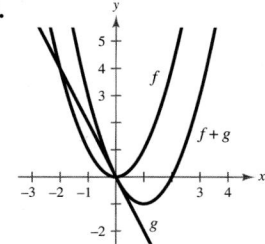

29.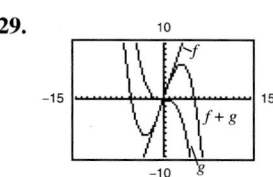
  $f(x), g(x)$

31. $T = \frac{3}{4}x + \frac{1}{15}x^2$

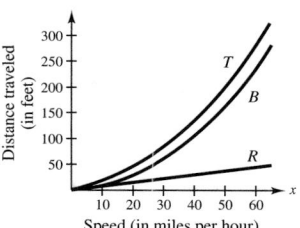

33. (a) $y_1 = 1.344t^2 - 9.38t + 163.6$
    $y_2 = 17.34t + 238.5$
    $y_3 = 3.38t + 28.1$
  (b) $y_1 + y_2 + y_3 = 1.344t^2 + 11.34t + 430.2$; the total amount spent on health services and supplies
  (c)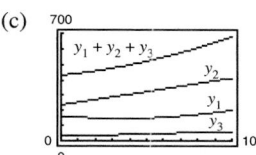
  (d) 2003: \$804.76 billion; 2005: \$902.70 billion

35. (a) $(x - 1)^2$ (b) $x^2 - 1$ (c) x^4

37. (a) x (b) x (c) $\sqrt[3]{\sqrt[3]{x - 1} - 1}$

39. (a) $\sqrt{x^2 + 4}$ (b) $x + 4$
  Domain of f and $g \circ f$: $x \geq -4$;
  Domain of g and $f \circ g$: all real numbers

41. (a) $x + 1$ (b) $\sqrt{x^2 + 1}$
  Domain of f and $g \circ f$: all real numbers
  Domain of g and $f \circ g$: $x \geq 0$

43. (a) $|x + 6|$ (b) $|x| + 6$
  Domain of $f, g, f \circ g,$ and $g \circ f$: all real numbers

45. (a) $\dfrac{1}{x + 3}$ (b) $\dfrac{1}{x} + 3$
  Domain of f and $g \circ f$: all real numbers $x \neq 0$
  Domain of g: all real numbers
  Domain of $f \circ g$: all real numbers $x \neq -3$

47. (a) 3 (b) 0 **49.** (a) 0 (b) 4

51. Answers will vary. Sample answer:
  $f(x) = x^2,\ g(x) = 2x + 1$

53. Answers will vary. Sample answer:
  $f(x) = \sqrt[3]{x},\ g(x) = x^2 - 4$

55. Answers will vary. Sample answer:
  $f(x) = \dfrac{1}{x},\ g(x) = x + 2$

57. Answers will vary. Sample answer:

$$f(x) = \frac{x + 3}{4 + x}, \quad g(x) = -x^2$$

59. (a) $r(x) = \dfrac{x}{2}$ (b) $A(r) = \pi r^2$

(c) $(A \circ r)(x) = \pi\left(\dfrac{x}{2}\right)^2$; $(A \circ r)(x)$ represents the area of the circular base of the tank on the square foundation with side length x.

61. False. $(f \circ g)(x) = 6x + 1$ and $(g \circ f)(x) = 6x + 6$

63. $g(f(x))$ represents 3 percent of an amount over \$500,000.

65. Odd

Section P.10   *(page 109)*

1. c   **2.** b   **3.** a   **4.** d   **5.** $f^{-1}(x) = \frac{1}{6}x$

7. $f^{-1}(x) = x - 9$ **9.** $f^{-1}(x) = \dfrac{x - 1}{3}$

11. $f^{-1}(x) = x^3$

13. (a) $f(g(x)) = f\left(\dfrac{x}{2}\right) = 2\left(\dfrac{x}{2}\right) = x$

$g(f(x)) = g(2x) = \dfrac{(2x)}{2} = x$

(b)

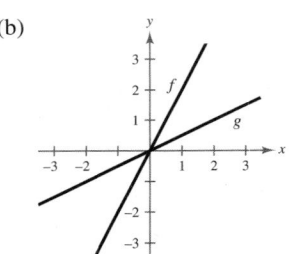

15. (a) $f(g(x)) = f\left(\dfrac{x - 1}{7}\right) = 7\left(\dfrac{x - 1}{7}\right) + 1 = x$

$g(f(x)) = g(7x + 1) = \dfrac{(7x + 1) - 1}{7} = x$

(b)

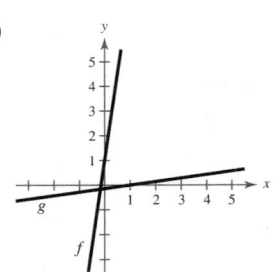

17. (a) $f(g(x)) = f\left(\sqrt[3]{8x}\right) = \dfrac{\left(\sqrt[3]{8x}\right)^3}{8} = x$

$g(f(x)) = g\left(\dfrac{x^3}{8}\right) = \sqrt[3]{8\left(\dfrac{x^3}{8}\right)} = x$

(b)

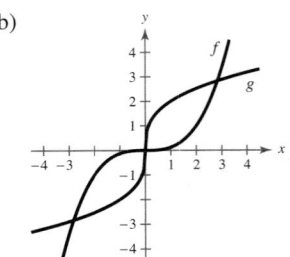

19. (a) $f(g(x)) = f(x^2 + 4), \; x \geq 0$

$\qquad = \sqrt{(x^2 + 4) - 4} = x$

$g(f(x)) = g\left(\sqrt{x - 4}\right)$

$\qquad = \left(\sqrt{x - 4}\right)^2 + 4 = x$

(b)

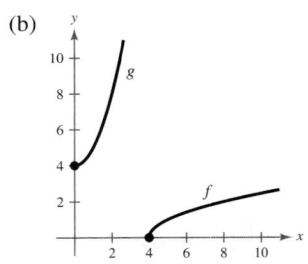

21. (a) $f(g(x)) = f\left(\sqrt{9 - x}\right), \; x \leq 9$

$\qquad = 9 - \left(\sqrt{9 - x}\right)^2 = x$

$g(f(x)) = g(9 - x^2), \; x \geq 0$

$\qquad = \sqrt{9 - (9 - x^2)} = x$

(b)

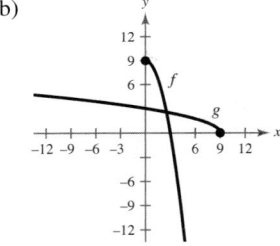

23. (a) $f(g(x)) = f\left(-\dfrac{5x + 1}{x - 1}\right) = \dfrac{-\left(\dfrac{5x + 1}{x - 1}\right) - 1}{-\left(\dfrac{5x + 1}{x - 1}\right) + 5}$

$\qquad = \dfrac{-5x - 1 - x + 1}{-5x - 1 + 5x - 5} = x$

$g(f(x)) = g\left(\dfrac{x - 1}{x + 5}\right) = \dfrac{-5\left(\dfrac{x - 1}{x + 5}\right) - 1}{\dfrac{x - 1}{x + 5} - 1}$

$\qquad = \dfrac{-5x + 5 - x - 5}{x - 1 - x - 5} = x$

(b)

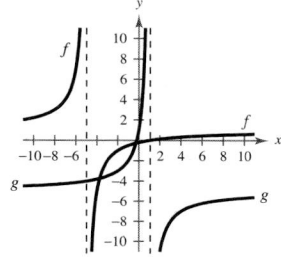

25. No

27.

x	-2	0	2	4	6	8
$f^{-1}(x)$	-2	-1	0	1	2	3

29. Yes **31.** No

33. 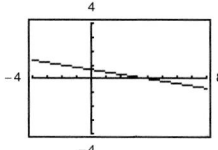 **35.**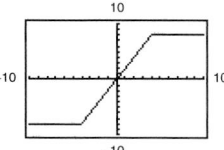

The function has an inverse. The function does not have an inverse.

37.

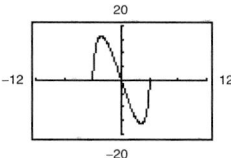

The function does not have an inverse.

39. $f^{-1}(x) = \dfrac{x+3}{2}$ **41.** $f^{-1}(x) = \sqrt[5]{x+2}$

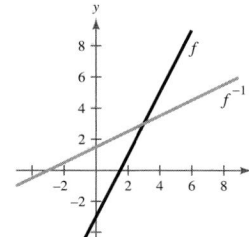 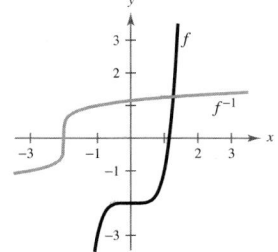

43. $f^{-1}(x) = x^2,\ x \geq 0$

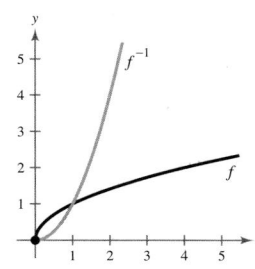

45. $f^{-1}(x) = \sqrt{4-x^2},\ 0 \leq x \leq 2$

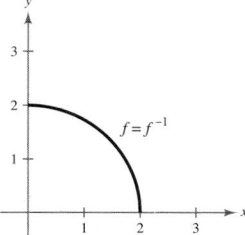

47. $f^{-1}(x) = \dfrac{4}{x}$ **49.** $f^{-1}(x) = \dfrac{2x+1}{x-1}$

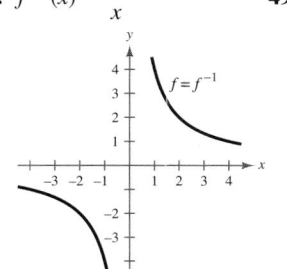 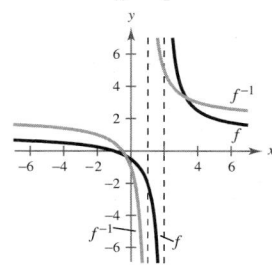

51. $f^{-1}(x) = x^3 + 1$ **53.** $f^{-1}(x) = \dfrac{5x-4}{6-4x}$

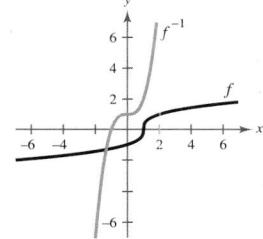 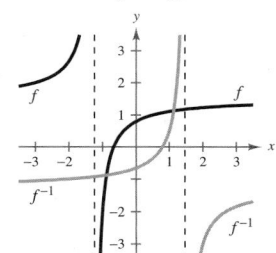

55. No inverse **57.** $g^{-1}(x) = 8x$ **59.** No inverse

61. $f^{-1}(x) = \sqrt{x} - 3$ **63.** No inverse

65. No inverse **67.** $f^{-1}(x) = \dfrac{x^2-3}{2},\ x \geq 0$ **69.** 32

71. 600 **73.** $2\sqrt[3]{x+3}$ **75.** $\dfrac{x+1}{2}$ **77.** $\dfrac{x+1}{2}$

79. (a) 9

(b) f^{-1} yields the year for a given number of households.

(c) $y = 1266.54x + 92{,}255.54$

(d) $f^{-1} = \dfrac{x - 92{,}255.54}{1266.54}$ (e) 15

81. (a) Yes

(b) f^{-1} yields the year for a given number of miles traveled by motor vehicles.

(c) 8

(d) No. $f(t)$ would not pass the Horizontal Line Test.

83. (a) $y = \sqrt{\dfrac{x - 245.50}{0.03}}$, $245.5 < x < 545.5$

$x =$ degrees Fahrenheit; $y = \%$ load

(b) 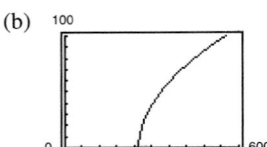 (c) $0 < x < 92.11$

85. False. $f(x) = x^2$ has no inverse.

87.

x	1	3	4	6
y	1	2	6	7

x	1	2	6	7
$f^{-1}(x)$	1	3	4	6

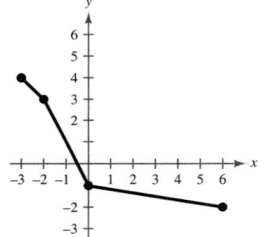

89.

x	-2	-1	3	4
y	6	0	-2	-3

x	-3	-2	0	6
$f^{-1}(x)$	4	3	-1	-2

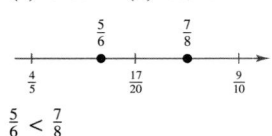

91. $k = \frac{1}{4}$

Review Exercises   *(page 115)*

1. (a) 11   (b) 11, -14

(c) 11, -14, $-\frac{8}{9}$, $\frac{5}{2}$, 0.4   (d) $\sqrt{6}$

3. (a) $0.8\overline{3}$ (b) 0.875

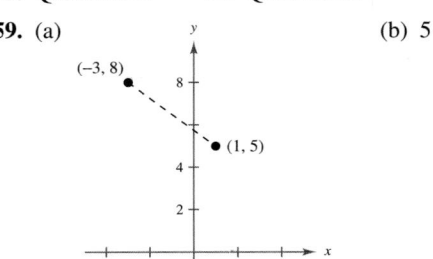

$\frac{5}{6} < \frac{7}{8}$

5. The set consists of all real numbers less than or equal to 7.

7. 155   **9.** $|x - 25| \le 10$ **11.** (a) -1 (b) -3

13. Associative Property of Addition

15. Additive Identity Property   **17.** -11 **19.** $\frac{1}{12}$

21. -144 **23.** Identity   **25.** 20   **27.** $-\frac{1}{2}$

29. $-\frac{60}{7}$ **31.** 9   **33.** $-\frac{5}{2}, 3$ **35.** $\pm\sqrt{2}$

37. $-4 \pm 3\sqrt{2}$ **39.** $6 \pm \sqrt{6}$ **41.** $-\dfrac{5}{4} \pm \dfrac{\sqrt{241}}{4}$

43. $0, \frac{12}{5}$ **45.** $\pm\sqrt{2}, \pm\sqrt{3}$ **47.** 5   **49.** No solution

51. $-124, 126$ **53.** $-5, 15$

55. Quadrant IV   **57.** Quadrant IV

59. (a) 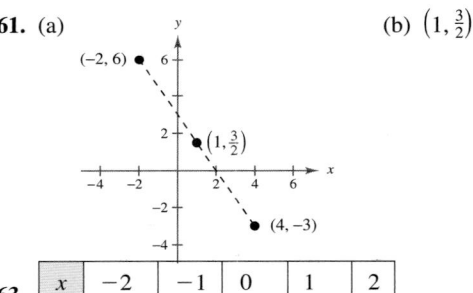 (b) 5

61. (a) (b) $\left(1, \frac{3}{2}\right)$

63.

x	-2	-1	0	1	2
y	-11	-8	-5	-2	1

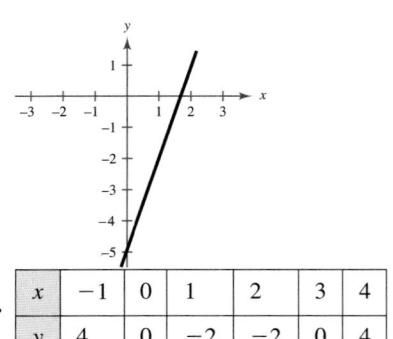

65.

x	-1	0	1	2	3	4
y	4	0	-2	-2	0	4

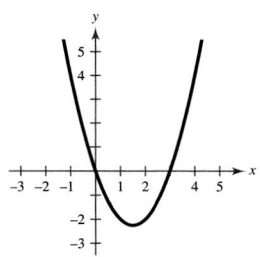

67.

x	−4	0	1	4	5
y	3	2.24	2	1	0

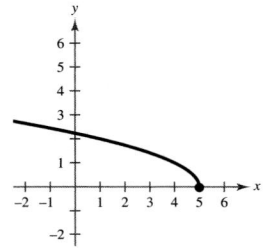

69. x-intercept: $\left(\frac{9}{2}, 0\right)$

y-intercept: $(0, -9)$

71. x-intercepts: $(0, 0), (3, 0), (-3, 0)$

y-intercept: $(0, 0)$

73. y-axis symmetry **75.** No symmetry

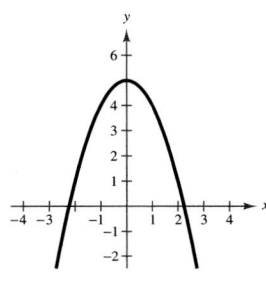 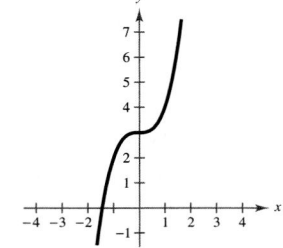

77. x-axis symmetry

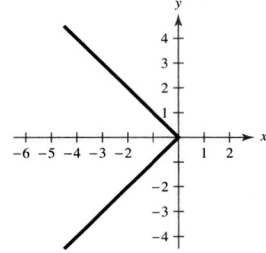

79. Center: $(0, 0)$; Radius: 3 **81.** Center: $(-2, 0)$; Radius: 4

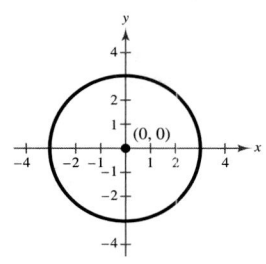 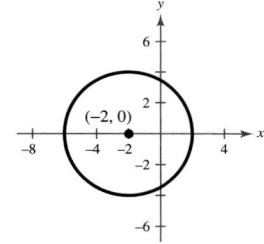

83. $(x - 2)^2 + (y + 3)^2 = 13$

85. **87.**

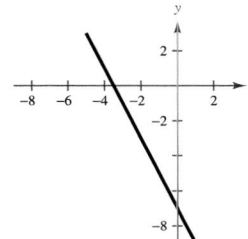 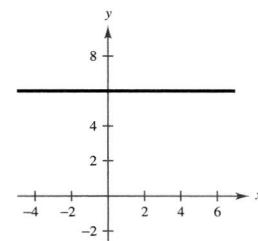

89. **91.**

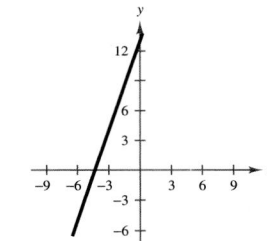 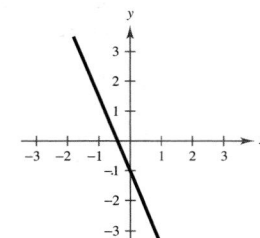

93. **95.**

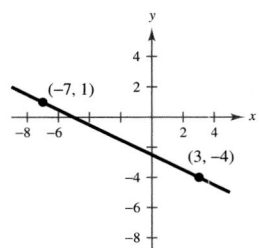 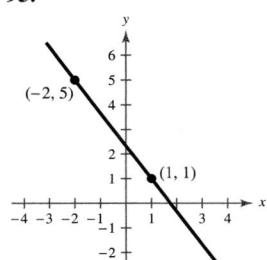

$m = -\frac{1}{2}$ $m = -\frac{4}{3}$

97.

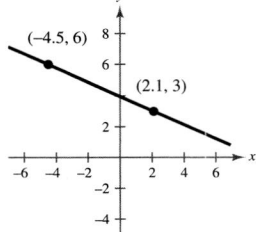

$m = -\frac{5}{11}$

99. $x = 0$ **101.** $4x + 3y - 8 = 0$

103. $4x + y + 10 = 0$

105. $3x - 2y - 10 = 0$ **107.** $x + 2y - 4 = 0$

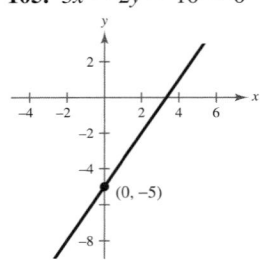   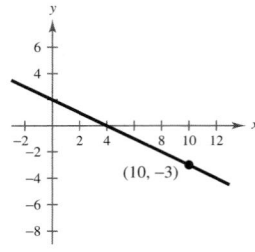

109. (a) $5x - 4y - 23 = 0$ (b) $4x + 5y - 2 = 0$

111. (a) $x = 4$ (b) $y = -1$

113. $V = 850t + 9100,\ 4 \le t \le 9$

115. $210,000

117. (a) Not a function, because 20 in the domain corresponds to two values in the range

(b) A function, because each input value has exactly one output value

(c) A function, because each input value has exactly one output value

(d) Not a function, because 30 in A is not matched with any element in B

119. No   **121.** Yes

123. (a) -3 (b) -1 (c) 2   (d) 6

125. $-5 \le x \le 5$ **127.** All real numbers $s \ne 3$

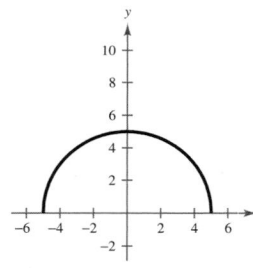   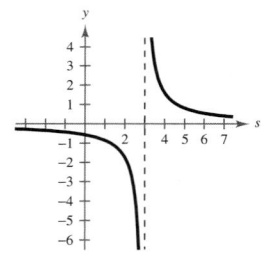

129. All real numbers $x \ne 3, -2$

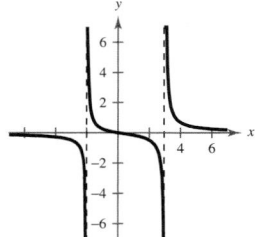

131. $4x + 2h + 3,\ h \ne 0$

133. (a) 16 feet per second   (b) 1.5 seconds

(c) -16 feet per second

135. (a) $A = x(12 - x)$ (b) $0 < x < 12$

137. Function   **139.** Not a function   **141.** $-1, \frac{1}{5}$

143. $\frac{7}{3}, 3$

145. Increasing on $(0, \infty)$

Decreasing on $(-\infty, -1)$

Constant on $[-1, 0]$

147. Neither even nor odd   **149.** Odd

151. $f(x) = -3x$ **153.** $f(x) = \frac{5}{3}x + \frac{10}{3}$

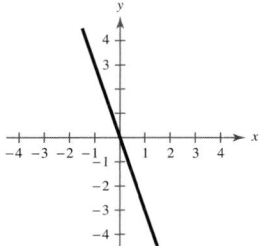   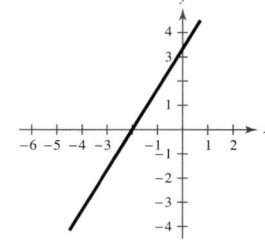

155.   **157.**

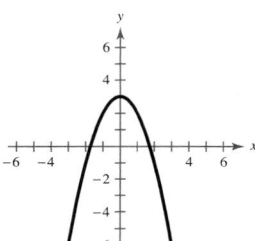   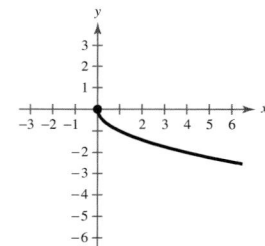

159.   **161.**

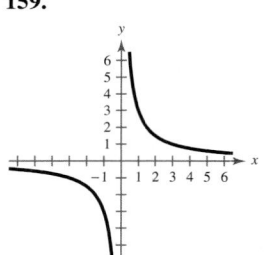   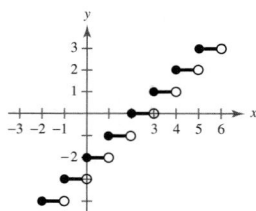

163.   **165.** $y = (x + 4)^3 + 4$

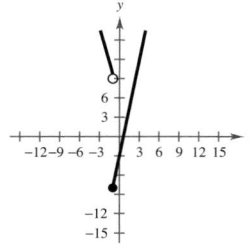

167. Vertical shift of nine units downward

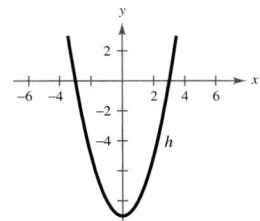

169. Horizontal shift of seven units to the right

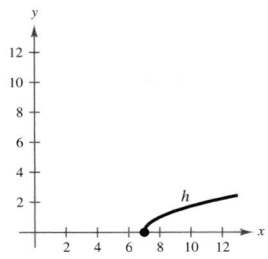

177. Horizontal shift of nine units to the right and vertical stretch

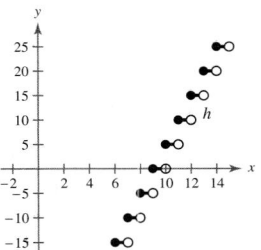

171. Reflection in the x-axis, horizontal shift of three units to the left, and vertical shift of one unit upward

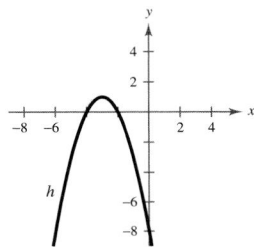

179. Reflection in the x-axis, vertical stretch, and horizontal shift of four units to the right

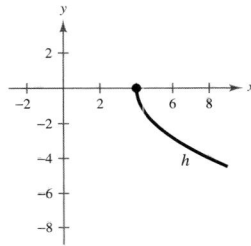

173. Reflection in the x-axis and vertical shift of six units upward

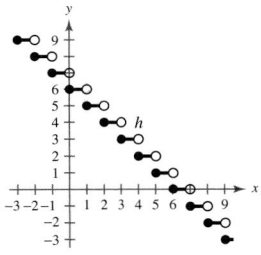

175. Reflections in the x-axis and the y-axis, horizontal shift of four units to the right, and vertical shift of six units upward

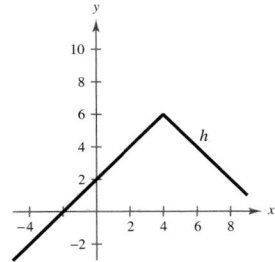

181. (a) $x^2 + 2x + 2$ (b) $x^2 - 2x + 4$

    (c) $2x^3 - x^2 + 6x - 3$ (d) $\dfrac{x^2 + 3}{2x - 1};\quad x \neq \dfrac{1}{2}$

183. (a) $x - \frac{8}{3}$ (b) $x - 8$

    Domain of $f, g, f \circ g,$ and $g \circ f$: all real numbers

185. $f(x) = x^3, g(x) = 6x - 5$

187. $y_1 = 0.207t^2 + 8.65t + 14.2$

    $y_2 = 1.414t^2 - 7.28t + 146.9$

189. $f^{-1}(x) = x + 7$

    $f(f^{-1}(x)) = x + 7 - 7 = x$

    $f^{-1}(f(x)) = x - 7 + 7 = x$

191. The function has an inverse.

193. The function has an inverse.

195. The function has an inverse.

197. (a) $f^{-1}(x) = 2x + 6$

    (b)

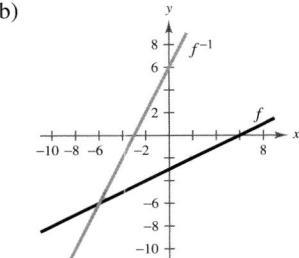

    (c) $f^{-1}(f(x)) = 2\left(\frac{1}{2}x - 3\right) + 6 = x - 6 + 6 = x$

    $f(f^{-1}(x)) = \frac{1}{2}(2x + 6) - 3 = x + 3 - 3 = x$

199. (a) $f^{-1}(x) = x^2 - 1, \; x \geq 0$

(b)

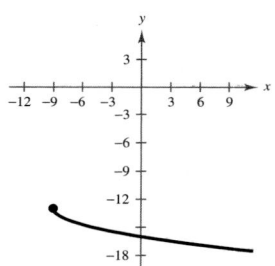

(c) $f^{-1}(f(x)) = f^{-1}\left(\sqrt{x + 1}\right)$
$= (x + 1) - 1$
$= x$

$f(f^{-1}(x)) = f(x^2 - 1), \; x \geq 0$
$= \sqrt{x^2 - 1 + 1}$
$= x$

201. $x \geq 4; \; f^{-1}(x) = \sqrt{\dfrac{x}{2} + 4}$

203. False. The graph is reflected in the x-axis, shifted 9 units to the left, and then shifted 13 units downward.

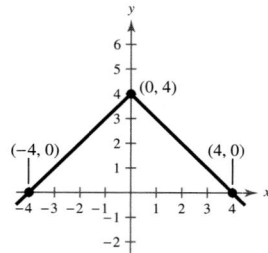

205. A function from a set A to a set B is a relation that assigns to each element x in the set A exactly one element y in the set B.

Chapter Test   *(page 120)*

1. $-\dfrac{10}{3} > -|-4|$ **2.** 9.15

3. Additive Identity Property

4. y-axis symmetry

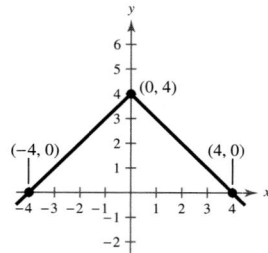

5. y-axis symmetry

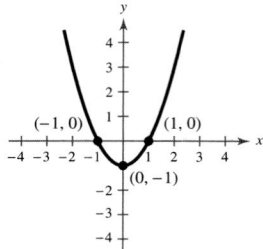

6. Center: $(3, 0)$
Radius: 3

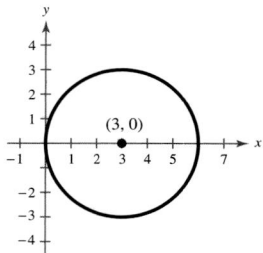

7. $\dfrac{128}{11}$ **8.** $-4, 5$ **9.** 4

10. $-2, \dfrac{8}{3}$

11. (a) $4x - 7y + 44 = 0$ (b) $7x + 4y - 53 = 0$

12. (a) -9 (b) 1   (c) $|x - 4| - 15$

13. $-10 \leq x \leq 10$

14. (a)

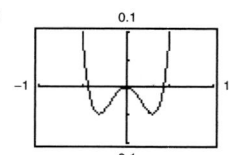

(b) Increasing on $(-0.31, 0), (0.31, \infty)$
Decreasing on $(-\infty, -0.31), (0, 0.31)$

(c) Even

15. (a)

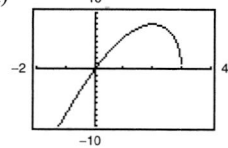

(b) Increasing on $(-\infty, 2)$
Decreasing on $(2, 3)$

(c) Neither even nor odd

16. (a)

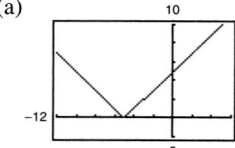

(b) Increasing on $(-5, \infty)$
Decreasing on $(-\infty, -5)$

(c) Neither even nor odd

17.

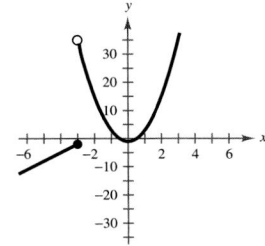

18. Reflection in the x-axis and vertical shift of $y = x^3$

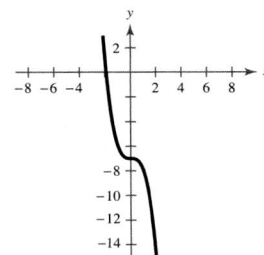

19. Reflection in the x-axis, horizontal shift, and vertical shift of $y = \sqrt{x}$

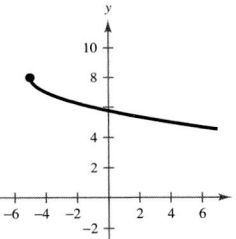

20. Vertical shrink, horizontal shift, and vertical shift of $y = |x|$

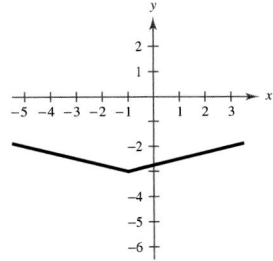

21. (a) $2x^2 - 4x - 2$ (b) $4x^2 + 4x - 12$

(c) $-3x^4 - 12x^3 + 22x^2 + 28x - 35$

(d) $\dfrac{3x^2 - 7}{-x^2 - 4x + 5}$, $x \neq 1, -5$

(e) $3x^4 + 24x^3 + 18x^2 - 120x + 68$

(f) $-9x^4 + 30x^2 - 16$

22. (a) $\dfrac{1 + 2x^{3/2}}{x}$, $x > 0$ (b) $\dfrac{1 - 2x^{3/2}}{x}$, $x > 0$

(c) $\dfrac{2\sqrt{x}}{x}$, $x > 0$ (d) $\dfrac{1}{2x^{3/2}}$, $x > 0$

(e) $\dfrac{\sqrt{x}}{2x}$, $x > 0$ (f) $\dfrac{2\sqrt{x}}{x}$, $x > 0$

23. $f^{-1}(x) = \sqrt[3]{x - 8}$ **24.** No inverse

25. $f^{-1}(x) = \left(\tfrac{1}{3}x\right)^{2/3}$, $x \geq 0$ **26.** $\$153$

Problem Solving   *(page 122)*

1. (a) $W_1 = 2000 + 0.07S$ (b) $W_2 = 2300 + 0.05S$

(c)

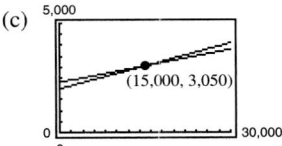

Both jobs pay the same monthly salary if sales equal $\$15,000$.

(d) No. Job 1 would pay $\$3400$ and job 2 would pay $\$3300$.

3. (a) The function will be even.

(b) The function will be odd if the two functions are not equal.

(c) The function will be neither even nor odd.

5. $f(x) = a_{2n}x^{2n} + a_{2n-2}x^{2n-2} + \cdots + a_2x^2 + a_0$

$f(-x) = a_{2n}(-x)^{2n} + a_{2n-2}(-x)^{2n-2}$

$+ \cdots + a_2(-x)^2 + a_0$

$= f(x)$

7. (a) $81\frac{2}{3}$ hours   (b) $25\frac{5}{7}$ miles per hour

(c) $y = \dfrac{-180}{7}x + 3400$

Domain: $0 \leq x \leq \frac{1190}{9}$

Range: $0 \leq y \leq 3400$

(d)

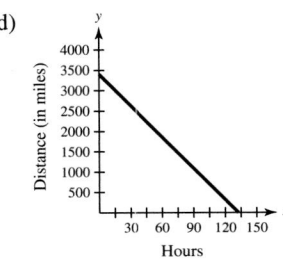

9. (a) $T = \dfrac{1}{2}\sqrt{4 + x^2} + \dfrac{1}{4}\sqrt{x^2 - 6x + 10}$

(b) $0 \le x \le 3$

(c)

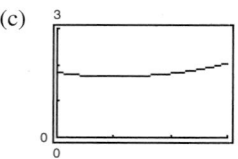

(d) $x = 1$

(e) The distance $x = 1$ yields a time of 1.68 hours.

11. (a) Domain: all real numbers $x \neq 1$

Range: all real numbers

(b) $f(f(x)) = \dfrac{x - 1}{x}$

Domain: all real numbers $x \neq 0, 1$

(c) $f(f(f(x))) = x$

The graph is not a line because there are holes at $x = 0$ and $x = 1$.

13. (a)
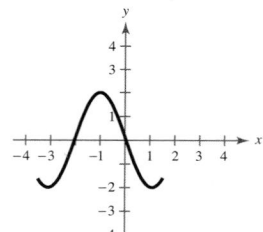

(b)
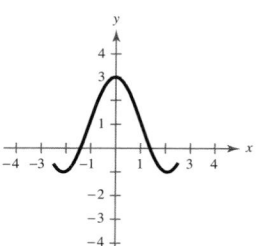

(c)
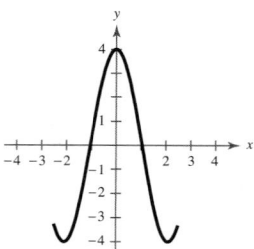

(d)
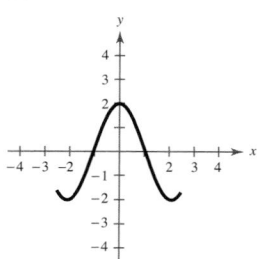

(e)
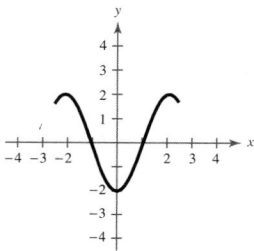

(f)
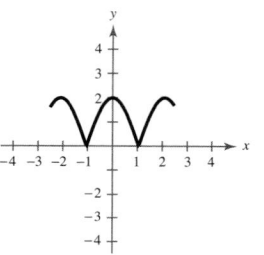

(g)

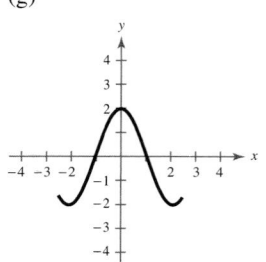

Chapter 1

Section 1.1 *(page 133)*

1. 2 radians **3.** -3 radians **5.** 1 radian

7. (a) Quadrant I (b) Quadrant III

9. (a) Quadrant IV (b) Quadrant III

11. (a) Quadrant III (b) Quadrant II

13. (a)

(b)

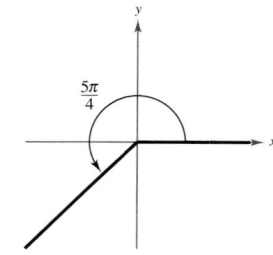

15. (a)

(b)

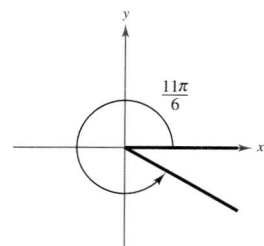

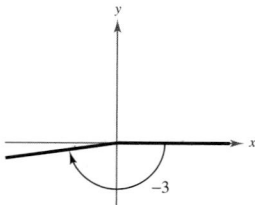

17. (a) $\dfrac{13\pi}{6}, \; -\dfrac{11\pi}{6}$ (b) $\dfrac{17\pi}{6}, \; -\dfrac{7\pi}{6}$

19. (a) $\dfrac{8\pi}{3}, \; -\dfrac{4\pi}{3}$ (b) $\dfrac{25\pi}{12}, \; -\dfrac{23\pi}{12}$

21. (a) Complement: $\dfrac{\pi}{6}$; Supplement: $\dfrac{2\pi}{3}$

(b) Complement: none; Supplement: $\dfrac{\pi}{4}$

23. (a) Complement: $\dfrac{\pi}{2} - 1 \approx 0.57$;

Supplement: $\pi - 1 \approx 2.14$

(b) Complement: none; Supplement: $\pi - 2 \approx 1.14$

25. (a) $\dfrac{\pi}{6}$ (b) $\dfrac{5\pi}{6}$ **27.** (a) $-\dfrac{\pi}{9}$ (b) $-\dfrac{4\pi}{3}$

29. 2.007 **31.** -3.776 **33.** 9.285 **35.** -0.014

37. (a) $270°$ (b) $210°$ **39.** (a) $420°$ (b) $-66°$

41. $25.714°$ **43.** $337.500°$ **45.** $-756.000°$

47. $-114.592°$ **49.** $210°$ **51.** $-60°$ **53.** $165°$

55. (a) Quadrant II (b) Quadrant IV

57. (a) Quadrant III (b) Quadrant I

59. (a) (b)

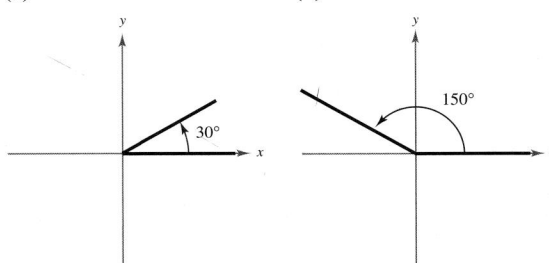

61. (a) (b)

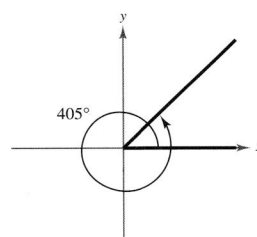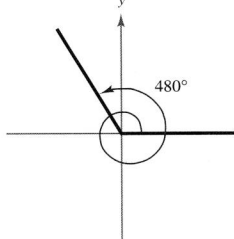

63. (a) $405°, -315°$ (b) $324°, -396°$

65. (a) $600°, -120°$ (b) $180°, -540°$

67. (a) Complement: $72°$; Supplement: $162°$

(b) Complement: none; Supplement: $65°$

69. (a) Complement: $11°$; Supplement: $101°$

(b) Complement: none; Supplement: $30°$

71. (a) $54.75°$ (b) $-128.5°$

73. (a) $85.308°$ (b) $330.007°$

75. (a) $240° 36'$ (b) $-145° 48'$

77. (a) $2° 30'$ (b) $-3° 34' 48''$

79. $\frac{6}{5}$ radians **81.** $\frac{32}{7}$ radians

83. $\frac{2}{9}$ radian **85.** $\frac{50}{29}$ radians

87. 15π inches ≈ 47.12 inches **89.** 3 meters

91. 591.7 miles **93.** 1141.0 miles

95. 0.071 radian $\approx 4.04°$ **97.** $\frac{5}{12}$ radian

99. (a) 728.3 revolutions per minute

(b) 4576 radians per minute

101. (a) $\dfrac{14\pi}{3}$ feet per second; ≈ 10 miles per hour

(b) $d = \dfrac{7\pi}{7920}n$ (c) $d = \dfrac{7\pi}{7920}t$

(d) The functions are both linear.

103. False. A measurement of 4π radians corresponds to two complete revolutions from the initial to the terminal side of an angle.

105. False. The terminal side of the angle lies on the x-axis.

107. Increases. The linear velocity is proportional to the radius.

109. The arc length increases. If θ is constant, the length of the arc is proportional to the radius ($s = r\theta$).

111. **113.**

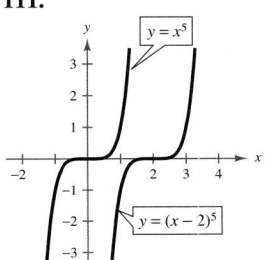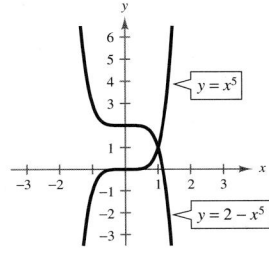

115. $\dfrac{\sqrt{2}}{2}$ **117.** $\sqrt{2}$ **119.** $2\sqrt{10}$ **121.** $12\sqrt{2}$

Section 1.2 *(page 142)*

1. $\sin\theta = \frac{15}{17}$ **3.** $\sin\theta = -\frac{5}{13}$

$\cos\theta = -\frac{8}{17}$ $\cos\theta = \frac{12}{13}$

$\tan\theta = -\frac{15}{8}$ $\tan\theta = -\frac{5}{12}$

$\csc\theta = \frac{17}{15}$ $\csc\theta = -\frac{13}{5}$

$\sec\theta = -\frac{17}{8}$ $\sec\theta = \frac{13}{12}$

$\cot\theta = -\frac{8}{15}$ $\cot\theta = -\frac{12}{5}$

5. $\left(\dfrac{\sqrt{2}}{2}, \dfrac{\sqrt{2}}{2}\right)$ **7.** $\left(-\dfrac{\sqrt{3}}{2}, -\dfrac{1}{2}\right)$

9. $\left(-\dfrac{1}{2}, -\dfrac{\sqrt{3}}{2}\right)$ **11.** $(0, -1)$

13. $\sin\dfrac{\pi}{4} = \dfrac{\sqrt{2}}{2}$ **15** $\sin\left(-\dfrac{\pi}{6}\right) = -\dfrac{1}{2}$

$\cos\dfrac{\pi}{4} = \dfrac{\sqrt{2}}{2}$ $\cos\left(-\dfrac{\pi}{6}\right) = \dfrac{\sqrt{3}}{2}$

$\tan\dfrac{\pi}{4} = 1$ $\tan\left(-\dfrac{\pi}{6}\right) = -\dfrac{\sqrt{3}}{3}$

17. $\sin\left(-\dfrac{7\pi}{4}\right) = \dfrac{\sqrt{2}}{2}$

$\cos\left(-\dfrac{7\pi}{4}\right) = \dfrac{\sqrt{2}}{2}$

$\tan\left(-\dfrac{7\pi}{4}\right) = 1$

19 $\sin\dfrac{11\pi}{6} = -\dfrac{1}{2}$

$\cos\dfrac{11\pi}{6} = \dfrac{\sqrt{3}}{2}$

$\tan\dfrac{11\pi}{6} = -\dfrac{\sqrt{3}}{3}$

21. $\sin\left(-\dfrac{3\pi}{2}\right) = 1$

$\cos\left(-\dfrac{3\pi}{2}\right) = 0$

$\tan\left(-\dfrac{3\pi}{2}\right)$ is undefined.

23. $\sin\dfrac{3\pi}{4} = \dfrac{\sqrt{2}}{2}$ $\qquad$ $\csc\dfrac{3\pi}{4} = \sqrt{2}$

$\cos\dfrac{3\pi}{4} = -\dfrac{\sqrt{2}}{2}$ $\qquad$ $\sec\dfrac{3\pi}{4} = -\sqrt{2}$

$\tan\dfrac{3\pi}{4} = -1$ $\qquad$ $\cot\dfrac{3\pi}{4} = -1$

25. $\sin\left(-\dfrac{\pi}{2}\right) = -1$ $\qquad$ $\csc\left(-\dfrac{\pi}{2}\right) = -1$

$\cos\left(-\dfrac{\pi}{2}\right) = 0$ $\qquad$ $\sec\left(-\dfrac{\pi}{2}\right)$ is undefined.

$\tan\left(-\dfrac{\pi}{2}\right)$ is undefined. $\quad$ $\cot\left(-\dfrac{\pi}{2}\right) = 0$

27. $\sin\dfrac{4\pi}{3} = -\dfrac{\sqrt{3}}{2}$ $\qquad$ $\csc\dfrac{4\pi}{3} = -\dfrac{2\sqrt{3}}{3}$

$\cos\dfrac{4\pi}{3} = -\dfrac{1}{2}$ $\qquad$ $\sec\dfrac{4\pi}{3} = -2$

$\tan\dfrac{4\pi}{3} = \sqrt{3}$ $\qquad$ $\cot\dfrac{4\pi}{3} = \dfrac{\sqrt{3}}{3}$

29. $\sin 5\pi = \sin \pi = 0$ $\qquad$ **31.** $\cos\dfrac{8\pi}{3} = \cos\dfrac{2\pi}{3} = -\dfrac{1}{2}$

33. $\cos\left(-\dfrac{15\pi}{2}\right) = \cos\dfrac{\pi}{2} = 0$

35. $\sin\left(-\dfrac{9\pi}{4}\right) = \sin\dfrac{7\pi}{4} = -\dfrac{\sqrt{2}}{2}$

37. (a) $-\dfrac{1}{3}$ (b) -3 $\qquad$ **39.** (a) $-\dfrac{1}{5}$ (b) -5

41. (a) $\dfrac{4}{5}$ (b) $-\dfrac{4}{5}$ $\qquad$ **43.** 0.7071 $\qquad$ **45.** 1.0378

47. -0.1288 $\qquad$ **49.** 1.3940 $\qquad$ **51.** -1.4486

53. (a) -1 (b) -0.4

55. (a) 0.25 or 2.89 (b) 1.82 or 4.46

57. (a)

t	0	$\frac{1}{2}$	1	$\frac{3}{2}$	2
y	0	1.1481	2.1213	2.7716	3

(b) $y = 3$ when $t = 2$. $\qquad$ (c) $t = 4$

59. False. $\sin(-t) = -\sin t$ means that the function is odd, not that the sine of a negative angle is a negative number.

61. (a) y-axis symmetry $\qquad$ (b) $\sin t_1 = \sin(\pi - t_1)$

(c) $\cos(\pi - t_1) = -\cos t_1$

63. $\sin(0.25) + \sin(0.75) = 0.9290 \neq \sin 1 = 0.8415$

65. $f^{-1}(x) = \frac{2}{3}(x + 1)$ $\qquad$ **67.** $f^{-1}(x) = \sqrt{x^2 + 4}, \quad x \geq 0$

Section 1.3 $\quad$ (page 151)

1. $\sin \theta = \frac{3}{5}$ $\qquad$ $\csc \theta = \frac{5}{3}$

$\cos \theta = \frac{4}{5}$ $\qquad$ $\sec \theta = \frac{5}{4}$

$\tan \theta = \frac{3}{4}$ $\qquad$ $\cot \theta = \frac{4}{3}$

3. $\sin \theta = \frac{9}{41}$ $\qquad$ $\csc \theta = \frac{41}{9}$

$\cos \theta = \frac{40}{41}$ $\qquad$ $\sec \theta = \frac{41}{40}$

$\tan \theta = \frac{9}{40}$ $\qquad$ $\cot \theta = \frac{40}{9}$

5. $\sin \theta = \dfrac{1}{3}$ $\qquad$ $\csc \theta = 3$

$\cos \theta = \dfrac{2\sqrt{2}}{3}$ $\qquad$ $\sec \theta = \dfrac{3\sqrt{2}}{4}$

$\tan \theta = \dfrac{\sqrt{2}}{4}$ $\qquad$ $\cot \theta = 2\sqrt{2}$

The triangles are similar, and corresponding sides are proportional.

7. $\sin \theta = \frac{3}{5}$ $\qquad$ $\csc \theta = \frac{5}{3}$

$\cos \theta = \frac{4}{5}$ $\qquad$ $\sec \theta = \frac{5}{4}$

$\tan \theta = \frac{3}{4}$ $\qquad$ $\cot \theta = \frac{4}{3}$

The triangles are similar, and corresponding sides are proportional.

9.

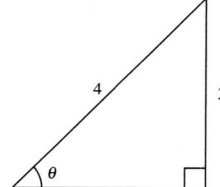

$\cos \theta = \dfrac{\sqrt{7}}{4}$ $\qquad$ $\sec \theta = \dfrac{4\sqrt{7}}{7}$

$\tan \theta = \dfrac{3\sqrt{7}}{7}$ $\qquad$ $\cot \theta = \dfrac{\sqrt{7}}{3}$

$\csc \theta = \dfrac{4}{3}$

11.

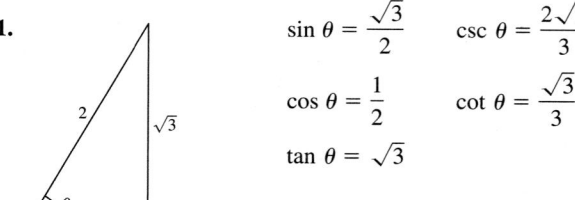

$\sin \theta = \dfrac{\sqrt{3}}{2}$ $\qquad$ $\csc \theta = \dfrac{2\sqrt{3}}{3}$

$\cos \theta = \dfrac{1}{2}$ $\qquad$ $\cot \theta = \dfrac{\sqrt{3}}{3}$

$\tan \theta = \sqrt{3}$

13.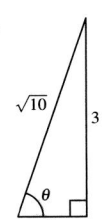

$$\sin \theta = \frac{3\sqrt{10}}{10} \qquad \sec \theta = \sqrt{10}$$

$$\cos \theta = \frac{\sqrt{10}}{10} \qquad \cot \theta = \frac{1}{3}$$

$$\csc \theta = \frac{\sqrt{10}}{3}$$

15.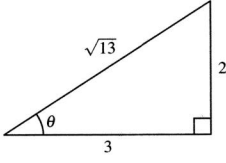

$$\sin \theta = \frac{2\sqrt{13}}{13}$$

$$\cos \theta = \frac{3\sqrt{13}}{13}$$

$$\tan \theta = \frac{2}{3}$$

$$\csc \theta = \frac{\sqrt{13}}{2}$$

$$\sec \theta = \frac{\sqrt{13}}{3}$$

17. (a) $\sqrt{3}$ (b) $\frac{1}{2}$ (c) $\frac{\sqrt{3}}{2}$ (d) $\frac{\sqrt{3}}{3}$

19. (a) $\frac{2\sqrt{13}}{13}$ (b) $\frac{3\sqrt{13}}{13}$ (c) $\frac{2}{3}$ (d) $\frac{\sqrt{13}}{2}$

21. (a) 3 (b) $\frac{2\sqrt{2}}{3}$ (c) $\frac{\sqrt{2}}{4}$ (d) $\frac{1}{3}$

23. (a) $\frac{1}{2}$ (b) 2 (c) $\sqrt{3}$

25. (a) $\frac{\sqrt{2}}{2}$ (b) $\frac{\sqrt{3}}{2}$ (c) $\frac{\sqrt{3}}{3}$

27. (a) 0.1736 (b) 0.1736

29. (a) 0.2815 (b) 3.5523

31. (a) 1.3499 (b) 1.3432

33. (a) 5.0273 (b) 0.1989

35. (a) 1.8527 (b) 0.9817

37. (a) $30° = \frac{\pi}{6}$ (b) $30° = \frac{\pi}{6}$

39. (a) $60° = \frac{\pi}{3}$ (b) $45° = \frac{\pi}{4}$

41. (a) $60° = \frac{\pi}{3}$ (b) $45° = \frac{\pi}{4}$

43. (a) $0.83° \approx 0.015$ (b) $27° \approx 0.474$

45. (a) $0.72° \approx 0.012$ (b) $67° \approx 1.169$

47–55. Answers will vary. **57.** $30\sqrt{3}$ **59.** $\frac{32\sqrt{3}}{3}$

61. (a)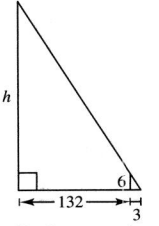

Not drawn to scale

(b) $\cos \theta = \frac{6}{3} = \frac{h}{135}$

(c) 270 feet

63. (a)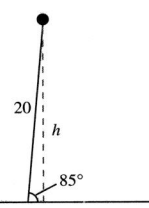

(b) $\sin 85° = \frac{h}{20}$ (c) 19.9 meters

(d) The side of the triangle labeled h will become shorter.

(e)

Angle, θ	80°	70°	60°	50°
Height	19.7	18.8	17.3	15.3

Angle, θ	40°	30°	20°	10°
Height	12.9	10.0	6.8	3.5

(f) As $\theta \to 0°$, $h \to 0$.

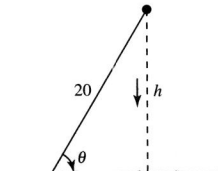

65. 137.6 feet **67.** $(x_1, y_1) = (28\sqrt{3}, 28)$
$(x_2, y_2) = (28, 28\sqrt{3})$

69. $\sin 20° \approx 0.34$
$\cos 20° \approx 0.94$
$\tan 20° \approx 0.36$
$\csc 20° \approx 2.92$
$\sec 20° \approx 1.06$
$\cot 20° \approx 2.75$

71. True, $\csc x = \frac{1}{\sin x}$. **73.** False, $\frac{\sqrt{2}}{2} + \frac{\sqrt{2}}{2} \neq 1$.

75. False, $1.7321 \neq 0.0349$.

77. Corresponding sides of similar triangles are proportional.

79. (a)

θ	0.1	0.2	0.3	0.4	0.5
$\sin \theta$	0.0998	0.1987	0.2955	0.3894	0.4794

(b) As θ approaches 0, $\sin \theta$ approaches θ.

81. $\dfrac{x}{x-2}$, $x \neq \pm 6$ **83.** $\dfrac{2(x^2 - 5x - 10)}{(x-2)(x+2)^2}$ **85.** $x = \dfrac{2}{3}$

Section 1.4   *(page 161)*

1. (a) $\sin \theta = \dfrac{3}{5}$ (b) $\sin \theta = -\dfrac{15}{17}$
$\cos \theta = \dfrac{4}{5}$ $\cos \theta = \dfrac{8}{17}$
$\tan \theta = \dfrac{3}{4}$ $\tan \theta = -\dfrac{15}{8}$
$\csc \theta = \dfrac{5}{3}$ $\csc \theta = -\dfrac{17}{15}$
$\sec \theta = \dfrac{5}{4}$ $\sec \theta = \dfrac{17}{8}$
$\cot \theta = \dfrac{4}{3}$ $\cot \theta = -\dfrac{8}{15}$

3. (a) $\sin \theta = -\dfrac{1}{2}$ (b) $\sin \theta = \dfrac{\sqrt{17}}{17}$
$\cos \theta = -\dfrac{\sqrt{3}}{2}$ $\cos \theta = -\dfrac{4\sqrt{17}}{17}$
$\tan \theta = \dfrac{\sqrt{3}}{3}$ $\tan \theta = -\dfrac{1}{4}$
$\csc \theta = -2$ $\csc \theta = \sqrt{17}$
$\sec \theta = -\dfrac{2\sqrt{3}}{3}$ $\sec \theta = -\dfrac{\sqrt{17}}{4}$
$\cot \theta = \sqrt{3}$ $\cot \theta = -4$

5. $\sin \theta = \dfrac{24}{25}$ $\csc \theta = \dfrac{25}{24}$
$\cos \theta = \dfrac{7}{25}$ $\sec \theta = \dfrac{25}{7}$
$\tan \theta = \dfrac{24}{7}$ $\cot \theta = \dfrac{7}{24}$

7. $\sin \theta = \dfrac{5\sqrt{29}}{29}$ $\csc \theta = \dfrac{\sqrt{29}}{5}$
$\cos \theta = -\dfrac{2\sqrt{29}}{29}$ $\sec \theta = -\dfrac{\sqrt{29}}{2}$
$\tan \theta = -\dfrac{5}{2}$ $\cot \theta = -\dfrac{2}{5}$

9.
$\sin \theta \approx \dfrac{68\sqrt{5849}}{5849} \approx 0.9$ $\csc \theta = \dfrac{\sqrt{5849}}{68} \approx 1.1$

$\cos \theta \approx -\dfrac{35\sqrt{5849}}{5849} \approx -0.5$ $\sec \theta = -\dfrac{\sqrt{5849}}{35} \approx -2.2$

$\tan \theta \approx -\dfrac{68}{35} \approx -1.9$ $\cot \theta \approx -\dfrac{35}{68} \approx -0.5$

11. Quadrant III   **13.** Quadrant II

15. $\sin \theta = \dfrac{3}{5}$ $\csc \theta = \dfrac{5}{3}$
$\cos \theta = -\dfrac{4}{5}$ $\sec \theta = -\dfrac{5}{4}$
$\tan \theta = -\dfrac{3}{4}$ $\cot \theta = -\dfrac{4}{3}$

17. $\sin \theta = -\dfrac{15}{17}$ $\csc \theta = -\dfrac{17}{15}$
$\cos \theta = \dfrac{8}{17}$ $\sec \theta = \dfrac{17}{8}$
$\tan \theta = -\dfrac{15}{8}$ $\cot \theta = -\dfrac{8}{15}$

19. $\sin \theta = -\dfrac{\sqrt{10}}{10}$ $\csc \theta = -\sqrt{10}$
$\cos \theta = \dfrac{3\sqrt{10}}{10}$ $\sec \theta = \dfrac{\sqrt{10}}{3}$
$\tan \theta = -\dfrac{1}{3}$ $\cot \theta = -3$

21. $\sin \theta = \dfrac{\sqrt{3}}{2}$ $\csc \theta = \dfrac{2\sqrt{3}}{3}$
$\cos \theta = -\dfrac{1}{2}$ $\sec \theta = -2$
$\tan \theta = -\sqrt{3}$ $\cot \theta = -\dfrac{\sqrt{3}}{3}$

23. $\sin \theta = 0$ $\csc \theta$ is undefined.
$\cos \theta = -1$ $\sec \theta = -1$
$\tan \theta = 0$ $\cot \theta$ is undefined.

25. $\sin \theta = \dfrac{\sqrt{2}}{2}$ $\csc \theta = \sqrt{2}$
$\cos \theta = -\dfrac{\sqrt{2}}{2}$ $\sec \theta = -\sqrt{2}$
$\tan \theta = -1$ $\cot \theta = -1$

27. $\sin \theta = -\dfrac{2\sqrt{5}}{5}$ $\csc \theta = -\dfrac{\sqrt{5}}{2}$
$\cos \theta = -\dfrac{\sqrt{5}}{5}$ $\sec \theta = -\sqrt{5}$
$\tan \theta = 2$ $\cot \theta = \dfrac{1}{2}$

29. -1 **31.** Undefined

33. Undefined   **35.** Undefined

37. $\theta' = 23°$ **39.** $\theta' = 65°$

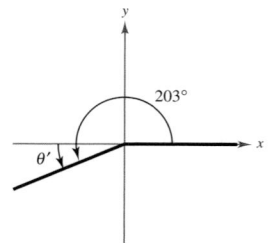

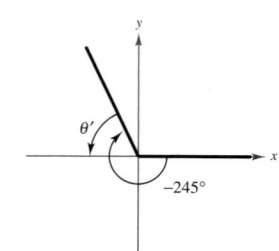

41. $\theta' = \dfrac{\pi}{3}$

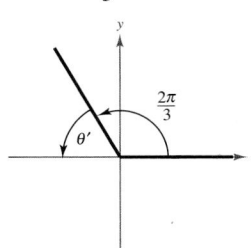

43. $\theta' = 3.5 - \pi$

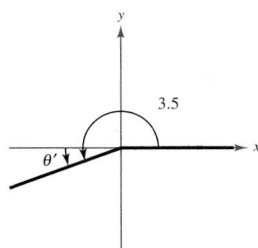

45. $\sin 225° = -\dfrac{\sqrt{2}}{2}$

$\cos 225° = -\dfrac{\sqrt{2}}{2}$

$\tan 225° = 1$

47. $\sin 750° = \dfrac{1}{2}$

$\cos 750° = \dfrac{\sqrt{3}}{2}$

$\tan 750° = \dfrac{\sqrt{3}}{3}$

49. $\sin(-150°) = -\dfrac{1}{2}$

$\cos(-150°) = -\dfrac{\sqrt{3}}{2}$

$\tan(-150°) = \dfrac{\sqrt{3}}{3}$

51. $\sin \dfrac{4\pi}{3} = -\dfrac{\sqrt{3}}{2}$

$\cos \dfrac{4\pi}{3} = -\dfrac{1}{2}$

$\tan \dfrac{4\pi}{3} = \sqrt{3}$

53. $\sin\left(-\dfrac{\pi}{6}\right) = -\dfrac{1}{2}$

$\cos\left(-\dfrac{\pi}{6}\right) = \dfrac{\sqrt{3}}{2}$

$\tan\left(-\dfrac{\pi}{6}\right) = -\dfrac{\sqrt{3}}{3}$

55. $\sin \dfrac{11\pi}{4} = \dfrac{\sqrt{2}}{2}$

$\cos \dfrac{11\pi}{4} = -\dfrac{\sqrt{2}}{2}$

$\tan \dfrac{11\pi}{4} = -1$

57. $\sin\left(-\dfrac{3\pi}{2}\right) = 1$

$\cos\left(-\dfrac{3\pi}{2}\right) = 0$

$\tan\left(-\dfrac{3\pi}{2}\right)$ is undefined.

59. 0.1736 **61.** -0.3420

63. 4.6373 **65.** 0.3640 **67.** -0.6052

69. (a) $30° = \dfrac{\pi}{6}$, $150° = \dfrac{5\pi}{6}$ (b) $210° = \dfrac{7\pi}{6}$, $330° = \dfrac{11\pi}{6}$

71. (a) $60° = \dfrac{\pi}{3}$, $120° = \dfrac{2\pi}{3}$ (b) $135° = \dfrac{3\pi}{4}$, $315° = \dfrac{7\pi}{4}$

73. (a) $45° = \dfrac{\pi}{4}$, $225° = \dfrac{5\pi}{4}$ (b) $150° = \dfrac{5\pi}{6}$, $330° = \dfrac{11\pi}{6}$

75. $54.99°$, $125.01°$ **77.** $115.89°$, $244.11°$

79. $0.175, 6.109$ **81.** $0.873, 4.014$ **83.** $1.955, 4.328$

85. $\dfrac{4}{5}$ **87.** $-\dfrac{\sqrt{13}}{2}$ **89.** $\dfrac{8}{5}$

91. (a) $N = 22.66 \sin(0.51t - 2.12) + 54.58$

$F = 37.18 \sin(0.51t - 1.91) + 26.17$

(b) February: $N = 34°$, $F = -3°$

March: $N = 42°$, $F = 12°$

May: $N = 64°$, $F = 48°$

June: $N = 73°$, $F = 60°$

August: $N = 76°$, $F = 57°$

September: $N = 69°$, $F = 43°$

November: $N = 47°$, $F = 6°$

(c) Answers will vary.

93. (a) 2 centimeters

(b) 0.14 centimeter

(c) -1.98 centimeters

95. False. In each of the four quadrants, the signs of the secant function and cosine function will be the same, because these functions are reciprocals of each other.

97. As θ increases from $0°$ to $90°$, x decreases from 12 cm to 0 cm and y increases from 0 cm to 12 cm. Therefore, $\sin \theta = y/12$ increases from 0 to 1 and $\cos \theta = x/12$ decreases from 1 to 0. Thus, $\tan \theta = y/x$ and increases without bound. When $\theta = 90°$, the tangent is undefined.

99.

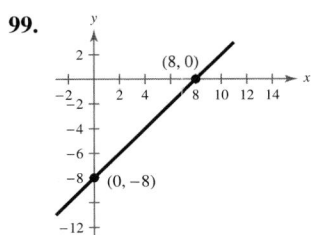

x-intercept: $(8, 0)$

y-intercept: $(0, -8)$

Domain: All real numbers

101.

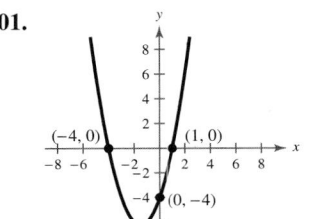

x-intercepts:

$(1, 0), (-4, 0)$

y-intercept: $(0, -4)$

Domain: All real numbers

103.

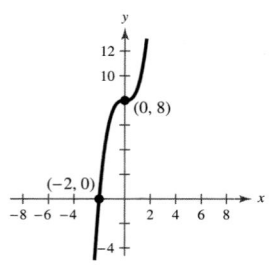

x-intercept: $(-2, 0)$

y-intercept: $(0, 8)$

Domain: All real numbers

105.

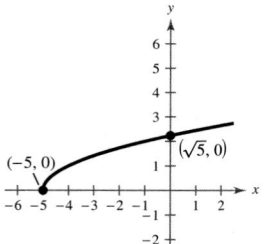

x-intercept: $(-5, 0)$

y-intercept: $(0, \sqrt{5})$

Domain: All real numbers,
   $x \geq -5$

Section 1.5   (page 171)

1. Period: π **3.** Period: 4π **5.** Period: 6

Amplitude: 3   Amplitude: $\frac{5}{2}$ Amplitude: $\frac{1}{2}$

7. Period: 2π

Amplitude: 2

9. Period: $\dfrac{\pi}{5}$

Amplitude: 3

11. Period: 3π **13.** Period: 1

Amplitude: $\frac{1}{2}$ Amplitude: $\frac{1}{4}$

15. g is a shift of f π units to the right.

17. g is a reflection of f in the x-axis.

19. The period of f is twice the period of g.

21. g is a shift of f three units upward.

23. The graph of g has twice the amplitude of the graph of f.

25. The graph of g is a horizontal shift of the graph of f π units to the right.

27.

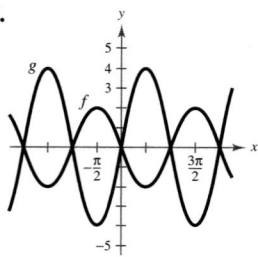

29.

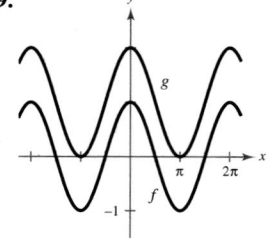

31.

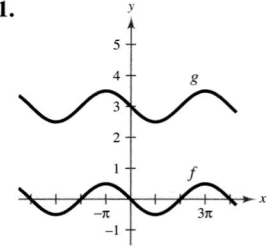

33.

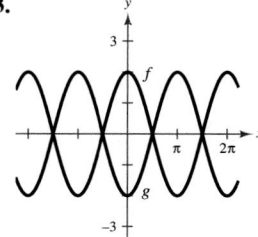

35.

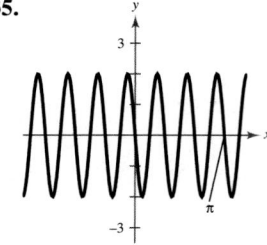

37.

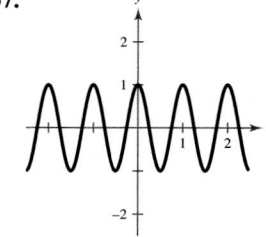

39.

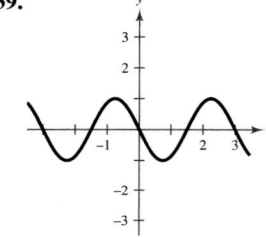

41.

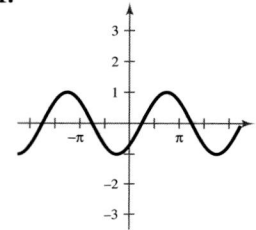

43.

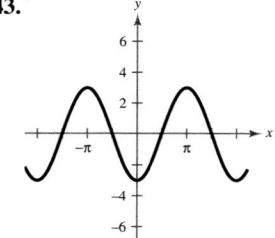

45.

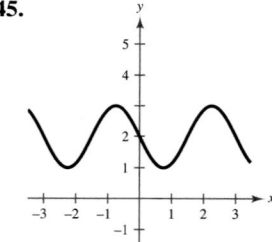

47.

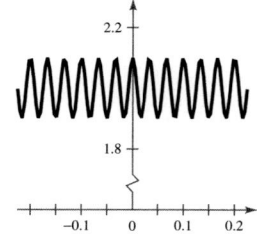

49.

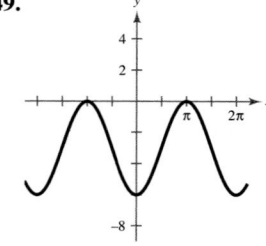

51.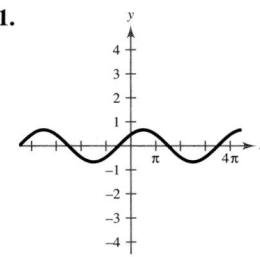

53.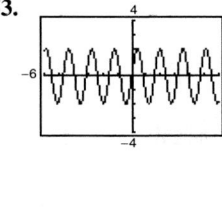

55.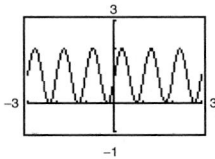

57.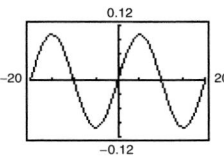

59. $a = 2, d = 1$ **61.** $a = -4, d = 4$

63. $a = -3, b = 2, c = 0$ **65.** $a = 2, b = 1, c = -\dfrac{\pi}{4}$

67.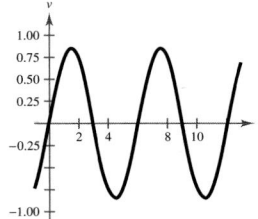

$x = -\dfrac{\pi}{6}, -\dfrac{5\pi}{6}, \dfrac{7\pi}{6}, \dfrac{11\pi}{6}$

69. (a) 6 seconds   (b) 10 cycles per minute

(c)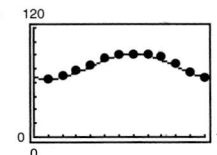

71. (a) $\dfrac{1}{440}$ second   (b) 440 cycles per second

73. (a) $C(t) = 56.35 + 27.35 \cos\left(\dfrac{\pi t}{6} - 3.67\right)$

(b)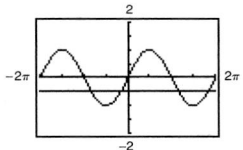

The model is a good fit.

(c)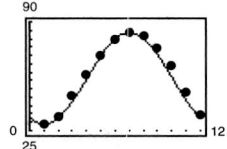

The model is a good fit.

(d) Tallahassee: 77.60°; Chicago: 56.35°
The constant term gives annual average temperature.

(e) 12. Yes. One full period is 1 year.

(f) Chicago; amplitude; The greater the amplitude, the greater the variability in the temperature.

75. (a) 365. Yes. One year is 365 days.

(b) 30.3 gallons; the constant term

(c)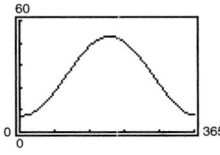

$124 < t < 252$

77. False. The function $y = \frac{1}{2} \cos 2x$ has an amplitude that is one-half that of $y = \cos x$. For $y = a \cos bx$, the amplitude is $|a|$.

79. 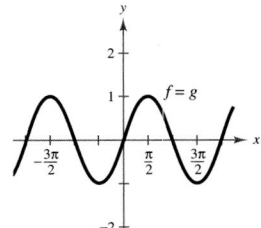   Conjecture:

$\sin x = \cos\left(x - \dfrac{\pi}{2}\right)$

81.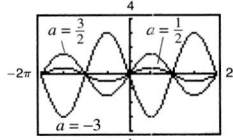

Amplitude changes

83.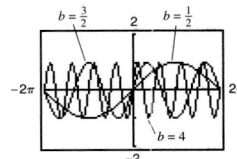

Period changes

85. (a)

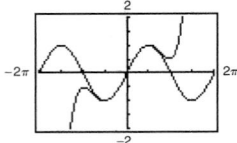

The graphs appear to coincide from $-\dfrac{\pi}{2}$ to $\dfrac{\pi}{2}$.

(b)

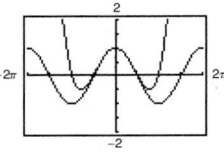

The graphs appear to coincide from $-\dfrac{\pi}{2}$ to $\dfrac{\pi}{2}$.

(c) $-\dfrac{x^7}{7!}, \ -\dfrac{x^6}{6!}$

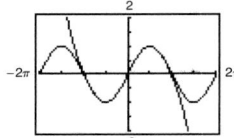

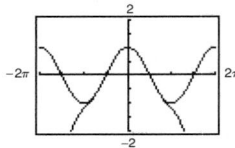

The accuracy increased.

87. $\dfrac{4}{x(1-x)}$ **89.** $\dfrac{3x-2}{x(x-1)}$

91. All real numbers $x \neq 11$ **93.** $-9 \le x \le 9$

Section 1.6   *(page 182)*

1. e, π **2.** c, 2π **3.** a, 1   **4.** d, 2π

5. f, 4   **6.** b, 4

7.

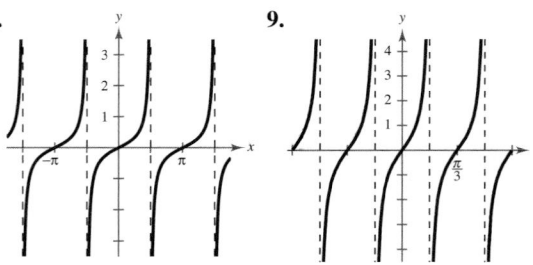

9.

11.

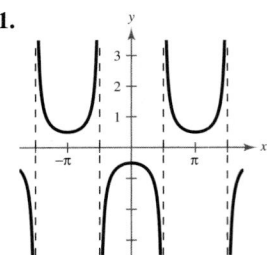

13.

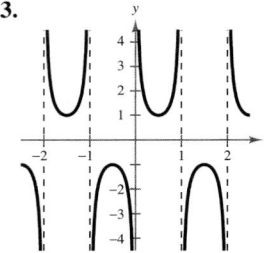

15.

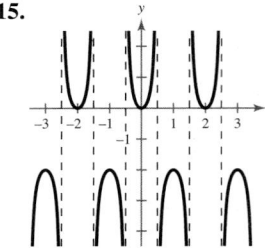

17.

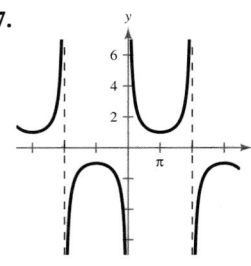

19.

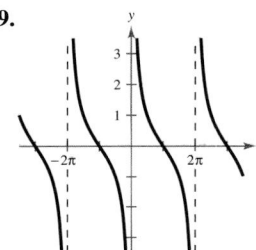

21.

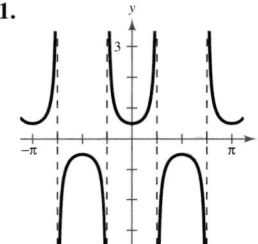

23.

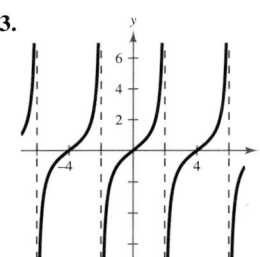

25.

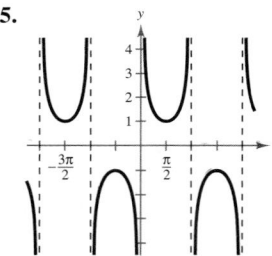

27.

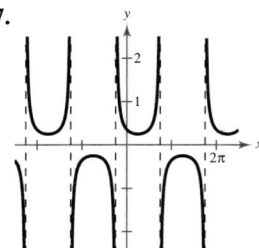

29.

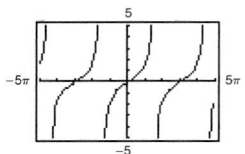

31.

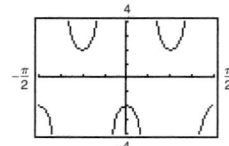

33.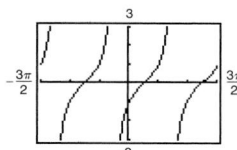

35.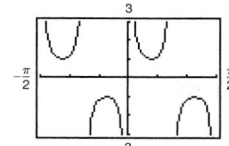

37.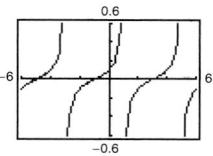

39. $-\dfrac{7\pi}{4}, -\dfrac{3\pi}{4}, \dfrac{\pi}{4}, \dfrac{5\pi}{4}$

41. $-\dfrac{4\pi}{3}, -\dfrac{\pi}{3}, \dfrac{2\pi}{3}, \dfrac{5\pi}{3}$

43. $-\dfrac{4\pi}{3}, -\dfrac{2\pi}{3}, \dfrac{2\pi}{3}, \dfrac{4\pi}{3}$

45. $-\dfrac{7\pi}{4}, -\dfrac{5\pi}{4}, \dfrac{\pi}{4}, \dfrac{3\pi}{4}$

47. Even

49. (a)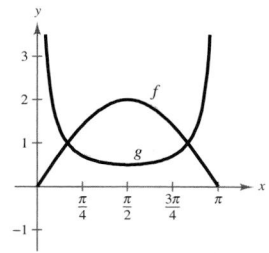

(b) $\dfrac{\pi}{6} < x < \dfrac{5\pi}{6}$

(c) f approaches 0 and g approaches $+\infty$ because the cosecant is the reciprocal of the sine.

51.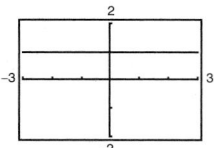

The expressions are equivalent except that when $x = 0$, y_1 is undefined.

53.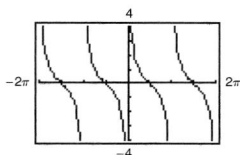

The expressions are equivalent.

55. d, $f \to 0$ as $x \to 0$.   **56.** a, $f \to 0$ as $x \to 0$.

57. b, $g \to 0$ as $x \to 0$.   **58.** c, $g \to 0$ as $x \to 0$.

59.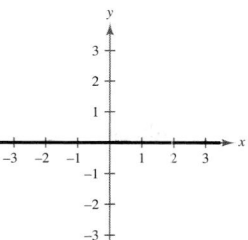

The functions are equal.

61.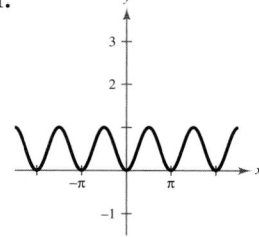

The functions are equal.

63.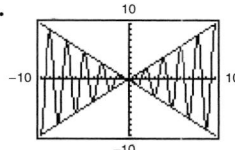

As $x \to \infty$, $f(x)$ oscillates.

65.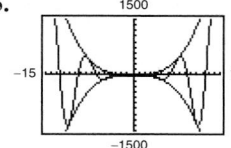

As $x \to \infty$, $g(x)$ oscillates.

67.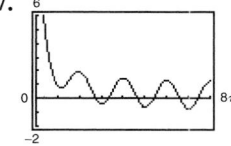

As $x \to 0$, $y \to \infty$.

69.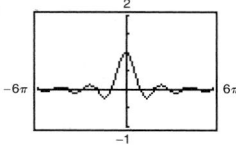

As $x \to 0$, $g(x) \to 1$.

71.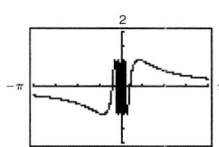

As $x \to 0$, $f(x)$ oscillates between 1 and -1.

73. $d = 7 \cot x$

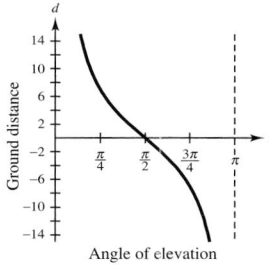

75. (a)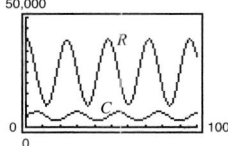

(b) As the predator population increases, the number of prey decreases. When the number of prey is small, the number of predators decreases.

(c) C: 24 months, R: 24 months; Answers will vary.

77. (a) 12

(b) Summer; Winter

(c) 1 month

79. True. For a given value of x, the y-coordinate of csc x is the reciprocal of the y-coordinate of sin x.

81. As x approaches $\pi/2$ from the left, f approaches ∞. As x approaches $\pi/2$ from the right, f approaches $-\infty$.

83. (a)

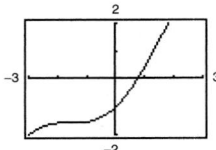

0.7391

(b) 1, 0.5403, 0.8576, 0.6543, 0.7935, 0.7014, 0.7640, 0.7221, 0.7504, 0.7314, . . . ; 0.7391

85.

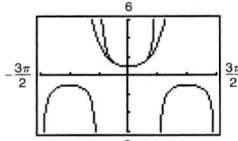

The graphs appear to coincide on the interval $-1.1 \le x \le 1.1$.

87. ± 8 **89.** $\frac{3}{2}$ **91.** $3 \pm \sqrt{5}$ **93.** $-\frac{10}{3}, 5$

Section 1.7 *(page 192)*

1. $\dfrac{\pi}{6}$ **3.** $\dfrac{\pi}{3}$ **5.** $\dfrac{\pi}{6}$ **7.** $\dfrac{5\pi}{6}$ **9.** $-\dfrac{\pi}{3}$ **11.** $\dfrac{2\pi}{3}$

13. $\dfrac{\pi}{3}$ **15.** 0   **17.** 1.29   **19.** -0.85 **21.** -1.25

23. 0.32   **25.** 1.99   **27.** 0.74   **29.** 0.85

31. 1.29   **33.** $-\dfrac{\pi}{3}, -\dfrac{\sqrt{3}}{3}, 1$

35.

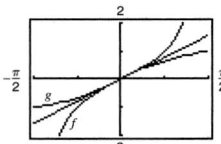

37. $\theta = \arctan \dfrac{x}{4}$

39. $\theta = \arcsin \dfrac{x+2}{5}$ **41.** $\theta = \arccos \dfrac{x+3}{2x}$ **43.** 0.3

45. -0.1 **47.** 0   **49.** $\dfrac{3}{5}$ **51.** $\dfrac{\sqrt{5}}{5}$ **53.** $\dfrac{12}{13}$

55. $\dfrac{\sqrt{34}}{5}$ **57.** $\dfrac{\sqrt{5}}{3}$ **59.** $\dfrac{1}{x}$ **61.** $\sqrt{1-4x^2}$

63. $\sqrt{1-x^2}$ **65.** $\dfrac{\sqrt{9-x^2}}{x}$ **67.** $\dfrac{\sqrt{x^2+2}}{x}$

69.

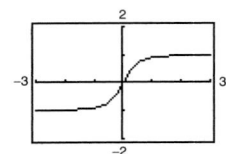

Asymptotes: $y = \pm 1$

71. $\dfrac{9}{\sqrt{x^2+81}}, x > 0;\ \dfrac{-9}{\sqrt{x^2+81}}, x < 0$

73. $\dfrac{|x-1|}{\sqrt{x^2-2x+10}}$

75.

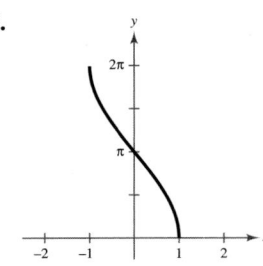

77.

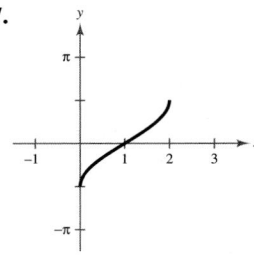

79.

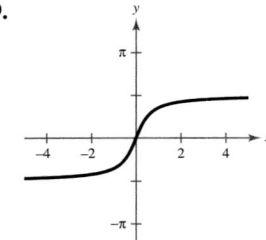

81.

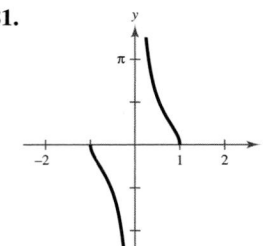

83.

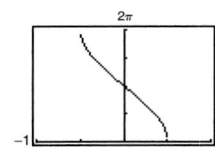

85.

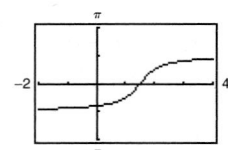

87.

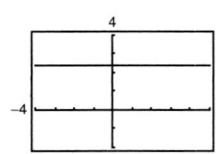

89. $3\sqrt{2} \sin\left(2t + \dfrac{\pi}{4}\right)$

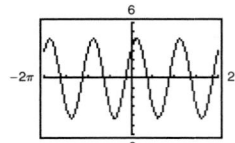

The graph implies that the identity is true.

91. (a) $\theta = \arcsin \dfrac{5}{s}$ (b) 0.13, 0.25

93. (a)

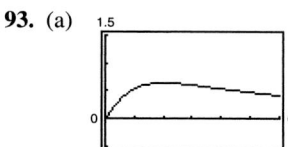

(b) 2 feet

(c) $\beta = 0$; As x increases, β approaches 0.

95. (a) $\theta \approx 26.0°$ (b) 24.4 feet

97. (a) $\theta = \arctan \dfrac{x}{20}$ (b) 14.0°, 31.0°

99. False. $\dfrac{5\pi}{4}$ is not in the range of the arctangent.

101. Domain: $(-\infty, -1] \cup [1, \infty)$

Range: $[0, \pi/2) \cup (\pi/2, \pi]$

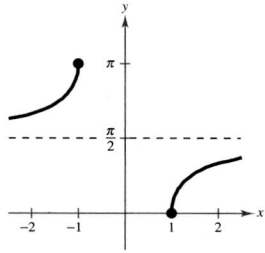

103. (a) $\dfrac{\pi}{4}$ (b) 0   (c) $\dfrac{5\pi}{6}$ (d) $\dfrac{\pi}{6}$

105.

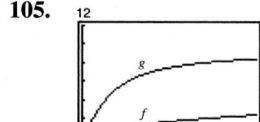

As x increases to infinity, g approaches 3π, but f has no maximum.

$a \approx 87.54$

107–111. Answers will vary.

113.   **115.**

![triangle with sides 4, 3, √7 and angle θ]   ![triangle with sides 6, 5, √11 and angle θ]

117. Eight people

Section 1.8   *(page 202)*

1. $a \approx 3.64$
   $c \approx 10.64$
   $B = 70°$

3. $a \approx 8.26$
   $c \approx 25.38$
   $A = 19°$

5. $c \approx 11.66$
   $A \approx 30.96°$
   $B \approx 59.04°$

7. $a \approx 49.48$
   $A \approx 72.08°$
   $B \approx 17.92°$

9. $a \approx 91.34$
   $b \approx 420.70$
   $B = 77°45'$

11. 2.56 inches

13. 19.99 inches   **15.** 107.2 feet   **17.** 19.7 feet

19. (a)

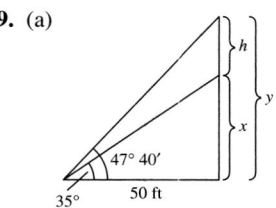

(b) $h = 50(\tan 47° 40' - \tan 35°)$ (c) 19.9 feet

21. 2236.8 feet

23. (a)

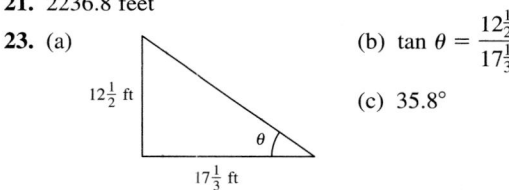

(b) $\tan \theta = \dfrac{12\frac{1}{2}}{17\frac{1}{3}}$

(c) 35.8°

25. 2.06°   **27.** 0.73 mile

29. 554 miles north; 709 miles east   **31.** 5.46 kilometers

33. 1933.3 feet   **35.** ≈ 3.23 miles or $\approx 17{,}054$ feet

37. 78.7°   **39.** 35.3°   **41.** 29.4 inches

43. $y = \sqrt{3}r$ **45.** $a \approx 12.2, b \approx 7$

47. (a) 4   (b) 4   (c) $\frac{1}{16}$

49. (a) $\frac{1}{16}$ (b) 60   (c) $\frac{1}{120}$ **51.** $d = 4\sin(\pi t)$

53. $d = 3\cos\left(\dfrac{4\pi t}{3}\right)$ **55.** $\omega = 528\pi$

57. (a)

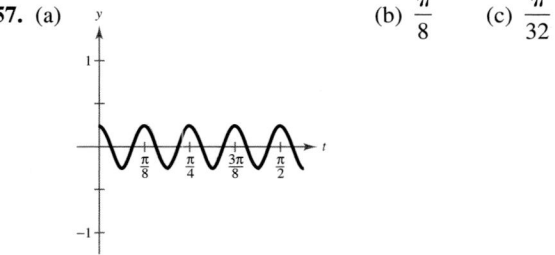

(b) $\dfrac{\pi}{8}$ (c) $\dfrac{\pi}{32}$

59. False. One period is the time for one complete cycle of the motion.

61. (a) and (b)

Base 1	Base 2	Altitude	Area
8	$8 + 16 \cos 10°$	$8 \sin 10°$	22.1
8	$8 + 16 \cos 20°$	$8 \sin 20°$	42.5
8	$8 + 16 \cos 30°$	$8 \sin 30°$	59.7
8	$8 + 16 \cos 40°$	$8 \sin 40°$	72.7
8	$8 + 16 \cos 50°$	$8 \sin 50°$	80.5
8	$8 + 16 \cos 60°$	$8 \sin 60°$	83.1
8	$8 + 16 \cos 70°$	$8 \sin 70°$	80.7

≈ 83.1 square feet when $\theta = 60°$

(c) $A = 64(1 + \cos \theta)(\sin \theta)$

(d)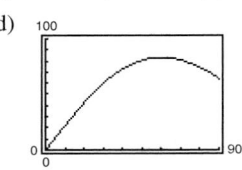

≈ 83.1 square feet when $\theta = 60°$

63. (a)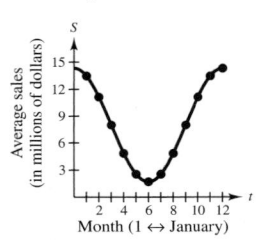

(b) $S = 8 + 6.3 \cos\left(\dfrac{\pi}{6}t\right)$ or $S = 8 + 6.3 \sin\left(\dfrac{\pi}{6}t + \dfrac{\pi}{2}\right)$

The model is a good fit.

(c) 12. Yes, sales of outerwear are seasonal.

(d) Maximum displacement from average sales of $8 million

65.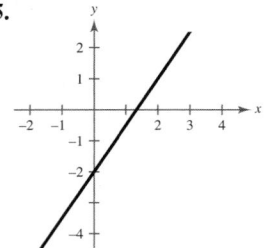

67.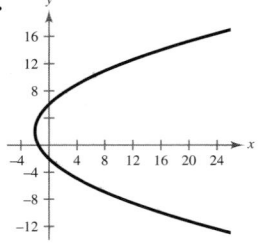

69.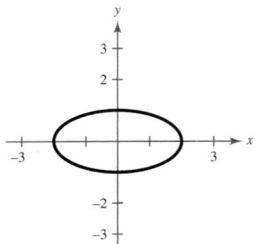

71.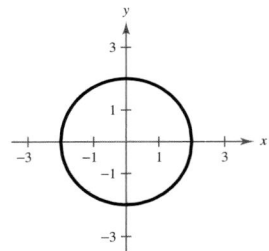

Review Exercises (page 208)

1. 0.5 radian **3.** 4.5 radians

5.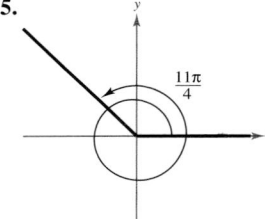

$\dfrac{3\pi}{4}, -\dfrac{5\pi}{4}$

7.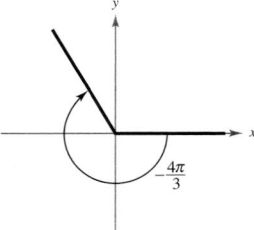

$\dfrac{2\pi}{3}, -\dfrac{10\pi}{3}$

9.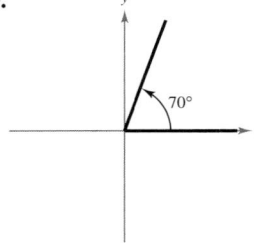

$430°, -290°$

11.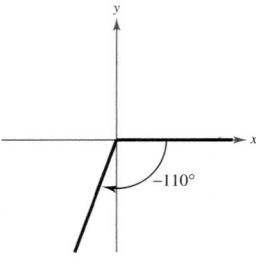

$250°, -470°$

13. $128.57°$ **15.** $-200.54°$

17. 8.3776 **19.** -0.5890

21. (a) $66\frac{2}{3}\pi$ radians per minute

(b) 400π inches per minute

23. $\left(-\dfrac{1}{2}, \dfrac{\sqrt{3}}{2}\right)$ **25.** $\left(-\dfrac{\sqrt{3}}{2}, \dfrac{1}{2}\right)$

27. $\sin \dfrac{7\pi}{6} = -\dfrac{1}{2}$ $\qquad$ $\csc \dfrac{7\pi}{6} = -2$

$\cos \dfrac{7\pi}{6} = -\dfrac{\sqrt{3}}{2}$ $\qquad$ $\sec \dfrac{7\pi}{6} = -\dfrac{2\sqrt{3}}{3}$

$\tan \dfrac{7\pi}{6} = \dfrac{\sqrt{3}}{3}$ $\qquad$ $\cot \dfrac{7\pi}{6} = \sqrt{3}$

29. $\sin\left(-\dfrac{2\pi}{3}\right) = -\dfrac{\sqrt{3}}{2}$ $\quad$ $\csc\left(-\dfrac{2\pi}{3}\right) = -\dfrac{2\sqrt{3}}{3}$

$\cos\left(-\dfrac{2\pi}{3}\right) = -\dfrac{1}{2}$ $\quad$ $\sec\left(-\dfrac{2\pi}{3}\right) = -2$

$\tan\left(-\dfrac{2\pi}{3}\right) = \sqrt{3}$ $\quad$ $\cot\left(-\dfrac{2\pi}{3}\right) = \dfrac{\sqrt{3}}{3}$

31. $\sin \dfrac{11\pi}{4} = \sin \dfrac{3\pi}{4} = \dfrac{\sqrt{2}}{2}$

33. $\sin\left(-\dfrac{17\pi}{6}\right) = \sin \dfrac{7\pi}{6} = -\dfrac{1}{2}$

35. -75.31 $\qquad$ **37.** 3.24

39. $\sin \theta = \dfrac{4\sqrt{41}}{41}$ $\qquad$ **41.** $\sin \theta = \dfrac{\sqrt{3}}{2}$

$\cos \theta = \dfrac{5\sqrt{41}}{41}$ $\qquad$ $\cos \theta = \dfrac{1}{2}$

$\tan \theta = \dfrac{4}{5}$ $\qquad$ $\tan \theta = \sqrt{3}$

$\csc \theta = \dfrac{\sqrt{41}}{4}$ $\qquad$ $\csc \theta = \dfrac{2\sqrt{3}}{3}$

$\sec \theta = \dfrac{\sqrt{41}}{5}$ $\qquad$ $\sec \theta = 2$

$\cot \theta = \dfrac{5}{4}$ $\qquad$ $\cot \theta = \dfrac{\sqrt{3}}{3}$

43. (a) 3 (b) $\dfrac{2\sqrt{2}}{3}$ (c) $\dfrac{3\sqrt{2}}{4}$ (d) $\dfrac{\sqrt{2}}{4}$

45. (a) $\dfrac{1}{4}$ (b) $\dfrac{\sqrt{15}}{4}$ (c) $\dfrac{4\sqrt{15}}{15}$ (d) $\dfrac{\sqrt{15}}{15}$

47. 0.65 $\quad$ **49.** 0.56 $\quad$ **51.** 3.67 $\quad$ **53.** 71.3 meters

55. $\sin \theta = \frac{4}{5}$ $\qquad$ $\csc \theta = \frac{5}{4}$

$\cos \theta = \frac{3}{5}$ $\qquad$ $\sec \theta = \frac{5}{3}$

$\tan \theta = \frac{4}{3}$ $\qquad$ $\cot \theta = \frac{3}{4}$

57. $\sin \theta = \dfrac{15\sqrt{241}}{241}$ $\qquad$ $\csc \theta = \dfrac{\sqrt{241}}{15}$

$\cos \theta = \dfrac{4\sqrt{241}}{241}$ $\qquad$ $\sec \theta = \dfrac{\sqrt{241}}{4}$

$\tan \theta = \dfrac{15}{4}$ $\qquad$ $\cot \theta = \dfrac{4}{15}$

59. $\sin \theta \approx 1$ $\qquad$ $\csc \theta \approx 1$

$\cos \theta \approx -0.1$ $\qquad$ $\sec \theta \approx -9$

$\tan \theta \approx -9$ $\qquad$ $\cot \theta \approx -0.1$

61. $\sin \theta = \dfrac{4\sqrt{17}}{17}$ $\qquad$ $\csc \theta = \dfrac{\sqrt{17}}{4}$

$\cos \theta = \dfrac{\sqrt{17}}{17}$ $\qquad$ $\sec \theta = \sqrt{17}$

$\tan \theta = 4$ $\qquad$ $\cot \theta = \dfrac{1}{4}$

63. $\sin \theta = -\dfrac{\sqrt{11}}{6}$ $\qquad$ **65.** $\cos \theta = -\dfrac{\sqrt{55}}{8}$

$\cos \theta = \dfrac{5}{6}$ $\qquad$ $\tan \theta = -\dfrac{3\sqrt{55}}{55}$

$\tan \theta = -\dfrac{\sqrt{11}}{5}$ $\qquad$ $\csc \theta = \dfrac{8}{3}$

$\csc \theta = -\dfrac{6\sqrt{11}}{11}$ $\qquad$ $\sec \theta = -\dfrac{8\sqrt{55}}{55}$

$\cot \theta = -\dfrac{5\sqrt{11}}{11}$ $\qquad$ $\cot \theta = -\dfrac{\sqrt{55}}{3}$

67. $\sin \theta = \dfrac{\sqrt{21}}{5}$

$\tan \theta = -\dfrac{\sqrt{21}}{2}$

$\csc \theta = \dfrac{5\sqrt{21}}{21}$

$\sec \theta = -\dfrac{5}{2}$

$\cot \theta = -\dfrac{2\sqrt{21}}{21}$

69. $\sin \dfrac{\pi}{3} = \dfrac{\sqrt{3}}{2}$; $\cos \dfrac{\pi}{3} = \dfrac{1}{2}$; $\tan \dfrac{\pi}{3} = \sqrt{3}$

71. $\sin\left(-\dfrac{7\pi}{3}\right) = -\dfrac{\sqrt{3}}{2}$; $\cos\left(-\dfrac{7\pi}{3}\right) = \dfrac{1}{2}$;

$\tan\left(-\dfrac{7\pi}{3}\right) = -\sqrt{3}$

73. $\sin 495° = \dfrac{\sqrt{2}}{2}$; $\cos 495° = -\dfrac{\sqrt{2}}{2}$; $\tan 495° = -1$

75. $\sin(-240°) = \dfrac{\sqrt{3}}{2}$; $\cos(-240°) = -\dfrac{1}{2}$;

$\tan(-240°) = -\sqrt{3}$

77. -0.76 $\quad$ **79.** 0.06 $\quad$ **81.** 3.24

83.

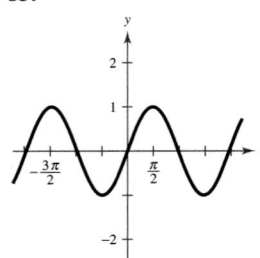

85.

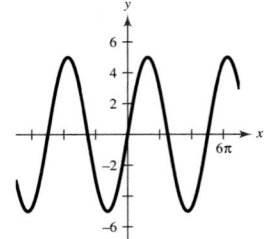

87.

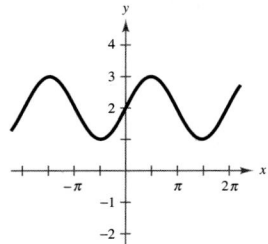

89.

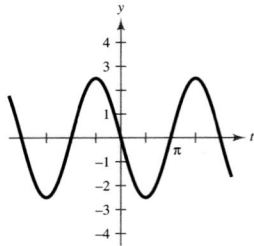

91. (a) $y = 2 \sin 528\pi x$ (b) 264 cycles per second

93.

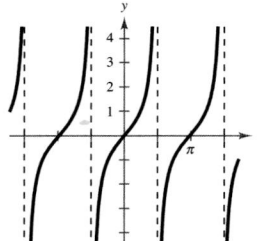

95.

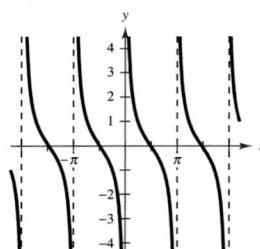

97.

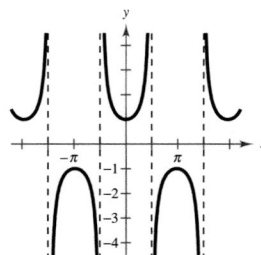

99.

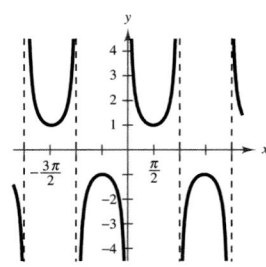

101.

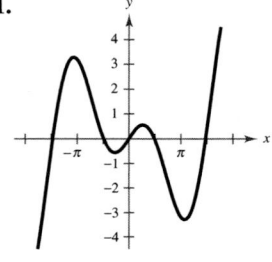

103. $-\dfrac{\pi}{6}$ **105.** 0.41

107. -0.46 **109.** $\dfrac{\pi}{6}$ **111.** π **113.** 1.24

115. 0.12 **117.** 1.40 **119.** -0.98 **121.** 0.72

123. 0 **125.** $\frac{4}{5}$ **127.** $\frac{13}{5}$ **129.** 66.8°

131. 1221 miles, 85.6°

133. False. The sine or cosine function is often useful for modeling simple harmonic motion.

135. False. For each θ there corresponds exactly one value of y.

137. d; The period is 2π and the amplitude is 3.

138. a; The period is 2π and, because $a < 0$, the graph is reflected in the x-axis.

139. b; The period is 2 and the amplitude is 2.

140. c; The period is 4π and the amplitude is 2.

141. The function is undefined because $\sec \theta = 1/\cos \theta$.

143. The ranges of the other four trigonometric functions are $(-\infty, \infty)$ or $(-\infty, -1] \cup [1, \infty)$.

145. Answers will vary.

Chapter Test *(page 212)*

1. (a)

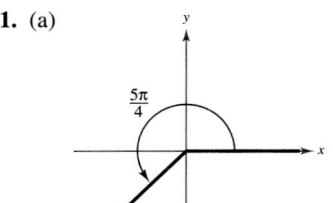

(b) $\dfrac{13\pi}{4}, \ -\dfrac{3\pi}{4}$

(c) 225°

2. 3000 radians per minute

3. $\sin \theta = \dfrac{3\sqrt{10}}{10}$ $\csc \theta = \dfrac{\sqrt{10}}{3}$

$\cos \theta = -\dfrac{\sqrt{10}}{10}$ $\sec \theta = -\sqrt{10}$

$\tan \theta = -3$ $\cot \theta = -\dfrac{1}{3}$

4. For $0 \le \theta < \dfrac{\pi}{2}$:   For $\pi \le \theta < \dfrac{3\pi}{2}$:

$\sin \theta = \dfrac{3\sqrt{13}}{13}$ $\sin \theta = -\dfrac{3\sqrt{13}}{13}$

$\cos \theta = \dfrac{2\sqrt{13}}{13}$ $\cos \theta = -\dfrac{2\sqrt{13}}{13}$

$\csc \theta = \dfrac{\sqrt{13}}{3}$ $\csc \theta = -\dfrac{\sqrt{13}}{3}$

$\sec \theta = \dfrac{\sqrt{13}}{2}$ $\sec \theta = -\dfrac{\sqrt{13}}{2}$

$\cot \theta = \dfrac{2}{3}$ $\cot \theta = \dfrac{2}{3}$

5. $\theta' = 70°$

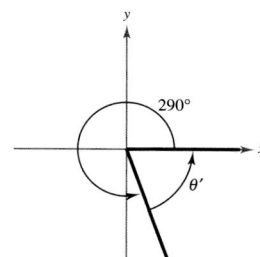

6. Quadrant III   **7.** 150°, 210°   **8.** 1.33, 1.81

9. $\sin \theta = -\frac{4}{5}$ **10.** $\sin \theta = \frac{15}{17}$

$\tan \theta = -\frac{4}{3}$ $\cos \theta = -\frac{8}{17}$

$\csc \theta = -\frac{5}{4}$ $\tan \theta = -\frac{15}{8}$

$\sec \theta = \frac{5}{3}$ $\csc \theta = \frac{17}{15}$

$\cot \theta = -\frac{3}{4}$ $\cot \theta = -\frac{8}{15}$

11. 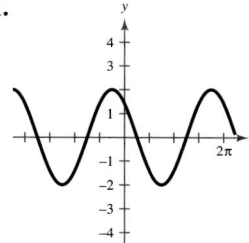   **12.**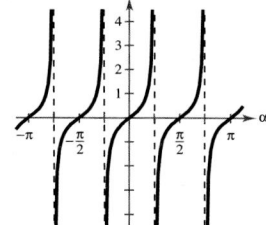

13. 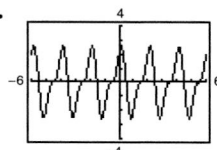   **14.**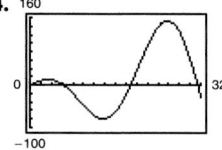

Period: 2   Not periodic

15. $a = -2, b = \frac{1}{2}, c = -\frac{\pi}{4}$ **16.** $\frac{\sqrt{5}}{2}$

17. 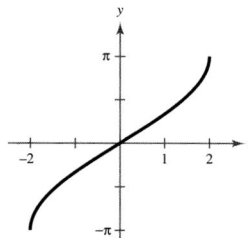   **18.** 310.1°

19. $d = -6 \cos \pi t$

Problem Solving   (page 214)

1. (a) $\frac{11\pi}{2}$ radians or 990°   (b) ≈816.42 feet

3. (a) 4767 feet   (b) 3705 feet

(c) $\tan 63° = \dfrac{w + 3705}{3000}$,

$w = 2183$ feet

5. (a)

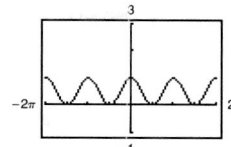

Even

(b)

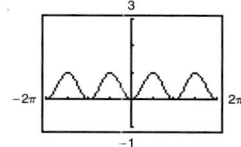

Even

7. $h = 51 - 50 \sin\left(8\pi t + \dfrac{\pi}{2}\right)$

9. (a)

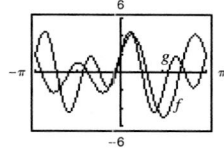

(b) Period of f: 2π

Period of g: π

(c) Yes, because the sine and cosine functions are periodic.

11. (a) Equal; two-period shift

(b) Not equal; $f\left(t + \frac{1}{2}c\right)$ is a horizontal translation and $f\left(\frac{1}{2}t\right)$ is a period change.

(c) Not equal; For example, $\sin\left[\frac{1}{2}(\pi + 2\pi)\right] \neq \sin\left(\frac{1}{2}\pi\right)$.

13. (a)

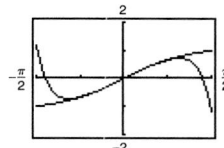

The approximation is accurate over the interval $-1 \le x \le 1$.

(b) $\dfrac{x^9}{9}$

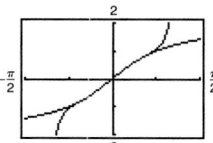

The accuracy improved.

Chapter 2

Section 2.1 *(page 223)*

1. $\tan x = -\sqrt{3}$

$\csc x = \dfrac{2\sqrt{3}}{3}$

$\sec x = -2$

$\cot x = -\dfrac{\sqrt{3}}{3}$

3. $\cos \theta = \dfrac{\sqrt{2}}{2}$

$\tan \theta = -1$

$\csc \theta = -\sqrt{2}$

$\cot \theta = -1$

5. $\sin x = -\dfrac{5}{13}$

$\cos x = -\dfrac{12}{13}$

$\csc x = -\dfrac{13}{5}$

$\cot x = \dfrac{12}{5}$

7. $\sin \phi = -\dfrac{\sqrt{5}}{3}$

$\cos \phi = \dfrac{2}{3}$

$\tan \phi = -\dfrac{\sqrt{5}}{2}$

$\cot \phi = -\dfrac{2\sqrt{5}}{5}$

9. $\sin x = \dfrac{1}{3}$

$\cos x = -\dfrac{2\sqrt{2}}{3}$

$\csc x = 3$

$\sec x = -\dfrac{3\sqrt{2}}{4}$

$\cot x = -2\sqrt{2}$

11. $\sin \theta = -\dfrac{2\sqrt{5}}{5}$

$\cos \theta = -\dfrac{\sqrt{5}}{5}$

$\csc \theta = -\dfrac{\sqrt{5}}{2}$

$\sec \theta = -\sqrt{5}$

$\cot \theta = \dfrac{1}{2}$

13. $\cos \theta = 0$

$\tan \theta$ is undefined.

$\csc \theta = -1$

$\sec \theta$ is undefined.

15. d   **16.** a   **17.** b   **18.** f   **19.** e   **20.** c

21. b   **22.** c   **23.** f   **24.** a   **25.** e   **26.** d

27. $\csc \theta$ **29.** $\cos^2 \phi$ **31.** $\cos x$ **33.** $\sin^2 x$

35. 1   **37.** $\tan x$ **39.** $1 + \sin y$ **41.** $\sec \beta$

43. $\cos u + \sin u$ **45.** $\sin^2 x$ **47.** $\sin^2 x \tan^2 x$

49. $\sec x + 1$ **51.** $\sec^4 x$ **53.** $\sin^2 x - \cos^2 x$

55. $\cot^2 x(\csc x - 1)$ **57.** $1 + 2\sin x \cos x$

59. $4\cot^2 x$ **61.** $2\csc^2 x$ **63.** $2\sec x$

65. $1 + \cos y$ **67.** $3(\sec x + \tan x)$

69.

x	0.2	0.4	0.6	0.8	1.0
y_1	0.1987	0.3894	0.5646	0.7174	0.8415
y_2	0.1987	0.3894	0.5646	0.7174	0.8415

x	1.2	1.4
y_1	0.9320	0.9854
y_2	0.9320	0.9854

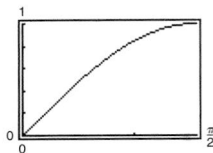

$y_1 = y_2$

71.

x	0.2	0.4	0.6	0.8	1.0
y_1	1.2230	1.5085	1.8958	2.4650	3.4082
y_2	1.2230	1.5085	1.8958	2.4650	3.4082

x	1.2	1.4
y_1	5.3319	11.6814
y_2	5.3319	11.6814

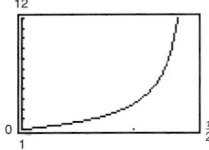

$y_1 = y_2$

73. $\csc x$ **75.** $\tan x$ **77.** $3\sin \theta$ **79.** $3\tan \theta$

81. $5\sec \theta$ **83.** $3\cos \theta = 3; \sin \theta = 0; \cos \theta = 1$

85. $4\sin \theta = 2\sqrt{2}; \sin \theta = \dfrac{\sqrt{2}}{2}; \cos \theta = \dfrac{\sqrt{2}}{2}$

87. $0 \le \theta \le \pi$ **89.** $0 \le \theta < \dfrac{\pi}{2}, \dfrac{3\pi}{2} < \theta < 2\pi$

91. (a) $\csc^2 132° - \cot^2 132° \approx 1.8107 - 0.8107 = 1$

(b) $\csc^2 \dfrac{2\pi}{7} - \cot^2 \dfrac{2\pi}{7} \approx 1.6360 - 0.6360 = 1$

93. (a) $\cos(90° - 80°) = \sin 80° \approx 0.9848$

(b) $\cos\left(\dfrac{\pi}{2} - 0.8\right) = \sin 0.8 \approx 0.7174$

95. (a) $\cot 25° = \dfrac{\cos 25°}{\sin 25°} \approx 2.1445$

(b) $\cot \dfrac{\pi}{8} = \dfrac{\cos \pi/8}{\sin \pi/8} \approx 2.4142$

97. (a) $\tan(90° - 5°) = \cot 5° \approx 11.4301$

(b) $\tan\left(\dfrac{\pi}{2} - \dfrac{11\pi}{12}\right) = \cot \dfrac{11\pi}{12} \approx -3.7321$

99. $\mu = \tan \theta$

101. False. The sign of a value of a trigonometric function depends on which quadrant the angle lies in.

103. $1, 1$ **105.** $\infty, 0$

107. Not an identity because $\cos \theta = \pm\sqrt{1 - \sin^2 \theta}$

109. Not an identity because $\dfrac{\sin k\theta}{\cos k\theta} = \tan k\theta$

111. Identity because $\sin \theta \cdot \dfrac{1}{\sin \theta} = 1$

113. Answers will vary. For example:

$$\sin^2 \theta + \cos^2 \theta = \frac{y^2}{r^2} + \frac{x^2}{r^2} = \frac{x^2 + y^2}{r^2} = \frac{r^2}{r^2} = 1$$

115. $x - 25$

117. $\dfrac{x^2 + 6x - 8}{(x + 5)(x - 8)}$ **119.** $\dfrac{-5x^2 + 8x + 28}{(x + 2)(x - 2)(x + 4)}$

Section 2.2   *(page 231)*

1–39. Answers will vary.

41. Identity   **43.** Not an identity   **45.** Identity

47. Identity   **49.** Not an identity   **51.** Identity

53. 1   **55.** 2   **57.** Answers will vary.

59. False. An identity is an equation that is true for all real values of θ.

61. The equation is not an identity because $\sin \theta = \pm\sqrt{1 - \cos^2 \theta}$.

Possible answer: $\dfrac{7\pi}{4}$

63. $3 \pm \sqrt{3}i$ **65.** $-1 \pm \sqrt{3}i$

Section 2.3   *(page 240)*

1–5. Answers will vary.   **7.** $\dfrac{2\pi}{3} + 2n\pi, \dfrac{4\pi}{3} + 2n\pi$

9. $\dfrac{\pi}{3} + 2n\pi, \dfrac{2\pi}{3} + 2n\pi$ **11.** $\dfrac{\pi}{6} + n\pi, \dfrac{5\pi}{6} + n\pi$

13. $n\pi, \dfrac{3\pi}{2} + 2n\pi$ **15.** $\dfrac{\pi}{3} + n\pi, \dfrac{2\pi}{3} + n\pi$

17. $\dfrac{\pi}{8} + \dfrac{n\pi}{2}, \dfrac{3\pi}{8} + \dfrac{n\pi}{2}$ **19.** $\dfrac{n\pi}{3}, \dfrac{\pi}{4} + n\pi$

21. $0, \dfrac{\pi}{2}, \pi, \dfrac{3\pi}{2}$ **23.** $0, \pi, \dfrac{\pi}{6}, \dfrac{5\pi}{6}, \dfrac{7\pi}{6}, \dfrac{11\pi}{6}$

25. $\dfrac{\pi}{3}, \dfrac{5\pi}{3}, \pi$ **27.** No solution   **29.** $\dfrac{\pi}{3}, \dfrac{5\pi}{3}, \pi$

31. $\dfrac{\pi}{6}, \dfrac{5\pi}{6}, \dfrac{7\pi}{6}, \dfrac{11\pi}{6}$ **33.** $\dfrac{\pi}{6} + n\pi, \dfrac{5\pi}{6} + n\pi$

35. $\dfrac{\pi}{12} + \dfrac{n\pi}{3}$ **37.** $\dfrac{\pi}{2} + 4n\pi, \dfrac{7\pi}{2} + 4n\pi$

39. $-1, 3$ **41.** ± 2

43. $\dfrac{2}{3}, \dfrac{3}{2}; \ 0.8411 + 2n\pi, 5.4421 + 2n\pi$

45. $2.6779, 5.8195$ **47.** $1.0472, 5.2360$

49. $0.8603, 3.4256$ **51.** $0, 2.6779, 3.1416, 5.8195$

53. $0.9828, 1.7682, 4.1244, 4.9098$

55. $0.3398, 0.8481, 2.2935, 2.8018$

57. $1.9357, 2.7767, 5.0773, 5.9183$

59. $\dfrac{\pi}{4}, \dfrac{5\pi}{4}, \arctan 5, \arctan 5 + \pi$ **61.** $\dfrac{\pi}{3}, \dfrac{5\pi}{3}$

63. (a)

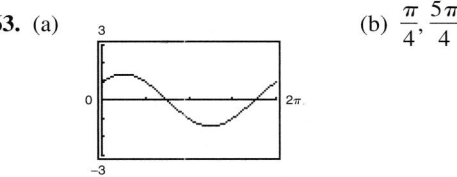

(b) $\dfrac{\pi}{4}, \dfrac{5\pi}{4}$

Maximum: $\left(\dfrac{\pi}{4}, \sqrt{2}\right)$

Minimum: $\left(\dfrac{5\pi}{4}, -\sqrt{2}\right)$

65. 1

67. (a) All real numbers $x \neq 0$

(b) y-axis symmetry; Horizontal asymptote: $y = 1$

(c) Oscillates

(d) Infinitely many solutions

(e) Yes, 0.6366

69. 0.04 second, 0.43 second, 0.83 second

71. February, March, and April   **73.** 1.9°

75. (a)

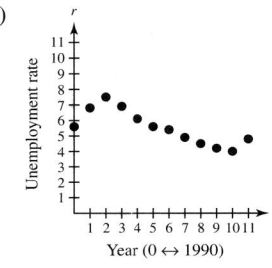

Year (0 ↔ 1990)

(b) Model (1)

(c) The constant term, 5.51

(d) ≈ 13 years

(e) 2004

77. True. The first equation has a smaller period than the second equation, so it will have more solutions in the interval $[0, 2\pi)$.

79. 1   **81.** $C = 24°$

$$a \approx 54.8$$
$$b \approx 50.1$$

83. $\sin 390° = \dfrac{1}{2}$ **85.** $\sin(-1845°) = -\dfrac{\sqrt{2}}{2}$

$\cos 390° = \dfrac{\sqrt{3}}{2}$ $\cos(-1845°) = \dfrac{\sqrt{2}}{2}$

$\tan 390° = \dfrac{\sqrt{3}}{3}$ $\tan(-1845°) = -1$

87. $1.36°$

Section 2.4 *(page 248)*

1. (a) $\dfrac{\sqrt{2} - \sqrt{6}}{4}$ (b) $\dfrac{\sqrt{2} + 1}{2}$

3. (a) $\dfrac{1}{2}$ (b) $\dfrac{-\sqrt{3} - 1}{2}$

5. (a) $\dfrac{-\sqrt{2} - \sqrt{6}}{4}$ (b) $\dfrac{-1 + \sqrt{2}}{2}$

7. $\sin 105° = \dfrac{\sqrt{2}}{4}(\sqrt{3} + 1)$

$\cos 105° = \dfrac{\sqrt{2}}{4}(1 - \sqrt{3})$

$\tan 105° = -2 - \sqrt{3}$

9. $\sin 195° = \dfrac{\sqrt{2}}{4}(1 - \sqrt{3})$

$\cos 195° = -\dfrac{\sqrt{2}}{4}(\sqrt{3} + 1)$

$\tan 195° = 2 - \sqrt{3}$

11. $\sin \dfrac{11\pi}{12} = \dfrac{\sqrt{2}}{4}(\sqrt{3} - 1)$

$\cos \dfrac{11\pi}{12} = -\dfrac{\sqrt{2}}{4}(\sqrt{3} + 1)$

$\tan \dfrac{11\pi}{12} = -2 + \sqrt{3}$

13. $\sin \dfrac{17\pi}{12} = -\dfrac{\sqrt{2}}{4}(\sqrt{3} + 1)$

$\cos \dfrac{17\pi}{12} = \dfrac{\sqrt{2}}{4}(1 - \sqrt{3})$

$\tan \dfrac{17\pi}{12} = 2 + \sqrt{3}$

15. $\sin 285° = -\dfrac{\sqrt{2}}{4}(\sqrt{3} + 1)$

$\cos 285° = \dfrac{\sqrt{2}}{4}(\sqrt{3} - 1)$

$\tan 285° = -(2 + \sqrt{3})$

17. $\sin(-165°) = -\dfrac{\sqrt{2}}{4}(\sqrt{3} - 1)$

$\cos(-165°) = -\dfrac{\sqrt{2}}{4}(1 + \sqrt{3})$

$\tan(-165°) = 2 - \sqrt{3}$

19. $\sin \dfrac{13\pi}{12} = \dfrac{\sqrt{2}}{4}(1 - \sqrt{3})$

$\cos \dfrac{13\pi}{12} = -\dfrac{\sqrt{2}}{4}(1 + \sqrt{3})$

$\tan \dfrac{13\pi}{12} = 2 - \sqrt{3}$

21. $\sin\left(-\dfrac{13\pi}{12}\right) = \dfrac{\sqrt{2}}{4}(\sqrt{3} - 1)$

$\cos\left(-\dfrac{13\pi}{12}\right) = -\dfrac{\sqrt{2}}{4}(\sqrt{3} + 1)$

$\tan\left(-\dfrac{13\pi}{12}\right) = -2 + \sqrt{3}$

23. $\cos 40°$ **25.** $\tan 239°$ **27.** $\sin 1.8$ **29.** $\tan 3x$

31. $-\dfrac{\sqrt{3}}{2}$ **33.** $\dfrac{\sqrt{3}}{2}$ **35.** -1 **37.** $-\dfrac{63}{65}$

39. $\dfrac{16}{65}$ **41.** $-\dfrac{63}{16}$ **43.** $\dfrac{65}{56}$ **45.** $\dfrac{3}{5}$ **47.** $-\dfrac{44}{117}$

49. $\dfrac{5}{3}$ **51.** 1 **53.** 0 **55–63.** Answers will vary.

65. $-\sin x$ **67.** $-\cos \theta$ **69.** $\dfrac{\pi}{2}$ **71.** $\dfrac{5\pi}{4}, \dfrac{7\pi}{4}$

73. $\dfrac{\pi}{4}, \dfrac{7\pi}{4}$ **75.** (a) $y = \dfrac{5}{12}\sin(2t + 0.6435)$

(b) $\dfrac{5}{12}$ feet   (c) $\dfrac{1}{\pi}$ cycle per second

77. False. $\sin(u \pm v) = \sin u \cos v \pm \cos u \sin v$

79. False.

$$\cos\left(x - \dfrac{\pi}{2}\right) = \cos x \cos \dfrac{\pi}{2} + \sin x \sin \dfrac{\pi}{2} = \sin x$$

81. Answers will vary.   **83.** Answers will vary.

85. (a) $\sqrt{2}\sin\left(\theta + \dfrac{\pi}{4}\right)$ (b) $\sqrt{2}\cos\left(\theta - \dfrac{\pi}{4}\right)$

87. (a) $13\sin(3\theta + 0.3948)$ (b) $13\cos(3\theta - 1.1760)$

89. $2\cos \theta$ **91.** $15°$

93.

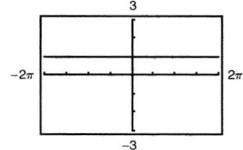

$$\sin^2\left(\theta + \frac{\pi}{4}\right) + \sin^2\left(\theta - \frac{\pi}{4}\right) = 1$$

95. Answers will vary.   **97.** $f^{-1}(x) = \dfrac{x + 15}{5}$

99. Because f is not one-to-one, f^{-1} does not exist.

Section 2.5   *(page 258)*

1. $\dfrac{\sqrt{17}}{17}$ **3.** $\dfrac{15}{17}$ **5.** $\dfrac{8}{15}$ **7.** $\dfrac{17}{8}$

9. $0, \dfrac{\pi}{3}, \pi, \dfrac{5\pi}{3}$ **11.** $\dfrac{\pi}{12}, \dfrac{5\pi}{12}, \dfrac{13\pi}{12}, \dfrac{17\pi}{12}$

13. $0, \dfrac{2\pi}{3}, \dfrac{4\pi}{3}$ **15.** $\dfrac{\pi}{2}, \dfrac{\pi}{6}, \dfrac{5\pi}{6}, \dfrac{7\pi}{6}, \dfrac{3\pi}{2}, \dfrac{11\pi}{6}$

17. $0, \dfrac{\pi}{2}, \pi, \dfrac{3\pi}{2}$ **19.** $3 \sin 2x$ **21.** $4 \cos 2x$

23. $\sin 2u = \dfrac{24}{25}$ **25.** $\sin 2u = \dfrac{24}{25}$

$\cos 2u = -\dfrac{7}{25}$ $\cos 2u = \dfrac{7}{25}$

$\tan 2u = -\dfrac{24}{7}$ $\tan 2u = \dfrac{24}{7}$

27. $\sin 2u = -\dfrac{4\sqrt{21}}{25}$

$\cos 2u = -\dfrac{17}{25}$

$\tan 2u = \dfrac{4\sqrt{21}}{17}$

29. $\frac{1}{8}(3 + 4\cos 2x + \cos 4x)$ **31.** $\frac{1}{8}(1 - \cos 4x)$

33. $\frac{1}{16}(1 + \cos 2x - \cos 4x - \cos 2x \cos 4x)$

35. $\dfrac{4\sqrt{17}}{17}$ **37.** $\dfrac{1}{4}$ **39.** $\sqrt{17}$

41. $\sin 75° = \frac{1}{2}\sqrt{2 + \sqrt{3}}$

$\cos 75° = \frac{1}{2}\sqrt{2 - \sqrt{3}}$

$\tan 75° = 2 + \sqrt{3}$

43. $\sin 112° 30' = \frac{1}{2}\sqrt{2 + \sqrt{2}}$

$\cos 112° 30' = -\frac{1}{2}\sqrt{2 - \sqrt{2}}$

$\tan 112° 30' = -1 - \sqrt{2}$

45. $\sin\dfrac{\pi}{8} = \frac{1}{2}\sqrt{2 - \sqrt{2}}$ **47.** $\sin\dfrac{3\pi}{8} = \frac{1}{2}\sqrt{2 + \sqrt{2}}$

$\cos\dfrac{\pi}{8} = \frac{1}{2}\sqrt{2 + \sqrt{2}}$ $\cos\dfrac{3\pi}{8} = \frac{1}{2}\sqrt{2 - \sqrt{2}}$

$\tan\dfrac{\pi}{8} = \sqrt{2} - 1$ $\tan\dfrac{3\pi}{8} = \sqrt{2} + 1$

49. $\sin\dfrac{u}{2} = \dfrac{5\sqrt{26}}{26}$ **51.** $\sin\dfrac{u}{2} = \sqrt{\dfrac{89 - 8\sqrt{89}}{178}}$

$\cos\dfrac{u}{2} = \dfrac{\sqrt{26}}{26}$ $\cos\dfrac{u}{2} = -\sqrt{\dfrac{89 + 8\sqrt{89}}{178}}$

$\tan\dfrac{u}{2} = 5$ $\tan\dfrac{u}{2} = \dfrac{8 - \sqrt{89}}{5}$

53. $\sin\dfrac{u}{2} = \dfrac{3\sqrt{10}}{10}$ **55.** $|\sin 3x|$ **57.** $-|\tan 4x|$

$\cos\dfrac{u}{2} = -\dfrac{\sqrt{10}}{10}$

$\tan\dfrac{u}{2} = -3$

59. π **61.** $\dfrac{\pi}{3}, \pi, \dfrac{5\pi}{3}$

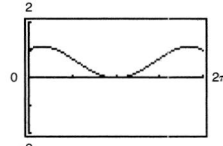   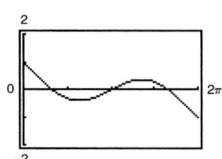

63. $3\left(\sin\dfrac{\pi}{2} + \sin 0\right)$

65. $\frac{1}{2}(\sin 10\theta + \sin 2\theta)$ **67.** $\frac{5}{2}(\cos 8\beta + \cos 2\beta)$

69. $\frac{1}{2}(\cos 2y - \cos 2x)$ **71.** $\frac{1}{2}(\sin 2\theta + \sin 2\pi)$

73. $5(\cos 60° + \cos 90°)$ **75.** $2 \sin 45° \cos 15°$

77. $-2 \sin\dfrac{\pi}{2} \sin\dfrac{\pi}{4}$ **79.** $2 \cos 4\theta \sin \theta$

81. $2 \cos 4x \cos 2x$ **83.** $2 \cos \alpha \sin \beta$

85. $-2 \sin\theta \sin\dfrac{\pi}{2}$

87. $0, \dfrac{\pi}{4}, \dfrac{\pi}{2}, \dfrac{3\pi}{4}, \pi, \dfrac{5\pi}{4}, \dfrac{3\pi}{2}, \dfrac{7\pi}{4}$

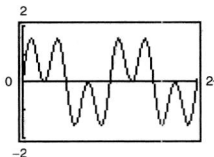

89. $\dfrac{\pi}{6}, \dfrac{5\pi}{6}$

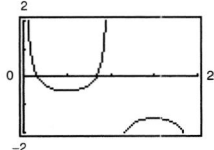

91. $\dfrac{25}{169}$ **93.** $\dfrac{4}{13}$ **95–109.** Answers will vary.

111.

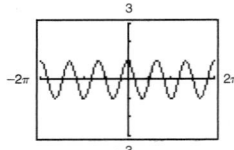

113.

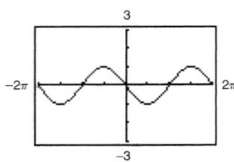

115.

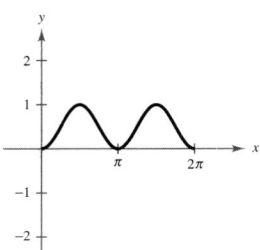

117. $2x\sqrt{1-x^2}$

119. (a) $A = 100 \sin\dfrac{\theta}{2}\cos\dfrac{\theta}{2}$

    (b) $A = 50\sin\theta$

    The area is maximum when $\theta = \pi/2$.

121. (a) π (b) 0.4482

    (c) 760 miles per hour; 3420 miles per hour

    (d) $\theta = 2\sin^{-1}\left(\dfrac{1}{m}\right)$

123. False. For $u < 0$,

$$\begin{aligned}
\sin 2u &= -\sin(-2u)\\
&= -2\sin(-u)\cos(-u)\\
&= -2(-\sin u)\cos u\\
&= 2\sin u \cos u.
\end{aligned}$$

125. (a)

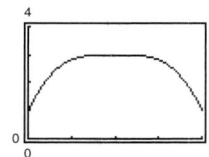

    Maximum: $(\pi, 3)$

    (b) π

127. (a) $\frac{1}{4}(3 + \cos 4x)$

    (b) $2\cos^4 x - 2\cos^2 x + 1$

    (c) $1 - 2\sin^2 x \cos^2 x$

    (d) $1 - \frac{1}{2}\sin^2 2x$

    (e) No. There is often more than one way to rewrite a trigonometric expression.

129. September: $235,000   **131.** ≈ 127 feet

    October: $272,600

Review Exercises *(page 263)*

1. $\sec x$ **3.** $\cos x$ **5.** $\cot x$

7. $\tan x = \dfrac{3}{4}$

   $\csc x = \dfrac{5}{3}$

   $\sec x = \dfrac{5}{4}$

   $\cot x = \dfrac{4}{3}$

11. $\sin^2 x$ **13.** 1   **15.** $\cot\theta$ **17.** $\cot^2 x$

19. $\sec x + 2\sin x$ **21.** $-2\tan^2\theta$

23–31. Answers will vary.

33. $\dfrac{\pi}{3} + 2n\pi, \dfrac{2\pi}{3} + 2n\pi$ **35.** $\dfrac{\pi}{6} + n\pi$

37. $\dfrac{\pi}{3} + n\pi, \dfrac{2\pi}{3} + n\pi$ **39.** $0, \dfrac{2\pi}{3}, \dfrac{4\pi}{3}$

41. $0, \dfrac{\pi}{2}, \pi$ **43.** $\dfrac{\pi}{8}, \dfrac{3\pi}{8}, \dfrac{9\pi}{8}, \dfrac{11\pi}{8}$

45. $0, \dfrac{\pi}{8}, \dfrac{3\pi}{8}, \dfrac{5\pi}{8}, \dfrac{7\pi}{8}, \dfrac{9\pi}{8}, \dfrac{11\pi}{8}, \dfrac{13\pi}{8}, \dfrac{15\pi}{8}$ **47.** $0, \pi$

49. $\arctan(-4) + \pi, \arctan(-4) + 2\pi, \arctan 3,$

   $\pi + \arctan 3$

51. $\sin 285° = -\dfrac{\sqrt{2}}{4}\left(\sqrt{3} + 1\right)$

   $\cos 285° = \dfrac{\sqrt{2}}{4}\left(\sqrt{3} - 1\right)$

   $\tan 285° = -2 - \sqrt{3}$

53. $\sin\dfrac{25\pi}{12} = \dfrac{\sqrt{2}}{4}\left(\sqrt{3} - 1\right)$

   $\cos\dfrac{25\pi}{12} = \dfrac{\sqrt{2}}{4}\left(\sqrt{3} + 1\right)$

   $\tan\dfrac{25\pi}{12} = 2 - \sqrt{3}$

55. $\sin 15°$ **57.** $\tan 35°$ **59.** $-\frac{3}{52}\left(5 + 4\sqrt{7}\right)$

61. $\frac{1}{52}\left(5\sqrt{7} + 36\right)$ **63.** $\frac{1}{52}\left(5\sqrt{7} - 36\right)$

65. $\dfrac{\pi}{4}, \dfrac{7\pi}{4}$ **67.** $\dfrac{\pi}{6}, \dfrac{11\pi}{6}$

69.

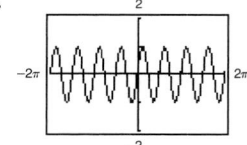

71. $\sin 2u = \dfrac{24}{25}$

   $\cos 2u = -\dfrac{7}{25}$

   $\tan 2u = -\dfrac{24}{7}$

73. $\theta = 15°$ or $\dfrac{\pi}{12}$ **75.** $\dfrac{1 - \cos 4x}{1 + \cos 4x}$

77. $\dfrac{3 - 4 \cos 2x + \cos 4x}{4(1 + \cos 2x)}$

79. $\sin(-75°) = -\frac{1}{2}\sqrt{2 + \sqrt{3}}$
$\cos(-75°) = \frac{1}{2}\sqrt{2 - \sqrt{3}}$
$\tan(-75°) = -2 - \sqrt{3}$

81. $\sin \dfrac{19\pi}{12} = -\dfrac{1}{2}\sqrt{2 + \sqrt{3}}$ **83.** $-|\cos 5x|$

$\cos \dfrac{19\pi}{12} = \dfrac{1}{2}\sqrt{2 - \sqrt{3}}$

$\tan \dfrac{19\pi}{12} = -2 - \sqrt{3}$

85. $\sin \dfrac{u}{2} = \dfrac{\sqrt{10}}{10}$ **87.** $\dfrac{1}{2}\sin \dfrac{\pi}{3}$ **89.** $\dfrac{1}{2}(\cos 2\theta + \cos 8\theta)$

$\cos \dfrac{u}{2} = \dfrac{3\sqrt{10}}{10}$

$\tan \dfrac{u}{2} = \dfrac{1}{3}$

91. $2 \sin 75° \cos 15°$ **93.** $-2 \sin x \sin \dfrac{\pi}{6}$

95. (a) $y = \dfrac{1}{2}\sqrt{10} \sin\left(8t - \arctan \dfrac{1}{3}\right)$

(b) $\dfrac{1}{2}\sqrt{10}$ feet (c) $\dfrac{4}{\pi}$ cycles per second

97. False. Using the sum and difference formula, $\sin(x + y) = \sin x \cos y + \cos x \sin y$.

99. True by the product-to-sum formula

101. No. For an equation to be an identity, the equation must be true for all real numbers x. $\sin \theta = \frac{1}{2}$ has an infinite number of solutions but is not an identity.

103. $y_1 = y_2 + 1$

105. $-1.8431, 2.1758, 3.9903, 8.8935, 9.8820$

Chapter Test *(page 266)*

1. $\sin \theta = -\dfrac{3\sqrt{13}}{13}$ **2.** 1 **3.** 1 **4.** $\csc \theta \sec \theta$

$\cos \theta = -\dfrac{2\sqrt{13}}{13}$

$\csc \theta = -\dfrac{\sqrt{13}}{3}$

$\sec \theta = -\dfrac{\sqrt{13}}{2}$

$\cot \theta = \dfrac{2}{3}$

5. $\theta = 0, \dfrac{\pi}{2} < \theta \le \pi, \dfrac{3\pi}{2} < \theta < 2\pi$

6.

$y_1 = y_2$

7–12. Answers will vary.

13. $\dfrac{1}{16}\left(\dfrac{10 - 15 \cos 2x + 6 \cos 4x - \cos 6x}{1 + \cos 2x}\right)$ **14.** $\tan 2\theta$

15. $2(\sin 6\theta + \sin 2\theta)$ **16.** $-2 \cos \dfrac{7\theta}{2} \sin \dfrac{\theta}{2}$

17. $0, \dfrac{3\pi}{4}, \pi, \dfrac{7\pi}{4}$ **18.** $\dfrac{\pi}{6}, \dfrac{\pi}{2}, \dfrac{5\pi}{6}, \dfrac{3\pi}{2}$

19. $\dfrac{\pi}{6}, \dfrac{5\pi}{6}, \dfrac{7\pi}{6}, \dfrac{11\pi}{6}$ **20.** $\dfrac{\pi}{6}, \dfrac{5\pi}{6}, \dfrac{3\pi}{2}$

21. $-2.938, -2.663, 1.170$

22. $\dfrac{\sqrt{2} - \sqrt{6}}{4}$ **23.** $\sin 2u = \dfrac{4}{5}, \tan 2u = -\dfrac{4}{3}$

24. Day 123 to day 223

Problem Solving *(page 270)*

1. (a) $\cos \theta = \pm\sqrt{1 - \sin^2 \theta}$

$\tan \theta = \pm\dfrac{\sin \theta}{\sqrt{1 - \sin^2 \theta}}$

$\cot \theta = \pm\dfrac{\sqrt{1 - \sin^2 \theta}}{\sin \theta}$

$\sec \theta = \pm\dfrac{1}{\sqrt{1 - \sin^2 \theta}}$

$\csc \theta = \dfrac{1}{\sin \theta}$

(b) $\sin \theta = \pm\sqrt{1 - \cos^2 \theta}$

$\tan \theta = \pm\dfrac{\sqrt{1 - \cos^2 \theta}}{\cos \theta}$

$\csc \theta = \pm\dfrac{1}{\sqrt{1 - \cos^2 \theta}}$

$\sec \theta = \dfrac{1}{\cos \theta}$

$\cot \theta = \pm\dfrac{\cos \theta}{\sqrt{1 - \cos^2 \theta}}$

3. Answers will vary. **5.** $y = \frac{1}{64} v^2 \sin^2 \theta$

7. $\sin\dfrac{\theta}{2} = \sqrt{\dfrac{1-\cos\theta}{2}}$

$\cos\dfrac{\theta}{2} = \sqrt{\dfrac{1+\cos\theta}{2}}$

$\tan\dfrac{\theta}{2} = \dfrac{\sin\theta}{1+\cos\theta}$

9. (a)

(b) $t = 91$, $t = 274$; Spring Equinox and Fall Equinox

(c) Seward; The amplitudes: 6.4 and 1.9

(d) 365.2 days

11. (a) $\dfrac{\pi}{6} \le x \le \dfrac{5\pi}{6}$

(b) $\dfrac{2\pi}{3} \le x \le \dfrac{4\pi}{3}$

(c) $\dfrac{\pi}{2} < x < \pi,\ \dfrac{3\pi}{2} < x < 2\pi$

(d) $0 \le x \le \dfrac{\pi}{4},\ \dfrac{5\pi}{4} \le x \le 2\pi$

13. (a) $\sin(u + v + w)$

$\quad = \sin u \cos v \cos w - \sin u \sin v \sin w$
$\qquad + \cos u \sin v \cos w + \cos u \cos v \sin w$

(b) $\tan(u + v + w)$

$\quad = \dfrac{\tan u + \tan v + \tan w - \tan u \tan v \tan w}{1 - \tan u \tan v - \tan u \tan w - \tan v \tan w}$

15. (a)

(b) 233.3 times per second

Chapter 3

Section 3.1   *(page 280)*

1. $C = 105°$, $b \approx 28.28$, $c \approx 38.64$

3. $C = 120°$, $b \approx 4.75$, $c \approx 7.17$

5. $B \approx 21.55°$, $C \approx 122.45°$, $c \approx 11.49$

7. $B = 60.9°$, $b \approx 19.32$, $c \approx 6.36$

9. $B = 42°4'$, $a \approx 22.05$, $b \approx 14.88$

11. $A \approx 10°11'$, $C \approx 154°19'$, $c \approx 11.03$

13. $A \approx 25.57°$, $B \approx 9.43°$, $a \approx 10.53$

15. $B \approx 18°13'$, $C \approx 51°32'$, $c \approx 40.06$

17. $C = 83°$, $a \approx 0.62$, $b \approx 0.51$ **19.** No solution

21. No solution   **23.** No solution

25. (a) $b \le 5$, $b = \dfrac{5}{\sin 36°}$

(b) $5 < b < \dfrac{5}{\sin 36°}$

(c) $b > \dfrac{5}{\sin 36°}$

27. (a) $b \le 10.8$, $b = \dfrac{10.8}{\sin 10°}$

(b) $10.8 < b < \dfrac{10.8}{\sin 10°}$

(c) $b > \dfrac{10.8}{\sin 10°}$

29. 10.4   **31.** 1675.2   **33.** 3204.5   **35.** 15.3 meters

37. 16.1°   **39.** 77 meters

41. (a)

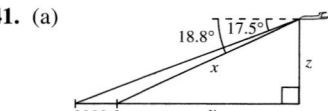

Not drawn to scale

(b) 22.6 miles

(c) 21.4 miles

(d) 7.3 miles

43. 3.2 miles

45. True. If an angle of a triangle is obtuse (greater than 90°), than the other two angles must be acute and therefore less than 90°. The triangle is oblique.

47. (a) $\alpha = \arcsin(0.5 \sin \beta)$

(b)

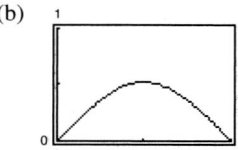

Domain: $0 < \beta < \pi$

Range: $0 < \alpha < \dfrac{\pi}{6}$

(c) $c = \dfrac{18 \sin[\pi - \beta - \arcsin(0.5 \sin \beta)]}{\sin \beta}$

(d)

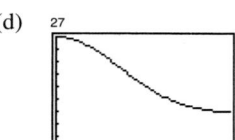

Domain: $0 < \beta < \pi$

Range: $9 < c < 27$

(e)

β	0.4	0.8	1.2	1.6
α	0.1960	0.3669	0.4848	0.5234
c	25.95	23.07	19.19	15.33

β	2.0	2.4	2.8
α	0.4720	0.3445	0.1683
c	12.29	10.31	9.27

As β increases from 0 to π, α increases and then decreases, and c decreases from 27 to 9.

49. $\cos x$　　**51.** $\sin^2 x$　　**53.** $3(\sin 11\theta + \sin 5\theta)$

Section 3.2　(page 287)

1. $A \approx 23.07°, B \approx 34.05°, C \approx 122.88°$

3. $B \approx 23.79°, C \approx 126.21°, a \approx 18.59$

5. $A \approx 31.99°, B \approx 42.38°, C \approx 105.63°$

7. $A \approx 92.94°, B \approx 43.53°, C \approx 43.53°$

9. $B \approx 13.45°, C \approx 31.55°, a \approx 12.16$

11. $A \approx 141°45', C \approx 27°40', b \approx 11.87$

13. $A = 27°10', C = 27°10', b \approx 56.94$

15. $A \approx 33.80°, B \approx 103.20°, c \approx 0.54$

	a	b	c	d	θ	ϕ
17.	5	8	12.07	5.69	45°	135°
19.	10	14	20	13.86	68.2°	111.8°
21.	15	16.96	25	20	77.2°	102.8°

23. 16.25　　**25.** 10.44　　**27.** 52.11

29.

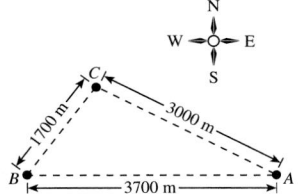

N 37.1° E, S 63.1° E

31. 373.3 meters　　**33.** 72.3°　　**35.** 43.3 miles

37. (a) N 58.4° W　(b) S 81.5° W　　**39.** 63.7 feet

41. 24.2 miles　　**43.** $\overline{PQ} \approx 9.4, \overline{QS} = 5, \overline{RS} \approx 12.8$

45.

d (inches)	9	10	12	13	14
θ (degrees)	60.9°	69.5°	88.0°	98.2°	109.6°
s (inches)	20.88	20.28	18.99	18.28	17.48

d (inches)	15	16
θ (degrees)	122.9°	139.8°
s (inches)	16.55	15.37

47. 46,837.5 square feet

49. False. For s to be the average of the lengths of the three sides of the triangle, s would be equal to $(a + b + c)/3$.

51. False. The three side lengths do not form a triangle.

53. (a) 570.60　　(b) 5909.2　　(c) 177.09

55. Answers will vary.　　**57.** $-\dfrac{\pi}{2}$　　**59.** $\dfrac{\pi}{3}$

61. $-\dfrac{\pi}{3}$　　**63.** $\dfrac{1}{\sqrt{1 - 4x^2}}$　　**65.** $\dfrac{1}{x - 2}$

67. $\cos \theta = 1$
$\sec \theta = 1$
$\csc \theta$ is undefined.

69. $\tan \theta = -\dfrac{\sqrt{3}}{3}$

$\sec \theta = \dfrac{2\sqrt{3}}{3}$

$\csc \theta = -2$

71. $-2 \sin \dfrac{7\pi}{12} \sin \dfrac{\pi}{4}$

Section 3.3　(page 300)

1. $\mathbf{v} = \langle 3, 2 \rangle; \|\mathbf{v}\| = \sqrt{13}$　　**3.** $\mathbf{v} = \langle -3, 2 \rangle; \|\mathbf{v}\| = \sqrt{13}$

5. $\mathbf{v} = \langle 0, 5 \rangle; \|\mathbf{v}\| = 5$　　**7.** $\mathbf{v} = \langle 16, 7 \rangle; \|\mathbf{v}\| = \sqrt{305}$

9. $\mathbf{v} = \langle 8, 6 \rangle; \|\mathbf{v}\| = 10$　　**11.** $\mathbf{v} = \langle -9, -12 \rangle; \|\mathbf{v}\| = 15$

13.

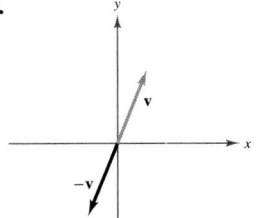

15.

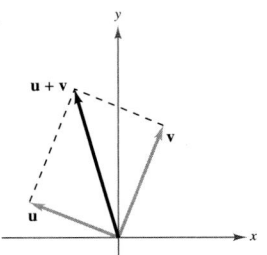

17.

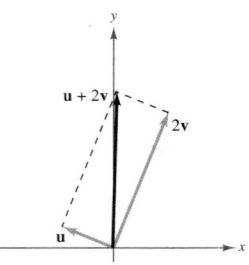

19. (a) $\langle 3, 4 \rangle$ (b) $\langle 1, -2 \rangle$

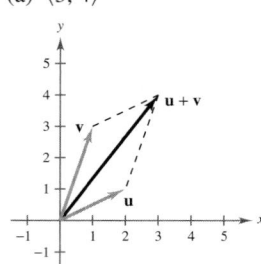
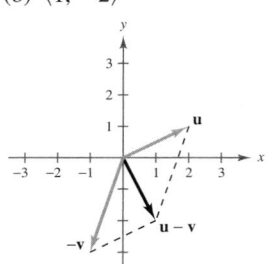

(c) $\langle 1, -7 \rangle$

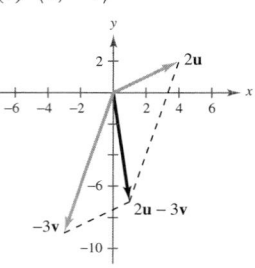

21. (a) $\langle -5, 3 \rangle$ (b) $\langle -5, 3 \rangle$

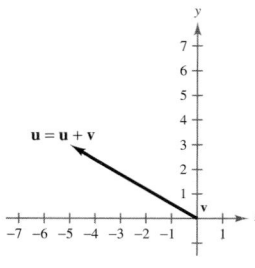
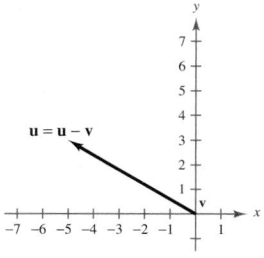

(c) $\langle -10, 6 \rangle$

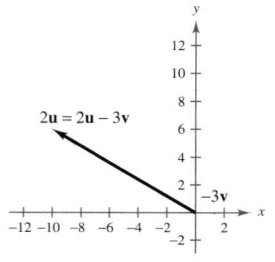

23. (a) $3\mathbf{i} - 2\mathbf{j}$ (b) $-\mathbf{i} + 4\mathbf{j}$

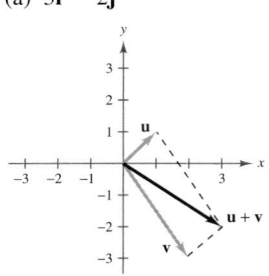
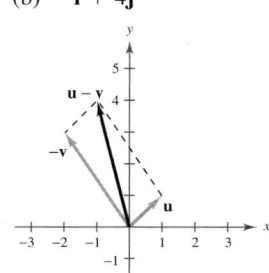

(c) $-4\mathbf{i} + 11\mathbf{j}$

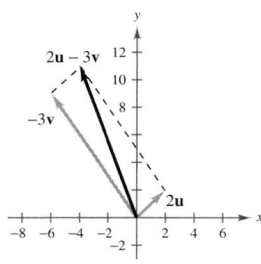

25. (a) $2\mathbf{i} + \mathbf{j}$ (b) $2\mathbf{i} - \mathbf{j}$

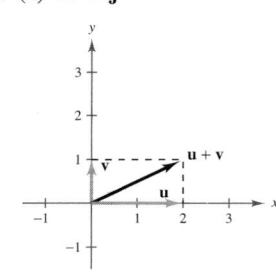
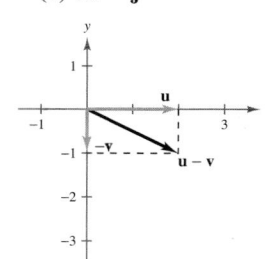

(c) $4\mathbf{i} - 3\mathbf{j}$

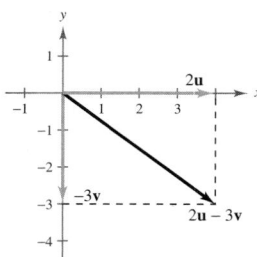

27. $\langle 1, 0 \rangle$ **29.** $\left\langle -\dfrac{\sqrt{2}}{2}, \dfrac{\sqrt{2}}{2} \right\rangle$ **31.** $\dfrac{3\sqrt{10}}{10}\mathbf{i} - \dfrac{\sqrt{10}}{10}\mathbf{j}$

33. $\mathbf{j}$ **35.** $\dfrac{\sqrt{5}}{5}\mathbf{i} - \dfrac{2\sqrt{5}}{5}\mathbf{j}$ **37.** $\left\langle \dfrac{5\sqrt{2}}{2}, \dfrac{5\sqrt{2}}{2} \right\rangle$

39. $\left\langle \dfrac{18\sqrt{29}}{29}, \dfrac{45\sqrt{29}}{2} \right\rangle$

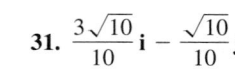

41. $\mathbf{v} = \left\langle 3, -\frac{3}{2} \right\rangle$

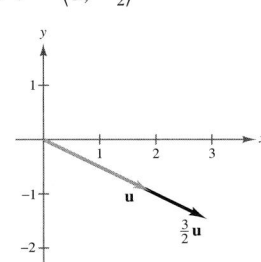

43. $\mathbf{v} = \langle 4, 3 \rangle$

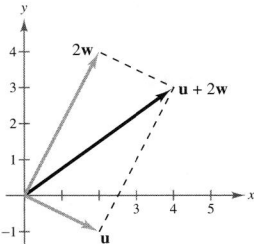

45. $\mathbf{v} = \left\langle \frac{7}{2}, -\frac{1}{2} \right\rangle$

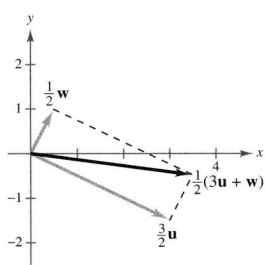

47. $\|\mathbf{v}\| = 3$; $\theta = 60°$ **49.** $\|\mathbf{v}\| = 6\sqrt{2}$; $\theta = 315°$

51. $\mathbf{v} = \langle 3, 0 \rangle$ **53.** $\mathbf{v} = \left\langle -\frac{7\sqrt{3}}{4}, \frac{7}{4} \right\rangle$

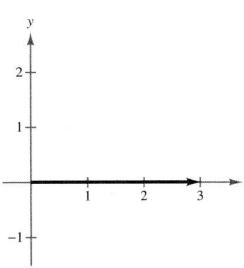

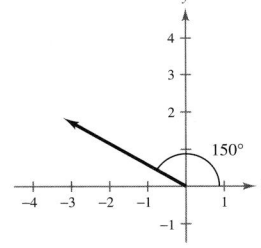

55. $\mathbf{v} = \left\langle -\frac{3\sqrt{6}}{2}, \frac{3\sqrt{2}}{2} \right\rangle$ **57.** $\mathbf{v} = \left\langle \frac{\sqrt{10}}{5}, \frac{3\sqrt{10}}{5} \right\rangle$

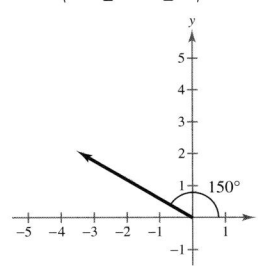

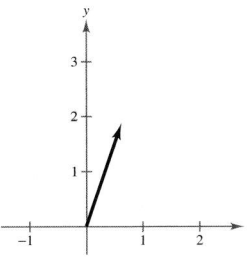

59. $\langle 5, 5 \rangle$ **61.** $\left\langle 10\sqrt{2} - 50, 10\sqrt{2} \right\rangle$ **63.** $90°$

65. $63.4°$ **67.** $62.7°$ **69.** $12.8°$; 398.32 newtons

71. $71.3°$; 228.5 pounds

73. Vertical component: $70 \sin 35° \approx 40.15$ feet per second

Horizontal component: $70 \cos 35° \approx 57.34$ feet per second

75. $T_{AC} \approx 1758.8$ pounds   **77.** 3154.4 pounds

$T_{BC} \approx 1305.4$ pounds

79. N 21.4° E; 138.7 kilometers per hour

81. 1928.4 foot-pounds   **83.** True. See Example 1.

85. (a) $0°$ (b) $180°$

(c) No. The magnitude is at most equal to the sum when the angle between the vectors is $0°$.

87. Answers will vary.   **89.** $\langle 1, 3 \rangle$ or $\langle -1, -3 \rangle$

91. $8 \tan \theta$ **93.** $6 \sec \theta$

95. $\dfrac{\pi}{2} + n\pi, \pi + 2n\pi$ **97.** $n\pi, \dfrac{\pi}{6} + 2n\pi, \dfrac{11\pi}{6} + 2n\pi$

Section 3.4   *(page 311)*

1. -9 **3.** 6   **5.** 8; scalar   **7.** $\langle -6, 8 \rangle$; vector

9. 13   **11.** $5\sqrt{41}$ **13.** 6   **15.** $90°$ **17.** $143.13°$

19. $60.26°$ **21.** $90°$ **23.** $\dfrac{5\pi}{12}$

25. $26.57°, 63.43°, 90°$ **27.** $41.63°, 53.13°, 85.24°$

29. -20 **31.** Parallel   **33.** Neither   **35.** Orthogonal

37. $\frac{1}{37}\langle 84, 14 \rangle, \frac{1}{37}\langle -10, 60 \rangle$ **39.** $\frac{45}{229}\langle 2, 15 \rangle, \frac{6}{229}\langle -15, 2 \rangle$

41. $\langle -5, 3 \rangle, \langle 5, -3 \rangle$ **43.** $\frac{2}{3}\mathbf{i} + \frac{1}{2}\mathbf{j}, -\frac{2}{3}\mathbf{i} - \frac{1}{2}\mathbf{j}$ **45.** 32

47. (a) $58,762.50

This value gives the total revenue that can be earned by selling all of the units.

(b) $1.05\mathbf{v}$

49. 735 newton-meters   **51.** 779.4 foot-pounds

53. False. Work is represented by a scalar.

55. (a) $\theta = \dfrac{\pi}{2}$ (b) $0 \le \theta < \dfrac{\pi}{2}$ (c) $\dfrac{\pi}{2} < \theta \le \pi$

57. Answers will vary.   **59.** $12\sqrt{7}$

61. $0, \dfrac{\pi}{6}, \pi, \dfrac{11\pi}{6}$ **63.** $0, \pi$ **65.** $-\dfrac{253}{325}$ **67.** $\dfrac{204}{325}$

Review Exercises   *(page 314)*

1. $C = 74°, b \approx 13.19, c \approx 13.41$

3. $A = 26°, a \approx 24.89, c \approx 56.23$

5. $C = 66°, a \approx 2.53, b \approx 9.11$

7. $B = 108°, a \approx 11.76, c \approx 21.49$

9. $A \approx 20.41°, C \approx 9.59°, a \approx 20.92$

11. $B \approx 39.48°, C \approx 65.52°, c \approx 48.24$

13. 7.9   **15.** 33.5   **17.** 31.1 meters   **19.** 31.01 feet

21. $A \approx 29.69°, B \approx 52.41°, C \approx 97.90°$

23. $A \approx 29.92°, B \approx 86.18°, C \approx 63.90°$

25. $A = 35°, C = 35°, b \approx 6.55$

27. $A \approx 45.76°, B \approx 91.24°, c \approx 21.42$

29. 615.1 meters **31.** 9.80 **33.** 8.36

35.

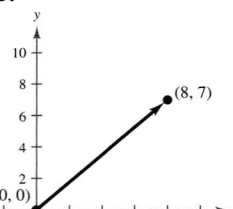

37.

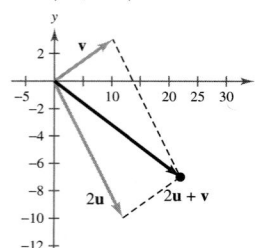

39. $\langle 7, -5 \rangle$ **41.** $\langle 7, -7 \rangle$ **43.** $\langle -4, 4\sqrt{3} \rangle$

45. $\langle 22, -7 \rangle$ **47.** $\langle 30, 9 \rangle$

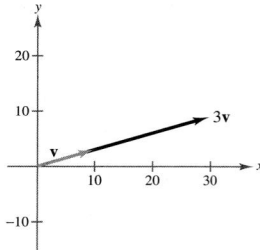

49. $-3\mathbf{i} + 4\mathbf{j}$ **51.** $6\mathbf{i} + 4\mathbf{j}$

53. $10\sqrt{2}(\cos 135° \, \mathbf{i} + \sin 135° \, \mathbf{j})$ **55.** $\|\mathbf{v}\| = 7; \theta = 60°$

57. $\|\mathbf{v}\| = \sqrt{41}; \theta = 38.7°$ **59.** $\|\mathbf{v}\| = 3\sqrt{2}; \theta = 225°$

61. The resultant force is 133.92 pounds and 5.6° from the 85-pound force.

63. 422.30 miles per hour; 130.4° **65.** 45 **67.** −2

69. 50; scalar **71.** $\langle 6, -8 \rangle$; vector **73.** $\dfrac{11\pi}{12}$

75. 160.5° **77.** Orthogonal **79.** Neither

81. $-\frac{13}{17}\langle 4, 1 \rangle, \frac{16}{17}\langle -1, 4 \rangle$ **83.** $\frac{5}{2}\langle -1, 1 \rangle, \frac{9}{2}\langle 1, 1 \rangle$ **85.** 48

87. True. sin 90° is defined in the Law of Sines.

89. True. By definition, $\mathbf{u} = \dfrac{\mathbf{v}}{\|\mathbf{v}\|}$, so $\mathbf{v} = \|\mathbf{v}\|\mathbf{u}$

91. $\dfrac{a}{\sin A} = \dfrac{b}{\sin B} = \dfrac{c}{\sin C}$

93. Direction and magnitude

95. a; The angle between the vectors is acute.

97. The diagonal of the parallelogram with $\mathbf{u}$ and $\mathbf{v}$ as its adjacent sides

Chapter Test *(page 317)*

1. $C = 88°, b \approx 27.81, c \approx 29.98$

2. $A = 43°, b \approx 25.75, c \approx 14.45$

3. Two solutions:

$B \approx 29.12°, C \approx 126.88°, c \approx 22.03$

$B \approx 150.88°, C \approx 5.12°, c \approx 2.46$

4. No solution **5.** $A \approx 39.96°, C \approx 40.04°, c \approx 15.02$

6. $A \approx 23.43°, B \approx 33.57°, c \approx 86.46$

7. 2052.5 square meters **8.** 606.3 miles; 29.1°

9. $\langle 14, -23 \rangle$ **10.** $\left\langle \dfrac{18\sqrt{34}}{17}, -\dfrac{30\sqrt{34}}{17} \right\rangle$

11. $\langle -4, 6 \rangle$ **12.** $\langle 10, 4 \rangle$

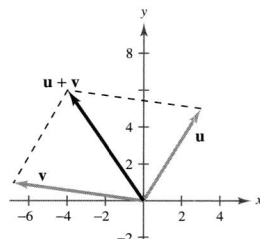 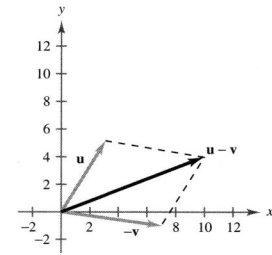

13. $\langle 36, 22 \rangle$ **14.** $\langle \frac{4}{5}, -\frac{3}{5} \rangle$

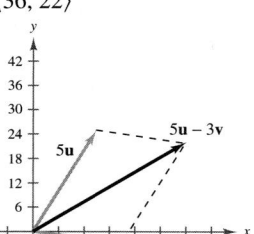

15. 14.9°; 250.15 pounds **16.** 135° **17.** No

18. $\dfrac{37}{26}\langle 5, 1 \rangle; \dfrac{29}{26}\langle -1, 5 \rangle$

Cumulative Test for Chapters 1–3 *(page 318)*

1. (a)

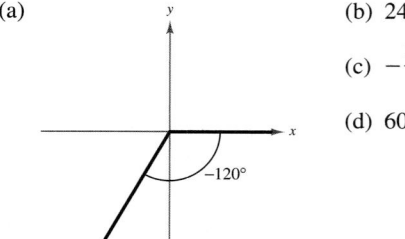

(b) 240°

(c) $-\dfrac{2\pi}{3}$

(d) 60°

(e) $\sin(-120°) = -\dfrac{\sqrt{3}}{2}$ $\qquad \csc(-120°) = -\dfrac{2\sqrt{3}}{3}$

$\cos(-120°) = -\dfrac{1}{2}$ $\qquad \sec(-120°) = -2$

$\tan(-120°) = \sqrt{3}$ $\qquad \cot(-120°) = \dfrac{\sqrt{3}}{3}$

2. $134.6°$ **3.** $\frac{3}{5}$

4. Period: 2; Amplitude: 2 **5.** Period: π

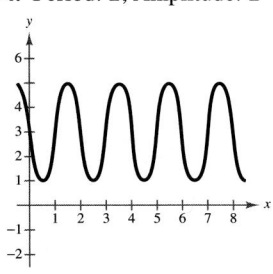 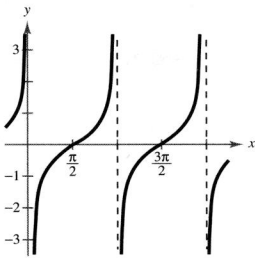

6. $a = -3, b = \pi, c = 0$

7.

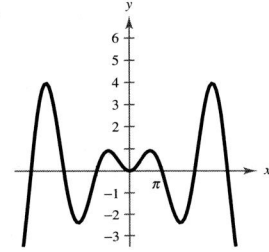

8. 6.7 **9.** $\frac{3}{4}$ **10.** $\sqrt{1 - 4x^2}$ **11.** 1

12. $2 \tan \theta$ **13–15.** Answers will vary.

16. $\frac{\pi}{3}, \frac{\pi}{2}, \frac{3\pi}{2}, \frac{5\pi}{3}$ **17.** $\frac{\pi}{6}, \frac{5\pi}{6}, \frac{7\pi}{6}, \frac{11\pi}{6}$ **18.** $\frac{3\pi}{2}$

19. $\frac{16}{63}$ **20.** $\frac{4}{3}$ **21.** $\frac{\sqrt{5}}{5}, \frac{2\sqrt{5}}{5}$ **22.** $\frac{5}{2}\left(\sin \frac{5\pi}{2} - \sin \pi\right)$

23. $B \approx 26.39°, C \approx 123.61°, c \approx 15.0$

24. $B \approx 52.48°, C \approx 97.52°, a \approx 5.04$

25. $B = 60°, a \approx 5.77, c \approx 11.55$

26. $A = 26.38°, B \approx 62.72°, C \approx 90.90°$

27. 36.4 square inches **28.** 85.2 square inches

29. $3\mathbf{i} + 5\mathbf{j}$ **30.** -5 **31.** $-\frac{1}{13}\langle 1, 5 \rangle; \frac{21}{13}\langle 5, -1 \rangle$

32. 5 feet

33. ≈ 500 revolutions per minute; ≈ 20 minutes

34. $22.6°$ **35.** $d = 4 \cos \frac{\pi}{4}t$

36. $32.6°$; 543.9 kilometers per hour

Problem Solving *(page 324)*

1. 2.01 feet

3. (a) Station A: 27.45 miles; Station B: 53.03 miles

(b) 11.03 miles; S 21.7° E

5. (a) (i) $\sqrt{2}$ (ii) $\sqrt{5}$ (iii) 1

(iv) 1 (v) 1 (vi) 1

(b) (i) 1 (ii) $3\sqrt{2}$ (iii) $\sqrt{13}$

(iv) 1 (v) 1 (vi) 1

(c) (i) $\frac{\sqrt{5}}{2}$ (ii) $\sqrt{13}$ (iii) $\frac{\sqrt{85}}{2}$

(iv) 1 (v) 1 (vi) 1

(d) (i) $2\sqrt{5}$ (ii) $5\sqrt{2}$ (iii) $5\sqrt{2}$

(iv) 1 (v) 1 (vi) 1

7. $\mathbf{w} = \frac{1}{2}(\mathbf{u} + \mathbf{v}); \mathbf{w} = \frac{1}{2}(\mathbf{v} - \mathbf{u})$

9. (a)

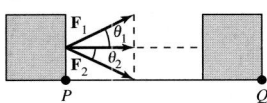

The amount of work done by $\mathbf{F}_1$ is equal to the amount of work done by $\mathbf{F}_2$.

(b)

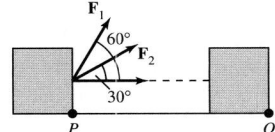

The amount of work done by $\mathbf{F}_2$ is $\sqrt{3}$ times as great as the amount of work done by $\mathbf{F}_1$.

Chapter 4

Section 4.1 *(page 333)*

1. $a = -10, b = 6$ **3.** $a = 6, b = 5$ **5.** $4 + 3i$

7. $2 - 3\sqrt{3}i$ **9.** $5\sqrt{3}i$ **11.** 8 **13.** $-1 - 6i$

15. $0.3i$ **17.** $11 - i$ **19.** 4 **21.** $3 - 3\sqrt{2}i$

23. $-14 + 20i$ **25.** $\frac{1}{6} + \frac{7}{6}i$ **27.** $-2\sqrt{3}$ **29.** -10

31. $5 + i$ **33.** $12 + 30i$ **35.** 24 **37.** $-9 + 40i$

39. -10 **41.** $6 - 3i, 45$ **43.** $-1 + \sqrt{5}i, 6$

45. $-2\sqrt{5}i, 20$ **47.** $\sqrt{8}, 8$ **49.** $-5i$

51. $\frac{8}{41} + \frac{10}{41}i$ **53.** $\frac{4}{5} + \frac{3}{5}i$ **55.** $-5 - 6i$

57. $-\frac{120}{1681} - \frac{27}{1681}i$ **59.** $-\frac{1}{2} - \frac{5}{2}i$ **61.** $\frac{62}{949} + \frac{297}{949}i$

63. $1 \pm i$ **65.** $-2 \pm \frac{1}{2}i$ **67.** $-\frac{3}{2}, -\frac{5}{2}$

69. $2 \pm \sqrt{2}i$ **71.** $\frac{5}{7} \pm \frac{5\sqrt{15}}{7}$ **73.** $-1 + 6i$

75. $-5i$ **77.** $-375\sqrt{3}i$ **79.** i

81. (a) 8 (b) 8 (c) 8

83. (a) 1 (b) i (c) -1 (d) $-i$

85. False. If the complex number is real, the number equals its conjugate.

87. False.

$$i^{44} + i^{150} - i^{74} - i^{109} + i^{61} = 1 - 1 + 1 - i + i = 1$$

89. Answers will vary. **91.** $-x^2 - 3x + 12$

93. $3x^2 + \frac{23}{2}x - 2$ **95.** -31 **97.** $\frac{27}{2}$

99. $a = \dfrac{\sqrt{3V\pi b}}{2\pi b}$ **101.** 1 liter

Section 4.2 *(page 340)*

1. Three solutions **3.** Four solutions

5. No real solutions **7.** Two real solutions

9. Two real solutions **11.** No real solutions

13. $\pm\sqrt{5}$ **15.** $-5 \pm \sqrt{6}$ **17.** 4 **19.** $-1 \pm 2i$

21. $\frac{1}{2} \pm i$ **23.** $20 \pm 2\sqrt{215}$ **25.** $-3 \pm 2\sqrt{2}i$

27.

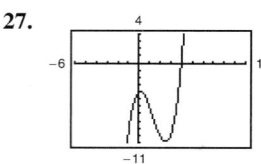

$4, \pm i$; The number of real zeros and the number of x-intercepts is the same.

29.

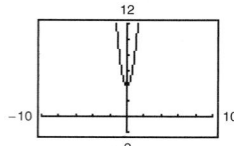

$\pm\sqrt{2}i$; The number of real zeros and the number of x-intercepts is the same.

31. $\pm5i$; $(x + 5i)(x - 5i)$

33. $2 \pm \sqrt{3}$; $\left(x - 2 - \sqrt{3}\right)\left(x - 2 + \sqrt{3}\right)$

35. $\pm3, \pm3i$; $(x + 3)(x - 3)(x + 3i)(x - 3i)$

37. $1 \pm i$; $(z - 1 + i)(z - 1 - i)$

39. $-3, \pm\sqrt{3}$; $(x + 3)\left(x + \sqrt{3}\right)\left(x - \sqrt{3}\right)$

41. $4, \pm4i$; $(x - 4)(x + 4i)(x - 4i)$

43. $\frac{1}{2}, \pm3\sqrt{2}i$; $(2x - 1)\left(x + 3\sqrt{2}i\right)\left(x - 3\sqrt{2}i\right)$

45. $0, 4, \pm6i$; $x(x - 4)(x + 6i)(x - 6i)$

47. $\pm i, \pm3i$; $(x + i)(x - i)(x + 3i)(x - 3i)$

49. $-\frac{3}{2}, \pm5i$ **51.** $\pm2i, 1, -\frac{1}{2}$ **53.** $-3 \pm i, \frac{1}{4}$

55. $2, -3 \pm \sqrt{2}i, 1$ **57.** $x^3 - x^2 + 25x - 25$

59. $x^3 + 4x^2 - 31x - 174$

61. $3x^4 - 17x^3 + 25x^2 + 23x - 22$

63. No. Setting $h = 64$ and solving the resulting equation yields imaginary roots.

65. False. The most complex zeros it can have is two, and the Linear Factorization Theorem guarantees that there are three linear factors, so one zero must be real.

67. r_1, r_2, r_3 **69.** $5 + r_1, 5 + r_2, 5 + r_3$

71. The zeros cannot be determined.

73. (a) $x^2 + b$ (b) $x^2 - 2ax + a^2 + b^2$

75. $-11 + 9i$ **77.** $20 + 40i$

79.

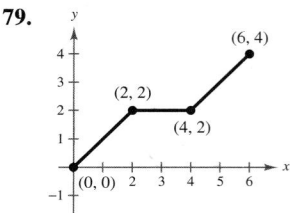

81.

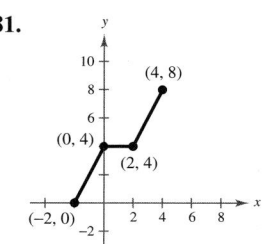

83.

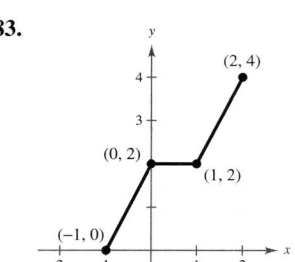

Section 4.3 *(page 347)*

1.

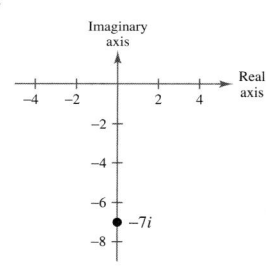

7

3.

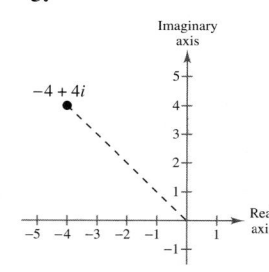

$4\sqrt{2}$

5.

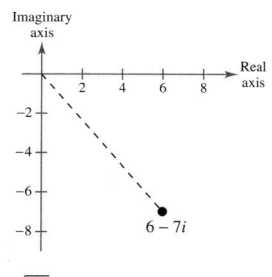

$\sqrt{85}$

7. $3\left(\cos \dfrac{\pi}{2} + i \sin \dfrac{\pi}{2}\right)$ **9.** $\sqrt{10}\left(\cos 5.96 + i \sin 5.96\right)$

11.

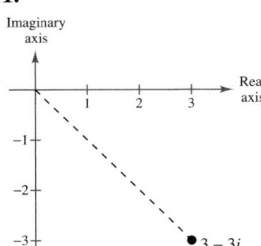

$$3\sqrt{2}\left(\cos\frac{7\pi}{4} + i\sin\frac{7\pi}{4}\right)$$

13.

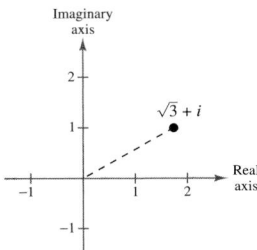

$$2\left(\cos\frac{\pi}{6} + i\sin\frac{\pi}{6}\right)$$

15.

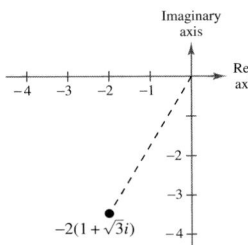

$$4\left(\cos\frac{4\pi}{3} + i\sin\frac{4\pi}{3}\right)$$

17.

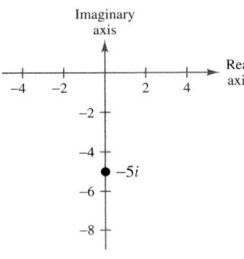

$$5\left(\cos\frac{3\pi}{2} + i\sin\frac{3\pi}{2}\right)$$

19.

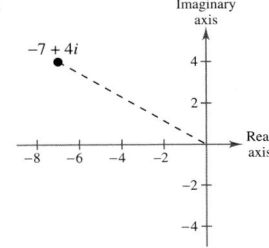

$$\sqrt{65}\left(\cos 2.62 + i\sin 2.62\right)$$

21.

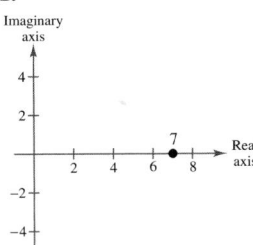

$$7(\cos 0 + i\sin 0)$$

23.

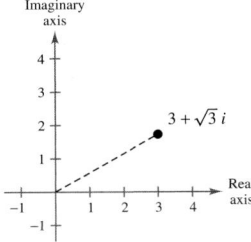

$$2\sqrt{3}\left(\cos\frac{\pi}{6} + i\sin\frac{\pi}{6}\right)$$

25.

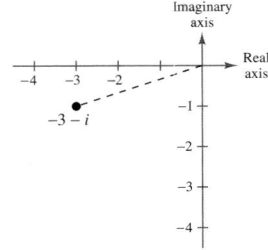

$$\sqrt{10}\left(\cos 3.46 + i\sin 3.46\right)$$

27. $5.39(\cos 0.38 + i\sin 0.38)$

29. $3.16(\cos 2.82 + i\sin 2.82)$

31. $8.19\left(\cos 5.26 + i\sin 5.26\right)$

33. $11.79(\cos 3.97 + i\sin 3.97)$

35.

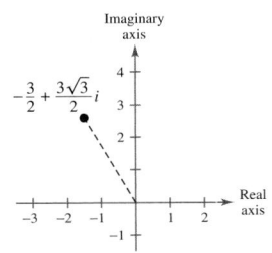

$$-\frac{3}{2} + \frac{3\sqrt{3}}{2}i$$

37.

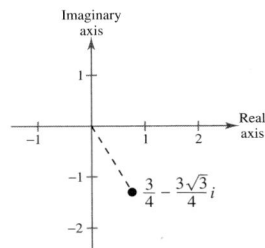

$$\frac{3}{4} - \frac{3\sqrt{3}}{4}i$$

39.

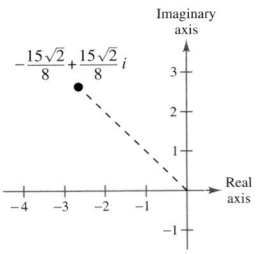

$$-\frac{15\sqrt{2}}{8} + \frac{15\sqrt{2}}{8}i$$

41.

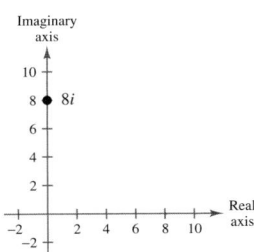

$$8i$$

43.

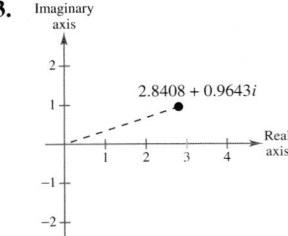

$$2.8408 + 0.9643i$$

45. $4.6985 + 1.7101i$ **47.** $-2.9044 + 0.7511i$

49. $12\left(\cos \dfrac{\pi}{3} + i \sin \dfrac{\pi}{3}\right)$ **51.** $\dfrac{10}{9}(\cos 200° + i \sin 200°)$

53. $0.27(\cos 150° + i \sin 150°)$ **55.** $\cos 30° + i \sin 30°$

57. $\cos \dfrac{2\pi}{3} + i \sin \dfrac{2\pi}{3}$ **59.** $4(\cos 302° + i \sin 302°)$

61. (a) $\left[2\sqrt{2}\left(\cos \dfrac{\pi}{4} + i \sin \dfrac{\pi}{4}\right)\right]\left[\sqrt{2}\left(\cos \dfrac{7\pi}{4} + i \sin \dfrac{7\pi}{4}\right)\right]$

(b) $4(\cos 0 + i \sin 0) = 4$

(c) 4

63. (a) $2\left(\cos \dfrac{3\pi}{2} + i \sin \dfrac{3\pi}{2}\right)\left[\sqrt{2}\left(\cos \dfrac{\pi}{4} + i \sin \dfrac{\pi}{4}\right)\right]$

(b) $2\sqrt{2}\left(\cos \dfrac{7\pi}{4} + i \sin \dfrac{7\pi}{4}\right) = 2 - 2i$

(c) $-2i - 2i^2 = -2i + 2 = 2 - 2i$

65. (a) $[5(\cos 0.93 + i \sin 0.93)] \div \left[2\left(\cos \dfrac{5\pi}{3} + i \sin \dfrac{5\pi}{3}\right)\right]$

(b) $\dfrac{5}{2}(\cos 1.97 + i \sin 1.97) = -0.982 + 2.299i$

(c) $\approx -0.982 + 2.299i$

67. (a) $[5(\cos 0 + i \sin 0)] \div \left[\sqrt{13}(\cos 0.98 + i \sin 0.98)\right]$

(b) $\dfrac{5}{\sqrt{13}}([\cos 5.30 + i \sin 5.30) \approx 0.769 - 1.154i$

(c) $\dfrac{10}{13} - \dfrac{15}{13}i \approx 0.769 - 1.154i$

69.

71.

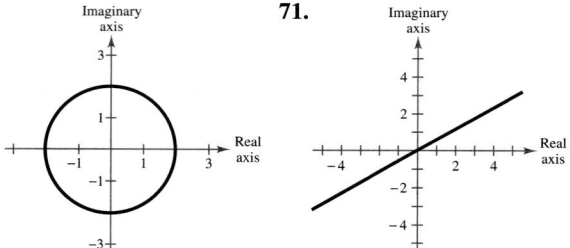

73. True, by the definition of the absolute value of a complex number.

75. Answers will vary.

77. (a) r^2 (b) $\cos 2\theta + i \sin 2\theta$

79. $B = 68°, b \approx 19.80, c \approx 21.36$

81. $B = 60°, a \approx 65.01, c \approx 130.02$

83. $B = 47°45', a \approx 7.53, b \approx 8.29$

85. $16; 2$ **87.** $\frac{1}{16}; 0$

Section 4.4 *(page 353)*

1. $-4 - 4i$ **3.** $-32i$ **5.** $-128\sqrt{3} - 128i$

7. $\dfrac{125}{2} + \dfrac{125\sqrt{3}}{2}i$ **9.** -1

11. $608.0204 + 144.6936i$

13. $-597 - 122i$ **15.** $-43\sqrt{5} + 4i$

17. $\dfrac{81}{2} + \dfrac{81\sqrt{3}}{2}i$ **19.** $32.3525 - 120.7407i$

21. $32i$ **23.** 27

25. (a) $\sqrt{5}(\cos 60° + i \sin 60°)$

$\sqrt{5}(\cos 240° + i \sin 240°)$

(b)

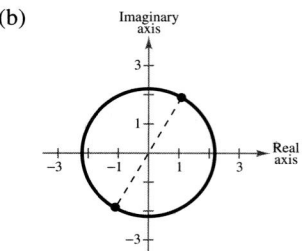

(c) $\dfrac{\sqrt{5}}{2} + \dfrac{\sqrt{15}}{2}i, \ -\dfrac{\sqrt{5}}{2} - \dfrac{\sqrt{15}}{2}i$

27. (a) $2\left(\cos \dfrac{2\pi}{9} + i \sin \dfrac{2\pi}{9}\right)$

$2\left(\cos \dfrac{8\pi}{9} + i \sin \dfrac{8\pi}{9}\right)$

$2\left(\cos \dfrac{14\pi}{9} + i \sin \dfrac{14\pi}{9}\right)$

(b)

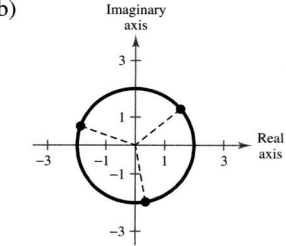

(c) $1.5321 + 1.2856i, -1.8794 + 0.6840i,$

$0.3473 - 1.9696i$

29. (a) $3\left(\cos \dfrac{\pi}{30} + i \sin \dfrac{\pi}{30}\right)$

$3\left(\cos \dfrac{13\pi}{30} + i \sin \dfrac{13\pi}{30}\right)$

$3\left(\cos \dfrac{5\pi}{6} + i \sin \dfrac{5\pi}{6}\right)$

$3\left(\cos \dfrac{37\pi}{30} + i \sin \dfrac{37\pi}{30}\right)$

$3\left(\cos \dfrac{49\pi}{30} + i \sin \dfrac{49\pi}{30}\right)$

(b)

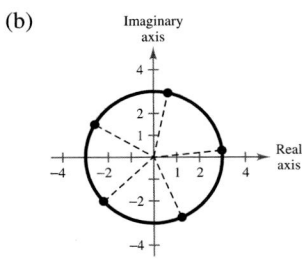

(c) $2.9836 + 0.3136i, 0.6237 + 2.9344i,$
$-2.5981 + 1.5i, -2.2294 - 2.0074i, 1.2202 - 2.7406i$

31. (a) $5\left(\cos \dfrac{3\pi}{4} + i \sin \dfrac{3\pi}{4}\right)$

$5\left(\cos \dfrac{7\pi}{4} + i \sin \dfrac{7\pi}{4}\right)$

(b)

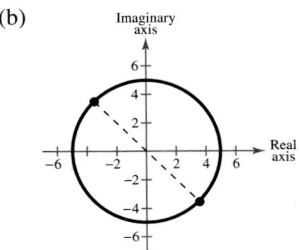

(c) $-\dfrac{5\sqrt{2}}{2} + \dfrac{5\sqrt{2}}{2}i, \dfrac{5\sqrt{2}}{2} - \dfrac{5\sqrt{2}}{2}i$

33. (a) $3\left(\cos \dfrac{\pi}{8} + i \sin \dfrac{\pi}{8}\right)$

$3\left(\cos \dfrac{5\pi}{8} + i \sin \dfrac{5\pi}{8}\right)$

$3\left(\cos \dfrac{9\pi}{8} + i \sin \dfrac{9\pi}{8}\right)$

$3\left(\cos \dfrac{13\pi}{8} + i \sin \dfrac{13\pi}{8}\right)$

(b)

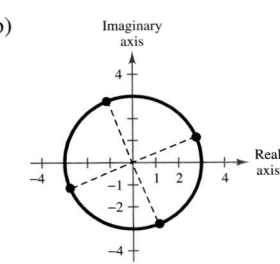

(c) $2.7716 + 1.1481i, -1.1481 + 2.7716i,$
$-2.7716 - 1.1481i, 1.1481 - 2.7716i$

35. (a) $5\left(\cos \dfrac{4\pi}{9} + i \sin \dfrac{4\pi}{9}\right)$

$5\left(\cos \dfrac{10\pi}{9} + i \sin \dfrac{10\pi}{9}\right)$

$5\left(\cos \dfrac{16\pi}{9} + i \sin \dfrac{16\pi}{9}\right)$

(b)

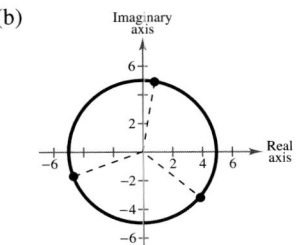

(c) $0.8682 + 4.9240i, -4.6985 - 1.7101i,$
$3.8302 - 3.2140i$

37. (a) $2(\cos 0 + i \sin 0)$

$2\left(\cos \dfrac{\pi}{2} + i \sin \dfrac{\pi}{2}\right)$

$2(\cos \pi + i \sin \pi)$

$2\left(\cos \dfrac{3\pi}{2} + i \sin \dfrac{3\pi}{2}\right)$

(b)

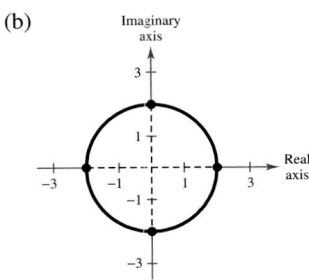

(c) $2, 2i, -2, -2i$

39. (a) $\cos 0 + i \sin 0$

$\cos \dfrac{2\pi}{5} + i \sin \dfrac{2\pi}{5}$

$\cos \dfrac{4\pi}{5} + i \sin \dfrac{4\pi}{5}$

$\cos \dfrac{6\pi}{5} + i \sin \dfrac{6\pi}{5}$

$\cos \dfrac{8\pi}{5} + i \sin \dfrac{8\pi}{5}$

(b)

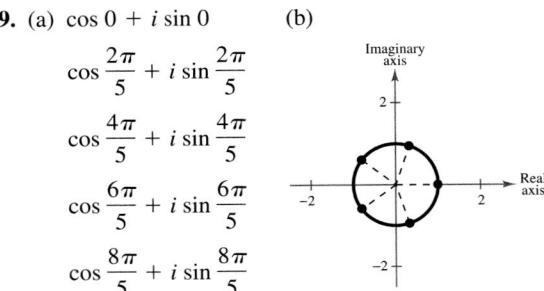

(c) $1, 0.3090 + 0.9511i, -0.8090 + 0.5878i,$
$-0.8090 - 0.5878i, 0.3090 - 0.9511i$

41. (a) $5\left(\cos\dfrac{\pi}{3} + i\sin\dfrac{\pi}{3}\right)$

$5(\cos\pi + i\sin\pi)$

$5\left(\cos\dfrac{5\pi}{3} + i\sin\dfrac{5\pi}{3}\right)$

(b)

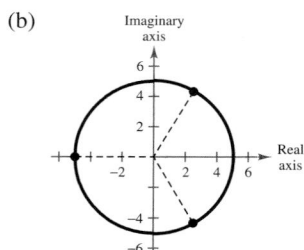

(c) $\dfrac{5}{2} + \dfrac{5\sqrt{3}}{2}i,\ -5,\ \dfrac{5}{2} - \dfrac{5\sqrt{3}}{2}i$

43. (a) $2\sqrt[5]{4\sqrt{2}}\left(\cos\dfrac{3\pi}{20} + i\sin\dfrac{3\pi}{20}\right)$

$2\sqrt[5]{4\sqrt{2}}\left(\cos\dfrac{11\pi}{20} + i\sin\dfrac{11\pi}{20}\right)$

$2\sqrt[5]{4\sqrt{2}}\left(\cos\dfrac{19\pi}{20} + i\sin\dfrac{19\pi}{20}\right)$

$2\sqrt[5]{4\sqrt{2}}\left(\cos\dfrac{27\pi}{20} + i\sin\dfrac{27\pi}{20}\right)$

$2\sqrt[5]{4\sqrt{2}}\left(\cos\dfrac{7\pi}{4} + i\sin\dfrac{7\pi}{4}\right)$

(b)

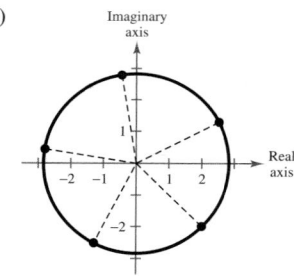

(c) $2.5201 + 1.2841i,\ -0.4425 + 2.7936i,$

$-2.7936 + 0.4425i,\ -1.2841 - 2.5201i,\ 2 - 2i$

45. $\cos\dfrac{3\pi}{8} + i\sin\dfrac{3\pi}{8}$

$\cos\dfrac{7\pi}{8} + i\sin\dfrac{7\pi}{8}$

$\cos\dfrac{11\pi}{8} + i\sin\dfrac{11\pi}{8}$

$\cos\dfrac{15\pi}{8} + i\sin\dfrac{15\pi}{8}$

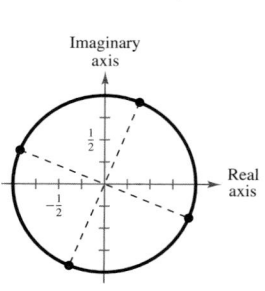

47. $\cos\dfrac{\pi}{6} + i\sin\dfrac{\pi}{6}$

$\cos\dfrac{\pi}{2} + i\sin\dfrac{\pi}{2}$

$\cos\dfrac{5\pi}{6} + i\sin\dfrac{5\pi}{6}$

$\cos\dfrac{7\pi}{6} + i\sin\dfrac{7\pi}{6}$

$\cos\dfrac{3\pi}{2} + i\sin\dfrac{3\pi}{2}$

$\cos\dfrac{11\pi}{6} + i\sin\dfrac{11\pi}{6}$

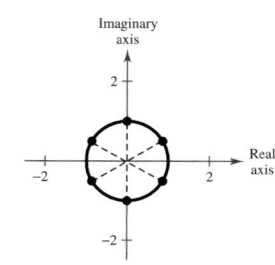

49. $3\left(\cos\dfrac{\pi}{5} + i\sin\dfrac{\pi}{5}\right)$

$3\left(\cos\dfrac{3\pi}{5} + i\sin\dfrac{3\pi}{5}\right)$

$3(\cos\pi + i\sin\pi)$

$3\left(\cos\dfrac{7\pi}{5} + i\sin\dfrac{7\pi}{5}\right)$

$3\left(\cos\dfrac{9\pi}{5} + i\sin\dfrac{9\pi}{5}\right)$

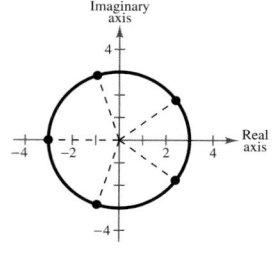

51. $2(\cos 0 + i\sin 0)$

$2\left(\cos\dfrac{2\pi}{5} + i\sin\dfrac{2\pi}{5}\right)$

$2\left(\cos\dfrac{4\pi}{5} + i\sin\dfrac{4\pi}{5}\right)$

$2\left(\cos\dfrac{6\pi}{5} + i\sin\dfrac{6\pi}{5}\right)$

$2\left(\cos\dfrac{8\pi}{5} + i\sin\dfrac{8\pi}{5}\right)$

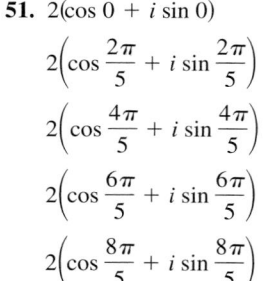

53. $2\left(\cos\dfrac{3\pi}{8} + i\sin\dfrac{3\pi}{8}\right)$

$2\left(\cos\dfrac{7\pi}{8} + i\sin\dfrac{7\pi}{8}\right)$

$2\left(\cos\dfrac{11\pi}{8} + i\sin\dfrac{11\pi}{8}\right)$

$2\left(\cos\dfrac{15\pi}{8} + i\sin\dfrac{15\pi}{8}\right)$

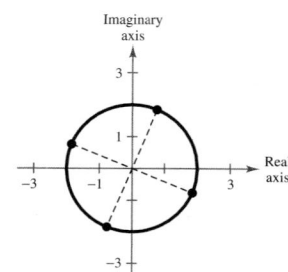

55. $2\left(\cos\dfrac{\pi}{8} + i\sin\dfrac{\pi}{8}\right)$

$2\left(\cos\dfrac{5\pi}{8} + i\sin\dfrac{5\pi}{8}\right)$

$2\left(\cos\dfrac{9\pi}{8} + i\sin\dfrac{9\pi}{8}\right)$

$2\left(\cos\dfrac{13\pi}{8} + i\sin\dfrac{13\pi}{8}\right)$

57. $\sqrt[6]{2}\left(\cos\dfrac{7\pi}{12} + i\sin\dfrac{7\pi}{12}\right)$

$\sqrt[6]{2}\left(\cos\dfrac{5\pi}{4} + i\sin\dfrac{5\pi}{4}\right)$

$\sqrt[6]{2}\left(\cos\dfrac{23\pi}{12} + i\sin\dfrac{23\pi}{12}\right)$

59. $\sqrt[12]{2}\left(\cos\dfrac{5\pi}{24} + i\sin\dfrac{5\pi}{24}\right)$

$\sqrt[12]{2}\left(\cos\dfrac{13\pi}{24} + i\sin\dfrac{13\pi}{24}\right)$

$\sqrt[12]{2}\left(\cos\dfrac{7\pi}{8} + i\sin\dfrac{7\pi}{8}\right)$

$\sqrt[12]{2}\left(\cos\dfrac{29\pi}{24} + i\sin\dfrac{29\pi}{24}\right)$

$\sqrt[12]{2}\left(\cos\dfrac{37\pi}{24} + i\sin\dfrac{37\pi}{24}\right)$

$\sqrt[12]{2}\left(\cos\dfrac{15\pi}{8} + i\sin\dfrac{15\pi}{8}\right)$

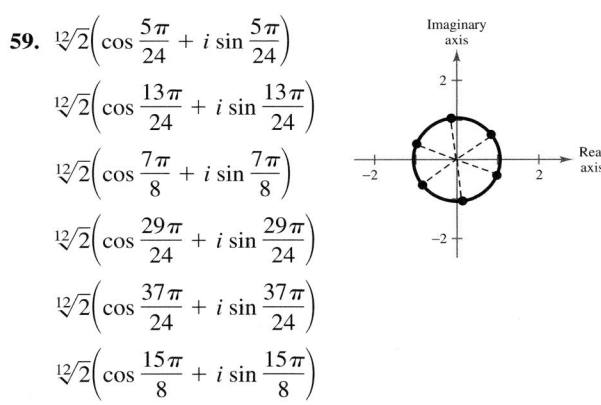

61. False. They are equally spaced along the circle centered at the origin with radius $\sqrt[n]{r}$.

63. Answers will vary.

65. (a) $2(\cos 30° + i\sin 30°)$

$2(\cos 150° + i\sin 150°)$

$2(\cos 270° + i\sin 270°)$

(b) $8i$

67. $\dfrac{\sqrt{34}}{136}(5\sqrt{7} - 9)$ **69.** $\dfrac{3\sqrt{34}}{136}(\sqrt{7} - 5)$

71. $\dfrac{120 + 51\sqrt{7}}{47}$ **73.** $\dfrac{15}{8}$ **75.** $\dfrac{1}{2}\sqrt{\dfrac{4 - \sqrt{7}}{2}}$

77. $\langle 1, 0\rangle$ **79.** $\dfrac{12}{13}\mathbf{i} - \dfrac{5}{13}\mathbf{j}$

81. 5   **83.** 41   **85.** $\sqrt{493}$ **87.** $\sqrt{205}$

Review Exercises   *(page 356)*

1. $6 + 2i$ **3.** $-1 + 3i$ **5.** $3 + 7i$

7. $40 + 65i$ **9.** $-4 - 46i$ **11.** $\dfrac{23}{17} + \dfrac{10}{17}i$

13. $\dfrac{21}{13} - \dfrac{1}{13}i$ **15.** $\pm\dfrac{\sqrt{3}}{3}i$ **17.** $1 \pm 3i$

19. Five solutions   **21.** Four solutions

23. Two real solutions   **25.** No real solutions

27. $0, 2$ **29.** $-4 \pm \sqrt{6}$

31. $-\dfrac{1}{2} \pm \dfrac{\sqrt{5}}{2}i$ **33.** $\dfrac{3}{2}, \pm 5i$ **35.** $\pm\dfrac{\sqrt{5}}{2}, \pm\sqrt{2}i$

37. $-7, 2; (x + 7)(x - 2)^2$

39. $-5, 1 \pm 2i; (x + 5)(x - 1 - 2i)(x - 1 + 2i)$

41. $-\dfrac{1}{2}, 5 \pm 3i; (2x + 1)(x - 5 - 3i)(x - 5 + 3i)$

43. $-2, 3, -3 \pm \sqrt{5}i;$

$(x - 3)(x + 2)(x + 3 - \sqrt{5}i)(x + 3 + \sqrt{5}i)$

45. $12x^4 - 19x^3 + 9x - 2$ **47.** $x^3 - 7x^2 + 13x - 3$

49. $3x^4 - 14x^3 + 17x^2 - 42x + 24$ **51.** $x^4 + 27x^2 + 50$

53. Yes. A price of \$95.41 or \$119.59 per unit would yield a profit of 9 million dollars.

55.   **57.**

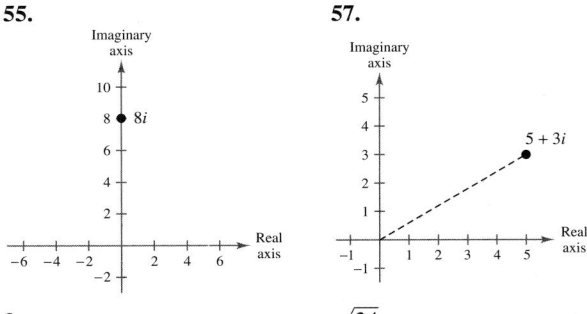

8 $\sqrt{34}$

59. $5\sqrt{2}\left(\cos\dfrac{7\pi}{4} + i\sin\dfrac{7\pi}{4}\right)$

61. $6\left(\cos\dfrac{5\pi}{6} + i\sin\dfrac{5\pi}{6}\right)$

63. (a) $z_1 = 4\left(\cos\dfrac{11\pi}{6} + i\sin\dfrac{11\pi}{6}\right)$

$z_2 = 10\left(\cos\dfrac{3\pi}{2} + i\sin\dfrac{3\pi}{2}\right)$

(b) $z_1 z_2 = 40\left(\cos\dfrac{10\pi}{3} + i\sin\dfrac{10\pi}{3}\right)$

$\dfrac{z_1}{z_2} = \dfrac{2}{5}\left(\cos\dfrac{\pi}{3} + i\sin\dfrac{\pi}{3}\right)$

65. $\dfrac{625}{2} + \dfrac{625\sqrt{3}}{2}i$ **67.** $2035 - 828i$

69. $3\left(\cos \dfrac{\pi}{4} + i \sin \dfrac{\pi}{4}\right)$

$3\left(\cos \dfrac{7\pi}{12} + i \sin \dfrac{7\pi}{12}\right)$

$3\left(\cos \dfrac{11\pi}{12} + i \sin \dfrac{11\pi}{12}\right)$

$3\left(\cos \dfrac{5\pi}{4} + i \sin \dfrac{5\pi}{4}\right)$

$3\left(\cos \dfrac{19\pi}{12} + i \sin \dfrac{19\pi}{12}\right)$

$3\left(\cos \dfrac{23\pi}{12} + i \sin \dfrac{23\pi}{12}\right)$

71. $3\left(\cos \dfrac{\pi}{4} + i \sin \dfrac{\pi}{4}\right) = \dfrac{3\sqrt{2}}{2} + \dfrac{3\sqrt{2}}{2}i$

$3\left(\cos \dfrac{3\pi}{4} + i \sin \dfrac{3\pi}{4}\right) = -\dfrac{3\sqrt{2}}{2} + \dfrac{3\sqrt{2}}{2}i$

$3\left(\cos \dfrac{5\pi}{4} + i \sin \dfrac{5\pi}{4}\right) = -\dfrac{3\sqrt{2}}{2} - \dfrac{3\sqrt{2}}{2}i$

$3\left(\cos \dfrac{7\pi}{4} + i \sin \dfrac{7\pi}{4}\right) = \dfrac{3\sqrt{2}}{2} - \dfrac{3\sqrt{2}}{2}i$

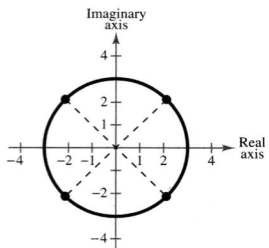

73. $2\left(\cos \dfrac{\pi}{2} + i \sin \dfrac{\pi}{2}\right) = 2i$

$2\left(\cos \dfrac{7\pi}{6} + i \sin \dfrac{7\pi}{6}\right) = -\sqrt{3} - i$

$2\left(\cos \dfrac{11\pi}{6} + i \sin \dfrac{11\pi}{6}\right) = \sqrt{3} - i$

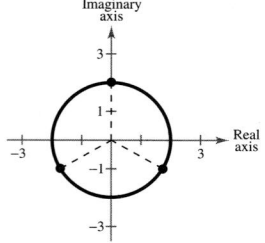

75. False.

$\sqrt{-18}\sqrt{-2} = 3\sqrt{2}i\,\sqrt{2}i$ and $\sqrt{(-18)(-2)} = \sqrt{36}$

$\qquad\qquad\quad = 3\sqrt{4}i^2 \qquad\qquad\qquad\qquad = 6$

$\qquad\qquad\quad = 6i^2$

$\qquad\qquad\quad = -6$

77. False. A fourth-degree polynomial can have at most four zeros, and complex zeros occur in conjugate pairs.

79. (a) $4(\cos 60° + i \sin 60°)$

$\qquad 4(\cos 180° + i \sin 180°)$

$\qquad 4(\cos 300° + i \sin 300°)$

(b) -64

81. $z_1 z_2 = -4, \dfrac{z_1}{z_2} = -i$

Chapter Test   *(page 358)*

1. $-3 + 9i$ **2.** $-3 + 5i$ **3.** $-32 + 24i$

4. 7 **5.** $2 - i$ **6.** $\dfrac{1}{2} \pm \dfrac{\sqrt{5}}{2}i$

7. Five solutions   **8.** Four solutions

9. $6, \pm\sqrt{5}i$ **10.** $\pm\sqrt{6}, \pm 2i$

11. $\pm 2, \pm\sqrt{2}i; (x + 2)(x - 2)\left(x + \sqrt{2}i\right)\left(x - \sqrt{2}i\right)$

12. $\dfrac{3}{2}, 2 \pm i; (2v - 3)(v - 2 - i)(v - 2 + i)$

13. $x^4 - 9x^3 + 28x^2 - 30x$

14. $x^4 - 8x^3 + 28x^2 - 60x + 63$

15. No. If $a + bi$ is a zero, its conjugate $a - bi$ is also a zero.

16. $5\sqrt{2}\left(\cos \dfrac{7\pi}{4} + i \sin \dfrac{7\pi}{4}\right)$

17. $-3 + 3\sqrt{3}i$ **18.** $-\dfrac{6561}{2} - \dfrac{6561\sqrt{3}}{2}i$ **19.** $5832i$

20. $4\sqrt[4]{2}\left(\cos \dfrac{\pi}{12} + i \sin \dfrac{\pi}{12}\right)$

$4\sqrt[4]{2}\left(\cos \dfrac{7\pi}{12} + i \sin \dfrac{7\pi}{12}\right)$

$4\sqrt[4]{2}\left(\cos \dfrac{13\pi}{12} + i \sin \dfrac{13\pi}{12}\right)$

$4\sqrt[4]{2}\left(\cos \dfrac{19\pi}{12} + i \sin \dfrac{19\pi}{12}\right)$

21. $3\left(\cos \dfrac{\pi}{6} + i \sin \dfrac{\pi}{6}\right)$

$3\left(\cos \dfrac{5\pi}{6} + i \sin \dfrac{5\pi}{6}\right)$

$3\left(\cos \dfrac{3\pi}{2} + i \sin \dfrac{3\pi}{2}\right)$

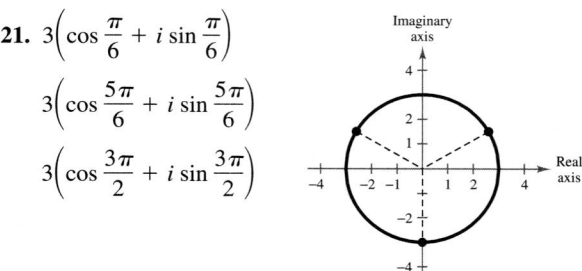

Problem Solving   *(page 360)*

1. (a) $z^3 = 8$ for all three complex numbers.

(b) $z^3 = 27$ for all three complex numbers.

(c) The complex numbers are the cube roots of z^3.

3. $(a + bi)(a - bi)$

$= a^2 + abi - abi - b^2i^2$

$= a^2 + b^2$

5. (a) $k > 1$ (b) $k < 1$

7. (a) No   (b) Yes

9. (a) $-2, 1, 4$

(b) The graph touches the x-axis at $x = 1$.

(c) The least possible degree of the function is 4, because there are at least four real zeros (1 is repeated) and a function can have at most the number of real zeros equal to the degree of the function. The product of two odd powers and an even power will always be an even degree polynomial.

(d) Positive. From the information in the table, it can be concluded that the graph will eventually rise to the left and rise to the right.

(e) $f(x) = x^4 - 4x^3 - 3x^2 + 14x - 8$

(f)

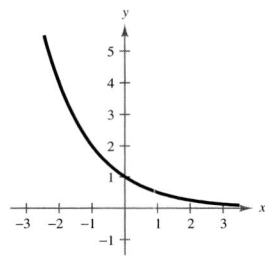

11. (a)

Function	Zeros	Sum of zeros	Product of zeros
$f_1(x)$	$2, 3$	5	6
$f_2(x)$	$-3, 1, 2$	0	-6
$f_3(x)$	$-3, 1, \pm 2i$	-2	-12
$f_4(x)$	$0, 2, -3, 2, \pm\sqrt{3}$	3	0

(b) The sum of the zeros is equal to the opposite of the coefficient of the $(n-1)^{\text{th}}$ term.

(c) The product of the zeros is equal to the constant term if the function is of an even degree and to the opposite of the constant term if the function is of an odd degree.

13. Answers will vary.; circle

Chapter 5

Section 5.1   *(page 372)*

1. 946.852   **3.** 0.006   **5.** 0.472

7. d   **8.** c   **9.** a   **10.** b

11. Shift the graph of four units to the right.

13. Shift the graph of five units upward.

15. Reflect f in the x-axis and shift four units to the left.

17. Reflect f in the x-axis and shift five units upward.

19.

x	-2	-1	0	1	2
$f(x)$	4	2	1	0.5	0.25

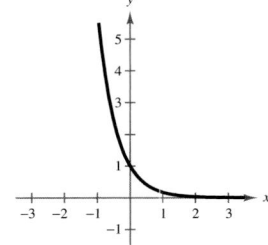

21.

x	-2	-1	0	1	2
$f(x)$	36	6	1	0.167	0.028

23.

x	-2	-1	0	1	2
$f(x)$	0.125	0.25	0.5	1	2

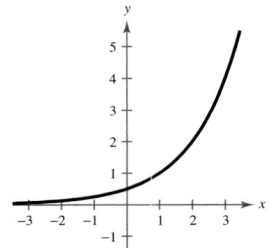

25.

x	−2	−1	0	1	2
f(x)	0.135	0.368	1	2.718	7.389

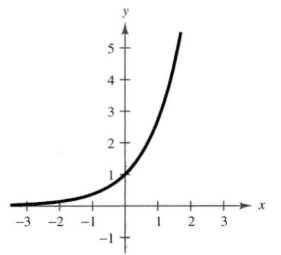

27.

x	−8	−7	−6	−5	−4
f(x)	0.055	0.149	0.406	1.104	3

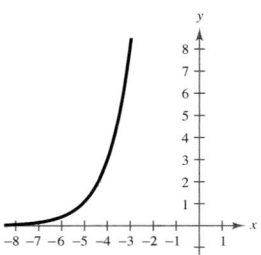

29.

x	−2	−1	0	1	2
f(x)	4.037	4.100	4.271	4.736	6

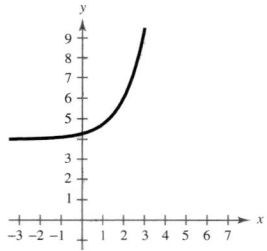

31.

x	−1	0	1	2	3
f(x)	3.004	3.016	3.063	3.25	4

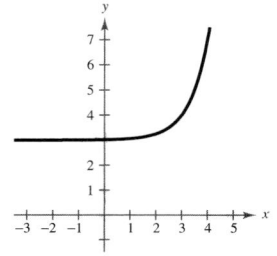

33.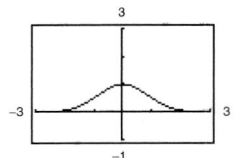

35.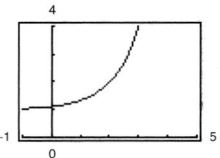

37.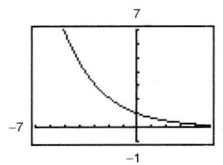

39.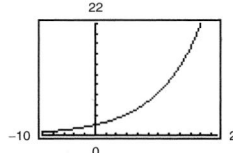

41.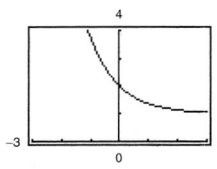

43.

n	1	2	4
A	\$5397.31	\$5477.81	\$5520.10

n	12	365	Continuous
A	\$5549.10	\$5563.36	\$5563.85

45.

n	1	2	4
A	\$11,652.39	\$12,002.55	\$12,188.60

n	12	365	Continuous
A	\$12,317.01	\$12,380.41	\$12,382.58

47.

t	10	20
A	\$26,706.49	\$59,436.39

t	30	40	50
A	\$132,278.12	\$294,390.36	\$655,177.80

49.

t	10	20
A	\$22,986.49	\$44,031.56

t	30	40	50
A	\$84,344.25	\$161,564.86	\$309,484.08

51. \$222,822.57

53. \$35.45   **55.** (a) 100   (b) 300   (c) 900

57. (a) 25 grams (b) 16.30 grams

(c)

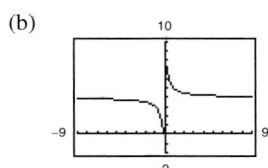

59. (a)

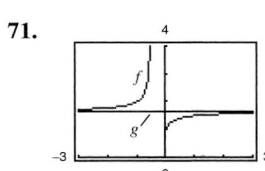

(b)

x	0	25	50	75	100
y	13	45	82	96	99

(c) 63.1% (d) 38.2

61. True. As $x \to -\infty$, $f(x) \to -2$ but never reaches -2.

63. $f(x) = h(x)$ **65.** $f(x) = g(x) = h(x)$

67. (a) $x < 0$ (b) $x > 0$

69. (a)

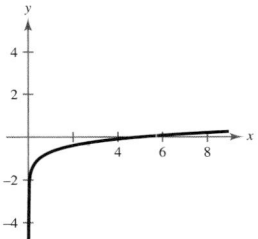

Horizontal asymptotes: $y = 0$, $y = 8$

(b)

Horizontal asymptote: $y = 4$

Vertical asymptote: $x = 0$

71.

As $x \to \infty$, $f(x) \to g(x)$.

As $x \to -\infty$, $f(x) \to g(x)$.

73. c, d **75.** $y = \frac{1}{7}(2x + 14)$ **77.** $y = \pm\sqrt{25 - x^2}$

Section 5.2 *(page 382)*

1. $4^3 = 64$ **3.** $7^{-2} = \frac{1}{49}$ **5.** $32^{2/5} = 4$

7. $e^{-0.693\ldots} = \frac{1}{2}$ **9.** $\log_5 125 = 3$ **11.** $\log_{81} 3 = \frac{1}{4}$

13. $\log_6 \frac{1}{36} = -2$ **15.** $\ln 20.0855\ldots = 3$

17. $\ln 4 = x$ **19.** 4 **21.** 0 **23.** 3 **25.** 2

27. -0.097 **29.** 2.913 **31.** -0.575 **33.** c

34. f **35.** d **36.** e **37.** b **38.** a

39. Domain: $(0, \infty)$

x-intercept: $(1, 0)$

Vertical asymptote: $x = 0$

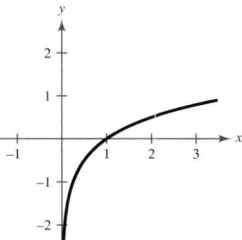

41. Domain: $(0, \infty)$

x-intercept: $(9, 0)$

Vertical asymptote: $x = 0$

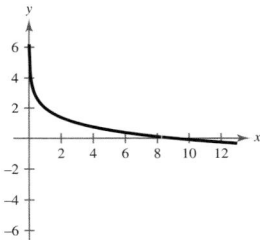

43. Domain: $(-2, \infty)$

x-intercept: $(-1, 0)$

Vertical asymptote: $x = -2$

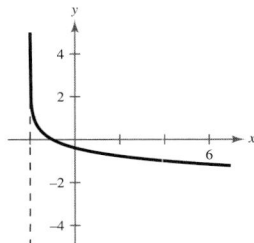

45. Domain: $(0, \infty)$

x-intercept: $(5, 0)$

Vertical asymptote: $x = 0$

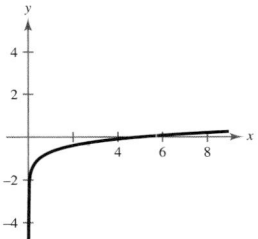

47. Domain: $(2, \infty)$

x-intercept: $(3, 0)$

Vertical asymptote: $x = 2$

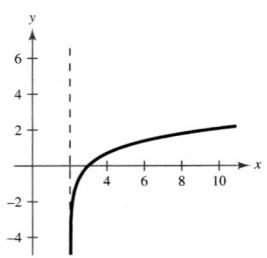

49. Domain: $(-\infty, 0)$

x-intercept: $(-1, 0)$

Vertical asymptote: $x = 0$

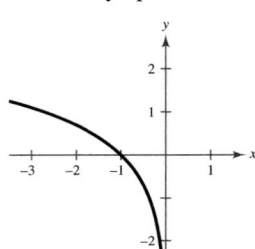

51.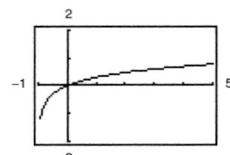

53.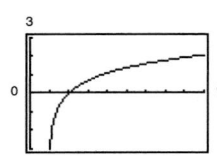

55.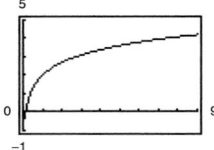

57. (a) 30 years; 20 years   (b) \$396,234; \$301,123.20

(c) \$246,234; \$151,123.20

(d) $x = 1000$; The monthly payment must be greater than \$1000.

59. (a)

r	0.005	0.01	0.015	0.02	0.025	0.03
t	138.6	69.3	46.2	34.7	27.7	23.1

(b)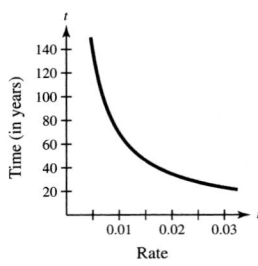

(c) Answers will vary.

61. (a)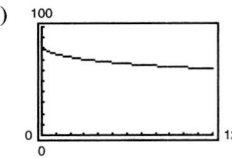

(b) 80   (c) 68.1   (d) 62.3

63. False. Reflecting $g(x)$ about the line $y = x$ will determine the graph of $f(x)$.

65.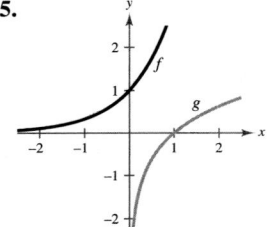

$g = f^{-1}$

67.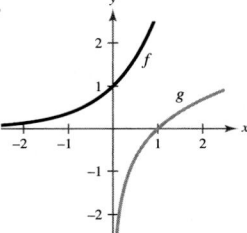

$g = f^{-1}$

69. (a)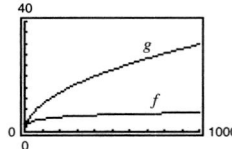

$g(x)$; The natural log function grows at a slower rate than the square root function.

(b)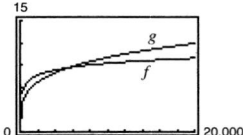

$g(x)$; The natural log function grows at a slower rate than the fourth root function.

71. $(0, \infty)$ **73.** $3 < x < 4$

75. (a)

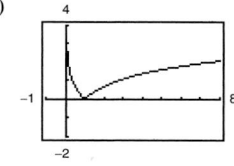

(b) Increasing: $(1, \infty)$

Decreasing: $(0, 1)$

(c) Relative minimum: $(1, 0)$

77. $83.95 + 37.50t$

Section 5.3 *(page 389)*

1. (a) $\dfrac{\log_{10} x}{\log_{10} 5}$ (b) $\dfrac{\ln x}{\ln 5}$ **3.** (a) $\dfrac{\log_{10} x}{\log_{10} \frac{1}{5}}$ (b) $\dfrac{\ln x}{\ln \frac{1}{5}}$

5. (a) $\dfrac{\log_{10} \frac{3}{10}}{\log_{10} x}$ (b) $\dfrac{\ln \frac{3}{10}}{\ln x}$ **7.** (a) $\dfrac{\log_{10} x}{\log_{10} 2.6}$ (b) $\dfrac{\ln x}{\ln 2.6}$

9. 1.771 **11.** -2.000 **13.** -0.417 **15.** 2.633

17. $\log_4 5 + \log_4 x$ **19.** $4 \log_8 x$ **21.** $1 - \log_5 x$

23. $\frac{1}{2} \ln z$ **25.** $\ln x + \ln y + 2 \ln z$

27. $\ln z + 2 \ln(z - 1)$ **29.** $\frac{1}{2} \log_2(a - 1) - \log_2 9$

31. $\frac{1}{3} \ln x - \frac{1}{3} \ln y$ **33.** $4 \ln x + \frac{1}{2} \ln y - 5 \ln z$

35. $2 \log_5 x - 2 \log_5 y - 3 \log_5 z$

37. $\frac{3}{4} \ln x + \frac{1}{4} \ln(x^2 + 3)$ **39.** $\ln 3x$ **41.** $\log_4 \dfrac{z}{y}$

43. $\log_2(x + 4)^2$ **45.** $\log_3 \sqrt[4]{5x}$

47. $\ln \dfrac{x}{(x + 1)^3}$ **49.** $\log_{10} \dfrac{xz^3}{y^2}$ **51.** $\ln \dfrac{x}{(x^2 - 4)^4}$

53. $\ln \sqrt[3]{\dfrac{x(x + 3)^2}{x^2 - 1}}$ **55.** $\log_8 \dfrac{\sqrt[3]{y(y + 4)^2}}{y - 1}$

57. $\log_2 \frac{32}{4} = \log_2 32 - \log_2 4$; Property 2 **59.** 2

61. $\frac{3}{4}$ **63.** 2.4 **65.** -9 is not in the domain of $\log_3 x$.

67. 4.5 **69.** $-\frac{1}{2}$ **71.** 7 **73.** 2

75. $\frac{3}{2}$ **77.** $-3 - \log_5 2$ **79.** $6 + \ln 5$

81. (a) 90 (b) 77 (c) 73 (d) 9 months

(e) $90 - \log_{10}(t + 1)^{15}$

(f)

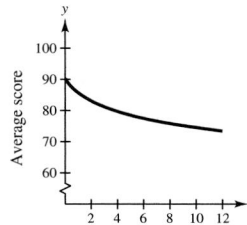

83. False. $\ln 1 = 0$ **85.** False. $\ln(x - 2) \neq \ln x - \ln 2$

87. False. $u = v^2$ **89.** Answers will vary.

91. $f(x) = \dfrac{\log_{10} x}{\log_{10} 2} = \dfrac{\ln x}{\ln 2}$ **93.** $f(x) = \dfrac{\log_{10} x}{\log_{10} \frac{1}{2}} = \dfrac{\ln x}{\ln \frac{1}{2}}$

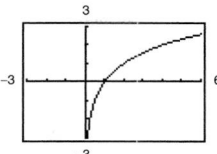 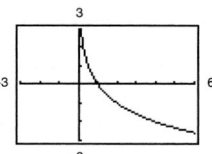

95. $f(x) = \dfrac{\log_{10} x}{\log_{10} 11.8} = \dfrac{\ln x}{\ln 11.8}$

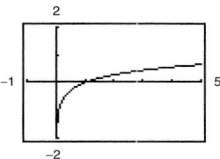

97. $f(x) = h(x)$; Property 2

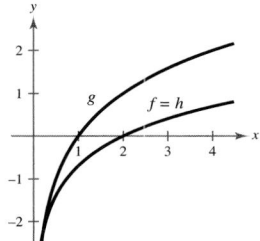

99. $\dfrac{3x^4}{2y^3}$, $x \neq 0$ **101.** 1, $x \neq 0$, $y \neq 0$

103. $-1, \dfrac{1}{3}$ **105.** $\dfrac{-1 \pm \sqrt{97}}{6}$

Section 5.4 *(page 398)*

1. (a) Yes (b) No

3. (a) No (b) Yes (c) Yes, approximate

5. (a) Yes, approximate (b) No (c) Yes **7.** 2

9. 4 **11.** -2 **13.** -5 **15.** 3 **17.** 2

19. $\ln 2 \approx 0.693$ **21.** $e^{-1} \approx 0.368$ **23.** 64

25. $\frac{1}{10}$ **27.** $(3, 8)$ **29.** $(9, 2)$

31. $\dfrac{\ln 5}{\ln 3} \approx 1.465$ **33.** $\ln 5 \approx 1.609$

35. $\ln 28 \approx 3.332$ **37.** $\dfrac{\ln 80}{2 \ln 3} \approx 1.994$ **39.** 2

41. 4 **43.** $3 - \dfrac{\ln 565}{\ln 2} \approx -6.142$

45. $\dfrac{1}{3} \log_{10}\left(\dfrac{3}{2}\right) \approx 0.059$ **47.** $1 + \dfrac{\ln 7}{\ln 5} \approx 2.209$

49. $\dfrac{\ln 12}{3} \approx 0.828$ **51.** $-\ln \dfrac{3}{5} \approx 0.511$ **53.** 0

55. $\dfrac{\ln \frac{8}{3}}{3 \ln 2} + \dfrac{1}{3} \approx 0.805$ **57.** $\ln 5 \approx 1.609$

59. $\ln 4 \approx 1.386$ **61.** $2 \ln 75 \approx 8.635$

63. $\frac{1}{2} \ln 1498 \approx 3.656$ **65.** $\dfrac{\ln 4}{365 \ln\left(1 + \frac{0.065}{365}\right)} \approx 21.330$

67. $\dfrac{\ln 2}{12 \ln\left(1 + \frac{0.10}{12}\right)} \approx 6.960$

69. 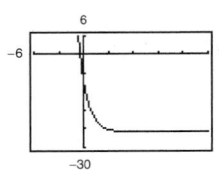   **71.**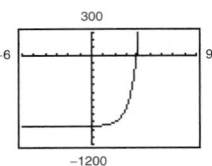

-0.427 3.847

73. 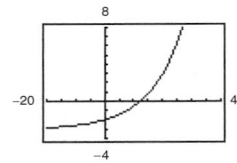   **75.**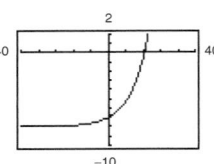

12.207   16.636

77. $e^{-3} \approx 0.050$ **79.** $\dfrac{e^{2.4}}{2} \approx 5.512$ **81.** 1,000,000

83. $2(3^{11/6}) \approx 14.988$ **85.** $\dfrac{e^{10/3}}{5} \approx 5.606$

87. $e^2 - 2 \approx 5.389$ **89.** $e^{-2/3} \approx 0.513$

91. No solution   **93.** $1 + \sqrt{1 + e} \approx 2.928$

95. No solution   **97.** 7   **99.** $\dfrac{-1 + \sqrt{17}}{2} \approx 1.562$

101. 2   **103.** $\dfrac{725 + 125\sqrt{33}}{8} \approx 180.384$

105. 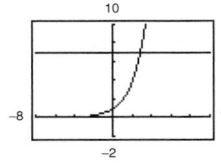   **107.**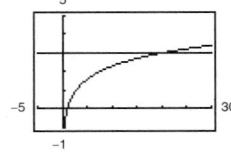

2.807   20.086

109. 8.2 years   **111.** 12.9 years

113. (a) 1426 units   (b) 1498 units

115. (a)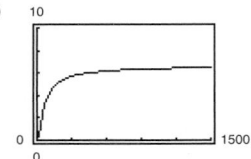

(b) $V = 6.7$; The yield will approach 6.7 million cubic feet per acre.

(c) 29.3 years

117. (a) $y = 100$ and $y = 0$; The range falls between 0% and 100%.

(b) Males: 69.71 inches   Females: 64.51 inches

119. (a)

x	0.2	0.4	0.6	0.8	1.0
y	162.6	78.5	52.5	40.5	33.9

(b)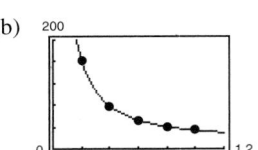

The model appears to fit the data well.

(c) 1.2 meters

(d) No. According to the model, when the number of g's is less than 23, x is between 2.276 meters and 4.404 meters, which isn't realistic in most vehicles.

121. $\log_b uv = \log_b u + \log_b v$

True by Property 1 in Section 5.3.

123. $\log_b(u - v) = \log_b u - \log_b v$

False.

$1.95 \approx \log_{10}(100 - 10) \neq \log_{10} 100 - \log_{10} 10 = 1$

125. Yes. See Exercise 95.

127. Yes. Time to double: $t = \dfrac{\ln 2}{r}$;

Time to quadruple: $t = \dfrac{\ln 4}{r} = 2\left(\dfrac{\ln 2}{r}\right)$

129. $4|x|y^2\sqrt{3y}$ **131.** $5\sqrt[3]{3}$ **133.** $M = kp^3$

135. $d = kab$ **137.** 1.226   **139.** -5.595

Section 5.5   (page 409)

1. c   **2.** e   **3.** b

4. a   **5.** d   **6.** f

Initial Investment	Annual % Rate	Time to Double	Amount After 10 years
7. $1000	12%	5.78 yr	$3320.12
9. $750	8.9438%	7.75 yr	$1834.37
11. $500	11.0%	6.3 yr	$1505.00
13. $6376.28	4.5%	15.4 yr	$10,000.00

15. $112,087.09

17. (a) 6.642 years   (b) 6.330 years

(c) 6.302 years   (d) 6.301 years

19.

r	2%	4%	6%	8%	10%	12%
t	54.93	27.47	18.31	13.73	10.99	9.16

21.

r	2%	4%	6%	8%	10%	12%
t	55.48	28.01	18.85	14.27	11.53	9.69

23.

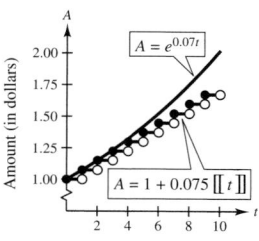

Continuous compounding

Isotope	Half-life (years)	Initial Quantity	Amount After 1000 Years
25. ^{226}Ra	1620	10 g	6.52 g
27. ^{14}C	5730	2.26 g	2 g
29. ^{239}Pu	24,360	2.16 g	2.1 g

31. $y = e^{0.7675x}$ **33.** $y = 5e^{-0.4024x}$ **35.** 2003

37. $k = 0.0274$; 720,738   **39.** 3.15 hours   **41.** 95.8%

43. (a) $V = -4500t + 22,000$ (b) $V = 22,000e^{-0.263t}$

(c)

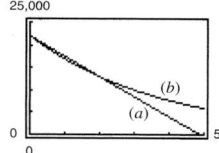

Exponential

(d) 1 year: Straight-line, $17,500;

Exponential, $16,912

3 years: Straight-line, $8500;

Exponential, $9995

(e) The value decreases $4500 per year.

45. (a) $S(t) = 100(1 - e^{-0.1625t})$

(b)

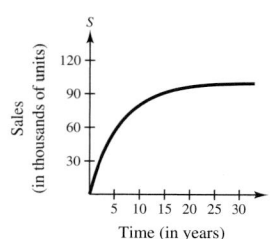

(c) 55,625

47. (a) $S = 10(1 - e^{-0.0575x})$ (b) 3314 units

49. (a) $N = 30(1 - e^{-0.050t})$ (b) 36 days

51. (a) 7.91   (b) 7.68   (c) 5.40

53. (a) 20 decibels   (b) 70 decibels

(c) 95 decibels   (d) 120 decibels

55. 95%   **57.** 4.64   **59.** 1.58×10^{-6} moles per liter

61. 10^7 **63.** 3:00 A.M.

65. (a)

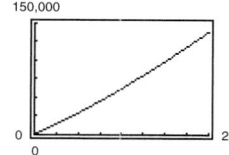

(b) $\approx$ 21 years; Yes

67. False. The domain can be the set of real numbers for a logistic growth function.

69. False. The graph of $f(x)$ is the graph of $g(x)$ shifted upward five units.

71. (a) Logarithmic   (b) Logistic   (c) Exponential

(d) Linear   (e) None of the above   (f) Exponential

73.

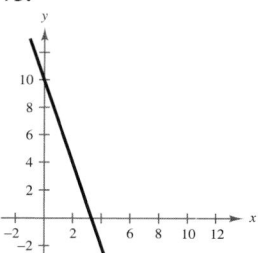

75.

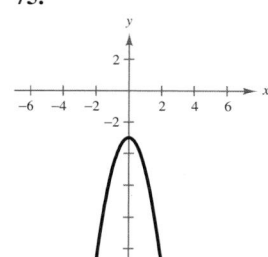

77.

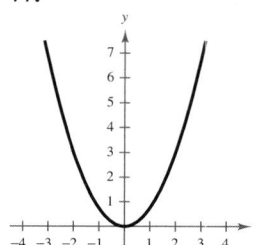

79.

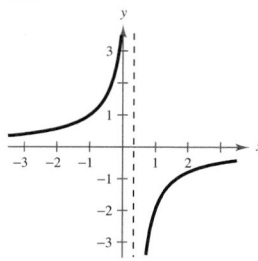

81.

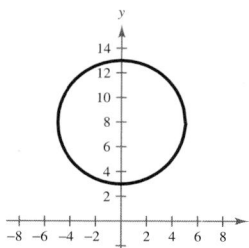

83.

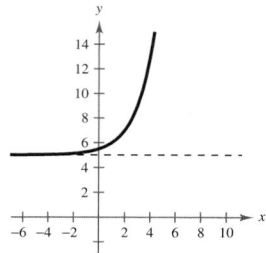

85.

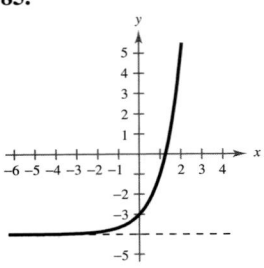

Review Exercises   *(page 416)*

1. 76.699   **3.** 0.337   **5.** 1201.845   **7.** c

8. d   **9.** a   **10.** b

11. Shift the graph of *f* one unit to the right.

13. Reflect *f* in the *x*-axis and shift two units to the left.

15.

x	−1	0	1	2	3
f(*x*)	8	5	4.25	4.063	4.016

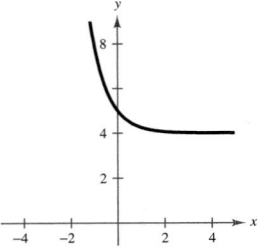

17.

x	−2	−1	0	1	2
f(*x*)	−0.377	−1	−2.65	−7.023	−18.61

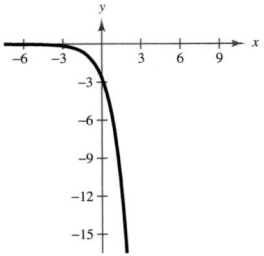

19.

x	−1	0	1	2	3
f(*x*)	4.008	4.04	4.2	5	9

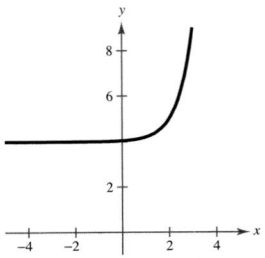

21.

x	−2	−1	0	1	2
f(*x*)	3.25	3.5	4	5	7

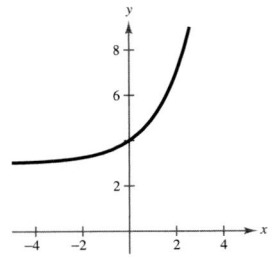

23. 2980.958   **25.** 0.183

27.

x	−2	−1	0	1	2
h(*x*)	2.72	1.65	1	0.61	0.37

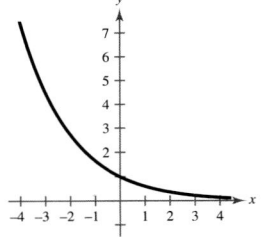

29.

x	−3	−2	−1	0	1
f(*x*)	0.37	1	2.72	7.39	20.09

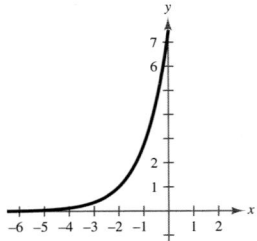

31.

n	1	2	4	12
A	\$6569.98	\$6635.43	\$6669.46	\$6692.64

n	365	Continuous
A	\$6704.00	\$6704.39

33. (a) 0.154   (b) 0.487   (c) 0.811

35. (a) \$1,069,047.14   (b) 7.9 years

37. $\log_4 64 = 3$ **39.** 3   **41.** -3

43. Domain: $(0, \infty)$ **45.** Domain: $(0, \infty)$

x-intercept: $(1, 0)$ x-intercept: $(3, 0)$

Vertical asymptote: $x = 0$ Vertical asymptote: $x = 0$

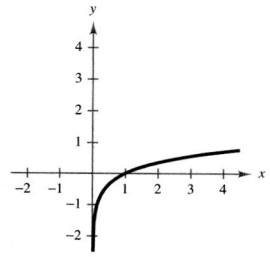 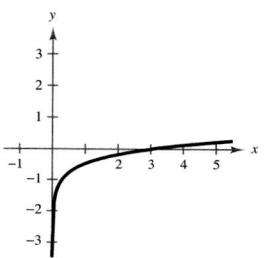

47. Domain: $(-5, \infty)$

x-intercept: $(9995, 0)$

Vertical asymptote: $x = -5$

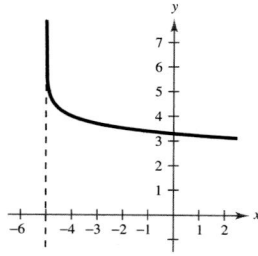

49. 3.118   **51.** -12 **53.** 2.034

55. Domain: $(0, \infty)$ **57.** Domain: $(-\infty, 0), (0, \infty)$

x-intercept: $(e^{-3}, 0)$ x-intercept: $(\pm 1, 0)$

Vertical asymptote: $x = 0$ Vertical asymptote: $x = 0$

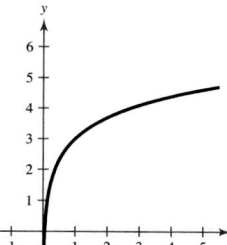 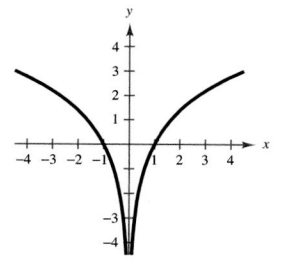

59. 53.4 inches   **61.** 1.585   **63.** -2.322

65. $1 + 2 \log_5 |x|$ **67.** $1 + \log_3 2 - \frac{1}{3} \log_3 x$

69. $2 \ln x + 2 \ln y + \ln z$ **71.** $\ln(x + 3) - \ln x - \ln y$

73. $\log_2 5x$ **75.** $\ln \dfrac{x}{\sqrt[4]{y}}$ **77.** $\log_8 y^7 \sqrt[3]{x + 4}$

79. $\ln \dfrac{\sqrt{|2x - 1|}}{(x + 1)^2}$

81. (a) $0 \le h < 18{,}000$

(b)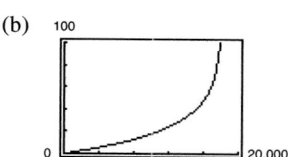

Vertical asymptote: $h = 18{,}000$

(c) The plane is climbing at a slower rate, so the time required increases.

(d) 5.46 minutes

83. 3   **85.** -3 **87.** $\ln 3 \approx 1.099$ **89.** 16

91. $e^4 \approx 54.598$

93. $\ln 12 \approx 2.485$ **95.** $-\dfrac{\ln 44}{5} \approx -0.757$

97. $\dfrac{\ln 22}{\ln 2} \approx 4.459$ **99.** $\dfrac{\ln 17}{\ln 5} \approx 1.760$

101. $\ln 2 \approx 0.693, \ln 5 \approx 1.609$

103. 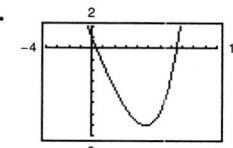   **105.**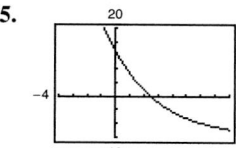

0.39, 7.48   2.45

107. $\frac{1}{3} e^{8.2} \approx 1213.650$ **109.** $\frac{1}{4} e^{7.5} \approx 452.011$

111. $3e^2 \approx 22.167$ **113.** $e^4 - 1 \approx 53.598$

115. No solution   **117.** 0.900

119. 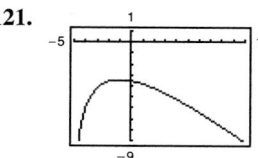   **121.**

1.64   No solution

123. 15.2 years   **125.** e   **126.** b   **127.** f

128. d   **129.** a   **130.** c   **131.** 2004

133. (a) 13.8629%   (b) \$11,486.98   **135.** $y = 2e^{0.1014x}$

137. (a) 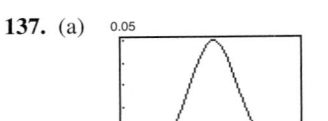   (b) 71

139. $10^{-3.5}$ watt per square centimeter

141. True by the inverse properties

143. $b < d < a < c$

    b and d are negative.

    a and c are positive.

Chapter Test   *(page 420)*

1. 1123.690   **2.** 687.291   **3.** 0.497   **4.** 22.198

5.

x	-1	$-\frac{1}{2}$	0	$\frac{1}{2}$	1
$f(x)$	10	3.162	1	0.316	0.1

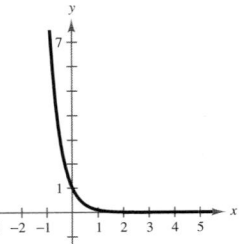

6.

x	-1	0	1	2	3
$f(x)$	-0.005	-0.028	-0.167	-1	-6

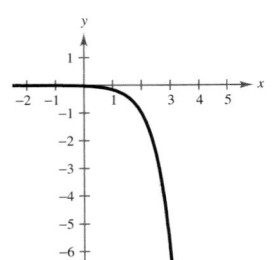

7.

x	-1	$-\frac{1}{2}$	0	$\frac{1}{2}$	1
$f(x)$	0.865	0.632	0	-1.718	-6.389

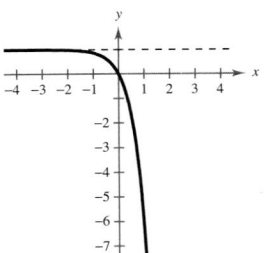

8. (a) -0.89 (b) 9.2

9.

x	$\frac{1}{2}$	1	$\frac{3}{2}$	2	4
$f(x)$	-5.699	-6	-6.176	-6.301	-6.602

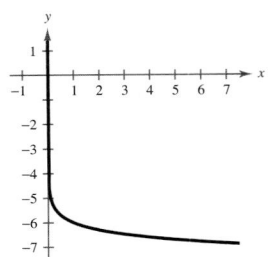

Vertical asymptote: $x = 0$

10.

x	5	7	9	11	13
$f(x)$	0	1.099	1.609	1.946	2.197

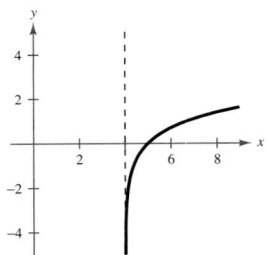

Vertical asymptote: $x = 4$

11.

x	-5	-3	-1	0	1
$f(x)$	1	2.099	2.609	2.792	2.946

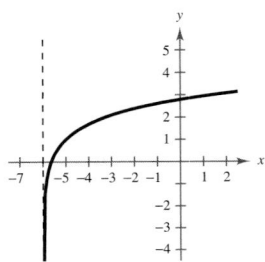

Vertical asymptote: $x = -6$

12. 1.945   **13.** 0.115   **14.** 1.328

15. $\log_2 3 + 4 \log_2 |a|$ **16.** $\ln 5 + \frac{1}{2}\ln x - \ln 6$

17. $\log_3 13y$ **18.** $\ln \dfrac{x^4}{y^4}$ **19.** $\dfrac{\ln 197}{4} \approx 1.321$

20. $\frac{800}{501} \approx 1.597$ **21.** $y = 2745e^{0.1570x}$ **22.** 55%

23. (a)

x	$\frac{1}{4}$	1	2	4	5	6
H	58.720	75.332	86.828	103.43	110.59	117.38

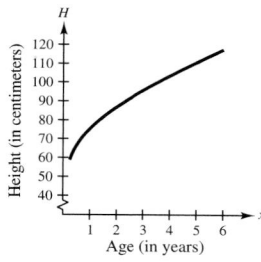

(b) 103 centimeters; 103.43 centimeters

Problem Solving   *(page 422)*

1.

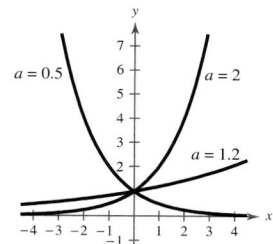

$y = 0.5^x$ and $y = 1.2^x$.

$0 \le a \le 1.44$.

3. As $x \to \infty$, the graph of e^x increases at a greater rate than the graph of x^n.

5. Answers will vary.

7. (a)

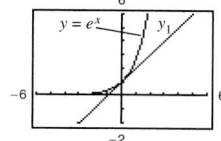

(b)

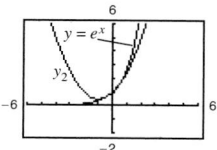

(c)

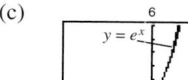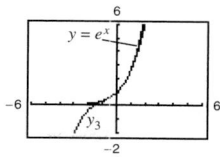

9.

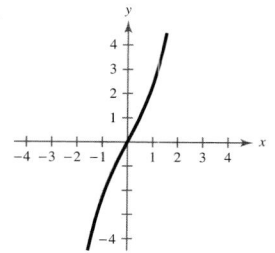

$$f^{-1}(x) = \ln\left(\frac{x + \sqrt{x^2 + 4}}{2}\right)$$

11. c   **13.** $t = \dfrac{\ln c_1 - \ln c_2}{\left(\dfrac{1}{k_2} - \dfrac{1}{k_1}\right)\ln \dfrac{1}{2}}$

15. (a) $y_1 = 252{,}606(1.0310)^t$

(b) $y_2 = 400.88t^2 - 1464.6t + 291{,}782$

(c)

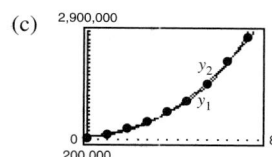

(d) The exponential model is a better fit. No, because the model is rapidly approaching infinity.

17. $1, e^2$

19. $y_4 = (x - 1) - \frac{1}{2}(x - 1)^2 + \frac{1}{3}(x - 1)^3 - \frac{1}{4}(x - 1)^4$

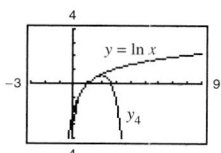

The pattern implies that

$$\ln x = (x - 1) - \tfrac{1}{2}(x - 1)^2 + \tfrac{1}{3}(x - 1)^3 - \cdots$$

21.

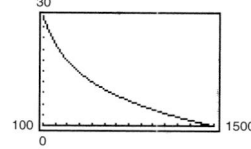

17.7 cubic feet per minute

Chapter 6

Section 6.1 (page 430)

1. $\dfrac{\sqrt{3}}{3}$ **3.** -1 **5.** $\sqrt{3}$ **7.** 3.2236

9. $\dfrac{3\pi}{4}$ radians, 135° **11.** $\dfrac{\pi}{4}$ radian, 45°

13. 0.6435 radian, 36.9° **15.** 1.0517 radians, 60.3°

17. 2.1112 radians, 121.0° **19.** 1.2490 radians, 71.6°

21. 2.1112 radians, 121.0° **23.** 1.1071 radians, 63.4°

25. 0.1974 radian, 11.3° **27.** 1.4289 radians, 81.9°

29. 0.9273 radian, 53.1° **31.** 0.8187 radian, 46.9°

33. (2, 1): 42.3°; (4, 4): 78.7°; (6, 2): 59.0°

35. $(-4, -1)$: 11.9°; (3, 2): 21.8°; (1, 0): 146.3° **37.** 0

39. $\dfrac{7}{5}$ **41.** 7 **43.** $\dfrac{8\sqrt{37}}{37} \approx 1.3152$

45. (a) 4 (b) 8 **47.** (a) $\dfrac{35\sqrt{37}}{74}$ (b) $\dfrac{35}{8}$

49. $2\sqrt{2}$ **51.** 0.1003, 1054 feet

53. (a) 18.4° (b) 15.8 meters

55. $\alpha \approx 33.69°$; $\beta \approx 56.31°$

57. True. The inclination of a line is related to its slope by $m = \tan\theta$. If the angle is greater than $\pi/2$ but less than π, then the angle is in the second quadrant, where the tangent function is negative.

59. (a) $d = \dfrac{4}{\sqrt{m^2 + 1}}$

(b)

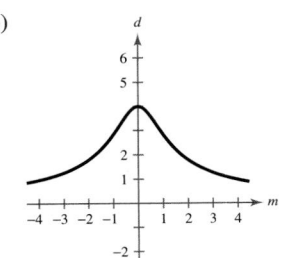

(c) $m = 0$

(d) $d = 0$. As the line approaches the vertical, the distance approaches 0.

61. x-intercept: (7, 0)

y-intercept: (0, 49)

63. x-intercepts: $\left(5 \pm \sqrt{5}, 0\right)$

y-intercept: (0, 20)

65. x-intercepts: $\left(\dfrac{7 \pm \sqrt{53}}{2}, 0\right)$

y-intercept: (0, −1)

67. $f(x) = 3\left(x + \frac{1}{3}\right)^2 - \frac{49}{3}$

Vertex: $\left(-\frac{1}{3}, -\frac{49}{3}\right)$

69. $f(x) = 5\left(x + \frac{17}{5}\right)^2 - \frac{324}{5}$

Vertex: $\left(-\frac{17}{5}, -\frac{324}{5}\right)$

71. $f(x) = 6\left(x - \frac{1}{12}\right)^2 - \frac{289}{24}$

Vertex: $\left(\frac{1}{12}, -\frac{289}{24}\right)$

73. **75.**

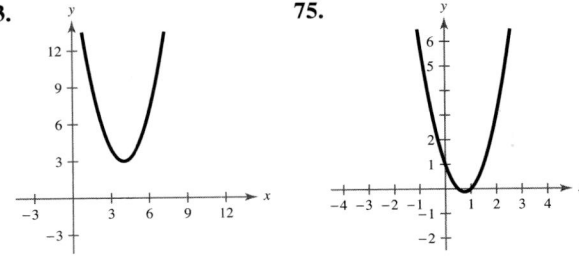

Section 6.2 (page 438)

1. A circle is formed when a plane intersects the top or bottom half of a double-napped cone and is perpendicular to the axis of the cone.

3. A parabola is formed when a plane intersects the top or bottom half of a double-napped cone, is parallel to the side of the cone, and does not intersect the vertex.

5. e **6.** b **7.** d **8.** f **9.** a **10.** c

11. Vertex: (0, 0)

Focus: $\left(0, \frac{1}{2}\right)$

Directrix: $y = -\frac{1}{2}$

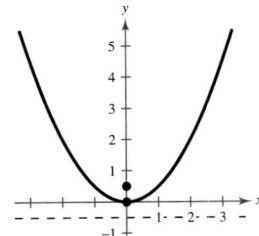

13. Vertex: (0, 0)

Focus: $\left(-\frac{3}{2}, 0\right)$

Directrix: $x = \frac{3}{2}$

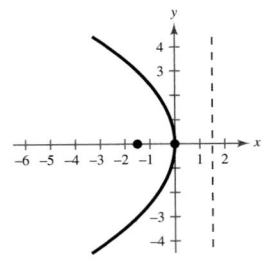

15. Vertex: (0, 0)

Focus: $\left(0, -\frac{3}{2}\right)$

Directrix: $y = \frac{3}{2}$

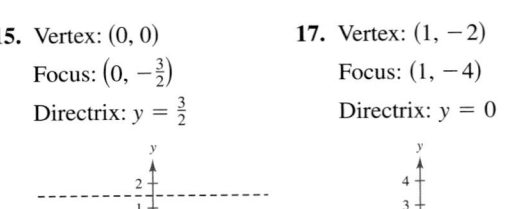

17. Vertex: (1, −2)

Focus: (1, −4)

Directrix: $y = 0$

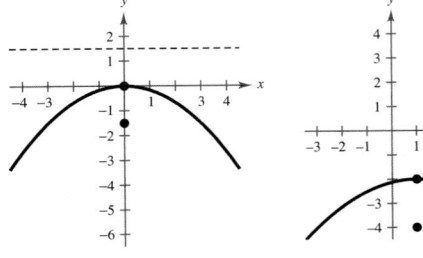

19. Vertex: $\left(-\frac{3}{2}, 2\right)$
Focus: $\left(-\frac{3}{2}, 3\right)$
Directrix: $y = 1$

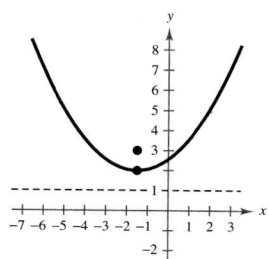

21. Vertex: $(1, 1)$
Focus: $(1, 2)$
Directrix: $y = 0$

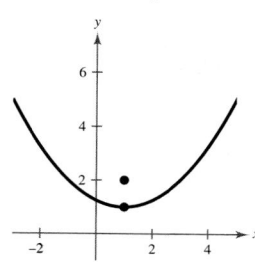

23. Vertex: $(-2, -3)$
Focus: $(-4, -3)$
Directrix: $x = 0$

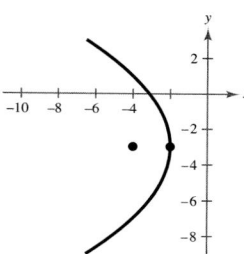

25. Vertex: $(-2, 1)$
Focus: $\left(-2, -\frac{1}{2}\right)$
Directrix: $y = \frac{5}{2}$

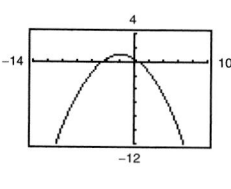

27. Vertex: $\left(\frac{1}{4}, -\frac{1}{2}\right)$
Focus: $\left(0, -\frac{1}{2}\right)$
Directrix: $x = \frac{1}{2}$

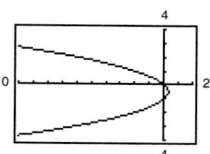

29.

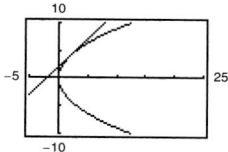

$(2, 4)$

31. $x^2 = \frac{3}{2}y$ **33.** $x^2 = -6y$ **35.** $y^2 = -8x$
37. $x^2 = 4y$ **39.** $y^2 = -8x$ **41.** $y^2 = 9x$
43. $(x - 3)^2 = -(y - 1)$ **45.** $y^2 = 4(x + 4)$
47. $(y - 2)^2 = -8(x - 5)$ **49.** $x^2 = 8(y - 4)$
51. $(y - 2)^2 = 8x$ **53.** $y = \sqrt{6(x + 1)} + 3$
55. $4x - y - 8 = 0; (2, 0)$ **57.** $4x - y + 2 = 0; \left(-\frac{1}{2}, 0\right)$

59.

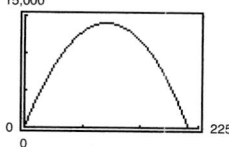

$x = 106$ units
61. $y = \frac{1}{18}x^2$ **63.** (a) $y = -\frac{1}{640}x^2$ (b) 8 feet
65. (a) $17,500\sqrt{2}$ miles per hour
(b) $x^2 = -16,400(y - 4100)$
67. (a) $x^2 = -64(y - 75)$
(b) 69.3 feet
69. False. If the graph crossed the directrix, there would exist points nearer the directrix than the focus.
71. (a)

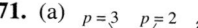
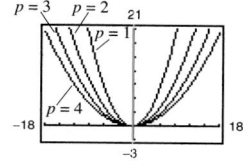

As p increases, the graph becomes wider.
(b) $(0, 1), (0, 2), (0, 3), (0, 4)$
(c) $4, 8, 12, 16; 4|p|$
(d) Easy way to determine two additional points on the graph
73. $m = \frac{x_1}{2p}$ **75.** $y = x^3 - 7x^2 + 17x - 15$
77. $\frac{1}{2}, -\frac{5}{3}, \pm 2$ **79.** $B \approx 23.67°, C \approx 121.33°, c \approx 14.89$
81. $C = 89°, a \approx 1.93, b \approx 2.33$
83. $A \approx 16.39°, B \approx 23.77°, C \approx 139.84°$
85. $B \approx 24.62°, C \approx 90.38°, a \approx 10.88$

Section 6.3 *(page 448)*

1. b **2.** c **3.** d **4.** f **5.** a **6.** e
7. Center: $(0, 0)$
Vertices: $(\pm 5, 0)$
Foci: $(\pm 3, 0)$
Eccentricity: $\frac{3}{5}$

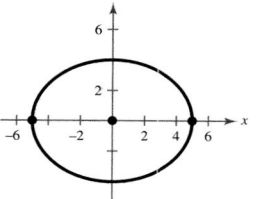

9. Center: $(0, 0)$
Vertices: $(0, \pm 3)$
Foci: $(0, \pm 2)$
Eccentricity: $\frac{2}{3}$

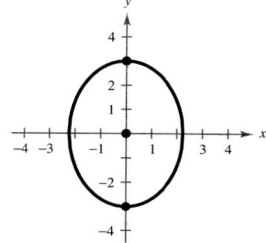

11. Center: $(-3, 5)$

Vertices: $(-3, 10), (-3, 0)$

Foci: $(-3, 8), (-3, 2)$

Eccentricity: $\frac{3}{5}$

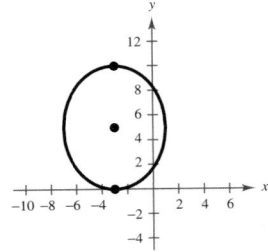

13. Center: $(-5, 1)$

Vertices: $\left(-\frac{7}{2}, 1\right), \left(-\frac{13}{2}, 1\right)$

Foci: $\left(-5 \pm \frac{\sqrt{5}}{2}, 1\right)$

Eccentricity: $\frac{\sqrt{5}}{3}$

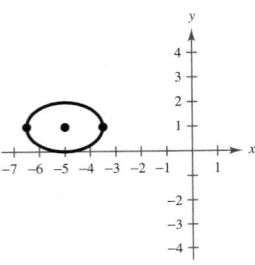

15. Center: $(-2, 3)$

Vertices: $(-2, 6), (-2, 0)$

Foci: $\left(-2, 3 \pm \sqrt{5}\right)$

Eccentricity: $\frac{\sqrt{5}}{3}$

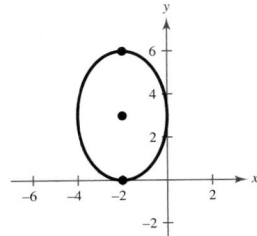

17. Center: $(4, 3)$

Vertices: $(14, 3), (-6, 3)$

Foci: $\left(4 \pm 4\sqrt{5}, 3\right)$

Eccentricity: $\frac{2\sqrt{5}}{5}$

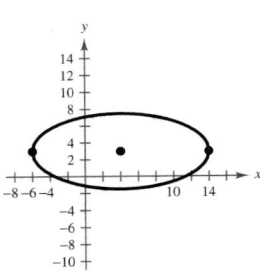

19. Center: $\left(-\frac{3}{2}, \frac{5}{2}\right)$

Vertices: $\left(-\frac{3}{2}, \frac{5}{2} \pm 2\sqrt{3}\right)$

Foci: $\left(-\frac{3}{2}, \frac{5}{2} \pm 2\sqrt{2}\right)$

Eccentricity: $\frac{\sqrt{6}}{3}$

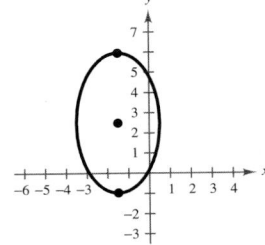

21. Center: $(1, -1)$

Vertices: $\left(\frac{9}{4}, -1\right), \left(-\frac{1}{4}, -1\right)$

Foci: $\left(\frac{7}{4}, -1\right), \left(\frac{1}{4}, -1\right)$

Eccentricity: $\frac{3}{5}$

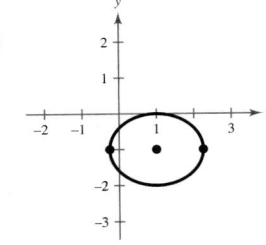

23.

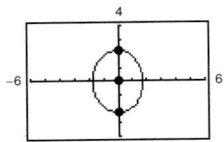

Center: $(0, 0)$

Vertices: $\left(0, \pm\sqrt{5}\right)$

Foci: $\left(0, \pm\sqrt{2}\right)$

25.

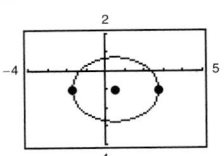

Center: $\left(\frac{1}{2}, -1\right)$

Vertices: $\left(\frac{1}{2} \pm \sqrt{5}, -1\right)$

Foci: $\left(\frac{1}{2} \pm \sqrt{2}, -1\right)$

27. $\dfrac{x^2}{4} + \dfrac{y^2}{16} = 1$ **29.** $\dfrac{x^2}{36} + \dfrac{y^2}{32} = 1$

31. $\dfrac{x^2}{36} + \dfrac{y^2}{11} = 1$ **33.** $\dfrac{21x^2}{400} + \dfrac{y^2}{25} = 1$

35. $\dfrac{(x-2)^2}{1} + \dfrac{(y-3)^2}{9} = 1$

37. $\dfrac{(x+2)^2}{16} + \dfrac{(y-3)^2}{9} = 1$

39. $\dfrac{(x-2)^2}{4} + \dfrac{(y-4)^2}{1} = 1$ **41.** $\dfrac{x^2}{48} + \dfrac{(y-4)^2}{64} = 1$

43. $\dfrac{x^2}{16} + \dfrac{(y-4)^2}{12} = 1$ **45.** $\dfrac{(x-2)^2}{4} + \dfrac{(y-2)^2}{1} = 1$

47. $\dfrac{x^2}{25} + \dfrac{y^2}{16} = 1$

49. (a)

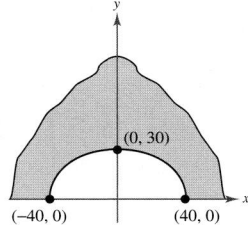

(b) $\dfrac{x^2}{1600} + \dfrac{y^2}{900} = 1$ (c) 14.5 feet

51. 40   **53.** $e \approx 0.052$

55.

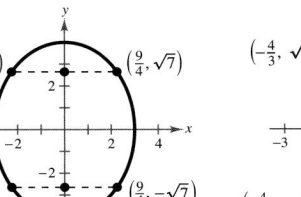

57.

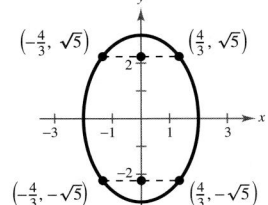

59. False. The equation of an ellipse is second degree in x and y.

61. (a) $A = \pi a(20 - a)$ (b) $\dfrac{x^2}{196} + \dfrac{y^2}{36} = 1$

(c)

a	8	9	10	11	12	13
A	301.6	311.0	314.2	311.0	301.6	285.9

$a = 10$, circle

(d)

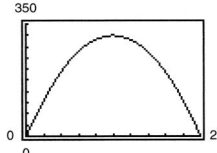

The shape of an ellipse with a maximum area is a circle.

63. 0, 10   **65.** $0, \pm 1$

67. (a) 37   (b) -28 (c) $5x - 43$

69. (a) -9 (b) $2\sqrt{7} - 9$ (c) $-9 + 3\sqrt{2}i$

Section 6.4   (page 458)

1. b   **2.** c   **3.** a   **4.** d

5. Center: $(0, 0)$

   Vertices: $(\pm 1, 0)$

   Foci: $\left(\pm \sqrt{2}, 0\right)$

   Asymptotes: $y = \pm x$

7. Center: $(0, 0)$

   Vertices: $(0, \pm 5)$

   Foci: $\left(0, \pm \sqrt{106}\right)$

   Asymptotes: $y = \pm \dfrac{5}{9}x$

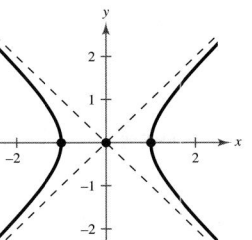

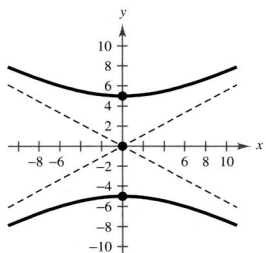

9. Center: $(1, -2)$

   Vertices: $(3, -2), (-1, -2)$

   Foci: $\left(1 \pm \sqrt{5}, -2\right)$

   Asymptotes: $y = -2 \pm \dfrac{1}{2}(x - 1)$

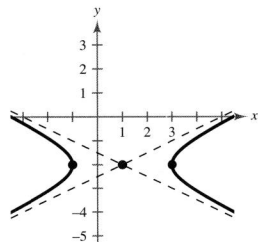

11. Center: $(2, -6)$

   Vertices: $\left(2, -\dfrac{17}{3}\right), \left(2, -\dfrac{19}{3}\right)$

   Foci: $\left(2, -6 \pm \dfrac{\sqrt{13}}{6}\right)$

   Asymptotes: $y = -6 \pm \dfrac{2}{3}(x - 2)$

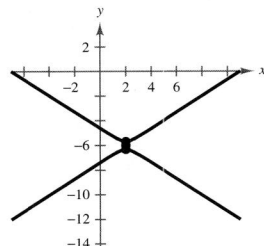

13. Center: $(2, -3)$

   Vertices: $(3, -3), (1, -3)$

   Foci: $\left(2 \pm \sqrt{10}, -3\right)$

   Asymptotes:

   $y = -3 \pm 3(x - 2)$

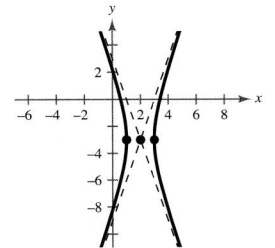

15. The graph of this equation is two lines intersecting at $(-1, -3)$.

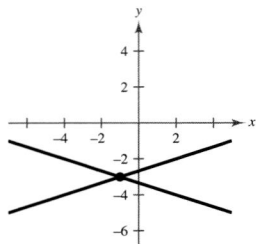

17. Center: $(0, 0)$

Vertices: $\left(\pm\sqrt{3}, 0\right)$

Foci: $\left(\pm\sqrt{5}, 0\right)$

Asymptotes: $y = \pm\dfrac{\sqrt{6}}{3}x$

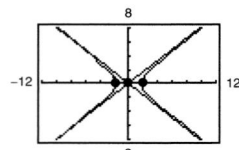

19. Center: $(1, -3)$

Vertices: $\left(1, -3 \pm \sqrt{2}\right)$

Foci: $\left(1, -3 \pm 2\sqrt{5}\right)$

Asymptotes: $y = -3 \pm \frac{1}{3}(x - 1)$

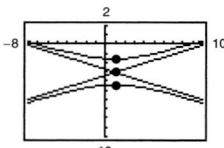

21. $\dfrac{y^2}{4} - \dfrac{x^2}{12} = 1$ **23.** $\dfrac{x^2}{1} - \dfrac{y^2}{25} = 1$

25. $\dfrac{17y^2}{1024} - \dfrac{17x^2}{64} = 1$ **27.** $\dfrac{(x-4)^2}{4} - \dfrac{y^2}{12} = 1$

29. $\dfrac{(y-5)^2}{16} - \dfrac{(x-4)^2}{9} = 1$ **31.** $\dfrac{y^2}{9} - \dfrac{4(x-2)^2}{9} = 1$

33. $\dfrac{(y-2)^2}{4} - \dfrac{x^2}{4} = 1$ **35.** $\dfrac{(x-2)^2}{1} - \dfrac{(y-2)^2}{1} = 1$

37. $\dfrac{(x-3)^2}{9} - \dfrac{(y-2)^2}{4} = 1$

39. (a) $x \approx 110.3$ miles   (b) 57.0 miles

(c) 0.00129 second

(d) The ship is at the position $(144.2, 60)$.

41. Circle   **43.** Hyperbola   **45.** Ellipse   **47.** Parabola

49. True. For a hyperbola, $c^2 = a^2 + b^2$. The larger the ratio of b to a, the larger the eccentricity of the hyperbola, $e = c/a$.

51. Answers will vary.   **53.** $x(x + 4)(x - 4)$

55. $2x(x - 6)^2$

57.

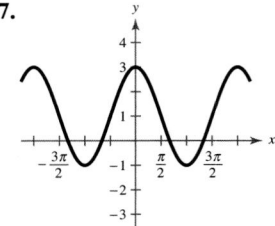

59.

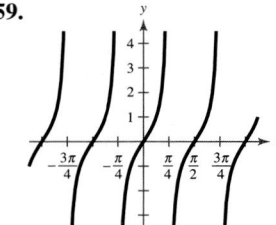

Section 6.5   (page 466)

1. $(3, 0)$ **3.** $\left(\dfrac{3 + \sqrt{3}}{2}, \dfrac{3\sqrt{3} - 1}{2}\right)$

5. $\left(\dfrac{3\sqrt{2}}{2}, -\dfrac{\sqrt{2}}{2}\right)$

7. $\dfrac{(y')^2}{2} - \dfrac{(x')^2}{2} = 1$ **9.** $y' = \pm\dfrac{\sqrt{2}}{2}$

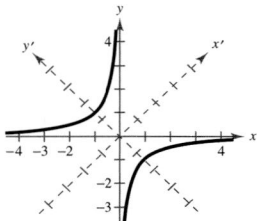
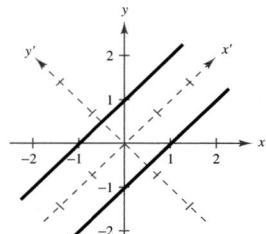

11. $\dfrac{\left(x' - 3\sqrt{2}\right)^2}{16} - \dfrac{\left(y' - \sqrt{2}\right)^2}{16} = 1$

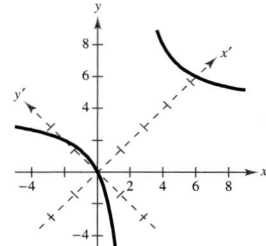

13. $\dfrac{(x')^2}{6} + \dfrac{(y')^2}{\frac{3}{2}} = 1$ **15.** $x' = -(y')^2$

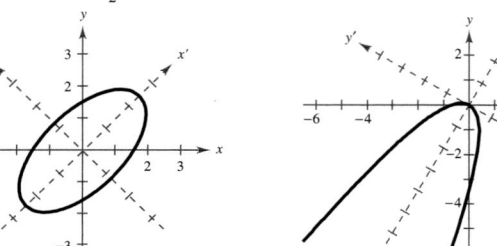
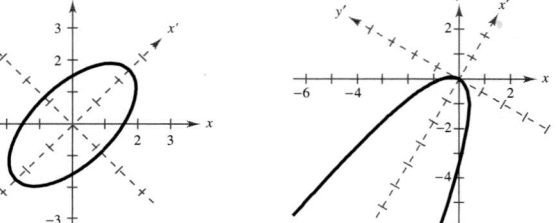

17. $y' = \frac{1}{6}(x')^2 - \frac{1}{3}x'$

19.

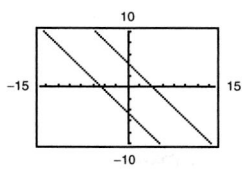

$\theta = 45°$

21.

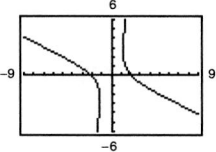

$\theta \approx 26.57°$

23.

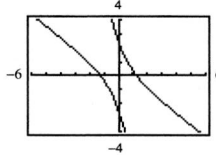

$\theta \approx 31.72°$

25.

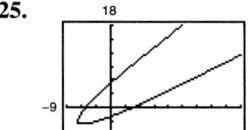

$\theta \approx 33.69°$

27. e   **28.** f   **29.** b
30. a   **31.** d   **32.** c

33. Parabola

$$y = \frac{(8x - 5) \pm \sqrt{(8x - 5)^2 - 4(16x^2 - 10x)}}{2}$$

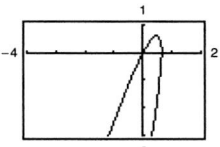

35. Ellipse

$$y = \frac{6x \pm \sqrt{36x^2 - 28(12x^2 - 45)}}{14}$$

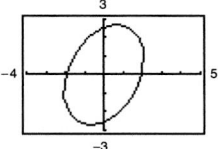

37. Hyperbola

$$y = \frac{6x \pm \sqrt{36x^2 + 20(x^2 + 4x - 22)}}{-10}$$

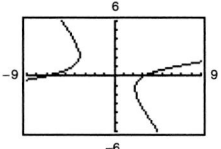

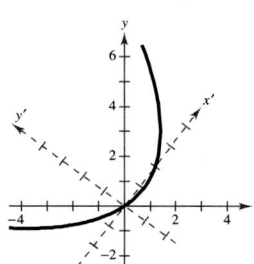

39. Parabola

$$y = \frac{-(4x - 1) \pm \sqrt{(4x - 1)^2 - 16(x^2 - 5x - 3)}}{8}$$

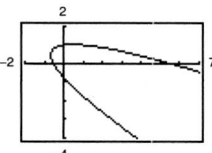

41.

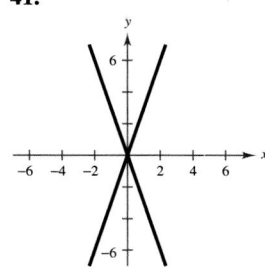

43.

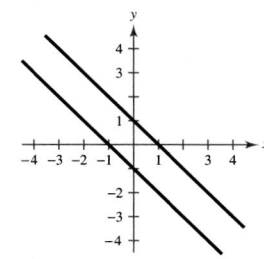

45. $(2, 2), (2, 4)$ **47.** $(-8, 12)$ **49.** $(0, 8), (12, 8)$
51. $(0, 4)$ **53.** $\left(1, \sqrt{3}\right), \left(1, -\sqrt{3}\right)$
55. No solution   **57.** $\left(0, \frac{3}{2}\right), (-3, 0)$

59. True. The graph of the equation can be classified by finding the discriminant. For a graph to be a hyperbola, the discriminant must be greater than zero. If $k \geq \frac{1}{4}$, then the discriminant would be less than or equal to zero.

61. Answers will vary.

63.

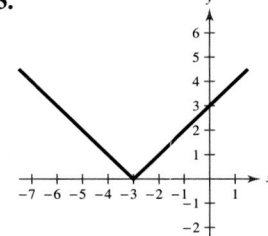

65.

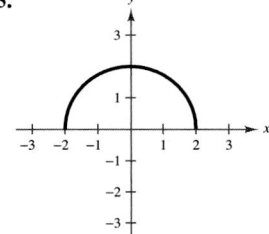

67.

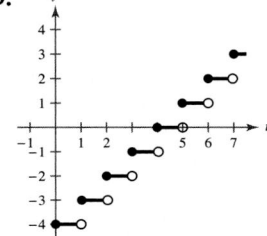

69.

71. 45.11   **73.** 48.60   **75.** 64.95 feet

Section 6.6 *(page 473)*

1. (a)

t	0	1	2	3	4
x	0	1	$\sqrt{2}$	$\sqrt{3}$	2
y	3	2	1	0	-1

(b)

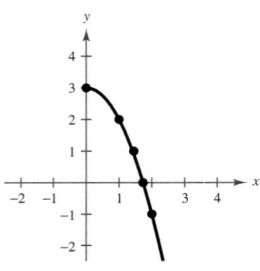

(c) $y = 3 - x^2$

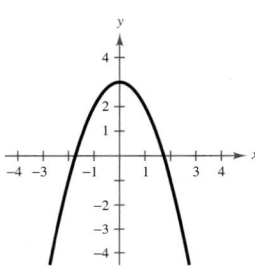

The graph of the rectangular equation shows the entire parabola rather than just the right half.

3.

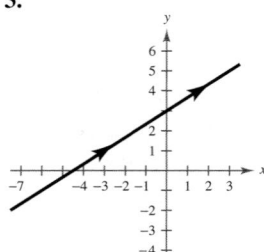

5.

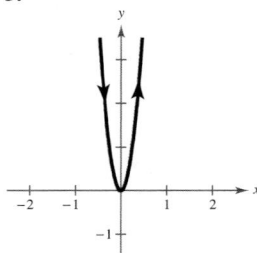

7.

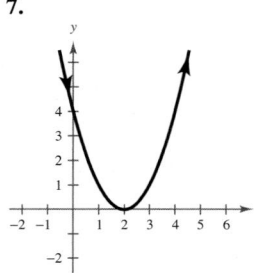

9.

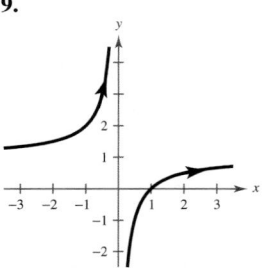

11.

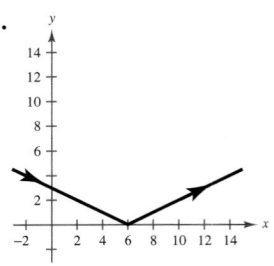

13.

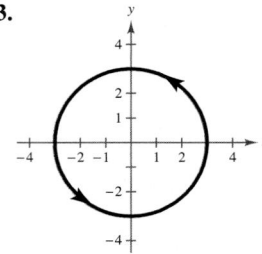

15.

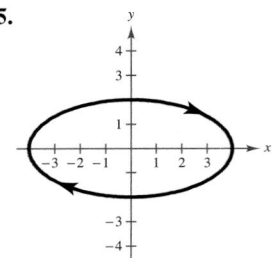

17.

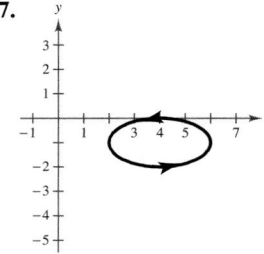

19.

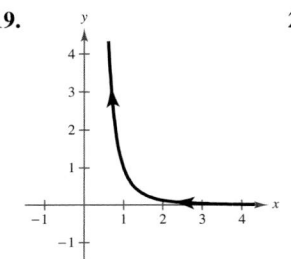

21.

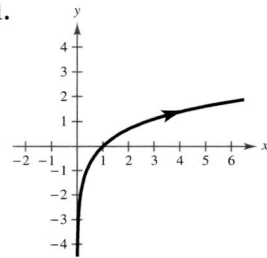

23. Each curve represents a portion of the line $y = 2x + 1$.

	Domain	*Orientation*
(a)	$(-\infty, \infty)$	Left to right
(b)	$[-1, 1]$	Depends on θ
(c)	$(0, \infty)$	Right to left
(d)	$(0, \infty)$	Left to right

25. $y - y_1 = m(x - x_1)$ **27.** $\dfrac{(x - h)^2}{a^2} + \dfrac{(y - k)^2}{b^2} = 1$

29. $x = 6t$ **31.** $x = 3 + 4\cos\theta$ **33.** $x = 4\cos\theta$
$\quad\;\; y = -3t$ $\qquad\qquad\; y = 2 + 4\sin\theta$ $\qquad\; y = \sqrt{7}\sin\theta$

35. $x = 4\sec\theta$ **37.** (a) $x = t,\; y = 3t - 2$
$\quad\;\; y = 3\tan\theta$ $\qquad\quad$ (b) $x = -t + 2,\; y = -3t + 4$

39. (a) $x = t,\, y = t^2$
$\qquad$ (b) $x = -t + 2,\, y = t^2 - 4t + 4$

41. (a) $x = t,\, y = t^2 + 1$
$\qquad$ (b) $x = -t + 2,\, y = t^2 - 4t + 5$

43. (a) $x = t,\, y = \dfrac{1}{t}$

$\qquad$ (b) $x = -t + 2,\, y = -\dfrac{1}{t - 2}$

13. Maximum: $|r| = 20$ when $\theta = \dfrac{3\pi}{2}$

Zero: $r = 0$ when $\theta = \dfrac{\pi}{2}$

15. Maximum: $|r| = 4$ when $\theta = 0,\ \dfrac{\pi}{3},\ \dfrac{2\pi}{3}$

Zero: $r = 0$ when $\theta = \dfrac{\pi}{6},\ \dfrac{\pi}{2},\ \dfrac{5\pi}{6}$

17.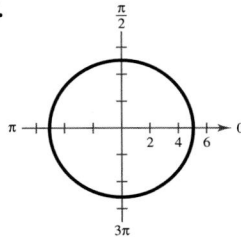

19.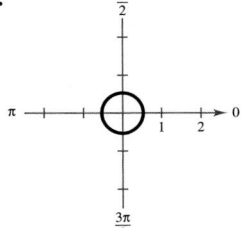

21.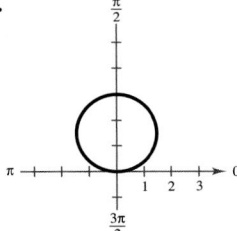

23.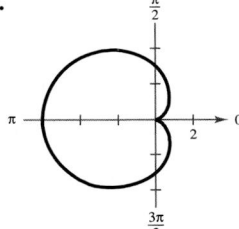

25.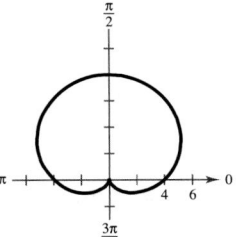

27.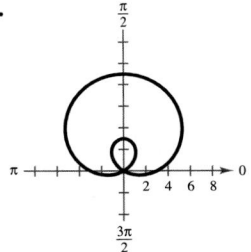

29.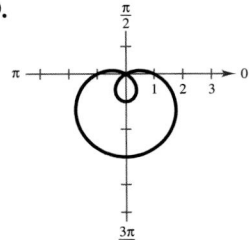

31.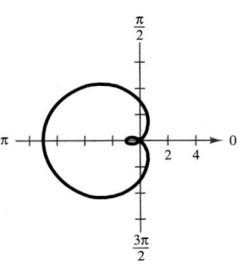

33.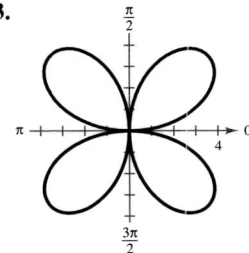

35.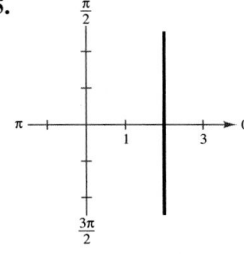

37.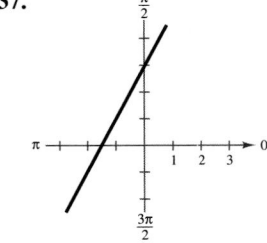

39.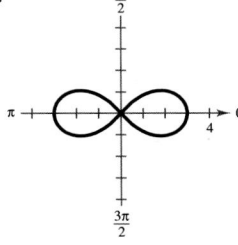

41.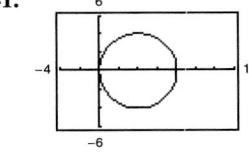

43.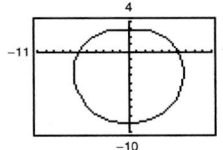

45.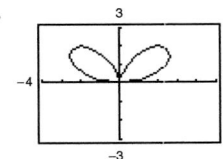

47.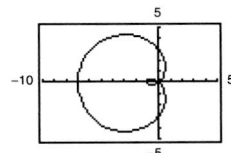

$0 \le \theta < 2\pi$

49.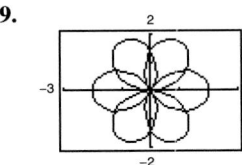

51.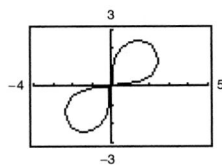

$0 \le \theta < 4\pi$ $0 \le \theta < \pi$

53.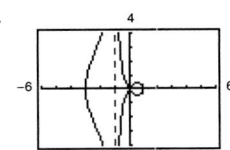

55.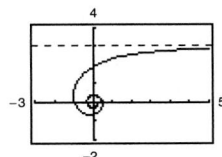

57. True. For a graph to have polar axis symmetry, replace $(r,\ \theta)$ by $(r,\ -\theta)$ or $(-r,\ \pi - \theta)$.

59. (a)

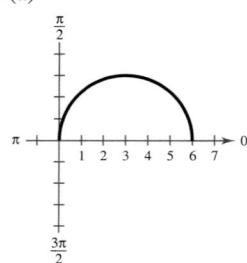

Upper half of circle

(b)

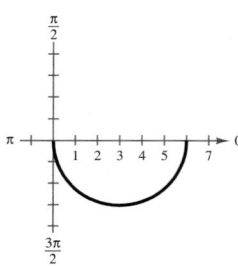

Lower half of circle

(c)

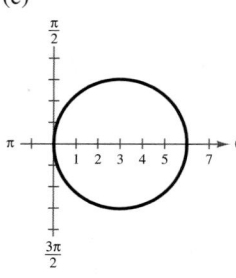

Full circle

(d)

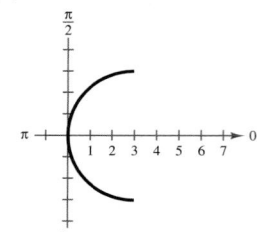

Left half of circle

61. Answers will vary.

63. (a) $r = 2 - \sin\left(\theta - \dfrac{\pi}{4}\right)$ (b) $r = 2 + \cos\theta$

 (c) $r = 2 + \sin\theta$ (d) $r = 2 - \cos\theta$

65. (a)

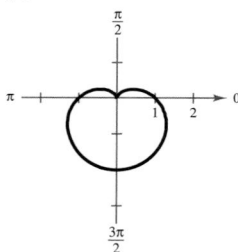

(b)

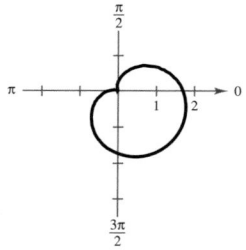

67.

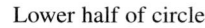

| $k = 0$, circle |
| $k = 1$, limaçon |
| $k = 2$, cardioid |
| $k = 3$, limaçon |

69. $\dfrac{(x + 1)^2}{9} + \dfrac{(y - 2)^2}{4} = 1$

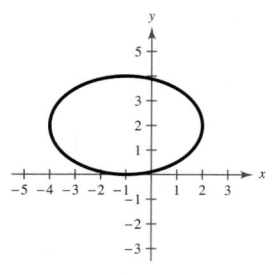

Section 6.9   *(page 494)*

1. $e = 1$: $r = \dfrac{4}{1 + \cos\theta}$, parabola

 $e = 0.5$: $r = \dfrac{2}{1 + 0.5\cos\theta}$, ellipse

 $e = 1.5$: $r = \dfrac{6}{1 + 1.5\cos\theta}$, hyperbola

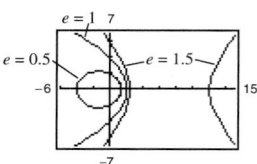

3. $e = 1$: $r = \dfrac{4}{1 - \sin\theta}$, parabola

 $e = 0.5$: $r = \dfrac{2}{1 - 0.5\sin\theta}$, ellipse

 $e = 1.5$: $r = \dfrac{6}{1 - 1.5\sin\theta}$, hyperbola

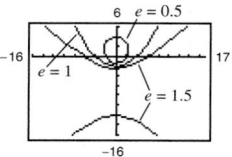

5. f **6.** c **7.** d **8.** e **9.** a **10.** b

11. Parabola **13.** Parabola

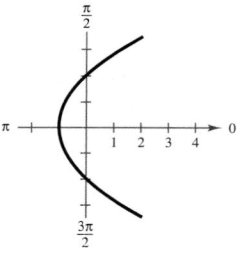

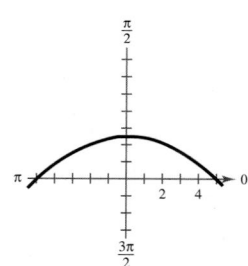

15. Ellipse **17.** Ellipse

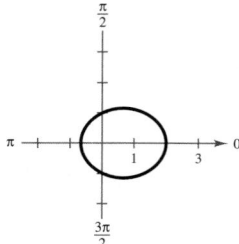

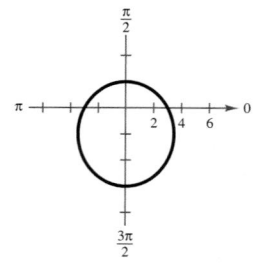

19. Hyperbola

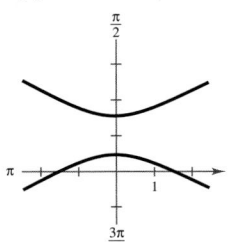

21. Hyperbola

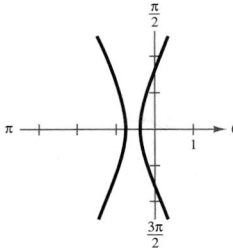

23. Ellipse

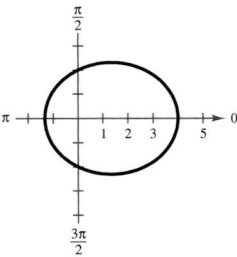

25.

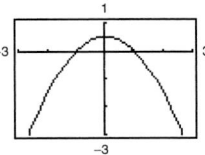

Parabola

27.

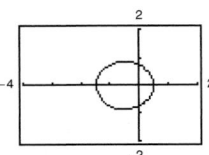

Ellipse

29.

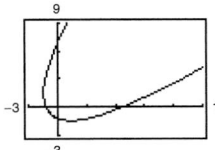

31.

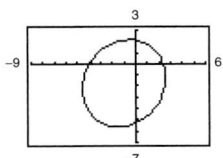

33. $r = \dfrac{1}{1 - \cos \theta}$ **35.** $r = \dfrac{1}{2 + \sin \theta}$

37. $r = \dfrac{2}{1 + 2 \cos \theta}$ **39.** $r = \dfrac{2}{1 - \sin \theta}$

41. $r = \dfrac{10}{1 - \cos \theta}$ **43.** $r = \dfrac{10}{3 + 2 \cos \theta}$

45. $r = \dfrac{20}{3 - 2 \cos \theta}$ **47.** $r = \dfrac{9}{4 - 5 \sin \theta}$

49. Answers will vary.

51. $r = \dfrac{9.2934 \times 10^7}{1 - 0.0167 \cos \theta}$

Perihelion: 9.1408×10^7 miles

Aphelion: 9.4512×10^7 miles

53. $r = \dfrac{5.5368 \times 10^9}{1 - 0.2481 \cos \theta}$

Perihelion: 4.4362×10^9 kilometers

Aphelion: 7.3638×10^9 kilometers

55. $r = \dfrac{1.3977 \times 10^8}{1 - 0.0934 \cos \theta}$

Perihelion: 1.2783×10^8 miles

Aphelion: 1.5417×10^8 miles

57. (a) $r = \dfrac{8200}{1 + \sin \theta}$

(b)

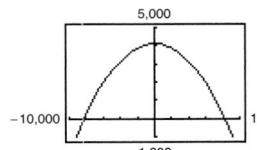

(c) 1467 miles

(d) 394 miles

59. True. The graphs represent the same hyperbola.

61. Answers will vary.   **63.** $r^2 = \dfrac{24{,}336}{169 - 25 \cos^2 \theta}$

65. $r^2 = \dfrac{144}{25 \cos^2 \theta - 9}$ **67.** $r^2 = \dfrac{144}{25 \cos^2 \theta - 16}$

69. $\dfrac{\pi}{6} + n\pi$ **71.** $\dfrac{\pi}{3} + n\pi, \dfrac{2\pi}{3} + n\pi$

73. $\dfrac{\pi}{2} + n\pi$ **75.** $\dfrac{\sqrt{2}}{10}$ **77.** $\dfrac{7\sqrt{2}}{10}$

79. $\sin 2u = -\dfrac{24}{25}$

$\cos 2u = -\dfrac{7}{25}$

$\tan 2u = \dfrac{24}{7}$

Review Exercises   *(page 498)*

1. $\dfrac{\pi}{4}$ radian, $45°$ **3.** 1.1071 radians, $63.43°$

5. 0.4424 radian, $25.35°$ **7.** 0.6588 radian, $37.75°$

9. $2\sqrt{2}$ **11.** Hyperbola   **13.** $(x - 4)^2 = -8(y - 2)$

15. $(y - 2)^2 = 12x$ **17.** $y = -2x + 2; (1, 0)$

19. $8\sqrt{6}$ meters   **21.** $\dfrac{(x - 2)^2}{25} + \dfrac{y^2}{21} = 1$

23. $\dfrac{(x - 2)^2}{4} + (y - 1)^2 = 1$

25. The foci occur 3 feet from the center of the arch on a line connecting the tops of the pillars.

27. Center: $(-2, 1)$

Vertices: $(-2, 11), (-2, -9)$

Foci: $\left(-2, 1 \pm \sqrt{19}\right)$

Eccentricity: $\dfrac{\sqrt{19}}{10}$

29. Center: $(1, -4)$

Vertices: $(1, 0), (1, -8)$

Foci: $\left(1, -4 \pm \sqrt{7}\right)$

Eccentricity: $\dfrac{\sqrt{7}}{4}$

31. $y^2 - \dfrac{x^2}{8} = 1$ **33.** $\dfrac{5(x-4)^2}{16} - \dfrac{5y^2}{64} = 1$

35. Center: $(3, -5)$
Vertices: $(7, -5), (-1, -5)$
Foci: $\left(3 \pm 2\sqrt{5}, -5\right)$
Asymptotes: $y = -5 \pm \frac{1}{2}(x - 3)$

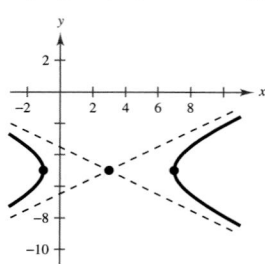

37. Center: $(1, -1)$
Vertices: $(5, -1), (-3, -1)$
Foci: $(6, -1), (-4, -1)$
Asymptotes: $y = -1 \pm \frac{3}{4}(x - 1)$

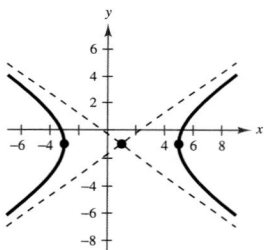

39. 72 miles   **41.** Hyperbola

43. $\dfrac{(x')^2}{8} - \dfrac{(y')^2}{8} = 1$ **45.** $\dfrac{(x')^2}{3} + \dfrac{(y')^2}{2} = 1$

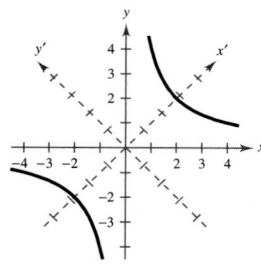   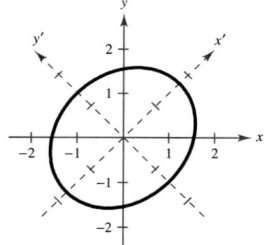

47. Parabola
$$y = \dfrac{24x + 40 \pm \sqrt{(24x + 40)^2 - 36(16x^2 - 30x)}}{18}$$

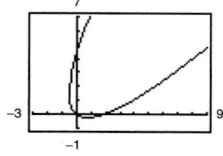

49. Parabola
$$y = \dfrac{-\left(2x - 2\sqrt{2}\right) \pm \sqrt{\left(2x - 2\sqrt{2}\right)^2 - 4\left(x^2 + 2\sqrt{2}x + 2\right)}}{2}$$

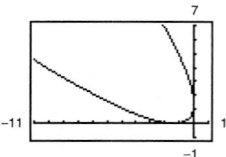

51. $x = 3, y = 0$ **53.** $x = \dfrac{3\sqrt{3}}{2}, y = \dfrac{1}{2}$

55.

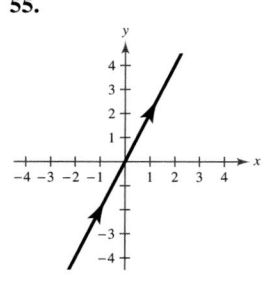

$y = 2x$

57.

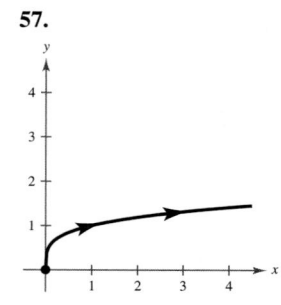

$y = \sqrt[4]{x}$

59.

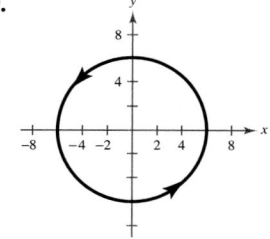

$x^2 + y^2 = 36$

61. $x = 5 + 6\cos\theta$
$y = 4 + 6\sin\theta$

63. $x = 3\tan\theta$
$y = 4\sec\theta$

65.

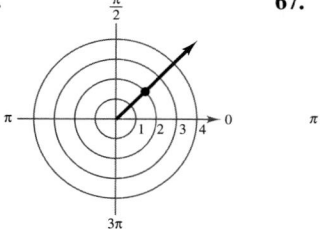

$\left(2, \dfrac{9\pi}{4}\right), \left(-2, \dfrac{5\pi}{4}\right)$

67.

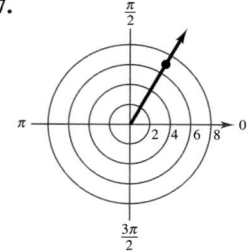

$(7, 1.05), (-7, 10.47)$

69. $\left(-\dfrac{1}{2}, -\dfrac{\sqrt{3}}{2}\right)$ **71.** $\left(-\dfrac{3\sqrt{2}}{2}, \dfrac{3\sqrt{2}}{2}\right)$

73. $\left(2, \dfrac{\pi}{2}\right)$ **75.** $\left(2\sqrt{13}, 0.9828\right)$

77. $r = 6 \sin \theta$ **79.** $r^2 = 10 \csc 2\theta$

81. $x^2 + y^2 = 3x$ **83.** $x^2 + y^2 = y^{2/3}$

85. Symmetry: $\theta = \dfrac{\pi}{2}$, polar axis, pole

Maximum value of $|r|$: $|r| = 4$ when $\theta = 0, \dfrac{\pi}{2}, \pi, \dfrac{3\pi}{2}$

No zeros of r

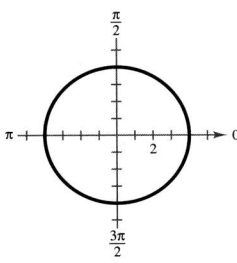

87. Symmetry: $\theta = \dfrac{\pi}{2}$, polar axis and the pole

Maximum value of $|r|$: $|r| = 4$ when $\theta = \dfrac{\pi}{4}, \dfrac{3\pi}{4}, \dfrac{5\pi}{4}, \dfrac{7\pi}{4}$

Zeros of r: $r = 0$ when $\theta = 0, \dfrac{\pi}{2}, \pi, \dfrac{3\pi}{2}$

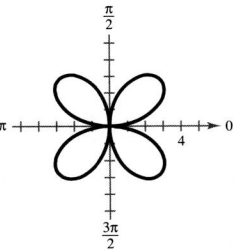

89. Symmetry: polar axis

Maximum value of $|r|$: $|r| = 4$ when $\theta = 0$

Zeros of r: $r = 0$ when $\theta = \pi$

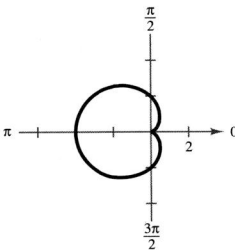

91. Symmetry: $\theta = \dfrac{\pi}{2}$

Maximum value of $|r|$: $|r| = 8$ when $\theta = \dfrac{\pi}{2}$

Zeros of r: $r = 0$ when $\theta = 3.4814, 5.9433$

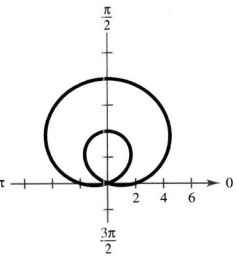

93. Symmetry: $\theta = \dfrac{\pi}{2}$, polar axis and the pole

Maximum value of $|r|$: $|r| = 3$ when $\theta = 0, \dfrac{\pi}{2}, \pi, \dfrac{3\pi}{2}$

Zeros of r: $r = 0$ when $\theta = \dfrac{\pi}{4}, \dfrac{3\pi}{4}, \dfrac{5\pi}{4}, \dfrac{7\pi}{4}$

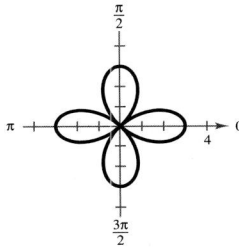

95. Limaçon **97.** Rose curve

99. Hyperbola **101.** Ellipse

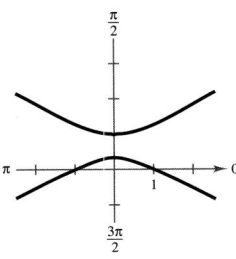 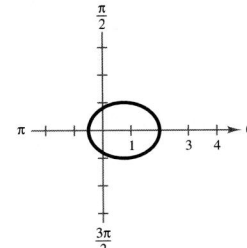

103. $r = \dfrac{4}{1 - \cos \theta}$ **105.** $r = \dfrac{5}{3 - 2 \cos \theta}$

107. $r = \dfrac{7977.2}{1 - 0.937 \cos \theta}$; 11,008.8 miles

109. False. When classifying an equation of the form $Ax^2 + Bxy + Cy^2 + Dx + Ey + F = 0$, its graph can be determined by its discriminant. For a graph to be a parabola, its discriminant, $B^2 - 4AC$, must equal zero. So, if $B = 0$, then A or C equals 0.

111. False. The following are two sets of parametric equations for the line.

$x = t$, $y = 3 - 2t$

$x = 3t$, $y = 3 - 6t$

113. 5. The ellipse becomes more circular and approaches a circle of radius 5.

115. (a) The speed would double.

(b) The elliptical orbit would be flatter; the length of the major axis would be greater.

117. (a) The graphs are the same.

(b) The graphs are the same.

Chapter Test   *(page 502)*

1. 0.2783 radian, 15.9°   **2.** 0.8330 radian, 47.7°

3. $\dfrac{7\sqrt{2}}{2}$

4. Parabola: $y^2 = 4(x - 1)$

Vertex: $(1, 0)$

Focus: $(2, 0)$

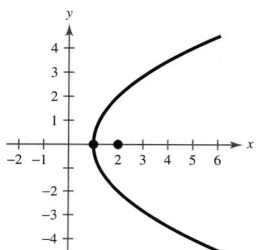

5. Hyperbola: $\dfrac{(x - 2)^2}{4} - y^2 = 1$

Center: $(2, 0)$

Vertices: $(0, 0), (4, 0)$

Foci: $\left(2 \pm \sqrt{5}, 0\right)$

Asymptotes: $y = \pm\tfrac{1}{2}(x - 2)$

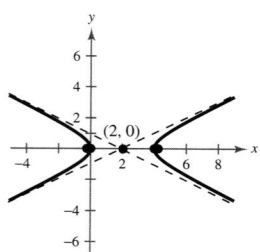

6. Ellipse: $\dfrac{(x + 3)^2}{16} + \dfrac{(y - 1)^2}{9} = 1$

Center: $(-3, 1)$

Vertices: $(1, 1), (-7, 1)$

Foci: $\left(-3 \pm \sqrt{7}, 1\right)$

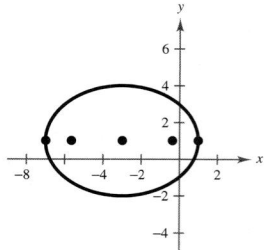

7. Circle: $(x - 2)^2 + (y - 1)^2 = \tfrac{1}{2}$

Center: $(2, 1)$

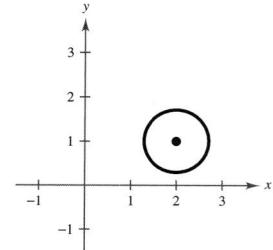

8. $(x - 3)^2 = \dfrac{3}{2}(y + 2)$ **9.** $\dfrac{5(y - 2)^2}{4} - \dfrac{5x^2}{16} = 1$

10. (a) $45°$

(b)

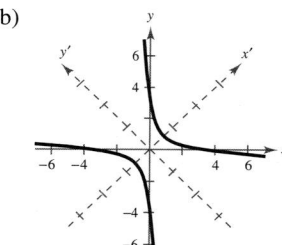

11.

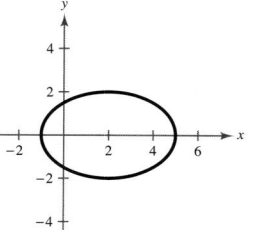

$\dfrac{(x - 2)^2}{9} + \dfrac{y^2}{4} = 1$

12. $x = 6 + 4t$

$y = 4 + 7t$

13. $\left(\sqrt{3}, -1\right)$

14. $\left(2\sqrt{2}, \dfrac{7\pi}{4}\right), \left(-2\sqrt{2}, \dfrac{3\pi}{4}\right), \left(2\sqrt{2}, -\dfrac{\pi}{4}\right)$

15. $r = 4 \sin \theta$

16.

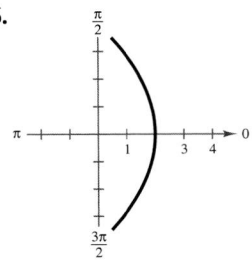

Parabola

17.

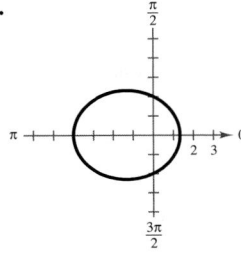

Ellipse

18.

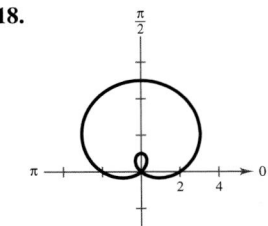

Limaçon

19.

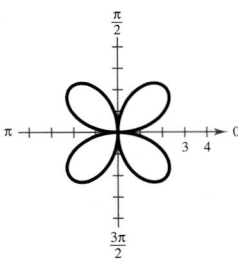

Rose curve

20. Slope: 0.1511; Change in elevation: 789.03 feet

21. No; Yes

Cumulative Test for Chapters 4–6
(page 503)

1. $3 - 5i$ **2.** $-2 - 3i$ **3.** $5 - 12i$ **4.** 4

5. $\frac{8}{5} + \frac{4}{5}i$ **6.** $-2, \pm 2i$ **7.** $-7, 0, 3$

8. $x^4 + 3x^3 - 11x^2 + 9x + 70$

9. $2\sqrt{2}\left(\cos\frac{3\pi}{4} + i\sin\frac{3\pi}{4}\right)$ **10.** $-12\sqrt{3} + 12i$

11. $-8 + 8\sqrt{3}i$ **12.** -64

13. $\cos 0 + i\sin 0$

$\cos\frac{2\pi}{3} + i\sin\frac{2\pi}{3}$

$\cos\frac{4\pi}{3} + i\sin\frac{4\pi}{3}$

14. $3\left(\cos\frac{\pi}{8} + i\sin\frac{\pi}{8}\right)$

$3\left(\cos\frac{5\pi}{8} + i\sin\frac{5\pi}{8}\right)$

$3\left(\cos\frac{9\pi}{8} + i\sin\frac{9\pi}{8}\right)$

$3\left(\cos\frac{13\pi}{8} + i\sin\frac{13\pi}{8}\right)$

15. Reflect f in the x-axis and y-axis, and shift f three units to the right.

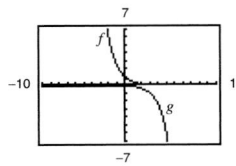

16. Reflect f in the x-axis, and shift f four units upward.

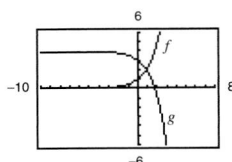

17. 1.991   **18.** -0.067 **19.** 1.717   **20.** 0.281

21. 0.302   **22.** -1.733 **23.** -4.087

24. $\ln(x + 4) + \ln(x - 4) - 4\ln x, \ x > 4$

25. $\ln\frac{x^2}{\sqrt{x + 5}}, \ x > 0$ **26.** $\frac{\ln 12}{2} \approx 1.242$

27. $\frac{\ln 9}{\ln 4} + 5 \approx 6.585$ **28.** $\frac{64}{5} = 12.8$

29. $\frac{1}{2}e^8 \approx 1490.479$

30.

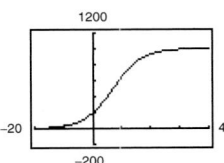

Horizontal asymptotes: $y = 0, \ y = 1000$

31. 6.3 hours   **32.** $81.87°$ **33.** $\frac{11\sqrt{5}}{5}$

34. Ellipse; $\frac{(x - 2)^2}{4} + \frac{(y + 1)^2}{9} = 1$

Center: $(2, -1)$

Vertices: $(2, 2), (2, -4)$

Foci: $\left(2, -1 \pm \sqrt{5}\right)$

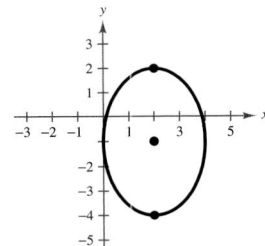

35. Hyperbola; $x^2 - \frac{y^2}{4} = 1$

Center: $(0, 0)$

Vertices: $(1, 0), (-1, 0)$

Foci: $\left(\sqrt{5}, 0\right), \left(-\sqrt{5}, 0\right)$

Asymptotes: $y = \pm 2x$

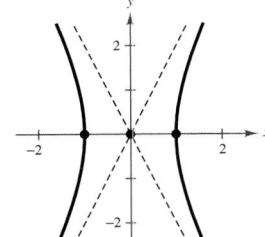

36. Circle; $(x + 1)^2 + (y - 3)^2 = 22$

Center: $(-1, 3)$

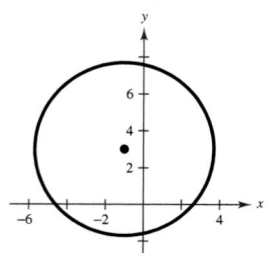

37. Parabola; $y^2 = -2(x + 1)$

Vertex: $(-1, 0)$

Focus: $\left(-\frac{3}{2}, 0\right)$

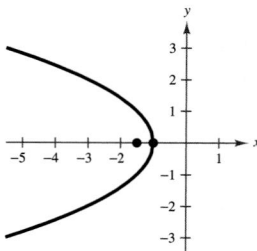

38. $x^2 - 4x + y^2 + 8y - 48 = 0$

39. $5y^2 - 4x^2 - 30y + 25 = 0$

40. (a) $45°$

(b)

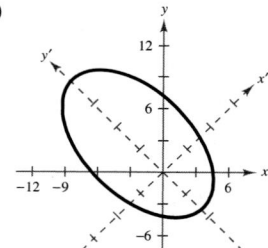

41.

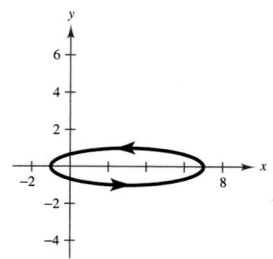

$\dfrac{(x - 3)^2}{16} + y^2 = 1$

42. $x = 3 - 6t$
   $y = -2 + 6t$

43. $r = 6 \sin \theta$

44.

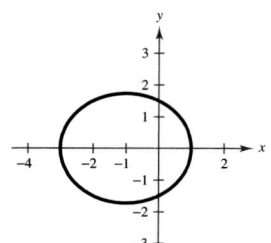

45.

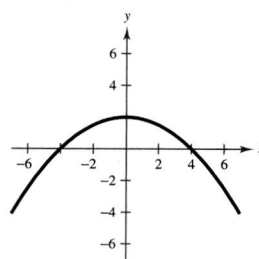

46. (a) iii   (b) i   (c) ii

Problem Solving   *(page 508)*

1. (a) 1.2016 radians

(b) 2420 feet, 5971 feet

3. $y^2 = 4p(x + p)$

5. (a) The sum of the distances from the two islands the tour boat can travel is constant.

(b) Island 1: $(-6, 0)$;
   Island 2: $(6, 0)$

(c) 20 miles; Vertex: $(10, 0)$ (d) $\dfrac{x^2}{100} + \dfrac{y^2}{64} = 1$

7. Answers will vary.

9. Answers will vary. For example:

$x = \cos(-t)$

$y = 2 \sin(-t)$

11. (a) $y^2 = x^2 \left(\dfrac{1 - x}{1 + x} \right)$ (b) $\cos 2\theta \sec \theta$

(c)

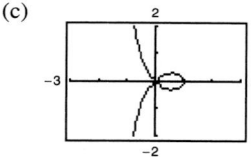

13. Circle

15.

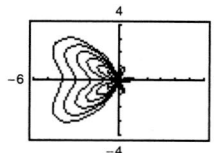 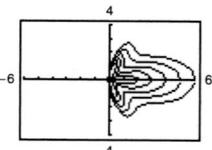

For $n \geq 1$, a bell is produced.

For $n \leq -1$, a heart is produced.

For $n = 0$, a rose curve is produced.

Instructor's Answers

Chapter P

Section P.1   *(page 9)*

2. (a) $12, 5$ (b) $-7, 0, -3, 12, 5$
  (c) $-7, -\frac{7}{3}, 0, 3.12, \frac{5}{4}, -3, 12, 5$ (d) $\sqrt{5}$

4. (a) 4 (b) $-75, 4$
  (c) $0.7575, -4.63, -75, 4$
  (d) $2.3030030003 \ldots, \sqrt{10}$

6. (a) $25, \sqrt{9}, 7, 13$ (b) $25, -17, \sqrt{9}, 7, 13$
  (c) $25, -17, -\frac{12}{5}, \sqrt{9}, 3.12, 7, -11.1, 13$ (d) $\frac{1}{2}\pi$

8. $0.\overline{3}$ **10.** $0.\overline{54}$ **12.** $-6 < -2.5$

14.
-3.5

$-3.5 < 1$

16.
$\frac{16}{3}$

$1 < \frac{16}{3}$

18.
$-\frac{8}{7}$ $-\frac{3}{7}$

$-\frac{8}{7} < -\frac{3}{7}$

20. $x \geq -2$ denotes the set of all real numbers greater than or equal to -2. Unbounded

22. $x > 3$ denotes the set of all real numbers greater than 3. Unbounded

24. $x < 2$ denotes the set of all real numbers less than 2. Unbounded

26. $0 \leq x \leq 5$ denotes the set of all real numbers greater than or equal to zero and less than or equal to 5. Bounded

28. $0 < x \leq 6$ denotes the set of positive real numbers less than or equal to 6. Bounded

30. $-6 \leq y < 0$ **32.** $y \leq 25$

34. $-3 \leq k < 5$ **36.** $2.5\% \leq r \leq 5\%$

38. This interval consists of all real numbers greater than or equal to -5 and less than or equal to 7.

40. This interval consists of all real numbers less than or equal to 4.

42. 0 **44.** 3 **46.** -6 **48.** -9 **50.** 1

52. $|-4| = |4|$ **54.** $-|-6| < |-6|$

56. $-(-2) > -2$ **58.** $\frac{5}{2}$ **60.** 51

62. $\frac{5}{2}$ **64.** 14.99

66. $|\$9772 - \$9400| = \$372 < \$500$
  $0.05(\$9400) = \$470$

  Because the difference between the actual expenses and the budget is less than $500 and less than 5% of the budgeted amount, there is compliance with the "budget variance test."

68. $|\$2613 - \$2575| = \$38 < \$500$
  $0.05(\$2575) = \$128.75$

  Because the difference between the actual expenses and the budget is less than $500 and less than 5% of the budgeted amount, there is compliance with the "budget variance test."

70.

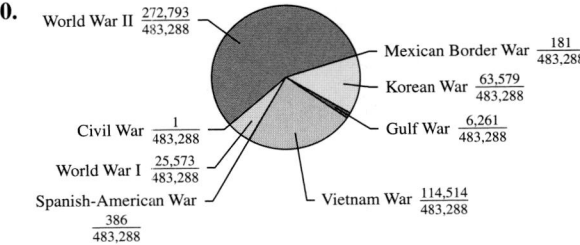

World War II $\frac{272,793}{483,288}$
Mexican Border War $\frac{181}{483,288}$
Korean War $\frac{63,579}{483,288}$
Gulf War $\frac{6,261}{483,288}$
Vietnam War $\frac{114,514}{483,288}$
Spanish-American War $\frac{386}{483,288}$
World War I $\frac{25,573}{483,288}$
Civil War $\frac{1}{483,288}$

72. $|326 - 351| = 25$ miles   **74.** $|48 - 82| = 34°$

76. $|x + 10| \geq 6$ **78.** $|y - a| \leq 2$

80. $6x^3$ and $-5x$ are the terms; 6 and -5 are the coefficients.

82. $3\sqrt{3}x^2$ and 1 are the terms; $3\sqrt{3}$ is the coefficient.

84. $3x^4$ and $-x^2/4$ are the terms; 3 and $-\frac{1}{4}$ are the coefficients.

86. (a) 30 (b) -12 **88.** (a) -10 (b) 0

90. (a) $\frac{1}{2}$ (b) Division by 0 is undefined.

92. Multiplicative Inverse Property

94. Additive Inverse Property

96. Additive Identity Property

98. Associative Property of Addition

100. $\frac{1}{7}(7 \cdot 12) = \left(\frac{1}{7} \cdot 7\right)12$ Associative Property of Multiplication

          $= 1 \cdot 12$ Multiplicative Inverse Property

          $= 12$ Multiplicative Identity Property

102. $\frac{2}{7}$ **104.** $\frac{59}{66}$ **106.** -3 **108.** $\frac{5x}{27}$

110. (a)

n	1	10	100	10,000	100,000
$5/n$	5	0.5	0.05	0.0005	0.00005

     (b) The value of $5/n$ approaches 0 as n increases without bound.

112. False. The denominators cannot be added when adding fractions.

114. Yes. y is nonnegative if $y \geq 0$. y is positive if $y > 0$.

116.

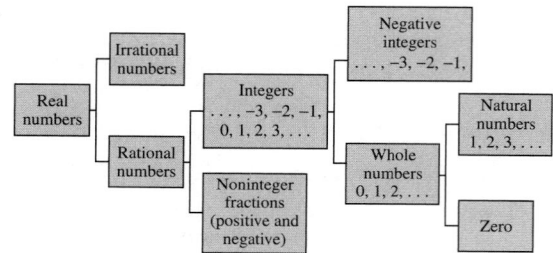

118. (a) Positive   (b) Positive

Section P.2   (page 21)

2. Conditional equation   **4.** Identity

6. Identity   **8.** Identity   **10.** Conditional equation

12. -12 **14.** 3   **16.** -5 **18.** $-\frac{5}{8}$

20. All real numbers   **22.** -5 **24.** 6

26. 50   **28.** No solution. The x-terms sum to zero.

30. $\frac{1}{2}$ **32.** 0   **34.** $\frac{9}{7}$ **36.** $\frac{5}{3}$

38. $\frac{11}{6}$ **40.** $-\frac{13}{3}$ **42.** $\frac{7}{4}$

44. No solution. The solution is extraneous.   **46.** $\frac{1}{5}$

48. No solution. The x-terms sum to zero.

50. $x^2 - 16x = 0$ **52.** $3x^2 + 42x + 134 = 0$

54. $4x^2 - 2x + 1 = 0$ **56.** $-\frac{1}{3}, \frac{1}{3}$ **58.** 9, 1

60. $-\frac{3}{2}$ **62.** $-\frac{3}{2}, 11$ **64.** 2, 6   **66.** $-8, 16$

68. $-a - b, -a + b$ **70.** $\pm 13; \pm 13.00$

72. $\pm 4\sqrt{2}; \pm 5.66$ **74.** $\pm 2; \pm 2.00$

76. $-18, -8; -18.00, -8.00$

78. $5 \pm \sqrt{30}; 10.48, -0.48$

80. $-\frac{7}{4} \pm \frac{\sqrt{11}}{2}; -0.09, -3.41$ **82.** $-\frac{9}{2}; -4.50$

84. $0, -4$ **86.** $3, -1$ **88.** $-4 \pm \sqrt{2}$

90. $\frac{2}{3} \pm \sqrt{2}$ **92.** $\frac{11}{2}, -\frac{9}{2}$ **94.** $1, -\frac{1}{2}$

96. $\frac{3}{5}, \frac{1}{5}$ **98.** $5 \pm \sqrt{3}$ **100.** $-3 \pm \sqrt{13}$

102. $\frac{1}{2} \pm \frac{\sqrt{5}}{2}$ **104.** $\frac{5}{4} \pm \frac{\sqrt{3}}{4}$ **106.** $-\frac{1}{3} \pm \frac{\sqrt{11}}{6}$

108. $\frac{5}{4} \pm \frac{\sqrt{5}}{2}$ **110.** $-\frac{3}{2} \pm \frac{\sqrt{13}}{2}$ **112.** $-\frac{8}{5} \pm \frac{\sqrt{3}}{5}$

114. $-7 \pm \sqrt{13}$ **116.** $\frac{686 \pm 196\sqrt{6}}{25}$

118. $1.400, -0.150$ **120.** $2.137, 18.063$

122. $0.672, -0.968$ **124.** $-2.995, 2.971$

126. $0, -3$ **128.** 7   **130.** $-\frac{3}{2} \pm \sqrt{3}$

132. $-\frac{b}{a}, \frac{b}{a}$ **134.** ± 1 **136.** $0, \pm\frac{5}{2}$ **138.** ± 2

140. $\frac{8}{3}$ **142.** $0, \frac{4}{3}$ **144.** -2 **146.** ± 2 **148.** ± 2

150. $\pm\frac{\sqrt{7}}{6}$ **152.** $-\sqrt[3]{2}, -1$ **154.** $\frac{9}{16}$ **156.** -4

158. $\frac{124}{3}$ **160.** $3, -2$ **162.** No solution   **164.** 1

166. $-29, 25$ **168.** $\frac{1 \pm 5\sqrt{5}}{2}$ **170.** $0, 1, \frac{3}{5}$

172. $2, -12$ **174.** $-3, 1$ **176.** $\frac{3}{4}, -1$ **178.** ± 1

180. $\frac{5}{3}, -3$ **182.** $-6, -3, 3$ **184.** $10, -1$

186. 23,437.5 miles

188. (a)   (b) $w^2 + 14w - 1632 = 0$

          w (c) $w = 34$ feet

                             $l = 48$ feet

         $w + 14$

190. $\frac{20\sqrt{3}}{3} \approx 11.55$ inches

192. 26,250 passengers   **194.** 270,900 units

196. False. If both sides of the equation are graphed, you can see that they intersect, which means that the equation has a real solution.

198. Yes. Dividing by x loses the solution $x = 0$. The student should have subtracted $15x$ from both sides to make the right side of the equation equal to zero. Factoring out an x shows that there are two solutions, $x = 0$ and $x = 6$.

200. Remove symbols of grouping, combine like terms, simplify fractions.

Add (or subtract) the same quantity to (from) each side of the equation.

Multiply (or divide) each side of the equation by the same nonzero quantity.

Interchange the two sides of the equation.

202. (a) $x = 0, -\dfrac{b}{a}$ (b) $x = 0, 1$

204. Isolate the absolute value by subtracting x from each side of the equation. The expression inside the absolute value signs can be positive or negative, so two separate equations must be solved. The first equation assumes the case $x - a = |x - a|$. The second equation assumes the case that $-(x - a) = |x - a|$.

206. Isolate the radical by subtracting x from each side of the equation. Square each side and solve the resulting equation.

Section P.3   *(page 34)*

2. $A: \left(\frac{3}{2}, -4\right)$, $B: (0, -2)$, $C: \left(-3, \frac{5}{2}\right)$, $D: (-6, 0)$

4. $(4, -8)$ **6.** Quadrant III   **8.** Quadrant I

10. Quadrant I or IV   **12.** Quadrant III

14. Quadrant II or IV

16.

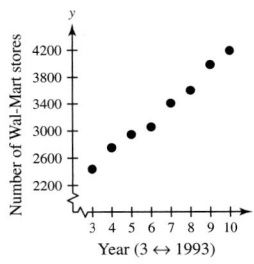

18. (a) $5, 12, 13$ (b) $5^2 + 12^3 = 13^2$

20. (a) $4, 7, \sqrt{65}$ (b) $4^2 + 7^2 = \left(\sqrt{65}\right)^2$

22. (a)

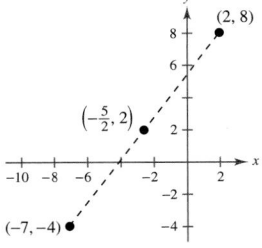

   (b) 15
   (c) $\left(-\frac{5}{2}, 2\right)$

24. (a)

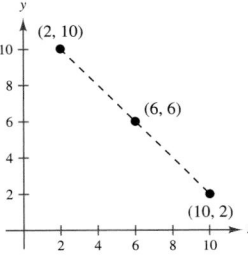

   (b) $8\sqrt{2}$
   (c) $(6, 6)$

26. (a)

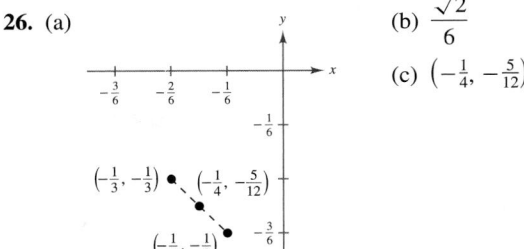

   (b) $\dfrac{\sqrt{2}}{6}$
   (c) $\left(-\frac{1}{4}, -\frac{5}{12}\right)$

28. (a)

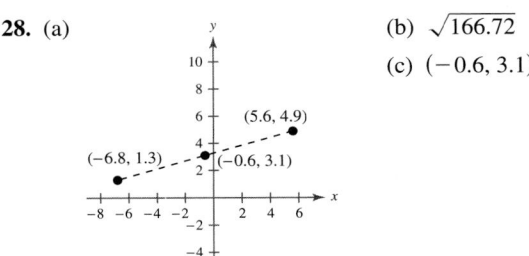

   (b) $\sqrt{166.72}$
   (c) $(-0.6, 3.1)$

30.

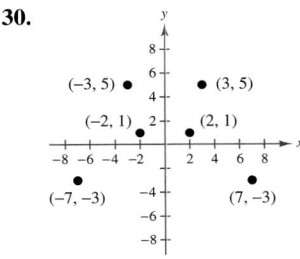

(a) The point is reflected through the y-axis.

(b) The point is reflected through the x-axis.

(c) The point is reflected through the origin.

32. $30\sqrt{41} \approx 192$ kilometers

34. \$1748.55 million

36. (a) Yes   (b) No   **38.** (a) Yes   (b) No

40.

x	-1	0	1	2	3
y	4	0	-2	-2	0
(x, y)	$(-1, 4)$	$(0, 0)$	$(1, -2)$	$(2, -2)$	$(3, 0)$

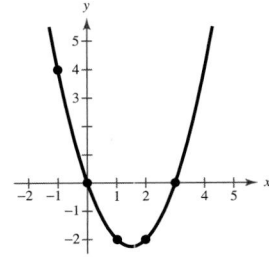

CHAPTER P

42. x-intercept: $(-3, 0)$
y-intercept: $(0, 9)$

44. x-intercept: $\left(\frac{8}{3}, 0\right)$
y-intercept: $(0, 8)$

46. x-intercept: $\left(\frac{1}{2}, 0\right)$
No y-intercept

48. x-intercept: $(-10, 0)$
y-intercept: $(0, -10)$

50. x-intercepts: $(\pm\sqrt{5}, 0)$
y-intercept: $(0, -25)$

52. x-intercept: $(-1, 0)$
y-intercepts: $(0, \pm 1)$

54. x-axis symmetry

56. y-axis symmetry

58. y-axis symmetry

60. Origin symmetry

62.

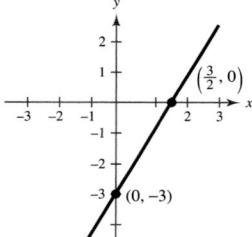

64.

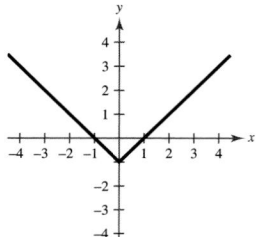

66.

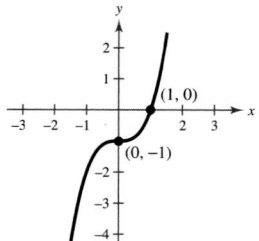

68.

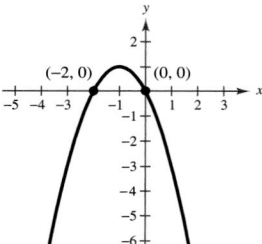

70.

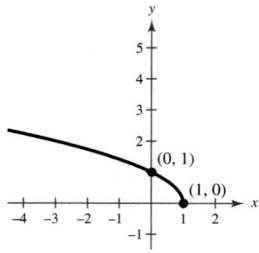

72.

74.

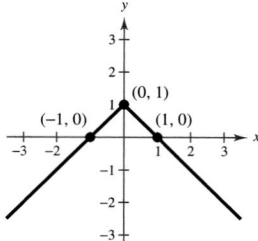

76.

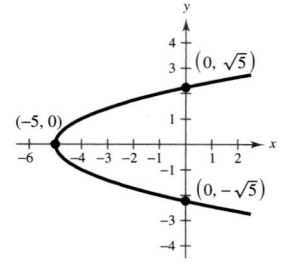

78. $(x + 7)^2 + (y + 4)^2 = 49$

80. $(x - 3)^2 + (y + 2)^2 = 25$ **82.** $x^2 + y^2 = 17$

84. Center: $(0, 0)$; Radius: 4

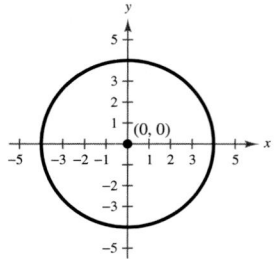

86. Center: $(0, 1)$; Radius: 1

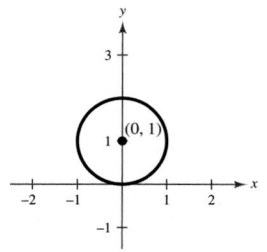

88. Center: $(2, -1)$; Radius: $\sqrt{3}$

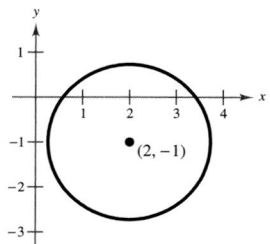

90.

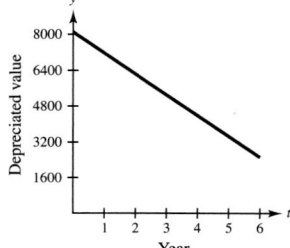

92. (a)

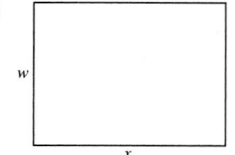

(b) Answers will vary.

(c)

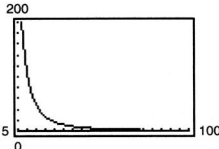

(d) $x = 5.5$, $w = 5.5$

94.

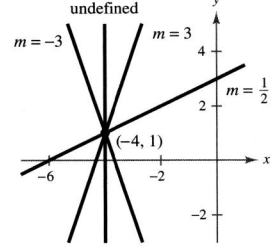

3.938 ohms

96. True. Two sides of the triangle have lengths of $\sqrt{149}$, and the third side has a length of $\sqrt{18}$.

98. True. All linear equations of the form $y = mx + b$, which excludes vertical lines, cross the y-axis one time.

100. No. It depends on the magnitudes of the quantity measured.

102. The viewing window is incorrect. Change the viewing window. Examples will vary.

Section P.4   *(page 47)*

2. (a) L_2 (b) L_1 (c) L_3

4.

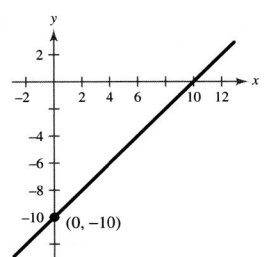

6. $\frac{8}{3}$ **8.** -1

10. $m = 1$; y-intercept: $(0, -10)$

12. $m = -\frac{3}{2}$; y-intercept: $(0, 6)$

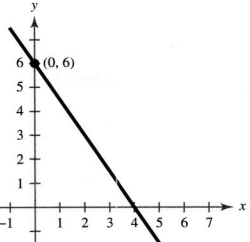

14. $m = 0$; y-intercept: $\left(0, -\frac{5}{3}\right)$

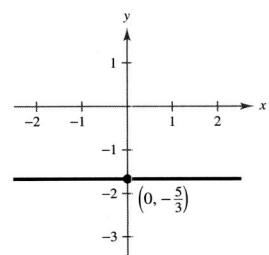

16. $m = -\frac{2}{3}$; y-intercept: $(0, 3)$

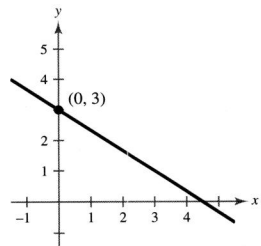

18. $m = 0$; y-intercept: $(0, -4)$

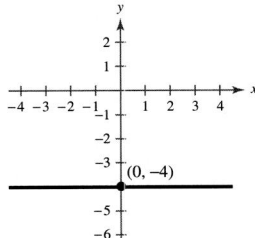

20. m is undefined. There is no y-intercept.

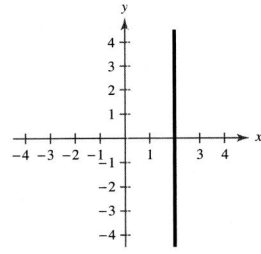

CHAPTER P

22.
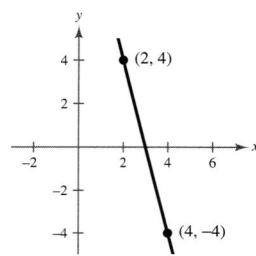
$m = -4$

24.
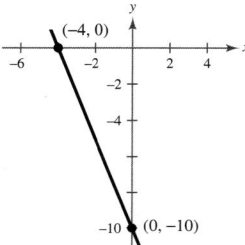
$m = -\frac{5}{2}$

26.
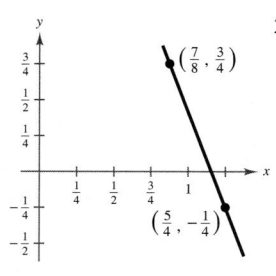
$m = -\frac{8}{3}$

28.
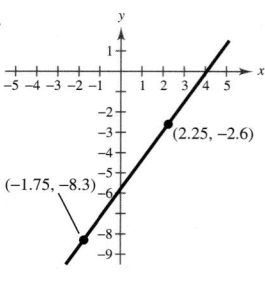
$m = 1.425$

30. Answers will vary. Sample answer:
$(-4, 0), (-4, 3), (-4, 5)$

32. Answers will vary. Sample answer:
$(0, 4), (9, -5), (11, -7)$

34. Answers will vary. Sample answer:
$(-4, -1), (-2, -1), (0, -1)$

36. Answers will vary. Sample answer:
$(-2, -5), (1, -11), (3, -15)$

38. Answers will vary. Sample answer:
$(-3, -5), (1, -7), (5, -9)$

40. Neither parallel nor perpendicular   **42.** Perpendicular

44. (a) Revenues increasing $400 per day.

(b) Revenues increasing $100 per day.

(c) No change in revenues.

46. (a) Greatest increase: 1998–1999
Smallest increase: 1996–1997

(b) 12.92

(c) Each year, the net profit increases by $12.92 million.

48. 12 feet   **50.** $V = 4.5t + 142.5$

51. b; The slope is -20, which represents the decrease in the amount of the loan each week.

52. c; The slope is 2, which represents the hourly wage per unit produced.

53. a; The slope is 0.32, which represents the increase in travel cost for each mile driven.

54. d; The slope is -100, which represents the decrease in the value of the word processor each year.

56. $y = -x + 10$
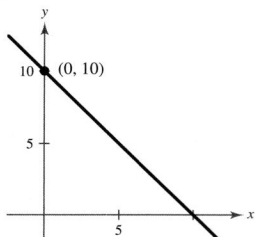

58. $y = 4x$
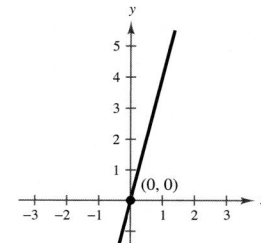

60. $y = \frac{3}{4}x - \frac{7}{2}$
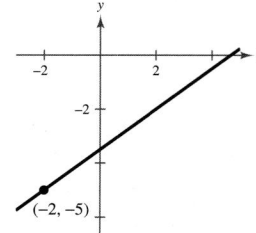

62. $x = -10$
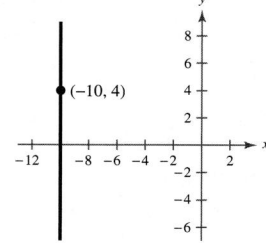

64. $y = \frac{3}{2}$
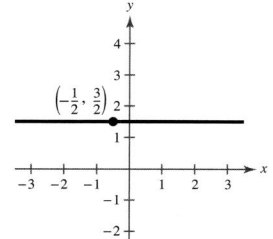

66. $y = -2.5x - 2.75$
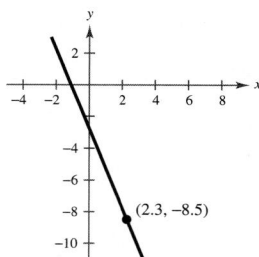

68. $y = \frac{7}{8}x - \frac{1}{2}$
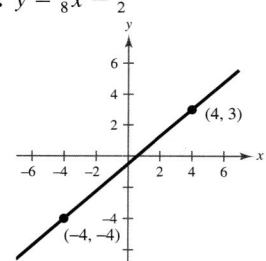

70. $y = 4$
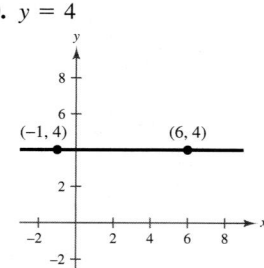

72. $y = -\frac{1}{3}x + \frac{4}{3}$
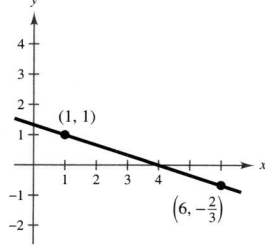

74. $y = -\frac{3}{25}x + \frac{159}{100}$
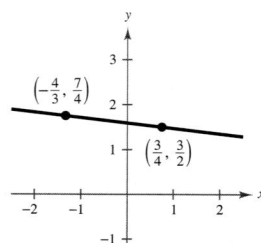

76. $y = 0.3x - 1.8$ **78.** $y = -2$

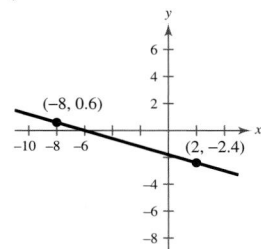

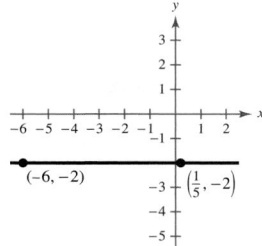

80. $x = 1.5$

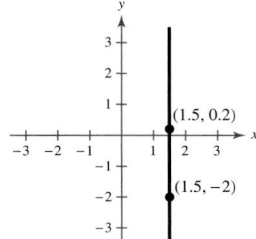

82. $4x - 3y + 12 = 0$ **84.** $3x - y - 2 = 0$

86. $x + y - 1 = 0$

88. (a) $y = -x - 1$ (b) $y = x + 5$

90. (a) $y = -\frac{5}{3}x + \frac{53}{24}$ (b) $y = \frac{3}{5}x + \frac{9}{40}$

92. (a) $y = -2$ (b) $x = 4$

94. (a) $x = -5$ (b) $y = 1$

96. (a) $y = -3x - 13.1$ (b) $y = \frac{1}{3}x - 0.1$

98. Line (a) is parallel to line (c). Line (b) is perpendicular to line (a) and line (c).

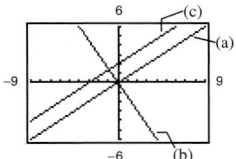

100. Line (a) is parallel to line (b). Line (c) is perpendicular to line (a) and line (b).

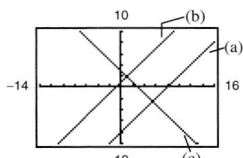

102. $5x + 13y + 2 = 0$ **104.** $128x + 168y + 39 = 0$

106. $y = -31.75t + 3927$; 2005: 3641 stores, 2010: 3483 stores

108. 28,765   **110.** $V = 25,000 - 2300t$

112. $W = 0.75x + 11.50$

114. (a) $x = -\frac{1}{15}p + \frac{266}{3}$ (b) 45   (c) 49

116. $W = 0.07S + 2500$

118. Answers will vary. Sample answer: $y = 199t - 13$

120. (a) and (b)

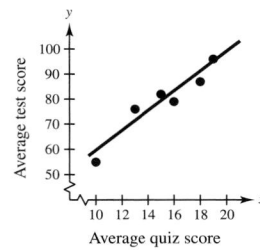

     (c) $y = 4x + 19$ (d) 87

     (e) Vertical shift four units upward

122. False. The slope of the first line is $\frac{2}{7}$ and the slope of the second line is $-\frac{11}{7}$.

124. The slope of a vertical line is undefined because division by zero is undefined.

126. The line with a slope of -4 is steeper. The slope with the greatest magnitude corresponds to the steepest line.

128. No. The slopes of two perpendicular lines have opposite signs (assume that neither line is vertical or horizontal).

Section P.5   *(page 61)*

2. No   **4.** Yes

6. No, the input values of 0 and 1 each have two different output values.

8. Yes, it does not matter that each input value has the same output value.

10. (a) Not a function, because the element c in A corresponds to two elements, 2 and 3, in B.

     (b) Function

     (c) Not a function from A to B. (It is instead a function from B to A.)

     (d) Function

12. 11 million   **14.** Not a function   **16.** Not a function

18. Not a function   **20.** Function   **22.** Not a function

24. (a) 7   (b) 0   (c) $1 - 3s$

26. (a) 0   (b) -0.75 (c) $x^2 + 2x$

28. (a) 2   (b) 5   (c) $\sqrt{x} + 2$

30. (a) $\dfrac{11}{4}$ (b) Undefined   (c) $\dfrac{2x^2 + 3}{x^2}$

32. (a) 6   (b) 6   (c) $x^2 + 4$

34. (a) 6   (b) 3   (c) 10   **36.** (a) 19   (b) 17   (c) 0

38.

x	3	4	5	6	7
$g(x)$	0	1	$\sqrt{2}$	$\sqrt{3}$	2

40.

s	0	1	$\frac{3}{2}$	$\frac{5}{2}$	4
$f(s)$	-1	-1	-1	1	1

42.

x	1	2	3	4	5
$h(x)$	8	5	0	1	2

44. $-\frac{1}{5}$ **46.** $\pm 2\sqrt{3}$ **48.** 3, 5   **50.** 1, ± 2

52. 0, ± 2 **54.** 4   **56.** All real numbers

58. All real numbers $y \neq -5$ **60.** All real numbers

62. $x \leq -3, x \geq 0$ **64.** All real numbers $x \neq 0, 2$

66. $x > -6$ **68.** $x < -3, x > 3$

70. $\{(-2, 1), (-1, -2), (0, -3), (1, -2), (2, 1)\}$

72. $\{(-2, 1), (-1, 0), (0, 1), (1, 2), (2, 3)\}$

74. $f(x) = cx; c = \frac{1}{4}$ **76.** $h(x) = c\sqrt{|x|}; c = 3$

78. $-(5 + h), h \neq 0$ **80.** $8x + 4h - 2, h \neq 0$

82. $\frac{1}{t - 2}, t \neq 1$ **84.** $\frac{x^{2/3} - 4}{x - 8}$ **86.** $A = \frac{C^2}{4\pi}$

88. (a) The maximum profit is \$3375.

(b)

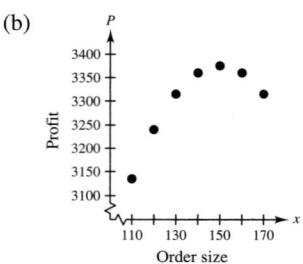

Yes, P is a function of x.

(c) $P = 45x - 0.15x^2, x > 100$

90. $A = 2xy = 2x\sqrt{36 - x^2}, \ 0 < x < 6$

92. (a) $V = x^2 y = x^2(108 - 4x)$

$= 108x^2 - 4x^3, 0 < x < 27$

(b)

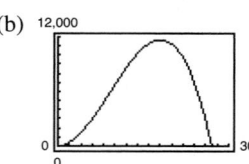

(c) 18 inches $\times$ 18 inches $\times$ 36 inches

94. (a) $C = 6000 + 0.95x$ (b) $\overline{C} = \dfrac{C}{x} = \dfrac{6000}{x} + 0.95$

96. (a)

y	5	10	20
$F(y)$	26,474.08	149,760.00	847,170.49

y	30	40
$F(y)$	2,334,527.36	4,792,320.00

The deeper the water, the greater the force.

(b) 21 feet   (c) 21.37 feet

98. Yes, the ball will be at a height of 6 feet.

100. False. The range is $[-1, \infty)$.

102. The domain is the set of inputs of the function, and the range is the set of outputs.

Section P.6   *(page 73)*

2. Domain: $(-\infty, \infty)$ **4.** Domain: $(-\infty, 1), (1, \infty)$

Range: $[0, \infty)$ Range: $-1, 1$

6. (a) 4   (b) 4   (c) 2   (d) 0

8. (a) 0   (b) 1   (c) 2   (d) 3

10. Function   **12.** Not a function   **14.** Not a function

16. $-8, \frac{2}{3}$ **18.** 2, 7   **20.** $\pm 3, 4$ **22.** $0, \pm\frac{5}{3}$ **24.** $-\frac{2}{3}$

26.

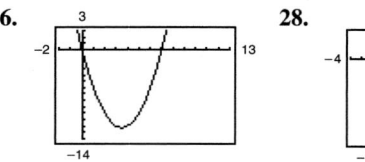

0, 7

28.

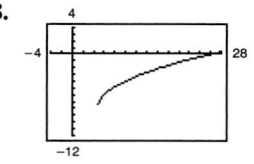

26

30.

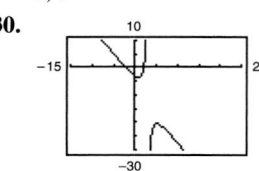

$\pm\dfrac{3\sqrt{2}}{2}$

32. Increasing on $(2, \infty)$; Decreasing on $(-\infty, 2)$

34. Increasing on $(1, \infty)$; Decreasing on $(-\infty, -1)$

36. Increasing on $(-\infty, -1), (0, \infty)$

Decreasing on $(-1, 0)$

38. Increasing on $(-\infty, -2), (0, \infty)$

Decreasing on $(-2, -1), (-1, 0)$

40. (a) 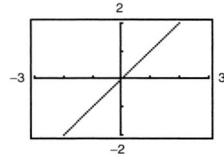   Increasing on $(-\infty, \infty)$

(b)

x	-2	-1	0	1	2
$g(x)$	-2	-1	0	1	2

42. (a) 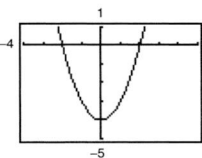   Increasing on $(0, \infty)$;

Decreasing on $(-\infty, 0)$

(b)

x	-2	-1	0	1	2
$h(x)$	0	-3	-4	-3	0

44. (a)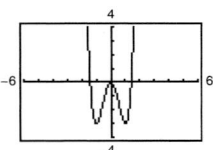

Increasing on $(-1, 0)$, $(1, \infty)$

Decreasing on $(-\infty, -1)$, $(0, 1)$

(b)

x	-2	-1	0	1	2
$f(x)$	24	-3	0	-3	24

46. (a)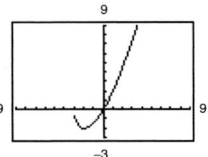

Increasing on $(-2, \infty)$

Decreasing on $(-3, -2)$

(b)

x	-3	-2	-1	0	1
$f(x)$	0	-2	-1.414	0	2

48. (a)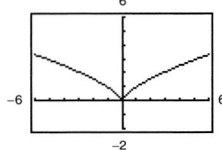

Decreasing on $(-\infty, 0)$; Increasing on $(0, \infty)$

(b)

x	-2	-1	0	1	2
$f(x)$	1.59	1	0	1	1.59

50. 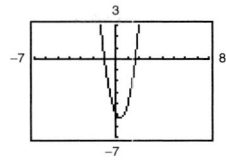   Relative minimum: $\left(\frac{1}{3}, -\frac{16}{3}\right)$

52.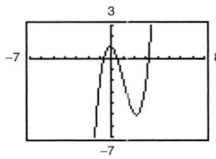

Relative maximum: $(-0.15, 1.08)$

Relative minimum: $(2.15, -5.08)$

54. 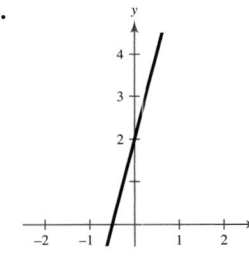   **56.**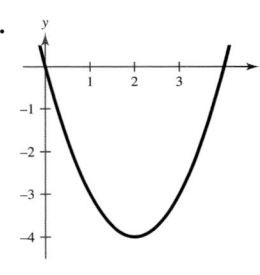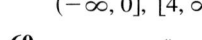

$\left[-\frac{1}{2}, \infty\right)$ $(-\infty, 0]$, $[4, \infty)$

58. 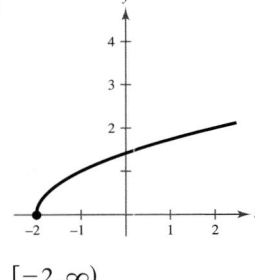   **60.**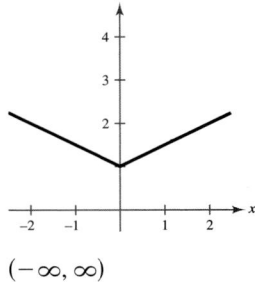

$[-2, \infty)$ $(-\infty, \infty)$

62. Neither even nor odd   **64.** Odd   **66.** Even

68. $h = 3 - 4x + x^2$ **70.** $h = 2 - \sqrt[3]{x}$

72. $L = 2 - \sqrt[3]{2y}$ **74.** $L = \dfrac{2}{y}$

76. (a) and (e)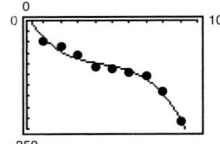

(b) y is a function of x.

(c) $y = -1.3767x^3 + 18.156x^2 - 89.40x + 16.5$

(d) $1 \le x \le 9$ (f) 1996; 1992

78. (a) $A = 64 - 2x^2$, $0 \le x \le 4$

(b)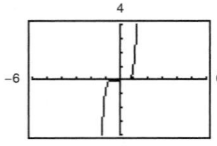

$32 \le A \le 64$

(c) Square with sides of $4\sqrt{2}$ meters

80. False. An odd function is symmetric with respect to the origin, so its domain must include negative values.

82. Yes. For each value of y there corresponds one and only one value of x.

84. (a) $\left(\frac{5}{3}, -7\right)$ (b) $\left(\frac{5}{3}, 7\right)$

86. (a) $(-5, -1)$ (b) $(-5, 1)$

88. Both graphs will pass through the origin. $y = x^7$ will be symmetric with respect to the origin, and $y = x^8$ will be symmetric with respect to the y-axis.

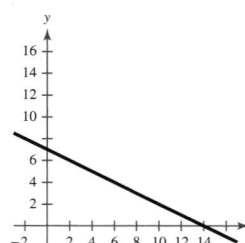

Section P.7   *(page 82)*

2. $f(x) = \frac{5}{2}x - \frac{1}{2}$

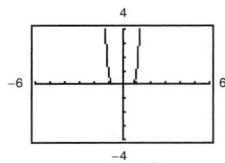

4. $f(x) = 5x - 6$

6. $f(x) = -\frac{1}{2}x + 7$

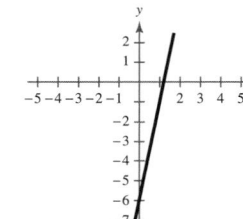

8. $f(x) = \frac{3}{4}x - 8$

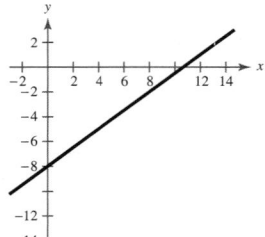

10.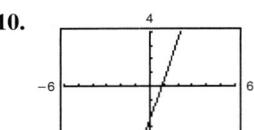

12.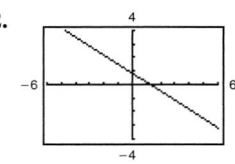

14.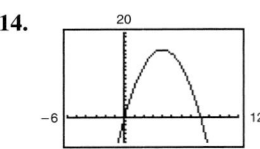

16.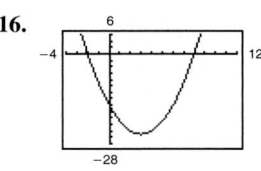

18.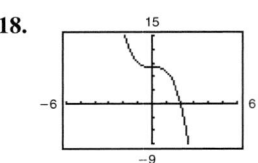

20.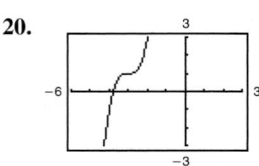

22.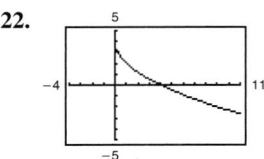

24.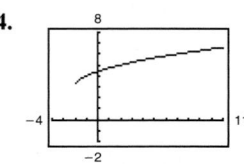

26.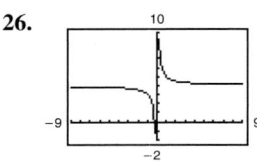

28.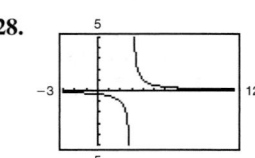

30. (a) -6 (b) 0 (c) 18 (d) 6

32. (a) 7 (b) -1 (c) 31 (d) 11

34. (a) 8 (b) 2 (c) 6 (d) 13

36. (a) -22 (b) -85 (c) 6 (d) -29

38.   **40.**

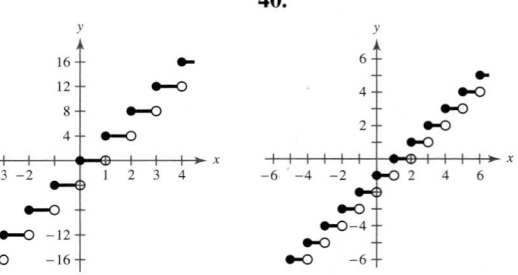

42.

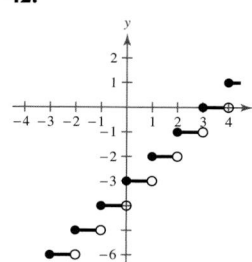

44.

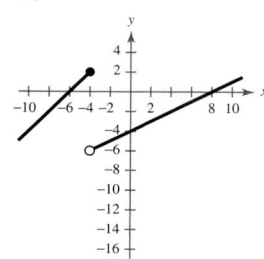

(b)

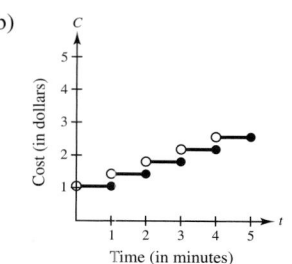

Time (in minutes)

$7.89

66. (a) $C = 9.8 + 2.5[\![x]\!]$

(b)

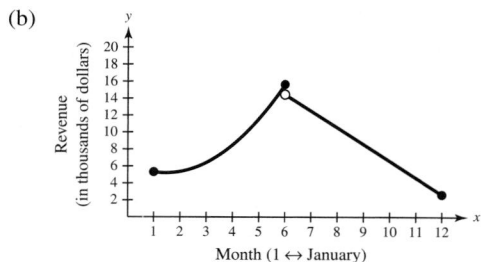

Weight (in pounds)

46.

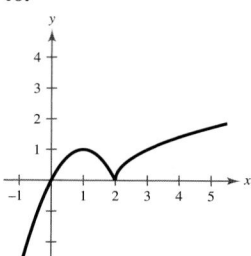

48.

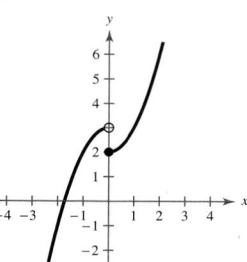

68. (a) $f(x) = \begin{cases} 0.505x^2 - 1.47x + 6.3, & 1 \le x \le 6 \\ -1.97x + 26.3, & 6 < x \le 12 \end{cases}$

Answers will vary. Sample answer: The domain is determined by inspection of a graph of the data with the two models.

(b)

50.

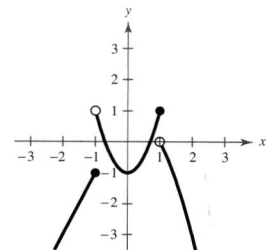

52.

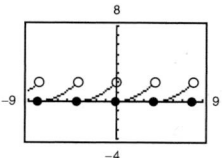

Domain: $(-\infty, \infty)$

Range: $[0, 2)$

Sawtooth pattern

(c) $f(5) = 11.575, f(11) = 4.63$; These values represent the revenue for the months of May and November, respectively.

(d) These values are quite close to the actual data values.

54. $f(x) = \sqrt{x}; g(x) = 1 + \sqrt{x + 2}$

56. $f(x) = \dfrac{1}{x}; g(x) = \dfrac{1}{x} - 2$

58. $f(x) = x^2; g(x) = 1 - (x + 2)^2$

60. $f(x) = [\![x]\!]; g(x) = [\![x - 1]\!]$

62. $f(x) = x^3; g(x) = -x^3$

64. (a) C_2 is the appropriate model, because the cost does not increase until after the next minute of conversation has started.

70. False. A piecewise-defined function is a function that is defined by two or more equations over a specified domain. That domain may or may not include x- and y-intercepts.

72. $f(x) = \begin{cases} -\frac{4}{3}x + 6, & 0 \le x \le 3 \\ -\frac{2}{5}x + \frac{16}{5}, & 3 < x \le 8 \end{cases}$

CHAPTER P

Section P.8   *(page 90)*

2. (a)

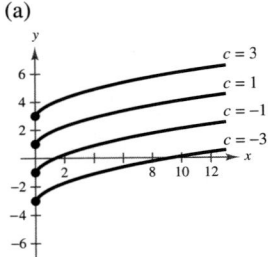

(b)

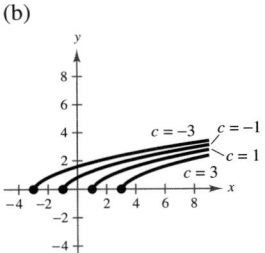

(c)

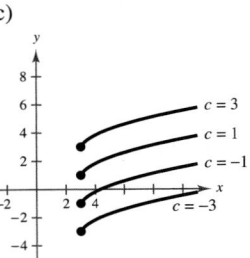

4. (a)

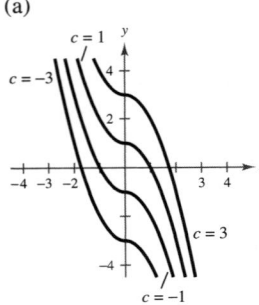

(b)

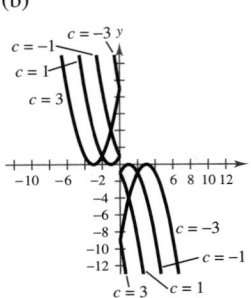

6. (a)

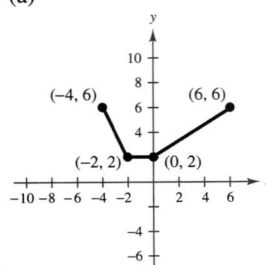

(b)

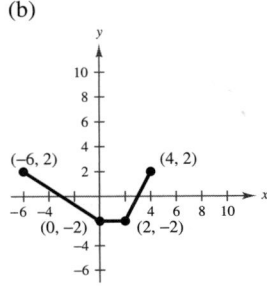

(c)

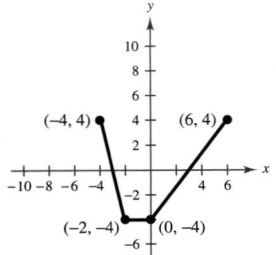

(d)

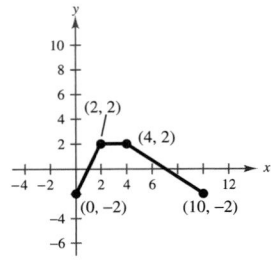

(e)

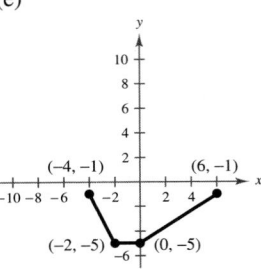

(f)

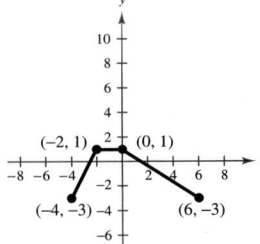

(g)

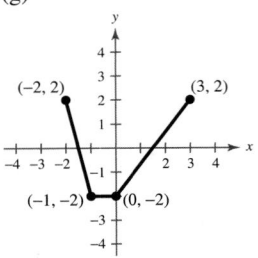

8. (a)

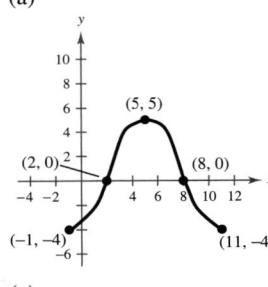

(b)

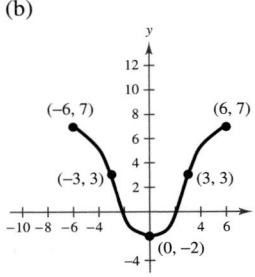

(c)

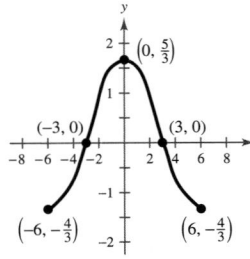

(d)

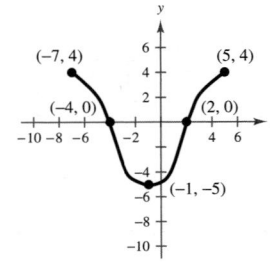

(e)

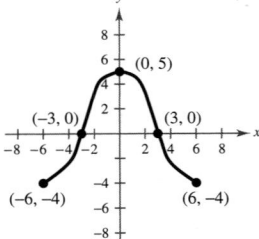

(f)

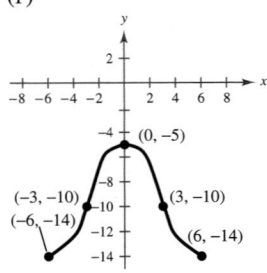

(g)

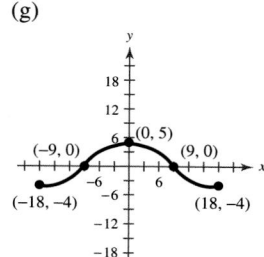

10. (a) $y = 1 - x^3$ (b) $y = (x - 1)^3 + 1$

(c) $y = -(x + 3)^3 - 1$ (d) $y = (x - 10)^3 - 4$

12. (a) $y = \sqrt{x} - 3$ (b) $y = \sqrt{x + 1} - 7$

(c) $y = -\sqrt{x - 5} + 5$ (d) $y = -\sqrt{-x + 3} - 4$

14. Vertical shrink of $y = x$; $y = \frac{1}{2}x$

16. Vertical shift of $y = [\![x]\!]$; $y = [\![x]\!] + 4$

18. Horizontal shift of $y = |x|$; $y = |x + 2|$

20. Horizontal shift eight units to the right of $f(x) = x^2$

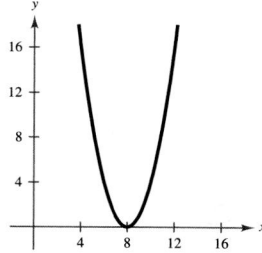

22. Reflection in the x-axis, and vertical shift one unit downward, of $f(x) = x^3$

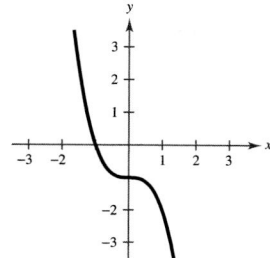

24. Reflection in the x-axis, horizontal shift ten units to the left, and vertical shift five units upward, of $f(x) = x^2$

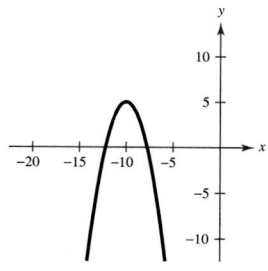

26. Horizontal shift three units to the left, and vertical shift ten units downward, of $f(x) = x^3$

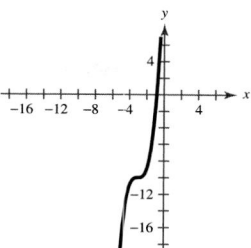

28. Reflection in the x-axis, horizontal shift five units to the left, and vertical shift six units upward, of $f(x) = |x|$

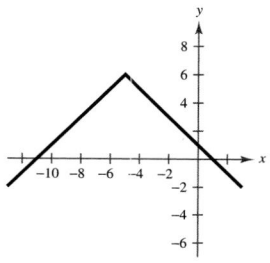

30. Reflection in the y-axis, horizontal shift three units to the right, and vertical shift nine units upward, of $f(x) = |x|$

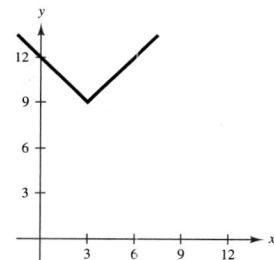

32. Horizontal shift five units to the left, and vertical stretch of two, of $f(x) = [\![x]\!]$

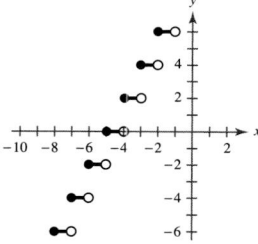

CHAPTER P

34. Horizontal shift four units to the left, and vertical shift eight units upward, of $f(x) = \sqrt{x}$

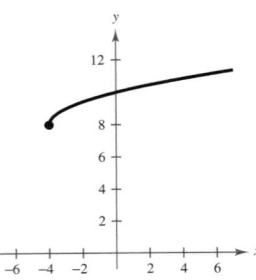

36. Reflection in the x-axis, horizontal shift one unit to the left, and vertical shift six units downward, of $f(x) = \sqrt{x}$

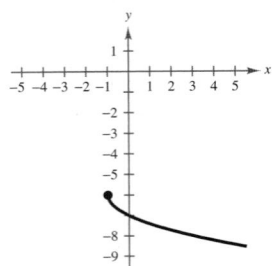

38. Vertical shift one unit upward, and horizontal shrink of $\frac{1}{3}$, of $f(x) = \sqrt{x}$

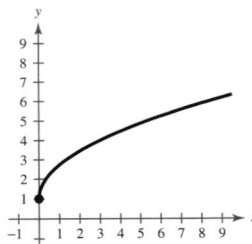

40. $f(x) = -(x+3)^2 + 7$ **42.** $f(x) = (-x-6)^3 - 6$

44. $f(x) = |x+1| - 7$ **46.** $f(x) = -\sqrt{-x} - 9$

48. (a) $y = \frac{1}{4}x^3$ (b) $y = -2x^3$

50. (a) $y = 8\sqrt{x}$ (b) $y = -\frac{1}{4}\sqrt{x}$

52. Vertical stretch of $y = |x|$; $y = 6|x|$

54. Horizontal stretch of $y = [\![x]\!]$; $y = [\![\frac{1}{2}x]\!]$

56. Reflection in the x-axis, vertical shift two units downward, and vertical stretch, of $y = |x|$;

$y = -2|x| - 2$

58. $y = |x+4| - 2$ **60.** $y = (x-2)^2 + 4$

62. (a)

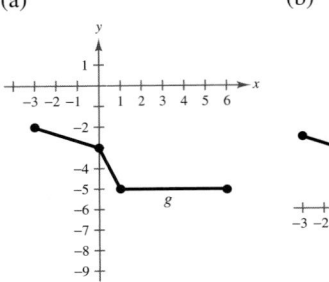

(b)

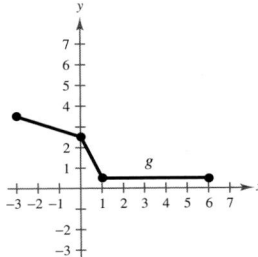

(c)

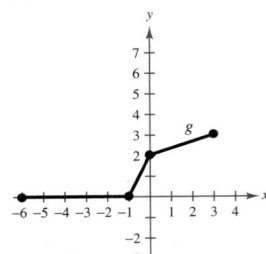

(d)

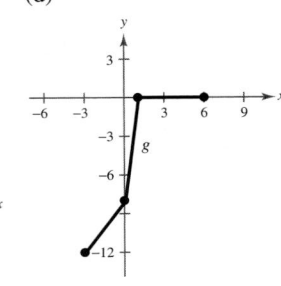

(e)

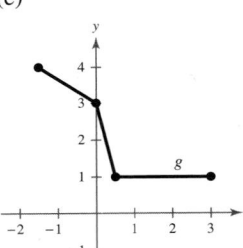

(f)

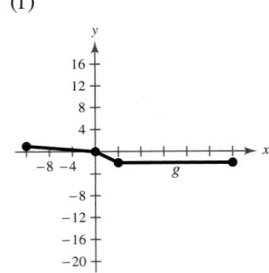

64. (a) Horizontal shift of 14.979 units to the left and vertical shrink of 0.0037

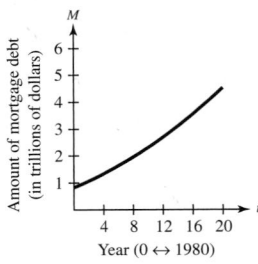

(b) $f(t) = 0.0037(t + 24.979)^2$. The graph was shifted 10 units to the left.

66. False. The point $(-2, -61)$ will lie on the transformation.

68. If you consider the x-axis to be a mirror, the graph of $y = -f(x)$ is the mirror image of the graph of $y = f(x)$.

70. Answers will vary.

Section P.9   *(page 100)*

2.

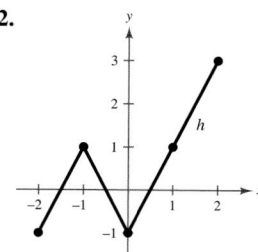

4.

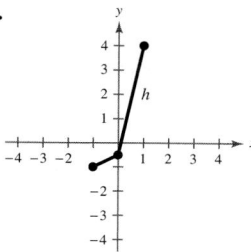

(e) $T(t) = \begin{cases} 60, & 0 \le t \le 6 \\ 12t - 12, & 6 < t < 7 \\ 72, & 7 \le t < 20 \\ -12t + 312, & 20 < t < 21 \\ 60, & 21 \le t \le 24 \end{cases}$

36. (a) $20 - 3x$ (b) $-3x$ (c) $9x + 20$

38. (a) $\dfrac{1}{x^3}$ (b) $\dfrac{1}{x^3}$ (c) x^9

40. (a) $\sqrt[3]{x^3 - 4}$ (b) $x - 4$

Domain of $f, g, f \circ g,$ and $g \circ f$: all real numbers

42. (a) x^4 (b) x^4

Domain of $f, g, f \circ g,$ and $g \circ f$: all real numbers

44. (a) $|-1 - x|$ (b) $3 - |x - 4|$

Domain of $f, g, f \circ g,$ and $g \circ f$: all real numbers

46. (a) $\dfrac{3}{x^2 + 2x}$ (b) $\dfrac{x^2 + 2}{x^2 - 1}$

Domain of f and $g \circ f$: all real numbers $x \ne \pm 1$

Domain of g: all real numbers

Domain of $f \circ g$: all real numbers $x \ne 0, -2$

48. (a) -1 (b) 0 **50** (a) 2 (b) 2

52. $f(x) = x^3,\ g(x) = 1 - x$

54. $f(x) = \sqrt{x},\ g(x) = 9 - x$

56. $f(x) = \dfrac{4}{x^2},\ g(x) = 5x + 2$

58. $f(x) = \dfrac{27x + 6\sqrt[3]{x}}{10 - 27x},\ g(x) = x^3$

60. $(A \circ r)(t) = 0.36\pi t^2$; $(A \circ r)(t)$ represents the area of the circle at time t.

62. True. The range of g must be a subset of the domain of f for $(f \circ g)(x)$ to be defined.

64. Answers will vary.

6. (a) $x - 3$ (b) $3x - 7$

(c) $-2x^2 + 9x - 10$ (d) $\dfrac{2x - 5}{2 - x};\ x \ne 2$

8. (a) $2x - 1$ (b) $2x - 9$

(c) $8x - 20$ (d) $\frac{1}{2}x - \frac{5}{4};\ -\infty < x < \infty$

10. (a) $\sqrt{x^2 - 4} + \dfrac{x^2}{x^2 + 1}$ (b) $\sqrt{x^2 - 4} - \dfrac{x^2}{x^2 + 1}$

(c) $\dfrac{x^2\sqrt{x^2 - 4}}{x^2 + 1}$ (d) $\dfrac{(x^2 + 1)\sqrt{x^2 - 4}}{x^2};\ |x| \ge 2$

12. (a) $\dfrac{x^4 + x^3 + x}{x + 1}$ (b) $\dfrac{-x^4 - x^3 + x}{x + 1}$

(c) $\dfrac{x^4}{x + 1}$ (d) $\dfrac{1}{x^2(x + 1)};\ x \ne 0, -1$

14. 7 **16.** -1 **18.** $t^2 - 3t - 1$ **20.** -370

22. $-\frac{1}{4}$ **24.** 43

26.

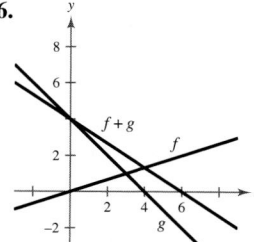

28.

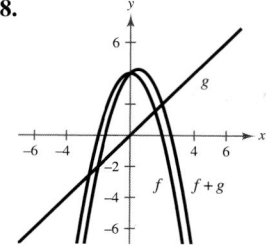

30.

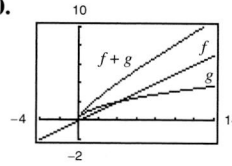

$g(x),\ f(x)$

32. $R_T = 734 - 7.22t - 0.8t^2$

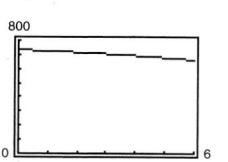

34. (a) For each time t there corresponds one and only one temperature T.

(b) $60°,\ 72°$

(c) All the temperature changes occur 1 hour later.

(d) The temperature is decreased by 1 degree.

Section P.10   *(page 109)*

1. c   **2.** b   **3.** a   **4.** d   **6.** $f^{-1}(x) = 3x$

8. $f^{-1}(x) = x + 4$ **10.** $f^{-1}(x) = 5x + 1$

12. $f^{-1}(x) = \sqrt[5]{x}$

14. (a) $f(g(x)) = f(x + 5) = (x + 5) - 5 = x$

$g(f(x)) = g(x - 5) = (x - 5) + 5 = x$

(b)

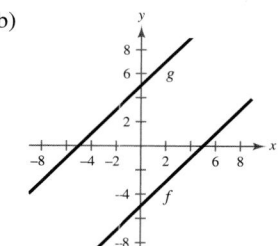

16. (a) $f(g(x)) = f\left(\dfrac{3 - x}{4}\right) = 3 - 4\left(\dfrac{3 - x}{4}\right) = x$

$g(f(x)) = g(3 - 4x) = \dfrac{3 - (3 - 4x)}{4} = x$

(b)

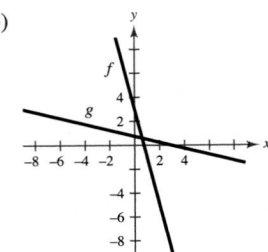

18. (a) $f(g(x)) = f\left(\dfrac{1}{x}\right) = \dfrac{1}{1/x} = x$

$g(f(x)) = g\left(\dfrac{1}{x}\right) = \dfrac{1}{1/x} = x$

(b)

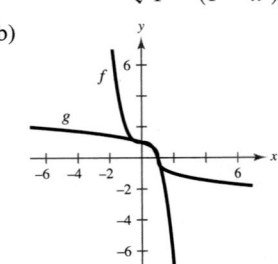

20. (a) $f(g(x)) = f\left(\sqrt[3]{1 - x}\right)$

$= 1 - \left(\sqrt[3]{1 - x}\right)^3 = x$

$g(f(x)) = g(1 - x^3)$

$= \sqrt[3]{1 - (1 - x^3)} = x$

(b)

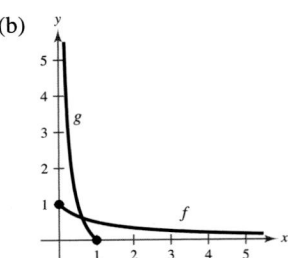

22. (a) $f(g(x)) = f\left(\dfrac{1 - x}{x}\right), \quad 0 < x \le 1$

$= \dfrac{1}{1 + (1 - x)/x} = \dfrac{x}{x + 1 - x} = x$

$g(f(x)) = g\left(\dfrac{1}{1 + x}\right), \quad x \ge 0$

$= \dfrac{1 - 1/(1 + x)}{1/(1 + x)} = \dfrac{1 + x - 1}{1} = x$

(b)

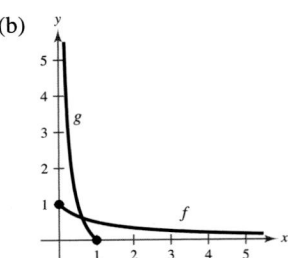

24. (a) $f(g(x)) = f\left(\dfrac{2x + 3}{x - 1}\right) = \dfrac{\left(\dfrac{2x + 3}{x - 1}\right) + 3}{\left(\dfrac{2x + 3}{x - 1}\right) - 2}$

$= \dfrac{2x + 3 + 3x - 3}{2x + 3 - 2x + 2} = x$

(b) $g(f(x)) = g\left(\dfrac{x + 3}{x - 2}\right) = \dfrac{2\left(\dfrac{x + 3}{x - 2}\right) + 3}{\left(\dfrac{x + 3}{x - 2}\right) - 1}$

$= \dfrac{2x + 6 + 3x - 6}{x + 3 - x + 2} = x$

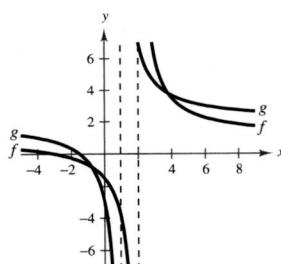

26. Yes

28.

x	-10	-7	-4	-1	2	5
$f^{-1}(x)$	-3	-2	-1	0	1	2

30. No   **32.** Yes

34.

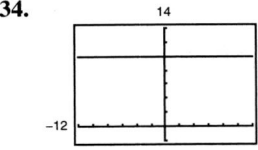

The function does not have an inverse.

36.

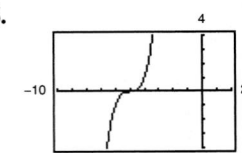

The function has an inverse.

38.

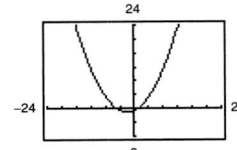

The function does not have an inverse.

40. $f^{-1}(x) = \dfrac{x - 1}{3}$

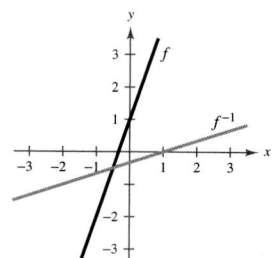

42. $f^{-1}(x) = \sqrt[3]{x - 1}$

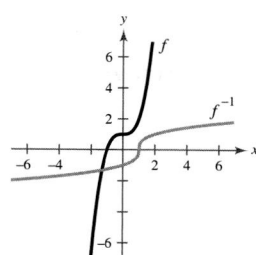

54. $f^{-1}(x) = \dfrac{-6x - 4}{2x - 8}$

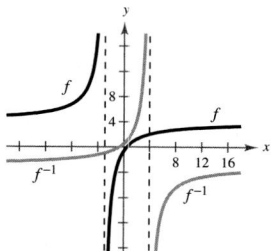

44. $f^{-1}(x) = \sqrt{x}$

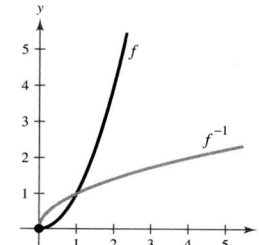

56. No inverse   **58.** $f^{-1}(x) = \dfrac{x - 5}{3}$

60. $f^{-1}(x) = \dfrac{5x - 4}{3}$ **62.** No inverse

64. No inverse   **66.** $f^{-1}(x) = 2 - x, \quad x \geq 0$

68. $f^{-1}(x) = x^2 + 2, \quad x \geq 0$ **70.** 0   **72.** $-\sqrt[9]{4}$

74. $2\sqrt[3]{x + 3}$ **76.** $\dfrac{x - 3}{2}$ **78.** $\dfrac{x - 3}{2}$

80. (a) Yes

(b) The inverse, f^{-1}, yields the year for a given per capita consumption of bottled water.

(c) 8

46. $f^{-1}(x) = -\sqrt{x + 2}$ **48.** $f^{-1}(x) = -\dfrac{2}{x}$

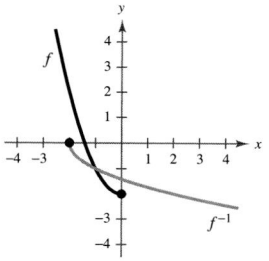

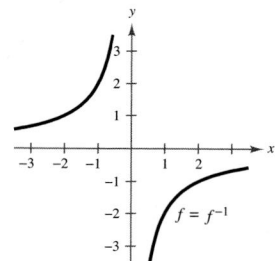

82. (a) $y = \dfrac{x - 8}{0.75}$

(b) $y =$ number of units produced; $x =$ hourly wage

(c) 19 units

84. (a) $y = \dfrac{80 - x}{0.35}$

$x =$ total cost; $y =$ number of pounds of the less expensive ground beef

(b) $62.5 \leq x \leq 80$ (c) 20 pounds

86. True. If $f(x) = x - 6$ and $f^{-1}(x) = x + 6$, then the y-intercept of f is $(0, -6)$ and the x-intercept of f^{-1} is $(-6, 0)$.

50. $f^{-1}(x) = \dfrac{-2x - 3}{x - 1}$ **52.** $f^{-1}(x) = x^{5/3}$

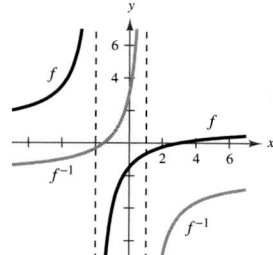

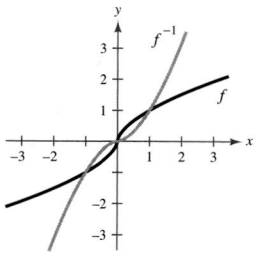

88.

x	-2	-1	1	3
y	-5	-2	2	3

x	-5	-2	2	3
$f^{-1}(x)$	-2	-1	1	3

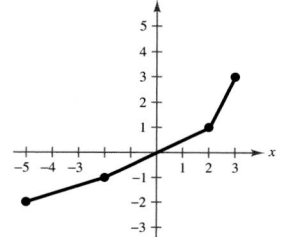

CHAPTER P

90.

x	-4	-2	0	3
y	3	4	0	-1

The graph of f does not pass the Horizontal Line Test, so $f^{-1}(x)$ does not exist.

92. $k = -\frac{1}{2}$

Review Exercises   *(page 115)*

2. (a) None   (b) $-22, 0$

(c) $-22, -\frac{10}{3}, 0, 5.2, \frac{3}{7}$ (d) $\sqrt{15}$

4. (a) 0.36 (b) $0.\overline{714285}$

$\frac{9}{25} < \frac{5}{7}$

6. The set consists of all real numbers greater than 1.

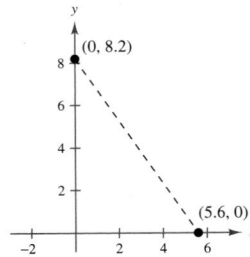

8. 106   **10.** $|z + 16| > 8$

12. (a) $\frac{1}{2}$ (b) Division by 0 is undefined.

14. Commutative Property of Multiplication

16. Multiplicative Inverse Property

18. -1 **20.** 2   **22.** -100

24. Conditional equation

26. $-\frac{9}{2}$ **28.** $-\frac{17}{3}$ **30.** 18   **32.** $\frac{11}{3}$

34. $-\frac{7}{2}, 4$ **36.** $\pm\frac{5}{4}$ **38.** $8 \pm \sqrt{15}$

40. $-3 \pm 2\sqrt{3}$ **42.** $\frac{1}{2} \pm \frac{\sqrt{249}}{6}$ **44.** $0, \frac{3}{2}$

46. $0, -3, \pm\frac{2}{3}$ **48.** 66   **50.** $\frac{38 + 5\sqrt{3}}{24}$ **52.** 79

54. $2, -5$ **56.** Quadrant I or II   **58.** Quadrant I or III

60. (a)   (b) $\sqrt{98.6} \approx 9.9$

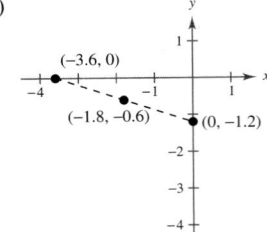

62. (a)   (b) $(-1.8, -0.6)$

64.

x	-4	-2	0	2	4
y	4	3	2	1	0

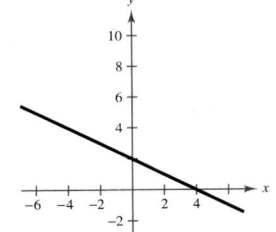

66.

x	-2	-1	0	1	2	3
y	1	-6	-9	-8	-3	6

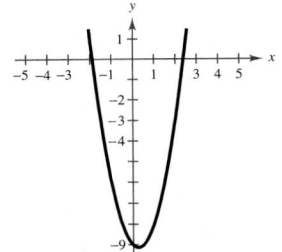

68.

x	-2	-1	0	2	4
y	0	1	1.41	2	2.45

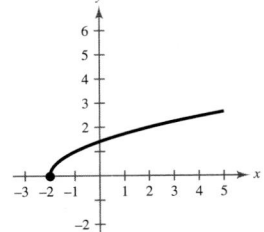

70. x-intercepts: $(2, 0)$, $(6, 0)$

y-intercept: $(0, -12)$

72. No x-intercept

y-intercept: $(0, 5)$

74. x-axis, y-axis, and origin symmetry

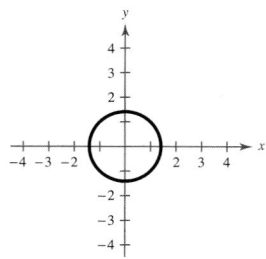

76. No symmetry

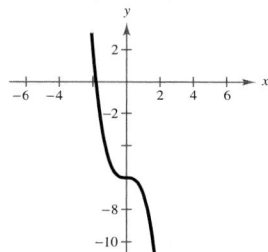

78. y-axis symmetry

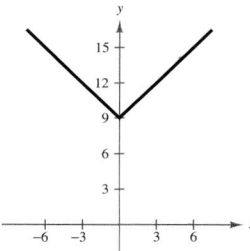

80. Center: $(0, 8)$; Radius: 9

82. Center: $\left(-4, \frac{3}{2}\right)$; Radius: 10

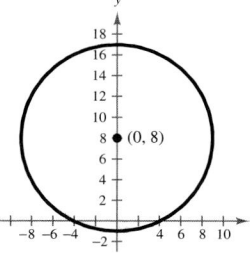

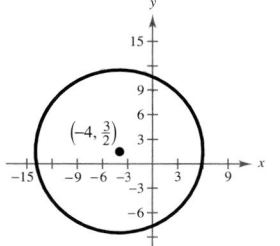

84. $(x - 1)^2 + \left(y + \frac{13}{2}\right)^2 = \frac{85}{4}$

86.

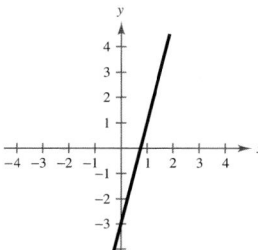

88.

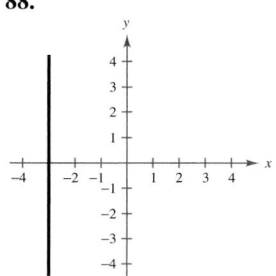

90.

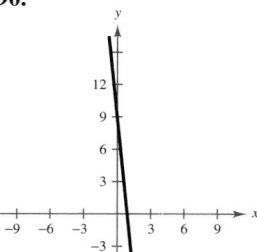

92.

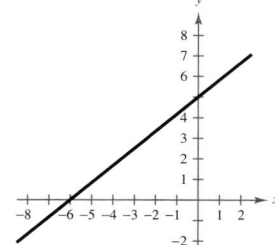

94.

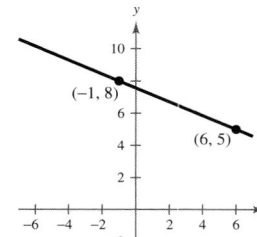

$m = -\frac{3}{7}$

96.

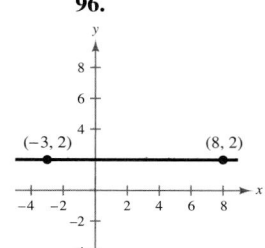

$m = 0$

98.

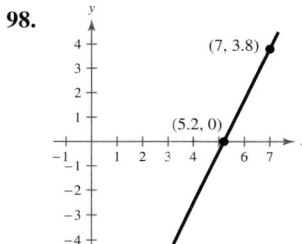

$m = \frac{19}{9}$

100. $3x - 2y + 4 = 0$

102. $x + 5y - 1 = 0$

104. $5x + 4y - 1 = 0$

106. $y - 6 = 0$

108. $x + 8 = 0$

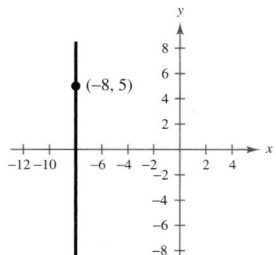

110. (a) $2x + 3y + 7 = 0$ (b) $3x - 2y + 30 = 0$

112. (a) $y = 5$ (b) $x = -2$

114. $V = 5.15t + 52.35$, $4 \le t \le 9$

116. (a) $V = 3.75t + 66.25$

(b)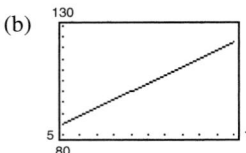

(c) $103.75

118. (a) Not a function, because u in the domain corresponds to two values in the range

(b) A function, because each input value has exactly one output value.

(c) A function, because each input value has exactly one output value.

(d) Not a function, because w in the domain corresponds to two values in the range and because u in A is not matched with any element in B

120. Yes   **122.** No

124. (a) 2   (b) $\dfrac{2}{13}$ (c) $\dfrac{4}{t^2 + 1}$ (d) 4

126. All real numbers

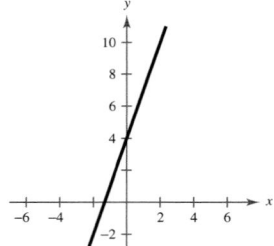

128. $x \le -8, x \ge 0$

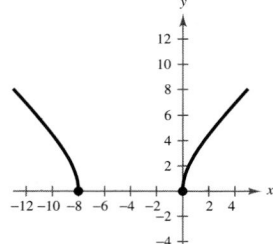

130. All real numbers

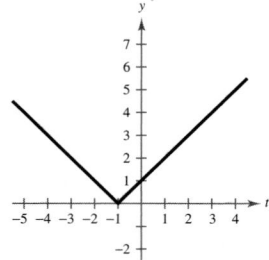

132. $3x^2 + 3xh + h^2 - 10x - 5h + 1, \ h \ne 0$

134. (a) $C(x) = 5.35x + 16{,}000$

(b) $P(x) = R - C = 2.85x - 16{,}000$

136. (a) $f(x) = x + 0.4(50 - x) = 0.6x + 20$

(b) Domain: $0 \le x \le 50$; Range: $20 \le y \le 50$

(c) $8\frac{1}{3}$ liters

138. Function   **140.** Not a function   **142.** $-\frac{3}{8}$

144. $1, \pm 5$

146. Increasing on $(-2, 0)$ and $(2, \infty)$

Decreasing on $(-\infty, -2)$ and $(0, 2)$

148. Even   **150.** Even

152. $f(x) = -\frac{3}{4}x - 5$ **154.** $f(x) = \frac{7}{8}x + \frac{217}{80}$

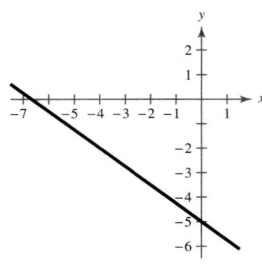   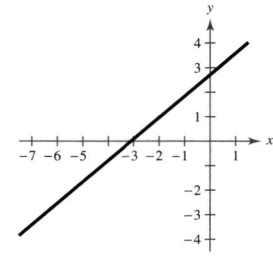

156.   **158.**

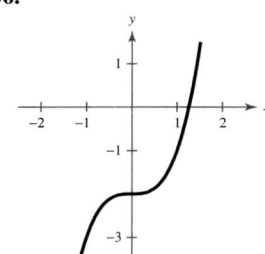   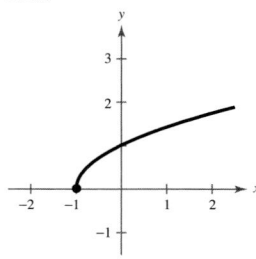

160.   **162.**

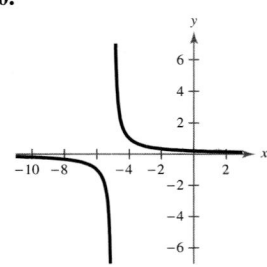   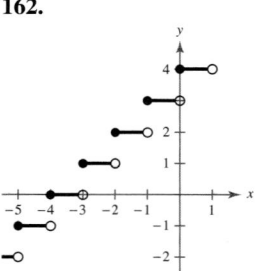

164.   **166.** $y = \sqrt{x}$

168. Horizontal shift of two units to the right and vertical shift of two units upward

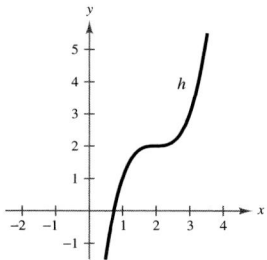

170. Horizontal shift of three units to the left and vertical shift of five units downward

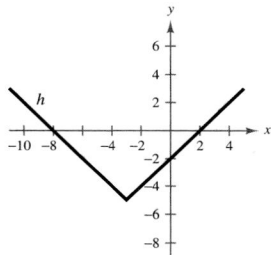

172. Reflection in the x-axis, horizontal shift of five units to the right, and vertical shift of five units downward

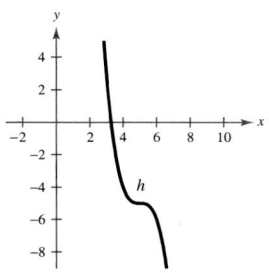

174. Reflection in the x-axis, horizontal shift of one unit to the left, and vertical shift of nine units upward

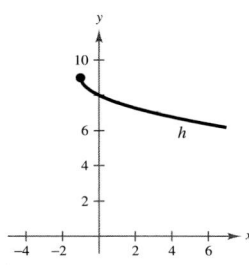

176. Reflection in the x-axis, horizontal shift of one unit to the left, and vertical shift of three units downward

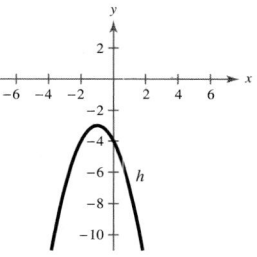

178. Reflection in the x-axis and vertical shrink

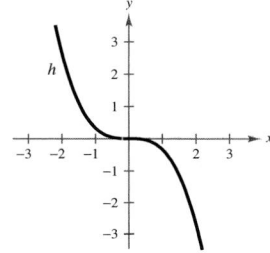

180. Vertical shrink and vertical shift of one unit downward

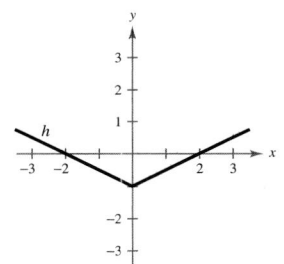

182. (a) $x^2 - 4 + \sqrt{3 - x}$ (b) $x^2 - 4 - \sqrt{3 - x}$

(c) $(x^2 - 4)\sqrt{3 - x}$ (d) $\dfrac{x^2 - 4}{\sqrt{3 - x}}$; $x < 3$

184. (a) $x + 3$ (b) $\sqrt[3]{x^3 + 3}$

Domain of $f, g, f \circ g,$ and $g \circ f$: all real numbers

186. $f(x) = \sqrt[3]{x},\ g(x) = x + 2$

188.

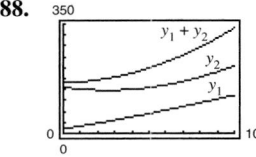

$\approx \$546.4$ billion

190. $f^{-1}(x) = x - 5$

$f(f^{-1}(x)) = x - 5 + 5 = x$

$f^{-1}(f(x)) = x + 5 - 5 = x$

CHAPTER P

192. The function does not have an inverse.

194. The function does not have an inverse.

196. The function does have an inverse.

198. (a) $f^{-1}(x) = \dfrac{x+7}{5}$

(b)

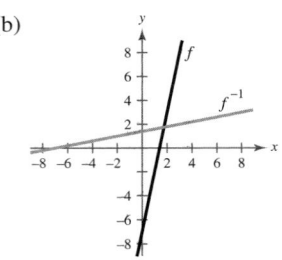

(c) $f^{-1}(f(x)) = f^{-1}(5x - 7)$

$$= \dfrac{5x - 7 + 7}{5}$$

$$= x$$

$$f(f^{-1}(x)) = f\left(\dfrac{x+7}{5}\right)$$

$$= 5\left(\dfrac{x+7}{5}\right) - 7$$

$$= x$$

200. (a) $f^{-1}(x) = \sqrt[3]{x-2}$

(b)

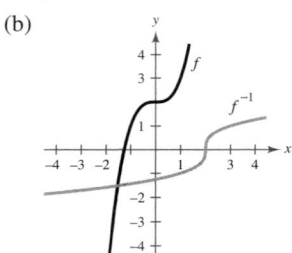

(c) $f^{-1}(f(x)) = f^{-1}(x^3 + 2)$

$$= \sqrt[3]{x^3 + 2 - 2}$$

$$= x$$

$$f(f^{-1}(x)) = f\left(\sqrt[3]{x-2}\right)$$

$$= \left(\sqrt[3]{x-2}\right)^3 + 2$$

$$= x$$

202. $x \geq 2$; $f^{-1}(x) = x + 2, x \geq 0$

204. True, by the definition of inverse function

Problem Solving   *(page 122)*

2. Mapping numbers onto letters is not a function since each number corresponds to three letters.

Mapping letters onto numbers is a function since every letter is assigned exactly one number.

4. $f(x) = x$ $\qquad\qquad\qquad$ $g(x) = -x$

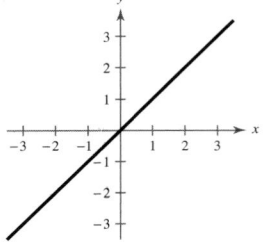 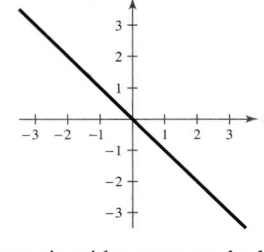

Both graphs are already symmetric with respect to the line $y = x$.

General formula: $y = -x + c, c \geq 0$

6. $(6, 8)$

$$f(x) = \begin{cases} \frac{12}{7}x - \frac{16}{7}, & 2.5 \leq x \leq 6 \\ -\frac{12}{7}x + \frac{128}{7}, & 6 < x \leq 9.5 \end{cases}$$

8. (a) $(f \circ g)(x) = 4x + 24$.   (b) $(f \circ g)^{-1}(x) = \frac{1}{4}x - 6$

(c) $f^{-1}(x) = \frac{1}{4}x; g^{-1}(x) = x - 6$

(d) $(g^{-1} \circ f^{-1})(x) = \frac{1}{4}x - 6$

(e) $(f \circ g)(x) = 8x^3 + 1; (f \circ g)^{-1}(x) = \frac{1}{2}\sqrt[3]{x - 1};$

$f^{-1}(x) = \sqrt[3]{x - 1}; g^{-1}(x) = \frac{1}{2}x;$

$(g^{-1} \circ f^{-1})(x) = \frac{1}{2}\sqrt[3]{x - 1}$

(f) Answers will vary.

(g) $(f \circ g)^{-1}(x) = (g^{-1} \circ f^{-1})(x)$

10. (a) $\qquad\qquad\qquad\qquad$ (b)

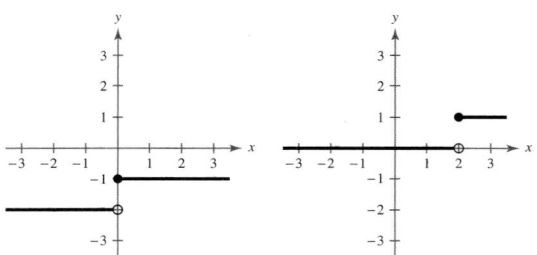

(c) $\qquad\qquad\qquad\qquad$ (d)

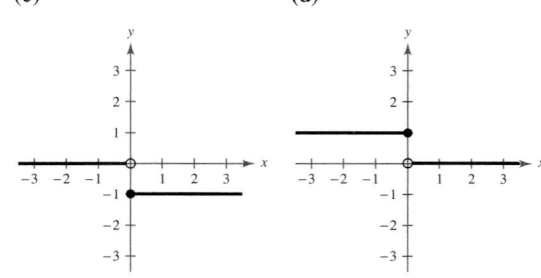

(e)   (f)

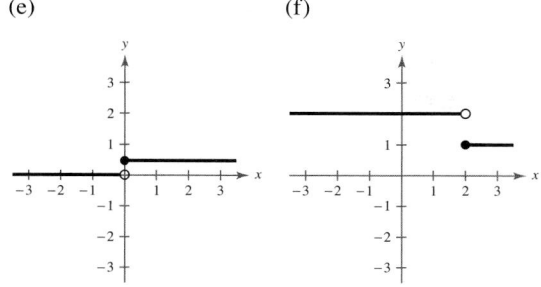

12. Answers will vary. Sample answer:

Let $f(x) = x^2$, $g(x) = x + 1$, $h(x) = 2x + 3$.

$(f \circ (g \circ h))(x) = f(g(2x + 3))$

$\qquad\qquad\qquad = f(2x + 4)$

$\qquad\qquad\qquad = 4x^2 + 16x + 16$

$((f \circ g) \circ h)(x) = f(x + 1) \circ h(x)$

$\qquad\qquad\qquad = x^2 + 2x + 1 \circ h(x)$

$\qquad\qquad\qquad = 4x^2 + 16x + 16$

14. (a)

x	-4	-2	0	4
$f(f^{-1}(x))$	-4	-2	0	4

(b)

x	-3	-2	0	1
$(f + f^{-1})(x)$	5	1	-3	-5

(c)

x	-3	-2	0	1
$(f \circ f^{-1})(x)$	4	0	2	6

x	-4	-3	0	4
$|f^{-1}(x)|$	2	1	1	3

Chapter 1

Section 1.1   *(page 133)*

2. 5.5 radians   **4.** -4 radians   **6.** 6.5 radians

8. (a) Quadrant III   (b) Quadrant III

10. (a) Quadrant IV   (b) Quadrant II

12. (a) Quadrant IV   (b) Quadrant II

14. (a)   (b)

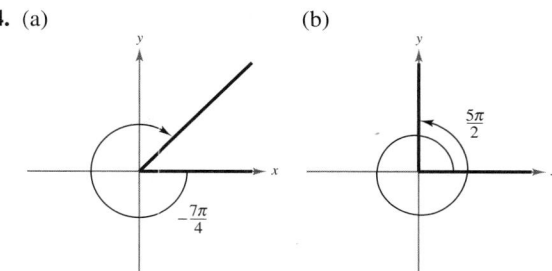

16. (a)   (b)

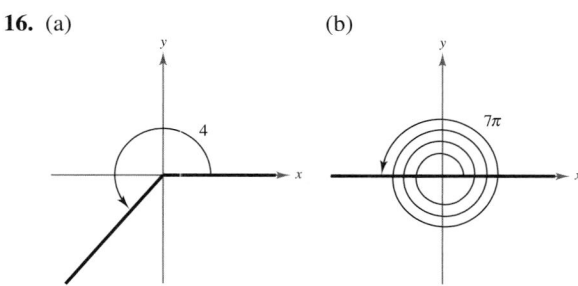

18. (a) $\dfrac{19\pi}{6}$, $-\dfrac{5\pi}{6}$ (b) $\dfrac{\pi}{6}$, $-\dfrac{23\pi}{6}$

20. (a) $\dfrac{7\pi}{4}$, $-\dfrac{\pi}{4}$ (b) $\dfrac{28\pi}{15}$, $-\dfrac{32\pi}{15}$

22. (a) Complement: $\dfrac{5\pi}{12}$; Supplement: $\dfrac{11\pi}{12}$

    (b) Complement: none; Supplement: $\dfrac{\pi}{12}$

24. (a) Complement: none; Supplement: $\pi - 3 \approx 0.14$

    (b) Complement: $\dfrac{\pi}{2} - 1.5 \approx 0.07$;

       Supplement: $\pi - 1.5 \approx 1.64$

26. (a) $\dfrac{7\pi}{4}$ (b) $\dfrac{2\pi}{3}$ **28.** (a) $-\dfrac{3\pi}{2}$ (b) $\dfrac{4\pi}{5}$

30. 1.525   **32.** -0.842 **34.** 6.021   **36.** 0.009

38. (a) $-105°$ (b) $20°$ **40.** (a) $330°$ (b) $408°$

42. $81.818°$ **44.** $1170.000°$ **46.** $864.000°$

48. $-32.659°$ **50.** $120°$ **52.** $-330°$ **54.** $10°$

56. (a) Quadrant I   (b) Quadrant III

58. (a) Quadrant II   (b) Quadrant IV

60. (a)   (b)

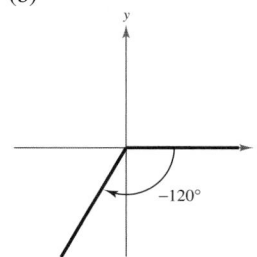

62. (a)   (b)

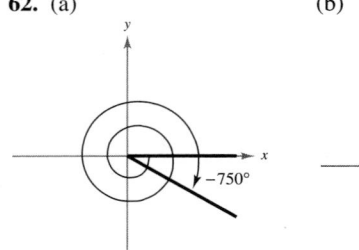
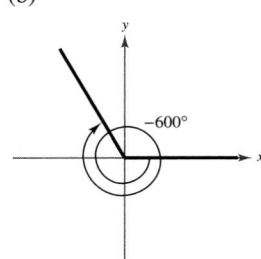

64. (a) $480°, -240°$ (b) $300°, -60°$

66. (a) $300°, -60°$ (b) $590°, -130°$

68. (a) Complement: $87°$; Supplement: $177°$

(b) Complement: $26°$; Supplement: $116°$

70. (a) Complement: none; Supplement: $50°$

(b) Complement: none; Supplement: $10°$

72. (a) $245.167°$ (b) $2.2°$

74. (a) $-135.01°$ (b) $-408.272°$

76. (a) $-345°7'12''$ (b) $0°27'$

78. (a) $-0°21'18''$ (b) $0°47'11.4''$

80. $\frac{29}{10}$ radians   **82.** $-\frac{4}{5}$ radian

84. $\frac{4}{7}$ radian   **86.** 2 radians   **88.** 3π feet ≈ 9.42 feet

90. 5π centimeters ≈ 15.71 centimeters

92. 686.4 miles   **94.** 4045.7 miles

96. 0.063 radian $\approx 3.59°$ **98.** $275°$

100. (a) 3400π radians per minute; 1700π radians per minute

(b) 850 revolutions per minute

102. 20.16π inches per second

104. True. Let α and β represent coterminal angles, and let n represent an integer.

$\alpha = \beta + n(360°)$

$\alpha - \beta = n(360°)$

106. (a) The vertex is at the origin and the initial side is on the positive x-axis.

(b) Clockwise rotation of the terminal side

(c) Two angles in standard position where the terminal sides coincide

(d) The magnitude of the angle is between $90°$ and $180°$.

108. Radian. 1 radian $\approx 57.3°$ **110.** Answers will vary.

112.

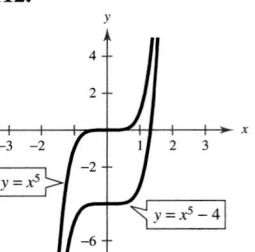

114.

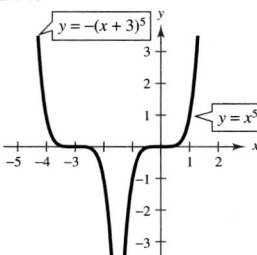

116. $\dfrac{2\sqrt{3}}{3}$ **118.** $\dfrac{5\sqrt{2}}{4}$ **120.** $6\sqrt{13}$ **122.** $4\sqrt{13}$

Section 1.2   *(page 142)*

2. $\sin\theta = \frac{5}{13}$ $\csc\theta = \frac{13}{5}$

$\cos\theta = \frac{12}{13}$ $\sec\theta = \frac{13}{12}$

$\tan\theta = \frac{5}{12}$ $\cot\theta = \frac{12}{5}$

4. $\sin\theta = -\frac{3}{5}$ $\csc\theta = -\frac{5}{3}$

$\cos\theta = -\frac{4}{5}$ $\sec\theta = -\frac{5}{4}$

$\tan\theta = \frac{3}{4}$ $\cot\theta = \frac{4}{3}$

6. $\left(\dfrac{1}{2}, \dfrac{\sqrt{3}}{2}\right)$ **8.** $\left(-\dfrac{\sqrt{2}}{2}, -\dfrac{\sqrt{2}}{2}\right)$

10. $\left(\dfrac{1}{2}, -\dfrac{\sqrt{3}}{2}\right)$ **12.** $(-1, 0)$

14. $\sin\dfrac{\pi}{3} = \dfrac{\sqrt{3}}{2}$ **16.** $\sin\left(-\dfrac{\pi}{4}\right) = -\dfrac{\sqrt{2}}{2}$

$\cos\dfrac{\pi}{3} = \dfrac{1}{2}$ $\cos\left(-\dfrac{\pi}{4}\right) = \dfrac{\sqrt{2}}{2}$

$\tan\dfrac{\pi}{3} = \sqrt{3}$ $\tan\left(-\dfrac{\pi}{4}\right) = -1$

18. $\sin\left(-\dfrac{4\pi}{3}\right) = \dfrac{\sqrt{3}}{2}$ **20.** $\sin\dfrac{5\pi}{3} = -\dfrac{\sqrt{3}}{2}$

$\cos\left(-\dfrac{4\pi}{3}\right) = -\dfrac{1}{2}$ $\cos\dfrac{5\pi}{3} = \dfrac{1}{2}$

$\tan\left(-\dfrac{4\pi}{3}\right) = -\sqrt{3}$ $\tan\dfrac{5\pi}{3} = -\sqrt{3}$

22. $\sin(-2\pi) = 0$

$\cos(-2\pi) = 1$

$\tan(-2\pi) = 0$

24. $\sin\dfrac{5\pi}{6} = \dfrac{1}{2}$

$\cos\dfrac{5\pi}{6} = -\dfrac{\sqrt{3}}{2}$

$\tan\dfrac{5\pi}{6} = -\dfrac{\sqrt{3}}{3}$

$\csc\dfrac{5\pi}{6} = 2$

$\sec\dfrac{5\pi}{6} = -\dfrac{2\sqrt{3}}{3}$

$\cot\dfrac{5\pi}{6} = -\sqrt{3}$

26. $\sin\dfrac{3\pi}{2} = -1$

$\cos\dfrac{3\pi}{2} = 0$

$\tan\dfrac{3\pi}{2}$ is undefined.

$\csc\dfrac{3\pi}{2} = -1$

$\sec\dfrac{3\pi}{2}$ is undefined.

$\cot\dfrac{3\pi}{2} = 0$

28. $\sin\dfrac{7\pi}{4} = -\dfrac{\sqrt{2}}{2}$

$\cos\dfrac{7\pi}{4} = \dfrac{\sqrt{2}}{2}$

$\tan\dfrac{7\pi}{4} = -1$

$\csc\dfrac{7\pi}{4} = -\sqrt{2}$

$\sec\dfrac{7\pi}{4} = \sqrt{2}$

$\cot\dfrac{7\pi}{4} = -1$

30. $\cos 5\pi = \cos \pi = -1$

32. $\sin\dfrac{9\pi}{4} = \sin\dfrac{\pi}{4} = \dfrac{\sqrt{2}}{2}$

34. $\sin\dfrac{19\pi}{6} = \sin\dfrac{7\pi}{6} = -\dfrac{1}{2}$

36. $\cos\left(-\dfrac{8\pi}{3}\right) = \cos\dfrac{4\pi}{3} = -\dfrac{1}{2}$

38. (a) $-\frac{3}{8}$ (b) $-\frac{8}{3}$ **40.** (a) $-\frac{3}{4}$ (b) $-\frac{4}{3}$

42. (a) $-\frac{4}{5}$ (b) $-\frac{4}{5}$ **44.** 1.7321   **46.** 0.6421

48. -0.8011 **50.** -4.4014 **52.** -0.7833

54. (a) 0.7   (b) -0.8

56. (a) 4.0 or 5.4   (b) 0.7 or 5.6

58. (a) 0.25 foot   (b) 0.02 foot   (c) -0.25 foot

60. True. The tangent function has a period of π.

62. Answers will vary.   **64.** Odd

66. $f^{-1}(x) = \sqrt[3]{4(x-1)}$ **68.** $f^{-1}(x) = \dfrac{2(2x+1)}{x-1}$

Section 1.3   *(page 157)*

2. $\sin\theta = \frac{5}{13}$ $\csc\theta = \frac{13}{5}$

$\cos\theta = \frac{12}{13}$ $\sec\theta = \frac{13}{12}$

$\tan\theta = \frac{5}{12}$ $\cot\theta = \frac{12}{5}$

4. $\sin\theta = \dfrac{\sqrt{2}}{2}$ $\csc\theta = \sqrt{2}$

$\cos\theta = \dfrac{\sqrt{2}}{2}$ $\sec\theta = \sqrt{2}$

$\tan\theta = 1$ $\cot\theta = 1$

6. $\sin\theta = \frac{8}{17}$ $\csc\theta = \frac{17}{8}$

$\cos\theta = \frac{15}{17}$ $\sec\theta = \frac{17}{15}$

$\tan\theta = \frac{8}{15}$ $\cot\theta = \frac{15}{8}$

The triangles are similar, and corresponding sides are proportional.

8. $\sin\theta = \dfrac{\sqrt{5}}{5}$ $\csc\theta = \sqrt{5}$

$\cos\theta = \dfrac{2\sqrt{5}}{5}$ $\sec\theta = \dfrac{\sqrt{5}}{2}$

$\tan\theta = \dfrac{1}{2}$ $\cot\theta = 2$

The triangles are similar, and corresponding sides are proportional.

10.

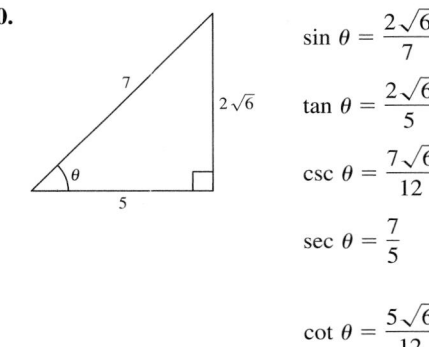

$\sin\theta = \dfrac{2\sqrt{6}}{7}$

$\tan\theta = \dfrac{2\sqrt{6}}{5}$

$\csc\theta = \dfrac{7\sqrt{6}}{12}$

$\sec\theta = \dfrac{7}{5}$

$\cot\theta = \dfrac{5\sqrt{6}}{12}$

12.

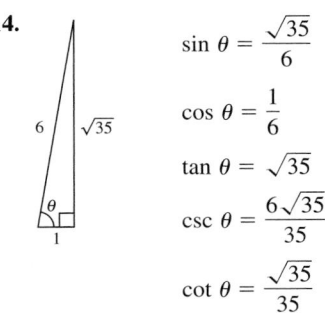

$\sin\theta = \dfrac{\sqrt{26}}{26}$ $\csc\theta = \sqrt{26}$

$\cos\theta = \dfrac{5\sqrt{26}}{26}$ $\sec\theta = \dfrac{\sqrt{26}}{5}$

$\tan\theta = \dfrac{1}{5}$

14.

$\sin\theta = \dfrac{\sqrt{35}}{6}$

$\cos\theta = \dfrac{1}{6}$

$\tan\theta = \sqrt{35}$

$\csc\theta = \dfrac{6\sqrt{35}}{35}$

$\cot\theta = \dfrac{\sqrt{35}}{35}$

16.

$$\sin \theta = \frac{4}{17} \qquad \sec \theta = \frac{17\sqrt{273}}{273}$$

$$\cos \theta = \frac{\sqrt{273}}{17} \qquad \cot \theta = \frac{\sqrt{273}}{4}$$

$$\tan \theta = \frac{4\sqrt{273}}{273}$$

18. (a) 2   (b) $\dfrac{\sqrt{3}}{3}$ (c) $\dfrac{\sqrt{3}}{2}$ (d) $\sqrt{3}$

20. (a) $\dfrac{1}{5}$ (b) $\dfrac{\sqrt{6}}{12}$ (c) $2\sqrt{6}$ (d) $\dfrac{2\sqrt{6}}{5}$

22. (a) $\dfrac{1}{5}$ (b) $\dfrac{\sqrt{26}}{26}$ (c) $\dfrac{1}{5}$ (d) $\dfrac{\sqrt{26}}{5}$

24. (a) 1   (b) $\dfrac{\sqrt{2}}{2}$ (c) $\sqrt{2}$

26. (a) $\dfrac{\sqrt{3}}{2}$ (b) 1   (c) $\dfrac{2\sqrt{3}}{3}$

28. (a) 0.4348   (b) 0.4348

30. (a) 0.9598   (b) 0.9609

32. (a) 0.9964   (b) 1.0036

34. (a) 1.7946   (b) 0.5572

36. (a) 2.6695   (b) 0.0699

38. (a) $45° = \dfrac{\pi}{4}$ (b) $45° = \dfrac{\pi}{4}$

40. (a) $60° = \dfrac{\pi}{3}$ (b) $60° = \dfrac{\pi}{3}$

42. (a) $60° = \dfrac{\pi}{3}$ (b) $45° = \dfrac{\pi}{4}$

44. (a) $10° \approx 0.175$ (b) $29° \approx 0.506$

46. (a) $22° \approx 0.384$ (b) $68° \approx 1.187$

48–56. Answers will vary.

58. $9\sqrt{3}$ **60.** $20\sqrt{2}$

62. 18 feet

64. (a)

(b) $\sin \theta = \dfrac{3\frac{1}{3}}{20}$ (c) $9.59°$

66. 1.3 miles

68. 6.57 centimeters

70. $\sin 75° \approx 0.97$ $\csc 75° \approx 1.04$

   $\cos 75° \approx 0.26$ $\sec 75° \approx 3.86$

   $\tan 75° \approx 3.73$ $\cot 75° \approx 0.27$

72. True, $\sec x = \csc(90 - x)$.

74. True, $\cot^2 \theta - \csc^2 \theta = -1$ for all θ.

76. False, $\tan 25° \neq (\tan 5°)(\tan 5°)$.

78. Yes. Tan θ is equal to opp/adj. You can find the value of the hypotenuse by the Pythagorean Theorem, then you can find sec θ, which is equal to hyp/adj.

80. (a)

θ	0°	18°	36°	54°	72°	90°
$\sin \theta$	0	0.3090	0.5878	0.8090	0.9511	1
$\cos \theta$	0	0.9511	0.8090	0.5878	0.3090	0

(b) Increasing function   (c) Decreasing function

(d) As the angle increases, the length of the side opposite the angle increases relative to the length of the hypotenuse and the length of the side adjacent to the angle decreases relative to the length of the hypotenuse. Thus, the sine increases and the cosine decreases.

82. $\dfrac{2t + 3}{4 - t}$, $t \neq \pm\dfrac{3}{2}$, -4 **84.** $\dfrac{1}{4}$, $x \neq 0, 12$

86. $x = \frac{2}{3}$

Section 1.4   (page 161)

2. (a) $\sin \theta = -\dfrac{5}{13}$ $\csc \theta = -\dfrac{13}{5}$

   $\cos \theta = -\dfrac{12}{13}$ $\sec \theta = -\dfrac{13}{12}$

   $\tan \theta = \dfrac{5}{12}$ $\cot \theta = \dfrac{12}{5}$

(b) $\sin \theta = \dfrac{\sqrt{2}}{2}$ $\csc \theta = \sqrt{2}$

   $\cos \theta = -\dfrac{\sqrt{2}}{2}$ $\sec \theta = -\sqrt{2}$

   $\tan \theta = -1$ $\cot \theta = -1$

4. (a) $\sin \theta = \dfrac{\sqrt{10}}{10}$ $\csc \theta = \sqrt{10}$

   $\cos \theta = \dfrac{3\sqrt{10}}{10}$ $\sec \theta = \dfrac{\sqrt{10}}{3}$

   $\tan \theta = \dfrac{1}{3}$ $\cot \theta = 3$

(b) $\sin \theta = -\dfrac{\sqrt{2}}{2}$ $\csc \theta = -\sqrt{2}$

   $\cos \theta = \dfrac{\sqrt{2}}{2}$ $\sec \theta = \sqrt{2}$

   $\tan \theta = -1$ $\cot \theta = -1$

6. $\sin \theta = \frac{15}{17}$ $\csc \theta = \frac{17}{15}$

   $\cos \theta = \frac{8}{17}$ $\sec \theta = \frac{17}{8}$

   $\tan \theta = \frac{15}{8}$ $\cot \theta = \frac{8}{15}$

8. $\sin \theta = -\dfrac{2\sqrt{29}}{29}$ $\csc \theta = -\dfrac{\sqrt{29}}{2}$

   $\cos \theta = -\dfrac{5\sqrt{29}}{29}$ $\sec \theta = -\dfrac{\sqrt{29}}{5}$

   $\tan \theta = \dfrac{2}{5}$ $\cot \theta = \dfrac{5}{2}$

10. $\sin \theta = -\dfrac{31\sqrt{1157}}{1157}$ $\csc \theta = -\dfrac{\sqrt{1157}}{31}$

   $\cos \theta = \dfrac{14\sqrt{1157}}{1157}$ $\sec \theta = \dfrac{\sqrt{1157}}{14}$

   $\tan \theta = -\dfrac{31}{14}$ $\cot \theta = -\dfrac{14}{31}$

12. Quadrant I   **14.** Quadrant IV

16. $\sin \theta = -\frac{3}{5}$ $\csc \theta = -\frac{5}{3}$

   $\cos \theta = -\frac{4}{5}$ $\sec \theta = -\frac{5}{4}$

   $\tan \theta = \frac{3}{4}$ $\cot \theta = \frac{4}{3}$

18. $\sin \theta = -\frac{15}{17}$ $\csc \theta = -\frac{17}{15}$

   $\cos \theta = \frac{8}{17}$ $\sec \theta = \frac{17}{8}$

   $\tan \theta = -\frac{15}{8}$ $\cot \theta = -\frac{8}{15}$

20. $\sin \theta = \dfrac{1}{4}$ $\csc \theta = 4$

   $\cos \theta = -\dfrac{\sqrt{15}}{4}$ $\sec \theta = -\dfrac{4\sqrt{15}}{15}$

   $\tan \theta = -\dfrac{\sqrt{15}}{15}$ $\cot \theta = -\sqrt{15}$

22. $\sin \theta = 0$ $\csc \theta$ is undefined.

   $\cos \theta = -1$ $\sec \theta = -1$

   $\tan \theta = 0$ $\cot \theta$ is undefined.

24. $\sin \theta = -1$ $\csc \theta = -1$

   $\cos \theta = 0$ $\sec \theta$ is undefined.

   $\tan \theta$ is undefined. $\cot \theta = 0$

26. $\sin \theta = -\dfrac{\sqrt{10}}{10}$ $\csc \theta = -\sqrt{10}$

   $\cos \theta = -\dfrac{3\sqrt{10}}{10}$ $\sec \theta = -\dfrac{\sqrt{10}}{3}$

   $\tan \theta = \dfrac{1}{3}$ $\cot \theta = 3$

28. $\sin \theta = -\frac{4}{5}$ $\csc \theta = -\frac{5}{4}$

   $\cos \theta = \frac{3}{5}$ $\sec \theta = \frac{5}{3}$

   $\tan \theta = -\frac{4}{3}$ $\cot \theta = -\frac{3}{4}$

30. 0 **32.** -1 **34.** 0 **36.** 0

38. $\theta' = 51°$ **40.** $\theta' = 35°$

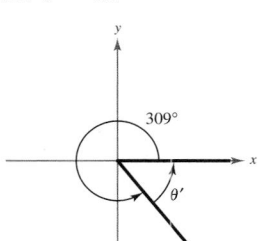   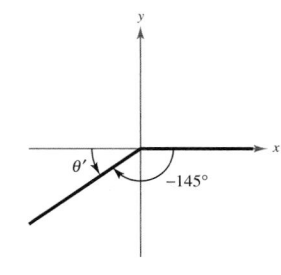

42. $\theta' = \dfrac{\pi}{4}$ **44.** $\theta' = \dfrac{\pi}{3}$

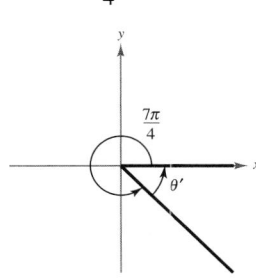   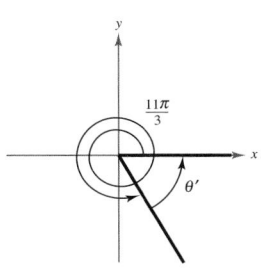

46. $\sin 300° = -\dfrac{\sqrt{3}}{2}$ **48.** $\sin(-405°) = -\dfrac{\sqrt{2}}{2}$

   $\cos 300° = \dfrac{1}{2}$ $\cos(-405°) = \dfrac{\sqrt{2}}{2}$

   $\tan 300° = -\sqrt{3}$ $\tan(-405°) = -1$

50. $\sin(-840°) = -\dfrac{\sqrt{3}}{2}$

   $\cos(-840°) = -\dfrac{1}{2}$

   $\tan(-840°) = \sqrt{3}$

52. $\sin \dfrac{\pi}{4} = \dfrac{\sqrt{2}}{2}$ **54.** $\sin\left(-\dfrac{\pi}{2}\right) = -1$

   $\cos \dfrac{\pi}{4} = \dfrac{\sqrt{2}}{2}$ $\cos\left(-\dfrac{\pi}{2}\right) = 0$

   $\tan \dfrac{\pi}{4} = 1$ $\tan\left(-\dfrac{\pi}{2}\right)$ is undefined.

56. $\sin \dfrac{10\pi}{3} = -\dfrac{\sqrt{3}}{2}$ **58.** $\sin\left(-\dfrac{25\pi}{4}\right) = -\dfrac{\sqrt{2}}{2}$

   $\cos \dfrac{10\pi}{3} = -\dfrac{1}{2}$ $\cos\left(-\dfrac{25\pi}{4}\right) = \dfrac{\sqrt{2}}{2}$

   $\tan \dfrac{10\pi}{3} = \sqrt{3}$ $\tan\left(-\dfrac{25\pi}{4}\right) = -1$

60. -1.4142 **62.** 2.0000 **64.** 0.2245

66. -0.3640 **68.** 0.6052

CHAPTER 1

70. (a) $45° = \dfrac{\pi}{4}$, $315° = \dfrac{7\pi}{4}$ (b) $135° = \dfrac{3\pi}{4}$, $225° = \dfrac{5\pi}{4}$

72. (a) $60° = \dfrac{\pi}{3}$, $300° = \dfrac{5\pi}{3}$ (b) $120° = \dfrac{2\pi}{3}$, $240° = \dfrac{4\pi}{3}$

74. (a) $60° = \dfrac{\pi}{3}$, $120° = \dfrac{2\pi}{3}$ (b) $240° = \dfrac{4\pi}{3}$, $300° = \dfrac{5\pi}{3}$

76. $29.00°$, $331.00°$ **78.** $220.65°$, $319.35°$

80. 0.018, 3.124 **82.** 0.175, 3.316 **84.** 1.898, 4.385

86. $\dfrac{\sqrt{10}}{10}$ **88.** $-\sqrt{3}$ **90.** $\dfrac{\sqrt{65}}{4}$

92. (a) $26{,}134$ units (b) $31{,}438$ units (c) $21{,}452$ units
(d) $26{,}756$ units

94. (a) 12 miles (b) 6 miles (c) $4\sqrt{3} \approx 6.9$ miles

96. False. For θ in Quadrant II, $\theta' = 180° - \theta$. For θ in Quadrant III, $\theta' = \theta - 180°$. For θ in Quadrant IV, $\theta' = 360° - \theta$.

98. Determine the trigonometric function of the reference angle and prefix the appropriate sign.

100.

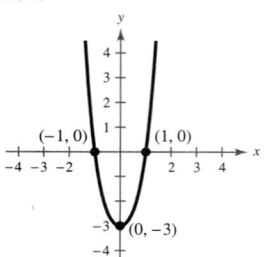

x-intercept: $\left(\dfrac{6}{7}, 0\right)$
y-intercept: $(0, 6)$
Domain: All real numbers

102.

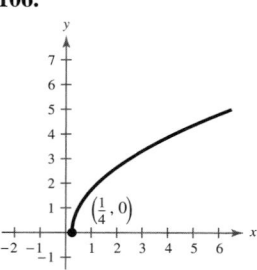

Intercept: $(0, 0)$
x-intercept: $\left(\dfrac{5}{2}, 0\right)$
Domain: All real numbers

104.

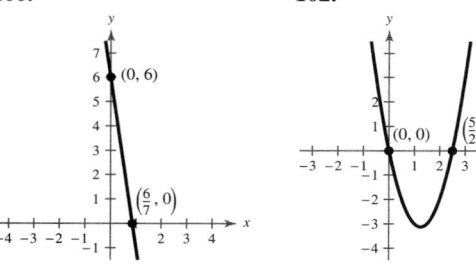

x-intercepts: $(1, 0)$, $(-1, 0)$
y-intercept: $(0, -3)$
Domain: All real numbers

106.

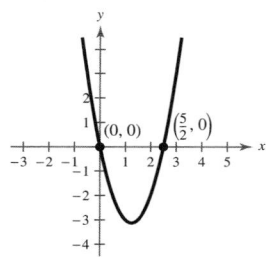

x-intercept: $\left(\dfrac{1}{4}, 0\right)$
Domain: All real numbers
where $x \geq \dfrac{1}{4}$

Section 1.5   *(page 171)*

2. Period: $\dfrac{2\pi}{3}$
Amplitude: 2

4. Period: 6π
Amplitude: 3

6. Period: 4
Amplitude: $\dfrac{3}{2}$

8. Period: 3π
Amplitude: 1

10. Period: $\dfrac{\pi}{4}$
Amplitude: $\dfrac{1}{3}$

12. Period: 8π
Amplitude: $\dfrac{5}{2}$

14. Period: 20
Amplitude: $\dfrac{2}{3}$

16. g is a shift of f π units to the left.

18. g is a reflection of f in the x-axis and in the y-axis.

20. The period of g is one-third the period of f.

22. g is a shift of f two units downward.

24. The period of g is $\dfrac{1}{3}$ the period of f.

26. g is a shift of f two units upward.

28.

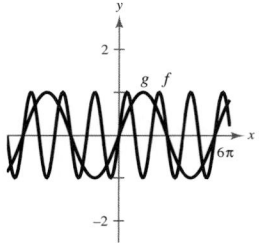

30.

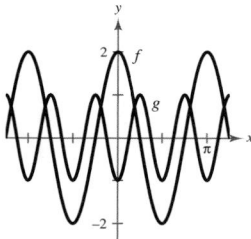

32.

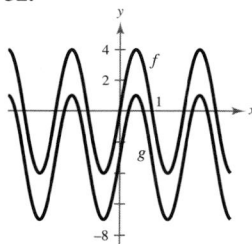

34.

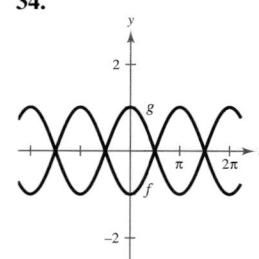

36.

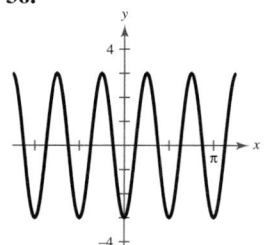

38.

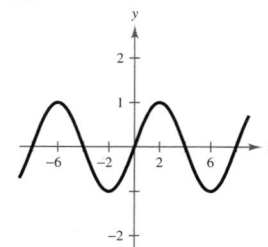

40.

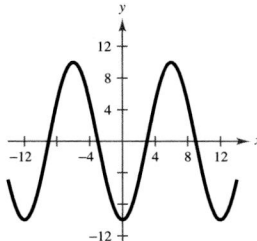

42.

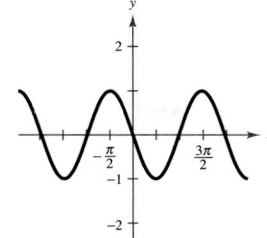

44.

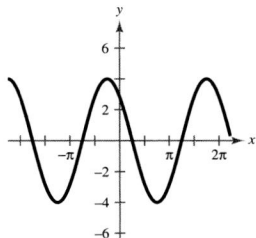

46.

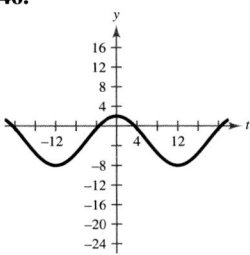

48.

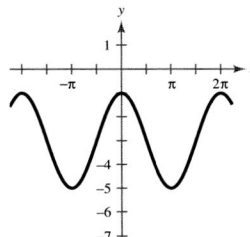

50.

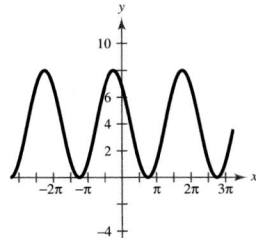

52.

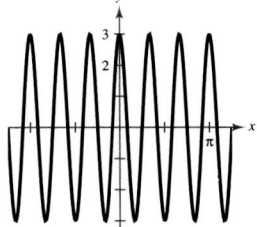

54.

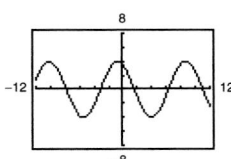

56.

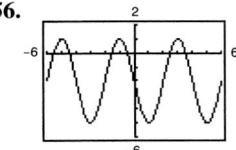

58.

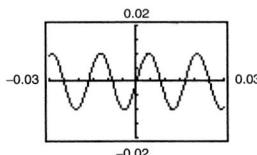

60. $a = 2, d = -1$ **62.** $a = -1, d = -3$

64. $a = 2, b = \dfrac{1}{2}, c = 0$ **66.** $a = 2, b = \dfrac{\pi}{2}, c = -\dfrac{\pi}{2}$

68.

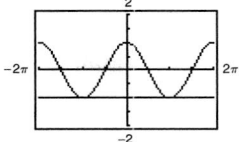

$x = \pi, -\pi$

70. (a) 4 seconds (b) 15 cycles per minute

(c)

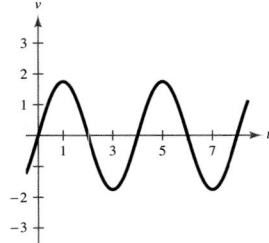

72. (a) $\frac{6}{5}$ seconds (b) 50 heartbeats per minute

74. (a) and (c)

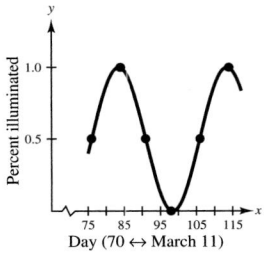

The model is a good fit.

(b) $y = \dfrac{1}{2} + \dfrac{1}{2} \sin\left[\dfrac{\pi}{15}(x - 76)\right]$ (d) 30 days (e) 0.0

76. False. The graph of $f(x) = \sin(x + 2\pi)$ translates the graph of $f(x) = \sin x$ exactly one period to the left so that the two graphs look identical.

78. True. Because $\cos x = \sin\left(x + \dfrac{\pi}{2}\right)$, $y = -\cos x$ is a reflection in the x-axis of $y = \sin\left(x + \dfrac{\pi}{2}\right)$.

80.

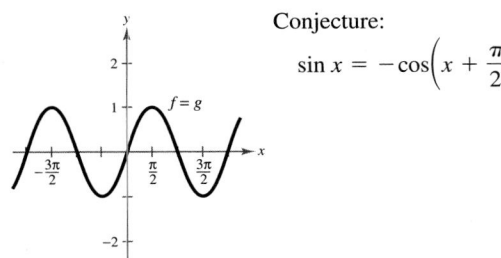

Conjecture:
$$\sin x = -\cos\left(x + \dfrac{\pi}{2}\right)$$

CHAPTER 1

82.

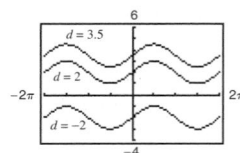

84.

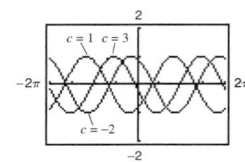

Vertical translations   Horizontal shifts

86. (a) 0.4794, 0.4794   (b) 0.8417, 0.8415   (c) 0.5, 0.5

(d) 0.8776, 0.8776   (e) 0.5417, 0.5403

(f) 0.7074, 0.7071

The error increases as x moves farther away from 0.

88. $\dfrac{-20}{(x+5)(x-5)}$ **90.** $\dfrac{3x-5}{2(x-5)}$

92. All real numbers $x \geq 3, x \neq 8$

94. All real numbers

Section 1.6   *(page 182)*

1. e, π **2.** c, 2π **3.** a, 1

4. d, 2π **5.** f, 4   **6.** b, 4

8.

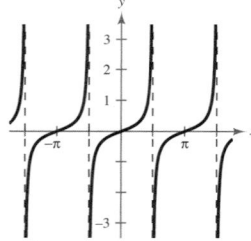

10.

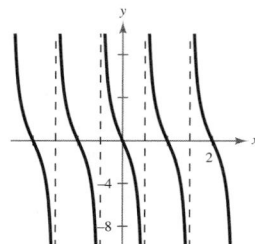

12.

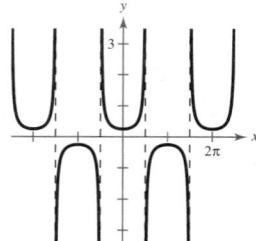

14.

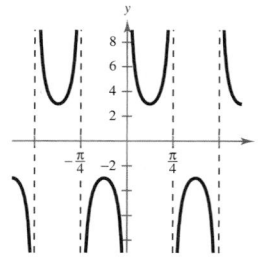

16.

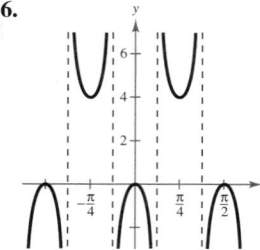

18.

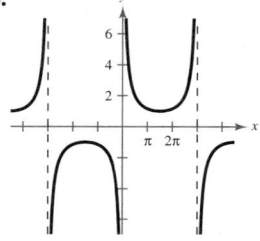

20.

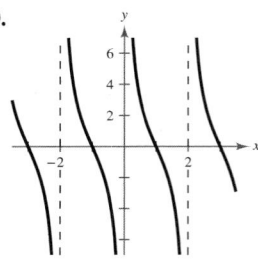

22.

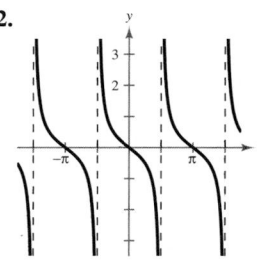

24.

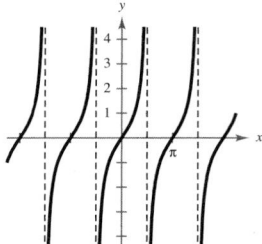

26.

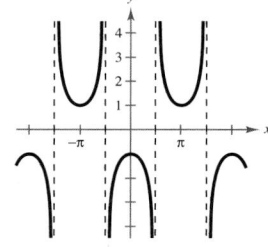

28.

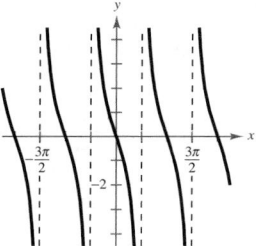

30.

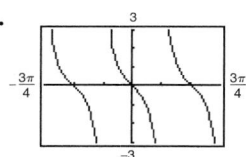

32.

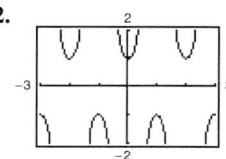

34.

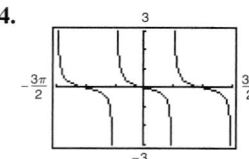

36.

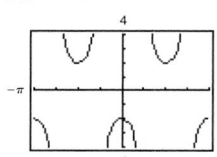

38.

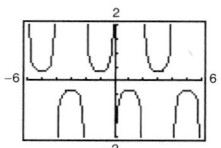

40. $-\dfrac{5\pi}{3}, -\dfrac{2\pi}{3}, \dfrac{\pi}{3}, \dfrac{4\pi}{3}$ **42.** $-\dfrac{7\pi}{4}, -\dfrac{3\pi}{4}, \dfrac{\pi}{4}, \dfrac{5\pi}{4}$

44. $-\dfrac{5\pi}{3}, -\dfrac{\pi}{3}, \dfrac{\pi}{3}, \dfrac{5\pi}{3}$ **46.** $-\dfrac{2\pi}{3}, -\dfrac{\pi}{3}, \dfrac{4\pi}{3}, \dfrac{5\pi}{3}$

48. Odd

50. (a) 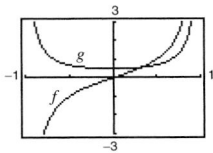   (b) $\left(-1, \dfrac{1}{3}\right)$

    (c) $\left(-1, \dfrac{1}{3}\right)$; The intervals are the same.

52.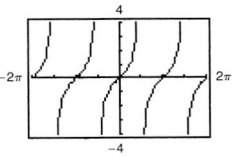

The expressions are equivalent.

54.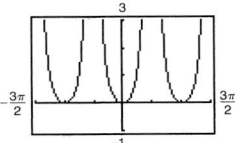

The expressions are equivalent.

55. d, $f \to 0$ as $x \to 0$.   **56.** a, $f \to 0$ as $x \to 0$.

57. b, $g \to 0$ as $x \to 0$.   **58.** c, $g \to 0$ as $x \to 0$.

60. 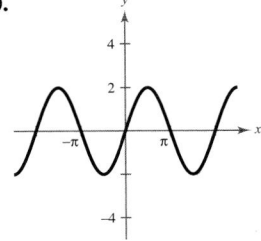   **62.**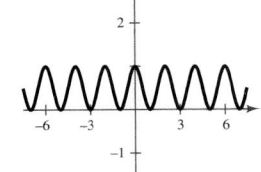

  The functions are equal.   The functions are equal.

64. 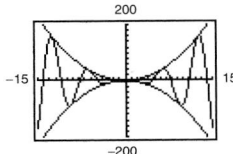   **66.**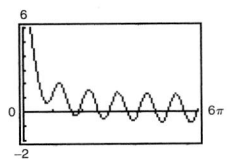

As $x \to \infty, f(x)$ oscillates.   As $x \to \infty, h(x)$ oscillates.

68. 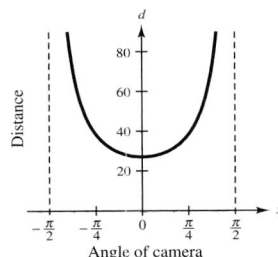   **70.**

As $x \to 0, y \to \infty$.   As $x \to 0, f(x) \to 0$.

72.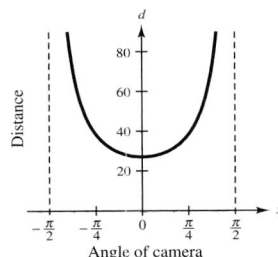

As $x \to 0, h(x)$ oscillates.

74. $d = 27 \sec x$

76.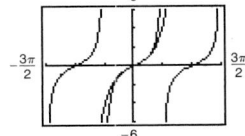

78. (a)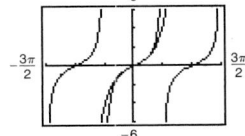

(b) y oscillates as t increases.

80. True. $y = \sec x$ is equal to $y = 1/\cos x$, and if the reciprocal of $y = \sin x$ is translated $\pi/2$ units to the left, then

$$\dfrac{1}{\sin\left(x + \dfrac{\pi}{2}\right)} = \dfrac{1}{\cos x} = \sec x.$$

82. As x approaches π from the left, f approaches ∞. As x approaches π from the right, f approaches $-\infty$.

84.

The graphs appear to coincide on the interval $-1.1 \le x \le 1.1$.

86. (a)

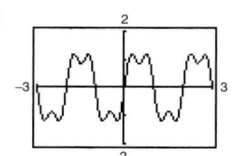

(b) $y_3 = \dfrac{4}{\pi}\left[\sin(\pi x) + \dfrac{1}{3}\sin(3\pi x) + \dfrac{1}{5}\sin(5\pi x)\right.$

$\left. + \dfrac{1}{7}\sin(7\pi x)\right]$

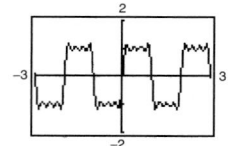

(c) $y_4 = \dfrac{4}{\pi}\left[\sin(\pi x) + \dfrac{1}{3}\sin(3\pi x) + \dfrac{1}{5}\sin(5\pi x)\right.$

$\left. + \dfrac{1}{7}\sin(7\pi x) + \dfrac{1}{9}\sin(9\pi x)\right]$

88. $5 \pm 2\sqrt{2}$ **90.** $-\frac{1}{3}, -1$ **92.** $3, -1$ **94.** $\frac{11}{8}$

Section 1.7   *(page 192)*

2. 0   **4.** $\dfrac{\pi}{2}$ **6.** $-\dfrac{\pi}{4}$ **8.** $-\dfrac{\pi}{4}$ **10.** $\dfrac{\pi}{3}$ **12.** $\dfrac{\pi}{4}$

14. $-\dfrac{\pi}{6}$ **16.** 0   **18.** 0.47   **20.** 2.35   **22.** 1.50

24. 1.31   **26.** -0.13 **28.** 1.23   **30.** 1.91

32. -1.50 **34.** $\pi, \dfrac{2\pi}{3}, \dfrac{\sqrt{3}}{2}$

36.

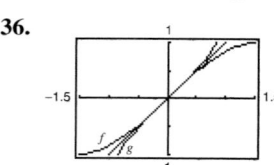

38. $\theta = \arccos\dfrac{4}{x}$

40. $\theta = \arctan\dfrac{x+1}{10}$ **42.** $\theta = \arctan\dfrac{1}{x+1}, \; x \neq 1$

44. 25   **46.** -0.2 **48.** $\dfrac{\pi}{2}$ **50.** $\dfrac{5}{3}$ **52.** $\dfrac{2\sqrt{5}}{5}$

54. $-\dfrac{13}{5}$ **56.** $-\dfrac{3\sqrt{7}}{7}$ **58.** $\dfrac{8}{5}$ **60.** $\dfrac{x}{\sqrt{x^2+1}}$

62. $\sqrt{9x^2+1}$ **64.** $\dfrac{1}{\sqrt{2x-x^2}}$ **66.** x

68. $\dfrac{\sqrt{r^2-(x-h)^2}}{r}$

70.

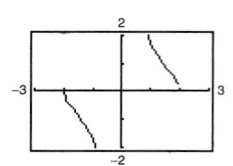

Asymptote: $x = 0$

72. $\dfrac{x}{6}$ **74.** $\dfrac{\sqrt{4x - x^2}}{x - 2}$

76.

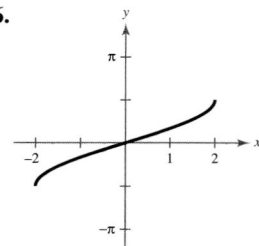

78.

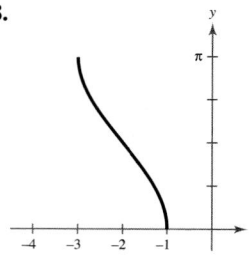

80.

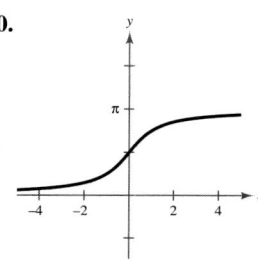

82.

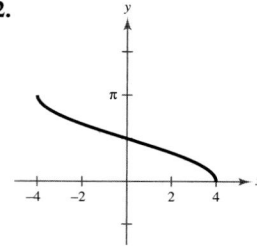

84.

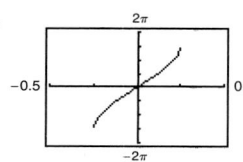

86.

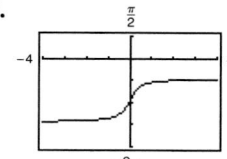

88.

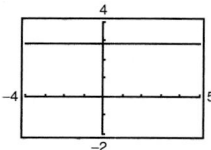

90. $5\sin\left(\pi t + \arctan\frac{4}{3}\right)$

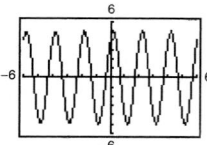

The graph implies that the identity is true.

92. (a) $\theta = \arctan\dfrac{s}{750}$ (b) $21.8°, 58.0°$

94. (a) $\theta = 32.9°$ (b) 12.94 feet

96. (a) $\theta = \arctan\dfrac{6}{x}$ (b) $40.6°, 80.5°$

98. False. $\dfrac{5\pi}{6}$ is not in the range of the arcsine.

100. Domain: $(-\infty, \infty)$

Range: $(0, \pi)$

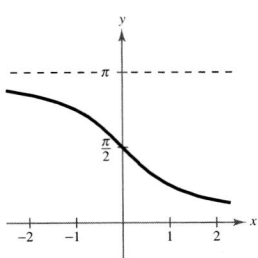

102. Domain: $(-\infty, -1] \cup [1, \infty)$

Range: $[-\pi/2, 0) \cup (0, \pi/2]$

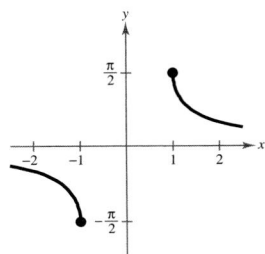

104. (a) $\dfrac{\pi}{4}$ (b) $\dfrac{\pi}{2}$ (c) 1.25 (d) 2.03

106. (a) $f \circ f^{-1}$

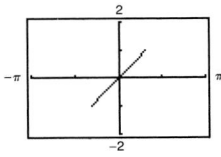

$f^{-1} \circ f$

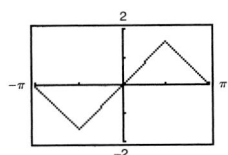

(b) The domains and ranges of the functions are restricted. The graphs of $f \circ f^{-1}$ and $f^{-1} \circ f$ differ because of the domains and ranges of f and f^{-1}.

108–112. Answers will vary.

114. 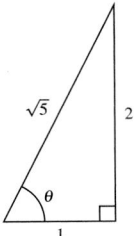   **116.**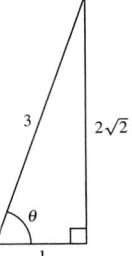

118. 3 miles per hour

Section 1.8   *(page 202)*

2. $a \approx 8.82$ **4.** $b \approx 274.27$ **6.** $b \approx 24.49$

$b \approx 12.14$ $c \approx 277.24$ $A \approx 45.58°$

$A = 36°$ $B = 81.6°$ $B \approx 44.42°$

8. $a \approx 9.36$ **10.** $b \approx 30.73$

$A \approx 81.97°$ $c \approx 33.85$

$B \approx 8.03°$ $A = 24°48'$

12. 1.62 meters   **14.** 2.80 feet   **16.** 1648.5 feet

18. 81.2 feet   **20.** 123.5 feet

22. 56.3°   **24.** 75.97°

26. (a) 5099 feet   (b) 117.7 seconds

28. 6.8°; 2516.3 feet

30. (a) N 58° E   (b) 68.82 meters

32. N 56.3° W   **34.** 11.8 kilometers

36. ≈ 1.025 miles or ≈ 5412 feet   **38.** 52.1°

40. 54.7°   **42.** 25 inches   **44.** 9.06 centimeters

46. $a \approx 21.6$ feet, $b \approx 7.2$ feet, $c \approx 13$ feet

48. (a) $\frac{1}{2}$ (b) 10   (c) $\frac{1}{40}$

50. (a) $\dfrac{1}{64}$ (b) 396   (c) $\dfrac{1}{792}$

52. $d = 3 \sin\left(\dfrac{\pi t}{3}\right)$

54. $d = 2 \cos\left(\dfrac{\pi t}{5}\right)$ **56.** $d = \dfrac{7}{4} \cos \dfrac{\pi t}{5}$

58. False. The tower is leaning, so it is not perfectly vertical and does not form a right angle with the ground.

60. (a)

θ	L_1	L_2	$L_1 + L_2$
0.1	$\dfrac{2}{\sin 0.1}$	$\dfrac{3}{\cos 0.1}$	23.0
0.2	$\dfrac{2}{\sin 0.2}$	$\dfrac{3}{\cos 0.2}$	13.1
0.3	$\dfrac{2}{\sin 0.3}$	$\dfrac{3}{\cos 0.3}$	9.9
0.4	$\dfrac{2}{\sin 0.4}$	$\dfrac{3}{\cos 0.4}$	8.4

(b)

θ	L_1	L_2	$L_1 + L_2$
0.5	$\dfrac{2}{\sin 0.5}$	$\dfrac{3}{\cos 0.5}$	7.6
0.6	$\dfrac{2}{\sin 0.6}$	$\dfrac{3}{\cos 0.6}$	7.2
0.7	$\dfrac{2}{\sin 0.7}$	$\dfrac{3}{\cos 0.7}$	7.0
0.8	$\dfrac{2}{\sin 0.8}$	$\dfrac{3}{\cos 0.8}$	7.1

7.0 (minimum length)

(c) $L = L_1 + L_2 = \dfrac{2}{\sin \theta} + \dfrac{3}{\cos \theta}$

(d)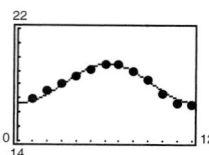

7.0 (minimum length)

62. (a)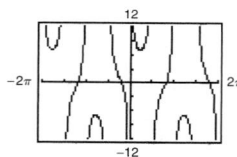

(b) 12. Yes. One period is 1 year.

(c) 1.41. 1.41 represents the maximum change in time from the average time ($d = 18.09$) of sunset.

64. No. N 24° E means 24 degrees east of north.

66.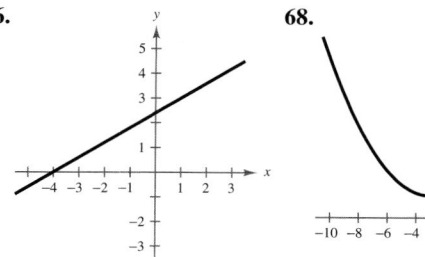

68.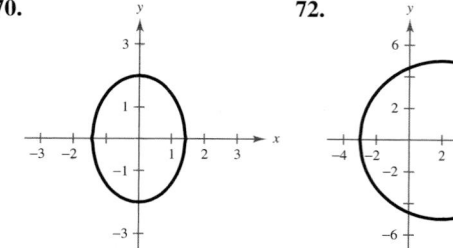

70.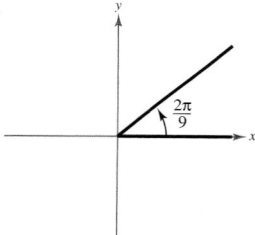

72.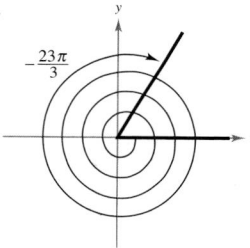

Review Exercises   *(page 208)*

2. 2 radians   **4.** -3.5 radians

6.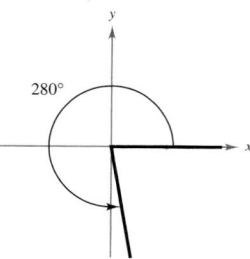

8.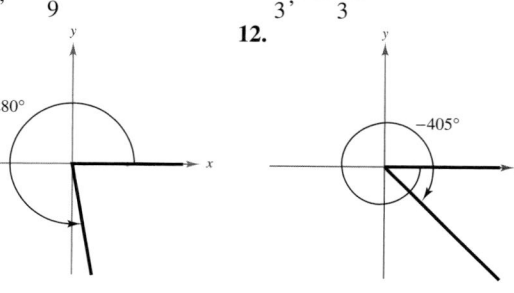

$\dfrac{20\pi}{9}, -\dfrac{16\pi}{9}$ $\dfrac{\pi}{3}, -\dfrac{17\pi}{3}$

10.

$280°$

$640°, -80°$

12.

$-405°$

$315°, -45°$

14. $-330.00°$ **16.** $326.59°$ **18.** -2.2253

20. 3.4432 **22.** 212.1 inches per second

24. $\left(-\dfrac{\sqrt{2}}{2}, \dfrac{\sqrt{2}}{2}\right)$ **26.** $\left(-\dfrac{1}{2}, \dfrac{\sqrt{3}}{2}\right)$

28. $\sin \dfrac{\pi}{4} = \dfrac{\sqrt{2}}{2}$ $\csc \dfrac{\pi}{4} = \sqrt{2}$

$\cos \dfrac{\pi}{4} = \dfrac{\sqrt{2}}{2}$ $\sec \dfrac{\pi}{4} = \sqrt{2}$

$\tan \dfrac{\pi}{4} = 1$ $\cot \dfrac{\pi}{4} = 1$

30. $\sin 2\pi = 0$ $\csc 2\pi$ is undefined.

$\cos 2\pi = 1$ $\sec 2\pi = 1$

$\tan 2\pi = 0$ $\csc 2\pi$ is undefined.

32. $\cos 4\pi = \cos 0 = 1$

34. $\cos\left(-\dfrac{13\pi}{3}\right) = \cos \dfrac{5\pi}{3} = \dfrac{1}{2}$ **36.** -1.14 **38.** -0.34

40. $\sin \theta = \dfrac{\sqrt{2}}{2}$ $\csc \theta = \sqrt{2}$

$\cos \theta = \dfrac{\sqrt{2}}{2}$ $\sec \theta = \sqrt{2}$

$\tan \theta = 1$ $\cot \theta = 1$

42. $\sin \theta = \dfrac{5}{9}$ $\csc \theta = \dfrac{9}{5}$

$\cos \theta = \dfrac{2\sqrt{14}}{9}$ $\sec \theta = \dfrac{9\sqrt{14}}{28}$

$\tan \theta = \dfrac{5\sqrt{14}}{28}$ $\cot \theta = \dfrac{2\sqrt{14}}{5}$

44. (a) $\dfrac{1}{4}$ (b) $\sqrt{17}$ (c) $\dfrac{\sqrt{17}}{17}$ (d) $\dfrac{\sqrt{17}}{4}$

46. (a) $\dfrac{1}{5}$ (b) $2\sqrt{6}$ (c) $\dfrac{\sqrt{6}}{12}$ (d) 5   **48.** 5.24

50. 5.39   **52.** 0.20   **54.** 19.5 feet

56. $\sin \theta = -\dfrac{4}{5}$ $\csc \theta = -\dfrac{5}{4}$

$\cos \theta = \dfrac{3}{5}$ $\sec \theta = \dfrac{5}{3}$

$\tan \theta = -\dfrac{4}{3}$ $\cot \theta = -\dfrac{3}{4}$

58. $\sin \theta = -\dfrac{\sqrt{26}}{26}$ $\csc \theta = -\sqrt{26}$

$\cos \theta = -\dfrac{5\sqrt{26}}{26}$ $\sec \theta = -\dfrac{\sqrt{26}}{5}$

$\tan \theta = \dfrac{1}{5}$ $\cot \theta = 5$

60. $\sin \theta = 0.8$ $\csc \theta = 1.25$

$\cos \theta = 0.6$ $\sec \theta \approx 1.67$

$\tan \theta \approx 1.33$ $\cot \theta = 0.75$

62. $\sin \theta = -\dfrac{3\sqrt{13}}{13}$ $\csc \theta = -\dfrac{\sqrt{13}}{3}$

$\cos \theta = -\dfrac{2\sqrt{13}}{13}$ $\sec \theta = -\dfrac{\sqrt{13}}{2}$

$\tan \theta = \dfrac{3}{2}$ $\cot \theta = \dfrac{2}{3}$

64. $\sin \theta = \dfrac{2}{3}$ **66.** $\sin \theta = -\dfrac{5\sqrt{41}}{41}$

$\cos \theta = -\dfrac{\sqrt{5}}{3}$ $\cos \theta = -\dfrac{4\sqrt{41}}{41}$

$\tan \theta = -\dfrac{2\sqrt{5}}{5}$ $\csc \theta = -\dfrac{\sqrt{41}}{5}$

$\sec \theta = -\dfrac{3\sqrt{5}}{5}$ $\sec \theta = -\dfrac{\sqrt{41}}{4}$

$\cot \theta = -\dfrac{\sqrt{5}}{2}$ $\cot \theta = \dfrac{4}{5}$

68. $\cos \theta = \dfrac{\sqrt{3}}{2}$ $\sec \theta = \dfrac{2\sqrt{3}}{3}$

$\tan \theta = -\dfrac{\sqrt{3}}{3}$ $\cot \theta = -\sqrt{3}$

$\csc \theta = -2$

70. $\sin \dfrac{\pi}{4} = \dfrac{\sqrt{2}}{2}$; $\cos \dfrac{\pi}{4} = \dfrac{\sqrt{2}}{2}$; $\tan \dfrac{\pi}{4} = 1$

72. $\sin\left(-\dfrac{5\pi}{4}\right) = \dfrac{\sqrt{2}}{2}$; $\cos\left(-\dfrac{5\pi}{4}\right) = -\dfrac{\sqrt{2}}{2}$;

$\tan\left(-\dfrac{5\pi}{4}\right) = -1$

74. $\sin(-150°) = -\dfrac{1}{2}$; $\cos(-150°) = -\dfrac{\sqrt{3}}{2}$;

$\tan(-150°) = \dfrac{\sqrt{3}}{3}$

76. $\sin 315° = -\dfrac{\sqrt{2}}{2}$; $\cos 315° = \dfrac{\sqrt{2}}{2}$; $\tan 315° = -1$

78. -0.14 **80.** 0.09   **82.** 4.38

84.

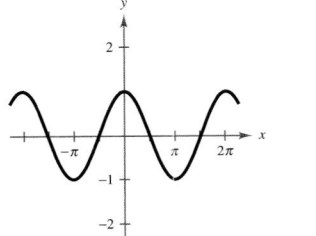

86.

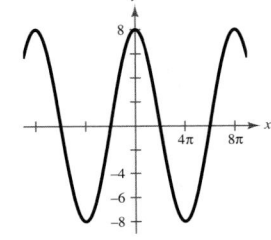

88.

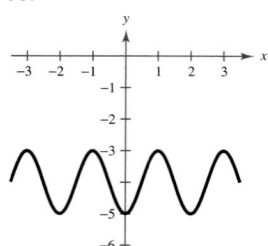

90.

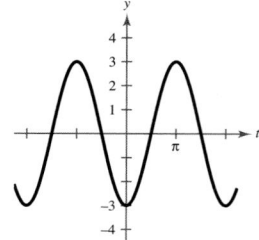

92. $y = 2 \cos 528\pi x$

94.

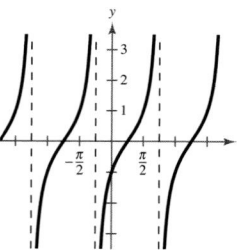

96.

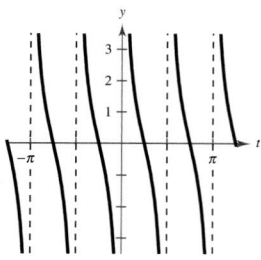

98.

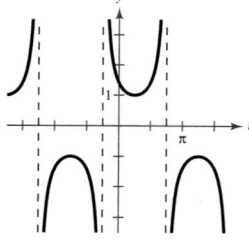

100.

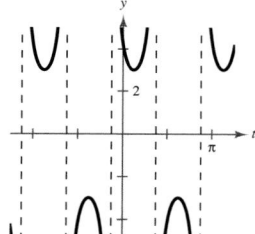

102.

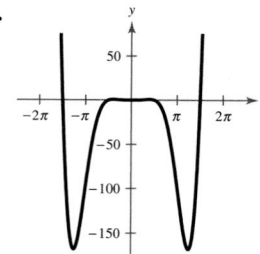

104. $-\dfrac{\pi}{2}$ **106.** 0.21

108. 1.10 **110.** $\dfrac{\pi}{4}$

112. $\dfrac{\pi}{6}$ **114.** 2.66

116. 1.17 **118.** 1.56

120. 1.45 **122.** 0.25

124. π **126.** $\frac{4}{3}$

128. $-\frac{5}{12}$

130. 9.6 feet **132.** $d = 0.75 \cos\left(\dfrac{2\pi t}{3}\right)$

134. False. The inverse sine, $y = \arcsin x$, cannot be defined as a function over any interval that is greater than the interval defined as $-\pi/2 \le x \le \pi/2$.

136. False. $3\pi/4$ is not in the range of the arctangent function.

137. d; The period is 2π and the amplitude is 3.

138. a; The period is 2π and, because $a < 0$, the graph is reflected in the x-axis.

139. b; The period is 2 and the amplitude is 2.

140. c; The period is 4π and the amplitude is 2.

142. (a)

θ	0.1	0.4	0.7
$\tan\left(\theta - \dfrac{\pi}{2}\right)$	-9.9666	-2.3652	-1.1872
$-\cot \theta$	-9.9666	-2.3652	-1.1872

θ	1.0	1.3
$\tan\left(\theta - \dfrac{\pi}{2}\right)$	-0.6421	-0.2776
$-\cot \theta$	-0.6421	-0.2776

(b) $\tan\left(\theta - \dfrac{\pi}{2}\right) = -\cot \theta$

144. (a) $A = 0.4r^2,\ r > 0$

$s = 0.8r,\ r > 0$

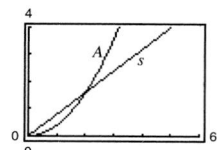

The area function increases at a greater rate than the arc length.

(b) $A = 50\theta,\ \theta > 0$

$s = 10\theta,\ \theta > 0$

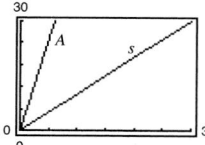

Problem Solving *(page 214)*

2. Gear 1: $270°$, $\dfrac{3\pi}{2}$ radians

Gear 2: $\approx 332.3°$, $\dfrac{24\pi}{13}$ radians

Gear 3: $\approx 392.7°$, $\dfrac{24\pi}{11}$ radians

Gear 4: $450°$, $\dfrac{5\pi}{2}$ radians

Gear 5: $\approx 454.7°$, $\dfrac{48\pi}{19}$ radians

4. (a) Answers will vary.   (b) The ratios are equal.

   (c) No; No

   (d) Yes, because all six trigonometric functions are ratios of right triangles.

6. (a) $h(x)$ is even.   (b) $h(x)$ is even.

8. (a)   (b)

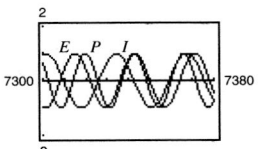 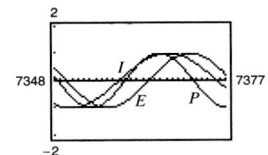

   (c) $P(7369) = 0.631$

     $E(7369) = 0.901$

     $I(7369) = 0.945$

10. (a) $3.35, 7.35$ (b) -0.65

   (c) Yes. There is a difference of nine periods between the values.

12. (a) $40.5°$ (b) $x \approx 1.71$ feet; $y \approx 3.46$ feet

   (c) ≈ 1.75 feet

   (d) As you move closer to the rock, d must get smaller and smaller. The angles θ_1 and θ_2 will decrease along with the distance y, so d will decrease.

Chapter 2

Section 2.1   *(page 223)*

2. $\sin x = -\dfrac{1}{2}$ **4.** $\sin \theta = \dfrac{3}{5}$

   $\csc x = -2$ $\cos \theta = \dfrac{4}{5}$

   $\sec x = -\dfrac{2\sqrt{3}}{3}$ $\sec \theta = \dfrac{5}{4}$

   $\cot x = \sqrt{3}$ $\cot \theta = \dfrac{4}{3}$

6. $\cos \phi = -\dfrac{3\sqrt{10}}{10}$ **8.** $\sin x = \dfrac{3}{5}$

   $\tan \phi = -\dfrac{1}{3}$ $\tan x = \dfrac{3}{4}$

   $\csc \phi = \sqrt{10}$ $\csc x = \dfrac{5}{3}$

   $\sec \phi = -\dfrac{\sqrt{10}}{3}$ $\sec x = \dfrac{5}{4}$

                        $\cot x = \dfrac{4}{3}$

10. $\sin x = \dfrac{\sqrt{15}}{4}$ **12.** $\sin \theta = -\dfrac{1}{5}$

   $\cos x = \dfrac{1}{4}$ $\cos \theta = -\dfrac{2\sqrt{6}}{5}$

   $\tan x = \sqrt{15}$ $\tan \theta = \dfrac{\sqrt{6}}{12}$

   $\csc x = \dfrac{4\sqrt{15}}{15}$ $\sec \theta = -\dfrac{5\sqrt{6}}{12}$

   $\cot x = \dfrac{\sqrt{15}}{15}$ $\cot \theta = 2\sqrt{6}$

14. $\sin \theta = 1$

   $\cos \theta = 0$

   $\csc \theta = 1$

   $\sec \theta$ is undefined.

   $\cot \theta = 0$

15. d   **16.** a   **17.** b   **18.** f   **19.** e   **20.** c

21. b   **22.** c   **23.** f   **24.** a   **25.** e   **26.** d

28. $\sin \beta$ **30.** 1   **32.** $\cot \theta$ **34.** $\cos^2 x$

36. $\sin^2 \theta$ **38.** $\sin x$ **40.** $\sec t$ **42.** $2 \sec \phi$

44. $\csc \theta \sec \theta$ **46.** $\cos^2 x$ **48.** 1   **50.** $\cos x + 2$

52. $\sin^4 x$ **54.** $\sec^2 x + \tan^2 x$ **56.** $\tan^2 x(\sec x - 1)$

58. -1 **60.** $9 \cos^2 x$ **62.** $-2 \cot^2 x$ **64.** $-\cot x$

66. $5(\sec x - \tan x)$ **68.** $\tan^4 x(\csc x - 1)$

70.

x	0.2	0.4	0.6	0.8	1.0
y_1	0.0403	0.1646	0.3863	0.7386	1.3105
y_2	0.0403	0.1646	0.3863	0.7386	1.3105

x	1.2	1.4
y_1	2.3973	5.7135
y_2	2.3973	5.7135

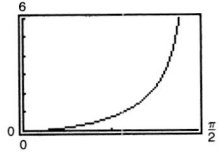

$y_1 = y_2$

72.

x	0.2	0.4	0.6	0.8	1.0
y_1	0.0428	0.2107	0.6871	2.1841	8.3087
y_2	0.0428	0.2107	0.6871	2.1841	8.3087

x	1.2	1.4
y_1	50.3869	1163.6143
y_2	50.3869	1163.6143

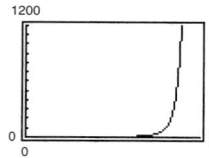

$y_1 = y_2$

74. $\cot x$ **76.** $\sec \theta$ **78.** $8 \sin \theta$ **80.** $2 \tan \theta$

82. $10 \sec \theta$ **84.** $6 \cos \theta = 3; \sin \theta = \pm\dfrac{\sqrt{3}}{2}; \cos \theta = \dfrac{1}{2}$

86. $10 \sin \theta = -5\sqrt{3}; \sin \theta = -\dfrac{\sqrt{3}}{2}; \cos \theta = \dfrac{1}{2}$

88. $\dfrac{\pi}{2} \le \theta \le \dfrac{3\pi}{2}$ **90.** $0 < \theta < \pi$

92. (a) $\tan^2 346° + 1 = \sec^2 346° \approx 1.0622$

(b) $\tan^2 3.1 + 1 = \sec^2 3.1 \approx 1.0017$

94. (a) $\sin(-250°) = -\sin 250° \approx 0.9397$

(b) $\sin\left(-\frac{1}{2}\right) = -\sin\left(\frac{1}{2}\right) \approx -0.4794$

96. (a) $1 + \cot^2 240° = \csc^2 240° \approx 1.3333$

(b) $1 + \cot^2 2.2 = \csc^2 2.2 \approx 1.5298$

98. (a) $\cos(-125°) = \cos 125° \approx -0.5736$

(b) $\cos\left(-\dfrac{5\pi}{6}\right) = \cos \dfrac{5\pi}{6} \approx -0.8660$

100. $\csc x \cot x - \cos x$

$= \dfrac{1}{\sin x} \cdot \dfrac{\cos x}{\sin x} - \cos x$

$= \dfrac{\cos x}{\sin^2 x} - \cos x$

$= \dfrac{\cos x - \cos x \, \sin^2 x}{\sin^2 x}$

$= \dfrac{\cos x(1 - \sin^2 x)}{\sin^2 x}$

$= \dfrac{\cos x \cos^2 x}{\sin^2 x}$

$= \cos x \cot^2 x$

102. False. A cofunction identity can be used to transform a tangent function so that it can be represented by a cotangent function.

104. 1, 1   **106.** 0, $-\infty$

108. Not an identity because $\cot \theta = \pm\sqrt{\csc^2 \theta - 1}$

110. Not an identity because $\dfrac{1}{5 \cos \theta} \neq \dfrac{5}{\cos \theta}$

112. Not an identity because θ and ϕ may not be equal

114. $\sin^2 \theta + \cos^2 \theta = 1$

$\dfrac{\sin^2 \theta + \cos^2 \theta}{\cos^2 \theta} = \dfrac{1}{\cos^2 \theta}$

$\tan^2 \theta + 1 = \sec^2 \theta$

$1 + \tan^2 \theta = \sec^2 \theta$

$\sin^2\theta + \cos^2\theta = 1$

$\dfrac{\sin^2 \theta + \cos^2 \theta}{\sin^2 \theta} = \dfrac{1}{\sin^2 \theta}$

$1 + \cot^2 \theta = \csc^2 \theta$

Answers will vary for discussion.

116. $4z + 12\sqrt{z} + 9$

118. $\dfrac{3(2x + 1)}{x - 4}$ **120.** $\dfrac{x(x^2 + 5x + 1)}{x^2 - 25}$

Section 2.2   *(page 231)*

2–40. Answers will vary.

42. Identity   **44.** Not an identity   **46.** Identity

48. Identity   **50.** Not an identity   **52.** Identity

54. 1   **56.** 2

58. (a) Answers will vary.

(b)

θ	10°	20°	30°	40°	50°
s	28.36	13.74	8.66	5.96	4.20

θ	60°	70°	80°	90°
s	2.89	1.82	0.88	0

(c) Greatest: 10°; Least: 90°

(d) Noon

60. True. An identity is an equation that is true for all real values in the domain of the variable.

62. The equation is not an identity because $\tan \theta = \pm\sqrt{\sec^2 \theta - 1}$.

Possible answer: $\dfrac{3\pi}{4}$

64. $\dfrac{-5 \pm \sqrt{3}i}{2}$ **66.** $\dfrac{1 \pm \sqrt{5}i}{4}$

Section 2.3   *(page 240)*

2–6. Answers will vary.   **8.** $\dfrac{7\pi}{6} + 2n\pi, \dfrac{11\pi}{6} + 2n\pi$

10. $\dfrac{2\pi}{3} + n\pi$ **12.** $\dfrac{\pi}{3} + n\pi, \dfrac{2\pi}{3} + n\pi$

14. $\dfrac{\pi}{6} + n\pi, \dfrac{5\pi}{6} + n\pi, \dfrac{\pi}{3} + n\pi, \dfrac{2\pi}{3} + n\pi$

16. $\dfrac{\pi}{3} + n\pi, \dfrac{2\pi}{3} + n\pi$ **18.** $\dfrac{\pi}{9} + \dfrac{n\pi}{3}, \dfrac{2\pi}{9} + \dfrac{n\pi}{3}$

20. $\dfrac{\pi}{4} + \dfrac{n\pi}{2}, \dfrac{2\pi}{3} + 2n\pi, \dfrac{4\pi}{3} + 2n\pi$

22. $0, \pi$ **24.** $\dfrac{\pi}{2}, \dfrac{3\pi}{2}, \dfrac{2\pi}{3}, \dfrac{4\pi}{3}$ **26.** $\dfrac{\pi}{3}, \dfrac{5\pi}{3}$ **28.** 0

30. $\dfrac{7\pi}{6}, \dfrac{3\pi}{2}, \dfrac{11\pi}{6}$ **32.** $\dfrac{\pi}{3}, \dfrac{5\pi}{3}$ **34.** $\dfrac{2\pi}{3} + n\pi, \dfrac{5\pi}{6} + n\pi$

36. $\dfrac{\pi}{12} + \dfrac{n\pi}{2}, \dfrac{5\pi}{12} + \dfrac{n\pi}{2}$ **38.** $\dfrac{8\pi}{3} + 4n\pi, \dfrac{10\pi}{3} + 4n\pi$

40. $\dfrac{3}{4} + n$ **42.** ± 2

44. $y = -5, 4$; No solution to the trigonometric equation

46. $0.7854, 2.3562, 3.6652, 3.9270, 5.4978, 5.7596$

48. $0.5236, 2.6180$ **50.** 4.9172

52. $0.5153, 2.7259, 3.6569, 5.8675$

54. $0.5236, 0.7297, 2.4119, 2.6180$

56. $0.5880, 2.0344, 3.7296, 5.1760$ **58.** $1.7794, 4.5038$

60. $\arctan(-2) + \pi, \arctan(-2) + 2\pi, \dfrac{\pi}{4}, \dfrac{5\pi}{4}$ **62.** $\dfrac{\pi}{6}, \dfrac{5\pi}{6}$

64. (a)

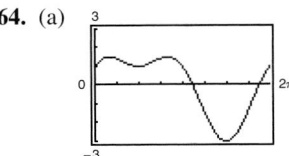

(b)

$\dfrac{\pi}{6} \approx 0.5236, \dfrac{\pi}{2} \approx 1.5708, \dfrac{5\pi}{6} \approx 2.6180, \dfrac{3\pi}{2} \approx 4.7124$

Maximum: $(0.5236, 1.5)$

Maximum: $(2.6180, 1.5)$

Minimum: $(1.5708, 1)$

Minimum: $(4.7124, -3)$

66. 0.7391

68. (a) All real numbers $x \neq 0$

(b) y-axis symmetry

(c) y approaches 1.

(d) Four solutions: $\pm\pi, \pm 2\pi$

70.

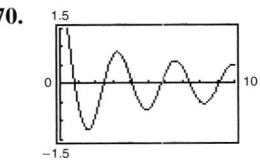

1.91 seconds

72. $36.9°, 53.1°$

74. (a)

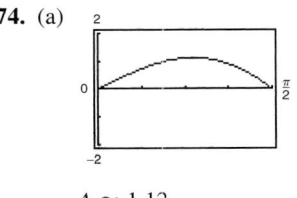

$A \approx 1.12$

(b) $0.6 < x < 1.1$

76. (a) $\dfrac{10}{3}$

(b)

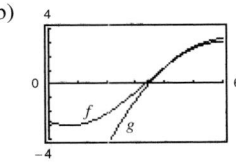

For $3.5 \leq x \leq 6$, the approximation appears to be good.

(c) $3.46, 8.81$

The value 3.46 is close to $\dfrac{10}{3} = 3.\overline{3}$, to the zero of f in the interval $[0, 6]$.

78. False. There is no value of x for which $\sin x = 3.4$.

80. 3

82. $C = 19°$

$a \approx 15.4$

$c \approx 5.0$

84. $\sin 570° = -\dfrac{1}{2}$ **86.** $\sin(-1410°) = \dfrac{1}{2}$

$\cos 570° = -\dfrac{\sqrt{3}}{2}$ $\cos(-1410°) = \dfrac{\sqrt{3}}{2}$

$\tan 570° = \dfrac{\sqrt{3}}{3}$ $\tan(-1410°) = \dfrac{\sqrt{3}}{3}$

88. 30 feet

Section 2.4   (page 248)

2. (a) $-\dfrac{\sqrt{6} + \sqrt{2}}{4}$ (b) $\dfrac{\sqrt{2} + 1}{2}$

4. (a) 0 (b) $\dfrac{\sqrt{3} - 1}{2}$

6. (a) $\dfrac{\sqrt{6} + \sqrt{2}}{4}$ (b) $\dfrac{\sqrt{2} - \sqrt{3}}{2}$

8. $\sin 165° = \dfrac{\sqrt{2}}{4}(\sqrt{3} - 1)$

$\cos 165° = -\dfrac{\sqrt{2}}{4}(\sqrt{3} + 1)$

$\tan 165° = -2 + \sqrt{3}$

10. $\sin 255° = -\dfrac{\sqrt{2}}{4}(\sqrt{3} + 1)$

$\cos 255° = \dfrac{\sqrt{2}}{4}(1 - \sqrt{3})$

$\tan 255° = 2 + \sqrt{3}$

12. $\sin \dfrac{7\pi}{12} = \dfrac{\sqrt{2}}{4}(\sqrt{3} + 1)$

$\cos \dfrac{7\pi}{12} = \dfrac{\sqrt{2}}{4}(1 - \sqrt{3})$

$\tan \dfrac{7\pi}{12} = -2 - \sqrt{3}$

14. $\sin\left(-\dfrac{\pi}{12}\right) = \dfrac{\sqrt{2}}{4}(1 - \sqrt{3})$

$\cos\left(-\dfrac{\pi}{12}\right) = \dfrac{\sqrt{2}}{4}(\sqrt{3} + 1)$

$\tan\left(-\dfrac{\pi}{12}\right) = -2 + \sqrt{3}$

16. $\sin(-105°) = -\dfrac{\sqrt{2}}{4}(\sqrt{3} + 1)$

$\cos(-105°) = \dfrac{\sqrt{2}}{4}(1 - \sqrt{3})$

$\tan(-105°) = 2 + \sqrt{3}$

18. $\sin 15° = \dfrac{\sqrt{2}}{4}(\sqrt{3} - 1)$

$\cos 15° = \dfrac{\sqrt{2}}{4}(1 + \sqrt{3})$

$\tan 15° = 2 - \sqrt{3}$

20. $\sin\left(-\dfrac{7\pi}{12}\right) = -\dfrac{\sqrt{2}}{4}(\sqrt{3} + 1)$

$\cos\left(-\dfrac{7\pi}{12}\right) = \dfrac{\sqrt{2}}{4}(1 - \sqrt{3})$

$\tan\left(-\dfrac{7\pi}{12}\right) = 2 + \sqrt{3}$

22. $\sin \dfrac{5\pi}{12} = \dfrac{\sqrt{2}}{4}(1 + \sqrt{3})$

$\cos \dfrac{5\pi}{12} = \dfrac{\sqrt{2}}{4}(\sqrt{3} - 1)$

$\tan \dfrac{5\pi}{12} = \sqrt{3} + 2$

24. $\sin 190°$ **26.** $\tan 80°$ **28.** $\cos \dfrac{12\pi}{35}$

30. $\cos(3x - 2y)$ **32.** $\dfrac{\sqrt{2}}{2}$ **34.** $\dfrac{\sqrt{2}}{2}$ **36.** $\dfrac{\sqrt{3}}{3}$

38. $\frac{56}{65}$ **40.** $-\frac{33}{65}$ **42.** $\frac{65}{33}$ **44.** $-\frac{16}{63}$ **46.** $\frac{4}{5}$

48. $\dfrac{117}{44}$ **50.** $-\dfrac{125}{44}$ **52.** $\dfrac{2x^2 - \sqrt{1 - x^2}}{\sqrt{4x^2 + 1}}$

54. $\dfrac{x + x\sqrt{1 - x^2}}{\sqrt{x^2 + 1}}$ **56–64.** Answers will vary.

66. $-\cos x$ **68.** $\tan \theta$ **70.** $\dfrac{\pi}{3}, \dfrac{5\pi}{3}$

72. $0, \dfrac{\pi}{3}, \pi, \dfrac{5\pi}{3}$ **74.** $0, \pi$ **76.** Answers will vary.

78. False. $\cos(u \pm v) = \cos u \cos v \mp \sin u \sin v$

80. True.

$\sin\left(x - \dfrac{\pi}{2}\right) = \sin x \cos \dfrac{\pi}{2} - \cos x \sin \dfrac{\pi}{2} = -\cos x$

82. Answers will vary.   **84.** Answers will vary.

86. (a) $5 \sin(2\theta + 0.9273)$ (b) $5 \cos(2\theta - 0.6435)$

88. (a) $\sqrt{2} \sin\left(2\theta - \dfrac{\pi}{4}\right)$ (b) $\sqrt{2} \cos\left(2\theta + \dfrac{\pi}{4}\right)$

90. $-\dfrac{5\sqrt{2}}{2}\sin \theta - \dfrac{5\sqrt{2}}{2}\cos \theta$ **92.** $15°$ **94.** $u + v = w$

96. Answers will vary.   **98.** $f^{-1}(x) = -8x + 7$

100. $f^{-1}(x) = x^2 + 16$

Section 2.5   (page 258)

2. $\frac{1}{4}$ **4.** $\frac{8}{17}$ **6.** $\frac{17}{15}$ **8.** $\frac{15}{8}$

10. $\dfrac{\pi}{2}, \dfrac{7\pi}{6}, \dfrac{3\pi}{2}, \dfrac{11\pi}{6}$ **12.** $\dfrac{\pi}{2}, \dfrac{3\pi}{2}, \dfrac{\pi}{4}, \dfrac{3\pi}{4}, \dfrac{5\pi}{4}, \dfrac{7\pi}{4}$

14. $\dfrac{\pi}{2}, \dfrac{7\pi}{6}, \dfrac{11\pi}{6}$ **16.** $\dfrac{\pi}{6}, \dfrac{\pi}{2}, \dfrac{5\pi}{6}, \dfrac{3\pi}{2}$

18. $0, \dfrac{\pi}{4}, \dfrac{\pi}{2}, \dfrac{3\pi}{4}, \pi, \dfrac{5\pi}{4}, \dfrac{3\pi}{2}, \dfrac{7\pi}{4}$

20. $3 \cos 2x$ **22.** $\cos 2x$

24. $\sin 2u = -\dfrac{4\sqrt{5}}{9}$

$\cos 2u = -\dfrac{1}{9}$

$\tan 2u = 4\sqrt{5}$

26. $\sin 2u = -\dfrac{8}{17}$

$\cos 2u = \dfrac{15}{17}$

$\tan 2u = -\dfrac{8}{15}$

28. $\sin 2u = -\dfrac{4\sqrt{2}}{9}$

$\cos 2u = \dfrac{7}{9}$

$\tan 2u = -\dfrac{4\sqrt{2}}{7}$

30. $\frac{1}{128}(35 - 48\cos 2x + 28\cos 4x - 16\cos 2x\cos 4x + \cos 8x)$

32. $\frac{1}{128}(3 - 4\cos 4x + \cos 8x)$

34. $\frac{1}{16}(1 - \cos 2x - \cos 4x + \cos 2x\cos 4x)$

36. $\dfrac{\sqrt{17}}{17}$ **38.** $\dfrac{\sqrt{17}}{4}$ **40.** 4

42. $\sin 165° = \frac{1}{2}\sqrt{2 - \sqrt{3}}$

$\cos 165° = -\frac{1}{2}\sqrt{2 + \sqrt{3}}$

$\tan 165° = \sqrt{3} - 2$

44. $\sin 67°\,30' = \frac{1}{2}\sqrt{2 + \sqrt{2}}$

$\cos 67°\,30' = \frac{1}{2}\sqrt{2 - \sqrt{2}}$

$\tan 67°\,30' = 1 + \sqrt{2}$

46. $\sin\dfrac{\pi}{12} = \frac{1}{2}\sqrt{2 - \sqrt{3}}$

$\cos\dfrac{\pi}{12} = \frac{1}{2}\sqrt{2 + \sqrt{3}}$

$\tan\dfrac{\pi}{12} = 2 - \sqrt{3}$

48. $\sin\dfrac{7\pi}{12} = \frac{1}{2}\sqrt{2 + \sqrt{3}}$

$\cos\dfrac{7\pi}{12} = -\frac{1}{2}\sqrt{2 - \sqrt{3}}$

$\tan\dfrac{7\pi}{12} = -2 - \sqrt{3}$

50. $\sin\dfrac{u}{2} = \dfrac{\sqrt{5}}{5}$

$\cos\dfrac{u}{2} = \dfrac{2\sqrt{5}}{5}$

$\tan\dfrac{u}{2} = \dfrac{1}{2}$

52. $\sin\dfrac{u}{2} = \frac{1}{2}\sqrt{\dfrac{10 + 3\sqrt{10}}{5}}$

$\cos\dfrac{u}{2} = -\frac{1}{2}\sqrt{\dfrac{10 - 3\sqrt{10}}{5}}$

$\tan\dfrac{u}{2} = -3 - \sqrt{10}$

54. $\sin\dfrac{u}{2} = \dfrac{3\sqrt{14}}{14}$

$\cos\dfrac{u}{2} = \dfrac{\sqrt{70}}{14}$

$\tan\dfrac{u}{2} = \dfrac{3\sqrt{5}}{5}$

56. $|\cos 2x|$ **58.** $-\left|\sin\!\left(\dfrac{x-1}{2}\right)\right|$

60. $0, \dfrac{\pi}{3}, \dfrac{5\pi}{3}$

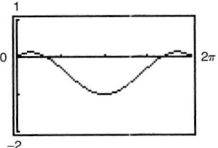

62. $0, \dfrac{\pi}{2}, \dfrac{3\pi}{2}$

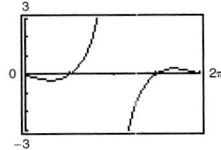

64. $2\left(\sin\dfrac{7\pi}{6} + \sin\dfrac{\pi}{2}\right)$

66. $\frac{3}{2}[\cos(\alpha) - \cos 5\alpha]$ **68.** $\frac{1}{2}(\cos 2\theta + \cos 6\theta)$

70. $\frac{1}{2}(\sin 2x + \sin 2y)$ **72.** $\frac{1}{2}(\cos 2\pi - \cos 2\theta)$

74. $3(\sin 60° + \sin 30°)$ **76.** $2\cos 75°\cos 45°$

78. $2\cos\pi\sin\dfrac{\pi}{4}$ **80.** $2\sin 2\theta\cos\theta$

82. $2\sin 3x\cos 2x$ **84.** $2\cos(\phi + \pi)\cos\pi$

86. $2\sin x\cos\dfrac{\pi}{2}$

88. $0, \dfrac{\pi}{4}, \dfrac{\pi}{2}, \dfrac{3\pi}{4}, \pi, \dfrac{5\pi}{4}, \dfrac{3\pi}{2}, \dfrac{7\pi}{4}$

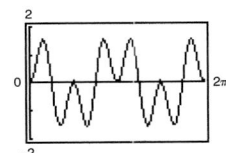

90. $0, \dfrac{\pi}{2}, \pi, \dfrac{3\pi}{2}, \dfrac{\pi}{4}, \dfrac{3\pi}{4}, \dfrac{5\pi}{4}, \dfrac{7\pi}{4}$

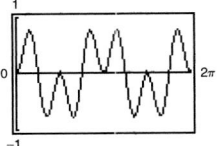

92. $\dfrac{144}{169}$ **94.** $\dfrac{36}{65}$

96–110. Answers will vary.

112.

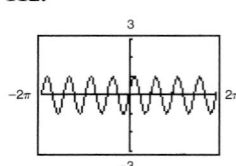

114.

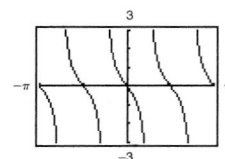

116.

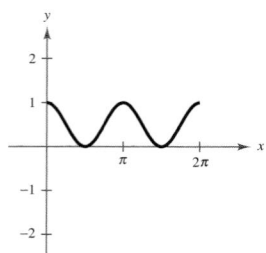

118. $2x^2 - 1$

120. $r = \frac{1}{16} v_0{}^2 \sin\theta \cos\theta$ **122.** $x = 2r(1 - \cos\theta)$

124. False. $\sin\dfrac{u}{2} = \sqrt{\dfrac{1 - \cos u}{2}}$ when $\dfrac{u}{2}$ is in the second quadrant.

126. (a)

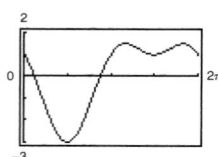

Minima: $\left(\dfrac{\pi}{2}, -3\right)$, $\left(\dfrac{3\pi}{2}, 1\right)$

Maxima: $\left(\dfrac{7\pi}{6}, \dfrac{3}{2}\right)$, $\left(\dfrac{11\pi}{6}, \dfrac{3}{2}\right)$

(b) $\dfrac{\pi}{2}, \dfrac{7\pi}{6}, \dfrac{3\pi}{2}, \dfrac{11\pi}{6}$

128. (a)

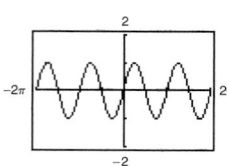

(b) $g(x) = \sin 2x$

(c) Answers will vary.

130. ≈ 15.7 gallons

Review Exercises *(page 263)*

2. $\csc x$ **4.** $\cot x$ **6.** $|\sec x|$

8. $\sin\theta = \dfrac{2\sqrt{13}}{13}$ **10.** $\cos\theta = \dfrac{1}{9}$

$\cos\theta = \dfrac{3\sqrt{13}}{13}$ $\tan\theta = 4\sqrt{5}$

$\csc\theta = \dfrac{\sqrt{13}}{2}$ $\csc\theta = \dfrac{9\sqrt{5}}{20}$

$\cot\theta = \dfrac{3}{2}$ $\sec\theta = 9$

   $\cot\theta = \dfrac{\sqrt{5}}{20}$

12. $\sec\theta \csc\theta$ **14.** $\cos^2 x$ **16.** $\tan u \sec u$ **18.** 1

20. $2\tan^2 x - 2\sec x \tan x + 1$ **22.** $1 + \sin x$

24–32. Answers will vary.

34. $\dfrac{\pi}{3} + 2n\pi, \dfrac{5\pi}{3} + 2n\pi$ **36.** $\dfrac{\pi}{3} + 2n\pi, \dfrac{5\pi}{3} + 2n\pi$

38. $\dfrac{\pi}{6} + n\pi, \dfrac{5\pi}{6} + n\pi$ **40.** $\dfrac{\pi}{6}, \dfrac{\pi}{2}, \dfrac{5\pi}{6}$ **42.** 0

44. $0, \dfrac{\pi}{3}, \dfrac{2\pi}{3}, \pi, \dfrac{4\pi}{3}, \dfrac{5\pi}{3}$ **46.** No solution   **48.** $\dfrac{\pi}{2}, \dfrac{3\pi}{2}$

50. $\dfrac{3\pi}{4}, \dfrac{7\pi}{4}, \arctan(-5) + \pi, \arctan(-5) + 2\pi$

52. $\sin 345° = \dfrac{\sqrt{2}}{4}\left(1 - \sqrt{3}\right)$

   $\cos 345° = \dfrac{\sqrt{2}}{4}\left(1 + \sqrt{3}\right)$

   $\tan 345° = -2 + \sqrt{3}$

54. $\sin\dfrac{19\pi}{12} = -\dfrac{\sqrt{2}}{4}\left(\sqrt{3} + 1\right)$

   $\cos\dfrac{19\pi}{12} = \dfrac{\sqrt{2}}{4}\left(\sqrt{3} - 1\right)$

   $\tan\dfrac{19\pi}{12} = -2 - \sqrt{3}$

56. $\cos 165°$ **58.** $\tan(-47°)$ **60.** $\dfrac{960 + 507\sqrt{7}}{1121}$

62. $\dfrac{12\sqrt{7} - 15}{52}$ **64.** $\dfrac{-960 + 507\sqrt{7}}{1121}$ **66.** $\dfrac{3\pi}{2}$

68. $0, \pi$

70.

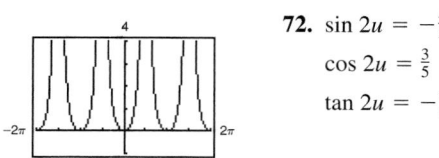

72. $\sin 2u = -\dfrac{4}{5}$

   $\cos 2u = \dfrac{3}{5}$

   $\tan 2u = -\dfrac{4}{3}$

74. $\theta \approx 40°$ **76.** $\dfrac{1 + \cos 6x}{2}$ **78.** $\dfrac{1 - \cos 2x}{2}$

80. $\sin 15° = \frac{1}{2}\sqrt{2 - \sqrt{3}}$

$\cos 15° = \frac{1}{2}\sqrt{2 + \sqrt{3}}$

$\tan 15° = 2 - \sqrt{3}$

82. $\sin\left(-\frac{17\pi}{12}\right) = \frac{1}{2}\sqrt{2 + \sqrt{3}}$

$\cos\left(-\frac{17\pi}{12}\right) = -\frac{1}{2}\sqrt{2 - \sqrt{3}}$

$\tan\left(-\frac{17\pi}{12}\right) = -2 - \sqrt{3}$

84. $\tan 3x$

86. (a) $V = \sin\frac{\theta}{2}\cos\frac{\theta}{2}$ cubic meters

(b) $V = \frac{1}{2}\sin\theta$ cubic meters

Volume is maximum when $\theta = \pi/2$.

88. $3[\cos(-30°) - \cos 60°]$ **90.** $2(\sin 5\alpha + \sin \alpha)$

92. $2\cos\frac{5\theta}{2}\cos\frac{\theta}{2}$ **94.** $2\cos x \sin\frac{\pi}{4}$

96. False. If $\frac{\pi}{2} < \theta < \pi$, then $\cos\frac{\theta}{2} > 0$. The sign of $\cos\frac{\theta}{2}$

depends on the quadrant in which $\frac{\theta}{2}$ lies.

98. True. $4\sin(-x)\cos(-x) = 4(-\sin x)\cos x$

$= -4\sin x \cos x$

$= -2(2\sin x \cos x)$

$= -2\sin 2x$

100. Reciprocal identities: $\sin\theta = \frac{1}{\csc\theta}$, $\cos\theta = \frac{1}{\sec\theta}$,

$\tan\theta = \frac{1}{\cot\theta}$, $\csc\theta = \frac{1}{\sin\theta}$, $\sec\theta = \frac{1}{\cos\theta}$,

$\cot\theta = \frac{1}{\tan\theta}$

Quotient identities: $\tan\theta = \frac{\sin\theta}{\cos\theta}$, $\cot\theta = \frac{\cos\theta}{\sin\theta}$

Pythagorean identities: $\sin^2\theta + \cos^2\theta = 1$,

$1 + \tan^2\theta = \sec^2\theta$, $1 + \cot^2\theta = \csc^2\theta$

102. $-1 \le \sin x \le 1$ for all x **104.** $y_1 = 1 - y_2$

106. $-3.1395, -2.0000, -0.4378, 2.0000$

Problem Solving   *(page 270)*

2. Answers will vary.

4. (a) $p_1(t) = \sin(524\,\pi t)$

$p_2(t) = \frac{1}{2}\sin(1048\,\pi t)$

$p_3(t) = \frac{1}{3}\sin(1572\,\pi t)$

$p_5(t) = \frac{1}{5}\sin(2620\,\pi t)$

$p_6(t) = \frac{1}{6}\sin(3144\,\pi t)$

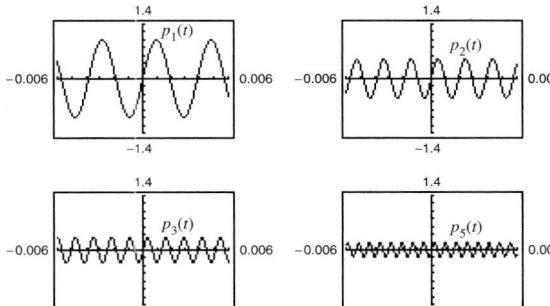

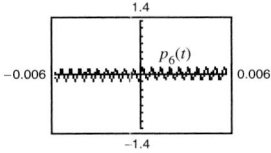

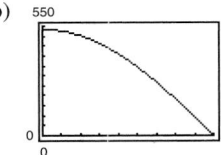

(b) p_1: $\frac{1}{262}$ p_5: $\frac{1}{1310}$

p_2: $\frac{1}{524}$ p_6: $\frac{1}{1572}$

p_3: $\frac{1}{786}$

p is periodic with period $\frac{1}{262}$.

(c) $(0, 0)$, $(0.00096, 0)$, $(0.00191, 0)$

$(0.00285, 0)$, $(0.00382, 0)$

(d) Maximum: $(0.00406, 1.19524)$

Minimum: $(0.00357, -1.19525)$

6. Answers will vary.

8. (a) $F = \dfrac{0.6\,W\cos\theta}{\sin 12°}$

(b)

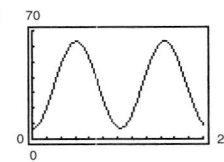

(c) Maximum: $\theta = 0°$

Minimum: $\theta = 90°$

10. (a) High tides: 6:12 A.M., 6:36 P.M.

Low tides: 12:00 A.M., 12:24 P.M.

(b) The water depth never falls below 7 feet.

(c)

CHAPTER 2

12. (a) $n = \dfrac{1}{2}\left(\cot \dfrac{\theta}{2} + \sqrt{3}\right)$

    (b) $\theta \approx 76.5°$

14. (a) $\cos 3\theta = \cos \theta - 4 \sin^2 \theta \cos \theta$

    (b) $\cos 4\theta = \cos^4 \theta - 6 \sin^2 \theta \cos^2 \theta + \sin^4 \theta$

Chapter 3

Section 3.1   *(page 280)*

2. $A = 35°, a \approx 11.88, b \approx 13.31$

4. $A = 35°, a \approx 36.50, b \approx 11.05$

6. Two solutions:

    $B \approx 45.79°, C \approx 74.21°, b \approx 7.45$

    $B \approx 14.21°, C \approx 105.79°, b \approx 2.55$

8. $B = 101.1°, a \approx 1.35, b \approx 3.23$

10. $C = 166°5', a \approx 3.30, c \approx 8.05$

12. $A \approx 174°41', C \approx 2°34', a \approx 11.99$

14. $B \approx 75.48°, C \approx 4.52°, b \approx 122.87$

16. $A \approx 44°14', B \approx 50°26', b \approx 38.67$

18. $A = 48°, b \approx 2.29, c \approx 4.73$

20. Two solutions:

    $B \approx 72.21°, C \approx 49.79°, c \approx 10.27$

    $B \approx 107.79°, C \approx 14.21°, c \approx 3.30$

22. $B \approx 36.82°, C \approx 67.18°, c \approx 32.30$

24. $B \approx 48.74°, C \approx 21.26°, c \approx 48.23$

26. (a) $b \le 10, b = \dfrac{10}{\sin 60°}$ (b) $10 < b < \dfrac{10}{\sin 60°}$

    (c) $b > \dfrac{10}{\sin 60°}$

28. (a) $b \le 315.6, b = \dfrac{315.6}{\sin 88°}$ (b) $315.6 < b < \dfrac{315.6}{\sin 88°}$

    (c) $b > \dfrac{315.6}{\sin 88°}$

30. 474.9   **32.** 4.5   **34.** 159.3

36. (a)

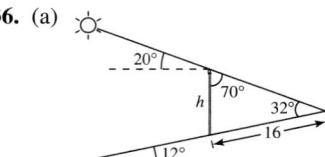

    (b) $\dfrac{16}{\sin 70°} = \dfrac{h}{\sin 32°}$

    (c) 9 meters

38. 240°

40. (a)

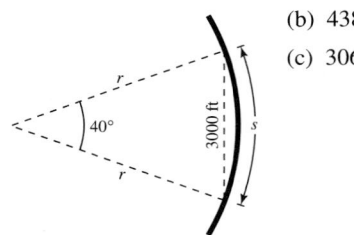

    (b) 4385.71 feet

    (c) 3061.80 feet

42. From Pine Knob: 42.4 kilometers

    From Colt Station: 15.5 kilometers

44. (a) $\alpha \approx 4.39°$ (b) $\beta = 85.61° - \theta$

    (c) $\dfrac{d}{\sin(85.61° - \theta)} = \dfrac{58.36}{\sin \theta}$

    (d)

θ	10°	20°	30°	40°	50°	60°
d	325.54	155.41	96.32	64.88	44.36	29.13

46. False. Two angles and one side determine a unique triangle, while two sides and one opposite angle do not necessarily determine a unique triangle.

48. (a) $A = 20\left(15 \sin \dfrac{3\theta}{2} - 4 \sin \dfrac{\theta}{2} - 6 \sin \theta\right)$

    (b)

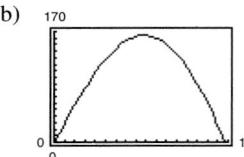

    (c) Domain: $0 \le \theta \le 1.6690$

    The area would increase and the domain would increase in length.

50. $\tan x$ **52.** $\sec^2 x$ **54.** $\sin 7\theta - \sin 3\theta$

Section 3.2   *(page 287)*

2. $A \approx 61.22°, B \approx 19.19°, C \approx 99.59°$

4. $A \approx 53.73°, B \approx 21.27°, c \approx 11.98$

6. $A \approx 39.35°, B \approx 16.74°, C \approx 123.91°$

8. $A \approx 86.68°, B \approx 31.82°, C \approx 61.50°$

10. $B \approx 16.54°, C \approx 108.46°, a \approx 8.64$

12. $A \approx 37°6', C \approx 67°34', b \approx 9.94$

14. $A \approx 157°2', B \approx 7°43', c \approx 4.21$

16. $A \approx 23.69°, B \approx 53.31°, c \approx 0.91$

a	b	c	d	θ	ϕ
18. 25	35	52.20	31.22	60°	120°
20. 40	60	63.25	80	104.47°	75.53°
22. 35.18	25	50	35	68.75°	111.25°

24. 54   **26.** 1350.2   **28.** 0.6

30.

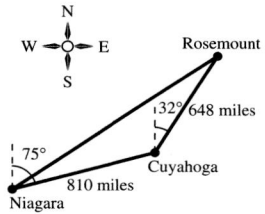

1357.8 miles, 56°

32. 41.2°, 52.9°   **34.** 127.2°   **36.** 131.1 feet, 118.6 feet

38. (a) N 59.7° E   (b) N 72.8° E   **40.** 103.9 feet

42. 3.8 miles

44. (a) $x^2 - 3x \cos \theta - 46.75 = 0$

(b) $x = \frac{1}{2}\left(3 \cos \theta + \sqrt{9 \cos^2 \theta + 187}\right)$

(c) 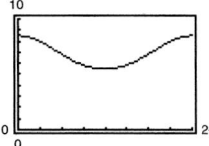   (d) 8.5 inches

46. 3.95 feet   **48.** 6577.8 square meters

50. False. To solve an SSA triangle, the Law of Sines is needed.

52. Answers will vary.   **54.** 405.3 feet

56. Answers will vary.   **58.** $\frac{\pi}{2}$ **60.** $-\frac{\pi}{3}$ **62.** $\frac{5\pi}{6}$

64. $\frac{\sqrt{1 - 9x^2}}{3x}$ **66.** $\frac{\sqrt{4 - (x - 1)^2}}{2}$

68. $\sin \theta = -\frac{\sqrt{2}}{2}$ **70.** $12 = 6 \sec \theta$

$\sec \theta = \sqrt{2}$ $\sec \theta = 2$

$\csc \theta = -\sqrt{2}$ $\csc \theta = \pm\frac{2\sqrt{3}}{3}$

72. $2 \cos x \sin \left(-\frac{\pi}{2}\right)$

Section 3.3   *(page 300)*

2. $\mathbf{v} = \langle -4, -2 \rangle$; $\|\mathbf{v}\| = 2\sqrt{5}$ **4.** $\mathbf{v} = \langle 4, 6 \rangle$; $\|\mathbf{v}\| = 2\sqrt{13}$

6. $\mathbf{v} = \langle 7, 0 \rangle$; $\|\mathbf{v}\| = 7$ **8.** $\mathbf{v} = \langle 8, -8 \rangle$; $\|\mathbf{v}\| = 8\sqrt{2}$

10. $\mathbf{v} = \langle 12, 29 \rangle$; $\|\mathbf{v}\| = \sqrt{985}$

12. $\mathbf{v} = \langle 7, -24 \rangle$; $\|\mathbf{v}\| = 25$

14.

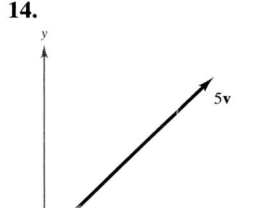

16.

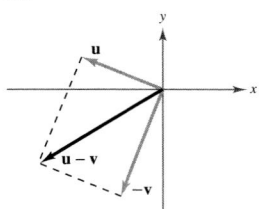

18.

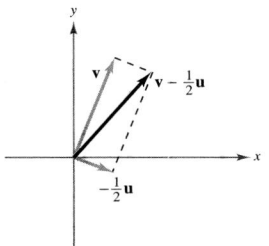

20. (a) $\langle 6, 3 \rangle$

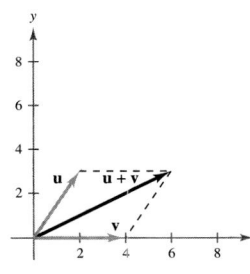

(b) $\langle -2, 3 \rangle$

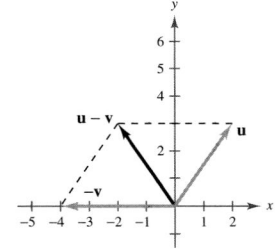

(c) $\langle -8, 6 \rangle$

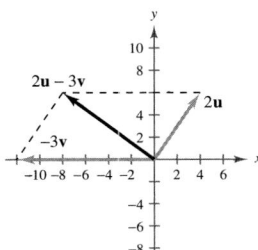

22. (a) $\langle 2, 1 \rangle$

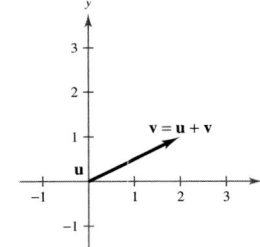

(b) $\langle -2, -1 \rangle$

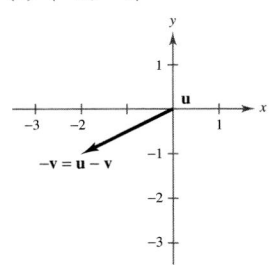

(c) $\langle -6, -3 \rangle$

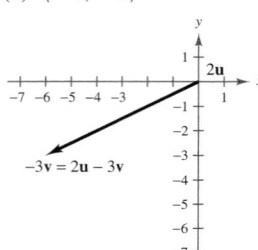

24. (a) $-3\mathbf{i} + 3\mathbf{j}$

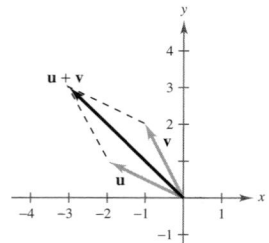

(b) $-\mathbf{i} - \mathbf{j}$

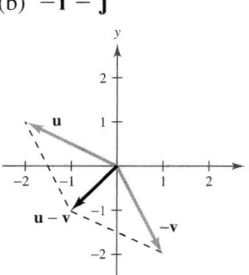

(c) $-\mathbf{i} - 4\mathbf{j}$

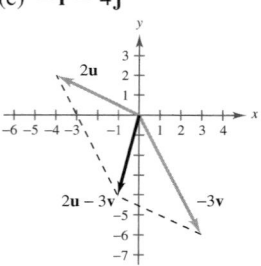

26. (a) $2\mathbf{i} + 3\mathbf{j}$

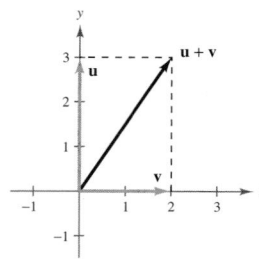

(b) $-2\mathbf{i} + 3\mathbf{j}$

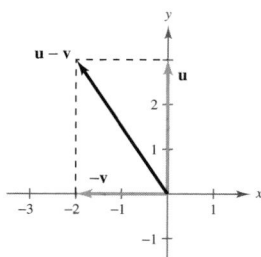

(c) $-6\mathbf{i} + 6\mathbf{j}$

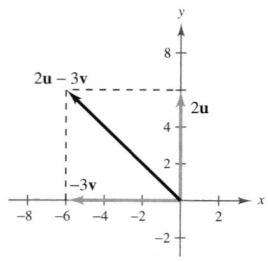

28. $\langle 0, -1 \rangle$ **30.** $\left\langle \dfrac{5}{13}, -\dfrac{12}{13} \right\rangle$ **32.** $\dfrac{\sqrt{2}}{2}\mathbf{i} + \dfrac{\sqrt{2}}{2}\mathbf{j}$

34. $-\mathbf{i}$ **36.** $-\dfrac{3\sqrt{58}}{58}\mathbf{i} + \dfrac{7\sqrt{58}}{58}\mathbf{j}$ **38.** $\langle -3\sqrt{2}, 3\sqrt{2} \rangle$

40. $\langle -10, 0 \rangle$

42. $\mathbf{v} = \left\langle \frac{3}{4}, \frac{3}{2} \right\rangle$ **44.** $\mathbf{v} = \langle -1, 3 \rangle$

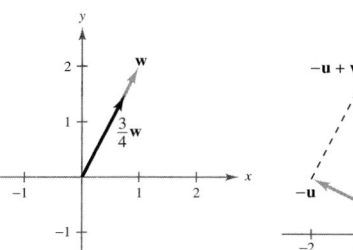

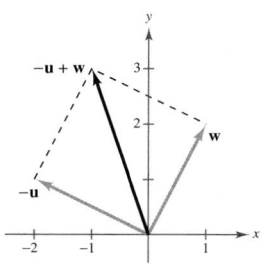

46. $\mathbf{v} = \langle 0, -5 \rangle$

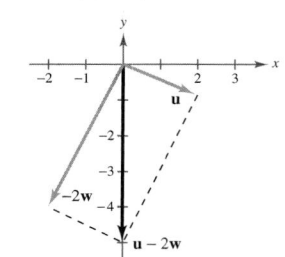

48. $\|\mathbf{v}\| = 8$; $\theta = 135°$ **50.** $\|\mathbf{v}\| = \sqrt{41}$; $\theta = 141.3°$

52. $\mathbf{v} = \left\langle \dfrac{\sqrt{2}}{2}, \dfrac{\sqrt{2}}{2} \right\rangle$ **54.** $\mathbf{v} = \left\langle \dfrac{5\sqrt{2}}{4}, \dfrac{5\sqrt{2}}{4} \right\rangle$

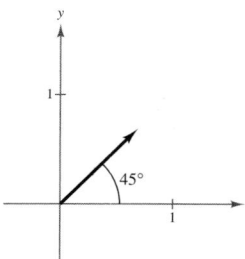

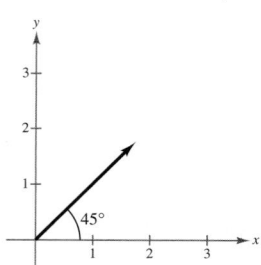

56. $\mathbf{v} = \langle 0, 4\sqrt{3} \rangle$ **58.** $\mathbf{v} = \left\langle \frac{9}{5}, \frac{12}{5} \right\rangle$

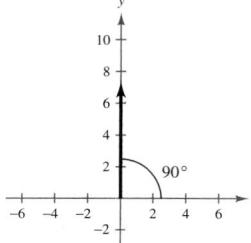

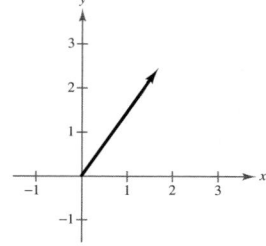

60. $\langle 2, 4 + 2\sqrt{3} \rangle$ **62.** $\langle 33.04, 53.19 \rangle$ **64.** $78.7°$

66. 90°   **68.** 47.4°   **70.** 8.7°; 2396.2 newtons

72. 37.5°; 58.6 pounds

74. Vertical component: 1200 sin 6° ≈ 125.4 feet per second

Horizontal component:

1200 cos 6° ≈ 1193.4 feet per second

76. $T_{AC} \approx 3611.1$ pounds

$T_{BC} \approx 2169.5$ pounds

78.

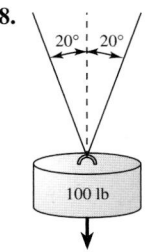

53.2 pounds

80. (a)

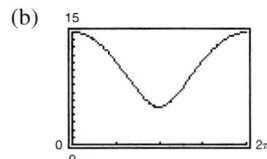

(b) $60\langle \cos 45°, \sin 45° \rangle$

(c) $580\langle \cos 118°, \sin 118° \rangle$

(d) 600.3 miles per hour

(e) 337.5°

82. $\sqrt{2}$ pounds; 1 pound

84. True, by the definition of a unit vector.

86. (a) $5\sqrt{5 + 4 \cos \theta}$

(b)

(c) Range: [5, 15]

Maximum is 15 when $\theta = 0$.

Minimum is 5 when $\theta = \pi$.

(d) The magnitudes of $\mathbf{F}_1$ and $\mathbf{F}_2$ are not the same.

88. Answers will vary.   **90.** $\langle 10, 50 \rangle$ or $\langle -10, -50 \rangle$

92. $8 \cos \theta$ **94.** $125 \tan^3 \theta$

96. $n\pi, \dfrac{5\pi}{4} + 2n\pi, \dfrac{7\pi}{4} + 2n\pi$

98. $\dfrac{\pi}{2} + n\pi, \dfrac{5\pi}{4} + 2n\pi, \dfrac{7\pi}{4} + 2n\pi$

Section 3.4   (page 311)

2. 9   **4.** 13   **6.** 8; scalar   **8.** 6; scalar   **10.** $2\sqrt{5}$

12. 20   **14.** 21   **16.** 33.69°   **18.** 7.13°

20. 53.13°   **22.** 93.18°   **24.** $\dfrac{\pi}{4}$

26. 41.40°, 64.15°, 74.45°   **28.** 21.80°, 41.63°, 116.57°

30. $12,500\sqrt{3}$ **32.** Neither   **34.** Neither

36. Orthogonal   **38.** $\langle 0, 0 \rangle, \langle 4, 2 \rangle$

40. $\frac{14}{17}\langle -4, -1 \rangle, \frac{5}{17}\langle 1, -4 \rangle$ **42.** $\langle 3, 8 \rangle, \langle -3, -8 \rangle$

44. $3\mathbf{i} - \frac{5}{2}\mathbf{j}, -3\mathbf{i} + \frac{5}{2}\mathbf{j}$ **46.** 14

48. (a) Force $= 30,000 \sin d$

(b)

d	0°	1°	2°	3°	4°	5°
Force	0	523.6	1047.0	1570.1	2092.7	2614.7

d	6°	7°	8°	9°	10°
Force	3135.9	3656.1	4175.2	4693.0	5209.4

(c) 29,885.8 pounds

50. 12,000 foot-pounds

52. 1,048,514.62 kilogram-meters (10,282,651.78 newton-meters)

54. True. cos 90° = 0

56. (a) $\mathbf{u}$ and $\mathbf{v}$ are parallel.   (b) $\mathbf{u}$ and $\mathbf{v}$ are orthogonal.

58. Answers will vary.   **60.** $12\sqrt{14}$

62. $\dfrac{\pi}{2}, \dfrac{5\pi}{4}, \dfrac{3\pi}{2}, \dfrac{7\pi}{4}$ **64.** $\dfrac{7\pi}{6}, \dfrac{3\pi}{2}, \dfrac{11\pi}{6}$

66. $-\dfrac{323}{325}$ **68.** $-\dfrac{253}{204}$

Review Exercises   (page 314)

2. $C = 37°, b \approx 38.90, c \approx 27.31$

4. $A = 150°, a \approx 48.24, b \approx 16.75$

6. $C = 40°, a \approx 162.42, b \approx 115.29$

8. $A = 80°, b \approx 334.95, c \approx 219.04$

10. No solution

12. Two solutions:

$A \approx 40.92°, C \approx 114.08°, c \approx 8.64$

$A \approx 139.08°, C \approx 15.92°, c \approx 2.60$

CHAPTER 3

14. 15.8   **16.** 44.1   **18.** 4.8   **20.** 586.4 feet

22. $A \approx 53.13°, B \approx 36.87°, C = 90°$

24. $A \approx 101.47°, B \approx 31.73°, C \approx 46.80°$

26. $A \approx 9.90°, C \approx 20.10°, b \approx 29.09$

28. $B \approx 35.20°, C \approx 82.80°, a \approx 17.37$

30.

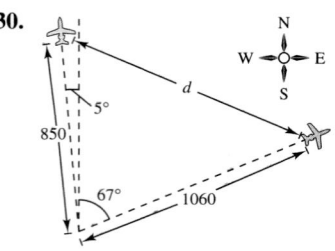

1135.5 miles

32. 36.98   **34.** 242.63

36.

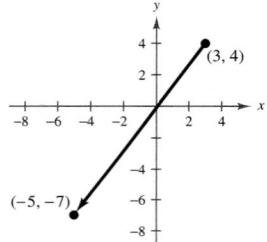

38.

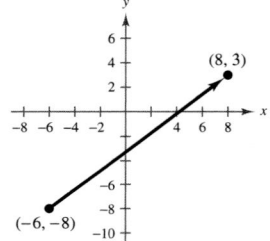

40. $\langle 6, \frac{5}{2} \rangle$ **42.** $\langle 14, 4 \rangle$ **44.** $\left\langle -\frac{\sqrt{2}}{4}, -\frac{\sqrt{2}}{4} \right\rangle$

46. $\langle -26, -35 \rangle$ **48.** $\langle 5, \frac{3}{2} \rangle$

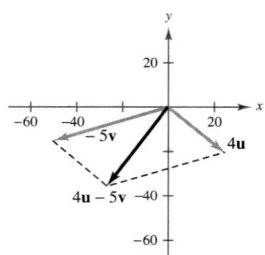

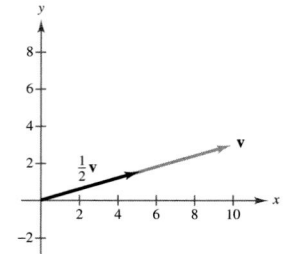

50. $-6\mathbf{i} - 8\mathbf{j}$ **52.** $7\mathbf{i} - 16\mathbf{j}$

54. $\sqrt{17}(\cos 346° \, \mathbf{i} + \sin 346° \, \mathbf{j})$ **56.** $\|\mathbf{v}\| = 3; \theta = 150°$

58. $\|\mathbf{v}\| = \sqrt{65}; \theta = 119.7°$ **60.** $\|\mathbf{v}\| = \sqrt{65}; \theta = 352.9°$

62. 180 pounds each

64. 740.5 kilometers per hour; 32.1°   **66.** -140

68. -136 **70.** 5; scalar   **72.** -6; scalar

74. 105°   **76.** 22.4°   **78.** Parallel   **80.** Orthogonal

82. $\langle 5, 0 \rangle, \langle 0, 6 \rangle$ **84.** $\frac{25}{29}\langle -5, 2 \rangle, \frac{19}{29}\langle 2, 5 \rangle$ **86.** -132

88. False. There may be no solution, one solution, or two solutions.

90. False. If $\mathbf{v} = a\mathbf{i} + b\mathbf{j} = \mathbf{0}$, then $a = b = 0$.

92. $a^2 = b^2 + c^2 - 2bc \cos A,$ **94.** A and C
$b^2 = a^2 + c^2 - 2ac \cos B,$
$c^2 = a^2 + b^2 - 2ab \cos C$

96. If $k > 0$, the direction is the same and the magnitude is k times as great.

If $k < 0$, the result is a vector in the opposite direction and the magnitude is $|k|$ times as great.

Problem Solving   *(page 324)*

2. S 22.09° E; 1025.88 yards

4. (a)

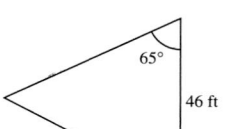

(b) 50.5 feet

(c) 22 bags

6. (a) $\mathbf{u} = \langle 0, -120 \rangle, \mathbf{v} = \langle 40, 0 \rangle$

(b)

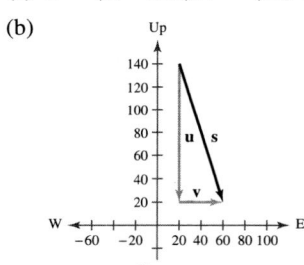

(c) 126.5 miles per hour; The magnitude gives the velocity of the skydiver's fall.

(d) 108.43°

(e)

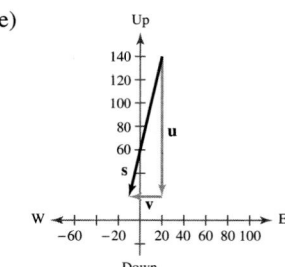

123.7 miles per hour

8. $\mathbf{u} \cdot \mathbf{v} = 0$ and $\mathbf{u} \cdot \mathbf{w} = 0$

$\mathbf{u} \cdot (c\mathbf{v} + d\mathbf{w})$

$= \mathbf{u} \cdot c\mathbf{v} + \mathbf{u} \cdot d\mathbf{w}$

$= c(\mathbf{u} \cdot \mathbf{v}) + d(\mathbf{u} \cdot \mathbf{w})$

$= 0$

10. (a)

θ	0.5°	1.0°	1.5°
$\|\mathbf{v}\|\sin\theta$	0.873	1.745	2.618
$\|\mathbf{v}\|\cos\theta$	99.996	99.985	99.966

θ	2.0°	2.5°	3.0°
$\|\mathbf{v}\|\sin\theta$	3.490	4.362	5.234
$\|\mathbf{v}\|\cos\theta$	99.939	99.905	99.863

    (b) No. Find the square root of the sum of the squares of the vertical and horizontal components.

    (c) (i) 150 miles per hour

        (ii) 150 miles per hour

Chapter 4

Section 4.1   *(page 333)*

2. $a = 13,\ b = 4$ **4.** $a = 0,\ b = -\frac{5}{2}$ **6.** $3 + 4i$

8. $1 + 2\sqrt{2}i$ **10.** $2i$ **12.** 45 **14.** $4 + 2i$

16. $0.02i$ **18.** $8 + 4i$ **20.** $-3 - 11i$ **22.** 4

24. $17 + 18i$ **26.** $-4.2 + 7.5i$ **28.** $-5\sqrt{2}$

30. -75 **32.** $6 - 22i$ **34.** $32 - 72i$

36. $\left(21 + 5\sqrt{2}\right) + \left(7\sqrt{5} - 3\sqrt{10}\right)i$

38. $-5 - 12i$ **40.** $-8i$ **42.** $7 + 12i,\ 193$

44. $-3 - \sqrt{2}i,\ 11$ **46.** $-\sqrt{15}i,\ 15$

48. $1 + \sqrt{8},\ 9 + 4\sqrt{2}$ **50.** $7i$ **52.** $\frac{5}{2} + \frac{5}{2}i$

54. $4 + i$ **56.** $8 - 4i$ **58.** $\frac{60}{169} - \frac{25}{169}i$ **60.** $\frac{12}{5} + \frac{9}{5}i$

62. $\frac{5}{17} - \frac{20}{17}i$ **64.** $-3 \pm i$ **66.** $\frac{1}{3} \pm 2i$

68. $\frac{1}{8} \pm \frac{\sqrt{11}}{8}i$ **70.** $\frac{3}{7} \pm \frac{\sqrt{34}}{14}i$ **72.** $\frac{1}{3} \pm \frac{\sqrt{23}}{3}i$

74. $-4 + 2i$ **76.** i **78.** -8 **80.** $\frac{1}{8}i$

82. (a) 16   (b) 16   (c) 16   (d) 16

84. (a) $z_1 = 9 + 16i,\ z_2 = 20 - 10i$

    (b) $z = \dfrac{11{,}240}{877} + \dfrac{4630}{877}i$

86. True. $x^4 - x^2 + 14 = 56$

$$\left(-i\sqrt{6}\right)^4 - \left(-i\sqrt{6}\right)^2 + 14 \overset{?}{=} 56$$

$$36 + 6 + 14 \overset{?}{=} 56$$

$$56 = 56$$

88. $\sqrt{-6}\sqrt{-6} = \sqrt{6}i\sqrt{6}i = 6i^2 = -6$

90. Answers will vary.   **92.** $x^3 + x^2 + 2x - 6$

94. $4x^2 - 20x + 25$ **96.** 14   **98.** $-\frac{4}{3}$

100. $r = \dfrac{\sqrt{\alpha m_1 m_2 F}}{F}$

Section 4.2   *(page 340)*

2. Six solutions   **4.** Five solutions

6. Two real solutions   **8.** No real solutions

10. Two real solutions   **12.** No real solutions

14. $\pm\dfrac{\sqrt{3}}{3}$ **16.** $-3, 5$ **18.** $-\dfrac{1}{2}$ **20.** $8 \pm \sqrt{118}$

22. $\dfrac{1}{2} \pm \sqrt{5}i$ **24.** $\dfrac{75 \pm 25\sqrt{7}}{2}$ **26.** $1 \pm \sqrt{6}$

28.

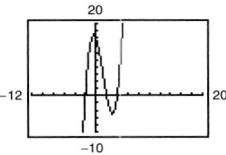

$\pm 2,\ 4$; The number of real zeros and the number of x-intercepts is the same.

30.

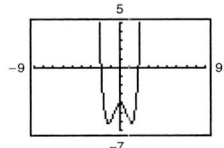

$\pm 2, \pm i$; The number of real zeros and the number of x-intercepts is the same.

32. $\dfrac{1 \pm \sqrt{223}i}{2};\ \left(x - \dfrac{1 - \sqrt{223}i}{2}\right)\left(x - \dfrac{1 + \sqrt{223}i}{2}\right)$

34. $-5 \pm \sqrt{2};\ \left(x + 5 + \sqrt{2}\right)\left(x + 5 - \sqrt{2}\right)$

36. $\pm 5, \pm 5i;\ (y + 5)(y - 5)(y + 5i)(y - 5i)$

38. $3 \pm \sqrt{19};\ \left(x - 3 + \sqrt{19}\right)\left(x - 3 - \sqrt{19}\right)$

40. $8, \pm 2\sqrt{3};\ (x - 8)\left(x + 2\sqrt{3}\right)\left(x - 2\sqrt{3}\right)$

42. $-5, \pm\sqrt{2}i;\ (x + 5)\left(x + \sqrt{2}i\right)\left(x - \sqrt{2}i\right)$

44. $-\frac{3}{4}, \pm 2\sqrt{6}i;\ (4x + 3)\left(x + 2\sqrt{6}i\right)\left(x - 2\sqrt{6}i\right)$

46. $0, -1, \pm 10i;\ x(x + 1)(x + 10i)(x - 10i)$

48. $\pm 2i, \pm 5i;\ (x + 2i)(x - 2i)(x + 5i)(x - 5i)$

50. $-1, \pm 3i$ **52.** $-3, 5 \pm 2i$ **54.** $-\frac{2}{3}, 1 \pm \sqrt{3}i$

56. $-2, -1 \pm 3i$ **58.** $x^3 - 4x^2 + 9x - 36$

60. $x^3 - 10x^2 + 33x - 34$

62. $x^4 + 8x^3 + 9x^2 - 10x + 100$

64. (a) $R = -265.193t^2 + 5621.38t - 19{,}147.5$

    (b)

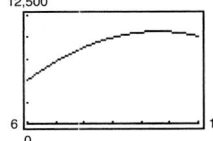

    (c) The revenue never reached \$11 billion.

    (d) The revenue never reached \$11 billion.

66. False. f does not have real coefficients.

68. r_1, r_2, r_3 **70.** $\dfrac{r_1}{2}, \dfrac{r_2}{2}, \dfrac{r_3}{2}$ **72.** $-r_1, -r_2, -r_3$

74. (a) Not correct because f has $(0, 0)$ as an intercept.

(b) Not correct because the function must be at least a fourth-degree polynomial.

(c) Correct function

(d) Not correct because k has $(-1, 0)$ as an intercept.

76. $12 + 11i$ **78.** 106

80.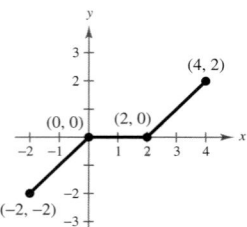

82.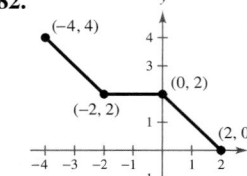

84.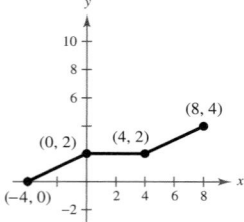

Section 4.3 (page 347)

2.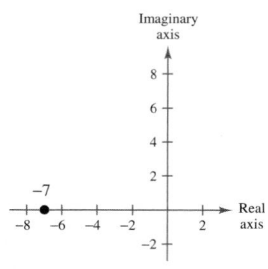

7

4.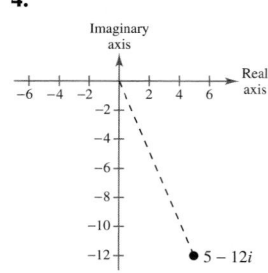

13

6.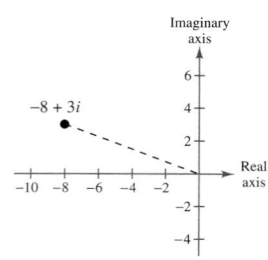

$\sqrt{73}$

8. $2(\cos \pi + i \sin \pi)$ **10.** $2\left(\cos \dfrac{2\pi}{3} + i \sin \dfrac{2\pi}{3}\right)$

12.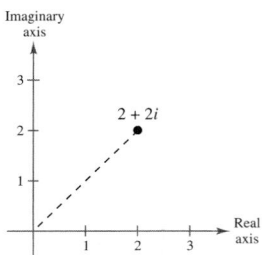

$2\sqrt{2}\left(\cos \dfrac{\pi}{4} + i \sin \dfrac{\pi}{4}\right)$

14.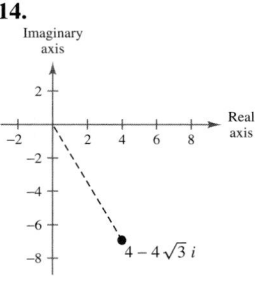

$8\left(\cos \dfrac{5\pi}{3} + i \sin \dfrac{5\pi}{3}\right)$

16.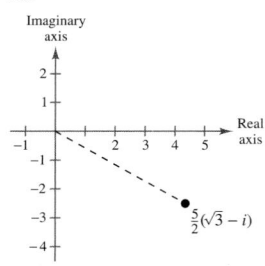

$5\left(\cos \dfrac{11\pi}{6} + i \sin \dfrac{11\pi}{6}\right)$

18.

$4\left(\cos \dfrac{\pi}{2} + i \sin \dfrac{\pi}{2}\right)$

20.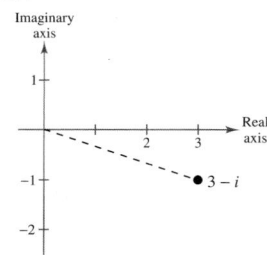

$\sqrt{10}(\cos 5.96 + i \sin 5.96)$

22.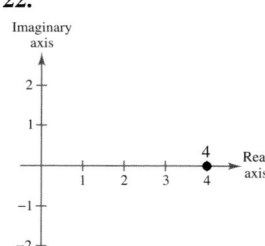

$4(\cos 0 + i \sin 0)$

24.

$3(\cos 5.94 + i \sin 5.94)$

26.

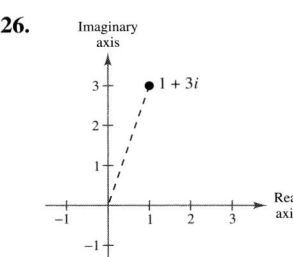

$\sqrt{10}(\cos 1.25 + i \sin 1.25)$

28. $8.54(\cos 0.36 + i \sin 0.36)$

30. $5.10(\cos 3.34 + i \sin 3.34)$

32. $9.80(\cos 5.86 + i \sin 5.86)$

34. $11(\cos 3.75 + i \sin 3.75)$

36.

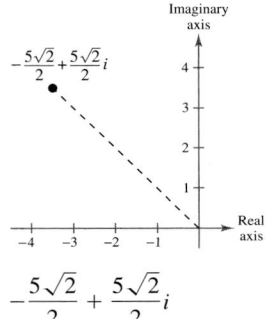

$-\dfrac{5\sqrt{2}}{2} + \dfrac{5\sqrt{2}}{2}i$

38.

$-\dfrac{\sqrt{2}}{8} - \dfrac{\sqrt{2}}{8}i$

40.

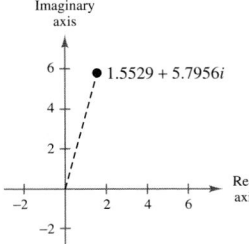

$1.5529 + 5.7956i$

42.

7

44.

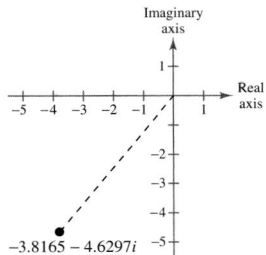

$-3.8165 - 4.6297i$

46. $3.0902 + 9.5106i$

48. $4.7693 + 7.6324i$

50. $3\left(\cos \dfrac{13\pi}{12} + i \sin \dfrac{13\pi}{12}\right)$

52. $0.4(\cos 40° + i \sin 40°)$ **54.** $\cos 25° + i \sin 25°$

56. $\frac{1}{2}(\cos 80° + i \sin 80°)$ **58.** $\frac{5}{4}(\cos 2.2 + i \sin 2.2)$

60. $\frac{6}{7}(\cos 300° + i \sin 300°)$

62. (a) $\left[2\left(\cos \dfrac{\pi}{6} + i \sin \dfrac{\pi}{6}\right)\right]\left[\sqrt{2}\left(\cos \dfrac{\pi}{4} + i \sin \dfrac{\pi}{4}\right)\right]$

(b) $2\sqrt{2}\left(\cos \dfrac{5\pi}{12} + i \sin \dfrac{5\pi}{12}\right) \approx 0.732 + 2.732i$

(c) $(\sqrt{3} - 1) + (\sqrt{3} + 1)i \approx 0.732 + 2.732i$

64. (a) $[4(\cos 0 + i \sin 0)]\left[2\left(\cos \dfrac{5\pi}{3} + i \sin \dfrac{5\pi}{3}\right)\right]$

(b) $8\left(\cos \dfrac{5\pi}{3} + i \sin \dfrac{5\pi}{3}\right) = 4 - 4\sqrt{3}i$

(c) $4 - 4\sqrt{3}i$

66. (a) $\left[2\left(\cos \dfrac{\pi}{3} + i \sin \dfrac{\pi}{3}\right)\right] \div$

$[3\sqrt{5}(\cos 5.82 + i \sin 5.82)]$

(b) $\dfrac{2\sqrt{5}}{15}(\cos 1.51 + i \sin 1.51) \approx 0.018 + 0.298i$

(c) $\approx 0.018 + 0.298i$

68. (a) $\left[4\left(\cos \dfrac{\pi}{2} + i \sin \dfrac{\pi}{2}\right)\right] \div [2\sqrt{5}(\cos 2.68 + i \sin 2.68)]$

(b) $\dfrac{2}{\sqrt{5}}(\cos 5.18 + i \sin 5.18) \approx 0.403 - 0.798i$

(c) $\dfrac{2}{5} - \dfrac{4}{5}i = 0.400 - 0.800i$

70.

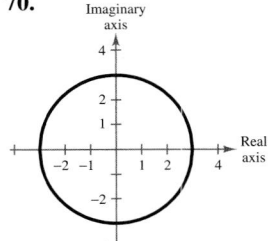

72.

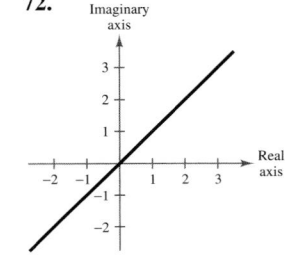

74. True. $z_1 z_2 = r_1 r_2 [\cos(\theta_1 + \theta_2) + i \sin(\theta_1 + \theta_2)] = 0$ if and only if $r_1 = 0$ and/or $r_2 = 0$.

76. Answers will vary.   **78.** Answers will vary.

80. $A = 24°, b \approx 75.24, c \approx 82.36$

82. $A = 84°, a \approx 2009.43, c \approx 2020.50$

84. $A = 8°30', a \approx 1.01, b \approx 6.73$

86. $\frac{1}{8}; \frac{1}{24}$ **88.** $\frac{1}{12}; 0$

Section 4.4   *(page 353)*

2. $-512i$ **4.** $-239 + 28,560i$ **6.** -32

8. $-\dfrac{81}{2} - \dfrac{81\sqrt{3}}{2}i$ **10.** 256   **12.** 1

14. $15{,}939 + 18{,}460i$ **16.** $-47 - 8\sqrt{3}i$

18. $44.4539 + 252.1108i$ **20.** $44.4539 + 252.1108i$

22. $-32\sqrt{2} + 32\sqrt{2}i$ **24.** $62.8930 + 234.7200i$

26. (a) $4(\cos 30° + i \sin 30°)$

$4(\cos 210° + i \sin 210°)$

(b)

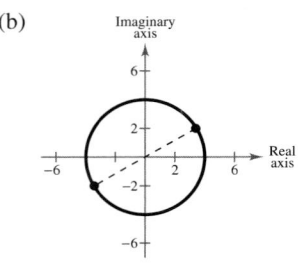

(c) $2\sqrt{3} + 2i, -2\sqrt{3} - 2i$

28. (a) $4\left(\cos \dfrac{\pi}{9} + i \sin \dfrac{\pi}{9}\right)$

$4\left(\cos \dfrac{7\pi}{9} + i \sin \dfrac{7\pi}{9}\right)$

$4\left(\cos \dfrac{13\pi}{9} + i \sin \dfrac{13\pi}{9}\right)$

(b)

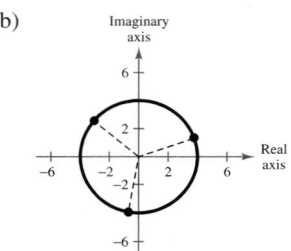

(c) $3.7588 + 1.3681i, -3.0642 + 2.5712i,$

$-0.6946 - 3.9392i$

30. (a) $2\left(\cos \dfrac{\pi}{6} + i \sin \dfrac{\pi}{6}\right)$

$2\left(\cos \dfrac{17\pi}{30} + i \sin \dfrac{17\pi}{30}\right)$

$2\left(\cos \dfrac{29\pi}{30} + i \sin \dfrac{29\pi}{30}\right)$

$2\left(\cos \dfrac{41\pi}{30} + i \sin \dfrac{41\pi}{30}\right)$

$2\left(\cos \dfrac{53\pi}{30} + i \sin \dfrac{53\pi}{30}\right)$

(b)

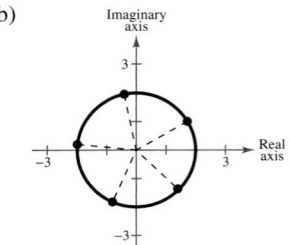

(c) $\sqrt{3} + i, -0.4158 + 1.9563i, -1.9890 + 0.2091i,$

$-0.8135 - 1.8271i, 1.4863 - 1.3383i$

32. (a) $6\left(\cos \dfrac{3\pi}{4} + i \sin \dfrac{3\pi}{4}\right)$

$6\left(\cos \dfrac{7\pi}{4} + i \sin \dfrac{7\pi}{4}\right)$

(b)

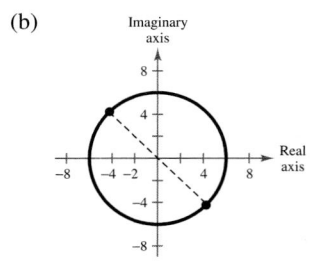

(c) $-3\sqrt{2} + 3\sqrt{2}i, 3\sqrt{2} - 3\sqrt{2}i$

34. (a) $5\left(\cos \dfrac{\pi}{8} + i \sin \dfrac{\pi}{8}\right)$

$5\left(\cos \dfrac{5\pi}{8} + i \sin \dfrac{5\pi}{8}\right)$

$5\left(\cos \dfrac{9\pi}{8} + i \sin \dfrac{9\pi}{8}\right)$

$5\left(\cos \dfrac{13\pi}{8} + i \sin \dfrac{13\pi}{8}\right)$

(b)

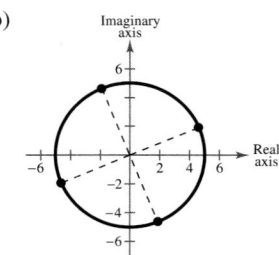

(c) $4.6194 + 1.9134i, -1.9134 + 4.6194i,$

$-4.6194 - 1.9134i, 1.9134 - 4.6194i$

36. (a) $2\left(\cos\dfrac{\pi}{4} + i\sin\dfrac{\pi}{4}\right)$

$2\left(\cos\dfrac{11\pi}{12} + i\sin\dfrac{11\pi}{12}\right)$

$2\left(\cos\dfrac{19\pi}{12} + i\sin\dfrac{19\pi}{12}\right)$

(b)

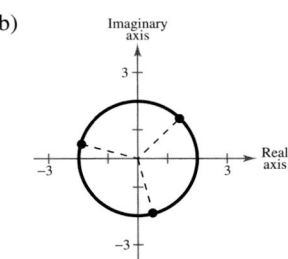

(c) $\sqrt{2} + \sqrt{2}i,\ -1.9319 + 0.5176i,\ 0.5176 - 1.9319i$

38. (a) $\cos\dfrac{\pi}{8} + i\sin\dfrac{\pi}{8}$

$\cos\dfrac{5\pi}{8} + i\sin\dfrac{5\pi}{8}$

$\cos\dfrac{9\pi}{8} + i\sin\dfrac{9\pi}{8}$

$\cos\dfrac{13\pi}{8} + i\sin\dfrac{13\pi}{8}$

(b)

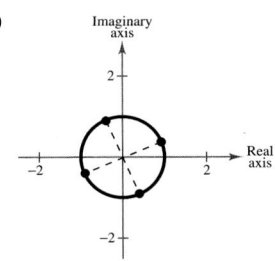

(c) $0.9239 + 0.3827i,\ -0.3827 + 0.9239i,$
$-0.9239 - 0.3827i,\ 0.3827 - 0.9239i$

40. (a) $10(\cos 0 + i\sin 0)$

$10\left(\cos\dfrac{2\pi}{3} + i\sin\dfrac{2\pi}{3}\right)$

$10\left(\cos\dfrac{4\pi}{3} + i\sin\dfrac{4\pi}{3}\right)$

(b)

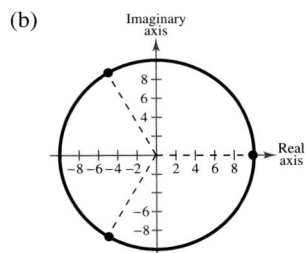

(c) $10,\ -5 + 5\sqrt{3}i,\ -5 - 5\sqrt{3}i$

42. (a) $\sqrt{2}\left(\cos\dfrac{\pi}{4} + i\sin\dfrac{\pi}{4}\right)$

$\sqrt{2}\left(\cos\dfrac{3\pi}{4} + i\sin\dfrac{3\pi}{4}\right)$

$\sqrt{2}\left(\cos\dfrac{5\pi}{4} + i\sin\dfrac{5\pi}{4}\right)$

$\sqrt{2}\left(\cos\dfrac{7\pi}{4} + i\sin\dfrac{7\pi}{4}\right)$

(b)

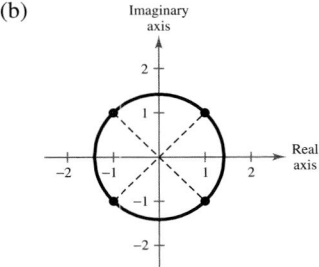

(c) $1 + i,\ -1 + i,\ -1 - i,\ 1 - i$

44. (a) $2\left(\cos\dfrac{\pi}{12} + i\sin\dfrac{\pi}{12}\right)$

$2\left(\cos\dfrac{5\pi}{12} + i\sin\dfrac{5\pi}{12}\right)$

$2\left(\cos\dfrac{3\pi}{4} + i\sin\dfrac{3\pi}{4}\right)$

$2\left(\cos\dfrac{13\pi}{12} + i\sin\dfrac{13\pi}{12}\right)$

$2\left(\cos\dfrac{17\pi}{12} + i\sin\dfrac{17\pi}{12}\right)$

$2\left(\cos\dfrac{7\pi}{4} + i\sin\dfrac{7\pi}{4}\right)$

(b)

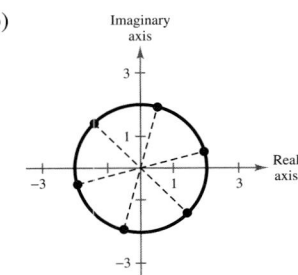

(c) $1.9319 + 0.5176i,\ 0.5176 + 1.9319i,$
$-\sqrt{2} + \sqrt{2}i,\ -1.9319 - 0.5176i,$
$-0.5176 - 1.9319i,\ \sqrt{2} - \sqrt{2}i$

CHAPTER 4

46. $\cos \dfrac{\pi}{6} + i \sin \dfrac{\pi}{6}$

$\cos \dfrac{5\pi}{6} + i \sin \dfrac{5\pi}{6}$

$\cos \dfrac{3\pi}{2} + i \sin \dfrac{3\pi}{2}$

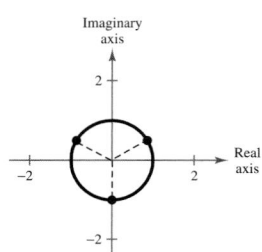

48. $\cos \dfrac{\pi}{3} + i \sin \dfrac{\pi}{3}$

$\cos \pi + i \sin \pi$

$\cos \dfrac{5\pi}{3} + i \sin \dfrac{5\pi}{3}$

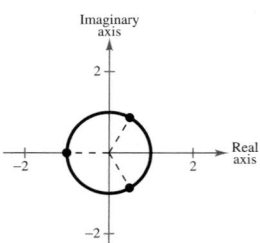

50. $5\left(\cos \dfrac{\pi}{3} + i \sin \dfrac{\pi}{3}\right)$

$5(\cos \pi + i \sin \pi)$

$5\left(\cos \dfrac{5\pi}{3} + i \sin \dfrac{5\pi}{3}\right)$

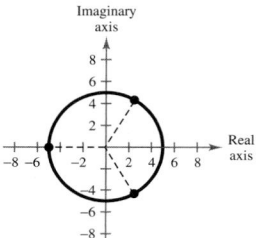

52. $3(\cos 0 + i \sin 0)$

$3\left(\cos \dfrac{2\pi}{3} + i \sin \dfrac{2\pi}{3}\right)$

$3\left(\cos \dfrac{4\pi}{3} + i \sin \dfrac{4\pi}{3}\right)$

54. $3\left(\cos \dfrac{\pi}{2} + i \sin \dfrac{\pi}{2}\right)$

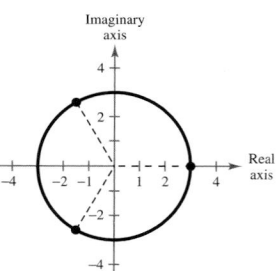

$3\left(\cos \dfrac{7\pi}{6} + i \sin \dfrac{7\pi}{6}\right)$

$3\left(\cos \dfrac{11\pi}{6} + i \sin \dfrac{11\pi}{6}\right)$

56. $2\left(\cos \dfrac{\pi}{4} + i \sin \dfrac{\pi}{4}\right)$

$2\left(\cos \dfrac{7\pi}{12} + i \sin \dfrac{7\pi}{12}\right)$

$2\left(\cos \dfrac{11\pi}{12} + i \sin \dfrac{11\pi}{12}\right)$

$2\left(\cos \dfrac{5\pi}{4} + i \sin \dfrac{5\pi}{4}\right)$

$2\left(\cos \dfrac{19\pi}{12} + i \sin \dfrac{19\pi}{12}\right)$

$2\left(\cos \dfrac{23\pi}{12} + i \sin \dfrac{23\pi}{12}\right)$

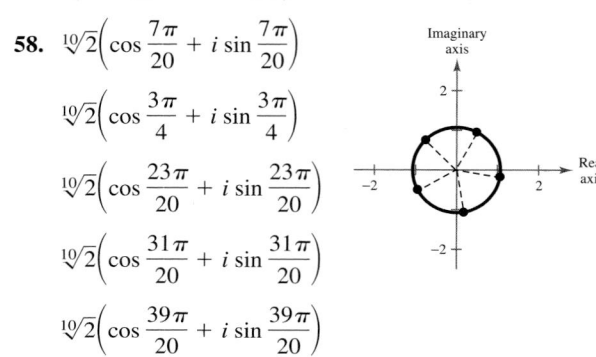

58. $\sqrt[10]{2}\left(\cos \dfrac{7\pi}{20} + i \sin \dfrac{7\pi}{20}\right)$

$\sqrt[10]{2}\left(\cos \dfrac{3\pi}{4} + i \sin \dfrac{3\pi}{4}\right)$

$\sqrt[10]{2}\left(\cos \dfrac{23\pi}{20} + i \sin \dfrac{23\pi}{20}\right)$

$\sqrt[10]{2}\left(\cos \dfrac{31\pi}{20} + i \sin \dfrac{31\pi}{20}\right)$

$\sqrt[10]{2}\left(\cos \dfrac{39\pi}{20} + i \sin \dfrac{39\pi}{20}\right)$

60. $\sqrt[8]{2}\left(\cos \dfrac{5\pi}{16} + i \sin \dfrac{5\pi}{16}\right)$

$\sqrt[8]{2}\left(\cos \dfrac{13\pi}{16} + i \sin \dfrac{13\pi}{16}\right)$

$\sqrt[8]{2}\left(\cos \dfrac{21\pi}{16} + i \sin \dfrac{21\pi}{16}\right)$

$\sqrt[8]{2}\left(\cos \dfrac{29\pi}{16} + i \sin \dfrac{29\pi}{16}\right)$

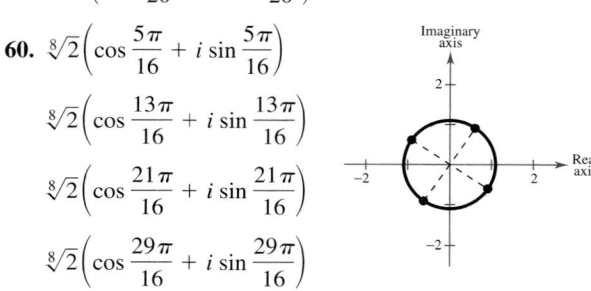

62. False. The complex number needs to be converted to trigonometric form before using DeMoivre's Theorem.

$$\left(4 + \sqrt{6}\,i\right)^8 \approx \left[\sqrt{22}\,(\cos 0.55 + i \sin 0.55)\right]^8$$

64. Answers will vary.

66. (a) $3(\cos 45° + i \sin 45°)$

$3(\cos 135° + i \sin 135°)$

$3(\cos 225° + i \sin 225°)$

$3(\cos 315° + i \sin 315°)$

(b) -81

68. $\dfrac{3\sqrt{34}}{136}\left(\sqrt{7} + 5\right)$ **70.** $\dfrac{\sqrt{34}}{136}\left(5\sqrt{7} + 9\right)$

72. $\dfrac{51\sqrt{7} - 120}{47}$ **74.** $\dfrac{3\sqrt{7}}{8}$ **76.** $\pm \dfrac{1}{2}\sqrt{\dfrac{34 + 5\sqrt{34}}{17}}$

78. $\left\langle -\dfrac{3\sqrt{58}}{58}, \dfrac{7\sqrt{58}}{58} \right\rangle$ **80.** j   **82.** $\sqrt{74}$

84. 13   **86.** $4\sqrt{17}$ **88.** $8\sqrt{13}$

Review Exercises *(page 356)*

2. $3 - 5i$ **4.** $-1 - 5i$ **6.** $-\sqrt{2}\,i$

8. $17 + 28i$ **10.** $9 + 20i$ **12.** $\frac{17}{26} + \frac{7}{26}i$

14. $\dfrac{9}{85} + \dfrac{83}{85}i$ **16.** $\pm\frac{1}{2}i$ **18.** $-\dfrac{1}{4} \pm \dfrac{\sqrt{71}}{4}i$

20. Six solutions   **22.** Three solutions

24. One repeated solution   **26.** One repeated solution

28. $0, 6$ **30.** $2 \pm \sqrt{7}$ **32.** $\dfrac{-5 \pm \sqrt{7}\,i}{4}$

34. $\frac{1}{4}, \pm 4\sqrt{2}\,i$ **36.** $\pm 5i, \pm \dfrac{\sqrt{5}}{5}i$

38. $-2, -\frac{3}{5}, \frac{1}{2}; (x + 2)(5x + 3)(2x - 1)$

40. $4, 2 \pm 3i; (x - 4)(x - 2 - 3i)(x - 2 + 3i)$

42. $\frac{4}{5}, \pm 2i; (5x - 4)(x - 2i)(x + 2i)$

44. $2 \pm \sqrt{3}\,i, 1 \pm \sqrt{2}\,i;$
$\left(x - 2 - \sqrt{3}\,i\right)\left(x - 2 + \sqrt{3}\,i\right)$
$\left(x - 1 - \sqrt{2}\,i\right)\left(x - 1 + \sqrt{2}\,i\right)$

46. $x^4 - 6x^3 + 5x^2 + 24x - 36$ **48.** $x^3 - 7x^2 + 9x + 5$

50. $x^4 - x^3 - 3x^2 + 17x - 30$ **52.** $x^4 + 20x^2 + 64$

54. No. There is no price for which the profit will reach $9 million.

56.

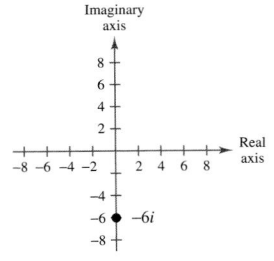

6

58.

$2\sqrt{29}$

60. $13(\cos 1.176 + i \sin 1.176)$ **62.** $9(\cos \pi + i \sin \pi)$

64. (a) $z_1 = 3\sqrt{2}\left(\cos \dfrac{5\pi}{4} + i \sin \dfrac{5\pi}{4}\right)$

$z_2 = 4\left(\cos \dfrac{\pi}{6} + i \sin \dfrac{\pi}{6}\right)$

(b) $z_1 z_2 = 12\sqrt{2}\left(\cos \dfrac{17\pi}{12} + i \sin \dfrac{17\pi}{12}\right)$

$\dfrac{z_1}{z_2} = \dfrac{3\sqrt{2}}{4}\left(\cos \dfrac{13\pi}{12} + i \sin \dfrac{13\pi}{12}\right)$

66. $-16 - 16\sqrt{3}\,i$ **68.** 16   **70.** $\pm 4, \pm 4i$

72. $2(\cos 0 + i \sin 0) = 2$

$2\left(\cos \dfrac{2\pi}{5} + i \sin \dfrac{2\pi}{5}\right) = 0.6180 + 1.9021i$

$2\left(\cos \dfrac{4\pi}{5} + i \sin \dfrac{4\pi}{5}\right) = -1.6180 + 1.1756i$

$2\left(\cos \dfrac{6\pi}{5} + i \sin \dfrac{6\pi}{5}\right) = -1.6180 - 1.1756i$

$2\left(\cos \dfrac{8\pi}{5} + i \sin \dfrac{8\pi}{5}\right) = 0.6180 - 1.9021i$

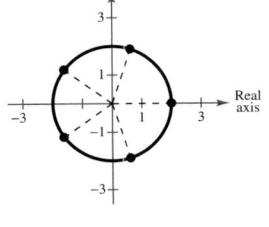

74. $\cos 0 + i \sin 0 = 1$

$\cos \dfrac{\pi}{2} + i \sin \dfrac{\pi}{2} = i$

$\cos \dfrac{2\pi}{3} + i \sin \dfrac{2\pi}{3} = -\dfrac{1}{2} + \dfrac{\sqrt{3}}{2}i$

$\cos \dfrac{4\pi}{3} + i \sin \dfrac{4\pi}{3} = -\dfrac{1}{2} - \dfrac{\sqrt{3}}{2}i$

$\cos \dfrac{3\pi}{2} + i \sin \dfrac{3\pi}{2} = -i$

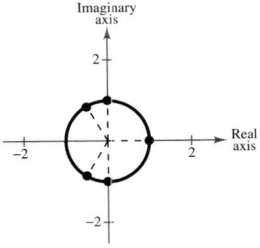

76. False. The equation has no real solution. The solutions are:
$\dfrac{717}{650} \pm \dfrac{\sqrt{3311}\,i}{650}.$

78. Answers will vary.

80. (a) $4(\cos 60° + i \sin 60°)$ (b) $-128 - 128\sqrt{3}\,i$
$4(\cos 150° + i \sin 150°)$
$4(\cos 240° + i \sin 240°)$
$4(\cos 330° + i \sin 330°)$

82. (a) 3

(b) On the circle 120°, 210°, and 300° from the positive real axis

Problem Solving   *(page 360)*

2. (a) $\frac{1}{2} - \frac{1}{2}i$ (b) $\frac{3}{10} + \frac{1}{10}i$ (c) $-\frac{1}{34} - \frac{2}{17}i$

4. Answers will vary.

6. (a) $0 < k < 4$ (b) $k < 0$ (c) $k > 4$

8. $f(x) = -2x^3 + 3x^2 + 11x - 6$

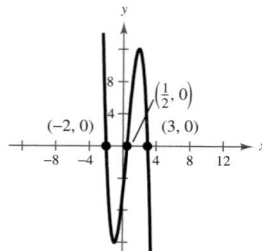

(Equations and graphs will vary.) There are infinitely many possible functions for f.

10. (a) $-2, 0, 2$

(b) The graph touches the x-axis at $x = 0$.

(c) The least possible degree of the function is 4, because there are four real zeros (0 is repeated) and a function can have at most the number of real zeros equal to the degree of the function. The degree cannot be odd because there is an even number of real zeros and complex solutions occur in conjugate pairs, so the degree must be even.

(d) Negative. From the information in the table, it can be concluded that the graph will eventually fall to the left and fall to the right.

(e) $f(x) = -x^4 + 4x^2$

(f)

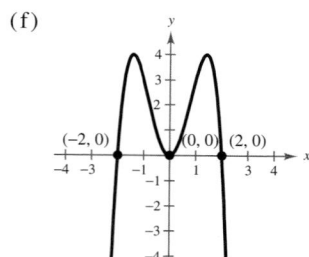

12. (a) $1 + i, 3 + i$

(b) $1 - i, 2 + 3i$

(c) $1 + i, -\frac{7}{2} + 3i$

(d) $4 + 5i, -\frac{1}{3} - \frac{1}{3}i$

14. Answers will vary.

Chapter 5

Section 5.1   *(page 372)*

2. 3.488   **4.** 1767.767   **6.** 24.533

7. d   **8.** c   **9.** a   **10.** b

12. Shift the graph of f one unit upward.

14. Reflect the graph of f in the y-axis and shift three units to the right.

16. Reflect the graph of f in the x-axis and y-axis and shift six units to the right.

18. Reflect the graph of f in the x-axis and y-axis and shift eight units upward.

20.

x	-2	-1	0	1	2
$f(x)$	0.25	0.5	1	2	4

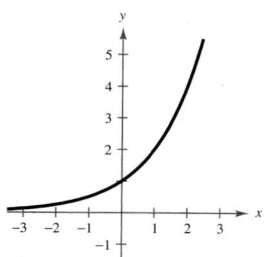

22.

x	-2	-1	0	1	2
$f(x)$	0.028	0.167	1	6	36

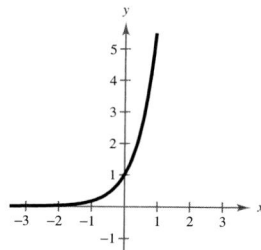

24.

x	-4	-3	-2	-1	0
$f(x)$	0.111	0.333	1	3	9

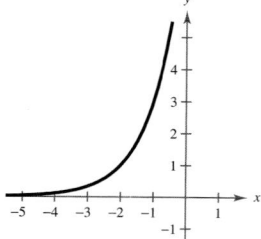

26.

x	-2	-1	0	1	2
$f(x)$	7.389	2.718	1	0.368	0.135

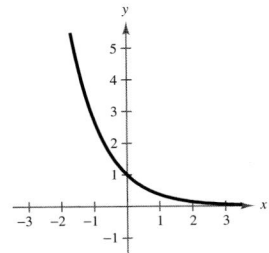

28.

x	-2	-1	0	1	2
$f(x)$	5.437	3.297	2	1.213	0.736

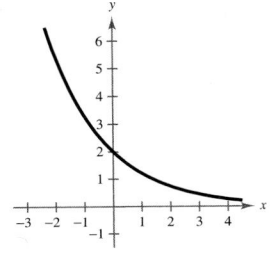

30.

x	0	2	4	5	6
$f(x)$	2.007	2.050	2.368	3	4.718

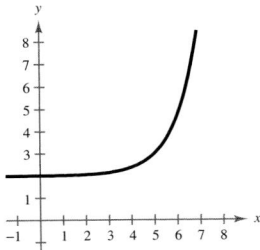

32.

x	-1	0	1	3	4
$f(x)$	-3.004	-3.016	-3.063	-4	-7

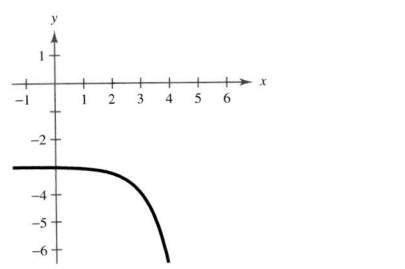

34. 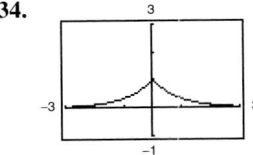　　**36.**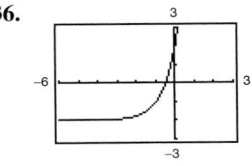

38. 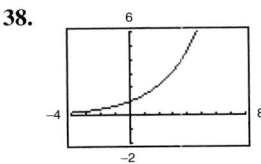　　**40.**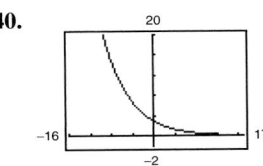

42.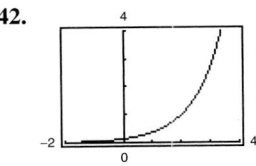

44.

n	1	2	4	12
A	\$1790.85	\$1806.11	\$1814.02	\$1819.40

n	365	Continuous
A	\$1822.03	\$1822.12

46.

n	1	2	4
A	\$10,285.72	\$10,640.89	\$10,828.46

n	12	365	Continuous
A	\$10,957.45	\$11,021.00	\$11,023.18

48.

t	10	20
A	\$21,865.43	\$39,841.40

t	30	40	50
A	\$72,595.77	\$132,278.12	\$241,026.44

50.

t	10	20
A	\$25,404.00	\$53,780.27

t	30	40	50
A	\$113,852.83	\$241,026.44	\$510,252.98

52. \$212,605.41

CHAPTER 5

54. (a)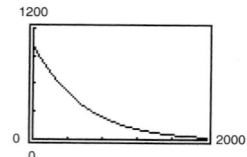

(b) $421.12

(c) $350

56. (a) 3351   (b) 4492

58. (a) 10 grams   (b) 7.85 grams

(c)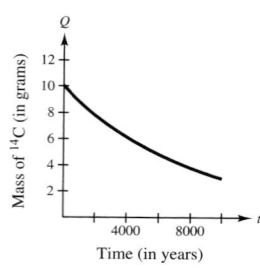

60. (a)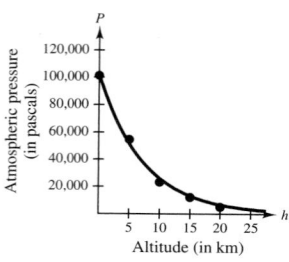

(b) 34,190 pascals

62. False. e is an irrational number.

64. $g(x) = h(x)$ **66.** None are equal.

68. (a) $x > 0$ (b) $x < 0$

70. (a)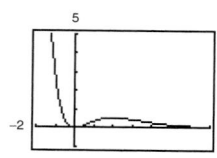

Decreasing: $(-\infty, 0), (2, \infty)$

Increasing: $(0, 2)$

Relative maximum: $(2, 4e^{-2})$

Relative minimum: $(0, 0)$

(b)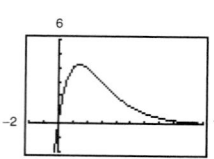

Decreasing: $(1.44, \infty)$

Increasing: $(-\infty, 1.44)$

Relative maximum: $(1.44, 4.25)$

72. The value of $f(x)$ approaches e^r.

x	1	10	100	200	500
$[1 + (1/x)]^x$	2	2.5937	2.7048	2.7115	2.7156

x	1000	10,000
$[1 + (1/x)]^x$	2.7170	2.718

74. $1 < \sqrt{2} < 2$ **76.** $y = \frac{1}{3}(4 - x^2)$

$2^1 < 2^{\sqrt{2}} < 2^2$

78. $y = x - 2$ and $y = -(x - 2)$, $x \geq 2$

Section 5.2   *(page 382)*

2. $3^4 = 81$ **4.** $10^{-3} = \frac{1}{1000}$ **6.** $16^{3/4} = 8$

8. $e^{1.386\cdots} = 4$ **10.** $\log_8 64 = 2$ **12.** $\log_9 27 = \frac{3}{2}$

14. $\log_{10} 0.001 = -3$ **16.** $\ln 1.6487\ldots = \frac{1}{2}$

18. $\log_u w = v$ **20.** $\frac{1}{2}$ **22.** 1   **24.** -2

26. -3 **28.** 1.097   **30.** -3.418 **32.** 0.693

33. c   **34.** f   **35.** d   **36.** e   **37.** b   **38.** a

40. Domain: $(0, \infty)$

x-intercept: $(1, 0)$

Vertical asymptote: $x = 0$

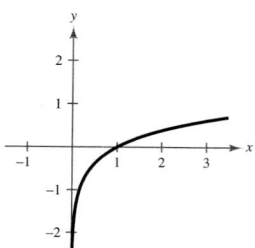

42. Domain: $(3, \infty)$

x-intercept: $(4, 0)$

Vertical asymptote: $x = 3$

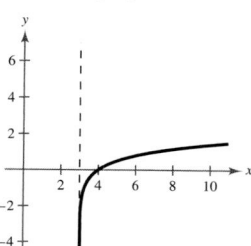

44. Domain: $(1, \infty)$

x-intercept: $\left(\frac{626}{625}, 0\right)$

Vertical asymptote: $x = 1$

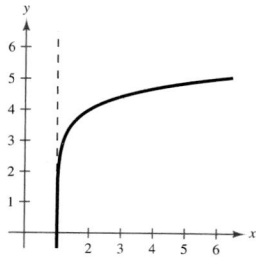

46. Domain: $(-\infty, 0)$

x-intercept: $(-1, 0)$

Vertical asymptote: $x = 0$

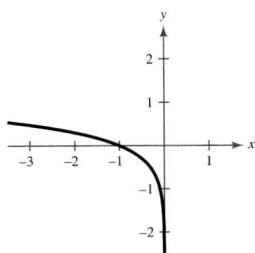

48. Domain: $(-1, \infty)$

x-intercept: $(0, 0)$

Vertical asymptote: $x = -1$

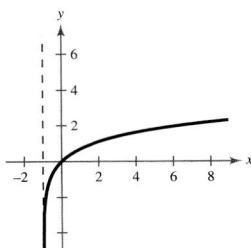

50. Domain: $(-\infty, 3)$

x-intercept: $(2, 0)$

Vertical asymptote: $x = 3$

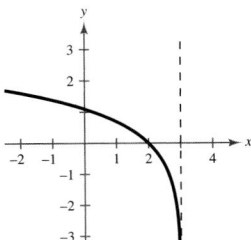

52. 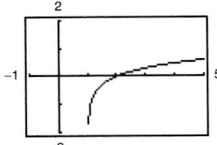 **54.**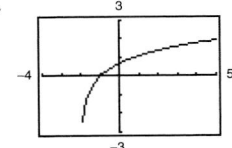

56.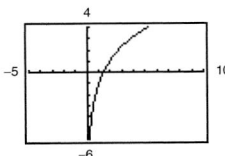

58. (a)

K	1	2	4	6	8	10	12
t	0	7.3	14.6	18.9	21.9	24.2	26.2

The number of years required to multiply the original investment by K increases with K. However, the larger the value of K, the fewer the years required to increase the value of the investment by an additional multiple of the original investment.

(b)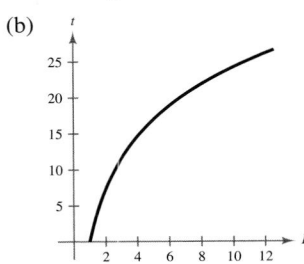

60. (a) 120 decibels

(b) 100 decibels

(c) No, the difference results from the logarithmic relationship between intensity and number of decibels.

62. (a)

x	1	5	10	10^2
$f(x)$	0	0.322	0.230	0.046

x	10^4	10^6
$f(x)$	0.00092	0.0000138

(b) 0

(c)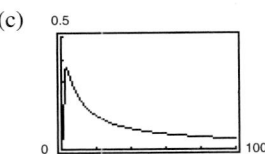

64. True. $\log_3 27 = 3 \implies 3^3 = 27$

66.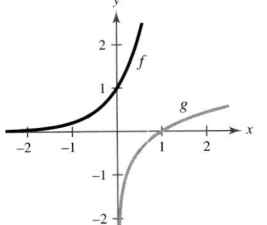

$g = f^{-1}$

68.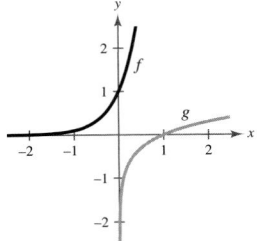

$g = f^{-1}$

70. (a) False   (b) True   (c) True   (d) False

72. $f^{-1}(x) = 10^x$ **74.** Answers will vary.

76. (a)

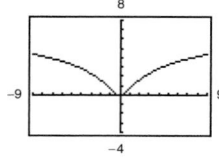

(b) Increasing: $(0, \infty)$

Decreasing: $(-\infty, 0)$

(c) Relative minimum: $(0, 0)$

78. $w^2 + 10w$

Section 5.3   *(page 389)*

2. (a) $\dfrac{\log_{10} x}{\log_{10} 3}$ (b) $\dfrac{\ln x}{\ln 3}$ **4.** (a) $\dfrac{\log_{10} x}{\log_{10} \frac{1}{3}}$ (b) $\dfrac{\ln x}{\ln \frac{1}{3}}$

6. (a) $\dfrac{\log_{10} \frac{3}{4}}{\log_{10} x}$ (b) $\dfrac{\ln \frac{3}{4}}{\ln x}$ **8.** (a) $\dfrac{\log_{10} x}{\log_{10} 7.1}$ (b) $\dfrac{\ln x}{\ln 7.1}$

10. 0.712   **12.** -1.161 **14.** -0.694 **16.** -3.823

18. $\log_3 10 + \log_3 z$ **20.** $\log_{10} y - \log_{10} 2$

22. $-3 \log_6 z$ **24.** $\frac{1}{3} \ln t$

26. $\log_{10} 4 + 2 \log_{10} x + \log_{10} y$

28. $\ln(x + 1) + \ln(x - 1) - 3 \ln x$

30. $\ln 6 - \frac{1}{2} \ln(x^2 + 1)$ **32.** $\frac{1}{2}(2 \ln x - 3 \ln y)$

34. $\frac{1}{2} \log_2 x + 4 \log_2 y - 4 \log_2 z$

36. $\log_{10} x + 4 \log_{10} y - 5 \log_{10} z$

38. $\ln x + \dfrac{1}{2} \ln(x + 2)$ **40.** $\ln yt$ **42.** $\log_5 \dfrac{8}{t}$

44. $\log_7(z - 2)^{2/3}$ **46.** $\log_6 \dfrac{1}{16x^4}$ **48.** $\ln 64(z - 4)^5$

50. $\log_3 \dfrac{x^3 y^4}{z^4}$ **52.** $\ln \dfrac{z^4(z + 5)^4}{(z - 5)^2}$ **54.** $\ln\left(\dfrac{x^3}{x^2 - 1}\right)^2$

56. $\log_4 \left[x^6(x - 1)\sqrt{x + 1}\right]$

58. $\log_7 \sqrt{70} = \frac{1}{2}(\log_7 7 + \log_7 10)$

$\qquad = \frac{1}{2} + \log_7 \sqrt{10}$; Properties 1 and 3

60. -3 **62.** $\frac{1}{3}$ **64.** -0.8

66. -16 is not in the domain of $\log_2 x$.

68. 12   **70.** $\frac{3}{4}$ **72.** 7   **74.** 3   **76.** $4 + 4 \log_2 3$

78. $\log_{10} 3 - 2$ **80.** $\ln 6 - 2$

82. $\beta = 10(\log_{10} I + 12)$; 60 decibels

84. True; Property 1   **86.** False. $f(\sqrt{x}) = \frac{1}{2} f(x)$

88. True   **90.** Answers will vary.

92. $f(x) = \dfrac{\log_{10} x}{\log_{10} 4} = \dfrac{\ln x}{\ln 4}$ **94.** $f(x) = \dfrac{\log_{10} x}{\log_{10} \frac{1}{4}} = \dfrac{\ln x}{\ln \frac{1}{4}}$

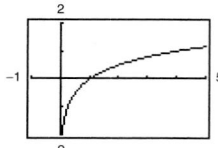 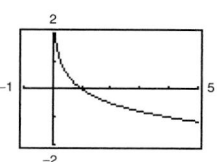

96. $f(x) = \dfrac{\log_{10} x}{\log_{10} 12.4} = \dfrac{\ln x}{\ln 12.4}$

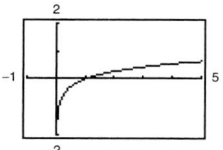

98. $\ln 2 \approx 0.6931$ $\ln 10 \approx 2.3025$

$\ln 3 \approx 1.0986$ $\ln 12 \approx 2.4848$

$\ln 4 \approx 1.3862$ $\ln 15 \approx 2.7080$

$\ln 5 \approx 1.6094$ $\ln 16 \approx 2.7724$

$\ln 6 \approx 1.7917$ $\ln 18 \approx 2.8903$

$\ln 8 \approx 2.0793$ $\ln 20 \approx 2.9956$

$\ln 9 \approx 2.1972$

100. $\dfrac{27y^3}{8x^6}$ **102.** $\dfrac{(xy)^2}{x + y}$ **104.** $1, \dfrac{1}{4}$ **106.** $\dfrac{1 \pm \sqrt{31}}{2}$

Section 5.4   *(page 398)*

2. (a) No   (b) No

4. (a) Yes   (b) Yes, approximate   (c) No

6. (a) Yes   (b) Yes, approximate   (c) No

8. 5   **10.** 6   **12.** $\frac{2}{3}$ **14.** -3

16. 2   **18.** 5   **20.** $\ln 4 \approx 1.386$

22. $e^{-7} \approx 0.000912$ **24.** 0.008   **26.** 100

28. $\left(\dfrac{2}{3}, 9\right)$ **30.** $(5, 0)$ **32.** $\dfrac{\ln 16}{\ln 5} \approx 1.723$

34. $\ln \dfrac{91}{4} \approx 3.125$ **36.** $\dfrac{\ln 37}{\ln 6} \approx 2.015$

38. $\dfrac{\ln 3000}{5 \ln 6} \approx 0.894$ **40.** $-\dfrac{\ln(0.10)}{3 \ln 4} \approx 0.554$ **42.** 8

44. $\dfrac{-\ln 64 - \ln 431}{\ln 8} \approx -4.917$

46. $6 + \log_{10} \frac{7}{5} \approx 6.146$ **48.** $6 - \dfrac{\ln 5}{\ln 3} \approx 4.535$

50. $\dfrac{\ln 50}{2} \approx 1.956$　**52.** $-\dfrac{1}{4}\ln\dfrac{3}{40} \approx 0.648$

54. $\ln\dfrac{25}{3} \approx 2.120$　**56.** $3 - \dfrac{\ln\frac{7}{2}}{2\ln 4} \approx 2.548$

58. $\ln 2 \approx 0.693$; $\ln 3 \approx 1.099$　**60.** No solution

62. $\ln 7 \approx 1.946$　**64.** $\dfrac{\ln 31}{6} \approx 0.572$

66. $\dfrac{\ln 21}{9\ln 3.938225} \approx 0.247$　**68.** $\dfrac{\ln 30}{3\ln\left(16 - \frac{0.878}{26}\right)} \approx 0.409$

70.
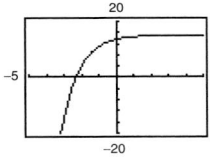
-2.322

72.
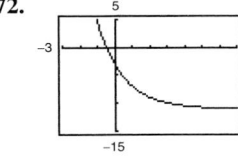
-0.478

74.
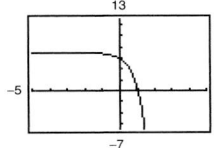
1.081

76.
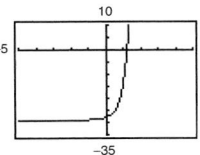
1.236

78. $e^2 \approx 7.389$　**80.** $\dfrac{e}{4} \approx 0.680$　**82.** $\dfrac{100}{3} \approx 33.333$

84. $10^{11/5} + 2 \approx 160.489$　**86.** $e^{7/2} \approx 33.115$

88. $e^{10} + 8 = 22{,}034.466$　**90.** $e^{-4/3} \approx 0.264$

92. $\dfrac{-1 + \sqrt{1 + 4e}}{2} \approx 1.223$　**94.** $\dfrac{-3 + \sqrt{9 + 4e}}{2} \approx 0.729$

96. $\dfrac{3 + \sqrt{13}}{2} \approx 3.303$　**98.** no solution

100. 2　**102.** 9

104. $\dfrac{1225 + 125\sqrt{73}}{2} \approx 1146.500$

106.
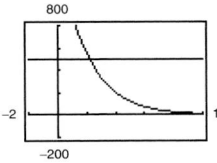
2.197

108.
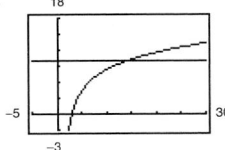
14.182

110. 5.8 years　**112.** 9.2 years

114. (a) 303 units　(b) 528 units

116. 12.76 inches

118. (a)
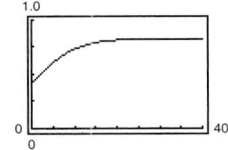

(b) Horizontal asymptotes: $P = 0$, $P = 0.83$

The proportion of correct responses will approach 0.83 as the number of trials increases.

(c) ≈ 5 trials

120. (a) $T = 20$; Room temperature　(b) ≈ 0.81 hour

122. $\log_b(u + v) = (\log_b u)(\log_b v)$

False.

$2.04 \approx \log_{10}(10 + 100) \neq (\log_{10}10)(\log_{10}100) = 2$

124. $\log_b \dfrac{u}{v} = \log_b u - \log_b v$

True by Property 2 in Section 5.3.

126. For $rt < \ln 2$ years, double the amount you invest. For $rt > \ln 2$ years, double your interest rate or double the number of years, because either of these will double the exponent in the exponential function.

128. Answers will vary.

130. $4\sqrt{2} - 10$　**132.** $\dfrac{1}{2}\sqrt{10} + 1$　**134.** $t = \dfrac{k}{s^3}$

136. $x = \dfrac{k}{b - 3}$　**138.** 1.262　**140.** 1.486

Section 5.5　(page 409)

1. c　**2.** e　**3.** b　**4.** a　**5.** d　**6.** f

Initial Investment	Annual % Rate	Time to Double	Amount After 10 Years
8. $20,000	10.5%	6.60 yr	$57,153.02
10. $10,000	5.7762%	12 yr	$17,817.97
12. $600	34.66%	2 yr	$19,205.00
14. $8986.58	8%	8.66 yr	$20,000.00

16. $4214.16

18. (a) 6.94 years　(b) 6.63 years

(c) 6.602 years　(d) 6.601 years

20.
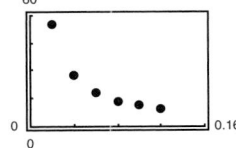

Use PwrReg: $t = 1.099r^{-1}$

22.

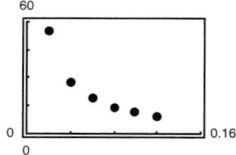

Use PwrReg: $t = 1.222r^{-1}$

24.

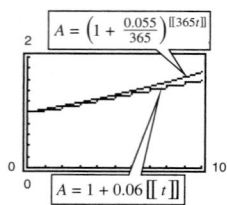

Daily compounding

Isotope	Half-life (years)	Initial Quantity	Amount After 1000 Years
26. ^{226}Ra	1620	2.30 g	1.5 g
28. ^{14}C	5730	3 g	2.66 g
30. ^{239}Pu	24,360	0.41 g	0.4 g

32. $y = \frac{1}{2}e^{0.5756x}$ **34.** $y = e^{-0.4621x}$

36. (a) Canada: $y = 31.3e^{0.00915t}$; 41.2 million

China: $y = 1261.8e^{0.00743t}$; 1576.9 million

Italy: $y = 57.6e^{-0.00035t}$; 57.0 million

United Kingdom: $y = 59.5e^{0.00183t}$; 62.9 million

United States: $y = 275.6e^{0.00852t}$; 355.9 million

(b) b; The greater the rate of growth, the greater the value of b.

(c) b determines whether the population is increasing ($b > 0$) or decreasing ($b < 0$).

38. $k = 0.0132$; 291,677   **40.** 61.16 hours

42. 15,683 years

44. (a) $V = -750t + 2000$ (b) $V = 2000e^{-0.6931t}$

(c)

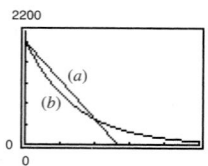

Exponential

(d) 1 year:
Straight-line, $1250; Exponential, $1000
3 years:
Straight-line, $-$250; Exponential, $250

(e) The value decreases $750 per year.

46. (a) $k \approx .053$ (b) 267,450 units

48. $496,880

50. (a) 203 animals

(b) 13 months

(c)

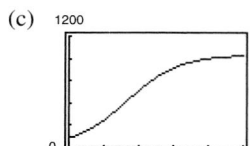

Horizontal asymptotes: $y = 0$, $y = 1000$. The population size will approach 1000 as time increases.

52. (a) 158,489,319   (b) 5,011,872   (c) 50,118,723

54. (a) 30 decibels   (b) 85 decibels

(c) 90 decibels   (d) 115 decibels

56. 97%   **58.** 4.95

60. $10^{-3.2} \approx 6.3 \times 10^{-4}$ moles per liter   **62.** 10

64. (a)

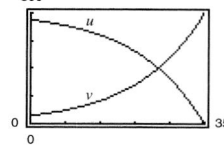

(b) Interest; $t \approx 26$ years

(c)

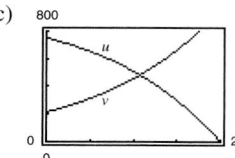

Interest; $t \approx 11$ years; The interest is still the majority of the monthly payment in the early years, but now the principal and interest are nearly equal when $t \approx 11$ years.

66. (a) $t_3 = 0.2729s - 6.0143$

$t_4 = 1.5385e^{0.02913s}$ or $t_4 = 1.5385(1.0296)^s$

(b)

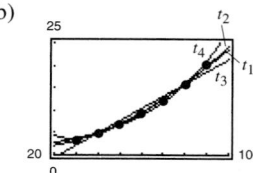

(c)

s	30	40	50	60	70	80	90
t_1	3.6	4.6	6.7	9.4	12.5	15.9	19.6
t_2	3.3	4.9	7.0	9.5	12.5	15.9	19.9
t_3	2.2	4.9	7.6	10.4	13.1	15.8	18.5
t_4	3.7	4.9	6.6	8.9	11.9	15.9	21.2

(d) Model t_1: Sum = 2.0

Model t_2: Sum = 1.1

Model t_3: Sum = 5.6

Model t_4: Sum = 2.6

The quadratic (model t_2) fits best.

68. False. A logistic growth function never has an x-intercept.

70. True. The graph of a Gaussian model will never have an x-intercept.

72. Answers will vary.

74.

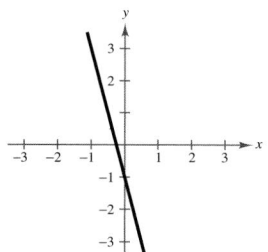

76.

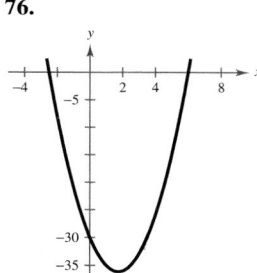

78.

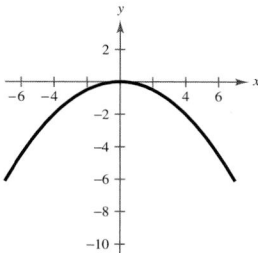

80.

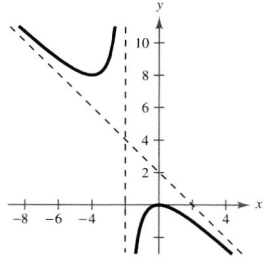

82.

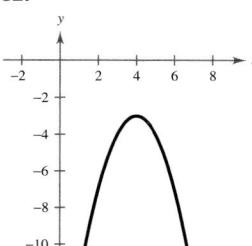

84.

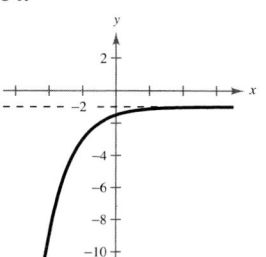

86.

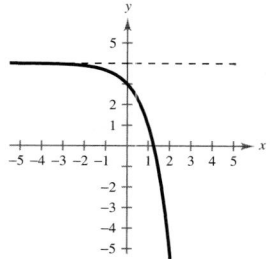

Review Exercises   *(page 416)*

2. -3.863 **4.** 4.181   **6.** 0.002   **7.** c   **8.** d

9. a   **10.** b   **12.** Shift the graph of f three units downward.

14. Reflect f in the x-axis and shift f eight units upward.

16.

x	-2	-1	0	1	2
$f(x)$	-3.063	-3.25	-4	-7	-19

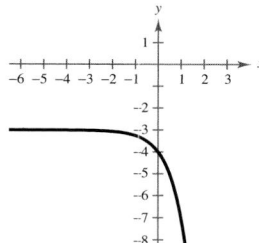

18.

x	-3	-1	0	1	3
$f(x)$	0.020	0.142	0.377	1	7.023

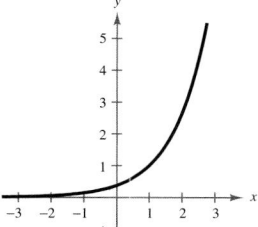

20.

x	0	5	6	7	8	9
$f(x)$	-4.984	-4.5	-4	-3	-1	3

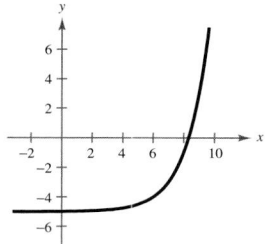

CHAPTER 5

22.

x	-3	-2	-1	0	2
$f(x)$	3	-4	-4.875	-4.984	-5

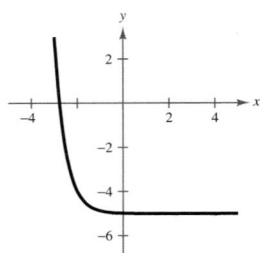

24. 1.868   **26.** 1.320

28.

x	-2	-1	0	1	2
$h(x)$	-0.72	0.35	1	1.39	1.63

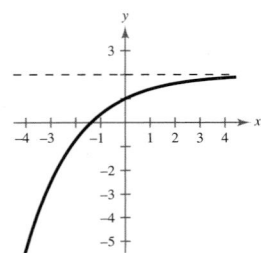

30.

t	$\frac{1}{2}$	1	2	3	4
$s(t)$	0.07	0.54	1.47	2.05	2.43

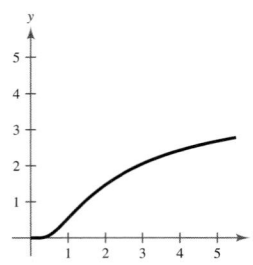

32.

n	1	2	4	12
A	\$8643.88	\$8799.58	\$8880.43	\$8935.49

n	365	Continuous
A	\$8962.46	\$8963.38

34. (a)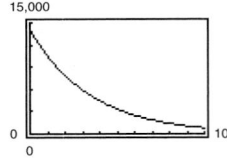

(b) \$7875   (c) At the beginning. Yes

36. (a) 100 grams   (b) 58.7 grams

(c)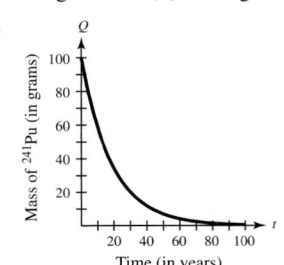

38. $\log_{25} 125 = \frac{3}{2}$ **40.** $\frac{1}{2}$ **42.** -1

44. Domain: $(0, \infty)$

x-intercept: $(1, 0)$

Vertical asymptote: $x = 0$

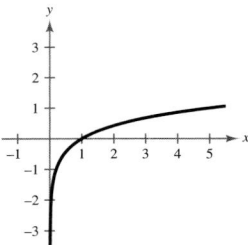

46. Domain: $(0, \infty)$

x-intercept: $(10^{-6}, 0)$

Vertical asymptote: $x = 0$

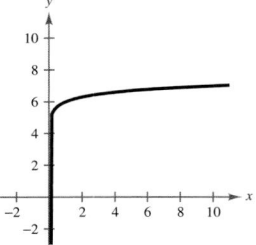

48. Domain: $(3, \infty)$

x-intercept: $(3.1, 0)$

Vertical asymptote: $x = 3$

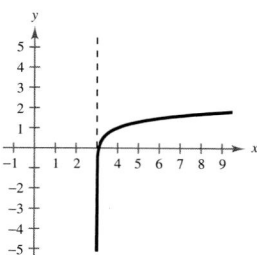

50. -0.020 **52.** 7 **54.** -1.530

56. Domain: $(3, \infty)$

x-intercept: $(4, 0)$

Vertical asymptote: $x = 3$

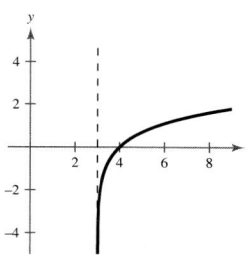

58. Domain: $(0, \infty)$

x-intercept: $(1, 0)$

Vertical asymptote: $x = 0$

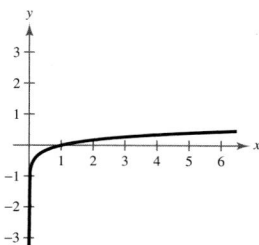

60 27.16 miles **62.** 2.132 **64.** -1.159

66. $\log_{10} 7 + 4 \log_{10} |x|$ **68.** $\frac{1}{2} \log_7 x - \log_7 4$

70. $\ln 3 + \ln x + 2 \ln y$ **72.** $2 \ln(y - 1) - \ln 16$

74. $\log_6 \dfrac{y}{z^2}$ **76.** $\ln x^3(x + 1)^2$

78. $\log_{10} \dfrac{1}{x^2(x + 6)^5}$ **80.** $\ln \dfrac{(x - 2)^5}{x^3(x + 2)}$

82. The score decreased 10 points over the four-month period.

84. 6 **86.** -2 **88.** $\ln 6$ **90.** $\frac{1}{6}$ **92.** e^{-3}

94. $\dfrac{\ln 25}{3} \approx 1.073$ **96.** $\frac{1}{3}(\ln 40 - 2) \approx 0.563$

98. $\dfrac{\ln 20}{\ln 6} \approx 1.672$ **100.** $\dfrac{\ln 95}{\ln 12} \approx 1.833$

102. $\ln 2 \approx 0.693$, $\ln 4 \approx 1.386$

104.

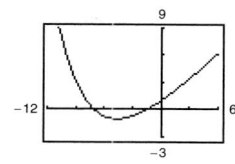

106.

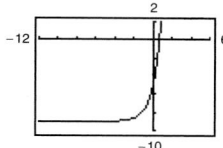

$-1.53, -7.04$ 0.68

108. $\frac{1}{5}e^{7.2} \approx 267.886$ **110.** $\frac{1}{3}e^{15/4} \approx 14.174$

112. $e^6 - 8 \approx 395.429$ **114.** $5e^4 \approx 272.991$

116. $-2 + \sqrt{6} \approx 0.449$ **118.** -104

120.

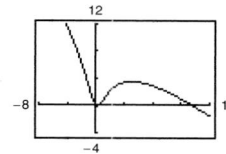

122.

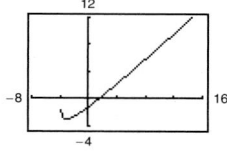

$0, 0.42, 13.63$ $-3.99, 1.48$

124. (a) 1151 units (b) 1325 units **125.** e **126.** b

127. f **128.** d **129.** a **130.** c **132.** 98.6%

134. 9.8 hours **136.** $y = \frac{1}{2}e^{0.4605x}$

138. (a) 7.7 weeks (b) 13.3 weeks

140. (a) 251,188,643 (b) 7,079,458 (c) 1,258,925,412

142. False. $\ln x + \ln y = \ln(xy) \neq \ln(x + y)$

Problem Solving *(page 422)*

2.

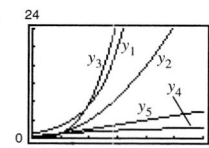

y_1 increases at the fastest rate.

4. Answers will vary.

6. Answers will vary.

8. $y_4 = 1 + \dfrac{x}{1!} + \dfrac{x^2}{2!} + \dfrac{x^3}{3!} + \dfrac{x^4}{4!}$

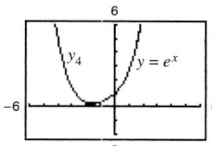

The pattern implies that

$$e^x = 1 + \frac{x}{1!} + \frac{x^2}{2!} + \frac{x^3}{3!} + \cdots$$

CHAPTER 5

10. $f^{-1}(x) = \dfrac{\ln\left(\dfrac{x+1}{x-1}\right)}{\ln a}$

12. (a) The upper graph represents the investment compounded annually and the lower graph represents simple interest.

(b) $y_1 = 500(1.07)^t$

$y_2 = 35t + 500$

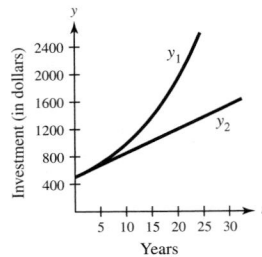

14. $B = 500a^{\left(\frac{\ln 2/5}{\ln a}\right)t}$ **16.** Answers will vary.

18. (a)

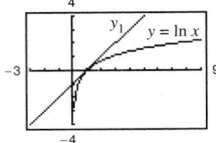

(b)

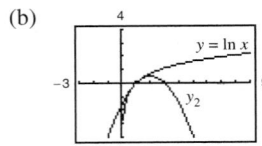

(c)

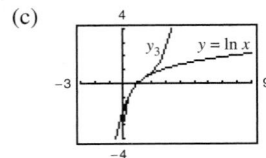

20. Slope $= \ln b$; y-intercept: $(0, \ln a)$

Slope $= b$; y-intercept: $(0, \ln a)$

22. (a) 15 cubic feet per minute

(b) 382.0 cubic feet

(c) 382.0 square feet

Chapter 6

Section 6.1   *(page 430)*

2. 1   **4.** $-\sqrt{3}$ **6.** $-\dfrac{\sqrt{3}}{3}$ **8.** -0.2677

10. 2.0344 radians, 116.6°   **12.** 1.1071 radians, 63.4°

14. 1.9513 radians, 111.8°   **16.** 0.6023 radian, 34.5°

18. 2.0344 radians, 116.6°   **20.** 2.4669 radians, 141.3°

22. $\dfrac{\pi}{4}$ radian, 45°   **24.** $\dfrac{\pi}{4}$ radian, 45°

26. 1.1071 radians, 63.4°   **28.** 1.4109 radians, 80.8°

30. 1.0808 radians, 61.9°   **32.** 1.0240 radians, 58.7°

34. $(-3, 2)$: 35.8°; $(1, 3)$: 94.4°; $(2, 0)$: 49.8°

36. $(-3, 4)$: 32.5°; $(2, 1)$: 16.9°; $(-2, 2)$: 130.6°

38. $\dfrac{4\sqrt{5}}{5} \approx 1.7889$ **40.** $\dfrac{5\sqrt{2}}{2} \approx 3.5355$ **42.** 4

44. $9\sqrt{2} \approx 12.7279$ **46.** (a) $\dfrac{33\sqrt{29}}{29}$ (b) $\dfrac{31}{2}$

48. (a) $\dfrac{31\sqrt{389}}{389}$ (b) $\dfrac{31}{2}$ **50.** $\dfrac{9}{5}$

52. 0.2027, 1049 feet   **54.** 31.0°

56. (a) 0.6167 radian, 35.3°   (b) 518.5 feet

(c) $y = 0.709x$

(d)

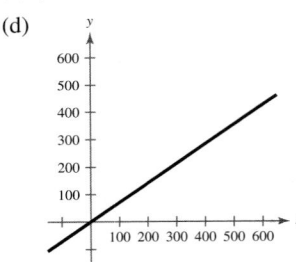

58. False. Substitute $\tan \theta_1$ and $\tan \theta_2$ for m_1 and m_2 in the formula for the angle between two lines.

60. (a) $d = \dfrac{3|m+1|}{\sqrt{m^2+1}}$

(b)

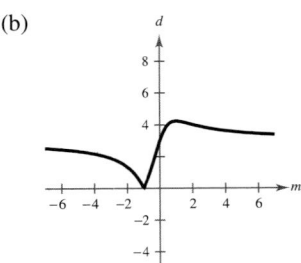

(c) $m = 1$ (d) Yes. $m = -1$

(e) $d = 3$. As the line approaches the vertical, the distance approaches 3.

62. x-intercept: $(-9, 0)$

y-intercept: $(0, 81)$

64. No x-intercepts

y-intercept: $(0, 133)$

66. x-intercepts: $(-11, 0), (2, 0)$

y-intercept: $(0, -22)$

68. $f(x) = 2\left(x - \frac{1}{4}\right)^2 - \frac{169}{8}$

Vertex: $\left(\frac{1}{4}, -\frac{169}{8}\right)$

70. $f(x) = -(x + 4)^2 + 1$

Vertex: $(-4, 1)$

72. $f(x) = -8\left(x + \frac{17}{8}\right)^2 + \frac{121}{8}$

Vertex: $\left(-\frac{17}{8}, \frac{121}{8}\right)$

74.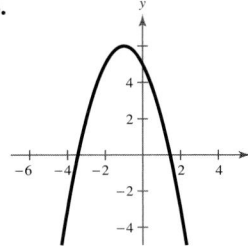

76.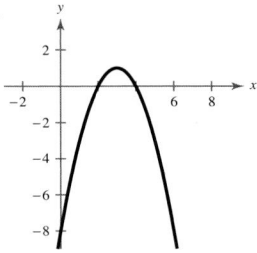

Section 6.2   *(page 438)*

2. An ellipse is formed when a plane intersects only the top or bottom half of a double-napped cone but is not parallel or perpendicular to the axis of the cone, is not parallel to the side of the cone, and does not intersect the vertex.

4. A hyperbola is formed when a plane intersects both halves of a double-napped cone, is parallel to the axis of the cone, and does not intersect the vertex.

5. e   **6.** b   **7.** d   **8.** f   **9.** a   **10.** c

12. Vertex: $(0, 0)$

Focus: $\left(0, -\frac{1}{8}\right)$

Directrix: $y = \frac{1}{8}$

14. Vertex: $(0, 0)$

Focus: $\left(\frac{3}{4}, 0\right)$

Directrix: $x = -\frac{3}{4}$

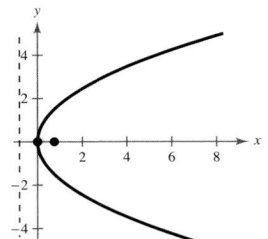

16. Vertex: $(0, 0)$

Focus: $\left(-\frac{1}{4}, 0\right)$

Directrix: $x = \frac{1}{4}$

18. Vertex: $(-5, 1)$

Focus: $\left(-\frac{21}{4}, 1\right)$

Directrix: $x = -\frac{19}{4}$

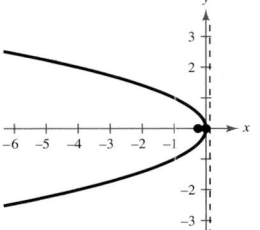
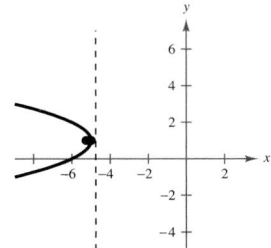

20. Vertex: $\left(-\frac{1}{2}, 1\right)$

Focus: $\left(-\frac{1}{2}, 2\right)$

Directrix: $y = 0$

22. Vertex: $(8, -1)$

Focus: $(9, -1)$

Directrix: $x = 7$

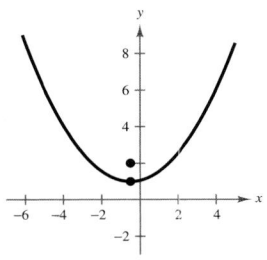

24. Vertex: $(-1, 2)$

Focus: $(0, 2)$

Directrix: $x = -2$

26. Vertex: $(1, -1)$

Focus: $(1, -3)$

Directrix: $y = 1$

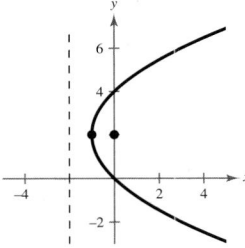
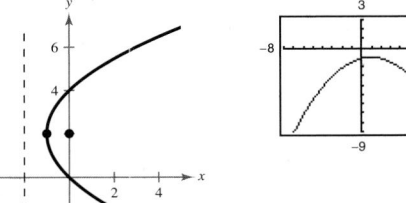

28. Vertex: $(-1, 0)$

Focus: $(0, 0)$

Directrix: $x = -2$

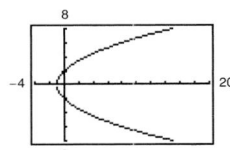

CHAPTER 6

30. 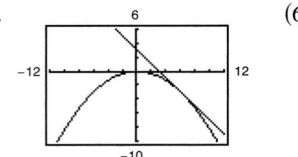 $(6, -3)$

32. $y^2 = -18x$ **34.** $y^2 = 8x$ **36.** $x^2 = -8y$

38. $x^2 = -12y$ **40.** $y^2 = 12x$ **42.** $x^2 = -3y$

44. $(y - 3)^2 = -2(x - 5)$ **46.** $(x - 3)^2 = 3(y + 3)$

48. $(x + 1)^2 = -8(y - 2)$ **50.** $(y - 1)^2 = -12(x + 2)$

52. $x^2 = -16(y - 4)$ **54.** $y = -\sqrt{2(x - 4)} - 1$

56. $6x + 2y + 9 = 0; \left(-\frac{3}{2}, 0\right)$

58. $8x + y - 8 = 0; (1, 0)$

60. 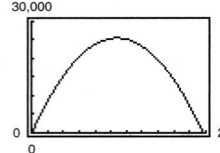 $x = 135$

62. (a)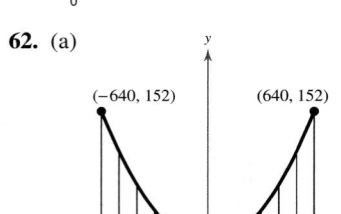

(b) $y = \dfrac{19x^2}{51,200}$

(c)

Distance, x	0	250	400	500	1000
Height, y	0	23.19	59.38	92.77	371.09

64. $y^2 = 640x$

66. (a)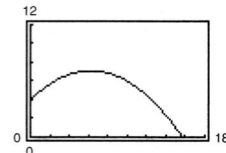

(b) Highest point: $(6.25, 7.125)$

Range: 15.69 feet

68. 41,329.5 feet

70. True. The directrix of a parabola is perpendicular to the axis of the parabola.

72. (a) $\dfrac{64\sqrt{2}}{3} \approx 30.17$

(b) As p approaches zero, the parabola becomes narrower and narrower, thus the area becomes smaller and smaller.

74. Answers will vary. **76.** $\frac{3}{2}, \pm 5i$

78.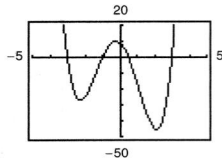

$\frac{1}{2}, -1, \pm 3$

80. $A \approx 96.37°, C \approx 29.63°, a \approx 22.11$

82. $A = 50°, b \approx 10.87, c \approx 24.07$

84. $A \approx 43.53°, B \approx 19.42°, C \approx 117.05°$

86. $A \approx 41.85°, C \approx 67.15°, b \approx 29.76$

Section 6.3 *(page 448)*

1. b **2.** c **3.** d **4.** f **5.** a **6.** e

8. Center: $(0, 0)$
Vertices: $(0, \pm 12)$
Foci: $\left(0, \pm 3\sqrt{7}\right)$
Eccentricity: $\dfrac{\sqrt{7}}{4}$

10. Center: $(0, 0)$
Vertices: $(\pm 8, 0)$
Foci: $(\pm 6, 0)$
Eccentricity: $\dfrac{3}{4}$

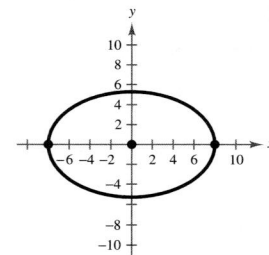

12. Center: $(4, -3)$
Vertices: $(4, 1), (4, -7)$
Foci: $(4, -1), (4, -5)$
Eccentricity: $\frac{1}{2}$

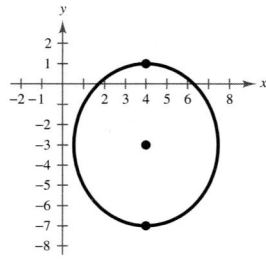

14. Center: $(-2, -4)$
Vertices: $(-3, -4), (-1, -4)$
Foci: $\left(\dfrac{-4 \pm \sqrt{3}}{2}, -4\right)$
Eccentricity: $\dfrac{\sqrt{3}}{2}$

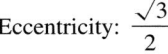
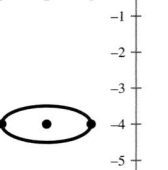

16. Center: $(3, -5)$

Vertices: $(3, 1), (3, -11)$

Foci: $\left(3, -5 \pm 2\sqrt{5}\right)$

Eccentricity: $\dfrac{\sqrt{5}}{3}$

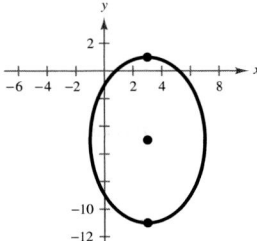

18. Center: $(-3, 1)$

Vertices: $(-3, 7), (-3, -5)$

Foci: $\left(-3, 1 \pm 2\sqrt{6}\right)$

Eccentricity: $\dfrac{\sqrt{6}}{3}$

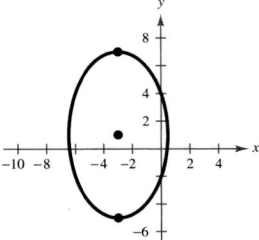

20. Center: $\left(3, -\dfrac{5}{2}\right)$

Vertices: $\left(9, -\dfrac{5}{2}\right), \left(-3, -\dfrac{5}{2}\right)$

Foci: $\left(3 \pm 3\sqrt{3}, -\dfrac{5}{2}\right)$

Eccentricity: $\dfrac{\sqrt{3}}{2}$

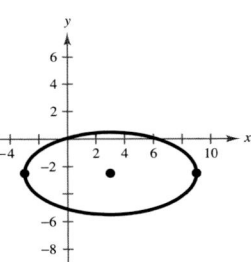

22. Center: $(2, 1)$

Vertices: $\left(\dfrac{7}{3}, 1\right), \left(\dfrac{5}{3}, 1\right)$

Foci: $\left(\dfrac{34}{15}, 1\right), \left(\dfrac{26}{15}, 1\right)$

Eccentricity: $\dfrac{4}{5}$

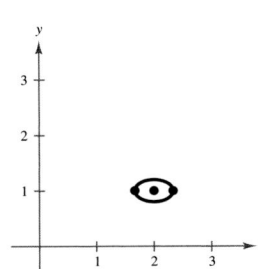

24.

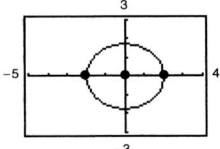

Center: $(0, 0)$

Vertices: $(\pm 2, 0)$

Foci: $(\pm 1, 0)$

26.

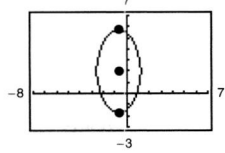

Center: $\left(-\dfrac{2}{3}, 2\right)$

Vertices: $\left(-\dfrac{2}{3}, 2 \pm \dfrac{2\sqrt{31}}{3}\right)$

Foci: $\left(-\dfrac{2}{3}, 2 \pm \dfrac{\sqrt{93}}{3}\right)$

28. $\dfrac{x^2}{4} + \dfrac{4y^2}{9} = 1$ **30.** $\dfrac{x^2}{48} + \dfrac{y^2}{64} = 1$

32. $\dfrac{x^2}{16} + \dfrac{y^2}{12} = 1$ **34.** $\dfrac{x^2}{4} + \dfrac{y^2}{16} = 1$

36. $\dfrac{(x-4)^2}{9} + \dfrac{y^2}{16} = 1$ **38.** $\dfrac{(x-2)^2}{4} + \dfrac{(y+1)^2}{1} = 1$

40. $\dfrac{(x-2)^2}{16} + \dfrac{y^2}{12} = 1$ **42.** $\dfrac{(x-2)^2}{1} + \dfrac{4(y+1)^2}{9} = 1$

44. $\dfrac{(x-3)^2}{36} + \dfrac{(y-2)^2}{32} = 1$

46. $\dfrac{(x-5)^2}{16} + \dfrac{(y-6)^2}{36} = 1$ **48.** $\dfrac{x^2}{48} + \dfrac{y^2}{64} = 1$

50. Positions: $\left(\pm\sqrt{5}, 0\right)$; Length of string: 6 feet

52. (a) $\dfrac{x^2}{327.25} + \dfrac{y^2}{19.34} = 1$

(b)

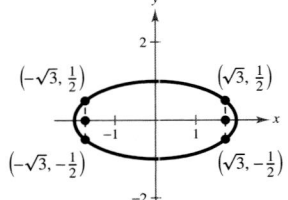

(c) Apogee: 35.64 astronomical units

Perigee: 0.54 astronomical units

54. Answers will vary.

56.   **58.**

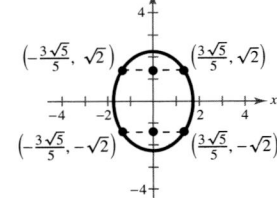

60. True. As the eccentricity of an ellipse approaches 1, the graph of an ellipse is elongated and the foci are close to the vertices. As the eccentricity approaches 0, the graph of an ellipse resembles that of a circle.

62. (a) $2a$

(b) The sum of the distances from the two fixed points is constant.

64. $-5, 15$ **66.** $\frac{5}{4}$

68. (a) -24 (b) 144 (c) $x^2 - 18x + 56$

70. (a) -1 (b) 15 (c) 18

Section 6.4   *(page 458)*

1. b   **2.** c   **3.** a   **4.** d

6. Center: $(0, 0)$

Vertices: $(\pm 3, 0)$

Foci: $\left(\pm \sqrt{34}, 0\right)$

Asymptotes: $y = \pm \frac{5}{3}x$

8. Center: $(0, 0)$

Vertices: $(\pm 6, 0)$

Foci: $\left(\pm 2\sqrt{10}, 0\right)$

Asymptotes: $y = \pm \frac{1}{3}x$

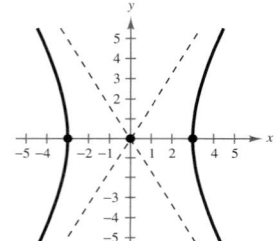

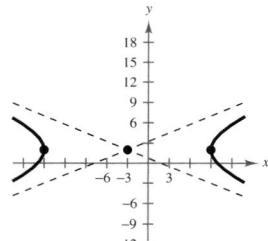

10. Center: $(-3, 2)$

Vertices: $(9, 2), (-15, 2)$

Foci: $(10, 2), (-16, 2)$

Asymptotes:

$y = 2 \pm \frac{5}{12}(x + 3)$

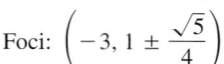

12. Center: $(-3, 1)$

Vertices: $\left(-3, \frac{3}{2}\right), \left(-3, \frac{1}{2}\right)$

Foci: $\left(-3, 1 \pm \frac{\sqrt{5}}{4}\right)$

Asymptotes:

$y = 1 \pm 2(x + 3)$

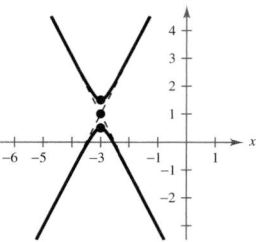

14. Center: $(0, 2)$

Vertices: $(\pm 6, 2)$

Foci: $\left(\pm 2\sqrt{10}, 2\right)$

Asymptotes: $y = 2 \pm \frac{1}{3}x$

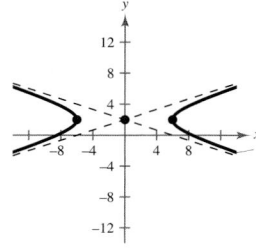

16. The graph of this equation is two lines intersecting at $(1, -2)$.

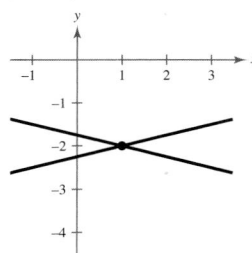

18. Center: $(0, 0)$

Vertices: $\left(0, \pm \sqrt{3}\right)$

Foci: $(0, \pm 3)$

Asymptotes: $y = \pm \frac{\sqrt{2}}{2}x$

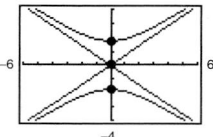

20. Center: $(-3, 5)$

Vertices: $\left(-\frac{10}{3}, 5\right), \left(-\frac{8}{3}, 5\right)$

Foci: $\left(-3 \pm \frac{\sqrt{10}}{3}, 5\right)$

Asymptotes: $y = 5 \pm 3(x + 3)$

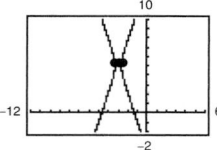

22. $\dfrac{x^2}{16} - \dfrac{y^2}{20} = 1$ **24.** $\dfrac{y^2}{9} - \dfrac{x^2}{1} = 1$

26. $\dfrac{x^2}{64} - \dfrac{y^2}{36} = 1$ **28.** $\dfrac{y^2}{9} - \dfrac{(x - 2)^2}{27} = 1$

30. $\dfrac{x^2}{4} - \dfrac{(y - 1)^2}{5} = 1$ **32.** $\dfrac{x^2}{4} - \dfrac{7(y - 1)^2}{12} = 1$

34. $\dfrac{y^2}{4} - \dfrac{(x - 1)^2}{4} = 1$ **36.** $\dfrac{(y - 3)^2}{9} - \dfrac{(x - 3)^2}{9} = 1$

38. $\dfrac{(y - 2)^2}{4} - \dfrac{(x - 3)^2}{9} = 1$

40. $\left(12(\sqrt{5}-1),0\right) \approx (14.83,0)$ **42.** Ellipse

44. Parabola   **46.** Hyperbola   **48.** Circle

50. False. For the trivial solution of two intersecting lines to occur, the standard form of the equation of the hyperbola would be equal to zero,

$$\frac{(x-h)^2}{a^2} - \frac{(y-k)^2}{b^2} = 0 \text{ or } \frac{(y-k)^2}{a^2} - \frac{(x-h)^2}{b^2} = 0.$$

52. The extended diagonals of the central rectangle are asymptotes of the hyperbola.

54. $(x+7)^2$ **56.** $x(3x+2)(2x-5)$

58. 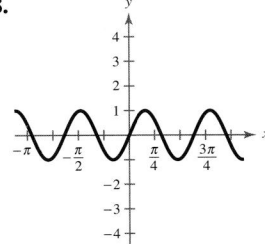   **60.**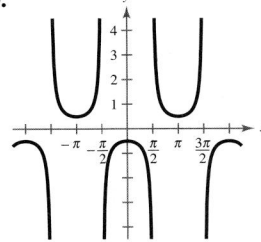

Section 6.5   (page 496)

2. $\left(3\sqrt{2},0\right)$ **4.** $\left(\dfrac{3+\sqrt{3}}{2}, \dfrac{1-3\sqrt{3}}{2}\right)$

6. $\left(\sqrt{3}+2, 2\sqrt{3}-1\right)$

8. $\dfrac{(x')^2}{4} - \dfrac{(y')^2}{4} = 1$

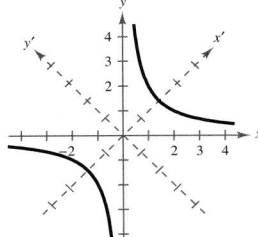

10. $\dfrac{\left[y'+\left(3\sqrt{2}/2\right)\right]^2}{10} - \dfrac{\left[x'-\left(\sqrt{2}/2\right)\right]^2}{10} = 1$

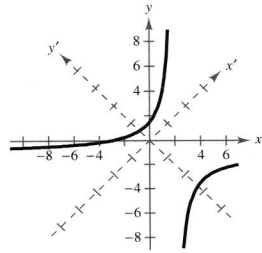

12. $\dfrac{(x')^2}{4} - \dfrac{(y')^2}{4} = 1$ **14.** $(x')^2 + \dfrac{(y')^2}{4} = 1$

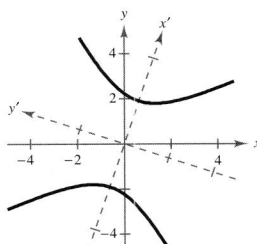   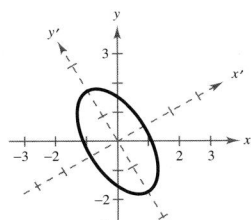

16. $(y')^2 = 4(x'-1)$ **18.** $y' = \frac{1}{4}(x')^2$

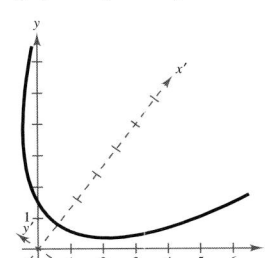   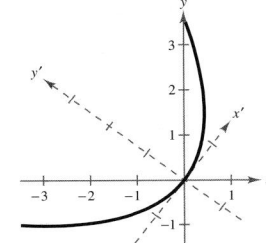

20.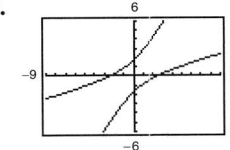

$\theta \approx 37.98°$

22.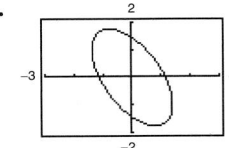

$\theta \approx 33.69°$

24.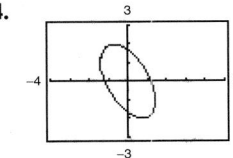

$\theta \approx 28.15°$

26.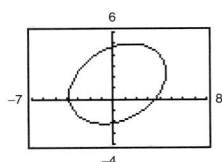

$\theta \approx 31.72°$

27. e   **28.** f   **29.** b   **30.** a   **31.** d   **32.** c

34. Hyperbola

$$y = \frac{4x \pm \sqrt{16x^2 + 8(x^2-6)}}{-4}$$

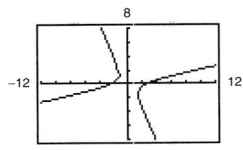

36. Ellipse

$$y = \frac{-(4x - 4) \pm \sqrt{(4x - 4)^2 - 20(2x^2 + 3x - 20)}}{10}$$

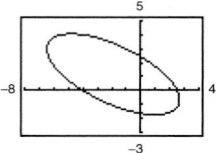

38. Parabola

$$y = \frac{60x - 9 \pm \sqrt{(60x - 9)^2 - 3600x^2}}{50}$$

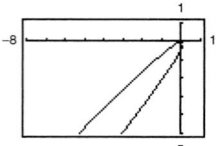

40. Ellipse

$$y = \frac{-(x + 1) \pm \sqrt{(x + 1)^2 - 16(x^2 + x - 4)}}{8}$$

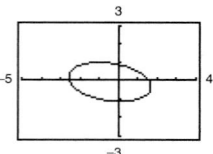

42. 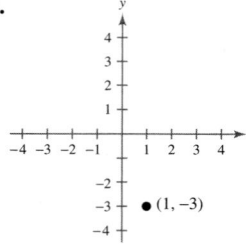   **44.**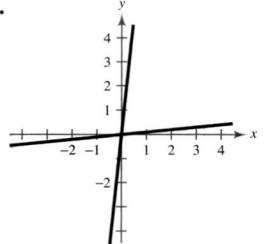

46. $(-7, 0), (-1, 0)$ **48.** $(14, -8), (6, -8)$

50. $(1, 0)$ **52.** $(-2, 8), \left(5, 8 \pm 4\sqrt{21}\right)$

54. $(\pm 3, 2)$ **56.** $(1, 1), (3, 1)$

58. $\left(\frac{1}{6}(3 + \sqrt{30}), \frac{1}{6}(3 - \sqrt{30})\right), \left(\frac{1}{6}(3 - \sqrt{30}), \frac{1}{6}(3 + \sqrt{30})\right)$

60. False. The coefficients of the new equation after it has been rotated are obtained by making the substitutions $x = x' \cos \theta - y' \sin \theta$ and $y = x' \sin \theta + y' \cos \theta$.

62. Major axis: 4; Minor axis: 2

64. 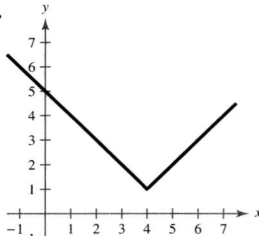   **66.**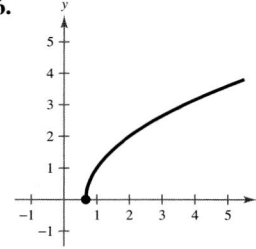

68. 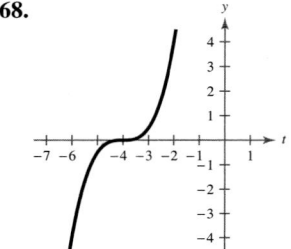   **70.**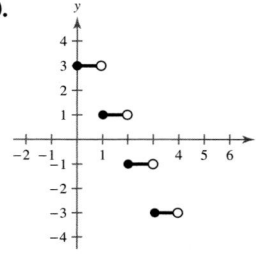

72. 187.94   **74.** 310.39

76. 107 miles south; 54 miles west

Section 6.6   (page 473)

2. (a)

θ	$-\dfrac{\pi}{2}$	$-\dfrac{\pi}{4}$	0	$\dfrac{\pi}{4}$	$\dfrac{\pi}{2}$
x	0	2	4	2	0
y	-2	$-\sqrt{2}$	0	$\sqrt{2}$	2

(b)

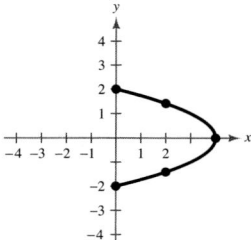

(c) $x = -y^2 + 4$

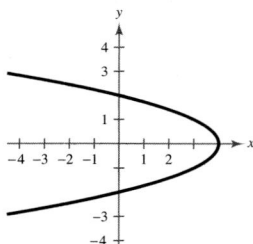

The graph of the rectangular equation continues the graph into the second and third quadrants.

4.

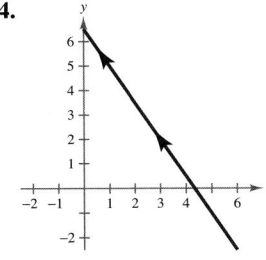

6.

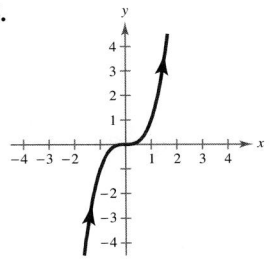

8.

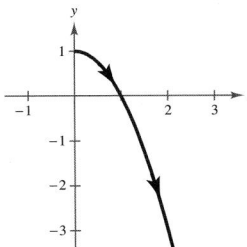

10.

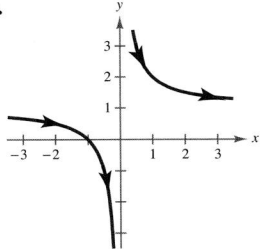

12.

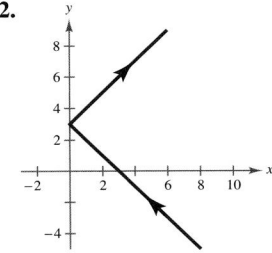

14.

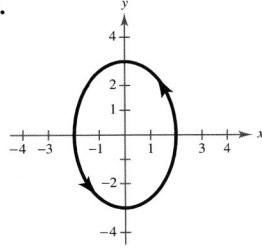

16.

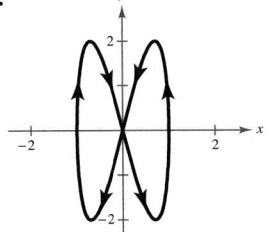

18.

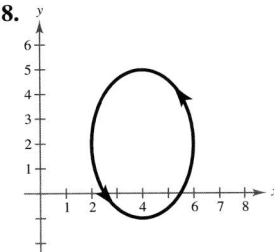

20.

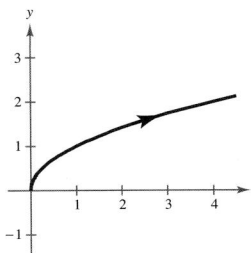

22.

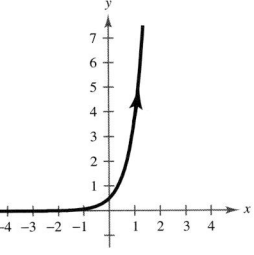

24. Each curve represents a portion of the parabola $y = x^2 - 1$.

Domain	Orientation
(a) $(-\infty, \infty)$	Left to right
(b) $[0, \infty)$	Depends on t
(c) $[-1, 1]$	Depends on t
(d) $(0, \infty)$	Left to right

26. $(x - h)^2 + (y - k)^2 = r^2$

28. $\dfrac{(x - h)^2}{a^2} - \dfrac{(y - k)^2}{b^2} = 1$

30. $x = 2 + 4t$
$\quad\ y = 3 - 6t$

32. $x = -3 + 5 \cos \theta$
$\quad\ y = 2 + 5 \sin \theta$

34. $x = 4 + 4 \cos \theta$
$\quad\ y = 2 + 5 \sin \theta$

36. $x = 2 \sec \theta$
$\quad\ y = 2\sqrt{3} \tan \theta$

38. (a) $x = t, y = \frac{1}{3}(t + 2)$
$\quad$ (b) $x = -t + 2, y = -\frac{1}{3}(t - 4)$

40. (a) $x = t, y = t^3$
$\quad$ (b) $x = -t + 2, y = (-t + 2)^3$

42. (a) $x = t, y = 2 - t$
$\quad$ (b) $x = -t + 2, y = t$

44. (a) $x = t, y = \dfrac{1}{2t}$

$\quad$ (b) $x = -t + 2, y = \dfrac{1}{-2t + 4}$

46.

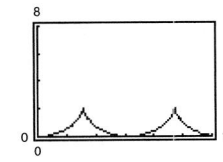

48.

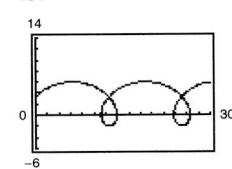

50.

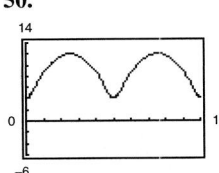

52.

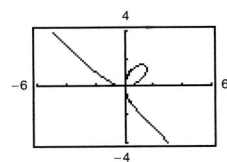

53. b
$\quad$ Domain: $[-2, 2]$
$\quad$ Range: $[-1, 1]$

54. c
$\quad$ Domain: $[-4, 4]$
$\quad$ Range: $[-6, 6]$

55. d
$\quad$ Domain: $(-\infty, \infty)$
$\quad$ Range: $(-\infty, \infty)$

56. a
$\quad$ Domain: $(-\infty, \infty)$
$\quad$ Range: $[-2, 2]$

58. (a)

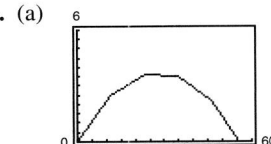

$\quad$ Maximum height: 3.8 feet
$\quad$ Range: 56.3 feet

CHAPTER 6

(b)

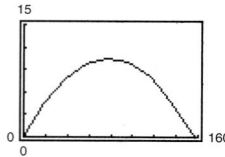

Maximum height: 10.5 feet

Range: 156.3 feet

(c)

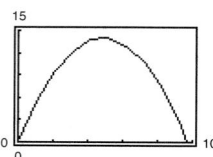

Maximum height: 14.1 feet

Range: 97.4 feet

(d)

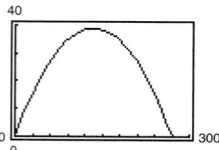

Maximum height: 39.1 feet

Range: 270.6 feet

60. (a) $x = (240 \cos 10°)t$

$y = 5 + (240 \sin 10°)t - 16t^2$

(b) 643 feet

(c)

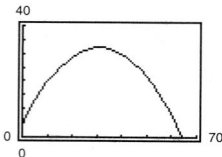

32.1 feet

(d) 2.72 seconds

62. (a) $h = 7, v_0 = 40, \theta = 45°$

$x = (40 \cos 45°)t$

$y = 7 + (40 \sin 45°)t - 16t^2$

(b)

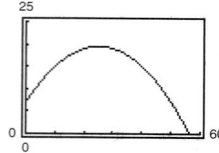

(c) Maximum height: 19.5 feet

Range: 56.2 feet

64. $x = 3 \cos \theta - \cos 3\theta$
$y = 3 \sin \theta - \sin 3\theta$

66. False. $y = x$ for $x \geq 0$

68. $\theta' = 50°$

70. $\theta' = \dfrac{\pi}{6}$

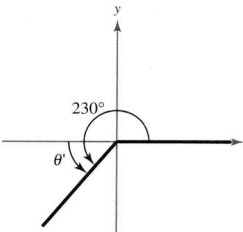

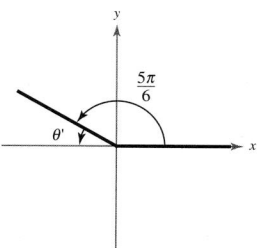

72.

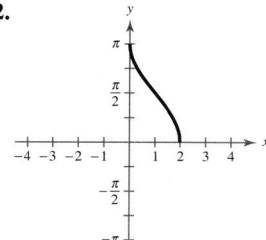

74.

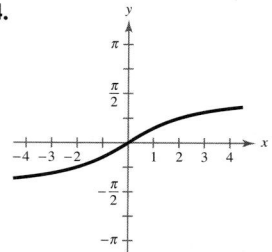

Section 6.7   *(page 480)*

2.

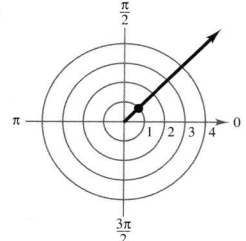

$\left(-1, \dfrac{5\pi}{4}\right), \left(1, \dfrac{\pi}{4}\right)$

4.

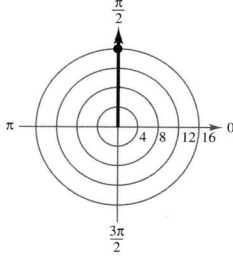

$\left(16, \dfrac{\pi}{2}\right), \left(-16, \dfrac{3\pi}{2}\right)$

6.

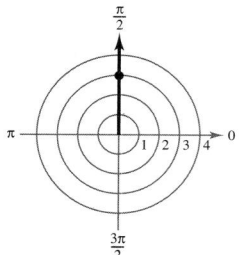

$(-3, 4.7132), (3, 1.5716)$

8.

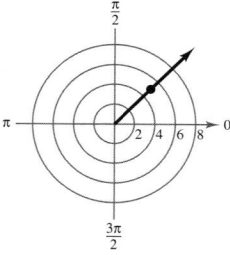

$(-5, 3.9232), (5, 0.7816)$

10. $(0, -3)$ **12.** $(0, 0)$ **14.** $\left(\sqrt{3}, 1\right)$

16. $(-7.7258, -2.8940)$ **18.** $\left(3\sqrt{2}, \dfrac{5\pi}{4}\right)$

20. $\left(5, \dfrac{3\pi}{2}\right)$ **22.** $\left(\sqrt{10}, 5.9614\right)$ **24.** $\left(2, \dfrac{11\pi}{6}\right)$

26. $(13, 1.1760)$ **28.** $\left(\sqrt{29}, 2.7611\right)$ **30.** $\left(6, \dfrac{\pi}{4}\right)$

32. $(2.3049, 0.7086)$ **34.** $r = 4$ **36.** $\theta = \dfrac{\pi}{4}$

38. $r = 4a \sec \theta$ **40.** $r = \dfrac{2}{3 \cos \theta + 5 \sin \theta}$

42. $r^2 = \frac{1}{2} \sec \theta \csc \theta = \csc 2\theta$

44. $r^2 = 9 \cos 2\theta$ **46.** $r = 3a$ **48.** $r = 2a \sin \theta$

50. $x^2 + y^2 - 2x = 0$ **52.** $\sqrt{3}x + y = 0$

54. $x^2 + y^2 = 100$ **56.** $x = -3$

58. $(x^2 + y^2)^2 = 2xy$ **60.** $(x^2 + y^2)^3 = 9(x^2 - y^2)^2$

62. $y^2 = 2x + 1$ **64.** $2x - 3y = 6$

66. $x^2 + y^2 = 64$ **68.** $x + y = 0$

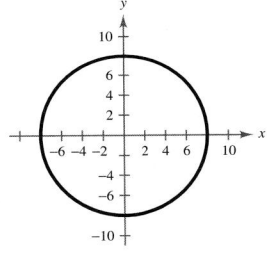   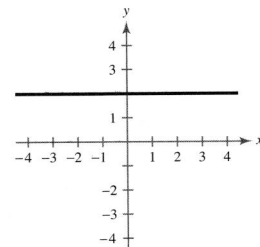

70. $y - 2 = 0$

72. False. If $r_1 = -r_2$, then (r_1, θ) and (r_2, θ) are different points.

74. $\left(x - \frac{1}{2}\right)^2 + \left(y - \frac{3}{2}\right)^2 = \frac{5}{2}$; circle

76. (a) Horizontal: x-coordinate changes

   Vertical: y-coordinate changes

   (b) Horizontal: r and θ both change

   Vertical: r and θ both change

   (c) Unlike r and θ, x and y measure horizontal and vertical changes, respectively.

78. $\frac{1}{4} + \frac{1}{2} \log_4 x - \log_4 y$ **80.** $\ln 5 + 2 \ln x + \ln(x^2 + 1)$

82. $\log_5 a(x + 1)^8$ **84.** $\ln \dfrac{6y}{x - 3}$

Section 6.8   *(page 488)*

2. Cardioid   **4.** Lemniscate   **6.** Circle

8. Polar axis   **10.** Polar axis   **12.** Pole

14. Maximum: $|r| = 18$ when $\theta = 0$

   Zeros: $r = 0$ when $\theta = \dfrac{2\pi}{3}, \dfrac{4\pi}{3}$

16. Maximum: $|r| = 3$ when $\theta = \dfrac{\pi}{4}, \dfrac{3\pi}{4}, \dfrac{5\pi}{4}, \dfrac{7\pi}{4}$

   Zeros: $r = 0$ when $\theta = 0, \dfrac{\pi}{2}, \pi, \dfrac{3\pi}{2}$

18.

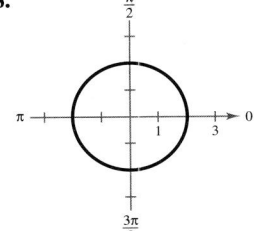

20.

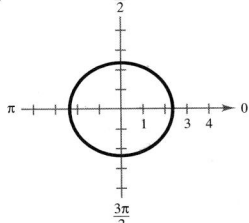

22.

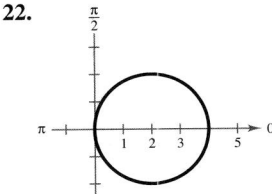

24.

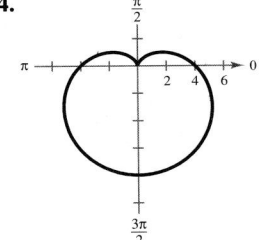

26.

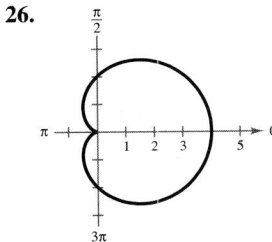

28.

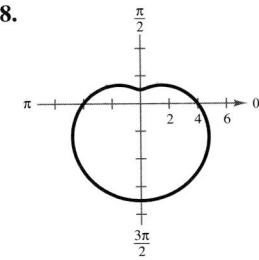

30.

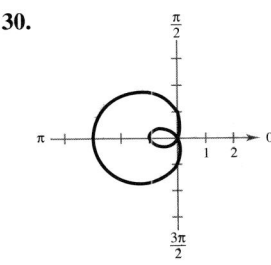

32.

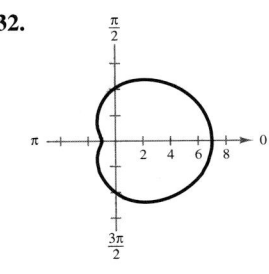

34.

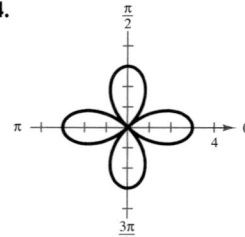

36.

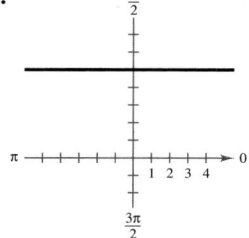

38.

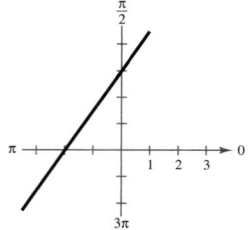

40.

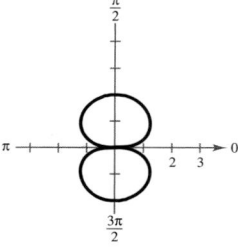

42.

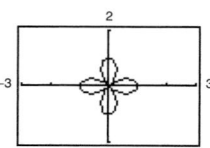

44.

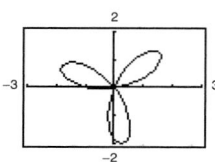

46.

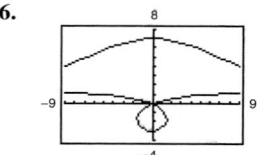

48.

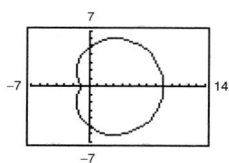

$0 \le \theta < 2\pi$

50.

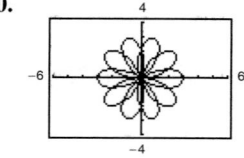

$0 \le \theta < 4\pi$

52.

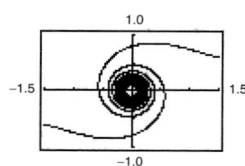

$0 \le \theta < \infty$

54.

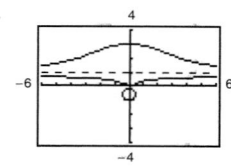

56.

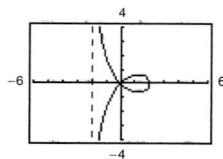

58. False. For a graph symmetric with respect to the pole, one portion of the graph coincides with the other portion when rotated π radians about the pole.

60. (a)

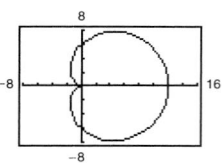

(b)

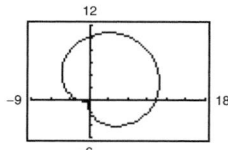

(c)

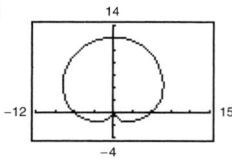

The angle ϕ controls rotation of the axis of symmetry.
$r = 6(1 + \sin \theta)$

62. Answers will vary.

64. (a) $r = 4 \sin\left(\theta - \dfrac{\pi}{6}\right) \cos\left(\theta - \dfrac{\pi}{6}\right)$

(b) $r = -4 \sin \theta \cos \theta$

(c) $r = 4 \sin\left(\theta - \dfrac{2\pi}{3}\right) \cos\left(\theta - \dfrac{2\pi}{3}\right)$

(d) $r = 4 \sin \theta \cos \theta$

66. (a)

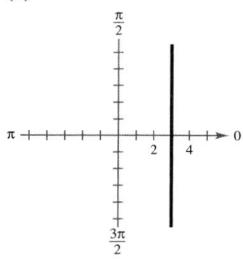

(b)

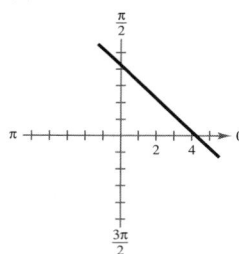

(c)

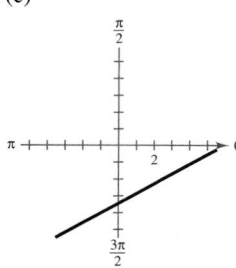

(d)

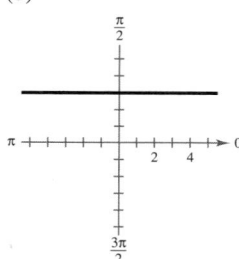

68. (a)

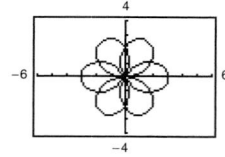

$0 \le \theta < 4\pi$

(b)

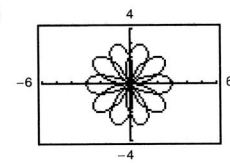

$$0 \le \theta < 4\pi$$

(c) Yes. Explanations will vary.

70. $\dfrac{(x-3)^2}{7} + \dfrac{(y+1)^2}{16} = 1$

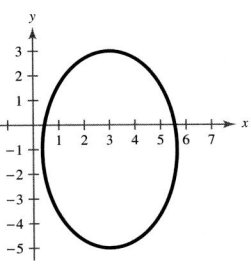

Section 6.9 *(page 494)*

2. $e = 1$: $r = \dfrac{4}{1 - \cos\theta}$, parabola

$e = 0.5$: $r = \dfrac{2}{1 - 0.5\cos\theta}$, ellipse

$e = 1.5$: $r = \dfrac{6}{1 - 1.5\cos\theta}$, hyperbola

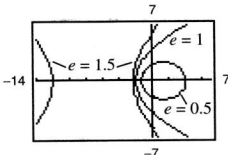

4. $e = 1$: $r = \dfrac{4}{1 + \sin\theta}$, parabola

$e = 0.5$: $r = \dfrac{2}{1 + 0.5\sin\theta}$, ellipse

$e = 1.5$: $r = \dfrac{6}{1 + 1.5\sin\theta}$, hyperbola

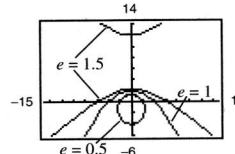

5. f   **6.** c   **7.** d   **8.** e   **9.** a   **10.** b

12. Parabola

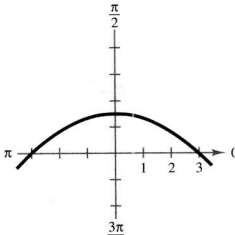

14. Parabola

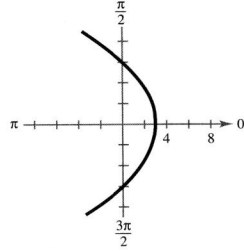

16. Ellipse

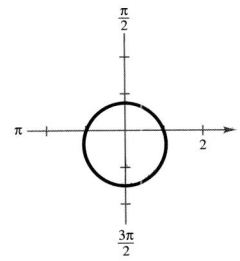

18. Ellipse

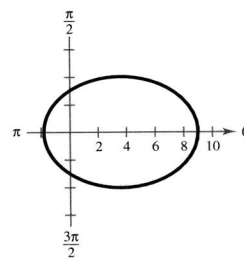

20. Hyperbola

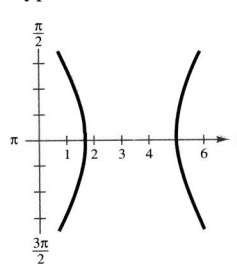

22. Hyperbola

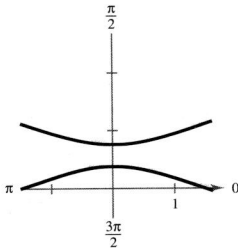

24. Hyperbola

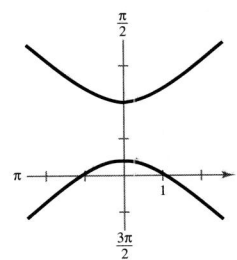

26.

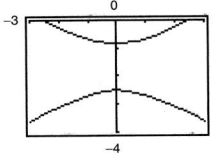

Hyperbola

28.

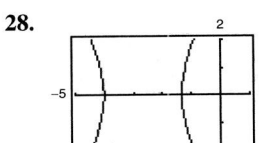

Hyperbola

30.

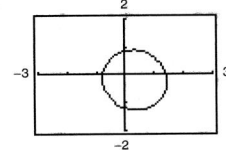

32.

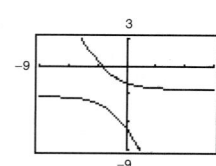

34. $r = \dfrac{2}{1 - \sin\theta}$

36. $r = \dfrac{9}{4 - 3\sin\theta}$

38. $r = \dfrac{3}{2 - 3\cos\theta}$

40. $r = \dfrac{12}{1 + \cos\theta}$ **42.** $r = \dfrac{20}{1 + \sin\theta}$

44. $r = \dfrac{8}{3 + \sin\theta}$ **46.** $r = \dfrac{16}{3 + 5\cos\theta}$

48. $r = \dfrac{8}{3 + 5\sin\theta}$ **50.** Answers will vary.

52. $r = \dfrac{1.4248 \times 10^9}{1 - 0.0543\cos\theta}$

Perihelion: 1.3514×10^9 kilometers

Aphelion: 1.5066×10^9 kilometers

54. $r = \dfrac{3.4459 \times 10^7}{1 - 0.2056\cos\theta}$

Perihelion: 2.8583×10^7 miles

Aphelion: 4.3377×10^7 miles

56. $r = \dfrac{7.7658 \times 10^8}{1 - 0.0484\cos\theta}$

Perihelion: 7.4073×10^8 kilometers

Aphelion: 8.1607×10^8 kilometers

58. If e remains fixed and p changes, then the lengths of both the major axis and the minor axis change. For example, graph

$r = \dfrac{5}{1 - \frac{2}{3}\cos\theta}$, with $e = \frac{2}{3}$ and $p = \frac{15}{2}$, and graph

$r = \dfrac{6}{1 - \frac{2}{3}\cos\theta}$, with $e = \frac{2}{3}$ and $p = 9$, on the same set of coordinate axes.

60. False. The graph has a horizontal directrix below the pole.

62. Answers will vary. **64.** $r^2 = \dfrac{400}{25 - 9\cos^2\theta}$

66. $r^2 = \dfrac{36}{8\cos^2\theta - 9}$ **68.** $r^2 = \dfrac{225}{25 - 16\cos^2\theta}$

70. $\dfrac{\pi}{3} + 2n\pi, \dfrac{5\pi}{3} + 2n\pi$ **72.** $\dfrac{\pi}{3} + n\pi, \dfrac{2\pi}{3} + n\pi$

74. $\dfrac{\pi}{3} + 2n\pi, \dfrac{5\pi}{3} + 2n\pi$ **76.** $-\dfrac{7\sqrt{2}}{10}$ **78.** $\dfrac{\sqrt{2}}{10}$

80. $\sin 2u = -\dfrac{\sqrt{3}}{2}$

$\cos 2u = -\dfrac{1}{2}$

$\tan 2u = \sqrt{3}$

Review Exercises *(page 498)*

2. 2.6012 radians, 149.04° **4.** 0.7086 radian, 40.60°

6. 0.4424 radian, 25.35° **8.** 1.4309 radians, 81.98°

10. $\dfrac{6\sqrt{5}}{5}$ **12.** Parabola **14.** $y^2 = -8(x - 2)$

16. $(x - 2)^2 = 8(y - 2)$ **18.** $y = 4x + 8; (-2, 0)$

20. $y^2 = 6x$ **22.** $\dfrac{(x - 2)^2}{3} + \dfrac{(y - 2)^2}{4} = 1$

24. $\dfrac{(x + 4)^2}{4} + \dfrac{(y - 5)^2}{36} = 1$

26. Longest distance: 36 feet

Shortest distance: 28 feet

Distance between foci: $16\sqrt{2}$ feet

28. Center: $(5, -3)$ **30.** Center: $(-2, 3)$

Vertices: $(5, 3), (5, -9)$ Vertices: $(3, 3), (-7, 3)$

Foci: $\left(5, -3 \pm \sqrt{35}\right)$ Foci: $\left(-2 \pm \sqrt{21}, 3\right)$

Eccentricity: $\dfrac{\sqrt{35}}{6}$ Eccentricity: $\dfrac{\sqrt{21}}{5}$

32. $\dfrac{x^2}{4} - \dfrac{(y - 2)^2}{12} = 1$ **34.** $\dfrac{5y^2}{16} - \dfrac{5(x - 3)^2}{4} = 1$

36. Center: $(0, 1)$

Vertices: $(0, 3), (0, -1)$

Foci: $\left(0, 1 \pm \sqrt{5}\right)$

Asymptotes: $y = 1 \pm 2x$

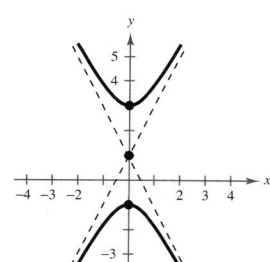

38. Center: $(-1, -3)$

Vertices: $(-1, -1), (-1, -5)$

Foci: $\left(-1, -3 \pm \sqrt{29}\right)$

Asymptotes: $y = -3 \pm \frac{2}{5}(x + 1)$

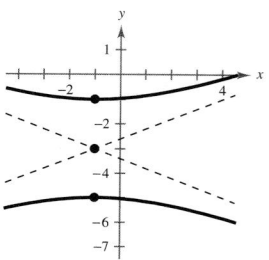

40. $\dfrac{576x^2}{25} - \dfrac{576y^2}{2279} = 1, \dfrac{64(x-1)^2}{25} - \dfrac{64y^2}{39} = 1$

42. Parabola

44. $\dfrac{(x')^2}{1/4} - \dfrac{(y')^2}{1/6} = 1$ **46.** $y' = -4(x')^2 - 8x'$

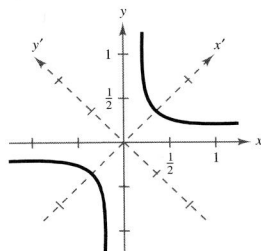 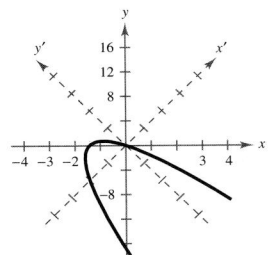

48. Ellipse

$y = \dfrac{8x \pm \sqrt{64x^2 - 28(13x^2 - 45)}}{14}$

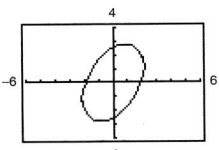

50. Hyperbola

$y = \dfrac{10x \pm \sqrt{100x^2 - 4(x^2 + 1)}}{2}$

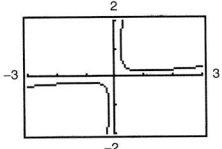

52. $x = \dfrac{3}{2}, y = \dfrac{3}{2}$ **54.** $x = \dfrac{3\sqrt{2}}{2}, y = 1$

56.   **58.**

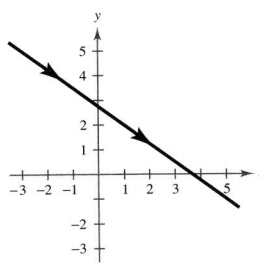

$y = -\dfrac{3}{4}x + \dfrac{11}{4}$ $y = (x-4)^2$

60.

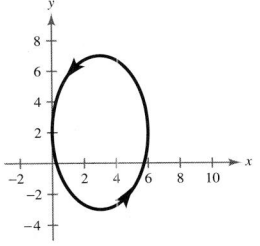

$\dfrac{(x-3)^2}{9} + \dfrac{(y-2)^2}{25} = 1$

62. $x = -3 + 4\cos\theta$ **64.** Answers will vary.
$\quad\ \ y = 4 + 3\sin\theta$

66.   **68.**

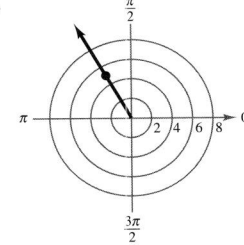 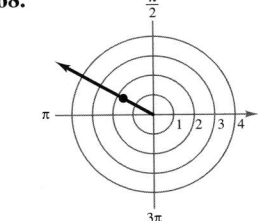

$\left(-5, \dfrac{5\pi}{3}\right), \left(5, \dfrac{2\pi}{3}\right)$ $\left(\sqrt{3}, 8.90\right), \left(-\sqrt{3}, 5.76\right)$

70. $\left(-\sqrt{2}, -\sqrt{2}\right)$ **72.** $(0,0)$ **74.** $\left(\sqrt{10}, \dfrac{3\pi}{4}\right)$

76. $(5, 5.3559)$ **78.** $r = 4\cos\theta$ **80.** $r^2 = -4\csc 2\theta$

82. $x^2 + y^2 = 8y$ **84.** $(x^2 + y^2)^2 = x^2 - y^2$

86. Symmetry: $\theta = \dfrac{\pi}{2}$, polar axis, pole

Maximum value of $|r|$: $|r| = 11$ when $\theta = 0, \dfrac{\pi}{2}, \pi, \dfrac{3\pi}{2}$

No zeros of r

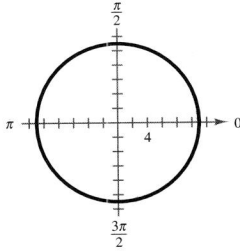

88. Symmetry: polar axis

Maximum value of $|r|$: $|r| = 1$ when $\theta = 0, \dfrac{2\pi}{5}, \dfrac{4\pi}{5}, \dfrac{6\pi}{5},$ $\dfrac{8\pi}{5}$

Zeros of r: $r = 0$ when $\theta = \dfrac{\pi}{10}, \dfrac{3\pi}{10}, \dfrac{5\pi}{10}, \dfrac{7\pi}{10}, \dfrac{9\pi}{10}$

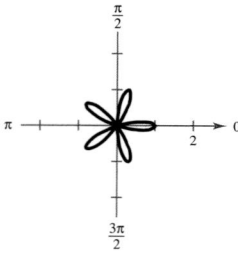

90. Symmetry: polar axis

Maximum value of $|r|$: $|r| = 7$ when $\theta = \pi$

Zeros of r: $r = 0$ when $\theta = \arccos \frac{3}{4}, 2\pi - \arccos \frac{3}{4}$

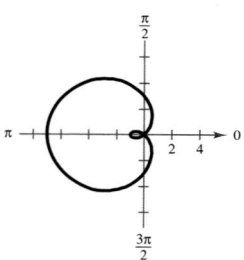

92. Symmetry: polar axis

Maximum value of $|r|$: $|r| = 10$ when $\theta = \pi$

Zero of r: $r = 0$ when $\theta = 0$

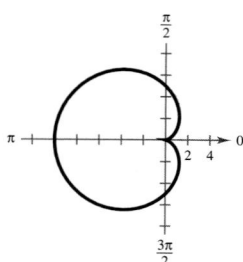

94. Symmetry: $\theta = \dfrac{\pi}{2}$, polar axis

Maximum value of $|r|$: $|r| = 1$ when $\theta = 0, \dfrac{\pi}{2}, \pi, \dfrac{3\pi}{2}$

Zeros of r: $r = 0$ when $\theta = \dfrac{\pi}{4}, \dfrac{3\pi}{4}, \dfrac{5\pi}{4}, \dfrac{7\pi}{4}$

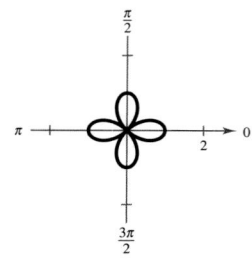

96. Limaçon   **98.** Lemniscate
100. Parabola   **102.** Hyperbola

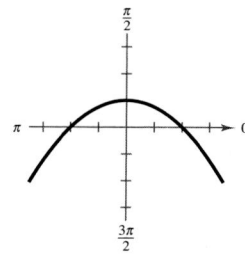   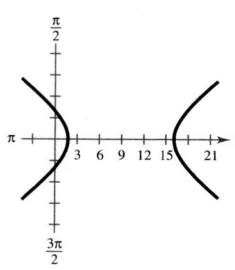

104. $r = \dfrac{4}{1 + \sin \theta}$ **106.** $r = \dfrac{7}{3 + 4 \cos \theta}$

108. $r = \dfrac{12,000,000}{1 + \sin \theta}$; 89,600,000 miles

110. False. The equation of a hyperbola is a second-degree equation.

112. False. $(2, \pi/4), (-2, 5\pi/4),$ and $(2, 9\pi/4)$ all represent the same point.

114. Yes. The orientation would be reversed.

116. (a) Symmetric to the pole
   (b) Symmetric to the polar axis
   (c) Symmetric to $\theta = \pi/2$

Problem Solving *(page 508)*

2. (a) $\dfrac{x^2}{2352.25} + \dfrac{y^2}{529} = 1$

   (b) ≈ 85.4 feet   (c) 1115.5π square feet

4. $A = \dfrac{4 a^2 b^2}{a^2 + b^2}$ **6.** $\dfrac{(x - 6)^2}{9} - \dfrac{(y - 2)^2}{7} = 1$

8. (a) The first set of parametric equations models projectile motion along a straight line. The second set of parametric equations models projectile motion of an object launched at a height of h units above the ground that will eventually fall back to the ground.

   (b) $y = (\tan \theta)x$; $y = h + x \tan \theta - \dfrac{16x^2 \sec^2 \theta}{v_0^2}$

(c) In the first case, the path of the moving object is not affected by changing the velocity because eliminating the parameter removes v_0.

10. (a)

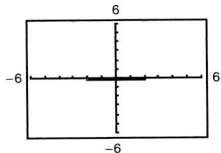

The graph is a line between -2 and 2 on the x-axis.

(b)

The graph is a three-sided figure with counter-clockwise orientation.

(c)

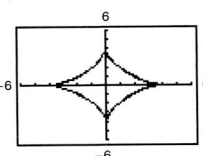

The graph is a four-sided figure with counter-clockwise orientation.

(d)

The graph is a 10-sided figure with counter-clockwise orientation.

(e)

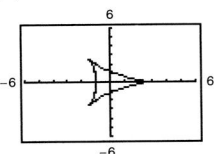

The graph is a three-sided figure with clockwise orientation.

(f)

The graph is a four-sided figure with clockwise orientation.

12.

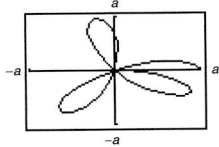

$r = a\cos(3.5\theta)$ $r = a\sin(2.63\theta)$

The graphs are rose curves with partially completed petals.

14. (a) No. Because of the exponential, the graph will continue to trace the butterfly curve at larger values of r.

(b) $r \approx 4.1$. This value will increase if θ is increased.

Explorations

Chapter P

(page 5)

a. 6   **b.** 1   **c.** 3   **d.** 3

Absolute value expressions are never negative.

(page 38)

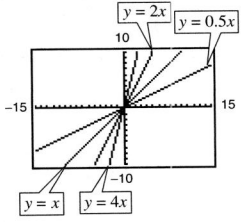

The line $y = 4x$.

The line $y = -4x$.

As $|m|$ increases, the line rises or falls faster.

(page 45)

$$d_1 = \sqrt{1 + m_1^2}, d_2 = \sqrt{1 + m_2^2}; m_1 = -\frac{1}{m_2}$$

(page 72)

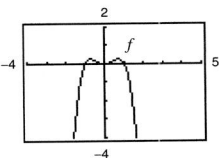

Even   Neither

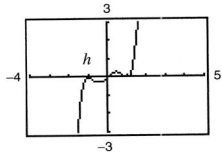

Odd   Even

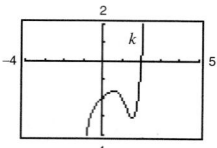

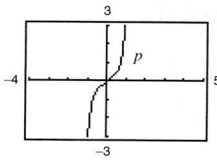

Neither   Odd

Equations of odd functions contain only odd powers of x. Equations of even functions contain only even powers of x. Odd functions have all variables raised to odd powers and even functions have all variables raised to even powers. A function that has variables raised to even and odd powers is neither odd nor even.

(page 86)

a.

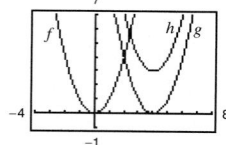

g is a right shift of four units. h is a right shift of four units and an upward shift of three units.

b.

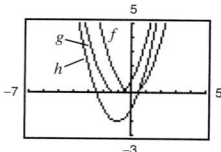

g is a left shift of one unit. h is a left shift of one unit and a downward shift of two units.

c.

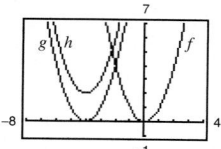

g is a left shift of four units. h is a left shift of four units and an upward shift of two units.

(page 87)

No. $g(x) = -x^4 - 2$. Yes. $h(x) = -(x - 3)^4$.

(page 99)

Option (b), because option (a) takes a 10% discount on a smaller amount. Option (b) represents $f(g(x))$; Option (a) represents $g(f(x))$.

(page 104)

x	-10	0	7	45
$f(f^{-1}(x))$	-10	0	7	45
$f^{-1}(f(x))$	-10	0	7	45

The functions are inverses of each other.

(page 107)

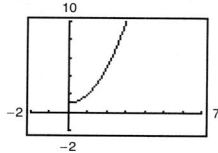

There is an inverse function $f^{-1}(x) = \sqrt{x - 1}$ since the domain of f is equal to the range of f^{-1} and the range of f is equal to the domain of f^{-1}.

Chapter 1

(page 140)

1.

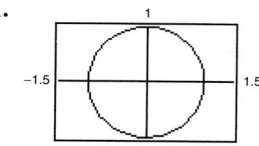

The graph is a circle.

2. The t-values represent the central angle in radians. The x- and y-values represent the location in the coordinate plane.

3. $-1 \le x \le 1, -1 \le y \le 1$

(page 167)

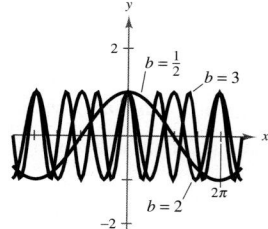

b affects the period of the graph.

$b = \frac{1}{2}$: $\frac{1}{2}$ cycle;

$b = 2$: 2 cycles;

$b = 3$: 3 cycles

(page 168)

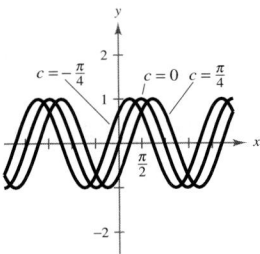

c shifts the graph horizontally.

Chapter 2

(page 237)

Yes. Preferences will vary.

(page 244)

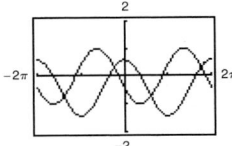

The graphs are different. No, it is not true.

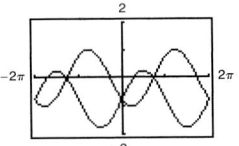

The graphs are different. No, it is not true.

Chapter 3

(page 284)

Pythagorean Theorem.

The Pythagorean Theorem is just a special case of the more general Law of Cosines.

Chapter 4

(page 330)

$i, -1, -i, 1, i, -1, -i, 1$;

The pattern repeats the first four results. Divide the exponent by 4.

If the remainder is 1, the result is i.

If the remainder is 2, the result is -1.

If the remainder is 3, the result is $-i$.

If the remainder is 0, the result is 1.

(page 350)

The given equation can be written as

$$x^4 = -16 = 16(\cos \pi + i \sin \pi)$$

which means that you can solve the equation by finding the four fourth roots of -16. Each of these roots has the form

$$\sqrt[4]{16}\left(\cos \frac{\pi + 2\pi k}{4} + i \sin \frac{\pi + 2\pi k}{4}\right).$$

Finally, using $k = 0, 1, 2,$ and 3, you obtain the roots

$$2\left(\cos \frac{\pi}{4} + i \sin \frac{\pi}{4}\right) = 2\left(\frac{\sqrt{2}}{2} + \frac{\sqrt{2}}{2}i\right) = \sqrt{2} + \sqrt{2}i$$

$$2\left(\cos \frac{3\pi}{4} + i \sin \frac{3\pi}{4}\right) = 2\left(-\frac{\sqrt{2}}{2} + \frac{\sqrt{2}}{2}i\right) = -\sqrt{2} + \sqrt{2}i$$

$$2\left(\cos \frac{5\pi}{4} + i \sin \frac{5\pi}{4}\right) = 2\left(-\frac{\sqrt{2}}{2} - \frac{\sqrt{2}}{2}i\right) = -\sqrt{2} - \sqrt{2}i$$

$$2\left(\cos \frac{7\pi}{4} + i \sin \frac{7\pi}{4}\right) = 2\left(\frac{\sqrt{2}}{2} - \frac{\sqrt{2}}{2}i\right) = \sqrt{2} - \sqrt{2}i.$$

Chapter 5

(page 366)

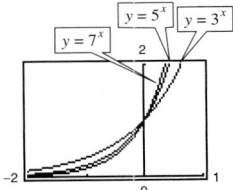

The graphs rise to the right and intersect at $(0, 1)$. $y = 3^x$. $y = 7^x$. $y = 7^x$. $y = 3^x$.

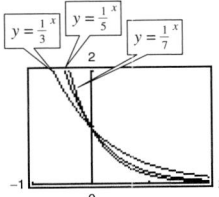

The graphs fall to the right and intersect at $(0, 1)$. $y = \frac{1}{7}x$. $y = \frac{1}{3}x$. $y = \frac{1}{3}x$. $y = \frac{1}{7}x$. For positive values of x, the greater the value of a, the more steeply the graph rises. The smaller the value of a, the more sharply the graph falls.

(page 369)
$A = \$5466.09$, $A = \$5466.35$, $A = \$5466.36$, $A = \$5466.38$.

No. Explanations will vary.

(page 376)

x	-2	-1	0	1	2
$f(x) = 10^x$	$\frac{1}{100}$	$\frac{1}{10}$	1	10	100

x	$\frac{1}{100}$	$\frac{1}{10}$	1	10	100
$f(x) = \log_{10} x$	-2	-1	0	1	2

The domain of $f(x) = 10^x$ is equal to the range of $f(x) = \log_{10} x$ and vice versa. $f(x) = 10^x$ and $f(x) = \log_{10} x$ are inverses of each other.

(page 387)

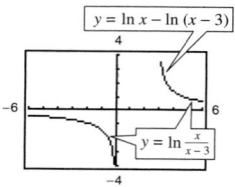

No; the domain of the first graph is $(3, \infty)$ and the domain of the second graph is $(-\infty, 0) \cup (3, \infty)$.

(page 396)

a. 7%   **b.** 7.251%   **c.** 7.186%   **d.** 7.45%

Savings plan (d) will have the greatest effective yield. Savings plan (d) will have the highest balance after 5 years.

Chapter 6

(page 470)

t should be greater than -1. The upper bound of t varies, but one possibility is $t = 20$.

(page 477)

a. Yes. $\theta \approx 3.927$, $x \approx -2.121$, $y \approx -2.121$

b. Yes. Answers and explanations will vary.

Technology

Chapter P

(page 45)

The lines appear perpendicular with the square setting.

(page 57)

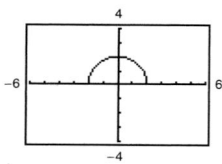

Domain: $[-2, 2]$

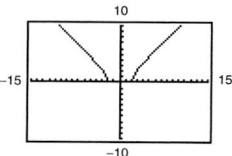

Domain: $(-\infty, -2] \cup [2, \infty)$

Yes, for -2 and 2.

Chapter 1

(page 165)

No graph is visible. Try $-\pi \le x \le \pi$ and $-0.5 \le y \le 0.5$ as a viewing window.

Chapter 5

(page 403)

$P = 53403(1.013)^t$. The models are very similar.

Chapter 6

(page 472)

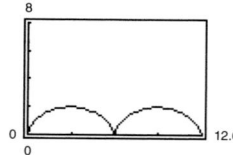

Writing About Mathematics

Chapter P

Section P.5   (page 60)

Answers will vary.

Section P.9   (page 99)

a.

x	1	2	3	4	5	6
$g(x)$	0	1	2	0	1	-1

b.

x	1	2	3	4	5	6
$h(x)$	2	1	-3	-1	0	5

Section P.10   (page 108)

a. The function does not have an inverse, because the retail price cannot be determined exactly because the sales tax is rounded.

b. The function does have an inverse because $f(f^{-1}(x)) = x$ and $f^{-1}(f(x)) = x$.

Chapter 1

Section 1.6   (page 181)

$h(x) = x + \sin x$

(a) $f(x) = x$ and $g(x) = \sin x$

(b) Tables will vary. One example is:

x	-3	-2	-1
$f(x) = x$	-3	-2	-1
$g(x) = \sin x$	-0.14112	-0.90930	-0.84147
$h(x) = f(x) + g(x)$	-3.14112	-2.90930	-1.84147

x	0	1	2
$f(x) = x$	0	1	2
$g(x) = \sin x$	0	0.84147	0.90930
$h(x) = f(x) + g(x)$	0	1.84147	2.90930

x	3
$f(x) = x$	3
$g(x) = \sin x$	0.14112
$h(x) = f(x) + g(x)$	3.14112

(c) A graph of h can be formed by adding the graphs of f and g point by point.

$h(x) = \cos x - \sin 3x$

(a) $f(x) = \cos x$ and $g(x) = \sin 3x$

(b) Tables will vary. One example is:

x	-3	-2	-1
$f(x) = \cos x$	-0.98999	-0.41615	0.54030
$g(x) = \sin 3x$	-0.41212	0.27942	-0.14112
$h(x) = f(x) - g(x)$	-0.57787	-0.69557	0.68142

x	0	1	2
$f(x) = \cos x$	1	0.54030	-0.41615
$g(x) = \sin 3x$	0	0.14112	-0.27942
$h(x) = f(x) - g(x)$	1	0.39918	-0.13673

x	3
$f(x) = \cos x$	-0.98999
$g(x) = \sin 3x$	0.41212
$h(x) = f(x) - g(x)$	-1.40211

(c) A graph of h can be formed by subtracting the graph of g from the graph of f point by point.

Examples will vary. One example is $f(x) = \sin x$ and

$g(x) = \cos\left(x + \dfrac{\pi}{2}\right).$

Section 1.8   (page 201)

(a) Amplitude modulation

(b) Frequency modulation

Explanations will vary.

Chapter 2

Section 2.2   (page 230)

This particular set of range settings actually shows the statement's left side and right side as coinciding, when an algebraic verification shows that they are not equal (using a different set of range settings will confirm this). A trigonometric identity should always be verified algebraically first, and then graphically. Showing that the graphs of the left and right sides of a statement appear to coincide is not strong enough proof to conclude that the statement is an identity.

Section 2.3 *(page 239)*

Equations a and b. Either $-2 \le \sqrt{b^2 - 4c} + b \le 2$ or $-2 \le \sqrt{b^2 - 4c} - b \le 2$ must be true.

Section 2.5 *(page 257)*

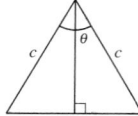

Let c be the length of the two equal sides of an isosceles triangle, and let θ be the angle between the two equal sides.

$$\text{Area} = \left(\frac{1}{2} \text{ base}\right)(\text{height})$$

$$= \left(c \sin \frac{\theta}{2}\right)\left(c \cos \frac{\theta}{2}\right)$$

$$= c^2 \sqrt{\frac{1 - \cos \theta}{2}} \sqrt{\frac{1 + \cos \theta}{2}}$$

$$= c^2 \sqrt{\frac{1 - \cos^2\theta}{4}}$$

$$= \frac{1}{2}c^2 \sqrt{\sin^2 \theta}$$

$$= \frac{1}{2}c^2 \sin \theta$$

Examples will vary.

Chapter 3

Section 3.1 *(page 279)*

Yes

(a) $A = 40°$, $a \approx 12.86$, $b \approx 15.32$

(b) $A = 40°$, $b \approx 11.92$, $c \approx 15.56$

It is probably easier to use the right triangle definitions of sine, cosine, and tangent to solve the triangle.

Section 3.2 *(page 286)*

(a) Area $= \frac{1}{2}(2)(4) \sin 50° \approx 3.064$ square feet

(b) Area $= \sqrt{\frac{9}{2}(\frac{9}{2} - 2)(\frac{9}{2} - 3)(\frac{9}{2} - 4)} \approx 2.905$ square feet

(c) Area $= \frac{1}{2}(2)(4) = 4$ square feet

(d) Area $= \sqrt{6(6 - 3)(6 - 4)(6 - 5)} = 6$ square feet

Chapter 4

Section 4.2 *(page 339)*

Answers will vary.

Section 4.3 *(page 346)*

To approximate the product of two complex numbers graphically, add the two angles.

(a) $\left(\cos \frac{\pi}{3} + i \sin \frac{\pi}{3}\right)\left(\cos \frac{2\pi}{3} + i \sin \frac{2\pi}{3}\right) = -1$

(b) $\left(\cos \frac{\pi}{6} + i \sin \frac{\pi}{6}\right)\left(\cos \frac{\pi}{3} + i \sin \frac{\pi}{3}\right) = i$

Section 4.4 *(page 352)*

$e^{a+bi} = e^a(\cos b + i \sin b)$

Let $a = 0$ and $b = \pi$.

$$e^{0 + \pi i} = e^0(\cos \pi + i \sin \pi)$$

$$e^{\pi i} = -1$$

$$e^{\pi i} + 1 = 0$$

Chapter 5

Section 5.1 *(page 371)*

$g(x)$ can be generated by f_5.

$h(x)$ can be generated by f_2 or f_3.

All six functions have $(0, 8)$ as their y-intercept.

The functions f_2 and f_3 are the same because

$$f_2(x) = 8\left(\tfrac{1}{2}\right)^x = 2^3(2)^{-x} = 2^{(3-x)} = \left(\tfrac{1}{2}\right)^{-(3-x)}$$

$$= \left(\tfrac{1}{2}\right)^{(x-3)} = f_3(x).$$

The functions f_1 and f_6 are the same because

$f_1(x) = 2^{(x+3)} = 2^x 2^3 = (8)2^x = f_6(x)$.

Examples of exponential functions with y-intercepts of $(0, -3)$ will vary. Possibilities include $f(x) = -3(2^x)$ and $f(x) = \left(\tfrac{1}{3}\right)^x - 4$.

Section 5.2 *(page 381)*

Approximately 11 months. Explanations will vary.

Section 5.4 *(page 397)*

a.

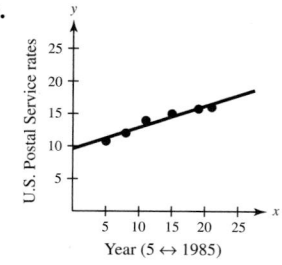

Linear model: $y = 0.329x + 9.58$

2004

b.

ln x	1.6094	2.0794	2.3979	2.7081	2.9444	3.0445
ln y	2.3749	2.4849	2.6355	2.7081	2.7568	2.7726

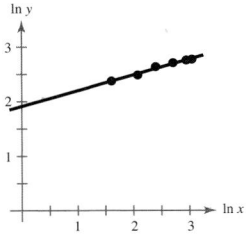

Logarithmic model: $\ln y = 0.28888 \ln x + 1.9103$

2007

c.

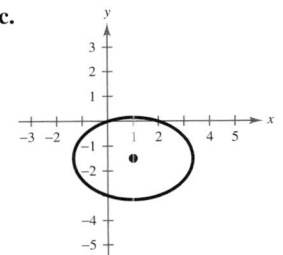

$y = 6.7551x^{0.28888}$ seems to fit the data better and will probably better predict future shipments.

Section 5.5 (page 408)

The quadratic model is better.

Chapter 6

Section 6.1 (page 429)

The angle of inclination is the positive angle made with the x-axis measured counterclockwise. Therefore, there are two cases: (1) a line with positive slope, implying $\theta < 90°$, and (2) a line with negative slope, implying $90° < \theta < 180°$. The angle between two lines is, by definition, the smaller of the two angles formed by the intersecting lines. The two angles formed are supplementary; therefore, one angle must be acute—i.e., less than 90°—or, if the lines are perpendicular, both angles equal 90°.

False. The inclination is the *positive angle measured counterclockwise* from the x-axis, not necessarily the angle between the line and the x-axis.

Section 6.2 (page 437)

When a television signal hits the antenna dish, the parabolic shape of the dish reflects the signal inward to the focus of the parabola. The signal is then transferred to the receiving equipment.

Section 6.3 (page 447)

a. Answers will vary.

b.

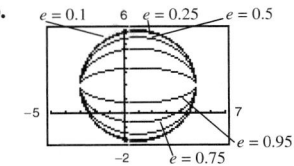

As e approaches 0, the shape of the ellipse approaches the shape of a circle.

c. When $e = 0$, the graph of the equation is circular. The equation of the ellipse is

$$\frac{(x - h)^2}{a^2} + \frac{(y - k)^2}{a^2} = 1,$$

otherwise known as the equation of a circle.

Section 6.4 (page 457)

a.   **b.**

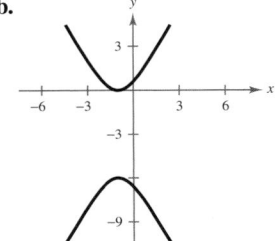

c.   **d.**

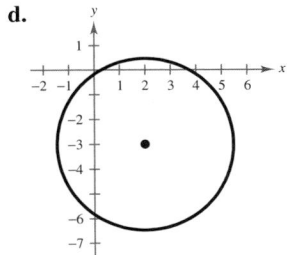

Discussions will vary.

Section 6.5 (page 465)

$f(x) = 1/x$

$y = 1/x$

$xy = 1$

$xy - 1 = 0 \Rightarrow B = 1$ and $F = -1$.

So, $B^2 - 4AC = (1)^2 - 4(0)(0) = 1 > 0$.

Therefore, the graph of $f(x) = 1/x$ is a hyperbola.

WRITING ABOUT MATHEMATICS

Index of Applications

U.S. Demographics

Index

A

Absolute value
  of a complex number, 342
  properties of, 5
  of a real number, 5
Acute angle, 127
Addition
  of a complex number, 329
  vector, 293
    properties of, 295
    resultant of, 293
Additive identity
  for a complex number, 329
  for a real number, 7
Additive inverse, 6
  for a complex number, 329
  for a real number, 7
Adjacent side of a right triangle, 144
Algebraic expression, 6
  evaluate, 6
  term of, 6
Algebraic function, 364
Algebraic tests for symmetry, 32
Alternative definition of conic, 490
Alternative form of Law of Cosines, 283, 321
Amplitude
  modulation, 201
  of sine and cosine curves, 166
Angle, 126
  acute, 127
  between two lines, 427
  between two vectors, 305, 323
  central, 127
  complementary, 129
  conversions between radians and degrees, 130
  coterminal, 126
  degree, 129
  of depression, 149
  of elevation, 149
  initial side, 126
  measure of, 127
  negative, 126
  obtuse, 127
  positive, 126
  radian, 127
  reference, 157
  standard position, 126
  supplementary, 129
  terminal side, 126

vertex, 126
Angular speed, 131
Arc length, 131
Arccosine function, 188
Arcsine function, 186, 188
Arctangent function, 188
Area, of an oblique triangle, 278
Argument, of a complex number, 343
Arithmetic combination, 95
Associative Property of Addition
  for complex numbers, 330
  for real numbers, 7
Associative Property of Multiplication
  for complex numbers, 330
  for real numbers, 7
Astronomical unit, 493
Asymptotes, of a hyperbola, 453
Axis (axes)
  imaginary, 342
  of a parabola, 434
  polar, 476
  real, 342
  rotation of, 460

B

Base, natural, 368
Basic Rules of Algebra, 6, 7
Bearings, 198
Bell-shaped curve, 406
Book value, 46
Bounded intervals, 3
Branches of a hyperbola, 451
Butterfly curve, 509

C

Cardioid, 486
Cartesian plane, 25
Center
  of an ellipse, 442
  of a hyperbola, 451
Central angle of a circle, 127
Change-of-base formula, 385
Characteristics of a function from set A to set B, 53
Circle, 33, 486
  central angle, 127
  classifying
    by discriminant, 464
    by general equation, 457
  standard form of the equation of, 33
  unit, 137

Classification of conics by the discriminant, 464
Classifying a conic from its general equation, 457
Coefficient, of a variable term, 6
Cofunction identities, 218
Common logarithmic function, 376
Commutative Property of Addition
  for complex numbers, 330
  for real numbers, 7
Commutative Property of Multiplication
  for complex numbers, 330
  for real numbers, 7
Complementary angles, 129
Completing the square, 5
Complex conjugates, 331
Complex number(s), 328
  absolute value of, 342
  addition of, 329
  additive identity, 329
  additive inverse, 329
  argument of, 343
  Associative Property of Addition, 330
  Associative Property of Multiplication, 330
  Commutative Property of Addition, 330
  Commutative Property of Multiplication, 330
  Distributive Property, 330
  equality of, 328
  imaginary part of, 328
  modulus of, 343
  nth root of, 350, 351
  nth roots of unity, 352
  polar form, 343
  product of two, 345
  quotient of two, 345
  real part of, 328
  standard form of, 328
  subtraction of, 329
  trigonometric form of, 343
Complex plane, 342
  imaginary axis, 342
  real axis, 342
Complex solutions occur in conjugate pairs, 337
Component form of a vector **v**, 292
Components, vector, 307, 308
Composite number, 8
Composition, 97
Compound interest

Definition of the Six Trigonometric Functions

Right triangle definitions, where $0 < \theta < \pi/2$.

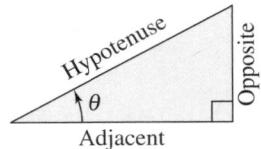

$$\sin\theta = \frac{\text{opp.}}{\text{hyp.}} \qquad \csc\theta = \frac{\text{hyp.}}{\text{opp.}}$$

$$\cos\theta = \frac{\text{adj.}}{\text{hyp.}} \qquad \sec\theta = \frac{\text{hyp.}}{\text{adj.}}$$

$$\tan\theta = \frac{\text{opp.}}{\text{adj.}} \qquad \cot\theta = \frac{\text{adj.}}{\text{opp.}}$$

Circular function definitions, where θ is any angle.

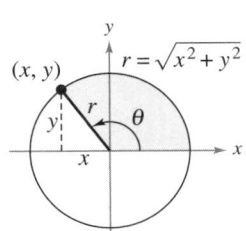

$$\sin\theta = \frac{y}{r} \qquad \csc\theta = \frac{r}{y}$$

$$\cos\theta = \frac{x}{r} \qquad \sec\theta = \frac{r}{x}$$

$$\tan\theta = \frac{y}{x} \qquad \cot\theta = \frac{x}{y}$$

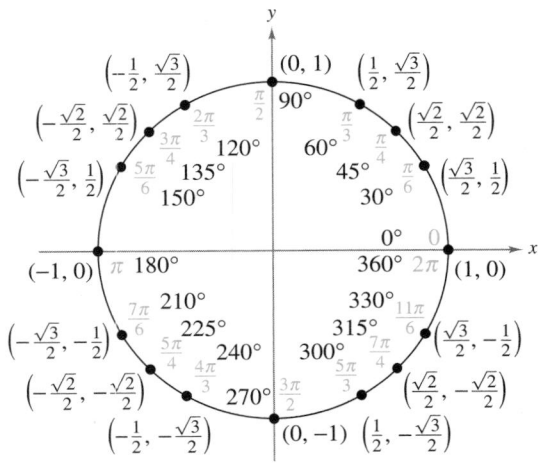

Reciprocal Identities

$$\sin u = \frac{1}{\csc u} \qquad \cos u = \frac{1}{\sec u} \qquad \tan u = \frac{1}{\cot u}$$

$$\csc u = \frac{1}{\sin u} \qquad \sec u = \frac{1}{\cos u} \qquad \cot u = \frac{1}{\tan u}$$

Quotient Identities

$$\tan u = \frac{\sin u}{\cos u} \qquad \cot u = \frac{\cos u}{\sin u}$$

Pythagorean Identities

$$\sin^2 u + \cos^2 u = 1$$

$$1 + \tan^2 u = \sec^2 u \qquad 1 + \cot^2 u = \csc^2 u$$

Cofunction Identities

$$\sin\left(\frac{\pi}{2} - u\right) = \cos u \qquad \cot\left(\frac{\pi}{2} - u\right) = \tan u$$

$$\cos\left(\frac{\pi}{2} - u\right) = \sin u \qquad \sec\left(\frac{\pi}{2} - u\right) = \csc u$$

$$\tan\left(\frac{\pi}{2} - u\right) = \cot u \qquad \csc\left(\frac{\pi}{2} - u\right) = \sec u$$

Even/Odd Identities

$$\sin(-u) = -\sin u \qquad \cot(-u) = -\cot u$$

$$\cos(-u) = \cos u \qquad \sec(-u) = \sec u$$

$$\tan(-u) = -\tan u \qquad \csc(-u) = -\csc u$$

Sum and Difference Formulas

$$\sin(u \pm v) = \sin u \cos v \pm \cos u \sin v$$

$$\cos(u \pm v) = \cos u \cos v \mp \sin u \sin v$$

$$\tan(u \pm v) = \frac{\tan u \pm \tan v}{1 \mp \tan u \tan v}$$

Double-Angle Formulas

$$\sin 2u = 2\sin u \cos u$$

$$\cos 2u = \cos^2 u - \sin^2 u = 2\cos^2 u - 1 = 1 - 2\sin^2 u$$

$$\tan 2u = \frac{2\tan u}{1 - \tan^2 u}$$

Power-Reducing Formulas

$$\sin^2 u = \frac{1 - \cos 2u}{2}$$

$$\cos^2 u = \frac{1 + \cos 2u}{2}$$

$$\tan^2 u = \frac{1 - \cos 2u}{1 + \cos 2u}$$

Sum-to-Product Formulas

$$\sin u + \sin v = 2\sin\left(\frac{u + v}{2}\right)\cos\left(\frac{u - v}{2}\right)$$

$$\sin u - \sin v = 2\cos\left(\frac{u + v}{2}\right)\sin\left(\frac{u - v}{2}\right)$$

$$\cos u + \cos v = 2\cos\left(\frac{u + v}{2}\right)\cos\left(\frac{u - v}{2}\right)$$

$$\cos u - \cos v = -2\sin\left(\frac{u + v}{2}\right)\sin\left(\frac{u - v}{2}\right)$$

Product-to-Sum Formulas

$$\sin u \sin v = \frac{1}{2}[\cos(u - v) - \cos(u + v)]$$

$$\cos u \cos v = \frac{1}{2}[\cos(u - v) + \cos(u + v)]$$

$$\sin u \cos v = \frac{1}{2}[\sin(u + v) + \sin(u - v)]$$

$$\cos u \sin v = \frac{1}{2}[\sin(u + v) - \sin(u - v)]$$

FORMULAS FROM GEOMETRY

Triangle:

$h = a \sin \theta$

$\text{Area} = \dfrac{1}{2}bh$

(Laws of Cosines)

$c^2 = a^2 + b^2 - 2ab \cos \theta$

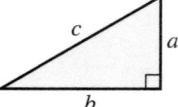

Right Triangle:

(Pythagorean Theorem)

$c^2 = a^2 + b^2$

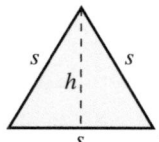

Equilateral Triangle:

$h = \dfrac{\sqrt{3}s}{2}$

$\text{Area} = \dfrac{\sqrt{3}s^2}{4}$

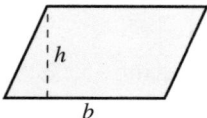

Parallelogram:

$\text{Area} = bh$

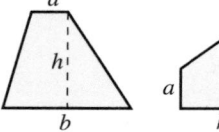

Trapezoid:

$\text{Area} = \dfrac{h}{2}(a + b)$

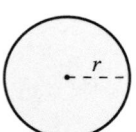

Circle:

$\text{Area} = \pi r^2$

$\text{Circumference} = 2\pi r$

Sector of Circle:

(θ in radians)

$\text{Area} = \dfrac{\theta r^2}{2}$

$s = r\theta$

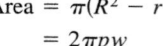

Circular Ring:

(p = average radius,

w = width of ring)

$\text{Area} = \pi(R^2 - r^2)$

$= 2\pi pw$

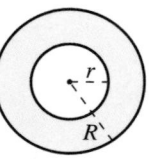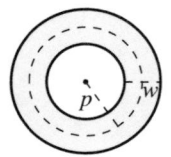

Sector of Circular Ring:

(p = average radius,

w = width of ring,

θ in radians)

$\text{Area} = \theta pw$

Ellipse:

$\text{Area} = \pi ab$

$\text{Circumference} \approx 2\pi \sqrt{\dfrac{a^2 + b^2}{2}}$

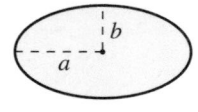

Cone:

(A = area of base)

$\text{Volume} = \dfrac{Ah}{3}$

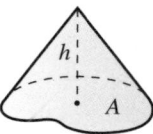

Right Circular Cone:

$\text{Volume} = \dfrac{\pi r^2 h}{3}$

$\text{Lateral Surface Area} = \pi r \sqrt{r^2 + h^2}$

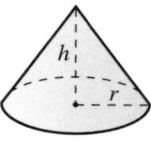

Frustum of Right Circular Cone:

$\text{Volume} = \dfrac{\pi(r^2 + rR + R^2)h}{3}$

$\text{Lateral Surface Area} = \pi s(R + r)$

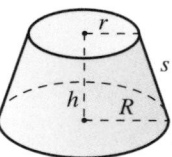

Right Circular Cylinder:

$\text{Volume} = \pi r^2 h$

$\text{Lateral Surface Area} = 2\pi rh$

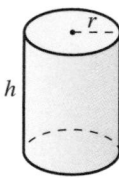

Sphere:

$\text{Volume} = \dfrac{4}{3}\pi r^3$

$\text{Surface Area} = 4\pi r^2$

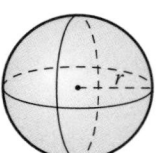

Wedge:

(A = area of upper face,

B = area of base)

$A = B \sec \theta$

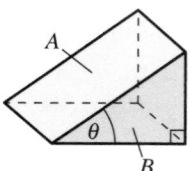